## of the elements

*Top 3*
*3 anions*

*+3 cation*

| | | | IIIA | IVA | VA | VIA | VIIA | 18 |
| | | | IIIB | IVB | VB | VIB | VIIB | 2 He 4.00260 |
| | | | 13 | 14 | 15 | 16 | 17 | |
| | IB | IIB | 5 no ion B 10.81 | 6 C 12.011 | 7 N 14.0067 | 8 O 15.9994 | 9 F 18.9984 | 10 Ne 20.179 |
| | IB | IIB | 13 Al 26.9815 | 14 Si 28.0855 | 15 P 30.9738 | 16 S 32.06 | 17 Cl 35.453 | 18 Ar 39.948 |
| 10 | 11 | 12 | | | | | | |
| 28 Ni 58.69 | 29 Cu 63.546 | 30 Zn 65.39 | 31 Ga 69.72 | 32 Ge 72.59 | 33 As 74.9216 | 34 Se 78.96 | 35 Br 79.904 | 36 Kr 83.80 |
| 46 Pd 106.42 | 47 Ag 107.868 | 48 Cd 112.41 | 49 In 114.82 | 50 Sn 118.71 | 51 Sb 121.75 | 52 Te 127.60 | 53 I 126.905 | 54 Xe 131.29 |
| 78 Pt 195.08 | 79 Au 196.967 | 80 Hg 200.59 | 81 Tl 204.383 | 82 Pb 207.2 | 83 Bi 208.980 | 84 Po (209) | 85 At (210) | 86 Rn (222) |

*Transitional elements*

*+4 cation & +2 cation*

*Representative Elements form one type of ion*

| 63 Eu 151.96 | 64 Gd 157.25 | 65 Tb 158.925 | 66 Dy 162.50 | 67 Ho 164.930 | 68 Er 167.26 | 69 Tm 168.934 | 70 Yb 173.04 | 71 Lu 174.967 |

| 95 Am (243) | 96 Cm (247) | 97 Bk (247) | 98 Cf (251) | 99 Es (252) | 100 Fm (257) | 101 Md (258) | 102 No (259) | 103 Lr (260) |

not found naturally on Earth

## Physical Constants

| | |
|---|---|
| Avogadro's number | $N_A = 6.022045 \times 10^{23} \text{ mol}^{-1}$ |
| Bohr radius | $a_o = 0.5291771 \text{ Å} = 5.291771 \times 10^{-11} \text{ m}$ |
| Boltzmann constant | $k = 1.38066 \times 10^{-23} \text{ J K}^{-1}$ |
| Coulomb's law constant | $\alpha = 1.1126501 \times 10^{-10} \text{ C}^2 \text{ N}^{-1} \text{ m}^{-2}$ |
| Electron charge | $e = 1.602189 \times 10^{-19} \text{ C}$ |
| Electron mass | $m_e = 9.10953 \times 10^{-31} \text{ kg}$ |
| Faraday's constant | $F = 96{,}485 \text{ C mol}^{-1}$ |
| Gas constant | $R = 8.3144 \text{ J mol}^{-1} \text{ K}^{-1}$ |
| | $= 0.082057 \text{ L atm mol}^{-1} \text{ K}^{-1}$ |
| Gravitational acceleration | $g = 9.80665 \text{ m s}^{-2}$ |
| Ideal gas molar volume at STP | $\overline{V}_{id} = 22.4138 \text{ L mol}^{-1}$ |
| Neutron mass | $m_n = 1.674954 \times 10^{-27} \text{ kg}$ |
| Planck's constant | $h = 6.62618 \times 10^{-34} \text{ J s}$ |
| Proton mass | $m_p = 1.672648 \times 10^{-27} \text{ kg}$ |
| Rydberg constant | $\mathcal{R} = 1.09737318 \times 10^7 \text{ m}^{-1}$ |
| Speed of light (in vacuum) | $c = 2.9979246 \times 10^8 \text{ m s}^{-1}$ |

## Conversion Factors

1 electron volt (eV) = $1.602189 \times 10^{-19}$ J
1 calorie = 4.184 J
1 u = $1.660566 \times 10^{-27}$ kg = 931.502 MeV
1 L atm = 101.325 J

# Modern University Chemistry

# MODERN UNIVERSITY CHEMISTRY
## SECOND EDITION

## NORBERT T. PORILE
Purdue University

McGraw-Hill, Inc.
College Custom Series

New York  St. Louis  San Francisco  Auckland  Bogotá
Caracas  Lisbon  London  Madrid  Mexico  Milan  Montreal
New Delhi  Paris  San Juan  Singapore  Sydney  Tokyo  Toronto

McGraw Hill's **College Custom Series** consists of products that are produced from camera-ready copy. Peer review, class testing, and accuracy are primarily the responsibility of the author.

This book is printed on acid-free paper.

MODERN UNIVERSITY CHEMISTRY

Copyright © 1993 by McGraw-Hill, Inc. All rights reserved. Printed in the United States of America. Except as permitted under the United States Copyright Act of 1976, no part of this publication may be reproduced or distributed in any form or by any means, or stored in a data base retrieval system, without prior written permission of the publisher.

1 2 3 4 5 6 7 8 9 0   MAL MAL   9 0 9 8 7 6 5 4 3

ISBN 0-07-050639-6

Editor: J.D. Ice

Cover Design: Amy L. Agurkis

Printer/Binder: Malloy Lithographing, Inc.

# Preface

*Modern University Chemistry* is intended for a first-year course for students who have had introductory high school courses in the physical sciences and enough mathematics to be taking calculus concurrently. I envision this textbook as being useful in courses taken by science and engineering students as well as for students in other areas who desire a strong background in chemistry.

The distinguishing feature of this book is its emphasis on chemistry as a quantitative experimental science. My approach is to present the experimental facts, the quantitative models based on these facts, and the more general theories that serve to unify the subject. In a textbook that stresses the quantitative aspects of a subject, it is easy, but undesirable, to reach beyond its confines and introduce relationships whose validity cannot be demonstrated. I have avoided this practice by either deriving all quantitative relationships or at least discussing their origin. The level of the mathematics used is dictated by the nature of each topic. Calculus, for example, is used in the chapters on thermodynamics because this makes it possible to derive all the important quantitative relationships.

There is no single generally accepted organizational scheme for a general chemistry text. Some instructors prefer to cover macroscopic phenomena before turning to the underlying microscopic behavior, while others prefer the opposite approach; some instructors prefer to integrate descriptive chemistry as fully as possible, while others prefer to treat it separately. In certain cases, the need to permit students to move between different general chemistry courses determines the sequence of topics; in others, the coordination between laboratory and lecture topics is of importance. I have written this textbook in a way that lends itself to diverse topical sequences and have, in fact, used the material in various sequences in my own course.

The basic scheme I follow is to develop macroscopic and microscopic chemistry in parallel. Thus, the first three chapters are introductory and examine the structure of matter, stoichiometry, and properties of gases. Chapter 4 describes the molecular basis of gas behavior and leads to a detailed consideration of atomic structure and chemical periodicity in the next two chapters. Chapters 7 through 9 deal with various aspects of chemical equilibrium; in the following four chapters, thermodynamics and its applications to phase equilibria, solutions, and electrochemistry are considered. After a discussion of chemical kinetics in Chapter 14, I return to microscopic chemistry: Chapters 15 through 18 describe chemical bonding, molecular structure, and intermolecular interactions. The separation of the chapters on bonding from those on atomic structure makes it necessary for the student, when studying chemical bonding, to review the introduction to quantum mechanics presented in the chapters on atomic structure. I believe that a second exposure to these abstract concepts is highly beneficial. The textbook continues with an examination of the representative elements, the solid state, and transition metals (Chapters 19 through 21), and concludes with organic and biochemistry and nuclear chemistry (Chapters 22 and 23). The five appendices present units and conversion factors, a review of the mathematics used, and numerical data tables. Chapter 1 includes an appendix on significant figures in numerical problems.

The book is readily adaptable to diverse sequences of topics. Instructors who wish to cover macroscopic chemistry before microscopic chemistry can defer Chapters 5 and 6 until after Chapter 14. On the other hand, those who wish to present microscopic chemistry first can cover Chapters 15 through 17 after Chapter 6. Descriptive chemistry is incorporated throughout the chapters. However, the acid–base chemistry and redox chemistry of the representative elements are treated separately in Chapter 19. This material can be integrated more fully by covering the sections on acid–base chemistry immediately after Chapter 8, which describes acid–base equilibria, and by covering the sections on redox chemistry after Chapter 13 (electrochemistry).

The design and organization of this textbook enhances the student's study and retention of the subject. Numbered sections and lettered subsections signal individual topics. Important terms are printed in boldface when first introduced and defined. Every chapter includes examples illustrating the application of the pertinent chemical principles. Students should be encouraged to work these examples, since mastering the art of problem-solving is an important aspect of chemistry. Each chapter concludes with many problems of varying difficulty. Wherever appropriate, problems that require the use of calculus are included, marked by an asterisk (*). Answers to most numerical problems are given at the back of the book. A complete solutions manual is available to instructors.

I have generally used SI units throughout this book. I have not been overly rigid about this, however, and have used several non-SI units because of their convenience or their wide use in the scientific literature. These include the standard atmosphere and the torr (mm Hg) as units of pressure, the angstrom as a unit of length in atoms and molecules, and the electron volt as a unit of energy of single particles. I have taken a conservative approach to the current controversy over the periodic table group numbers and have used the conventional American designations. However, both the old and the new IUPAC conventions are mentioned briefly. I have generally avoided the IUPAC provisional names and symbols for newly discovered elements; instead, I refer to these elements by their atomic numbers.

This textbook developed from my involvement during the past decade with the Honors General Chemistry course at Purdue University. My understanding of the subject has been clarified and sharpened as a result of my interaction with the outstanding students in this course. It is my pleasure to acknowledge their contributions. Special thanks are due to the class of 1985–86, which used a draft version of this book and made numerous suggestions for its improvement.

In writing this book, I have profited from numerous discussions with my colleagues at Purdue University. I particularly thank George Bodner, Robert Grimley, John Grutzner, Jurgen Honig, David McMillin, and William Robinson. I am very appreciative of the detailed comments on the manuscript by the following reviewers: Ruth H. Aranow (Johns Hopkins University), Mario E. Baur (University of California at Los Angeles), J. Aaron Bertrand (Georgia Institute of Technology), Luther K. Brice Jr. (Virginia Polytechnic Institute), Alan Campion (University of Texas), Leigh B. Clark (University of California, San Diego), Richard F. Fenske (University of Wisconsin), Robert E. Frost (State University of New York at Albany), Patrick L. Jones (The Ohio State University), Darl H. McDaniel (Univeristy of Cincinnati), Clyde Metz (College of Charleston), Peter B. Moore (Yale University), Norman H. Nachtrieb (University of Chicago), Richard P. Schmitt (Texas A & M University), Peter E. Siska (University of Pittsburgh), Richard M. Stratt (Brown University), Bradford Wayland (University of Pennsylvania), D. T. Zajicek (University of Massachusetts), J. J. Zuckerman (University of Oklahoma), and Steven S. Zumdahl (University of Illinois). Clyde Metz also independently worked out all the examples and end-of-chapter problems.

I am grateful to all the people at Academic Press and Harcourt Brace Jovanovich who worked on this book, particularly to my acquisitions editors, Jeff Holtmeier and Don Schumacher, who supported and encouraged me throughout. I am grateful to Mary Castellion, who did an excellent job as developmental editor. I also wish to thank Gail Shively for ably and cheerfully typing the manuscript in its several drafts.

Part of this book was written while I was the recipient of a Senior U.S. Scientist Award from the Alexander Von Humboldt Foundation at Philipps University, Marburg, Federal Republic of Germany. I wish to thank the Humboldt Foundation for its support and the staff of the Kernchemie, Marburg, for their hospitality.

Finally, I am grateful to my family for their constant support and encouragement during the years I devoted to this book. My wife Miriam helped me with numerous questions of English usage as well as with the proofreading. My son Jim read portions of the text, providing me with numerous helpful suggestions from his perspective as an undergraduate chemistry major.

Norbert T. Porile
West Lafayette

# Chapter Titles

1. The Basic Constituents of Matter  1
2. Chemical Reactions  33
3. Properties of Gases  65
4. Molecular Basis of Gas Properties  91
5. The Quantized Atom  123
6. Multielectron Atoms and the Periodic Table  175
7. Chemical Equilibrium  221
8. Acids and Bases  249
9. Ionic Equilibria  298
10. Energy in Chemical Processes—The First Law of Thermodynamics  331
11. Spontaneity in Chemical Processes—The Second Law of Thermodynamics  370
12. Phase Equilibria  410
13. Electrochemistry  456
14. Kinetics—The Rates and Mechanisms of Chemical Reactions  499
15. Lewis Structures and Chemical Bonding  552
16. Theories of Chemical Bonding  588
17. Molecular Structure and Chemical Bonding  628
18. Intermolecular Interactions  658
19. The Representative Elements  693
20. Structure and Bonding in Solids  738
21. The Transition Elements and Their Coordination Compounds  785
22. The Chemistry of Carbon  828
23. Nuclear Chemistry  876

Appendices  918

Answers to Selected Problems  950

Index  954

# Contents

## 1 The Basic Constituents of Matter   1

### Atoms and Elements

   1.1  The Elements   2
   1.2  Atoms   3
   1.3  The Structure of the Atom   4
   1.4  Atomic Masses   7
        a.  The Atomic Mass Scale   7
        b.  Atomic Weights of the Elements   8
        c.  Mass and Weight; Density   9

### Molecules, Ions, and Compounds

   1.5  Molecular Compounds   11
        a.  Molecules   11
        b.  Compounds and Mixtures   13
   1.6  Determination of Atomic and Molecular Weights   14
   1.7  Ionic Compounds   16
        a.  Ions   16
        b.  Nomenclature   17
   1.8  The Mole and Avogadro's Number   20
   1.9  Conversion between Mass and Moles   21

### Conclusion   25

### Appendix: Experimental Uncertainties and Significant Figures

    1A.1  Uncertainties in Experimentally Determined Quantities   25
    1A.2  Propagation of Errors and Significant Figures   27
    1A.3  Significant Figures in Numerical Problems   28

## 2 Chemical Reactions   33

### Chemical Equations and Stoichiometry

    2.1  Chemical Equations and the Conservation of Mass   33
        a.  The Conservation of Mass in Chemical Reactions   33
        b.  Balanced Chemical Equations   35
    2.2  Stoichiometry of Chemical Reactions   35
        a.  The Meaning of Chemical Equations and Stoichiometric Calculations   35
        b.  The Limiting Reagent and Incomplete Reactions   38
    2.3  Chemical Reactions in Aqueous Solution and the Conservation of Charge   40
        a.  Ions in Solution   40
        b.  Molarity of Solutions   42
        c.  The Conservation of Charge and Net Ionic Equations   45
    2.4  Conservation of Energy in Chemical Reactions   47
    2.5  A Classification of Chemical Reactions   48
        a.  Decomposition Reactions   48
        b.  Combination Reactions   48
        c.  Displacement Reactions   49
        d.  Double Displacement Reactions   49

### Oxidation-Reduction Reactions

    2.6  Oxidation and Reduction   51
    2.7  Oxidation States   53
    2.8  Balancing Redox Equations   55
        a.  The Half-Reaction Method   55
        b.  The Oxidation Number Method   58
    2.9  Disproportionation   59
    2.10  Redox Titrations and Stoichiometry   60

### Conclusion   62

## 3 Properties of Gases   65

### The Ideal Gas

    3.1  Avogadro's Law and Gas Stoichiometry   66
    3.2  Pressure of a Gas   68
    3.3  The Gas Laws   70
        a.  Dependence of Volume on Pressure   70
        b.  Dependence of Volume on Temperature   72
        c.  The Combined Gas Law   73
    3.4  The Ideal Gas Equation of State   74
    3.5  Dalton's Law of Partial Pressures   77

### The Atmosphere

- 3.6 Properties of the Atmosphere    82
- 3.7 The Special Role of Ozone    85
- 3.8 Air Pollution    86
  - a. Photochemical Smog    86
  - b. Acid Rain    87
  - c. The Greenhouse Effect    87

### Conclusion    88

## 4   Molecular Basis of Gas Properties    91

### The Kinetic Theory of Gases

- 4.1 Assumptions of the Kinetic Theory    92
- 4.2 Predictions of the Kinetic Theory    94
  - a. Derivation of Boyle's Law    94
  - b. Temperature and Kinetic Energy of Motion    98
- 4.3 Molecular Speeds    99
- 4.4 Effusion and Diffusion of Gases    101
- 4.5 Molecular Collisions    103
- 4.6 Distribution of Molecular Speeds    106

### Real Gases

- 4.7 Departures from Ideal Gas Behavior    110
- 4.8 The Critical Point and Condensation    113
  - a. The Shape of P-V Isotherms    113
  - b. The Law of Corresponding States    115
- 4.9 The van der Waals Equation of State    117

### Conclusion    121

## 5   The Quantized Atom    123

### The Wave-Particle Duality

- 5.1 The Structure of the Atom    124
  - a. The Charge/Mass Ratio of the Electron    124
  - b. The Charge of the Electron    125
  - c. The Scattering of Alpha Particles by Atoms    127
- 5.2 Wave Properties of Light    129
- 5.3 The Electromagnetic Spectrum    133
- 5.4 Corpuscular Properties of Electromagnetic Radiation    135
  - a. Black-Body Radiation    135
  - b. The Photoelectric Effect    137
  - c. The Compton Effect    140
- 5.5 Atomic Spectra    141
  - a. Line Spectra of Atoms    141
  - b. The Spectrum of Atomic Hydrogen    143
- 5.6 Wave Properties of Matter    148
- 5.7 The Uncertainty Principle    150

### Quantum Mechanics of the Hydrogen Atom

    5.8   The Schrödinger Wave Equation   153
    5.9   The Particle in a Box   154
    5.10  The Wave Equation for the Hydrogen Atom   161
    5.11  The Quantum Numbers   163
    5.12  Solutions of the Wave Equation   165
         a.   The Radial Functions   165
         b.   The Angular Functions   169

### Conclusion   171

## 6 Multielectron Atoms and the Periodic Table   175

### The Electronic Structure of Atoms

    6.1   Historical Development of the Periodic Table   175
         a.   Some Early Classification Schemes   175
         b.   The Periodic Table of Mendeleev   177
    6.2   Quantum Mechanics of Multielectron Atoms   179
         a.   Splitting of the Energy Levels   179
         b.   Factors Determining Orbital Filling   182
    6.3   Electronic Configuration of the Elements   184
    6.4   The Periodic Table and the Electronic Structure of the Elements   190

### Periodic Properties

    6.5   Ionization Energies   192
         a.   The First Ionization Energy   192
         b.   Higher Ionization Energies   197
         c.   X-Rays and Photoelectron Spectroscopy   199
    6.6   Electron Affinity   204
    6.7   Electronegativity   207
    6.8   The Sizes of Atoms and Ions   208
         a.   Atomic Volume   208
         b.   Atomic Radii   209
         c.   Ionic Radii   209
    6.9   Metals and Nonmetals   213
         a.   Classification of the Elements   213
         b.   Trends in Properties of the Elements   215
    6.10  Oxidation Numbers and Chemical Bonds   216

### Conclusion   218

## 7 Chemical Equilibrium   221

    7.1   The Nature of Chemical Equilibrium   222
    7.2   The Equilibrium Constant   224
         a.   Definition of the Equilibrium Constant   224
         b.   Equilibrium Constant for Reactions Involving Gases   226
         c.   Properties of the Equilibrium Constant   227
         d.   Magnitude of Equilibrium Constants   229

7.3 The Reaction Quotient  230
7.4 Solving Equilibrium Problems  231
7.5 Dissociation Equilibria  237
7.6 Factors Affecting the Position of Equilibrium  240
   a. The Effect of Concentration Changes  241
   b. The Effect of Pressure  243
   c. The Effect of Temperature on Equilibrium  244
7.7 Chemical Equilibrium for Real Gases  245

**Conclusion**  246

# 8 Acids and Bases  249

## Properties of Acids and Bases

8.1 The Arrhenius Model and its Drawbacks  250
   a. The Arrhenius Model and the Conductivity of Solutions  250
   b. Drawbacks of the Arrhenius Model  251
8.2 The Brønsted-Lowry Model  252
8.3 The Lewis Model  254
8.4 Relative Strengths of Acids and Bases  255
   a. Acid Ionization Constants  255
   b. The Ion Product of Water and pH of Aqueous Solutions  257
   c. Ionization Constants of Bases  259

## Equilibria Involving Acids and Bases

8.5 Weak Acids and Bases  261
8.6 Salts of Weak Acids or Bases  265
8.7 The Ionization of Water  266
8.8 Buffer Solutions  269
8.9 Acid-Base Titrations  276
   a. Some Fundamental Aspects  276
   b. pH Indicators  278
   c. Titration Curves for Strong Acids or Bases  280
   d. Titration of a Weak Acid or Base  281
   e. Graphical Representation of Concentration Changes  285
8.10 Polyprotic Acids  286
   a. Multistep Equilibria  286
   b. Equilibria of Amphiprotic Species  291

**Conclusion**  295

# 9 Ionic Equilibria  298

## Solubility Equilibria

9.1 The Solubility of Salts  298
9.2 The Solubility Product  300
9.3 The Common Ion Effect  304
9.4 Selective Precipitation  306

### Equilibria of Metal-Ligand Complexes

    9.5   Metal-Ligand Complexes   307
    9.6   Complex Ion Equilibria   307

### Effect of Other Equilibria on Solubility

    9.7   Complex Ion Formation and Solubility   312
        a.   The Effect of Complexing Ligands   312
        b.   Identification of Insoluble Metal Chlorides   314
        c.   Complex Formation Involving a Common Ion   316
    9.8   The Effect of pH on the Solubility of Salts of Weak Acids   319
        a.   The Effect of pH on Solubility Equilibria   319
        b.   Sulfide Separations in Qualitative Analysis   320
    9.9   Effect of Hydrolysis on Solubility   322
    9.10   The Solubility of Metal Hydroxides   326

### Conclusion   328

## 10   Energy in Chemical Processes — The First Law of Thermodynamics   331

### Fundamental Aspects of the First Law of Thermodynamics

    10.1   The Transfer of Energy between a System and its Surroundings   331
    10.2   Work   335
        a.   Mechanical Work   335
        b.   Work Associated with Expansion or Compression of a Gas   336
        c.   Expansion Against a Variable External Pressure   337
        d.   Work in Reversible Processes   339
    10.3   Heat   342
        a.   The Mechanical Equivalent of Heat   342
        b.   Heat Capacity   343
    10.4   The First Law of Thermodynamics   344
        a.   Statement of the Law   344
        b.   State Functions   345
    10.5   Constant Volume and Constant Pressure Processes   350
        a.   Constant Volume Processes   350
        b.   Constant Pressure Processes   352
        c.   Relationship between Constant Pressure and Constant Volume Processes   353
    10.6   Molecular Interpretation of the Thermodynamic Functions   357

### Thermochemistry

    10.7   Thermochemical Equations   358
    10.8   Hess's Law: The Indirect Determination of Reaction Enthalpies   360
    10.9   Standard Enthalpies of Formation   362
    10.10   Combustion Reactions and Natural Fuels   364
    10.11   Temperature Dependence of Reaction Enthalpies   366

### Conclusion   367

## 11 Spontaneity in Chemical Processes — The Second Law of Thermodynamics  370

### Entropy and Spontaneity

11.1 Spontaneous Processes  370
11.2 Entropy and the Second Law of Thermodynamics  374
11.3 Molecular Interpretation of Entropy  378
11.4 Entropy Changes in Chemical Processes  381
 a. Dependence of Entropy on Temperature  382
 b. Entropy and Changes of Phase  382
11.5 Third Law Entropies  385
 a. Evaluation of Third Law Entropies  385
 b. Standard Entropies of Substances  386
 c. Entropy Changes in Chemical Reactions  389

### Free Energy and Chemical Equilibrium

11.6 The Free Energy Function  390
11.7 Free Energy Changes in Chemical Reactions  392
 a. The Makeup of the Free Energy Change  392
 b. Standard Free Energies of Formation  393
11.8 Free Energy and the Equilibrium Constant  394
11.9 Temperature Dependence of K and $\triangle G$  400
 a. Variation of the Equilibrium Constant with Temperature  400
 b. Variation of $\triangle G$ with Temperature  403

### Conclusion  406

## 12 Phase Equilibria  410

### Phase Equilibria of Pure Substances

12.1 Changes of Phase  410
 a. The Molecular Viewpoint  410
 b. The Thermodynamic Viewpoint  413
12.2 Vapor Pressure  416
12.3 Phase Diagrams  422
 a. The Shape of a Phase Diagram  422
 b. The Phase Rule  425

### Phase Equilibria of Solutions

12.4 Ideal Solutions  426
 a. Retrospective View of Solutions  426
 b. Properties of Ideal Solutions  426
12.5 Phase Diagrams of Ideal Solutions  429
 a. The Liquid-Vapor Phase Diagram  429
 b. The Phase Rule for a Two-Component System  432
 c. Distillation  433
12.6 Non-Ideal Solutions and their Phase Diagrams  434
 a. The Nature of Non-Ideal Solutions  434
 b. Henry's Law  435
 c. Liquid-Vapor Phase Diagrams  438

12.7 Solid-Liquid Phase Equilibria   439
   a. Phase Equilibrium Diagrams   439
   b. Thermal Analysis   442
12.8 Colligative Properties   444
   a. Vapor Pressure of a Solution of a Nonvolatile Solute   444
   b. Boiling Point Elevation and Freezing Point Depression   444
   c. Osmotic Pressure   448
12.9 Colloids   451

**Conclusion   452**

# 13 Electrochemistry   456

## Fundamentals of Electrochemistry

13.1 Galvanic Cells   456
   a. The Operation of a Cell   456
   b. The Cell EMF   460
   c. Types of Electrodes   461
13.2 Cell EMF and the Reaction Free Energy   464
13.3 The Nernst Equation   467
13.4 Concentration Cells   468
13.5 Standard Electrode Potentials   470
   a. Determination of Standard Electrode Potentials   470
   b. The EMF Series   472
   c. Standard Electrode Potentials of Elements with Three or More Oxidation States   475
13.6 Determination of Equilibrium Constants from Standard Reduction Potentials   477

## Applications of Electrochemistry

13.7 Batteries   480
   a. Primary Cells   480
   b. Storage Cells   482
   c. Fuel Cells   484
13.8 Electrochemical Determination of pH   485
13.9 Corrosion   486
13.10 Electrolysis and Its Applications   488
   a. Electrolytic Cells   488
   b. Preparation of Active Metals   491
   c. Electrolysis of Aqueous Solutions   493
   d. Electrolytic Refining and Electroplating of Metals   494

**Conclusion   495**

# 14 Kinetics — The Rates and Mechanisms of Chemical Reactions   499

## Empirical Rate Laws

14.1 An Overview of Kinetics   500
14.2 The Rate Law   501
   a. The Rate of a Reaction   501
   b. Types of Rate Laws   503

        c. The Initial Rate Method for the Determination of the Rate Law   505
- 14.3 Integrated Rate Laws   508
  - a. First Order Kinetics   508
  - b. Second Order Kinetics   511
  - c. The Isolation Method   514

### Reaction Mechanisms

- 14.4 Elementary Reactions   517
  - a. Molecularity and Reaction Order   517
  - b. Elementary Reactions Attaining Equilibrium   519
- 14.5 Understanding Reaction Mechanisms   519
  - a. The Rate-Determining Step   520
  - b. The Steady State Approximation   521
- 14.6 Chain Reactions   526

### The Rates of Elementary Reactions

- 14.7 The Rate of Molecular Collisions   528
- 14.8 Temperature Dependence of Reaction Rates and the Magnitude of Rate Constants   530
  - a. The Arrhenius Equation   530
  - b. The Basis of the Arrhenius Equation   531
  - c. Orientation Effects   533
- 14.9 The Activated Complex   534
- 14.10 Rates and Mechanisms of Chemical Reactions in Solution   540
- 14.11 Catalysis   543

### Conclusion   546

## 15 Lewis Structures and Chemical Bonding   552

### Lewis Structures

- 15.1 An Overview of Chemical Bonding   553
- 15.2 Simple Lewis Structures   554
  - a. Bonded Atoms with Complete Valence Shells   554
  - b. Bonded Atoms with Incomplete Valence Shells   557
  - c. Bonded Atoms with Expanded Valence Shells   558
  - d. Ionic Compounds   560
- 15.3 Coordinate Covalent Bonds   560
  - a. The Nature of Coordinate Covalent Bonds   560
  - b. Lewis Acids and Bases and Coordinate Covalent Bonds   562
- 15.4 Multiple Bonds   562
- 15.5 The Distribution of Electronic Charge   564
  - a. Formal Charge   564
  - b. Oxidation Number   566
- 15.6 Resonance   567

### Molecular Geometry

- 15.7 The Shapes of Molecules   571
- 15.8 Dipole Moments   573
  - a. Diatomic Molecules   573
  - b. Polyatomic Molecules   575

15.9 The VSEPR Model  578

**Conclusion**  585

# 16 Theories of Chemical Bonding  588

## Valence Bond Theory

16.1 The Hydrogen Molecule  588
16.2 Hybridization of Atomic Orbitals  592
16.3 Ionic Character of Covalent Bonds  596

## Molecular Orbital Theory

16.4 Molecular Orbitals  599
16.5 Sigma and Pi Orbitals  604
16.6 Homonuclear Diatomic Molecules  608
    a. Electronic Configuration  608
    b. Bond Properties  612
16.7 Heteronuclear Diatomic Molecules  613
16.8 Bonding in Simple Hydrocarbons  617
    a. Localized Bonding  617
    b. Delocalized Bonding  620
16.9 Three-Center Bonds in Boron Hydrides  623

**Conclusion**  625

# 17 Molecular Structure and Chemical Bonding  628

17.1 Bond Energies  628
    a. Diatomic Molecules  628
    b. Polyatomic Molecules  630
    c. Bond Energies and Reaction Enthalpies  632
17.2 Rotational Spectra and Bond Lengths  634
    a. Molecular Spectroscopy  634
    b. Rotational Spectra of Diatomic Molecules  637
    c. Bond Lengths  640
17.3 Vibrational Spectra of Diatomic Molecules and Bond Force Constants  641
17.4 Electronic Spectra  647
    a. Transitions Involving Pi Electrons  647
    b. Deexcitation of Electronic States  650
17.5 The Laser  653

**Conclusion**  655

# 18 Intermolecular Interactions  658

## Interactions Between Molecules or Ions

18.1 The Ion Pair Interaction  658
18.2 Van der Waals Forces  664
    a. The Interaction between Instantaneous and Induced Dipoles  664
    b. Dipole-Dipole Interactions  665

        c. Repulsive Interactions   666
        d. The Lennard-Jones Potential   666
  18.3  The Hydrogen Bond   670
  18.4  Experimental Consequences of Hydrogen Bonding   673
        a. Properties of the Hydrides of the Nonmetals   673
        b. Azeotropic Solutions   676

## Water

  18.5  Hydrogen Bonding in Ice and Water   677
  18.6  Solvent Properties of Water   680
        a. The Hydration of Ions   680
        b. The Dielectric Constant of Water   681
        c. The Hydrogen Ion in Aqueous Solution   683
        d. Hydrates   683
        e. The Dissolution of Ionic Solids   685
        f. The Solubility of Molecular Compounds   686
  18.7  Natural Waters   688
        a. The Hydrosphere   688
        b. Hard Water   689

## Conclusion   690

# 19 The Representative Elements   693

## Acid-Base Chemistry

  19.1  Binary Compounds with Hydrogen   694
        a. The Ionic Hydrides   694
        b. The Basicity of Ammonia   695
        c. The Hydrohalic Acids   697
  19.2  Binary Oxides   699
  19.3  Hydroxides and Oxoacids   704
        a. Acidic versus Basic Behavior   704
        b. The Common Oxoacids   704
        c. Trends in the Strength of Oxoacids   706

## Redox Chemistry

  19.4  Common Oxidation States of the Representative Elements   709
  19.5  The Metallic Representative Elements   711
        a. The s-Block Elements and Their Reactions   711
        b. Diagonal Relationships   713
        c. Reactivity of the Alkali and Alkaline Earth Metals in Solution   714
        d. The p-Block Metals   717
  19.6  The Redox Chemistry of Nitrogen and the Use of Electrode Potential Diagrams   718
  19.7  The Redox Chemistry of Oxygen and Sulfur   724
        a. Oxygen   724
        b. Sulfur   725
  19.8  The Halogens   727
        a. The Elements   727
        b. Redox Chemistry   728

        c. The Halides and Pseudohalides   730
        d. Interhalogen Compounds   732
19.9  The Noble Gases   733

**Conclusion**   734

# 20  Structure and Bonding in Solids   738

## Some Properties of Solids

20.1  Crystalline and Amorphous Solids   738
20.2  The Heat Capacities of Solids   740
20.3  Crystal Systems   742
        a. Lattices and Unit Cells   742
        b. The Cubic Crystal System   745

## Metals

20.4  The Structure of Metals   748
20.5  The Electron Sea Model of Metallic Bonding   753
20.6  The Band Theory of Solids   754
        a. The Molecular Orbital Approach to Metallic Bonding   754
        b. Metals, Insulators, and Semiconductors   757
        c. Doped Semiconductors   758

## Ionic Crystals

20.7  Some Common Ionic Crystal Structures   761
20.8  The Packing of Ions in Crystals   764
        a. Interstitial Sites   764
        b. The Radius-Ratio Rule   766
20.9  Defects in Crystals   768
20.10  The Stability of Ionic Crystals   770
        a. Some Properties of Ionic Crystals   770
        b. The Crystal Lattice Energy   771
        c. The Born-Haber Cycle   774

## Giant Molecules

20.11  Network Covalent Solids   776
20.12  The Structure of Silicates   779

**Conclusion**   781

# 21  The Transition Elements and Their Coordination Compounds   785

## Transition Element Chemistry

21.1  General Properties   785
        a. Electronic Configuration   785
        b. The Elements   786
        c. Oxidation States   788
21.2  Redox Chemistry of Transition Elements with Few Oxidation States   789
        a. The Iron Triad   789

       b. The Coinage Metals   791
       c. The Post-Transition Elements   793
       d. The Scandium Family and the Lanthanides   793
21.3 Redox Chemistry of Transition Elements with Multiple Oxidation States   794

## Coordination Compounds

21.4 Fundamentals of Coordination Chemistry   796
       a. Some Basic Definitions   796
       b. The Nomenclature of Coordination Compounds   798
21.5 Structure and Isomerism   800
       a. The Geometry of Metal Complexes   800
       b. Structural Isomerism   801
       c. Stereoisomerism   802
       d. Optical Stereoisomerism   803
21.6 Color and Magnetism of Complexes   806
21.7 Crystal Field Theory   808
       a. Bonding in Coordination Complexes   808
       b. Octahedral Crystal Field Splitting   809
       c. Absorption Spectra and the Crystal Field   813
       d. Tetrahedral and Square-Planar Complexes   817
21.8 Metal Clusters   820
21.9 Reactions of Coordination Complexes   821
21.10 Some Uses of Coordination Compounds   823

**Conclusion   825**

# 22 The Chemistry of Carbon   828

## Organic Chemistry

22.1 Hydrocarbons   829
       a. The Alkanes   830
       b. Nomenclature   832
       c. Reactions of Alkanes   834
       d. Alkenes and Alkynes   835
       e. Aromatic Hydrocarbons   838
22.2 The Chemistry of Functional Groups   840
       a. Alcohols and Ethers   840
       b. Aldehydes and Ketones   842
       c. Carboxylic Acids and Esters   843
       d. Redox Reactions Involving Functional Groups   846
22.3 Optical Stereoisomerism   848
22.4 Mechanisms of Organic Reactions   850
       a. Reactive Intermediates   850
       b. Nucleophilic Substitution in Alkyl Halides   851
       c. Alkene Addition Reactions   854

## Biochemistry

22.5 Carbohydrates   856
       a. Monosaccharides   856
       b. Polysaccharides   858
       c. Photosynthesis and Metabolism of Carbohydrates   859

22.6 Proteins and Nucleic Acids    861
    a. Amino Acids    861
    b. Protein Structure    864
    c. The Structure of DNA    867

**Conclusion    872**

# 23 Nuclear Chemistry    876

## Nuclear Properties

23.1 The Atomic Nucleus    876
    a. Composition    877
    b. Nuclear Radii    877
    c. Mass and Binding Energy    878
23.2 Nuclear Stability    880
    a. Binding Energy per Nucleon    880
    b. The Liquid Drop Model of the Nucleus    881
23.3 Beta Decay    884
    a. Relative Stabilities of Isobaric Nuclides    884
    b. $\beta^-$ Decay    886
    c. Positron Decay and Electron Capture Decay    888
    d. The Valley of Nuclear Stability    891
23.4 Alpha Decay    893
    a. Energetics of Alpha Decay    893
    b. The Effect of the Coulomb Barrier    894
    c. The Decay Series of the Naturally Occurring Heavy Element Isotopes    895

## Nuclear Reactions

23.5 Fundamentals of Nuclear Reactions    896
23.6 Types of Nuclear Reactions    898
    a. The (n, $\gamma$) Reaction    898
    b. Nuclear Fission    900
    c. Reactions Induced by Charged Particles    901
23.7 Artificially Produced Elements    903

## Applications of Nuclear Chemistry

23.8 Radioactive Decay    907
23.9 Age Determinations by Radioactivity Measurements    909
23.10 Radiocarbon Dating    910
23.11 Isotopic Tracers in Chemistry    910
23.12 Nuclear Magnetic Resonance Spectroscopy    912

**Conclusion    914**

# Appendices    918

## Appendix 1 Units and Conversions    918

## Appendix 2 Mathematical Operations    921

A2.1 Exponentials    921
A2.2 Logarithms    922

A2.3 The Quadratic Formula  923
A2.4 Mathematical Approximations  923
A2.5 Differentiation  924
A2.6 Integration  927

## Appendix 3 Equilibrium Constants  931

## Appendix 4 Thermodynamic Data Tables  935

## Appendix 5 Standard Reduction Potentials at 25°C  946

## Answers to Selected Problems  950

## Index  954

# 1
# The Basic Constituents of Matter

Chemistry is the science dealing with the composition, properties, and structure of substances, the reactions by which they are converted into other substances, and the energy changes that accompany these transformations. This broad definition suggests that chemistry occupies a central place among the sciences. A knowledge of chemistry is indeed very useful in a variety of fields. For example, chemistry plays a significant role in modern medicine, where it is employed in the synthesis of potent new drugs and the development of new materials for prosthetic devices and surgical procedures. Biochemistry—the chemistry of biological systems—provides the foundation for many of these applications.

A totally different example of the uses of chemistry is in its applications to art. Chemistry has been used in the preservation and restoration of old masterpieces, the detection of art forgeries, and the development of new materials for the creation of art works. The combination of chemistry and other sciences has given rise to many new specialties such as geochemistry, the study of the chemical aspects of geology, and radiochemistry, the applications of nuclear physics to chemistry.

While much of current chemical research involves such interdisciplinary fields, the traditional areas of chemistry continue to offer numerous opportunities for creative work. Organic chemistry—the chemistry of carbon—has long been a flourishing field. The use of computers in the development of novel approaches to the synthesis of complex new molecules is one interesting recent development. Inorganic chemistry—the chemistry of the elements other than carbon—has extended its scope in recent years. The incorporation of atoms of various elements into organic and biological molecules has led to the development of organometallic chemistry and bioinorganic chemistry, respectively. Physical chemistry deals with the structure and properties of matter and with the theories that attempt to explain the experimental observations. The use of lasers to probe the behavior of molecules has become an area of intense research activity in physical chemistry. Analytical chemistry deals with the identification and quantitative assay of substances. The use of computers and modern electronics is leading to increasingly sensitive methods of analysis.

## Atoms and Elements

One of the prominent features of the world we live in is the enormous variety of its constituent matter. Substances come in all sorts of textures and colors; some exist as solids, others as liquids, and still others as gases. Moreover, variables such as temperature and pressure affect the physical state and other properties of materials. In spite of this diversity, underlying regularities have been sought since antiquity. As long ago as the fifth century B.C. the Greek philosopher Empedocles postulated that all matter was composed in varying proportions of earth, fire, air, and water. This postulate was based on a generalization of observations of the combustion of wood, a common chemical process both then and now. Empedocles noted that burning wood emits smoke (i.e., air) and flames (i.e., fire) and is eventually consumed, resulting in the formation of ashes (i.e., earth). If a cold surface is held near the flames water vapor condenses, thereby accounting for the fourth constituent. These ideas gradually became dissociated from their origin in observations and were given a more mystical connotation by the medieval alchemists. Nonetheless, the idea that all matter is composed of a few basic constituents is one that has survived in one form or another throughout the ages.

Modern experimental techniques, which allow the study of the microscopic structure of matter, have revealed that there is indeed an underlying uniformity to the diversity revealed by our senses. All substances are made up of atoms, there being only about 100 different kinds of atoms. The determination of the constitution of matter is one of the great triumphs of modern science and one in which chemistry has played a central role.

### 1.1 The Elements

The **elements** are the basic substances that make up the vast assortment of known materials. Elements cannot be decomposed into other substances or converted into other elements by chemical means. Although elements can be transformed into each other by means of the techniques of nuclear physics (Section 23.5), such transformations do not involve chemical processes.

Over 100 elements are known at present. Their names and chemical symbols are listed on the inside of the back cover of this book. The symbols are one- or two-letter abbreviations of the names in English or Latin. The symbol for sodium, for example, is Na after the Latin *natrium*. Approximately 90 of these elements occur naturally on Earth and the rest have been created artificially. The abundance of the elements found on Earth varies widely. Table 1.1 lists the elements most abundant in the parts of the Earth whose composi-

**TABLE 1.1 Elemental Composition of the Outer Part of the Earth (Crust, Surface, and Atmosphere)**

| Element | Percentage of Total Mass |
|---|---|
| Oxygen | 49.20% |
| Silicon | 25.67 |
| Aluminum | 7.50 |
| Iron | 4.71 |
| Calcium | 3.39 |
| Sodium | 2.63 |
| Potassium | 2.40 |
| Magnesium | 1.93 |
| Hydrogen | 0.87 |
| Titanium | 0.58 |
| Chlorine | 0.19 |
| Phosphorus | 0.11 |
| Manganese | 0.09 |
| Carbon | 0.08 |
| Sulfur | 0.06 |
| Barium | 0.04 |
| Nitrogen | 0.03 |
| Fluorine | 0.03 |
| Strontium | 0.02 |
| All others | 0.47 |

tion is well known: the surface, the subsurface layer known as the crust, and the atmosphere. Only eleven elements account for about 99% of the total mass of these outermost regions, which in turn make up only about 2% of the mass of the planet. Figure 1.1 shows the most abundant elements found in different parts of the Earth. The inner core is thought to consist mostly of iron and nickel, and the mantle, which extends outward from the core, is believed to be made primarily of silicon and oxygen in the form of silicate minerals. Most of the elements are actually not found in the elemental state in nature; rather, they occur in chemical combination with other elements.

## 1.2 Atoms

Elements are made up of submicroscopic constituents called **atoms,** which are the smallest particles characteristic of a specific element. The concept of the atom has a long history. The term was introduced by the Greek philosopher Democritus nearly 2500 years ago. In seeking to explain the nearly endless variety of substances known to mankind he introduced as a unifying principle the idea that matter was composed of small indivisible particles called atoms. This idea remained untested until the beginning of the nineteenth century when John Dalton (1766–1844) proposed the notion that atoms had specific properties, such as mass. All atoms of a given element were postulated to have the same mass, different from that of the atoms of any other element. This postulate is subject to experimental verification and has been found, in essence, to be correct.

# 4  The Basic Constituents of Matter  1.3

**FIGURE 1.1** The elemental composition of the Earth. Only the most abundant elements are listed.

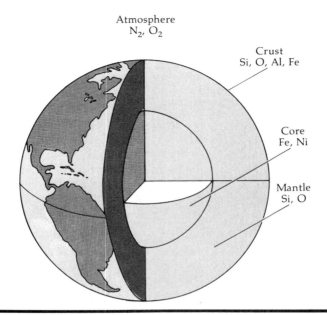

Let us examine some of the properties of atoms. As already noted, atoms are exceedingly small. Atomic radii, when expressed in meters, are in the range of $1-3 \times 10^{-10}$ m. In order to obtain a feeling for the size of atoms, consider that a barely visible grain of sand has a radius of $10^{-4}$ m, roughly a million times larger than an atom. To simplify the notation and eliminate powers of ten, atomic radii have traditionally been expressed in angstroms, where

$$1 \text{ Å} = 10^{-10} \text{ m}$$

(See Appendix 1 for a discussion of the units used in this book. In most instances we use the internationally adopted SI system of units. However, we retain the angstrom as the unit in which atomic and molecular distances will be expressed.)

The incredibly small size of atoms has, until recently, made it impossible to "see" an isolated atom even with the aid of the most sophisticated instruments. Although this fact has not precluded the universal acceptance of the reality of atoms, it is nonetheless gratifying that microscopy has finally reached the stage where individual atoms can be sufficiently magnified to render them visible. In 1976, Albert Crewe (b. 1927) and his co-workers at the University of Chicago developed a special type of electron microscope in which a beam of electrons was scattered off uranium atoms deposited on a thin graphite foil. The resulting image magnified the uranium atoms by a factor of over five million, which is just sufficient to render them visible. Figure 1.2 shows the results of this experiment.

## 1.3 The Structure of the Atom

Atoms are complex particles. Figure 1.3 shows a schematic diagram of the structure of two simple atoms. Over 99.9% of the mass of an atom is concentrated in its **nucleus**, which only occupies a minute fraction of the volume. Nu-

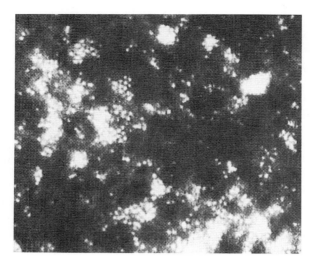

**FIGURE 1.2** Individual uranium atoms (and clusters of atoms) photographed with a scanning electron microscope. The blurred image is in part due to the thermal motion of the atoms. The carbon atoms of the graphite on which the uranium atoms are supported are too small to be resolved.

clear radii are over 10,000 times smaller than atomic radii and range from about $10^{-15}$ m to $10^{-14}$ m. Nuclei contain two kinds of particles of nearly equal mass: protons and neutrons. **Protons** are positively charged while **neutrons** are uncharged. High-energy physics tells us that protons and neutrons, too, are composite particles and are made up of quarks and gluons. However, the latter play no discernible role in the chemical behavior of atoms and need not concern us further.

The relatively vast region of the atom surrounding the nucleus contains **electrons.** These particles are nearly 2000 times lighter than protons and carry a

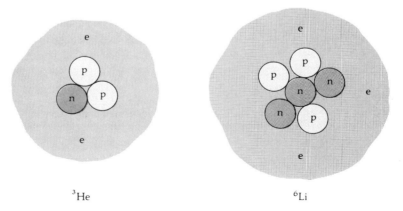

**FIGURE 1.3** Schematic diagram of the structure of the $^3$He and $^6$Li atoms. The diagram is *not* drawn to scale; the atomic nuclei are too large by many orders of magnitude.

negative charge that is equal in magnitude to that of protons. Table 1.2 summarizes the basic properties of these atomic constituents.

The number of protons in the nucleus of an atom is known as the **atomic number** and is usually designated Z. The atomic number determines the identity of an element. For example, carbon has $Z = 6$, and iron has $Z = 26$. The atomic numbers of the elements are listed in the table inside the back cover of this book. The number of electrons in an atom is equal to the number of protons and, thus, to the atomic number. The charge on a proton and the charge on an electron are equal in magnitude but opposite in sign, thus accounting for the observed electrical neutrality of ordinary matter. Although the chemical properties of an element are determined by the electrons in its atoms rather than by the protons, the atomic number is nonetheless defined by the proton number. The reason is that atoms can gain or lose electrons in chemical processes whereas the number of protons remains unchanged.

Atoms of a given element differing in the number of neutrons are called **isotopes.** The sum of the number of protons and neutrons in the nucleus of a particular isotope is equal to the **mass number** $A$. Since the neutron has mass but does not have charge, isotopes of an element differ in mass and mass number, but not in atomic number. The symbol of an isotope consists of the chemical symbol for the element in question preceded by the mass number written as a superscript, e.g., $^2H$, $^{12}C$, and $^{20}Ne$. The atomic number of the element is sometimes included as a subscript, as in $^{20}_{10}Ne$. (Since the chemical symbol and the atomic number are essentially synonymous, this inclusion is largely superfluous.) Most elements as they occur in nature consist of one or more stable isotopes; radioactive, or unstable, isotopes are also known for all the elements (Section 23.3). For example, hydrogen consists of the stable isotopes $^1H$ and $^2H$ (known as deuterium). The one known radioactive isotope of hydrogen is $^3H$ (tritium).

---

**Example 1.1** The isotope of uranium used as fuel in nuclear reactors has $A = 235$. Write the symbol for this isotope and indicate the number of protons, neutrons, and electrons in an atom.

The symbol is $^{235}U$. The atomic number of uranium is $Z = 92$. One atom of this uranium isotope therefore contains 92 protons, 92 electrons and $235 - 92 = 143$ neutrons. ∎

---

**TABLE 1.2 Mass and Charge of Atomic Constituents**

| Particle | Mass (kg) | Mass (u) | Charge[a] (C) |
|---|---|---|---|
| Proton | $1.67264 \times 10^{-27}$ | 1.00728 | $1.60219 \times 10^{-19}$ |
| Neutron | $1.67495 \times 10^{-27}$ | 1.00867 | 0 |
| Electron | $9.10953 \times 10^{-31}$ | $5.48580 \times 10^{-4}$ | $-1.60219 \times 10^{-19}$ |

[a] C is the symbol for coulomb, the SI unit of electrical charge. One coulomb is the quantity of electricity transported in 1 second by a current of 1 ampere.

## 1.4 Atomic Masses

**a. The Atomic Mass Scale**  The masses of individual atoms are exceedingly small, in the range of $10^{-27}$ to $10^{-25}$ kg. It is customary to express atomic masses in units in which such large negative powers of ten do not appear. These units are called atomic mass units, abbreviated u (formerly amu). The atomic mass scale is defined by setting the mass of the most abundant isotope of carbon, $^{12}$C, equal to exactly 12 atomic mass units:

Mass of $^{12}$C = 12 u

The atomic mass scale is a relative scale in which the masses of all other atoms are expressed relative to that of $^{12}$C. For example, the statement that the mass of $^{1}$H is 1.0078 u means that an $^{1}$H atom has a mass 1.0078/12 as large as that of a $^{12}$C atom.

The mass of an atom is slightly smaller than the sum of the masses of its constituent protons, neutrons, and electrons. The difference arises from the fact that when the individual particles combine to form an atom, some of their mass is converted to energy. (The relation between energy and mass is given by Einstein's famous formula $E = mc^2$, where $E$ is the energy corresponding to a mass $m$, and $c$ is the speed of light.)

---

**Example 1.2**  Calculate the mass lost when a $^{12}$C atom is assembled from protons, neutrons, and electrons.

A $^{12}$C atom contains 6 protons, 6 electrons, and 6 neutrons. Using the masses listed in Table 1.2 we obtain for the sum of the constituent masses

Protons:   $6 \times 1.00728$ u       = 6.04368 u

Neutrons: $6 \times 1.00867$ u       = 6.05202 u

Electrons: $6 \times 5.48580 \times 10^{-4}$ u = 0.00329148 u

Total mass of constituents:          12.09899 u

Since, by definition of the atomic mass scale, the mass of $^{12}$C is exactly 12 u, the mass loss is 0.09899 u, that is, nearly 1% of the total mass. (The Appendix at the end of this chapter should be consulted regarding the appropriate number of digits appearing in the final answer to this and other calculations.) ∎

Because the masses of the proton and neutron are both approximately 1 u, and also because the mass lost in assembling an atom is only about 1%, the masses of atoms are numerically very close to their mass numbers. We have, for example,

$^{19}$F    $A = 19$   $m = 18.99840$ u

$^{63}$Cu   $A = 63$   $m = 62.92959$ u

$^{107}$Ag  $A = 107$  $m = 106.90508$ u

$^{238}$U   $A = 238$  $m = 238.0508$ u

As implied by the large number of significant figures, the relative masses of isotopes can be accurately measured (Section 1.6).

***b. Atomic Weights of the Elements*** The atomic mass of an element, or its **atomic weight**, as it is more commonly known, is one of the important properties characterizing an element. The atomic weights of the elements are listed in the table inside the back cover of this book. A look at the tabulated values indicates that while, in analogy to individual isotopes, some elements have near-integral atomic weights, others do not. For example, the atomic weight of chlorine is 35.453 u and that of copper is 63.54 u, neither one of which is close to an integral value. The occurrence of more than one stable isotope of these elements accounts for their fractional atomic weights. Terrestrial copper, for instance, consists of a mixture of 69.09% $^{63}$Cu and 30.91% $^{65}$Cu. The atomic weight of copper is a weighted average of the atomic weights of the individual isotopes in which the isotopic abundances are the weighting factors:

$$\text{mass (Cu)} = [\text{fractional abundance } (^{63}\text{Cu}) \times \text{mass } (^{63}\text{Cu})]$$
$$+ [\text{fractional abundance } (^{65}\text{Cu}) \times \text{mass } (^{65}\text{Cu})]$$
$$= (0.6909 \times 62.926 \text{ u}) + (0.3091 \times 64.928 \text{ u})$$
$$= 63.54 \text{ u}$$

The experimentally determined atomic weights reflect this type of weighted average for all elements that have more than one naturally occurring isotope. Tin, for instance, has ten stable isotopes, the most of any element. Table 1.3 summarizes the isotopic data for this element. The averaging procedure used to relate the atomic weight of an element to those of the isotopes of which it is comprised can be summarized by the formula

$$\text{Atomic weight} = \sum_i (\text{Fractional Abundance}_i)(\text{Mass}_i) \tag{1.1}$$

(The symbol $\Sigma$ stands for summation and indicates that the quantities following it are to be summed over all $i$ values, where $i$ is an integer. In the case of

**TABLE 1.3 Abundances and Masses of the Stable Isotopes of Tin**[a]

| Isotope | Abundance (%) | Mass (u) |
|---|---|---|
| $^{112}$Sn | 0.96 | 111.90481 |
| $^{114}$Sn | 0.66 | 113.90276 |
| $^{115}$Sn | 0.35 | 114.90335 |
| $^{116}$Sn | 14.30 | 115.90174 |
| $^{117}$Sn | 7.61 | 116.90294 |
| $^{118}$Sn | 24.03 | 117.90160 |
| $^{119}$Sn | 8.58 | 118.90330 |
| $^{120}$Sn | 32.85 | 119.90219 |
| $^{122}$Sn | 4.72 | 121.90343 |
| $^{124}$Sn | 5.94 | 123.90526 |
| | 100.00% | |

[a] Atomic weight of Sn = 118.69 u.

tin, for example, $i$ takes on the values 1, 2, ..., 10, corresponding to each of the ten isotopes.)

The isotopic distribution in a terrestrial sample of an element is nearly constant (to within 0.01–0.1%) regardless of the source of the material; therefore, the atomic weights of the elements are considered to be constant on Earth. This lack of variation indicates that all the isotopes of a given element have behaved in the same way in response to the chemical and physical changes that have occurred since the formation of the Earth. The absence of significant separation and redistribution of isotopes is a reflection of the insensitivity of most processes to isotopic mass. The separation of isotopes requires rather elaborate procedures. It is therefore not surprising that the observation of large isotopic variations, or anomalies, is an indication of the occurrence of a fairly unusual natural phenomenon.

An interesting example of an isotopic anomaly was discovered in uranium extracted in 1972 from the Oklo mine in the African country of Gabon. Normal uranium consists primarily of $^{238}U$ but also contains 0.72% $^{235}U$, the isotope which undergoes nuclear fission. (In fission the nucleus of an atom splits into two nuclei of approximately half the mass with the liberation of a large amount of energy, Section 23.6). When uranium samples from Oklo were isotopically analyzed, it was discovered that they contained only about half the expected amount of $^{235}U$. Further investigation of Oklo samples revealed that isotopes formed as products of the fission of $^{235}U$ also exhibited isotopic anomalies, indicating that the missing $^{235}U$ had undergone fission. It appears that conditions at Oklo some 1–2 billion years ago favored the spontaneous occurrence of nuclear fission on a rather grand scale. While not all known isotopic anomalies have such a spectacular origin, they generally are the result of some unusual process.

*c. Mass and Weight; Density* The use of the term atomic weight for the mass of an atom is an example of the common tendency to use the terms *mass* and *weight* interchangeably. The two are actually different, though closely related, quantities. The mass of a body is a measure of the amount of matter in the body and is independent of its location. The weight of a body is the force exerted on it by the gravitational attraction of the Earth and is measured with a balance (Figure 1.4). According to Newton's second law of motion, this force is

**FIGURE 1.4** A modern single-pan analytical balance. The mass of a sample can be determined to within $\pm 10^{-4}$ g with such a balance.

given by the product of the mass of the object and the acceleration due to gravity, $g$,

$$w = mg \qquad (1.2)$$

The average value of $g$ at sea level is 9.807 m s$^{-2}$. Consequently, a given object has approximately the same weight anywhere on Earth. However, in outer space or on other celestial bodies, the value of $g$ can be considerably different than it is on Earth. An object of a certain mass will therefore have a different weight than it does on Earth. For example, an object that weighs 100 kg on Earth would weigh 38.1 kg on Mars and 253 kg on Jupiter.

The mass of a particular substance is not a characteristic property of that substance. A small amount of the substance has a smaller mass than a large amount. A property that depends on the amount of material present is called an **extensive property**. In addition to its mass, the volume of a substance is a familiar extensive property.

In contrast to extensive properties, an **intensive property** is independent of the amount of material present. The **density** of a substance, defined as the ratio of its mass to its volume,

$$\rho = m/V \qquad (1.3)$$

is a familiar intensive property. Table 1.4 lists the densities of several elements. The densities of solid or liquid substances are customarily expressed in g cm$^{-3}$

**TABLE 1.4  Densities of Some Elements at 0°C or 20°C and a Pressure of 1 Atmosphere**

| Solid Elements | Density at 20°C (g cm$^{-3}$) |
|---|---|
| Aluminum | 2.6989 |
| Calcium | 1.55 |
| Copper | 8.96 |
| Gold | 19.32 |
| Iron | 7.874 |
| Lead | 11.35 |
| Magnesium | 1.738 |
| Potassium | 0.862 |
| Sodium | 0.971 |

| Liquid Elements | Density at 20°C (g cm$^{-3}$) |
|---|---|
| Bromine | 3.12 |
| Mercury | 13.546 |

| Gaseous Elements | Density at 0°C (g L$^{-1}$) |
|---|---|
| Fluorine | 1.696 |
| Helium | 0.1785 |
| Hydrogen | 0.08988 |
| Nitrogen | 1.2506 |
| Oxygen | 1.429 |

(grams per cubic centimeter) while those of gases are given in g L$^{-1}$ (grams per liter). Intensive properties such as density generally depend on the values of certain experimental variables, such as temperature and pressure, and these must be specified, as is done in Table 1.4.

The atomic weight of an element is another example of an intensive property. While this might seem surprising since mass and weight are extensive properties, we must keep in mind that the atomic weight is the mass of a single atom, and so is independent of the number of atoms.

---

**Example 1.3** What is the volume of a piece of gold containing $1.00 \times 10^{23}$ atoms? Assume a temperature of 20°C and atmospheric pressure. Use the relation 1 u = $1.66 \times 10^{-24}$ g.

The atomic weight of gold is 196.97 u. Therefore, the mass of $1.00 \times 10^{23}$ atoms is

$$m = (1.00 \times 10^{23})(196.97 \text{ u})(1.66 \times 10^{-24} \text{ g u}^{-1})$$
$$= 32.70 \text{ g}$$

The density of gold at 20°C is 19.32 g cm$^{-3}$. The volume is

$$V = \frac{m}{\rho} = \frac{32.70 \text{ g}}{19.32 \text{ g cm}^{-3}}$$
$$= 1.69 \text{ cm}^3 \quad \blacksquare$$

---

## Molecules, Ions, and Compounds

### 1.5 Molecular Compounds

***a. Molecules*** A **molecule** may be defined as a group of atoms held together by specific forces which are strong enough so that the atoms act as a single unit. The atoms in a particular molecule may be atoms of the same element, e.g., H$_2$, O$_2$, S$_4$, or those of different elements, such as H$_2$O, or CH$_4$.

The forces between bound atoms are called **chemical bonds.** Much of modern chemistry has been concerned with the elucidation of chemical bonding (see Chapters 15–18). At this stage it suffices to say that two atoms are held together in a molecule by one or more **covalent bonds,** which are specific types of chemical bonds resulting from a pair of electrons shared between the atoms in question. The shared electrons are the outermost atomic electrons, known as the **valence electrons.**

Figure 1.5 shows several different ways of representing some simple molecules. To be specific, let us examine the entries for ammonia in order to understand the significance of the various representations. The first entry, NH$_3$, is the molecular formula. This formula merely indicates the number of atoms of each element present in a molecule of the compound. It says nothing about how the atoms are joined to each other.

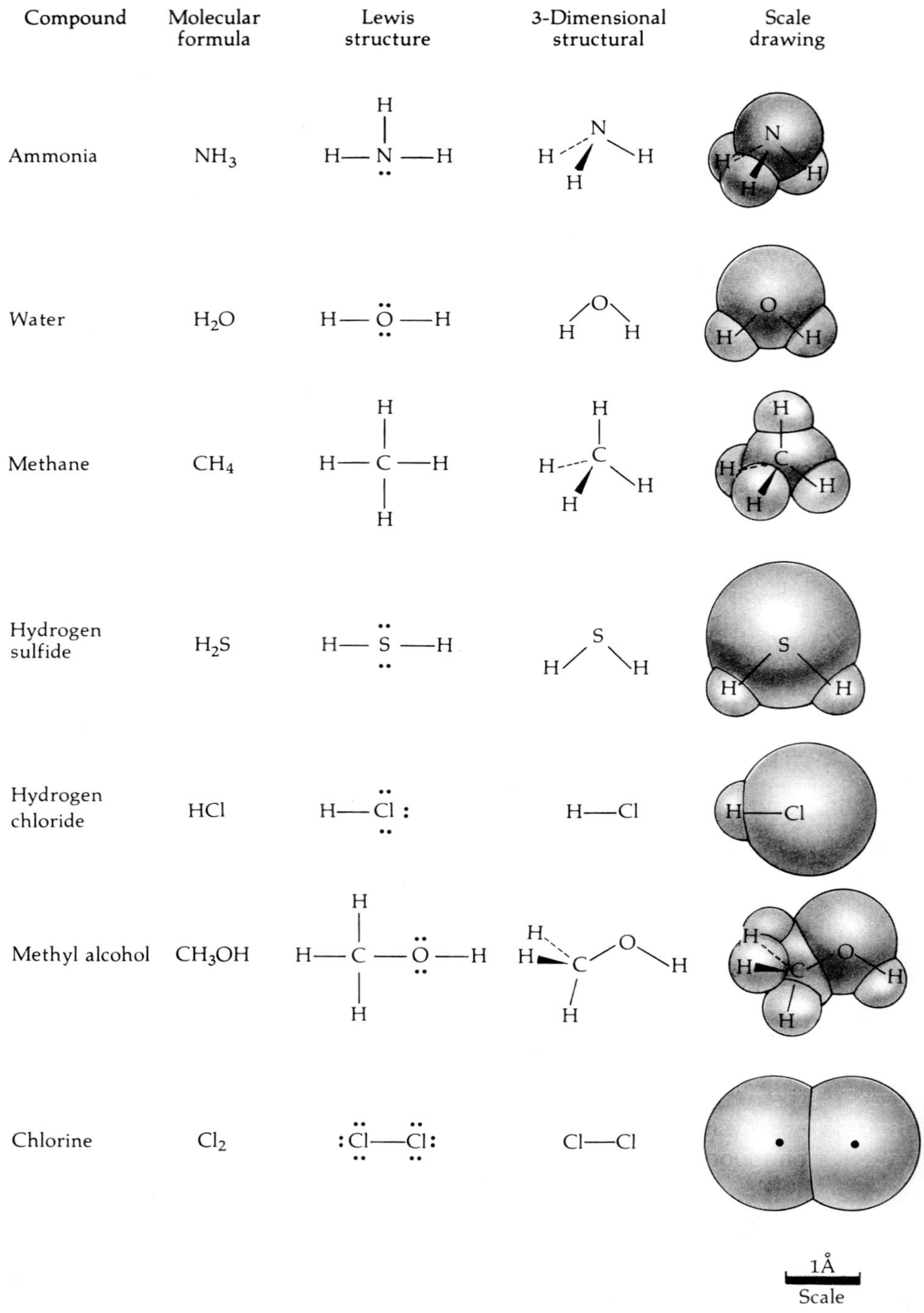

## 12  The Basic Constituents of Matter  1.5

| Compound | Molecular formula | Lewis structure | 3-Dimensional structural | Scale drawing |
|---|---|---|---|---|
| Ammonia | $NH_3$ | H—N(H)(H) with lone pair | tetrahedral N with 3 H | space-filling model |
| Water | $H_2O$ | H—Ö—H with two lone pairs | bent O with 2 H | space-filling model |
| Methane | $CH_4$ | H—C(H)(H)—H | tetrahedral C with 4 H | space-filling model |
| Hydrogen sulfide | $H_2S$ | H—S̈—H with two lone pairs | bent S with 2 H | space-filling model |
| Hydrogen chloride | HCl | H—Cl̈: with three lone pairs | H—Cl | space-filling model |
| Methyl alcohol | $CH_3OH$ | H—C(H)(H)—Ö—H | tetrahedral C with O—H | space-filling model |
| Chlorine | $Cl_2$ | :C̈l—C̈l: | Cl—Cl | space-filling model |

1Å Scale

**FIGURE 1.5**  Different representations of the molecules of some simple compounds.

The second entry, which can be written in several ways,

$$H \overset{..}{\underset{..}{N}} H \qquad H:\overset{..}{N}:H \qquad H—\overset{|}{\underset{|}{N}}—H \quad \text{and} \quad \overset{H}{\underset{H \quad H}{\diagdown N \diagdown}}$$

is known as a **Lewis formula.** Such formulas represent valence electrons as dots and show explicitly the way in which the various atoms are held together. The first structure distinguishes between the electrons originating in each atom by the use of different symbols. One can tell at a glance that the nitrogen atom contributes five valence electrons, two of which are unshared. The second structure dispenses with this distinction and the third, which is the one shown in Figure 1.5, represents a shared electron pair by a short line segment. The final representation takes note of the fact that covalent bonds are highly directional and make specific angles with respect to each other. This fact is conveyed, albeit imperfectly, by the two-dimensional projection of the three-dimensional structure.

The third entry in Figure 1.5 conveys more precise information about the three-dimensional structure of the molecule by showing the direction in which the various bonds point. The structure of ammonia is known to be pyramidal, with the nitrogen atom located at the apex of the pyramid. This description is conveyed by the structural formula:

$$H \overset{\text{-- N}}{\underset{H}{\diagup | \diagdown H}}$$

where the tapered bond is directed out of the plane of the page with the wide end pointing towards the observer, and the dashed line stands for an out-of-plane bond directed away from the observer and into the page. The final structure in the figure shows a drawing of the molecule in which the sizes of the atoms are shown to scale.

**b. Compounds and Mixtures** A **compound** is a substance containing two or more elements chemically combined with each other and therefore present in definite relative amounts. A molecular compound is made up of identical molecules just as an element is made up of identical atoms. The presence of identical constituents endows a compound with a distinctive set of properties. As is well known, water has a density of 1.000 g cm$^{-3}$ at 4°C, a melting point of 0.00°C and a boiling point of 100.00°C at a pressure of 1 atmosphere.

A compound may be contrasted with a **mixture,** which consists of two or more substances present in varying relative amounts and retaining their own identity. Since each component of a mixture has a distinctive set of **physical properties,** that is, properties that do not involve a chemical change, a mixture can be separated on the basis of differences in physical properties. Among the physical properties that can be used for this purpose are solubility, melting or boiling points, and magnetic properties. Figure 1.6 illustrates the separation of a mixture of iron and sulfur. By contrast, when this mixture is heated, the elements react to form the compound iron(II) sulfide, FeS. Such a compound has a fixed set of physical properties and, therefore, cannot be separated into its constituent elements by physical means. To effect such a separation, the

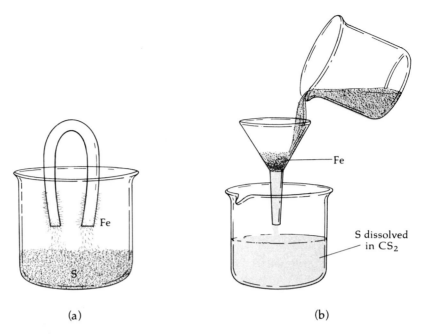

**FIGURE 1.6** The separation of a mixture of iron filings and powdered sulfur by two different methods. (a) Iron is magnetic and can be removed from the mixture with a magnet. Sulfur is nonmagnetic and is not affected by the magnet. (b) Sulfur is soluble in liquid carbon disulfide and iron is not. Iron can be separated from the solution by filtration and sulfur can be recovered by evaporation of the carbon disulfide.

---

**chemical properties** of the substance—the changes that occur as a result of chemical reaction—must be exploited.

## 1.6 Determination of Atomic and Molecular Weights

The relative atomic weights of many of the elements were determined during the nineteenth century. The determination of this fundamental property was stimulated by the atomic theory of Dalton. Experiments on the composition of compounds showed that the elements making up a compound are present in fixed proportions by mass (**law of definite proportions**). For example, hydrogen fluoride invariably consists of 95% fluorine and 5% hydrogen by mass. Dalton postulated that elements were made of atoms, that atoms of different elements differed in mass, and that compounds are formed by the combination of atoms of different elements. This theory readily accounted for the law of definite proportions. The composition of hydrogen fluoride, for instance, could be understood if its formula were HF and the atomic weights of fluorine and hydrogen stood in the ratio of 19 to 1.

An important corollary of Dalton's theory, the **law of multiple proportions,** states that if elements A and B combine to form more than one compound, e.g., AB and $AB_2$, the weights of B combining with a given weight of A are in the ratio of small integers. This law is a direct consequence of atomic theory; in this example, the ratio of the masses of element B in $AB_2$ and AB is just two.

The experimental verification of the law of multiple proportions provided strong support for the validity of the atomic theory. However, Dalton was unable to arrive at a consistent set of relative atomic weights on the basis of only weight composition data. The work of Amadeo Avogadro (1776–1856) and Stanislao Cannizaro (1826–1910) on the volumes of gaseous substances (Section 3.1) eventually led to the correct atomic weight scale.

The **molecular weight** of a compound, that is, the mass of a single molecule, is just the sum of the masses of its constituent atoms. The development of the atomic weight scale permitted the characterization of compounds by their molecular weights. For example, the molecular weight of methyl alcohol (methanol), $CH_3OH$, is obtained by combining the atomic weights of one carbon, one oxygen, and four hydrogen atoms:

$$C = 12.011 \text{ u}$$
$$O = 15.999 \text{ u}$$
$$4H = \underline{4 \times 1.008 \text{ u}}$$
$$CH_3OH = 32.042 \text{ u}$$

Modern values of atomic and molecular weights are obtained by **mass spectrometry.** The most common form of this technique is based on the response of a charged particle to the presence of electric and magnetic fields. An electric field accelerates a charged particle to a certain energy and a magnetic field deflects the moving particle in a way that depends on its mass (Section 5.1). The operation of a mass spectrometer is illustrated in Figure 1.7.

Atomic and molecular weights, as well as isotopic abundances can be de-

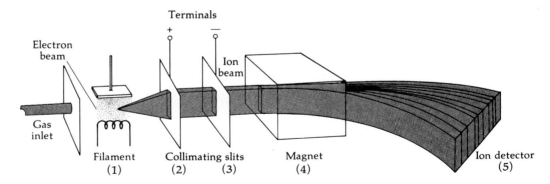

**FIGURE 1.7** Operation of a mass spectrometer. The mass measurement involves the following sequential steps: (1) ionization of the gaseous substance, e.g., by electron bombardment (see Section 1.7); (2) acceleration of the positively charged ions toward the negative terminal; (3) collimation of the ion beam by means of slits; (4) passage of the ion beam into a highly evacuated region where it is bent into a circular trajectory by a magnetic field; and (5) detection of the ions. Particles with the same ionic-charge-to-mass ratio ($ne/m$), where $n$ is a small integer, strike the ion detector at the same location. The trajectories of particles with large ($ne/m$) values are bent more than those of particles with small ($ne/m$) values. The spectrometer, therefore, disperses the ions at the detector in order of decreasing ($ne/m$) value from left to right.

termined to high accuracy with mass spectrometers. Figure 1.8 shows the distribution of magnesium isotopes obtained in 1920 by Arthur Dempster (1886–1950), one of the pioneers in mass spectrometry. The atomic weights of the three stable isotopes are determined from the exact position of the peaks and the isotopic abundances are obtained from the areas under the peaks. Individual isotopic masses can be measured to an accuracy of about eight significant figures. The mass of $^{24}$Mg, for example, is 23.985045 u. Since natural variations in the isotopic composition of the elements are normally in the range of 0.01% to 0.1%, the tabulated atomic weights of most elements are somewhat less accurate than those of their constituent isotopes.

## 1.7 Ionic Compounds

*a. Ions* **Ions** are atoms that have gained or lost electrons and have thereby become charged. The loss of electrons, known as **ionization,** results in the formation of positively charged ions which are known as **cations** (e.g., Na$^+$, Mg$^{2+}$, in which one or two electrons, respectively, have been removed). The gain of electrons leads to the formation of negatively charged ions called **anions** (e.g., Cl$^-$, S$^{2-}$). While atoms can be converted to ions by such diverse techniques as bombardment by electrons, exposure to light, and heating, the most familiar way is through chemical reactions (Chapter 2). For example, the

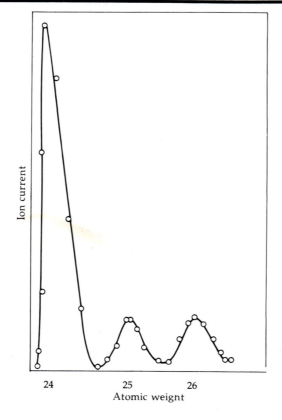

**FIGURE 1.8** Mass spectrum of natural magnesium, from an early (1920) measurement by Dempster. The resolution (i.e., separating power) of modern mass spectrometers is much higher and the mass peaks are correspondingly narrower.

reaction of metallic sodium with chlorine gas forms sodium chloride—the familiar table salt—which is an ionic compound consisting of $Na^+$ and $Cl^-$ ions.

Oppositely charged ions are held together in an ionic compound by **ionic bonds,** which result from the electrostatic attraction between particles of opposite charge, known as the coulombic force. The magnitude of the attractive force $F$ between a cation and an anion is proportional to the product of their charges $q_1$ and $q_2$ and inversely proportional to the square of the distance $r$ between them (**Coulomb's law**):

$$F = q_1 q_2 / \alpha r^2 \tag{1.4}$$

In the SI system, the proportionality constant $\alpha$ has the value

$$\alpha = 1.113 \times 10^{-10} \, C^2 \, N^{-1} \, m^{-2}$$

so that the force has units of newtons (N) if the charges are expressed in coulombs and the distance in meters. (See Appendix 1 for the definition of a newton).

When a number of oppositely charged ions are brought into proximity of each other, as in an ionic compound, a mutual electrostatic interaction of a magnitude determined by Coulomb's law arises among all the ions. Ionic compounds are consequently very stable substances as evidenced, for example, by their high melting points. Ionic compounds commonly exist in the form of crystals in which cations and anions, occupy fixed positions in a regular repeating array. Each cation is closely surrounded by a number of anions and vice versa. Figure 1.9 shows the ionic structure of a sodium chloride crystal, a structure displayed by a number of ionic compounds. Although it is customary to write the formula of this compound as NaCl, it is apparent that there is no one discrete molecule of this type in the crystal. Due to this difference from molecular compounds, the term **formula weight** is sometimes used to denote the mass of a formula unit.

**b. Nomenclature** Many of the substances of interest in inorganic chemistry are ionic compounds. Their names are based on those of their constituent ions according to an internationally accepted convention. Table 1.5 lists and names many of the simple ions formed by the elements. Cations are named by adding the word *ion* to the name of the element. When an element forms more than

**FIGURE 1.9** The structure of an NaCl crystal.

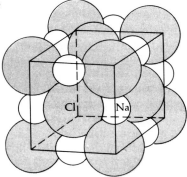

NaCl Crystals

**TABLE 1.5 Common Simple Ions of the Elements**

| | | Cations | | | |
|---|---|---|---|---|---|
| +1 | | +2 | | +3 | |
| Hydrogen | H$^+$ | Beryllium | Be$^{2+}$ | Aluminum | Al$^{3+}$ |
| Lithium | Li$^+$ | Magnesium | Mg$^{2+}$ | Gallium | Ga$^{3+}$ |
| Sodium | Na$^+$ | Calcium | Ca$^{2+}$ | Indium | In$^{3+}$ |
| Potassium | K$^+$ | Strontium | Sr$^{2+}$ | Thallium(III)$^c$ | Tl$^{3+}$ |
| Rubidium | Rb$^+$ | Barium | Ba$^{2+}$ | (Thallic) | |
| Cesium | Cs$^+$ | Manganese(II)$^a$ | Mn$^{2+}$ | Scandium | Sc$^{3+}$ |
| Copper(I)$^a$ | Cu$^+$ | (Manganous) | | Yttrium | Y$^{3+}$ |
| (Cuprous) | | Iron(II)$^a$ | Fe$^{2+}$ | Lanthanum | La$^{3+}$ |
| Silver | Ag$^+$ | (Ferrous) | | Chromium(III)$^c$ | Cr$^{3+}$ |
| Thallium(I)$^a$ | Tl$^+$ | Nickel | Ni$^{2+}$ | (Chromic) | |
| (Thallous) | | Copper(II)$^c$ | Cu$^{2+}$ | Iron(III)$^c$ | Fe$^{3+}$ |
| Mercury(I)$^b$ | Hg$_2^{2+}$ | (Cupric) | | (Ferric) | |
| (Mercurous) | | Zinc | Zn$^{2+}$ | Cobalt | Co$^{3+}$ |
| | | Cadmium | Cd$^{2+}$ | | |
| | | Mercury(II)$^c$ | Hg$^{2+}$ | | |
| | | (Mercuric) | | | |
| | | Tin(II)$^a$ | Sn$^{2+}$ | | |
| | | (Stannous) | | | |
| | | Lead | Pb$^{2+}$ | | |

| | | Anions | | | |
|---|---|---|---|---|---|
| −1 | | −2 | | −3 | |
| Fluoride | F$^-$ | Oxide | O$^{2-}$ | Nitride | N$^{3-}$ |
| Chloride | Cl$^-$ | Sulfide | S$^{2-}$ | Phosphide | P$^{3-}$ |
| Bromide | Br$^-$ | | | | |
| Iodide | I$^-$ | | | | |

$^a$ If an element forms more than one simple cation, the Roman numeral after the name gives the ionic charge; in a still widely used older nomenclature, the suffix -ous designates the ion having the lower charge.
$^b$ Mercury(I) exists as a dimer.
$^c$ The suffix -ic is used for the ion with higher charge.

one cation, the magnitude of the ionic charge is indicated by a Roman numeral following the name of the element, as in iron(II) and iron(III), for example. Anions are named by replacing the ending of the element name by -ide.

In addition to simple ions, many elements also form polyatomic ions, in which they combine with other species, often with oxygen to form **oxoanions.** Table 1.6 gives the names and formulas of some common polyatomic ions. The use of specific prefixes and suffixes to identify closely related ions, such as the various oxoanions of chlorine, is described in the table. Figure 1.10 shows the structures of a few of these ions. The nitrate ion, $NO_3^-$, for instance, has three oxygen atoms located at the corners of an equilateral triangle in whose center is a nitrogen atom, the overall charge of the ion being −1. Polyatomic ions retain their identity in many chemical reactions and form ionic compounds in the same way that simple ions do.

Ionic compounds are named by giving the name of the cation followed by that of the anion. The formula of a compound can be written by recognizing that electrical neutrality must be maintained. Thus, the number of cations and

## TABLE 1.6 Some Common Polyatomic Ions

*Memorize all*

| Cations | | | | Anions | | | | | |
|---|---|---|---|---|---|---|---|---|---|
| **+1** | | **+2** | | **−1** | | **−2** | | **−3** | |
| Ammonium | $NH_4^+$ | Vanadyl | $VO^{2+}$ | Hydroxide | $OH^-$ | Carbonate | $CO_3^{2-}$ | Phosphate | $PO_4^{3-}$ |
| Nitryl | $NO_2^+$ | Uranyl | $UO_2^{2+}$ | Nitrite[a] | $NO_2^-$ | Sulfite[a] | $SO_3^{2-}$ | Arsenite[a] | $AsO_3^{3-}$ |
| Nitrosyl | $NO^+$ | | | Nitrate[a] | $NO_3^-$ | Sulfate[a] | $SO_4^{2-}$ | Arsenate[a] | $AsO_4^{3-}$ |
| | | | | Hypochlorite[b] | $ClO^-$ | Thiosulfate | $S_2O_3^{2-}$ | Borate | $BO_3^{3-}$ |
| | | | | Chlorite[b] | $ClO_2^-$ | Chromate | $CrO_4^{2-}$ | | |
| | | | | Chlorate | $ClO_3^-$ | Dichromate | $Cr_2O_7^{2-}$ | | |
| | | | | Perchlorate[b] | $ClO_4^-$ | Oxalate | $C_2O_4^{2-}$ | | |
| | | | | Cyanide | $CN^-$ | Peroxide | $O_2^{2-}$ | | |
| | | | | Thiocyanate | $SCN^-$ | Hydrogen phosphate[c] | $HPO_4^{2-}$ | | |
| | | | | Acetate | $CH_3COO^-$ | | | | |
| | | | | Permanganate | $MnO_4^-$ | | | | |
| | | | | Hydrogen carbonate[c] | $HCO_3^-$ | | | | |
| | | | | Hydrogen sulfite[c] | $HSO_3^-$ | | | | |
| | | | | Hydrogen sulfate[c] | $HSO_4^-$ | | | | |
| | | | | Dihydrogen phosphate[c] | $H_2PO_4^-$ | | | | |

*need not know

[a] When an element forms two oxoanions, the ion with the smaller number of oxygens takes the suffix *-ite* and the ion with the larger number of oxygens the suffix *-ate*.

[b] When an element forms more than two oxoanions the prefixes *hypo-* and *per-* are used to designate the ions with the smallest and largest number of oxygen atoms, respectively.

[c] When two or more anions differ in the number of hydrogen atoms they contain they are distinguished by preceding the name of the anion by hydrogen (one H atom) or dihydrogen (two H atoms). The prefix *bi-* to denote the presence of hydrogen is part of an older but still commonly used nomenclature.

anions per formula unit is such that the total positive charge just equals the total negative charge.

**Example 1.4** Name the following compounds: (a) CuI, (b) CuCl$_2$, (c) NaHSO$_4$, (d) Sc$_2$O$_3$, (e) Fe(OH)$_3$, and (f) NH$_4$HSO$_3$.

The systematic names are given first, followed by the older names in parenthesis.

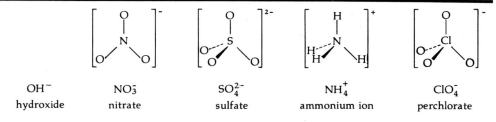

**FIGURE 1.10** The structure of some common polyatomic ions.

(a) CuI, copper(I) iodide (cuprous iodide)
(b) CuCl$_2$, copper(II) chloride (cupric chloride)
(c) NaHSO$_4$, sodium hydrogen sulfate (sodium bisulfate)
(d) Sc$_2$O$_3$, scandium oxide
(e) Fe(OH)$_3$, iron(III) hydroxide (ferric hydroxide)
(f) NH$_4$HSO$_3$, ammonium hydrogen sulfite (ammonium bisulfite) ∎

---

**Example 1.5** Write the formulas of the following compounds: (a) sodium hydrogen carbonate (sodium bicarbonate), (b) aluminum nitrate, (c) potassium hydrogen phosphate, (d) iron(III) chromate, and (e) copper(II) sulfate pentahydrate.

(a) The ions in sodium hydrogen carbonate are Na$^+$ and HCO$_3^-$. Since the charges are equal, the formula unit contains one ion of each type. The formula is NaHCO$_3$.

(b) The ions in aluminum nitrate are Al$^{3+}$ and NO$_3^-$. Three NO$_3^-$ ions are required to balance the charge of one Al$^{3+}$ ion. The formula is Al(NO$_3$)$_3$.

(c) The ions in potassium hydrogen phosphate are K$^+$ and HPO$_4^{2-}$. Two K$^+$ ions are needed for each HPO$_4^{2-}$ ion and the formula is K$_2$HPO$_4$.

(d) The ions in iron(III) chromate are Fe$^{3+}$ and CrO$_4^{2-}$. Charge balance requires that there be three CrO$_4^{2-}$ ions for each two Fe$^{3+}$ ions. The formula is Fe$_2$(CrO$_4$)$_3$.

(e) The ions in copper(II) sulfate are Cu$^{2+}$ and SO$_4^{2-}$. A pentahydrate contains five water molecules. The formula is CuSO$_4$·5H$_2$O. ∎

## 1.8 The Mole and Avogadro's Number

We have so far focused on single atoms and molecules and expressed their atomic and molecular weights in atomic mass units. However, chemists usually deal with a sufficiently large number of molecules that their collective mass can be determined by weighing and consequently is better expressed in grams or some multiple thereof. It is convenient to scale up from the molecular to the laboratory level and introduce a quantity called the mole, which always contains the same number of constituents, be they atoms, molecules, ions, or some other species. This procedure has the advantage that the relation between the masses of the constituents is preserved when we scale up to the macroscopic level. For example, we know from the atomic weights that a C atom weighs about six times more than an H$_2$ molecule. If we now consider, say 10$^{23}$ C atoms and an equal number of H$_2$ molecules, the former must still weigh six times more than the latter; however, the masses are now sufficiently large that they can be determined with a balance.

A **mole** of a substance is the amount containing Avogadro's number of constituents. **Avogadro's number** is equal to the number of atoms in exactly 12 grams of $^{12}$C. The numerical value of this important constant depends on the choice of 12 u for the mass of a $^{12}$C atom and on the amount of mass taken to be equal to one gram, and must be determined experimentally. The value of Avogadro's number is

$$N_A = 6.022045 \times 10^{23} \text{ mol}^{-1}$$

where mol is the customary abbreviation for mole.

This definition of the mole has an important implication for the relationship between the masses of macroscopic amounts of substances. We note that the mass of one $^{12}C$ atom expressed in atomic mass units (12 u) is numerically equal to the mass of a mole of $^{12}C$ atoms expressed in grams (12 g). Since a mole of any substance contains $N_A$ atoms or molecules, this same relationship between the mass of a single constituent expressed in atomic mass units and that of a mole expressed in grams must hold; that is, *the mass of one mole of any substance in grams is numerically equal to the atomic or molecular weight of its constituents in atomic mass units.* A few examples indicate the universal applicability of this most fundamental of concepts:

The molecular weight of nitrogen dioxide, $NO_2$, is 46.00 u. Therefore, 1 mol of $NO_2$ has a mass of 46.00 g and contains $6.022 \times 10^{23}$ $NO_2$ molecules.

The atomic weight of gold is 196.97 u. Therefore, 1 mol of gold has mass of 196.97 g and contains $6.0220 \times 10^{23}$ atoms of gold.

The atomic weight of the electron is $5.486 \times 10^{-4}$ u. Therefore, 1 mol of electrons has a mass of $5.486 \times 10^{-4}$ g = 0.5486 mg and consists of $6.022 \times 10^{23}$ electrons.

The mass of a golf ball expressed in atomic mass units is nearly $10^{26}$ u. Therefore, 1 mol of golf balls has a mass of close to $10^{26}$ g. This value is comparable to the mass of the Moon!

The preceding examples indicate that the mole concept can be reasonably applied to any entity whose mass is conveniently expressed in atomic mass units, e.g., atoms and molecules. However, the concept is not useful for macroscopic objects such as golf balls both because their mass in atomic mass units is so large and because the number encountered in any situation is many orders of magnitude smaller than Avogadro's number.

The molecular weight of a substance has the same numerical value in two different units: atomic mass units per molecule and grams per mole. Due to the possible confusion between the two units, the term **molar mass** has been introduced for the mass in grams of one mole, leaving molecular weight to stand for the mass of a single molecule. Since it is usually obvious from the context whether a mole or a single molecule is under consideration, we shall continue to use the term molecular weight in both contexts.

The numerical value of Avogadro's number yields a conversion factor between atomic mass units and grams. Consider a constituent with an atomic weight of $\overline{A}$ atomic mass units. Avogadro's number of these constituents weigh $\overline{A}$ grams. Consequently,

$$(6.022045 \times 10^{23})(\overline{A} \text{ u}) = \overline{A} \text{ g}$$

and cancelling the common factor $\overline{A}$,

$$6.022045 \times 10^{23} \text{ u} = 1 \text{ g}$$

$$1 \text{ u} = 1.660565 \times 10^{-24} \text{ g}$$

## 1.9 Conversion between Mass and Moles

One of the most fundamental relationships in chemistry is that between the formula of a substance and its composition by weight. The formula states the number of atoms of each element present in a molecule or, equivalently, the number of moles of each element present in one mole of the substance. On the other hand, the composition by weight is the information obtainable in the laboratory. For example, the formula of water, $H_2O$, indicates that one molecule is made up of two hydrogen atoms and one oxygen atom or that one mole of water contains two moles of hydrogen atoms and one mole of oxygen atoms. The experimental information indicates that water contains eight grams of oxygen for each gram of hydrogen and that the molecular weight is 18 g $mol^{-1}$. The two statements are actually equivalent: the first expresses the answer in terms of the basic constituents of matter, and the second expresses the answer in terms of the results of laboratory measurements. Knowing the one we can obtain the other, and vice versa. The procedure involves the conversion between mass expressed in the units in which it is actually measured (e.g., grams) and the number of moles. This conversion is one of the most basic chemical calculations and one that must be mastered by every student of the subject. The following examples illustrate various aspects of this procedure.

---

**Example 1.6** The density of liquid mercury at room temperature is 13.595 g $cm^{-3}$. A container with a capacity of 20.00 mL is filled with mercury. How many kilograms of mercury are in the container? How many moles? How many atoms of mercury are present?

The density of mercury can be used to obtain the mass from the volume:

$$\rho = m/V$$

Therefore,

$$m = \rho V = (13.595 \text{ g cm}^{-3})(20.00 \text{ mL})(1 \text{ cm}^3 \text{ mL}^{-1})(10^{-3} \text{ kg g}^{-1})$$
$$= 0.2719 \text{ kg}$$

Recognizing that the atomic weight of mercury (200.59) is equal to the number of grams in one mole, we obtain for the number of moles ($n$) of mercury

$$n = (0.2719 \text{ kg})(10^3 \text{ g kg}^{-1})\left(\frac{1}{200.59 \text{ g mol}^{-1}}\right) = 1.356 \text{ mol}$$

Finally, Avogadro's number is used to obtain the number of Hg atoms:

$$\text{Number of atoms} = (1.356 \text{ mol})(6.022 \times 10^{23} \text{ atoms mol}^{-1})$$
$$= 8.166 \times 10^{23} \text{ atoms} \quad \blacksquare$$

All three of the calculations in Example 1.6 have been performed by a method in which the original quantity is multiplied by a series of factors whose purpose is to convert it, one step at a time, into the desired quantity. Let us examine this approach more closely by focusing on the calculation of the number of moles given the mass in kilograms. The first step is to express the mass in grams by multiplication by the appropriate unit conversion factor, $10^3$ g $kg^{-1}$. Next, multiplication by the reciprocal of the atomic weight, expressed in grams per mole converts the mass to the desired number of moles. Note that each factor has been labelled with its units. When a given unit appears to the same power in both the numerator and denominator it is cancelled. In the calculation in question, kg cancels kg$^{-1}$ and g cancels g$^{-1}$, thus leaving the final answer in mol. Since this is the correct unit of the answer, the procedure serves as a check that the problem has been correctly formulated.

**Example 1.7** What is the elemental composition (in grams) of 15.0 g of sodium sulfate, Na$_2$SO$_4$?

The formula tells us that one mole of the compound contains 2 mol of Na, 1 mol of S, and 4 mol of O atoms. The molecular weight $\overline{M}$ of Na$_2$SO$_4$ is obtained from the atomic weights $\overline{A}$ of the constituent elements:

$$\overline{M}(Na_2SO_4) = \frac{(2 \text{ mol Na})(\overline{A} \text{ Na})}{(\text{mol Na}_2SO_4)} + \frac{(1 \text{ mol S})(\overline{A} \text{ S})}{(\text{mol Na}_2SO_4)} + \frac{(4 \text{ mol O})(\overline{A} \text{ O})}{(\text{mol Na}_2SO_4)}$$

$$= (2 \times 22.99 \text{ g mol}^{-1}) + (32.06 \text{ g mol}^{-1}) + (4 \times 16.00 \text{ g mol}^{-1})$$

$$= 142.04 \text{ g mol}^{-1}$$

The number of moles of Na$_2$SO$_4$ is

$$n = \frac{15.0 \text{ g}}{142.04 \text{ g mol}^{-1}} = 0.1056 \text{ mol}$$

The elemental composition in grams is obtained by reversing the preceding conversion from grams to moles:

$$m_{Na} = (0.1056 \text{ mol Na}_2SO_4)\left(\frac{2 \text{ mol Na}}{\text{mol Na}_2SO_4}\right)(22.99 \text{ g mol}^{-1}) = 4.86 \text{ g}$$

$$m_{S} = (0.1056 \text{ mol Na}_2SO_4)\left(\frac{1 \text{ mol S}}{\text{mol Na}_2SO_4}\right)(32.06 \text{ g mol}^{-1}) = 3.39 \text{ g}$$

$$m_{O} = (0.1056 \text{ mol Na}_2SO_4)\left(\frac{4 \text{ mol O}}{\text{mol Na}_2SO_4}\right)(16.00 \text{ g mol}^{-1}) = 6.76 \text{ g} \quad \blacksquare$$

The following example illustrates a more difficult type of problem involving a mixture of compounds.

**Example 1.8** A 100.0 g mixture of sodium chloride (NaCl), sodium sulfate (Na$_2$SO$_4$), and sodium nitrate (NaNO$_3$) is analyzed and found to contain 30.57 g of sodium and 42.89 g of oxygen. What is the composition of the mixture?

This problem involves three unknown masses: $m_{NaCl}$, $m_{Na_2SO_4}$, and $m_{NaNO_3}$. Let $x = m_{NaCl}$ and $y = m_{Na_2SO_4}$. Since the total mass is 100.0 g we can express the mass of sodium nitrate as

$$m_{NaNO_3} = 100.0 - x - y$$

We need two independent equations to solve for the unknowns $x$ and $y$. We are given the total mass of sodium and that of oxygen. The total mass of each of these elements can be expressed in terms of the unknown masses of the three compounds as follows:

$$m_{Na} = 30.57 \text{ g} = \left(\frac{22.99 \text{ g mol}^{-1} \text{ Na}}{58.44 \text{ g mol}^{-1} \text{ NaCl}}\right)(x \text{ g}) + \left(\frac{22.99 \text{ g mol}^{-1} \text{ Na}}{142.04 \text{ g mol}^{-1} \text{ Na}_2\text{SO}_4}\right)$$

$$\left(\frac{2 \text{ mol Na}}{\text{mol Na}_2\text{SO}_4}\right)(y \text{ g}) + \left(\frac{22.99 \text{ g mol}^{-1} \text{ Na}}{85.00 \text{ g mol}^{-1} \text{ NaNO}_3}\right)[(100.0 - x - y)\text{g}]$$

and

$$m_O = 42.89 \text{ g} = \left(\frac{16.00 \text{ g mol}^{-1} \text{ O}}{142.04 \text{ g mol}^{-1} \text{ Na}_2\text{SO}_4}\right)\left(\frac{4 \text{ mol O}}{\text{mol Na}_2\text{SO}_4}\right)(y \text{ g})$$

$$+ \left(\frac{16.00 \text{ g mol}^{-1} \text{ O}}{85.00 \text{ g mol}^{-1} \text{ NaNO}_3}\right)\left(\frac{3 \text{ mol O}}{\text{mol NaNO}_3}\right)[(100.0 - x - y)\text{g}]$$

The terms multiplying the unknown masses are the fractions of the total mass of each compound present as either sodium or oxygen, as obtained from the atomic weights of these elements and the molecular weights and formulas of the compounds. The equations reduce to

$$0.1229x + 0.0532y = 3.52$$

and

$$-0.5647x - 0.1141y = -13.58$$

respectively. Eliminating $y$ by multiplying the first equation by $0.1141/0.0532$ and adding the two equations, we obtain $x = 20.0$. Substitution of this value into either equation gives $y = 19.9$.

The mixture consists of 20.0g NaCl, 19.9g Na$_2$SO$_4$, and 60.1g NaNO$_3$. ■

The formula of a compound can be established by a procedure similar to that in Example 1.7. Chemical analysis yields the percent composition of a compound by weight. Knowledge of the atomic weights of the constituent elements enables us to determine the molar composition of the compound. From this information, the simplest formula consistent with the experimentally determined composition, commonly known as the **empirical formula,** can be found. The actual **molecular formula** of the compound—the formula which represents the actual number of atoms in each molecule—may be more complex than the empirical formula. However, the relative numbers of atoms of different elements must be the same in both formulas. For example, if the empirical formula of a compound is CH$_2$, the actual formula is some multiple in which the 2:1 hydrogen to carbon ratio is preserved, for instance, CH$_2$, C$_2$H$_4$, or C$_3$H$_6$.

---

**Example 1.9** THIP is a promising new analgesic compound. Its percent composition by weight is 51.4% C, 5.71% H, 20.0% N, and 22.9% O. Determine the empirical formula of this compound.

It is convenient to assume that we are dealing with a 100 g sample since the given weight percent values are then equal to the actual masses of the elements in the compound. Division of these masses by the appropriate atomic weights yields the relative number of moles present:

$$C \ \frac{51.4 \text{ g}}{12.0 \text{ g mol}^{-1}} = 4.28 \text{ mol} \qquad H \ \frac{5.71 \text{ g}}{1.01 \text{ g mol}^{-1}} = 5.66 \text{ mol}$$

$$N \ \frac{20.0 \text{ g}}{14.0 \text{ g mol}^{-1}} = 1.43 \text{ mol} \qquad O \ \frac{22.9 \text{ g}}{16.0 \text{ g mol}^{-1}} = 1.43 \text{ mol}$$

Since a molecule must contain an integral number of atoms the results must be converted to integers without, at the same time, changing the relative values. This procedure is most conveniently carried out by dividing the relative numbers of moles by the smallest value, 1.43:

$$C \ \frac{4.28}{1.43} = 3 \qquad H \ \frac{5.66}{1.43} = 4 \qquad N \ \frac{1.43}{1.43} = 1 \qquad O \ \frac{1.43}{1.43} = 1$$

The empirical formula is found to be $C_3H_4NO$. ■

In order to determine the actual formula from the empirical formula the molecular weight of the compound must be known. The molecular weight of THIP, for instance, turns out to be 140 u, which is twice the sum of the atomic weights of the atoms in the empirical formula. The molecular formula of this substance is therefore $C_6H_8N_2O_2$.

## Conclusion

In this chapter we have examined some of the properties of the basic constituents of matter—atoms, ions, and molecules. Atomic and molecular masses or weights, which are among the most fundamental of these properties, can be determined by mass spectrometry. Since the masses of the individual constituents are so small, they are customarily expressed in atomic mass units. To scale up from single atoms and molecules to macroscopic samples of elements and compounds, the mole, the fundamental unit used in dealing with such sizeable amounts of substances, has been introduced. The mole concept permits us to count molecules or ions by weighing the compounds that contain them; the conversion between mass and number of moles undoubtedly is the most frequently performed calculation in chemistry. Several common applications of this procedure, including the determination of empirical and molecular formulas and the composition of compounds have been examined.

# Appendix: Experimental Uncertainties and Significant Figures

## 1A.1 Uncertainties in Experimentally Determined Quantities

Experimentally measured quantities are subject to error; they cannot be determined with perfect accuracy. The validity of this statement is illustrated in Table 1A.1 which lists the currently accepted values of some of the important

**TABLE 1A.1 Experimental Uncertainties in the Values of Some Important Constants**

| Quantity | Numerical Value | Uncertainty | Units |
|---|---|---|---|
| Avogadro's number | $6.022045 \times 10^{23}$ | $0.000031 \times 10^{23}$ | $mol^{-1}$ |
| Electron charge | $1.6021892 \times 10^{-19}$ | $0.0000046 \times 10^{-19}$ | C |
| Electron mass | $9.109534 \times 10^{-31}$ | $0.000047 \times 10^{-31}$ | kg |
| Proton mass | $1.6726485 \times 10^{-27}$ | $0.0000086 \times 10^{-27}$ | kg |
| Neutron mass | $1.6749543 \times 10^{-27}$ | $0.0000086 \times 10^{-27}$ | kg |

constants introduced in this chapter along with their experimental uncertainties. In each case the last two digits of the numerical value have an uncertainty of the indicated magnitude. For example, Avogadro's number is known to within 31 parts in 6,022,045 corresponding to an uncertainty of approximately $5 \times 10^{-4}\%$. The other constants are known to a comparable degree of accuracy. While it is indeed remarkable that the important constants of nature have been determined with such high accuracy, perfect knowledge of their magnitude is unattainable.

The uncertainty in an experimentally determined quantity is composed of both **random** and **systematic** errors. Random errors are those that vary in an irreproducible manner from one measurement of a quantity to another. For example, if a particular object is weighed on a balance a number of times, the weights will, in general, be slightly different. The differences reflect the occurrence of random errors. These can be due to such factors as the limited precision of the balance, which makes it necessary to determine the last digit in the weight by interpolation between two marks on a scale, to the effect of fluctuations in such external factors as temperature and humidity, which affect the performance of the balance, or to other variable factors. Due to their unpredictability, random errors can be either positive or negative. One measurement might yield a larger value than the "true" weight while another might give a smaller value.

Systematic errors, by contrast, always have the same sign. The various weight determinations might, for instance, underestimate the true weight because of an incorrect calibration of the balance, a failure to correct for the buoyancy of air, or because of some other factor which remains constant during the experiment and affects the results.

Both random and systematic errors generally contribute to the experimental uncertainty in a measured quantity. The magnitude of the random error is readily determined: a number of determinations of the quantity of interest are made and the mean value is computed. The magnitude of the deviations between the individual determinations and the mean determines the random error. If the individual results cluster closely around the mean, the random error is smaller than if the spread in values is large. Statistical methods have been developed which permit a straightforward determination of the magnitude of the random error.

By contrast, systematic errors are difficult to determine. Since they always affect a particular measurement in the same way, repeated trials do not reveal the presence of a systematic error. Sometimes, a change in instruments can reveal the occurrence of systematic errors. For example an object can be weighed on a number of different balances. Neglecting the random errors, which can be evaluated in the preceding manner, the differences are a measure of the systematic errors. In general, however, systematic errors are rather more difficult to determine, and the scientific literature contains numerous examples of measurements which were later found to contain large unsuspected systematic errors.

Measurements having a small random error yield results of high **precision,** meaning that the results tend to be reproducible. This does not necessarily mean that the results have high **accuracy.** An accurate set of measurements of a quantity yields a mean value close to the true value and, thus, one having a small systematic error. Clearly, the most desirable situation is to obtain data which are both accurate and precise since the overall uncertainty will then be minimized. The distinction between accuracy and precision and that between the systematic and random errors which they reflect is graphically illustrated in Figure 1A.1.

## 1A.2 Propagation of Errors and Significant Figures

The determination of some property of interest frequently cannot be performed directly but is based, instead, on the measurement of some related quantities, whose values are algebraically combined to yield the desired result. The density of a substance, for example, is usually determined by measurement of its mass and volume, and the values are then combined according to the relation $\rho = m/V$. The uncertainty in the density is related to those in the mass and volume by means of standard statistical methods for the propagation of errors. While this subject is beyond the scope of this book, it is nonetheless necessary to understand how the precision of the measured quantities relates to that of the derived result.

It is helpful to examine this relation in terms of a concrete example. Suppose we wish to measure the density of a liquid in the manner just described. The volume of a sample is measured with a graduated cylinder as 37.3 ± 0.1

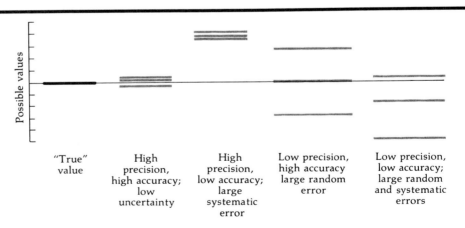

**FIGURE 1A.1** Distinction between accuracy and precision in experimental measurements.

mL. The mass is determined with an analytical balance as $36.115 \pm 0.003$ g. The density then is

$$\rho = \frac{36.115 \pm 0.003 \text{ g}}{37.3 \pm 0.1 \text{ mL}} = 0.96823 \pm 0.00260 \text{ g mL}^{-1}$$

where the error of $\pm 0.00260$ g mL$^{-1}$ has been found by statistical methods. The relative uncertainty in the result, that is, the uncertainty expressed as a percent of the actual value (0.269%), is essentially equal to the percent uncertainty in the volume; the mass is so much more accurately determined than the volume that the effect of its uncertainty is practically negligible.

We have given the same number of digits for the density as those in the mass. In view of the fact that the third digit after the decimal point is already uncertain ($8 \pm 2$), the succeeding digits are so uncertain that it is almost meaningless to include them. With practically no loss in accuracy, the density can be given as

$$\rho = 0.968 \pm 0.003 \text{ g mL}^{-1}$$

Note that the number of digits in this result is such that only the last digit is uncertain. When this prescription for writing a quantity and its associated uncertainty is followed, the digits constituting the result are termed **significant figures.** In the preceding illustration the density is known to three significant figures. The following additional examples further illustrate the concept of significant figures:

$1.230 \pm 0.002$—this number has four significant figures.

$0.00468 \pm 0.00003$—this number has three significant figures. The leading three zeros do not contribute to the precision of the number. This can be clearly seen by writing the number in exponential form, $4.68 \pm 0.03 \times 10^{-3}$.

$1250 \pm 25$—this number has three significant figures. The final zero is not significant since the preceding digit is already uncertain.

## 1A.3 Significant Figures in Numerical Problems

So long as the experimental uncertainties in the quantities used to obtain the value of some derived quantity are known, the latter can be written in terms of the correct number of significant figures. The precision of the result properly reflects that of the input data. Numerical problems, such as those found in this book, generally do not give the uncertainties associated with experimental data. A somewhat different procedure, therefore, must be used to determine the correct number of significant figures that the answer to the problem should have. We shall assume that all given data have the same numerical uncertainty in the last significant figure and that all preceding digits are known exactly. The actual magnitude of the uncertainty is immaterial; all that matters is that it be the same for all the given numbers. The following list of quantities can be understood to have the meaning given in the second column below where, to be specific, we assume that the uncertainty in the last significant figure is $\pm 1$:

1.45 g         means   $1.45 \pm 0.01$ g
0.003 mL       means   $0.003 \pm 0.001$ mL or $3 \pm 1 \times 10^{-3}$ mL

$1.60 \times 10^4$ m  means  $1.60 \pm 0.01 \times 10^4$ m
16,000 m  means  $16,000 \pm 1$ m, or $1.6000 \pm 0.0001 \times 10^4$ m

The following approximate rules can then be used to determine the appropriate number of significant figures in the answer to a problem, thereby ensuring that the implied uncertainty in the result is commensurate with that in the input data:

1. In addition or subtraction, the result has the same number of decimal places (i.e., digits to the right of the decimal point) as the number having the fewest decimal places. This rule ensures that the absolute uncertainty in the answer is comparable to that in the least precisely known quantity. For example, $6.5 + 0.34$ is expressed as 6.8.
2. In multiplication or division, the result has the same number of significant figures as those in the number having the fewest significant figures. The relative uncertainty in the result will then be comparable to that in the least precisely known input value. For example, $0.381 \times 1.5 = 0.57$.
3. When taking the logarithm of a number $L = \log A$, the number of decimal places in $L$ is equal to the number of significant figures in $A$. For example, $\log(3.5 \times 10^4) = 4.54$. Conversely, the number of significant figures in the antilogarithm of a number, $A = \text{antilog } L$ or $A = 10^L$, is equal to the number of decimal places in $L$.
4. Exact numbers, such as integers, are treated as if they had infinitely many significant figures.
5. The digits in the answer are reduced to the desired number by a procedure known as **rounding off.** The following steps are involved:
   a. Discard the unwanted digits.
   b. If the first of the discarded digits is less than 5, the last retained digit is left unchanged. For example, 1.5438 rounded off to three significant figures is 1.54.
   c. If the first of the discarded digits is equal to or greater than 5, the last retained digit is increased by one. For example, 0.54501 rounded off to two significant figures is 0.55 while 1.606 rounded off to three significant figures is 1.61. Most calculators automatically round off numbers according to these rules
6. In performing computations, it is wise to avoid rounding off until the final answer is obtained since premature rounding off can result in an erroneous result. Calculators automatically do not round off until the final answer is obtained.

The following examples illustrate the application of these rules.

**Example 1A.1** Different samples of a liquid, whose volumes are 20.1, 1.45, 0.317, and 11.3 mL are combined. What is the final volume?

Adding the volumes we have

20.1  mL
 1.45
 0.317
11.3
―――――
33.167 mL

The least precise entries in this list have one decimal place. The answer must therefore be rounded off to 33.2 mL. ■

**Example 1A.2** Determine the density of a liquid from the following data: The weight of a beaker containing 10.07 mL liquid is 29.98 g, and the weight of the empty beaker is 20.01 g.

$$\rho = \frac{m}{V} = \left(\frac{29.98 \text{ g} - 20.01 \text{ g}}{10.07 \text{ mL}}\right)\left(\frac{1 \text{ mL}}{\text{cm}^3}\right)$$

$$= \frac{9.97 \text{ g}}{10.07 \text{ cm}^3} = 0.99007 \text{ g cm}^{-3} = 0.990 \text{ g cm}^{-3}$$

Note that the result contains only three significant figures even though each input value is known to four figures. One digit is lost in subtraction because the two numbers are comparable in magnitude. ■

## Problems

1. How many protons, neutrons, and electrons does each of the following atoms or ions contain: $^{24}$Na, $^{36}$Cl, $^{39}$Ar, $^{53}$Fe$^{2+}$, $^{64}$Cu$^+$, $^{78}$Br$^-$?

2. The mass of $^{19}$F is 18.99840 u. How much mass is converted to energy when a $^{19}$F atom is assembled from its constituent protons, neutrons, and electrons? What is the mass decrease expressed in percent?

3. Nitrogen consists primarily of $^{14}$N (99.63%) with a small amount of $^{15}$N (0.37%) present. The masses of these isotopes are 14.0031 u and 15.0001 u, respectively. Calculate the atomic weight of nitrogen.

4. Iron has four stable isotopes whose masses and natural abundances are as follows:

   | Mass | Abundance |
   | --- | --- |
   | 53.9396 u | 5.8% |
   | 55.9499 u | 91.8% |
   | 56.9354 u | 2.1% |
   | 57.9333 u | 0.3% |

   Determine the number of protons and neutrons and write the nuclidic symbol for each isotope. Calculate the atomic weight of iron.

5. Chlorine consists of a mixture of $^{35}$Cl (34.97 u) and $^{37}$Cl (36.97 u). Given the atomic weight of chlorine listed on the inside back cover of this book, determine the natural abundance of the two isotopes of chlorine.

6. Prior to the adoption of $^{12}$C as the standard defining the atomic mass scale, two different atomic mass scales were in use: the physical scale ($^{16}$O = 16 u) and the chemical scale (natural oxygen = 16 u). Given that the masses (on the $^{12}$C scale) and abundances of the naturally occurring isotopes of oxygen are $^{16}$O, 15.99491 u, 99.759%; $^{17}$O, 16.99913 u, 0.0374%; $^{18}$O, 17.99916 u, 0.2039%; determine the relation between the three scales. Arrange the three mass scales in order of increasing value of the atomic mass unit.

7. A certain element has three naturally occurring isotopes whose atomic weights are $(A - 4)$, $A$, and $(A + 2)$. If the atomic weight of the element is just $A$, what are the relative abundances of the three isotopes expressed in terms of $x$, the abundance of the lightest isotope? For each isotope, determine the range of allowed percent abundances consistent with the given data.

8. The three oxides of manganese contain 63.2%, 69.6%, and 77.4% manganese by weight, respectively. Show that these values are consistent with the law of multiple proportions. Using the known atomic weight of oxygen, determine two possible values for the atomic weight of manganese. For each value, give the simplest formula for each of these oxides.

9. A certain metal forms two chlorides which contain 47.26% and 64.19% of the metal by weight. (a) Show that these compounds are consistent with the law of multiple proportions. (b) What are the simplest formulas of the compounds and what are

the corresponding atomic weight and identity of the metal? (c) What other chemically plausible formulas are possible and to what atomic weights and elements do they correspond?

10. Oxygen ($O_2$) and methanol ($CH_3OH$) both have molecular weights of approximately 32 g mol$^{-1}$. However, the difference between their molecular weights is sufficiently large to be determined by mass spectrometry. By how many parts per thousand do they differ in molecular weight? Which species is heavier?

11. Calculate the molecular weights of $CaCO_3$, $NH_4NO_3$, and $H_2SO_4$. Give the names of these compounds.

12. $Na^+$ and $Cl^-$ ions are brought to a distance of 3.00 Å of each other. What is the magnitude of the force between them?

13. Write the formulas of the following compounds: Copper(I) cyanide, barium perchlorate, manganese(II) nitrite, iron(III) hydrogen sulfate, scandium carbonate, aluminum sulfate, ammonium dichromate, mercury(II) phosphate, tin(II) hydrogen phosphate, potassium hypochlorite.

14. Give the names of the following compounds: $K_2Cr_2O_7$, $NaClO_2$, $Ba(ClO_4)_2$, $Ca(NO_2)_2$, $LiHCrO_4$, $Na_2S_2O_3$, $Sr_3(PO_4)_2$, $MgHPO_4$, $Cu(OH)_2$, $NH_4NO_3$.

15. How many moles are present in each of the following?
    (a) 200 g of KCl
    (b) 1.00 kg of $Al_2O_3$
    (c) $3.05 \times 10^{21}$ molecules of $CH_4$
    (d) 18.5 mL of liquid n-octane ($C_8H_{18}$) (density = 0.703 g cm$^{-3}$)

16. What is the mass of each of the following samples?
    (a) 2.50 mol of carbon monoxide
    (b) 1.75 mol of carbon dioxide
    (c) $4.70 \times 10^{22}$ molecules of nitrogen dioxide
    (d) 9.70 mmol of sulfur dioxide
    What fraction of the mass of each of these compounds is oxygen? What fraction of the atoms in each compound are oxygen atoms?

17. How many atoms are present in each of the following samples?
    (a) 0.386 mol of $H_2O$
    (b) 5.48 g of $SiO_2$
    (c) 15.3 mL of $C_2H_5OH$ (density = 0.789 g cm$^{-3}$)
    (d) a mixture of 2.45 g of $Fe_2O_3$ and 1.81 g of NiO

18. If Avogadro's number is defined as the number of atoms in 1 mol of $^{12}C$ and the old physical mass scale is used (mass of $^{16}O$ = 16 u), what is its numerical value? (See problem 6 for the mass of $^{16}O$.)

19. Sodium borohydride ($NaBH_4$) is a white solid with a density of 1.074 g cm$^{-3}$. Evaluate the following quantities for this compound: (a) volume occupied by 1.50 mol, (b) mass of 755 mmol, (c) mass containing $2.50 \times 10^{23}$ H atoms, and (d) mass containing as many hydrogen atoms as 50.0 g of water.

20. The compound adiponitrile ($C_6N_2H_8$) has a density of 0.9676 g cm$^{-3}$. Determine for this compound (a) the volume occupied by 2.75 mol, (b) the mass containing $4.50 \times 10^{23}$ C atoms, (c) the mass of carbon in 83.5 g, and (d) the mass of carbon in a sample containing as many N atoms as 25.0 g of ammonia.

21. Copper(II) nitrate hexahydrate ($Cu(NO_3)_2 \cdot 6H_2O$) is an ionic solid with a density of 2.074 g cm$^{-3}$. When heated to 26.4°C it is converted to the trihydrate, $Cu(NO_3)_2 \cdot 3H_2O$. Determine for the hexahydrate the (a) formula weight, (b) number of moles of copper in 83.5 g of the compound, (c) mass of copper in 100 g of the compound, and (d) number of water molecules given off when 15.0 mL of the compound is heated to 26.4°C.

22. A 14.31 g sample of hydrated magnesium sulfate ($MgSO_4 \cdot xH_2O$) is heated to 400°C to remove the water of crystallization and 6.99 g of anhydrous magnesium sulfate ($MgSO_4$) is left. What is the value of $x$?

23. When strontium bromide ($SrBr_2$) is heated in a stream of chlorine gas, it is completely converted to strontium chloride ($SrCl_2$). Assuming no losses, how many grams of $SrCl_2$ can be obtained from 14.3 g of $SrBr_2$?

24. As in problem 23, barium bromide can be converted to barium chloride by the action of chlorine gas; 2.50 g of $BaBr_2$ yields 1.75 g of $BaCl_2$. On the basis of these data and the atomic weights of bromine and chlorine, what is the atomic weight of barium?

25. An organic compound was found to contain 74.03% carbon, 8.70% hydrogen, and 17.27% nitrogen by weight. What is its empirical formula?

26. Carbohydrates are compounds containing carbon, hydrogen, and oxygen in which there are 2 hydrogen atoms for each oxygen atom. A certain carbohydrate contains 40.0% carbon by weight. What is its empirical formula? Knowing that the molecular weight is close to 180, what is the actual formula? What is the accurate molecular weight of the compound?

27. An oxide of phosphorus contains 43.6% phosphorus by weight, and its molecular weight is approximately 280. What is the formula of the oxide? Determine an accurate value of the molecular weight.

28. A compound containing carbon, hydrogen, and

chlorine consists of 37.2% carbon by weight; 5.00 g of the compound contain $2.33 \times 10^{23}$ H atoms. What is its empirical formula?

29. A certain compound has a molecular weight of approximately 80 and contains 43.5% of chlorine and 39.3% of oxygen by weight. A third element accounts for the remaining weight. Determine the identity of this element, the molecular formula, and an accurate value of the molecular weight of the compound. Note that more than one choice of a third element exists and the incorrect choices must be rejected.

30. What mass of each of the constituent elements is present in the following?
    (a) 10.0 g of $K_2Cr_2O_7$
    (b) 10.0 g of $H_2SO_4 \cdot 2H_2O$
    (c) 10.0 mL of $C_3H_5O_2Cl$ (density = 1.259 g cm$^{-3}$)

31. A certain element forms an oxide with the empirical formula $M_2O_3$. The oxide contains 68.4% of the element by weight. What are the atomic weight and identity of this element?

32. A certain element forms a chloride containing 44.7% chlorine by weight, which has the formula $MCl_n$, where $n$ is a positive integer. What are the possible atomic weights of this element?

33. A fertilizer consisting of a mixture of $KNO_3$ and $(NH_4)_3PO_4$ contains 62.0% of its total nitrogen in the nitrate and the remainder in the ammonium compound. What is the percent by weight of each compound in the fertilizer?

34. A mixture of $NaNO_3$ and $Na_2SO_4$ contains 30.05% sodium by weight. What is the percent by weight of each compound?

35. A mixture of $AgNO_3$, $Ag_2SO_4$, and $AgClO_4$ contains 60.05% silver and 27.05% oxygen by weight. What is the composition by weight of the mixture?

# 2
# Chemical Reactions

In this chapter we introduce a topic of central importance in chemistry: chemical reactions. Chemical reactions are involved in virtually all areas of chemistry since they constitute the means by which new compounds are synthesized and their chemical properties determined. In addition, chemical reactions have important practical applications. The chemical industry uses many different types of reactions to produce a variety of useful products such as pharmaceuticals, fertilizers, and plastics, to name a few. Chemical reactions have also shaped the geological development of the Earth, and their continuing occurrence in nature is manifested in such phenomena as the corrosion of metals, the fixation of atmospheric nitrogen by plants, and the destruction of forests by fire.

We focus here on the quantitative aspects of chemical reactions, by means of which the amounts of products obtainable from given amounts of reactants can be calculated. The fundamental importance of these stoichiometric calculations cannot be overemphasized.

## Chemical Equations and Stoichiometry

### 2.1 Balanced Chemical Equations and the Conservation of Mass

*a. The Conservation of Mass in Chemical Reactions* One of the first chemists to make quantitative measurements of the masses of reactants and products of chemical reactions was Antoine Lavoisier (1743–1794). Among the many reactions he studied were the decomposition of mercury(II) oxide at elevated temperatures and its reverse, the reaction of mercury with oxygen. On the basis of mass and volume measurements (Figure 2.1), Lavoisier concluded that the mass of the reactants was equal to that of the products of the reaction. Generalizing from these and similar observations, he proposed that mass must be conserved in chemical reactions. No exceptions to this generalization have been found and the law of mass conservation has become one of the most fundamental statements about chemical phenomena.

Lavoisier's work preceded the development of the atomic theory of matter. Once this theory was developed, the conservation of mass was seen to be equivalent to the conservation of the number of atoms of each element participating in a reaction. *Chemical reactions do not change the number of atoms of a particular element but only affect their state of combination with atoms of the same or other elements.* For example, the decomposition of mercury(II) oxide can be represented by the equation

$$2HgO(s) \rightarrow 2Hg(l) + O_2(g) \tag{2.1}$$

(It is customary to indicate the phase of each reactant and product. The designations s, l, and g stand for solid, liquid, and gas, respectively.) Equation 2.1 clearly shows that the reaction does not change the number of mercury and oxygen atoms present. Since the atomic weight of an element is independent of its chemical form, the mass does not change either. Thus, the equation is con-

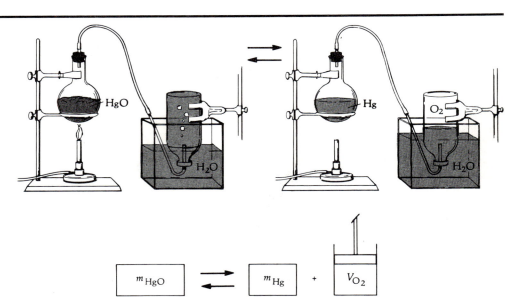

**FIGURE 2.1** Lavoisier's demonstration of the conservation of mass in the decomposition of mercury(II) oxide. The masses of mercury(II) oxide and the mercury formed from its decomposition by heat are measured; a decrease in mass is noted. The volume of oxygen given off in the reaction is measured. When the same mass of mercury is allowed to react with an equal volume of oxygen, the mass of the resulting mercury(II) oxide exceeds that of mercury by an amount equal to that lost in the decomposition reaction.

sistent with the experimental fact that 433.2 g of the oxide yields 401.2 g of mercury and 32.0 g of oxygen.

**b. Balanced Chemical Equations** A balanced chemical equation representing a reaction is one in which the number of atoms of each element participating in the reaction remains unchanged, and which therefore is consistent with the conservation of mass. (See Section 2.3c for an additional condition that must be obeyed by balanced equations involving ions.) Balancing a chemical equation is, in essence, a matter of bookkeeping. When the equation is simple, as in Equation 2.1, the process is nearly self-evident. We must resort to more complicated reactions in order to examine the balancing process in further detail.

---

Example 2.1  Solid dimethylhydrazine, $(CH_3)_2N_2H_2$, and liquefied nitrogen tetroxide, $N_2O_4$, together constitute a rocket fuel since the reaction between them forms gaseous carbon dioxide, nitrogen, and water vapor, which are propelled as exhaust gases. Write a balanced equation for this reaction.

We start by writing a skeleton equation in which we simply identify reactants and products:

$$(CH_3)_2N_2H_2 + N_2O_4 \rightarrow CO_2 + N_2 + H_2O$$

This equation is reasonably complicated since there are two reactants and three products, involving four different elements. It is helpful to see whether any of these elements originate exclusively in one compound and end up exclusively in another. The relation between the number of molecules of the compounds in question is then particularly easy to determine. Proceeding in this fashion we note that dimethylhydrazine is the sole source of carbon and hydrogen and that these elements exclusively form carbon dioxide and water, respectively. Since each molecule of $(CH_3)_2N_2H_2$ contains two C atoms and eight H atoms we obtain as a first step:

$$(CH_3)_2N_2H_2 + N_2O_4 \rightarrow 2CO_2 + N_2 + 4H_2O$$

The resulting eight atoms of oxygen must originate in $N_2O_4$ and we therefore need two molecules of this compound:

$$(CH_3)_2N_2H_2 + 2N_2O_4 \rightarrow 2CO_2 + N_2 + 4H_2O$$

The equation is now balanced with respect to every element except nitrogen. Since there are six N atoms on the left side there must be six on the right and the stoichiometric (i.e., numerical) coefficient of $N_2$ must be three. The balanced equation is

$$(CH_3)_2N_2H_2(s) + 2N_2O_4(l) \rightarrow 2CO_2(g) + 3N_2(g) + 4H_2O(g)$$

as can be verified by doing the bookkeeping on the number of atoms of each element present. ■

## 2.2 Stoichiometry of Chemical Reactions

*a. The Meaning of Chemical Equations and Stoichiometric Calculations* A balanced chemical equation implicitly contains a number of quantitative relations concerning the amounts of products obtainable from given amounts of reactants, known as the **stoichiometry** of the reaction. The reaction of nitrogen with oxygen to form nitric oxide, which occurs at high temperatures, may be used to explore the type of information conveyed by the chemical equation:

$$N_2(g) + O_2(g) \rightarrow 2NO(g) \tag{2.2}$$

On the most fundamental level, the equation indicates that 1 molecule of $N_2$ reacts with 1 molecule of $O_2$ to form 2 molecules of NO (Figure 2.2a). Since we do not deal with individual molecules in the laboratory, we can scale up from molecules to moles and state that 1 mol of $N_2$ reacting with 1 mol of $O_2$ yields 2 mol of NO (Figure 2.2b). This statement, while perfectly correct, is unnecessarily restrictive. The most general meaning of Equation 2.2 is that 2 mol of NO are formed for each mole of $N_2$ or $O_2$ that react (Figure 2.2c). The actual amounts of material undergoing reaction need not be fixed; only the relative amounts are determined by the reaction stoichiometry. Since the stoichiometric coefficients in a balanced equation determine the ratio of moles of products and reactants, Equation 2.2 could be written equally well as

$$\tfrac{1}{2}N_2(g) + \tfrac{1}{2}O_2(g) \rightarrow NO(g)$$

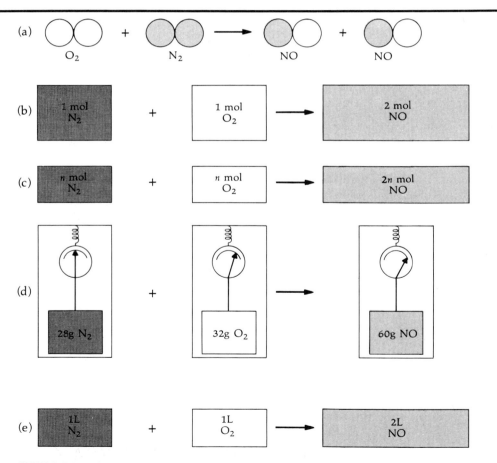

**FIGURE 2.2** The various meanings of a balanced chemical equation.

Note, however, that this equation makes no sense if we assume that it applies to individual molecules since there is no such entity as half a molecule. For this reason equations with fractional stoichiometric coefficients are sometimes avoided.

Since a given compound is characterized by its molecular weight, a chemical equation implies a definite mass relationship between reactants and products. This relationship follows directly from the numerical equality of the mass of a mole (in units of g mol$^{-1}$) and that of a molecule (in units of u) considered in Section 1.8. Taking note of the fact that the molecular weights of $N_2$, $O_2$, and NO are 28 u, 32 u, and 30 u, respectively, the balanced equation implies that 28 g of $N_2$ will react with 32 g of $O_2$ to form $2 \times 30 = 60$ g of NO (Figure 2.2d) or, more generally, that $\frac{60}{28}$ and $\frac{60}{32}$ grams of NO will be obtained, respectively, for each gram of $N_2$ or $O_2$ that reacts.

## Example 2.2

The reaction of nitrogen dioxide with water

$$3NO_2(g) + H_2O(l) \rightarrow NO(g) + 2HNO_3(l)$$

is one of the steps in the production of nitric acid. What mass of nitric acid is produced when 350 g of $NO_2$ reacts?

The fundamental point that must be kept in mind in solving any stoichiometry problem is that the stoichiometric coefficients in the balanced equation relate the number of *moles* of reactants and products involved in the reaction. Since a mole of one substance generally has a different mass than a mole of another, the coefficients do *not* relate the relative number of grams. For example, in this equation it would be incorrect to say that 2 g of $HNO_3$ are formed for every 3 g of $NO_2$ that react; the correct statement is that 2 moles of $HNO_3$ are formed for every 3 moles of $NO_2$ that react. In order to relate the masses of the two species we must therefore proceed via the number of moles, using the appropriate molecular weights to make the conversion from mass to moles and vice versa.

The molecular weights are $\overline{M}_{NO_2} = 46.01$ g mol$^{-1}$ and $\overline{M}_{HNO_3} = 63.02$ g mol$^{-1}$. The mass of $HNO_3$ is

$$m_{HNO_3} = \left(\frac{350 \text{ g NO}_2}{46.01 \text{ g NO}_2 \text{ mol}^{-1}}\right)\left(\frac{2 \text{ mol HNO}_3}{3 \text{ mol NO}_2}\right)(63.02 \text{ g HNO}_3 \text{ mol}^{-1})$$

$$= 320 \text{ g}$$

Note that the sequence of steps involves the following successive conversions:

$$\text{mass (NO}_2) \rightarrow \text{moles (NO}_2) \rightarrow \text{moles (HNO}_3) \rightarrow \text{mass (HNO}_3)$$

The conversion of the number of moles of $NO_2$ to the number of moles of $HNO_3$ is a key step in this sequence and one that, if forgotten, will lead to an erroneous answer. As shown in the equation for $m_{HNO_3}$, this conversion involves the ratio of the appropriate stoichiometric coefficients. ∎

The following problem involving the reaction of a mixture of compounds illustrates a more complicated stoichiometric calculation.

**Example 2.3** A mixture of magnesium and calcium carbonates weighing 4.15 g decomposes upon heating to 2.04 g of the corresponding oxides. What mass of magnesium carbonate was present in the mixture?

This problem is complicated by the fact that it involves the simultaneous occurrence of two reactions (Δ is the symbol for heat):

$$MgCO_3(s) \xrightarrow{\Delta} MgO(s) + CO_2(g)$$
$$CaCO_3(s) \xrightarrow{\Delta} CaO(s) + CO_2(g)$$

The following two mass relations can be written on the basis of the given data:

$$m_{MgCO_3} = y \text{ g}, \quad m_{CaCO_3} = (4.15 - y) \text{ g}$$

where $y$ is the unknown we want to determine, and

$$m_{MgO} + m_{CaO} = 2.04 \text{ g}$$

The second relation suggests an approach to the solution of the problem: Express the masses of the oxides in terms of those of the corresponding carbonates and solve the resulting equation for the unknown $y$. The procedure developed in Example 2.2 may be adapted for this purpose. For example, the mass of MgO can be expressed in terms of that of $MgCO_3$ by the relation

$$m_{MgO} = \left(\frac{(m_{MgCO_3})}{(\overline{M}_{MgCO_3})}\right)\left(\frac{(n_{MgO})}{(n_{MgCO_3})}\right)(\overline{M}_{MgO})$$

A similar expression can be written for the mass of CaO. Adding the two expressions and inserting the appropriate numerical values we obtain

$$\frac{(y \text{ g } MgCO_3)}{(84.31 \text{ g mol}^{-1} MgCO_3)}\left(\frac{1 \text{ mol } MgO}{1 \text{ mol } MgCO_3}\right)(40.31 \text{ g mol}^{-1} MgO)$$
$$+ \frac{[(4.15 - y)\text{g } CaCO_3]}{(100.08 \text{ g mol}^{-1} CaCO_3)}\left(\frac{1 \text{ mol } CaO}{1 \text{ mol } CaCO_3}\right)(56.08 \text{ g mol}^{-1} CaO) = 2.04 \text{ g}$$

Solving for $y$ we obtain

$$y = 3.5 \text{ g}$$

indicating that the mixture contained 3.5 g of $MgCO_3$. (Note that one significant digit is lost in the subtraction step.) ∎

As illustrated in Examples 2.2 and 2.3, stoichiometric calculations can be used to infer the mass of a product from that of a reactant undergoing a chemical reaction. When a chemical reaction involves gaseous substances, as in Equation 2.2, similar information about their volumes can be inferred from the equation. While stoichiometry involving gases is examined in Section 3.1, it is worth pointing out here that the volume of a gas is proportional to the number of moles of gas, provided the temperature and pressure remain constant. Thus, Equation 2.2 implies that 2 L of NO are formed for each liter of $N_2$ or $O_2$ that undergo reaction (Figure 2.2e).

**b. The Limiting Reagent and Incomplete Reactions** When two substances undergo a chemical reaction they are usually not present in stoichiometric amounts, that is, in the amounts consistent with the balanced equation. Instead, one of them may be present in excess. When the reaction is over, some of this reagent remains unreacted. The reagent present in insufficient amount to react with the entire amount of the other reagent is called the **limiting reagent**. The limiting reagent determines the maximum yield of a reaction product that is obtainable.

Limiting reagents can be used to optimize certain reactions. If, in a particular reaction, one of the reacting species is more valuable than the other, it is sensible to use a limited amount of the expensive reagent and an excess of the inexpensive one. For example, a given amount of silver bromide, a component of photographic film, can be made least expensively in the reaction of silver nitrate and sodium bromide solutions when the more expensive silver nitrate is used as the limiting reagent.

In certain instances the outcome of a chemical reaction can be determined by which of two reagents is present in limited amount. In the reaction of carbon with oxygen, for example, the product is carbon dioxide ($CO_2$) when oxygen is present in excess whereas carbon monoxide (CO) is formed when oxygen is the limiting reagent.

---

**Example 2.4** What mass of carbon monoxide is formed when 30.0 g of carbon reacts with 30.0 g of oxygen?

The number of moles of each reactant is evaluated first:

$$n_C = \frac{30.0 \text{ g}}{12.0 \text{ g mol}^{-1}} = 2.50 \text{ mol}$$

$$n_{O_2} = \frac{30.0 \text{ g}}{32.0 \text{ g mol}^{-1}} = 0.938 \text{ mol}$$

The balanced equation is

$$2C(s) + O_2(g) \rightarrow 2CO(g)$$

indicating that 2 mol of carbon react for each mole of oxygen. Therefore, 1.876 mol of carbon will react with all the oxygen present. Since there are 2.50 mol of carbon, oxygen is the limiting reagent. The mass of CO is obtained from the number of moles of oxygen:

$$m_{CO} = (0.938 \text{ mol } O_2) \frac{(2 \text{ mol CO})}{(\text{mol } O_2)} (28.0 \text{ g mol}^{-1}) = 52.5 \text{ g} \quad \blacksquare$$

While the limiting reagent determines the *maximum* yield of a reaction product, the *actual* yield may be significantly lower. The reasons for this are varied. Many reactions, particularly those between organic reagents, do not go to completion and a fraction of the reacting species do not react. In some instances, the separation of a particular product from a mixture of products is difficult and some of the substance may be lost in the separation process. In yet other instances, a reaction might be quenched before completion in order to avoid possible additional reactions between the products. The actual yield of a reaction product is frequently expressed as a percentage of the maximum possible yield.

**Example 2.5**  When 10.0 g of hydrogen is heated with 10.0 g of iodine, 9.05 g of hydrogen iodide is obtained. What is the percentage yield of hydrogen iodide?

As in Example 2.4, we first obtain the number of moles of each reactant:

$$n_{H_2} = \frac{10.0 \text{ g}}{2.02 \text{ g mol}^{-1}} = 4.95 \text{ mol}$$

$$n_{I_2} = \frac{10.0 \text{ g}}{254 \text{ g mol}^{-1}} = 0.0394 \text{ mol}$$

The balanced equation is

$$H_2(g) + I_2(s) \rightarrow 2HI(g)$$

Since $H_2$ and $I_2$ react in equimolar amounts and there are fewer moles of $I_2$, iodine is the limiting reagent. Complete conversion of $I_2$ to HI would produce 0.0788 mol HI corresponding to a mass of

$$m_{HI} = (0.0788 \text{ mol})(128 \text{ g mol}^{-1}) = 10.1 \text{ g}$$

Since the actual mass of HI obtained in the reaction is 9.05 g, the percentage yield is

$$\frac{9.05 \text{ g}}{10.1 \text{ g}} \times 100 = 89.6\% \quad \blacksquare$$

## 2.3 Chemical Reactions in Aqueous Solution and the Conservation of Charge

**a. Ions in Solution**  A **solution** is a homogenous mixture consisting of a **solvent,** the substance present in highest concentration (i.e., amount in a given volume) and one or more **solutes,** the substances present in lesser concentration. Solutions are of great importance in chemistry because they are an excellent medium for the occurrence of chemical reactions. **Aqueous solutions,** in which the solvent is water, are particularly common. We shall deal here with aqueous solutions in which the solute is present in the form of ions.

Figure 2.3 depicts the changes that occur in the dissolution of a crystalline ionic compound. The ions in the solid are held rigidly in place by their mutual electrostatic interaction (Section 20.10b). The strength of this interaction is evident from the high melting points of most ionic compounds. Nonetheless, many of these compounds are readily soluble in water. As discussed in detail in Section 18.6e, water is an excellent solvent for ionic compounds because it is a polar substance. If two atoms held together by a covalent bond differ in their attraction for electrons, the bond between them is said to be **polar.** The atom with a greater tendency to attract electrons acquires a slight negative charge while the other atom acquires a slight positive charge. Water is polar because oxygen exerts a stronger attraction for electrons than hydrogen (Figure 2.3a). [The tendency of an atom to attract electrons from the atoms to which it is bonded is called its **electronegativity.** See Section 6.7.]

When an ionic compound is dissolved in water, the positively charged cations attract the slightly negative oxygen atoms of water while the negatively charged anions attract the hydrogens (Figure 2.3b). The net effect is that the ions in solution become **hydrated,** that is, closely surrounded by electrostati-

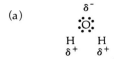

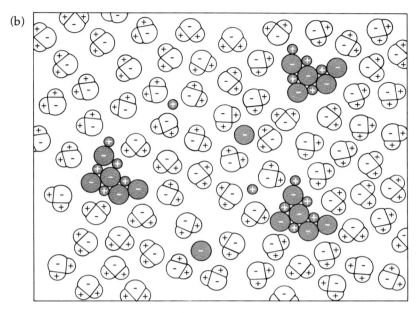

**FIGURE 2.3** The dissolution of an ionic compound in water. (a) Water is polar because the oxygen atom exerts a greater attraction for the bonding electrons than the hydrogen atoms. The oxygen atom therefore acquires a slight negative charge, indicated by the symbol $\delta^-$, while the hydrogens acquire a slight positive charge $\delta^+$. (b) When an ionic compound is dissolved in water, the oppositely charged ions surrounding a given ion are replaced by water molecules oriented so that the oxygen ends are closest to the cations while the hydrogen ends are closest to the anions. The ions are said to be hydrated.

cally attracted water molecules. If the interaction between the ions and the water molecules is sufficiently strong to overcome the mutual attraction between the oppositely charged ions in the crystal, the ionic compound will be soluble. Sodium chloride is a familiar example. It is customary to indicate that a chemical reaction involves hydrated ions by writing the symbol (aq) after each ion in the equation for the reaction, e.g., Na$^+$(aq) or Cl$^-$(aq).

Three types of substances dissolve in water to form ions: acids, bases, and salts. An **acid** can be defined as a substance that increases the concentration of hydrogen ions when added to water; a **base** is a substance that increases the concentration of hydroxide ions when added to water. (More complete definitions of acids and bases are given in Sections 8.1–8.3.) Acids and bases are among the most widely used chemical reagents. Table 2.1 lists some common acids and bases. The list is subdivided into *strong* acids and bases, which are completely ionized in dilute aqueous solutions, and *weak* acids and

**TABLE 2.1 Some Acids, Bases, and Salts**

| Acids | | Bases | | Salts | |
|---|---|---|---|---|---|
| **Strong Acids** | | **Strong Bases** | | | |
| Perchloric | HClO$_4$ | Lithium hydroxide | LiOH | Lithium perchlorate | LiClO$_4$ |
| Nitric | HNO$_3$ | Sodium hydroxide | NaOH | Sodium nitrate | NaNO$_3$ |
| Sulfuric | H$_2$SO$_4$ | Potassium hydroxide | KOH | Potassium chloride | KCl |
| Hydrochloric | HCl | | | Ammonium bromide | NH$_4$Br |
| Hydrobromic | HBr | | | Calcium fluoride | CaF$_2$ |
| | | | | Sodium acetate | NaOAc |
| **Weak Acids** | | **Weak Bases** | | | |
| Sulfurous | H$_2$SO$_3$ | Ammonia | NH$_3$ | | |
| Nitrous | HNO$_2$ | | | | |
| Hydrofluoric | HF | | | | |
| Formic | HCOOH | | | | |
| Acetic | CH$_3$COOH (HOAc) | | | | |
| Phosphoric | H$_3$PO$_4$ | | | | |

bases, which are only slightly ionized in dilute aqueous solution, and are present mostly in molecular form.

A **salt** is an ionic compound whose cation is not H$^+$ and whose anion is not OH$^-$. As we shall see, salts are formed in the reaction of an acid and a base, the other product being water. Most ionic compounds are salts; Table 2.1 lists some familiar examples.

The presence of ions in aqueous solutions can be inferred from the observation that solutions of acids, bases, or salts conduct electricity. An electric current consists of charged particles in motion. The electricity we are familiar with in daily life involves a flow of electrons. On the other hand, electrical conduction in aqueous solution involves the motion of ions, as may be demonstrated by means of the simple apparatus illustrated in Figure 2.4. Substances whose aqueous solutions contain ions and consequently conduct electricity are called **electrolytes.** This definition indicates that electrolytes are either acids, bases, or salts. Electrolytes can be subdivided into strong and weak electrolytes on the same basis as strong or weak acids and bases. While many acids and bases are not completely ionized in dilute solutions and are therefore weak electrolytes, the vast majority of salts are strong electrolytes. The most notable exceptions are a few salts of mercury, such as mercury(II) chloride (HgCl$_2$) and mercury(II) cyanide, (Hg(CN)$_2$), which are weak electrolytes.

***b. Molarity of Solutions*** In order to deal with the stoichiometry of chemical reactions occurring in solution, we must first specify the composition of a solution. Depending on the particular context, one or another of various sets of units may be used. For the present, we shall confine ourselves to the most common unit, molarity. The **molarity** of a solution is defined as the number of moles of solute in 1 L of solution. The units of molarity are moles per liter (mol L$^{-1}$), usually symbolized as $M$. It is important to note that the molarity expresses the amount of solute in a given volume of *solution;* the amount of solvent present is not stated explicitly. Moreover, since the unit consists of a ratio

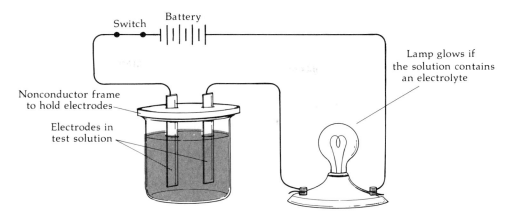

**FIGURE 2.4** Apparatus for demonstrating the electrical conductivity of a solution. If the solution contains an electrolyte, an electric current can flow through the metallic plates (electrodes) inserted in the solution, as evidenced by the glowing light bulb. If the solution does not contain an electrolyte, the bulb does not glow.

of moles to volume, it is an intensive property, that is, independent of total amount. A unit having these properties is a measure of **concentration.** The concentration of a species is symbolized by the chemical formula of that species written within brackets. For example, the equation $[SO_4^{2-}] = 0.5\ M$ means that the concentration of sulfate ion is 0.5 mol L$^{-1}$.

Suppose we have a 1.0 $M$ solution of NaCl in water. This statement tells us that 1 L of the solution contains 1 mol of NaCl or, equivalently, 58.4 g of this salt. Knowing that NaCl dissociates completely in water to yield Na$^+$ and Cl$^-$ ions, we can also state that the solution is 1.0 $M$ in Na$^+$ and 1.0 $M$ in Cl$^-$ ions. Similarly, a 0.1 $M$ solution of Na$_2$SO$_4$ contains 0.1 mol of this salt per liter of solution. The concentration of SO$_4^{2-}$ ions also is 0.1 $M$; however, that of Na$^+$ ions is 0.2 $M$ since the salt contains 2 Na$^+$ ions per formula unit.

Molarity is a very convenient unit to work with in the laboratory because solutions of known molarity can be prepared readily: The mass of solute is determined by weighing and the solution volume is measured in a calibrated container of the type illustrated in Figure 2.5. Since the volume of a liquid changes with temperature, the molarity will also be dependent on this variable. The effect is generally small, but for experimental work of the highest accuracy the temperature at which the concentration of a given solution is measured must be specified.

The ability to use the concept of molarity is of fundamental importance in chemistry. The following examples illustrate various aspects of its use.

**Example 2.6** A sample of BaCl$_2$ weighing 10.00 g is dissolved in a small amount of water and the solution is diluted to 100.0 mL. What is the molarity of the resulting solution?

The number of moles of BaCl$_2$ is

$$n = \frac{10.00\ \text{g}}{208.1\ \text{g mol}^{-1}} = 0.04805\ \text{mol}$$

44  Chemical Reactions  2.3

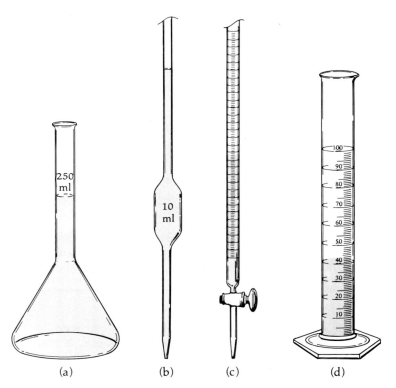

**FIGURE 2.5** Containers used for determination of the volume of liquids: (a) volumetric flask—used for the preparation of a specified volume of a liquid; (b) transfer pipet—used for the transfer of a specified volume of a liquid; (c) buret—used for the delivery of an accurately determined volume of a liquid; and (d) graduated cylinder—used for determination of the volume of a liquid.

and the molarity of the solution is

$$\frac{0.04805 \text{ mol}}{(100.0 \text{ mL})(0.001 \text{ L mL}^{-1})} = 0.4805 \, M \quad \blacksquare$$

Example 2.7   What is the concentration of the solution made by the addition of 10.0 mL of 0.100 $M$ HCl to 50.0 mL of 0.0250 $M$ HCl?

The combination of two solutions poses a problem because molarities are not additive. We can, however, determine the number of moles of HCl in each solution, and thus, the total number of moles. We will also assume that the volume of the combined solution is equal to the sum of the separate volumes. (As discussed in Section 12.4 this is frequently a valid assumption.)

We obtain for the total number of moles of HCl:

$$n_{HCl} = (0.100 \text{ mol L}^{-1})(0.0100 \text{ L}) + (0.0250 \text{ mol L}^{-1})(0.0500 \text{ L})$$
$$= 2.25 \times 10^{-3} \text{ mol}$$

Since the total volume is 60.0 mL, the molarity of the resulting solution is

$$\frac{2.25 \times 10^{-3} \text{ mol}}{0.0600 \text{ L}} = 0.0375 \; M$$

This type of calculation is also used to solve problems involving the dilution of a solution with water. ∎

**Example 2.8** The density of an aqueous solution of sulfuric acid, 50.0% $H_2SO_4$ by weight, is 1.395 g cm$^{-3}$. What is the molarity?

Knowing the density we can calculate the mass of a given volume of solution, say 1 L. We obtain

$$m = \rho V = (1.395 \text{ g cm}^{-3})(1.000 \text{ L})(10^3 \text{ cm}^3 \text{ L}^{-1}) = 1395 \text{ g}$$

Since the solution is 50.0% $H_2SO_4$ by weight, we have

$$m_{H_2SO_4} = (0.500)(1395 \text{ g}) = 697.5 \text{ g}$$

Dividing by the molecular weight (98.08 g mol$^{-1}$) gives the number of moles:

$$n_{H_2SO_4} = \frac{697.5 \text{ g}}{98.08 \text{ g mol}^{-1}} = 7.11 \text{ mol}$$

We have assumed the volume to be exactly 1 L. The solution is therefore 7.11 M. ∎

Commercially available acids are frequently labelled with the percent by weight and density but not the molarity. The preceding procedure can be followed to obtain this more useful measure of concentration.

**c. The Conservation of Charge and Net Ionic Equations** Experiments indicate that normal matter is electrically neutral. To be sure, it is indeed possible to induce a charge on a body. For example, a glass rod rubbed with silk acquires a positive charge. However, highly charged bodies are inherently unstable and tend to give up their charge. The occurrence of lightning is one familiar example of this phenomenon.

Chemical reactions do not generate or destroy electric charge. Therefore, chemical equations for reactions involving ions must conserve ionic charge. Consider, for example, the reaction of aluminum with hydrochloric acid:

$$Al(s) + 2H^+(aq) + 2Cl^-(aq) \rightarrow Al^{3+}(aq) + H_2(g) + 2Cl^-(aq)$$

When written in this fashion the equation conserves the number of atoms of each element. However the equation is not properly balanced because it does not represent the conservation of charge. The net charge of the reactants is zero while that of the products is +1. If this equation were correct as written, then the reaction of 1 mol of aluminum would result in a solution having a net positive charge of nearly $10^5$ C. Such a strongly charged solution would be highly unstable because of the electrostatic repulsion between particles of like charge.

The conservation of charge can be represented by the use of suitable stoichiometric coefficients:

$$2Al(s) + 6H^+(aq) + 6Cl^-(aq) \rightarrow 2Al^{3+}(aq) + 3H_2(g) + 6Cl^-(aq)$$

## 46 Chemical Reactions 2.3

Each side of the equation now has a net charge of zero as well as the same number of atoms of each element. The equation is therefore balanced.

Examination of the preceding equation shows the presence of six $Cl^-$ ions on each side. Clearly, the $Cl^-$ ions do not participate in the reaction and are merely included to make each side of the equation electrically neutral. Ions that do not participate in a reaction are said to be **spectator ions**. It is often customary to delete such species from the balanced equation even though this deletion necessarily leads to a net, though equal, charge on both sides:

$$2Al(s) + 6H^+(aq) \rightarrow 2Al^{3+}(aq) + 3H_2(g)$$

Equations of this type are called **net ionic equations**.

The following example illustrates the stoichiometry of chemical reactions occurring in aqueous solution.

---

**Example 2.9** The addition of 20.0 mL of 0.100 $M$ NaOH solution to 40.0 mL of 0.0200 $M$ $Cu(NO_3)_2$ results in the precipitation of copper(II) hydroxide, $Cu(OH)_2$. Calculate the concentration of ions present in the resulting solution.

The reaction can be written as

$$Cu^{2+} + 2NO_3^- + 2Na^+ + 2OH^- \rightarrow Cu(OH)_2(s) + 2Na^+ + 2NO_3^-$$

and the net ionic equation is

$$Cu^{2+} + 2OH^- \rightarrow Cu(OH)_2(s)$$

(Since all ions are hydrated in aqueous solution, we will frequently omit the designation (aq) in equations for ionic reactions in aqueous solution.) Let us first compute the concentration of the $Na^+$ and $NO_3^-$ spectator ions. Even though these ions do not react, their concentrations nonetheless change because of the change in volume. We obtain the new concentrations by first calculating the number of moles present and then dividing by the final volume:

$$n_{Na^+} = (0.100 \text{ mol L}^{-1} \text{ Na}^+)(0.0200 \text{ L}) = 0.00200 \text{ mol}$$

$$[Na^+] = \frac{0.00200 \text{ mol Na}^+}{0.0200 \text{ L} + 0.0400 \text{ L}} = 0.0333 \text{ } M$$

Similarly,

$$n_{NO_3^-} = (0.0200 \text{ mol L}^{-1} \text{ Cu(NO}_3)_2) \left( \frac{2 \text{ mol NO}_3^-}{\text{mol Cu(NO}_3)_2} \right)(0.0400 \text{ L})$$

$$= 0.00160 \text{ mol}$$

$$[NO_3^-] = \frac{0.00160 \text{ mol NO}_3^-}{0.0600 \text{ L}} = 0.0267 \text{ } M$$

If $Cu^{2+}$ and $OH^-$ had been added in stoichiometric amounts, both of these ions would have been essentially completely removed from the solution. It is easy to show, however, that this is not the case. The number of moles of these ions present initially is

$$n_{Cu^{2+}} = (0.0200 \text{ mol L}^{-1} \text{ Cu}^{2+})(0.0400 \text{ L}) = 8.00 \times 10^{-4} \text{ mol}$$

and

$n_{OH^-} = n_{Na^+} = 0.00200$ mol

Since the formula of copper(II) hydroxide is Cu(OH)$_2$, the number of moles of OH$^-$ reacting with $8.00 \times 10^{-4}$ mol of copper is

$$2 \times 8.00 \times 10^{-4} \text{ mol} = 0.00160 \text{ mol}$$

indicating that Cu$^{2+}$ is the limiting reagent. The amount of unreacted OH$^-$ is

$$n_{OH^-} = 0.00200 \text{ mol} - 0.00160 \text{ mol} = 4.0 \times 10^{-4} \text{ mol}$$

and the concentration is

$$[OH^-] = \frac{4.0 \times 10^{-4} \text{ mol OH}^-}{0.0600 \text{ L}} = 0.0067 \ M$$

After the reaction is complete, the solution contains $0.0333 \ M$ Na$^+$, $0.0267 \ M$ NO$_3^-$, and $0.0067 \ M$ OH$^-$. ■

## 2.4 Conservation of Energy in Chemical Reactions

It has been experimentally verified that energy is conserved in all known processes. This statement is as true for chemical reactions as it is for mechanical processes, such as the motion of a pendulum. Energy appears in a number of forms: gravitational energy, kinetic energy, thermal energy, and others. While one form of energy can be converted into another, the sum total of all the various forms remains constant. As noted in Section 1.4, the observation of the enormous amounts of energy liberated in nuclear transformations has indicated that mass, too, must be regarded as a form of energy. The extension of the law of conservation of energy to account for the loss of mass in nuclear processes indicates that the law can be applied with confidence to newly discovered phenomena.

Chemical reactions that liberate energy are said to be **exothermic** while those that absorb energy are called **endothermic.** Since energy is conserved in chemical reactions, the energy given off in an exothermic reaction must come from somewhere. The source of this energy is the molecules of the reacting species. Molecules contain various types of energy, known collectively as their **internal energy** (Section 10.6). (Mass changes in chemical reactions associated with the conversion between mass and energy are so small as to be negligible.) In an exothermic reaction, the internal energy of the products must be smaller than that of the reactants, the difference corresponding to the liberated energy. The opposite holds true in an endothermic reaction. The study of the energy changes in chemical reactions is known as **thermochemistry** (Chapter 10).

Chemical equations summarize the stoichiometric relationships between the species participating in a chemical reaction and, as discussed in Section 10.7, also are involved in the evaluation of energy changes. While chemical equations thus convey much valuable information about the reactions they represent, they tell us far from everything. A balanced chemical equation says nothing about the rate at which a particular reaction occurs. The rates of chemical reactions are immensely variable. Some reactions occur in nanoseconds or less; others occur so slowly that the reaction is only perceptible on a geological time scale. (The rates of chemical reactions are discussed in Chapter 14.) Even more fundamentally, a balanced equation gives no assurance that the reaction it describes will, in fact, occur as written. For example, several plausible reactions between nitrogen and oxygen can be written:

$$N_2 + O_2 \rightarrow 2NO$$
$$2N_2 + O_2 \rightarrow 2N_2O$$
$$N_2 + 2O_2 \rightarrow 2NO_2$$

It is impossible to predict just on the basis of the equations which of these reactions will occur for a particular set of experimental conditions. This question can only be answered by experiment.

## 2.5 A Classification of Chemical Reactions

The number of known chemical reactions is legion. In order to make some sense of the great variety of possible reactions it is customary to divide them into a few broad categories. Although there is no single universally accepted classification scheme, the one presented here has the advantages of simplicity and widespread application.

*a. Decomposition Reactions* A **decomposition** reaction is one in which a single compound decomposes into two or more other substances. This type of reaction usually occurs when the compound to be decomposed is heated to a sufficiently high temperature. The compound in question is frequently a solid and one or more of the products are usually gases. The thermal decompositions of metal carbonates and hydroxides to the corresponding oxides are typical, for example,

$$CaCO_3(s) \xrightarrow{\Delta} CaO(s) + CO_2(g)$$
$$2Al(OH)_3(s) \xrightarrow{\Delta} Al_2O_3(s) + 3H_2O(g)$$

The decomposition of potassium chlorate in the presence of manganese dioxide,

$$2KClO_3(s) \xrightarrow[MnO_2]{\Delta} 2KCl(s) + 3O_2(g)$$

illustrated in Figure 2.6, has been commonly used for the generation of oxygen in the laboratory.

*b. Combination Reactions* A **combination** reaction is the reverse of decomposition: Two substances combine with each other to form a third. **Combustion** provides a good illustration of this process. This term is applied to fast reactions that are accompanied by the evolution of light and heat. The burning of elements in air with the formation of oxides is a common example of combustion:

$$2Mg(s) + O_2(g) \rightarrow 2MgO(s)$$
$$S(s) + O_2(g) \rightarrow SO_2(g)$$

Other types of combination reactions are those involving acidic or basic oxides, which react with water to form acids and bases, respectively,

$$SO_2(g) + H_2O(l) \rightarrow H_2SO_3(aq)$$
$$Na_2O(s) + H_2O(l) \rightarrow 2Na^+(aq) + 2OH^-(aq)$$

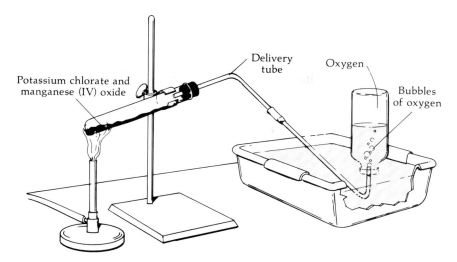

**FIGURE 2.6** Laboratory preparation of oxygen by the decomposition of potassium chlorate in the presence of manganese dioxide.

and with each other to form salts, for example,

$$CaO(s) + CO_2(g) \rightarrow CaCO_3(s)$$

***c. Displacement Reactions*** As the name implies, a **displacement** reaction involves the displacement of one species by another in a particular compound. Such a reaction provides a measure of the relative activity of the two species in question since the more reactive one will usually displace the other. The formation of an insoluble substance or the evolution of a gas removes the product of a displacement reaction from the reaction site and drives the reaction to completion.

Displacement reactions provide a useful qualitative measure of the activity of different metals (Section 6.9) since a more reactive metal will displace a less reactive one from a solution of one of its salts. For example, copper is more reactive than silver and so displaces the latter from a silver nitrate solution:

$$Cu(s) + 2AgNO_3(aq) \rightarrow Cu(NO_3)_2(aq) + 2Ag(s)$$

The liberation of hydrogen in the reactions of many metals with acids is another example of a displacement reaction. The reaction of magnesium with hydrochloric acid,

$$Mg(s) + 2HCl(aq) \rightarrow MgCl_2(aq) + H_2(g)$$

has been commonly used for the laboratory preparation of hydrogen gas (Figure 2.7). Note that the preceding equation is not written in the form of a net ionic equation in order to identify the acid used in the reaction. The corresponding net ionic equation would be

$$Mg(s) + 2H^+(aq) \rightarrow Mg^{2+}(aq) + H_2(g)$$

***d. Double Displacement Reactions*** **Double displacement** (sometimes called metathesis) involves a reaction in which an exchange of partners occurs be-

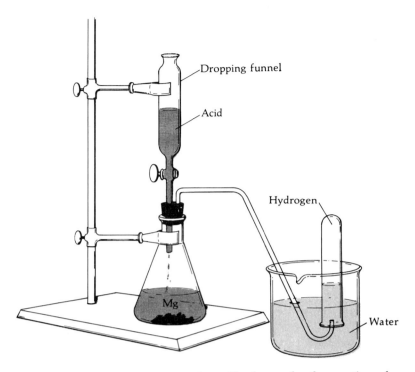

**FIGURE 2.7** Laboratory preparation of hydrogen by the reaction of magnesium with hydrochloric acid.

tween the reacting species. This type of reaction commonly involves aqueous solutions of ionic compounds. The exchange of ions results in a reaction when either an insoluble compound, a gas, or an essentially un-ionized substance such as water is formed.

The following are some examples of double displacement in which the reaction is driven to completion by the formation of an insoluble compound:

$$2Na^+ + 2OH^- + Fe^{2+} + 2NO_3^- \rightarrow Fe(OH)_2(s) + 2Na^+ + 2NO_3^-$$

$$2K^+ + CO_3^{2-} + Ca^{2+} + 2NO_3^- \rightarrow CaCO_3(s) + 2K^+ + 2NO_3^-$$

Note that ions such as $Na^+$ and $NO_3^-$ (in the first reaction) do not participate in the reaction and merely play the role of spectator ions. Their inclusion in the equation serves to emphasize the exchange aspect of this type of reaction, which could also be written

$$Fe^{2+} + 2OH^- \rightarrow Fe(OH)_2(s)$$

It is important to emphasize that in order for double displacement to occur, the exchange of partners must remove one of the compounds from solution. If this removal does not occur, the mixing of two ionic compounds in aqueous solution does not result in a reaction. For example, when sodium chloride is added to potassium nitrate in solution, no discernible reaction occurs because all the four possible salts (NaCl, KNO₃, NaNO₃, and KCl) are soluble and there can be no reaction between them.

In addition to being driven by the formation of an insoluble salt, double displacement reactions can also occur as a result of gas formation. The reaction of many metal sulfides with acid results in the evolution of $H_2S$, for example,

$$ZnS(s) + 2HCl(aq) \rightarrow H_2S(g) + ZnCl_2(aq)$$

The reaction between a strong acid and a strong base is called **neutralization** and is one of the most frequently encountered chemical reactions. A typical example,

$$Na^+ + OH^- + H^+ + Cl^- \rightarrow H_2O(l) + Na^+ + Cl^-$$

shows that neutralization involves double displacement, in which the formation of water drives the reaction to completion and the spectator ions form a salt solution. Since water is nearly completely un-ionized, the $H^+$ and $OH^-$ ions from which it was formed have been effectively removed from the solution just as if an insoluble compound or a gas had been formed. In contrast to other double displacement reactions, the net reaction involved in neutralization is the same regardless of the identity of the acid and base:

$$H^+ + OH^- \rightarrow H_2O(l)$$

Only the salts left in solution will differ. Acids and bases are important reagents since they react with a variety of substances to cause chemical transformations that would otherwise be difficult to achieve. The dissolution of metals with the liberation of hydrogen is just one example.

**Example 2.10** Classify the following reactions by means of the scheme outlined in this section:

(a) $2NO(g) + O_2(g) \rightarrow 2NO_2(g)$
This is a combination reaction.

(b) $Br_2(aq) + 2I^- \rightarrow I_2(aq) + 2Br^-$
Bromine displaces iodide from a solution of one of its salts, e.g., KI.

(c) $Na_2SO_3(aq) + 2HCl(aq) \rightarrow 2NaCl(aq) + SO_2(g) + H_2O(l)$
Double displacement occurs with the formation of NaCl(aq) and $H_2SO_3$(aq). However, this acid decomposes to $SO_2$(g) and $H_2O$(l).

(d) $2MnO_4^- + 3H_2S(g) \rightarrow 3S(s) + 2MnO_2(s) + 2OH^- + 2H_2O(l)$
This reaction does not fit within the simple classification scheme outlined in this section. ■

# Oxidation–Reduction Reactions

## 2.6 Oxidation and Reduction

The term "oxidation" originally was applied to reactions of substances with oxygen. Many of the elements undergo this reaction, for example,

$$2Ca(s) + O_2(g) \rightarrow 2CaO(s) \tag{2.3}$$

The resulting oxide is an ionic compound in which calcium is present in the form of calcium ions ($Ca^{2+}$), and oxygen is present as oxide ions ($O^{2-}$). Evidently, the reaction involves a transfer of two electrons from calcium to oxygen.

Calcium can undergo a similar loss of electrons in its reaction with many other substances, for example,

$$Ca(s) + Cl_2(g) \rightarrow CaCl_2(s)$$

in which the product consists of $Ca^{2+}$ and $Cl^-$ ions, and

$$Ca(s) + 2HCl(aq) \rightarrow CaCl_2(aq) + H_2(g)$$

in which the products include $Ca^{2+}$ and $Cl^-$ ions in solution. Since all the above reactions involve the loss of electrons by a calcium atom, the concept of oxidation can be generalized to include them all. We say that *a substance undergoes* **oxidation** *when it loses electrons.*

The electrons lost by the calcium atom in the preceding examples are gained by one of the other species, such as oxygen or chlorine. *A substance is said to undergo* **reduction** *when it gains electrons.* In view of the conservation of charge in chemical reactions, oxidation and reduction always occur together. The term **redox reactions** is frequently used to describe reactions involving oxidation–reduction. It is customary to call a substance that causes some other substance to be oxidized an **oxidizing agent.** Similarly, a substance that brings about the reduction of another substance is called a **reducing agent.** For example, in the reaction represented by Equation 2.3, $O_2$ acts as an oxidizing agent because it oxidizes Ca, while Ca acts as a reducing agent because it reduces $O_2$. It is apparent that in a redox reaction the oxidizing agent is always reduced while the reducing agent is always oxidized.

The occurrence of electron transfer is more apparent in some redox reactions than in others. The reduction of $Ni^{2+}$ by magnesium metal is an example of a reaction obviously involving electron transfer:

$$\overset{\overset{\displaystyle 2e^-}{\nearrow\quad\searrow}}{Mg\ +\ Ni^{2+}} \rightarrow Mg^{2+} + Ni$$

On the other hand, the reaction between hypochlorite and nitrite ions in aqueous solution,

$$ClO^- + NO_2^- \rightarrow NO_3^- + Cl^- \tag{2.4}$$

cannot be interpreted as a redox reaction in such a direct manner. All four of the participating species have ionic charges of $-1$, so that it is not at all evident that electron gain and loss actually occur. Some redox reactions do not, in fact, even require the transfer of electrons. For example, the reaction between methane and oxygen,

$$2CH_4 + O_2 \rightarrow 2CH_3OH$$

involves only uncharged molecules and there is no electron transfer. We might conjecture that the reaction involves oxidation–reduction since one of the reactants is oxygen, but the validity of this conjecture must be established in some other manner.

## 2.7 Oxidation States

In order to facilitate the analysis of redox reactions, **oxidation states,** also called **oxidation numbers,** are assigned to the atoms in molecules or polyatomic ions. With their aid, the occurrence of oxidation and reduction in reactions where it might not be apparent is readily established. When oxidation–reduction does occur, oxidation numbers are useful in balancing the equations. In addition to these uses, which depend on their function as an electron bookkeeping device, oxidation numbers are also important in the systematization of chemical properties (see Section 6.10 and Chapter 19).

Oxidation states have a simple meaning in the case of monatomic ions: They are just equal to the ionic charge. For example, the oxidation states of $Cu^{2+}$, $H^+$, $F^-$, and $O^{2-}$ are +2, +1, −1, and −2, respectively.

The extension of the concept to polyatomic species is less straightforward. Consider for example, the oxidation states of the hydrogen and oxygen atoms making up a molecule of water. As noted in Section 2.3a, we expect the electrons in the molecule to be concentrated in the vicinity of the oxygen atom. The hydrogen atoms therefore acquire a slight positive charge while the oxygen atom becomes somewhat negatively charged. However, the actual charges are not easily determined. Moreover, the actual charge on a hydrogen atom in a different molecule, such as HCl or $H_2S$, will differ somewhat from the value in $H_2O$. The oxidation numbers consequently do not represent the actual charges on atoms present in polyatomic species. Rather, they are arbitrarily assigned on the basis of the following specific rules:

1. The oxidation number of atoms in an uncombined element, such as $H_2$, S, or $Cl_2$, is zero.
2. The oxidation number of a monatomic ion is equal to its ionic charge.
3. The oxidation number of hydrogen is +1 in all its compounds except for those with the metals. Hydrogen combines with the alkali (Li, Na, K, Rb, Cs) and alkaline earth (Be, Mg, Ca, Sr, Ba) metals to form hydrides, examples of which are LiH and $MgH_2$, in which the oxidation number of hydrogen is −1.
4. The oxidation number of fluorine in all its compounds is −1, those of the alkali metals in their compounds are +1, and those of the alkaline earth metals in their compounds are +2.
5. The oxidation number of oxygen is −2 in all its compounds excepting (a) those containing an O–O bond (i.e., peroxides), such as $H_2O_2$, and superoxides, such as $KO_2$, where its oxidation state is −1 and $-\frac{1}{2}$, respectively, and (b) compounds with fluorine, where rule 4 takes precedence.
6. The oxidation numbers of all other elements in polyatomic ions or molecules are chosen so that the sum of the oxidation numbers of all the atoms is equal to the ionic charge or to zero, respectively.

As an illustration of these rules let us assign oxidation numbers to the atoms in potassium permanganate, $KMnO_4$. The oxidation number of O is −2 and that of K (one of the alkali elements) is +1. Since the overall charge is zero, the oxidation number of Mn must be +7, since

$$0 \quad -(+1) - 4 \times (-2) = +7$$
$$(KMnO_4) \quad (K) \qquad (O) \quad (Mn)$$

In this equality, the oxidation numbers of one potassium atom and four oxygen atoms are subtracted from zero (the charge on KMnO$_4$) to give the oxidation number of Mn. The same procedure may be used in the case of a polyatomic ion, $Cr_2O_7^{2-}$, for example. Once again, the oxidation state of oxygen is $-2$. Since the overall charge is $-2$, the oxidation state of Cr must be $+6$, since

$$\frac{-2 - (7)(-2)}{2} = +6$$

$$(Cr_2O_7^{2-})(O) \quad (Cr)$$

where division by 2 takes into account the fact that there are two chromium atoms in each dichromate ion.

Having seen how oxidation numbers are assigned, we are now ready to apply them to redox reactions. To do so, we first generalize the definitions of oxidation and reduction given in Section 2.6 as follows: *An atom is oxidized when its oxidation number increases and it is reduced when its oxidation number decreases.* Rules 1 and 2 for the assignment of oxidation numbers indicate that this generalization is consistent with the earlier definitions in terms of electron gain and loss. For example, when iron is oxidized from Fe to $Fe^{2+}$, it loses two electrons and at the same time its oxidation number increases from 0 to $+2$.

Let us now reexamine the reactions represented by Equations 2.4 and 2.5 in order to determine the occurrence of oxidation and reduction. In Equation 2.4,

$$ClO^- + NO_2^- \rightarrow NO_3^- + Cl^-$$

chlorine is present first as $ClO^-$ and then as $Cl^-$. According to the rules, the oxidation number of chlorine decreases from $+1$ to $-1$, indicating the occurrence of reduction. Consequently, nitrogen must be oxidized, as can be ascertained by noting that its oxidation number increases from $+3$ in $NO_2^-$ to $+5$ in $NO_3^-$. In Equation 2.5,

$$2CH_4(g) + O_2(g) \rightarrow 2CH_3OH(l)$$

oxygen changes from the free element to $CH_3OH$ and its oxidation number decreases from 0 to $-2$, indicating that the element is reduced. Carbon is initially present as $CH_4$, where its oxidation number is $-4$, but the conversion to $CH_3OH$ increases the oxidation number to $-2$, thus showing the occurrence of oxidation. Note that hydrogen is present in the $+1$ state in both compounds and is therefore neither oxidized nor reduced.

The various meanings of oxidation and reduction are summarized in Figure 2.8 in terms of a specific example.

---

| Zn | + | $Cu^{2+}$ | $\longrightarrow$ | $Zn^{2+} + Cu$ |
|:---:|:---:|:---:|:---:|:---:|
| Undergoes oxidation | | Undergoes reduction | | |
| Oxidation number increases | | Oxidation number decreases | | |
| Loses electrons | | Gains electrons | | |
| Reducing agent | | Oxidizing agent | | |

**FIGURE 2.8** The various meanings of oxidation and reduction.

## 2.8 Balancing Redox Equations

The equations for chemical reactions involving oxidation–reduction can, in principle, be balanced by adjusting the various stoichiometric coefficients until the number of atoms of each element and the total ionic charge are properly conserved. For a simple redox reaction such as that between zinc and copper(II) ions,

$$Zn(s) + Cu^{2+} \rightarrow Zn^{2+} + Cu(s)$$

the procedure is trivial since just writing the reactants and products automatically results in a balanced equation. Redox equations, however, are frequently rather complicated. Consider, for example, the oxidation of $Fe^{2+}$ to $Fe^{3+}$ in acidic solution by $MnO_4^-$, resulting in the formation of $Mn^{2+}$. The balanced equation for this reaction is

$$MnO_4^- + 5Fe^{2+} + 8H^+ \rightarrow Mn^{2+} + 5Fe^{3+} + 4H_2O$$

While this result can be obtained by trial and error, the process can be both lengthy and frustrating. Perhaps even more importantly, such an approach does not shed any light on the redox features of the reaction. Clearly, a more systematic approach is desirable and several have been developed.

**a. The Half-Reaction Method** In this approach, a redox reaction is divided into **half-reactions,** one involving oxidation and the other involving reduction. Electron loss or gain are explicitly shown in a half-reaction. For example, the reaction of zinc with copper(II) ions can be represented by the following two half-reactions:

$$Zn \rightarrow Zn^{2+} + 2e^- \quad \text{oxidation}$$

$$Cu^{2+} + 2e^- \rightarrow Cu \quad \text{reduction}$$

Each half-reaction can be separately balanced and the results combined so that the electrons cancel to yield an overall balanced equation.

This author has found the following procedure using half-reactions *and* oxidation numbers to be particularly reliable in yielding correctly balanced redox equations in even the most complicated cases:

1. Identify the species undergoing oxidation and reduction by determining changes in oxidation numbers. Then write skeleton half-reactions for these processes, including enough electrons to balance the change in oxidation number per atom oxidized or reduced. For example, if $Cl^-$ is oxidized to $Cl_2$, for an oxidation number change of +1, write

$$Cl^- \rightarrow Cl_2 + e^-$$

while if $MnO_4^-$ is reduced to $Mn^{2+}$, write

$$MnO_4^- + 5e^- \rightarrow Mn^{2+}$$

2. Balance each half-reaction, first with respect to charge and next with respect to number of atoms of each element. Reactions occurring in acidic media require the addition of $H^+$ to the side of the equation deficient in positive charge and $H_2O$ to the other side. Reactions occurring in basic media require the addition of $OH^-$ to the side deficient in negative charge and $H_2O$ to balance the number of atoms.

3. Multiply the half-reactions by the smallest coefficients needed to make the number of electrons involved in them equal to each other. Add the half-reactions; the electrons will cancel. If there are common terms on both sides of the equation (e.g., $H_2O$), then cancel them.

The following examples illustrate this procedure.

---

**Example 2.11** Balance the following equation:

$$Cr_2O_7^{2-} + Br^- \rightarrow Cr^{3+} + Br_2(aq) \quad \text{(acid)}$$

1. We can tell by inspection that $Br^-$ (oxidation number = $-1$) is oxidized to $Br_2$ (oxidation number = 0). Therefore, $Cr_2O_7^{2-}$ must be reduced. This is confirmed by determining the oxidation number of Cr in $Cr_2O_7^{2-}$ (+6) and noting the decrease to +3 in $Cr^{3+}$.

   The skeleton half-reactions are

   $$Br^- \rightarrow Br_2 + e^- \quad \text{(oxidation)}$$

   $$Cr_2O_7^{2-} + 3e^- \rightarrow Cr^{3+} \quad \text{(reduction)}$$

   Note that the number of electrons in each skeleton half-reaction is just equal to the change in oxidation number per atom.

2. We next balance each half-reaction.
   Oxidation: The half-reaction is balanced by noting that a $Br_2$ molecule requires two $Br^-$ ions for its formation and that two electrons are lost in the process:

   $$2Br^- \rightarrow Br_2 + 2e^-$$

   Reduction: First balance the species being reduced (i.e., Cr); the reduction of 2 chromium atoms from the +6 to the +3 state requires six electrons and 2 $Cr^{3+}$ ions:

   $$Cr_2O_7^{2-} + 6e^- \rightarrow 2Cr^{3+}$$

   Recalling that the reaction takes place in acidic medium, balance the charge by the addition of $H^+$. Since the charge on the left side is $-8$ and that on the right is +6, $14H^+$ must be added to the left:

   $$Cr_2O_7^{2-} + 14H^+ + 6e^- \rightarrow 2Cr^{3+}$$

   The $14H^+$ will combine with the 7 oxygen atoms in $Cr_2O_7^{2-}$ to form 7 $H_2O$ molecules whose inclusion on the right side completes the balancing of this half-reaction:

   $$Cr_2O_7^{2-} + 14H^+ + 6e^- \rightarrow 2Cr^{3+} + 7H_2O$$

3. In order for the electrons to cancel on addition of the two half-reactions, the first of them must be multiplied by 3:

$$6Br^- \rightarrow 3Br_2 + 6e^-$$

$$\underline{Cr_2O_7^{2-} + 14H^+ + 6e^- \rightarrow 2Cr^{3+} + 7H_2O}$$

$$Cr_2O_7^{2-} + 6Br^- + 14H^+ \rightarrow 2Cr^{3+} + 3Br_2 + 7H_2O$$

There being no common terms to cancel, the equation is balanced, as can be ascertained. ∎

---

**Example 2.12** Write a balanced equation for the reaction

$$Fe(OH)_3(s) + ClO^- \rightarrow FeO_4^{2-} + Cl^-$$

in basic solution.

1. The oxidation number of Fe is +3 in $Fe(OH)_3$ and +6 in $FeO_4^{2-}$. The oxidation half-reaction in its skeleton form is

    $$Fe(OH)_3 \rightarrow FeO_4^{2-} + 3e^-$$

    The oxidation number of Cl is +1 in $ClO^-$ and $-1$ in $Cl^-$. The reduction half-reaction is therefore:

    $$ClO^- + 2e^- \rightarrow Cl^-$$

2. We next balance each half-reaction.
    Oxidation: The equation is already balanced with respect to Fe and has the appropriate number of electrons. Balancing charge, we note that the total charge on the left is 0 while on the right it is $-5$. Since the reaction takes place in basic medium the charge must be balanced by the addition of $OH^-$:

    $$Fe(OH)_3 + 5\,OH^- \rightarrow FeO_4^{2-} + 3e^-$$

    The 8 hydrogen atoms on the left side of the equation form 4 $H_2O$ molecules which must be added to the right:

    $$Fe(OH)_3 + 5\,OH^- \rightarrow FeO_4^{2-} + 4H_2O + 3e^-$$

    This procedure also balances oxygen and the half-reaction is thus balanced.
    Reduction: The skeleton equation contains the correct number of Cl and $e^-$. Two $OH^-$ ions must be added to the right side to balance the $-3$ charge on the left side:

    $$ClO^- + 2e^- \rightarrow Cl^- + 2OH^-$$

    The two hydrogen atoms on the right side are balanced by addition of one $H_2O$ molecule to the left side, thus resulting in the following balanced half-reaction:

    $$ClO^- + H_2O + 2e^- \rightarrow Cl^- + 2OH^-$$

3. In order to cancel the electrons upon addition, the coefficients of the oxidation half-reaction must be multiplied by 2 and those of the reduction half-reaction by 3:

    $$2Fe(OH)_3 + 10OH^- \rightarrow 2FeO_4^{2-} + 8H_2O + 6e^-$$
    $$3ClO^- + 3H_2O + 6e^- \rightarrow 3Cl^- + 6OH^-$$

    ---

    $$2Fe(OH)_3 + 3ClO^- + 10OH^- + 3H_2O \rightarrow 2FeO_4^{2-} + 3Cl^- + 8H_2O + 6OH^-$$

    Note that $OH^-$ and $H_2O$ appear on both sides of the equation and the smaller number of each species must be eliminated:

$$2Fe(OH)_3 + 3ClO^- + 4OH^- \rightarrow 2FeO_4^{2-} + 3Cl^- + 5H_2O$$

The equation is now balanced, as can be checked by counting atoms and charges. ■

The method illustrated in Examples 2.11 and 2.12 combines the use of half-reactions with that of oxidation numbers. It is also possible to balance redox equations by the half-reaction method without the use of oxidation numbers. The following procedure, illustrated by means of the reaction balanced in Example 2.11, may be used.

1. Separate the reaction into half-reactions written without electrons. The oxidation of bromide by dichromate in acid would be written as

$$Cr_2O_7^{2-} \rightarrow Cr^{3+}$$

$$Br^- \rightarrow Br_2$$

2. Balance the number of atoms in each half-reaction by addition of $H^+$ and $H_2O$ or $OH^-$ and $H_2O$, as appropriate to the conditions of the problem. The first half-reaction is balanced by adding $H^+$ to the left side and $H_2O$ to the right side and equalizing the number of Cr atoms:

$$Cr_2O_7^{2-} + 14H^+ \rightarrow 2Cr^{3+} + 7H_2O$$

The second half-reaction requires only that the number of Br atoms be balanced:

$$2Br^- \rightarrow Br_2$$

3. Balance the charge by the addition of electrons. Since the net charge on the left side of the first half-reaction is +12 and that on the right side is +6, six electrons must be added to the left side:

$$Cr_2O_7^{2-} + 14H^+ + 6e^- \rightarrow 2Cr^{3+} + 7H_2O$$

The second half-reaction requires the addition of two electrons to the right side:

$$2Br^- \rightarrow Br_2 + 2e^-$$

4. Combine the two half-reactions following the procedure outlined in step 3:

$$Cr_2O_7^{2-} + 6Br^- + 14H^+ \rightarrow 2Cr^{3+} + 3Br_2 + 7H_2O$$

**b. The Oxidation Number Method** In this method, the reaction is balanced on the basis of the changes in oxidation numbers without separation into half-reactions. We illustrate the procedure to be followed with the oxidation of nitrite ion by chlorine to form nitrate and chloride ions in basic solution.

1. Determine the oxidation numbers of the species undergoing oxidation and reduction:

$$NO_2^- + Cl_2 \rightarrow NO_3^- + Cl^-$$
$$\phantom{NO_2^-}+3 \phantom{++} 0 \phantom{+++} +5 \phantom{++} -1$$

2. Balance the changes in oxidation numbers. The oxidation number of nitrogen increases from +3 to +5 corresponding to the loss of two electrons.

These electrons are gained by Cl atoms. Since each Cl atom gains one electron, two Cl atoms must be reduced:

$$NO_2^- + Cl_2 \rightarrow NO_3^- + 2Cl^-$$

3. Balance the charges. Since the reaction takes place in basic solution, hydroxide ions must be added. The charge on the left side is −1 and that on the right side is −3 indicating that two OH⁻ ions must be added to the left:

$$NO_2^- + Cl_2 + 2OH^- \rightarrow NO_3^- + 2Cl^-$$

4. Balance the number of atoms. Only hydrogen and oxygen atoms remain unbalanced. The addition of one H₂O to the right side balances the overall equation,

$$NO_2^- + Cl_2 + 2OH^- \rightarrow NO_3^- + 2Cl^- + H_2O$$

as can be ascertained.

The choice between the three methods for balancing redox equations is, to some extent, a matter of personal preference. The half-reaction method is simpler to use for complicated equations, particularly in conjunction with oxidation numbers, as illustrated in Examples 2.11 and 2.12. In addition, this method has some more fundamental advantages to recommend it. Redox reactions are intimately connected with electrochemistry (Chapter 13), the subject dealing with the chemical generation or utilization of electricity. The explicit inclusion of the number of transferred electrons in the balanced half-reactions permits the establishment of a quantitative relation between electric current and amount of material undergoing oxidation or reduction. Furthermore, the use of half-reactions provides a convenient way to summarize the reactions of oxidizing and reducing agents. Table 2.2 lists some common oxidizing and reducing agents and shows the half-reactions they undergo as they are reduced and oxidized, respectively.

## 2.9 Disproportionation

**Disproportionation** is a special type of redox reaction in which the same element is both oxidized and reduced. In order for disproportionation to be possible, the element in question must have at least three different possible oxidations states, that in the reactant, and one higher and one lower oxidation state. The reaction of chlorine with water,

$$Cl_2(g) + H_2O(l) \rightarrow HCl(aq) + HOCl(aq)$$
$$\phantom{Cl_2(g) + H_2O(l) \rightarrow } 0 \phantom{aaaaaaa} -1 \phantom{aaa} +1$$

and that of nitrite ion with acid,

$$3NO_2^- + 2H^+ \rightarrow 2NO(g) + NO_3^- + H_2O(l)$$
$$\phantom{3NO_2^-} +3 \phantom{aaaaaa} +2 \phantom{aa} +5$$

are examples of disproportionation, as can be seen from the oxidation numbers of chlorine and nitrogen in the various species present. Some additional examples of compounds that undergo disproportionation are sulfites, manganates ($MnO_4^{2-}$), and copper(I) salts.

TABLE 2.2 Some Common Oxidizing and Reducing Agents and Their Half-reactions

| Oxidizing agent | Half-reaction |
|---|---|
| $MnO_4^-$ (acidic solution) | $MnO_4^- + 8H^+ + 5e^- \rightarrow Mn^{2+} + 4H_2O$ |
| $MnO_4^-$ (basic solution) (permanganate ion) | $MnO_4^- + 2H_2O + 3e^- \rightarrow MnO_2 + 4OH^-$ |
| $Cr_2O_7^{2-}$ (acidic solution) (dichromate ion) | $Cr_2O_7^{2-} + 14H^+ + 6e^- \rightarrow 2Cr^{3+} + 7H_2O$ |
| $BiO_3^-$ (acidic solution) (bismuthate ion) | $BiO_3^- + 6H^+ + 2e^- \rightarrow Bi^{3+} + 3H_2O$ |
| $H_2O_2(aq)$ (hydrogen peroxide) | $H_2O_2 + 2H^+ + 2e^- \rightarrow 2H_2O$ |
| $Cl_2(g)$ | $Cl_2 + 2e^- \rightarrow 2Cl^-$ |
| $Br_2(l)$ | $Br_2 + 2e^- \rightarrow 2Br^-$ |
| $Ce^{4+}$ | $Ce^{4+} + e^- \rightarrow Ce^{3+}$ |
| $O_2(g)$ | $O_2 + 4e^- \rightarrow 2O^{2-}$ (as oxide, etc.) |
| $H_2SO_4$(conc) | $H_2SO_4 \rightarrow SO_2, S, H_2S$ (depending on reaction) |
| $HNO_3$(conc)[a] | $NO_3^- + 2H^+ + e^- \rightarrow NO_2 + H_2O$ |
| $HNO_3$(dil)[a] | $NO_3^- + 4H^+ + 3e^- \rightarrow NO + 2H_2O$ |

| Reducing Agent | Half-reaction |
|---|---|
| Mg(s) | $Mg \rightarrow Mg^{2+} + 2e^-$ |
| Al(s) | $Al \rightarrow Al^{3+} + 3e^-$ |
| Zn(s) | $Zn \rightarrow Zn^{2+} + 2e^-$ |
| $SO_3^{2-}$ (sulfite ion) | $SO_3^{2-} + H_2O \rightarrow SO_4^{2-} + 2H^+ + 2e^-$ |
| $Cr^{2+}$ | $Cr^{2+} \rightarrow Cr^{3+} + e^-$ |

[a] Other possible products include $NH_4^+$, $HNO_2$, etc., depending on the identity of the reducing agent and the acid concentration.

## 2.10 Redox Titrations and Stoichiometry

The concentration of many species capable of undergoing oxidation or reduction can be determined by **redox titration.** This procedure involves the addition of an accurately measurable volume of oxidizing (or reducing) agent of known concentration to a solution of a reducing (or oxidizing) agent of unknown concentration. The oxidizing agent is dispensed until the **equivalence point** is reached, at which point the oxidizing agent has been added in the correct stoichiometric amount to just oxidize all the reducing agent present. Under appropriate conditions, the equivalence point is marked by a change in the color of the solution and can be observed readily. A solution of potassium permanganate ($KMnO_4$) is frequently used in redox titrations because its intense purple color provides an accurate visual indication of the equivalence point. (The point at which the change in color is observed is called the **endpoint.**) Figure 2.9 illustrates the titration of oxalic acid ($H_2C_2O_4$) with $KMnO_4$ solution. An additional requirement is that the redox reaction occur rapidly, so that the reaction is complete by the time the reagent has been added. Many redox reactions satisfy this requirement. The following example illustrates the quantitative aspects of a redox titration.

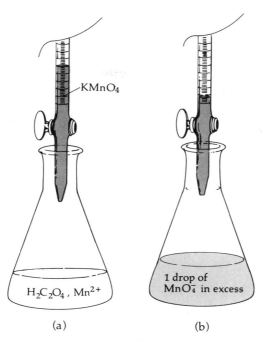

**FIGURE 2.9** Determination of the concentration of an oxalic acid solution by titration with potassium permangante. (a) Before the equivalence point, $MnO_4^-$ is reduced to $Mn^{2+}$. Since both $H_2C_2O_4$ and $Mn^{2+}$ are colorless, the resulting solution is colorless. (b) At the equivalence point, because of the intense purple color of $KMnO_4$, the addition of just a single drop of permanganate which remains unreduced produces a readily visible purple color whose appearance marks the equivalence point.

**Example 2.13** When 25.0 mL of oxalic acid in acidic solution is titrated with 0.103 $M$ $KMnO_4$, the equivalence point is reached after the addition of 12.3 mL of permanganate solution. What was the concentration of oxalic acid in the original solution? Oxalic acid is oxidized to carbon dioxide.

The first step is to determine the stoichiometric relation between the reactants. We can balance the equation following one of the procedures described in Section 2.8:

$$2MnO_4^- + 5H_2C_2O_4(aq) + 6H^+ \rightarrow 2Mn^{2+} + 10CO_2(g) + 8H_2O(l)$$

The number of moles of $KMnO_4$ can be obtained from the molarity and the added volume; the stoichiometric coefficients permit the conversion of this number to that of moles of oxalic acid:

$$n_{H_2C_2O_4} = (0.103 \text{ mol L}^{-1} \text{ } KMnO_4)(0.0123 \text{ L } KMnO_4)\left(\frac{5 \text{ mol } H_2C_2O_4}{2 \text{ mol } KMnO_4}\right)$$

$$= 3.167 \times 10^{-3} \text{ mol}$$

The concentration of the oxalic acid in solution is now obtained by dividing by the volume:

$$[H_2C_2O_4] = \frac{3.167 \times 10^{-3} \text{ mol}}{0.0250 \text{ L}} = 0.127 \, M \quad \blacksquare$$

## Conclusion

The conservation of mass and charge in chemical reactions provide the experimental basis for balanced chemical equations and stoichiometry. Balancing chemical equations is essentially a matter of bookkeeping. However, special balancing techniques are available for oxidation–reduction reactions. Stoichiometric calculations permit us to relate the masses of products to those of reactants. Such calculations are basic to all areas of chemistry. The ability to perform calculations of this type can best be acquired through practice in working stoichiometry problems.

## Problems

1. What mass of phosphoric acid ($H_3PO_4$) is needed to completely convert 80.7 g of MgO into $Mg_3(PO_4)_2$?

2. Laughing gas, $N_2O$, has been used as an anesthetic in dentistry. It can be made by the thermal decomposition of ammonium nitrate. How many grams of $N_2O$ can be obtained from 18.5 g of $NH_4NO_3$?

3. The reaction of vanadium(II) oxide with iron(III) oxide results in the formation of vanadium(V) oxide and iron(II) oxide. Assuming that no other products are formed, calculate the mass of $V_2O_5$ obtainable when 7.83 g of VO reacts with an excess of $Fe_2O_3$.

4. Hydrogen gas is produced when calcium metal is added to water. What mass of hydrogen is produced from the reaction of 2.05 g of Ca with an excess of water?

5. Cobalt(II) chloride can be converted to $Co(NH_3)_6Cl_3$ by a series of reactions. What mass of this compound can be obtained from the complete conversion of 5.30 g of $CoCl_2$?

6. A 4.52 g sample of calcium carbonate was heated until all the carbon dioxide had evolved. What was the mass of the remaining solid?

7. The air oxidation of iron pyrite ($FeS_2$) to iron(III) oxide in the presence of water is believed to be a source of the dilute sulfuric acid found in underground mine water. What mass of $FeS_2$ is required for the formation of exactly 1 mol of $H^+$?

8. Silver combines with sulfur to form black silver sulfide ($Ag_2S$), the substance that causes silver to tarnish when exposed to various compounds containing sulfur. What mass of $Ag_2S$ is obtained from the complete reaction of 5.00 g of silver with 10.0 g of sulfur?

9. What mass of silver chloride will be precipitated when a solution containing 2.34 g of silver nitrate is added to a solution containing 5.05 g of sodium chloride? Will any sodium chloride remain in the solution? If so, how much?

10. Zinc and iodine combine when heated to form zinc iodide ($ZnI_2$). A mixture of 25.0 g of zinc and 26.3 g of iodine is heated in a closed tube until the combination reaction has occurred completely. What substances will be present in the tube and what are their masses?

11. Hexane ($C_6H_{14}$) is a liquid whose density is 0.660 g $cm^{-3}$. How many grams of $CO_2$ can be obtained from the combustion of 30.0 mL of hexane? Assume that $CO_2$ is the only product containing carbon formed.

12. $PbO_2$ and $BaO_2$ evolve oxygen upon thermal decomposition, forming PbO and BaO, respectively. A mixture of $PbO_2$ and $BaO_2$ weighing 17.1 g yielded 15.7 g of PbO and BaO upon decomposition. What mass of $BaO_2$ was present in the mixture?

13. An 8.27 g sample of a mixture of $MgCl_2$ and KCl in aqueous solution was treated with enough $Na_2CO_3$ to precipitate all the magnesium as $MgCO_3$. Thermal decomposition of the carbonate yielded 2.43 g of MgO. How many grams of $MgCl_2$ were present in the original sample?

14. An 8.37 g mixture of copper(I) oxide ($Cu_2O$) and copper(II) oxide (CuO) is reduced to the metal and

7.13 g of copper is obtained. What percentage by weight of the mixture was copper(I) oxide?

15. A mixture of $Mn_2O_3$ and $MnO_2$ weighing 4.73 g is heated, thereby converting each of these oxides into $Mn_3O_4$. If 4.35 g of $Mn_3O_4$ is obtained, what was the mass of $MnO_2$ originally present?

16. A sample of iron weighing 10.0 g is oxidized to iron(III) oxide by oxygen gas generated in the thermal decomposition of potassium chlorate. Only enough oxygen is generated to oxidize some of the metal, resulting in a mixture of Fe and $Fe_2O_3$ weighing 11.5 g. What mass of $KClO_3$ was decomposed?

17. A mixture of CuO and $Ag_2O$ weighing 15.4 g is heated in a stream of hydrogen until complete reduction to the metals (with the evolution of $H_2O$ vapor) has been achieved. If 1.78 g of $H_2O$ is produced, what weight of $Ag_2O$ was present in the initial mixture? What fraction by weight of the final mixture of metals is silver?

18. The displacement of silver by nickel is studied by adding 15.0 g of Ni metal to a beaker containing $AgNO_3$ solution. The beaker is dropped before all the nickel has reacted and the solution is lost. However, the solid metals are completely recovered and found to weigh 25.0 g. How much nickel and silver are present?

19. When 4.703 g of sodium metal reacted with excess of oxygen, 5.563 g of sodium oxide ($Na_2O$) were obtained. What is the percentage yield of $Na_2O$?

20. The rusting of iron in air results in the formation of iron(III) oxide, $Fe_2O_3$. An iron sample weighing 13.263 g is exposed to oxygen and 2.381 g of $O_2$ react. What mass of $Fe_2O_3$ is formed? What percentage of the iron rusts?

21. What is the molarity of a solution made by dissolving 7.45 g of potassium iodide in water and adjusting the volume to 50.0 mL?

22. What mass of sodium nitrate is contained in 25.0 mL of a 0.150 M solution?

23. The density of 65% nitric acid is 1.40 g cm$^{-3}$. What volume of nitric acid is needed to prepare 100 mL of a 0.20 M solution?

24. Concentrated hydrochloric acid contains 37.50% HCl by weight and has a density of 1.205 g cm$^{-3}$. How much water must be added to 50.00 mL of concentrated hydrochloric acid to make a 0.1000 M solution? Assume no volume change upon mixing.

25. What is the concentration of a solution made by mixing 25.0 mL of 0.20 M sodium sulfate with 50.0 mL of 0.050 M sodium sulfate? Assume no volume change upon mixing. Assuming that dissociation is complete, what are the concentrations of the $Na^+$ and $SO_4^{2-}$ ions?

26. What mass of $BaSO_4$ is obtained when 100.0 mL of 0.105 M $Na_2SO_4$ is added to 50.0 mL of 0.0850 M $Ba(NO_3)_2$? What are the concentrations of the ions remaining in solution?

27. Excess NaI was added to 25.0 mL of $AgNO_3$ solution and 3.50 g of AgI precipitate was formed. What was the concentration of the $AgNO_3$ solution?

28. It took 31.2 mL of an acidified solution of $Fe^{2+}$ ion to reduce 0.385 g of $KMnO_4$. What was the concentration of the $Fe^{2+}$ solution?

29. Oxalic acid ($H_2C_2O_4$) can act as an acid and also as a reducing agent, being oxidized to $CO_2$ by $KMnO_4$ in acidic solution. A 10.0 mL sample of an oxalic acid solution requires 21.4 mL of 0.165 M NaOH solution to react completely. What volume of 0.136 M $KMnO_4$ solution would be required to react completely with a second 10.0 mL aliquot (i.e., sample) of the same oxalic acid solution?

30. What volume of 0.163 M KOH solution is required to react completely with 25.0 mL of 0.135 M $H_2SO_4$?

31. It takes 18.3 mL of 0.102 M NaOH to react completely with 25.0 mL HCl solution. What is the concentration of the HCl solution?

32. What are the concentrations of the ions remaining in solution when 10.0 mL of 0.205 M $HNO_3$ is added to 15.0 mL of 0.235 M NaOH?

33. A strong base liberates ammonia from ammonium salts as a consequence of the decomposition of $NH_4OH$ formed in the double displacement reaction. To 25.0 mL of $NH_4Cl$ solution, was added 50.0 mL of 1.00 M NaOH and all the $NH_3$ was expelled by heating. The remaining solution was titrated with HCl and 46.4 mL of 0.805 M HCl was required for complete reaction. What was the concentration of the $NH_4Cl$ solution? What mass of $NH_3$ was obtained?

34. Balance each of the following equations. Indicate whether a redox reaction occurs; classify each reaction in one of the various categories introduced in this chapter, e.g., decomposition, displacement.

(a) $Al_2O_3(s) + HCl(g) \xrightarrow{\Delta} AlCl_3(s) + H_2O(g)$
(b) $F_2(g) + H_2O(l) \rightarrow HF(aq) + O_2(g)$
(c) $Fe_2O_3(s) + H_2O(l) \rightarrow Fe(OH)_3(s)$
(d) $H_2S(g) + SO_2(g) \rightarrow S(s) + H_2O(l)$
(e) $BaCO_3(s) + H^+ \rightarrow Ba^{2+} + CO_2(g) + H_2O(l)$
(f) $HCO_3^- \rightarrow CO_2(g) + CO_3^{2-} + H_2O(l)$
(g) $Al(s) + H^+ \rightarrow Al^{3+} + H_2(g)$
(h) $K_2CrO_4(aq) + AgNO_3(aq) \rightarrow KNO_3(aq) + Ag_2CrO_4(s)$
(i) $P_4O_{10}(s) + H_2O(l) \rightarrow H_3PO_4(aq)$
(j) $(NH_4)_2S(s) \xrightarrow{\Delta} NH_3(g) + H_2S(g)$
(k) $Ca_3(PO_4)_2(s) + SiO_2(s) \rightarrow CaSiO_3(s) + P_4O_{10}(g)$
(l) $SiBr_4(l) + H_2O(l) \rightarrow H_2SiO_3(s) + HBr(aq)$

(m) $SO_2(g) + O_2(g) \rightarrow SO_3(g)$
(n) $Pb(NO_3)_2(s) \xrightarrow{\Delta} PbO(s) + NO_2(g) + O_2(g)$
(o) $AuCl_3(s) \xrightarrow{\Delta} Au(s) + Cl_2(g)$
(p) $H_2O_2(l) \xrightarrow{\Delta} H_2O(l) + O_2(g)$

35. Complete and balance the following equations for reactions which occur in acidic solution:
    (a) $MnO_4^{2-} \rightarrow MnO_2 + MnO_4^-$
    (b) $ClO_3^- + As_2S_3 \rightarrow Cl^- + H_2AsO_4^- + HSO_4^-$
    (c) $V^{2+} + H_2O_2 \rightarrow H_4VO_4^+$
    (d) $PH_4^+ + Cr_2O_7^{2-} \rightarrow P + Cr^{3+}$
    (e) $MnO_4^- + Cl^- \rightarrow Mn^{2+} + HClO$
    (f) $NH_4^+ + MnO_4^- \rightarrow NO_3^- + Mn^{2+}$
    (g) $NO_3^- + I_2 \rightarrow IO_3^- + NO_2$
    (h) $CuS + NO_3^- \rightarrow Cu^{2+} + HSO_4^- + NO$
    (i) $Ag + NO_3^- \rightarrow Ag^+ + NO$

36. Complete and balance the following equations for reactions which occur in basic solution:
    (a) $PbO_2 + Cl^- \rightarrow ClO^- + Pb(OH)_3^-$
    (b) $ClO_2 \rightarrow ClO_2^- + ClO_3^-$
    (c) $Mn(CN)_6^{4-} + O_2 \rightarrow Mn(CN)_6^{3-}$
    (d) $HO_2^- + Cr(OH)_3^- \rightarrow CrO_4^{2-}$
    (e) $MnO_4^- + Fe_3O_4 \rightarrow Fe_2O_3 + MnO_2$
    (f) $CrO_4^{2-} + CN^- \rightarrow CNO^- + Cr(OH)_3$
    (g) $As_2S_3 + NO_3^- \rightarrow HAsO_4^{2-} + S + NO_2$
    (h) $MnO_4^- + IO_3^- \rightarrow MnO_2 + IO_4^-$
    (i) $Cl_2 \rightarrow ClO_3^- + Cl^-$
    (j) $S^{2-} + Cl_2 \rightarrow SO_4^{2-} + Cl^-$

# 3
# Properties of Gases

All gaseous substances behave in a remarkably uniform manner. The relationships among their pressures, molar volumes, and temperatures are nearly independent of their identity for a wide range of conditions. Many experimental observations can consequently be summarized by a few empirical laws. Since gases consist of large numbers of atoms or molecules, it makes sense to suppose that these laws must reflect some aspects of molecular behavior. As we shall see in Chapter 4, a simple model based on a few plausible assumptions about the motion of molecules can be used to account for the empirical laws. In this manner, the properties of gases serve to illustrate the connection between the macroscopic phenomena observed in the laboratory and the behavior of molecules. This connection is fundamental to all of chemistry.

From a more practical viewpoint, many substances used in both laboratory and industrial applications are gases under the conditions of their use. Several of the light (i.e., low atomic weight) elements as well as many of their compounds normally occur as gases, and two of them, nitrogen and oxygen, are the principal components of the atmosphere. Chemical and physical changes occurring in the atmosphere have a substantial bearing on long-term environmental conditions on the surface of the Earth. Thus, an understanding of the behavior of gases is of general importance.

# The Ideal Gas

## 3.1 Avogadro's Law and Gas Stoichiometry

While many of the early experiments on the combining weights of substances were performed with solids, experiments with gases played a crucial role in placing the atomic theory of matter on a firm footing. In the early years of the nineteenth century, Joseph Gay-Lussac (1778–1850) performed a series of experiments on the volumes of gases undergoing chemical reaction. He found that the volumes of gaseous reactants and products stood in the ratio of small integers provided the volumes were measured at the same temperature and pressure. For example, in the reaction of hydrogen and chlorine to form hydrogen chloride, 2 liters of HCl gas were formed for each liter of $H_2$ and each liter of $Cl_2$ that reacted.

Gay-Lussac's results were interpreted by Avogadro, who advanced the hypothesis that *equal volumes of different gases at the same temperature and pressure contain equal numbers of molecules*. The relative volumes of gaseous reactants and products observed by Gay-Lussac could then be interpreted as corresponding to the relative numbers of molecules undergoing reaction. In the reaction between hydrogen and chlorine, for example, the relative volumes correspond to the stoichiometric coefficients in the equation

$$H_2(g) + Cl_2(g) \rightarrow 2HCl(g)$$

If 2 liters of HCl are formed for each liter of $H_2$ or $Cl_2$ that reacts, then 2 molecules of HCl must be formed for each molecule of $H_2$ or $Cl_2$ that reacts, and 2 moles of HCl are formed for each mole of $H_2$ or $Cl_2$ that reacts. Thus, in addition to the relationship between the masses of reactants and products implicit in a balanced equation, there is an even more direct relationship between the volumes of gaseous species measured at the same temperature and pressure (see Figure 2.2). In other words, the molecules in a gas can be "counted" not only by weighing the gas but also by measuring its volume.

If the volume of 1 mole of various gases is measured at the same temperature and pressure, then according to Avogadro's law, the same value should be obtained for each gas. Table 3.1 lists the molar volumes of several gases at 0°C and a pressure of 1 atmosphere, conditions known as *standard temperature and*

**TABLE 3.1 Molar Volume of Gases at Standard Temperature and Pressure (STP)**

| Gas | Molar Volume (L) |
|---|---|
| $H_2$ | 22.43 |
| $N_2$ | 22.40 |
| $O_2$ | 22.39 |
| $CO_2$ | 22.26 |
| $NH_3$ | 22.08 |
| $Cl_2$ | 22.06 |

*pressure* (STP). While the values are in close agreement, they do differ from each other by up to a few percent. This difference can practically be eliminated by reducing the pressure. It is found that as the pressure approaches zero, the volumes of all gases approach a value of 22.414 L at 0°C. Avogadro's law is an example of a limiting law since it becomes exact only in the limit of low pressure. Nonetheless, Avogadro's law is very useful in stoichiometric calculations involving gases at ordinary pressures. A molar volume of 22.4 L at STP may be used in such calculations, as illustrated in the following example.

---

**Example 3.1**   An evacuated 1.00 L gas bulb was filled with 1.34 g of an unknown hydrocarbon (compound of carbon and hydrogen). The sample was then burned in excess oxygen thereby completely converting it to carbon dioxide and 2.41 g water. What is the molecular formula of the hydrocarbon?

The mass of hydrogen in the unknown sample can be obtained from the mass of water:

$$m_H = \frac{(2.41 \text{ g H}_2\text{O})}{(18.0 \text{ g mol}^{-1})} \frac{(2 \text{ mol H})}{(\text{mol H}_2\text{O})} (1.01 \text{ g mol}^{-1}) = 0.270 \text{ g}$$

The mass of carbon in the sample is obtained by subtracting the mass of hydrogen from that of the sample:

$$m_C = 1.34 \text{ g} - 0.270 \text{ g} = 1.07 \text{ g}$$

Knowing the composition of the hydrocarbon, its empirical formula may be obtained (see Example 1.9):

$$n_C = \frac{1.07 \text{ g}}{12.0 \text{ g mol}^{-1}} = 0.0892 \text{ mol}$$

$$n_H = \frac{0.270 \text{ g}}{1.01 \text{ g mol}^{-1}} = 0.270 \text{ mol}$$

Therefore, the ratio of H to C atoms in the hydrocarbon is

$$\frac{0.270 \text{ mol H}}{0.0892 \text{ mol C}} = 3.03 = 3$$

The empirical formula is therefore $CH_3$ and the molecular formula can be written as $(CH_3)_x$. To evaluate $x$ we note that the volume of the hydrocarbon at STP is 1.00 L. Using Avogadro's law we obtain the corresponding number of moles as

$$n_{(CH_3)_x} = \frac{1.00 \text{ L}}{22.4 \text{ L mol}^{-1}} = 0.0446 \text{ mol}$$

The empirical formula permits us to express the mass of 0.0446 mol in terms of $x$. Since the mass of the hydrocarbon is known, we can solve for $x$:

$$m_{(CH_3)_x} = (0.0446 \text{ mol})(15.0x \text{ g mol}^{-1}) = 1.34 \text{ g}$$

$$x = 2.00$$

The molecular formula of the hydrocarbon is $C_2H_6$.   ■

## 3.2 Pressure of a Gas

A characteristic feature of gases is that they expand to fill the entire volume of their containers. If the volume of the container is adjustable, as in the illustration in Figure 3.1, the volume of a gas is found to be strongly dependent on the applied pressure. A doubling of the pressure results in a halving of the volume. Solids and liquids, by contrast, are nearly incompressible. For example, the volume of liquid water is reduced by only about 0.005% when the externally applied pressure is doubled. This difference is not surprising since the molecules in a gas are much farther apart than those in a liquid or solid (Section 4.1). Due to this characteristic feature of gases, pressure is an important variable when dealing with this state of matter.

The pressure on a gas may be defined with reference to Figure 3.1. When the container depicted in this figure is filled with a gas and a force $F$ is applied on the piston, a pressure is exerted on the gas. The **pressure** is the force per unit cross-sectional area:

$$P = F/A \tag{3.1}$$

The units of pressure are in turn determined by those of the applied force. We recall from Section 1.4 that force is defined by Newton's second law,

$$F = ma \tag{3.2}$$

$$(1 \text{ newton}) = (1 \text{ kg})(1 \text{ m s}^{-2})$$

which states that a force exerted on a body of mass $m$ produces an acceleration $a$. The corresponding SI unit of pressure is the pascal (Pa), so that one pascal equals one newton per square meter (N m$^{-2}$).

It should be noted that the pressure resulting from the force applied to the piston in Figure 3.1 is the external pressure acting on the gas in the container. In turn, the gas pushes back against the piston. If the external pressure is significantly larger than the pressure exerted by the gas, the piston moves inward and the gas is compressed. This process continues until the external and gas pressures become equal.

The most familiar type of pressure is that exerted by the atmosphere (Section 3.6). The gases making up the atmosphere occupy the space surrounding the Earth up to an altitude of approximately 80 km. The gravitational attraction of the Earth causes these gases to exert a force, and thus a pressure, on the matter beneath. Since the density of atmospheric gases decreases with increasing altitude, atmospheric pressure also decreases with altitude.

The pressure exerted by the atmosphere can be measured with a mercury barometer, a simple device invented by Evangelista Torricelli (1608–1647). As shown in Figure 3.2, a tube closed at one end is filled with mercury and inverted into a dish of mercury exposed to the atmosphere. Some of the mercury will run out of the tube but enough will remain so that the pressure it exerts on the surface of the pool of mercury is equal to that exerted by the atmosphere.

It is customary to express pressure in units of atmospheres. An **atmosphere** is defined as the pressure that will support a column of mercury to a height of exactly 760 mm at sea level and a temperature of 0°C. (Because the Earth is not exactly spherical, the force of gravity at sea level varies slightly with location. An average "sea level" can be defined as the place where the acceleration due to gravity is

$$g = 9.80665 \text{ m s}^{-2}$$

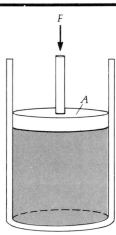

**FIGURE 3.1** A container of adjustable volume. An external force $F$ applied to a piston of cross-sectional area $A$ exerts a pressure $P$ on the gas in the container.

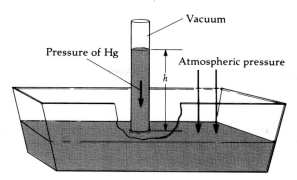

**FIGURE 3.2** Operation of a barometer. The pressure exerted by the atmosphere on the pool of mercury is equal to the force exerted per unit area by a column of mercury of height $h$. The space above the column of mercury is essentially a vacuum, containing only a small amount of mercury vapor.

---

A freely falling body subject to this acceleration increases its speed by 9.80665 m s$^{-1}$ in each second of fall.) It has been customary in the past to call a "millimeter of mercury" a *torr*, after Torricelli. Therefore,

$$1 \text{ atm} = 760 \text{ mm Hg} = 760 \text{ torr}$$

Pressure is commonly expressed in all three units.

The factor for the conversion of pressure in atmospheres to pressure in pascals, the SI unit, may be obtained by means of the following calculation. As in Figure 3.2, we consider a column of mercury of height $h$ and cross-sectional area $A$. The force exerted by this column on the surface of the mercury pool is given by Equation 3.2, where the acceleration is that due to gravity,

$$a = g = 9.80665 \text{ m s}^{-2}$$

(see Equation 1.2). It is convenient to relate the force to the density of mercury. This relation can be obtained by noting that the volume $V$ of the cylindrical mercury column is equal to the product of its cross-sectional area $A$ and height $h$. Thus, the density $\rho$ is

$$\rho = m/V = m/Ah \tag{3.3}$$

Solving this equation for $m$ and substituting into Equation 3.2 for the force, we obtain

$$F = \rho A g h \tag{3.4}$$

from which the pressure is immediately obtained by means of Equation 3.1:

$$P = \rho g h \tag{3.5}$$

The desired conversion factor between atmospheres and pascals is obtained by numerical substitution into this expression. The height of mercury corresponding to 1 atmosphere is 760 mm and the density at 0°C is 13.595 g cm$^{-3}$. Thus,

$$1 \text{ atm} = (13.595 \text{ g cm}^{-3})(10^{-3} \text{ kg g}^{-1})(10^2 \text{ cm m}^{-1})^3(9.80665 \text{ m s}^{-2})$$
$$(760 \text{ mm})(10^{-3} \text{ m mm}^{-1})(1 \text{ N/kg m s}^{-2})$$

$$1 \text{ atm} = 1.0132 \times 10^5 \text{ N m}^{-2} = 1.0132 \times 10^5 \text{ Pa}$$

**Example 3.2** What is the height of a column of water supported by exactly one atmosphere?

According to Equation 3.5, the height of a column of liquid supported by the atmosphere is inversely proportional to its density. Relating the height and density of water to the corresponding quantities for mercury, we obtain

$$h_{H_2O} \times \rho_{H_2O} = h_{Hg} \times \rho_{Hg}$$

and substituting the appropriate numerical values,

$$h_{H_2O} = \frac{(13.59 \text{ g cm}^{-3} \text{ Hg})(0.760 \text{ m Hg})}{(1.00 \text{ g cm}^{-3} \text{ H}_2\text{O})} = 10.3 \text{ m}$$

A 10-m high barometer would be very unwieldy, which is why it is customary to use mercury, a very dense liquid, as the working fluid in a barometer. ∎

The various scientific units used to express pressure are summarized in Table 3.2. In addition to the units discussed so far, the table also lists the bar, a multiple of the pascal which has nearly the same magnitude as one atmosphere. The kilobar is a commonly used unit for high pressures.

## 3.3 The Gas Laws

Experiments performed in the last three centuries have revealed that the dependence of the volume of a gas on pressure and temperature is virtually independent of the identity of the gas provided that the pressure is sufficiently low. The empirical laws derived from these observations serve as the foundation for a more complete understanding of gas behavior.

*a. Dependence of Volume on Pressure*  Robert Boyle (1627–1691), one of the first scientists to perform quantitative measurements, investigated the relationship between the volume of a gas and its pressure using the apparatus illustrated in Figure 3.3. The same relationship was obtained for a number of different gases, namely, that the volume of a fixed amount of gas is inversely proportional to its pressure provided that the temperature $T$ remains constant **(Boyle's law)**. This relationship can be expressed by the following equations:

$$PV = \text{constant} \qquad \text{(at constant } T, n\text{)} \tag{3.6a}$$

$$P = \text{constant} \times \frac{1}{V} \qquad \text{(at constant } T, n\text{)} \tag{3.6b}$$

**TABLE 3.2 Scientific Units Used to Express Pressure**

| Unit (Symbol) | Conversion Factor |
|---|---|
| pascal (Pa) | 1 N m$^{-2}$ (SI unit of pressure) |
| atmosphere (atm) | 1.01325 × 10$^5$ Pa |
| mm Hg or torr (torr) | (1/760) atm |
| bar (bar) | 10$^5$ Pa |

Properties of Gases 3.3   71

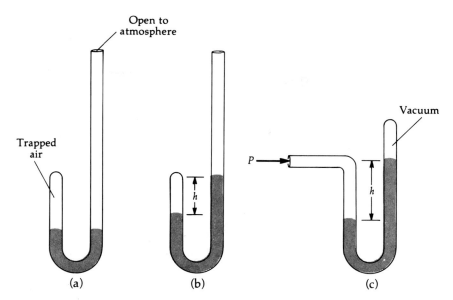

**FIGURE 3.3** Boyle's apparatus for the study of the relationship between the pressure and volume of a gas. (a) The volume of a certain amount of trapped air at atmospheric pressure is proportional to the height of the trapped air column. (b) When additional mercury is poured into the container, the pressure on the gas increases and its volume decreases. The new pressure is equal to the difference in height $h$ between the mercury levels in the two arms plus atmospheric pressure. (c) The *manometer*, an adaptation of Boyle's apparatus, may be used to measure the pressure of a gas in the laboratory. The open arm is attached to the gas container. The difference between the mercury levels in the two arms is equal to the gas pressure.

Boyle's law is shown graphically in Figure 3.4. The plot of pressure versus volume (Equation 3.6a) yields a curve with the shape of a hyperbola (Figure 3.4a). A plot of Equation 3.6b ($P$ versus $V^{-1}$) shows a linear dependence (Figure 3.4b). Since a straight line is easier to recognize than a hyperbola, this plot makes Boyle's law more obvious visually. Yet another graphical illustration is given by a plot of $PV$ versus $P$ (Figure 3.4c). Equation 3.6a indicates that at constant temperature $PV$ is constant and, therefore, independent of $P$. The plot in question consequently yields a horizontal line. Deviations from Boyle's law ob-

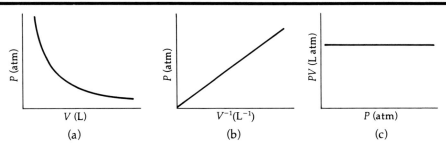

**FIGURE 3.4** Boyle's law: (a) pressure $P$ versus volume $V$, (b) $P$ versus $V^{-1}$, and (c) $PV$ versus $P$.

served at high pressures are readily seen in such a plot as departures from the horizontal.

**b. Dependence of Volume on Temperature** All substances generally expand when heated. However, gases expand substantially more than solids or liquids. The volume of liquid water at room temperature, for example, increases by approximately 0.03% per degree, while that of a gas maintained at constant pressure increases by about 0.3% per degree. The expansion of many substances, including that of all gases, is essentially linear over a broad range of temperatures. This property serves as the basis of the familiar method of measuring temperature. A linear temperature scale requires two reference temperatures. The widely used Celsius scale uses the normal ($P = 1$ atm) freezing and boiling points of water for this purpose. These temperatures are defined as 0°C and 100°C, respectively. A 1 Celsius degree temperature interval is therefore equal to $\frac{1}{100}$ of the difference between the temperatures of the normal boiling and freezing points of water.

The linear relationship between the volume of a gas at constant pressure and the temperature, known as the **law of Charles and Gay-Lussac,** was established experimentally in the eighteenth century. Figure 3.5 shows some typical data obtained for several values of the gas pressure. Since gases condense to liquids at sufficiently low temperatures there is a lower temperature limit below which data cannot be obtained. However, if the lines in the figure are extrapolated to $V = 0$, it is found that they intersect the abscissa at a temperature of about −273°C, *regardless of the identity of the gas.* This striking result indicates that −273°C (actually −273.15°C) is the lowest possible temperature, since lower values would result in physically impossible negative volumes. This temperature is therefore an *absolute zero* as opposed to the arbitrary zero corresponding to the freezing point of water.

The preceding considerations suggest that a different temperature scale, in which the lowest possible temperature is set equal to zero degrees, should be useful. This scale was introduced by William Thomson, Lord Kelvin

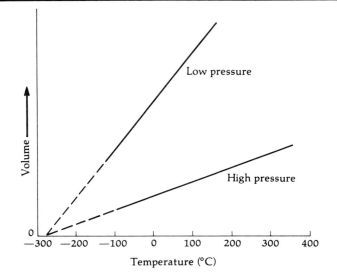

**FIGURE 3.5** Variation of the volume of a gas with temperature at two different pressures. The dashed lines show the extrapolation to $V = 0$.

(1824–1907) and is known as the Kelvin or absolute temperature scale. The **absolute temperature** $T$ is measured in kelvins (K) and is related to the Celsius scale as follows:

$$T(K) = T(°C) + 273.15°C \tag{3.7}$$

Note that a given temperature change is identical on the two scales. For instance, the difference between the normal boiling and freezing points of water is 100°C or 100 K. The law of Charles and Gay-Lussac takes on a particularly simple form when the temperature is expressed in kelvins:

$$V = \text{constant} \times T \quad \text{(at constant } P, n\text{)} \tag{3.8}$$

**c. The Combined Gas Law** The laws of Boyle and Charles may be combined to express the variations of the pressure, volume, and temperature of a gas with each other. Since $PV$ equals a constant, and $V$ is proportional to $T$, the combined gas law for a given number of moles may be written as

$$PV = \text{constant} \times T \tag{3.9}$$

or for two particular sets of values of these properties, designated 1 and 2,

$$\frac{P_1 V_1}{T_1} = \frac{P_2 V_2}{T_2} \tag{3.10}$$

The combined gas law is shown graphically in Figure 3.6 as a plot of $P$ versus $V$ at different $T$. Such curves are known as **isotherms,** meaning that the temperature along each curve remains constant. As is true for Avogadro's law, the laws of Boyle and Charles (and, hence, the combined gas law) are strictly valid only in the limit of low pressure. However, they provide a reasonably accurate description of the behavior of most gases at pressures up to a few atmospheres at room temperature or higher.

The combined gas law, as expressed in Equation 3.10, permits a quantitative determination of any one of the three properties of interest if the other two

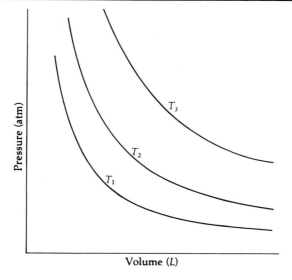

**FIGURE 3.6** The combined gas law is shown as a series of $(P,V)$ isotherms at temperatures $T_1 < T_2 < T_3$.

74  Properties of Gases  3.4

are known and if all three are known for another set of conditions. A variety of problems, all of which are really only variants of each other, may be attacked by making use of this equation.

**Example 3.3**  A balloon carrying instruments to monitor pollution of the upper atmosphere is filled with helium gas at 25°C and a pressure of 1.00 atm. The volume of the balloon under these conditions is $1.75 \times 10^4$ L. When the balloon reaches its highest altitude the pressure drops to 0.605 atm, and the temperature drops to −35°C. What is the volume of the balloon under these conditions?

Equation 3.10 can be used to solve for the unknown volume. Designating the upper atmosphere conditions by the subscript 2, we have after rearrangement:

$$V_2 = V_1 \times \frac{T_2}{T_1} \times \frac{P_1}{P_2}$$

$$V_2 = (1.75 \times 10^4 \text{ L})\left(\frac{238 \text{ K}}{298 \text{ K}}\right)\left(\frac{1.00 \text{ atm}}{0.605 \text{ atm}}\right) = 2.31 \times 10^4 \text{ L}$$

Note that the temperatures must first be converted to kelvins. This is *always* the case in problems involving the use of the gas laws. Note, too, that the change in volume arises from two separate factors, the decrease in temperature and the decrease in pressure. Since $V$ is directly proportional to $T$ and inversely proportional to $P$, the decrease in $T$ leads to a concomitant decrease in $V$, while the decrease in $P$ has the opposite effect. If the relations between $V$ and $P$ and between $V$ and $T$ are kept in mind when setting up a problem of this type, the chances of accidentally reversing the ratios of the pressures or temperatures will be minimized.  ∎

## 3.4  The Ideal Gas Equation of State

The preceding gas laws tacitly assume that the amount of gas present does not change as the pressure and temperature are varied. If gas is added or removed from a container at constant temperature, the pressure or volume must change. If the container has a fixed volume, an increase in the amount of gas present increases the pressure; if, as in Figure 3.1, the volume is allowed to vary, an increase in the amount of gas leads to an increase in volume provided that constant pressure is maintained. Equation 3.9 can be modified to show explicitly the dependence on the number of moles of gas, $n$:

$$\frac{PV}{T} = n \times \text{constant} \tag{3.11}$$

Recall that Avogadro's law (Section 3.1) indicated that at a given temperature and pressure, the same volumes of different gases contain the same number of molecules and, thus, of moles. Consequently, the constant in Equation 3.11 must have the same numerical value for all gases. It is customary to designate this **gas constant** as $R$ and to write Equation 3.11 as

$$PV = nRT \tag{3.12}$$

This equation is known as the ideal gas equation of state or more simply, as the **ideal gas law;** a gas that obeys this relation is called an **ideal gas.**

An **equation of state** is a statement of the relationships among various properties of a system, such as temperature and pressure. Equation 3.12 shows how any one of the properties of an ideal gas is related to the others. Note that this equation does not contain any constants which depend on the identity of the gas. In the limit of low pressures, the ideal gas law applies to *all* gases. The various laws considered in the preceding section are seen to be special cases of the ideal gas law. For example, Equation 3.12 reduces to Boyle's law when the temperature and number of moles remain constant.

The value of the gas constant $R$ is obtained most directly from the fact that the molar volume at STP of those gases which obey Equation 3.12 at these conditions is 22.414 L. Substituting into Equation 3.12 we obtain

$$R = \frac{(1 \text{ atm})(22.414 \text{ L mol}^{-1})}{(273.15 \text{ K})} = 0.082057 \text{ L atm mol}^{-1} \text{ K}^{-1}$$

This result can be expressed in SI units by means of the previously derived conversion factor between atmospheres and pascals:

$$R = (0.082057 \text{ L atm mol}^{-1} \text{ K}^{-1})(10^{-3} \text{ m}^3 \text{ L}^{-1})(1.01325 \times 10^5 \text{ Pa atm}^{-1})$$
$$\times (1 \text{J/Pa m}^3) = 8.3144 \text{ J mol}^{-1} \text{ K}^{-1}$$

The last factor in this expression, involving the replacement of a (pascal × cubic meter) by a joule, requires some amplification. The equivalence of these two units can be obtained from the following dimensional analysis, in which the pascal, the newton, and the joule are expressed in terms of the fundamental units of mass, length, and time:

$$1 \text{ Pa m}^3 = [(1 \text{ N m}^{-2})(\text{m}^3)][1(\text{kg m s}^{-2})\text{N}^{-1}][1 \text{ J}(\text{kg m}^2\text{s}^{-2})^{-1}]$$
$$= 1 \text{ J}$$

The individual unit conversion factors are explained in Appendix 1. Note that the product of pressure and volume, $PV$, which can be expressed in units of Pa m$^3$, has the dimensions of energy.

---

**Example 3.4** The instrument balloon of Example 3.3 had a volume of $1.75 \times 10^4$ L when the H$_2$ pressure at 25°C was 1.00 atm. How many moles of H$_2$ are needed to achieve these conditions? The hydrogen is obtained from a standard gas cylinder, pressurized when full to 15,180 kPa, with an internal volume of 43.8 L. How many cylinders are needed to fill the balloon?

The first part of this problem merely requires substitution of the known $P$, $V$, and $T$ into the ideal gas law (Equation 3.12), which is then solved for the remaining variable, $n$. Note that $R$ must be expressed in L atm mol$^{-1}$ K$^{-1}$ for proper cancellation of units:

$$n = \frac{PV}{RT} = \frac{(1.00 \text{ atm})(1.75 \times 10^4 \text{ L})}{(0.08206 \text{ L atm mol}^{-1} \text{ K}^{-1})(298.15 \text{ K})} = 715 \text{ mol}$$

The second part of the problem deals with the practical question of how many gas cylinders are needed to inflate the balloon. The procedure is essentially the same as in the first part, except that the unknown quantity is the volume occupied by 715 moles at the indicated pressure and temperature. Note, however, that the pressure is expressed in SI units, so that $R$ must be expressed in J mol$^{-1}$ K$^{-1}$. Solving for $V$, we obtain

$$V = \frac{nRT}{P} = \frac{(715 \text{ mol})(8.314 \text{ J mol}^{-1}\text{ K}^{-1})(298.15 \text{ K})(1 \text{ Pa m}^3/\text{J})}{(1.5180 \times 10^7 \text{ Pa})} = 0.117 \text{ m}^3$$

$$= (0.117 \text{ m}^3)(10^3 \text{ L m}^{-3}) = 117 \text{ L}$$

Since one cylinder has a volume of 43.8 L, three cylinders will be required, the third cylinder being only partially depleted. ∎

Although Equation 3.12 is the usually encountered form of the ideal gas law, other variants are of use in certain applications. An expression relating the molecular weight of a gas to other gas properties can be derived from Equation 3.12. This expression forms the basis of an experimental technique for the determination of the molecular weight of gases.

This ideal gas law may be written as

$$P = nRT/V$$

Multiplying both sides by the molecular weight of the gas, $\overline{M}$, gives

$$P\overline{M} = n\overline{M}RT/V = mRT/V \tag{3.13}$$

where the mass $m$ is equal to the product of the molecular weight and the number of moles $n$. Since the density $\rho$ is equal to the ratio of mass to volume, we obtain

$$P\overline{M} = \rho RT \tag{3.14}$$

indicating that at constant temperature and pressure the density of a gas is proportional to its molecular weight.

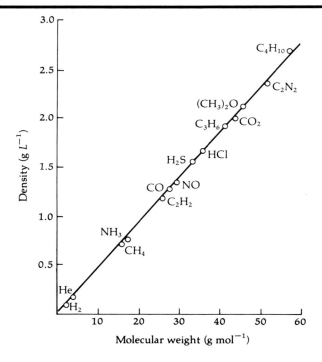

**FIGURE 3.7** Density of gases at 0°C and atmospheric pressure (STP) as a function of molecular weight.

Equation 3.14 may be used to demonstrate the validity of Avogadro's law. The direct proportionality between the density and molecular weight of gases at a given temperature and pressure will be observed only if the gas constant $R$ is indeed the same for all gases. As noted earlier, this condition is equivalent to Avogadro's law. Figure 3.7 shows a plot of the density of various gases at STP as a function of their molecular weight. The linear variation confirms the general validity of Equation 3.14 and, thus, of Avogadro's law.

**Example 3.5** A glass bulb weighs 30.0000 g when evacuated, 130.0000 g when filled with water at 25°C, and 30.0357 g when filled with a certain gas at 25°C and 150.0 torr pressure. The density of water at 25°C is 0.99567 g cm$^{-3}$. What is the molecular weight of the gas?

The first two weight measurements together with the density of water permit a determination of the volume of the bulb:

$$m_{H_2O} = 130.0000 \text{ g} - 30.0000 \text{ g} = 100.0000 \text{ g}$$

$$V = \frac{m_{H_2O}}{\rho_{H_2O}} = \frac{100.0000 \text{ g}}{0.99567 \text{ g cm}^{-3}} = 100.43 \text{ cm}^3$$

The density of the gas is next obtained from its weight and volume:

$$m_{gas} = 30.0357 \text{ g} - 30.0000 \text{ g} = 0.0357 \text{ g}$$

$$\rho_{gas} = \frac{0.0357 \text{ g}}{100.43 \text{ cm}^3} = 3.555 \times 10^{-4} \text{ g cm}^{-3}$$

Solving Equation 3.14 for the molecular weight and inserting the known data, we obtain

$$\overline{M} = \frac{\rho RT}{P}$$

$$= \frac{(3.555 \times 10^{-4} \text{ g cm}^{-3})(10^3 \text{ cm}^3 \text{ L}^{-1})(0.08206 \text{ L atm mol}^{-1} \text{ K}^{-1})(298.15 \text{ K})}{(150.0 \text{ torr})[(1/760.0) \text{ atm torr}^{-1}]}$$

$$= 44.1 \text{ g mol}^{-1} \quad \blacksquare$$

## 3.5 Dalton's Law of Partial Pressures

Our discussion of gases has so far been restricted to single substances. It has been found experimentally that mixtures of gases also obey the ideal gas law, a result which is not surprising in view of the fact that this law does not depend on the identity of a gas. Thus, it should not matter whether a single gas or a mixture of gases is present.

If a container is filled with a mixture of gases, each gas will fill the entire volume and exert a certain pressure, known as its **partial pressure**. **Dalton's law** summarizes the results of experimental observations by stating that *the total pressure of a gas mixture is equal to the sum of the partial pressures of the individual components*. If there are $j$ components in the mixture we have

$$P = P_1 + P_2 + \cdots + P_j = \sum_{i=1}^{j} P_i \tag{3.15}$$

Since each component obeys the ideal gas law, Equation 3.12 can be separately applied to each one. The partial pressure of the $i$th component is

$$P_i = \frac{n_i RT}{V} \tag{3.16}$$

Dividing $P_i$ by the total pressure $P$ and substituting for $P$ the quantities given by the ideal gas law, we obtain

$$\frac{P_i}{P} = \frac{n_i RT/V}{nRT/V} = \frac{n_i}{n} \tag{3.17}$$

which is customarily written as follows (an alternative way of stating Dalton's law):

$$P_i = \frac{n_i P}{n} = X_i P \tag{3.18}$$

where the quantity $X_i$, the ratio of the number of moles of the $i$th component to the total number of moles of gas present, is called the **mole fraction** of component $i$. This form of Dalton's law indicates that the partial pressure of a gas may be obtained from the total pressure and the composition of the mixture expressed in mole fractions.

---

**Example 3.6** A sample of solid ammonium sulfide, $(NH_4)_2S$, is placed in a closed bulb which is then filled with 0.0150 mol of inert helium gas. The pressure of helium at 25°C is 145 mm Hg. The bulb is then heated to 325°C and all the ammonium sulfide decomposes to form gaseous ammonia and hydrogen sulfide. The total gas pressure at this temperature is 875 mm Hg. Assuming that the volume of the ammonium sulfide is negligible compared to that of the bulb determine (a) the partial pressures of the three gases at 325°C, and (b) the mass of ammonium sulfide present initially. Assume ideal gas behavior.

(a) According to Dalton's law, the total pressure at 325°C is equal to the sum of the partial pressures,

$$P = 875 \text{ mm Hg} = P_{He} + P_{NH_3} + P_{H_2S}$$

The pressure of He at 325°C can be obtained by means of the combined gas law, Equation 3.10. Since the volume and the amount of He present remain constant, we have

$$P_{He} = P_2 = P_1 \frac{T_2}{T_1}$$

$$= (145 \text{ mm Hg}) \frac{(598 \text{ K})}{(298 \text{ K})} = 291 \text{ mm Hg}$$

The decomposition reaction is

$$(NH_4)_2S \xrightarrow{\Delta} 2NH_3 + H_2S$$

indicating that 2 moles of $NH_3$ are formed for each mole of $H_2S$. Since the gases occupy the same volume and are at the same temperature, the partial pressure of $NH_3$ must be twice that of $H_2S$,

$$P_{NH_3} = 2P_{H_2S}$$

Substituting into Dalton's law, we have

$$P = 875 \text{ mm Hg} = 291 \text{ mm Hg} + 2P_{H_2S} + P_{H_2S}$$

$$P_{H_2S} = 194.7 \text{ mm Hg}$$

Finally,

$$P_{NH_3} = 2P_{H_2S} = 389 \text{ mm Hg}$$

The partial pressures of He, H$_2$S, and NH$_3$ are 291, 195, and 389 mm Hg, respectively.

(b) The amount of (NH$_4$)$_2$S present initially can be obtained from the reaction stoichiometry and the number of moles of product formed. The latter can, in turn, be found from the partial pressures and the ideal gas law:

$$n_{(NH_4)_2S} = n_{H_2S} = \frac{P_{H_2S} V}{RT}$$

The volume is not known explicitly but can be obtained from the data for He and the ideal gas law, $V = n_{He}RT/P_{He}$. Combining these two relations we obtain

$$n_{(NH_4)_2S} = \frac{(P_{H_2S})(n_{He})}{P_{He}}$$

Note that $P_{He}$ must be evaluated at 325°C for the temperatures to cancel:

$$n_{(NH_4)_2S} = \frac{(195 \text{ mm Hg})(0.0150 \text{ mol})}{(291 \text{ mm Hg})} = 0.0101 \text{ mol}$$

The corresponding mass is obtained from the formula weight,

$$m_{(NH_4)_2S} = (0.0101 \text{ mol})(68.1 \text{ g mol}^{-1}) = 0.688 \text{ g} \quad \blacksquare$$

A common laboratory application of Dalton's law is the study of the stoichiometry of reactions in which gaseous products are formed. As illustrated in Figure 3.8, gases can be conveniently collected by displacement of water. The

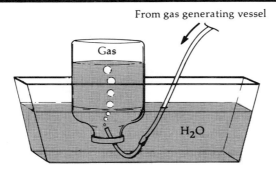

**FIGURE 3.8** The collection over water of a gas generated in a chemical reaction, illustrating an experiment in reaction stoichiometry. When the water level is the same inside and outside the collection vessel, the total gas pressure in the collection vessel is equal to the outside pressure, i.e., atmospheric pressure. The vapor pressure of water at the measured water temperature must be subtracted from the measured atmospheric pressure to obtain the partial pressure of the generated gas.

## 80  Properties of Gases  3.5

**TABLE 3.3  The Vapor Pressure of Water Between 18 and 30°C**

| T (°C) | $P_{H_2O}$ (mm Hg) | T (°C) | $P_{H_2O}$ (mm Hg) |
|---|---|---|---|
| 18 | 15.477 | 25 | 23.756 |
| 19 | 16.477 | 26 | 25.209 |
| 20 | 17.535 | 27 | 26.739 |
| 21 | 18.650 | 28 | 28.349 |
| 22 | 19.827 | 29 | 30.043 |
| 23 | 21.068 | 30 | 31.824 |
| 24 | 22.377 |   |   |

collected gas consists of the gaseous reaction product and water vapor. The pressure exerted by the water vapor is known as the vapor pressure of water (Section 12.2). The vapor pressure of a liquid depends only on the temperature and is independent of the amount of liquid present. Table 3.3 lists the vapor pressure of water at room temperatures. The partial pressure of the gaseous reaction product can be obtained by subtracting the vapor pressure of water at the measured temperature from the total gas pressure.

---

**Example 3.7**  In an experiment in which hydrogen is generated in the reaction of magnesium with dilute hydrochloric acid, the gas is collected at 27°C in a bottle inverted in a trough of water. When the level of water in the originally full bottle is the same as that in the trough, the volume of the collected gas is 910 mL. How many moles of $H_2$ have been collected and what is the mole fraction of $H_2$ in the gas? How many grams of magnesium reacted? The barometric pressure during the experiment was measured as 746 torr.

The number of moles of $H_2$ can be determined by means of the ideal gas law, after first subtracting the vapor pressure of water from the total gas pressure. Since the level of water inside the bottle is the same as that in the trough, the total gas pressure is equal to atmospheric pressure. Therefore,

$$P_{H_2} = P - P_{H_2O} = 746 \text{ mm Hg} - 26.7 \text{ mm Hg} = 719.3 \text{ mm Hg}$$

$$n_{H_2} = \frac{P_{H_2}V}{RT}$$

$$= \frac{(719.3 \text{ mm Hg})[(1/760) \text{ atm mm Hg}^{-1}](0.910 \text{ L})}{(0.08206 \text{ L atm mol}^{-1} \text{ K}^{-1})(300 \text{ K})} = 0.0350 \text{ mol}$$

The mole fraction of hydrogen is

$$X_{H_2} = \frac{P_{H_2}}{P}$$

$$= \frac{719.3 \text{ mm Hg}}{746 \text{ mm Hg}} = 0.964$$

The mass of magnesium can be determined from the number of moles of hydrogen and the reaction stoichiometry. The balanced equation for the reaction is

$$Mg(s) + 2HCl(aq) \rightarrow H_2(g) + Mg^{2+} + 2Cl^-$$

indicating that $n_{Mg} = n_{H_2} = 0.0350$. The mass of magnesium that reacted is

$$m_{Mg} = (0.0350 \text{ mol})(24.31 \text{ g mol}^{-1}) = 0.851 \text{ g} \blacksquare$$

The following example illustrates the use of the mole fraction concept in conjunction with the ideal gas law.

---

**Example 3.8** A natural gas mixture, consisting of methane ($CH_4$) and acetylene ($C_2H_2$) in unknown proportion, occupies a certain volume at a total pressure of 105 mm Hg. The gases are burned completely in excess oxygen forming carbon dioxide and water. The pressure of carbon dioxide is measured at the same temperature and volume as the original mixture and found to be 197 mm Hg. Determine the mole fractions of methane and acetylene.

Letting $n_1$ and $n_2$ be the number of moles of $CH_4$ and $C_2H_2$, respectively, we can write the following expression for the total pressure before the reaction:

$$P = 105 \text{ mm Hg} = n_1 RT/V + n_2 RT/V$$

Similarly, after the reaction we have

$$P = P_{CO_2} = 197 \text{ mm Hg} = n_{CO_2} RT/V$$

The number of moles of $CO_2$ formed in the reaction can be related readily to the number of moles of each reactant. Since all the carbon in the reactants ends up as carbon dioxide, conservation of mass requires that each mole of $CH_4$ yield 1 mole of $CO_2$ and each mole of $C_2H_2$ yield 2 moles of $CO_2$. The total number of moles of $CO_2$ formed is therefore $n_{CO_2} = n_1 + 2n_2$. Substituting this result into the expression for $P_{CO_2}$ gives

$$P_{CO_2} = 197 \text{ mm Hg} = (n_1 + 2n_2)(RT/V)$$

The equation for the total pressure before reaction can be written as

$$105 \text{ mm Hg} = (n_1 + n_2)(RT/V)$$

Dividing the one equation by the other we obtain

$$\frac{197}{105} = \frac{n_1 + 2n_2}{n_1 + n_2}$$

This equation simplifies to

$$1.88 n_1 + 1.88 n_2 = n_1 + 2n_2$$

from which we obtain

$$n_1 = 0.14 n_2$$

The mole fractions are

$$X_{C_2H_2} = \frac{n_2}{n_{tot}} = \frac{n_2}{n_1 + n_2} \quad \text{and} \quad X_{CH_4} = \frac{n_1}{n_{tot}} = \frac{n_1}{n_1 + n_2} = 1 - X_{C_2H_2}$$

Substituting for $n_1$ we obtain

$$X_{C_2H_2} = \frac{n_2}{0.14\,n_2 + n_2} = \frac{1}{1.14} = 0.88 \quad \text{and} \quad X_{CH_4} = 1 - 0.88 = 0.12$$

The composition of the gas mixture is 12 mol% methane and 88 mol% acetylene. ∎

# The Atmosphere

No discussion of gases can be complete without consideration of the atmosphere, the mixture of gases surrounding the Earth which makes life on this planet possible. In this section we briefly examine the properties of the atmosphere and some of the sources of pollution which threaten the quality of the air we breathe. Since air is a mixture, its properties can be examined by means of Dalton's law of partial pressures.

## 3.6 Properties of the Atmosphere

The two chief constituents of the atmosphere are nitrogen and oxygen, which together make up 99% of air by volume (or mole fraction). As indicated in Table 3.4, a large number of different substances account for the remaining 1%. Many of the rare components enter the atmosphere as the result of either natural processes on Earth, such as decay of vegetation, or as the products of man-made pollution. Their abundance is consequently quite variable, as shown.

TABLE 3.4 Average Composition of Dry Air

| Substance | Mole percent |
|---|---|
| Nitrogen ($N_2$) | 78.08 |
| Oxygen ($O_2$) | 20.94 |
| Argon (Ar) | 0.934 |
| Carbon dioxide ($CO_2$) | 0.032 |
| Neon (Ne) | $1.8 \times 10^{-3}$ |
| Helium (He) | $5.2 \times 10^{-4}$ |
| Methane ($CH_4$) | $1.6 \times 10^{-4}$ |
| Krypton (Kr) | $1.1 \times 10^{-4}$ |
| Hydrogen ($H_2$) | $5 \times 10^{-5}$ |
| Xenon (Xe) | $9 \times 10^{-6}$ |
| Carbon monoxide (CO) | $10^{-6} - 10^{-5}$ |
| Nitric oxide (NO) | $10^{-6} - 10^{-4}$ |
| Nitrogen dioxide ($NO_2$) | $10^{-6} - 10^{-4}$ |
| Nitrous oxide ($N_2O$) | $2-4 \times 10^{-7}$ |
| Ammonia ($NH_3$) | $<10^{-4}$ |
| Formaldehyde (HCHO) | $<10^{-5}$ |
| Ozone ($O_3$) | $0-5 \times 10^{-5}$ |

**Example 3.9** What are the partial pressures at STP of the components of dry air? What is the composition by weight? Calculate the density of dry air under these conditions. Assume that air consists of only $N_2$(78.0 mol%), $O_2$(21.0 mol%), and Ar(1.0 mol%) and that the ideal gas law applies.

The partial pressures may be obtained by means of Dalton's law (Equation 3.18), $P_i = X_i P$. When $P = 760$ torr, we obtain

$$P_{N_2} = 0.780 \times 760 \text{ torr} = 593 \text{ torr}$$

$$P_{O_2} = 0.210 \times 760 \text{ torr} = 160 \text{ torr}$$

$$P_{Ar} = 0.010 \times 760 \text{ torr} = 7.6 \text{ torr}$$

To obtain the composition by weight, consider 1 mole of air. The number of moles of the constituents are then numerically equal to their mole fractions and the corresponding masses are obtained by means of the molecular weights:

$$m_{N_2} = X_{N_2} \times \overline{M}_{N_2} = 0.780 \times 28.01 \text{ g mol}^{-1} = 21.85 \text{ g mol}^{-1}$$

$$m_{O_2} = X_{O_2} \times \overline{M}_{O_2} = 0.210 \times 32.00 \text{ g mol}^{-1} = 6.72 \text{ g mol}^{-1}$$

$$m_{Ar} = X_{Ar} \times \overline{M}_{Ar} = 0.010 \times 39.95 \text{ g mol}^{-1} = 0.40 \text{ g mol}^{-1}$$

$$\sum = 28.97 \text{ g mol}^{-1}$$

The sum of the masses is the mass of one mole of air, that is, the average molecular weight. The weight percentages are given by the ratios of the masses of the components to the total mass:

$$\text{wt.\% } N_2 = \frac{21.85 \text{ g mol}^{-1}}{28.97 \text{ g mol}^{-1}} \times 100 = 75.4\%$$

The corresponding values for $O_2$ and Ar are 23.2% and 1.4%, respectively.

Since we have just calculated the average molecular weight of air, the most direct way of obtaining the average density is by means of the ideal gas law in the form of Equation 3.14, $P\overline{M} = \rho RT$, where, for a mixture, $\overline{M}$ and $\rho$ are the average values of the molecular weight and density, respectively. Solving for $\rho$ we obtain

$$\rho = \frac{(1.00 \text{ atm})(28.97 \text{ g mol}^{-1})}{(0.08206 \text{ L atm mol}^{-1} \text{ K}^{-1})(273.15 \text{ K})} = 1.29 \text{ g L}^{-1}$$

which is in excellent agreement with the experimental value at STP, 1.29 g $L^{-1}$. ∎

The pressure exerted by the atmosphere decreases with increasing altitude for the simple reason that the mass of air present above a given elevation decreases as the altitude increases. On the assumption that air is an ideal gas, and that the temperature, molecular weight, and acceleration due to gravity, $g$, are independent of the altitude $h$, the variation of atmospheric pressure with altitude is given by (see Problem 32) the **barometric formula,**

$$P_h = P_0 e^{-\overline{M}gh/RT} \tag{3.19}$$

where $P_h$ is the pressure at altitude $h$ and $P_0$ is the pressure when $h = 0$, i.e.,

**84** Properties of Gases 3.6

**FIGURE 3.9** Exponential decrease of atmospheric pressure with altitude. The pressure decreases by a factor of 2 for each 6 km change in altitude.

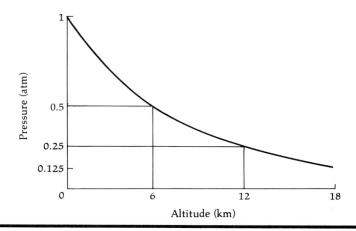

1 atm. Figure 3.9 displays the exponential decrease of pressure with increasing altitude.

The temperature of the atmosphere is actually not independent of altitude, as assumed in the barometric formula, but displays the oscillations shown in Figure 3.10. The two minima and the intervening maximum in this temperature profile delineate the limits of the four regions into which the atmosphere is commonly divided: the troposphere, stratosphere, mesosphere, and thermosphere. In addition to the main components which are present in

**FIGURE 3.10** Variation of average atmospheric temperature with altitude. The atmosphere is divided into four regions by the minima and maxima in the temperature–altitude curve. Some of the minor constituents of the atmosphere present in enhanced concentration in each of the regions are shown.

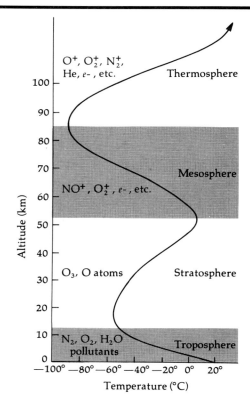

essentially constant concentration throughout, various substances are concentrated in different regions of the atmosphere, where they serve some specific functions, as we shall now see.

## 3.7 The Special Role of Ozone

Ozone ($O_3$) is a rare **allotrope** of oxygen, that is, a different form of the element. At room temperature it is a gas with a characteristic pungent odor; it condenses to a blue liquid at $-11°C$. Ozone is generally prepared by passing oxygen through an electric discharge:

$$3O_2(g) \rightarrow 2O_3(g)$$

Ozone is found in the stratosphere; its concentration reaches a maximum of approximately 10 ppm (parts per million) at an altitude of 25–30 km. The gas is formed by the action of ultraviolet radiation (Section 5.3) emanating from the sun. The radiation provides the energy to split an $O_2$ molecule into its atoms; the latter then react with molecular oxygen ($h\nu$ is the symbol for radiation, see Section 5.4):

$$O_2 \xrightarrow{h\nu} 2O$$
$$O + O_2 \longrightarrow O_3$$

Reactions initiated by radiation are called **photochemical reactions.**

In addition to being formed in the presence of ultraviolet radiation, ozone can also absorb it. Since this type of radiation has harmful effects (e.g., it causes skin cancer), the stratospheric ozone layer plays a beneficial role as a filter of ultraviolet radiation. However, owing to its great chemical reactivity, ozone is particularly susceptible to destruction by various pollutants that diffuse into the stratosphere after being released into the air at ground level. The most harmful agents are the chlorofluorocarbons and nitric oxide.

The chlorofluorocarbons, commonly known as *Freons*, are compounds of carbon, chlorine, and fluorine. The gases have been used as propellants in aerosol sprays and as coolants in refrigeration equipment. Their reaction with ozone in the stratosphere involves, as a first step, the photochemical release of chlorine atoms:

$$CF_2Cl_2 \xrightarrow{h\nu} CF_2Cl + Cl$$
$$CFCl_3 \xrightarrow{h\nu} CFCl_2 + Cl$$

The chlorine atoms are highly reactive and can destroy ozone in amounts that are out of proportion to their relatively small number. This happens because the reactions involved in the destruction of ozone regenerate chlorine; a given chlorine atom can therefore react successively with many ozone molecules:

$$Cl + O_3 \rightarrow ClO + O_2$$
$$ClO + O \rightarrow Cl + O_2$$

Reactions in which one of the reactants is also a product are known as **chain reactions.** Owing to the effectiveness of the *Freons* in reducing the concentration of ozone, their use as aerosol propellants has been banned in the United States.

The action of nitric oxide, formed in the combustion of gasoline (Section 3.8), also involves a chain reaction mechanism:

$$NO + O_3 \rightarrow NO_2 + O_2$$

$$NO_2 + O \rightarrow NO + O_2$$

Free oxygen atoms for both of these chain reactions are available in the stratosphere from the photochemical decomposition of molecular oxygen.

## 3.8 Air Pollution

The injection of ever increasing amounts of gaseous waste products into the atmosphere has led to increasingly noticeable deleterious effects. We shall briefly examine three effects that have recently received considerable attention.

*a. Photochemical Smog* **Photochemical smog** involves the formation of a noxious, hazy layer of pollution as a result of complex chemical reactions primarily involving automobile exhaust fumes. The problem occurs in specific locations, as in the city of Los Angeles, which have a sufficiently high automobile concentration, as well as the topographical and atmospheric conditions that lend themselves to smog formation.

The three essential ingredients of photochemical smog are oxides of nitrogen, hydrocarbons, and sunlight. Nitric oxide is produced at high temperatures by the oxidation of atmospheric nitrogen:

$$N_2(g) + O_2(g) \rightarrow 2NO(g)$$

Nitric oxide in turn is further oxidized to nitrogen dioxide:

$$2NO(g) + O_2(g) \rightarrow 2NO_2(g)$$

The internal combustion engine of automobiles is the most important source of nitric oxide. At the same time, incomplete combustion and evaporation of gasoline releases volatile hydrocarbons into the air. While these substances are not overly harmful by themselves, the action of sunlight produces some very reactive species which convert the gases into smog.

The first step in smog formation is the photochemical formation of atomic oxygen from nitrogen dioxide:

$$NO_2 \xrightarrow{h\nu} NO + O$$

Atomic oxygen reacts with $O_2$ to form ozone:

$$O_2 + O \rightarrow O_3$$

and with hydrocarbons to form highly reactive intermediates containing unpaired electrons, known as **free radicals:**

$$O + \text{hydrocarbons} \rightarrow R\cdot + RCO\cdot$$

The symbol R stands for the hydrocarbon part of a molecule (Section 22.2) (e.g., $CH_3CH_2$) and the dot represents an unpaired electron. Free radicals can undergo a variety of reactions many of which are involved in the formation of the irritating substances making up photochemical smog. These substances include various organic compounds, ozone, and nitrogen dioxide.

In order to reduce the occurrence of smog, the emission of nitric oxide and hydrocarbons by automobiles must be substantially reduced. The incorporation of catalytic converters in automobile exhaust systems has been very helpful in this respect. These devices reduce the concentration of unburned hydrocarbons in the exhaust and also promote decomposition of NO to the elements.

**b. Acid Rain** A second kind of atmospheric pollution involves the formation of **acid rain.** This problem is associated with the increasing use of fossil fuels having a relatively high sulfur content. For example, U.S. coal mined east of the Mississippi contains up to 6% sulfur. When this material undergoes combustion, sulfur is oxidized to sulfur dioxide which, in turn, is oxidized to sulfur trioxide. Sulfur trioxide readily dissolves in rainwater droplets with the formation of sulfuric acid, the main active constituent of acid rain. The overall process may be represented by the following three equations:

$$S(s) + O_2(g) \rightarrow SO_2(g)$$
$$2SO_2(g) + O_2(g) \rightarrow 2SO_3(g)$$
$$SO_3(g) + H_2O(l) \rightarrow H_2SO_4(aq)$$

Although acid rain has been known for a long time in northern Europe, it has been observed only recently in North America. Its effects include a decline in the population of lakewater fish, and the gradual destruction of forests. One of the characteristics of the process is that the effects can be seen many hundreds of miles from the pollution source as a result of the incorporation of the sulfur oxides in high-altitude air currents. This fact makes it difficult to determine the precise source of the pollution. Experiments are currently being performed to more precisely establish the connection between source and effect. Small amounts of distinctive tracer gases whose concentration can be determined down to 1 part in $10^{15}$ are released from specific industrial areas. The progress of the tracer plume across the continent can be tracked using both aircraft and ground sampling stations. Ultimately, a solution of this problem will require the removal of $SO_2$ from the combustion gases right at the power plant. The economic aspects of this problem are easily as important as the technological ones.

**c. The Greenhouse Effect** As a last example of atmospheric pollution we examine a possible occurrence of a climatic change resulting from increases in the concentration of carbon dioxide. The combustion of *all* fossil fuels results in the formation of $CO_2$. It is estimated that the present concentration of $CO_2$, about 300 ppm (Table 3.4), will double in approximately 100 years. In contrast to the other pollutants considered so far, the effect of $CO_2$ is not the result of its chemical reactions but, rather, of its interaction with different forms of radiation. Carbon dioxide is transparent to the visible light incident upon the surface of the Earth and so does not affect incoming radiation. However, some of the radiant energy absorbed by the Earth is reradiated in the form of infrared radiation (Section 5.3). Carbon dioxide acts by absorbing infrared radiation and preventing its escape into space. As a consequence, the average temperature at the surface of the Earth is expected to rise as the $CO_2$ concentration rises. In this respect $CO_2$ acts in a similar way as the glass enclosure of a greenhouse, and the warming of the Earth's surface due to $CO_2$ and other absorbing gases in the atmosphere is referred to as the **greenhouse effect.**

The best current estimates are that a doubling of the $CO_2$ concentration will lead to an increase in average Earth temperature of between 1.5 and 4.5°C. The increase is likely to be larger than average in the Earth's polar regions. While a change of a few degrees might not appear to be large, it should be noted that the temperature changes associated with the coming and going of the ice ages of the past million years amounted to only 6 or 7°C. One possible consequence of the predicted temperature change could be the melting of the polar ice caps. The resulting increase in sea level might then flood many of the Earth's coastal regions. Substantial changes in climate, including changes in the distribution of rainfall, also are predicted. While the various predictions are subject to great uncertainties, the potential impact of the greenhouse effect is so large that it would be totally irresponsible to ignore the problem.

## Conclusion

All gases behave in essentially the same manner at low pressures. These observations form the basis of the ideal gas law, which describes the relationship between the pressure, volume, temperature, and number of moles of either a single gas or a mixture of gases (Dalton's law). The ideal gas law provides the means by which the stoichiometric calculations described in Chapter 2 can be extended to reactions involving gases. As we shall see in Chapter 4, the ideal gas law reflects certain common features of molecular behavior which are independent of the identity of the molecules.

## Problems

(Assume that all gases obey the ideal gas law in working these problems.)

1. Incomplete combustion of 1.00 L of ethane ($C_2H_6$) with 3.15 L of oxygen, with both volumes measured at the same temperature and pressure, results in the formation of a mixture of gaseous CO, $CO_2$, and $H_2O$. What fraction of the number of moles of carbon appears in the form of CO?

2. An equimolar mixture of hydrogen and oxygen gas is sparked to initiate the formation of water vapor. Assuming that the reaction goes to completion, what is the final volume of the gas mixture relative to the initial volume if both volumes are measured at the same temperature and pressure?

3. An equimolar mixture of nitrogen and hydrogen gases is heated to produce ammonia. The final volume of the resulting gas mixture ($N_2$, $H_2$, and $NH_3$) measured at the same temperature and pressure is 78.5% of the initial volume. What fraction of the hydrogen and nitrogen reacted? What is the percentage yield of ammonia?

4. What volume of ammonia at STP is produced from 25.0 g of $H_2$ reacting with excess $N_2$?

5. A 3.16 g sample of an unknown gas occupies a volume of 863 mL at STP. What is the molecular weight of the gas?

6. When 0.334 g of a gaseous hydrocarbon was completely burned in excess oxygen, 1.05 g of $CO_2$ was obtained. (a) What is the empirical formula of the hydrocarbon? (b) The density of the hydrocarbon at STP is 1.25 g $L^{-1}$. What is the molecular formula and exact molecular weight of this compound?

7. A 2.50 L cylinder of neon at 25°C and 2.00 atm is compressed and heated to a final volume of 1.00 L and final temperature of 300°C. What is the final pressure?

8. A cylinder containing helium at a pressure of $1.40 \times 10^4$ kPa at 20°C is in danger of exploding if the pressure rises above $3.00 \times 10^4$ kPa. What is the maximum temperature at which this cylinder could be safely stored?

9. What is the pressure exerted by 2.50 kg of nitrogen in a volume of 15.0 L at 25°C?

10. How many moles of gas are there in a sample contained in a 2.00 L vessel at a pressure of 17.5 mm Hg and a temperature of 120°C?

11. Ethylene ($C_2H_4$) produced in a chemical plant is to be stored in large tanks. What is the volume of a tank required to store 1.0 ton (short) of ethylene at a pressure of 10 atm and a temperature of 20°C?

12. How many moles of air are present in an empty room with a volume of $1.0 \times 10^3$ m$^3$ at a temperature of 20°C if the atmospheric pressure is 756 mm Hg?

13. What weight of helium at STP will occupy the same volume that 40.0 g of oxygen occupies at 20°C and 700 torr?

14. Liquid gallium can be used as the fluid in a high-temperature barometer. How high is a column of gallium supported by a pressure of 1.00 atm? The density of liquid gallium is 6.1 g cm$^{-3}$.

15. A tank is filled with 2500 g of neon at 10°C and 15.0 atm. The tank is then heated to 30°C and a valve, which allows the gas to escape, is opened. Assuming that the external pressure is 1.0 atm and the temperature is kept at 30°C, what is the weight of neon that escapes from the cylinder?

16. The minimum temperature of the stratosphere, at an altitude of about 15 km, is −55°C. The atmospheric pressure at this altitude is 85 mm Hg. How many moles of air per liter are there at this altitude?

17. Suppose that 1.50 L of air is inhaled at 1.00 atm and −15°C. If one holds one's breath long enough for the air to warm up to body temperature (37°C), what would be the resulting air pressure if there were no change in volume? Assume that air is not soluble in body fluids.

18. Two glass bulbs, each of which has a volume of 1.00 L, are connected by a thin tube of negligible volume. A valve in the tube can either open or close the two bulbs to each other. The bulbs each contain 0.100 mol of $N_2$ at 20°C. While the valve is open, bulb A is cooled with liquid nitrogen to −196°C while bulb B is maintained at 20°C. After equilibrium is established, the valve is closed and bulb A is warmed back to 20°C. Taking note of the fact that at equilibrium the pressure of a gas is the same everywhere in a container, calculate (a) the initial pressure in each bulb and (b) the final pressure in each bulb.

19. The pressure of a gas in a certain container is 500 mm Hg. A certain amount of the gas is withdrawn and found to occupy 1.75 mL at a pressure of 760 mm Hg. The pressure of the gas remaining in the container was 400 mm Hg. Assuming that all measurements were made at the same temperature, calculate the volume of the container.

20. One of the oxides of nitrogen is a gas with a density of 1.33 g L$^{-1}$ at 150°C and a pressure of 763 mm Hg. What is the molecular weight of the oxide and what is its formula?

21. A certain gas has a density of 0.48 g L$^{-1}$ at 20°C and a pressure of 200 mm Hg. What is the molecular weight of the gas? If the gas contains 63.64% nitrogen by weight, how many atoms of nitrogen are there in one molecule of the gas? What is the molecular weight to four significant figures?

22. A certain compound contains 49.0% carbon, 2.7% hydrogen, and 48.3% chlorine by weight. Its density at 120°C and a pressure of 250 mm Hg is 1.50 g L$^{-1}$. What is the molecular formula?

23. A mixture of ethylene ($C_2H_4$) and propane ($C_3H_8$) occupied a certain volume at a total pressure of 125 torr. The gas was burned in excess oxygen resulting in the formation of $CO_2$ and $H_2O$. The $CO_2$ gas was collected and its pressure was found to be 280 torr in the same volume and at the same temperature as the original mixture. What was the mole fraction of ethylene?

24. Nitrogen dioxide actually exists as a mixture of gaseous $NO_2$ and $N_2O_4$ at ordinary conditions. At 50°C and 1.20 atm total pressure, the density of this gas mixture is 2.40 g L$^{-1}$. What is the partial pressure of each substance?

25. A 100.0 L vessel contains an equal mixture by weight of neon and argon at 25°C and 1.50 atm total pressure. The vessel is heated to 100°C. Determine (a) the total pressure at 100°C, (b) the partial pressure of each gas at 25°C and at 100°C, and (c) the density at 25°C and at 100°C.

26. Carbon monoxide is a pollutant emitted in automobile exhausts. If the concentration of CO is 9.15 mg m$^{-3}$ at 20°C and atmospheric pressure, determine (a) the partial pressure and (b) the mole fraction of CO in the air.

27. A sealed glass bulb with a volume of 600 mL contains $O_2$ at a pressure of 400 torr and a temperature of 25°C; a second bulb of equal volume contains $N_2$ at a pressure of 800 torr and a temperature of 10°C. The two bulbs are connected so that the gases can mix completely and are brought to a final temperature of 20°C. What is the final pressure?

28. A 200 mL sample of oxygen at a pressure of 500 torr is mixed with a 100 mL sample of nitrogen at a pressure of 700 torr in a container whose volume is 250 mL. A constant temperature of 25°C is maintained during mixing. Thereafter, the mixture is heated to

100°C. Determine the partial pressure of each gas at (a) 25°C and (b) 100°C.

29. A 0.631 g sample of carbon is completely burned in a vessel containing 425.0 mL of oxygen at a pressure of 3.75 atm and a temperature of 20.0°C. Assuming that $CO_2$ is the only combustion product and that the temperature goes up by 2.0°C as a result of the reaction, calculate the mole fraction of each gas and the final pressure. Correct your answer for the fact that the total volume available to the gas increases as a result of the consumption of solid carbon, whose density is 2.0 g cm$^{-3}$.

30. A sample of oxygen gas is collected over water at 22°C in a volume of 500 mL. The total gas pressure is found to be 745 mm Hg. How many moles of $O_2$ were collected?

31. At an altitude of 20 km, the density of air is 0.093 g L$^{-1}$ and the temperature is 218 K. Assuming that air consists of 80 mol% $N_2$ and 20 mol% $O_2$, calculate the pressure exerted by the atmosphere at this altitude.

32.* (This problem requires use of calculus.) The barometric formula (Equation 3.19) can be derived by starting with the fact that the change in pressure $dP$ for an infinitesimal change in altitude $dh$ is $dP = -\rho g\, dh$, where $\rho$ is the gas density (see Equation 3.5). Obtain the barometric formula by integrating this expression for the conditions described in the text (i.e., ideal gas and constant temperature).

33. According to the barometric formula, the composition of the atmosphere varies with altitude. The heavier molecules tend to sink to the bottom of the atmosphere because $exp-(\overline{M}gh)$ depends more strongly on height when $\overline{M}$ is large. If the mole fraction of helium at sea level is $5.2 \times 10^{-6}$, what is the mole fraction of He at an altitude of (a) 10 km (e.g., Mt. Everest) and (b) 100 km? Assume a constant temperature of −40°C and an average molecular weight of air of 29 g mol$^{-1}$.

# 4

# Molecular Basis of Gas Properties

It is reasonable to suppose that the relationships among the properties of a gas embodied in the ideal gas law reflect the behavior of the molecules in the gas. In this chapter we examine a theoretical model of gases—the kinetic theory—based on a number of specific assumptions about the molecules in a gas. We shall see that the model is able to predict Boyle's law and other macroscopic properties of gases. The agreement between the predicted behavior and the experimental facts summarized in the ideal gas law is an indication that the assumptions about molecular behavior which underlie the model may be reasonable.

A good theoretical model is not only able to account for the experimental observations but, in addition, usually provides a real insight into the phenomena under consideration and often reveals hitherto unsuspected relationships between various properties. As we shall see, the kinetic theory of gases offers a much more satisfactory explanation of temperature than that of relative "hotness." Furthermore, the theory yields a quantitative measure of the average speed of the molecules in a gas, a quantity that at first sight bears no relation to the ideal gas law. Finally, the deviations from ideal gas behavior observed at high pressures (and also at low temperatures) can be ascribed to the breakdown of specific assumptions about molecular properties. Modifications of these assumptions lead to an improved description of gas behavior, as we shall see.

The study of gases offers an excellent illustration of the interplay between experiment and theory in science. Experimental observations of gas properties are summarized in a few empirical laws such as Boyle's law. A theory is developed to account for the observations. The theory usually suggests additional experiments which sometimes confirm the theory but often suggest certain modifications which can be tested further. This continuous interplay can be exceedingly fruitful in the development of new scientific knowledge and a more profound understanding of nature.

# The Kinetic Theory of Gases

## 4.1 Assumptions of the Kinetic Theory

The kinetic theory of gas behavior was developed in the nineteenth century chiefly by Rudolf Clausius (1822–1888), James Clerk Maxwell (1831–1879), and Ludwig Boltzmann (1844–1906). The theory is based on the following assumptions about the nature of gases on the molecular level:

1. *A gas consists of a large number of molecules that are so small compared to the average distance between them that their size is negligible.* This assumption consists of two related parts: one deals with the number of molecules present, and the other deals with their size. The first part requires little discussion as it is quite obvious that any macroscopic system contains an enormous number of molecules.

    The validity of the second part can be examined with reference to the properties of an ideal gas. The molar volume at STP is 22.4 L. This volume can be apportioned among all the molecules present. The volume corresponding to a single molecule is

    $$\frac{(22.4 \text{ L mol}^{-1})(10^{-3} \text{ m}^3 \text{ L}^{-1})}{(6.0 \times 10^{23} \text{ mol}^{-1})} = 3.7 \times 10^{-26} \text{ m}^3$$

    For simplicity, let us assume that each molecule is located inside a cubic region of space whose volume has this value. As shown in Figure 4.1a, the side of this cube corresponds to the average distance between two molecules. This distance is

    $$(3.7 \times 10^{-26} \text{ m}^3)^{1/3}(10^9 \text{ nm m}^{-1}) = 3.3 \text{ nm}$$

    Since molecular diameters are only about 0.2–0.4 nm (2–4 Å), the assumption is reasonably good at atmospheric pressure. Figure 4.1b illustrates the effect of increasing pressure on the distance between the molecules. It is apparent that the molecules occupy an increasingly larger fraction of the volume as the pressure increases. Since the size of a molecule depends on its identity, the validity of the assumption at a given pressure depends on the gas. The one general statement that can be made is that the size of molecules is negligible at sufficiently low pressures.

    The temperature of the gas is also of importance. In view of the proportionality, at a given pressure, between the volume of a gas and temperature (Charles' law), molecules occupy an increasingly large fraction of the available volume as the temperature decreases. Thus, unless the pressure is low enough, the temperature cannot be too low if the assumption is to be valid.

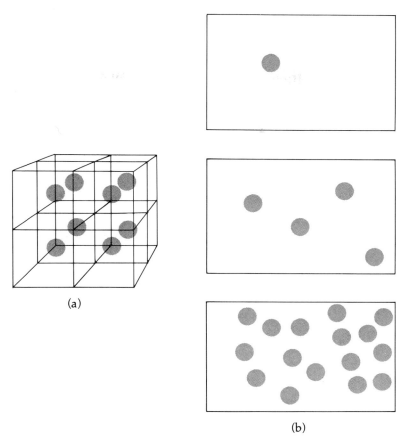

**FIGURE 4.1** Size of molecules of a gas compared to the distance between them. (a) Division of the volume of a gas at STP into a set of cubes each of which contains a single molecule. A side of the cube is equal to the average distance between neighboring molecules. (b) Scale drawing of the number of argon atoms in a 2 nm thick slab of gas at room temperature and a pressure of $10^{-2}$ atm (top), 1 atm (middle), and 100 atm (bottom).

2. *The molecules in a gas are in continuous random motion.* This assumption is obvious on the basis of our daily experience. For instance, when a foul smelling gas such as hydrogen sulfide is released at some particular location, the odor can very soon be detected at a distance. Evidently, the molecules of the gas in question have traveled from one location to the other. The motion of the molecules is random, meaning that there is no preferred direction to their movement. On a windless day, the odor spreads in all directions from the source.
3. *The forces between molecules can be neglected.* All gases condense to liquids at sufficiently low temperatures, thus showing that there are forces between molecules. Intermolecular forces in liquids and solids account for the cohesiveness of these states of matter—liquids and solids occupy a certain volume independent of the size of their container. This is not so for gases. A

gas will expand to occupy the entire volume in which it is confined indicating that the forces holding the molecules together can be neglected. This characteristic difference between gases and condensed phases can be attributed to the much greater distance between the molecules in a gas and those in a liquid or solid, as evidenced by the much smaller density of gases (see Table 1.4).

As noted in the discussion of assumption 1, the distance between the molecules in a gas decreases as the pressure increases or the temperature decreases. The importance of intermolecular forces can therefore be expected to increase under these conditions. Both assumptions are valid under the same conditions, that is, as long as the pressure is not too high and the temperature is not too low.

4. *Molecules make elastic collisions with the container walls and with each other.* Because of their random motion, the molecules in a gas must eventually collide with the walls of their container and with each other. These collisions are **elastic,** meaning that the molecules lose no energy when they collide with the wall and lose no net energy when they collide with each other, i.e., the sum of their energies remains constant. If this were not so, the pressure of a gas, which is the result of the collisions of molecules with the container walls, would gradually decrease. However, so long as the temperature and volume of a given amount of gas remain constant, the gas pressure remains unchanged.

## 4.2 Predictions of the Kinetic Theory

*a. Derivation of Boyle's Law* The kinetic theory may be used to derive Boyle's law on the basis of the preceding assumptions. As the molecules randomly move about in their container, they periodically collide with the walls. Each time a molecule strikes a wall and rebounds, it exerts a momentary force on the wall. The ratio of this force to the area of the wall is the momentary pressure due to this collision. The number of molecules striking a container wall every second is usually very large. It is this continuous bombardment of the walls that gives rise to a measurable pressure. The evaluation of the pressure involves a computation of the force exerted on the wall by the impact of many molecules.

We begin by reviewing a few basic concepts of physics. The terms velocity and speed are frequently used synonymously; here we must distinguish between them. The **velocity** of a particle is the distance traversed per unit time *in a specific direction*. The **speed** of a particle is equal to the magnitude of its velocity but conveys no information about the direction of motion. Both quantities are expressed in the same units, i.e., meters per second (m s$^{-1}$). The distinction is further illustrated in Figure 4.2. The **momentum** of a particle is equal to the product of its mass and velocity,

$$p = mv \qquad (4.1)$$

In view of the definition of velocity, it is apparent that the momentum of a particle is directed in a specific direction. By contrast, the **kinetic energy** of a particle,

$$\epsilon = \tfrac{1}{2}mv^2 \qquad (4.2)$$

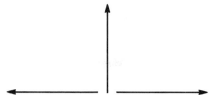

**FIGURE 4.2** Difference between velocity and speed. Three particles traverse the same distance (given by the length of the arrows) in a given time. These particles have the same speed, but since they move in different directions, they differ in velocity.

has no direction. When the velocity is squared, the directional information is lost. This fact is most readily seen by considering two particles moving with identical speeds in opposite directions; their velocities are $+v$ and $-v$. When squared, both of these quantities are equal to $v^2$. Thus, knowing only $v^2$ it is impossible to ascertain the direction of motion.

We now proceed to evaluate the pressure exerted by a gas. Consider the collisions that a molecule undergoes with the container walls. We assume that the molecule has velocity $v$, with components along the $x$, $y$, and $z$ axes designated $v_x$, $v_y$, and $v_z$, respectively. It is mathematically convenient to pick the directions of this coordinate system so that the axes are perpendicular to the walls of the container, as shown in Figure 4.3. The advantage of this choice is that in a collision with a wall only one of the velocity components is affected, namely, the one directed perpendicular to the wall in question. For instance, only $v_x$ is affected in a collision with the wall drawn in color. Therefore, although the molecule will collide with all the walls and consequently travel successively in many different directions, we need only concern ourselves with the velocity component $v_x$ so long as we are solely interested in collisions with the wall in question. This results in a major mathematical simplification since we have to consider collisions in only one rather than in three dimensions.

When a molecule collides with the wall perpendicular to the $x$ axis, its velocity component $v_x$ changes to $-v_x$ while its other velocity components remain unchanged. If the particle has mass $m$, its component of momentum along the $x$ axis similarly changes from $mv_x$ to $-mv_x$. The magnitude of the momentum

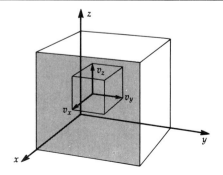

**FIGURE 4.3** Resolution of the velocity of a molecule confined to a rectangular box into its components. The velocity component $v_x$ is directed perpendicular to the wall drawn in color.

change is $2m|v_x|$ (the symbol $|\ |$ signifies the absolute value of the quantity bracketed by it, that is, the value irrespective of sign).

The total number of collisions with the wall occurring in a time interval $\Delta t$ is equal to the number of molecules able to reach the wall in that time. The distance a molecule of speed $|v_x|$ can travel in a time $\Delta t$ is $|v_x|\Delta t$. Suppose, for a moment, that all the molecules in the container have the same value of $|v_x|$. Then, all the molecules located within a distance $|v_x|\Delta t$ of the wall will strike the wall provided they are moving towards it. Figure 4.4 shows that all these molecules lie in a region of the container with length $|v_x|\Delta t$ and cross-sectional area $A$, whose volume is $|v_x|\Delta t\, A$. If there are $N^*$ molecules per unit volume, the total number of molecules in the volume of interest will be $N^*|v_x|\Delta t\, A$. On average, half the molecules will be moving towards the wall and half will be moving away, so that the average number of collisions with the wall occurring in the time interval $\Delta t$ is

$$\text{number of collisions} = \tfrac{1}{2} N^* |v_x| \Delta t\, A \tag{4.3}$$

Since the momentum change in a single collision is $2m|v_x|$, the total momentum change $\Delta p$ is

momentum change = (number of collisions)(momentum change per collision)

$$\Delta p = (\tfrac{1}{2} N^* |v_x| \Delta t\, A)(2m|v_x|) \tag{4.4}$$
$$= N^* m A v_x^2\, \Delta t$$

(The quantity $v_x^2$ is necessarily positive. Thus, the absolute value symbol is not needed.) We know from physics that the force exerted by an object is equal to the rate of change of its momentum,

$$F = \Delta p / \Delta t$$
$$1\ \text{newton} = 1\ \text{kg m s}^{-2} \tag{4.5}$$

Therefore,

$$F = N^* m A v_x^2 \tag{4.6}$$

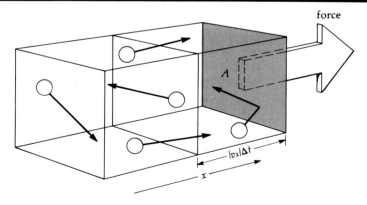

**FIGURE 4.4** A gas is confined to a rectangular container. The area of the wall shown in color is designated $A$. The force exerted by the gas on this wall arises from the collisions with the wall of molecules lying within a distance $|v_x|\Delta t$ of it in a time $\Delta t$. The pressure on the wall is equal to the force divided by the area $A$.

Furthermore, we have seen that pressure is force per unit area (see Equation 3.1). Since the wall of the container has area $A$, the pressure exerted by the molecules in the gas is

$$\text{pressure} = \frac{F}{A} = N^* m v_x^2 \tag{4.7}$$

In obtaining Equation 4.7, we have made the unnecessarily restrictive assumption that all the molecules in a gas have the same value of $|v_x|$. Owing to the occurrence of molecular collisions, the molecules constantly change their speeds. Thus, there is a distribution in molecular speeds (Section 4.6). This can be taken into account by replacing $v_x^2$ in Equation 4.7 by its average value, $\langle v_x^2 \rangle$, giving

$$P = N^* m \langle v_x^2 \rangle \tag{4.8}$$

We have so far restricted the discussion to the component of motion along the $x$ axis. Since the molecules are in random motion, all directions are equivalent and the average of $v_x^2$ is the same as the average of the corresponding quantities in the $y$ and $z$ directions:

$$\langle v_x^2 \rangle = \langle v_y^2 \rangle = \langle v_z^2 \rangle \tag{4.9}$$

Making use of the properties of right triangles, as detailed in Figure 4.5, we can obtain the speed of a molecule from the relation

$$v^2 = v_x^2 + v_y^2 + v_z^2 \tag{4.10}$$

In combination with Equation 4.9, this relation indicates that the **mean square speed** is

$$\langle v^2 \rangle = \langle v_x^2 \rangle + \langle v_y^2 \rangle + \langle v_z^2 \rangle = 3 \langle v_x^2 \rangle \tag{4.11}$$

Substituting this result into Equation 4.8 for the pressure yields

$$P = \tfrac{1}{3} N^* m \langle v^2 \rangle \tag{4.12}$$

Finally, the number of molecules per unit volume, $N^*$, can be replaced by $N/V$, where $N$ is the actual number of molecules present and $V$ is the volume of the

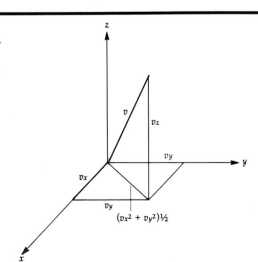

**FIGURE 4.5** Relation between $v$ and its components $v_x$, $v_y$, and $v_z$. The hypotenuse of the right triangle whose sides have length $v_x$ and $v_y$ is $(v_x^2 + v_y^2)^{1/2}$. The length of the hypotenuse of the right triangle whose sides are $(v_x^2 + v_y^2)^{1/2}$ and $v_z$ is $v$; thus, $v^2 = v_x^2 + v_y^2 + v_z^2$.

container. Expressing the number of molecules in terms of Avogadro's number, $N = nN_A$, where $n$ is the number of moles, yields the end result of the derivation:

$$PV = \tfrac{1}{3} n N_A m \langle v^2 \rangle \tag{4.13}$$

Equation 4.13 is an important result of the kinetic theory. Comparison with Boyle's law (Equation 3.6),

$$PV = \text{constant} \ (n, \ T \ \text{constant})$$

indicates that the two are identical provided that the quantities on the right side of Equation 4.13 are constant at a given temperature and number of moles. The only quantity for which this constancy is not obvious is the mean square speed. We can therefore conclude that the kinetic theory predicts Boyle's law provided that the mean square speed of the gas molecules is constant at a given temperature.

Before exploring the consequences of this provision it is necessary to examine in more detail one aspect of molecular motion neglected in this discussion. The preceding statement that all molecules with speed $|v_x|$ located within a distance $|v_x| \Delta t$ of the wall will strike the wall within a time $\Delta t$ if they are moving toward it is tantamount to neglecting collisions between molecules. As we shall see, at ordinary pressures a given molecule collides with other molecules at a high rate. The effect of these collisions is exceedingly complex. The velocity of a given molecule generally changes as a result of each collision, and its path, illustrated in Figure 4.6, is reminiscent of a drunkard's progress across a room and is known as a "random walk." Consequently, a particular molecule is unlikely to reach the wall in the requisite time. Fortunately, this complication is more apparent than real. Owing to the large number of molecules present and to their random motion, the effect of intermolecular collisions will, on average, cancel out. That is, for each molecule whose velocity in the $x$ direction changes from $v_x$ to $-v_x$ as the result of a collision, another molecule will, on average, experience an equal but opposite change. The validity of Equation 4.13 is not affected by the occurence of intermolecular collisions.

**b. Temperature and Kinetic Energy of Motion**  The result of the kinetic theory, Equation 4.13, may be expressed in terms of the average kinetic energy of the molecules in a gas, $\bar{\epsilon}$, which is given by

$$\bar{\epsilon} = \tfrac{1}{2} m \langle v^2 \rangle \tag{4.14}$$

Substituting into Equation 4.13, we obtain

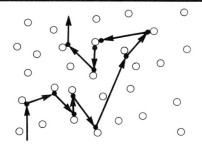

**FIGURE 4.6**  The path traced by a molecule experiencing successive intermolecular collisions is called a random walk.

$$PV = \tfrac{2}{3} n N_A \bar{\epsilon} \tag{4.15}$$

or, designating the average total kinetic energy of one mole of gas as $E$,

$$PV = \tfrac{2}{3} nE \tag{4.16}$$

Comparing this result to the experimentally determined ideal gas law, $PV = nRT$, we obtain the remarkably simple result that

$$E = \tfrac{3}{2} RT \tag{4.17}$$

In other words, temperature can be understood as a measure of the translational kinetic energy of motion of gas molecules. An increase in the temperature of a gas corresponds to an increase in the average kinetic energy and speed of its molecules while a decrease in temperature has the opposite effect. In this view, absolute zero is the temperature at which all motion ceases. Thus, the kinetic theory explains the origin of a familiar yet somewhat obscure property, temperature, on the basis of molecular behavior.

If we divide both sides of Equation 4.17 by Avogadro's number, we obtain

$$\bar{\epsilon} = E/N_A = \tfrac{3}{2} kT \tag{4.18}$$

where

$$k = \frac{R}{N_A} = \frac{(8.314 \text{ J mol}^{-1} \text{ K}^{-1})}{(6.022 \times 10^{23} \text{ mol}^{-1})} = 1.381 \times 10^{-23} \text{ J K}^{-1}$$

The constant $k$, the gas constant per molecule, is known as the **Boltzmann constant.**

With the results developed in this section we have attained one of the major goals of this chapter: to show how empirical laws about the macroscopic properties of gases can be understood by means of a model dealing with the behavior of molecules. Two important assumptions, namely, that the size of molecules as well as the forces between them are negligibly small, lead to a result that is consistent with the experimental ideal gas law. These assumptions are valid at low pressures and moderate or high temperatures, when molecules are far apart and most of the space they occupy is void. Therefore, kinetic theory indicates that gases should become ideal at sufficiently low pressures and at moderate temperatures, as is indeed observed experimentally. By implication, the model indicates that at high pressures and/or low temperatures, gases should deviate from ideal behavior in a manner that manifests the effects of molecular size and intermolecular forces. As we shall see, the actual behavior of real gases can be understood by a simple extension of the model. Before we turn to this subject it is worth exploring some additional aspects of kinetic theory. The success of this theory in correctly explaining the conditions for ideal gas behavior suggests that additional predictions of the theory should also be valid.

## 4.3 Molecular Speeds

The close connection between temperature and molecular motion makes it possible to obtain numerical estimates of molecular speeds. Equations 4.14 and 4.18 may be combined to give

$$\bar{\epsilon} = \tfrac{1}{2} m \langle v^2 \rangle = \tfrac{3}{2} kT$$

**TABLE 4.1 Evaluation of the Mean and Root Mean Square Values of Molecular Speeds**

| Mean Speed | Root Mean Square Speed |
|---|---|
| $v = 1 \times 10^3$ m s$^{-1}$ | $v^2 = 1 \times 10^6$ m$^2$ s$^{-2}$ |
| $2 \times 10^3$ | $4 \times 10^6$ |
| $3 \times 10^3$ | $9 \times 10^6$ |
| $\sum = 6 \times 10^3$ | $\sum = 14 \times 10^6$ |
| $\bar{v} = \dfrac{6 \times 10^3}{3} = 2 \times 10^3$ m s$^{-1}$ | $v_{rms} = \langle v^2 \rangle^{1/2} = \left(\dfrac{14 \times 10^6}{3}\right)^{1/2} = 2.16 \times 10^3$ m s$^{-1}$ |

Solving for the speed, we obtain

$$v_{rms} = \langle v^2 \rangle^{1/2} = \left(\frac{3kT}{m}\right)^{1/2} \tag{4.19}$$

The quantity $\langle v^2 \rangle^{1/2}$ is expressed by the less cumbersome symbol $v_{rms}$ and is called the **root mean square speed**—the square root of the mean square speed. Alternatively, the root mean square speed may be expressed in terms of the molecular weight instead of the mass of a single molecule by multiplying both numerator and denominator in Equation 4.19 by Avogadro's number and replacing $kN_A$ by the gas constant $R$:

$$v_{rms} = \left(\frac{3RT}{\overline{M}}\right)^{1/2} \tag{4.20}$$

The distinction between the root mean square speed and the mean speed is illustrated by the data in Table 4.1. For a given gas at a particular temperature, the root mean square speed exceeds the mean speed $\bar{v}$ by approximately 8%. We ignore here the numerical difference between these two kinds of averages.

---

**Example 4.1** Calculate the root mean square speed of an $O_2$ molecule at 20°C.

The solution is obtained by substitution into Equation 4.20:

$$v_{rms} = \left(\frac{3RT}{\overline{M}}\right)^{1/2}$$

$$= \left[\frac{(3)(8.314 \text{ J mol}^{-1} \text{ K}^{-1})(293 \text{ K})(1 \text{ kg m}^2 \text{ s}^{-2}/\text{J})}{(32.0 \text{ g mol}^{-1})(10^{-3} \text{ kg g}^{-1})}\right]^{1/2}$$

$$= 478 \text{ m s}^{-1}$$

The large magnitude of molecular speeds becomes even more apparent when it is realized that this value corresponds to a speed somewhat in excess of 1000 miles per hour. ∎

In spite of these very large speeds, the net distance traversed by a molecule at atmospheric pressure is much smaller than that traversed in the same time by, say, an airplane. Collisions with other molecules result in the tortuous kind of path already illustrated in Figure 4.6. It is apparent that the net distance covered is much shorter than the actual path length.

Equation 4.20 indicates that two factors determine the mean speed of a molecule: the temperature and the molecular weight. At a given temperature the mean speed is inversely proportional to the square root of the molecular weight. Thus, the mean speeds of molecules with molecular weight $\overline{M}_1$ and $\overline{M}_2$ are related as follows:

$$\frac{v_{rms_1}}{v_{rms_2}} = \left(\frac{\overline{M}_2}{\overline{M}_1}\right)^{1/2} \quad (4.21)$$

For example, the mean speed of a hydrogen molecule ($\overline{M} = 2$ u) is four times greater than that of an oxygen molecule ($\overline{M} = 32$ u) at the same temperature.

## 4.4 Effusion and Diffusion of Gases

The flow of gas from a region of high pressure to one of low pressure makes the mass dependence of the mean molecular speed directly observable. Consider a partitioned container, as illustrated in Figure 4.7, in which gas is separated from an evacuated region by a wall with a pinhole. If this hole is sufficiently small, gas molecules will leak out of the container only if their random motion happens to carry them through the hole. The process by which molecules escape from a container under these conditions is called **effusion.** Since the effusion rate is determined by the probability that the molecules enter the hole in the container wall, it is possible to understand this process in terms of the kinetic theory.

The rate $Z_{eff}$ at which molecules confined to a volume $V$ effuse through a hole of area $A$ can be obtained from Equation 4.3 for the number of collisions made by molecules with the wall of a container. The result is

$$Z_{eff} = \tfrac{1}{4} n N_A \overline{v} A / V \quad (4.22)$$

where the units of $Z_{eff}$ are molecules per second. The factor of $\tfrac{1}{4}$ comes from a more detailed analysis in which molecules are allowed to approach the hole from all directions. Since the mean speed is inversely proportional to the square root of the mass (Equation 4.21), the ratio of effusion rates of two gases at the same temperature and pressure is

$$\frac{Z_{eff_2}}{Z_{eff_1}} = \left(\frac{\overline{M}_1}{\overline{M}_2}\right)^{1/2} \quad (4.23)$$

For instance, hydrogen ($\overline{M} = 2$ u) will effuse four times as fast as oxygen ($\overline{M} = 32$ u) through a given orifice.

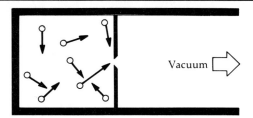

**FIGURE 4.7** Representation of molecular effusion. Molecules in the gas container incident on the small pinhole in the wall are able to pass through and escape into the evacuated region. If the wall is replaced by a porous barrier the molecules are transported into the vacuum by diffusion.

Thomas Graham (1805–1869) observed that the rates of effusion of gases are inversely proportional to their densities. Since the density of a gas is proportional to its molecular weight (Equation 3.14), Graham's law is equivalent to Equation 4.23 and is explained by kinetic theory.

If the opening in the container sketched in Figure 4.7 is replaced by a porous barrier, the passage of gas molecules into a vacuum will occur at a faster rate. The porous wall prevents the bulk flow of gas but permits the occurrence of **diffusion**. This process is more complex than effusion because the transport process involves molecular collisions inside the porous wall. Nonetheless, Equation 4.23 for the mass dependence of the rate of diffusion is still applicable.

Because of its dependence on molecular weight, diffusion can be used to separate the isotopes of an element. The most important application of diffusion to isotope separation involves the element uranium. Natural uranium consists of 99.3% $^{238}$U and 0.7% $^{235}$U. Only the latter undergoes nuclear fission, an important source of energy (Section 23.6). In order to make fission practical, the abundance of $^{235}$U in uranium must be increased to a few percent. The most widely used enrichment process involves diffusion of gaseous uranium hexafluoride. The molecular weights of $^{235}$UF$_6$ and $^{238}$UF$_6$ are 349.0 u and 352.0 u, respectively. According to Equation 4.23, the ratio of diffusion rates is $(352.0/349.0)^{1/2} = 1.004$, which obviously produces a very small enrichment. However, if the resulting material is passed through a second porous wall the overall enrichment will be increased to $1.004 \times 1.004 = 1.008$. This result indicates that sequential diffusion can be used to provide the desired enrichment. For example, a 400-stage diffusion process leads to an enrichment factor of $(1.004)^{400} = 5$ and produces uranium with the desired $^{235}$U abundance.

---

**Example 4.2** Compare gaseous $^1$H$_2$ ($\overline{M} = 2.016$ u) and $^2$H$_2$ ($\overline{M} = 4.028$ u) with respect to (a) the average molar translational kinetic energy at 25.0°C, (b) the relative mean molecular speeds at 100°C, and (c) the relative rates of effusion at 100°C and atmospheric pressure.

(a) The average molar translational kinetic energy of a gas is

$$E = \tfrac{3}{2}RT$$

independent of the identity of the gas. The value for either gas is

$$E = (\tfrac{3}{2})(8.314 \text{ J mol}^{-1}\text{ K}^{-1})(298.2 \text{ K}) = 3719 \text{ J mol}^{-1}$$

(b) The relative mean molecular speeds of any two gases at the same temperature are inversely proportional to the square roots of their molecular weights, Equation 4.31. Thus,

$$\frac{v_{rms}(^1H_2)}{v_{rms}(^2H_2)} = \left(\frac{\overline{M}_{2H_2}}{\overline{M}_{1H_2}}\right)^{1/2} = \left(\frac{4.028}{2.016}\right)^{1/2} = 1.414$$

The actual numerical value for $^2$H$_2$ is

$$v_{rms} = \left(\frac{3RT}{\overline{M}}\right)^{1/2}$$

$$= \left[\frac{(3)(8.314 \text{ J mol}^{-1}\text{ K}^{-1})(373.2 \text{ K})(1 \text{ kg m}^2 \text{ s}^{-2}/\text{J})}{(4.028 \text{ g mol}^{-1})(10^{-3} \text{ kg g}^{-1})}\right]^{1/2} = 1520 \text{ m s}^{-1}$$

(c) The relative rates of effusion of two gases, given by Equation 4.23, depend on the molecular weights in exactly the same way as their mean speeds. The answer is therefore the same as that to part (b):

$$\frac{Z_{eff}(^1H_2)}{Z_{eff}(^2H_2)} = 1.414 \quad \blacksquare$$

## 4.5 Molecular Collisions

Molecules are constantly in motion and as a result often collide with each other. Kinetic theory may be used to derive an expression for the number of collisions per unit time that a given molecule makes with other molecules, the so-called **collision frequency** $Z_1$. The frequency of molecular collisions is of importance to an understanding of chemical reactions because, before two molecules can react with each other, they must first collide.

What are the factors that determine $Z_1$? We might expect this quantity to be proportional to the mean molecular speed $\bar{v}$ since a swift molecule will encounter more molecules in a given time than a slow one. The collision frequency should also be related to the target area presented by a molecule when viewed head on, known as the cross-sectional area $\pi(d/2)^2$, where $d$ is the molecular diameter. This follows since large molecules confined to a certain volume are more likely to collide than small ones, much as two trucks passing each other on a narrow road are more likely to sideswipe each other than two sportcars. Finally, $Z_1$ should be proportional to the number of molecules per unit volume, $N^*$. Figure 4.8 shows graphically how these factors affect the collision frequency. The expression obtained from kinetic theory but not derived here is

$$Z_1 = \sqrt{2}\ \pi d^2 \bar{v} N^*$$
$$s^{-1} = (m^2)(m\ s^{-1})(m^{-3}) \tag{4.24}$$

where the units of the various quantities in this expression show that the equation is dimensionally correct.

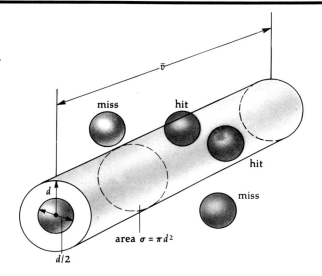

**FIGURE 4.8** A molecule with speed $\bar{v}$ can collide in one second with all molecules contained in a cylinder of volume $\pi d^2 \bar{v}$, where $d$ is the molecular diameter.

A property closely related to the collision frequency is the average distance traversed between collisions, known as the **mean free path** $\Lambda$. The path traced by a molecule in a gas has already been illustrated in Figure 4.6. The mean free path is just the average of the various line segments that constitute the path. The mean free path is given by the distance traversed by a molecule in 1 second divided by the number of collisions experienced in that time. Since in 1 second a molecule traverses a distance numerically equal to its speed,

$$\Lambda = \frac{\bar{v}}{Z_1} = \frac{\bar{v}}{\sqrt{2}\,\pi d^2 \bar{v} N^*} = \frac{1}{\sqrt{2}\,\pi d^2 N^*} \tag{4.25}$$

Equations 4.24 and 4.25 indicate that the values of the collision frequency and the mean free path are determined by those of the mean molecular speed $\bar{v}$, the molecular diameter $d$, and the number of molecules per unit volume $N^*$. The value of $\bar{v}$ has been derived from kinetic theory ($v_{rms}$ may be used) and that of $N^*$ can be obtained from the ideal gas law. Molecular diameters have been measured by means of various techniques and their values are well known. Thus, the collision frequency and mean free path can be calculated. Since these quantities are not readily amenable to experimental measurement, kinetic theory provides the means by which the dynamics of molecular behavior in the gas phase can be examined.

---

**Example 4.3** Calculate the collision frequency and mean free path at 25°C of an $O_2$ molecule (a) at atmospheric pressure and (b) in a good laboratory vacuum, $1.00 \times 10^{-6}$ torr. The collision diameter of an $O_2$ molecule is 3.57 Å.

We begin by evaluating $N^*$, the number of molecules per unit volume. This quantity can be expressed as

$$\underset{\text{(molecules m}^{-3})}{N^*} = \underset{\text{(molecules mol}^{-1})}{N_A} \times \underset{\text{(mol m}^{-3})}{n/V}$$

where the units show that this relation is dimensionally correct. Rewriting the ideal gas law as

$$\frac{n}{V} = \frac{P}{RT}$$

and substituting into the preceding expression gives

$$N^* = \frac{N_A P}{RT}$$

We can now evaluate $N^*$ numerically.

(a) When $P = 1.00$ atm:

$$N^* = \frac{(6.022 \times 10^{23} \text{ mol}^{-1})(1.00 \text{ atm})}{(0.08206 \text{ L atm mol}^{-1} \text{ K}^{-1})(298 \text{ K})(10^{-3} \text{ m}^3 \text{ L}^{-1})}$$

$$= 2.46 \times 10^{25} \text{ m}^{-3}$$

(b) When $P = 1.00 \times 10^{-6}$ torr, the only quantity in the expression for $N^*$ that is different for (b) than for (a) is the pressure. Therefore, to minimize computation, the result obtained for (a) can be used:

$$N^* = (2.46 \times 10^{25} \text{ m}^{-3}) \frac{(1.00 \times 10^{-6} \text{ torr})[(1/760) \text{ atm torr}^{-1}]}{(1.00 \text{ atm})}$$

$$= 3.24 \times 10^{16} \text{ m}^{-3}$$

The mean velocity (we use $v_{rms}$ for $\bar{v}$) is obtained by means of the previously derived relation (Equation 4.20):

$$\bar{v} = \left(\frac{3RT}{\overline{M}}\right)^{1/2}$$

$$= \left[\frac{(3)(8.314 \text{ J mol}^{-1} \text{ K}^{-1})(298 \text{ K})(1 \text{ kg m}^2 \text{ s}^{-2}/\text{J})}{(32.0 \text{ g mol}^{-1})(10^{-3} \text{ kg g}^{-1})}\right]^{1/2}$$

$$= 482 \text{ m s}^{-1}$$

which is independent of pressure. Substituting the values of $N^*$ and $\bar{v}$ into Equations 4.24 and 4.25 for the collision frequency and mean free path, respectively, we obtain:

(a) At atmospheric pressure:

$$Z_1 = \sqrt{2}\ \pi d^2 \bar{v} N^*$$

$$= \sqrt{2}\ \pi[(3.57 \text{ Å})(10^{-10} \text{ m Å}^{-1})]^2 (482 \text{ m s}^{-1})(2.46 \times 10^{25} \text{ m}^{-3})$$

$$= 6.71 \times 10^9 \text{ s}^{-1}$$

and

$$\Lambda = \frac{1}{\sqrt{2}\ \pi d^2 N^*}$$

$$= [\sqrt{2}\ \pi(3.57 \times 10^{-10} \text{ m})^2 (2.46 \times 10^{25} \text{ m}^{-3})]^{-1}$$

$$= 7.18 \times 10^{-8} \text{ m}$$

(b) At a pressure of $1.00 \times 10^{-6}$ torr:

$$Z_1 = \sqrt{2}\ \pi(3.57 \times 10^{-10} \text{ m})^2 (482 \text{ m s}^{-1})(3.24 \times 10^{16} \text{ m}^{-3})$$

$$= 8.84 \text{ s}^{-1}$$

and

$$\Lambda = [\sqrt{2}\ \pi(3.57 \times 10^{-10} \text{ m})^2 (3.24 \times 10^{16} \text{ m}^{-3})]^{-1}$$

$$= 54.5 \text{ m}$$

We find that at atmospheric pressure an oxygen molecule undergoes billions of collisions per second while its mean free path is but the smallest fraction of the length of any ordinary laboratory gas container. On the other hand, the situation is much less hectic in a laboratory vacuum. An oxygen molecule, on the average, makes fewer than ten collisions per second in a vacuum and its mean free path is large compared to the dimensions of an average container. It is clear that under these conditions a molecule will make many collisions with the walls of the container before it collides with another molecule, whereas the opposite is true at atmospheric pressure. ∎

## 4.6 Distribution of Molecular Speeds

Suppose all the molecules in a certain gas were moving at the same speed at a given moment. Would this condition be maintained for any length of time? From what we have learned so far about molecular collisions this cannot be the case. Molecules undergo a huge number of collisions and generally change their speed after each one. As a result, the molecules in a gas will move with a range of speeds. Is it possible to determine how the molecules are distributed in speed? At first sight this appears to be a hopeless task because the speed of a given molecule will change so quickly. Even if we could measure the speed of individual molecules, the information would be out of date in an instant. Fortunately, since the number of molecules in any gas sample is very large, it becomes possible to treat the problem in a statistical manner and predict how many molecules will have a particular speed. While we will be unable to identify which molecules in a gas have that speed at a given point in time, this drawback is of no consequence since one molecule is like another. What matters is the number of molecules having a given speed, not their identity. This situation is analogous to that involved in the use of actuarial tables by insurance companies. If a life insurance company has a sufficiently large number of customers, accurate predictions can be made as to the number of expected deaths during a given year. However, the identity of the deceased individuals cannot be determined in advance.

The distribution of molecular speeds is known as the **Maxwell–Boltzmann distribution** and can be written as an expression for the fraction of molecules with speed $v$ as a function of $v$. When written in this form, the distribution is independent of the actual number of molecules. For example, if 5% of the molecules in a gas have speeds ranging from $v_1$ to $v_2$, this will be so whether there are $10^{23}$ molecules in the gas or $10^{24}$ molecules. With $N(v)$ the number of molecules with speed $v$ and $N_{tot}$ the total number of molecules, the fraction of molecules with speed $v$ is

$$\frac{N(v)}{N_{tot}} = 4\pi \left(\frac{m}{2\pi kT}\right)^{3/2} v^2 \exp(-mv^2/2kT) \tag{4.26}$$

where $m$ is the mass of a molecule, $T$ is the absolute temperature, $k$ is the Boltzmann constant, and exp (or $e$) = 2.7... is the base of natural logarithms. The derivation of this expression is based on the second assumption of the kinetic theory, namely, that the molecules in a gas are in continuous random motion. Random motion can be analyzed by probability theory and this approach, which will not be developed here, leads to the result in Equation 4.26.

Since the Maxwell–Boltzmann distribution is fairly complicated, it is helpful to break it down into the separate factors which determine the dependence of $N(v)$ on $v$. Two such factors may be noted: the exponential term, $\exp(-mv^2/2kT)$, and the term preceding the exponential (a *preexponential* term), which is equal to the product of several constants and $v^2$. The exponent in the first term consists of the negative of the ratio of the kinetic energy of a molecule $mv^2/2$ to a thermal energy $kT$, which is closely related to the mean kinetic energy of translational motion (Equation 4.18).

Figure 4.9 shows the separate behavior of these two factors. As already noted, the preexponential term varies as $v^2$ and so increases quadratically with

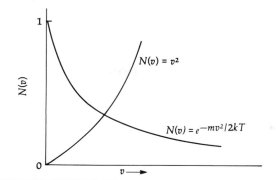

**FIGURE 4.9** The two speed-dependent terms whose product constitutes the Maxwell–Boltzmann distribution.

$v$ from a value of zero when $v = 0$. The exponential term displays an opposite dependence on $v$: when $v = 0$ the exponent is equal to zero and $e^{-0} = 1$. As $v$ increases, the function decreases continuously and approaches the $v$ axis asymptotically as $v$ approaches infinity (i.e., $e^{-\infty} = 0$). The combined effect of these two terms leads to the distribution of molecular speeds illustrated in Figure 4.10. The following are some of the interesting features of this distribution:

1. The function initially increases with $v$ from a value of zero when $v = 0$, goes through a peak, and then decreases, reflecting the combined effect of the two functions plotted in Figure 4.9. The magnitude of the speed at the peak (i.e., the value of the most probable speed $v_{mp}$) can be determined by means of differential calculus. The result is

$$v_{mp} = \left(\frac{2kT}{m}\right)^{1/2} \qquad (4.27)$$

(See Appendix A2.5 for a derivation of this equation.) When this value is compared with that of the root mean square speed (Equation 4.19), it is found that the latter is some 22% larger.

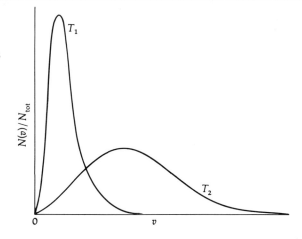

**FIGURE 4.10** Maxwell–Boltzmann distribution of molecular speeds at temperatures $T_1$ and $T_2$, where $T_2 > T_1$.

2. While a substantial fraction of the molecules have speeds not too different from the most probable speed, the distribution is not symmetrical about its maximum and extends to indefinitely high speeds. (As discussed in Section 14.8a, it is precisely the few molecules with speeds far greater than the most probable value that undergo chemical reaction.)
3. The distribution of speeds broadens with increasing temperature and the peak shifts to higher values of $v$. This is a consequence of the fact that the ratio $v^2/T$ appears in the exponential term. An increase in temperature requires a larger speed to yield the same value of $N(v)$.

The following examples illustrate some of these points in a quantitative way.

**Example 4.4** The most probable speed of $N_2$ molecules at 20°C is 417 m s$^{-1}$. What is the ratio of the number of molecules with speed of 1251 m s$^{-1}$ to the number with a speed of 417 m s$^{-1}$?

To solve this problem we must evaluate the function $N(v)/N_{tot}$ at the two values of $v$ and take the ratio. The procedure becomes somewhat simpler if we determine the ratio directly because certain common terms will cancel. Equation 4.26 evaluated at temperature $T$ for two values of $v$ ($v_2$ and $v_1$) can be written as

$$\frac{N(v_2)/N_{tot}}{N(v_1)/N_{tot}} = \frac{[4\pi(m/2\pi kT)^{3/2}](v_2^2)\exp(-mv_2^2/2kT)}{[4\pi(m/2\pi kT)^{3/2}](v_1^2)\exp(-mv_1^2/2kT)}$$

and canceling common factors we obtain

$$\frac{N(v_2)}{N(v_1)} = \frac{(v_2^2)\exp(-mv_2^2/2kT)}{(v_1^2)\exp(-mv_1^2/2kT)}$$

$$= \frac{v_2^2}{v_1^2}\exp[-\tfrac{m}{2kT}(v_2^2 - v_1^2)]$$

We are now ready to make the numerical substitutions. As noted in Section 3.6, the ratio of the mass of a molecule to the Boltzmann constant, $m/k$, which appears in the preceding equation, can be replaced by the ratio of the molecular weight to the gas constant, $\overline{M}/R$. This procedure circumvents the necessity of converting known molecular weights to masses of single molecules. Substituting the given values, we obtain

$$\frac{N(1251 \text{ m s}^{-1})}{N(417 \text{ m s}^{-1})} = \frac{(1251 \text{ m s}^{-1})^2}{(417 \text{ m s}^{-1})^2}$$

$$\times \exp\left\{\frac{-(28.0 \text{ g mol}^{-1})(10^{-3} \text{ kg g}^{-1})[(1251 \text{ m s}^{-1})^2 - (417 \text{ m s}^{-1})^2]}{(2)(8.314 \text{ J mol}^{-1} \text{ K}^{-1})(293 \text{ K})(1 \text{ kg m}^2 \text{ s}^{-2}/\text{J})}\right\}$$

$$= 3.03 \times 10^{-3}$$

This result shows that the number of molecules having a speed three times as large as the most probable value is some 300 times smaller than the number having the most probable speed. ∎

**Example 4.5** Compare the number of $N_2$ molecules with a speed of 1251 m s$^{-1}$ at 45°C to the number with that speed at 20°C.

We proceed in the same manner as in the preceding example, except that now $v_2 = v_1$ but $T_2 \neq T_1$. We obtain for the ratio

$$\frac{N(T_2)}{N(T_1)} = \left(\frac{T_1}{T_2}\right)^{3/2} \exp\left[-\frac{mv^2}{2k}\left(\frac{1}{T_2} - \frac{1}{T_1}\right)\right]$$

and making the numerical substitutions,

$$\frac{N(318\text{ K})}{N(293\text{ K})} = \left(\frac{293\text{ K}}{318\text{ K}}\right)^{3/2} \exp\left[-\frac{(28.0\text{ g mol}^{-1})(10^{-3}\text{ kg g}^{-1})(1251\text{ m s}^{-1})^2}{(2)(8.314\text{ J mol}^{-1}\text{ K}^{-1})(1\text{ kg m}^2\text{ s}^{-2}/\text{J})}\left(\frac{1}{318\text{ K}} - \frac{1}{293\text{ K}}\right)\right]$$
$$= 1.79$$

This example shows that a relatively small increase in temperature leads to a substantial increase in the number of very fast molecules. This fact accounts for the large increase with temperature of the rate at which chemical reactions occur (Section 14.8a). ∎

The Maxwell–Boltzmann distribution has been experimentally verified by measurement of the number of molecules as a function of their speed. The method, which is illustrated in Figure 4.11, involves a determination of the time it takes molecules to traverse a known distance, and is known as a **time-of-flight** measurement. The speed, of course, is equal to the ratio of the distance to the elapsed time.

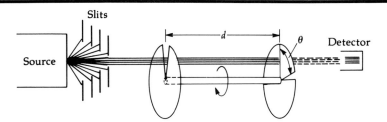

**FIGURE 4.11** Time-of-flight determination of the Maxwell–Boltzmann distribution of molecular speeds. Molecules effuse through a hole from a reservoir at temperature $T$. Collimating slits form the molecules into a well-defined beam. The beam encounters two rotating discs separated by a distance $d$. Each disc has a wedge-shaped opening which allows molecules to pass through. The two wedges are displaced from each other by an angle $\theta$, thereby permitting only molecules of a certain speed to reach the detector. The time of flight can be determined from the rotational speed of the wheels and the angle $\theta$. Different values of the rotational speed and of $\theta$ allow molecules of different speed to reach the detector. The distribution is obtained from the number of molecules detected as a function of their speed.

# Real Gases

Real gases obey the ideal gas law at temperatures well above their condensation points and at sufficiently low pressures. As conditions depart from these limits, deviations from ideal behavior may be observed. In this section we examine the nature of these deviations and attempt to understand them both macroscopically and in terms of the underlying molecular behavior.

## 4.7 Departures from Ideal Gas Behavior

The fact that real gases deviate from the ideal gas law is already apparent from the data in Table 3.1, which showed that the molar volumes of various gases at STP are not exactly equal to the ideal gas value of 22.414 L. A more useful measure of departures from ideal behavior is provided by the **compressibility factor,** which is defined as the ratio of the actual volume of 1 mole of gas, $V$, to the ideal gas volume at the same temperature and pressure, $V_{id}$,

$$Z = V/V_{id} \tag{4.28}$$

Since according to the ideal gas law $V_{id} = RT/P$, we can also write

$$Z = PV/RT \tag{4.29}$$

The compressibility factor of an ideal gas is unity by definition. Real gases display values of $Z$ which, depending on the temperature and pressure, can be larger or smaller than unity. Figure 4.12 shows the variation of $Z$ with pressure at 0°C for a number of gases. In some instances $Z$ initially decreases with pressure, goes through a minimum, and then increases; in other instances $Z$ in-

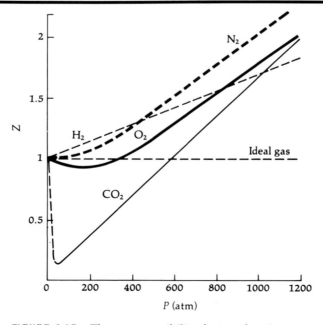

**FIGURE 4.12** The compressibility factor of various gases at 0°C.

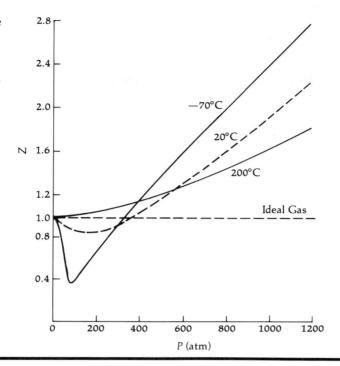

**FIGURE 4.13** Pressure dependence of the compressibility factor of methane at three different temperatures.

creases with pressure from the outset. Note that the deviations from ideal behavior can be rather large, in some cases exceeding a factor of 2. However, the values of Z for all gases converge to unity as the pressure approaches zero, thereby confirming that ideal behavior is observed when this condition is met.

The pressure dependence of the compressibility factor evolves in a distinct manner with increasing temperature. As an example, the results for methane are illustrated in Figure 4.13. At relatively low temperatures Z initially decreases with increasing pressure but this trend reverses itself at high pressures. On the other hand, at high temperatures the initial decrease in Z does not occur. Similar results are obtained for all gases although the changes in the behavior of Z occur at different temperatures. The data in Figure 4.12 can thus be understood as showing that 0°C is a "low" temperature for some gases and a "high" temperature for others. The pressure scale in Figures 4.12 and 4.13 extends to rather large values because this is where deviations from the ideal gas law become most severe. The ideal gas law actually works well at ordinary pressures.

In order to understand the behavior of the compressibility factor, it is necessary to examine the interaction between the molecules of a real gas in more detail. We already know that when molecules are far apart (low pressure, Figure 4.14a) the interactions between them are negligibly small, and ideal behavior is observed. When a gas is compressed to the point that the average distance between molecules is the equivalent of a few molecular diameters (moderate pressure, Figure 4.14b), the attractive intermolecular forces begin to be felt. A gas in which attractive forces are important is more readily compressed than an ideal gas due to the mutual attraction between the molecules. Thus, we might expect its volume to be smaller than the ideal gas volume and

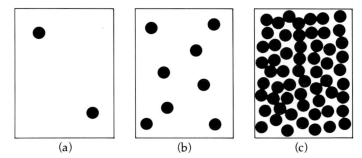

**FIGURE 4.14** Intermolecular interactions and the distance between molecules. (a) When the molecules in a gas are far apart the gas behaves ideally (low pressure). (b) When the distance between molecules is a few molecular diameters attractive forces cause the gas to be more readily compressed than an ideal gas (moderate pressure). (c) When the molecules are brought into close contact they repel each other and the gas is less compressible than an ideal gas (high pressure).

its compressibility factor to become less than unity, as is indeed observed for many gases. As the gas is compressed beyond this stage (high pressure, Figure 4.14c), the molecules are squeezed into the same space. Under these conditions the intermolecular forces become repulsive and the gas becomes more difficult to compress than an ideal gas. The volume of the gas is now larger than the ideal gas volume at the same temperature and pressure, and the compressibility factor therefore becomes larger than unity.

The interaction between two molecules is customarily shown by a plot of the intermolecular **potential energy**—the energy which the molecules have in the presence of a force (e.g., the Coulomb force) by virtue of their relative position—as a function of the distance between their centers. We know from physics that the force between two molecules separated by a distance $d$ is equal to the negative of the rate of change with respect to distance of the potential energy $U$ evaluated at $d$,

$$F = -\Delta U/\Delta d \qquad (4.30)$$

If the slope of the potential energy curve, $\Delta U/\Delta d$, is positive (i.e., $U$ increases with $d$) the force between the molecules is attractive, but if the slope of the potential energy curve is negative, the force between the molecules is repulsive. With these relationships in mind, let us examine the potential energy between two molecules, shown in Figure 4.15.

Starting at very small values of the intermolecular distance, the potential energy $U$ decreases with increasing distance $d$. The slope of the curve is negative and the molecules repel each other. Figure 14.14c shows a typical configuration of gas molecules corresponding to this potential energy. As the distance increases, the potential energy goes through a minimum and then increases with increasing distance—the slope of the curve becomes positive. The force between the molecules then becomes attractive and the molecular configuration is shown in Figure 14.14b. With increasing distance beyond the minimum the curve gradually flattens out and the attractive force between the molecules asymptotically decreases to zero in the limit of large separation. Figure 14.14a shows the corresponding molecular configuration.

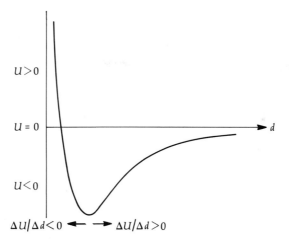

**FIGURE 4.15** Intermolecular potential energy as a function of separation distance. The distance is expressed in units of the molecular diameter. When the potential energy curve has negative slope ($\Delta U/\Delta d < 0$) the molecules repel each other, and when the slope is positive ($\Delta U/\Delta d > 0$) the molecules attract each other. The horizontal line at $U = 0$ corresponds to an ideal gas, for which there is no interaction between the molecules.

The potential energy curve of Figure 14.15 provides another representation of the difference between an ideal and a real gas. Since there are no molecular interactions in an ideal gas, the potential energy is zero at all separations. A real gas approaches this limit for moderately large separation distances but deviates from this behavior in the manner described above for small distances.

## 4.8 The Critical Point and Condensation

*a. The Shape of P–V Isotherms* The most radical departure from ideal behavior occurs when a gas condenses to form a liquid. This process occurs when a gas is cooled and compressed. The average speed of the molecules in a gas is reduced by cooling while the average distance between the molecules is reduced by both cooling and compression. Eventually, a point is reached when the motion of the molecules is sufficiently slow and the distance between them sufficiently small that the attractive forces overcome the random thermal motion of the molecules and condensation occurs. This sequence of events is illustrated in Figure 4.16, which extends a series of isotherms, such as that shown in Figure 3.5, to lower temperatures. The data refer to carbon dioxide, but similar behavior is shown by all gases, although the numerical values of $P$, $T$, and $V$ vary from one substance to another.

At the highest indicated temperature Boyle's law is obeyed, as evidenced by the hyperbolic shape of the isotherm. With decreasing temperature, increasingly drastic changes in the shape of the isotherms may be noted. At a temperature of 31°C and below, the isotherms exhibit sharp discontinuities, becoming horizontal over a certain range of volumes, and nearly vertical for small volumes.

It is instructive to examine in detail one particular isotherm exhibiting these discontinuities. Consider a container of $CO_2$ maintained at 13°C, one of whose walls is a movable piston. At relatively low pressures this substance ex-

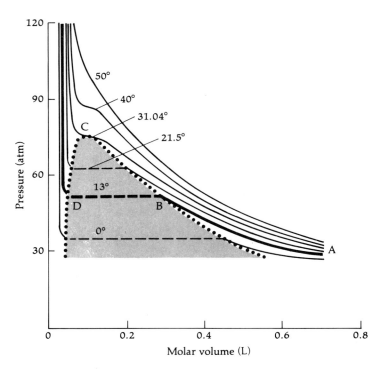

**FIGURE 4.16** Isotherms of $CO_2$ in the vicinity of the critical point (point C). Condensation occurs at any combination of $P$ and $V$ within the shaded area. The isotherm for 13°C (in color) and points A, B, and D on this isotherm are discussed in detail in the text.

---

ists as a gas (point A on the isotherm). As the piston is pushed in, the volume decreases and the pressure of the gas increases until point B is reached. At this point the behavior of the system changes drastically from that observed at lower pressures. Further compression reduces the volume of the container until point D is reached. However, the pressure remains constant as the volume decreases. If we were to look inside the container we would see a dramatic change in the appearance of the carbon dioxide. At point B the first droplets of liquid carbon dioxide would be seen. As the system is compressed further, more of the carbon dioxide condenses until point D is reached, when all the carbon dioxide in the container is present as a liquid. Further compression requires a large increase in pressure but leads to only a small reduction in volume; the isotherm becomes nearly vertical. The substance being compressed is now liquid carbon dioxide. As noted in Section 3.2, liquids are nearly incompressible—a large increase in pressure leads to a very small decrease in volume.

The horizontal segments of the isotherms become narrower as the temperature increases and define a conical region which is shaded in Figure 4.16. To the right of this region, a substance exists as a gas. To the left, it exists as a liquid. Both phases coexist in the shaded region. The apex of the shaded region, designated C in Figure 4.16, is of special significance. The isotherm passing through this point corresponds to the highest temperature at which it is possible to condense a gas to a liquid. The random thermal motion of the

**TABLE 4.2  Critical Point Parameters of Some Gases**

| Gas | $T_c$ (K) | $P_c$ (atm) | $V_c$ (L mol$^{-1}$) | $T_{bp}^a$ (K) | $T_c/T_{bp}$ | $Z_c = P_cV_c/RT_c$ |
|---|---|---|---|---|---|---|
| He | 5.3 | 2.26 | 0.0576 | 4.3 | 1.2 | 0.30 |
| H$_2$ | 33.3 | 12.8 | 0.0650 | 20.4 | 1.6 | 0.30 |
| N$_2$ | 126.1 | 33.5 | 0.0900 | 77.4 | 1.6 | 0.29 |
| CO | 134.0 | 35.0 | 0.0900 | 81.7 | 1.6 | 0.29 |
| Ar | 150.7 | 48.0 | 0.0771 | 87.5 | 1.7 | 0.30 |
| O$_2$ | 153.4 | 49.7 | 0.0744 | 90.2 | 1.7 | 0.29 |
| CH$_4$ | 190.2 | 45.6 | 0.0988 | 109.2 | 1.7 | 0.29 |
| CO$_2$ | 304.3 | 73.0 | 0.0957 | 194.7 | 1.6 | 0.28 |
| NH$_3$ | 405.6 | 111.5 | 0.0724 | 239.8 | 1.7 | 0.24 |
| n-C$_5$H$_{12}$ | 470.3 | 33.0 | 0.3102 | 309.3 | 1.5 | 0.27 |
| CH$_3$OH | 513.1 | 78.5 | 0.1177 | 338.3 | 1.5 | 0.22 |
| C$_6$H$_6$ | 561.6 | 47.9 | 0.2564 | 353.3 | 1.6 | 0.27 |
| H$_2$O | 647.2 | 217.7 | 0.0450 | 373.2 | 1.7 | 0.18 |

$^a$ Normal boiling point.

molecules becomes so violent above this point that the attractive forces are unable to bring about condensation even when the molecules are squeezed together. This point is known as the **critical point,** and it is characterized by a particular pressure $P_c$, molar volume $V_c$, and temperature $T_c$. Note that the critical isotherm changes from concave to convex at the critical point; in mathematical language, the curve has a point of inflection at C. Table 4.2 summarizes the critical parameters for a number of substances. The critical pressure is one or two orders of magnitude higher than atmospheric pressure while the critical temperature is generally 50–70% higher than the normal boiling point (in kelvins). For example, the critical point of water occurs at 647 K and a pressure of 217.7 atm.

***b. The Law of Corresponding States***  In view of the relation between the critical point and condensation it is perhaps not surprising that the previously noted departures from ideal behavior are closely linked to the paramaters defining the critical point. It turns out that the P–V–T data for all gases can be systematized into a single plot of the type illustrated in Figure 4.13 provided these variables are expressed in terms of the corresponding critical point values. In order to do so, we define the ratios

$$P_R = P/P_c, \quad T_R = T/T_c, \quad V_R = V/V_c$$

These ratios, or **reduced variables,** are dimensionless quantities that can be obtained by dividing experimental P–V–T values by the corresponding tabulated critical point values. For example, at a pressure of 1.00 atm, water is at a reduced pressure of 1.00 atm/217.7 atm = 0.00459, while carbon dioxide is at a reduced pressure of 1.00 atm/73.0 atm = 0.0137.

A plot of the compressibility factor Z versus $P_R$ for different values of $T_R$ is shown in Figure 4.17 for a number of different gases. The uniformity of behavior brought out by the use of the reduced variables is striking, particularly when compared with Figure 4.12, where similar data were plotted without using reduced variables. The observation that real gases in the same state of reduced volume and temperature exert approximately the same reduced pressure

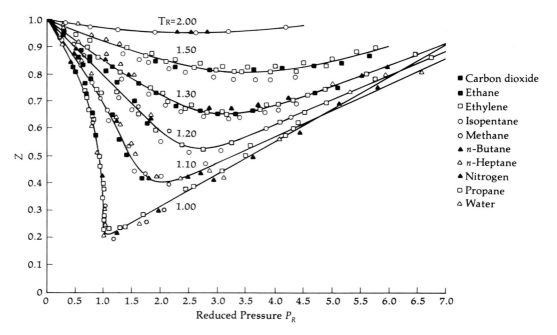

**FIGURE 4.17** Compressibility factor of various gases as a function of reduced pressure and temperature, illustrating the law of corresponding states.

is known as the **law of corresponding states.** The critical point parameters provide a further illustration of the validity of this law. Since gases at their respective critical points are by definition in corresponding states ($P_R = V_R = T_R = 1.0$), their values of the compressibility factor, $P_c V_c / RT_c$ should be approximately equal. Table 4.2 shows that this is indeed so.

The data illustrated in Figure 4.17 are of value in calculations involving real gases. For example, many industrial processes involving gases are carried out at pressures and temperatures where large deviations from ideal behavior can be expected. A better estimate of the P–V–T relationships expected under these conditions can be obtained from the reduced variable data.

Example 4.6  Use the data in Figure 4.17 and Table 4.2 to estimate the volume of 1 mole of $CO_2$ at a pressure of 73 atm and a temperature of 365 K. Compare this volume with that found from the ideal gas law.

From Table 4.2 we obtain $P_c = 73.0$ atm and $T_c = 304.3$ K. The reduced pressure and temperature are

$$P_R = \frac{P}{P_c} = \frac{73 \text{ atm}}{73 \text{ atm}} = 1.0 \qquad T_R = \frac{T}{T_c} = \frac{365 \text{ K}}{304.3 \text{ K}} = 1.20$$

Figure 4.17 indicates that at these values of $P_R$ and $T_R$, the compressibility factor is $Z \approx 0.80$. We then obtain $V$ from the definition of $Z$,

$$Z = PV/RT$$
$$V = ZRT/P$$

$$= \frac{(0.80)(0.08206 \text{ L atm mol}^{-1} \text{ K}^{-1})(365 \text{ K})}{73 \text{ atm}}$$

$$= 0.33 \text{ L mol}^{-1}$$

The ideal gas volume is simply

$$V = \frac{RT}{P} = \frac{(0.08206 \text{ L atm mol}^{-1} \text{ K}^{-1})(365 \text{ K})}{73 \text{ atm}}$$

$$= 0.41 \text{ L mol}^{-1}$$

The actual volume of $CO_2$ under these conditions is some 20% smaller than the ideal volume. ∎

## 4.9 The van der Waals Equation of State

The deviations from ideal gas behavior discussed in the preceding sections reflect the fact that molecules interact with each other. We have already noted in Section 4.7 that at moderate intermolecular distances attractive forces between molecules manifest themselves. On the other hand, at very short distances, the finite volume of molecules causes them to repel each other. The ideal gas law can be modified readily to take these two effects into account. These modifications were first proposed by Johannes van der Waals (1837–1923) in the derivation of the equation that bears his name.

Since molecules have a finite volume, some fraction of the space in which a gas is confined will, in effect, be occupied. The free volume in which the molecules can move is therefore smaller than the physical volume of their container (Figure 4.18a). The volume $V$ appearing in the ideal gas law is consequently reduced by a term $nb$, where $b$ is a measure of the volume of Avogadro's number of molecules:

$$P(V - nb) = nRT \quad (4.31)$$

The effect of the attractive forces is incorporated into the ideal gas law in an equally direct manner. The attraction that a molecule exerts on its neighbors

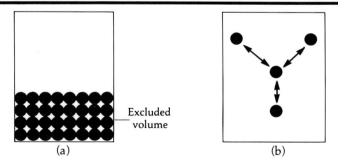

**FIGURE 4.18** Corrections to the ideal gas law in the van der Waals equation. (a) The excluded volume. The molecules occupy space and therefore the actual volume in which they can move is smaller than the container volume. (b) The attractive forces between the molecules reduce the force with which they strike the container walls and, therefore, reduce the gas pressure. The effect of one molecule on another is proportional to $n/V$; since the second molecule has a similar effect on the first molecule, the reduction in pressure is proportional to $(n/V)^2$.

tends to draw them closer and so mimics the effect of a decrease in the pressure exerted by the gas. The effect of one molecule in causing this reduction in pressure is proportional to the number of nearby molecules on which it can act. If there are $n$ moles of gas present in a volume $V$, this number is proportional to $n/V$. Since each of the neighboring molecules in turn interacts with the first molecule and with its other neighbors, the total reduction in pressure is proportional to $(n/V) \times (n/V)$, or $(n/V)^2$ (Figure 4.18b). The gas pressure given by Equation 4.31 is therefore reduced by an amount $an^2/V^2$, where $a$ is a proportionality constant:

$$P = \frac{nRT}{V - nb} - \frac{an^2}{V^2} \tag{4.32}$$

or

$$\left(P + \frac{an^2}{V^2}\right)(V - nb) = nRT \tag{4.33}$$

Equation 4.33 is known as the **van der Waals equation of state.** The two correction terms by which it differs from the ideal gas equation of state, $an^2/V^2$ and $nb$, are often called the **internal pressure** and the **excluded volume,** respectively. The attractive and repulsive intermolecular forces which give rise to these correction terms are collectively known as **van der Waals forces.**

The van der Waals equation is characterized by the presence of two coefficients, $a$ and $b$. These constants vary from one gas to another; some typical values are tabulated in Table 4.3. They are obtained by fitting the equation to experimental P–V–T data and adjusting their values so as to obtain the best fit. Equation 4.32 shows that the van der Waals equation can be solved readily for $P$, while Equation 4.33 indicates that the equation can also be easily solved for $T$. The solutions for $V$ and $n$ are more difficult to obtain since the equation is cubic in each of these variables (i.e., it includes terms up to $V^3$ or $n^3$).

**TABLE 4.3 van der Waals Constants for Gases**

| Gas | $a$ (L$^2$ atm mol$^{-2}$) | $b$ (L mol$^{-1}$) |
|---|---|---|
| He | 0.03412 | 0.02370 |
| Ne | 0.2107 | 0.01709 |
| Ar | 1.345 | 0.03219 |
| Kr | 2.318 | 0.03978 |
| Xe | 4.194 | 0.05105 |
| H$_2$ | 0.2444 | 0.02661 |
| N$_2$ | 1.390 | 0.03913 |
| O$_2$ | 1.360 | 0.03183 |
| Cl$_2$ | 6.493 | 0.05622 |
| CO$_2$ | 3.592 | 0.04267 |
| NH$_3$ | 4.170 | 0.03707 |
| H$_2$O | 5.464 | 0.03049 |
| CH$_4$ | 2.253 | 0.04278 |
| CH$_3$OH | 9.523 | 0.06702 |
| C$_6$H$_6$ | 18.00 | 0.115 |

**Example 4.7** In an industrial process, methane must be heated to 1000°C at constant volume. If 5000 mol of the gas are introduced into a 1000 L vessel at 25°C, what is the final pressure exerted at 1000°C? Assume that methane acts (a) as a van der Waals gas and (b) as an ideal gas.

(a) The pressure of a van der Waals gas is given directly by Equation 4.32:

$$P = \frac{nRT}{V - nb} - \frac{an^2}{V^2}$$

Using the van der Waals constants for methane given in Table 4.3 we obtain

$$P = \frac{(5000 \text{ mol})(0.08206 \text{ L atm mol}^{-1} \text{ K}^{-1})(1273 \text{ K})}{1000 \text{ L} - (5000 \text{ mol})(0.04278 \text{ L mol}^{-1})}$$
$$- \frac{(2.253 \text{ L}^2 \text{ atm mol}^{-2})(5000 \text{ mol})^2}{(1000 \text{ L})^2}$$

= 608.1 atm

(b) For an ideal gas,

$$P = \frac{nRT}{V}$$

$$= \frac{(5000 \text{ mol})(0.08206 \text{ L atm mol}^{-1} \text{ K}^{-1})(1273 \text{ K})}{1000 \text{ L}}$$

= 522.3 atm

The comparison shows that the pressure calculated by means of the van der Waals equation is, in this particular instance, nearly 20% higher than that obtained for an ideal gas. The difference could be of practical importance since it gives a more accurate indication of the pressures that the confinement vessel must withstand. ∎

How good is the van der Waals equation? One indication of its validity is given by the data in Table 4.4, in which the experimental and calculated molar volumes of $CO_2$ are compared at various pressures. We see that the van der Waals pressure closely agrees with the experimental value up to $P = 100$ atm, although a significant discrepancy is seen for $P = 200$ atm. Nonetheless, the agreement is much better than that obtained with the ideal gas law.

**TABLE 4.4 Experimental and Calculated Molar Volumes of $CO_2$ at 320 K**

| P (atm) | V (L mol$^{-1}$) Experimental | van der Waals | Ideal Gas |
|---|---|---|---|
| 1 | 26.2 | 26.2 | 26.3 |
| 10 | 2.52 | 2.53 | 2.63 |
| 40 | 0.54 | 0.55 | 0.66 |
| 100 | 0.098 | 0.099 | 0.26 |
| 200 | 0.056 | 0.072 | 0.13 |

**FIGURE 4.19** Isotherms of $CO_2$ as calculated by means of the van der Waals equation.

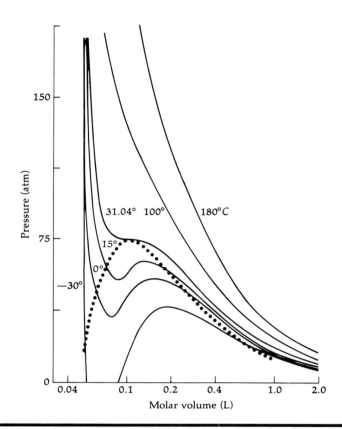

A more graphical view of the van der Waals equation is given in Figure 4.19, which shows a number of calculated isotherms for $CO_2$. The curves are directly comparable to the experimental isotherms shown in Figure 4.16. At high temperatures the isotherms are similar to ideal gas isotherms except for a somewhat steeper rise at high pressures, which is also exhibited by the data. This behavior can be understood by noting that at high $T$, the first term in Equation 4.32 is much larger than the second. Under these conditions the van der Waals equation (i.e., Equation 4.32) reduces to

$$P = \frac{nRT}{V - nb} \qquad (4.34)$$

For a given value of $V$, $P$ is larger than predicted by the ideal gas law, and the isotherms are correspondingly steeper.

At low temperatures, the van der Waals isotherms show the occurrence of oscillations. These occur in the region where condensation takes place and correspond to the horizontal portions of the experimental isotherms. The oscillations are obviously unrealistic; however, we cannot expect a simple equation to fit the observed discontinuities. It is rather remarkable that as the critical point is approached the minima and maxima of the oscillations come closer and that when $T_c$ is reached, they coincide. The critical isotherm is consequently characterized by a point of inflection, just as is the experimental critical isotherm. Since the location of an inflection point can be determined by means

## Conclusion

In this chapter we have seen how the kinetic theory of gases, which is based on several assumptions about molecular behavior, can account for the macroscopic properties of gases embodied in the ideal gas law. The consequences of the theory are far reaching. The theory explains the meaning of temperature and predicts the magnitude of the mean speed of molecules as well as the distribution of speeds. Several assumptions of the kinetic theory are approximations which break down at high pressures and low temperatures. The failure of the ideal gas equation under these conditions can be traced back to these assumptions. Some reasonable modifications lead to the van der Waals equation, which fits experimental P–V–T data for gases over a much broader range of conditions than the ideal gas law. The success of the kinetic theory in predicting macroscopic properties of gases on the basis of a molecular model was one of the chief reasons for the early acceptance of the existence of atoms and molecules by the scientific community.

of differential calculus, the location of the critical point can be predicted by means of the van der Waals equation. Considering that the van der Waals equation has only two independent constants ($a$ and $b$), it does a very good, though not perfect, job of fitting the P–V–T data for real gases.

## Problems

1. How many air molecules are there in a 1.0 L vessel which has been evacuated to a pressure of $8.0 \times 10^{-7}$ torr at 20°C?

2. A 1.5 L bulb contains 0.500 mol of neon gas at 25°C. (a) What is the total translational kinetic energy? (b) Neon is monatomic and possesses only translational kinetic energy. How much energy must be expended in heating the gas to 150°C?

3. Calculate the root mean square speed of a nitrogen molecule at 25°C; determine the total translational kinetic energy per mole of nitrogen at this temperature.

4. The root mean square speed of the molecules of a certain gaseous oxide at 10°C is 502 m s$^{-1}$. What is the molecular formula?

5. Calculate the ratio of the root mean square speeds of gaseous molecules of $D_2O$ and $H_2O$ (D = deuterium).

6. It takes 28.0 L of air to fill a certain automobile tire to a pressure of 1600 torr at 30°C. (a) What is the effusion rate through a 1 mm diameter pinhole? For simplicity, assume that air consists of pure nitrogen. (b) Assuming that the effusion rate and volume remain constant, determine the length of time it would take for half of the air to escape.

7. The effusion rate of a certain gas at $-106°C$ is three times as high as that of neon at 25°C. What is the molecular weight of the gas? Assume that both gases are at the same pressure and occupy the same volume.

8. Calculate the number of collisions experienced per second and the mean free path of a nitrogen molecule at 20°C and a pressure of (a) 1.0 atm, and (b) 0.010 torr. The collision diameter of $N_2$ is 3.68 Å.

9. Although interstellar space is nearly empty it nonetheless contains various atoms, molecules, and other species in extremely low concentrations. The most abundant species by far is atomic hydrogen, whose density is $2.0 \times 10^{-28}$ g cm$^{-3}$. Assuming a collision diameter of 1.06 Å and a temperature of 4.0 K calculate (a) the H–H collision frequency, (b) the mean free path of a hydrogen atom, and (c) the average number of years elapsed between collisions.

10. Recognizing that an intermolecular collision in-

volves *two* molecules, derive an expression for the total number of intermolecular collisions occurring per second in a unit volume. Use this expression to determine the total number of collisions per second occurring in 1.0 m³ of $N_2$ gas at a pressure of 1.0 atm at 20°C. The collision diameter of $N_2$ is 3.68 Å.

11. Calculate the ratio of the number of Ne atoms in a sample having a speed of 2500 m s⁻¹ at 30°C to the number having a speed of 250 m s⁻¹ at this temperature.

12. Calculate the ratio of the number of $N_2$ molecules having a speed of $3.0 \times 10^3$ m s⁻¹ at 100°C to the number having this speed at 25°C. Assume the total number of molecules is the same at both temperatures.

13. Determine the ratio of the number of molecules in a gas having a speed ten times larger than the root mean square speed to the number having a speed equal to the root mean square speed. Explain why this ratio is independent of temperature.

14. When $\frac{1}{2}mv^2 \ll kT$, the Maxwell–Boltzmann distribution reduces to

$$\frac{N(v)}{N_{tot}} = 4\pi \left(\frac{m}{2\pi kT}\right)^{3/2} v^2 \left(1 - \frac{mv^2}{2kT}\right)$$

Derive this expression by recognizing that $e^{-x} \approx 1 - x$ when $x \ll 1$. Sketch this function and comment on how it compares in shape with the Maxwell-Boltzmann distribution. What are the minimum and maximum values of the speed allowed by this approximation?

15.* Derive the value of the most probable speed of molecules obeying the distribution given in Problem 14.

16.* The average speed can be obtained from the relation $\bar{v} = \int f(v)v \, dv / \int f(v) \, dv$, where $f(v)$ is the distribution function and the integration is performed over all allowed values of $v$. What is the average speed of the distribution function of Problem 14?

17.* Derive the value of the most probable speed of molecules obeying the Maxwell-Boltzmann distribution.

18. A 5.00 L container is filled with 85.0 g of $NH_3$ at 40°C. Calculate the pressure exerted by the gas using (a) the ideal gas law and (b) the van der Waals equation.

19. Assuming that the van der Waals equation gives the correct pressure of methane at 50°C, determine the percentage error made in using the ideal gas law to calculate the pressure exerted by 25.0 mol of methane confined to a 1.75 L container at this temperature.

20. What is the volume occupied by 1.00 mol of Ar at 1000 atm and 0°C? Treat argon as a van der Waals gas. Compare with the ideal gas value.

21. The pressure of methane in a 1000 L vessel at 25°C is 100.0 atm. The vessel is heated to 1000°C. What is the pressure assuming the gas is (a) a van der Waals gas and (b) an ideal gas.

22.* At the critical point of a gas, the critical isotherm has a point of inflection. At such a point, the first and second derivatives of the function are zero (i.e., $dP/dV = 0$ and $d^2P/dV^2 = 0$). Applying these relations to the van der Waals equation for one mole of a gas derive relations for the constants $a$ and $b$ in terms of the critical point parameters $P_c$ and $T_c$. Using these relations and the data in Table 4.2, evaluate the van der Waals constants for $O_2$. Compare them with the values listed in Table 4.3.

23. The results obtained in Problem 22 may be rewritten to express the critical point parameters in terms of the van der Waals constants:

$$P_c = a/27b^2, \qquad V_c = 3b, \qquad T_c = 8a/27bR$$

By inserting these results into the van der Waals equation (for 1 mole of gas) expressed in terms of the reduced variables $P_R$, $V_R$, and $T_R$, show that this equation is consistent with the law of corresponding states, i.e., show that the van der Waals equation expressed in reduced variables does not involve the constants $a$ and $b$ and is thus the same for all real gases.

24. Using the relations between the critical point parameters and the van der Waals constants given in Problem 23, obtain the value for the compressibility factor at the critical point $Z_c$. Compare it with the data in Table 4.2.

25. When the attractive and repulsive forces between molecules just balance each other, a real gas mimics the behavior of an ideal gas: its compressibility factor is unity over a broad range of pressures. The temperature at which this happens is called the Boyle temperature. Derive an expression for the Boyle temperature in terms of the van der Waals constants. [The derivation requires that the quantity $1/(1 - b/V)$ appearing in the van der Waals expression for the compressibility factor be replaced by $(1 + b/V)$, which can be justified on the basis of a power series expansion, i.e., $1/(1 - x) = 1 + x$ when $x \ll 1$ (see Appendix A2.4).] Use your expression to determine the Boyle temperature for $O_2$.

# 5
# The Quantized Atom

The detailed structure of the atom was discovered in a series of brilliant experiments performed during a few decades spanning the end of the nineteenth and the beginning of the twentieth century, a period that must rank as one of the most creative in the annals of science. The results of these experiments had a profound effect on modern chemistry and physics since they could not be understood within the framework of classical physics. As developed by Isaac Newton (1642–1727) and the many others who followed him, classical physics held that moving bodies follow definite trajectories in response to the forces acting on them. The energy of a body can be changed continuously by acceleration, and such a body can also be brought to rest and its properties examined. The laws of classical physics are familiar to all of us since we see their manifestation in a variety of common macroscopic phenomena, such as the motion of an automobile or the flight of a spacecraft.

However, experience has shown that the laws of classical physics cannot explain experimental results in the atomic and molecular domain. Instead, the behavior of atoms and molecules is described by quantum mechanics. This radically different theory, developed in the 1920s, has become one of the cornerstones of modern chemistry and physics. The results of quantum mechanics, indeed its very language, are strange and unfamiliar. This should come as no great surprise since the world of atoms and molecules that it describes lies outside the realm of daily experience. It is best to approach the subject with an open mind as we trace its development in this chapter. After all, we must measure the validity of a theory by its success in correlating, explaining, and predicting the results of experiments, rather than by the ease with which it can be discussed in familiar terms.

# The Wave—Particle Duality

## 5.1 The Structure of the Atom

The nuclear model of the atom, briefly summarized in Section 1.3, was developed as a result of a number of highly original experiments performed near the beginning of the twentieth century. In this section we examine the three major experiments which led to the modern concept of the atom.

*a. The Charge/Mass Ratio of the Electron*  The most important constituent of the atom from the perspective of a chemist is the electron. The first quantitative determination of its properties was performed in 1897 by J. J. Thomson (1856–1940) in an experiment in which he measured the ratio of the charge of the electron to its mass, $e/m$. It had been known for a number of years that when a high voltage is applied between metal electrodes in an evacuated tube, negatively charged particles are emitted from the negative electrode (the cathode). These so-called "cathode rays" are accelerated towards the positive electrode (the anode), and their tracks are marked by the light emitted by the residual gas through which they pass.

Using an apparatus such as that illustrated in Figure 5.1, Thomson measured the response of cathode rays (i.e., electrons) to electric and magnetic fields. In the absence of either type of field, the rays, which are formed into a narrow beam by the collimating hole in the anode, strike a zinc sulfide screen

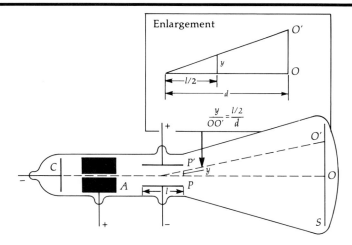

**FIGURE 5.1**  A cathode ray tube for the determination of $e/m$ for an electron. Electrons ejected from the cathode (plate C) in an evacuated container are accelerated towards the anode A and collimated into a beam by the hole through the anode. A zinc sulfide screen S at the end of the tube fluoresces (i.e., emits light) at the point struck by the electron beam (point O). When an electric field is applied between plates P and P', the electrons are deflected towards the positive plate by a distance $y$, as evidenced by the shift in fluorescence from point O to O'. A magnetic field (not shown) of appropriate magnitude applied perpendicular to the plane of the paper shifts the beam direction back until the electrons once again strike point O. The insert shows the relation between $y$, $l$, $OO'$, and $d$, the distance between the center of the electric field and the zinc sulfide screen.

located at the end of the tube at point O. If an electric field $E$ is applied between plates P and P', the cathode rays are deflected towards the positive plate P' by a force, $F = eE$, where $e$ is the charge on the electron. This force can be related to the acceleration by Newton's second law of motion:

$$F = eE = ma \tag{5.1}$$

In order to evaluate $e/m$ from Equation 5.1, the acceleration produced by the electric field must be determined. Elementary mechanics allows us to relate the acceleration to the upward deflection $y$ produced by the electric field during the time $t$ that the electrons spend between the plates:

$$y = \tfrac{1}{2}at^2 \tag{5.2}$$

As shown in Figure 5.1, $y$ can be obtained from the displacement of the electron point of impact from $O$ to $O'$. In turn, the elapsed time $t$ is given by the ratio of the length of the plates $l$ to the velocity of the electrons, $v$:

$$t = l/v \tag{5.3}$$

The velocity of the electrons can be determined with the aid of a magnetic field applied at right angles to the electric field and to their direction of motion, and extending over the same distance $l$. If the intensity of the magnetic field is $H$, the cathode rays will experience an additional force $F'$, given by

$$F' = Hev \tag{5.4}$$

The direction of the deflection due to the magnetic field will be downward; the field can be adjusted so that the cathode rays do not deviate from their original path, as evidenced by the position of the spot on the screen. This condition occurs when the forces due to the electric and magnetic fields are just equal, in which case

$$eE = Hev \tag{5.5}$$

from which we obtain

$$v = E/H \tag{5.6}$$

Combining this result with Equations 5.3, 5.2, and 5.1 gives:

$$e/m = 2yE/l^2H^2 \tag{5.7}$$

All the quantities on the right side of this equation can be determined experimentally, thereby yielding a value of $e/m$ for which the presently accepted value is $1.7588 \times 10^{11}$ C kg$^{-1}$.

**b. The Charge of the Electron**  The full significance of Thomson's result could not be appreciated until either the charge or mass of the electron was determined in a separate experiment. In combination with the value of their ratio, either quantity would serve as a determination of the other. A comparison of the electronic charge and mass with that of a simple ion, such as H$^+$, would then be of value in determining the relationship between an electron and an atom. The measurement of the electron charge by Robert Millikan (1868–1953) about a decade after Thomson's experiment provided this information.

Millikan's experiment was performed with an apparatus such as that depicted in Figure 5.2. Small oil drops are sprayed by means of an atomizer into an air-filled box. These drops drift downward under the combined effect of the

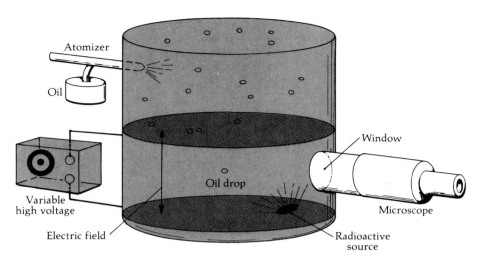

**FIGURE 5.2** Apparatus for measuring the charge on the electron. Oil drops are sprayed into the container by an atomizer. Their motion in the electric field between the two plates is observed with a microscope. A radioactive source ionizes the air; collisions of the drops with ions cause them to become charged.

gravitational force, $mg$, and a resisting force due to the viscosity of air. This resisting force causes the velocity of a drop to attain a constant value $v$:

$$v = mg/k \tag{5.8}$$

where $k$ is a constant proportional to the viscosity of air and the radius of the drop.

An electric field $E$ is applied along the vertical direction as shown in the figure. So long as the oil drop is uncharged, the electric field has no effect on it. However, collisions with gaseous ions produced by the action on air of X rays or radioactive substances occasionally cause the drops to become charged. A drop that has acquired a negative charge $q$ will be attracted upward towards the positive electrode with a force equal to the difference between the electrical force $qE$ and the gravitational force. The net velocity $v'$ is then given by the expression

$$v' = (qE - mg)/k \tag{5.9}$$

The velocity of the drop can be determined by measurement of the distance traversed in a given time interval. The quantity $k$ and the mass of the drop, $m$, can be evaluated in subsidiary experiments, and the field strength $E$ is known. Consequently, the magnitude of the charge on the drop can be determined. It was found that while the drops could acquire charges of different magnitude, all the values were multiples of $1.602 \times 10^{-19}$ C. One could therefore deduce that this value corresponded to the charge of an electron.

The combination of the results of Thomson's and Millikan's experiments indicated that the mass of the electron was $9.1 \times 10^{-31}$ kg. This value was nearly 2000 times smaller than that of a hydrogen atom, the lightest known atom. Since electrons were known to be ejected from matter, which is made up of atoms, it could be concluded that the electron must be a constituent of the atom.

*c. The Scattering of α Particles by Atoms* The experiments described earlier demonstrated that atoms, already known to be electrically neutral, must consist of negatively charged electrons and some much more massive positively charged constituent. However, it was not known how the various particles were arranged. Thomson had proposed a model of the atom based on the not unreasonable assumption that the volume of the positive part of the atom was proportional to its mass. Since most of the mass resided in this positive constituent, Thomson pictured an atom as a positively charged sphere in which the electrons were embedded in a way which minimized their repulsion from each other. It remained for Ernest Rutherford (1871–1937) and his collaborators to perform the last of the three experiments by means of which the structure of the atom was established.

The phenomenon of radioactivity (Chapter 23) had been discovered at the end of the nineteenth century. Among the particles emitted in the decay of naturally occurring radioactive isotopes are **α particles,** which were shown to be very energetic, doubly ionized helium atoms. These α particles could be collimated, shot at thin metal foils, and the manner in which they interacted with the atoms of the target element could be observed. Rutherford and co-workers performed such experiments, as illustrated in Figure 5.3.

The results of an α-particle scattering experiment are shown in Figure 5.4 as a plot of the angular distribution of the α particles passing through a thin gold foil. While most of the α particles emerged very close to their initial direction of motion (i.e., at $\theta \approx 0°$), a small fraction were scattered through large angles, a few of them actually bouncing off backwards at close to 180°. Could these results be reconciled with Thomson's model of the atom? We would expect the α particle to be deflected owing to the Coulomb force between it and the charged constituents of the atom. However, since in Thomson's model the positive charge was uniformly spread over the entire atom, an α particle would experience little if any deflection as it passed through the positive charge. Fur-

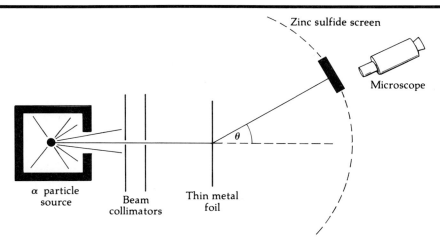

**FIGURE 5.3** Rutherford's α-particle scattering experiment. The α particles emitted by polonium are collimated into a beam and allowed to strike a thin metal foil. The number of α particles scattered through a particular angle $\theta$ is measured by a zinc sulfide screen which can be moved into any one of the indicated locations.

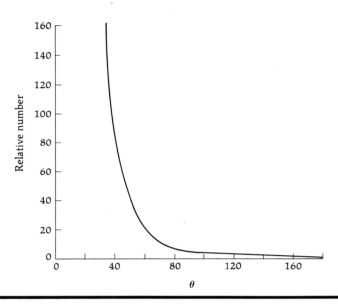

**FIGURE 5.4** Results of a Rutherford $\alpha$-particle scattering experiment. The plot gives the relative number of $\alpha$ particles observed by the detector at angle $\theta$ as a function of $\theta$. The angle $\theta$ is measured with respect to the original direction of the beam of $\alpha$ particles.

thermore, because an $\alpha$ particle is so much more massive than an electron, the effect of the electrons on the motion of the $\alpha$ particle should also be negligible, much like the effect of a hailstorm on the trajectory of a cannonball fired into it. We might, at most, expect a very slight deviation after the traversal of several atoms (Figure 5.5a). Clearly, the experimental results were at variance with Thomson's model.

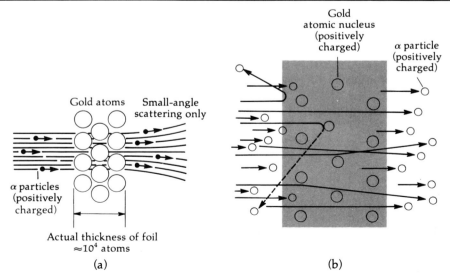

**FIGURE 5.5** (a) $\alpha$ particles incident on a "Thomson" atom. Only small-angle scattering can occur; (b) $\alpha$ particles incident on a "Rutherford" atom. Most of the $\alpha$ particles miss the nucleus and are nearly undeflected; the occasional $\alpha$ particle that makes a head-on collision (shown in color) is scattered through a large angle.

Rutherford realized that a totally different model of the atom was required: the positive charge of the atom had to be concentrated in a tiny "nucleus" containing most of the mass. Most of the time the $\alpha$ particle would approach an atom on a trajectory that would not bring it close to the nucleus. Since the Coulomb force is inversely proportional to the square of the distance (Equation 1.4), the $\alpha$ particles would experience little deflection. Occasionally, however, an $\alpha$ particle would approach a nucleus on a "head-on" trajectory (Figure 5.5b). Since such a collision involves an $\alpha$ particle and a more massive body—the nucleus of a gold atom—the $\alpha$ particle would feel the effect: it would be scattered backward as a consequence of the Coulomb repulsion between the two positively charged bodies. The expected angular distribution could be calculated on the basis of Newton's laws of motion and the assumption that the interaction between the $\alpha$ particle and the nucleus obeyed Coulomb's law. The result indicated that the angular distribution should vary as $[\sin^4(\theta/2)]^{-1}$. This characteristic dependence on $\theta$ precisely matches the experimental angular distribution displayed in Figure 5.4.

Since the radius of an atom was known to be about $10^{-10}$ m, Rutherford's analysis showed that the nucleus had to be much smaller than this. Subsequent experiments with higher-energy $\alpha$ particles, which were energetic enough to penetrate the nucleus instead of merely bouncing off (as in Rutherford's experiment), showed that nuclear radii were approximately $10^{-14}$ m.

The three experiments described in this section show how the ideas of classical physics applied in a bold and imaginative manner to relatively simple experiments provided a quantitative description of the fundamental constituents of the atom. They undoubtedly rank among the greatest experiments of physical science. The nuclear model of the atom to which they led has formed the basis for much of modern chemistry and physics.

Although the interpretation of these experiments involved the application of such classical principles as Newton's laws of motion, the resulting model of the atom could not be reconciled with classical physics. Rutherford's model, in essence, required a massive, small nucleus at the center of a set of orbiting electrons. Two equally unpalatable choices about the motion of the electrons were possible according to the principles of classical physics. First, the electrons could be stationary. However, if this were the case, the attraction between the positively charged nucleus and the negatively charged electrons would cause the electrons to fall into the nucleus, thereby destroying the atom. Second, the electrons could revolve about the nucleus, much as the planets revolve about the sun. However, classical physics demands that charged particles in circular motion radiate energy. The resulting energy losses would cause the electrons to slow down and ultimately force them to fall into the nucleus just as if they had been stationary to begin with. The very fact that atoms are stable indicates that the arrangement and motion of the electrons about the nucleus must be far more complex than suggested by the Rutherford model. More detailed information about the electronic structure of atoms is obtained by means of experiments on the light emitted by atoms under certain conditions.

## 5.2 Wave Properties of Light

In order to explore the connection between atomic structure and light, we must first examine the properties of light waves. The phenomenon of wave motion is a familiar one. It is manifested, for example, by water waves in oceans or lakes, by the undulatory motion of a rope attached at one end and swung up

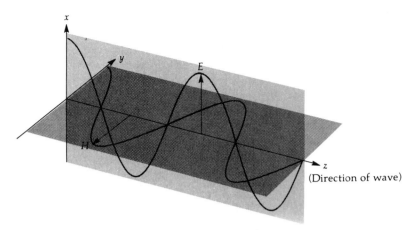

**FIGURE 5.6** An electromagnetic wave. The electric field $E$ oscillates in the $xz$ plane; the magnetic field $H$ oscillates in the $yz$ plane; the wave is propagated in the $z$ direction.

and down at the other, and by the transmission of sound through a medium. All these waves travel through some form of matter and consist of vibrations of the medium through which they move. In the absence of such a medium, this type of wave motion cannot occur, as can be deduced, for example, from the fact that sound cannot be transmitted through a vacuum. Light waves are different: they do not require the presence of matter for their propagation. The simple fact that stars are visible shows that this is indeed so.

A detailed understanding of the nature of light waves was obtained by James Clerk Maxwell (1831–1879). His theory showed that light is an **electromagnetic wave** generated by the oscillation of an electric charge. The oscillating motion produces electric and magnetic fields whose motion constitutes the light wave. These fields oscillate in a direction perpendicular to each other and to the direction of motion of the wave, as illustrated in Figure 5.6. A small charged body placed in the path of a light wave will oscillate back and forth in response to the oscillating electric field. Other types of experiments show that the magnetic field also oscillates in the manner shown in the figure.

An electromagnetic wave is characterized by several properties that can be discussed with reference to Figure 5.7. This figure is like a snapshot of the electric component of a wave. The **wavelength** $\lambda$ is the distance between two successive maxima, and the **amplitude** $A$ is the maximum value of the displacement. A charged body will experience a maximum displacement in one direction when the wave is at its maximum and a maximum displacement in the opposite direction when the wave is at its minimum. Electromagnetic waves

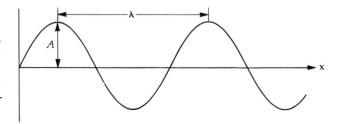

**FIGURE 5.7** A wave propagating through space can be characterized by its wavelength $\lambda$ and amplitude $A$. The shape of the wave is that of the sine function $A \sin(2\pi x/\lambda)$.

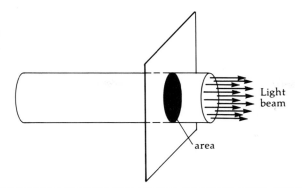

**FIGURE 5.8.** The intensity of a wave is the energy carried through a unit area in 1 second or, equivalently, the power per second.

transport energy; it is customary to express the energy in terms of the wave **intensity,** which is defined as the energy passing through a unit area per unit time (Figure 5.8). Intensity is measured in units of joules per second per square meter or, since energy per unit time is a measure of **power** (1 watt = 1 joule per second), intensity can also be expressed in units of watts per square meter. The intensity of a wave is proportional to the square of its amplitude.

All electromagnetic waves travel in vacuum at the same speed $c$, whose value is

$$c = 2.99792458 \times 10^8 \text{ m s}^{-1}$$

The number of wave maxima that move past a stationary point in 1 second is known as the **frequency** and is given by the expression

$$\nu = c/\lambda \qquad (5.10)$$

$$(\text{s}^{-1}) = (\text{m s}^{-1})/(\text{m})$$

It is customary to express frequency in units of hertz (Hz), where 1 Hz = 1 s$^{-1}$. One final quantity that is frequently used to characterize electromagnetic radiation emanating from atoms and molecules is the **wave number** $\bar{\nu}$, which is just the reciprocal of the wavelength:

$$\bar{\nu} = 1/\lambda \qquad (5.11)$$

It is customary to express $\bar{\nu}$ in units of cm$^{-1}$.

---

**Example 5.1** A 100 watt light bulb radiates light isotropically (i.e., uniformly in all directions) at a wavelength of 600 nm. Determine (a) the frequency of the radiation, (b) the wave number, (c) the time elapsed between the instant the bulb is switched on and the instant the light reaches a point 10.0 m away, and (d) the intensity of the light wave at that point.

(a) The frequency is given by Equation 5.10:

$$\nu = \frac{c}{\lambda} = \frac{2.998 \times 10^8 \text{ m s}^{-1}}{(600 \text{ nm})(10^{-9} \text{ m nm}^{-1})}$$

$$= 5.00 \times 10^{14} \text{ Hz}$$

(b) The wave number is just the reciprocal of the wavelength in cm$^{-1}$:

$$\bar{\nu} = \frac{1}{\lambda} = \frac{1}{(600 \text{ nm})(10^{-7} \text{ cm nm}^{-1})}$$

$$= 1.67 \times 10^4 \text{ cm}^{-1}$$

(c) The time of travel is equal to the ratio of the traversed distance to the speed:

$$t = \frac{d}{c} = \frac{10.0 \text{ m}}{2.998 \times 10^8 \text{ m s}^{-1}}$$

$$= 3.34 \times 10^{-8} \text{ s}$$

$$= 33.4 \text{ ns}$$

(d) The intensity of the wave is the power per unit area. Since the source emits isotropically, the total area on which the light falls at a distance of 10.0 m from the source is the surface of a sphere with a radius of 10.0 m:

$$\text{Intensity} = \frac{\text{Power}}{\text{Area}} = \frac{\text{Power}}{4\pi r^2}$$

$$= \frac{100 \text{ W}}{4\pi (10 \text{ m})^2}$$

$$= 7.96 \times 10^{-2} \text{ W m}^{-2}$$

where W is the symbol for watts. ∎

One of the characteristic features of wave motion is the occurrence of interference between two waves arriving at the same point at exactly the same time. Let us examine what happens when two waves of the same frequency

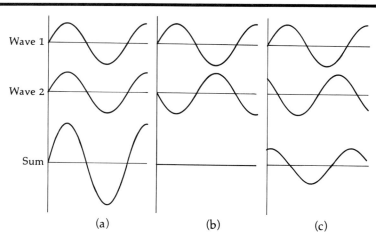

**FIGURE 5.9** Superposition of two waves of the same frequency. (a) The waves have equal amplitude and are in phase; they add to give a wave of twice the amplitude. (b) The waves are exactly out of phase and their amplitudes cancel; the intensity of the radiation vanishes. (c) The two waves are partly out of phase.

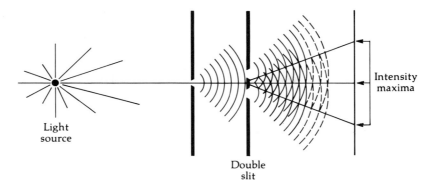

**FIGURE 5.10** Occurrence of diffraction in a double-slit experiment. The indicated rays perpendicular to the wave fronts mark the points where the two waves reinforce each other resulting in the occurrence of intensity maxima.

are superimposed. The result of this superposition is determined by the phase difference between the two waves. The **phase** of a wave specifies the position of its maximum displacement with respect to some fixed point. If two waves reach this point so that their maxima coincide in time, the waves are said to be *in phase*. As shown in Figure 5.9a, the result of this superposition is the addition of the wave amplitudes; if the two waves have equal amplitude the result is a wave of twice the amplitude. The two waves are said to undergo **constructive interference.** If the two waves arrive at a point so that the crest of one coincides with the trough of the other, the two waves are *out of phase* and undergo **destructive interference.** As may be seen in Figure 5.9b, the electric and magnetic fields of the two waves cancel and the net intensity is zero. The intermediate case (Figure 5.9c) involves two waves which are partly out of phase and undergo partial reinforcement. Note that in all three cases the amplitudes of the waves are additive; the frequency and wavelength remain unchanged.

Interference effects are readily observable in the double slit experiment illustrated in Figure 5.10. Light waves emitted by a source are collimated and allowed to impinge on two closely spaced slits. Each slit acts as a separate light source and two circular wave patterns are formed. The circles represent the positions of the wave fronts, that is, of successive crests of the waves. As shown in the figure, the two waves overlap and interfere with each other. Constructive interference occurs along the straight lines, the light "rays" that show the direction in which the waves travel. A screen intercepting the overlapping waves shows a characteristic alternation of light and dark bands corresponding, repetively, to the regions where constructive and destructive interference occurs. The result is called a **diffraction pattern** (Figure 5.10). The same phenomenon is observed when the slits are replaced by a regular array of atoms in a solid crystal and can be used to study crystal structure (Section 20.3a). The occurrence of diffraction is one of the characteristic properties of wave motion.

## 5.3 The Electromagnetic Spectrum

As shown in Example 5.1, the frequency of visible light is in the range of $10^{14}$ Hz. Maxwell's theory predicted that it should be possible to generate waves of lower frequency by slower oscillations of electric charges than those giving rise to visible light. It took some twenty years for this prediction to be confirmed by Heinrich Hertz's (1857–1894) discovery of radio waves. This discovery has led to the development of modern radio and television in one of the most successful practical applications of basic science. The presently known spectrum of electromagnetic radiation encompasses a range of wavelengths and frequencies extending over some 15 orders of magnitude. It is not surprising that there are drastic differences in the origin and form of the radiation over such a wide span of wavelengths.

The extent of the electromagnetic spectrum is displayed in Figure 5.11 and a more complete summary of the various spectral regions is given in Table 5.1. **Gamma (γ) rays,** electromagnetic radiation emitted by the atomic nucleus, have the shortest known wavelengths. Processes involving the inner shell electrons of atoms (i.e., *not* the valence electrons) give rise to the emission of **X rays.** The X rays emitted by the elements have characteristic wavelengths which can be used for atomic number assignment (Section 6.5c). The wavelengths of X rays are comparable to the sizes of atoms. This fact makes it possible to use X rays to probe the arrangement of atoms in crystalline solids by means of **X-ray diffraction** (Section 20.3a).

The broad region of the spectrum with wavelengths between approximately $10^{-8}$ m and $10^{-1}$ m is known as the **optical region.** It can be subdivided into the **ultraviolet** (UV) region, with wavelengths shorter than those of visible light, the **visible** (VIS) region, lying between approximately 400 and 700 nm, and the **infrared** (IR) region, with wavelengths longer than those of visible light. Ultraviolet and visible radiation are associated with various processes involving the outer electrons of atoms and molecules (Section 17.4). The infrared

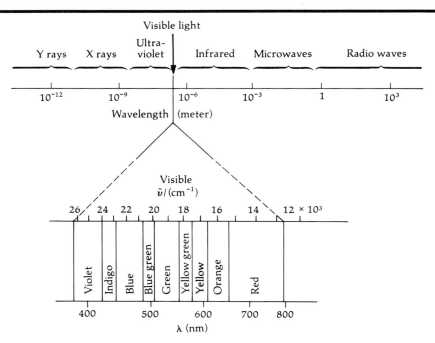

**FIGURE 5.11** Regions of the electromagnetic spectrum on a logarithmic scale of wavelengths. The visible region is shown in greater detail.

TABLE 5.1  The Electromagnetic Spectrum

| Radiation | Typical Wavelength (m) | Frequency (Hz) | Energy (eV) | Mode of Generation |
|---|---|---|---|---|
| γ rays | $10^{-12}$ | $3 \times 10^{20}$ | $10^6$ | Nuclear phenomena |
| X rays | $10^{-10}$ | $3 \times 10^{18}$ | $10^4$ | Transitions of inner atomic electrons |
| Ultraviolet | $2 \times 10^{-7}$ | $10^{15}$ | 4 | |
| Visible | | | | |
| violet | $5 \times 10^{-7}$ | | | Atomic and molecular transitions |
| ↓ | | $6 \times 10^{14}$ | 2.5 | |
| red | $7 \times 10^{-7}$ | | | |
| Near infrared | $10^{-6}$ | $3 \times 10^{14}$ | 1.2 | |
| Far infrared | $10^{-4}$ | $3 \times 10^{12}$ | 0.01 | |
| Microwaves | $10^{-1}$ | $3 \times 10^9$ | $10^{-5}$ | |
| Radio waves | | | | |
| FM, TV | 3 | $10^8$ | $10^{-7}$ | |
| AM | $3 \times 10^3$ | $10^5$ | $10^{-10}$ | Electrical oscillations |
| AC power | $5 \times 10^6$ | 60 | $10^{-13}$ | |

band is generally divided into the **near infrared** region, associated with the vibration of the atoms in a molecule, and the **far infrared** region, associated with rotational motion in molecules (Sections 17.2 and 17.3).

Beyond wavelengths of approximately 0.1 m lies the broad band of **radio waves,** including microwaves, FM radio waves, AM radio waves, and ac power. Radio waves have become of importance in chemistry as a consequence of the development of nuclear magnetic resonance spectroscopy (NMR), a powerful tool for the study of molecular structure (Section 23.12).

## 5.4  Corpuscular Properties of Electromagnetic Radiation

The discovery of radio waves and X rays confirmed Maxwell's wave theory of electromagnetic radiation. However, a number of experiments performed in the late nineteenth and early twentieth centuries indicated that the classical wave theory could not explain the emission and absorption of radiation by atoms or molecules. These experiments led to a drastic revision of the wave theory of light by showing that in some cases the behavior of light could only be explained by picturing light as a stream of particles, that is, by assuming that light has corpuscular properties. In this section we describe the three types of experiments that led to this conclusion and were instrumental in the development of quantum mechanics.

*a. Black-Body Radiation*  When a solid is heated it emits electromagnetic radiation. The intensity of the radiation varies with wavelength in the manner depicted in Figure 5.12a. The curves are those for an ideal emitter of radiation, a body capable of equally emitting and absorbing radiation at all frequencies. Such an emitter is called a *black body*.

Figure 5.12a shows that the intensity of the emitted radiation increases with the temperature of the black body. The distribution in wavelengths has a

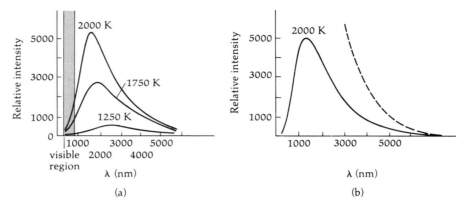

**FIGURE 5.12** Relative intensity of electromagnetic radiation emitted by a black body at the indicated temperatures as a function of wavelength. (a) Observed relative intensities. (b) Comparsion of observed relative intensity (solid curve) with the intensity predicted by classical theory at 2000 K (dashed curve).

characteristic shape: the emission probability increases with increasing wavelength, goes through a maximum, and then decreases. The wavelength at the peak is seen to shift to lower values with increasing temperature. This shift is in line with the common observation that a heated object first glows red, then yellow, and finally, at the highest temperatures, white. However, most of the emitted radiation lies in the infrared and, owing to the peak in the spectrum, the emission of short wavelength radiation is very improbable.

Attempts to explain the spectrum of black-body radiation by means of classical physics were based on the idea that the radiation originated in the oscillation, or vibration, of atoms in the solid radiator. Since, according to classical physics, atoms could possess an arbitrary amount of vibrational energy, radiation of all wavelengths should have been emitted. More detailed considerations indicated that the intensity of black-body radiation should increase continuously with decreasing wavelength, varying as $\lambda^{-4}$. As may be seen in Figure 5.12b, this classical intensity curve deviates strongly from the experimental spectrum at short wavelengths.

In 1900, Max Planck (1858–1947) proposed an explanation of the spectrum of black-body radiation which required a radically different assumption about the behavior of matter. He postulated that the energy of a vibrating atom in the black body was discrete, or *quantized*, and was related to the vibrational frequency $\nu$ by the expression

$$\epsilon = nh\nu \tag{5.12}$$

where $n$ is an integer and $h$ is a constant now known as **Planck's constant.** The significance of this assumption can be seen by considering an atom vibrating at some particular frequency, for example at $3.0 \times 10^{14}$ s$^{-1}$. According to Equation 5.12, the vibrational energy of this atom can be equal to only one of the following values: $(3.0 \times 10^{14}h)$ J, $(6.0 \times 10^{14}h)$ J, $(9.0 \times 10^{14}h)$ J, and so on. Intermediate energies [e.g., $(4 \times 10^{14}h)$ J] are forbidden. The quantization of energy invalidates the classical assumption that an atom can possess any arbitrary energy.

On the basis of this assumption, Planck derived the following expression for the intensity per unit wavelength interval of black-body radiation as a function of wavelength:

$$I = \left(\frac{2\pi hc^2}{\lambda^5}\right)\left(\frac{1}{e^{h\nu/kT} - 1}\right) \tag{5.13}$$

This expression gives an excellent fit to the spectra in Figure 5.12 provided $h$ has the value

$$h = 6.6262 \times 10^{-34} \text{ J s}$$

Planck's formula differs from the classical expression for the spectrum of black-body radiation by the exponential $e^{h\nu/kT}$, involving the ratio of the quantized vibrational energy to the thermal energy $kT$. This exponential term reduces the intensity at large values of $\nu$ (i.e., small $\lambda$) and accounts for the peak in the spectrum. Although Planck's constant is small, it is not infinitesimally small, as implied by classical physics. The difference has made it necessary to revise the laws of classical physics by recognizing that the energy of the vibrating atoms comprising the black body must be quantized.

**b. The Photoelectric Effect** Experiments performed in the late nineteenth century had demonstrated that light was capable of knocking electrons out of metals. As shown in Figure 5.13, a beam of light incident on the surface of a metal contained in a vacuum results in the flow of an electric current, a phenomenon known as the **photoelectric effect.** In principle, this effect was understandable within the framework of classical physics since light, being an electromagnetic wave, should be able to interact with a charged particle such as the electron. However, when examined more closely, the photoelectric effect turned out to be completely inconsistent with the wave theory of light.

The characteristic features of the photoelectric effect are shown in Figure 5.14. It was observed that, regardless of the light intensity, no electrons are ejected unless the light frequency is larger than some threshold value $\nu_0$ whose magnitude depends on the identity of the metal (Figure 5.14a). Furthermore, the kinetic energy of the ejected electrons was found to be a linear function of

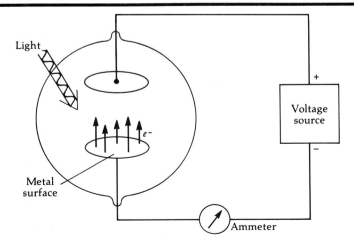

**FIGURE 5.13** The photoelectric effect. Light incident on a metal surface in an evacuated container knocks out electrons and causes an electric current to flow.

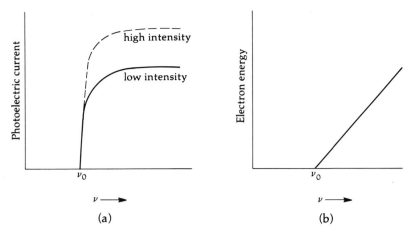

**FIGURE 5.14** Results of the photoelectric effect: (a) electric current as a function of frequency; the dashed curve shows the current observed for a greater light intensity. (b) Maximum kinetic energy of the electrons as a function of frequency.

the light frequency (Figure 5.14b). Finally, the onset of the photoelectric effect is independent of the light intensity. If light of a certain frequency could cause a photoelectric current to flow, the effect was observed even at very low intensities. The only effect of a higher intensity was to increase the magnitude of the current; higher intensity did not affect the threshold frequency or the kinetic energy of the electrons (Figure 5.14a).

According to the wave theory, the energy of a light wave is proportional to its intensity and independent of its frequency. Hence, the wave theory could explain neither the occurrence of a threshold frequency nor the dependence of the electron energy on frequency. On the other hand, the feature expected from the wave theory, namely, a dependence of the electron energy on light intensity, was not seen.

The photoelectric effect was explained in 1905 by Albert Einstein (1879–1955) by means of an extension of Planck's explanation of black-body radiation. Einstein postulated that light consisted of discrete particles, or **photons,** of energy $h\nu$,

$$\epsilon_\gamma = h\nu \tag{5.14}$$

where $\epsilon_\gamma$ is the photon energy. When a photon interacts with an electron in the metal surface, this energy is transferred to the electron. Some of the energy is used to overcome the forces binding the electron to the metal, and the remainder appears as the kinetic energy of the electron. If the energy with which the electron is bound to the metal, called the **work function,** is designated $\phi$, conservation of energy requires that

$$\tfrac{1}{2}mv^2 = h\nu - \phi \tag{5.15}$$

Thus, the kinetic energy of the electrons should be a linear function of the light frequency above a threshold $\nu_0$ determined by the condition $\phi = h\nu_0$. As already shown in Figure 5.14b, this is precisely what is observed experimentally. The idea that light can be regarded as a collection of photons was further confirmed by the dependence of the photoelectric current on light intensity (shown in Figure 5.14a). The observed increase in intensity means that we should associate light intensity with the number of photons arriving at a given point per unit time; the greater the number of photons impinging on the metal, the greater is the number of ejected electrons.

The application of Equation 5.15 to actual data permits a determination of a value of Planck's constant. The value of $h$ is just equal to the slope of the straight line obtained in a plot of electron kinetic energy versus light frequency (Figure 5.14b). The value of $h$ found in this way was the same as that obtained by Planck in his explanation of black-body radiation. This result was exceedingly important because it showed that quantization of energy occurred not only in the special case of oscillating atoms but was a much more general phenomenon: light itself was quantized.

The wave theory of light, which had been so successful in explaining the behavior of electromagnetic radiation, was incapable of explaining the interaction of radiation with electrons. In addition to its wave properties light also has corpuscular properties, that is, it behaves like particles termed photons.

**Example 5.2** Electrons are bound to the surface of cesium with an energy of $3.43 \times 10^{-19}$ J. What are the threshold frequency and wavelength necessary for the observation of the photoelectric effect in cesium?

At threshold, we have from Equations 5.14 and 5.15,

$$\phi = \epsilon_\gamma = h\nu_0$$

Therefore,

$$\nu_0 = \frac{\phi}{h} = \frac{3.43 \times 10^{-19} \text{ J}}{6.626 \times 10^{-34} \text{ J s}}$$

$$= 5.18 \times 10^{14} \text{ s}^{-1}$$

Since the speed of light equals the product of the frequency and wavelength,

$$\nu\lambda = c$$

we obtain for the corresponding threshold wavelength

$$\lambda = \frac{c}{\nu} = \frac{2.998 \times 10^8 \text{ m s}^{-1}}{5.18 \times 10^{14} \text{ s}^{-1}}$$

$$= 5.79 \times 10^{-7} \text{ m}$$

$$= 579 \text{ nm}$$

In other words, the photoelectric effect in cesium can be observed when yellow light (or light of shorter wavelength) is used. Light of longer wavelength (for example, red light) cannot eject electrons from this metal. ■

In dealing with individual atomic particles or photons, the joule is an inconveniently large energy unit. It is customary to express the energies of single particles in units of electron volts. An **electron volt** is defined as the energy acquired by a particle with the charge of an electron when it is accelerated through an electrical potential difference of exactly 1 V (volt). The relationship between these two units (see Appendix 1) is

$$1 \text{ eV} = 1.6021892 \times 10^{-19} \text{ J}$$

The energy of particles or photons is also commonly expressed in kilojoules per mole. The relationship between kilojoules per mole and electron volts is based on the fact that 1 mole contains $N_A$ particles:

$$1 \text{ eV/particle} = (1.6021892 \times 10^{-19} \text{ J/particle}) \times$$
$$(6.022045 \times 10^{23} \text{ particles/mol}) \times (10^{-3} \text{ kJ/J}) \quad (5.16)$$
$$= 96.48456 \text{ kJ mol}^{-1}$$

The "per particle" has been incorporated into the units in order to make clear the difference between the energy of a single particle and that of a mole (i.e., of $N_A$ particles). We shall generally not follow this practice.

Since the energy of a photon is proportional to its frequency and thus to its wave number, the latter also is a measure of photon energy. The energies of photons associated with various atomic or molecular processes are frequently reported in wave numbers.

---

**Example 5.3** (a) What is the energy of photons whose wavelength is 600 nm (orange light)? Express the answer in eV and in kJ/mol. (b) To how many wave numbers does this wavelength correspond?

(a) Combining Equations 5.14 and 5.10 yields the relation between photon energy and wavelength:

$$\epsilon_\gamma = \frac{hc}{\lambda}$$

$$= \frac{(6.626 \times 10^{-34} \text{ J s})(2.998 \times 10^8 \text{ m s}^{-1})}{(600 \text{ nm})(10^{-9} \text{ m nm}^{-1})(1.602 \times 10^{-19} \text{ J eV}^{-1})}$$

$$= 2.07 \text{ eV}$$

$$= (2.07 \text{ eV})(96.48 \text{ kJ mol}^{-1} \text{ eV}^{-1}) = 200 \text{ kJ mol}^{-1}$$

(b) The wave number is obtained most directly from its definition as the reciprocal of the wavelength (Equation 5.11):

$$\bar{\nu} = \frac{1}{\lambda}$$

$$= \frac{1}{(600 \text{ nm})(10^{-7} \text{ cm nm}^{-1})}$$

$$= 1.67 \times 10^4 \text{ cm}^{-1}$$

From the answers to (a) and (b) we see that 1 eV corresponds to just over 8000 cm$^{-1}$. ∎

**c. The Compton Effect** In the photoelectric effect a photon interacts with an electron by imparting all its energy to it and vanishing in the process. The corpuscular nature of light is even more clearly revealed in the **Compton effect,** in which a photon ejects an electron from an atom and in the process changes its direction of travel by an angle $\theta$, while its wavelength increases by an amount $\Delta\lambda$. This process, which is illustrated in Figure 5.15, can be understood as in-

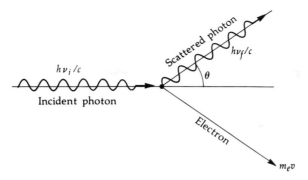

**FIGURE 5.15** Directions and momenta of a photon and an electron involved in Compton scattering. The wavy line represents a photon, and the straight line represents an electron. The initial and final momenta of the photon and the momentum of the electron are $h\nu_i/c$, $h\nu_f/c$, and $m_e v$, respectively. The photon is scattered through an angle $\theta$.

volving a collision between a photon of energy $h\nu$ and momentum $h\nu/c$ with an electron of mass $m_e$, in which both energy and momentum are conserved, much in the same way as in a collision between two molecules in a gas (Section 4.2). The analysis of the process leads to the following expression for the relation between the wavelength shift and the photon scattering angle:

$$\Delta\lambda = \frac{h(1 - \cos\theta)}{m_e c} \tag{5.17}$$

For example, the maximum shift in wavelength occurs when $\theta = 180°$, that is, when light bounces off in the backward direction. The value of $\Delta\lambda$ then is:

$$\Delta\lambda = \frac{h(1 - \cos 180°)}{m_e c} = \frac{2h}{m_e c}$$

$$= \frac{(2)(6.626 \times 10^{-34} \text{ J s})}{(9.110 \times 10^{-31} \text{ kg})(2.998 \times 10^8 \text{ m s}^{-1})}$$

$$= 4.85 \times 10^{-12} \text{ m}$$

$$= 4.85 \text{ pm}$$

The shift amounts to a change in the wavelength of an X ray of a few percent and is readily detectable in the scattering of X rays by matter. Thus, while some phenomena, such as interference and diffraction, manifest the wave properties of radiation, the photoelectric and Compton effects depend on its particle properties.

## 5.5 Atomic Spectra

*a. Line Spectra of Atoms* When gases are heated to high temperatures, or when an electric current is passed through them, electromagnetic radiation is emitted. The radiation, which usually consists of a complex mixture of wavelengths, can be dispersed into its individual components by means of a spectrometer (Figure 5.16). The results of such a measurement yield an **emission**

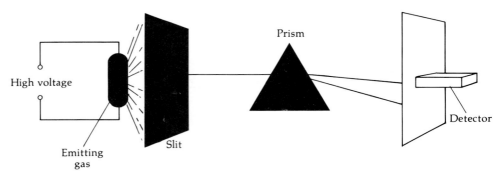

**FIGURE 5.16** A prism spectrometer for the measurement of visible light spectra. Light from electrically excited atoms is collimated onto a prism. Because the speed of light in matter is a function of wavelength, light of different color is refracted (i.e., bent) to a different extent. The resulting dispersion separates light according to wavelength. If the detector is a photographic plate rather than a photon counter, the instrument is called a spectrograph.

**spectrum,** a plot of the intensity of the emitted radiation as a function of its wavelength or some related variable. The spectra of atoms in the gas phase consist of a number of sharp lines, and the visible spectrum of atomic hydrogen is particularly simple, as shown in Figure 5.17.

The occurrence of atomic (and molecular) line spectra provides the most compelling evidence that the energies of atoms and molecules must be quantized. When gases are heated or subjected to an electric current, the atoms of the gas become more energetic. After a brief moment, this added energy can be reemitted in the form of photons, thereby giving rise to the observed emission spectrum. If each atom were able to accept an arbitrary amount of energy, the large number of atoms making up a normal gas sample would have a continuous distribution of energies. Consequently, the observed radiation spectrum would also be continuous. Since the observed spectra consist of sharp lines instead, it follows that the energies of atoms and molecules must be discrete rather than continuous. While we have already encountered this idea in our discussion of black-body radiation, the evidence given by atomic line spectra is far more direct.

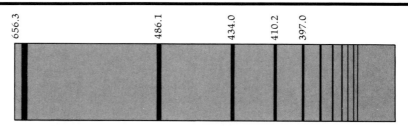

**FIGURE 5.17** The visible spectrum of atomic hydrogen. The wavelengths (in nanometers) of the lowest energy lines are given.

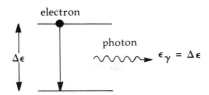

**FIGURE 5.18** Representation of an atom with two energy levels that have an energy difference $\Delta\epsilon$. When an electron in the excited state makes a transition to the ground state the energy is emitted in the form of radiation of energy $\epsilon_\gamma = \Delta\epsilon$.

Since atoms can only possess discrete amounts of energy, they are described as having energy levels. A given level is characterized by a particular energy. The lowest energy level—the normal state of the atom—is called the **ground state.** The levels with higher energies are called **excited states;** they are frequently referred to in sequential order (e.g., the first excited state, the second excited state, etc.). We know, of course, that atoms have a complex structure. Can the various levels be identified with one of the constituents of an atom? The states of present interest (i.e., those associated with the emission of optical radiation) refer to different states of atomic electrons. Molecules, too, have electronic states but for the present, we shall concentrate on the electronic states of atoms.

The preceding concepts are illustrated in Figure 5.18, which shows a simple energy level diagram, in which an atomic electron can occupy either the ground state or the first excited state. Suppose that, as a result of the absorption of some energy, an atomic electron has jumped from the ground state to the first excited state. At some point, the electron makes a transition back to the ground state. Since this state has less energy than the excited state, the atom must lose energy, which it can do by emitting a photon. Since energy must be conserved in this process, the energy lost by the atom, $\Delta\epsilon$, must be equal to the photon energy $\epsilon_\gamma$. In combination with Equations 5.14 and 5.10 we can then write the fundamental formula of spectroscopy,

$$\Delta\epsilon = \epsilon_\gamma = h\nu = hc/\lambda \tag{5.18}$$

which relates the observed wavelength of the emitted light to the energy difference between the initial and final states of the atom.

***b. The Spectrum of Atomic Hydrogen*** The concept of quantized energy levels for electrons in atoms was developed by Niels Bohr (1885–1962) in his explanation of the spectrum of atomic hydrogen. It is not surprising that the emission spectrum of hydrogen is relatively simple since an atom of this element has only a single electron. In addition to the lines in the visible region displayed in Figure 5.17, there are several series of lines in the infrared and ultraviolet regions. The following empirical formula gives the wave numbers of *all* the spectral lines:

$$\bar{\nu} = \mathcal{R}\left(\frac{1}{n_2^2} - \frac{1}{n_1^2}\right) \tag{5.19}$$

### TABLE 5.2 The Spectrum of Atomic Hydrogen

| Series | $n_2$ | $n_1$ | Region | $\lambda^a_{(n_2+1) \to n_2}$ (nm) |
|---|---|---|---|---|
| Lyman | 1 | 2, 3, ... | Ultraviolet | 121.6 |
| Balmer | 2 | 3, 4, ... | Visible | 656.3 |
| Paschen | 3 | 4, 5, ... | Infrared | 1875 |
| Brackett | 4 | 5, 6, ... | Infrared | 4051 |
| Pfund | 5 | 6, 7, ... | Infrared | 7458 |

[a] Wavelength for the transition between levels characterized by quantum numbers $n_2 + 1$ and $n_2$.

where $n_1$ and $n_2$ are integers such that $n_1 > n_2$, and $\mathcal{R}$ is called the **Rydberg constant**, with a value:

$$\mathcal{R} = 1.097 \times 10^5 \text{ cm}^{-1}$$

All lines belonging to a given series have a common value of $n_2$ and a range of values of $n_1$. For example, all the lines in the visible spectrum have wave numbers given by Equation 5.19 with $n_2 = 2$ and $n_1 = 3, 4, 5, \ldots$. These lines constitute the *Balmer series*. Other series have $n_2 = 1, 3, \ldots$ and appropriate values of $n_1$. Table 5.2 summarizes the properties of the various series and Figure 5.19 shows their location in the spectrum.

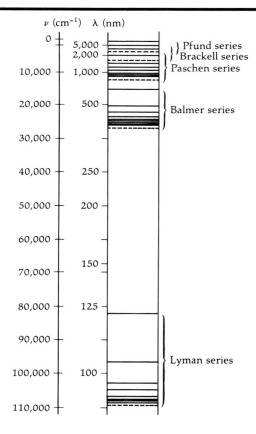

**FIGURE 5.19** Relative positions of the lines of the various spectral series of atomic hydrogen.

**Example 5.4**  Calculate the wavelength of the visible line in the hydrogen spectrum that has the second longest wavelength. What is the energy difference between the two levels of the atom involved in the transition giving rise to this line?

We use the Rydberg formula (Equation 5.19) to obtain the wave number of the line in question. In view of the reciprocal relation between $\lambda$ and $\bar{\nu}$, a long wavelength corresponds to a small wave number; thus, we want the line with the second smallest wave number. Since the visible lines have $n_2 = 2$, we want the line for which $n_1 = 4$. Substituting into Equation 5.19, we obtain

$$\bar{\nu} = \mathcal{R}\left(\frac{1}{n_2^2} - \frac{1}{n_1^2}\right)$$

$$= (1.097 \times 10^5 \text{ cm}^{-1})\left(\frac{1}{2^2} - \frac{1}{4^2}\right)$$

$$= 2.057 \times 10^4 \text{ cm}^{-1}$$

from which we get the wavelength as

$$\lambda = \frac{1}{\bar{\nu}}$$

$$= \frac{1}{(2.057 \times 10^4 \text{ cm}^{-1})(10^{-7} \text{ cm nm}^{-1})}$$

$$= 486.1 \text{ nm}$$

indicating that the line is blue–green. Note that the result agrees to at least four significant figures with the experimental value listed in Figure 5.17.

The energy difference between the two states of the hydrogen atom can be obtained by means of Equation 5.18:

$$\Delta\epsilon = \frac{hc}{\lambda}$$

$$= \frac{(6.626 \times 10^{-34} \text{ J s})(2.998 \times 10^8 \text{ m s}^{-1})}{(486.1 \text{ nm})(10^{-9} \text{ m nm}^{-1})(1.602 \times 10^{-19} \text{ J eV}^{-1})}$$

$$= 2.55 \text{ eV}$$

showing that an energy difference of 2.55 eV corresponds to the emission of blue–green light. ∎

Bohr's theory of the hydrogen atom involved a combination of the laws of classical physics with the quantization of energy of electrons orbiting the nucleus. He was able to derive Equation 5.19 and to show that the empirically determined Rydberg constant was given by the following combination of constants:

$$\mathcal{R} = 2\pi^2 m e^4/ch^3 = 1.097 \times 10^5 \text{ cm}^{-1} \tag{5.20}$$

where $m$ is the mass of the electron. Note that the theoretical value of the Rydberg constant is in excellent agreement with the experimental results. Figure 5.20 shows the energy level diagram of atomic hydrogen derived from the spectrum on the basis of Bohr's interpretation. Each horizontal line corresponds to one of the electronic energy levels of the atom. The levels are num-

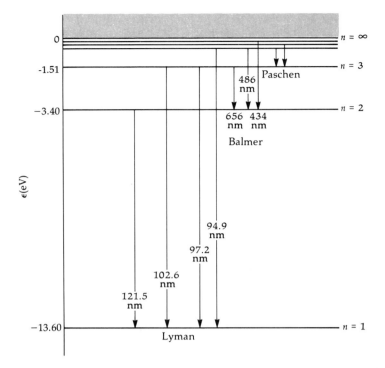

**FIGURE 5.20** Energy level diagram of the hydrogen atom. Each level is labelled by its quantum number $n$ and its energy relative to that of the ionized atom ($\epsilon = 0$) is given on the left. The vertical arrows show various electronic transitions between different levels and the spectral series to which they give rise. The transitions corresponding to the Lyman and Balmer series are labelled with the wavelength of the emitted radiation in nanometers.

bered sequentially in order of increasing energy by an integer $n$ called a **quantum number**. For example, the ground state is assigned $n = 1$, the first excited state, $n = 2$, and so on. It is customary to assign zero energy to the state in which the electron has been completely separated from the nucleus and both are at rest, that is, to the ionized atom. All the levels in which the electron is bound to the nucleus consequently have negative energy, as shown in Figure 5.20.

The actual energies of the electronic levels of the hydrogen atom were calculated by Bohr in the course of his derivation of the Rydberg formula (Equation 5.19). The result is

$$\epsilon_n = -hc\mathcal{R}/n^2 \tag{5.21}$$

where $\epsilon_n$ is the energy of the level with quantum number $n$. To see how the Rydberg formula follows from Equation 5.21 we write for the energy difference between levels characterized by quantum numbers $n_1$ and $n_2$

$$\Delta\epsilon = \epsilon_{n_1} - \epsilon_{n_2} = hc\mathcal{R}\left(\frac{1}{n_2^2} - \frac{1}{n_1^2}\right) \tag{5.22}$$

The Rydberg formula is obtained by substituting for $\Delta\epsilon$ from Equation 5.18:

$$\bar{\nu} = \frac{\Delta \epsilon}{hc} = \mathcal{R}\left(\frac{1}{n_2^2} - \frac{1}{n_1^2}\right)$$

When the atom is in its ground state ($n = 1$) it has the lowest possible energy, whose value according to Equation 5.21 is

$$\epsilon_1 = -hc\mathcal{R}$$
$$= \frac{-(6.626 \times 10^{-34} \text{ J s})(2.998 \times 10^8 \text{ m s}^{-1})(1.097 \times 10^5 \text{ cm}^{-1})(10^2 \text{ cm m}^{-1})}{(1.602 \times 10^{-19} \text{ J eV}^{-1})}$$
$$= -13.60 \text{ eV}$$

The energy required to remove an electron from an atom in its ground state is called the **ionization energy**. In view of the choice of energy zero, the ionization energy of the hydrogen atom, $I_H$, is given by the expression

$$I_H = 0 - (-hc\mathcal{R}) = hc\mathcal{R} = 13.56 \text{ eV} \tag{5.23}$$

Ionization energies can be determined experimentally. The experimental value for hydrogen is in excellent agreement with the value derived from the Bohr model.

---

**Example 5.5** The electron in a hydrogen atom is excited to the $n = 3$ level. Calculate the wavelengths of all the photons that can be emitted as the electron returns to the ground state.

A glance at Figure 5.20 shows that electronic transitions between the following levels of interest are possible:

$$n = 3 \rightarrow n = 1$$

or

$$n = 3 \rightarrow n = 2 \quad \text{followed by} \quad n = 2 \rightarrow n = 1$$

The wavelengths of the radiation emitted when the electron makes transitions between these levels are obtained by means of the Rydberg formula:

$$\lambda = (\bar{\nu})^{-1} = \frac{1}{\mathcal{R}\left(\frac{1}{n_2^2} - \frac{1}{n_1^2}\right)}$$

For the $n = 3 \rightarrow n = 1$ transition

$$\lambda = \frac{1}{(1.097 \times 10^5 \text{ cm}^{-1})\left(\frac{1}{1^2} - \frac{1}{3^2}\right)} (10^7 \text{ nm cm}^{-1})$$

$$= 102.6 \text{ nm}$$

For the $n = 3 \rightarrow n = 2$ transition

$$\lambda = \frac{1}{(1.097 \times 10^5 \text{ cm}^{-1})\left(\frac{1}{2^2} - \frac{1}{3^2}\right)} (10^7 \text{ nm cm}^{-1})$$

$$= 656.3 \text{ nm}$$

and for the $n = 2 \rightarrow n = 1$ transition

$$\lambda = \frac{1}{(1.097 \times 10^5 \text{ cm}^{-1})\left(\frac{1}{1^2} - \frac{1}{2^2}\right)} (10^7 \text{ nm cm}^{-1})$$

$$= 121.5 \text{ nm}$$

In order of increasing wavelength, the three lines observed in the deexcitation of the $n = 3$ state lie at 102.6, 121.5, and 656.3 nm. ■

The atomic spectra of the elements heavier than hydrogen are considerably more complex. It is impossible to write a simple relation, such as the Rydberg formula, that will fit the many spectral lines that are observed. As we shall see, the arrangement of the electrons in multielectron atoms can only be understood on the basis of quantum mechanics.

## 5.6 Wave Properties of Matter

Until the early 1920s, the idea that particles could exhibit wave properties, just as waves could exhibit particle properties, had received no serious consideration. This situation changed when Louis de Broglie (1892–1987) proposed in his celebrated 1924 doctoral thesis that particles in motion possessed a wavelength and could therefore act like waves.

De Broglie obtained an expression for the wavelength of a particle by drawing an analogy to the behavior of photons. We have seen that the relation between the energy of a photon and its wavelength is (Equations 5.14 and 5.10) $\epsilon_\gamma = hc/\lambda$. Since the energy and the momentum $p$ of light are related as

$$\epsilon_\gamma = pc \tag{5.24}$$

the relation between the momentum and wavelength of a photon is just

$$p = h/\lambda \tag{5.25}$$

De Broglie conjectured that this same relation would be obeyed by a particle of mass $m$, whose momentum is given by the familiar relation (Section 4.2)

$$p = mv$$

The **de Broglie wavelength** of a particle is given by

$$\lambda = h/p = h/mv \tag{5.26}$$

While this equation seems plausible enough, it represents a strikingly different view of the behavior of particles such as the electron, which until that time had been thought to possess only corpuscular properties.

---

**Example 5.6** What is the wavelength of a 100 eV electron?

In order to use Equation 5.26, which defines the wavelength, we must determine the speed of the electron. Using the relation between kinetic energy and speed, $\epsilon = \frac{1}{2}mv^2$, we obtain:

$$v = \left(\frac{2\epsilon}{m}\right)^{1/2}$$

$$= \left(\frac{(2)(100 \text{ eV})(1.602 \times 10^{-19} \text{ J eV}^{-1})}{(9.110 \times 10^{-31} \text{ kg})(1 \text{ J kg}^{-1} \text{ m}^{-2} \text{ s}^2)}\right)^{1/2}$$

$$= 5.930 \times 10^6 \text{ m s}^{-1}$$

Substituting into Equation 5.26, we get

$$\lambda = \frac{h}{mv}$$

$$= \frac{(6.626 \times 10^{-34} \text{ J s})(10^9 \text{ nm m}^{-1})}{(9.110 \times 10^{-31} \text{ kg})(5.930 \times 10^6 \text{ m s}^{-1})(1 \text{ J kg}^{-1} \text{ m}^{-2} \text{ s}^2)}$$

$$= 0.123 \text{ nm}$$

The result shows that a 100 eV electron has a wavelength comparable to the size of atoms. ■

Table 5.3 lists the wavelengths of some additional particles. Note that only particles of atomic dimensions have significantly large wavelengths. The concept is completely unimportant for macroscopic objects.

De Broglie's conjecture would undoubtedly have faded into oblivion had it not been for its spectacular confirmation by experiment. As noted in Example 5.6, the wavelength of a 100 eV electron is of atomic dimensions. If electrons can indeed act like waves, they should be diffracted by crystalline solids in the same way as X rays, which have comparable wavelengths (Section 5.3). In 1927, C. J. Davisson (1881–1958) and L. H. Germer (1896–1971) observed the diffraction of electrons by crystalline solids and thereby confirmed de Broglie's hypothesis. Figure 5.21 compares diffraction patterns obtained by the use of X rays and electrons. The close similarity of the two patterns is striking and shows that electrons (and other particles) are diffracted just like electromagnetic waves. The techniques of electron diffraction and neutron diffraction have become important tools for the study of molecular structure.

The results of modern physical experiments indicate that the concepts of particle and wave blend together when applied on an atomic scale. Under certain conditions, particles exhibit the properties of waves and waves exhibit those of particles. This wave–particle duality is a consequence of the use of concepts that we best understand from their application to macroscopic objects, to the world of atoms and photons. It should come as no surprise that entities of atomic size cannot be neatly categorized as being either particles or waves. As we shall see, the development of quantum mechanics permits us to deal with the wave–particle duality in a way that resolves these semantic difficulties.

**TABLE 5.3  The de Broglie Wavelength of Some Typical Particles**

| Particle | Mass (kg) | Velocity (m s$^{-1}$) | Wavelength (nm) |
|---|---|---|---|
| 100 eV electron | $9.1 \times 10^{-31}$ | $5.9 \times 10^6$ | 0.12 |
| 0.025 eV neutron | $1.67 \times 10^{-27}$ | $2.2 \times 10^3$ | 0.18 |
| H$_2$ molecule at 200°C | $3.3 \times 10^{-27}$ | $2.4 \times 10^3$ | 0.084 |
| Golf ball | $4.5 \times 10^{-2}$ | 30 | $4.9 \times 10^{-15}$ |

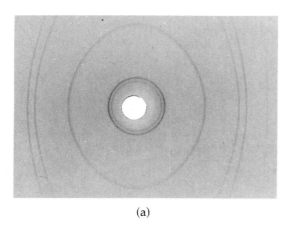 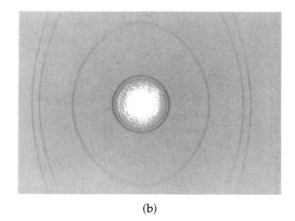

(a)          (b)

**FIGURE 5.21** Diffraction patterns obtained when (a) X rays and (b) electrons of comparable wavelength are incident on aluminum.

## 5.7 The Uncertainty Principle

The ability to predict the motion and position of macroscopic bodies is taken for granted. For example, the orbits of the planets and the moon can be calculated to high accuracy and the occurrence of eclipses predicted far into the future. Because of the wave nature of particles, however, it is impossible to make similar predictions about their motion. If the electron is described as a wave, then its position at a given time cannot be specified with complete accuracy because a wave cannot be localized in space.

The uncertainties associated with the wave nature of matter can be quantified by considering the interaction between the object under observation and the probe used to observe its motion. In dealing with macroscopic phenomena this interaction can either be ignored or else properly accounted for. For example, the flight of a baseball is in no way altered by the pitcher's observation of its trajectory. A different, and perhaps more insightful, example involves the measurement of the temperature of a beaker filled with hot water by means of a cold thermometer. While the insertion of the cold thermometer will lead to a small drop in temperature, the magnitude of the effect can be calculated. We can show that it is inherently impossible to make a similar correction for the effect of the measuring probe on the trajectory of an atomic particle such as the electron.

In order to calculate the trajectory of a particle we have to measure its position to know where it is and its momentum to know where it is going. The determination of the position and momentum of an electron can be made by the use of electromagnetic radiation of short wavelength, as in the Compton effect (Section 5.4c). According to the principles of optics, the position of an electron cannot be determined more closely than $\pm\lambda$, where $\lambda$ is the wavelength of the incident photons. The uncertainty in the position of the electron, $\Delta x$, is consequently approximately equal to $\lambda$. This fact suggests that the accuracy of the position measurement can be increased to whatever degree is desirable by the use of electromagnetic radiation of sufficiently short wavelength. However, we must now recognize that in measuring the position of an electron, a photon must collide with it. In this collision, some unknown fraction of the momentum of the photon will be transferred to the electron. We already know that the momentum of a photon of wavelength $\lambda$ is equal to $h/\lambda$ (Equation 5.25). The uncertainty in the momentum of the electron, $\Delta p$, is therefore comparable to the momentum of the photon, $\Delta p = h/\lambda$. The product of the uncertainties in the momentum and position of the electron is therefore

$$(\Delta p)(\Delta x) \approx (h/\lambda)(\lambda) \approx h$$

This argument is a simplified statement of **Heisenberg's uncertainty principle.** More detailed considerations indicate that the product of the uncertainties must be at least as large as $h/4\pi$

$$(\Delta p)(\Delta x) \geq h/4\pi \tag{5.27}$$

The uncertainty principle indicates that it is impossible to *simultaneously* measure both the momentum and the position of a particle so that the product of the uncertainties is less than $h/4\pi$. To be sure, the uncertainty in either quantity can be reduced as much as desired. As already noted, the uncertainty in the position of the electron can be minimized by reducing the wavelength of the radiation. However, this necessarily increases the photon momentum and therefore the uncertainty in the momentum of the electron. On the other hand, the uncertainty in the momentum of the electron can be minimized by the use of low-momentum photons. Since such photons have long wavelengths, the uncertainty in the position of the electron is correspondingly increased. As a result, the two uncertainties are inextricably coupled in a way that makes the determination of an accurate trajectory impossible. Because of the magnitude of Planck's constant, the uncertainty principle has a profound meaning for particles of atomic size but plays no role in macroscopic phenomena. The following example illustrates this point quantitatively.

**Example 5.7** (a) An electron travels with a speed of $2 \times 10^6$ m s$^{-1}$. Assuming that its position is measured to within 0.01 nm, or 10% of an atomic radius, what is the minimum uncertainty in its momentum? Compare the uncertainty in $p$ with the value of $p$. (b) Perform the same calculation for a 0.045 kg golf ball travelling at a speed of 30 m s$^{-1}$. Assume that the uncertainty in the position of the ball is equal to the wavelength of the light used to observe it, e.g., 500 nm.

(a) According to the uncertainty principle (Equation 5.27) the minimum uncertainty in the momentum, $\Delta p$, is:

$$\Delta p = \frac{h}{4\pi \Delta x}$$

$$= \frac{(6.63 \times 10^{-34} \text{ J s})}{(4\pi)(0.01 \text{ nm})(10^{-9} \text{ m nm}^{-1})(1 \text{ J kg}^{-1} \text{ m}^{-2} \text{ s}^2)}$$

$$= 5 \times 10^{-24} \text{ kg m s}^{-1}$$

The momentum of the electron is

$$p = mv$$
$$= (9.1 \times 10^{-31} \text{ kg})(2 \times 10^6 \text{ m s}^{-1})$$
$$= 2 \times 10^{-24} \text{ kg m s}^{-1}$$

The minimum uncertainty in the momentum, $5 \times 10^{-24}$ kg m s$^{-1}$, is more than twice as large as the momentum itself. Owing to this large uncertainty, we are unable to make any predictions about the position of the electron at a future time.

(b) The momentum of the golf ball is

$$p = (0.045 \text{ kg})(30 \text{ m s}^{-1}) = 1.4 \text{ kg m s}^{-1}$$

and the uncertainty in this quantity is

$$\Delta p = \frac{(6.63 \times 10^{-34} \text{ J s})}{(4\pi)(500 \text{ nm})(10^{-9} \text{ m nm}^{-1})(1 \text{ J kg}^{-1} \text{ m}^{-2} \text{ s}^2)}$$
$$= 1 \times 10^{-28} \text{ kg m s}^{-1}$$

This uncertainty in momentum is so small compared to the value of the momentum as to be completely undetectable. ∎

## Quantum Mechanics of the Hydrogen Atom

The experimental evidence accumulated by the 1920s had made it apparent that to describe matter on an atomic scale requires a different formalism than that used to describe macroscopic objects. Such a formalism must be able to account for the wave properties as well as the corpuscular properties of matter and for the fact that the energy levels of atoms and molecules are quantized (i.e., discrete) rather than continuous. Furthermore, owing to the limitations imposed by the uncertainty principle, the concept of the trajectory of a particle must be abandoned. For example, it makes no sense to attempt to calculate the precise distance between an electron and nucleus in an atom since this distance is indeterminate. All that we can aspire to know is the *probability* that an electron will be at some particular distance from the nucleus.

In this section we present an introduction to the quantum mechanical description of matter in which we focus on the hydrogen atom, the simplest of all atomic and molecular systems. In dealing with quantum mechanics we are faced with the inescapable fact that the formulation of the theory requires mathematics that is beyond the level of a first-year course. Therefore, we must concentrate on the results of the theory, which are not particularly difficult to understand, rather than on the manner in which these results are obtained. Nonetheless, it is imperative that we pay some attention to the relation between the formulation of the theory and its results, because otherwise the results may appear to be a collection of arbitrary statements and facts. Fortunately, there exists a simple model system for which the theory can be solved without advanced mathematics. By examining this system, the particle in a box (Section 5.9), we can see how the probabilistic description of the motion of a particle wave and the quantization of energy follow in a natural way from the form of the equations.

## 5.8 The Schrödinger Wave Equation

In 1926 Erwin Schrödinger (1887–1961) proposed an equation whose solution describes the quantum mechanical behavior of a particle. Schrödinger's equation represents the motion of a particle as being like that of a wave. Recall from Section 5.2 that an electromagnetic wave is characterized by an amplitude, whose value determines the distribution of the wave in space. Similarly, the Schrödinger equation characterizes a particle by an amplitude, or **wave function,** whose value determines the distribution of the particle wave in space. Schrödinger's equation is an imaginative postulate that permits the calculation of the wave function. For a particle free to move in one dimension (along the $x$ axis) the equation is

$$-\frac{\hbar^2}{2m}\frac{d^2\psi(x)}{dx^2} + V(x)\psi(x) = E\psi(x) \tag{5.28}$$

where $\hbar$ is the symbol for $h/2\pi$. The quantities that are known are the mass of the particle, $m$, and its potential energy at point $x$, $V(x)$. A solution of the equation yields the value of the wave function at point $x$, $\psi(x)$, and the total energy $E$ (kinetic + potential).

Equation 5.28 is an example of a *second-order differential equation* and involves the second derivative of $\psi(x)$ with respect to $x$ $(d^2\psi(x)/dx^2)$ (see Appendix A2.5). This quantity is the rate of change of $d\psi(x)/dx$ which, in turn, is the rate of change of $\psi(x)$ with respect to $x$. Differential equations of this type describe the motion of waves and are commonly encountered in classical wave theory. A good example of classical wave motion, illustrated in Figure 5.22, is the vibration of a string with fixed ends, such as a violin string. Vibrations of the string set up standing waves which are characterized by an amplitude that varies with position, as well as by the requirement that a half-integral number of wavelengths be accomodated along the length of the string. As shown in the figure, this second condition results from the fact that the string is fixed at its ends, a requirement known as a **boundary condition.** As in the case of electromagnetic radiation (see Figure 5.7), the intensity of the wave at any given point is given by the square of the amplitude. The wave function obtained as the solution to a simple Schrödinger equation closely resembles the relation that describes the amplitude of a vibrating string, as we will see in the next section.

As noted earlier, a solution of the Schrödinger equation yields the values of the wave function $\psi(x)$. In analogy to classical waves, the square of the wave function, like the square of the amplitude, corresponds to the intensity of the particle wave. In order to understand the physical meaning of the intensity of a particle wave, consider the intensity of a wave of energetic radiation, such as X rays. Such a wave can be understood as a flow of photons, all of the same energy, $\epsilon_\gamma = h\nu$. Recalling from Section 5.2 that the intensity of a wave is proportional to its energy, it follows that the intensity at a given small region of space is proportional to the number of photons passing through that region and thus, to the probability that the region in question will be struck by a photon. This line of reasoning provides us with a physical interpretation of the wave function: *the square of the wave function of a particle $\psi^2(x)$ is equal to the probability per unit length of finding the particle at $x$.* Instead of yielding the exact position of the particle, which would violate the uncertainty principle, the solution of the wave equation gives the probability of finding the particle as a function of its

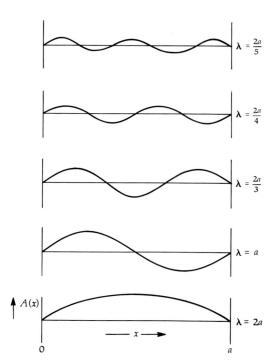

**FIGURE 5.22** Standing waves of motion in a vibrating violin string. Each curve represents a possible representation of the amplitude $A(x)$ as a function of the position $x$. The various waves are consistent with the requirement that $A(x) = 0$ when $x = 0$ and when $x = a$, where $a$ is the length of the string. This boundary condition arises because the string is tied down at its ends and results in the relations given for each wave between the wavelength $\lambda$ and the length of the string, $a$. The wave function of the Schrödinger equation for a particle in a box (Section 5.9) closely resembles the amplitude of a vibrating string.

position. According to quantum mechanics, this is the most that we can hope for.

## 5.9 The Particle in a Box

One of the simplest problems that can be solved by means of quantum mechanics deals with a particle confined to moving back and forth in a straight line between two impenetrable walls, the so-called particle in a box. Since the solution of the Schrödinger equation for this particle contains all the elements of the solutions to more complex problems, this readily solvable equation serves as a useful model for the more complicated cases. (The solution requires the use of calculus.)

We consider a particle of mass $m$ free to move along the $x$ axis between $x = 0$ and $x = L$. The potential energy of the particle $V(x)$ is taken to be zero at all points along its path except at the endpoints $x = 0$ and $x = L$, where, as a consequence of the presence of the impenetrable walls, it becomes infinite (Figure 5.23).

**FIGURE 5.23** Potential energy of a particle confined to a one-dimensional region by impenetrable walls.

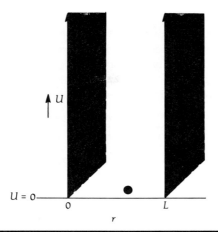

The Schrödinger equation for the region between $x = 0$ and $x = L$, where $V(x) = 0$, is

$$-\frac{\hbar^2}{2m}\frac{d^2\psi(x)}{dx^2} = E\psi(x) \quad (5.29)$$

This equation indicates that, except for the constants by which the terms are multiplied, the second derivative of the wave function, $d^2\psi(x)/dx^2$, is equal to the wave function $\psi(x)$. The sine function is one of the simplest functions that is equal (with appropriate multiplication by constants) to its second derivative, indicating that the solution to Equation 5.29 may be a sine function.

We can therefore assume that the solution of Equation 5.29 has the form

$$\psi(x) = A \sin kx \quad (5.30)$$

where $A$ and $k$ are constants. In order to check that this assumption is correct, we differentiate twice:

$$\frac{d\psi(x)}{dx} = kA \cos kx \quad (5.31)$$

$$\frac{d^2\psi(x)}{dx^2} = -k^2 A \sin kx \quad (5.32)$$

and substituting $\psi(x)$ for $A \sin kx$ (Equation 5.30) we obtain

$$\frac{d^2\psi(x)}{dx^2} = -k^2 \psi(x) \quad (5.33)$$

Comparing this equation with the wave equation (Equation 5.29), we see that they are identical provided the constant $k^2$ is set equal to $2mE/\hbar^2$. Making this substitution into our assumed solution (Equation 5.30), we obtain for the wave function of the particle in a box

$$\psi(x) = A \sin[(2mE)^{1/2} x/\hbar] \quad (5.34)$$

Up to this point we have not made any use of the boundary conditions, that is, of the fact that the potential energy becomes infinite at the walls. Equation 5.34 is therefore the wave function in one dimension of a free particle, one which has no constraints on its motion. The Schrödinger wave equation has a particularly simple interpretation for a free particle. Since the potential energy is in this instance zero, the total energy can be identified with the kinetic energy, $E = \frac{1}{2}mv^2$. Substituting this result into the argument of the sine function (i.e., the quantity whose sine is to be evaluated) in Equation 5.34, we obtain:

$$\frac{(2mE)^{1/2}x}{\hbar} = \frac{[2m(mv^2/2)]^{1/2}x}{\hbar} = \frac{2\pi mvx}{h} = \frac{2\pi x}{\lambda}$$

where $\hbar = h/2\pi$ and the quantity $h/mv$ has been replaced by the de Broglie wavelength $\lambda$. By introducing this result into Equation 5.34 we can express the wave function of a free particle as

$$\psi(x) = A \sin(2\pi x/\lambda) \tag{5.35}$$

which is just the equation describing an electromagnetic wave of wavelength $\lambda$ (Figure 5.7). This similarity allows us to interpret the Schrödinger equation for a free particle as the equation of motion of a particle wave whose de Broglie wavelength is $\lambda$.

A free particle has no constraints on its energy, that is, its energy can vary continuously. To see the origin of the quantization of energy we must now consider the fact that the particle is actually not free but is confined to the region between the walls. This means that the probability of finding the particle outside the walls is zero. Recalling that the probability is proportional to the square of the wave function allows us to conclude that the wave function itself must go to zero at the walls, i.e.,

$$\psi(0) = 0 \qquad \psi(L) = 0 \tag{5.36}$$

These are the boundary conditions.

The first condition, $\psi(0) = 0$, is automatically satisfied since setting $x = 0$ in Equation 5.34 yields $\psi = \sin(0) = 0$. The second condition requires closer examination. Evaluating Equation 5.34 for $x = L$ and imposing the boundary condition $\psi(L) = 0$ gives

$$\psi(L) = A \sin[(2mE)^{1/2}L/\hbar] = 0 \tag{5.37}$$

This equality can be satisfied by demanding that the sine function be zero. Since $\sin \theta = 0$ when $\theta = 0, \pi, 2\pi, \ldots$, the argument of the sine function must be some integral multiple of $\pi$, i.e.,

$$\frac{(2mE)^{1/2}L}{\hbar} = n\pi \qquad n = 1, 2, 3, \ldots \tag{5.38}$$

Note that the condition $n = 0$ is eliminated because it leads to the result $(2mE)^{1/2}/\hbar = 0$ and thus, to $\psi(x) = 0$ for all values of $x$. This is a physically impossible result since it means that the probability of finding the particle anywhere in the box is zero.

Equation 5.38 is obeyed for only certain values of the energy. These values are obtained by solving the equation for $E$:

$$E = n^2\pi^2\hbar^2/2mL^2 = n^2h^2/8mL^2 \qquad n = 1, 2, 3, \ldots \tag{5.39}$$

This result is important because it shows that the energy of the particle is quantized, with the allowed values determined by those of the quantum number $n$.

Evidently, the continuous energy of a free particle gives way to the quantized energy of a particle in a box as a consequence of the confinement.

In order to complete the solution of the Schrödinger equation for the particle in a box, we must evaluate the constant $A$ in Equation 5.34. The interpretation of $\psi^2(x)$ as the probability per unit length of finding the particle at $x$ provides us with a way to evaluate $A$. Since the particle is somewhere in the box, the sum of the probabilities evaluated at all values of $x$ must be unity, i.e., $\Sigma \psi^2(x) = 1$. If we attempt to perform this summation, we immediately realize that the number of different $x$ values is infinite because $x$ is a continuous variable. No matter how close two values of $x$ are to each other, it is always possible to pick a value of $x$ that has an intermediate value. Under these circumstances, the summation must be replaced by an integration:

$$\int_0^L \psi^2(x) dx = 1 \tag{5.40}$$

where $\psi^2(x)dx$ represents the probability of finding the particle in an infinitesimally wide interval $dx$ between $x$ and $x + dx$. Substituting for $\psi(x)$ from Equation 5.34, with $(2mE)^{1/2}/\hbar$ given by Equation 5.38, we write

$$A^2 \int_0^L \sin^2\left(\frac{n\pi x}{L}\right) dx = 1 \tag{5.41}$$

The value of the integral is $L/2$, so

$$A^2 L/2 = 1 \tag{5.42}$$

and, solving for $A$, we obtain

$$A = (2/L)^{1/2} \tag{5.43}$$

The constant $A$ is called a **normalization constant**; when the value given by Equation 5.43 is used, the wave function is said to be normalized to unity and $\psi^2(x)dx$ is the numerical probability of finding the particle between $x$ and $x + dx$.

Our final results for the wave function and the energy of a particle in a one-dimensional box are

$$\psi_n(x) = (2/L)^{1/2} \sin(n\pi x/L)$$
$$E_n = \frac{n^2 h^2}{8mL^2} \qquad n = 1, 2, 3, \ldots \tag{5.44}$$

where the allowed values of $\psi$ and $E$ have been labelled with the value of the quantum number $n$. A given value of $n$ determines the form of the wave function and the energy of the state it describes.

The solutions of the wave equation for the lowest values of $n$ are displayed in Figure 5.24. The particle has discrete translational energy levels reminiscent of the energy levels of an atom (Section 5.5). Note that the lowest energy that the particle can possess is not zero, as would be the case in classical mechanics, but $h^2/8mL^2$. This irreducible minimum energy is called the **zero-point energy**. Its origin can be understood with the aid of the uncertainty principle. Since the particle is confined to a region of length $L$, the uncertainty in its position can be no larger than $L$. Therefore, the uncertainty in its momentum must be approximately $\Delta p \approx \hbar/2L$. If the magnitude of the momentum is of

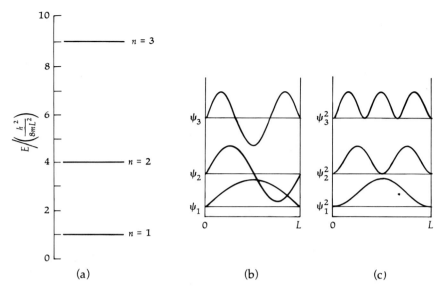

**FIGURE 5.24** Solutions of the Schrödinger wave equation for a particle confined to a one-dimensional box of length L. (a) The allowed energy levels, (b) the wave functions, and (c) the probability distributions.

the order of $\Delta p$, the kinetic energy ($p^2/2m$) must be roughly $\hbar^2/8mL^2$, which is of the order of magnitude of the zero-point energy.

The energy difference between neighboring levels is

$$E_{n+1} - E_n = \frac{(n+1)^2 h^2}{8mL^2} - \frac{n^2 h^2}{8mL^2} = \frac{(2n+1)h^2}{8mL^2} \tag{5.45}$$

Note that this energy difference is inversely proportional to the mass of the particle and to the square of the length of the container. As we shall see later, the energy difference between adjacent levels of a particle confined to a macroscopic container, i.e., one with very large L, is negligibly small. This result allows us to treat the translational kinetic energy of molecules confined to macroscopic containers as being continuous rather than quantized. Recall that this is precisely what was done in the derivation of the kinetic theory of gases (Section 4.2). Furthermore, the inverse dependence of the energy difference on the mass of the particle also indicates that the quantization of energy becomes completely negligible for macroscopic bodies. As a result, quantum mechanics reduces to classical mechanics for massive bodies or for essentially unconfined particles. It is only when these conditions are not met that the two aproaches differ.

Figure 5.24b shows the wave functions of the particle in a box. Note that the various curves are indentical to the standing waves which describe the motion of a vibrating string with fixed ends (Figure 5.22). The similarity comes from that in the boundary conditions, which in both cases require the wave amplitude to become zero at the ends. While all the solutions of the wave equation necessarily display this property, the wave functions for $n > 1$ alternate with increasing frequency between positive and negative values as $n$ increases. The points at which the wave functions pass through zero are called **nodes**. As shown in the figure, the number of nodes increases with the energy of the state; this result is of general validity in quantum mechanics. The bonding properties of electrons in molecules (Chapter 16) are, to a significant extent, determined by the location of the nodes in the electronic wave functions.

The probability distribution $\psi^2(x)$ is displayed in Figure 5.24c. For a given solution, the probability of finding the particle at a particular location varies widely. In its lowest translational energy level, the particle is most likely to be found away from the edges and towards the center of the box. This is not the case for the $n = 2$ state since the particle has zero probability of being at the point where the wave function has a node. Since the number of nodes increases with the quantum number $n$, the alternation between maxima and minima in the distribution also increases. For very large quantum numbers the maxima and minima lie so close to each other that, in effect, the particle has an equal probability of being anywhere in the box. Since this is precisely the condition which holds in a classical description, e.g., a particle in a macroscopic box, this condition is attained in the limit of large quantum numbers.

---

**Example 5.8** An electron is confined to a 1.50 nm linear molecule (i.e., 15.0 Å, about 7 atoms long). (a) What is the minimum energy of the electron? (b) What is the probability of finding the electron in the region of the molecule lying between (1) 0.100 nm and 0.101 nm and (2) 0.750 nm and 0.751 nm for $n = 1$? (c) What is the energy difference between the $n = 2$ and $n = 1$ states of the electron? What is the wavelength of the radiation emitted when the electron makes a transition between these states? (d) What is the energy difference between the $n = 2$ and $n = 1$ states of an $O_2$ molecule confined to a 1 m long container?

(a) The energy of a particle in a one-dimensional box is given by Equation 5.44, $E_n = n^2h^2/8mL^2$. The lowest energy corresponds to $n = 1$:

$$E = \frac{(1)^2(6.626 \times 10^{-34} \text{ J s})^2}{(8)(9.110 \times 10^{-31} \text{ kg})(1.50 \text{ nm} \times 10^{-9} \text{ m nm}^{-1})^2(1 \text{ J kg}^{-1} \text{ m}^{-2} \text{ s}^2)}$$

$$= 2.68 \times 10^{-20} \text{ J}$$

Recalling that the appropriate energy units for particles are either eV or kJ mol$^{-1}$, we obtain

$$E = \frac{2.68 \times 10^{-20} \text{ J}}{1.602 \times 10^{-19} \text{ J eV}^{-1}} = 0.167 \text{ eV}$$

$$= \frac{(2.68 \times 10^{-20} \text{ J})(6.022 \times 10^{23} \text{ mol}^{-1})}{(10^3 \text{ J kJ}^{-1})} = 16.1 \text{ kJ mol}^{-1}$$

(b) The probability of finding the electron in an infinitesimally wide interval between $x$ and $x + dx$ is $\psi^2(x)dx$. The given intervals are sufficiently narrow to permit the use of this expression. (Figure 5.25 shows a geometric interpretation of this statement.) Using Equation 5.44 for the wave function we write for $n = 1$:

$$\psi^2(x)dx = (2/L) \sin^2(\pi x/L) \, dx$$

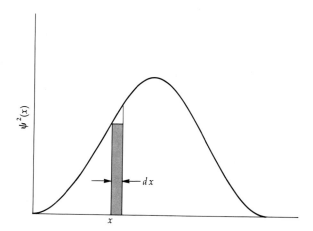

**FIGURE 5.25** The probability of finding a particle between $x$ and $x + dx$ is equal to $\psi^2(x)dx$. In Example 5.8b the infinitesimal interval $dx$ is replaced by a finite interval $\Delta x = 0.001$ nm. This is equivalent to replacing the area under the function $\psi^2(x)$ between the limits of the interval by that of a rectangle drawn through point $x$ as shown. The narrower the interval and the less rapidly the function changes in the interval, the better the approximation.

---

(1) Evaluating the probability at $x = 0.1005$ nm we obtain

$$\psi^2(x)dx = \left(\frac{2}{1.5 \text{ nm}}\right) \sin^2\left[\frac{(\pi)(0.1005 \text{ nm})}{1.50 \text{ nm}}\right](0.001 \text{ nm})$$

$$= 6 \times 10^{-5}$$

(To evaluate the sine function the calculator must be set to read the angle in radians; if this cannot be done, the angle must be converted from radians to degrees.)

(2) Performing a similar evaluation at $x = 0.7505$ nm we obtain

$$\psi^2(x)dx = \left(\frac{2}{1.5 \text{ nm}}\right) \sin^2\left[\frac{(\pi)(0.7505 \text{ nm})}{1.50 \text{ nm}}\right](0.001 \text{ nm})$$

$$= 1 \times 10^{-3}$$

Note that the probability of finding the electron in the middle of the molecule is much higher than that of finding it near either end, as already depicted in Figure 5.24c.

(c) The energy difference can be obtained from either Equation 5.44 or 5.45:

$$\Delta E = E_{n=2} - E_{n=1} = [(2)^2 - (1)^2]\frac{h^2}{8mL^2}$$

Using the result of part (a), $h^2/8mL^2 = 2.68 \times 10^{-20}$ J, we obtain

$$\Delta E = \frac{(3)(2.68 \times 10^{-20} \text{ J})}{(1.602 \times 10^{-19} \text{ J eV}^{-1})} = 0.502 \text{ eV} = 48.3 \text{ kJ mol}^{-1}$$

The relation between the energy difference of two states and the wavelength of the radiation emitted when a particle makes a transition between them (Equation 5.18) is $\Delta E = \epsilon_\gamma = hc/\lambda$. Solving for $\lambda$, we obtain

$$\lambda = \frac{hc}{\Delta E} = \frac{(6.626 \times 10^{-34} \text{ J s})(2.998 \times 10^8 \text{ m s}^{-1})}{(3)(2.68 \times 10^{-20} \text{ J})}$$

$$= 2.47 \times 10^{-6} \text{ m} = 2.47 \times 10^3 \text{ nm}$$

corresponding to emission in the infrared. The electron in a box can be used as a model for certain types of molecules, in which an electron is, in fact, free to move over the entire molecule. The electronic spectra of such molecules can be interpreted in the manner illustrated in this problem. For example, the wavelengths of the observed lines can be used to deduce the size of the molecule (Section 17.4a).

(d) This is essentially the same problem as in part (c) but applied to the macroscopic conditions considered in Chapter 4.

$$\Delta E = E_{n=2} - E_{n=1} = \frac{3h^2}{8mL^2}$$

$$= \frac{(3)(6.626 \times 10^{-34} \text{ J s})^2(10^{-3} \text{ kJ J}^{-1})(6.022 \times 10^{23} \text{ mol}^{-1})}{(8)(32 \text{ u})(1.67 \times 10^{-27} \text{ kg u}^{-1})(1 \text{ m})^2}$$

$$= 2 \times 10^{-21} \text{ kJ mol}^{-1}$$

This energy difference is some 22 orders of magnitude smaller than that between the comparable states of the electron in a molecule, 48 kJ mol$^{-1}$. The small magnitude of $\Delta E$ for gas molecules confined to a macroscopic container shows why energy quantization can be neglected for such systems. In these circumstances it is valid to treat energy by means of classical mechanics, i.e., as a continuous variable. ∎

## 5.10 The Wave Equation for the Hydrogen Atom

The hydrogen atom is the one atomic system for which an exact solution to the Schrödinger equation can be obtained. The resulting wave functions are important not only for what they tell us about the electronic structure of the hydrogen atom, but because they serve as a model of the structure of more complicated atoms and molecules.

The one-dimensional Schrödinger equation was given by Equation 5.28. In order to apply this equation to the hydrogen atom, it must first be generalized to the three dimensions in which a real atom exists. Thus, the second derivative of the wave function with respect to the coordinate $x$, $d^2\psi/dx^2$, must be replaced by a sum of three such second derivatives, one with respect to each of the coordinates, $x$, $y$, and $z$.

Next, the potential energy of the electron must be specified. The force acting on the electron is the Coulomb force between the oppositely charged electron (charge $= -e$) and nucleus (charge $= +e$) (see Equation 1.4). The potential energy associated with this force is

$$V(r) = -e^2/\alpha r \tag{5.46}$$

where $r$ is the distance between electron and nucleus and $\alpha$ is the same propor-

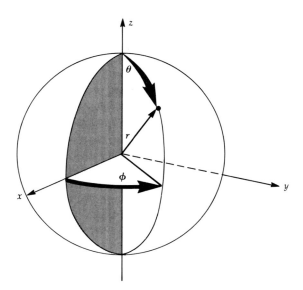

**FIGURE 5.26** Relation between rectilinear (or cartesian) coordinates $(x, y, z)$ and spherical polar coordinates $(r, \theta, \phi)$. Here $r$ is the radius, $\theta$ the polar angle or colatitude, and $\phi$ the azimuthal angle or longitude. The trigonometric relations between the two sets of coordinates are: $x = r \sin \theta \cos \phi$, $y = r \sin \theta \sin \phi$, $z = r \cos \theta$, and $r^2 = x^2 + y^2 + z^2$. In the application to the hydrogen atom, the nucleus is at the origin ($r = 0$) and the electron at the surface of a sphere of radius $r$.

tionality constant as in Equation 1.4. The hydrogen atom can be represented by a sphere; consequently $r$ can point in any direction. It is more convenient, in these circumstances, to replace the rectilinear coordinates $x$, $y$, and $z$ by spherical polar coordinates, $r$, $\theta$, and $\phi$, which are useful in dealing with spherically symmetric systems. The relation between these two sets of coordinates is shown in Figure 5.26.

The Schrödinger equation for the hydrogen atom can be written in terms of spherical polar coordinates as an equation involving the wave function $\psi(r, \theta, \phi)$, its second derivatives with respect to each of these coordinates, the potential energy $V(r)$, the total energy of the electron, $E$, and the mass of the electron, $m$. The process of solving the wave equation is facilitated by the fact that it is possible to write the wave function as the product of three functions of a single variable:

$$\psi(r, \theta, \psi) = R(r)\Theta(\theta)\Phi(\phi) \tag{5.47}$$

where $R(r)$ is called the **radial function,** and the product of the other two functions, $Y(\theta, \phi) = \Theta(\theta)\Phi(\phi)$ is called the **angular function.** This procedure permits the Schrödinger equation to be separated into three equations, each involving only one of these variables, which can then be solved. As in the case of the wave equation for the particle in a box, the solution of each equation depends on the value of a quantum number. These quantum numbers determine the allowed energies and the shapes of the wave functions of the electron. It is customary to call the wave function of an electron in an atom or molecule an **orbital**; the term is more general than "level," which refers to the quantized energy of an orbital but says nothing about its shape. An orbital may also be less precisely described as the region of space where the electron is found.

## 5.11 The Quantum Numbers

The radial equation can only be solved for integral values of a quantum number $n$, called the **principal quantum number**. The allowed values are

$$n = 1, 2, 3, \ldots$$

The energies of the orbitals depend only on the value of $n$. The allowed energies of the hydrogen atom (or of any other one-electron ion of atomic number Z) are given by the formula

$$E_n = -2\pi^2 me^4 Z^2/n^2 h^2 = -hc\mathcal{R}Z^2/n^2 \tag{5.48}$$

which, for $Z = 1$, is identical to the Bohr formula (Equation 5.21).

Two quantum numbers are obtained in the solution of the angular wave equations. The **azimuthal quantum number** $l$ determines the shape of the orbital as well as the angular momentum of the electron. According to classical mechanics, the angular momentum of a particle of mass $m$ travelling in a circular orbit of radius $r$ with velocity $v$ is $mvr$. In quantum mechanics, the angular momentum of a particle is quantized and its value is equal to $\hbar[l(l+1)]^{1/2}$. If the electron in the hydrogen atom possesses angular momentum it will also have kinetic energy of angular motion. Since this energy is necessarily limited by the total energy of the electron, which is determined by the value of the principal quantum number, it is not surprising that the allowed values of $l$ depend on those of $n$. For a solution of the wave equation corresponding to a given $n$, the allowed values of $l$ are:

$$l = 0, 1, 2, \ldots, n - 1$$

The second quantum number associated with the solution of the angular wave equations is the **magnetic quantum number** $m_l$, which determines the orientation of the orbital in space. Just like an electric current circulating in a loop, an electron with angular momentum generates a magnetic field. The magnetism of the atom is manifested when the atom is placed in an external magnetic field and depends on the value of $m_l$. Since the magnetism is associated with the angular momentum of the electron, it is not surprising that the allowed values of $m_l$ are determined by the value of $l$. For a given $l$, $m_l$ can assume the following $2l + 1$ values:

$$m_l = -l, -l + 1, \ldots, -1, 0, 1, \ldots, (l - 1), l$$

For example, when $l = 2$, $(2l + 1) = 5$ values of $m_l$ are allowed (i.e., $-2, -1, 0, 1,$ and $2$).

The three quantum numbers $n$, $l$, and $m_l$ are directly associated with the solutions to the three equations that comprise the Schrödinger equation for the hydrogen atom. In addition, a fourth quantum number must be introduced to take account of the fact that the electron itself acts as a small magnet and thus possesses an intrinsic angular momentum or spin. Experimental evidence based on atomic spectra indicates that the **spin quantum number** $m_s$ has only two possible values, $+\frac{1}{2}$ or $-\frac{1}{2}$. Table 5.4 summarizes the allowed values of the four quantum numbers.

## 5.11

### TABLE 5.4 Quantum Numbers of Atomic Electrons

| Name | Symbol | Allowed Values | Property Chiefly Determined |
|---|---|---|---|
| Principal | $n$ | 1, 2, 3, ... | Size and energy of orbital |
| Azimuthal | $l$ | 0, 1, 2, ..., $n-1$ | Shape of orbital |
| Magnetic | $m_l$ | $-l, -l+1, ..., 0, 1, ..., l$ | Orientation of orbital |
| Spin | $m_s$ | $\frac{1}{2}, -\frac{1}{2}$ | Spin of electron |

The orbitals obtained in the solution of the wave equation correspond to the various possible combinations of quantum numbers. Each orbital can be characterized by the values of the quantum numbers $n$ and $l$ associated with it. In order to avoid confusion between the similar numerical values of these two integers it is customary to express the values of $l$ by means of letters rather than numbers. Thus, an orbital with $l = 0$ is called an s orbital, one with $l = 1$ a p orbital, one with $l = 2$ a d orbital, and one with $l = 3$ an f orbital, where these designations have a historical origin. The complete designation of an orbital involves the numerical value of $n$ followed by the letter corresponding to the value of $l$; for example, an orbital with $n = 2$ and $l = 1$ is a 2p orbital.

In view of the correspondence between the orbitals and the various combinations of quantum numbers, a list of these allowed combinations facilitates the identification of the orbitals and their associated energy levels. For example, the ground state of the hydrogen atom corresponds to $n = 1$. Since $l$ can only have integral values ranging from 0 to $n - 1$, the only possible value of $l$ is $l = 0$. The value of $l$ restricts the values of $m_l$: the magnetic quantum number is restricted to integers ranging from $-l$ to $l$ and, in this particular instance, must be $m_l = 0$. Finally, irrespective of the values of $n$, $l$, and $m_l$, $m_s$ can be either $\frac{1}{2}$ or $-\frac{1}{2}$. We see that two possible combinations of quantum numbers correspond to the ground state of the hydrogen atom: $n$, $l$, $m_l$, and $m_s$ can either be 1, 0, 0, $\frac{1}{2}$ or 1, 0, 0, $-\frac{1}{2}$, respectively.

Table 5.5 summarizes the first few combinations of quantum numbers and identifies the orbitals and associated energy levels to which they correspond. Since the energy of the electron in a hydrogen atom is determined by the value of $n$, a number of different combinations of quantum numbers corre-

### TABLE 5.5 Hydrogen Atom Orbitals and Their Quantum Numbers

| Energy state | $n$ | $l$ | $m_l$ | $m_s$ | Orbital | Degeneracy |
|---|---|---|---|---|---|---|
| Ground state | 1 | 0 | 0 | $\pm\frac{1}{2}$ | 1s | 2 |
| 1st excited state | 2 | 0 | 0 | $\pm\frac{1}{2}$ | 2s | 2 ⎫ 8 |
|  | 2 | 1 | −1, 0, 1 | $\pm\frac{1}{2}$ | 2p | 6 ⎭ |
| 2nd excited state | 3 | 0 | 0 | $\pm\frac{1}{2}$ | 3s | 2 ⎫ |
|  | 3 | 1 | −1, 0, 1 | $\pm\frac{1}{2}$ | 3p | 6 ⎬ 18 |
|  | 3 | 2 | −2, −1, 0, 1, 2 | $\pm\frac{1}{2}$ | 3d | 10 ⎭ |
| 3rd excited state | 4 | 0 | 0 | $\pm\frac{1}{2}$ | 4s | 2 ⎫ |
|  | 4 | 1 | −1, 0, 1 | $\pm\frac{1}{2}$ | 4p | 6 ⎬ 32 |
|  | 4 | 2 | −2, −1, 0, 1, 2 | $\pm\frac{1}{2}$ | 4d | 10 |
|  | 4 | 3 | −3, −2, −1, 0, 1, 2, 3 | $\pm\frac{1}{2}$ | 4f | 14 ⎭ |

spond to the same energy level. We have seen that two combinations correspond to the ground state. Similarly, eight combinations correspond to the first excited state, eighteen to the second excited state, and so on. In general, the number of combinations of quantum numbers all corresponding to the same value of $n$ is $2n^2$. Since the energy of the electron depends only on the principal quantum number, it follows that an electron in any of the various orbitals characterized by the same value of $n$ has the same energy. For example, the electron in a 3s orbital has the same energy as one in any of the 3p orbitals or in any of the 3d orbitals. Different orbitals with the same energy are said to be **degenerate**. As we shall see, the electronic energy of atoms more complex than hydrogen depends on $l$ as well as $n$.

## 5.12 Solutions of the Wave Equations

As mentioned in Section 5.10, the wave function of the hydrogen atom can be separated into a radial and an angular part. Due to the complexity of these wave functions, it is convenient to examine the results separately. The radial function determines the size of the orbital and permits an evaluation of the probability of finding the electron as a function of its distance from the nucleus. The angular function determines the shape and orientation of the orbital. While this separation is convenient, it must be kept in mind that the probability per unit volume of finding the electron in some small region of space is given by the product of the squares of the radial and angular functions.

*a. The Radial Functions*  The radial part of the orbital, $R(r)$, depends on the values of $n$ and $l$ but is independent of $m_l$. The first few solutions are given in Table 5.6. The constant $a_0$, which appears in each of the solutions, is known as the **Bohr radius** and is a collection of fundamental constants with the value

$$a_0 = \alpha \hbar^2 / me^2 = 0.529 \text{ Å} \tag{5.49}$$

Recall from Section 5.8 that the square of the wave function is a measure of the probability per unit interval of finding the particle whose motion it describes at some particular location. For example, the probability of finding the particle in a box between $x$ and $x + dx$ is $\psi^2(x)\,dx$ (Section 5.9). Similarly, the probability of finding the electron in an infinitesimal interval $dr$ at a distance $r$ from the nucleus is $R^2(r)\,dr$. As can be seen in Figure 5.27a, this is the probability of finding the electron *along a particular line*. This is actually not the desired probability of locating the electron since, owing to the spherical symmetry of the atom, there are an infinite number of lines originating at the nucleus along which the orbital can extend. All lines of length $r$ have one feature in common, namely, they terminate on the surface of a sphere of radius $r$. Since the surface area of a sphere ($4\pi r^2$) increases with its radius, the probability of finding the electron at a distance $r$ *irrespective of direction* depends on this geometric factor as well as on the square of the wave function. The probability of finding the electron in a shell of thickness $dr$ at a distance $r$ from the nucleus is $4\pi r^2 R^2(r)\,dr$, where the factor $4\pi r^2 R^2(r)$ is known as the **radial probability function**. Figure 5.27b graphically illustrates the difference between this function and the square of the radial wave function.

The radial probability functions for the first few orbitals are graphed in Figure 5.28 and represented pictorially in Figure 5.29. The 1s orbital, which corresponds to the ground state of the hydrogen atom, has a particularly simple

## TABLE 5.6 First Few Hydrogenic Wave Functions[a]

**Radial Function $R(r)$**

$$R_{1s} = 2a^{-3/2} \exp(-r/a)$$

$$R_{2s} = (2a)^{-3/2}\left(2 - \frac{r}{a}\right) \exp(-r/2a)$$

$$R_{2p} = \frac{1}{\sqrt{3}}(2a)^{-3/2}\frac{r}{a} \exp(-r/2a)$$

$$R_{3s} = \frac{2}{27}(3a)^{-3/2}\left(27 - \frac{18r}{a} + \frac{2r^2}{a^2}\right) \exp(-r/3a)$$

$$R_{3p} = \frac{1}{81\sqrt{3}}\left(\frac{a}{2}\right)^{-3/2}\left(6\frac{r}{a} - \frac{r^2}{a^2}\right)\exp(-r/3a)$$

$$R_{3d} = \frac{1}{81\sqrt{15}}\left(\frac{a}{2}\right)^{-3/2}\frac{r^2}{a^2} \exp(-r/3a)$$

**Angular Function $Y(\theta, \phi)$**

$$Y_s = \left(\frac{1}{4\pi}\right)^{1/2}$$

$$Y_{d_{z^2}} = \left(\frac{5}{16\pi}\right)^{1/2}(3\cos^2\theta - 1)$$

$$Y_{p_z} = \left(\frac{3}{4\pi}\right)^{1/2}\cos\theta$$

$$Y_{d_{xz}} = \left(\frac{15}{4\pi}\right)^{1/2}\sin\theta\cos\theta\cos\phi$$

$$Y_{p_x} = \left(\frac{3}{4\pi}\right)^{1/2}\sin\theta\cos\phi$$

$$Y_{d_{yz}} = \left(\frac{15}{4\pi}\right)^{1/2}\sin\theta\cos\theta\sin\phi$$

$$Y_{p_y} = \left(\frac{3}{4\pi}\right)^{1/2}\sin\theta\sin\phi$$

$$Y_{d_{x^2-y^2}} = \left(\frac{15}{4\pi}\right)^{1/2}\sin^2\theta\cos 2\phi$$

$$Y_{d_{xy}} = \left(\frac{15}{4\pi}\right)^{1/2}\sin^2\theta\sin 2\phi$$

[a] For H, $a = a_0$; for a one-electron atom with atomic number Z, $a = a_0/Z$.

radial probability function. The probability increases sharply from a value of zero at the origin and peaks at the Bohr radius $a_0$. Beyond this maximum, the exponential decrease dominates, indicating that the probability of finding the electron at a greater distance from the nucleus decreases rapidly.

The orbitals for the excited states of the hydrogen atom (2s, 2p, etc.) exhibit a more complex radial dependence than the 1s orbital. The exponential term varies in a systematic way with $n$, the argument of the exponent changing from $r/a_0$, to $r/2a_0$, to $r/3a_0$ as $n$ increases from one to two to three (Table 5.6). The exponential decreases more slowly with increasing $n$, and the electron tends to be further from the nucleus as $n$ increases. A feature of the excited state orbitals is the appearance of subsidiary peaks close to the origin. As in the case of the wave functions for the particle in a box, these peaks are separated by nodes, points where the probability of finding the electron drops to zero. The number of nodes in the radial wave function varies in a systematic way

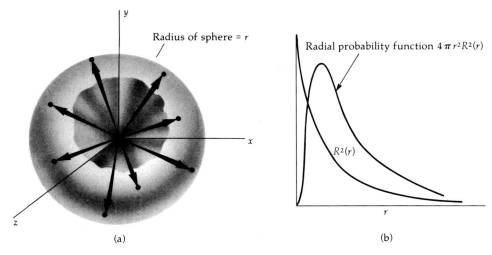

**FIGURE 5.27** (a) The probability of finding the electron in an infinitesimal interval $dr$ at $r$ along any one of the infinite number of radial lines is $R^2(r)\,dr$. The number of different ways an electron can be at a distance $r$ from the nucleus is equal to $4\pi r^2$, the surface area of a sphere of radius $r$. (b) The difference between $R^2(r)$ and the radial probability function, $4\pi r^2 R^2(r)$, for a 1s orbital.

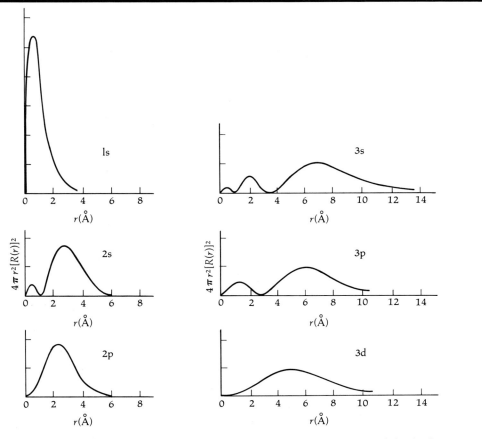

**FIGURE 5.28** Radial probability functions for the first few orbitals of the hydrogen atom. The same scale applies in all cases.

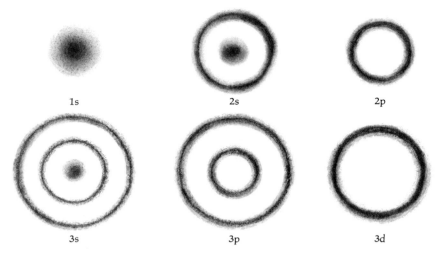

**FIGURE 5.29** Two-dimensional representation of the radial probability functions. The intensity of the shading is proportional to the probability of finding the electron at that location.

and can be obtained from the relation

$$\text{number of radial nodes} = n - l - 1 \tag{5.50}$$

For example, the 3s orbital has $3 - 0 - 1 = 2$ nodes, the 3p orbital has $3 - 1 - 1 = 1$ node, and the 3d orbital has $3 - 2 - 1 = 0$ nodes. The occurrence of these radial nodes and associated maxima means that an electron in an s orbital can approach the nucleus more closely than one in a p orbital and is said to be more "penetrating." While this difference does not affect the properties of the hydrogen atom, it accounts for one of the important features of the energy levels of multielectron atoms (Section 6.2).

**Example 5.9** What is the probability that an electron in the 1s orbital of the hydrogen atom will be four times farther from the nucleus than the most probable distance $a_0$?

To answer this question we need to evaluate the radial probability function for a 1s electron when $r = 4a_0$ and when $r = a_0$ and take the ratio. Using the 1s radial function in Table 5.6 we have,

$$4\pi r^2 R^2 = 4\pi(4a_0)^2(4a_0^{-3}) \exp[-2(4a_0/a_0)] \qquad r = 4a_0$$

$$4\pi r^2 R^2 = 4\pi a_0^2(4a_0^{-3}) \exp[-2(a_0/a_0)] \qquad r = a_0$$

and taking the ratio, we obtain after cancelling common factors:

$$\frac{\text{Probability } (r = 4a_0)}{\text{Probability } (r = a_0)} = \frac{(4)^2 \exp(-8)}{(1)^2 \exp(-2)}$$

$$= 16 \exp(-6) = 0.04$$

showing that the probability that the hydrogen is four times larger than its most likely size is only 4%. ∎

*b. The Angular Functions* The angular part of the wave function determines the shape of the orbital and its orientation in space. These factors are particularly important in determining the spatial properties of covalent bonds (Chapter 16). The angular functions are summarized in Table 5.6. In contrast to the radial functions, the angular functions are independent of the principal quantum number. It follows that all orbitals of a given type (e.g., s, p, or d) display the same angular behavior, which we now examine.

The angular function for an s orbital, $Y_s$, is independent of the angular coordinates $\theta$ and $\phi$. Thus, the orbital extends equally in all directions and is spherically symmetric. A simple visual representation of the shape of an orbital consists of a plot in a rectilinear coordinate system (in either two or three dimensions) of the dependence of the angular function on the polar angle $\theta$. It may be helpful to refer back to Figure 5.26, which shows the relation between $\theta$ and $\phi$ and the $x$, $y$, and $z$ axes. The representation of a spherically symmetric orbital is particularly simple in such a polar plot. Since $Y_s$ has the same value in all directions, it can be represented by a circle in a two-dimensional plot (Figure 5.30) or a sphere in a three-dimensional plot.

The probability of finding the electron in a given direction from the nucleus (at a fixed radius) is given by the square of the angular function. Since $Y_s$ is independent of $\theta$, so is $Y_s^2$, meaning that an s electron can be found with equal probability in all directions from the nucleus. Since $Y_s$ is represented by a sphere in a three-dimensional polar plot, $Y_s^2$ must also be spherical in such a representation (Figure 5.31).

The p orbitals have a more complex angular dependence than the s orbital. There are three p orbitals, corresponding to the three possible values of the magnetic quantum number, $m_l$ (when $l = 1$, $m_l = 0, \pm 1$). These orbitals depend on $\theta$ and $\phi$ and are not spherically symmetric. Let us consider $Y_{p_z}$, the first of the three angular functions for p orbitals in Table 5.6. This function is proportional to $\cos \theta$, where $\theta$ is the angle with respect to the z axis. Since $\cos \theta$ has its maximum value ($\cos \theta = 1$) when $\theta = 0$, the function attains its maximum value along the positive z axis. On the other hand, the function attains its most negative value along the negative z axis, for this is where $\theta = \pi$ and $\cos \pi = -1$. Since the z axis is perpendicular to the $xy$ plane, $\theta = \pi/2$, and $\cos \theta = 0$ anywhere in this plane. The angular wave function consequently has a node in the $xy$ plane. We can now see the reason for naming the orbital in question $p_z$: the wave function extends along the z axis.

The behavior of the $Y_{p_z}$ function is depicted graphically in Figure 5.30 as a plot of $\cos \theta$ versus $\theta$. The function consists of two tangent circles and is positive in sign along the positive z axis and negative in sign along the negative z axis. As noted earlier, the function attains its maximum magnitude along the z axis and has a node in the $xy$ plane (perpendicular to the plane of the paper). The angular probability distribution $Y_{p_z}^2$ is plotted in Figure 5.31. This is a plot of the function $\cos^2 \theta$ and consists of two pear-shaped lobes extending along the z axis. This representation makes the strong directional preference of the $p_z$ orbital even more apparent than does that of the angular function.

The angular dependence of the other two p orbitals can be examined in the same way. The $p_x$ function is proportional to $\sin \theta \cos \phi$. Since $\phi$ lies in the $xy$ plane and is measured from the $x$ axis, $\cos \phi = 0$ everywhere in the $yz$ plane (see Figure 5.26). Thus, the $p_x$ function has a node in the $yz$ plane and extends along the $x$ axis. Similarly, the $p_y$ function, which is proportional to $\sin \theta \sin \phi$, has a node in the $xz$ plane and extends along the $y$ axis. Figures 5.30 and 5.31

170   The Quantized Atom   5.12

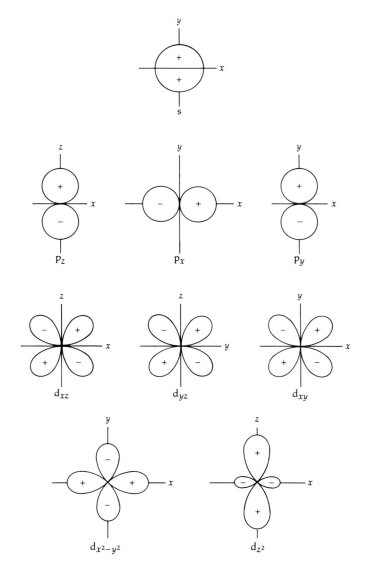

**FIGURE 5.30** Two-dimentional polar plots of the angular functions $Y(\theta, \phi)$ for s, p, and d orbitals of the hydrogen atom. The signs of the functions are marked.

show that the angular functions $Y_{p_x}$ and $Y_{p_y}$ and probability distributions $Y_{p_x}^2$ and $Y_{p_y}^2$ differ from those for the $p_z$ orbital only in their orientation; the shapes are identical.

The angular behavior of the five d orbitals (corresponding to $m_l = 0, \pm 1, \pm 2$) is more complicated, and we need not analyze these orbitals in detail. Figure 5.30 shows the shape of the angular functions. The orbital designations are based on the directions or planes in which the angular probability is highest.

To conclude this discussion, we must emphasize the fact that while the separation of the hydrogen atom wave functions into their radial and angular components facilitates the discussion of their properties, the actual functions

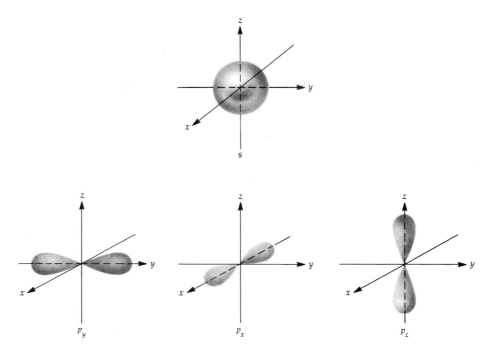

**FIGURE 5.31** Angular probability functions $Y^2(\theta, \phi)$ for s and p orbitals.

are the products of these components. Similarly the probability per unit volume of finding the electron at any point in space is given by the square of the overall wave function, $\psi^2(r, \theta, \phi)$. Figure 5.32 shows a cross-sectional profile of this distribution for selected orbitals. Note how the angular function modulates the behavior of the radial distribution by concentrating the probability of finding the electron along certain specific directions, which depend on the identity of the orbital. For example, s orbitals can be represented as a function of increasing $n$ by successively larger spherical shells, while p orbitals look like increasingly larger blobs concentrated on the appropriate axes.

The radial and angular wave functions can also be correlated by the number of nodes in each orbital. As in the case of the particle in a box, the total number of nodes in the wave function is $n - 1$. We have already noted that the radial wave function has $n - l - 1$ nodes. Therefore, the angular wave function has $l$ nodes. This result is evident in Figure 5.30 which shows, for example, that the various p orbitals each have one node.

# Conclusion

In this chapter we have seen how various experiments on the structure of atoms, the properties of electromagnetic radiation, and the interaction between atoms and radiation led to the development of quantum mechanics. The quantization of energy, the central result of these studies, was first revealed, albeit

# 172 Conclusion

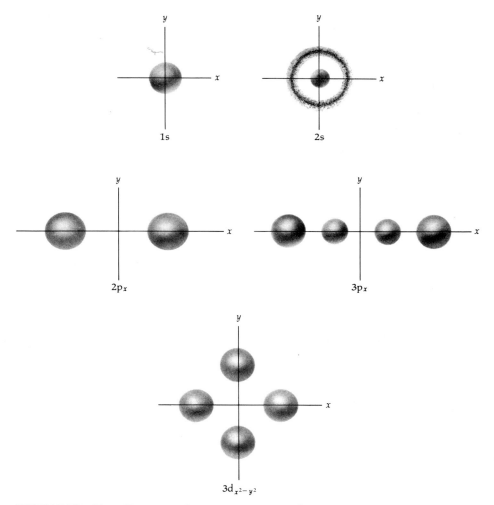

**FIGURE 5.32** Two-dimensional representation of $\psi^2(r, \theta, \phi)$, the overall electron probability distribution in the hydrogen atom. The distance from the origin corresponds to the distance from the center of the nucleus. The shaded areas correspond to the regions where the probability of finding the electron is highest.

indirectly, by Planck's analysis of the spectrum of black-body radiation. Einstein's explanation of the photoelectric effect showed that light itself is quantized in the form of photons, and Bohr's explanation of the spectrum of atomic hydrogen showed the connection between quantized energy levels and the energy of the emitted photons. Along with the studies that revealed the wave properties of particles, these experiments showed that a radically different theory of the behavior of atomic matter than the one provided by classical physics was needed. The presently accepted theoretical approach—quantum mechanics—involves the solution of the Schrödinger wave equation. A simple model system—the particle in a box—shows how the probability distribution and quantized energy levels are obtained. The detailed examination of this simple system makes it easier to understand the far more complex case of the hydro-

## Problems

1. (a) What is the frequency of electromagnetic radiation of wavelength 0.15 nm, 500 nm, 2.5 cm, and 200 m? In which region of the spectrum would each of these wavelengths be found? (b) What is the wavelength of a photon of frequency $10^{13}$ Hz, $10^{16}$ Hz, and $10^{19}$ Hz? Indicate in which region of the spectrum each of these wavelengths would be found.

2. A 150 W bulb radiates light of 550 nm wavelength isotropically. Determine (a) the time elapsed from the instant the light is turned on until an observer at a distance of 15 m sees it, (b) the intensity of the light at this point, and (c) the number of photons emitted by the bulb striking a 0.10 m² screen at a distance of 15 m in 1.0 s.

3. Electrons can be ejected from the surface of molybdenum by photons of wavelength 274 nm or less. (a) What is the work function of molybdenum (in kJ mol$^{-1}$)? (b) What is the maximum kinetic energy (in eV) of electrons ejected from molybdenum by 100 nm photons? What is the speed of these electrons?

4. The work function of tungsten is 439 kJ mol$^{-1}$. What is the maximum kinetic energy of electrons ejected by 150 nm photons?

5. The energy required to ionize lithium is 520 kJ mol$^{-1}$. What is the lowest frequency of light that can ionize a lithium atom, and what is its wavelength?

6. The human eye can perceive light when the radiant energy incident on it is at least $3.5 \times 10^{-17}$ J. For light of 500 nm wavelength, how many photons does this correspond to?

7. Use Planck's formula (Equation 5.13) to calculate the relative intensities of the radiation at 800 nm and 400 nm emitted by a black body at 2500 K.

8. The wavelength at which a black body emits with maximum intensity is $\lambda_{max} = hc/5kT$. What is the surface temperature of the sun, if we treat it as a black body emitting with highest intensity at 475 nm?

9. X rays of wavelength 70.8 pm (1 pm = $10^{-12}$ m) are scattered off a block of carbon. What is the energy (in eV) of the radiation detected at 90° to the X-ray beam?

10. Calculate the relative number of $\alpha$ particles striking a detector located at 15°, 90°, and 165° with respect to a beam of $\alpha$ particles incident on a thin metal foil. Assume that the angular distribution obeys Rutherford scattering.

11. Equation 5.19 for the wave numbers of the spectral lines of the hydrogen atom is also valid for other atoms with just one electron provided the Rydberg constant is multiplied by the square of the atomic number of the element in question, i.e., $\mathcal{R}$ is replaced by $\mathcal{R}Z^2$. With this fact in mind, consider a He$^+$ ion whose single electron has been excited to the $n = 3$ level. Calculate the wavelengths of all the possible optical lines associated with a transition back to the ground state.

12. Calculate the energy required to remove the electron from the following ions in their ground state: He$^+$, Li$^{2+}$, and Be$^{3+}$.

13. What are the wavelengths of the three Paschen lines of hydrogen having the lowest frequencies?

14. The energy of the ground state of the hydrogen atom is $-13.56$ eV. How many excited states of the atom lie in the energy interval $-0.5$ to $-2.0$ eV? What are their quantum numbers?

15. Derive an expression for the wave number of a transition between any two adjacent levels of the hydrogen atom. Give a simple approximation for your expression which becomes valid for large values of $n$. Use this approximation to calculate the wavelength of the radiation emitted when the electron makes a transition from the $n = 21$ to the $n = 20$ level. In which region of the spectrum does this wavelength lie?

16. The orange–yellow light of sodium vapor lamps is emitted when the valence electron of sodium makes a transition from an excited state back to the ground state of the atom. The light consists primarily of two wavelengths: 589.00 nm and 589.59 nm. What is the energy of each excited state relative to the ground state and what is the energy difference between the two states? Express the answers in electron volts.

17. A certain atom in an excited state can return to the ground state in two different ways: (a) via a direct transition in which a photon of wavelength $\lambda_1$, is emitted and (b) via an intermediate excited state,

reached by the emission of a photon of wavelength $\lambda_2$; this excited state in turn decays to the ground state by emitting a photon of wavelength $\lambda_3$. How are $\lambda_1$, $\lambda_2$, and $\lambda_3$ related? How are the corresponding wave numbers related?

18. The ionization energy of a certain element is 400 kJ mol$^{-1}$. However, when the atoms of this element are in their first excited state, the ionization energy is only 115 kJ mol$^{-1}$. What is the wavelength of the radiation emitted in a transition from the first excited state to the ground state?

19. Calculate the wavelength of (a) a 1.0 keV electron and (b) a 1.0 keV proton.

20. Calculate the wavelength of a neon atom whose speed is equal to the root mean square speed at 25°C.

21. What is the energy of an electron whose wavelength is equal to that of a 5.0 eV photon?

22. An electron and a photon have the same energy. Derive an expression for the wavelength of the electron in terms of that of the photon.

23. Thermal neutrons have speeds characteristic of the translational energies of gases at room temperature. What is the wavelength of thermal neutrons whose speed is equal to their root mean square speed at 300 K? What is the energy (in eV) of such a neutron?

24. (a) What is the de Broglie wavelength of a helium atom whose speed is equal to $v_{rms}$ at 35°C? (b) If the position of such an atom can be measured to within 1.0 nm, what is the uncertainty in its momentum? (c) How does this uncertainty compare with the value of the momentum?

25. (a) If the uncertainty in the position of a 500 eV electron is equal to its wavelength, what is the minimum uncertainty in the momentum of the electron? (b) What is the uncertainty in the position of the electron 1 second later?

26. Use the uncertainty principle to estimate the minimum kinetic energy of the electron in the hydrogen atom. Assume the uncertainty in position is equal to the diameter of the atom ($\approx 1.2$ Å) and that the minimum energy of the electron is equal to the uncertainty in the energy. Repeat this calculation for a proton confined to a nucleus 10 fm in diameter. Express the answers in electron volts.

27. An electron is confined to a 1.0 nm long box. (a) What is the wavelength of the radiation emitted in a transition from the $n = 3$ to the $n = 1$ state? (b) At what positions does the wave function for the $n = 3$ state have a node?

28. A nitrogen molecule is confined to a 10 cm long container. (a) For what value of the quantum number $n$ is the translational energy of the molecule equal to $\frac{1}{2}kT$, when $T = 27$°C? (b) What is the energy separation (as a fraction of the thermal energy $\frac{1}{2}kT$) between this level and its nearest neighbor?

29. An electron is confined to a 1.2 nm long box. How many energy levels are there between 7 and 14 eV?

30.* A particle of mass $m$ is confined to a box of length $L$. What is the probability that the particle is found in the middle third of the box (i.e., between $L/3$ and $2L/3$) for $n = 1$, 2, and 3?

31.* The average position of a particle in a box of length $L$ is given by the expression $\bar{x} = \int_0^L x\psi^2(x)\,dx$. What is the average position of the particle? Show that this result is independent of $n$.

32. What possible values of the quantum numbers $n$, $l$, $m_l$, $m_s$, can a 4f electron have?

33. From an examination of Figure 5.28, deduce what the shapes of the radial probability functions for the various $n = 4$ orbitals of the hydrogen atom must look like. Sketch these functions.

34. At what distance from the center of the nucleus in units of the Bohr radius $a_0$ do (a) the 2s wavefunction and (b) the 3p wave function of the hydrogen atom have a node?

35. Show that the 2s orbital is more penetrating than the 2p orbital by evaluating the ratio of the probabilities that an electron in these two orbitals will be at a distance $a_0/2$ from the nucleus of a hydrogen atom. [Evaluate the ratio of $\psi^2(2s)/((\psi^2(2p_z) + \psi^2(2p_x) + \psi^2(2p_y))$.]

36.* The wave function for a 1s orbital of an atom with atomic number $Z$ is

$$\psi = (Z^3/\pi a_0^3)^{1/2} \exp(-Zr/a_0)$$

Derive an expression for the most probable radius of a 1s electron. Evaluate the expression for the elements H–Ne. [Note that $\psi^2(r)$ must be multiplied by the area of a sphere of radius $r$ ($4\pi r^2$) in order to obtain the radial probability function.]

37. An electron occupies the following orbitals: (a) 1s, (b) 2p, (c) 3s, and (d) 3d. How many nodes are there in the radial wave function, in the angular function, and in the overall wave function? What is the magnitude of the angular momentum of the electron?

38. What is the number of different combinations of quantum numbers corresponding to the levels of the hydrogen atom characterized by energy $-hc\mathcal{R}/9$ and $-hc\mathcal{R}/25$? What is the degeneracy of these levels?

# 6

# Multielectron Atoms and the Periodic Table

Since the early days of chemistry, chemists have been struck by certain regularities in the chemical and physical properties of the elements and their compounds. Various attempts to systematize these regularities led to the development of the **periodic table**, a sequential arrangement of the elements which emphasizes the observed regularities. With an understanding of the structure of the atom came an understanding that many chemical and physical properties depend on how electrons are arranged in atoms. One of the successes of quantum mechanics is in allowing a connection to be made between the periodicity of observed properties on which the periodic table is based and the electronic structure of atoms.

Quantum mechanics provides the theoretical basis for the ordering of the elements. It turns out that the format of the periodic table is a direct result of the variations in the electronic energy levels available in multielectron atoms, as explained by quantum mechanics. The impressive outcome is a link between a powerful but abstract theory and the empirical regularities embodied in the periodic table. This joint contribution of experiment and theory to our understanding is an excellent illustration of how modern chemistry has developed.

## The Electronic Structure of Atoms

### 6.1 Historical Development of the Periodic Table

*a. Some Early Classification Attempts* Since chemistry was in its infancy, it has been known that certain elements exhibit similar chemical and physical properties. Over the years, many schemes for explaining and predicting such

TABLE 6.1  Dobereiner's Triads—the First Attempt At a Systematic Classification of the Elements

| Elements of triad | Atomic weight (u) | Atomic weight differences (u) |
|---|---|---|
| Lithium | 6.9 | 16.1 |
| Sodium | 23.0 | 16.1 |
| Potassium | 39.1 | |
| Calcium | 40.1 | 47.5 |
| Strontium | 87.6 | 49.7 |
| Barium | 137.3 | |
| Chlorine | 35.4 | 44.5 |
| Bromine | 79.9 | 47.0 |
| Iodine | 126.9 | |
| Sulfur | 32.1 | 46.9 |
| Selenium | 79.0 | 48.6 |
| Tellurium | 127.6 | |
| Chromium | 52.0 | 2.9 |
| Manganese | 54.9 | 0.9 |
| Iron | 55.8 | |

relationships have been proposed. The early ordering schemes were based on a sequencing of the elements according to their atomic weight, since the atomic number concept was not introduced until 1913 (Section 6.5c).

The first attempt at a systematic arrangement of the elements came in 1829, when J. W. Dobereiner (1780–1849) pointed out that certain groups of three elements with similar chemical properties differed by a constant number of atomic weight units. Sulfur, selenium, and tellurium form one such group of elements. Their similarity is manifested in various ways, for example, in the formation of similar ions with oxygen: $SO_4^{2-}$, $SeO_4^{2-}$, and $TeO_4^{2-}$. The atomic weights of sulfur and selenium and of selenium and tellurium differ by about 48 units, suggesting a recurrence in properties at such an interval. Table 6.1 lists some of Dobereiner's most well defined triads. Since there was no way in which different triads could be combined into a coherent whole, the scheme failed to win much acceptance.

Before a more sound classification system could be developed, reliable atomic weights had to be known for a reasonably large number of elements. These conditions were first met in the 1860s, by which time some 60 elements and their atomic weights were known. On the basis of this information, a more ambitious classification scheme was proposed by John Newlands (1838–1898) in 1863. He arranged the elements in order of increasing atomic weight in seven horizontal families. In this scheme, elements lying in a given row resembled each other, suggesting that a pair of similar elements was separated by six intervening elements, as shown in Figure 6.1. Newlands saw an analogy between this "law of octaves" and the seven-tone musical scale, in which there is a similar resemblance between the first and the eighth tone. While Newlands' scheme was largely dismissed as numerology, he deserves credit for detecting the recurrence of similar properties when elements are arranged in order of increasing atomic weight.

| No. | | No. | | No. | | No. | | No. | | No. | | No. | | No. | |
|---|---|---|---|---|---|---|---|---|---|---|---|---|---|---|---|
| H | 1 | F | 8 | Cl | 15 | Co & Ni | 22 | Br | 29 | Pd | 36 | I | 42 | Pt & Ir | 50 |
| Li | 2 | Na | 9 | K | 16 | Cu | 23 | Rb | 30 | Ag | 37 | Cs | 44 | Os | 51 |
| G | 3 | Mg | 10 | Ca | 17 | Zn | 24 | Sr | 31 | Cd | 38 | Ba & V | 45 | Hg | 52 |
| Bo | 4 | Al | 11 | Cr | 19 | Y | 25 | Ce & La | 33 | U | 40 | Ta | 46 | Tl | 53 |
| C | 5 | Si | 12 | Ti | 18 | In | 26 | Zr | 32 | Sn | 39 | W | 47 | Pb | 54 |
| N | 6 | P | 13 | Mn | 20 | As | 27 | Di & Mo | 34 | Sb | 41 | Nb | 48 | Ei | 55 |
| O | 7 | S | 14 | Fe | 21 | Se | 28 | Ro & Ru | 35 | Te | 43 | Au | 49 | Th | 56 |

**FIGURE 6.1** Octave arrangement of the elements by Newlands (1863). The elements are numbered in order of increasing atomic weight; similar elements supposedly recur at intervals of eight (e.g., Li, Na, and K).

***b. The Periodic Table of Mendeleev*** A substantially correct form of the periodic table was first proposed by Dmitri Mendeleev (1834–1907) and, independently, by Lothar Meyer (1830–1895) in 1869–1871. In Mendeleev's 1871 table (Figure 6.2) the elements were arranged in order of increasing atomic weight in twelve horizontal rows, thereby placing related elements in vertical columns, or groups, numbered I to VIII. Hydrogen was the sole occupant of the first row; the second and third rows each contained seven elements, ranging from lithium to fluorine, and from sodium to chlorine, respectively. If allowance is

| Row | Group I — $R_2O$ | Group II — RO | Group III — $R_2O_3$ | Group IV $RH_4$ $RO_2$ | Group V $RH_3$ $R_2O_5$ | Group VI $RH_2$ $RO_3$ | Group VII RH $R_2O_7$ | Group VIII — $RO_4$ |
|---|---|---|---|---|---|---|---|---|
| 1 | H = 1 | | | | | | | |
| 2 | Li = 7 | Be = 9.4 | B = 11 | C = 12 | N = 14 | O = 16 | F = 19 | |
| 3 | Na = 23 | Mg = 24 | Al = 27.3 | Si = 28 | P = 31 | S = 32 | Cl = 35.5 | |
| 4 | K = 39 | Ca = 40 | — = 44 | Ti = 48 | V = 51 | Cr = 52 | Mn = 55 | Fe = 56, Co = 59 Ni = 59, Cu = 63 |
| 5 | (Cu = 63) | Zn = 65 | — = 68 | — = 72 | As = 75 | Se = 78 | Br = 80 | |
| 6 | Rb = 85 | Sr = 87 | ?Yt = 88 | Zr = 90 | Nb = 94 | Mo = 96 | — = 100 | Ru = 104, Rh = 104, Pd = 106, Ag = 108 |
| 7 | (Ag = 108) | Cd = 112 | In = 113 | Sn = 118 | Sb = 122 | Te = 128 | I = 127 | |
| 8 | Cs = 133 | Ba = 137 | ?Di = 138 | ?Ce = 140 | | | | |
| 9 | | | | | | | | |
| 10 | | | ?Er = 178 | ?La = 180 | Ta = 182 | W = 184 | | Os = 195, Ir = 197 Pt = 198, Au = 199 |
| 11 | (Au = 199) | Hg = 200 | Tl = 204 | Pb = 207 | Bi = 208 | | | |
| 12 | | | | Th = 231 | | U = 240 | | |

**FIGURE 6.2** Mendeleev's 1871 version of the periodic table. The letter R in the column headings symbolizes the elements below in the formula of typical oxides and hydrides. The elements are arranged in order of increasing atomic weight. Elements in parenthesis indicate a continuation of a period from the preceding row (e.g., K to Br occupies two rows but is a single period). Note the empty spaces corresponding to atomic weights 44, 68, and 72, Mendeleev's predictions for then unknown scandium, gallium, and germanium.

made for the fact that the noble gases (He, Ne, etc.) were not discovered until some two decades later, the first three rows of Mendeleev's periodic table are identical to the modern version shown inside the front cover of this book.

Mendeleev also divided many of the groups into subgroups in an attempt to separate elements with different properties from each other. In effect, this procedure resulted in the institution of long periods for what are now known as the transition elements and made it unnecessary to place metals such as chromium and manganese in the same groups as nonmetals such as sulfur and chlorine. Note in Figure 6.2 that the long periods extend over two rows, thereby accounting for the large number of rows.

In arriving at his arrangement of the elements, Mendeleev had to exercise a certain amount of creative license. The properties of certain pairs of elements suggested that their sequence should be in reverse order from that indicated by their atomic weights. For example, Mendeleev decided that iodine should follow rather than precede tellurium, even though its atomic weight was lower, because this order would place it below chlorine and bromine, which it resembled. Similarly, because of its resemblance to silver, gold was placed in the table after platinum, in reverse order from that of their atomic weights. It was Mendeleev's genius to recognize that the preservation of the regularities was of primary importance and that more careful measurements of atomic weights would prove his order correct. This is indeed what happened in the case of gold and platinum. However, the atomic weight of iodine remains lower than that of tellurium. The subsequent realization that the elements are ordered according to atomic number rather than atomic weight eventually confirmed the validity of Mendeleev's bold procedure, since the atomic number of iodine is indeed larger than that of tellurium.

Mendeleev's most significant insight, and the one that led to universal acceptance of the periodic table, was his recognition of the occurrence of gaps in the table and the realization that they had to be filled by then unknown elements. For example, instead of placing titanium right after calcium in the fourth row, he left room for an unknown element, which, owing to its placement below boron, was given the provisional name *eka boron*. The properties of this element and its compounds could be predicted on the basis of its position in the table. Similarly, Mendeleev predicted the properties of *eka aluminum* and *eka silicon*, the unknown elements in the fifth row, between zinc and arsenic.

The discovery of these three elements in the ensuing two decades and the close similarity between their actual and predicted properties provided striking confirmation of the validity of Mendeleev's approach. Table 6.2 compares the predicted properties of *eka aluminum* and the actual properties of gallium, the first of these three elements to be discovered. This same predictive power of the periodic table has been used in recent years as a guide to the discovery of the transuranium elements (Section 23.7) and, most recently, has been used to predict the properties of the as yet unknown superheavy elements (atomic number $\geq 112$).

While Mendeleev's scheme was a remarkable advance in systematization, it was far from perfect, as careful examination of Figure 6.2 will demonstrate. For example, not enough of the rare earths (the elements La through Lu) were known to make it clear that their unique similarities required that they be grouped together. Instead, the known rare earths (e.g., Ce, Er) were scattered among the heavy elements. While some of these errors were corrected in the

**TABLE 6.2  Comparison of Mendeleev's Predictions for Eka Aluminum With the Actual Properties of Gallium**

| Eka aluminum (predicted in 1871) | Gallium (discovered in 1875) |
| --- | --- |
| Atomic weight 68 | Atomic weight 69.9 |
| Density ≈ 5.9 | Density = 5.93 |
| Melting point should be low | mp = 30.1°C |
| Formula of oxide, $Ea_2O_3$ | $Ga_2O_3$ |
| Should be found in aluminum ores | Found in aluminum ores |
| Metal should not be volatile | Low vapor pressure; bp = 2403°C |
| Metal should resist attack by common reagents | Pure metal does not react with water, air, or hydrochloric acid |

ensuing years, it was the discovery of the atomic number concept in 1913 that made it possible to order all the elements in correct sequence and to spot any remaining gaps in the periodic table. The reason for the particular form of the periodic table required by the observed properties of the elements became clear once quantum mechanics was developed and applied to the electronic structure of the atom.

The periodic table is undoubtedly the greatest single generalization in chemistry. Among the many properties that are systematized by the arrangement of the elements are their ionization energies, atomic and ionic sizes, oxidation numbers, metallic properties, and the acid–base character of the oxides. These regularities will be examined in detail in the course of the present and succeeding chapters.

## 6.2 Quantum Mechanics of Multielectron Atoms

*a. Splitting of the Energy Levels*   The quantum mechanical treatment of multielectron atoms is considerably more complicated than that of the hydrogen atom (Section 5.10). When an atom possesses more than one electron, the potential energy includes not only the energy associated with the attractive electron–nucleus interaction but also that associated with the mutually repulsive interactions among all the electrons. For example, the potential energy of the helium atom is

$$V(r_{10}, r_{20}, r_{12}) = -\frac{2e^2}{\alpha r_{10}} - \frac{2e^2}{\alpha r_{20}} + \frac{e^2}{\alpha r_{12}} \tag{6.1}$$

where $r_{10}$ and $r_{20}$ are the distances between the two electrons and the nucleus and $r_{12}$ is the distance between the two electrons. This expression is more complex than Equation 5.46, the corresponding expression for the hydrogen atom. It turns out, in fact, that the Schrödinger equation for atoms other than hydrogen can only be solved by making use of various approximations. Nonetheless, the advent of computers has made it possible to accurately calculate the orbitals and energy levels of complex atoms. These atomic orbitals describe how single electrons behave as a result of their interaction with the nucleus and the other electrons. The orbitals of multielectron atoms are qualitatively similar to those of the hydrogen atom and are labelled with the same quantum numbers used to describe the hydrogen atom orbitals.

**180** Multielectron Atoms and the Periodic Table 6.2

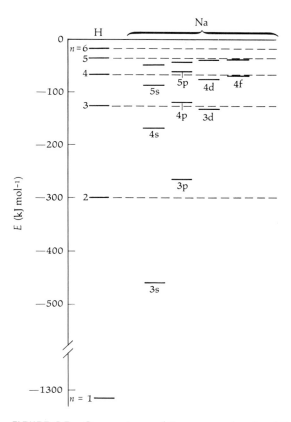

**FIGURE 6.3** Comparison of the energy levels of the hydrogen atom and the sodium atom. The energy required to remove an electron from a particular level to infinity (the ionization energy) corresponds to the difference between zero and the energy of the level in question. The energies of the 1s, 2s, and 2p levels in sodium lie below −1300 kJ mol$^{-1}$ and are not shown.

The presence of more than one electron in an atom has a major effect on the energy levels of the atom. Figure 6.3 shows a comparison between the energy levels of hydrogen and sodium, as derived from atomic spectra. In contrast to the energy levels of hydrogen, which depend only on the value of the principal quantum number $n$, the energies of the levels of sodium depend on both $n$ and $l$. As shown in the figure, the 3s level has a lower energy than the 3p level, which in turn has a lower energy than the 3d level, and so forth. While atoms of different elements exhibit varied patterns of energy levels, it is always the case that the energies depend on the values of both $n$ and $l$ such that $E_s < E_p < E_d < E_f$ for a given value of $n$. Only in the case of hydrogen are levels with different $l$ and the same $n$ virtually degenerate in energy.

To what factors can this difference between hydrogen and the other elements be attributed? The answer lies in the different penetration of the various orbitals, already mentioned in connection with Figure 5.28. Recall that, for a given value of $n$, an s orbital has a higher probability of being close to the nu-

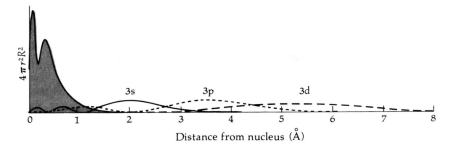

**FIGURE 6.4** The radial distribution function of the sodium atom in its ground and excited states. The individual curves show the distribution of the valence electron in the 3s orbital (ground state) and in the 3p and 3d orbitals (excited states). The shaded region represents the distribution of the ten core electrons.

cleus than a p orbital, which in turn has a higher probability of being close to the nucleus than a d orbital. While this difference in penetration does not affect the energy levels of the hydrogen atom, it does affect the levels of atoms of all other elements in the manner indicated earlier. To see why this is so, let us examine Figure 6.4, which shows the radial probability function for the ground state and excited states of the single valence electron of a sodium atom and that of the remaining ten electrons of this atom. The 10 nonvalence electrons form an inner core which screens the valence electron from the nucleus. The extent of such screening depends on the orbital occupied by the valence electron. When the electron is in the 3d orbital, corresponding to an excited state of sodium, the probability that it is outside the inner electron core is close to unity. Under these conditions, the 10 inner electrons effectively screen the valence electron from the nucleus. Since the nuclear charge of sodium is 11, the valence electron feels an effective nuclear charge of approximately $11 - 10 = 1$, comparable to that of hydrogen. Consequently, it is not surprising that the energy of the 3d level in sodium is very close to that of the $n = 3$ level in hydrogen (Figure 6.3).

Contrast this behavior with that of an electron in the 3s orbital, corresponding to the ground state of the sodium atom. Because of the subsidiary maxima in the radial probability distribution, a 3s electron has an appreciable probability of penetrating the inner electron core. Under these conditions, the core electrons are less efficient in screening the 3s electron from the nucleus, and the electron feels a larger effective nuclear charge. A 3s electron is therefore held more tightly than a 3d electron, and the energy of the 3s level is lower than that of the 3d level. The 3p orbital is seen to represent an intermediate situation.

The effect illustrated in Figure 6.4 occurs for every element except hydrogen and shows that the splitting of the energy levels with the same value of $n$ can be attributed to the combined effects of *penetration and screening*. Stated in another way, the level splitting arises from the different **effective nuclear charge** felt by electrons in different orbitals. It should be kept in mind, however, that while this explanation makes the splitting of the orbital energies understandable, it is only a highly approximate description of what is actually the result of the very complicated interactions of all the electrons in the atom with each other and with the nucleus.

***b. Factors Determining Orbital Filling***  The key question that has to be answered in considering the electronic structure of the elements is the order in which the atomic orbitals are filled with electrons. Since the chemical properties of the elements are determined by the electrons in the highest energy levels, it is important to identify the orbitals associated with these levels. In order to do so, the filling order of the orbitals must be known. Three principles guide us in this respect.

The first principle is known as the **Pauli exclusion principle.** It asserts that each electron in an atom has a unique set of values of the four quantum numbers $n$, $l$, $m_l$, and $m_s$. Furthermore, a given orbital can contain a maximum of two electrons, which must differ in the value of the spin quantum number, (i.e., $m_s = \frac{1}{2}$ or $m_s = -\frac{1}{2}$). The Pauli principle can be understood as the statement that each electron in an atom has its own quantum mechanical "space," which cannot be intruded upon by any other electron. The exclusion principle was proposed by Wolfgang Pauli (1900–1958) to account for some features of atomic spectra. It is the key to understanding electronic structure and the occurrence of chemical periodicity.

The role of the Pauli principle in explaining the electronic structure of atoms is illustrated in Figure 6.5, which shows the orbitals with $n = 2$. As already noted, the designations and numbers of orbitals of each type are identical to those for the hydrogen atom (Table 5.5). Thus, the available orbitals with $n = 2$ consist of one 2s orbital and three 2p orbitals. According to the Pauli principle each orbital can accommodate two electrons, and the $n = 2$ orbitals can therefore contain up to eight electrons as shown. Two electrons occupying the same orbital are said to be **paired**, or to constitute an **electron pair**, while a single electron occupying an orbital is said to be **unpaired.**

The second principle that explains the order in which the orbitals are filled is known as **Hund's rule** and states that when two electrons are placed in orbitals of equal energy, the atom has a lower energy and is therefore more sta-

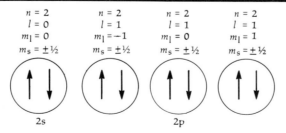

**FIGURE 6.5**  The role of the Pauli exclusion principle in determining the filling of the $n = 2$ orbitals. Each orbital is represented by a circle and an electron in an orbital is represented by an arrow. The opposite direction of the two arrows in a given circle corresponds to the fact that two electrons in the same orbital have opposite spins, i.e., $m_s = \frac{1}{2}$ and $-\frac{1}{2}$. The quantum numbers of the electrons are tabulated and it can be seen that no two electrons have identical quantum numbers.

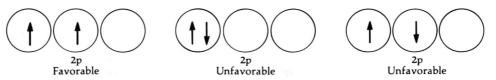

**FIGURE 6.6** Possible ways of placing two electrons in the three degenerate 2p orbitals. According to Hund's rule, the configuration shown at the left leads to a lower energy than those at the right and is, therefore, favored.

ble if the electrons have the same value of $m_s$. Since orbitals of a given energy have the same value of $n$ and the same value of $l$, the requirement that the electrons also have the same value of $m_s$ means that they must differ in $m_l$, otherwise the Pauli exclusion principle would be violated. As illustrated in Figure 6.6, the combined effect of the Pauli principle and Hund's rule is that before one of a set of degenerate orbitals is filled with a pair of electrons, all of them must acquire one electron apiece with a given $m_s$. Hund's rule provides a prescription by means of which the repulsion between two electrons is minimized, thereby resulting in a state of lower energy: when two electrons have parallel spin (i.e., same value of $m_s$), they must be placed in different degenerate orbitals. Owing to the different orientation of these orbitals (e.g., see Figure 5.30 showing the orientation of the $p_x$, $p_y$, and $p_z$ orbitals), the electrons occupy different regions of space and their repulsion is minimized.

The third principle which applies to the sequence in which the orbitals are filled is that the filling order must minimize the energy of the atom. Figure 6.7 shows the relative orbital energies for neutral atoms. These energies are

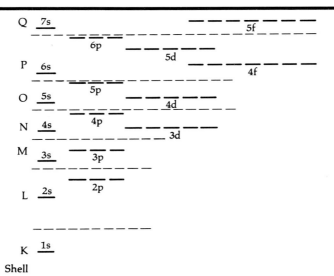

**FIGURE 6.7** Orbital energy diagram for neutral atoms. The energy scale is not linear; the difference in the energy of the lower orbitals (e.g., 1s and 2s) relative to that between the higher orbitals is much larger than indicated. The orbitals lying within adjacent dashed lines have comparable energy and belong to the same shell.

generally applicable to atoms for which the orbital in question is a valence orbital, and they vary slowly with atomic number. As can be seen in the figure, certain groups of orbitals have nearly the same energy and are separated from other orbitals by large energy gaps. These orbitals are said to belong to the same **shell**. For example, the 1s orbital belongs to the K shell, the 2s and 2p orbitals belong to the L shell, the 3s and 3p orbitals belong to the M shell, the 4s, 3d, and 4p orbitals belong to the N shell, and so on. Note that a given shell starts with an s orbital and ends with a p orbital with the same value of $n$ (the K shell is anomalous because it contains only the 1s orbital). Orbitals with the same values of $n$ and $l$ are said to form a **subshell**. For example, the 3p orbitals form a subshell, as do the 3d or the 4f orbitals. It is customary to refer to the shell containing the valence orbitals of an atom as the **valence shell** or **outer shell** and to the lower energy shells as the **inner shells**. The valence shell is of particular importance because the chemical behavior of the elements is, in essence, determined by the number of electrons in this shell. As shown in Figure 6.4, a high-energy (or "outer") orbital is, on average, farther from the nucleus than a low-energy (or "inner") orbital.

## 6.3 Electronic Configuration of the Elements

By applying the principles described in the preceding section, the electronic configurations of the elements can be built up one by one, a process known as **aufbau** (German for build-up). The procedure amounts to choosing the orbital placement of each successive electron based on the requirements that each electron has a unique set of quantum numbers, that the energy of the atom is minimized by first filling equivalent orbitals with unpaired electrons, and by filling lower energy orbitals before higher energy orbitals. Because of the splitting in the energy of orbitals with a given $n$ but different $l$, which is clearly in evidence in Figure 6.7, the pattern of orbital energies becomes rather complex for multielectron atoms. There is no problem for the first few orbitals, which are well separated in energy. The orbital energies increase in the order 1s, 2s, 2p, 3s, and 3p, and this is the order in which the orbitals are filled. The first complication arises with the filling of the 3d and 4s orbitals. Note that these orbitals have nearly the same energy. Under these circumstances, each exact orbital energy depends on the number of electrons with which an electron in a particular orbital can interact, and consequently varies with the atomic number.

If this particular complication is, for the moment, ignored, the electronic configurations of the elements are obtained by feeding up to two electrons into each orbital in order of increasing orbital energies. Degenerate orbitals are filled in a manner consistent with Hund's rule. Figure 6.8 shows a useful mnemonic diagram for the order in which the orbitals are filled based on the orbital energies shown in Figure 6.7

We start with the first element, hydrogen. The ground state of the hydrogen atom consists of one electron occupying the 1s orbital; the electronic configuration of the atom is said to be $1s^1$. The next element, in order of increasing atomic number, is helium; an atom of this element has two electrons. The atom attains its lowest energy when both electrons are placed in the 1s orbital, giving helium the electronic configuration of $1s^2$. Since the K shell contains only the 1s orbital, the shell is filled at this point, and helium has a **closed-shell** electronic configuration. A closed shell confers a high degree of

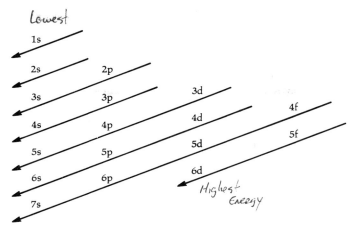

**FIGURE 6.8** Orbital filling sequence. Atomic orbitals are generally filled in the order indicated by the arrows, starting at the top. Deviations from this sequence are discussed in the text.

chemical inertness to the elements whose atoms possess this configuration; the elements with closed-shell configurations are known as the **noble gases**.

The next element, lithium, has the configuration

$$_3\text{Li} \quad 1s^2 2s^1$$

or, using a notation that becomes increasingly convenient for its brevity as we proceed to heavier elements,

$$_3\text{Li} \quad [\text{He}]2s^1$$

In this shorthand notation the chemical symbol of the noble gas in which a particular shell is filled replaces the string of characters designating its configuration. The 2s orbital is filled at beryllium,

$$_4\text{Be} \quad [\text{He}]2s^2$$

and the next six electrons fill the three 2p orbitals:

$$_5\text{B} \quad [\text{He}]2s^2 2p^1$$

C, N, O, F ↓

$$_{10}\text{Ne} \quad [\text{He}]2s^2 2p^6$$

As may be seen in Figure 6.7, the filling of the 2p orbitals completes the L shell. The eight elements from lithium through neon correspond to the filling of the L shell, with neon having a closed shell configuration.

**Example 6.1** How many unpaired electrons are there in an oxygen atom?

The number of unpaired electrons can be determined by the application of Hund's rule. The electronic configuration of oxygen is $[\text{He}]2s^2 2p^4$. Each of the first three p electrons occupies a different orbital and all three electrons have parallel spin (same value of $m_s$). The fourth electron must fill one of the orbitals, thereby leaving two unpaired electrons. The four electrons are placed in the equivalent p orbitals as follows:

⇅ ↑ ↑ ■

The filling of the M shell, which according to Figure 6.7 contains the 3s and 3p orbitals, is completely analogous to that of the L shell. The electronic configuration of the eight elements in which this shell is filling ranges from

$_{11}$Na    [Ne]3s$^1$

for sodium to

$_{18}$Ar    [Ne]3s$^2$3p$^6$

for argon, which has a closed shell configuration and completes the filling of the M shell. The electronic configuration of all the elements is given in Table 6.3.

---

**Example 6.2** (a) What is the configuration of the first excited electronic state of magnesium? (b) What is the electronic configuration of the Mg$^+$ ion?

(a) The electronic configuration of magnesium in its ground state is [Ne]3s$^2$. While the orbital energy level diagram in Figure 6.7 is only approximately valid, it is sufficiently accurate to permit us to conclude that the 3s orbital is followed in energy by the 3p orbitals. The first excited state is consequently formed by promoting a valence electron from the 3s to the 3p orbital. The resulting electronic configuration is [Ne]3s$^1$3p$^1$.

(b) In forming the ion, the most loosely bound electron, instead of being promoted to a higher energy orbital, as in the formation of an excited state, is actually removed from the atom. Thus, the electronic configuration of Mg$^+$ is [Ne]3s$^1$. ■

The electronic configuration of the elements corresponding to the filling of the first three shells (K, L, M) follows a systematic sequence. We now come to the N shell where, in contrast, the orbital filling sequence is more complex. As already noted, this complexity arises when two orbitals are close in energy causing the relative energies of these orbitals to depend on the total number of electrons present. A more complete version of Figure 6.7, one in which the relative orbital energies are plotted as a function of the total number of electrons present (i.e., as a function of atomic number) shows how the energies of such orbitals vary. Such a plot is shown in Figure 6.9.

We see that for low atomic numbers the 3d orbital lies well below the 4s orbital. Here, the orbital energy is primarily determined by the value of the principal quantum number. Up to atomic number 20, the 3d-orbital energy remains nearly constant with increasing atomic number. In contrast, the 4s-orbital energy and, to a lesser extent, the 4p-orbital energy decrease. This difference in behavior reflects the different penetration of the orbitals. As already noted in Figure 6.4, the 3d orbitals show little penetration of the inner core and their energy is essentially that expected for an atom with an effective nuclear charge close to +1. On the other hand, the 4s orbitals penetrate the inner core and, because they feel an increasing effective nuclear charge with increasing atomic number, their energy decreases. When the filling of the N shell com-

## TABLE 6.3 Electronic Configuration of the Elements

| Shell | Atomic number | Element | Electronic configuration | Shell | Atomic number | Element | Electronic configuration |
|---|---|---|---|---|---|---|---|
| K | 1 | H | $1s^1$ | P | 55 | Cs | $[Xe]6s^1$ |
|   | 2 | He | $1s^2$ |   | 56 | Ba | $[Xe]6s^2$ |
| L | 3 | Li | $[He]2s^1$ |   | 57 | La | $[Xe]5d^16s^2$ |
|   | 4 | Be | $[He]2s^2$ |   | 58 | Ce | $[Xe]4f^26s^2$ |
|   | 5 | B | $[He]2s^22p^1$ |   | 59 | Pr | $[Xe]4f^36s^2$ |
|   | 6 | C | $[He]2s^22p^2$ |   | 60 | Nd | $[Xe]4f^46s^2$ |
|   | 7 | N | $[He]2s^22p^3$ |   | 61 | Pm | $[Xe]4f^56s^2$ |
|   | 8 | O | $[He]2s^22p^4$ |   | 62 | Sm | $[Xe]4f^66s^2$ |
|   | 9 | F | $[He]2s^22p^5$ |   | 63 | Eu | $[Xe]4f^76s^2$ |
|   | 10 | Ne | $[He]2s^22p^6$ |   | 64 | Gd | $[Xe]4f^75d^16s^2$ |
| M | 11 | Na | $[Ne]3s^1$ |   | 65 | Tb | $[Xe]4f^96s^2$ |
|   | 12 | Mg | $[Ne]3s^2$ |   | 66 | Dy | $[Xe]4f^{10}6s^2$ |
|   | 13 | Al | $[Ne]3s^23p^1$ |   | 67 | Ho | $[Xe]4f^{11}6s^2$ |
|   | 14 | Si | $[Ne]3s^23p^2$ |   | 68 | Er | $[Xe]4f^{12}6s^2$ |
|   | 15 | P | $[Ne]3s^23p^3$ |   | 69 | Tm | $[Xe]4f^{13}6s^2$ |
|   | 16 | S | $[Ne]3s^23p^4$ |   | 70 | Yb | $[Xe]4f^{14}6s^2$ |
|   | 17 | Cl | $[Ne]3s^23p^5$ |   | 71 | Lu | $[Xe]4f^{14}5d^16s^2$ |
|   | 18 | Ar | $[Ne]3s^23p^6$ |   | 72 | Hf | $[Xe]4f^{14}5d^26s^2$ |
| N | 19 | K | $[Ar]4s^1$ |   | 73 | Ta | $[Xe]4f^{14}5d^36s^2$ |
|   | 20 | Ca | $[Ar]4s^2$ |   | 74 | W | $[Xe]4f^{14}5d^46s^2$ |
|   | 21 | Sc | $[Ar]3d^14s^2$ |   | 75 | Re | $[Xe]4f^{14}5d^56s^2$ |
|   | 22 | Ti | $[Ar]3d^24s^2$ |   | 76 | Os | $[Xe]4f^{14}5d^66s^2$ |
|   | 23 | V | $[Ar]3d^34s^2$ |   | 77 | Ir | $[Xe]4f^{14}5d^76s^2$ |
|   | 24 | Cr | $[Ar]3d^54s^1$ |   | 78 | Pt | $[Xe]4f^{14}5d^96s^1$ |
|   | 25 | Mn | $[Ar]3d^54s^2$ |   | 79 | Au | $[Xe]4f^{14}5d^{10}6s^1$ |
|   | 26 | Fe | $[Ar]3d^64s^2$ |   | 80 | Hg | $[Xe]4f^{14}5d^{10}6s^2$ |
|   | 27 | Co | $[Ar]3d^74s^2$ |   | 81 | Tl | $[Xe]4f^{14}5d^{10}6s^26p^1$ |
|   | 28 | Ni | $[Ar]3d^84s^2$ |   | 82 | Pb | $[Xe]4f^{14}5d^{10}6s^26p^2$ |
|   | 29 | Cu | $[Ar]3d^{10}4s^1$ |   | 83 | Bi | $[Xe]4f^{14}5d^{10}6s^26p^3$ |
|   | 30 | Zn | $[Ar]3d^{10}4s^2$ |   | 84 | Po | $[Xe]4f^{14}5d^{10}6s^26p^4$ |
|   | 31 | Ga | $[Ar]3d^{10}4s^24p^1$ |   | 85 | At | $[Xe]4f^{14}5d^{10}6s^26p^5$ |
|   | 32 | Ge | $[Ar]3d^{10}4s^24p^2$ |   | 86 | Rn | $[Xe]4f^{14}5d^{10}6s^26p^6$ |
|   | 33 | As | $[Ar]3d^{10}4s^24p^3$ | Q | 87 | Fr | $[Rn]7s^1$ |
|   | 34 | Se | $[Ar]3d^{10}4s^24p^4$ |   | 88 | Ra | $[Rn]7s^2$ |
|   | 35 | Br | $[Ar]3d^{10}4s^24p^5$ |   | 89 | Ac | $[Rn]6d^17s^2$ |
|   | 36 | Kr | $[Ar]3d^{10}4s^24p^6$ |   | 90 | Th | $[Rn]6d^27s^2$ |
| O | 37 | Rb | $[Kr]5s^1$ |   | 91 | Pa | $[Rn]5f^26d^17s^2$ |
|   | 38 | Sr | $[Kr]5s^2$ |   | 92 | U | $[Rn]5f^36d^17s^2$ |
|   | 39 | Y | $[Kr]4d^15s^2$ |   | 93 | Np | $[Rn]5f^46d^17s^2$ |
|   | 40 | Zr | $[Kr]4d^25s^2$ |   | 94 | Pu | $[Rn]5f^67s^2$ |
|   | 41 | Nb | $[Kr]4d^45s^1$ |   | 95 | Am | $[Rn]5f^77s^2$ |
|   | 42 | Mo | $[Kr]4d^55s^1$ |   | 96 | Cm | $[Rn]5f^76d^17s^2$ |
|   | 43 | Tc | $[Ar]4d^55s^2$ |   | 97 | Bk | $[Rn]5f^97s^2$ |
|   | 44 | Ru | $[Kr]4d^75s^1$ |   | 98 | Cf | $[Rn]5f^{10}7s^2$ |
|   | 45 | Rh | $[Kr]4d^85s^1$ |   | 99 | Es | $[Rn]5f^{11}7s^2$ |
|   | 46 | Pd | $[Kr]4d^{10}$ |   | 100 | Fm | $[Rn]5f^{12}7s^2$ |
|   | 47 | Ag | $[Kr]4d^{10}5s^1$ |   | 101 | Md | $[Rn]5f^{13}7s^2$ |
|   | 48 | Cd | $[Kr]4d^{10}5s^2$ |   | 102 | No | $[Rn]5f^{14}7s^2$ |
|   | 49 | In | $[Kr]4d^{10}5s^25p^1$ |   | 103 | Lr | $[Rn]5f^{14}6d^17s^2$ |
|   | 50 | Sn | $[Kr]4d^{10}5s^25p^2$ |   | 104 | — | $[Rn]5f^{14}6d^27s^2$ |
|   | 51 | Sb | $[Kr]4d^{10}5s^25p^3$ |   | 105 | — | $[Rn]5f^{14}6d^37s^2$ |
|   | 52 | Te | $[Kr]4d^{10}5s^25p^4$ |   | 106 | — | $[Rn]5f^{14}6d^47s^2$ |
|   | 53 | I | $[Kr]4d^{10}5s^25p^5$ |   | 107 | — | $[Rn]5f^{14}6d^57s^2$ |
|   | 54 | Xe | $[Kr]4d^{10}5s^25p^6$ |   | 108 | — | $[Rn]5f^{14}6d^67s^2$ |
|   |    |    |                       |   | 109 | — | $[Rn]5f^{14}6d^77s^2$ |

**188** Multielectron Atoms and the Periodic Table 6.3

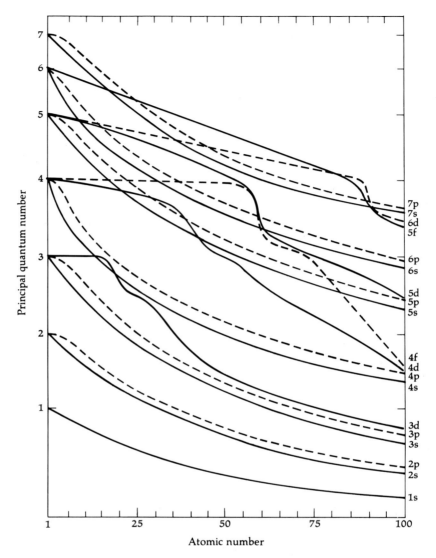

**FIGURE 6.9** Orbital energies in neutral atoms as a function of atomic number. As in Figure 6.7, the energies are not shown to scale.

---

mences at $Z = 19$, the atom has a lower energy when the 19th and 20th electrons occupy the 4s rather than the 3d orbital. As a result, the electronic configurations of potassium and calcium are

$$_{19}K \quad [Ar]4s^1$$

and

$$_{20}Ca \quad [Ar]4s^2$$

As electrons are added to the 4s orbital, the nuclear charge increases by two units. Since the 4s electrons are not effective in screening the 3d orbital from the nucleus because of their higher value of $n$, the 3d orbital suddenly feels an increased effective nuclear charge and its energy drops below that of the 4s orbital. Thus, in the next ten elements the five degenerate 3d orbitals are filled. The electronic configurations are given in Table 6.3 and range from

$_{21}$Sc    [Ar]3d$^1$4s$^2$

to

$_{30}$Zn    [Ar]3d$^{10}$4s$^2$

Slight anomalies in the electronic configuration of a few of these elements may be noted. For example, the configuration of chromium is [Ar]3d$^5$4s$^1$ instead of the expected [Ar]3d$^4$4s$^2$. Note that, according to the actual configuration, both 3d and 4s subshells are half filled. Evidently, half-filled shells confer extra stability to an atom. This effect can be understood in terms of Hund's rule: when the subshells are half-filled, the electrons are in different orbitals and their mutual repulsion is consequently minimized. A similar effect occurs at copper.

The ten elements from scandium to zinc constitute the first **transition series.** The complex variation in the energy of the various valence orbitals is responsible for the rich variety of transition element chemistry (Chapter 21). Note that in writing the electronic configuration of the transition elements (Table 6.3) the 3d electrons precede the 4s electrons in spite of the fact that the 4s orbital is filled before the 3d orbitals. This rather subtle switch in order is yet another manifestation of the fact that, when different valence orbitals have nearly the same energy, the total energy of the atom depends on the number of electrons present, as well as on the orbital energies. Experimental evidence for the electronic configurations of the transition elements given in Table 6.3 comes from the ions formed by these elements. For example, zinc readily forms a +2 ion, consistent with the electronic configuration

Zn$^{2+}$    [Ar]3d$^{10}$

Evidently, the 4s orbital is less stable than the 3d orbitals and electrons are removed from it more readily even though it is filled first.

After the 3d orbitals are filled, the next six electrons go into the 4p orbitals and the N shell closes at krypton, whose electronic configuration is

$_{36}$Kr    [Ar]3d$^{10}$4s$^2$4p$^6$

Note that in contrast to the preceding shells, the orbitals constituting the N shell can accommodate 18 electrons, corresponding to 18 different elements.

The filling of the O shell is completely analogous to that of the N shell. The electronic configuration of the first element, rubidium, is

$_{37}$Rb    [Kr]5s$^1$

and the shell is closed at xenon:

$_{54}$Xe    [Kr]4d$^{10}$5s$^2$5p$^6$

The filling of the P shell commences in the same manner as that of the O shell. The first two electrons fill the 6s orbital and the configurations of the corresponding elements, cesium and barium, are

$_{55}$Cs    [Xe]6s$^1$

$_{56}$Ba    [Xe]6s$^2$

Figure 6.9 shows that at this point the energies of both the 5d and 4f orbitals, which had been higher than that of the 6s orbital, suddenly drop. As in the case of the 3d orbitals, this drop occurs because the 6s electrons do not screen these orbitals with lower $n$ from the nucleus; the 5d and 4f orbitals consequently feel a greater effective nuclear charge. The configuration of lanthanum, the next element in the sequence, is

$$_{57}\text{La} \quad [\text{Xe}]5d^1 6s^2$$

and thereafter the seven 4f orbitals are filled, terminating at lutetium,

$$_{71}\text{Lu} \quad [\text{Xe}]4f^{14}5d^1 6s^2$$

Lanthanum plus the fourteen elements in which the 4f orbitals are filling are known as the **lanthanides.** Because the 4f orbitals drop below the valence shell as they are filled, the lanthanides have very similar chemical properties.

Once the 4f subshell has been filled, the filling of the 5d orbitals continues and the shell is finally closed by the filling of the 6p orbitals at the element radon, which has the electronic configuration

$$_{86}\text{Rn} \quad [\text{Xe}]4f^{14}5d^{10}6s^2 6p^6$$

Note that the filling of the P shell requires 32 electrons, the most of any of the shells considered so far.

The final shell to be filled is the Q shell and the order in which the orbitals are filled is essentially identical to that of the P shell. The elements associated with the filling of this shell are either naturally occurring unstable elements or artificially created unstable elements (Section 23.7). In the heaviest elements known to date the 6d orbitals are occupied; no elements with valence electrons in the 7p orbitals are as yet known.

**Example 6.3** The highly unstable element with $Z = 109$ was first made in 1982. What is its electronic configuration?

Element 109 has $109 - 86 = 23$ electrons in the Q shell. Using the configuration of the P shell elements as a guide, we can write the electronic configuration as

$$109 \quad [\text{Rn}]5f^{14}6d^7 7s^2 \quad \blacksquare$$

## 6.4 The Periodic Table and the Electronic Structure of the Elements

The periodic table arranges the elements in order of increasing atomic number in a series of horizontal rows, or **periods.** The arrangement reflects the order in which the various shells and subshells are filled. The periodic table in Figure 6.10 emphasizes the arrangement of the elements into distinct blocks associated with the filling of different subshells. In its modern form, the periodic table consists of seven periods each of which corresponds to the filling of one of the shells. For example, the first period, which is comprised of hydrogen and helium, corresponds to the filling of the K shell, the second period corresponds to the filling of the L shell, and so on. The number of elements in the various periods, 2, 8, 8, 18, etc., is determined by the number of orbitals belonging to the corresponding shells. Each period terminates with a noble gas element, whose atoms have a closed valence shell. The first three periods contain at most eight elements and are known as **short periods.** The remaining periods contain 18 or more elements and are known as **long periods.**

## Multielectron Atoms and the Periodic Table 6.4

|  | I A |  |  |  |  |  |  |  |  |  |  |  |  |  |  |  |  | VIII A, 0 |
|---|---|---|---|---|---|---|---|---|---|---|---|---|---|---|---|---|---|---|
| 1s | 1 H | II A |  |  |  |  |  |  |  |  |  |  | III A | IV A | V A | VI A | VII A | 2 He |
| 2s | 3 Li | 4 Be |  |  |  |  |  |  |  |  |  | 2p | 5 B | 6 C | 7 N | 8 O | 9 F | 10 Ne |
| 3s | 11 Na | 12 Mg |  | III B | IV B | V B | VI B | VII B | VIII |  | I B | II B | 3p | 13 Al | 14 Si | 15 P | 16 S | 17 Cl | 18 Ar |
| 4s | 19 K | 20 Ca | 3d | 21 Sc | 22 Ti | 23 V | 24 Cr | 25 Mn | 26 Fe | 27 Co | 28 Ni | 29 Cu | 30 Zn | 4p | 31 Ga | 32 Ge | 33 As | 34 Se | 35 Br | 36 Kr |
| 5s | 37 Rb | 38 Sr | 4d | 39 Y | 40 Zr | 41 Nb | 42 Mo | 43 Tc | 44 Ru | 45 Rh | 46 Pd | 47 Ag | 48 Cd | 5p | 49 In | 50 Sn | 51 Sb | 52 Te | 53 I | 54 Xe |
| 6s | 55 Cs | 56 Ba | 5d | 57- La | 72 Hf | 73 Ta | 74 W | 75 Re | 76 Os | 77 Ir | 78 Pt | 79 Au | 80 Hg | 6p | 81 Tl | 82 Pb | 83 Bi | 84 Po | 85 At | 86 Rn |
| 7s | 87 Fr | 88 Ra | 6d | 89- Ac | 104 | 105 | 106 | 107 | 108 | 109 |  |  |  | 7p |  |  |  |  |  |  |

| 4f | 58 Ce | 59 Pr | 60 Nd | 61 Pm | 62 Sm | 63 Eu | 64 Gd | 65 Tb | 66 Dy | 67 Ho | 68 Er | 69 Tm | 70 Yb | 71 Lu |
|---|---|---|---|---|---|---|---|---|---|---|---|---|---|---|
| 5f | 90 Th | 91 Pa | 92 U | 93 Np | 94 Pu | 95 Am | 96 Cm | 97 Bk | 98 Cf | 99 Es | 100 Fm | 101 Md | 102 No | 103 Lr |

**FIGURE 6.10** The periodic table emphasizing the separation into blocks associated with the filling of orbitals having different values of the azimuthal quantum number l.

The vertical division of the periodic table places elements of similar electronic configuration beneath one another in separate columns, or **groups.** For example, all the elements in the first group have electronic configuration $ns^1$ and differ only in the value of the principal quantum number $n$. The two groups arising from the filling of the s orbitals are placed on the left side of the table while the six groups associated with the filling of the p orbitals are placed on the right. These eight groups are referred to as the **representative elements.** Groups I and II are separated from the other six groups by the **transition elements,** which are formed by the filling of the d orbitals. The first transition series appears in the fourth period and corresponds to the filling of the 3d orbitals. There are four transition series, the last of which is incomplete owing to the nuclear instability of the heaviest elements. The third and fourth transition series are interrupted by the occurrence of the **inner transition series,** comprised of the **lanthanides** and **actinides,** elements formed by the filling of the f orbitals.

The various groups in the periodic table are labelled by a combination of Roman numerals and letters. The Roman numerals generally correspond to the number of valence electrons. For example, gallium is in group IIIA because its valence shell configuration is $4s^2 4p^1$. Similarly, scandium is in group IIIB because its valence shell configuration is $3d^1 4s^2$. As already noted, the $(n-1)d$ subshell is part of the valence shell of the transition elements but drops to lower energies for the subsequent representative elements in the same period.

At present, there is a distinct lack of worldwide agreement on the letter designations of the various groups. The labels used in this textbook reflect the common American practice: the representative element groups are labelled IA–VIIIA, from left to right (the noble gases, group VIIIA, are also labelled 0);

the transition element groups are labelled IIIB–VIIB, VIII, IB, and IIB, from left to right (see Figure 6.10). Representative and transition elements with the same number of valence electrons have the same Roman numeral but different letter designations. These designations serve to emphasize that there are both similarities and differences between representative and transition elements labelled with the same Roman numeral.

The International Union of Pure and Applied Chemistry (IUPAC) has recommended somewhat different designations: the representative element groups are designated IA, IIA, IIIB–VIIIB, while the transition element groups are designated IIIA–VIIIA, IB, and IIB. A scheme in which the A, B designations are omitted and the groups numbered sequentially from 1 to 18 is presently under consideration by the IUPAC.

Some of the representative element groups are universally known by their common names. Thus, the group IA elements (excepting hydrogen) are the **alkali metals,** the group IIA elements are the **alkaline earths,** the group VIA elements are the **chalcogens,** the group VIIA elements are the **halogens,** and the group VIIIA elements are the **noble gases.**

As already mentioned, the elements within each group resemble each other chemically, although gradual changes do occur. The representative elements display the broadest range of "horizontal" variations in chemical behavior from Group I to Group VIII. The combined effect of "vertical" and "horizontal" variations among these elements gives rise to characteristic trends and regularities in properties. In contrast to the representative elements, the transition elements resemble each other "horizontally" as well as "vertically." As already noted, the horizontal similarities reflect the decrease in the energy of the d and f orbitals being filled below that of the just-filled s orbitals.

# Periodic Properties

Among the many properties of the elements and their compounds, the greatest regularities are exhibited by those properties that depend on the behavior of isolated atoms of an element. In these circumstances, the regularities in properties can be directly attributed to regularities in electronic configuration. The ionization energy of the gaseous monatomic elements is a good example of such a property because the gaseous atoms act as essentially free and independent particles (Chapter 4). On the other hand, properties of bulk matter, such as density and melting point, depend on the interaction between an atom and its neighbors as well as on electronic configuration. Not surprisingly, these properties vary in a more irregular manner with atomic number. We shall begin by examining several properties that reflect the behavior of isolated atoms and then proceed to consider some of the bulk properties.

## 6.5 Ionization Energies

***a. The First Ionization Energy***  The first ionization energy $I_1$ is the minimum energy required to remove an electron from a gaseous atom in its ground state to form a gaseous ion, that is, the energy required for the process

$$M(g) \rightarrow M^+(g) + e^-$$

## Multielectron Atoms and the Periodic Table  6.5  193

**TABLE 6.4  Ionization Energies and Electron Affinities of the First 36 Elements**

| Element Z | Symbol | $I_1$ (kJ mol$^{-1}$) | $I_2$ (kJ mol$^{-1}$) | $I_3$ (kJ mol$^{-1}$) | $\varepsilon$ (kJ mol$^{-1}$) |
|---|---|---|---|---|---|
| 1  | H  | 1312  | —     | —                   | 72.8   |
| 2  | He | 2372  | 5250  | —                   | <0[a]  |
| 3  | Li | 520.2 | 7298  | $1.181 \times 10^4$ | 59.7   |
| 4  | Be | 899.4 | 1757  | $1.484 \times 10^4$ | <0     |
| 5  | B  | 800.6 | 2427  | 3660                | 26.8   |
| 6  | C  | 1086  | 2353  | 4620                | 121.9  |
| 7  | N  | 1402  | 2856  | 4578                | <0     |
| 8  | O  | 1314  | 3388  | 5300                | 141.1  |
| 9  | F  | 1681  | 3374  | 6050                | 328.0  |
| 10 | Ne | 2081  | 3952  | 6122                | <0     |
| 11 | Na | 495.6 | 4562  | 6912                | 52.9   |
| 12 | Mg | 737.7 | 1451  | 7733                | <0     |
| 13 | Al | 577.6 | 1817  | 2745                | 42.7   |
| 14 | Si | 786.4 | 1577  | 3232                | 133.6  |
| 15 | P  | 1012  | 1903  | 2912                | 72.0   |
| 16 | S  | 999.6 | 2251  | 3361                | 200.4  |
| 17 | Cl | 1251  | 2297  | 3822                | 348.8  |
| 18 | Ar | 1520  | 2666  | 3931                | <0     |
| 19 | K  | 418.8 | 3051  | 4411                | 48.4   |
| 20 | Ca | 589.8 | 1145  | 4912                | <0     |
| 21 | Sc | 631   | 1235  | 2389                | 18.2   |
| 22 | Ti | 658   | 1310  | 2653                | 7.7    |
| 23 | V  | 650   | 1414  | 2828                | 50.8   |
| 24 | Cr | 652.8 | 1496  | 2987                | 64.4   |
| 25 | Mn | 717.4 | 1509  | 3248                | <0     |
| 26 | Fe | 759.3 | 1561  | 2957                | 15.8   |
| 27 | Co | 758   | 1646  | 3232                | 63.9   |
| 28 | Ni | 736.7 | 1753  | 3393                | 111.6  |
| 29 | Cu | 745.4 | 1958  | 3554                | 118.5  |
| 30 | Zn | 906.4 | 1733  | 3833                | <0     |
| 31 | Ga | 578.8 | 1979  | 2963                | 30     |
| 32 | Ge | 762.1 | 1537  | 3302                | 116    |
| 33 | As | 947   | 1798  | 2736                | 78     |
| 34 | Se | 940.9 | 2045  | 2974                | 195.0  |
| 35 | Br | 1140  | 2103  | $3.5 \times 10^3$   | 324.6  |
| 36 | Kr | 1351  | 2350  | 3565                | <0     |

[a] A value $\varepsilon < 0$ indicates that the anion is unstable.

Since the electron that requires the least energy for its removal from an atom is the most loosely bound electron (i.e., the electron occupying the highest energy orbital), the first ionization energy is a measure of the relative stability of the atom with respect to ionization. The first ionization energies of the elements are tabulated in Table 6.4 and plotted as a function of atomic number in Figure 6.11.

The ionization energies (the designation "first" is usually omitted for brevity) display a striking periodicity that is closely correlated with the location of the elements in the periodic table. Starting with a given alkali element, the ionization energies generally increase across each period until the noble gas

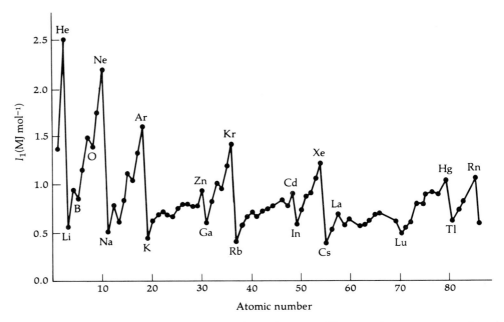

**FIGURE 6.11** First ionization energy $I_1$ of the elements. The energies are expressed in megajoules per mole (1 MJ = $10^6$ J).

which terminates the period is reached. At this point the ionization energy drops sharply and the trend is repeated in the following period. For example, the ionization energies of the elements in the second period increase from 520 kJ mol$^{-1}$ for lithium to 2081 kJ mol$^{-1}$ for neon. The value then plunges to 496 kJ mol$^{-1}$ for the next element, sodium, which begins the third period. Superimposed on this main trend are a number of irregularities, such as the decrease in ionization energy which occurs between beryllium and boron, and again between nitrogen and oxygen.

The ionization energy is a property of the isolated atom and consequently should be explainable by the same factors which determine electronic structure (Section 6.2). It is convenient to begin by recalling that the energy of an electron in a hydrogen-like atom (i.e., an atom with only one electron) is given by Equation 5.48. Since the zero of the energy scale corresponds to the ionized atom, the ionization energy has the same magnitude but opposite sign as the electron energy (see Section 5.5b). Owing to the effects of electron–electron repulsion, Equation 5.48 is not applicable to atoms with more than one electron. However, these effects can be approximated by replacing the actual nuclear charge by the effective nuclear charge, $Z_{\text{eff}}$, which represents by one single parameter the complex effects of electron–electron repulsion discussed in Section 6.2. Consequently we can use the expression

$$I_1 = hc\mathcal{R}(Z_{\text{eff}}^2/n^2) = (1312 \text{ kJ mol}^{-1})(Z_{\text{eff}}^2/n^2) \tag{6.2}$$

to represent the behavior of the ionization energies, where the numerical factor has been obtained by means of Equation 5.23 and the conversion from eV to kJ mol$^{-1}$ (Equation 5.16).

We begin by comparing the ionization energy of helium (2372 kJ mol$^{-1}$) with that of hydrogen (1312 kJ mol$^{-1}$). The increase in ionization energy is expected because of the greater attraction exerted by the doubly charged nucleus on the electrons. However, the repulsion between the two electrons has the opposite effect and tends to reduce the ionization energy. According to Equation 6.2, the ionization energy of helium is consistent with an effective nuclear charge of 1.34.

The large decrease in ionization energy observed between helium and lithium reflects the increase in the principal quantum number from 1 to 2. In the absence of electron repulsion, the increase in $n$ should be counteracted by that in $Z$. However, the 1s electrons are particularly effective in screening the 2s electron from the nucleus because of their high probability of being close to the nucleus (Figure 6.4). The ionization energy of lithium is consistent with an effective nuclear charge of only 1.26, close to the value of $3 - 2 = 1$ expected for perfect screening.

Focusing, now, on the trends in the second period, we note that the ionization energy of beryllium (899 kJ mol$^{-1}$) is greater than that of lithium, reflecting the increase in nuclear charge. It may be noted in Figure 6.11 that with the exception of two irregularities, the ionization energies increase throughout the period, indicating that the effect of the increasing nuclear charge is stronger than that of the increasing electron repulsion. The values of the effective nuclear charge, displayed in Figure 6.12, increase between lithium and neon in accordance with the experimental ionization energies.

The two irregularities observed in the second period are interesting because they lend experimental support to some of the results of the quantum mechanical analysis discussed in Section 6.2. First, consider the decrease in

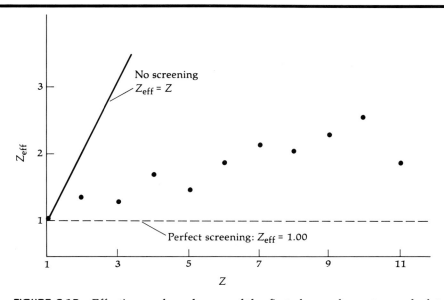

**FIGURE 6.12** Effective nuclear charges of the first eleven elements as calculated by means of Equation 6.2 and experimental ionization energies. The values of $Z_{eff}$ are closer to the values expected for perfect screening ($Z_{eff} = 1$) than to those for no screening ($Z_{eff} = Z$), thereby showing the importance of electron–electron repulsion.

ionization energy between beryllium and boron. The ionization process leads to the following changes in configurations:

Be → Be⁺   [He]2s² → [He]2s¹
B → B⁺    [He]2s²2p¹ → [He]2s²

Evidently, it requires less energy to remove a 2p electron than a 2s electron in spite of the increase in nuclear charge. This result is consistent with the shape of the radial probability functions: the p orbital is less penetrating than the s orbital and is therefore screened more effectively (Figure 6.4).

The second discontinuity occurs between nitrogen and oxygen, where once again the ionization energy decreases. The changes in configuration upon ionization for these elements are as follows:

N → N⁺   [He]2s²2p³ → [He]2s²2p²
O → O⁺   [He]2s²2p⁴ → [He]2s²2p³

As discussed earlier, Hund's rule indicates that the 2p orbitals of these elements are filled in the following manner:

N   ↑ ↑ ↑
O   ↑↓ ↑ ↑

The ionization of oxygen removes the first paired electron from the atom. The two electrons in this orbital are in close proximity and the repulsion between them is stronger than that between electrons in different p orbitals which, as already noted, tend to occupy different regions of space (Figure 5.31). Thus, the paired electron in oxygen is easier to remove than any of the electrons in nitrogen.

All of these features are repeated in the third period. As might have been expected, the behavior of the ionization energies in the fourth and succeeding periods is different because of the intervention of the transition series. The effective nuclear charge increases only slightly across these series because of the effectiveness of the $(n-1)d$ and $(n-2)f$ electrons in screening the $ns$ electrons from the nucleus. The increase in ionization energy across a particular transition series is correspondingly small.

In addition to the regularities observed across the periods, certain trends may be noted within the groups. Some of these trends are displayed in Figure 6.13. For the representative elements, the ionization energies decrease in succeeding periods with increasing principal quantum number (Equation 6.2). This decrease can be associated with the increase in the average distance between the valence electrons and the nucleus and the consequent decrease in the Coulomb force. The corresponding increase in nuclear charge, which affects the ionization energy in the opposite way, is counteracted by a similar increase in screening.

The transition elements in a given group display a different, and generally less regular, behavior. Of particular interest is the increase in ionization energy observed between the corresponding elements in the fifth and sixth periods. The large increase in nuclear charge associated with the filling of the 4f orbitals causes the valence electrons of the elements in the sixth period to be attracted more strongly to the nucleus, and the ionization energies are therefore correspondingly greater.

The regularities in the ionization energies of the elements manifest them-

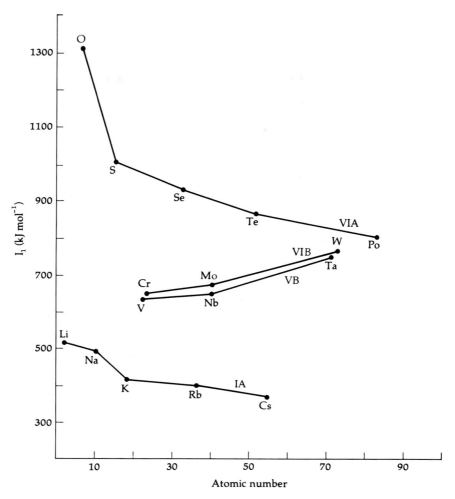

**FIGURE 6.13** The first ionization energies of elements in the same groups. Results are shown for Groups IA, VIA, VB, and VIB.

selves in corresponding regularities in chemical and physical properties, as we shall see. For example, the low ionization energies of the alkali elements are responsible for the strong tendency of these elements to react by forming singly charged cations. At the opposite extreme, the high ionization energies of the noble gases result in their low reactivity. However, these correlations cannot be pushed too far because the ionization energy is only one of a number of factors which determine the properties and reactions of matter in bulk.

**b. Higher Ionization Energies** A multielectron atom can lose more than one electron. The second ionization energy $I_2$ is the energy required for the process

$$M^+(g) \rightarrow M^{2+}(g) + e^-$$

Similarly, the third ionization energy $I_3$ is the energy that must be supplied for the process

$$M^{2+}(g) \rightarrow M^{3+}(g) + e^-$$

and so on. The second and third ionization energies of the elements are given in Table 6.4

We saw in Section 6.5a that the behavior of the first ionization energies reflects the combined effect of changes in nuclear charge, electron screening, and principal quantum number. A judicious examination of some of the higher ionization energies allows us to look at the isolated effect of these variables.

The effect of a change in nuclear charge can be seen most clearly in an examination of the ionization energies of **isoelectronic species**. These are atoms or ions with identical electronic configurations. As an illustration, consider the following processes:

$$_{11}Na \rightarrow {_{11}Na^+} + e^- \qquad I_1$$
$$_{12}Mg^+ \rightarrow {_{12}Mg^{2+}} + e^- \qquad I_2$$
$$_{13}Al^{2+} \rightarrow {_{13}Al^{3+}} + e^- \qquad I_3$$
$$_{14}Si^{3+} \rightarrow {_{14}Si^{4+}} + e^- \qquad I_4$$

All of these species are isoelectronic—they all have the [Ne]$3s^1$ configuration, and these processes can be represented by the following change in configuration:

$$[Ne]3s^1 \rightarrow [Ne]$$

Clearly, the only variable in the sequence is the nuclear charge. Its effect on the energy required to remove an electron is shown in Figure 6.14, where the ap-

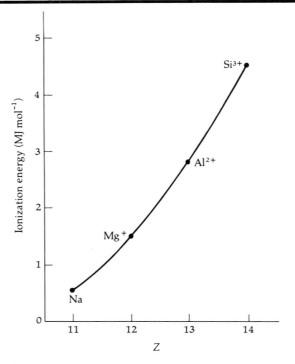

**FIGURE 6.14** Ionization energies of the isoelectronic species Na, Mg$^+$, Al$^{2+}$, and Si$^{3+}$. The data illustrate the effect of increasing nuclear charge.

propriate ionization energies are plotted versus atomic number. In the absence of competing factors, the effect of nuclear charge is clearly in evidence.

**Example 6.4** Use Equation 6.2 and the data in Table 6.4 to determine the effective nuclear charge felt by the 3s electron in Na, $Mg^+$, and $Al^{2+}$. Compare the differences between the effective nuclear charges with those between the actual nuclear charges of these species.

Equation 6.2 can be solved for the effective nuclear charge:

$$Z_{eff} = \frac{n(I \text{ kJ mol}^{-1})^{1/2}}{(1312 \text{ kJ mol}^{-1})^{1/2}}$$

The effective nuclear charge of $Al^{2+}$ is obtained by use of the third ionization energy of aluminum:

$$Z_{eff} = \frac{(3)(2745 \text{ kJ mol}^{-1})^{1/2}}{(1312 \text{ kJ mol}^{-1})^{1/2}} = 4.34$$

Similarly, the effective charge of $Mg^+$ is

$$Z_{eff} = \frac{(3)(1451 \text{ kJ mol}^{-1})^{1/2}}{(1312 \text{ kJ mol}^{-1})^{1/2}} = 3.15$$

while that of Na is $Z_{eff} = 1.84$. The differences are slightly larger than those between the actual nuclear charges of these elements. ∎

The effect of a change in principal quantum number can be seen most clearly by comparison of the energies required to remove successive valence shell electrons from an atom with that required to break the resulting closed shell. The successive ionization energies of sodium, magnesium, and aluminum are plotted in Figure 6.15. The common feature of all three plots is the sharp increase in energy required to remove an electron from the closed-shell [Ne] configuration. As a result, the removal of electrons beyond those in the valence shell does not occur in ordinary chemical reactions.

*c. X Rays and Photoelectron Spectroscopy* In 1895, while studying the effects of cathode rays (i.e., energetic electrons, see Section 5.1a) on matter, Wilhelm Roentgen (1845–1923) discovered the emission of a penetrating radiation, which he called X rays. As noted in Section 5.3, X rays are electromagnetic radiation of very short wavelength and are emitted as the result of electronic transitions of inner-shell electrons. A systematic investigation of the X rays emitted by different elements performed in 1913 by Henry Moseley (1887–1915) showed, for the first time, that the order of the elements in the periodic table is based on some fundamental property of the atom, which subsequently came to be known as the atomic number. Moseley found that the square root of the X-ray frequency was proportional to an integer that corresponded to the order of the elements in the periodic table. Moseley called this integer the atomic number and identified it with the positive charge of the nucleus. In this manner, the irregularities in the periodic table alluded to in Section 6.1, namely, the inversion between atomic weight and atomic number observed for certain pairs of elements, were accounted for. Atomic number, rather than atomic weight, was shown to be the characteristic property by which the elements are or-

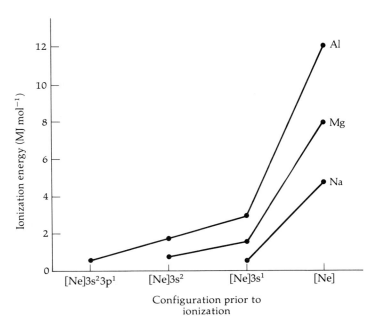

**FIGURE 6.15** Energy required to remove successive electrons from sodium, magnesium, and aluminum. The data illustrate the stability of a closed electron shell.

dered. Because the atomic number is an integer, unsuspected gaps in the periodic table could be readily determined. For example, Moseley observed a gap between neodymium and samarium, which was subsequently filled by the unstable element promethium (Z = 61). [More recently, the characteristic X rays emitted by newly created heavy elements (Section 23.7) have been used to confirm the atomic number assignments.]

Moseley's relation for the X-ray frequency can be written in the form

$$\nu = a(Z - b)^2 \tag{6.3}$$

where $a$ and $b$ are constants. Figure 6.16 shows the data for the elements of the fourth period. The dependence on the square of an effective charge $Z - b$ is similar to that of the ionization energy and for the same reason: the electrons screen each other from the nucleus, and the electronic energy levels (which determine the frequency of the emitted X rays) depend on an effective nuclear charge. The one striking difference between the behavior of the X-ray frequencies and that of the ionization energies considered so far is the absence of periodicity in the X-ray frequencies. In contrast to the repetitive pattern of the first ionization energies, the X-ray frequencies vary monotonically with atomic number. It is this very lack of periodicity that makes the X-ray frequencies a useful measure of atomic number. What is the reason for this difference? In order to answer this question we must first examine the relation between the X-ray frequencies and the ionization energies.

Figure 6.17 shows the process involved in X-ray emission. An energetic electron or photon is incident on an atom and ejects one of its K shell electrons, leaving the atom with a vacancy in the K shell. This is an unstable state and a

Multielectron Atoms and the Periodic Table  6.5  **201**

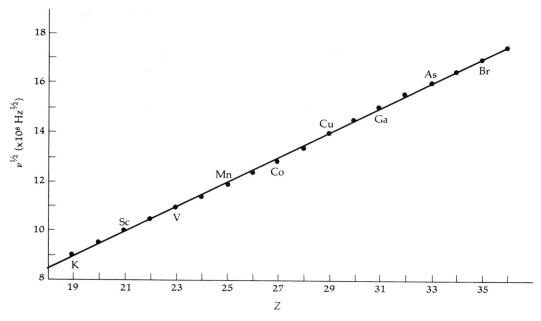

**FIGURE 6.16**  Relation between the square root of the X-ray frequency and atomic number for the elements of the fourth period. Note that the values of $\nu^{1/2}$ must be multiplied by $10^8$.

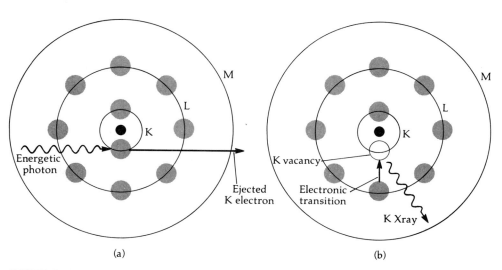

**FIGURE 6.17**  The emission of a K X ray. (a) An energetic electron or photon incident on an atom knocks out a K-shell electron (1s orbital) leaving a vacancy in the K shell. (b) An electron from a higher energy orbital makes a transition to the K shell and a K X ray carries off the difference in the energies of the two levels.

readjustment, in which one of the electrons in a higher energy orbital jumps down to fill the K-shell vacancy, occurs promptly. The emission of an X ray accompanies this electronic transition, much in the same way as a photon is emitted when the electron in the hydrogen atom makes a transition to a lower energy level (Section 5.5). Since the electron ends up in the K shell, the X ray is called a K X ray. If a vacancy had occurred in the L shell, the emission of an L X ray would have accompanied its filling. The K and L X rays in multielectron atoms are analogous to the Lyman and Balmer photons of the hydrogen atom, respectively. The L X rays have much lower energies than the K X rays and need not concern us further. It is the K X rays that were observed by Moseley and whose frequencies obey Equation 6.3.

The reason why the K X rays do not display a periodically repeating frequency now becomes apparent: their frequency is primarily determined by the ionization energy of the 1s electrons that make up the K shell. The ionization energy of these innermost electrons is very insensitive to the electronic configuration of the valence shell. Their principal quantum number, $n = 1$, is the same for all elements, as is the extent to which they are screened from the nucleus. The only variable is the nuclear charge, and the ionization energy of the 1s electrons increases with atomic number in the manner indicated in Equation 6.2.

---

**Example 6.5** The ionization energies of the 1s, 2p, and 3p electrons in a copper atom are $8.66 \times 10^5$, $9.08 \times 10^4$, and $7.14 \times 10^3$ kJ mol$^{-1}$, respectively. Calculate the energy, frequency, and wavelength of the K X ray accompanying (a) a 2p $\rightarrow$ 1s transition (called a $K_\alpha$ X ray) and (b) a 3p $\rightarrow$ 1s transition (called a $K_\beta$ X ray). For each transition indicate the change in electronic configuration.

(a) According to Equation 5.18, the energy of the X ray is just equal to the difference in energy of the initial state (designated i) and the final state (designated f)

$$\epsilon_\gamma = \Delta E = E_i - E_f$$

Since the ionization energy is just the negative of the energy of the state in question, we can also write

$$\epsilon_\gamma = I_f - I_i = I_{1s} - I_{2p}$$

We thus have:

$$\epsilon_{K_\alpha} = \frac{(8.66 \times 10^5 \text{ kJ mol}^{-1} - 9.08 \times 10^4 \text{ kJ mol}^{-1})(10^3 \text{ J kJ}^{-1})}{6.022 \times 10^{23} \text{ mol}^{-1}}$$

$$= 1.29 \times 10^{-15} \text{ J} = 8.04 \times 10^3 \text{ eV} = 8.04 \text{ keV}$$

The frequency and wavelength are obtained from the energy by means of Equation 5.18,

$$\nu = \frac{\epsilon}{h} = \frac{1.29 \times 10^{-15} \text{ J}}{6.626 \times 10^{-34} \text{ J s}}$$

$$= 1.95 \times 10^{18} \text{ s}^{-1}$$

and

$$\lambda = \frac{c}{\nu} = \frac{(2.998 \times 10^8 \text{ m s}^{-1})(10^9 \text{ nm m}^{-1})}{1.95 \times 10^{18} \text{ s}^{-1}}$$

$$= 0.154 \text{ nm}$$

The electronic configuration of a copper atom with a K-shell vacancy changes in the following way in a 2p → 1s transition:

$$1s^1 2s^2 2p^6 3s^2 3p^6 3d^{10} 4s^1 \rightarrow 1s^2 2s^2 2p^5 3s^2 3p^6 3d^{10} 4s^1$$

(b)  The energy of the $K_\beta$ X ray is

$$\epsilon_{K_\beta} = \frac{(8.66 \times 10^5 \text{ kJ mol}^{-1} - 7.14 \times 10^3 \text{ kJ mol}^{-1})(10^3 \text{ J kJ}^{-1})}{6.022 \times 10^{23} \text{ mol}^{-1}}$$

$$= 1.43 \times 10^{-15} \text{ J} = 8.90 \times 10^3 \text{ eV} = 8.90 \text{ keV}$$

The frequency of the $K_\beta$ X ray is

$$\nu = \frac{\epsilon}{h} = \frac{1.43 \times 10^{-15} \text{ J}}{6.626 \times 10^{-34} \text{ J s}}$$

$$= 2.15 \times 10^{18} \text{ s}^{-1}$$

and the wavelength is

$$\lambda = \frac{c}{\nu} = \frac{(2.998 \times 10^8 \text{ m s}^{-1})(10^9 \text{ nm m}^{-1})}{2.15 \times 10^{18} \text{ s}^{-1}}$$

$$= 0.139 \text{ nm}$$

The change in the electronic configuration of the copper atom is

$$1s^1 2s^2 2p^6 3s^2 3p^6 3d^{10} 4s^1 \rightarrow 1s^2 2s^2 2p^6 3s^2 3p^5 3d^{10} 4s^1$$

This example shows that the K X ray of copper (and of other elements) actually consists of a set of closely spaced X rays. The instrumentation available in Moseley's day did not have sufficiently high resolution to resolve the wavelengths, and, consequently, he did not have to deal with this additional complication. ∎

In discussing the X-ray emission process we have remarked that a vacancy in an electron shell is created when an electron is ejected from that shell and leaves the atom. The energy of the ejected electrons can be measured accurately by a technique called **photoelectron spectroscopy** (Figure 6.18), allowing the determination of the various ionization energies and, from these, the electronic energy levels. The process is analogous to the photoelectric effect (Section 5.4) except that, instead of being bound to the surface of a metal, the electrons are bound to an atom. The energy of the ejected electrons ($\frac{1}{2}mv^2$) is related to the energy of the incident photons ($h\nu$) by the expression:

$$h\nu = \tfrac{1}{2}mv^2 + I \tag{6.4}$$

which is identical to the equation for the photoelectric effect (Equation 5.15) with the ionization energy $I$ replacing the metal work function $\phi$.

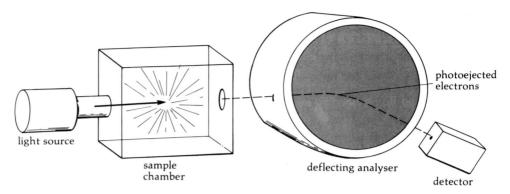

**FIGURE 6.18** A photoelectron spectrometer. Electromagnetic radiation of known frequency (visible light, UV, or X rays, depending on the application) strikes the sample to be analyzed in an evacuated chamber. The energy of the ejected photoelectrons is determined by electric or magnetic deflection (see Section 5.1). The number of ejected photoelectrons can be measured with a variety of detectors.

Since the inner-shell ionization energies of an element are unique to it, photoelectron spectroscopy can be used to establish quantitatively the presence of particular elements either as impurities or as constituents of a large variety of substances. In this important analytical application, the technique is known as electron spectroscopy for chemical analysis (**ESCA**). Photoelectron spectroscopy is also a powerful tool for the study of the electronic energy levels in molecules.

## 6.6 Electron Affinity

We have so far examined the removal of electrons from atoms, which results in the formation of cations. Electrons can also be added to atoms to form anions. The energy liberated when an electron is added to a gaseous atom to form a gaseous ion is called the **electron affinity** $\varepsilon$. Thus, it is the energy released in the process

$$X(g) + e^- \rightarrow X^-(g)$$

It is important to note that, for historical reasons, the sign of the electron affinity is defined *opposite* to the way in which the sign of the ionization energy is defined. A positive ionization energy means that energy must be supplied to remove an electron, while a positive electron affinity means that energy is liberated when an electron is added to an atom. In order to avoid any possible confusion, it may be helpful to think of the electron affinity as the ionization energy of the negative ion, i.e., as the energy that must be supplied for the process:

$$X^-(g) \rightarrow X(g) + e^-$$

When viewed in this fashion, a large positive value of the electron affinity means that the negative ion is very stable, just as a large ionization energy of an atom means that the atom is very stable.

The electron affinities of the elements are tabulated in Table 6.4. Most ele-

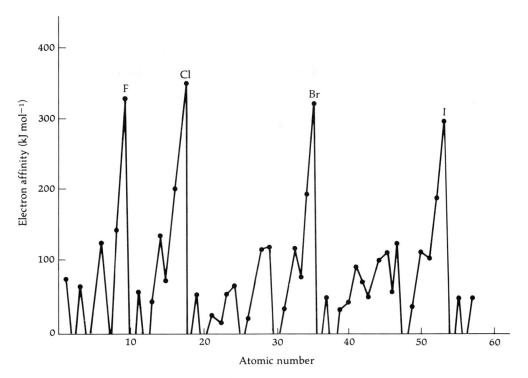

**FIGURE 6.19** Electron affinities of the elements.

ments have positive electron affinities, indicating that energy is released when an electron is added. A few elements, such as the alkaline earths and the noble gases, have negative electron affinities. The combination of large ionization energies and negative electron affinities indicates that it is energetically unfavorable for the noble gases to either gain or lose electrons, thereby accounting for their very limited tendency to undergo chemical reactions.

The dependence of the electron affinity on atomic number is displayed in Figure 6.19. The most striking regularities in the plotted values are the sharp peaks that occur at the halogens and the minima that occur at the noble gases and again at the alkaline earths. The halogen atoms have an unusually strong tendency to gain an electron. These elements have a single vacancy in the p subshell. The relatively high ionization energies of these elements indicate that the p electrons are strongly held by the atom, so it is not surprising that an extra p electron can be added with ease. The minima occur for elements with a complete p subshell (the noble gases) or with a complete s subshell (the alkaline earths). Evidently, the addition of an electron to atoms with these electronic configurations is energetically unfavorable.

While the addition of an electron to an atom of most elements releases energy, the addition of a second electron invariably requires an input of energy. The first extra electron effectively screens the second one from the nucleus and the net Coulomb force between the anion and the second electron is repulsive. Therefore, the second electron affinities of all elements are negative.

**Example 6.6** Examine the ionization energies and electron affinities tabulated in Table 6.4 in order to determine whether the reaction X(g) + X(g) → X⁺(g) + X⁻(g) is exothermic for any of the listed elements.

The reaction in question can be broken up into two separate processes:

| | |
|---|---|
| X(g) → X⁺(g) + e⁻ | Energy released = $-I_1$ |
| X(g) + e⁻ → X⁻(g) | Energy released = $\varepsilon$ |
| 2X(g) → X⁺(g) + X⁻(g) | Energy released = $\varepsilon - I_1$ |

Note that since $I_1$ is the energy absorbed by the atom in ionization, the energy released in this process is $-I_1$. In order for the reaction to be exothermic, there must be a net release of energy, or $\varepsilon - I_1 > 0$. The data in Table 6.4 indicate that none of the listed elements have first ionization energies and electron affinities that fulfill this requirement. For example, the result for chlorine, the element with the highest electron affinity, is

$$\varepsilon - I_1 = 348.8 \text{ kJ mol}^{-1} - 1251 \text{ kJ mol}^{-1}$$
$$= -902 \text{ kJ mol}^{-1} \quad \blacksquare$$

## 6.7 Electronegativity

The **electronegativity** of an atom is a measure of its tendency to attract electrons from atoms to which it is bonded. Electronegativity is an important property because it gives an indication of the types and properties of the chemical bonds that atoms form. Thus, when two bonded atoms have the same electronegativity, they have an equal tendency to attract electrons, and form a nonpolar covalent bond. At the other extreme, when the difference in electronegativity of two atoms is very large, the bond between them is so polar that it can best be described as ionic—the less electronegative (or more electropositive) atom transfers one or more electrons to the more electronegative atom. Many chemical properties of the elements and their compounds are correlated with their electronegativity, as we shall see. Consequently, it is possible to deduce chemical behavior from a knowledge of electronegativities.

Several different, but essentially equivalent, methods of determining the electronegativities of the elements have been proposed. The method proposed by R. S. Mulliken (1895–1986) is particularly direct. Since a large electronegativity indicates a strong tendency to attract and hold electrons, it should be related to both the first ionization energy and the electron affinity of an element. The electronegativity $\chi$ is defined as the average of these two quantities,

$$\chi = (I + \varepsilon)/2 \tag{6.5}$$

adjusted so that the electronegativity of fluorine, the most electronegative element, is 4.0.

The electronegativities of the representative elements are plotted in Figure 6.20. A characteristic periodicity may be noted. The electronegativities increase across a given period and decrease vertically in a given group. As a consequence of these two trends, the most electronegative elements are found in the upper right corner of the periodic table.

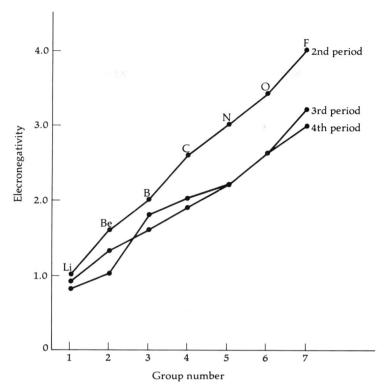

**FIGURE 6.20** Electronegativities of the representative elements.

---

**Example 6.7** Using the data in Table 6.4, determine the electronegativity of sodium.

The value of $\chi$ can be obtained by means of the tabulated first ionization energy and electron affinity, scaled to a value of 4.0 for fluorine. Applying Equation 6.5 to fluorine we have

$$\chi_F = \frac{I_F + \varepsilon_F}{2}$$

$$= \frac{1681 \text{ kJ mol}^{-1} + 328.0 \text{ kJ mol}^{-1}}{2}$$

$$= 1005 \text{ kJ mol}^{-1}$$

Adjusting this value to 4.0 yields a scaling factor of

$$4.0/1005 \text{ kJ mol}^{-1} = 3.98 \times 10^{-3} \text{ (kJ mol}^{-1})^{-1}$$

This factor must be used in Equation 6.5 to obtain the electronegativity of sodium or any other element:

$$\chi_{Na} = \left(\frac{I_{Na} + \varepsilon_{Na}}{2}\right)[3.98 \times 10^{-3} \text{ (kJ mol}^{-1})^{-1}]$$

$$= \left(\frac{495.6 \text{ kJ mol}^{-1} + 52.9 \text{ kJ mol}^{-1}}{2}\right)\left(\frac{3.98 \times 10^{-3}}{\text{kJ mol}^{-1}}\right)$$

$$= 1.1 \blacksquare$$

## 6.8 The Sizes of Atoms and Ions

Early views of atoms as very small spheres led naturally to the idea that atoms had a definite size and could be characterized by a radius. On the basis of our understanding of atomic structure, the radius of an atom can be defined as the distance of its valence electrons from the nucleus. However, the supposition that an atom has a definite radius is equivalent to the hypothesis that electrons are located at a well-defined distance from the nucleus. As we have seen in Section 5.7, the uncertainty principle indicates that particles such as the electron cannot be localized in this manner. Must we then abandon the idea that an atom has a certain size? The answer is no. The radial distribution functions (e.g., Figure 6.4) indicate that electrons have a high probability of being confined to within fairly narrow distances from the nucleus, although they are not, of course, at a unique distance. It is in this spirit that we must approach the concept of atomic size.

*a. Atomic Volume* A crude measure of the relative size of atoms is provided by the **atomic volume,** obtainable as the ratio of the atomic weight of an element (grams per mole) to its density (grams per cubic centimeter). Figure 6.21

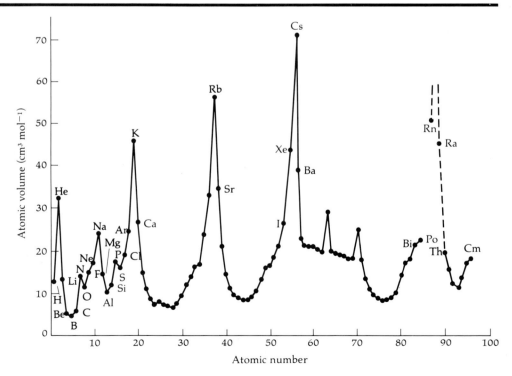

**FIGURE 6.21** Atomic volumes of the elements.

shows a plot of the atomic volume as a function of atomic number. Such a plot gives, at best, a qualitative indication of the relative size of different atoms owing to (1) the variation in how atoms of different elements are packed together in solids (Section 20.4), (2) the variation of density with temperature, and (3) the occurrence of different allotropic forms of many elements each generally characterized by a different density. Nonetheless, Figure 6.21 shows a definite periodicity in atomic volume and encourages us to search for a more quantitative measure.

***b. Atomic Radii*** The radii of isolated atoms cannot be measured directly. What can be done, instead, is to measure the mean internuclear distance between adjacent atoms in a metallic crystal or between the atoms in a covalently bonded diatomic molecule of an element (e.g., $Cl_2$). These internuclear distances can be determined by various techniques, notably for crystals by X-ray diffraction (Section 20.3a). The atomic radius can then be defined as half the internuclear distance between identical atoms. For example, the internuclear distance between Cl atoms in $Cl_2$ is 1.98 Å; therefore, the atomic radius is 0.99 Å.

In spite of the difficulties associated with the occurrence of different allotropic forms of many elements and with the difference in the types of bonds formed, a reasonably consistent set of atomic radii can be obtained. The results are listed in Table 6.5 and plotted as a function of atomic number in Figure 6.22. A very pronounced periodicity is readily apparent. In a given period, the atomic radii decrease by about a factor of 2 from the alkali elements to the halogens. Within a given group, the radii increase with atomic number, a trend that is particularly pronounced for the representative elements. For example, the radius of the alkali element atoms increases from 1.52 Å for lithium to 1.86 Å for sodium to 2.27 Å for potassium, and so on. Both trends are related to the changes in electronic configuration and can be explained in the same general way as the trends in the first ionization energies: in a given period, the effective nuclear charge increases from left to right. Consequently the attraction of the nucleus for the valence electrons generally increases and the atoms shrink in size. Within a given group, the increase in radius reflects the increase of the principal quantum number of the valence-shell electrons. Since the average distance between electron and nucleus increases with the value of $n$ (e.g., Figure 6.4), the observed trend follows.

Figure 6.22 also shows that the variation in the atomic radii across the transition series is substantially different from that observed for the representative elements in a given period. While the radii tend to decrease, this decrease is both much smaller and more irregular than that exhibited by atoms of the representative elements. The addition of electrons to the $(n-1)d$ or $(n-2)f$ orbitals, where they can screen the nuclear charge, results in only a slight increase in the effective nuclear charge felt by the $n$s electrons of these elements. The overall increase in nuclear charge resulting from the filling of the $(n-1)d$ and $(n-2)f$ orbitals is nonetheless large and has noticeable effects on the properties of the elements in succeeding periods. Some of these effects are discussed later, in connection with the radii of the common ions of the elements.

***c. Ionic Radii*** The radii of the common ions of the elements provide yet another indication of the relationship between size and electronic configuration, and they display some interesting regularities. Figure 6.23 shows a plot of the

**TABLE 6.5 Atomic Radii of the First 36 Elements**[a]

| Z | Element Symbol | r (Å) |
|---|---|---|
| 1 | H | 0.37 |
| 2 | He | (0.5)[b] |
| 3 | Li | 1.52 |
| 4 | Be | 1.11 |
| 5 | B | 0.80 |
| 6 | C | 0.77 |
| 7 | N | 0.74 |
| 8 | O | 0.74 |
| 9 | F | 0.71 |
| 10 | Ne | (0.65)[b] |
| 11 | Na | 1.86 |
| 12 | Mg | 1.60 |
| 13 | Al | 1.43 |
| 14 | Si | 1.18 |
| 15 | P | 1.10 |
| 16 | S | 1.03 |
| 17 | Cl | 0.99 |
| 18 | Ar | (0.95)[b] |
| 19 | K | 2.27 |
| 20 | Ca | 1.97 |
| 21 | Sc | 1.61 |
| 22 | Ti | 1.45 |
| 23 | V | 1.31 |
| 24 | Cr | 1.25 |
| 25 | Mn | 1.37 |
| 26 | Fe | 1.24 |
| 27 | Co | 1.25 |
| 28 | Ni | 1.25 |
| 29 | Cu | 1.28 |
| 30 | Zn | 1.33 |
| 31 | Ga | 1.22 |
| 32 | Ge | 1.23 |
| 33 | As | 1.25 |
| 34 | Se | 1.16 |
| 35 | Br | 1.14 |
| 36 | Kr | (1.10)[b] |

[a]The values are based on measurements of interatomic distances in either metallic or covalently bonded elements.
[b]Estimated values.

ionic radii of a number of ions. The various data points are connected in a way that emphasizes some of the important features. The near vertical lines connect the radii of isoelectronic ions. These ions, which have the same number of electrons, differ only in their nuclear charge. For example, the nuclear charge in the isoelectronic sequence that begins with $N^{3-}$ and ends with $Si^{4+}$ increases from 7 to 14. This increase in nuclear charge causes the ions to shrink dramatically in size: the ionic radii decrease from 1.71 Å for $N^{3-}$ to 0.41 Å for $Si^{4+}$.

The dashed lines in Figure 6.23 connect the ionic radii of elements in the same group. The same trend already noted for the atomic radii is apparent and

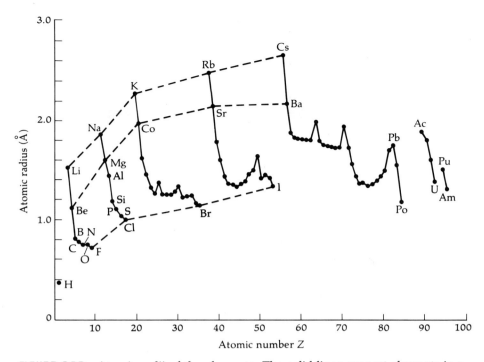

**FIGURE 6.22** Atomic radii of the elements. The solid lines connect elements in a given period; the dashed lines connect elements in a given group.

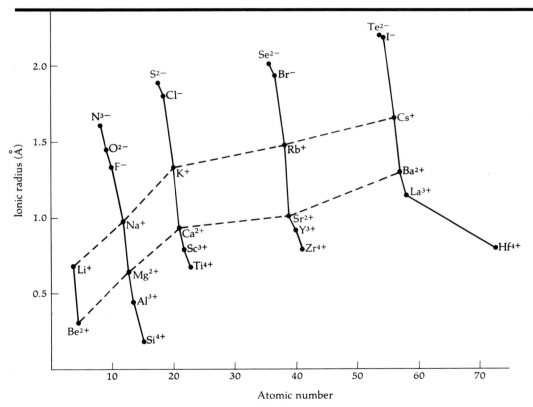

**FIGURE 6.23** Dependence of ionic radii on atomic number. The solid lines connect the points for isoelectronic ions and the dashed lines show the trends in a given group.

for the same reason: the radii increase with the value of the principal quantum number of the valence electrons. The similarity extends to some more detailed features. Note, in particular, the change in slope in the line connecting the ionic or atomic radii which occurs following the third period. For example, the radius of K$^+$ is some 40% larger than that of Na$^+$ while that of Rb$^+$ is only about 10% larger than that of K$^+$. This effect can be attributed to the occurrence of the transition series, which first intervene between, say, potassium and rubidium. Owing to the filling of the d or f orbitals, the increase in nuclear charge between successive members of the groups is larger than in the earlier periods. The Coulomb attraction between nucleus and valence electrons consequently increases and effectively counteracts the usually more dominant effect of the increase in principal quantum number of the valence electrons.

Another clear indication of the effect of increasing nuclear charge on ionic radii may be seen in an examination of the lanthanide ions. The electronic configuration of these elements generally is [Xe]4f$^{1-14}$6s$^2$, and they readily form +3 ions in which both 6s and one 4f electron are lost. Since the difference between succeeding lanthanides is in the filling of the f orbitals, the principal quantum number remains unchanged. As shown in Figure 6.24, the gradual increase in nuclear charge leads to a decrease in the size of the ions that is similar, though much less pronounced, than that shown for isoelectronic ions. A similar, albeit somewhat more irregular effect is displayed by the atomic radii of these elements (Figure 6.22). The decrease in size across the period for the lanthanum through lutetium atoms or ions is called the **lanthanide contraction.** As a result of this contraction, the ionic radii of elements immediately following the lanthanides are essentially equal to those of the preceding elements in

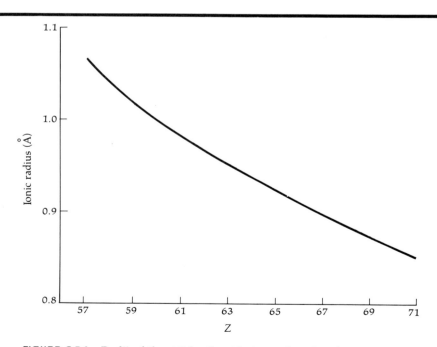

**FIGURE 6.24** Radii of the +3 lanthanide ions, showing the occurrence of the lanthanide contraction.

the same group. For instance, as may be noted in Figure 6.23, the ionic radius of hafnium, which immediately follows the lanthanides, is no larger than that of zirconium in spite of the increase in the principal quantum number of the valence shell. Since these elements belong to the same group and have the same ionic size, their chemical properties are unusually similar. The same effect may be noted for tantalum and niobium.

## 6.9 Metals and Nonmetals

*a. Classification of the Elements* The broadest classification of the elements is based on their metallic properties: elements can be categorized as metals, nonmetals, or semimetals (metalloids). **Metals** are good conductors of electricity and heat. With the exception of mercury, all metals are solids at room temperatures. (Gallium and cesium melt just below 30°C.) Solid metals are malleable (that is, they can be rolled into sheets) and ductile (they can be drawn into wires).

Although energy must be supplied to remove one or more electrons from the atom of any element (i.e., $I > 0$), metals have relatively low ionization energies. They consequently tend to lose one or more valence electrons in chemical reactions and become cations. The ease with which a metal can lose electrons is a major determinant of its reactivity. For example, sodium has a low ionization energy (495.6 kJ mol$^{-1}$) and reacts violently with cold water:

$$2Na(s) + 2H_2O \rightarrow 2Na^+ + 2OH^- + H_2(g)$$

On the other hand, copper ($I = 745.4$ kJ mol$^{-1}$) does not react with water.

**Nonmetals** are poor conductors of electricity and heat and variously form solids, liquids, or gases at room temperature. In contrast to the metals, solid nonmetals are either brittle or crumbly. Sulfur, for example, can readily be crumbled into a yellow powder.

Nonmetals are characterized by the ease with which they accept electrons to form anions, a property that is closely related to their electron affinity. The light halogens (fluorine and chlorine) are the most nonmetallic elements and undergo a variety of reactions in which they become anions, for example,

$$2K(s) + F_2(g) \rightarrow 2KF(s)$$

$$Cl_2(g) + H_2O \rightarrow H^+ + Cl^- + HClO(aq)$$

Because metallic and nonmetallic character depend on the ease with which atoms lose or gain electrons, this categorization of the elements is closely correlated with their electronegativity. The most reactive metals (e.g., potassium) have the lowest electronegativity while the most nonmetallic elements (e.g., fluorine) have the highest electronegativity. Metallic character generally decreases with increasing electronegativity while nonmetallic character increases. An electronegativity of 2.0 marks an approximate division between metals and nonmetals. **Semimetals**, which have intermediate properties, have electronegativities close to 2.0.

Since electronegativity increases from left to right and from bottom to top across the periodic table (Figure 6.20), the division of the periodic table into metals and nonmetals follows a diagonal. The division into metals, nonmetals, and semimetals, is depicted in Figure 6.25. This particular form of the periodic table, which separates the representative from the transition elements, has the

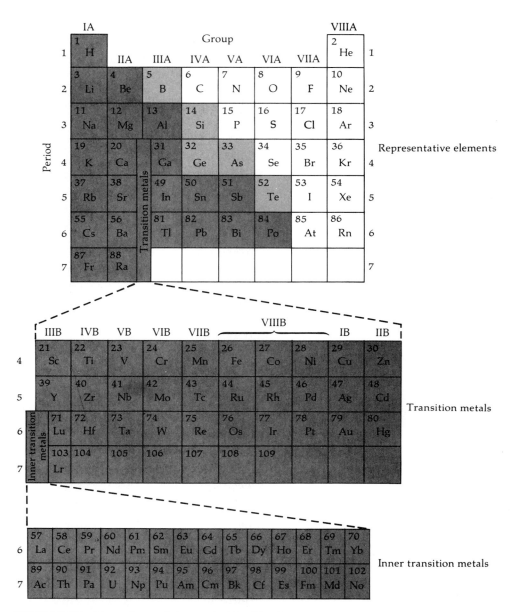

**FIGURE 6.25** Short form of the periodic table showing the categorization of the elements as metals (colored), semimetals (shaded), and nonmetals.

virtue of emphasizing the trends among the representative elements. The metals are concentrated in a triangle-like region on the left side of the representative block. In addition, all the transition and inner transition elements are metallic. The nonmetals lie in an inverted triangular region on the right side of the block of representative elements. The semimetals, which separate the metals from the nonmetals, lie along a diagonal extending from Group IIIA at the top to Group VIA near the bottom.

***b. Trends in Properties of the Elements*** Nearly 70 elements can be classified as metals. With the number being so large, it is not surprising that there are substantial differences in the behavior of different metals.

The alkali metals exhibit the greatest similarity among the various groups. They are all soft, low-melting metals of relatively low density, which react readily with oxygen and violently with water forming strongly basic solutions. In these as in other reactions, the metals readily lose an electron to form singly charged M$^+$ ions.

Although hydrogen is nominally placed in Group IA, it is very different from the alkali metals, the other elements in this group. In actuality, it does not belong in any of the groups of the periodic table. Hydrogen is the first element in the periodic table and this is what makes it unique. A hydrogen atom has just one valence electron and no inner electron shells. Thus, the hydrogen ion (H$^+$) is just a bare proton, and is several orders of magnitude smaller than any other ion.

Elemental hydrogen is a diatomic gas. Its boiling point (20.5 K) is lower than that of any other substance excepting helium. It is a colorless and odorless gas and is classified as a nonmetal. Hydrogen is the most abundant element in the universe and is the major constituent of the sun, where because of the extremely high temperature, it is dissociated into protons and electrons.

Hydrogen readily combines with most other elements; for example, it combines with oxygen to form water:

$$2H_2(g) + O_2(g) \rightarrow 2H_2O(l)$$

It reacts with the halogens to form hydrogen halides, HF, HCl, HBr, and HI. These gaseous compounds are very soluble in water and their aqueous solutions are the hydrohalic acids [e.g., hydrochloric acid, HCl(aq)]. Hydrogen also reacts with the alkali and most of the alkaline earth metals to form **hydrides**, in which hydrogen is present as the hydride ion, H$^-$. The reaction with sodium at elevated temperatures is typical:

$$2Na(s) + H_2(g) \rightarrow 2NaH(s)$$

The alkaline earths are distinctly different from the alkali metals. They have much higher melting points and densities and significantly lower reactivity. They readily form dipositive ions, M$^{2+}$, and their hydroxides, although not very soluble in water, form basic solutions.

The elements of Group IIIA exhibit considerably more variability than either the alkali or alkaline earth metals. The first element in the group, boron, is a semimetal rather than a metal and is characterized by a low electrical conductivity. It is the only member of the group that does not form a tripositive ion, M$^{3+}$, in its compounds. The other elements of Groups IIIA are all metallic and their reactivity generally increases from top to bottom. Aluminum is a very important structural metal due to its wide availability, low density, and natural resistance to corrosion.

The elements of Group IVA display a wide variability of properties and include a nonmetal (carbon), two semimetals (silicon and germanium), and two metals (tin and lead). Tin actually exists in various allotropic forms not all of which are metallic. Both tin and lead form a variety of compounds in which they display both +2 and +4 oxidation states.

The only metallic element in Group VA is bismuth and the remaining groups of representative elements contain only nonmetals or semimetals.

The transition metals are characterized by the filling of the d orbitals. The closeness of the $(n - 1)$d and $n$s orbitals makes electrons in both orbitals available for bonding and results in a variety of oxidation states. Manganese, for example, exists in every oxidation state from 0 to +7, although not all of them give rise to a large number of compounds. However, the multiplicity of oxidation states does not occur uniformly throughout the series (Section 21.1c). The elements of Group IIIB (scandium, yttrium, lanthanum, and actinium) primarily form +3 ions. The metals in Group IB (copper, silver, and gold) form +1 ions in many compounds, and the metals in Group IIB, sometimes called the post-transition elements, form +2 ions, such as $Zn^{2+}$, $Cd^{2+}$, and $Hg^{2+}$. Some of the other interesting features of the transition metals are their magnetic properties and the vivid colors displayed by many of their compounds (Chapter 21).

The transition elements display a much narrower range of properties than the representative elements. This is particularly true of the inner transition series—the lanthanides and the actinides. Since the elements in a given series differ only in the configuration of an f subshell and have the same valence shell configuration, they closely resemble each other. This is particularly true of the lanthanides, all of which are reactive metals that readily form $M^{3+}$ ions in their reactions.

The nonmetals (including the semimetals) constitute a majority of the representative elements and include the elements of fundamental importance in life processes, such as carbon, oxygen, nitrogen, and hydrogen (Chapter 22). While the metals all display some similar properties, the nonmetals are characterized by their diversity. Thus, they range from the semimetals, which verge on being metallic, to the halogens, which exhibit a strong tendency to acquire electrons, to the inert noble gases.

Of the nonmetals, the halogens are both the most reactive elements and the ones that most closely resemble each other chemically. All the elements form compounds with hydrogen, the hydrogen halides, in which they display the −1 oxidation state. All the halogens excepting fluorine also form oxides in which they have a positive oxidation number. The rich redox chemistry of the nonmetals is examined in more detail in Chapter 19.

## 6.10 Oxidation Numbers and Chemical Bonds

One of the earliest observed regularities in the chemical properties of the elements was the variation of the "combining power" of the elements in a given period. This regularity is illustrated in Figure 6.26, which shows the variation of the number of fluorine atoms combining with an atom of the various elements in the second and third periods to form their fluorides. The combining power increases linearly from one for lithium (e.g., LiF) to four for carbon (e.g., $CF_4$) and then decreases, again linearly. This trend repeats itself in the third period as well as among the representative elements of subsequent periods. Using the terminology introduced in Section 2.7, we can say that the oxidation numbers of the second-period elements in their fluorides, in which the oxidation number of fluorine is −1, increase from +1 for lithium to +4 for carbon and then decrease. The regularities in oxidation number provide a useful basis for the systematization of the chemical behavior of the elements (Chapter 19).

The regularities in the oxidation numbers displayed in Figure 6.26 do not, by themselves, say much about the nature of these compounds, that is, about

Multielectron Atoms and the Periodic Table  6.10  **217**

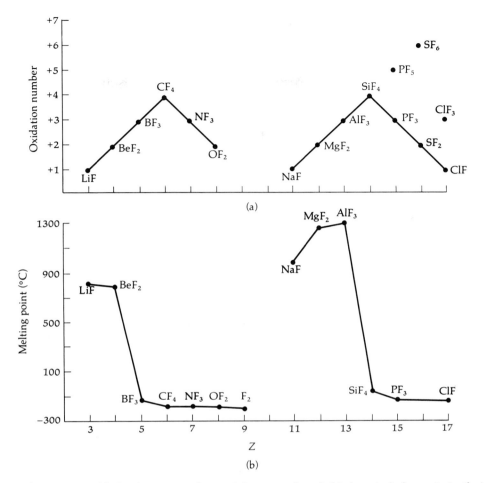

**FIGURE 6.26** (a) Oxidation numbers of the second and third period elements in their fluorides. The unconnected points represent additional oxidation numbers displayed by the elements in their fluorides. (The combining power is numerically equal to the oxidation number.) (b) Normal melting points of the fluorides shown in (a). The sublimation point, i.e., the temperature at which the solid is converted to a gas, is shown for compounds that do not exist as liquids at atmospheric pressure. Sulfur difluoride is only known in the gas phase; its melting point is therefore not given.

---

the nature of the bonds between fluorine and the various elements of the second or third period. For example both beryllium and oxygen display an oxidation number of +2 in $BeF_2$ and $OF_2$, respectively. In spite of this similarity, $BeF_2$ and $OF_2$ are vastly different compounds. One of the properties that is particularly revealing in this respect is the melting point, also shown in Figure 6.26. A clear dichotomy may be seen: LiF and $BeF_2$, as well as NaF, $MgF_2$, and $AlF_3$, have high melting points; all the other fluorides have very low melting points and, in fact, are already gaseous at room temperature.

The high melting points of compounds such as NaF, $MgF_2$, and $AlF_3$, show that these compounds are salts consisting of ions. The strong Coulomb attraction between the positively charged cations and negatively charged anions is responsible for the high temperatures required before the ionic bonds between neighboring ions in the solid crystal are broken. The formation of Na$^+$ and Mg$^{2+}$ ions occurs readily owing to the low ionization energies of these elements. This behavior is closely related to their metallic character. As already noted in Figure 6.25, sodium, magnesium, and aluminum are the only elements in the third period that are metals.

The low melting and boiling points of the remaining fluorides are suggestive of covalent rather than ionic bonding. These compounds exist as discrete molecules which can be represented by Lewis formulas such as

$$:\ddot{F}-B-\ddot{F}:$$
$$|$$
$$:\ddot{F}:$$

or

$$:\ddot{F}:$$
$$|$$
$$:\ddot{F}-C-\ddot{F}:$$
$$|$$
$$:\ddot{F}:$$

As already noted (Section 4.9), discrete molecules are held together in condensed phases by weak van der Waals forces. Owing to the weakness of these forces compared to the Coulomb force between oppositely charged ions, the covalent fluorides melt and boil at very low temperatures.

The occurrence of covalent and ionic bonding can be correlated with the difference in electronegativity of the atoms forming a bond. When this difference is large, the atom of the more electronegative element can actually pull one or more electrons away from the atom of the less electronegative element. The resulting ions are held together by an ionic bond. For example, sodium and fluorine, which form the solid salt NaF, differ in electronegativity by 3.1 units. On the other hand, silicon and fluorine, which form the gaseous compound SiF$_4$, differ in electronegativity by only 2.2 units.

## Conclusion

The application of the principles of quantum mechanics to multielectron atoms makes possible the determination of the electronic structure of the atoms of elements beyond hydrogen. The ordering of the atomic electrons in specific shells and subshells provides the theoretical foundation for the arrangement of the elements in the periodic table. Although the periodic table had been substantially developed prior to the advent of quantum mechanics on the basis of the observed regularities in the properties of the elements and their compounds, many of these regularities only become understandable on the basis of the electronic structure. This is particularly true of those properties that reflect the behavior of isolated gaseous atoms: ionization energies and electron affinities. Other properties examined in this chapter, such as metallic character, depend on both electronic structure and the interactions between an atom and its neighbors in a solid crystal. The interaction between neighboring atoms obscures some of the features attributable to the effects of electronic configuration but these effects are still observable. As we continue our study of the trends in various properties of the elements and their compounds we shall see that electronic structure is an important determinant of the observed features. However, intermolecular or interionic interactions are also of importance and in some cases overshadow the periodicity associated with similar electronic configurations. The occurrence of numerous exceptions to general rules of chemical behavior is an expected, if at times frustrating, result of the interplay between these different factors.

## Problems

1. Examine the atomic numbers and atomic weights of the stable elements and find all pairs of adjacent elements in which the element of higher atomic number has lower atomic weight. List the atomic numbers and atomic weights of the elements. Exclude elements heavier than bismuth.

2. The reversal of atomic number and atomic weight sequences considered in Problem 1 can be attributed to isotopic effects. Examine the stable isotopes of the elements showing this reversal (e.g., in the Handbook of Chemistry and Physics) and draw some generalizations about the common features of all of these cases.

3. How many electrons can occupy the following sets of orbitals: 2s, 3p, 3d, and 4f?

4. How many unpaired electrons do the following atoms or ions have: Be, C, Na$^+$, Al, Ti, Ti$^{2+}$, Fe$^{3+}$, Cu$^+$, Sr, and Xe?

5. Using only a periodic table for reference, write the electronic ground state configuration of each of the following: B, Be$^{2+}$, Si, Ar, Ti, Fe$^{2+}$, Sr, In$^{3+}$, and Xe.

6. How many electrons does each atom or ion in problem 5 have? How many of these are s electrons? How many are p electrons?

7. Identify the divalent cations with the following electronic configurations: [He]2s$^2$2p$^6$, [Ne]3s$^2$3p$^6$, [Ar]3d$^5$, [Ar]3d$^6$, [Ar]3d$^8$, [Xe]4f$^7$.

8. Indicate whether each of the following electronic configurations is associated with (a) a ground state atom of the listed element, (b) an excited state, (c) a cation formed by the element in question, (d) an anion formed by the element, or (e) a forbidden configuration.

    | | |
    |---|---|
    | Be | 1s$^2$2s$^2$ |
    | H | 1s$^2$ |
    | B | 1s$^2$2s$^2$2p$^2$ |
    | N | 1s$^2$2s$^2$2p$^2$3s$^1$ |
    | F | 1s$^2$2s$^2$2p$^4$2d$^1$ |
    | F | 1s$^2$2s$^2$2p$^6$ |

9. From the following list of atoms, molecules, and ions pick out those sets that are isoelectronic: H$^-$, NH$_3$, Na$^+$, He, Be$^{2+}$, H$^+$, CH$_4$, Li$^+$, F$^-$, HF, O$_2$.

10. Although all attempts to produce it have so far failed, element 114 is thought to be the most stable of the predicted superheavy elements. What is the electronic configuration of this element? In which group does it belong and which element in this group does it follow?

11. If a large number of superheavy elements are ever discovered, it may be possible to examine the properties of elements formed by the filling of the 5g orbitals, which are expected to commence at Z = 121. (a) How many elements will belong to this series? (b) Write the electronic configuration of the third element in this series. (c) What is the maximum number of unpaired electrons an element in this series can have?

12. Using the data in Table 6.4, calculate the effective nuclear charge felt by the valence electron of the first three alkali elements. Which is more constant, $Z_{eff}$ or $Z_{eff}/n$? Comment on the result.

13. The ionization energies of the elements in the third period generally increase between sodium and argon. Two exceptions to this trend occur: The ionization energy of aluminum is lower than that of magnesium and the ionization energy of sulfur is lower than that of phosphorus. Explain the reason for the occurrence of each of these exceptions.

14. The energy needed to remove an electron from a gaseous lithium atom is barely half as large as that needed to remove an electron from a gaseous beryllium atom, yet over four times as much energy is needed to remove an electron from Li$^+$ than from Be$^+$. Explain the contrast between the atoms and the ions.

15. Plot the ionization energies of K, Ca$^+$, and Sc$^{2+}$ versus atomic number. What does this plot show?

16. Plot the first three ionization energies of lithium and beryllium. What does this plot show?

17. Electrons with a velocity of $2.726 \times 10^7$ m s$^{-1}$ are ejected from iodine vapor when 584 pm light from a helium discharge lamp is incident. What is the ionization energy of iodine?

18. Use the data in Table 6.4 to calculate the effective nuclear charge felt by a 4s electron in (a) potassium and (b) calcium.

19. The ionization energies of the 1s, 2p, and 3p electrons in silver are $2.462 \times 10^6$, $3.400 \times 10^5$, and $5.81 \times 10^4$ kJ mol$^{-1}$, respectively. Calculate the energy and wavelength of (a) the $K_\alpha$ X ray of silver and (b) the $K_\beta$ X ray of silver.

20. For the electronic transitions accompanying the emission of a $K_\alpha$ and a $K_\beta$ X ray in silver (see problem 19) write the change in the electronic configuration of the silver atom.

21. The energy of the $K_\alpha$ X ray of strontium is 14.165 keV. What is the wavelength of the X ray?

22. The energies of the $K_\alpha$ and $K_\beta$ X rays of lanthanum are 33.442 keV and 37.8 keV, respectively. What is the difference in the ionization energies of the 2p and 3p electrons of lanthanum (expressed in kilojoules per mole)?

23. The energies of the $K_\alpha$ X rays of rubidium and strontium are 13.395 keV and 14.165 keV, respectively. On the basis of Moseley's formula, what is the effective nuclear charge of rubidium and strontium felt by the inner electrons in this transition?

24. The ionization energy of the 1s electron in iron is $6.8616 \times 10^5$ kJ mol$^{-1}$. What is the maximum wavelength of X rays able to eject the 1s electron from an iron atom?

25. The shortest wavelength X ray ($K_\beta$) emitted by lanthanum metal bombarded by energetic electrons has a wavelength of 32.0 pm; the corresponding wavelength of the X ray emitted by hafnium is 19.1 pm. The shortest wavelength of the X rays emitted by an unknown lanthanide metal was 24.8 pm. Identify the element.

26. Determine the maximum wavelength of light that can detach an electron from each of the following gaseous anions: C$^-$, F$^-$, S$^-$. (Measurements of this type using laser light of very precisely known wavelengths are, in fact, used to determine electron affinities.)

27. Explain why the electron affinities of the elements tend to increase from left to right in a given period. Why is the affinity of nitrogen zero while the elements on either side, carbon and oxygen, have substantial electron affinities?

28. Calculate, the energy required for the reaction Na(g) + Cl(g) → Na$^+$(g) + Cl$^-$(g).

29. Considering all the alkali halides that can be made from the elements listed in Table 6.4, is the reaction M(g) + X(g) → M$^+$(g) + X$^-$(g) exothermic for any M–X combination?

30. Plot the melting points of the representative elements in the second, third, and fourth periods (data can be found in Handbook of Chemistry and Physics) and comment on any periodic trends.

31. Describe the trend in atomic size expected for the alkaline earth elements. What factors account for this trend?

32. For each of the following pairs of ions, pick the one with larger radius: S$^{2-}$ and Cl$^-$, Cr$^{3+}$ and Co$^{3+}$, F$^-$ and Na$^+$, Tl$^+$ and Tl$^{3+}$, La$^{3+}$ and Lu$^{3+}$.

33. Among the elements with atomic number 6 to 16, pick (a) the one with highest ionization energy, (b) the one with lowest ionization energy, (c) the one with highest second ionization energy, (d) the metallic elements, (e) the nonmetals, (f) the semimetals, and (g) the one with largest atomic size.

34. For the unknown element eka francium predict: (a) the atomic number, (b) the electronic configuration, (c) the approximate value of the ionization energy, (d) metallic properties, (e) expected oxidation state, and (f) reactivity with air and water.

35. Which of the following chlorides are expected to be high-melting salts: KCl, SiCl$_4$, PCl$_5$, CaCl$_2$, CuCl$_2$?

# 7
# Chemical Equilibrium

The combination of two or more chemical substances can have one of several different results. Possibly the substances will have no effect at all upon each other and no reaction will occur. Perhaps the tendency for the substances to react will be so great that virtually all of the reactants will be converted to products; or it may be that some but not the entire amounts of reactants that are present will be converted to products. The latter case is the most likely. The mixture of reactants and products present when such a reaction is "over," that is, when no further change in the concentrations of reactants or products can be detected, has reached a state of chemical equilibrium. The concept of chemical equilibrium is widely applicable in all fields of chemistry and we now turn to an extended examination of this subject. In this chapter we introduce some fundamental aspects of this topic and examine chemical equilibrium in reactions of gases. The succeeding two chapters deal with equilibrium in reactions occurring in aqueous solution. Later, after we have learned some thermodynamics, we shall return to chemical equilibrium and examine the theoretical basis for the quantitative relationships introduced in these chapters.

Chemical equilibrium is also a subject of great practical importance. If it is known in advance that the equilibrium point of a certain reaction favors the formation of products, it may be worthwhile to invest considerable effort in developing conditions that will make the reaction occur both rapidly and with a minimum of undesirable side reactions. On the other hand, if the equilibrium mixture of a reaction consists primarily of the reactants, it is pointless to search for conditions that will substantially improve the product yield.

## 7.1 The Nature of Chemical Equilibrium

Consider the reaction between hydrogen and iodine at elevated temperatures:

$$H_2(g) + I_2(g) \rightarrow 2HI(g)$$

The reacting molecules will collide with each other at the enormously high frequency predicted by kinetic theory (Section 4.5). Occasionally, one of these collisions leads to a sequence of steps resulting in the formation of two HI molecules. Several things happen as the reaction proceeds. First, the concentrations of $H_2$ and $I_2$ gradually decrease as more and more of these molecules react. Owing to this decrease, the rates at which these molecules collide and react with each other also decrease. Second, the concentration of HI gradually increases. Therefore, HI molecules collide with each other with increasing frequency, and the rate of the reverse reaction,

$$HI(g) + HI(g) \rightarrow H_2(g) + I_2(g)$$

increases with time.

In Figure 7.1 we see that as the rate of the forward reaction between $H_2$ and $I_2$ gradually decreases, that of the reverse reaction gradually increases. Therefore, a time is eventually reached when the two rates become equal to each other. We also see that the concentrations of $H_2$ and $I_2$ decrease and then level off while that of HI increases and eventually attains a constant value. The changes in reaction rates and concentrations are coupled and define the occurrence of **chemical equilibrium**: *when the rates of the forward and reverse reactions become equal, the concentrations of reactants and products become constant and chemical equilibrium is reached.*

It is important to emphasize that although there is no further net change in the concentrations of the reacting species once equilibrium is attained, both the forward and the reverse reactions continue to occur. This is illustrated in Figure 7.1, which shows that the reaction rates do not decrease towards zero but level off at a finite value at equilibrium. Chemical equilibrium is a dynamic process.

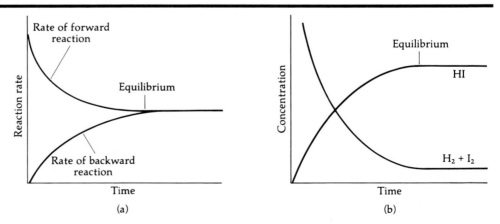

**FIGURE 7.1** Time dependence of (a) the forward and reverse reaction rates and of (b) the concentrations of reactants and products in the reaction $H_2(g) + I_2(g) \rightleftharpoons 2HI(g)$.

Clear evidence for the dynamic nature of equilibrium can be obtained by experiments with a radioactive isotope, or **radioisotope**, of one of the reacting species. Radioisotopes are useful because there is virtually no difference between their chemical properties and those of stable isotopes of the same element. Consequently, the radioisotope serves as a tracer for the behavior of the stable isotope under circumstances in which this behavior cannot otherwise be easily determined. If a small amount of iodine containing radioactive $^{131}$I is added to an equilibrium mixture of $H_2$, $I_2$, and HI, it is found that $^{131}$I is gradually incorporated into HI, thereby showing that the reaction continues to occur even though the system is at equilibrium.

In order for chemical equilibrium to be attainable, it is necessary for the reverse reaction to occur. In other words, equilibrium requires that a chemical reaction be **reversible.** The reversibility of a reaction is indicated by means of a double arrow in the equation, e.g.,

$$H_2(g) + I_2(g) \rightleftarrows 2HI(g)$$

This notation qualitatively indicates that significant amounts of both reactants and products are present at equilibrium.

An important consequence of the reversibility of a reaction is that equilibrium can be attained by starting with either reactants or products. As illustrated in Figure 7.2, at a given temperature and pressure the ratio of the HI to the $I_2$ or $H_2$ concentrations present at equilibrium is the same regardless of whether the reaction begins with a mixture of hydrogen and iodine or with hydrogen iodide.

Once a system has reached a state of equilibrium, this condition will be maintained indefinitely provided the system is left undisturbed. We might think that since a reaction occurs as a result of the random collisions of molecules, there is a chance that at a given moment the frequency of a particular type of collision might be altered significantly from its average value. For instance, if the collision frequency between HI molecules were suddenly to in-

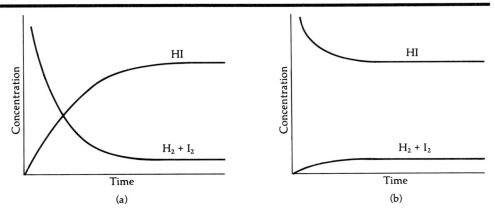

**FIGURE 7.2** The concentrations of reactants and products present at equilibrium are the same whether we start with (a) the reactants, e.g., $H_2(g) + I_2(g) \rightleftarrows 2HI(g)$ or (b) the products, e.g., $2HI(g) \rightleftarrows H_2(g) + I_2(g)$, provided that the conditions are identical.

crease due to a random fluctuation, the rate of the backward reaction would increase concomitantly and the system would move away from equilibrium. While such an effect can indeed occur when the number of molecules is small, the probability of fluctuations in the collision rate becomes negligibly small when the enormous number of molecules involved in a chemical reaction is considered. The only way the ratio of the concentrations of reactants and products present at equilibrium can be affected is by a change in one or more of the variables which affect the equilibrium point, such as the temperature or pressure (Section 7.6).

The results depicted in Figures 7.1 and 7.2 for the reaction of hydrogen and iodine show that, under the particular conditions of the experiment, the equilibrium mixture consists of approximately 85% HI and 15% $H_2 + I_2$. In qualitative terms, the position of chemical equilibrium in this reaction, or the equilibrium point, favors the formation of the product.

Different reactions generally differ with respect to the position of chemical equilibrium. For example, under the conditions just described for the formation of HI, the equilibrium mixture for the reaction

$$H_2(g) + Br_2(g) \rightleftharpoons 2HBr(g)$$

consists of only about $10^{-7}\%$ $H_2 + Br_2$. Since the products of a reaction are given on the right side of the chemical equation, we can characterize the equilibrium point of such a reaction by saying that it lies far to the right. From what we have learned so far about the nature of chemical equilibrium, it is clear that if this equation is written in the reverse direction:

$$2HBr(g) \rightleftharpoons H_2(g) + Br_2(g)$$

the position of equilibrium lies far to the left.

In our ensuing discussion in this chapter we will examine how the position of chemical equilibrium can be quantitatively determined for a given reaction. While we shall concentrate on reactions of gases, the results are generally applicable to all reactions. We shall not address here the equally important question of the time it takes for equilibrium to be established except to say that, depending on the reaction and the experimental conditions, this time can range from very small fractions of a second to times longer than the age of the Earth. The rate at which chemical reactions occur is the subject of the branch of chemistry called **kinetics** (Chapter 14).

## 7.2 The Equilibrium Constant

***a. Definition of the Equilibrium Constant*** Table 7.1 shows some data for the concentrations of $H_2$, $I_2$, and HI in the reaction

$$H_2(g) + I_2(g) \rightleftharpoons 2HI(g)$$

at 700 K. The first set of entries gives the initial concentrations, and the second set gives the equilibrium concentrations. Each line represents a different set of initial concentrations. In some cases the starting concentration of $H_2$ is greater than that of $I_2$; in other cases the reverse is true, and in the last two cases the species present initially is HI. The individual concentrations at equilibrium vary from experiment to experiment and depend on the initial concentrations. However, the quantity $[HI]^2_{eq}/[H_2]_{eq}[I_2]_{eq}$ has the same value (55) at 700 K, independent of the initial concentrations, provided equilibrium has been reached.

**TABLE 7.1  Equilibrium in the Reaction $H_2(g) + I_2(g) \rightleftharpoons 2HI(g)$ at 700 K**

| \multicolumn{3}{c}{Initial concentrations (mol L$^{-1}$)($10^3$)$^a$} ||| \multicolumn{3}{c}{Equilibrium concentrations (mol L$^{-1}$)($10^3$)$^a$} ||| $\dfrac{[HI]^2}{[H_2][I_2]}$ |
|---|---|---|---|---|---|---|
| $H_2$ | $I_2$ | HI | $H_2$ | $I_2$ | HI | |
| 11.11 | 9.946 | 0 | 2.869 | 1.706 | 16.48 | 55.5 |
| 11.28 | 7.508 | 0 | 4.505 | 0.7377 | 13.54 | 55.2 |
| 10.65 | 10.76 | 0 | 2.222 | 2.336 | 16.85 | 54.7 |
| 10.64 | 11.97 | 0 | 1.804 | 3.129 | 17.67 | 55.3 |
| 0 | 0 | 4.480 | 0.4714 | 0.4788 | 3.530 | 55.2 |
| 0 | 0 | 10.67 | 1.124 | 1.141 | 8.409 | 55.1 |

$^a$The concentrations are a factor of $10^3$ smaller than the tabulated values.

The preceding combination of concentrations constitutes an **equilibrium constant** for the reaction in question. In more general terms, if the equation for a chemical reaction is written in the form

$$aA + bB + \cdots \rightleftharpoons cC + dD + \cdots \tag{7.1}$$

the equilibrium constant is given by

$$K = \frac{[C]_{eq}^c[D]_{eq}^d \cdots}{[A]_{eq}^a[B]_{eq}^b \cdots} \tag{7.2}$$

In this expression, the equilibrium concentrations in moles per liter of the reaction products appear in the numerator and those of the reactants appear in the denominator. Each concentration is raised to a power equal to the stoichiometric coefficient of the species in question in the balanced equation. The subscript *eq* explicitly indicates that the concentrations must be evaluated at equilibrium. We shall usually omit this subscript in order to keep the notation simple.

Several points concerning Equation 7.2 deserve to be emphasized. First, the concentrations appearing in this expression are the equilibrium concentrations. The equilibrium constant is a constant, that is, independent of the initial concentrations, *only* if the equilibrium concentrations are used in Equation 7.2. Second, the value of the equilibrium constant of a particular reaction is a function of the temperature and does not change so long as the temperature remains invariant. Third, Equation 7.2 is strictly applicable only to ideal gases or ideal solutions (considered in Section 12.4). Some of the effects of non-ideal behavior are examined later in this chapter (Section 7.7)

One additional point about equilibrium constants is that only the concentrations of gases or of species in solution appear in Equation 7.2. The concentrations of pure liquids or solids participating in a chemical reaction are not included. As noted in Section 3.2, solids and liquids are nearly incompressible. Under ordinary conditions, the density of a solid or liquid at a given temperature is therefore constant and so, too, is its concentration. In view of this constancy it makes little sense to include the concentration of a pure solid or liquid in an equilibrium constant expression since, at a given temperature, this concentration always has the same value. Instead, the concentrations of pure solids and liquids are incorporated in the value of the equilibrium constant; the only concentrations appearing in the equilibrium expression are those of gases or of species in solution.

**Example 7.1** Express the equilibrium constants of the following reactions in terms of concentrations of reactants and products:

(a) $CaCO_3(s) \rightleftarrows CaO(s) + CO_2(g)$

(b) $N_2(g) + 3H_2(g) \rightleftarrows 2NH_3(g)$

(c) $2Hg(l) + O_2(g) \rightleftarrows 2HgO(s)$

(d) $AgCl(s) \rightleftarrows Ag^+ + Cl^-$

(e) $H^+ + OH^- \rightleftarrows H_2O$

(a) $K = [CO_2]$

(b) $K = \dfrac{[NH_3]^2}{[N_2][H_2]^3}$

(c) $K = \dfrac{1}{[O_2]}$

(d) $K = [Ag^+][Cl^-]$

(e) $K = \dfrac{1}{[H^+][OH^-]}$ ∎

***b. Equilibrium Constants for Reactions Involving Gases*** It is customary to express the amount of a gaseous substance present in a mixture of gases in terms of its partial pressure (Section 3.5). The equilibrium partial pressures of reacting gases can be related to a somewhat different form of the equilibrium constant by an expression analogous to Equation 7.2. For a chemical reaction written in the form of Equation 7.1, we define the equilibrium constant $K_P$ as

$$K_P = \frac{P_C^c P_D^d}{P_A^a P_B^b} \quad (7.3)$$

where $P$ is the equilibrium partial pressure in atmospheres of the substance identified by the subscript. As before, the superscripts are equal to the stoichiometric coefficients.

For a given reaction, $K$ and $K_P$ must be related since the concentrations and partial pressures whose ratios they express are related to each other. The relation between the two constants can be derived on the basis of the ideal gas law (Section 3.4). According to Equation 3.16, the partial pressure of the $i$th component of a mixture is given by

$$P_i = n_i RT/V = [i]RT$$

where $[i]$ is the molar concentration. Substituting for $P_i$ into Equation 7.3 we obtain:

$$K_P = \frac{[C]^c(RT)^c[D]^d(RT)^d}{[A]^a(RT)^a[B]^b(RT)^b} = \frac{[C]^c[D]^d}{[A]^a[B]^b}(RT)^{c+d-a-b}$$

and finally,

$$K_P = K(RT)^{\Delta n} \quad (7.4)$$

In other words, if there is no difference between the number of moles of reactants and products in the balanced equation for the reaction, the two constants are identical. Otherwise, they differ by the factor $(RT)^{\Delta n}$, where $\Delta n$ is the difference between the number of moles of gaseous products and gaseous reactants.

A dimensional analysis of the equilibrium constant expression indicates, at first sight, that the units of $K$ and $K_P$ depend on the reaction stoichiometry. For example, for the reaction

$$Zn(OH)_2(s) \rightleftharpoons Zn^{2+} + 2OH^-$$

$$K = [Zn^{2+}][OH^-]^2$$

and the units of $K$ are $(mol\ L^{-1})^3$. Similarly, for the reaction

$$N_2O_4(g) \rightleftharpoons 2NO_2(g)$$

$$K_P = \frac{P_{NO_2}^2}{P_{N_2O_4}}$$

and, with partial pressures expressed in atmospheres, the units of $K_P$ are atm.

In actuality, $K$ and $K_P$ are dimensionless quantities. The reason for this assertion will become clear in Section 11.8, where the equilibrium constant expression (Equation 7.3) is derived from thermodynamics. Anticipating the result of this derivation, it turns out that the partial pressures of reactants and products are divided by unit pressure (i.e., one atmosphere). Thus, Equation 7.3 should really be written as

$$K_P = \frac{(P_C/1\ \text{atm})^c (P_D/1\ \text{atm})^d}{(P_A/1\ \text{atm})^a (P_B/1\ \text{atm})^b} \tag{7.5}$$

indicating that $K_P$ is always dimensionless. The same statement may be made about the equilibrium constant $K$ since the reagent concentrations are similarly divided by unit molarity. It is, nonetheless, frequently convenient to retain the units of $K$ and $K_P$ since they facilitate the conversion between these two types of constants, as shown in the following example.

---

**Example 7.2** The equilibrium constant $K_P$ for the reaction

$$4HCl(g) + O_2(g) \rightleftharpoons 2Cl_2(g) + 2H_2O(g)$$

is 13.3 atm$^{-1}$ at 480°C. What is the value of the equilibrium constant $K$ at this temperature?

According to Equation 7.4 the two equilibrium constants are related by the expression

$$K_P = K(RT)^{\Delta n}$$

Substituting into this expression we obtain

$$13.3\ \text{atm}^{-1} = K[(0.08206\ \text{L atm mol}^{-1}\ \text{K}^{-1})(753\ \text{K})]^{-1}$$

$$K = 822\ (mol\ L^{-1})^{-1} \blacksquare$$

**c. Properties of the Equilibrium Constant** A number of relationships are implicit in the expression for the equilibrium constant (Equations 7.2 and 7.3).

The numerical value of the equilibrium constant of a reaction depends on the number of moles represented by the stoichiometric coefficients. Consider the following two representations of the reaction between hydrogen and nitrogen to make ammonia, corresponding to the formation of two moles and one mole of $NH_3$, respectively:

(a) $\quad N_2(g) + 3H_2(g) \rightleftarrows 2NH_3(g)$

(b) $\quad \tfrac{1}{2}N_2(g) + \tfrac{3}{2}H_2(g) \rightleftarrows NH_3(g)$

The corresponding values of the equilibrium constant $K_p$ are

$$K_p(a) = \frac{P_{NH_3}^2}{P_{N_2} P_{H_2}^3}$$

$$K_p(b) = \frac{P_{NH_3}}{P_{N_2}^{1/2} P_{H_2}^{3/2}}$$

It is apparent that $K_p(a) = K_p(b)^2$. This same relation also holds true if the equilibrium constant is expressed in terms of molar concentrations rather than partial pressures. In general, multiplying the stoichiometric coefficients of an equation by a factor $f$ changes the corresponding equilibrium constant from $K$ to $K^f$.

A special case of this relation is obtained by setting $f$ equal to $-1$. This corresponds to changing the direction of the reaction. The result is that the equilibrium constants for a forward reaction and the corresponding reverse reaction are reciprocals of each other. For instance, if we write the decomposition of $NH_3$ as

(c) $\quad 2NH_3(g) \rightleftarrows N_2(g) + 3H_2(g)$

we obtain

$$K_p(c) = \frac{P_{N_2} P_{H_2}^3}{P_{NH_3}^2}$$

indicating that $K_p(c) = K_p(a)^{-1}$.

When two equations are added to yield the equation for a third reaction, the corresponding equilibrium constants must be multiplied together to obtain the equilibrium constant of the new reaction. Consider, for example, the sequential oxidation of nitrogen:

(1) $\quad N_2(g) + O_2(g) \rightleftarrows 2NO(g) \qquad K_{p_1} = \dfrac{P_{NO}^2}{P_{N_2} P_{O_2}}$

(2) $\quad 2NO(g) + O_2(g) \rightleftarrows 2NO_2(g) \qquad K_{p_2} = \dfrac{P_{NO_2}^2}{P_{NO}^2 P_{O_2}}$

Adding these equations, we obtain

(3) $\quad N_2(g) + 2O_2(g) \rightleftarrows 2NO_2(g) \qquad K_{p_3} = \dfrac{P_{NO_2}^2}{P_{N_2} P_{O_2}^2}$

It is clear from the expressions for the equilibrium constants that

$$K_{p_3} = K_{p_1} K_{p_2}$$

**Example 7.3** The production of nitric acid from nitric oxide involves the following two sequential reactions:

(1) $\quad 2NO(g) + O_2(g) \rightleftarrows 2NO_2(g)$

(2) $\quad 3NO_2(g) + H_2O(l) \rightleftarrows 2H^+ + 2NO_3^- + NO(g)$

(a) Write an expression for the equilibrium constant of each reaction in terms of the concentrations of reactants and products. (b) Write an equation for the overall reaction and express its equilibrium constant in terms of $K_1$ and $K_2$.

(a) For reaction (1),

$$K_1 = \frac{[NO_2]^2}{[NO]^2[O_2]}$$

and for reaction (2),

$$K_2 = \frac{[H^+]^2[NO_3^-]^2[NO]}{[NO_2]^3}$$

Note that the constant concentration of water is not explicitly included in the equilibrium expression.

(b) Inspection of the two equations indicates that when the coefficients of the first are multiplied by $\frac{3}{2}$ and the two are added, the desired equation is obtained:

(1) $\quad 3NO(g) + \frac{3}{2}O_2(g) \rightleftarrows 3NO_2(g)$

(2) $\quad \underline{3NO_2(g) + H_2O(l) \rightleftarrows 2H^+ + 2NO_3^- + NO(g)}$
$\quad\quad\quad 2NO(g) + \frac{3}{2}O_2(g) + H_2O(l) \rightleftarrows 2H^+ + 2NO_3^-$

Applying the rules for the combination of equilibrium constants, we obtain

$$K = K_1^{3/2}K_2$$

as can be ascertained by writing out and combining the various equilibrium expressions. ∎

**d. Magnitude of Equilibrium Constants** Every chemical reaction can be characterized by an equilibrium constant the value of which depends only on the temperature. Table 7.2 lists the values of $K_p$ for a number of different reactions involving gases. Some reactions have equilibrium constants that are many orders of magnitude larger than unity; by contrast, others have equilibrium constants much smaller than unity. What does the magnitude of the equilibrium constant tell us about a reaction?

In view of the definition of the equilibrium constant, its value relative to unity must tell us something about the relative amounts of products and reactants present at equilibrium and, therefore, about the extent to which a particular reaction occurs. However, the relationship between the two is rather complicated. We have already noted that $K$ depends on the manner in which the equation of a chemical reaction is written. For example, if the stoichiometric coefficients are doubled, the value of the equilibrium constant is squared. However, the position of chemical equilibrium remains unchanged. Furthermore, as we shall presently see, the position of equilibrium can be altered by

**TABLE 7.2 Equilibrium Constants $K_p$ for Reactions Involving Gases at 25°C**

| Reaction | $K_p{}^a$ |
|---|---|
| $N_2(g) + O_2(g) \rightleftarrows 2NO(g)$ | $4.72 \times 10^{-31}$ |
| $PCl_5(g) \rightleftarrows PCl_3(g) + Cl_2(g)$ | $3.04 \times 10^{-7}$ |
| $N_2O_4(g) \rightleftarrows 2NO_2(g)$ | 0.148 |
| $H_2(g) + I_2(s) \rightleftarrows 2HI(g)$ | 0.254 |
| $N_2(g) + 3H_2(g) \rightleftarrows 2NH_3(g)$ | $5.81 \times 10^5$ |
| $2NO(g) + O_2(g) \rightleftarrows 2NO_2(g)$ | $2.23 \times 10^{12}$ |
| $H_2(g) + Br_2(l) \rightleftarrows 2HBr(g)$ | $5.35 \times 10^{18}$ |
| $2SO_2(g) + O_2(g) \rightleftarrows 2SO_3(g)$ | $6.77 \times 10^{24}$ |
| $H_2(g) + Cl_2(g) \rightleftarrows 2HCl(g)$ | $2.47 \times 10^{33}$ |
| $2C(s) + O_2(g) \rightleftarrows 2CO(g)$ | $1.16 \times 10^{48}$ |
| $C(s) + O_2(g) \rightleftarrows CO_2(g)$ | $1.23 \times 10^{69}$ |
| $2H_2(g) + O_2(g) \rightleftarrows 2H_2O(l)$ | $1.22 \times 10^{83}$ |
| $H_2(g) + F_2(g) \rightleftarrows 2HF(g)$ | $7.59 \times 10^{98}$ |

[a]The values apply to the reactions as written, with pressures expressed in atmospheres.

changes in the concentrations of reactants and products and, in some instances, by changes in the total pressure. In spite of these complications, it is generally true that when the equilibrium constant is much larger than unity, the position of equilibrium lies far to the right. On the other hand, when the equilibrium constant is much smaller than unity, the equilibrium point lies far to the left and the reverse reaction is favored. It is when the equilibrium constant lies within just a few orders of magnitude of unity that care must be taken in relating its value to the position of equilibrium.

Why is it that one reaction has a very large equilibrium constant and another does not? The answer to this very fundamental question must be sought in thermodynamics, the branch of chemistry that allows us to predict the spontaneity of various processes (Chapter 11).

## 7.3 The Reaction Quotient

When given amounts of possible reactants and/or products are combined, it is of interest to determine whether a reaction will occur, and if so, in which direction. Such questions are of practical as well as fundamental importance.

In order to determine in which direction a reaction will go, it is useful to introduce the concept of the **reaction quotient**. For the reaction $aA + bB \rightleftarrows cC + dD$ the reaction quotient $Q$ is defined as

$$Q = \frac{[C]^c[D]^d}{[A]^a[B]^b} \tag{7.6}$$

where the concentrations are evaluated at the particular experimental conditions of interest and are *not* necessarily the equilibrium conditions. It is obvious from the definition of the equilibrium constant that at equilibrium, $Q = K$. On the other hand, if the concentration of reactants is larger and that of products smaller than it is at equilibrium, then $Q < K$, while if the opposite condition holds, then $Q > K$. The comparison of the reaction quotient with the equilibrium constant shows the direction in which the reaction proceeds: if $Q < K$, the forward reaction occurs, while if $Q > K$, the reverse reaction occurs. Figure 7.3 shows the time dependence of $Q$ for these various cases. For a reaction in-

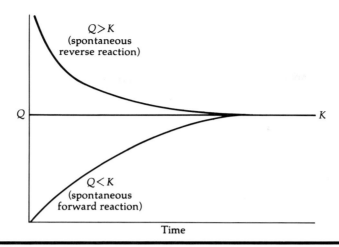

**FIGURE 7.3** Variation of reaction quotient $Q$ with time for $Q > K$, $Q < K$, and $Q = K$.

volving gases, it is more useful to evaluate the reaction quotient in terms of partial pressures since $Q$ can then be compared with $K_p$.

**Example 7.4** If nitrogen, hydrogen, and ammonia are introduced into an evacuated vessel at 25°C so that their partial pressures are 1.5 torr, 0.50 torr, and 1.2 atm, respectively, will more ammonia be formed?

According to Table 7.2, the equilibrium constant for the reaction

$$N_2(g) + 3H_2(g) \rightleftharpoons 2NH_3(g)$$

at 25°C is $K_p = 5.81 \times 10^5$ atm$^{-2}$. The reaction quotient is

$$Q = \frac{P_{NH_3}^2}{P_{N_2} P_{H_2}^3} = \frac{(1.2 \text{ atm})^2}{\left(\dfrac{1.5 \text{ torr}}{760 \text{ torr atm}^{-1}}\right)\left(\dfrac{0.50 \text{ torr}}{760 \text{ torr atm}^{-1}}\right)^3}$$

$$= 2.56 \times 10^{12} \text{ atm}^{-2}$$

Since $Q > K_p$, a net ammount of ammonia will actually decompose rather than be formed as the system proceeds to the equilibrium point. ∎

## 7.4 Solving Equilibrium Problems

Depending on the particular context, equilibrium calculations can be employed in a great many ways, for example, to estimate the extent of dissociation of the molecules of a gas into simpler molecules or atoms, to calculate the acidity of a solution, to predict the occurrence of precipitation, or to predict the maximum yield of products under a given set of conditions. The widespread use of equilibrium calculations in all areas of chemistry underlines the importance of mastering such calculations. Equilibrium calculations frequently involve the solution of complicated algebraic equations. In many instances, these equations can be simplified considerably by making judicious approximations, while still yielding a result of acceptable accuracy. To considerable measure, the art of solving equilibrium problems involves an understanding of these approximations.

**Example 7.5** The equilibrium constant for the reaction $2HBr(g) \rightleftarrows H_2(g) + Br_2(g)$ at 425°C is $1.8 \times 10^{-9}$. A 1.00 mol sample of HBr is heated to this temperature. How many moles of each gas are present at equilibrium?

This problem can be solved by relating to each other the number of moles of each species present at equilibrium by means of the reaction stoichiometry. Next, the equilibrium concentrations are expressed in terms of the equilibrium numbers of moles. The equilibrium concentrations are then inserted in the equilibrium constant expression and the equation is solved.

Let $x$ be the number of moles of HBr that react. Then $1.00 - x$ moles of HBr are present at equilibrium. Since half a mole apiece of $H_2$ and $Br_2$ are formed for each mole of HBr that reacts, the equilibrium number of moles of $H_2$ or $Br_2$ is $x/2$. It is helpful to summarize this information beneath the equation for the reaction:

$$2HBr(g) \rightleftarrows H_2(g) + Br_2(g)$$

Eq. no. of moles:  $1.00 - x$   $x/2$   $x/2$

The equilibrium concentrations are equal to the equilibrium numbers of moles divided by the volume $V$:

$$[HBr] = (1.00 - x)/V \qquad [H_2] = x/2V \qquad [Br_2] = x/2V$$

Expressing the equilibrium constant for the reaction in terms of the concentrations, we obtain

$$K = \frac{[H_2][Br_2]}{[HBr]^2}$$

$$= \frac{(x/2)(x/2)}{(1.00 - x)^2}$$

where a common factor of $V^2$ has been cancelled in the numerator and denominator. We are given that the equilibrium constant of the reaction is $1.8 \times 10^{-9}$. Since for this particular reaction $\Delta n = 0$, $K$ and $K_p$ are identical. Therefore,

$$\frac{(x/2)(x/2)}{(1.00 - x)^2} = 1.8 \times 10^{-9}$$

Simplifying, we obtain

$$\frac{x^2}{(4)(1.00 - x)^2} = 1.8 \times 10^{-9}$$

Taking the square root of both sides of the equation, we get

$$\frac{x}{(2)(1.00 - x)} = 4.24 \times 10^{-5}$$

$$x = 8.48 \times 10^{-5}$$

Therefore,

$$n_{H_2} = n_{Br_2} = x/2 = 4.2 \times 10^{-5} \text{ mol}$$

$$n_{HBr} = 1.00 - x = 1.00 \text{ mol} \quad \blacksquare$$

The next example requires the solution of a more complicated algebraic equation. The solution can be simplified considerably by making some reasonable approximations.

**Example 7.6** The equilibrium constant for the reaction $2IBr(g) \rightleftharpoons I_2(g) + Br_2(g)$ at 145°C is $1.15 \times 10^{-4}$. What are the equilibrium partial pressures of the three gases when IBr and $I_2$ at initial partial pressures of 1.10 atm and 0.0750 atm, respectively, are mixed at this temperature?

Since both reactant and product are present in the initial mixture, the first step is to determine the direction of the reaction, as illustrated in Example 7.4. A closer look at the problem indicates that the answer is obvious: since the initial partial pressure of $Br_2$ is zero, the reaction quotient $Q = 0$, and the reaction must proceed in the forward direction.

Since the gases are in the same container at constant temperature, Dalton's law (i.e., Equation 3.16) shows that the partial pressure of each gas is directly proportional to the number of moles of that gas, $P_i = n_i RT/V$. Therefore, the changes in the partial pressures as equilibrium is reached are directly proportional to the changes in the numbers of moles. The equilibrium partial pressures can consequently be related to each other by the reaction stoichiometry. Let $x$ be the equilibrium partial pressure of $Br_2$. The equilibrium partial pressure of $I_2$ must therefore also increase by $x$ while that of IBr must decrease by $2x$. The equilibrium pressures are summarized below the equation for the reaction:

$$2IBr(g) \rightleftharpoons I_2(g) + Br_2(g)$$

$P_i$ at eq. $(1.10 - 2x)$ $(0.0750 + x)$ $x$
(atm)

The equilibrium constant expression is

$$K_p = \frac{P_{I_2} P_{Br_2}}{P_{IBr}^2}$$

Substituting the appropriate values we get

$$\frac{x(0.0750 + x)}{(1.10 - 2x)^2} = 1.15 \times 10^{-4}$$

This is a quadratic equation in $x$ and can be solved by means of the quadratic formula (see Appendix A2.3). The results are $x = 0.00180$ or $x = -0.0773$. The second answer is physically impossible since $x$, which is the partial pressure of $Br_2$ at equilibrium, cannot be negative. Therefore, $x = 0.00180$.

While this method yields an exact solution, the procedure is fairly lengthy. An approximate solution, correct to within 5 or 10%, is frequently adequate and it is often possible to obtain such an approximate solution. As shown below, this procedure is much shorter than the quadratic formula.

Let us examine the preceding expression for the equilibrium constant. Since $K_p$ is very small, the equilibrium point lies to the left and very little IBr will be converted to $Br_2$. Therefore, $2x \ll 1.10$ and $x$ can be neglected in the denominator. It may also be true that $x$ is small compared to 0.0750, although this must be checked. On the assumption that this is the case, the equilibrium constant expression simplifies to

$$0.0750x/(1.10)^2 = 1.15 \times 10^{-4}$$

The solution is $x = 0.00186$. Since this result is indeed small compared to 0.0750, we were justified in making the approximation.

Whenever an approximate solution is obtained, is is necessary to check the result to see whether it is sufficiently accurate. This can be done by substituting the approximate value of $x$ wherever $x$ was neglected in the equilibrium expression and then solving the resulting equation for the unknown $x$ a second time. This procedure, which is known as the method of **successive approximations,** necessarily yields a more accurate answer because the first solution for $x$ is closer to the correct value than $x = 0$, the value assumed in the first approximation. In general, the error in the answer obtained in the second approximation is significantly smaller than the difference between the answers obtained in the first and second approximations.

Let us apply this successive approximation method to the preceding problem. Substituting $x = 0.00186$ into the equilibrium constant expression wherever $x$ had been neglected, we obtain

$$\frac{x(0.0750 + 0.00186)}{[1.10 - (2)(0.00186)]^2} = 1.15 \times 10^{-4}$$

The solution is $x = 0.00180$, which differs by some 3% from the first answer. Thus, the second approximation must differ from the true answer by substantially less than 3%. In fact, this answer is the same as that obtained by means of the quadratic formula.

Having solved for $x$, we can obtain the equilibrium partial pressures:

$P_{IBr} = 1.10 - 2x = 1.10$ atm

$P_{I_2} = 0.0750 + x = 0.0768$ atm

$P_{Br_2} = x = 0.00180$ atm ∎

Our next example illustrates the use of the approximation method under conditions where, at first sight, it does not appear to be applicable.

---

**Example 7.7** The partial pressures of $H_2$ and $I_2$ introduced into a container at 700 K are 0.10 and 0.20 atm, respectively. What are the equilibrium partial pressures of $H_2$, $I_2$, and HI, and what is the total pressure, given that the reaction $H_2(g) + I_2(g) \rightleftarrows 2HI(g)$ has $K_p = 55$?

The equilibrium partial pressures may be related to each other by means of the reaction stoichiometry in exactly the same manner as in Example 7.6. It is convenient to let $2x$ be the equilibrium partial pressure of HI as this eliminates fractional values of $x$. The equilibrium pressures are summarized:

$$\begin{array}{cccc} & H_2(g) & + \quad I_2(g) & \rightleftarrows 2HI(g) \\ P_i \text{ at eq.} & (0.10 - x) & (0.20 - x) & 2x \\ \text{(atm)} & & & \end{array}$$

The equilibrium expression is

$$K_p = \frac{P_{HI}^2}{P_{H_2}P_{I_2}} = \frac{(2x)^2}{(0.10-x)(0.20-x)} = \frac{4x^2}{(0.10-x)(0.20-x)} = 55$$

We now ask whether this quadratic equation can be solved by the approximation method developed in Example 7.6. A moment's thought indicates that this procedure will not work here. The equilibrium constant for the reaction is substantially larger than unity and most of the hydrogen present initially will consequently react. In these circumstances $x$ must be close to 0.10 and can most certainly not be neglected with respect to this quantity. This statement can readily be confirmed by neglecting $x$ with respect to 0.10 and 0.20 in the equilibrium expression. The resulting equation may be solved for $x$ and the answer is $x = 0.52$, which, of course, is actually larger than 0.2!

Since an approximation method will not work, the problem must be solved by use of the quadratic formula. Combining terms, we obtain

$$51x^2 - 17x + 1.1 = 0$$

The solutions of this equation are $x = 0.25$ and $x = 0.088$. The first solution is not physically meaningful because it exceeds the initial partial pressures of the reactants. Using the second value of $x$ we obtain for the equilibrium partial pressures

$$P_{H_2} = 0.01 \text{ atm} \qquad P_{I_2} = 0.11 \text{ atm} \qquad P_{HI} = 0.18 \text{ atm}$$

Since the total number of moles does not change in this reaction, the total pressure remains constant and is 0.30 atm. This result can be checked by summing the equilibrium partial pressures.

While $x$ cannot be neglected as it is defined here, the problem can be reformulated in a way that makes an approximate solution possible. Since the equilibrium constant is much larger than unity, practically all the hydrogen initially present reacts. The problem can therefore be reformulated by assuming that all but a small amount of hydrogen reacts and by designating the partial pressure of the remaining hydrogen as $x$. The equilibrium partial pressures then are

$$\begin{array}{cccc} & H_2(g) + & I_2(g) & \rightleftarrows & 2HI(g) \\ P_i \text{ at eq.} & x & (0.10 + x) & & (0.20 - 2x) \\ \text{(atm)} & & & & \end{array}$$

where $x$ is small compared to 0.10, the initial partial pressure of $H_2$. Note that since the initial partial pressure of $I_2$ is 0.20 atm, the equilibrium pressure of $I_2$ exceeds that of $H_2$ by 0.10 atm.

Using these values of the partial pressures the equilibrium expression is

$$K_p = \frac{(0.20 - 2x)^2}{x(0.10 + x)} = 55$$

Since $x \ll 0.1$ it can be neglected with respect to this quantity and the expression simplifies to

$$\frac{(0.20)^2}{0.10x} = 55$$

whose solution is $x = 0.0073$. The equilibrium partial pressures are

$$P_{H_2} = 0.0073 \text{ atm} \qquad P_{I_2} = 0.11 \text{ atm} \qquad P_{HI} = 0.19 \text{ atm}$$

in adequate agreement with the exact values. As indicated in Example 7.6, a second iteration would give even closer agreement. ∎

While the approximation illustrated in this example may appear to be merely a matter of convenience, we can readily show that for reactions with a somewhat more complex stoichiometry such an approximation is virtually a matter of necessity. Consider, for example, the reaction

$$N_2(g) + 3H_2(g) \rightleftarrows 2NH_3(g)$$

at low temperatures, where $K_p \gg 1$. Suppose that 1 atm of $N_2$ is mixed with 1 atm of $H_2$. If we let $x$ be the decrease in the partial pressure of $N_2$ from its initial value, the equilibrium expression becomes

$$K_p = \frac{P_{NH_3}^2}{P_{N_2} P_{H_2}^3} = \frac{(2x)^2}{(1-x)(1-3x)^3}$$

Since $K_p \gg 1$, $x$ cannot be neglected with respect to the initial pressures. The resulting equation contains terms in $x$ all the way up to $x^4$ and cannot be solved in a simple manner. By contrast, if we assume that the reaction goes essentially to completion and then let $3x$ be the small remaining partial pressure of $H_2$ at equilibrium, the resulting equilibrium expression can be solved in the manner illustrated in Example 7.7.

The procedure followed in making the approximations illustrated in the preceding two examples may be summarized as follows: *in order to make an approximate solution possible, the unknown must refer to the species present in smallest amount at equilibrium.* When $K_p \ll 1$, this species will be a reaction product while when $K_p \gg 1$, it will be a reactant. When $K_p \approx 1$, the choice is immaterial since only an exact solution of the equilibrium expression yields a satisfactory answer. More specifically, when $K_p \ll 1$, the actual unknown (which we have designated $x$) is either the decrease in the number of moles or in the partial pressure of a reactant, or equivalently, the number of moles or partial pressure of a product formed in the reaction. Similarly, when $K_p \gg 1$, the actual unknown is the remaining number of moles or the remaining partial pressure of a reactant. The reaction stoichiometry must, of course, be taken into account so that, as in Example 7.6, if $2x$ is the decrease in the partial pressure of a reactant, the partial pressure of the product is $x$. The unknown $x$ can then usually be neglected with respect to the initial number of moles or partial pressure of the reactant. The numerical value of $x$ obtained in the solution to the approximate equation serves as a check of the validity of this procedure. These same general considerations also apply to reactions occurring in solution, as we shall see in the next two chapters.

To conclude this section, we consider what is essentially the reverse of the problems solved so far: the determination of the equilibrium constant from data concerning the equilibrium state of the reaction in question.

---

**Example 7.8** A container is filled with carbon dioxide and heated to 1000 K; the gas pressure at this temperature is 0.50 atm. Graphite is added to the container and some of the carbon dioxide is converted to carbon monoxide. The equilibrium pressure at 1000 K is 0.80 atm. What is the value of $K_p$?

The reaction is $CO_2(g) + C(s) \rightleftarrows 2CO(g)$. Letting $x$ be the reduction in the $CO_2$ equilibrium pressure from the initial value, the partial pressures at equilibrium are

$$CO_2(g) + C(s) \rightleftarrows 2CO(g)$$

$P_i$ at eq. (atm): $0.50 - x$ , $2x$

The equilibrium expression is

$$K_p = \frac{P_{CO}^2}{P_{CO_2}} = \frac{(2x)^2}{0.50 - x}$$

The value of $x$ can be obtained from the total gas pressure at equilibrium:

$$P = P_{CO_2} + P_{CO} = (0.5 - x) + (2x) = 0.80 \text{ atm}$$

$$x = 0.30 \text{ atm}$$

Substituting into the equilibrium expression we get

$$K_p = \frac{(0.60 \text{ atm})^2}{0.20 \text{ atm}} = 1.8 \text{ atm} \quad \blacksquare$$

## 7.5 Dissociation Equilibria

Reactions involving the dissociation of a molecule frequently involve forward and reverse reactions having nearly comparable rates. Thus, chemical equilibrium is of importance in such reactions. The following are some typical examples of dissociation reactions:

$$N_2O_4(g) \rightleftarrows 2NO_2(g)$$

$$PCl_5(g) \rightleftarrows PCl_3(g) + Cl_2(g)$$

$$O_2(g) \rightleftarrows 2O(g)$$

$$NH_4Cl(s) \rightleftarrows NH_3(g) + HCl(g)$$

Dissociation reactions can be characterized by their **degree of dissociation** $f$, defined as the fraction of the number of moles of starting material that is dissociated at equilibrium.

When a substance dissociates, the total number of moles necessarily increases. Consequently, if the dissociation products are gases and the temperature and volume remain constant, the total pressure also increases. The degree of dissociation of a substance can be related to the total equilibrium pressure. Since the total pressure of a gas mixture is easily measured, the degree of dissociation can be obtained from such a measurement.

Consider a dissociation reaction of the type

$$A_2(g) \rightleftarrows 2A(g) \tag{7.7}$$

Of the preceding examples, the dissociation of $N_2O_4$ and of $O_2$ obey this stoichiometry. For each mole of $A_2$ present initially, $f$ moles dissociate. At equilibrium, $1 - f$ moles of $A_2$ and $2f$ moles of $A$ are present. The composition of the equilibrium mixture can be expressed in terms of the mole fractions of the two gases:

$$X_{A_2} = \frac{n_{A_2}}{n_{tot}} = \frac{1-f}{1+f}$$

$$X_A = \frac{n_A}{n_{tot}} = \frac{2f}{1+f}$$

where the total number of moles present at equilibrium is $1 + f$. Note that

$$X_{A_2} + X_A = \frac{1+f}{1+f} = 1$$

as expected from the definition of the mole fraction (Section 3.5). The mole fractions can be related to the partial pressures by means of Dalton's law (Equation 3.18), $P_i = X_i P$. We obtain

$$P_{A_2} = \frac{(1-f)P}{1+f} \qquad P_A = \frac{2fP}{1+f}$$

The equilibrium constant expression for the dissociation reaction is

$$K_p = \frac{P_A^2}{P_{A_2}}$$

Substituting for the partial pressures, we get

$$K_p = \frac{[2fP/(1+f)]^2}{(1-f)P/(1+f)}$$

which simplifies to

$$K_p = \frac{4f^2 P}{1-f^2} \tag{7.8}$$

Thus, a measurement of the equilibrium pressure and a knowledge of $K_p$ permit the determination of the degree of dissociation. However, it must be kept in mind that Equation 7.8 only applies to reactions obeying the stoichiometry of Equation 7.7. A different result will be obtained for a dissociation reaction with a different stoichiometry, as shown in the following example.

---

**Example 7.9** The equilibrium constant for the dissociation of gaseous $PCl_5$ into gaseous $PCl_3$ and $Cl_2$ at 450 K is $1.50 \times 10^{-2}$ atm. Exactly 1 mol of $PCl_5$ is heated to 450 K at a constant total pressure of 2.00 atm. (a) Derive an expression relating the equilibrium constant, the total pressure, and the degree of dissociation. (b) What is the degree of dissociation of $PCl_5$ and what are the equilibrium partial pressures for the given conditions?

(a) The equation for the dissociation of $PCl_5$ is $PCl_5(g) \rightleftarrows PCl_3(g) + Cl_2(g)$. For each mole of $PCl_5$ present initially, $f$ mol of $PCl_3$, $f$ mol of $Cl_2$, and $1 - f$ mol of $PCl_5$ will be present at equilibrium. The equilibrium mole fractions are therefore

$$X_{PCl_3} = \frac{f}{1+f} \qquad X_{Cl_2} = \frac{f}{1+f} \qquad X_{PCl_5} = \frac{1-f}{1+f}$$

The equilibrium partial pressures can be expressed in terms of the corresponding mole fractions and the total pressure by means of Dalton's law:

$$P_{PCl_3} = \frac{fP}{1+f} \qquad P_{Cl_2} = \frac{fP}{1+f} \qquad P_{PCl_5} = \frac{(1-f)P}{1+f}$$

The equilibrium constant expression is

$$K_p = \frac{P_{PCl_3} P_{Cl_2}}{P_{PCl_5}}$$

Substituting for the partial pressures, we obtain

$$K_p = \frac{[fP/(1+f)][(fP/(1+f)]}{(1-f)P/(1+f)}$$

which reduces to

$$K_p = \frac{f^2 P}{1-f^2}$$

(b) The degree of dissociation can be obtained by numerical substitution into this expression:

$$1.50 \times 10^{-2} = \frac{2.00 f^2}{1 - f^2}$$

$$2.015 f^2 = 1.50 \times 10^{-2}$$

$$f = 0.0863$$

Thus, 8.63% of $PCl_5$ dissociates under the given conditions.

The equilibrium partial pressures can be obtained by means of the relationships developed in (a):

$$P_{PCl_3} = P_{Cl_2} = \frac{fP}{1+f} = \frac{(0.0863)(2.00 \text{ atm})}{1 + 0.0863}$$

$$= 0.159 \text{ atm}$$

$$P_{PCl_5} = \frac{(1-f)P}{1+f} = \frac{(1 - 0.0863)(2.00 \text{ atm})}{1 + 0.0863}$$

$$= 1.68 \text{ atm} \quad \blacksquare$$

The final example of this section involves a heterogeneous dissociation equilibrium, that is, one involving substances present in two different phases.

**Example 7.10** The equilibrium constant of the reaction $NH_4HS(s) \rightleftarrows NH_3(g) + H_2S(g)$ at 20°C is $K_p = 0.110$ (atm)². (a) A 2.55 g sample of $NH_4HS$ is placed in an evacuated 1.00 L container at 25°C. What are the equilibrium partial pressures of $NH_3$ and $H_2S$? What are the equilibrium mass and the degree of dissociation of $NH_4HS$? (b) In a separate experiment, the container is filled with $NH_3$ and $H_2S$ gases of 25°C; the initial partial pressure of each gas is 0.100 atm. What are the equilibrium partial pressures and what is the mass of $NH_4HS$ at equilibrium?

(a) Since the initial pressures of $NH_3$ and $H_2S$ are zero, the reaction quotient,

$$Q = P_{NH_3} P_{H_2S} = 0$$

and $NH_4HS$ decomposes as the system moves towards equilibrium. The stoichiometry of the reaction indicates that $NH_3$ and $H_2S$ are formed in equimolar amounts; their partial pressures are therefore the same. The equilibrium constant expression is

$$K_p = 0.110 = P_{NH_3} P_{H_2S} = P_{NH_3}^2$$

and the solution is $P_{NH_3} = P_{H_2S} = 0.332$ atm.

The result must be taken as provisional because we do not know as yet whether the initial amount of $NH_4HS$ was large enough to yield the decomposition products at the indicated pressures. To check this point and at the same time answer the rest of the question we convert the partial pressures to the corresponding numbers of moles by means of Dalton's law:

$$P_i = n_i RT/V$$

$$n_i = \frac{P_i V}{RT} = \frac{(0.332 \text{ atm})(1.00 \text{ L})}{(0.08206 \text{ L atm mol}^{-1} \text{ K}^{-1})(298 \text{ K})}$$

$$= 0.0136 \text{ mol}$$

The initial mass of $NH_4HS$ corresponds to

$$n_{NH_4HS} = \frac{2.55 \text{ g}}{51.11 \text{ g mol}^{-1}} = 0.0499 \text{ mol}$$

indicating that there is enough $NH_4HS$ present to yield the desired amounts of products. Since 1 mol of $NH_4HS$ decomposes for each mole of $NH_3$ or $H_2S$ formed, the degree of dissociation is

$$f = \frac{0.0136 \text{ mol}}{0.0499 \text{ mol}} = 0.273$$

The mass of $NH_4HS$ present at equilibrium is

$$[(0.0499 - 0.0136) \text{ mol}](51.11 \text{ g mol}^{-1}) = 1.86 \text{ g}$$

(b) We first evaluate the reaction quotient:

$$Q = P_{NH_3} P_{H_2S} = (0.100 \text{ atm})(0.100 \text{ atm})$$

$$= 0.0100 \text{ (atm)}^2$$

Since $K_p = 0.110$ $(atm)^2$, we have $Q < K_p$, indicating that only the forward reaction can occur as the system moves towards equilibrium. Since the formation of $NH_4HS$ involves the reverse reaction, and this reaction cannot take place, $NH_4HS$ cannot be formed for the conditions of the problem. ∎

## 7.6 Factors Affecting the Position of Equilibrium

Once a chemical reaction reaches equilibrium, no further net changes in concentrations or partial pressures occur. However, owing to the dynamic nature of the equilibrium between forward and reverse reactions, the equilibrium point can be shifted by various changes in experimental conditions. The factors which can affect the position of equilibrium include changes in concentration or partial pressure of reactants or products, and changes in total pressure and temperature.

The effects on equilibrium of changes in the conditions that determine the equilibrium can be qualitatively understood by means of a general principle formulated in the late nineteenth century by Henri Le Chatelier (1850–1936). **Le Chatelier's principle** states that *a change in one of the variables governing the position of equilibrium shifts this position in the direction that tends to minimize the effect of the change.* With the help of this principle we can determine qualitatively the direction in which equilibrium will shift in response to a change in conditions. Such qualitative predictions are helpful as a guide to the quantitative analysis of equilibrium by means of the calculations described in Sections 7.4 and 7.5 and as a check of the correctness of such calculations.

***a. The Effect of Concentration Changes*** A change in the concentration or partial pressure of a reactant or product will shift the position of chemical equilibrium. According to Le Chatelier's principle, an increase in the concentration of a reactant causes a system to readjust by using up some of the added reactant, thereby shifting the position of equilibrium in the direction of product formation. Similarly, an increase in the concentration of a product causes some of the excess product to react and results in a concomitant shift in the position of equilibrium towards the reactants. Decreases in concentrations of reactants and products have exactly the opposite effects. The changes in concentration that result from the addition or removal of one of the substances from an equilibrium mixture of $H_2$, $I_2$, and HI are shown graphically in Figure 7.4. Compare this figure with Figure 7.1, which shows the undisturbed approach to equilibrium. Note that when HI is added to the equilibrium mixture, some but not all of the excess HI is consumed by the reverse reaction. If all the added HI reacted, the reaction quotient would be smaller than the equilibrium constant and the system would not be at equilibrium.

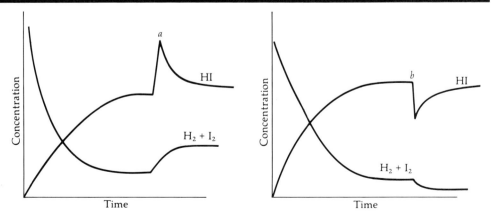

**FIGURE 7.4** Effect of changes in concentration on equilibrium in the reaction $H_2(g) + I_2(g) \rightleftarrows 2HI(g)$. The addition of product at point *a* shifts the equilibrium to the left, while the removal of product at point *b* shifts the equilibrium to the right.

**242** Chemical Equilibrium  7.6

The continuous removal of a reaction product from an equilibrium mixture is one of the standard ways of driving a reaction to completion, thereby increasing the yield of the desired product. In order to remove a reaction product, we take advantage of any major differences between physical or chemical properties of the product in question and those of the other species present. These differences include differences in solubility and volatility. Double displacement reactions (Section 2.5d) occur precisely because one of the reaction products is removed from the reaction mixture owing to its volatility, insolubility, etc.

---

**Example 7.11**  In Example 7.5 it was shown that when 1.00 mol HBr is heated to 425°C, a small amount decomposes to $H_2$ and $Br_2$; the equilibrium mixture was found to consist of $4.2 \times 10^{-5}$ mol $H_2$, $4.2 \times 10^{-5}$ mol $Br_2$, and 1.00 mol HBr. Suppose that when this mixture is passed through a certain solvent, 99.0% of the $H_2$ and $Br_2$ are removed but only 1.0% of the HBr is removed. The remaining gas mixture is reheated to 425°C. (a) Show that additional $H_2$ and $Br_2$ are formed. (b) Determine the composition of the new equilibrium mixture. The equilibrium constant of the reaction $2HBr(g) \rightleftarrows H_2(g) + Br_2(g)$ is $1.8 \times 10^{-9}$.

(a) The mixture of gases that have passed through the solvent consists of

$$n_{HBr} = (0.99)(1.00 \text{ mol}) = 0.99 \text{ mol}$$

$$n_{H_2} = (0.010)(4.2 \times 10^{-5} \text{ mol}) = 4.2 \times 10^{-7} \text{ mol}$$

$$n_{Br_2} = (0.010)(4.2 \times 10^{-5} \text{ mol}) = 4.2 \times 10^{-7} \text{ mol}$$

The reaction quotient for this mixture is

$$Q = \frac{[H_2][Br_2]}{[HBr]^2} = \frac{n_{H_2} n_{Br_2}}{n_{HBr}^2}$$

$$= \frac{(4.2 \times 10^{-7} \text{ mol})(4.2 \times 10^{-7} \text{ mol})}{(0.99 \text{ mol})^2}$$

$$= 1.8 \times 10^{-13}$$

Since this value is smaller than the equilibrium constant, additional HBr decomposes. This conclusion is also predictable by Le Chatelier's principle: the removal of a reaction product from an equilibrium mixture results in the formation of additional product.

(b) As in Example 7.5, let $x$ be the number of moles of HBr that react. We then have the following equilibrium mixture:

| | $2HBr(g)$ $\rightleftarrows$ | $H_2(g)$ | $+$ | $Br_2(g)$ |
|---|---|---|---|---|
| Eq. no. of moles | $(0.99 - x)$ | $\left(4.2 \times 10^{-7} + \dfrac{x}{2}\right)$ | | $\left(4.2 \times 10^{-7} + \dfrac{x}{2}\right)$ |

As in Example 7.5, we have

$$K = \frac{[H_2][Br_2]}{[HBr]^2} = \frac{n_{H_2} n_{Br_2}}{n_{HBr}^2}$$

Substituting for the number of moles, we obtain

$$\frac{[4.2 \times 10^{-7} + (x/2)]^2}{(0.99 - x)^2} = 1.8 \times 10^{-9}$$

Taking the square root of both sides of the equation we get

$$\frac{4.2 \times 10^{-7} + (x/2)}{0.99 - x} = 4.2 \times 10^{-5}$$

and the solution is $x = 8.3 \times 10^{-5}$. The composition of the equilibrium mixture is obtained by substituting this value of $x$ into the equilibrium number of moles. The result is

$$n_{\text{HBr}} = (0.99 - 8.3 \times 10^{-5}) \text{ mol} = 0.99 \text{ mol}$$

$$n_{\text{H}_2} = n_{\text{Br}_2} = [4.2 \times 10^{-7} + (8.3 \times 10^{-5})/2] \text{ mol} = 4.2 \times 10^{-5} \text{ mol} \quad \blacksquare$$

**b. The Effect of Pressure** An increase in the overall pressure of an equilibrium gas mixture, which can be achieved by decreasing the gas volume, increases the total concentration of gas molecules. According to Le Chatelier's principle, the system adjusts to this stress by reacting so as to decrease the number of gas molecules. The effect of pressure thus depends on whether the number of moles of gaseous substances changes as a result of the reaction. If $\Delta n = 0$, a change in $P$ has no effect on chemical equilibrium. [This statement is only true, however, so long as the pressure does not become so high that departures from ideal behavior become significant (Section 7.7).] If $\Delta n > 0$, that is, if the number of moles of gaseous products exceeds that of gaseous reactants, an increase in $P$ shifts the equilibrium towards the reactants while a decrease in $P$ has the opposite effect. On the other hand, if $\Delta n < 0$, the changes are exactly in the opposite direction.

As an example of the role of pressure, consider the dissociation equilibrium $N_2O_4(g) \rightleftarrows 2NO_2(g)$. Since $\Delta n > 0$, Le Chatelier's principle predicts that an increase in pressure shifts the equilibrium to the left and thereby decreases the degree of dissociation; the opposite effect results from a decrease in pressure. This prediction is born out by Equation 7.8, relating the degree of dissociation to the total pressure and to the equilibrium constant,

$$K_P = \frac{4f^2 P}{1 - f^2}$$

To see the relation between $P$ and $f$ as explicitly as possible, this expression may be rewritten in the form

$$f = \left(\frac{K_P}{K_P + 4P}\right)^{1/2} \tag{7.9}$$

Since $K_P$ is constant at a given temperature, it is apparent that an increase in $P$ requires that $f$ decrease while a decrease in $P$ has the opposite effect, in agreement with Le Chatelier's principle. The effect of a change in pressure can be achieved most directly by changing the volume occupied by the gas mixture: an expansion of the gas leads to a reduction in the total pressure and leads to an increase in the degree of dissociation while compression of the gas has the opposite effect.

The extent of a chemical reaction involving gases is frequently optimized by suitable adjustment of the gas pressure. The formation of ammonia from the elements is a good example. Although the equilibrium constant is large at low temperatures, the reaction proceeds at an exceedingly slow rate. At high temperatures, where the reaction occurs rapidly, the equilibrium constant is very small. The reaction is usually run at intermediate temperatures, where suitable catalysts can speed the reaction up to an acceptable rate, and where the yield can be optimized at high pressure. [A **catalyst** is a substance that speeds up the rate of a reaction without affecting the position of chemical equilibrium (Section 14.11).]

*c. The Effect of Temperature on Equilibrium*   Experimental studies have shown that the numerical values of equilibrium constants generally vary with temperature. Figure 7.5 displays the temperature dependence of the equilibrium constants of some typical reactions. The functional form of the dependence of $K_p$ on $T$ is considered in Section 11.9a; here we wish to examine the more general result that $K_p$ increases with $T$ for some reactions but decreases with $T$ for others. What factors determine this rather fundamental difference? Experience shows that the variation of $K$ or $K_p$ with $T$ is dependent on heat being absorbed

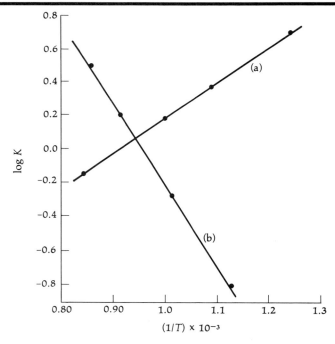

**FIGURE 7.5**   Temperature dependence of the equilibrium constants of the reactions (a) $CO_2(g) + H_2(g) \rightleftarrows CO(g) + H_2O(g)$ (an exothermic reaction) and (b) $SO_3(g) \rightleftarrows SO_2(g) + \frac{1}{2}O_2(g)$ (an endothermic reaction). The increase in log $K$ with $T^{-1}$ for (a) is equivalent to a decrease in $K$ with $T$; similarly, the decrease in log $K$ with $T^{-1}$ for (b) is equivalent to an increase in $K$ with $T$. The linear form of the plot is based on thermodynamics (see Chapter 11).

or released in the reaction: the equilibrium constant of an endothermic reaction (Section 2.4) increases with temperature while that of an exothermic reaction decreases with temperature.

These observations are readily understandable on the basis of Le Chatelier's principle. Heating causes an increase in temperature. An equilibrium system responds to this stress by a shift in the direction that absorbs some of the heat. Thus, for an endothermic reaction, the shift favors the formation of products because some of the added heat is thereby absorbed. For example, the reaction $N_2(g) + O_2(g) \rightleftharpoons 2NO(g)$ absorbs 90.4 kJ of heat per mole of NO. We see in Table 7.2 that the equilibrium constant at 298 K is only $4.7 \times 10^{-31}$; the equilibrium mixture consists virtually only of reactants, consistent with the well known fact that nitrogen and oxygen coexist in the atmosphere. If the temperature is increased by an order of magnitude to 3000 K, the equilibrium constant increases by nearly 29 orders of magnitude to 0.017 and the reaction can occur to a significant extent. As mentioned in Section 4.6, the rates of chemical reactions increase sharply with temperature. Thus, high temperatures make it possible to obtain the highest yields of products of endothermic reactions in the shortest possible time.

We can similarly see that, in the case of an exothermic reaction, Le Chatelier's principle predicts that heating must favor the formation of reactants, for this is the direction of reaction that absorbs heat. Thus, the equilibrium constants of exothermic reactions decrease with increasing temperature. For example, the reaction $2SO_2(g) + O_2(g) \rightleftharpoons 2SO_3(g)$, which is an intermediate step in the production of sulfuric acid, is exothermic, liberating 98.3 kJ of heat per mole of $SO_3$. The equilibrium constant is approximately $7 \times 10^{24}$ at room temperature, but the reaction occurs very slowly. The reaction does occur rapidly at 1300 K, but the equilibrium constant at this temperature is only about 0.1. The use of a platinum catalyst increases the rate of the reaction at moderate temperatures and makes the reaction viable. As noted earlier in this chapter, a catalyst does *not* change the position of chemical equilibrium.

## 7.7 Chemical Equilibrium for Real Gases

The equilibrium constant expressions (Equations 7.2 and 7.3) yield constant values of $K$ or $K_p$ so long as the species present at equilibrium behave ideally. We have seen (Section 4.7) that at sufficiently high pressures, gases do not obey the ideal gas law. Under these conditions the value of the equilibrium constant derived from the equilibrium partial pressures is no longer constant (as in Table 7.1) but changes with pressure. As an illustration of this effect, Table 7.3 summarizes the equilibrium pressures and resulting $K_p$ for the ammonia synthesis reaction as the total pressure increases from 10 to $10^3$ atm. Note that $K_p$ changes by nearly a factor of 4 over this interval, clearly showing a breakdown of Equation 7.3. Parenthetically, the observed deviation is in the direction of increasing the yield of ammonia at high pressure, now due to both the increase in $K_p$ as well as the already discussed effect of pressure.

In order to restore the concept of an equilibrium constant that remains invariant even when the reagents do not behave ideally, it is necessary to correct the pressures or concentrations for deviations from ideal behavior. The study of chemical equilibrium under non-ideal conditions constitutes an important topic in physical chemistry.

**TABLE 7.3** Pressure Dependence of Chemical Equilibrium in the Synthesis of Ammonia from the Elements at 450°C

| Total pressure (atm) | Equilibrium pressures (atm) |  |  | $K_p{}^a$ |
|---|---|---|---|---|
|  | $N_2$ | $H_2$ | $NH_3$ |  |
| 10 | 2.44 | 7.35 | 0.204 | 0.0066 |
| 50 | 11.3 | 34.1 | 4.58 | 0.0068 |
| 100 | 20.9 | 62.7 | 16.35 | 0.0072 |
| 300 | 48.4 | 145 | 106.6 | 0.0088 |
| 600 | 69.5 | 208 | 322 | 0.0129 |
| 1000 | 76.5 | 229 | 694 | 0.0231 |

$^a$The value of $K_p$ corresponds to the reaction $\frac{1}{2}N_2 + \frac{3}{2}H_2 \rightleftarrows NH_3$.

## Conclusion

When substances are combined, chemical reactions occur in the direction that leads to equilibrium—the point at which the rates of the forward and backward reactions become equal to each other and the concentrations or partial pressures (for gases) of reactants and products become constant. The equilibrium concentrations are determined by the value of the equilibrium constant and by the initial conditions. So long as a reaction involves ideal gases or ideal solutions, the equilibrium constant depends only on the temperature. The position of chemical equilibrium can be affected by changes in conditions, e.g., concentrations, pressure, and temperature. The direction in which the equilibrium point shifts is predicted by Le Chatelier's principle.

In this chapter we have not examined the question of why some reactions have large values of $K$ while others have small values. Rather, we have developed the methodology of obtaining equilibrium concentrations on the basis of the initial conditions and a given value of $K$. The ability to perform equilibrium calculations is of widespread importance in chemistry for it is often necessary to know the amounts of reactants and/or products present at equilibrium. The more fundamental question of why some reactions have large equilibrium constants while others have small ones can only be answered on the basis of thermodynamics (Chapter 11).

## Problems

1. Write equilibrium expressions in terms of $K_p$ for each of the following reactions:
   (a) $CO_2(g) + H_2(g) \rightleftarrows CO(g) + H_2O(g)$
   (b) $CO_2(g) + H_2(g) \rightleftarrows CO(g) + H_2O(l)$
   (c) $Br_2(l) + F_2(g) \rightleftarrows 2BrF(g)$
   (d) $4NH_3(g) + 5O_2(g) \rightleftarrows 4NO(g) + 6H_2O(g)$
   (e) $6CO_2(g) + 6H_2O(l) \rightleftarrows C_6H_{12}O_6(s) + 6O_2(g)$

2. The equilibrium constant for the reaction $H_2(g) + I_2(g) \rightleftarrows 2HI(g)$ at 700 K is 55.3. What is the equilibrium constant at this temperature for the reaction $HI(g) \rightleftarrows \frac{1}{2}H_2(g) + \frac{1}{2}I_2(g)$?

3. Express the equilibrium constant for the reaction $2C(s) + 3H_2(g) \rightleftarrows C_2H_6(g)$ in terms of the equilibrium constants for the following combustion reactions:

$2C_2H_6(g) + 7O_2(g) \rightleftarrows 4CO_2(g) + 6H_2O(l)$   $K_1$

$C(s) + O_2(g) \rightleftarrows CO_2(g)$   $K_2$

$2H_2(g) + O_2(g) \rightleftarrows 2H_2O(l)$   $K_3$

4. Nitrogen dioxide is formed in the oxidation of nitric oxide by oxygen: $2NO(g) + O_2(g) \rightleftharpoons 2NO_2(g)$. When NO is burned in oxygen at 700 K, the equilibrium partial pressures are $P_{NO} = 2.1 \times 10^{-2}$ atm, $P_{O_2} = 3.0 \times 10^{-3}$ atm, and $P_{NO_2} = 1.2$ atm. What is the equilibrium constant (a) for this reaction and (b) for the reaction $NO_2(g) \rightleftharpoons NO(g) + \frac{1}{2}O_2(g)$?

5. The equilibrium constant for the reaction $2NO_2(g) \rightleftharpoons N_2O_4(g)$ at 100° C is $1.2 \times 10^{-2}$ atm$^{-1}$. Will any of the following mixtures of these gases undergo further reaction at this temperature? If so, in which direction?
   (a) $P_{N_2O_4} = 0.002$ atm, $P_{NO_2} = 0.4$ atm.
   (b) $P_{N_2O_4} = 0.4$ atm, $P_{NO_2} = 0.002$ atm.
   (c) $P_{N_2O_4} = 0.02$ atm, $P_{NO_2} = 1$ atm.
   (d) $P_{N_2O_4} = 0$ atm, $P_{NO_2} = 0.0001$ atm.

6. The equilibrium constant for the reaction $2NO(g) + Br_2(g) \rightleftharpoons 2NOBr(g)$ at a certain temperature is 100 atm$^{-1}$. The three substances are introduced into a container so that their partial pressures if no reaction occurred would be $P_{NO} = 0.30$ atm, $P_{Br_2} = 0.10$ atm, and $P_{NOBr} = 1.10$ atm. In which direction, if any, does the reaction proceed?

7. Sulfuryl chloride partially dissociates into sulfur dioxide and chlorine: $SO_2Cl_2(g) \rightleftharpoons SO_2(g) + Cl_2(g)$. When the temperature is 300 K and the pressure 0.900 atm, 12.5% of the $SO_2Cl_2$ is dissociated at equilibrium. What are the values of $K_p$ and $K$?

8. When graphite is added to a container filled with 0.915 atm carbon dioxide at a certain temperature, carbon monoxide is formed and the pressure rises to 1.42 atm at equilibrium. What is the value of $K_p$ at this temperature?

9. The density of gaseous nitrosyl bromide stored in a vessel at 25°C and a pressure of 39.1 torr is 0.186 g L$^{-1}$. This value differs from that expected for the pure compound owing to its dissociation by the reaction $2NOBr(g) \rightleftharpoons 2NO(g) + Br_2(g)$. What is the equilibrium constant of this reaction?

10. It was found in a study of the reaction $3Fe(s) + 4H_2O(g) \rightleftharpoons Fe_3O_4(s) + 4H_2(g)$ at 1200 K that when the equilibrium partial pressure of water vapor was 15.0 torr, the total pressure of the equilibrium mixture was 36.3 torr. (a) What is the equilibrium constant of the reaction? (b) What fraction of $H_2O$ is reduced to $H_2$ at a total pressure of 1.00 atm when it is passed over Fe at 1200 K?

11. The water gas reaction, $C(s) + H_2O(g) \rightleftharpoons CO(g) + H_2(g)$, is one of the reactions that is useful for coal gasification since the reaction products burn exothermically in oxygen. At 1300 K, $K_p = 48$. What mass of carbon is gasified when 10.0 atm $H_2O$ vapor is introduced into a 100 L vessel containing coal?

12. Solid ammonium iodide dissociates by the reaction $NH_4I(s) \rightleftharpoons NH_3(g) + HI(g)$. In a particular experiment at 670 K excess $NH_4I$ is introduced into an evacuated container and the equilibrium pressure is found to be 0.920 atm. Eventually, 20% of the HI decomposes into $H_2$ and $I_2$. Calculate the final equilibrium pressure in the container.

13. The equilibrium constant for the reaction $FeO(s) + CO(g) \rightleftharpoons Fe(s) + CO_2(g)$ is 0.403 at 1000 K. Pure CO is passed over powdered FeO at this temperature so that equilibrium is reached. What is the composition of the exhaust gas?

14. The equilibrium constant of the reaction $N_2O_4(g) \rightleftharpoons 2NO_2(g)$ at 300 K is 0.14 atm. What pressure would result at equilibrium from placing 2.50 g $N_2O_4$ into an evacuated 1.50 L vessel at this temperature?

15. At a certain temperature, 1.00 mol of an equimolar mixture of gaseous $CO_2$ and $H_2$ is placed in an evacuated container, where the gases react to form CO and $H_2O$. When equilibrium is attained, 0.400 mol of each reactant is still present. At this point another 0.400 mol of each reactant is added. Find the composition of the new equilibrium state.

16. The partial pressures of an equilibrium mixture of $N_2O_4(g)$ and $NO_2(g)$ are $P_{N_2O_4} = 0.33$ atm and $P_{NO_2} = 1.2$ atm at a certain temperature. The volume of the container is doubled; find the partial pressures of the two gases when a new equilibrium is established.

17. The equilibrium constant for the reaction $CCl_4(g) \rightleftharpoons C(s) + 2Cl_2(g)$ at 700 K is 0.76 atm. Determine the initial pressure of carbon tetrachloride that will produce a total equilibrium pressure of 1.2 atm.

18. For the reaction $2SO_2(g) + O_2(g) \rightleftharpoons 2SO_3(g)$ the value of K at 1170 K is $3.20 \times 10^{-4}$ (mol L$^{-1}$)$^{-1}$. (a) Determine the value of $K_p$. (b) What is the equilibrium composition and total pressure if 2.20 mol of $SO_3$ is introduced into a 40.0 L flask at this temperature?

19. A sample of ammonia is placed in a vessel at 25°C. If 35% decomposes to the elements at equilibrium, what was the initial pressure of ammonia? Use the data in Table 7.2.

20. (a) What is the density of gaseous $N_2O_4$ in equilibrium with gaseous $NO_2$ at 25°C and a total pressure of 1.2 atm? (b) What is the apparent molecular weight of the mixture? Use the data in Table 7.2.

21. The equilibrium constant for the reaction $N_2(g) + O_2(g) \rightleftharpoons 2NO(g)$ is $1.5 \times 10^{-4}$ at 1000 K. (a) What is the equilibrium mole fraction of NO when a mixture of $N_2$ and $O_2$ in which the mole fraction of $N_2$ is 0.35 reacts at this temperature? At this same temperature $N_2$ and $O_2$ also undergo the reaction $N_2(g) + 2O_2(g) \rightleftharpoons 2NO_2(g)$ for which $K_p = 1.0 \times$

$10^{-5}$ atm$^{-1}$. (b) Assuming that this second reaction is the only reaction occurring when a mixture of $N_2$ and $O_2$ with this composition undergo reaction, what is the equilibrium mole fraction of $NO_2$ if the total pressure at equilibrium is 0.12 atm? (c) Assuming, now, that both reactions occur simultaneously, at what total equilibrium pressure will the yield of $NO_2$ be equal to that of NO?

22. At 5000 K and 1.000 atm, 83.0% of the oxygen molecules in a sample have dissociated to atomic oxygen. At what pressure will 95.0% of the molecules dissociate at this temperature?

23. The equilibrium constant for the reaction $CaCO_3(s) \rightleftharpoons CaO(s) + CO_2(g)$ at 1000 K is $5.98 \times 10^{-3}$ mol L$^{-1}$. If 0.200 mol of $CaCO_3$ is placed in an evacuated 15.0 L vessel at this temperature, determine (a) the number of moles of $CO_2$ present in the vessel at equilibrium, (b) the total equilibrium gas pressure, and (c) the fraction of the $CaCO_3$ that has decomposed when equilibrium is attained.

24. For the reaction $PCl_5(g) \rightleftharpoons PCl_3(g) + Cl_2(g)$ at 700 K, the equilibrium constant is 11.5 atm. Suppose that 2.450 g of $PCl_5$ is placed in an evacuated 500 mL bulb which is then heated to 700 K. (a) What would the pressure of $PCl_5$ be if it did not dissociate? (b) What is the partial pressure of $PCl_5$ at equilibrium? (c) What is the total pressure at equilibrium? (d) What is the degree of dissociation of $PCl_5$?

25. The equilibrium constant for the dissociation of gaseous $PCl_5$ into gaseous $PCl_3$ and $Cl_2$ at 450 K is $1.50 \times 10^{-2}$ atm. What are the equilibrium partial pressures of $PCl_5$, $PCl_3$, and $Cl_2$ if a mixture of 1.00 mol of $PCl_5$ and 1.50 mol of $PCl_3$ is heated to 450 K at a total pressure of 2.00 atm? What is the degree of dissociation of $PCl_5$ under these conditions?

26. Predict how the equilibrium yield of sugar produced by photosynthesis in the endothermic reaction $6CO_2(g) + 6H_2O(l) \rightleftharpoons C_6H_{12}O_6(s) + 6O_2(g)$ would be affected by (a) an increase in the partial pressure of $CO_2$, (b) an increase in the partial pressure of $O_2$, (c) the removal of half the sugar present, (d) an increase in the amount of water present, (e) the addition of a catalyst, (f) an increase in total pressure, and (g) an increase in temperature.

27. In the reaction $CaCO_3(s) \rightleftharpoons CaO(s) + CO_2(g)$ how will the amount of $CaCO_3$ present at equilibrium be affected by (a) the removal of $CO_2$ from the equilibrium system, (b) an increase in pressure, and (c) the addition of CaO.

28. For each of the following systems at equilibrium predict the effect of (a) heating the system at constant pressure and (b) decreasing the volume at constant temperature:

(1) $CO(g) + H_2O(g) \rightleftharpoons CO_2(g) + H_2(g)$
    (exothermic)

(2) $2SO_3(g) \rightleftharpoons 2SO_2(g) + O_2(g)$
    (endothermic)

(3) $2NO(g) + Cl_2(g) \rightleftharpoons 2NOCl(g)$
    (exothermic)

(4) $2N_2(g) + O_2(g) \rightleftharpoons 2N_2O(g)$
    (endothermic)

29. The synthesis of ammonia from the elements is an exothermic reaction. Predict the effect of each of the following changes on the yield of ammonia at equilibrium: (a) the temperature is increased at constant pressure, (b) the pressure is reduced at constant temperature, (c) the volume of the gas mixture is increased, (d) ammonia is removed from the equilibrium mixture, and (e) a metal oxide catalyst is added.

30. For each of the following reactions, (a) write an equilibrium expression for the constant $K_p$, (b) indicate the effect of a decrease in temperature at constant pressure, and (c) indicate the effect of a decrease in pressure at constant temperature:

(1) $2C(s) + O_2(g) \rightleftharpoons 2CO(g)$    (exothermic)
(2) $2P(s) + 3H_2(g) \rightleftharpoons 2PH_3(g)$    (endothermic)
(3) $S(s) + 3F_2(g) \rightleftharpoons SF_6(g)$    (exothermic)
(4) $2C(s) + H_2(g) \rightleftharpoons C_2H_2(g)$    (endothermic)

31. One of the processes that may be suitable for the production of a fuel from inaccessible coal deposits is the endothermic reaction $C(s) + 2H_2O(g) \rightleftharpoons CO_2(g) + 2H_2(g)$. How would each of the following changes affect the equilibrium yield of $H_2$? (a) An increase in temperature, (b) an increase in total pressure, (c) the addition of $CO_2$ to the equilibrium mixture, and (d) an increase in the gas volume.

32. The reaction of 1.0 mol of $N_2$ with 3.0 mol of $H_2$ to form $NH_3$ is studied at 450°C, where $K_p = 6.5 \times 10^{-3}$ (atm)$^{-2}$. The total pressure of the gas mixture is maintained at 2.0 atm. (a) How many moles of each species will be present at equilibrium? (b) Ammonia may be removed from the other gases present by taking advantage of its much greater solubility in water. The equilibrium mixture is bubbled through water and 75% of the ammonia but none of the other gases are removed. The resulting mixture is reheated to 450°C while maintaining $P = 2.0$ atm. How many moles of each substance are present when equilibrium is reestablished?

33. A mixture of $H_2$ and $N_2$ in a 3.0 : 1.0 mole ratio is passed over a catalyst at 450°C. What should the total pressure be if the mole fraction of $NH_3$ in the equilibrium mixture is to be 0.30? Use the data given in Problem 32.

# 8
# Acids and Bases

Acids and bases are among the most widely used reagents in chemistry. They are important ingredients in the synthesis of a large variety of chemicals both in the laboratory and in industrial production, they serve as catalysts for many reactions, and their ionization equilibria play a major role in such diverse areas as analytical separations and physiological processes. The properties of acids and bases are most clearly manifested in aqueous solution, where they form ions. Acid and base ionization equilibria serve as an excellent illustration of the principles of chemical equilibrium developed in Chapter 7.

## Properties of Acids and Bases

While acids and bases have been of general importance since the beginnings of chemistry, the meaning attached to these terms has evolved continuously. The early characterization of acids was phenomenological. It has been known since the time of Robert Boyle, some 300 years ago, that acids possess a number of distinctive properties: they dissolve many substances, they change the color of certain naturally occurring dyes (e.g., litmus changes from blue to red), they have a sour taste, and they lose these properties after being mixed with bases. Bases were also found to have some characteristic properties: they feel soapy, they have a bitter taste, they change the color of certain dyes but in a different way than acids do, and they lose these properties when mixed with acids.

The first theory of acids was proposed by Lavoisier, who believed that oxygen was responsible for their distinctive properties. This theory could not meet the test of further experimental evidence, which showed that acidic substances such as hydrochloric acid contained no oxygen. Furthermore, many substances that did contain oxygen, such as calcium oxide, did not possess acidic properties. The oxygen viewpoint was eventually replaced by the concept that all acids contain hydrogen, but in a form replaceable by metals, i.e., by means of displacement reactions with the liberation of hydrogen gas (Section 2.5). These ideas were placed on a firmer theoretical basis in 1887 by Svante Arrhenius (1859–1927), who developed the first modern theory of acids and bases.

## 8.1 The Arrhenius Model and Its Drawbacks

*a. The Arrhenius Model and the Conductivity of Solutions* Arrhenius' definitions of acids and bases are essentially those given in Section 2.3: they are substances which increase the concentration of hydrogen ions or hydroxide ions, respectively, when added to water. These ions are formed by the ionization of a molecular substance, for example,

$$HCl(g) \xrightarrow{H_2O} H^+ + Cl^-$$

or by the dissociation of an ionic substance, e.g.,

$$NaOH(s) \xrightarrow{H_2O} Na^+ + OH^-$$

As already noted in Section 2.3, solutions containing ions are electrolytes—they conduct electricity. The ability of a substance to conduct electricity is measured by its conductivity. Since the charge passing through a solution is carried by the ions present, the conductivity of a solution generally increases with its concentration. A very convenient measure of the conductivity of solutions is the **molar conductivity,** the conductivity per unit concentration, customarily expressed in units of $ohm^{-1}\ cm^2\ mol^{-1}$. A molar conductivity of $1\ ohm^{-1}\ cm^2\ mol^{-1}$ means that if a potential difference of 1 V is applied to opposite faces of a 1 cm cube containing a 1 $M$ solution, a current of 1 A will flow.

Figure 8.1 shows the dependence of the molar conductivity of some acids, bases, and salts on concentration. Most salts, as well as strong acids and bases, (e.g., HCl and NaOH) have molar conductivities in the range of 100–500 $ohm^{-1}\ cm^2\ mol^{-1}$, the actual values decreasing slowly with increasing concentration. This behavior is characteristic of **strong electrolytes,** substances that are extensively ionized in aqueous solution. Figure 8.1 also shows that weak acids and bases, such as acetic acid, have much smaller molar conductivities but that these increase sharply in the limit of low concentration. Substances displaying this characteristic behavior are termed **weak electrolytes,** meaning that they are only slightly ionized. The increasing ionization observed at very low concentrations can be explained on the basis of equilibrium considerations (see Section 8.5).

The Arrhenius theory successfully coupled the notion that acids and bases form ions in aqueous solution to their observed molar conductivities. The magnitude of the molar conductivity and its dependence on concentration provided a sound experimental basis for the division into strong and weak acids or bases. Furthermore, the decrease in the conductivity of acids and bases upon neutralization (Section 2.5d), for example, in the reaction

$$Na^+ + OH^- + H^+ + Cl^- \rightarrow Na^+ + Cl^- + H_2O$$

is consistent with the formation of essentially un-ionized water, a non–elec-

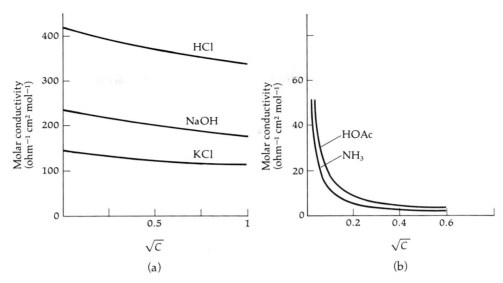

**FIGURE 8.1** Molar conductivity of solutions of (a) strong electrolytes and (b) weak electrolytes. The difference between strong and weak electrolytes is most clearly displayed by plots of molar conductivities versus the square root of concentration.

trolyte, and a resultant decrease in the concentration of ions. (Unless otherwise indicated, we shall assume that H$_2$O is present as a liquid and that ions are in aqueous solution.)

**b. Drawbacks of the Arrhenius Model** The Arrhenius model was able to account for many aspects of acid–base behavior and remains a useful, though overly simplified model. A number of experimental observations eventually indicated that a more general model of acids and bases was needed. For example, the behavior of aqueous ammonia as a weak base could not be explained correctly. Within the framework of the Arrhenius model, this behavior was understood as resulting from the ionization of "ammonium hydroxide,"

$$NH_4OH(aq) \rightleftarrows NH_4^+ + OH^-$$

Spectroscopic measurements indicate, however, that the species NH$_4$OH does not exist. The basic properties of aqueous ammonia are more properly represented by the reaction

$$NH_3(aq) + H_2O \rightleftarrows NH_4^+ + OH^-$$

in which a proton is transferred from water to ammonia. The concept of proton transfer can also be fruitfully applied to acid–base neutralization reactions that do not involve the formation of water, for instance to the following reaction occuring in liquid ammonia:

$$NH_4^+ + NH_2^- \rightleftarrows NH_3 + NH_3 \quad \text{(in liquid ammonia)}$$

which is analogous to the formation of water in the reaction of an acid and a base in aqueous solution.

The concept of proton transfer also permits a more accurate description of the behavior of acids in aqueous solution than does the Arrhenius ionization

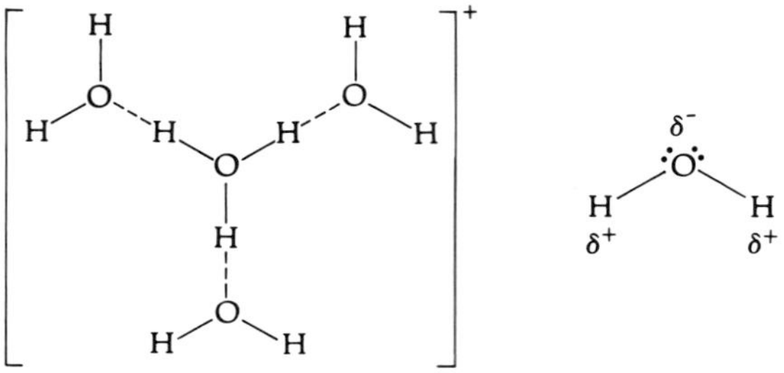

**FIGURE 8.2** A possible structure of the hydrated proton showing an $H_3O^+$ ion attached by hydrogen bonds (dashed lines) to three $H_2O$ molecules. The polar nature of the $H_2O$ molecule, which is responsible for this phenomenon, is illustrated at the right.

picture. For example, the behavior of aqueous nitric acid can be represented, according to these two models, by the equations

$$HNO_3(aq) + H_2O \rightleftarrows H_3O^+ + NO_3^- \tag{8.1}$$

and

$$HNO_3(aq) \rightleftarrows H^+ + NO_3^- \tag{8.2}$$

respectively. Equation 8.2 is unsatisfactory because it conveys the erroneous impression that hydrogen ions exist as separate entities in aqueous solution. A hydrogen ion is just a bare proton. As noted in Section 1.3, its diameter is only $10^{-15}$ m, compared to a diameter of $10^{-10}$ m of other ions. Such an exceedingly small charged particle cannot exist as a separate entity in aqueous solution owing to its strong Coulomb interaction with electrons of nearby atoms. What happens, instead, is that the proton firmly attaches itself to a molecule of water to form the **hydronium ion,** $H_3O^+$:

$$H^+ + H_2O \rightarrow H_3O^+$$

This is just the process described by Equation 8.1, indicating that proton transfer constitutes a more realistic description of acid behavior than proton ionization.

Like other ions, the hydronium ion itself undergoes further **hydration**, that is, it attaches several additional molecules of water. Figure 8.2 shows one of the common forms of the hydrated hydronium ion, which can be represented by the formula $H_9O_4^+$, where the hydronium ion is attached to the additional water molecules by hydrogen bonds (Section 18.6c). While a hydrogen ion in solution may be symbolized by either $H_3O^+$ or $H^+$, both designations should be recognized as representing clusters such as that depicted in Figure 8.2.

## 8.2 The Brønsted–Lowry Acid–Base Model

The notion that acid–base behavior involves proton transfer forms the basis of the Brønsted–Lowry model, developed in the 1920s. *A Brønsted acid is defined as a substance that can donate a proton and a Brønsted base as one that can accept a proton.* The ionization of HCl in water is viewed as involving the transfer of a proton from an acid to a base:

$$HCl + H_2O \rightleftharpoons H_3O^+ + Cl^-$$
(acid)  (base)

Similarly, the reaction of HCl with $NH_3$ involves the transfer of a proton from HCl to $NH_3$:

$$HCl + NH_3 \rightleftharpoons NH_4^+ + Cl^-$$
(acid)  (base)

The ionization of weak acid can be represented in exactly the same way:

$$HOAc + H_2O \rightleftharpoons H_3O^+ + OAc^-$$
(acid)  (base)

An important feature of the Brønsted–Lowry model becomes apparent if we focus on the reverse reaction associated with any of these examples. Clearly, if the forward reaction involves proton transfer, then so does the reverse reaction. In the reaction between $H_3O^+$ and $OAc^-$, for instance, a proton is transferred from the acid $H_3O^+$ to the base $OAc^-$. In the Brønsted–Lowry model, a proton transfer reaction always involves two acids and two bases, for example:

$$HCl + H_2O \rightarrow H_3O^+ + Cl^-$$
(acid 1) (base 2) (acid 2) (base 1)

or more generally:

$$HA + B \rightleftharpoons HB^+ + A^- \qquad (8.3)$$
(acid 1) (base 2) (acid 2) (base 1)

where bases $A^-$ and B can actually have positive or negative charges or be neutral. Acid 1 and base 1, or acid 2 and base 2, are called **conjugate acid–base pairs**; the acid member of such a pair contains one more $H^+$ than the base. According to the Brønsted–Lowry definitions, an acid–base reaction involves the participation of two conjugate acid–base pairs. In spite of the change from the Arrhenius viewpoint, the language of that model, e.g., dissociation of a base, continues to be used although the meaning is now different.

The following is a list of some common conjugate acid–base pairs:

*Acid*     *Base*
$HCl$      $Cl^-$
$NH_4^+$   $NH_3$
$H_3O^+$   $H_2O$
$H_2O$     $OH^-$
$H_2SO_4$  $HSO_4^-$
$HSO_4^-$  $SO_4^{2-}$

Note that the species classified as acids or bases can be ions or molecules. Note, too, that there are some species, such as water or hydrogen sulfate ion ($HSO_4^-$), that can either donate or accept a proton. Such species are said to be **amphiprotic**. We have already illustrated the role of water as a base. The reaction of water with ammonia or with carbonate ion ($CO_3^{2-}$) shows it in the role of an acid:

$$H_2O + NH_3 \rightleftharpoons NH_4^+ + OH^-$$
(acid 1) (base 2) (acid 2) (base 1)

$$H_2O + CO_3^{2-} \rightleftharpoons HCO_3^- + OH^-$$
(acid 1)  (base 2)  (acid 2)  (base 1)

Since water can act as both acid and base, it is not surprising that it undergoes a self-ionization process in which it exhibits its amphiprotism:

$$H_2O + H_2O \rightleftharpoons H_3O^+ + OH^-$$

## 8.3 The Lewis Model

A more general definition of acids and bases than either of the ones considered so far was proposed by G. N. Lewis (1875–1946). This definition is based on the electronic structure of the species in question. An acid is defined as a species containing an atom that can accept one or more electron pairs and is consequently an **electron-pair acceptor**. A base is correspondingly defined as a species containing an atom capable of donating one or more electron pairs and is an **electron-pair donor**.

Brønsted acids and bases are also Lewis acids and bases, respectively. To illustrate this equivalence, we must look at the Lewis formulas (Section 1.5) of these species. For example, the reaction of an acid with ammonia can be represented as

$$H^+ + :\!\!\underset{\underset{H}{|}}{\overset{\overset{H}{|}}{N}}\!\!-H \rightarrow \left[ \underset{\underset{H}{|}}{\overset{\overset{H}{|}}{H-N-H}} \right]^+$$

showing that the base, ammonia, donates an electron pair to the hydrogen ion, an acid. Similarly, the reaction

$$H^+ + OH^- \rightarrow H_2O$$

can be represented as

$$H^+ + [:\!\ddot{O}\!-H]^- \rightarrow H-\underset{\underset{H}{|}}{\ddot{O}:}$$

showing that the base OH⁻ donates an electron pair to the acid H⁺.

Although all Brønsted acids and bases are necessarily Lewis acids and bases, the converse is not necessarily true. The Lewis definition is more general and permits reactions such as

$$BF_3 + NH_3 \rightarrow BF_3NH_3$$

to be understood as acid–base reactions. In this reaction the base, ammonia, donates an electron pair to the acid, boron trifluoride:

$$\underset{\underset{F}{|}}{\overset{\overset{F}{|}}{F-B}} + :\!\!\underset{\underset{H}{|}}{\overset{\overset{H}{|}}{N}}\!\!-H \rightarrow \underset{\underset{F\quad H}{|\quad|}}{\overset{\overset{F\quad H}{|\quad|}}{F-B-N-H}}$$

Lewis acids and bases are considered further in connection with the Lewis model of covalent bonding (Section 15.3b).

## 8.4 Relative Strengths of Acids and Bases

**a. Acid Ionization Constants** The strength of an acid is measured by its tendency to transfer a proton to a base. The stronger this tendency, the greater will be the extent of the acid–base reaction at equilibrium. In order to compare the relative strengths of various acids, it is necessary to compare their reactions with the same base. When dealing with aqueous solutions, the obvious choice of a base is water. The quantitative measure of the strength of an acid HA is then given by the equilibrium constant for the reaction with water,

$$HA + H_2O \rightleftharpoons H_3O^+ + A^-$$

The equilibrium constant $K_a$, commonly known as the **acid ionization constant,** is related to the molar concentrations of the participating species by the expression

$$K_a = \frac{[H_3O^+][A^-]}{[HA]} \tag{8.4}$$

Note that the concentration of water is not included in the denominator. So long as an aqueous solution is not too concentrated, the concentration of water is constant, at approximately 55 M. Recall from Section 7.2 that, under these circumstances, the concentration of a reagent is not included in the equilibrium expression.

Table 8.1 lists a number of acids arranged in order of decreasing strength, that is, in order of decreasing value of the acid ionization constant. Each acid (HA) is listed along with its conjugate Brønsted base ($A^-$). Since the strengths of the acids decrease going down the list, the strengths of the conjugate bases must increase. For example, the equilibrium constant for the reaction

$$HF + H_2O \rightleftharpoons H_3O^+ + F^-$$

is less than unity. This means that $F^-$ has a stronger affinity for a proton than

**TABLE 8.1  Acid Ionization Constants In Aqueous Solution At 25°C**

| Acid | HA | $A^-$ | $K_a$ | $pK_a$ |
|---|---|---|---|---|
| Perchloric | $HClO_4$ | $ClO_4^-$ | Large | |
| Nitric | $HNO_3$ | $NO_3^-$ | Large | |
| Sulfuric | $H_2SO_4$ | $HSO_4^-$ | Large | |
| Hydrochloric | HCl | $Cl^-$ | Large | |
| Hydrobromic | HBr | $Br^-$ | Large | |
| Hydronium ion | $H_3O^+$ | $H_2O$ (solvent) | 1.00 | 0 |
| Chlorous | $HClO_2$ | $ClO_2^-$ | $1.1 \times 10^{-2}$ | 1.96 |
| Nitrous | $HNO_2$ | $NO_2^-$ | $7.2 \times 10^{-4}$ | 3.14 |
| Hydrofluoric | HF | $F^-$ | $6.5 \times 10^{-4}$ | 3.19 |
| Benzoic | $C_6H_5COOH$ | $C_6H_5COO^-$ | $6.6 \times 10^{-5}$ | 4.18 |
| Acetic | $CH_3COOH$ (HOAc) | $CH_3COO^-$ ($OAc^-$) | $1.76 \times 10^{-5}$ | 4.754 |
| Propionic | $CH_3CH_2COOH$ | $CH_3CH_2COO^-$ | $1.34 \times 10^{-5}$ | 4.873 |
| Hypochlorous | HClO | $ClO^-$ | $2.90 \times 10^{-8}$ | 7.538 |
| Periodic | $HIO_4$ | $IO_4^-$ | $5.6 \times 10^{-9}$ | 8.25 |
| Ammonium ion | $NH_4^+$ | $NH_3$ | $6.3 \times 10^{-10}$ | 9.25 |
| Hydrocyanic | HCN | $CN^-$ | $6.2 \times 10^{-10}$ | 9.21 |
| Water | $H_2O$ | $OH^-$ | $1.00 \times 10^{-14}$ | 14.000 |

H₂O; therefore, F⁻ is a stronger base than H₂O. Equivalently, H₃O⁺ is a stronger acid than HF. The strengths of an acid and its conjugate base are inversely related; a strong acid necessarily corresponding to a weak conjugate base and vice versa.

The value of $K_a$ for H₃O⁺, that is, the equilibrium constant for the reaction

$$H_3O^+ + H_2O \rightleftharpoons H_3O^+ + H_2O$$
(acid 1)   (base 2)   (acid 2)   (base 1)

is listed as unity, as required by the equivalence of reactants and products. Because water is present in great excess, any acid whose conjugate base is weaker than H₂O will be virtually completely ionized in aqueous solution. Such acids are termed **strong acids** and are characterized by $K_a \gg 1$. As an example, the strength of nitric acid is reflected in an equilibrium point for the reaction

$$HNO_3 + H_2O \rightleftharpoons H_3O^+ + NO_3^-$$

which lies so far to the right that the reverse reaction can be neglected. The number of strong acids commonly encountered is rather small. It is worthwhile to remember their identity:

Strong acids:
three of the hydrogen halides: HCl, HBr, and HI (but not HF)
perchloric acid, HClO₄
nitric acid, HNO₃
sulfuric acid, H₂SO₄ (ionization of the first proton)

Acids that are only partially ionized in aqueous solution and as a result have $K_a < 1$ are called **weak acids**. Table 8.1 lists a few examples, and a more complete compilation may be found in Appendix 3. The values of $K_a$ for weak acids are generally much smaller than unity and span many orders of magnitude. It is frequently convenient to express the values of $K_a$ in terms of their logarithms, thereby avoiding the use of exponentials. The quantity p$K_a$ is defined by the relation

$$pK_a = -\log K_a \tag{8.5}$$

where the logarithm is taken to the base 10. For example, if $K_a = 1.0 \times 10^{-10}$, then p$K_a$ = 10.00. The values of p$K_a$ of several weak acids are listed in Table 8.1.

All strong acids are completely ionized in water. The strong affinity of water for a proton drives the reaction for the ionization of a strong acid virtually to completion, e.g.,

$$HCl + H_2O \rightarrow H_3O^+ + Cl^-$$

However, strong acids differ in strength in a solvent that is more acidic than water, that is, one which has a weaker affinity for protons. Consider, for example, the ionization of the strong acids HClO₄, and HCl in anhydrous (no water present) acetic acid:

$$HClO_4 + HOAc \rightleftharpoons H_2OAc^+ + ClO_4^-$$

$$HCl + HOAc \rightleftharpoons H_2OAc^+ + Cl^-$$

While neither reaction goes to completion owing to the weak affinity of HOAc for a proton, the equilibrium point for the first of these reactions lies farther to the right than that for the second reaction, indicating that HClO₄ is a stronger acid than HCl. This difference cannot be seen in aqueous solution. Water is said to have a **leveling effect** on acids stronger than $H_3O^+$, meaning that in aqueous solution such acids are completely ionized and appear to be equally strong.

While an acidic solvent represses the ionization of an acid, a basic solvent enhances its ionization. In such a solvent even some weak acids are fully ionized. In liquid ammonia, for instance, even acetic acid is a "strong" acid, meaning that the equilibrium for the reaction

$$HOAc(l) + NH_3(l) \rightleftharpoons NH_4^+ + OAc^-$$

lies far to the right. It must therefore be realized that the strength of acids and bases depends to a large extent on the nature of the solvent. However, so long as we restrict ourselves to aqueous solutions, the division into strong and weak acids given in Table 8.1 is meaningful.

**b. The Ion Product of Water and pH of Aqueous Solutions**  Water is amphiprotic and can act as both an acid and a base:

$$H_2O + H_2O \rightleftharpoons H_3O^+ + OH^-$$
(acid 1)  (base 2)  (acid 2)  (base 1)

The equilibrium for this reaction strongly favors the reactants, so that water must be regarded as a very weak acid as well as a very weak base. Table 8.1 indicates that the acid ionization constant of water at 25°C is $1.00 \times 10^{-14}$. The equilibrium expression for this ionization reaction can be written as

$$[H_3O^+][OH^-] = K_w = 1.00 \times 10^{-14} \tag{8.6}$$

and the product of the concentrations is known as the **ion product** of water. This relation between the $H_3O^+$ and $OH^-$ concentrations holds for any dilute aqueous solution and consequently provides a direct method for the determination of the concentration of one of these species if the concentration of the other is known. We shall make frequent use of this important relationship in solving equilibrium problems.

The concentration of $H_3O^+$ is an important property of aqueous solutions and one that can vary by many orders of magnitude. The values of $[H_3O^+]$ can range from 12 $M$ in concentrated HCl to $10^{-15}$ $M$ in saturated NaOH solution. An aqueous solution is said to be **neutral** if $[H_3O^+] = [OH^-]$. According to Equation 8.6, this condition is met at 25°C when $[H_3O^+] = 1.00 \times 10^{-7}$ $M$. This value constitutes the dividing point between **acidic solutions,** for which $[H_3O^+] > 10^{-7}$ $M$ and **basic solutions,** for which $[H_3O^+] < 10^{-7}$ $M$ at 25°C.

As in the case of the acid ionization constant, the $H_3O^+$ concentrations of solutions generally vary over a broad range of negative powers of 10. A logarithmic concentration scale, similar to that represented by the p$K$ values (Equation 8.5), is convenient. This scale is called the **pH scale** and is defined by the relation

$$pH = -\log[H_3O^+] \tag{8.7}$$

For instance, the pH of a neutral solution, with $[H_3O^+] = 1 \times 10^{-7}$ $M$, is 7.0, that of a 1 $M$ HCl solution ($[H_3O^+] = 1$ $M$) is 0.0, and that of a saturated NaOH solution ($[H_3O^+] = 10^{-15}$ $M$) is 15. Since the pH scale is logarithmic, the pH in-

**FIGURE 8.3** The pH and [H₃O⁺] scales.

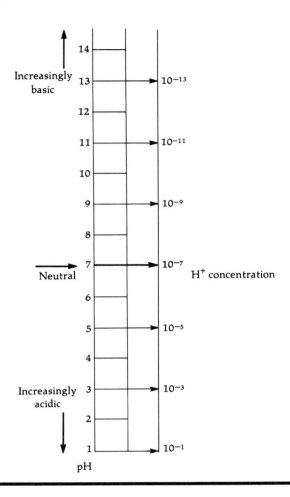

creases by one unit as the $H_3O^+$ concentration decreases by a factor of 10. Figure 8.3 illustrates the relation between the pH and $[H_3O^+]$ scales. A similar relationship holds between the $OH^-$ concentration and a logarithmic quantity called pOH. The ion product of water (Equation 8.6) can be expressed conveniently in terms of pH and pOH as

$$pH + pOH = 14.00 \tag{8.8}$$

This relation holds for any dilute aqueous solution at 25°C.

**Example 8.1** Determine the concentrations of $H_3O^+$ and $OH^-$, as well as the pH and pOH, of a 0.015 $M$ HCl solution.

Since HCl is a strong acid, it is virtually completely ionized; therefore, $[H_3O^+] = 0.015\ M$. Applying Equations 8.7 and 8.8,

$$pH = -\log[H_3O^+] = 1.82$$
$$pOH = 14.00 - pH = 12.18$$

Finally, [OH$^-$] can be obtained from either the relation

$$[H_3O^+][OH^-] = K_w = 1.00 \times 10^{-14}$$

or, equivalently, from

$$[OH^-] = 10^{-(pOH)}$$

By either method, we obtain [OH$^-$] = 6.6 × 10$^{-13}$ M. (See Section 1A.3 in Chapter 1 for the relationship between the number of significant figures in a quantity and that in its logarithm.) ∎

The pH values of some common solutions are listed in Table 8.2. A number of entries are worthy of comment. The pH of body fluids is seen to vary widely depending on the nature of the fluid. The acid–base equilibria that give rise to these pH values are of vital importance to a large variety of biochemical processes. From a different point of view, we take note of the entries for rain and acid rain. Ordinary rainwater is slightly acidic because it contains dissolved $CO_2$ (Section 8.10a). Acid rain is more acidic than ordinary rain primarily due to the presence of sulfuric acid (Section 3.8b).

**c. Ionization Constants of Bases** The strength of a base is determined by the ease with which it accepts a proton. The hydroxide ion (OH$^-$) readily accepts protons in aqueous solution. Thus, substances that dissociate completely to OH$^-$ and a cation when added to water, such as the alkali metal hydroxides, are strong bases, e.g.,

$$NaOH(s) \xrightarrow{H_2O} Na^+ + OH^-$$

$$KOH(s) \xrightarrow{H_2O} K^+ + OH^-$$

Note that the Arrhenius viewpoint adequately describes this process since the base OH$^-$ is directly formed by dissociation of the alkali hydroxide.

The behavior of weak bases can be understood more readily by means of the Brønsted–Lowry model. Consider, for instance, the reaction of ammonia

**TABLE 8.2  pH of Some Common Solutions**

| | |
|---|---|
| Gastric juice | ≈1 |
| Orange juice | 2.8 |
| Vinegar | 3.0 |
| Grapefruit juice | 3.2 |
| Wine | 3.5 |
| Tomato juice | 4.2 |
| Acid rain | 3–5 |
| Black coffee | 5.0 |
| Rain | 5.5–6 |
| Milk | 6.9 |
| Pure water at 25°C | 7.0 |
| Blood | 7.4 |
| Sea water | 8.0 |
| Baking soda solution (NaHCO$_3$) | 8.3 |
| Household ammonia | 11.9 |

with water:

$$NH_3 + H_2O \rightleftharpoons NH_4^+ + OH^-$$
(base 1)   (acid 2)   (acid 1)   (base 2)

The equilibrium point for this reaction lies far to the left indicating that $NH_3$ has a smaller affinity for protons than $OH^-$ and is, therefore, a weaker base. The quantitative measure of the strength of a weak base is the equilibrium constant for its reaction with water, that is, the base ionization constant. For the reaction of base B with $H_2O$,

$$B + H_2O \rightleftharpoons BH^+ + OH^-$$

the equilibrium constant $K_b$ is defined as

$$K_b = \frac{[BH^+][OH^-]}{[B]} \tag{8.9}$$

The Brønsted–Lowry model permits us to consider the proton transfer equilibria of acids and their conjugate bases from a unified viewpoint. An important relation between the value of $K_b$ and that of $K_a$ for the conjugate acid can be derived. Consider the transfer of a proton from acid HA to water:

$$HA + H_2O \rightleftharpoons H_3O^+ + A^-$$

We have noted that the equilibrium constant for this reaction is the acid ionization constant of HA:

$$K_a = \frac{[H_3O^+][A^-]}{[HA]}$$

Similarly, the transfer of a proton from water to base $A^-$,

$$H_2O + A^- \rightleftharpoons HA + OH^-$$

determines the ionization constant $K_b$ of base $A^-$,

$$K_b = \frac{[HA][OH^-]}{[A^-]}$$

Multiplying together the expressions for $K_a$ and $K_b$ we obtain:

$$K_a K_b = \left(\frac{[H_3O^+][A^-]}{[HA]}\right)\left(\frac{[HA][OH^-]}{[A^-]}\right)$$

which reduces to:

$$K_a K_b = [H_3O^+][OH^-] = K_w \tag{8.10}$$

In other words, *the product of the ionization constants of an acid and its conjugate base is equal to the ion product of water.*

Equation 8.10 allows $K_b$ for any base to be found from the value of $K_a$ for its conjugate acid, and vice versa. For example, the tabulated value of $K_a$ for $NH_4^+$, i.e., the equilibrium constant for the reaction

$$NH_4^+ + H_2O \rightleftharpoons H_3O^+ + NH_3$$

is $6.3 \times 10^{-10}$ (Table 8.1) indicating that the ammonium ion is a very weak acid.

TABLE 8.3  Ionization Constants of Some Weak Bases at 25°C

| Base | B | BH$^+$ | $K_b$ | $K_a = K_w/K_b$ |
|---|---|---|---|---|
| Dimethylamine | $(CH_3)_2NH$ | $(CH_3)_2NH_2^+$ | $5.9 \times 10^{-4}$ | |
| Methylamine | $CH_3NH_2$ | $CH_3NH_3^+$ | $3.9 \times 10^{-4}$ | |
| Trimethylamine | $(CH_3)_3N$ | $(CH_3)_3NH^+$ | $6.3 \times 10^{-5}$ | |
| Ammonia | $NH_3$ | $NH_4^+$ | $1.6 \times 10^{-5}$ | |
| Cyanide ion | $CN^-$ | HCN | $1.6 \times 10^{-5}$ | $6.2 \times 10^{-10}$ |
| Acetate ion | $OAc^-$ | HOAc | $5.68 \times 10^{-10}$ | $1.76 \times 10^{-5}$ |
| Fluoride ion | $F^-$ | HF | $1.5 \times 10^{-11}$ | $6.5 \times 10^{-4}$ |

The value of $K_b$ for $NH_3$ is

$$K_b = \frac{K_w}{K_a} = \frac{1.00 \times 10^{-14}}{6.3 \times 10^{-10}} = 1.6 \times 10^{-5}$$

In its reaction with water,

$$NH_3 + H_2O \rightleftharpoons NH_4^+ + OH^-$$

ammonia acts as a moderately weak base.

Table 8.3 lists the values of $K_b$ of some typical weak bases. The values of $K_a$ of some of the corresponding conjugate acids are also listed. It is seen that they agree with the values given in Table 8.1, as expected from Equation 8.10.

# Equilibria Involving Acids and Bases

In this section we apply the various principles and techniques developed in Chapter 7 to the solution of equilibrium problems involving acids, bases, and their salts. In addition to the obvious difference in the use of concentrations rather than partial pressures when dealing with a liquid solution, some more subtle differences will become apparent. Nonetheless, such general aspects of equilibrium as the use of the equilibrium constant expression, the application of the successive approximation method, and the usefulness of Le Chatelier's principle as a guide to the direction of change are equally applicable to acid–base equilibria (or any equilibria in aqueous solution) as to gaseous equilibria.

## 8.5 Weak Acids and Bases

Equilibrium problems involving weak acids or bases generally require the determination of the $H_3O^+$ or $OH^-$ concentrations, from which the pH of the solution can then be obtained by the procedure of Section 8.4b. The degree of ionization of a weak acid or base is another quantity of interest; it is completely analogous to the degree of dissociation of a gaseous substance (Section 7.5).

**Example 8.2**

Determine the pH of a 0.010 M HOAc solution. What is the degree of ionization of the acid?

We begin by writing the equation for the ionization of acetic acid and listing the equilibrium concentrations of the various species below the equation:

$$HOAc + H_2O \rightleftarrows H_3O^+ + OAc^-$$

Equilibrium concentration    $0.010 - x$     $x$     $x$

Note that the ionization constant of acetic acid (Table 8.1) is only $1.76 \times 10^{-5}$. Consequently, for reasonably concentrated solutions the equilibrium lies far to the left and only a small fraction of the acid ionizes. As in Section 7.4, we let the unknown $x$ refer to the species having the smallest equilibrium concentration. Substituting the equilibrium concentrations into the equilibrium constant expression (Equation 8.4) we obtain

$$K_a = \frac{[H_3O^+][OAc^-]}{[HOAc]} = \frac{x^2}{0.010 - x} = 1.76 \times 10^{-5}$$

As in Section 7.4, this equation can be solved for $x$ by using either the quadratic formula or the method of successive approximations. The result is $x = 4.1 \times 10^{-4}$ mol L$^{-1}$. Since $x$ is the concentration of $H_3O^+$, the pH of the solution is obtained by means of Equation 8.7:

$$pH = -\log[H_3O^+] = -\log(4.1 \times 10^{-4}) = 3.39$$

The degree of ionization $f$ is simply the fraction of the acid present in ionized form at equilibrium, that is, the ratio of the $H_3O^+$ concentration to the stoichiometric (i.e., total) acid concentration. Expressing the answer in percent, we obtain:

$$f = \frac{4.1 \times 10^{-4} \, M}{0.010 \, M} \times 100 = 4.1\% \quad \blacksquare$$

It is instructive to compare the pH of acetic acid obtained in the preceding example with that of a strong acid of the same molarity. The pH of $0.010 \, M$ HCl, for instance, is simply 2.00 (see Example 8.1). Thus, for the same molarity, the pH of a weak acid is higher than that of a strong acid, reflecting the fact that the weak acid has a lower $H_3O^+$ concentration.

How does the degree of ionization of a weak acid depend on its concentration? A qualitative answer to this question can be obtained by applying Le Chatelier's principle. A change in the concentration of a solution is analogous to a change in the pressure of a gas that can dissociate. Both quantities are a measure of the number of molecules of either gas or solute per unit volume. A decrease in the concentration of an acid increases its degree of ionization just as a decrease in gas pressure increases the dissociation of a gas. The validity of this statement can be confirmed by repeating the calculation of Example 8.2 for a more dilute solution, e.g., $0.0010 \, M$ HOAc. The equilibrium expression becomes:

$$\frac{x^2}{0.0010 - x} = 1.76 \times 10^{-5}$$

and the result is $x = 1.2 \times 10^{-4}$. The degree of ionization is

$$f = \frac{1.2 \times 10^{-4}}{0.0010} \times 100 = 12\%$$

compared to $f = 4.1\%$ for $0.010\ M$ HOAc. The increase in the degree of ionization is consequently in accord with Le Chatelier's principle.

Although the degree of ionization increases upon dilution, the *actual* $H_3O^+$ concentration decreases. For $0.0010\ M$ HOAc we obtain $[H_3O^+] = x = 1.2 \times 10^{-4}\ M$ compared to $[H_3O^+] = 4.1 \times 10^{-4}\ M$ for $0.010\ M$ HOAc. The contrasting behavior of these two quantities is understandable. Since there is less acid in a given volume of the dilute solution, the actual amount of acid that can ionize is smaller; however, the fraction that undergoes ionization is larger.

The variation of the degree of ionization of a weak acid with concentration is reflected in the behavior of its molar conductivity (Figure 8.1). As the concentration of the acid decreases, the number of ions per mole of acid increases, and the molar conductivity, which is a measure of this quantity, shows a concomitant increase.

The approach to the ionization of a weak acid illustrated in Example 8.2 may be applied to the general case of acid HA, which has ionization constant $K_a$, and is of stoichiometric concentration $C_0$ (i.e., $C_0$ includes both ionized and un-ionized acid). The equilibrium constant expression for the ionization reaction

$$HA + H_2O \rightleftharpoons H_3O^+ + A^-$$

is

$$K_a = \frac{[H_3O^+][A^-]}{[HA]}$$

The reaction stoichiometry indicates that

$$[H_3O^+] = [A^-] \tag{8.11}$$

(See Section 8.7 for the special case where the ionization of water is sufficiently important to negate this statement.) Furthermore, since species A is present either as HA or as $A^-$, mass balance requires that

$$C_0 = [HA] + [A^-] \tag{8.12}$$

Substituting for $[A^-]$ from Equation 8.11 gives

$$[HA] = C_0 - [H_3O^+] \tag{8.13}$$

Further substitution of Equations 8.13 and 8.11 into the equilibrium expression gives

$$K_a = \frac{[H_3O^+]^2}{C_0 - [H_3O^+]} \tag{8.14}$$

This quadratic equation may be solved for the unknown $H_3O^+$ concentration by means of the quadratic formula. The positive solution is

$$[H_3O^+] = \frac{-K_a + (K_a^2 + 4K_aC_0)^{1/2}}{2} \tag{8.15}$$

It is frequently possible to assume that the $H_3O^+$ concentration is much smaller than the stoichiometric concentration of acid, that is, $[H_3O^+] \ll C_0$. Example 8.2 and the discussion following it indicate that this assumption is most valid when $K_a$ is small and $C_0$ is large. When this assumption is made, Equation 8.14 becomes

$$K_a = \frac{[H_3O^+]^2}{C_0} \quad (8.16)$$

Solving for $[H_3O^+]$, we get the simple expression

$$[H_3O^+] = (K_a C_0)^{1/2} \quad (8.17)$$

from which the pH is obtained as

$$pH = \frac{pK_a - \log C_0}{2} \quad (8.18)$$

Comparison of the solutions to Equations (8.15) and (8.17) for different values of $K_a$ and $C_0$ indicates the conditions for which the simple approximation is valid. For example, for $K_a = 1 \times 10^{-4}$, the error made when using Equation 8.17 to evaluate $[H_3O^+]$ is 1.6% for $C_0 = 0.1\ M$, 5.3% for $C_0 = 0.01\ M$, and 17% for $C_0 = 0.001\ M$.

The pH of a weak acid solution can be determined from the known value of its ionization constant, and the reverse procedure can be used to determine the value of $K_a$, as shown in the following example.

---

**Example 8.3**  The pH of a 0.21 M solution of a newly synthesized monoprotic organic acid is 2.54. What is the value of $K_a$?

The equilibrium concentrations are tabulated below the species present at equilibrium:

$$HA\ +\ H_2O \rightleftarrows H_3O^+ + A^-$$

$$0.21 - x \qquad\qquad x \quad\ x$$

The equilibrium expression is

$$K_a = \frac{x^2}{0.21 - x}$$

where the value of $x$ may be obtained from the measured pH:

$$x = [H_3O^+] = 10^{(-pH)} = 10^{-2.54} = 2.88 \times 10^{-3}$$

Substituting into the equilibrium expression, we obtain:

$$K_a = \frac{(2.88 \times 10^{-3})^2}{0.21 - 2.88 \times 10^{-3}} = 4.0 \times 10^{-5} \quad \blacksquare$$

The ionization of weak bases involves the same principles as that of weak acids. The following example involving a weak base illustrates another aspect of ionization equilibria.

---

**Example 8.4**  How many moles of $NH_3$ are needed to prepare 100 mL of an ammonia solution of pH 11.00?

The reaction and equilibrium concentrations are

$$NH_3 + H_2O \rightleftharpoons NH_4^+ + OH^-$$
$$C_0 - x \qquad\qquad x \qquad x$$

Since we do not know the stoichiometric concentration of $NH_3$, we have labelled it $C_0$. This apparently leaves us with two unknowns, $C_0$ and $x$. We can obtain $x$ from the given pH:

$$pOH = 14.00 - pH = 14.00 - 11.00 = 3.00$$

$$[OH^-] = x = 10^{-(pOH)} = 10^{-3.00} = 1.0 \times 10^{-3}\ M$$

Substituting into the expression for the equilibrium constant we have

$$K_b = 1.6 \times 10^{-5} = \frac{[NH_4^+][OH^-]}{[NH_3]} = \frac{x^2}{C_0 - x}$$

Inserting the numerical value of $x$, we get

$$\frac{(1.0 \times 10^{-3})^2}{C_0 - 1.0 \times 10^{-3}} = 1.6 \times 10^{-5}$$

and the answer is $C_0 = 0.064\ M$. In order to prepare 100 mL of this solution $6.4 \times 10^{-3}$ mol of $NH_3$ are needed. ∎

## 8.6 Salts of Weak Acids or Bases

As noted in Section 2.3a, salts generally are strong electrolytes and dissociate completely in aqueous solution. When the salt is one whose component ions are those of a strong acid and a strong base (e.g., NaCl or $KClO_4$), the ions formed on dissociation of the salt do not react with water. The resulting solution is neutral and the pH is 7 at 25°C.

The situation is different when the salt consists of ions at least one of which is conjugate to either a weak acid or a weak base, for example, NaOAc or $NH_4Cl$. In this case, the ions formed when the salt is added to water react with water to form the conjugate weak acid or base. For example, when NaOAc is dissolved in water, the following equilibrium is established by the acetate ion:

$$OAc^- + H_2O \rightleftharpoons HOAc(aq) + OH^- \tag{8.19}$$

The reaction occurs because HOAc is a weak acid. The most direct evidence for the occurrence of this reaction is that a solution of NaOAc is alkaline—its pH is greater than 7. Similarly, when a salt of a weak base, such as $NH_4Cl$, is added to water, an acidic solution is formed owing to the occurrence of the reaction

$$NH_4^+ + H_2O \rightleftharpoons NH_3(aq) + H_3O^+ \tag{8.20}$$

which forms the weak base ammonia and liberates $H_3O^+$.

The reaction with water of a salt of either a weak acid or a weak base is known as **hydrolysis**. In spite of its distinguishing name, this reaction is just another acid–base reaction. In Equation 8.19, for example, $H_2O$ acts as an acid and transfers a proton to the base $OAc^-$. Similarly, in Equation 8.20, $NH_4^+$ acts as an acid and transfers a proton to $H_2O$, which in this instance acts as a base.

**Example 8.5** What is the pH of a 0.10 M NaOAc solution? To what extent does hydrolysis occur?

The equilibrium expression for the reaction

$$OAc^- + H_2O \rightleftharpoons HOAc + OH^-$$

is

$$K_b = \frac{[HOAc][OH^-]}{[OAc^-]}$$

The value of $K_b$ may be expressed in terms of $K_a$ for the conjugate acid,

$$K_a = \frac{[H_3O^+][OAc^-]}{[HOAc]}$$

by means of Equation 8.10,

$$K_b = \frac{K_w}{K_a} = \frac{1.00 \times 10^{-14}}{1.76 \times 10^{-5}} = 5.68 \times 10^{-10}$$

The problem can now be solved in the same way as Example 8.2. Letting $x$ be the concentration of $OH^-$, the equilibrium expression for the preceding reaction is

$$\frac{x^2}{0.10 - x} = 5.68 \times 10^{-10}$$

The solution to this equation is

$$x = [OH^-] = 7.54 \times 10^{-6} \, M$$

and the pH can be obtained by means of Equation 8.8:

$$pH = 14.00 - pOH = 14.00 - [-\log(7.54 \times 10^{-6})]$$
$$= 14.00 - 5.12 = 8.88$$

Since the pH is greater than 7, the solution is alkaline, as expected from earlier considerations.

The extent of the hydrolysis of NaOAc is given by the ratio of $OAc^-$ present as HOAc to the total amount in solution. The reaction stoichiometry indicates that

$$[HOAc] = x = 7.54 \times 10^{-6} \, M$$

The extent of the reaction, expressed in percent is

$$\frac{7.54 \times 10^{-6}}{0.10} \times 100 = 7.54 \times 10^{-3} \%$$

showing that very little of the $OAc^-$ undergoes hydrolysis in aqueous solution. ∎

## 8.7 The Ionization of Water

In solving equilibrium problems involving weak acids and bases, we have so far neglected the contribution to the $H_3O^+$ or $OH^-$ concentrations of the self-ionization of water. This neglect is justified in most situations by the fact that the $H_3O^+$ and $OH^-$ concentrations in water are only $10^{-7}$ M. In actuality, the contribution of the ionization of water to the $H_3O^+$ concentration of an acidic solution is even smaller than $10^{-7}$ M. According to Le Chatelier's principle, the presence of $H_3O^+$ ions in acidic solutions must repress the self-ionization of water, for $H_3O^+$ is a product of the self-ionization reaction. The contribution of the self-ionization of water to the $OH^-$ concentration of an alkaline solution is similarly reduced from $10^{-7}$ M.

To examine quantitatively this application of Le Chatelier's principle, let us determine the actual contribution from water to the $H_3O^+$ concentration in a $0.01$ M HOAc solution. The concentration of $H_3O^+$ resulting from the ionization of $0.01$ M HOAc has been evaluated in Example 8.2 as $4.1 \times 10^{-4}$ M. The $OH^-$ concentration in this solution can be obtained from the ion product of water (Equation 8.6):

$$[OH^-] = \frac{1.0 \times 10^{-14}}{[H_3O^+]} = \frac{1.0 \times 10^{-14}}{4.1 \times 10^{-4}} = 2.4 \times 10^{-11} \text{ M}$$

In an acidic solution, the only source of $OH^-$ is the ionization of water. Since the reaction stoichiometry is

$$2H_2O \rightleftharpoons H_3O^+ + OH^-$$

the concentration of $H_3O^+$ resulting from the ionization of water must consequently be equal to the $OH^-$ concentration, i.e.,

$$[H_3O^+]_{(\text{from } H_2O)} = 2.4 \times 10^{-11} \text{ M}$$

The presence of $0.01$ M HOAc is seen to have repressed the ionization of water by a factor of about $10^4$.

This estimate shows that we have been justified in neglecting the ionization of water in the various problems considered so far. Are there any circumstances when the ionization of water must be included explicitly in equilibrium calculations? The following illustration shows that at times it is indeed necessary to do so. Let us follow the procedure used in Example 8.2 to calculate the pH of a $10^{-7}$ M solution of HCN—this is a very dilute solution of an acid that is substantially weaker than HOAc. Using the notation of that example, we write

$$\frac{x^2}{1.0 \times 10^{-7} - x} = 6.2 \times 10^{-10}$$

Solving this equation by means of the quadratic formula, we obtain $x = [H_3O^+] = 7.6 \times 10^{-9}$ M, giving pH $= 8.12$. We have been led to the obviously incorrect result that a dilute solution of a weak acid has an alkaline pH! The source of the problem is clear: The ionization of water contributes more $H_3O^+$ ions than that of the dissolved HCN; the pH cannot exceed 7.0.

Let us generalize the analysis of weak acid ionization to take the ionization of water into account. The relevant equilibria are

$$HA + H_2O \rightleftharpoons H_3O^+ + A^-$$

$$2H_2O \rightleftharpoons H_3O^+ + OH^-$$

and the corresponding equilibrium expressions, which must be obeyed simultaneously, are

$$K_a = \frac{[H_3O^+][A^-]}{[HA]} \tag{8.21}$$

and

$$K_w = [H_3O^+][OH^-] \tag{8.22}$$

Since the number of unknowns (i.e., $[H_3O^+]$, $[A^-]$, $[HA]$, and $[OH^-]$) exceeds the number of equations, additional relationships must be sought in order to obtain a solution. The species A must be present as either HA or as $A^-$ and therefore the following mass balance relationship can be written:

$$[HA] + [A^-] = C_0 \tag{8.23}$$

where $C_0$ is the stoichiometric concentration of HA. We can also take advantage of the fact that the solution must be electrically neutral to write the following charge balance relationship:

$$[H_3O^+] = [A^-] + [OH^-] \tag{8.24}$$

This formulation of the simultaneous equilibrium of an acid and water involves four equations (Equations 8.21–8.24) and four unknowns. These equations can be combined appropriately to give a single equation involving the unknown $[H_3O^+]$ and the knowns $K_a$, $K_w$, and $C_0$. First, we express $[HA]$ in terms of $[A^-]$ by means of Equation 8.23,

$$[HA] = C_0 - [A^-]$$

next we express $[A^-]$ in terms of $[H_3O^+]$ and $[OH^-]$ by means of Equation 8.24,

$$[A^-] = [H_3O^+] - [OH^-]$$

and finally we express $[OH^-]$ in terms of $[H_3O^+]$ by means of Equation 8.22:

$$[OH^-] = K_w/[H_3O^+]$$

This procedure yields expressions for $[HA]$ and $[A^-]$ in terms of $[H_3O^+]$ and known quantities. Substitution into Equation 8.21 yields the desired result:

$$K_a = \frac{[H_3O^+]([H_3O^+] - K_w/[H_3O^+])}{C_0 - [H_3O^+] + K_w/[H_3O^+]} \tag{8.25}$$

It can be ascertained by multiplying out the terms that Equation 8.25 is cubic in $[H_3O^+]$. An equation of this type can be solved by a moderately lengthy successive approximation method. Instead of focusing on the exact solution, let us examine the conditions for which the exact solution is needed. Comparing Equation 8.25 with Equation 8.14 for the acid ionization equilibrium in which the ionization of water is not included,

$$K_a = \frac{[H_3O^+]^2}{C_0 - [H_3O^+]}$$

we see that Equation 8.25 reduces to Equation 8.14 when $K_w/[H_3O^+]$ is small compared to $[H_3O^+]$. Since $K_w = 10^{-14}$, this condition will be met whenever $[H_3O^+] > 10^{-6}\ M$, for then $K_w/[H_3O^+]$ is less than $10^{-8}\ M$. Only in the rare case when this condition is not met is the use of Equation 8.25 necessary. An

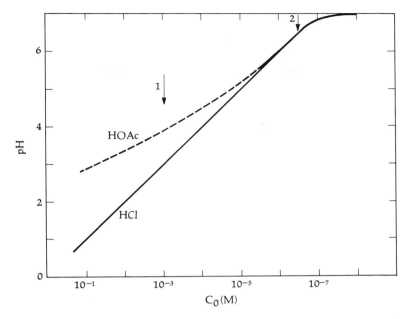

**FIGURE 8.4** Dependence of the pH of an acetic acid solution (dashed curve) and a hydrochloric acid solution on acid concentration. The arrows on the HOAc curve define the concentration intervals at which the various approximations used to calculate [$H_3O^+$] give results with an error of less than 7%. In the region to the left of arrow 1, Equation 8.17 may be used; in the region to the left of arrow 2, Equation 8.15 may be used. Only Equation 8.25 can be used at this level of accuracy to the right of arrow 2. Note that the concentration scale is logarithmic.

---

analogous approach may be followed to determine whether the ionization of water significantly affects the ionization equilibrium of a weak base.

Returning briefly to the case of the $10^{-7}$ M solution of HCN, we see that, since Equation 8.14 gave pH = 8.12, Equation 8.25 must indeed be used. The result is pH = 7.0, thus confirming the fact that in this instance the contribution of the acid to the $H_3O^+$ concentration is negligible compared to that of $H_2O$.

As a summary of this discussion of the ionization of weak acids, we show in Figure 8.4 the dependence of pH on acid concentration for a weak acid, HOAc, and a strong acid, HCl. For strong acids the relation is, of course, logarithmic so that a straight line is obtained on the log scale used in the graph. Only when the HCl concentration approaches $10^{-7}$ M does the ionization of water cause a deviation from linearity.

The shape of the pH curve for acetic acid is more complex. We have divided this curve into three regions, corresponding to the acid concentration intervals in which [$H_3O^+$] can be obtained with an error of less than, say, 7% by means of Equation 8.17, Equation 8.15, or Equation 8.25, the last yielding an exact answer at any concentration. We see that Equation 8.17, the simplest approximation, is satisfactory at the 7% level of accuracy so long as the acid concentration is greater than $10^{-3}$ M. In accord with Equation 8.18, which gives the variation of pH with acid concentration predicted by this approximation, the curve for HOAc is linear in this concentration regime. For more dilute solutions, Equation 8.15 must be used. Finally, when the acid concentration drops below $4 \times 10^{-7}$ $M$, the ionization of water must be taken into account by means of Equation 8.25.

## 8.8 Buffer Solutions

A solution that contains substantial amounts of both a weak acid and a salt of that acid (or of a weak base and a salt of that base) has the remarkable property that its pH is nearly unchanged by the addition of small amounts of another acid or base or by dilution. Such a solution is called a **buffer solution** or, simply, a buffer.

Buffers are of importance in a variety of processes which require a nearly constant pH. They are of supreme importance in physiological processes, where the pH of various body fluids must be maintained within a narrow range in order to ensure proper biochemical behavior. Many industrial processes require careful pH control, which can be achieved by the use of buffers. In the laboratory, buffers are useful in a variety of analytical applications. For example, the separation of certain elements by precipitation of their sulfides can be accomplished by fixing the pH at a particular value by means of a buffer (Section 9.8).

The ability of a buffer solution to maintain its pH can be understood by examining the effect of added acid or base on the ionization equilibrium. Consider, for example, the equilibrium established in an acetic acid–acetate buffer:

$$HOAc + H_2O \rightleftarrows H_3O^+ + OAc^-$$

where HOAc and $OAc^-$ have comparable concentrations. Suppose, now, that another acid is added. According to Le Chatelier's principle an increase in the $H_3O^+$ concentration shifts the equilibrium point to the left—the added $H_3O^+$ reacts with $OAc^-$ to form the un-ionized acid. The pH of the solution consequently remains nearly unchanged. On the other hand, if a base is added to the buffer, the resulting $OH^-$ reacts with $H_3O^+$ present in equilibrium with HOAc to form water. The decrease in $H_3O^+$ concentration shifts the equilibrium point to the right and more HOAc ionizes. Once again, the pH remains nearly the same. Because the $OAc^-$ and HOAc constituents of the buffer combine with added acid or base, a nearly constant $H_3O^+$ concentration is maintained.

Note that the preceding analysis has ignored the hydrolysis of $OAc^-$, that is, the reaction

$$OAc^- + H_2O \rightleftarrows HOAc + OH^-$$

Le Chatelier's principle tells us that this reaction is repressed by the presence of HOAc. Since $K_b$ of $OAc^-$ is very small to begin with (Table 8.3), the reaction is negligible and can be ignored.

These considerations also indicate why a buffer solution has to be moderately concentrated in order to be effective. If the number of moles of added acid or base approaches or exceeds the number of moles of buffer present, the capacity of the buffer to react with the added acid or base will be exceeded. Under these circumstances, the pH will change appreciably, as if the acid or base were added to a solution containing only HOAc or $OAc^-$, respectively.

**Example 8.6** A HOAc–OAc⁻ buffer is prepared by combining 100 mL of 0.0200 M HOAc with 100 mL of 0.0300 M OAc⁻. (a) What is the pH of the resultant solution? (b) What is the resulting pH of this buffer solution when 10.0 mL of 0.0100 M HCl is added? (c) Compare the change in pH evaluated in (b) to that resulting from the addition of 10.0 mL of 0.0100 M HCl to 200 mL of unbuffered 0.0100 M HOAc solution.

(a) Due to the increase in volume, the nominal concentrations of HOAc and OAc⁻ in the buffer solution are lower than those in the separate solutions. Since the volume is doubled, the concentrations must be halved to 0.010 M and 0.015 M, respectively. Knowing the stoichiometric concentrations, the equilibrium concentrations of the various species can be determined:

$$\text{HOAc} + \text{H}_2\text{O} \rightleftarrows \text{H}_3\text{O}^+ + \text{OAc}^-$$
$$0.010 - x \qquad \qquad x \qquad 0.015 + x$$

As is customary, we let $x$ refer to the species present in lowest concentration at equilibrium. Since HOAc is a weak acid this species is $H_3O^+$. Substituting into the equilibrium expression we obtain

$$K_a = 1.76 \times 10^{-5} = \frac{[H_3O^+][OAc^-]}{[HOAc]} = \frac{x(0.015 + x)}{0.010 - x}$$

Assuming that $x \ll 0.01$, the equation simplifies to

$$1.76 \times 10^{-5} = \frac{0.015x}{0.010}$$

and the solution is

$$[H_3O^+] = x = 1.17 \times 10^{-5} \, M$$

showing that the assumption $x \ll 0.01$ is indeed justified. The corresponding pH is 4.93.

(b) The increase in volume leads to changes in stoichiometric concentrations that must be computed first:

$$C_{\text{HOAc}} = (0.0100 \text{ mol L}^{-1}) \frac{(200 \text{ mL})}{(210 \text{ mL})} = 0.00952 \, M$$

$$C_{\text{OAc}^-} = (0.0150 \text{ mol L}^{-1}) \frac{(200 \text{ mL})}{(210 \text{ mL})} = 0.0143 \, M$$

$$C_{\text{H}_3\text{O}^+} = (0.0100 \text{ mol L}^{-1}) \frac{(10.0 \text{ mL})}{(210 \text{ mL})} = 4.76 \times 10^{-4} \, M$$

We assume that nearly all of the added $H_3O^+$ reacts with $OAc^-$ to form unionized HOAc. Letting $x$ be the $H_3O^+$ concentration at equilibrium, the equilibrium concentrations are

$$[H_3O^+] = x$$
$$[HOAc] = 0.00952 + 4.76 \times 10^{-4} - x = 1.000 \times 10^{-2} - x$$
$$[OAc^-] = 0.0143 - 4.76 \times 10^{-4} + x = 1.382 \times 10^{-2} + x$$

Substituting into the equilibrium expression we get

$$1.76 \times 10^{-5} = \frac{x(1.382 \times 10^{-2} + x)}{1.000 \times 10^{-2} - x}$$

As in part (a), we neglect $x$ with respect to terms whose magnitude is $10^{-2}$ and obtain

$$[H_3O^+] = x = (1.76 \times 10^{-5})\frac{(1.000 \times 10^{-2})}{(1.382 \times 10^{-2})} = 1.27 \times 10^{-5} \, M$$

and pH = 4.89. We see that the $H_3O^+$ concentration has increased by less than 10% and the pH has decreased by only 0.04 units.

(c) The addition of HCl represses the ionization of HOAc and the $H_3O^+$ ions in solution are essentially those resulting from the ionization of HCl. The $H_3O^+$ concentration of HCl changes only because of the increase in volume and can be determined without recourse to the equilibrium expression for HOAc. The result has already been obtained in (b):

$$[H_3O^+] = 4.76 \times 10^{-4} \, M$$

and pH = 3.32. The addition of HCl to HOAc increases the $H_3O^+$ concentration of the HOAc solution by a factor of 40 and decreases the pH by 1.6 units. Comparison of the change in pH with that evaluated in (b) illustrates dramatically the ability of a buffer solution to resist the effect of added acid on its pH. ■

The effect on the pH of a buffer resulting from the addition of base can be evaluated in the same manner as that illustrated in Example 8.6. While the addition of base causes an increase rather than a decrease in pH, the change is just as small as that resulting from the addition of acid. The pH of a buffer is also very insensitive to dilution, as illustrated in the following example.

---

**Example 8.7** The concentrations of $NH_3$ and $NH_4^+$ in a certain buffer solution are 0.12 $M$ and 0.16 $M$, respectively. The solution is diluted by a factor of 50. What is the change in pH?

The ionization reaction and equilibrium concentrations prior to dilution are

$$\begin{array}{cccc} NH_3 & + \; H_2O \rightleftharpoons & NH_4^+ & + \; OH^- \\ 0.12 - x & & 0.16 + x & x \end{array}$$

Since $NH_3$ is a weak base, the concentration of $OH^-$ is small compared to that of $NH_3$ and is designated as $x$. The equilibrium expression is

$$K_b = 1.6 \times 10^{-5} = \frac{[NH_4^+][OH^-]}{[NH_3]} = \frac{(0.16 + x)x}{0.12 - x}$$

If $x$ can be neglected with respect to 0.12, this expression reduces to

$$1.6 \times 10^{-5} = \frac{0.16x}{0.12}$$

$$[OH^-] = x = 1.2 \times 10^{-5} \, M$$

which can indeed be neglected with respect to 0.12. The pH is obtained from the $OH^-$ concentration in the manner illustrated in Example 8.1:

$$pH = 14.00 - pOH = 14.00 + \log(1.2 \times 10^{-5})$$
$$= 9.08$$

When the solution is diluted by a factor of 50 the stoichiometric concentrations of $NH_3$ and $NH_4^+$ decrease by this factor:

$$C_{NH_3} = \frac{0.12 \text{ mol L}^{-1}}{50} = 2.4 \times 10^{-3} \, M$$

$$C_{NH_4^+} = \frac{0.16 \text{ mol L}^{-1}}{50} = 3.2 \times 10^{-3} \, M$$

Letting $y$ be the new equilibrium concentration of $OH^-$ we obtain on substitution into the equilibrium expression:

$$1.6 \times 10^{-5} = \frac{(3.2 \times 10^{-3} + y)y}{2.4 \times 10^{-3} - y}$$

The answer obtained on the assumption that $y \ll 2.4 \times 10^{-3}$ is

$$[OH^-] = y = 1.2 \times 10^{-5}$$

Since this is the same value as that obtained before dilution, the pH remains unchanged. ∎

The effect of dilution on the pH of a buffer solution of a weak acid and its conjugate base is shown graphically in Figure 8.5. For the particular conditions illustrated, the pH remains constant even when the solution is diluted 1000-fold. By contrast, dilution of the weak acid alone, or of a strong acid, changes the pH by several units.

Let us now generalize the specific illustrations in Examples 8.6 and 8.7 and determine the pH range over which a particular buffer is effective. Starting with the equilibrium expression for weak acid HA,

$$K_a = \frac{[H_3O^+][A^-]}{[HA]}$$

we get for the $H_3O^+$ concentration

$$[H_3O^+] = \frac{[HA]K_a}{[A^-]}$$

Taking the logarithm of both sides of the equation, we obtain

$$\log[H_3O^+] = \log\frac{[HA]}{[A^-]} + \log K_a$$

The pH is obtained by multiplying through by $-1$ and introducing pH and $pK_a$:

$$pH = pK_a + \log\frac{[A^-]}{[HA]} \tag{8.26}$$

This equation is sometimes called the **Henderson–Hasselbach equation.** The

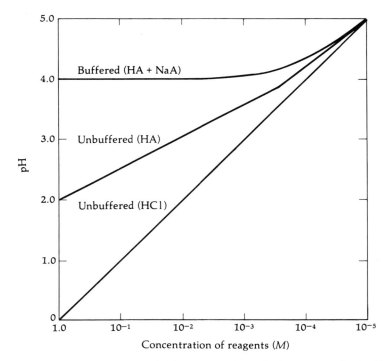

**FIGURE 8.5** The effect of dilution on the pH of an equimolar buffer solution of the weak acid HA ($K_a = 1.0 \times 10^{-4}$) and its conjugate base $A^-$. The effects on the pH of unbuffered HA and HCl are also shown. The initial nominal concentration of each solute is 1 $M$.

equation shows that for an equimolar acid–conjugate base solution, the pH is just equal to $pK_a$. If the $A^-$ concentration is larger than the HA concentration, the pH of the buffer is higher than $pK_a$, while if it is smaller, the pH is lower than $pK_a$. In practice, a buffer will maintain a nearly constant pH on dilution or on addition of small amounts of acid or base so long as the $[A^-]/[HA]$ concentration ratio lies between about 0.1 and 10 because of the slow variation of the logarithm term over this range of ratios. This means that a given acid–conjugate base couple will be an effective buffer over a range of 2 pH units centered at $pK_a$. For instance, since the $pK_a$ of HOAc is 4.75, an equimolar HOAc–$OAc^-$ buffer will resist changes in pH in the range of 3.75 to 5.75. If the concentrations of acid and conjugate base are unequal, the buffer will not be equally effective in protecting the pH from the effects of added acid or base. Figure 8.6 illustrates these points by showing the change in the pH of a 0.1 $M$ HOAc–$OAc^-$ buffer when small amounts of acid or base are added. It is seen that the pH varies slowly with the amount of added acid or base for approximately 1 pH unit on either side of $pK_a$ but begins to vary rapidly outside this interval.

Equation 8.26 indicates that different buffers will be effective at different pH, depending on the value of $K_a$. Figure 8.7 shows the pH range over which a number of common buffers are effective. The greatest constancy of pH is obtained in the middle of the indicated range, which corresponds to $pK_a$. Com-

Acids and Bases  8.8  275

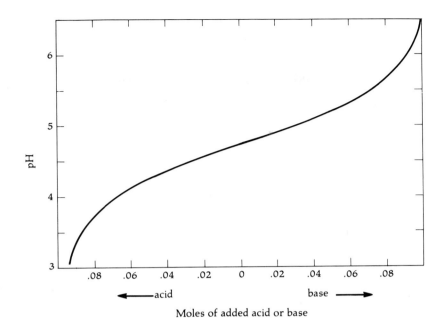

**FIGURE 8.6** The pH of a 0.1 M HOAc–OAc⁻ buffer as a function of the number of moles of added acid or base. The pH values were obtained assuming 1 L of buffer and neglecting changes in volume.

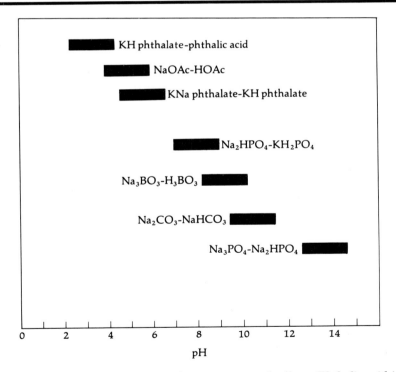

**FIGURE 8.7** The pH range of some common buffers. (Phthalic acid is a diprotic organic acid.)

mercial buffers consisting of a combination of several acid–conjugate base pairs are available. They have the virtue of being useful over a broad range of pH, from approximately 2 to 10.

**Example 8.8** Using the data in Table 8.1, design a buffer that will maintain a pH of 5.2.

Table 8.1 shows that propionic acid has $pK_a$ closest to the desired pH. The most effective value of the ratio of concentrations of propionate ion and propionic acid can be obtained by means of Equation 8.26 for the pH of a buffer:

$$pH = pK_a + \log \frac{[A^-]}{[HA]}$$

Substituting the values of pH and $pK_a$,

$$5.2 = 4.87 + \log \frac{[A^-]}{[HA]}$$

we get for the ratio of concentrations

$$\frac{[A^-]}{[HA]} = 2.14$$

A pH of 5.2 can be maintained by a propionic acid buffer in which $[A^-]/[HA] = 2.14$. If the propionic acid concentration is 0.100 $M$, the propionate ion concentration must therefore be 0.214 $M$. ∎

## 8.9 Acid–Base Titrations

In a titration, the amount or concentration of a particular species is determined by the addition of a known amount of a reagent capable of reacting rapidly and completely with the species in question. (Redox titrations were discussed in Section 2.10.) In this section we examine **acid–base titrations,** in which the concentration of a solution of an acid (or base) is determined by the addition of a known amount of base (or acid). Due to the widespread use of acids and bases, acid–base titrations are an important tool of analytical chemistry.

*a. Some Fundamental Aspects* Acid–base reactions were briefly examined in Section 2.5d as an example of double displacement reactions. The reaction of any strong acid with any strong base can be written as

$$H_3O^+ + OH^- \rightarrow 2H_2O$$

This reaction occurs rapidly and, as we have already noted in this chapter, goes practically to completion. These characteristics make it possible to determine to a high degree of accuracy the concentration of an acidic or basic solution by titration.

Figure 8.8 illustrates the titration of an acid of unknown concentration with a base of known concentration. In principle, the base is dispensed until an *equivalent* amount has been added, that is, until the amount of base is stoichiometrically equal to that of acid. For example, 1 mol of NaOH is equivalent to 1 mol of HCl but to only 0.5 mol of $H_2SO_4$. The addition of an equivalent amount of reagent marks the occurrence of the **equivalence point.** We sometimes say that, at the equivalence point, the base has "neutralized" the acid, or that

**FIGURE 8.8** Titration of an acid by a base. The base is dispensed from the buret until the endpoint is reached, as indicated by the change in color of the pH indicator. To minimize the titration error, the endpoint should occur at or extremely close to the equivalence point.

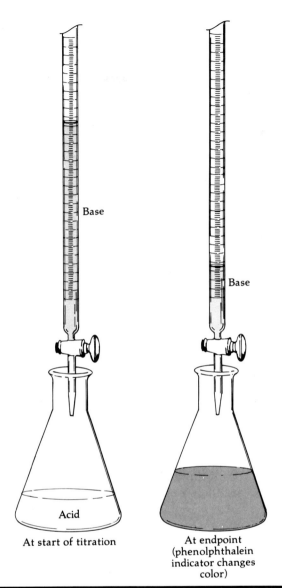

"neutralization" has taken place. This terminology is clear so long as it is realized that neutralization does *not* mean that the resulting solution is necessarily neutral, i.e., has pH of 7.0 at 25°C. Neutralization of an acid by a base, or vice versa, results in a neutral solution when both the acid and base are strong. However, when a weak acid or base is neutralized, the solution is generally not neutral due to the hydrolysis of the resulting salt. For example, at the equivalence point in the reaction of sodium hydroxide and acetic acid, the solution will be alkaline due to hydrolysis of the $OAc^-$ ion (see Equation 8.19).

The equivalence point of a titration is a theoretical concept. In practice, the titration is carried out until the **endpoint** is reached. This is the point at which some detectable physical change accompanying neutralization occurs. Ideally, the endpoint and the equivalence point should, of course, coincide.

This is not always possible and the difference constitutes a source of systematic error.

***b. pH Indicators*** A common method of endpoint detection in acid–base titrations involves the use of a supplementary compound that undergoes a change in color as a result of the change in pH which occurs in the vicinity of the equivalence point. Such a substance is called a **pH indicator.**

A pH indicator is a substance that changes color over some narrow range of pH values. Many organic dye molecules exhibit this property. These substances are weak acids or bases that have a very different color in the acid form than in the conjugate base form. For example, methyl orange is red in a solution of pH less than 3.0, and yellow in one with pH greater than 4.6. The change in color corresponds to an ionization equilibrium which can be represented by the general equation

$$HIn + H_2O \rightleftharpoons H_3O^+ + In^-$$

For methyl orange, HIn is red and $In^-$ is yellow. As in the case of any other weak acid, the equilibrium depends on the value of the acid ionization constant, designated $K_{In}$.

The pH range over which a given indicator changes color can be determined by applying Equation 8.26 to the equilibrium:

$$pH = pK_{In} + \log \frac{[In^-]}{[HIn]}$$

In general, the color of the acid will be dominant when the concentration of HIn is about a factor of ten larger than that of $In^-$ while the color of the base predominates when $[In^-]$ exceeds $[HIn]$ by a similar factor. The color change therefore occurs over a pH range of about two units, centered at $pK_{In}$. Figure

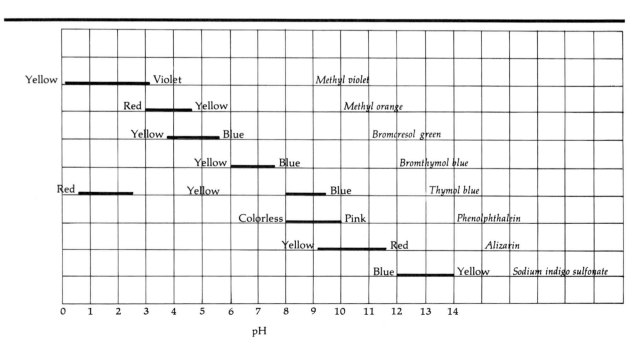

**FIGURE 8.9** pH indicators and their color changes.

8.9 shows the pH intervals at which various indicators change color. Combinations of indicators are available which undergo a continuous change of color over practically the entire range of pH.

Example 8.9  A solution of ammonia (approximate concentration 0.1 M) is titrated with HCl. (a) Which of the indicators in Figure 8.9 will give the most accurate endpoint? (b) Suppose that the only indicator available for this titration is phenolphthalein, which changes color at pH = 9.0. What is the error in the concentration of $NH_3$ resulting from the use of this indicator and the assumption that the endpoint corresponds to the equivalence point?

(a)  The reaction occurring in the titration is

$$NH_3 + H_3O^+ \rightarrow NH_4^+ + H_2O$$

At the equivalence point, all the $NH_3$ has been converted to $NH_4^+$ and no excess acid has been added. The resulting solution is acidic due to the hydrolysis of $NH_4^+$:

$$NH_4^+ + H_2O \rightleftharpoons NH_3 + H_3O^+$$

The pH can be determined in the manner illustrated in Section 8.6. Letting $x$ be the $H_3O^+$ concentration, the equilibrium expression can be written as

$$K_a = 6.3 \times 10^{-10} = \frac{[NH_3][H_3O^+]}{[NH_4^+]} = \frac{x^2}{0.1 - x}$$

The answer is

$$[H_3O^+] = x = 7.9 \times 10^{-6}$$

and the pH is 5.1. This pH lies in the range in which bromcresol green changes color and this indicator should therefore be used.

(b)  Let us first examine what happens in this titration. We start with a ≈0.1 M $NH_3$ solution. Using the approach developed in Section 8.5, the pH of this solution can be evaluated as 11.1. We now add HCl until the pH decreases to 9.0. We know from (a) that the pH at the equivalence point is 5.1. We can conclude that at a pH of 9.0 some, but not all, of the $NH_3$ will have been neutralized. The ammonia in the initial solution will therefore be present in part as $NH_3$ and in part as $NH_4^+$. The relative amounts of the two can be determined by noting that the ionization equilibrium of $NH_3$ must be obeyed. Letting $y$ be the concentration of $NH_4^+$ at pH = 9.0, we can write the equilibrium concentrations as

$$NH_3 + H_2O \rightleftharpoons NH_4^+ + OH^-$$
$$0.1 - y \qquad\qquad y \quad\; 1 \times 10^{-5}$$

The $OH^-$ concentration is obtained from the pH and the ion product of water (pOH = 14.0 − pH = 14.0 − 9.0 = 5.0, $[OH^-] = 1 \times 10^{-5}$). Substituting into the equilibrium expression, we obtain

$$K_b = 1.6 \times 10^{-5} = \frac{[NH_4^+][OH^-]}{[NH_3]} = \frac{1 \times 10^{-5} y}{0.1 - y}$$

Solving for $y$, we obtain

$$[NH_4^+] = y = 0.06\ M$$

indicating that 40% of the ammonia is still present as NH₃. The occurrence of a 40% error shows that the selection of a suitable pH indicator is important. ∎

***c. Titration Curves for Strong Acids or Bases*** An acid–base titration curve is a plot of the pH of an acidic (or basic) solution as a function of the amount of added base (or acid). A titration curve provides a single graphical representation of the manner in which the various factors which determine acid–base equilibria affect the pH. The titration curve also facilitates the choice of the best titration indicator and gives some insight into the accuracy attainable for a particular set of experimental conditions. In this section we examine the titration curves obtained in the titration of a strong acid by a strong base, or vice versa.

As an illustration of the procedure to be followed, let us calculate the titration curve for 100 mL of 0.10 $M$ HCl titrated with 0.10 $M$ NaOH. The pH of the solution changes in the course of the titration as a consequence of two separate factors: the neutralization reaction and the dilution resulting from the increase in solution volume. The effect of both factors must be evaluated. The calculation is conveniently divided into two parts, one dealing with the pH before the equivalence point and the other with the pH after the equivalence point. In the case of strong acids and bases, ionization equilibria do not come into play except at pH ≈ 7, where the ionization of water is manifested.

The calculation of a titration curve involves the decoupling of the effects of reaction and dilution. This is achieved by considering separately the change in the number of moles of acid or base and the change in volume. The concentrations of acid and base change as a consequence of both of these factors and cannot be used directly. The procedure for the region in which acid is present in excess involves a calculation of the number of moles of added base, the number of moles of unneutralized acid, and finally, the acid concentration and pH, all as a function of added base volume $V_{NaOH}$. The calculation is repeated for a sufficiently large number of points to permit a smooth curve to be drawn through them.

Applying this procedure to the titration that constitutes our illustration we have, in turn:

Original acid:    $n_{HCl} = 0.10$ mmol mL$^{-1}$ × 100 mL = 10 mmol
Added base:    $n_{NaOH} = 0.10$ mmol mL$^{-1}$ × $V_{NaOH}$ mL = $0.10 V_{NaOH}$ mmol
Net acid:    $n_{acid} = 10$ mmol $- 0.10 V_{NaOH}$ mmol
Total volume:    $V_{tot} = (100 + V_{NaOH})$ mL

$$[H_3O^+] = \frac{n_{acid}}{V_{tot}} = \frac{10 - (0.10 V_{NaOH})}{100 + V_{NaOH}} = \left(\frac{100 - V_{NaOH}}{100 + V_{NaOH}}\right)(0.10)\ M$$

Similarly, after the equivalence point we must determine how much base was added in excess of that required to neutralize the acid present. From the amount of base and the total volume we determine [OH⁻] and then the pH:

Added base:    $n_{NaOH} = 0.10$ mmol mL$^{-1}$ × $V_{NaOH}$ mL = $0.10 V_{NaOH}$ mmol
Net base:    $n_{base} = 0.10 V_{NaOH}$ mmol $- 10$ mmol
Total volume:    $V_{tot} = (100 + V_{NaOH})$ mL

$$[OH^-] = \frac{n_{base}}{V_{tot}} = \frac{(0.10\ V_{NaOH}) - 10}{100 + V_{NaOH}} = \left(\frac{V_{NaOH} - 100}{100 + V_{NaOH}}\right)(0.10)\ M$$

**TABLE 8.4 Titration of 100 mL of 0.10 M HCl with 0.10 M NaOH**

Before equivalence point:

| Volume of added NaOH $V_{NaOH}$ (mL) | $\dfrac{100 - V_{NaOH}}{100 + V_{NaOH}}$ | $[H_3O^+]$ | pH |
|---|---|---|---|
| 0     | 1.0                | 0.10               | 1.00 |
| 20    | 0.67               | 0.067              | 1.17 |
| 40    | 0.43               | 0.043              | 1.37 |
| 60    | 0.25               | 0.025              | 1.60 |
| 80    | 0.11               | 0.011              | 1.96 |
| 90    | 0.053              | 0.0053             | 2.28 |
| 99    | $5.0 \times 10^{-3}$ | $5.0 \times 10^{-4}$ | 3.30 |
| 99.9  | $5.0 \times 10^{-4}$ | $5.0 \times 10^{-5}$ | 4.30 |
| 99.99 | $5.0 \times 10^{-5}$ | $5.0 \times 10^{-6}$ | 5.30 |

After equivalence point:

| Volume of added NaOH $V_{NaOH}$ (mL) | $\dfrac{V_{NaOH} - 100}{V_{NaOH} + 100}$ | $[OH^-]$ | pH |
|---|---|---|---|
| 100.01 | $5.0 \times 10^{-5}$ | $5.0 \times 10^{-6}$ | 8.70 |
| 100.1  | $5.0 \times 10^{-4}$ | $5.0 \times 10^{-5}$ | 9.70 |
| 101    | $5.0 \times 10^{-3}$ | $5.0 \times 10^{-4}$ | 10.70 |
| 110    | 0.048 | 0.0048 | 11.70 |
| 120    | 0.091 | 0.0091 | 11.96 |
| 140    | 0.17  | 0.017  | 12.23 |
| 160    | 0.23  | 0.023  | 12.36 |
| 180    | 0.28  | 0.028  | 12.45 |
| 200    | 0.33  | 0.033  | 12.52 |

Table 8.4 gives a tabulation of the quantities entering into this calculation and of the calculated pH. The resulting titration curve is shown in Figure 8.10. The pH rises slowly with the amount of added base until the vicinity of the equivalence point is reached, where it rises sharply. In this region, a drop or two of added base will change the pH by 4 or 5 units. As a result, all indicators that change color somewhere within this range will yield the same endpoint and any of them can be used. Since both acid and base are strong, the pH at the equivalence point is 7.0 and the curve is symmetric about this point.

Figure 8.10 also includes a titration curve for 0.01 $M$ HCl titrated with 0.01 $M$ NaOH. The overall shape is similar to the curve for the 0.1 $M$ solution. However, since the solution is ten times more dilute, the pH before neutralization is higher while that after neutralization is lower than that for the more concentrated solution. Consequently, the steep increase in the vicinity of the equivalence point is confined to a narrower pH interval. This means that more care must be chosen in the selection of an endpoint indicator if an accurate result is to be obtained.

The curve for the titration of a strong base with a strong acid can be obtained by a similar analysis. The resulting curve is just the reverse of the one obtained for the titration of a strong acid: the pH decreases with added acid and shows the same behavior in the vicinity of pH = 7 as is displayed in Figure 8.10.

**d. Titration of a Weak Acid or Base** The titration of a weak acid by a strong

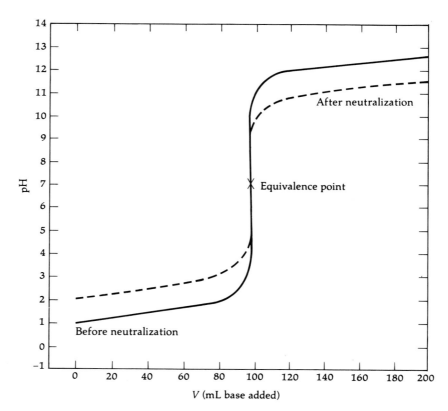

**FIGURE 8.10** Curve for the titration of HCl with NaOH. Solid curve, 100 mL of 0.1 M HCl titrated with 0.1 M NaOH; dashed curve, 100 mL of 0.01 M HCl titrated with 0.01 M NaOH.

base involves the reaction

$$HA + OH^- \rightleftarrows A^- + H_2O$$

and requires a considerably more complicated analysis than that outlined in Section 8.9c. The calculation of the titration curve is conveniently divided into four distinct parts. Since each of these parts corresponds to one or another of the equilibria considered in previous sections, it is convenient to recapitulate by restating the equilibria involved, using the titration of 100 mL of 0.10 M HOAc with 0.10 M NaOH for illustration, and defining $x$ as $[H_3O^+]$.

1. Before the addition of base, the pH is that of the weak acid HOAc. The equilibrium is

$$\begin{array}{cccc} HOAc & + H_2O \rightleftarrows & H_3O^+ & + OAc^- \\ 0.10 - x & & x & x \end{array}$$

$$K_a = \frac{x^2}{0.10 - x}$$

2. In the region of acid excess, the equilibrium is that for a weak acid and a salt of that acid. The pH is obtained as illustrated in Example 8.6. The equi-

librium is

$$HOAc + H_2O \rightleftharpoons H_3O^+ + OAc^-$$
$$C_a - x \qquad \qquad x \qquad C_b + x$$

$$K_a = \frac{x(C_b + x)}{C_a - x}$$

where $C_a$ is the remaining stoichiometric concentration of acid and $C_b$ is the concentration of added base. Note that when $C_b = C_a$, i.e., when half the initial acid has been neutralized, $pH = pK_a$.

3. At the equivalence point, the equilibrium is that for a salt of a weak acid (Section 8.6):

$$OAc^- + H_2O \rightleftharpoons HOAc + OH^-$$
$$0.05 - x \qquad \qquad x \qquad x$$

$$\frac{K_w}{K_a} = \frac{x^2}{0.05 - x}$$

Note that the concentration of $OAc^-$ is half that of the initial HOAc as a result of the doubling in volume.

4. Beyond the equivalence point the acid equilibrium no longer plays a role because the excess $OH^-$ represses the hydrolysis of $OAc^-$. The pH is determined by the concentration of excess NaOH, just as in the titration of a strong acid.

Figure 8.11 displays the titration curve resulting from this analysis. Let us examine it in detail and contrast it with the curve obtained for the titration of a

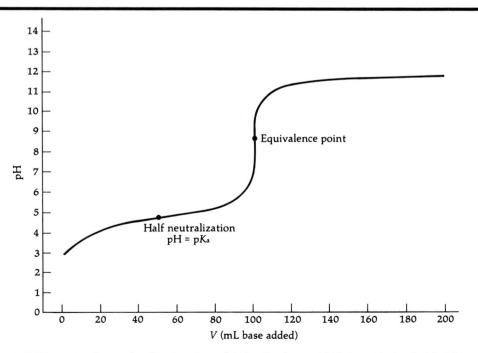

**FIGURE 8.11** Curve for the titration of 100 mL of 0.1 $M$ HOAc with 0.1 $M$ NaOH.

strong acid (i.e., Figure 8.10). We note that due to the lower H₃O⁺ concentration, the starting pH of the weak acid is higher. After an initial rise, the curve features a long plateau, reflecting the formation of the HOAc–OAc⁻ buffer. The pH in the middle of this region is 4.75, that is, equal to p$K_a$. As the equivalence point is approached, the pH increases more rapidly. Note, however, that the change in pH in the vicinity of the equivalence point is smaller than in the case of a strong acid. This result, coupled with the fact that the pH at the equivalence point is greater than 7 due to hydrolysis, indicates that we must be careful in the choice of indicator. An indicator that changes color at moderately low pH (e.g., methyl orange) would result in an erroneous endpoint. Accurate results require the use of one of the indicators that change color in the pH = 7–10 range.

The curve for the titration of a weak base with a strong acid can similarly be divided into four distinct regions corresponding to the dominant equilibrium processes. The titration curve is shaped approximately like an inverted titration curve for a weak acid, with the pH decreasing with the amount of added acid.

---

**Example 8.10** When 100.0 mL of an unknown weak acid is titrated with 0.100 $M$ NaOH, 38.5 mL of NaOH is needed to reach the endpoint. It is also observed that the pH of the solution is 4.21 when 24.2 mL of NaOH has been added. Assuming that the acid has the formula HA, determine its concentration in the initial solution. What is the value of $K_a$?

We assume that the endpoint occurs at the equivalence point. According to the reaction stoichiometry,

$$HA + OH^- \rightarrow A^- + H_2O$$

the number of moles of base added at the endpoint is equal to the number of moles of acid present initially:

$$n_{HA} = n_{NaOH} = (0.100 \text{ mol L}^{-1})(38.5 \text{ mL})(10^{-3} \text{ L mL}^{-1})$$
$$= 3.85 \times 10^{-3} \text{ mol}$$

The concentration of the acid solution is

$$C_{HA} = \frac{n_{HA}}{V_{HA}} = \frac{3.85 \times 10^{-3} \text{ mol}}{(100.0 \text{ mL})(10^{-3} \text{ L mL}^{-1})}$$
$$= 0.0385 \text{ } M$$

We have seen that the acid ionization constant affects the shape of the titration curve in the region up to the endpoint. Consequently, the value of $K_a$ can be determined from the pH at some known point on the titration curve. Before the equivalence point is reached, the equilibrium that must be considered is

$$HA + H_2O \rightleftharpoons H_3O^+ + A^-$$

We know that

$$[H_3O^+] = 10^{-pH} = 10^{-4.21} = 6.17 \times 10^{-5} \text{ } M$$

when 24.2 mL of NaOH has been added. This amount constitutes

$$\frac{24.2 \text{ mL}}{38.5 \text{ mL}} = 0.6286$$

of the total NaOH added. The stoichiometry of the neutralization reaction tells us that 0.6286 of the HA present initially will have been converted to $A^-$ at this point, thus leaving

$$1 - 0.6286 = 0.3714$$

of the acid still present in the form of HA. Therefore, the ratio of concentrations of $A^-$ and HA at this point in the titration is

$$\frac{[A^-]}{[HA]} = \frac{0.6286}{0.3714} = 1.693$$

Substituting into the equilibrium expression

$$K_a = \frac{[H_3O^+][A^-]}{[HA]}$$

we obtain

$$K_a = (6.17 \times 10^{-5})(1.693) = 1.04 \times 10^{-4} \quad \blacksquare$$

**e. Graphical Representation of Concentration Changes** We have noted that the concentrations of the various species present in a solution of a weak acid change in the course of a titration. First, the un-ionized acid HA predominates, next, HA and $A^-$ coexist in a buffer solution, and finally $A^-$ predominates. The changes in the concentrations of the various species as a function of pH are summarized graphically in Figure 8.12. This figure gives a plot of the loga-

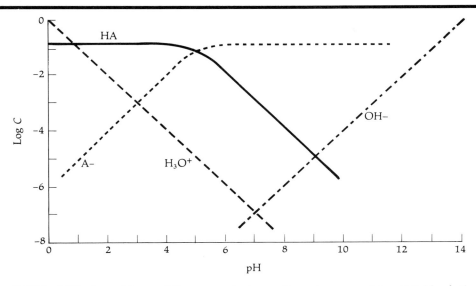

**FIGURE 8.12** Logarithms of the concentrations of species present in a 0.1 M solution of acid HA as a function of pH. Here $K_a$ is assumed to be $10^{-5}$.

rithms of the concentrations of the various species versus pH for an acid with a particular $K_a$ and a specific stoichiometric concentration.

The figure shows the variation of $[H_3O^+]$, $[OH^-]$, $[HA]$, and $[A^-]$ with pH. Since pH = $-\log[H_3O^+]$, the curves are, in effect, plots of the various concentrations versus the $H_3O^+$ concentration on a log–log scale. We see that the graph of $[H_3O^+]$ is a straight line of slope $-1$ and intercept 0 (i.e., $[H_3O^+]$ = 1 $M$ when pH = 0). Similarly, the graph of $[OH^-]$ is a straight line of slope $+1$ and intercept 0 when pH = 14. This behavior is just a consequence of the definition of pH. Of greater interest is the shape of the $[HA]$ and $[A^-]$ curves on this log–log plot. The HA concentration is virtually equal to the stoichiometric concentration at low pH because the high $H_3O^+$ concentration represses the ionization of the acid. With increasing pH, $[HA]$ gradually decreases and $[HA]$ and $[A^-]$ become equal when pH = $pK_a$. At higher pH, $[HA]$ decreases sharply as the continuing decrease in $[H_3O^+]$ induces further ionization. The concentration of $A^-$ shows the opposite dependence on pH since mass balance requires that $[HA]$ and $[A^-]$ always add up to the stoichiometric concentration of HA.

Figure 8.12 can be used to obtain quick semiquantitative answers to a variety of equilibrium problems involving the system in question:

What proportions of HA and NaA are required for a buffer with pH = 4.0 when the total stoichiometric concentration of these two species is 0.1 $M$? From the diagram we see that at pH = 4.0, $\log[HA] = -1.0$ and $\log[A^-] = -2.0$. Therefore,

$$\frac{[HA]}{[A^-]} = \frac{10^{-1}}{10^{-2}} = 10$$

What is the concentration of each species present in a 0.1 $M$ NaA solution? Since the reaction of importance is

$$A^- + H_2O \rightleftharpoons HA + OH^-$$

it follows that, to a good approximation, $[HA] = [OH^-]$. Figure 8.12 shows that these two concentrations are equal at pH = 9.0. At this point

$$[HA] = [OH^-] = 10^{-5} \ M$$
$$[H_3O^+] = 10^{-9} \ M$$

and

$$[A^-] = 10^{-1} \ M$$

Although the answers to problems of this type can be obtained by the algebraic techniques considered in this chapter, the use of Figure 8.12 provides a visual method which can be of value in more complex equilibria, as we shall see.

## 8.10 Polyprotic Acids

***a. Multistage Equilibria*** A **polyprotic acid** is one that can transfer more than one proton to a base. Examples of diprotic acids are $H_2SO_4$ and $H_2S$. The most common triprotic acid is phosphoric acid ($H_3PO_4$), which can undergo three successive proton transfer reactions in solution:

$$H_3PO_4 + H_2O \rightarrow H_3O^+ + H_2PO_4^- \qquad K_1 = 7.5 \times 10^{-3}$$

$$H_2PO_4^- + H_2O \rightarrow H_3O^+ + HPO_4^{2-} \quad K_2 = 6.6 \times 10^{-8}$$

$$HPO_4^{2-} + H_2O \rightarrow H_3O^+ + PO_4^{3-} \quad K_3 = 1 \times 10^{-12}$$

Each reaction is characterized by a different acid ionization constant. A solution of phosphoric acid is actually a mixture of three different acids, $H_3PO_4$, $H_2PO_4^-$, and $HPO_4^{2-}$, and their conjugate bases. In considering the equilibria of polyprotic acids we must address what at first sight appears to be a formidable task: the calculation of the concentrations of as many as six distinct species which simultaneously obey several equilibria. Fortunately, simplifications are possible because the successive ionization constants for a given acid generally have widely different values, i.e.,

$$K_1 \gg K_2 \gg K_3$$

A consequence of this difference is that no more than two of the successive species associated with the ionization of a polyprotic acid are present in significant concentrations at a particular pH. The equilibria are determined by only one, or at most two, of the acid ionization constants and the principles developed earlier in this chapter can be applied with only minor modifications.

Table 8.5 summarizes the acid ionization constants of some common polyprotic acids and shows that successive ionization constants of a given acid generally differ by a factor of $10^4$ or more. This is understandable, for if it is difficult to remove a proton from a neutral acid, it must be more so to remove one from a negatively charged ion.

We illustrate the evaluation of multistage ionization equilibria by considering the equilibria occurring in an aqueous solution of carbon dioxide. All natural waters contain some dissolved carbon dioxide. The equilibria involving this substance are important to the chemistry of natural waters and also play a significant role in such diverse processes as respiration and the manufacture of carbonated beverages. The equilibria occurring in aqueous $CO_2$ are

$$CO_2(aq) + 2H_2O \rightleftarrows H_3O^+ + HCO_3^- \tag{8.27}$$

$$HCO_3^- + H_2O \rightleftarrows H_3O^+ + CO_3^{2-} \tag{8.28}$$

**TABLE 8.5  Ionization Constants of Some Common Polyprotic Acids At 25°C**

| Acid | HA | A⁻ | K | | pK |
|---|---|---|---|---|---|
| Sulfuric | $H_2SO_4$ | $HSO_4^-$ | >1 | $(K_1)$ | |
| | $HSO_4^-$ | $SO_4^{2-}$ | $1.0 \times 10^{-2}$ | $(K_2)$ | 2.00 |
| Sulfurous | $H_2SO_3$ | $HSO_3^-$ | $1.43 \times 10^{-2}$ | $(K_1)$ | 1.845 |
| | $HSO_3^-$ | $SO_3^{2-}$ | $5.0 \times 10^{-8}$ | $(K_2)$ | 7.30 |
| Phosphoric | $H_3PO_4$ | $H_2PO_4^-$ | $7.5 \times 10^{-3}$ | $(K_1)$ | 2.12 |
| | $H_2PO_4^-$ | $HPO_4^{2-}$ | $6.6 \times 10^{-8}$ | $(K_2)$ | 7.18 |
| | $HPO_4^{2-}$ | $PO_4^{3-}$ | $1 \times 10^{-12}$ | $(K_3)$ | 12.0 |
| Carbon dioxide | $CO_2(aq)$ | $HCO_3^-$ | $4.5 \times 10^{-7}$ | $(K_1)$ | 6.35 |
| (Carbonic) | $HCO_3^-$ | $CO_3^{2-}$ | $4.8 \times 10^{-11}$ | $(K_2)$ | 10.32 |
| Hydrosulfuric | $H_2S$ | $HS^-$ | $1.0 \times 10^{-7}$ | $(K_1)$ | 7.00 |
| | $HS^-$ | $S^{2-}$ | $3 \times 10^{-13}$ | $(K_2)$ | 12.5 |

The formation of carbonic acid ($H_2CO_3$) in the reaction

$$CO_2(aq) + H_2O \rightleftarrows H_2CO_3(aq)$$

does not occur to a significant extent and can be ignored.

### Example 8.11

The concentration of $CO_2$ in a saturated solution at 25°C and atmospheric pressure is 0.040 $M$. Calculate the pH and the concentration of all species containing the carbonate group present in the solution.

Since $K_2 \ll K_1$ it is reasonable to treat the preceding two equilibria separately. This is an assumption whose validity must be checked. The first equilibrium yields the following equilibrium concentrations:

$$CO_2(aq) + 2H_2O \rightleftarrows H_3O^+ + HCO_3^-$$
$$0.040 - x \qquad\qquad x \qquad x$$

Inserting these values into the equilibrium expression, we obtain

$$K_1 = 4.5 \times 10^{-7} = \frac{[H_3O^+][HCO_3^-]}{[CO_2]} = \frac{x^2}{0.040 - x}$$

Solving for $x$, we obtain

$$x = [H_3O^+] = [HCO_3^-] = 1.3 \times 10^{-4} \, M$$

and the pH is 3.87. The concentration of $CO_2$ remains 0.040 $M$.

In obtaining these values we have not considered the effect of the second ionization step. This step will increase the concentration of $H_3O^+$ and decrease that of $HCO_3^-$. However, the change can be shown to be negligibly small because $K_2 \ll K_1$. The second step, however, must be considered in order to obtain the concentration of $CO_3^{2-}$. The evaluation of $[CO_3^{2-}]$ can, at the same time, be used to substantiate the assertion that the concentrations of $H_3O^+$ and $HCO_3^-$ are unaffected by the ionization of $HCO_3^-$. The second ionization step yields the following equilibrium concentrations:

$$HCO_3^- + H_2O \rightleftarrows H_3O^+ + CO_3^{2-}$$
$$1.3 \times 10^{-4} - x \qquad\qquad 1.3 \times 10^{-4} + x \qquad x$$

The equilibrium expression is

$$K_2 = 4.8 \times 10^{-11} = \frac{[H_3O^+][CO_3^{2-}]}{[HCO_3^-]} = \frac{x(1.3 \times 10^{-4} + x)}{1.3 \times 10^{-4} - x}$$

The solution, obtained by neglecting $x$ with respect to $1.3 \times 10^{-4}$, is

$$x = [CO_3^{2-}] = 4.8 \times 10^{-11} \, M$$

showing that the second ionization step is indeed completely negligible insofar as the concentrations of $H_3O^+$ and $HCO_3^-$ are concerned. ∎

The preceding example shows that in a moderately acidic solution (pH = 3.9) there is a decrease in the concentrations of carbonate species that result from successive ionization steps:

$$[CO_3^{2-}] \ll [HCO_3^-] \ll [CO_2]$$

This relationship must be strongly pH dependent. According to Le Chatelier's principle, if $[H_3O^+]$ is reduced by the addition of base, the equilibrium in the reactions given by Equations 8.27 and 8.28 is shifted to the right and the concentration of $CO_2$ decreases relative to those of $HCO_3^-$ and $CO_3^{2-}$. Let us determine how the concentrations of these three species vary with the pH of the solution.

The first equilibrium allows us to relate the concentrations of $CO_2$ and $HCO_3^-$ to each other:

$$K_1 = 4.5 \times 10^{-7} = \frac{[H_3O^+][HCO_3^-]}{[CO_2]}$$

Therefore,

$$\frac{[HCO_3^-]}{[CO_2]} = \frac{K_1}{[H_3O^+]}$$

This relation shows that $[HCO_3^-] = [CO_2]$ when $[H_3O^+] = K_1 = 4.5 \times 10^{-7}$, corresponding to pH = $pK_1$ = 6.35. At this pH, the solution consists of an equimolar $CO_2$–$HCO_3^-$ buffer. In more acidic solutions, $[CO_2]$ will be larger than $[HCO_3^-]$ while, conversely, the opposite holds true in more alkaline solutions.

Although the $HCO_3^-$ concentration has become substantial at pH = 6.35, the $CO_3^{2-}$ concentration is still very small. This can be seen by rewriting the expression for the second ionization equilibrium,

$$K_2 = \frac{[H_3O^+][CO_3^{2-}]}{[HCO_3^-]}$$

in terms of the ratio of $CO_3^{2-}$ and $HCO_3^-$ concentrations and evaluating this ratio at pH = 6.35:

$$\frac{[CO_3^{2-}]}{[HCO_3^-]} = \frac{K_2}{[H_3O^+]} = \frac{4.8 \times 10^{-11}}{4.5 \times 10^{-7}} = 1.1 \times 10^{-4}$$

However, as the pH is further increased, the second ionization step becomes increasingly important. In complete analogy to the first ionization step, the $CO_3^{2-}$ and $HCO_3^-$ concentrations become equal when $[H_3O^+] = K_2$, that is, at a pH of 10.32. Once again, the solution is buffered, the acid now being $HCO_3^-$ and the conjugate base being $CO_3^{2-}$.

The changes in the concentrations of the various carbonate species as a function of pH are conveniently displayed in a plot of log[C] versus pH. Figure 8.13 shows such a plot for 0.1 M $CO_2$. As demonstrated earlier, $CO_2$ predominates in acidic solution but its concentration rapidly decreases at pH greater than 6.4. Similarly, $HCO_3^-$ predominates between pH of 6.4 and 10.3, giving way to $CO_3^{2-}$ in more basic solution.

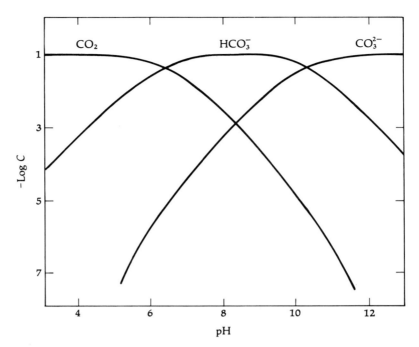

**FIGURE 8.13** Dependence of the concentration of carbonate species on pH for a 0.1 $M$ solution of $CO_2$.

---

**Example 8.12**  The pH of a 0.010 $M$ solution of $Na_2S$ is adjusted to 11.00. What are the concentrations of the various sulfur-containing species?

The equilibria involving $H_2S$ in aqueous solution are

$$H_2S(aq) + H_2O \rightleftarrows H_3O^+ + HS^- \qquad K_1 = 1.0 \times 10^{-7}$$

$$HS^- + H_2O \rightleftarrows H_3O^+ + S^{2-} \qquad K_2 = 3 \times 10^{-13}$$

We expect from the preceding considerations, given the values of $K_1$ and $K_2$, that at a pH of 11 the predominant species will be $HS^-$ and $S^{2-}$. Thus, we first focus on the second reaction and write the equilibrium concentrations:

$$\begin{array}{cccc} HS^- + H_2O \rightleftarrows & H_3O^+ & + & S^{2-} \\ x & 1.0 \times 10^{-11} & & 0.010 - x \end{array}$$

where the $H_3O^+$ concentration is obtained from the given pH. The equilibrium expression is

$$K_2 = 3 \times 10^{-13} = \frac{[H_3O^+][S^{2-}]}{[HS^-]} = \frac{(1.0 \times 10^{-11})(0.010 - x)}{x}$$

and the solution is

$$x = [HS^-] = 9.7 \times 10^{-3} \, M$$

$$0.010 - x = [S^{2-}] = 3 \times 10^{-4} \, M$$

The H₂S concentration can now be obtained by considering the first equilibrium:

$$H_2S + H_2O \rightleftharpoons H_3O^+ + HS^-$$

$$y \qquad\qquad 1.0 \times 10^{-11} \quad 9.7 \times 10^{-3} - y$$

Substituting into the equilibrium expression, we obtain

$$K_1 = 1.0 \times 10^{-7} = \frac{[H_3O^+][HS^-]}{[H_2S]} = \frac{(1.0 \times 10^{-11})(9.7 \times 10^{-3} - y)}{y}$$

$$y = [H_2S] = 9.7 \times 10^{-7} \, M$$

Note that the H₂S concentration is too small to affect the HS⁻ concentration obtained from the HS⁻–S²⁻ equilibrium. ∎

**b. Equilibria of Amphiprotic Species**  In this section we examine equilibria involving amphiprotic species, that is, species that can both donate and accept protons. This topic is intimately related to polyprotic acid equilibria for at least one of the species involved in the successive ionization of a polyprotic acid is amphiprotic. Let us consider the diprotic acid H₂A, with ionization constants $K_1$ and $K_2$, and focus on the acid–base equilibria of the species HA⁻. This ion is amphiprotic because it can undergo the following two reactions:

$$HA^- + H_2O \rightleftharpoons H_3O^+ + A^{2-} \qquad K = K_2 \qquad (8.29)$$

$$HA^- + H_2O \rightleftharpoons H_2A + OH^- \qquad K = K_w/K_1 \qquad (8.30)$$

In the first reaction HA⁻ acts as an acid, while in the second it acts as a base, with an equilibrium constant related to the ionization constant of the conjugate acid (H₂A) by Equation 8.10. For example, in an aqueous solution of CO₂, the amphiprotic species is the hydrogen carbonate ion ($HCO_3^-$). We are interested in the properties of a solution of a salt such as NaHA—its pH, the extent to which it undergoes the preceding reactions, and the dependence of these properties on concentration.

Equations 8.29 and 8.30 indicate that five species of unknown concentration are present in a solution of NaHA: HA⁻, H₂A, A²⁻, H₃O⁺, and OH⁻. The solution, of course, also contains Na⁺, but its concentration is equal to the stoichiometric concentration of the salt, $C_0$. In order to express the H₃O⁺ concentration in terms of known quantities, five equations relating the concentrations of these species are needed. The procedure followed in Section 8.7, in which the effect of the ionization of water on the pH of a weak acid was evaluated, must be followed here.

The first two equations involving the concentrations of these species are just the equilibrium expressions for the reactions represented by Equations 8.29 and 8.30:

$$K_2 = \frac{[H_3O^+][A^{2-}]}{[HA^-]} \qquad (8.31)$$

$$\frac{K_w}{K_1} = \frac{[H_2A][OH^-]}{[HA^-]} \qquad (8.32)$$

The ion product of water constitutes a third equilibrium expression:

$$[H_3O^+][OH^-] = K_w \tag{8.33}$$

The remaining two expressions are obtained from the mass and charge balance relationships. Since the species A must be present as $H_2A$, $HA^-$, or $A^{2-}$, the mass balance is

$$C_0 = [H_2A] + [HA^-] + [A^{2-}] \tag{8.34}$$

Charge balance requires that

$$[Na^+] + [H_3O^+] = [HA^-] + 2[A^{2-}] + [OH^-]$$

and, since the $Na^+$ concentration is just equal to the stoichiometric concentration of the salt,

$$C_0 + [H_3O^+] = [HA^-] + 2[A^{2-}] + [OH^-] \tag{8.35}$$

Note that the concentration of $A^{2-}$ has a coefficient of 2. Each $A^{2-}$ ion contributes twice as much charge as the other ions do because it has a charge of $-2$.

To express $[H_3O^+]$ in terms of known quantities, we proceed as follows. First, subtract Equations 8.34 from Equation 8.35:

$$[H_3O^+] = [A^{2-}] - [H_2A] + [OH^-] \tag{8.36}$$

Solving Equations 8.31 and 8.32 for $[A^{2-}]$ and $[H_2A]$, respectively, and substituting into Equation 8.36, we obtain

$$[H_3O^+] = \frac{K_2[HA^-]}{[H_3O^+]} - \frac{K_w[HA^-]}{K_1[OH^-]} + [OH^-] \tag{8.37}$$

Replacememt of $[OH^-]$ by $K_w/[H_3O^+]$ (from Equation 8.33) gives, after rearrangement,

$$[H_3O^+] = \left( \frac{K_2[HA^-] + K_w}{1 + [HA^-]/K_1} \right)^{1/2} \tag{8.38}$$

As we shall see in Example 8.13, the concentration of $HA^-$ is typically very close to its stoichiometric concentration $C_0$. Equation 8.38 can then be rewritten as

$$[H_3O^+] = \left( \frac{K_2 C_0 + K_w}{1 + C_0/K_1} \right)^{1/2} \tag{8.39}$$

which expresses the $H_3O^+$ concentration in terms of known quantities. Two additional approximations are frequently valid. If the stoichiometric concentration of NaHA is not too small and if $K_1$ is small, then $C_0/K_1 \gg 1$. Furthermore, the values of $K_2$ in Table 8.5 indicate that, for most acids, $K_2 C_0 \gg K_w$, provided that $C_0$ is not too small. With these approximations, Equation 8.39 reduces to the simple expression

$$[H_3O^+] = (K_1 K_2)^{1/2} \tag{8.40}$$

and the pH is

$$pH = \frac{pK_1 + pK_2}{2} \tag{8.41}$$

Equation 8.41 indicates that the pH of a solution of NaHA is equal to the average of the pK values of the corresponding diprotic acid $H_2A$ and, so long as the solution is not too dilute, is independent of the concentration of the solution. Although this is one of the characteristic properties of a buffer, a solution of a salt of this type is *not* a buffer. In fact, the addition of acid or base has a marked effect on the pH of an amphiprotic solution.

Equations 8.40 and 8.41 become invalid when the solution is very dilute. In the limit of infinite dilution, the pH must, of course, become that of water. Equation 8.39 shows that this is indeed so, since the equation reduces to $[H_3O^+] = K_w^{1/2}$ when terms involving the stoichiometric concentration, $C_0$, are neglected. The validity of Equations 8.40 and 8.41 for a given concentration depends on the actual values of $K_1$ and $K_2$ and can be determined by comparison of the calculated pH with that obtained by means of Equation 8.39.

**Example 8.13** Consider a 0.100 M $NaHCO_3$ solution. (a) What is the pH? (b) What are the concentrations of $CO_2$, $HCO_3^-$, and $CO_3^{2-}$?

(a) The pH may be calculated by means of Equation 8.41:

$$pH = \frac{pK_1 + pK_2}{2} = \frac{6.35 + 10.32}{2} = 8.34$$

(b) The $HCO_3^-$ ion can undergo the following three reactions:

(1) $HCO_3^- + H_2O \rightleftharpoons H_3O^+ + CO_3^{2-}$    $K_2 = 4.8 \times 10^{-11}$

(2) $HCO_3^- + H_2O \rightleftharpoons CO_2 + H_2O + OH^-$    $K_b = \frac{K_w}{K_1} = \frac{1.00 \times 10^{-14}}{4.5 \times 10^{-7}}$
$= 2.2 \times 10^{-8}$

(3) $2HCO_3^- \rightleftharpoons CO_2 + H_2O + CO_3^{2-}$    $K = \frac{K_2}{K_1} = \frac{4.8 \times 10^{-11}}{4.5 \times 10^{-7}}$
$= 1.1 \times 10^{-4}$

Reaction (3) has the largest equilibrium constant and is consequently the most important of the three reactions. We assume that it alone determines the various concentrations. Since its equilibrium constant is nonetheless small, let $x$ be the concentration of $CO_2$ or $CO_3^{2-}$. The equilibrium constant expression is

$$K = 1.1 \times 10^{-4} = \frac{x^2}{(0.100 - 2x)^2}$$

and the solution obtained by taking the square root of both sides is $x = 1.0 \times 10^{-3}$. The concentrations are

$$[CO_2] = [CO_3^{2-}] = x = 1.0 \times 10^{-3} \, M$$

and

$$[HCO_3^-] = 0.100 - 2x = 0.098 \, M$$

indicating that most of the hydrogen carbonate is present as $HCO_3^-$. ∎

Sodium hydrogen carbonate, commonly known as "bicarbonate of soda", is used as an antacid owing to its mildly alkaline pH and the release of $CO_2$ in its reaction with acid:

$$HCO_3^- + H_3O^+ \rightarrow CO_2(g) + 2H_2O$$

Sodium hydrogen carbonate is also a common ingredient of baking powders, where the release of $CO_2$ serves to leaven the dough.

To summarize the various equilibria which occur at different pH values in the $CO_2(aq)$–$HCO_3^-$–$CO_3^{2-}$ system, we show in Figure 8.14 the curve for the titration of $CO_2(aq)$ with NaOH solution. This rather complicated curve can best be understood as consisting of two successive titration curves of the type illustrated in Figure 8.11. The curve has been marked by several points to facilitate the discussion. Point 1 signals the beginning of the titration; the pH is that of a solution of $CO_2$, and may be evaluated as in Example 8.11. At point 2, half the $CO_2$ present has been converted to $HCO_3^-$, and the solution consists of an equimolar $CO_2$–$HCO_3^-$ buffer. The pH is just equal to $pK_1$ (Equation 8.26). Point 3 marks the first equivalence point: all the $CO_2$ has been converted to $HCO_3^-$. The pH may be obtained by means of Equation 8.41 as $\frac{1}{2}(pK_1 + pK_2)$. At point 4, half the $HCO_3^-$ has been converted to $CO_3^{2-}$. The solution is once again

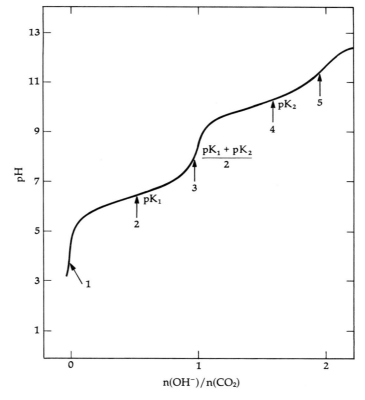

**FIGURE 8.14** Curve for the titration of $0.1\ M\ CO_2(aq)$ with NaOH. The abscissa, $n(OH^-)/n(CO_2)$, is the ratio of the number of moles of added base to that of $CO_2$ present initially. The numbers mark special points on the curve, as discussed in the text.

buffered and the pH is now equal to p$K_2$. The second equivalence point occurs at point 5. The solution now consists of $CO_3^{2-}$. Since the conjugate acid of this species, $HCO_3^-$, is very weak, $CO_3^{2-}$ hydrolyzes to form $HCO_3^-$ and $OH^-$. The pH can be determined as in Example 8.5.

## Conclusion

The concepts of chemical equilibrium developed in Chapter 7 have been applied in this chapter to solutions of acids and bases. The meaning of these terms has evolved with the development of chemistry. The Brønsted–Lowry definitions are particularly useful in the analysis of aqueous solutions. With their help, we have presented a unified approach to equilibria involving acids, bases, and their salts.

Our approach in this chapter has been to develop a number of general principles and to illustrate their use in specific situations. Some equilibrium problems may be solved by the use of simple approximations based on these principles while others may require a more complete treatment. The ability to choose the simplest approximation that will yield a valid solution is of general usefulness in chemistry since many chemical problems are sufficiently complicated that this is the only approach possible. This important skill can be fostered by practice in solving equilibrium problems.

## Problems

1. Write the conjugate bases for each of the following Brønsted acids: $HSO_4^-$, $HOCl$, $Fe(H_2O)_6^{3+}$.

2. Write the conjugate acids for each of the following Brønsted bases: $HSO_4^-$, $NH_2^-$, $PO_4^{3-}$.

3. Arrange the following bases in order of increasing strength: $OAc^-$, $NH_3$, $CN^-$, $H_2O$, $OH^-$, $ClO_4^-$.

4. What is the pH of each of the following solutions: $3.50 \times 10^{-3}$ M HCl, 0.0150 M NaOH, $2.70 \times 10^{-3}$ M NaOH?

5. What is the pH of a solution prepared by mixing 100 mL of 0.100 M NaOH with 150 mL of 0.120 M HCl?

6. What volume of 0.100 M NaOH solution must be added to 100 mL of 0.150 M HCl to increase the pH to 3.00?

7. Heavy water consists of deuterium oxide, $D_2O$ (D = $^2H$). The equilibrium constant for the ionization of pure $D_2O$ at 25°C is $1.35 \times 10^{-15}$. (a) If pD is defined analogously to pH, what is the pD of pure $D_2O$? (b) For what values of pD will a solution be acidic in this solvent system? (c) What is the relationship between pD and pOD?

8. How will the pH of pure water be affected by the addition of (a) $Na_2C_2O_4$, (b) HCN, (c) NaCN, (d) $NH_4Cl$, and (e) NaCl?

9. How will the pH of a solution of formic acid be affected by the addition of (a) $H_2O$, (b) $Na_2CO_3$, (c) NaCl, (d) HCl, and (e) $NH_4Cl$? Neglect the effects of dilution on (b) − (e).

10. Determine the concentrations of all species present in a 0.0250 M solution of HOAc. What is the pH of the solution? What is the degree of ionization?

11. What are the concentrations of all species present in 0.0325 M $NH_3$?

12. What is the concentration of $OAc^-$ in a solution prepared by mixing 50.0 mL of 0.150 M HOAc with 50.0 mL of 0.100 M $HNO_3$?

13. The weak acid HA has $K_a = 6.0 \times 10^{-10}$. (a) Write the equation for the ionization of HA in water. (b) For this process identify the acid, the base, the conjugate acid, and the conjugate base.

14. A solution of acetic acid has pH = 2.00. How many grams of acetic acid are there in 100 mL of this solution?

15. A 0.50 $M$ solution of a weak acid has the same pH as a 0.015 $M$ HCl solution. What are (a) $K_a$ and (b) the degree of ionization of this acid?

16. In aqueous solution, aniline hydrochloride, $C_6H_5NH_3Cl$, dissociates completely to $C_6H_5NH_3^+$ and $Cl^-$. In turn, $C_6H_5NH_3^+$ forms the weak base aniline ($C_6H_5NH_2$) by ionization. What is the pH of a 0.0030 $M$ solution of aniline hydrochloride?

17. The pH of a 0.10 $M$ solution of cyanic acid (HOCN) is 2.25. What is the pH of a 0.010 $M$ solution of HOCN?

18. The pH of a 0.055 $M$ solution of dichloracetic acid ($CHCl_2COOH$) is 1.53. (a) What is the value of $K_a$? (b) What is the degree of ionization of dichloracetic acid in this solution?

19. The pH of a 0.028 $M$ solution of a certain base is 11.95. Is this a strong base?

20. Equation 8.18 is used to estimate the pH of a 0.015 $M$ solution of nitrous acid. What is the percentage of error in this estimate?

21. When the concentration of a strong acid is not substantially higher than $10^{-7}$ $M$, the ionization of water must be taken into account. Derive an exact expression for the $H_3O^+$ concentration of such a solution. (The ion product of water and a charge balance relationship must be used.)

22. Use the equation derived in Problem 21 to determine the pH of a $1.00 \times 10^{-7}$ $M$ HCl solution.

23. The ionization constants of weak acids HA and HB are $K_1$ and $K_2$, respectively. A solution containing both acids is prepared in which the stoichiometric concentrations are $C_1$ and $C_2$, respectively. Using the charge balance relationship, derive an expression for $[H_3O^+]$ in terms of the given constants and stoichiometric concentrations. Assume that the contribution of the ionization of water is negligible and that the ionization of the acids proceeds to a very small extent.

24. What are the concentrations of all species present in a solution prepared by mixing 100 mL of 0.100 $M$ propionic acid with 0.120 $M$ benzoic acid? (See Problem 23.)

25. The degree of ionization of a weak electrolyte can be related to its molar conductivity $\Lambda$ by the formula $f = \Lambda/\Lambda_0$, where $\Lambda_0$ is the molar conductivity extrapolated to the limit of infinite dilution. (a) Derive an expression for the ionization constant of a weak acid in terms of $\Lambda$, $\Lambda_0$, and the stoichiometric concentration $C_0$. (b) Show that when $f \ll 1$, the molar conductivity of a weak acid varies as $C_0^{-1/2}$.

26. With the help of Le Chatelier's principle, predict the effect of the following changes on the extent of hydrolysis of sodium cyanide solution: (a) The solution is diluted with water; (b) NaOH is added; (c) HCl is added; and (d) KCl is added. Neglect volume changes in (b)–(d).

27. What is the hydroxide ion concentration of a 0.15 $M$ solution of sodium hypochlorite (NaOCl)?

28. Phenol ($C_6H_5OH$) is a weak acid. The pH of a 0.010 $M$ solution of sodium phenolate ($NaOC_6H_5$) is 11.0. What is the acid ionization constant of phenol?

29. How many grams of sodium acetate must be added to 500 mL of 0.180 $M$ HOAc to increase the pH to 4.35? Neglect any change in volume.

30. To 100 mL of a buffer solution with the stoichiometric composition 0.0600 $M$ HOAc and 0.0800 $M$ NaOAc, is added 7.00 mg of NaOH. What is the pH of the solution (a) before addition of NaOH and (b) after dissolution of NaOH? (c) What weight of NaOH would have to be added to the buffer solution to increase its pH to 6.10?

31. What is the change in the pH of 1.00 L of a solution of $NH_3$ and $NH_4Cl$, each present at a stoichiometric concentration of 0.500 $M$, when (a) the solution is diluted to 10.0 L; (b) 100 mL of 0.100 $M$ KOH is added; and (c) 100 mL of 0.250 $M$ HCl is added?

32. What volume of 0.1050 $M$ NaOH is required to neutralize 50.0 mL of a solution of HOAc whose pH is 2.770? What is the pH of the resulting solution?

33. A 0.1 $M$ HOAc solution is titrated with 0.1 $M$ NaOH. (a) Which of the indicators in Figure 8.9 gives the most accurate endpoint? (b) What is the error in the concentration of HOAc if the endpoint is determined by the use of bromthymol blue ($pK_a = 7.10$)?

34. The indicator methyl orange is a weak acid whose $pK_a$ is 3.46. Over what pH range does this indicator change from 95% HIn to 95% $In^-$?

35. The indicator $\alpha$-naphthyl red is effective for the determination of endpoints in the pH range 3.7–5.0. What is the value of $K_{in}$?

36. Using an NaOH solution of known concentration, how could you determine whether an unknown acid is monoprotic or diprotic?

37. An unknown monoprotic acid is titrated with NaOH. After 12.6 mL of base is added, the pH is 4.65. An additional 21.0 mL of base is required to reach the equivalence point. What is the value of $K_a$?

38. For the weak base trimethylamine, $(CH_3)_3N$, determine (a) the pH of a 0.20 $M$ solution; (b) the pH of a 0.10 $M$ solution of $(CH_3)_3NHCl$. (c) Using the answers to (a) and (b), as well as calculations for two or more additional points, sketch the curve for the titration of 0.20 $M$ $(CH_3)_3N$ with 0.20 $M$ HCl.

39. What are the concentrations of all species present in a solution prepared by adding 100 mL of 0.100 M HCl to 200 mL of 0.250 M Na$_2$CO$_3$ before the escape of CO$_2$ gas?

40. To prepare a buffer with pH = 10.35, what mass of NaHCO$_3$ must be dissolved in 500 mL of 0.250 M Na$_2$CO$_3$?

41. What is the concentration of all species present in a 0.25 M solution of sulfuric acid?

42. A 0.3150 g sample of Na$_2$CO$_3$ is dissolved in water and the solution is titrated with HCl solution to a methyl orange endpoint. If 45.08 mL of acid must be added, what is the molarity of the HCl solution?

43. What is the pH of the solution made by mixing the following solutions: (a) 25.0 mL of 0.100 M NH$_4$Cl with 35.0 mL of 0.0750 M NH$_3$; (b) 30.0 mL of 0.0400 M CO$_2$(aq) with 10.0 mL of 0.100 M NaHCO$_3$; and (c) 30.0 mL of 0.100 M HCl with 30.0 mL of 0.200 M NaOAc?

44. The molecular weight of a certain carbonate, designated MeCO$_3$, can be determined by adding excess hydrochloric acid to the carbonate and then "back titrating" the remaining excess acid with sodium hydroxide. (a) Write chemical equations for these reactions. (b) In a particular experiment, 75.00 mL of 0.500 M HCl was added to a 0.906 g sample of MeCO$_3$. The excess acid was neutralized by the addition of 21.24 mL of 1.000 M NaOH. Determine the molecular weight of the carbonate.

45. What is the pH of a 1.00 M solution of (a) H$_3$PO$_4$, (b) NaH$_2$PO$_4$, (c) Na$_2$HPO$_4$, and (d) Na$_3$PO$_4$? For which of these solutions is the pH essentially independent of concentration? Explain your answer.

46. The pH of a 0.150 M sodium oxalate solution is adjusted to 5.00. What are the concentrations of all species present in the solution?

47. What volumes of 1.00 M Na$_3$PO$_4$ and Na$_2$HPO$_4$ solutions must be mixed to prepare exactly 1 L of a buffer solution with pH = 12.5?

48. The principal reaction occurring in a solution of an amphiprotic ion HA$^-$ is

$$2HA^- \rightleftarrows H_2A + A^{2-}$$

in which HA$^-$ acts as both a weak acid and a weak base. A similar reaction occurs in a solution of a salt of a weak acid and a weak base, e.g., ammonium acetate. (a) Derive expressions analogous to Equations 8.39 and 8.40 for [H$_3$O$^+$] in a solution of a salt of a weak acid and a weak base. (b) Derive an expression for the fraction of the salt present in dissociated form at equilibrium. (Hint: Write the equilibrium constant expression for the reaction and express K in terms of $K_a$ and $K_b$. Use mass and charge balance to relate the concentrations.)

49. Using the expressions derived in Problem 48 determine the pH and the extent of hydrolysis of (a) 0.10 M NH$_4$OAc and (b) 0.10 M NH$_4$CN.

# 9

# Ionic Equilibria

The dissolution of sparingly soluble salts provides an excellent illustration of the interplay between different types of equilibrium processes. A particular anion, for example, might form an insoluble salt with a certain cation and, at the same time, be conjugate to a weak acid. Under these circumstances, the solubility of the salt will depend on the pH of the solution. The operation of simultaneous equilibria is widely exploited in analytical chemistry. For example, the precipitation of insoluble metal sulfides at different pH can be used to separate and identify a large number of metal ions. In this final chapter on chemical equilibrium, the principles developed in Chapters 7 and 8 are brought to bear on various aspects of the solubility of metal salts.

## Solubility Equilibria

### 9.1 The Solubility of Salts

The **solubility** of a substance is the amount that will dissolve at a particular temperature in a given amount of solvent at equilibrium. Solubility is commonly expressed as the number of moles per liter or, alternatively, as the number of grams of substance per 100 mL of solvent. Salts differ greatly in their solubility in water. An *insoluble* or *sparingly soluble* salt has a solubility of less than approximately $10^{-4}$ mol L$^{-1}$.

The dissolution of a salt in water is a rather complicated process. In a

**TABLE 9.1   The Solubility of Metal Salts in Water at 25°C**

| General Rule | Exceptions |
|---|---|
| 1. Carbonates ($CO_3^{2-}$), phosphates ($PO_4^{3-}$), sulfides ($S^{2-}$), cyanides ($CN^-$), sulfites ($SO_3^{2-}$), and oxalates ($C_2O_4^{2-}$) are insoluble | Group IA and $NH_4^+$ |
| 2. Metal oxides and hydroxides are insoluble | Group IA, $Ba^{2+}$ |
| 3. Sulfates ($SO_4^{2-}$) are soluble | $CaSO_4$, $SrSO_4$, $BaSO_4$, $PbSO_4$ |
| 4. Chlorides ($Cl^-$) are soluble | $AgCl$, $Hg_2Cl_2$, $PbCl_2$, $CuCl$ |
| 5. Bromides ($Br^-$) and iodides ($I^-$) are soluble | $AgX$, $Hg_2X_2$, $PbX_2$, $CuX$ (X = Br or I), $HgI_2$, $BiI_3$ |
| 6. Nitrates ($NO_3^-$), nitrites ($NO_2^-$), acetates ($OAc^-$), chlorates ($ClO_3^-$), perchlorates ($ClO_4^-$), and permanganates ($MnO_4^-$) are soluble | |

crystalline solid, the ions constituting the salt are arranged in an ordered array of alternating positive and negative ions (Section 20.7). When the salt is dissolved, the orderly arrangement is destroyed, and the individual ions are surrounded (hydrated) by several molecules of water much in the same way as is the hydronium ion (Section 8.1b). Although the specific factors that determine solubility can be evaluated (see Section 18.6), it is difficult to predict with certainty whether a particular salt will be soluble or not. It is customary, instead, to develop solubility rules based on experiment.

Table 9.1 lists a number of rules concerning the solubility of metal salts. Each rule generally has some exceptions, which are also listed. We see from these rules that the salts of many singly charged cations, e.g., alkali metal ions and the ammonium ion, are soluble. Similarly, many singly charged anions form soluble salts, e.g., nitrates and perchlorates. These results suggest that the electrostatic interaction between ions is one of the important factors governing solubility. The attractive force between single charged ions in a crystal is weaker than that between multiply charged ions and such salts consequently dissolve more readily. However, this cannot be the only important factor, as is demonstrated by the insolubility of most hydroxides as well as several of the halides.

Table 9.1 also shows that many of the salts of multiply charged anions are insoluble, e.g., carbonates, sulfides, and phosphates. While this once again suggests the importance of the electrostatic interaction, the solubility of most metal sulfates shows that here, too, other factors are operative.

The solubility rules in Table 9.1 are applicable at room temperature. The effect of temperature on solubility can be understood with the help of Le Chatelier's principle. Applying the considerations of the effect of temperature on equilibria (Section 7.6) to solubility, we can conclude that solubility should increase with temperature if the dissolution is endothermic and decrease with temperature if it is exothermic. Owing to the attraction between oppositely charged ions, energy must always be supplied to separate the ions of a crystal. On the other hand, the hydration of ions generally liberates energy. Thus, the relative magnitudes of these two energies determine whether the reaction is exothermic or endothermic. The dissolution of most salts is endothermic and their solubilities consequently increase with temperature.

Figure 9.1 displays the variation of solubility with temperature for a number of salts. Although most of the curves do indeed have a positive slope, there

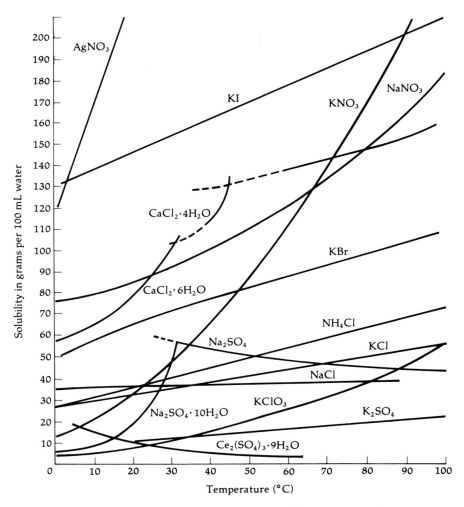

**FIGURE 9.1** Temperature dependence of the solubility in water of various salts.

is considerable variation in the rate of increase in solubility. Two important exceptions to the general trend are $CaSO_4$ and anhydrous $Na_2SO_4$, both of which exhibit decreasing solubilities with increasing temperature. Note, by contrast, that the solubility of sodium sulfate decahydrate, $Na_2SO_4 \cdot 10\,H_2O$, which is the stable form of this salt at low temperature, increases with temperature. The difference in the behavior of anhydrous and hydrated salts such as sodium sulfate is understandable: the anhydrous salt becomes hydrated upon dissolution with the liberation of hydration energy, whereas the hydrate is already hydrated in the solid.

## 9.2 The Solubility Product

A solution is said to be **saturated** when it contains as much of a particular substance as will dissolve in it at a given temperature at equilibrium. The dissolution of a salt in its saturated solution is an equilibrium process and can be rep-

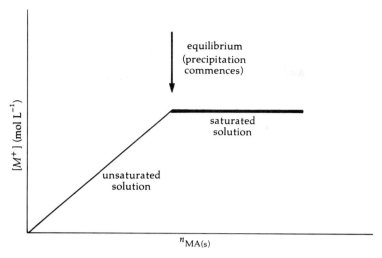

**FIGURE 9.2** Approach to equilibrium in the dissolution of salt MA in water shown in a plot of ionic concentration versus moles of added salt. The solubility product expression is applicable when [M$^+$] becomes constant.

---

resented by equations such as

$$AgCl(s) \underset{}{\overset{H_2O}{\rightleftharpoons}} Ag^+ + Cl^-$$

or

$$Mg_3(PO_4)_2(s) \underset{}{\overset{H_2O}{\rightleftharpoons}} 3Mg^{2+} + 2PO_4^{3-}$$

The equilibrium constant for the dissolution is called the **solubility product** and is customarily designated as $K_{sp}$. Recall from Section 7.2 that the concentration of a solid is not included in the equilibrium constant expression. The solubility product is therefore equal to the appropriate product of ionic concentrations at equilibrium. For example, the solubility products of AgCl and Mg$_3$(PO$_4$)$_2$ are given by the expressions

$$K_{sp} = [Ag^+][Cl^-]$$

and

$$K_{sp} = [Mg^{2+}]^3[PO_4^{3-}]^2$$

respectively.

The approach to equilibrium in the dissolution of a sparingly soluble salt is shown in Figure 9.2. This figure illustrates the important point that, so long as a solution of a salt is unsaturated, there is no equilibrium between the salt and the solution: the concentration of the ions increases as more salt is dissolved. It is only when the solution becomes saturated that the ionic concentrations become constant, equilibrium is established, and the solubility product expression applies.

Table 9.2 lists the solubility product constants at 25°C of a number of relatively insoluble salts. A more complete list is given in Appendix 3. These data may be used to illustrate the concept of solubility.

**TABLE 9.2 Solubility Product Constants of Sparingly Soluble Salts in Aqueous Solutions at 25°C**

| Substance | $K_{sp}$ |
|---|---|
| $PbCl_2$ | $1.6 \times 10^{-5}$ |
| $AgCl$ | $1.8 \times 10^{-10}$ |
| $AgBr$ | $5.0 \times 10^{-13}$ |
| $AgI$ | $8.3 \times 10^{-17}$ |
| $CuI$ | $1.1 \times 10^{-12}$ |
| $SrSO_4$ | $2.8 \times 10^{-7}$ |
| $BaSO_4$ | $1.1 \times 10^{-10}$ |
| $Ca_3(PO_4)_2$ | $1 \times 10^{-26}$ |
| $ZnS$ | $1.1 \times 10^{-21}$ |
| $CdS$ | $2 \times 10^{-28}$ |
| $CuS$ | $6 \times 10^{-36}$ |
| $Ag_2S$ | $7.1 \times 10^{-50}$ |
| $Mg(OH)_2$ | $7.1 \times 10^{-12}$ |
| $Zn(OH)_2$ | $3.3 \times 10^{-13}$ |
| $Fe(OH)_2$ | $8 \times 10^{-16}$ |
| $Fe(OH)_3$ | $6.3 \times 10^{-38}$ |

---

**Example 9.1**   What is the solubility of silver chloride in water at 25°C? Express the answer in mol $L^{-1}$ and in grams per 100 mL.

As in all equilibrium problems we write the equation for the process of interest and, beneath each species, its equilibrium concentration:

$$AgCl(s) \rightleftharpoons Ag^+ + Cl^-$$
$$\phantom{AgCl(s) \rightleftharpoons\ } x \quad\ \ x$$

The solubility product expression is

$$K_{sp} = 1.8 \times 10^{-10} = [Ag^+][Cl^-] = x^2$$

and the solution is

$$x = [Ag^+] = [Cl^-] = 1.3 \times 10^{-5} \text{ mol } L^{-1}$$

The stoichiometry of the reaction indicates that one mole of $Ag^+$ and one mole of $Cl^-$ are formed for each mole of AgCl that dissolves. Therefore, the solubility $s$ of silver chloride, which is given by the amount of AgCl dissolving in a given amount of water, is equal to the concentration of either ion:

$$s = x = 1.3 \times 10^{-5} \text{ mol } L^{-1}$$
$$= (1.3 \times 10^{-5} \text{ mol } L^{-1})(143.32 \text{ g mol}^{-1})[10^{-1} \text{ L}(100 \text{ mL})^{-1}]$$
$$= 1.9 \times 10^{-4} \text{ g}/100 \text{ mL} \quad \blacksquare$$

The difference between the solubility product and the solubility of a salt deserves to be emphasized. The solubility product is an equilibrium constant. Like all equilibrium constants, it is a constant for a particular salt at a given temperature provided only that the solution in equilibrium with the salt is sufficiently dilute to be ideal. On the other hand, the solubility of a salt at a given temperature can be changed by changing the composition of the solution, as we shall see. The two quantities are related, of course, and either one can be obtained if the other is known.

**Example 9.2**   The solubility of lead iodide in water at 25°C is $1.21 \times 10^{-3}$ mol L$^{-1}$. What is the solubility product of lead iodide at this temperature?

The dissolution reaction is

$$PbI_2(s) \xrightarrow{H_2O} Pb^{2+} + 2I^-$$

The stoichiometry of this process indicates that one mole of $Pb^{2+}$ and two moles of $I^-$ are formed for each mole of $PbI_2$ that dissolves. If the solubility of $PbI_2$ is designated $s$, the ionic concentrations are consequently

$$[Pb^{2+}] = s \qquad [I^-] = 2s$$

The solubility product is

$$K_{sp} = [Pb^{2+}][I^-]^2 = (s)(2s)^2 = 4s^3$$

and substituting the numerical value of $s$ we get

$$K_{sp} = (4)(1.21 \times 10^{-3})^3 = 7.09 \times 10^{-9} \qquad \blacksquare$$

Table 9.2 indicates that solubility products vary widely in magnitude. The solubility of salts in water also varies widely. The relative solubility of different salts may be inferred directly from the values of their solubility products provided that the salts have the same stoichiometry. For example, AgCl has a larger value of $K_{sp}$ than CuI. Since both salts have the same stoichiometry, i.e., they are 1 : 1 electrolytes, AgCl will be more soluble in water than CuI. Such a conclusion cannot be drawn directly for salts of different stoichiometry. For example, AgCl and $PbI_2$ have nearly equal values of $K_{sp}$, and yet $PbI_2$ is about 100 times more soluble in water than AgCl. In order to compare the solubilities of salts of different stoichiometries, a calculation of the type illustrated in Example 9.1 must be performed.

In addition to permitting an evaluation of the solubilities of salts, solubility product constants permit a determination of whether an insoluble salt will precipitate when its constituent ions are brought together in solution. This information is useful in qualitative analysis (Sections 9.7 and 9.8) since various metal ions can be identified by the characteristic precipitates they form on addition of appropriate reagents.

In order to determine whether precipitation will occur we must determine the direction of the reaction as the system moves towards equilibrium. As for other equilibrium reactions, this involves a comparison of the reaction quotient with the equilibrium constant (Section 7.3). If $Q < K_{sp}$, the solution is **unsaturated**; the direction of the reaction favors the dissolution of additional salt rather than the precipitation of the salt. If $Q > K_{sp}$ the solution is **supersaturated** and the salt will precipitate. It is worth noting, however, that precipitation does not always occur immediately. It is sometimes necessary to either vigorously stir a supersaturated solution or to let it stand for some time before a visible precipitate is formed. If $Q = K_{sp}$, the solution is saturated. The addition of any additional amount of either constituent ion results in precipitation.

**Example 9.3** 100 mL of 0.020 $M$ Ba(NO$_3$)$_2$ is added to 100 mL of 0.010 $M$ NaF. Will BaF$_2$ precipitate from this solution? Neglect the hydrolysis of F$^-$.

The solubility product for BaF$_2$ can be written as

$$K_{sp} = [\text{Ba}^{2+}][\text{F}^-]^2 = 1.0 \times 10^{-6}$$

Taking into account the dilution which occurs when the two solutions are mixed, we have

$$[\text{Ba}^{2+}] = 0.010 \ M$$
$$[\text{F}^-] = 0.0050 \ M$$

Therefore,

$$Q = [\text{Ba}^{2+}][\text{F}^-]^2 = (0.010)(0.0050)^2 = 2.5 \times 10^{-7}$$

Since $Q$ is less than $K_{sp}$, precipitation will not occur. ∎

It is worth noting that hydrolysis affects the solubility of a salt whenever one of its constituent ions is conjugate to either a weak acid or a weak base. Such effects have been ignored in this example. They are considered in Section 9.9.

## 9.3 The Common Ion Effect

We have so far considered the solubility of salts in pure water. Suppose, instead, that we dissolve a salt in a solution containing one of its ions, e.g., PbI$_2$ in a solution containing I$^-$ ions. Since the concentration of one of the constituent ions of the salt is now larger than it is in water, the equilibrium point must, according to Le Chatelier's principle, shift to the left. The solubility of the salt is consequently lower than it is in water. This phenomenon is known as the **common ion effect**. The effect is completely analogous to the reduction in the ionization of a weak acid in the presence of one of its salts (Section 8.8).

**Example 9.4** Compare the solubility of PbI$_2$ in 0.10 $M$ NaI solution with that in pure water at 25°C.

The equilibrium can be represented by the equation

$$\text{PbI}_2(s) \rightleftharpoons \text{Pb}^{2+} + 2\text{I}^-$$
$$\phantom{\text{PbI}_2(s) \rightleftharpoons } x \qquad (0.10 + 2x)$$

where the equilibrium concentrations in 0.10 $M$ NaI solution are given below the equation. The solubility product is

$$K_{sp} = 7.09 \times 10^{-9} = [\text{Pb}^{2+}][\text{I}^-]^2 = x(0.10 + 2x)^2$$

Since $K_{sp} \ll 1$, $2x$ can be neglected with respect to 0.10 in a first approximation. The expression then simplifies to

$$x(0.10)^2 = 7.09 \times 10^{-9}$$
$$x = 7.09 \times 10^{-7}$$

The approximation obviously is adequate. The solubility of PbI$_2$ is equal to the concentration of Pb$^{2+}$ and is therefore $7.09 \times 10^{-7}$ mol L$^{-1}$.

The solubility of PbI$_2$ in water is $1.2 \times 10^{-3}$ mol L$^{-1}$ (See Example 9.2). Thus, the presence of 0.10 $M$ I$^-$ decreases the solubility of PbI$_2$ by three orders of magnitude. ∎

The effect of a common ion on the solubility of a salt is illustrated graphically in Figure 9.3. As expected from the preceding considerations, the solubility decreases with increasing concentration of the common ion. A practical application of the common ion effect is the removal of small amounts of precious metals from their solutions. For example, the recovery of silver from exposed photographic film can be increased by precipitation of AgCl from a dilute NaCl solution.

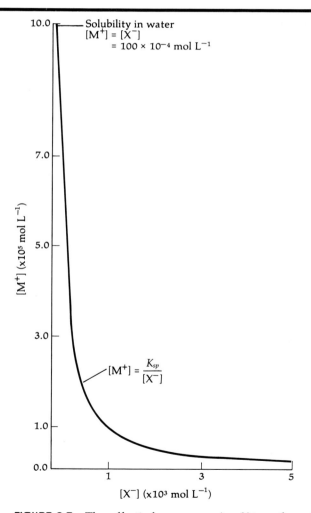

**FIGURE 9.3** The effect of a common ion X$^-$ on the solubility of salt MX. The value of $K_{sp}$ for MX is assumed to be $1 \times 10^{-8}$.

## 9.4 Selective Precipitation

The values of $K_{sp}$ vary over an exceedingly broad range. Even apparently similar compounds display large differences in $K_{sp}$. For instance, the solubility products of the alkaline earth sulfates, $BaSO_4$ and $SrSO_4$, differ by nearly three orders of magnitude. It is possible to take advantage of this difference in solubility to selectively precipitate one salt in the presence of the other and thus effect a separation. The following example shows how well $Ba^{2+}$ and $Sr^{2+}$ can be separated from each other by selective precipitation.

---

**Example 9.5** A solution of $Na_2SO_4$ is added dropwise to a solution that is 0.010 $M$ in $Ba^{2+}$ and $Sr^{2+}$. Assuming that the change in volume due to the addition of sulfate is negligible, (a) determine the concentration of $SO_4^{2-}$ for which one of the above ions (state which one) first precipitates out of solution as the sulfate. (b) Determine the $SO_4^{2-}$ concentration at which $SrSO_4$ begins to precipitate. What is the residual concentration of $Ba^{2+}$ at this point? The values of $K_{sp}$ are $K_{sp} = 1.1 \times 10^{-10}$ for $BaSO_4$ and $K_{sp} = 2.8 \times 10^{-7}$ for $SrSO_4$. Neglect the hydrolysis of $SO_4^{2-}$.

(a) Since both salts have the same stoichiometry, the one with the smaller $K_{sp}$ ($BaSO_4$) is less soluble and will precipitate first. Precipitation first occurs when the reaction quotient exceeds the solubility product:

$$Q = [Ba^{2+}][SO_4^{2-}] > 1.1 \times 10^{-10}$$

Since

$$[Ba^{2+}] = 0.010 \text{ mol L}^{-1}$$

it follows that

$$[SO_4^{2-}] > 1.1 \times 10^{-10}/0.010$$

$$> 1.1 \times 10^{-8} \text{ mol L}^{-1}$$

Let us check whether $SrSO_4$ will precipitate under these conditions. The reaction quotient is

$$Q = [Sr^{2+}][SO_4^{2-}] = (0.010)(1.1 \times 10^{-8}) = 1.1 \times 10^{-10}$$

Since this value is smaller than $K_{sp}$, $SrSO_4$ will not precipitate.

(b) The reaction quotient becomes equal to $K_{sp}$ of $SrSO_4$ when

$$Q = [Sr^{2+}][SO_4^{2-}] = 2.8 \times 10^{-7}$$

Therefore, since $[Sr^{2+}] = 0.010$ mol L$^{-1}$

$$[SO_4^{2-}] = 2.8 \times 10^{-5} \text{ mol L}^{-1}$$

At this point, the residual $Ba^{2+}$ concentration is given by the solubility product expression for this salt:

$$[Ba^{2+}][SO_4^{2-}] = 1.1 \times 10^{-10}$$

$$[Ba^{2+}] = \frac{1.1 \times 10^{-10} \text{ (mol L}^{-1})^2}{2.8 \times 10^{-5} \text{ mol L}^{-1}} = 3.9 \times 10^{-6} \text{ mol L}^{-1}$$

Expressed as a percentage of the original $Ba^{2+}$ present we find that

$$\frac{3.9 \times 10^{-6}}{0.01} \times 100 = 0.039\%$$

is left in the solution at the time $SrSO_4$ first starts to precipitate. These calculations show that careful addition of sulfate can be used to perform a virtually complete separation of these two ions. ∎

# Equilibria of Metal–Ligand Complexes

## 9.5 Metal–Ligand Complexes

Metal ions in aqueous solution are hydrated and can be represented by formulas such as $Cr(H_2O)_6^{3+}$ or $Na(H_2O)_4^{+}$. If certain other ions or molecules are also present, the possibility exists that these species will displace one or more of the water molecules to form a **complex ion** consisting of the metal ion and its coordinated (i.e., attached) ligands. The term **ligand** refers to the species that are coordinated to the metal ion. For instance, $Cr^{3+}$ is hydrated by six $H_2O$ ligands in aqueous solution. The addition of $Cl^-$ to the solution can lead to the displacement of up to all six $H_2O$ molecules to form a hexachloro complex ion of $Cr^{3+}$ by means of the following displacement reactions:

$$Cr(H_2O)_6^{3+} + Cl^- \rightleftarrows Cr(H_2O)_5Cl^{2+} + H_2O$$

$$Cr(H_2O)_5Cl^{2+} + Cl^- \rightleftarrows Cr(H_2O)_4Cl_2^{+} + H_2O$$

$$\vdots$$

$$Cr(H_2O)Cl_5^{2-} + Cl^- \rightleftarrows CrCl_6^{3-} + H_2O$$

The subject of coordination chemistry, the chemistry of the compounds comprised of metal–ligand complexes, is treated in Chapter 21. In this section we focus on only one aspect of this subject: the equilibria involved in the formation of complex ions. Complex ion equilibria have an intimate connection with solubility equilibria since many insoluble salts can be dissolved as soluble complex ions by the addition of suitable ligands.

## 9.6 Complex Ion Equilibria

A characteristic feature of complex ion equilibria is that the metal ion is distributed among a number of species, ranging from the fully hydrated ion (e.g., $Cr(H_2O)_6^{3+}$) to the complex resulting from the complete displacement of the water molecules (e.g., $CrCl_6^{3-}$). The relative concentrations of the various species depend on the values of the equilibrium constants for the successive displacement reactions and also on the overall concentration of the ligand. For example, for the ammonia complex of the copper(II) ion, the stepwise reactions and corresponding equilibrium constants are as follows:

$$Cu(H_2O)_4^{2+} + NH_3 \rightleftarrows Cu(H_2O)_3NH_3^{2+} + H_2O$$

$$K_1 = \frac{[Cu(NH_3)^{2+}]}{[Cu^{2+}][NH_3]} = 1.86 \times 10^4 \tag{9.1}$$

$$Cu(H_2O)_3(NH_3)^{2+} + NH_3 \rightleftarrows Cu(H_2O)_2(NH_3)_2^{2+} + H_2O$$

$$K_2 = \frac{[Cu(NH_3)_2^{2+}]}{[Cu(NH_3)^{2+}][NH_3]} = 3.88 \times 10^3 \tag{9.2}$$

$$Cu(H_2O)_2(NH_3)_2^{2+} + NH_3 \rightleftarrows Cu(H_2O)(NH_3)_3^{2+} + H_2O$$

$$K_3 = \frac{[Cu(NH_3)_3^{2+}]}{[Cu(NH_3)_2^{2+}][NH_3]} = 1.00 \times 10^3 \tag{9.3}$$

$$Cu(H_2O)(NH_3)_3^{2+} + NH_3 \rightleftarrows Cu(NH_3)_4^{2+} + H_2O$$

$$K_4 = \frac{[Cu(NH_3)_4^{2+}]}{[Cu(NH_3)_3^{2+}][NH_3]} = 1.55 \times 10^2 \tag{9.4}$$

(The hydration need not be included in the chemical formulas when representing concentrations of complex ions in equilibrium expressions.)

The equilibrium constants in Equations 9.1–9.4 are sometimes referred to as **formation constants** since the reactions involve the formation of the complex ions as products. Some tabulations give the values of the complex ion dissociation constants—the constants for the reverse reactions. The two types of constants are, of course, reciprocals of each other.

In order to obtain the distribution of copper among the various complexes we first note that the equilibrium expressions can be rewritten in the following manner to yield concentration ratios for the various species. The expression for $K_1$ yields the ratio of $Cu(NH_3)^{2+}$ and $Cu^{2+}$ concentrations:

$$\frac{[Cu(NH_3)^{2+}]}{[Cu^{2+}]} = K_1[NH_3] \tag{9.5}$$

Similar expressions yield the ratios of the concentrations of other complexes differing in composition by just one $NH_3$ molecule. For instance, the expression for $K_2$ yields the ratio

$$\frac{[Cu(NH_3)_2^{2+}]}{[Cu(NH_3)^{2+}]} = K_2[NH_3] \tag{9.6}$$

Note that these ratios are all proportional to the first power of the ammonia concentration. Additional concentration ratios can be obtained by combination of expressions such as Equations 9.5 and 9.6. For instance, the ratio of the concentration of species that differ by two $NH_3$ molecules (e.g., $Cu(NH_3)_2^{2+}$ and $Cu^{2+}$) is

$$\frac{[Cu(NH_3)_2^{2+}]}{[Cu^{2+}]} = \left(\frac{[Cu(NH_3)_2^{2+}]}{[Cu(NH_3)^{2+}]}\right)\left(\frac{[Cu(NH_3)^{2+}]}{[Cu^{2+}]}\right) = K_1 K_2 [NH_3]^2 \tag{9.7}$$

Note that this ratio is proportional to the square of the ammonia concentration. A simple extension of this procedure shows that the ratio of species differing by three ammonia molecules is proportional to $[NH_3]^3$:

$$\frac{[Cu(NH_3)_3^{2+}]}{[Cu^{2+}]} = K_1 K_2 K_3 [NH_3]^3 \tag{9.8}$$

The distribution of copper(II) is readily obtained with the help of the preceding ratios. For example, the fraction of copper(II) present as $Cu(NH_3)_2^{2+}$ is defined as

$$f_{Cu(NH_3)_2^{2+}} = \frac{[Cu(NH_3)_2^{2+}]}{[Cu^{2+}] + [Cu(NH_3)^{2+}] + [Cu(NH_3)_2^{2+}] + [Cu(NH_3)_3^{2+}] + [Cu(NH_3)_4^{2+}]}$$

Dividing the numerator and denominator by $[Cu^{2+}]$ we obtain

$$f_{Cu(NH_3)_2^{2+}} = \frac{[Cu(NH_3)_2^{2+}]/[Cu^{2+}]}{1 + \dfrac{[Cu(NH_3)^{2+}]}{[Cu^{2+}]} + \dfrac{[Cu(NH_3)_2^{2+}]}{[Cu^{2+}]} + \dfrac{[Cu(NH_3)_3^{2+}]}{[Cu^{2+}]} + \dfrac{[Cu(NH_3)_4^{2+}]}{[Cu^{2+}]}}$$

Substitution of the expressions for the individual concentration ratios then yields the desired result:

$$f_{Cu(NH_3)_2^{2+}} = \frac{K_1 K_2 [NH_3]^2}{1 + K_1[NH_3] + K_1 K_2 [NH_3]^2 + K_1 K_2 K_3 [NH_3]^3 + K_1 K_2 K_3 K_4 [NH_3]^4}$$

(9.9)

Similar expressions can be derived for the fractional concentrations of the other complex ions. Note that the denominator, which gives the sum of all the concentrations, is the same for all such expressions. Only the numerator changes.

The numerical values of the fractional concentrations can be obtained by substitution of the numerical values of the equilibrium constants and the free ammonia concentration. For example, when $[NH_3] = 0.001\ M$, the fraction of copper(II) present as $Cu(NH_3)_2^{2+}$ is obtained from Equation 9.9 as

$$f_{Cu(NH_3)_2^{2+}} = \frac{72.2}{1 + 18.6 + 72.2 + 72.2 + 11.2} = 0.412$$

The overall distribution at this ammonia concentration is obtained from the comparable expressions for all of the ions present:

$f_{Cu^{2+}} = 5.71 \times 10^{-3}$

$f_{Cu(NH_3)^{2+}} = 0.106$

$f_{Cu(NH_3)_2^{2+}} = 0.412$

$f_{Cu(NH_3)_3^{2+}} = 0.412$

$f_{Cu(NH_3)_4^{2+}} = 6.39 \times 10^{-2}$

Total = 1.00

Figure 9.4 shows the distribution of the various species as a function of the ammonia concentration. Note the characteristic manner in which each species is dominant over some particular range of $[NH_3]$ values and decreases rapidly in concentration outside this range. As might be expected, the species with high ammonia content increase in relative abundance with increasing ammonia concentration. At sufficiently high ammonia concentration, all the copper is present as the fully complexed $Cu(NH_3)_4^{2+}$ ion. This result is consistent with Le Chatelier's principle and is valid for any complex ion. Consequently, the overall constant for the formation of the fully complexed ion from the hydrated ion is of special importance. Since the reaction

$$Cu(H_2O)_4^{2+} + 4NH_3 \rightleftharpoons Cu(NH_3)_4^{2+} + 4H_2O$$

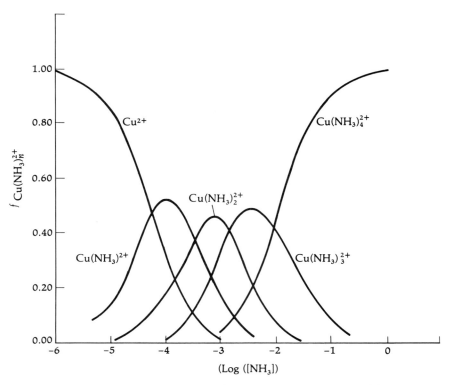

**FIGURE 9.4** Dependence of the relative concentrations of copper(II) ammonia complexes in aqueous solution at 25°C on the free ammonia concentration.

which represents this process, is obtained as the sum of the individual four reactions, the overall formation constant is the product of the four separate constants:

$$K = K_1 K_2 K_3 K_4 = 1.1 \times 10^{13} \tag{9.10}$$

The overall features of the results displayed in Figure 9.4 are pertinent to various aspects of the solution chemistry of metals. For example, the concentrations of many metal ions are determined by light absorption. While solutions of the hydrated ions may be colorless and thus nonabsorbing, many metals form colored complex ions that absorb light of a particular wavelength. Copper(II) can be assayed by this technique using the deeply blue colored $Cu(NH_3)_4^{2+}$ ion to absorb light (Figure 9.5). Figure 9.4 shows that a concentration of free ammonia greater than $\approx 0.3\ M$ is necessary to ensure that at least 98% of the copper is present in this particular form.

Table 9.3 lists the formation constants of a number of complex ions; a more complete list of complex dissociation constants is given in Appendix 3. These data can be used to calculate the concentration of free and complex ions in a solution of a metal to which some particular ligand has been added. The preceding discussion indicates, however, that the results are valid only when the ligand concentration is sufficiently high to ensure that the fully complexed ion is the dominant species.

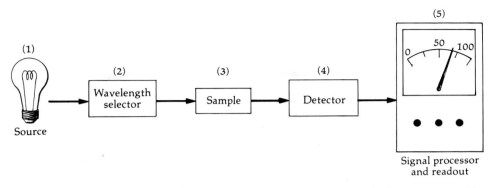

**FIGURE 9.5** Assay of $Cu(NH_3)_4^{2+}$ by absorption spectroscopy. Light from a source (1) passes through a wavelength selector (2) set to transmit light of the appropriate wavelength. This light is incident on a sample (3) containing $Cu(NH_3)_4^{2+}$, where it is absorbed. The fraction of the incident light transmitted to the light detector (4) depends on the concentration of $Cu(NH_3)_4^{2+}$. The signal from the detector is processed and read out in the processor and readout unit (5).

**TABLE 9.3 Complex Formation Constants at 25°C**

| Complex | $K_f$ |
|---|---|
| $Ag(NH_3)_2^+$ | $1.0 \times 10^8$ |
| $Zn(NH_3)_4^{2+}$ | $2.9 \times 10^9$ |
| $Cu(NH_3)_4^{2+}$ | $1.1 \times 10^{13}$ |
| $Cd(CN)_4^{2-}$ | $1.2 \times 10^{17}$ |
| $Ag(CN)_2^-$ | $1 \times 10^{22}$ |
| $Fe(CN)_6^{3-}$ | $7.7 \times 10^{43}$ |
| $Pb(OH)_3^-$ | $8 \times 10^{13}$ |
| $Zn(OH)_4^{2-}$ | $2 \times 10^{20}$ |

**Example 9.6** Enough $NH_3$ is added to a 0.100 M solution of $Cu^{2+}$ to make the initial concentration of $NH_3$ 1.000 M. What will be the equilibrium concentrations of $Cu^{2+}$ and $Cu(NH_3)_4^{2+}$?

The equilibrium constant for the reaction

$$Cu^{2+} + 4NH_3 \rightleftarrows Cu(NH_3)_4^{2+}$$

is $K_f = 1.1 \times 10^{13}$. Since this value is much larger than unity, the equilibrium point lies far to the right and nearly all the copper will be in the form of the complex ion. Recall from Section 7.4 that under these circumstances a simpler equilibrium expression is obtained by assuming that all the $Cu^{2+}$ present initially except for a small amount undergoes the preceding reaction. Let the equilibrium concentration of $Cu^{2+}$ be $x$, where $x$ is small compared to 0.100 M, the initial concentration of $Cu^{2+}$. The other equilibrium concentrations are

$$[Cu^{2+}] = x \qquad [NH_3] = 0.600 + 4x \qquad [Cu(NH_3)_4^{2+}] = 0.100 - x$$

The value for [NH$_3$] is obtained as follows: if 0.100 mol L$^{-1}$ of Cu$^{2+}$ were to react, then 0.400 mol L$^{-1}$ of NH$_3$ would also react leaving 0.600 mol L$^{-1}$ of unreacted NH$_3$. Since, in actuality, $x$ mol L$^{-1}$ of Cu$^{2+}$ does not react, $4x$ mol L$^{-1}$ of NH$_3$ cannot react either, giving a total NH$_3$ concentration of $(0.600 + 4x)$ mol L$^{-1}$.

Substituting into the expression for the equilibrium constant,

$$K_f = 1.1 \times 10^{13} = \frac{[Cu(NH_3)_4^{2+}]}{[Cu^{2+}][NH_3]^4}$$

we obtain

$$\frac{0.100 - x}{x(0.600 + 4x)^4} = 1.1 \times 10^{13}$$

Because $K_f$ is so large, $x$ can be neglected with respect to 0.100 so that the expression simplifies to

$$\frac{0.100}{x(0.600)^4} = 1.1 \times 10^{13}$$

$$x = [Cu^{2+}] = 7.0 \times 10^{-14} \text{ mol L}^{-1}$$

$$[Cu(NH_3)_4^{2+}] = 0.10 \text{ mol L}^{-1}$$

The small value of the Cu$^{2+}$ concentration confirms the validity of this approximation. ∎

If the existence of all the intermediate complexes ranging from Cu(NH$_3$)$^{2+}$ to Cu(NH$_3$)$_3^{2+}$ is considered by means of the previously developed equations, somewhat different answers are obtained. The concentration of Cu(NH$_3$)$_4^{2+}$ decreases slightly to 0.099 $M$, the remaining copper being present as these other species. The 1% difference is negligible for all but the most accurate work. Note that the procedure used in this example is only valid provided that the equilibrium concentration of NH$_3$ is large enough to ensure essentially complete formation of the fully complexed ion.

# Effect of Other Equilibria on Solubility

Many chemical reactions involve the simultaneous operation of two or more different equilibria. For example, solubility equilibria can also involve the operation of complex formation or acid ionization equilibria. Consequently, the solubility of a sparingly soluble salt can be affected markedly, under appropriate conditions, by the addition of a complexing ligand or by a change in pH. Many of the principles illustrated in this section serve as the basis for the identification and separation of metal ions.

## 9.7 Complex Ion Formation and Solubility

*a. The Effect of Complexing Ligands*   A complexing ligand increases the solubility of a relatively insoluble salt—it has the opposite effect of a common ion. Consider, for instance, the reaction of silver carbonate with ammonia. This reaction can be broken down into the following two steps:

(1) $\quad Ag_2CO_3(s) \xrightleftharpoons{H_2O} 2Ag^+ + CO_3^{2-} \quad K_{sp} = 8.1 \times 10^{-12}$

(2) $\quad Ag^+ + 2NH_3(aq) \rightleftharpoons Ag(NH_3)_2^+ \quad K_f = 1.0 \times 10^8$

The concentration of $Ag^+$ must simultaneously obey both equilibria. (The hydrolysis of $CO_3^{2-}$ gives rise to yet another equilibrium which we neglect at this stage.) Since $K_f$ is large, the second reaction reduces the concentration of free silver ion in solution. Le Chatelier's principle then tells us that the equilibrium point of the first reaction is shifted toward the ions in solution. The net effect is that the solubility of silver carbonate in a solution containing ammonia is much greater than it is in water. The overall reaction is the sum of the individual steps, with the second equation multiplied by two:

(3) $\quad Ag_2CO_3(s) + 4NH_3(aq) \rightleftharpoons 2Ag(NH_3)_2^+ + CO_3^{2-}$

The equilibrium constant for this reaction can be expressed in terms of those for reactions (1) and (2) in the manner discussed in Section 7.2 as

$$K = K_{sp}K_f^2$$

The particular relation between $K$ and $K_{sp}$ and $K_f$ depends on the stoichiometric relationship between the reactions in question. The following example illustrates a different stoichiometry.

---

**Example 9.7** What is the solubility of AgCl in 1.0 M $NH_3$?

The reaction can be divided into the following two steps:

(1) $\quad AgCl(s) \rightleftharpoons Ag^+ + Cl^- \quad K_{sp} = 1.8 \times 10^{-10}$

(2) $\quad Ag^+ + 2NH_3(aq) \rightleftharpoons Ag(NH_3)_2^+ \quad K_f = 1.0 \times 10^8$

Adding the two equations we obtain

(3) $\quad AgCl(s) + \underset{(1.0 - 2x)}{2NH_3(aq)} \rightleftharpoons \underset{x}{Ag(NH_3)_2^+} + \underset{x}{Cl^-} \quad K = K_{sp}K_f = 1.8 \times 10^{-2}$

where the equilibrium concentrations have been written below the equation in the usual fashion. Note that $Ag(NH_3)_2^+$ and $Cl^-$ have the same concentration. This will only be true if the concentration of $Ag^+$ is negligibly small compared to that of $Ag(NH_3)_2^+$. This assumption must be checked. However, we must first determine the concentration of $Ag(NH_3)_2^+$. Substituting into the expression for the equilibrium constant we obtain

$$K = 1.8 \times 10^{-2} = \frac{[Ag(NH_3)_2^+][Cl^-]}{[NH_3]^2} = \frac{x^2}{(1.0 - 2x)^2}$$

This equation can be solved readily by taking the square root of both sides:

$$\frac{x}{1.0 - 2x} = 0.13$$

$$x = 0.10$$

Since one complex ion is formed for each silver chloride ion pair that dissolves, the solubility is 0.10 M. By contrast, we note that the solubility of

AgCl in water is only $1.3 \times 10^{-5}$ mol L$^{-1}$ (Example 9.1). In essence, these numbers mean that silver chloride is insoluble in water but dissolves in 1.0 $M$ ammonia.

The concentration of free Ag$^+$ appears in the equilibria associated with the reactions represented by Equations (1) and (2) but not in the equilibrium associated with the overall reaction. The concentration of Ag$^+$ must satisfy the equilibria for both reactions in which this ion participates. Thus, either one can be used to determine its value. Using the dissolution of AgCl, we can write

$$K_{sp} = 1.8 \times 10^{-10} = [\text{Ag}^+][\text{Cl}^-] = [\text{Ag}^+](0.10)$$

$$[\text{Ag}^+] = 1.8 \times 10^{-9} \text{ mol L}^{-1}$$

This value is so small compared to that of [Ag(NH$_3$)$_2^+$] that the presence of free Ag$^+$ does not affect the calculated solubility of AgCl. If the values of $K_f$ and $K_{sp}$ were such that [Ag$^+$] were not negligibly small, a more complete analysis, analogous to that described in Section 8.7, would be necessary. Such an analysis requires that mass balance and charge balance relationships between all the species present be used. ∎

Silver(I) forms a relatively large number of insoluble salts and soluble complex ions. By arranging these species in order of decreasing Ag$^+$ equilibrium concentration it is possible to cycle through all the various precipitates and complexes from the most to the least soluble by the sequential addition of suitable reagents to a single solution of Ag$^+$. This procedure, which is illustrated in Figure 9.6 and is readily performed in the laboratory, constitutes a dramatic verification of the role of simultaneous equilibria. Note that there are several instances in this sequence in which one precipitate is directly transformed into another. For instance, the reaction

$$\text{AgOH(s)} + \text{Cl}^- \rightleftarrows \text{AgCl(s)} + \text{OH}^-$$

directly transforms black silver hydroxide into white silver chloride. If this transformation were to be attempted in the solid state, for instance by mixing silver hydroxide with sodium chloride, the reaction would be exceedingly slow. It occurs nearly instantaneously in solution because of the intervening equilibrium involving free silver ions,

$$\text{AgOH(s)} \rightleftarrows \text{Ag}^+ + \text{OH}^-$$

$$\text{Ag}^+ + \text{Cl}^- \rightleftarrows \text{AgCl(s)}$$

Since AgCl is the less soluble salt, its precipitation continously removes Ag$^+$ from the solution and so causes additional AgOH to dissolve.

**b. Identification of Insoluble Metal Chlorides** The insolubility of metal chlorides such as AgCl, Hg$_2$Cl$_2$, and PbCl$_2$ in water and the varied response of these salts to complexing agents serve as the basis of a procedure for the separation and identification of these metal ions. This procedure is one of the steps in **inorganic qualititative analysis,** in which the presence of specific cations and anions in a mixture in which any number of them may be present is determined on the basis of their chemical properties. A scheme for the identification of the insoluble metal chlorides is summarized in Figure 9.7.

Ionic Equilibria 9.7 **315**

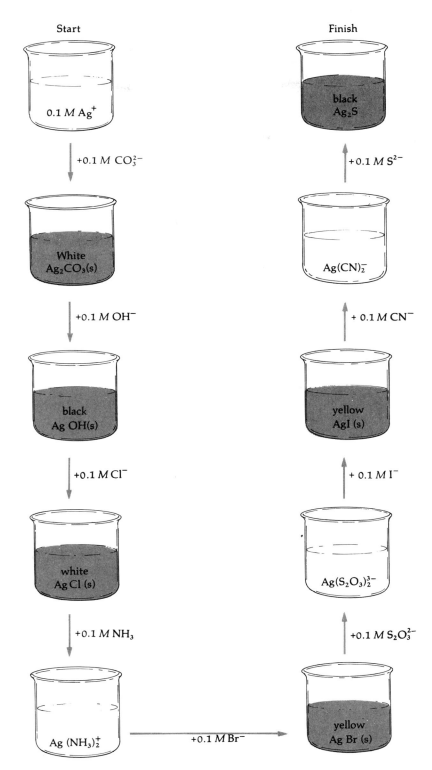

**FIGURE 9.6** The silver one-pot reactions.

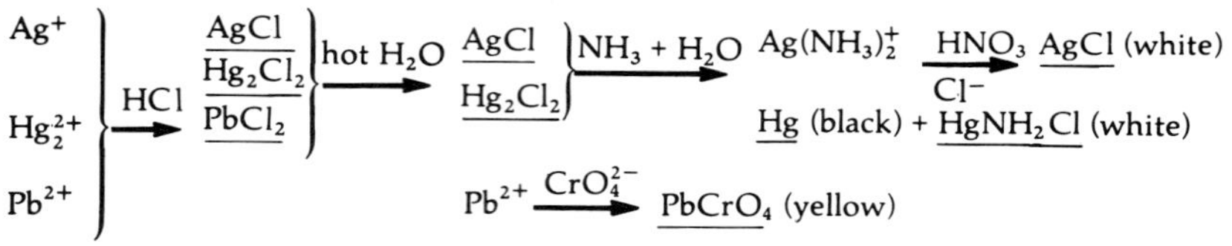

$$\begin{array}{c}
Ag^+ \\
Hg_2^{2+} \\
Pb^{2+}
\end{array} \Bigg\} \xrightarrow{HCl} \begin{array}{c} AgCl \\ Hg_2Cl_2 \\ PbCl_2 \end{array} \xrightarrow{hot\ H_2O} \begin{array}{c} AgCl \\ Hg_2Cl_2 \end{array} \xrightarrow{NH_3 + H_2O} \begin{array}{c} Ag(NH_3)_2^+ \xrightarrow{HNO_3} AgCl\ (white) \\ Cl^- \\ Hg\ (black) + HgNH_2Cl\ (white) \end{array}$$

$$Pb^{2+} \xrightarrow{CrO_4^{2-}} PbCrO_4\ (yellow)$$

**FIGURE 9.7** Flow diagram for the separation and identification of the insoluble metal chlorides. The various steps in the identification procedure are listed sequentially from left to right: Hydrochloric acid is added to a solution which may contain $Ag^+$, $Hg_2^{2+}$ and/or $Pb^{2+}$ ions, thus precipitating the metal chlorides. Lead chloride is redissolved in hot water and, after separation of any remaining insoluble chlorides, the presence of $Pb^{2+}$ is confirmed by precipitation of yellow $PbCrO_4$. Silver chloride is dissolved in ammonia and the presence of $Ag^+$ is confirmed by the reprecipitation of white AgCl. The addition of ammonia to $Hg_2Cl_2$ results in disproportionation of $Hg_2^{2+}$ with the formation of a characteristic gray–black precipitate.

---

The presence in solution of those metal ions that form insoluble metal chlorides may be determined by the addition of HCl and the observation of precipitate formation as a result of reactions such as

$$Hg_2^{2+} + 2Cl^- \rightleftarrows Hg_2Cl_2(s)$$

In order to identify the precipitate, we can take advantage of the difference in solubility of the metal chlorides in various reagents. Lead chloride is considerably more soluble than the other chlorides ($K_{sp} = 1.6 \times 10^{-5}$) and is even more soluble in hot water. Therefore, the addition of hot water to the mixed chlorides redissolves $PbCl_2$. Its presence in the solution, from which the other chlorides, if present, have been removed by centrifugation or filtration, can be confirmed by precipitation of the sparingly soluble yellow chromate:

$$Pb^{2+} + CrO_4^{2-} \rightleftarrows PbCrO_4(s)$$

Silver chloride and mercury(I) chloride can be distinguished from each other by their reaction with ammonia. Mercury(I) chloride undergoes disproportionation (Section 2.9) in an ammonia solution:

$$Hg_2Cl_2(s) + 2NH_3(aq) \rightarrow Hg(s) + HgNH_2Cl(s) + NH_4^+ + Cl^-$$

The precipitate has a characteristic gray-black color. On the other hand, AgCl dissolves in ammonia owing to the formation of the $Ag(NH_3)_2^+$ complex (Example 9.7). The presence of $Ag^+$ can be confirmed by reprecipitation of AgCl with nitric acid. The acid reacts with $NH_3$ to form $NH_4^+$, thereby leading to the dissociation of the complex and permitting free $Ag^+$ to precipitate as AgCl. The net reaction is the sum of the following three equilibria:

(1) $\quad Ag(NH_3)_2^+ \rightleftarrows Ag^+ + 2NH_3(aq) \qquad K_f^{-1}$

(2) $\quad 2NH_3(aq) + 2H_3O^+ \rightleftarrows 2NH_4^+ + 2H_2O \qquad (K_b/K_w)^2$

(3) $\quad \underline{Ag^+ + Cl^- \rightleftarrows AgCl(s) \qquad\qquad\qquad\qquad K_{sp}^{-1}}$

$\qquad Ag(NH_3)_2^+ + Cl^- + 2H_3O^+ \rightarrow AgCl(s) + 2NH_4^+ + 2H_2O$

**c. Complex Formation Involving a Common Ion** Many insoluble salts tend to react with one of their constituent ions to form soluble complexes. The solubil-

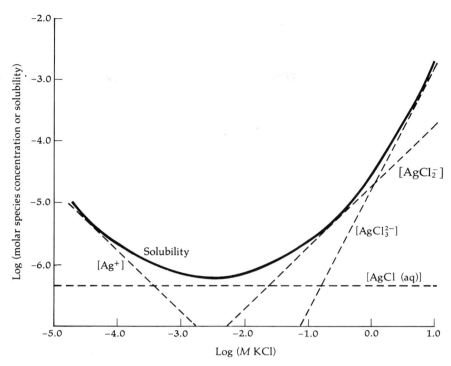

**FIGURE 9.8** Dependence of the solubility of AgCl on the concentration of Cl⁻. The solid curve shows the total concentration of dissolved AgCl. The dashed lines show the concentrations of the various species containing silver.

ity of AgCl in KCl solution, depicted in Figure 9.8, illustrates this phenomenon. The solubility of AgCl initially decreases with increasing KCl concentration. This decrease is due to the common ion effect and has already been illustrated in Figure 9.3. What is new in Figure 9.8 is the occurrence of a minimum in the solubility of AgCl when the concentration of KCl is approximately 0.003 $M$, followed by an increase in solubility at higher KCl concentrations.

The reactions responsible for the observed increase in solubility are

(1) $\quad AgCl(s) + Cl^- \rightleftarrows AgCl_2^- \qquad K_1 = 2.0 \times 10^{-5}$

(2) $\quad AgCl_2^- + Cl^- \rightleftarrows AgCl_3^{2-} \qquad K_2 = 1$

According to Le Chatelier's principle, an increase in the Cl⁻ concentration shifts the equilibrium towards the formation of the complex ions. To be sure, this tendency is counteracted by the common ion effect,

(3) $\quad AgCl(s) \rightleftarrows Ag^+ + Cl^- \qquad K_3 = K_{sp} = 1.8 \times 10^{-10}$

by which an increase in the Cl⁻ concentration shifts the equilibrium to the left. The overall solubility of AgCl is determined by the competition between all three equilibria and, for a particular salt, depends on the values of $K_1$, $K_2$, and $K_3$. A number of metal salts, including many metal hydroxides, display a similar behavior, as we shall see.

In addition to the various ionic equilibria represented by the preceding three equations, a small amount of undissociated AgCl is also present in solution. The equilibrium for the formation of this species can be represented by the equation

(4)  $AgCl(s) \rightleftarrows AgCl(aq)$    $K_4 = 4.7 \times 10^{-7}$

Owing to the small value of $K_4$, less than 3% of AgCl is present in undissociated form in a saturated aqueous solution. With the notable exception of $HgCl_2$, most salts behave in this manner. Equation (4) shows that the concentration of undissociated AgCl is independent of the $Cl^-$ concentration. It is consequently not surprising that when the $Cl^-$ concentration is such that the AgCl solubility is minimal, most of the dissolved AgCl is actually present in its undissociated form (Figure 9.8). While the contribution of AgCl(aq) to the solubility of AgCl can be ignored in an aqueous solution, this is obviously not the case for a 0.003 M KCl solution.

The solubility of a salt generally also increases when a strong electrolyte containing no ions in common with the salt is added to the solution. Figure 9.9 shows that the solubility of AgCl is 30% higher in 0.03 M $KNO_3$ than it is in water, while the solubility of $BaSO_4$ is over a factor of two higher. In these instances, the observed increase in solubility does not result from any additional equilibria. Rather, it is the result of the electrostatic interaction between the ions of the insoluble salt and those of the added electrolyte. As noted in Sec-

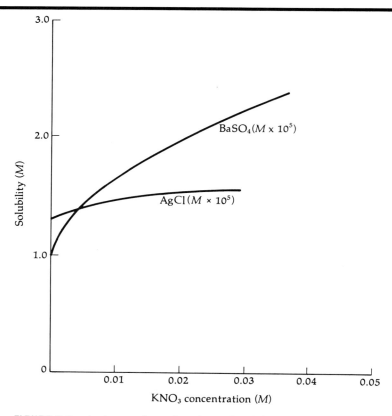

**FIGURE 9.9** A strong electrolyte (e.g., $KNO_3$) added to a saturated solution of a salt increases its solubility owing to the effect of electrostatic interactions.

tion 7.7, ionic concentrations can only be used in equilibrium expressions so long as the solutions are dilute.

## 9.8 The Effect of pH on the Solubility of Salts of Weak Acids

*a. The Effect of pH on Solubility Equilibria* The solubility of many salts can be affected by the pH of the solution. The occurrence of this effect is due to the presence of an anion with basic properties (i.e., its conjugate acid is weak) or to that of a cation with acidic properties (i.e., its conjugate base is weak). Under these circumstances acid–base ionization equilibria come into play and, as discussed in Chapter 8, these equilibria are pH dependent.

The solubility of metal sulfides provides an excellent illustration of the effect of pH on solubility, for many metal sulfides are insoluble. Aqueous hydrogen sulfide is a weak diprotic acid (Section 8.10) and undergoes ionization,

$$H_2S(aq) + H_2O \rightleftharpoons H_3O^+ + HS^- \qquad K_1 = 1.0 \times 10^{-7}$$

$$HS^- + H_2O \rightleftharpoons H_3O^+ + S^{2-} \qquad K_2 = 3 \times 10^{-13}$$

These equations may be added to give an equation for the overall ionization of hydrogen sulfide to form sulfide ion:

$$H_2S(aq) + 2H_2O \rightleftharpoons 2H_3O^+ + S^{2-}$$

The equilibrium expression for this process is

$$K = K_1 K_2 = 3 \times 10^{-20} = \frac{[H_3O^+]^2[S^{2-}]}{[H_2S]} \qquad (9.11)$$

In separations of metal sulfides, it is customary to saturate the solutions containing the metal ions with $H_2S$ gas, whose solubility at room temperature is approximately 0.1 mol $L^{-1}$. Owing to the small extent of ionization, the concentration of un-ionized $H_2S$ will consequently also be 0.1 mol $L^{-1}$. Thus, Equation 9.11 can be rewritten as

$$[S^{2-}] = \frac{3 \times 10^{-21}}{[H_3O^+]^2} \qquad (9.12)$$

showing that the sulfide ion concentration is inversely proportional to the square of the hydronium ion concentration. Since the hydronium ion concentration can be adjusted to a given value by means of a buffer, the sulfide ion concentration can be varied to suit the circumstances. For example, the precipitation of a highly insoluble metal sulfide, such as copper sulfide, requires only a low sulfide ion concentration in order for $Q$ to exceed $K_{sp}$ and can be achieved at a high hydronium ion concentration. On the other hand, the precipitation of a moderately insoluble sulfide, such as zinc sulfide, requires a higher sulfide ion concentration. It is consequently possible to precipitate copper sulfide from a solution of low pH without, at the same time, precipitating zinc sulfide.

---

**Example 9.8** Exactly 5 mL of 6.00 $M$ HCl are added to 100 mL of a solution containing 100 mg of $ZnSO_4$ and 100 mg of $CdSO_4$ and the solution is saturated with $H_2S$ at 25°C. (a) Which, if either, of the sulfides precipitates under these conditions? (b) What is the minimum pH at which zinc sulfide will precipitate from this solution?

(a) We first calculate the various concentrations:

$$[Zn^{2+}] = \frac{0.100 \text{ g}}{(161.5 \text{ g/mol})(0.105 \text{ L})} = 0.00590 \text{ mol L}^{-1}$$

$$[Cd^{2+}] = \frac{0.100 \text{ g}}{(209.5 \text{ g/mol})(0.105 \text{ L})} = 0.00457 \text{ mol L}^{-1}$$

$$[H_3O^+] = \frac{(6.00 \text{ mol/L})(5.00 \times 10^{-3} \text{ L})}{0.105 \text{ L}} = 0.286 \text{ mol L}^{-1}$$

Next, the $[S^{2-}]$ concentration is obtained from Equation 9.12 as

$$[S^{2-}] = \frac{3 \times 10^{-21}}{[H_3O^+]^2} = \frac{3 \times 10^{-21}}{(0.286)^2} = 4 \times 10^{-20} \text{ mol L}^{-1}$$

Finally, we evaluate the reaction quotients of the two sulfides and see whether they exceed the respective solubility products. For CdS,

$$Q = [Cd^{2+}][S^{2-}] = (0.00457)(4 \times 10^{-20}) = 2 \times 10^{-22}$$

Since $K_{sp} = 2 \times 10^{-28}$, CdS precipitates. For ZnS,

$$Q = [Zn^{2+}][S^{2-}] = (0.00590)(4 \times 10^{-20}) = 2 \times 10^{-22}$$

Since $K_{sp} = 1.1 \times 10^{-21}$, ZnS does not precipitate.

(b) Zinc sulfide will precipitate when the reaction quotient just exceeds the value of $K_{sp}$ of this salt. This condition will be met when

$$Q = [Zn^{2+}][S^{2-}] > 1.1 \times 10^{-21}$$

Therefore,

$$[S^{2-}] > \frac{1.1 \times 10^{-21}}{0.00590} > 1.9 \times 10^{-19} \text{ mol L}^{-1}$$

This sulfide ion concentration will be obtained when the $H_3O^+$ concentration satisfies Equation 9.12:

$$[H_3O^+]^2[S^{2-}] = 3 \times 10^{-21}$$

$$[H_3O^+] < \left(\frac{3 \times 10^{-21}}{1.9 \times 10^{-19}}\right)^{1/2} = 0.13 \text{ mol L}^{-1}$$

$$\text{pH} > 0.9 \quad \blacksquare$$

**b. Sulfide Separations in Qualitative Analysis** The existence of a large number of insoluble metal sulfides, coupled with the possibility of controlling precipitation by adjustments in pH, have made sulfide precipitation one of the most widely used procedures in qualitative analysis. Figure 9.10 summarizes a scheme that may be used for the identification of those metal ions whose sulfides are sufficiently insoluble to precipitate out of 0.3 M HCl. This group includes PbS, CuS, HgS (all colored black), $SnS_2$, CdS, $As_2S_3$ (all colored yellow), $Bi_2S_3$, and $Sb_2S_3$ (dark).

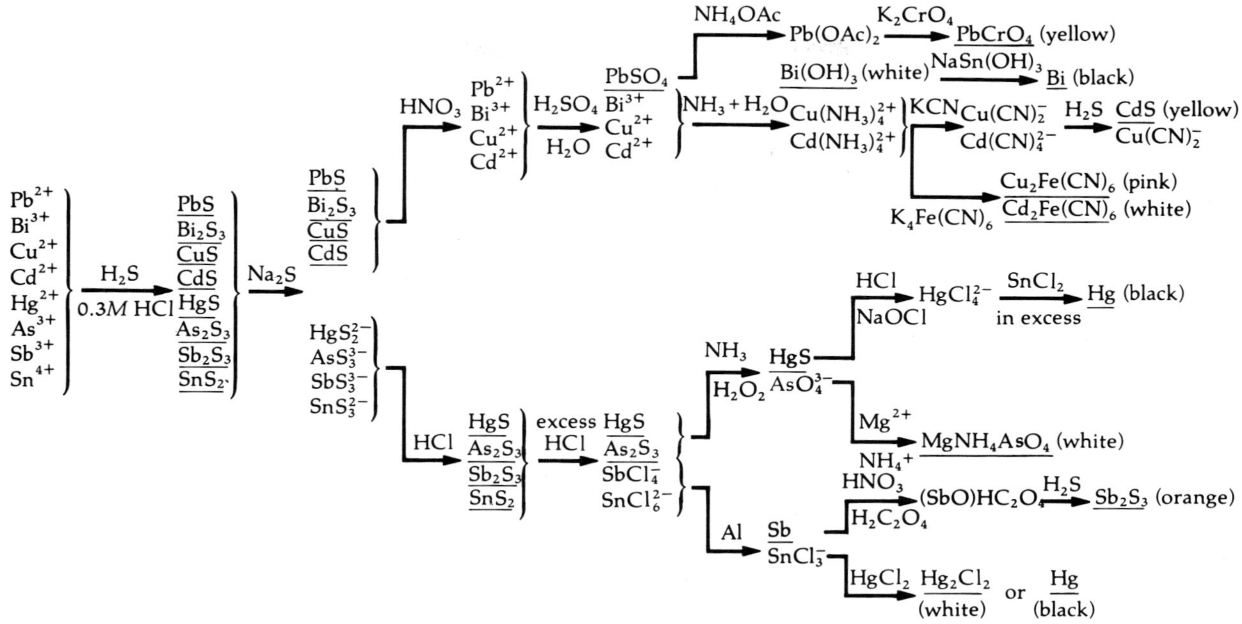

**FIGURE 9.10** Flow diagram for the separation and identification of the highly insoluble metal sulfides.

These sulfides can be further separated from each other by noting that some of them dissolve in an alkaline sulfide solution owing to the formation of complex ions with sulfide, of the type discussed in Section 9.7c. The concentration of sulfide ion in alkaline solution is sufficiently high for reactions of the following type to occur:

$$HgS(s) + S^{2-} \rightleftharpoons HgS_2^{2-} \quad \text{(thiomercurate ion)}$$

$$Sb_2S_3(s) + 3S^{2-} \rightleftharpoons 2SbS_3^{3-} \quad \text{(thioantimonite ion)}$$

In addition to these two sulfides, the sulfides of arsenic(III) and tin(IV) also undergo this type of reaction. The separation and identification of the individual ions is based on additional precipitation and complex ion formation reactions as summarized in Figure 9.10. Bismuth, for instance, can be identified by dissolving the insoluble sulfide in acid and precipitating white bismuth hydroxide with ammonia:

$$Bi^{3+} + 3NH_3(aq) + 3H_2O \rightarrow Bi(OH)_3 + 3NH_4^+$$

While copper hydroxide and cadium hydroxide follow the chemistry of bismuth up to this point, they redissolve in excess ammonia owing to the formation of $Cu(NH_3)_4^{2+}$ and $Cd(NH_3)_4^{2+}$. Bismuth does not form a complex ion of this type and remains behind as the solid hydroxide. Its identity can be confirmed by the characteristic black color to which the precipitate turns upon reduction to the elemental state by the addition of sodium stannite:

$$2Bi(OH)_3(s) + 3Sn(OH)_3^- + 3OH^- \rightarrow 2Bi(s) + 3Sn(OH)_6^{2-}$$

**322** Ionic Equilibria 9.9

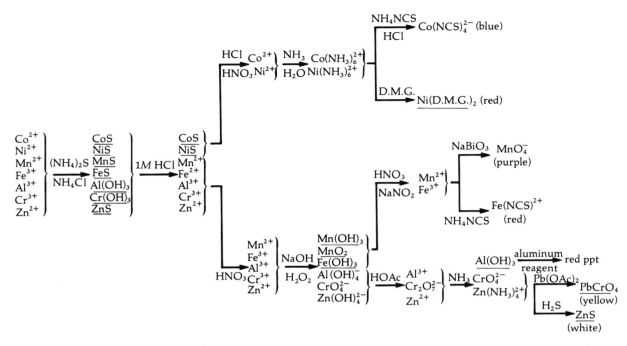

**FIGURE 9.11** Flow diagram for the separation and identification of the moderately insoluble metal sulfides.

---

The identification of the moderately insoluble sulfides is based on their precipitation from aqueous ammonia, the pH of which is sufficiently high to raise the sulfide ion concentration to the value needed for the reaction quotients to exceed the corresponding solubility product constants. As indicated in Figure 9.11, which summarizes the procedure, the ions precipitated in this fashion include $Co^{2+}$, $Ni^{2+}$, $Mn^{2+}$, $Fe^{3+}$, $Zn^{2+}$, $Al^{3+}$, and $Cr^{3+}$, the last two precipitating as the hydroxides. The further separation of the individual metal ions is based on a variety of reactions. Cobalt(II) and nickel(II), for example, are identified by the formation of the ammonia complexes, $Co(NH_3)_6^{2+}$ and $Ni(NH_3)_6^{2+}$, following the dissolution of the sulfides in a mixture of hydrochloric and nitric acids. The addition of dimethylglyoxime results in the formation of an insoluble, bright red complex compound with nickel. After its removal from the solution, the presence of cobalt(II) can be demonstrated by the formation of a blue complex with thiocyanate:

$$Co(NH_3)_6^{2+} + 4NCS^- \rightleftarrows Co(NCS)_4^{2-} + 6NH_3$$

The formation constant of this complex is greater than that of the ammonia complex, and the equilibrium consequently favors its formation.

## 9.9 Effect of Hydrolysis on Solubility

When a sparingly soluble salt of a weak acid or base is dissolved in water it undergoes hydrolysis (Section 8.6). Consider, for example, the dissolution of a relatively insoluble sulfide, such as lead sulfide. The following equilibria must be considered for the lead and sulfide ions:

(1) $PbS(s) \rightleftharpoons Pb^{2+} + S^{2-}$ $\quad K_{sp}$ $\qquad$ (9.13)

(2) $S^{2-} + H_2O \rightleftharpoons HS^- + OH^-$ $\quad K_w/K_2$ $\qquad$ (9.14)

(3) $HS^- + H_2O \rightleftharpoons H_2S(aq) + OH^-$ $\quad K_w/K_1$ $\qquad$ (9.15)

The second and third reactions involve the hydrolysis of the sulfide and hydrosulfide ions and lead to a reduction in the concentration of sulfide ion. According to Le Chatelier's principle, the equilibrium of the first reaction will consequently shift towards the formation of additional sulfide ion. Thus, the solubility of lead sulfide in water is greater than predicted by the value of $K_{sp}$ alone. The procedure described in Section 9.2 for the determination of the solubility of a salt must therefore be modified when one of the ions formed on dissolution can react with water.

The determination of the solubility in water of a salt of a weak acid such as $H_2S$ is somewhat more complicated than that of the solubility of such a salt in a solution of fixed pH. When the pH is fixed, the sulfide ion concentration is given by Equation 9.12, and the solubility of the salt can be obtained directly from the solubility product expression. However, when a salt such as lead sulfide is dissolved in water, the hydronium ion concentration is one of the unknowns and must be determined along with the concentration of sulfide ion.

The procedure to be followed is analogous to that described in connection with the effect of the ionization of water on equilibria involving weak acids (Section 8.7). A sufficiently large number of relationships between the unknown concentrations must be developed so that the number of equations is as large as the number of unknowns. In general, this requires the use of mass and charge balance relationships as well as explicit consideration of the ionization of water. Reasonable approximations generally must be made in order to permit the equations to be solved in a straightforward manner. Two such simplifying approximations are frequently valid: when the salt in question is very slightly soluble, the hydroxide ion concentration resulting from the reactions represented by Equations 9.14 and 9.15 is small compared to that resulting from the ionization of water. On the other hand, if the salt is slightly more soluble and if, furthermore, the anion is extensively hydrolyzed, the reverse approximation is valid—the concentration of hydroxide ion resulting from the ionization of water may be neglected with respect to that resulting from hydrolysis. The dissolution of lead sulfide illustrates the applicability of the first of these assumptions.

---

**Example 9.9** Taking the hydrolysis of the sulfide ion into account, determine the solubility of lead sulfide in water at 25°C. Compare the result with that obtained by neglecting hydrolysis.

We begin by writing all the equilibrium expressions for species involving sulfide, i.e., expressions for the processes represented by Equations 9.13–9.15:

(1) $[Pb^{2+}][S^{2-}] = K_{sp} = 1 \times 10^{-28}$

(2) $\dfrac{[HS^-][OH^-]}{[S^{2-}]} = \dfrac{K_w}{K_2} = \dfrac{1.0 \times 10^{-14}}{3 \times 10^{-13}} = 0.033$

(3) $\dfrac{[H_2S][OH^-]}{[HS^-]} = \dfrac{K_w}{K_1} = \dfrac{1.0 \times 10^{-14}}{1.0 \times 10^{-7}} = 1.0 \times 10^{-7}$

The unknowns in these expressions are $[Pb^{2+}]$, $[S^{2-}]$, $[HS^-]$, $[H_2S]$, and $[OH^-]$, with the solubility of lead sulfide equal to the concentration of $Pb^{2+}$. In addition, the concentration of $H_3O^+$ is unknown. Thus, there are six unknowns, indicating that three additional relationships are needed. The ion product of water provides one of these relationships:

(4) $[H_3O^+][OH^-] = K_w = 1.0 \times 10^{-14}$

In addition to the four equilibrium expressions, the mass balance expression is

(5) $[Pb^{2+}] = [S^{2-}] + [HS^-] + [H_2S]$

indicating that when one PbS ion pair dissolves, the sulfur must be distributed among these three species. Finally, the charge balance relationship is

(6) $2[Pb^{2+}] + [H_3O^+] = 2[S^{2-}] + [HS^-] + [OH^-]$

where the concentrations of the doubly charged ions are multiplied by 2 in order for the total positive and negative charges to balance.

Since we have six equations and six unknowns, a solution can be obtained. Solving such a large set of equations is not a pleasant task, however, and the process can be simplified considerably by making judicious approximations. Since the solubility of PbS is very small it is likely that the $OH^-$ concentration will not increase significantly as the precipitate dissolves. Consequently, we will assume tentatively that

$[OH^-] \approx [H_3O^+] = 1.0 \times 1.0^{-7}$

This assumption will be correct provided that in the charge balance [Equation (6)]

(7) $2[Pb^{2+}] \ll [H_3O^+]$

and

(8) $2[S^{2-}] + [HS^-] \ll [OH^-]$

The validity of these approximations must be checked once a provisional solution has been obtained.

Having made this approximation, we proceed by substituting $[OH^-] = 1.0 \times 10^{-7}$ into Equations (2) and (3), thereby obtaining from Equation (2)

(9) $[HS^-] = 3.3 \times 10^5[S^{2-}]$

and from Equation (3)

(10) $[H_2S] = [HS^-]$

Combining Equations (10) and (9) gives

(11) $[H_2S] = 3.3 \times 10^5[S^{2-}]$

We have now succeeded in expressing the concentrations of all species containing sulfur in terms of the concentration of the sulfide ion. Introducing the results into the mass balance [Equation (5)] gives

$$[Pb^{2+}] = [S^{2-}] + 3.3 \times 10^5[S^{2-}] + 3.3 \times 10^5[S^{2-}]$$

which can be used to express $[S^{2-}]$ in terms of $[Pb^{2+}]$:

(12) $\quad [S^{2-}] = 1.5 \times 10^{-6}[Pb^{2+}]$

Substituting Equation (12) into the solubility product expression [Equation (1)] we obtain

$$1.5 \times 10^{-6}[Pb^{2+}]^2 = 1 \times 10^{-28}$$

which can immediatly be solved to give

$$[Pb^{2+}] = 8 \times 10^{-12} \text{ mol L}^{-1}$$

for the solubility of PbS.

We must now check the validity of the approximations leading to this result. The first approximation [Equation (7)] is certainly valid, for the concentration of $Pb^{2+}$ is much smaller than $10^{-7}$ mol L$^{-1}$. The second approximation [Equation (8)] is equally valid, as can be ascertained by evaluating the concentrations of $S^{2-}$ and $HS^-$ from Equations (9) and (12) and the now known concentration of $Pb^{2+}$:

$$2[S^{2-}] + [HS^-] = 4 \times 10^{-12}$$

which is negligibly small compared to $1 \times 10^{-7}$.

Finally, we compare this result for the solubility of PbS with that obtained from the solubility product expression alone. The result is

$$K_{sp} = 1 \times 10^{-28} = [Pb^{2+}][S^{2-}] = [Pb^{2+}]^2$$

and

$$[Pb^{2+}] = 1 \times 10^{-14} \text{ mol L}^{-1}$$

We see that while the solubility of lead sulfide in water remains exceedingly small, hydrolysis leads to an 800-fold increase in solubility. ∎

We have so far considered only the effect of anion hydrolysis on solubility. Many hydrated metal cations hydrolyze in aqueous solution with the formation of acidic solutions. Cation hydrolysis occurs in reactions such as

$$Cu(H_2O)_4^{2+} + H_2O \rightleftharpoons Cu(H_2O)_3(OH)^+ + H_3O^+$$

and

$$Al(H_2O)_6^{3+} + H_2O \rightleftharpoons Al(H_2O)_5(OH)^{2+} + H_3O^+$$

where the cations are represented as hydrated ions. Evidently, these hydrated cations are stronger acids than water and can donate a proton to $H_2O$. This effect occurs because the electrons on the oxygen atoms of the attached water molecules are attracted to the positively charged metal ion, thereby increasing the strength of the cation–oxygen bond. As a result, the O—H bonds are weakened and the hydrated ion can more easily lose a proton to a water molecule. Cation hydrolysis is most important for ions of large charge density

**FIGURE 9.12** Hydrolysis of metal cations. (a) A metal ion of high charge density forms a strong bond with the oxygen atom of an attached water molecule. A proton can split off forming an acidic solution. (b) A metal ion of low charge density forms a weaker bond with oxygen and does not form an acidic solution.

(large positive charge, small radius), for such ions most strongly attract the electrons on the oxygen atoms (Figure 9.12). These properties are manifested by metal ions found toward the right in the periodic table (e.g., aluminum and many of the transition metal ions).

When a sparingly soluble salt of one of these cations (e.g., CuS) is dissolved in water, cation hydrolysis reduces the concentration of the cation and increases the solubility of the salt. In complete analogy to anion hydrolysis (e.g., Equations 9.13–9.15), the hydrolysis equilibrium expression as well as the solubility product expression must be satisfied by the concentrations of the various species present.

## 9.10 The Solubility of Metal Hydroxides

Many metal hydroxides are sparingly soluble in water. In addition to obeying the solubility equilibrium, the concentration of hydroxide ion must satisfy the ion product of water. Consider the ionization of sparingly soluble $M(OH)_2$. The equations representing these equilibria can be written as

$$M(OH)_2 \underset{}{\overset{H_2O}{\rightleftharpoons}} M^{2+} + 2OH^- \qquad K_{sp}$$

$$2H_2O \rightleftharpoons H_3O^+ + OH^- \qquad K_w$$

If the hydroxide is sufficiently soluble so that the $OH^-$ concentration is substantially higher than $10^{-7}$ mol $L^{-1}$, the concentration of $OH^-$ in water, the ionization of water can be neglected. In these circumstances, the solubility of the metal hydroxide is determined solely by its solubility product.

Suppose, though, that the metal hydroxide is sufficiently insoluble that the $OH^-$ concentration resulting from its dissolution is less than $10^{-7}$ mol $L^{-1}$. Due to the ionization of water, the actual concentration of $OH^-$ will be $10^{-7}$ mol $L^{-1}$. Since the concentration of $OH^-$ is larger than expected from the solubility equilibrium, $OH^-$ acts as a common ion and represses the solubility of the metal hydroxide. The exact form of the equilibrium expression can be derived in the manner shown in Example 9.9. To illustrate the magnitude of this effect, consider a metal hydroxide $M(OH)_2$ with $K_{sp} = 1.0 \times 10^{-26}$. Taking the ionization of water into account, we can calculate that the solubility is $1.0 \times 10^{-12}$ mol $L^{-1}$. If this equilibrium is ignored, the solubility is calculated incorrectly as $1.4 \times 10^{-9}$ mol $L^{-1}$. As noted earlier, the effect of the ionization of water decreases in importance as the value of $K_{sp}$ increases. For example, when $K_{sp} = 1.0 \times 10^{-18}$, the solubility is $6.3 \times 10^{-7}$ mol $L^{-1}$ regardless of whether this effect is considered.

When a solution of a sparingly soluble metal hydroxide is acidified, the solubility of the hydroxide increases. This effect is a consequence of the simultaneous action of the following two equilibria:

(1) $M(OH)_2(s) \xrightleftharpoons{H_2O} M^{2+} + 2OH^-$     $K_{sp}$

(2) $H_3O^+ + OH^- \rightleftharpoons 2H_2O$     $K_w^{-1}$

Since the equilibrium for the second reaction lies far to the right, the concentration of $OH^-$ in acidic solution is much smaller than it is in water. According to Le Chatelier's principle, the equilibrium of the dissolution reaction must consequently shift towards the formation of additional $OH^-$.

It might be expected that the addition of alkali to an aqueous solution of a sparingly soluble metal hydroxide would have the opposite effect than the addition of acid—the solubility should decrease because of the common ion effect. This is indeed the behavior displayed by many metal hydroxides. For certain metals, however, the addition of alkali, while first decreasing the solubility of the hydroxide, eventually leads to an increase in solubility. The behavior of $Zn(OH)_2$, illustrated in Figure 9.13, is typical of these cases. As the pH increases from 7 to approximately 9.5, the solubility of $Zn(OH)_2$ decreases. At this pH the solubility reaches its lowest value and thereafter increases with increasing pH.

Hydroxides or oxides that display this behavior are said to be **amphoteric,** meaning that they can act as either an acid or a base. For example, the behavior

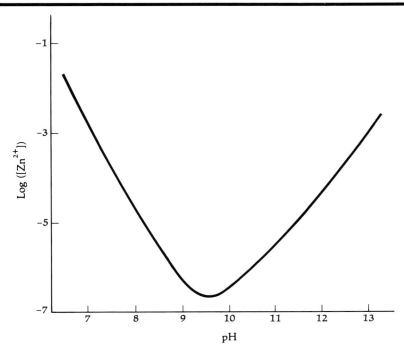

**FIGURE 9.13**  Solubility of $Zn(OH)_2$ at 25°C as a function of pH.

of $Zn(OH)_2$ as a base is manifested in its reaction with an acid

$$Zn(OH)_2(s) + 2H_3O^+ \rightleftharpoons Zn^{2+} + 4H_2O$$

while its behavior as an acid is seen in the reaction with a base

$$Zn(OH)_2(s) + 2OH^- \rightleftharpoons Zn(OH)_4^{2-}$$

The concept of amphoterism is similar to but more general than that of amphiprotism (Section 8.10), which refers to a substance that can either gain or lose a proton and thus act as either a Brønsted acid or base. All amphiprotic species are amphoteric. However, the reverse is not necessarily true since in certain non-aqueous solvents acid–base behavior does not necessarily involve proton transfer. For example, aluminium sulfite is amphoteric in liquid sulfur dioxide. It reacts with the acid $SO^{2+}$ by the reaction

$$Al_2(SO_3)_3(s) + 3SO^{2+} \rightleftharpoons 2Al^{3+} + 6SO_2$$

and with the base $SO_3^{2-}$ by the reaction

$$Al_2(SO_3)_3(s) + 3SO_3^{2-} \rightleftharpoons 2Al(SO_3)_3^{3-}$$

Note that neither reaction involves proton transfer.

In addition to $Zn(OH)_2$ (or ZnO) the amphoteric oxides or hydroxides include those of beryllium, aluminum, tin, and lead. This behavior is generally displayed by the elements lying near the boundary between the metals and nonmetals (Section 6.9).

## Conclusion

At first sight, solubility equilibria appear to be the simplest of all the various equilibrium processes considered so far, being governed by the relatively simple expression for the solubility product. A closer look indicates that these equilibria can actually be rather complicated, especially in those cases where the solubility is affected by the operation of additional equilibria involving one of the ions of a sparingly soluble salt. We have examined the effect on solubility equilibria of complex ion formation, pH, hydrolysis, and the ionization of water. The interplay between these various equilibria offers many possibilities for optimizing the separation of some particular metal ion from other metal ions and provides the means for the identification of metal ions.

## Problems

1. Express the solubility product of each of the following salts in terms of the solubility $s$ (expressed in mol $L^{-1}$): (a) $PbCl_2$, (b) $AgI$, (c) $Ca_3(PO_4)_2$, and (d) $Ag_3AsO_4$.

2. Using tabulated $K_{sp}$ values, calculate the equilibrium constant of the reaction

$$AgI(s) + Br^- \rightleftharpoons AgBr(s) + I^-$$

3. What is the molar solubility of $SrSO_4$ in (a) pure water and (b) 0.15 $M$ $Na_2SO_4$ solution at 25°C? Neglect the effect of hydrolysis.

4. The solubility of AgBr at 20°C is $8.4 \times 10^{-6}$ g of AgBr per 100 mL of water. What is the solubility product of AgBr at this temperature?

5. A concentrated $CrO_4^{2-}$ solution is added dropwise

to a solution containing 0.060 mol L$^{-1}$ Ba$^{2+}$ ions and 0.010 mol L$^{-1}$ Ag$^+$ ions. (a) Which ion precipitates first? (b) What percentage of this ion remains in solution when the other ion begins to precipitate?

6. Equal volumes of 0.10 $M$ AgNO$_3$ solution and 0.12 $M$ BaCl$_2$ solution are mixed. What are the concentrations of all ions present in the solution after equilibrium has been attained?

7. A 25 mL sample of a solution is to be tested for the presence of fluoride ion by addition of a drop (0.05 mL) of 0.5 $M$ Ca(NO$_3$)$_2$. What is the minimum concentration of F$^-$ that can be detected by the observation of a CaF$_2$ precipitate? Express your answer in grams per 100 mL. Neglect the effect of hydrolysis.

8. What are the equilibrium concentrations of Mg$^{2+}$ and PO$_4^{3-}$ ions when 20 mL of 0.10 $M$ Na$_3$PO$_4$ solution is added to 30 mL of 0.15 $M$ MgCl$_2$ solution? Neglect the effect of hydrolysis.

9. The solubility of silver carbonate in water at 27°C is 1.3 × 10$^{-4}$ mol L$^{-1}$. What is the solubility of silver carbonate in (a) 0.10 $M$ AgNO$_3$ and (b) 0.10 $M$ Na$_2$CO$_3$? Neglect hydrolysis effects.

10. What are the concentrations of Pb$^{2+}$ and Cl$^-$ ions in a solution made by mixing 50 mL of 0.0010 $M$ Pb(NO$_3$)$_2$ with 500 mL of 0.0020 $M$ NaCl? Does any PbCl$_2$ precipitate form?

11. A solution in which the free bromide concentration is 2.00 × 10$^{-5}$ mol L$^{-1}$ is in equilibrium with AgBr and AgI. What is the concentration of free iodide?

12. The complex formation constant of Ni(CN)$_4^{2-}$ is 1 × 10$^{30}$. What is the molar solubility of nickel carbonate in 0.30 $M$ NaCN? Neglect hydrolysis.

13. The solubility products of several sparingly soluble phosphates are listed below:

| Substance | $K_{sp}$ |
|---|---|
| Li$_3$PO$_4$ | 3 × 10$^{-13}$ |
| AlPO$_4$ | 5.8 × 10$^{-19}$ |
| Ba$_3$(PO$_4$)$_2$ | 3 × 10$^{-23}$ |
| BiPO$_4$ | 1.3 × 10$^{-23}$ |

Use these data to identify the phosphate with (a) the largest phosphate ion concentration in a saturated aqueous solution, (b) the smallest molar solubility in water, (c) the smallest cation concentration in a saturated aqueous solution, (d) the smallest molar solubility in a solution in which the PO$_4^{3-}$ concentration is 0.010 mol L$^{-1}$, and (e) the largest molar solubility in a solution in which the cation concentration is 0.05 mol L$^{-1}$. Neglect hydrolysis.

14. The solubility product constants of several sparingly soluble hydroxides are listed below:

| Hydroxide | $K_{sp}$ |
|---|---|
| BiO(OH) | 1 × 10$^{-12}$ |
| Pb(OH)$_2$ | 1.2 × 10$^{-15}$ |
| Cr(OH)$_3$ | 6 × 10$^{-31}$ |
| Sn(OH)$_4$ | 1 × 10$^{-57}$ |

Which of these hydroxides has (a) the smallest molar solubility in water, (b) the largest molar solubility in a solution with pH = 12, (c) the largest molar solubility in a solution that is 0.02 $M$ in the hydroxide cation, (d) the smallest cation concentration in a saturated solution, and (e) the highest pH in a saturated solution? Neglect effects associated with the ionization of water.

15. Establish whether either of the following separations are feasible in principle: (a) 0.20 $M$ Ag$^+$ from 0.050 $M$ Pb$^{2+}$ by the addition of Br$^-$ and (b) 0.10 $M$ Mg$^{2+}$ from 0.20 $M$ Cu$^{2+}$ by the addition of OH$^-$. Assume that a separation is feasible if the remaining concentration of one of the ions has been reduced to ≤1 × 10$^{-6}$ mol L$^{-1}$ before the second ion precipitates.

16. Enough ammonia is added to a 0.010 $M$ solution of Cu$^{2+}$ to make the stoichiometric concentration of NH$_3$ equal to 0.50 $M$. (a) What is the concentration of free Cu$^{2+}$ in the resulting solution? (b) What is the concentration of Cu(NH$_3$)$_4^{2+}$?

17. What must the concentration of free ammonia be in a solution containing copper(II) ions if the concentration of Cu$^{2+}$ is equal to that of Cu(NH$_3$)$_4^{2+}$? Under these conditions, which copper–ammonia complex is present in highest concentration?

18. What is the molar solubility of silver bromide in 0.20 $M$ NH$_3$?

19.* The molar solubility of a sparingly soluble salt MX is lowest at a certain concentration of X$^-$ due to the action of the following two equilibria:

$$MX(s) \rightleftarrows M^+ + X^- \qquad K_{sp}$$

$$M^+ + 2X^- \rightleftarrows MX_2^- \qquad K_f$$

Derive an expression for the concentration of X$^-$ for which the molar solubility of MX is lowest in terms of $K_{sp}$ and $K_f$.

20. The reaction of silver(I) with thiosulfate ion involves the following equilibria:

$$Ag^+ + S_2O_3^{2-} \rightleftarrows AgS_2O_3^- \qquad K_1 = 6.6 \times 10^8$$

$$AgS_2O_3^- + S_2O_3^{2-} \rightleftarrows Ag(S_2O_3)_2^{3-} \qquad K_2 = 4.4 \times 10^3$$

Determine the solubility of AgBr in 0.100 $M$ Na$_2$S$_2$O$_3$.

21. A separation of 0.010 $M$ Mn$^{2+}$ from 0.010 $M$ Fe$^{3+}$ is to be made by hydroxide precipitation. Determine

the pH range suitable for such a separation as follows: (a) the lowest pH is that at which 99.9% of $Fe^{3+}$ precipitates as $Fe(OH)_3$ and (b) the highest pH is 0.1 unit below that at which $Mn(OH)_2$ just precipitates.

22. A 0.10 $M$ solution of $Bi^{3+}$ is saturated with $H_2S$ at pH = 1.0. Will $Bi_2S_3$ precipitate under these conditions? If so, what is the remaining concentration of $Bi^{3+}$ in the solution?

23. Which of the following separations is feasible by control of the pH of a saturated $H_2S$ solution? (a) $Cu^{2+}$ and $Zn^{2+}$, (b) $Ni^{2+}$ and $Cd^{2+}$, (c) $Mn^{2+}$ and $Fe^{2+}$. Assume that the initial concentration of each ion is 0.10 $M$ and that the ion has been quantitatively removed from the solution if its residual concentration is $\leq 1.0 \times 10^{-6}$ $M$. If a separation is feasible, specify the pH range that could be employed.

24. Taking the hydrolysis of $S^{2-}$ into account, determine the molar solubility of CuS in water. How large is the effect of hydrolysis on the solubility?

25. Calculate the molar solubility of calcium oxalate in a solution with a fixed pH of 4.0. The relevant equilibrium constants are $CaC_2O_4$, $K_{sp} = 1 \times 10^{-9}$; $H_2C_2O_4$, $K_1 = 5.60 \times 10^{-2}$, $K_2 = 6.2 \times 10^{-5}$.

26. Determine the molar solubility of $Fe(OH)_3$ in water (a) by neglecting the ionization of water and (b) by taking the ionization of water into account.

27. Aluminum hydroxide is amphoteric and dissolves in alkaline solution owing to the reaction

$$Al(OH)_3(s) + OH^- \rightleftarrows Al(OH)_4^- \qquad K = 10$$

What volume of 1.5 $M$ NaOH must be added to 200 mL of water in which 500 mg of $Al(OH)_3$ is suspended in order to completely dissolve the precipitate?

28. The solubility product of magnesium hydroxide is $7.1 \times 10^{-12}$. (a) Below what value must the $OH^-$ concentration be maintained in order to prevent the precipitation of $Mg(OH)_2$ from a 0.050 $M$ solution of $Mg^{2+}$? (b) What is the ratio of $NH_4^+$ to $NH_3$ concentrations in a buffer that will just maintain this $OH^-$ concentration?

29. Lead hydroxide obeys the following equilibria in aqueous solution:

$$Pb(OH)_2(s) \rightleftarrows Pb^{2+} + 2OH^-$$
$$K_{sp} = 1.2 \times 10^{-15}$$

$$Pb(OH)_2(s) + OH^- \rightleftarrows Pb(OH)_3^-$$
$$K = 5.0 \times 10^{-2}$$

What are the minimum and the maximum pH values that will prevent the precipitation of $Pb(OH)_2$ from a 0.050 $M$ solution of $Pb^{2+}$?

30. Derive an exact expression for the solubility of a metal hydroxide MOH in water. Assume that the solubility is affected by $OH^-$ resulting from the ionization of water. To what does your expression reduce when $K_{sp} \ll K_w$, and when $K_w \ll K_{sp}$?

# 10

# Energy in Chemical Processes—The First Law of Thermodynamics

Thermodynamics is the branch of chemistry that deals with the energetics and spontaneity of chemical and physical processes. It is a far reaching subject of importance in all the sciences as well as in engineering. A few general laws distilled from a vast body of experimental observations provide the basis for relationships that expose the connection between different aspects of chemistry, for example, between equilibrium constants and energy changes. The approach of thermodynamics is macroscopic—the subject deals with the bulk properties of matter rather than with molecular behavior. By being independent of the atomic and molecular structure of matter, the laws of thermodynamics need not be modified as knowledge of molecular structure and interactions increases. These laws have acquired a permanence that stands in marked contrast to the continuous evolution of other theories.

## Fundamental Aspects of the First Law of Thermodynamics

### 10.1 The Transfer of Energy between a System and Its Surroundings

The transfer of energy is a familiar process. We see it in such diverse operations as the heating of the interior of a building, in which the energy stored in oil or natural gas is transformed to thermal energy, or the operation of an automobile, in which the expansion of the combustion products of gasoline is used to power the rotary motion of the wheels. The analysis of energy transfer processes is one of the central concerns of thermodynamics.

The transfer of energy takes place between a system and its surroundings. A **system** is the part of the universe that we are interested in studying. A chemical system might consist of some reagents undergoing a chemical reaction or a gas confined to a chamber. The rest of the universe, lying outside the boundaries of the system, is called the **surroundings.** A system is said to be in a definite **state** when each of its properties has a definite value. A system has many properties, such as temperature, pressure, volume, density, and chemical composition. It is usually necessary to specify the values of only a few of these properties in order to describe the state of the system because many properties are related to each other. For example, the density of an ideal gas is automatically specified if the pressure, temperature, and molecular weight of the gas are given (see Equation 3.14).

The processes by which a system interacts with its surroundings generally involve transfers of energy and mass between the two. A system is called **open** when mass can move across the boundary between system and surroundings. The study of open systems is an important topic in physical chemistry but is beyond the scope of this text. We are interested here in **closed systems,** which have boundaries that do not allow the transfer of mass. Closed systems can only interact with their surroundings by the transfer of energy across their boundaries. A closed system can also be **isolated,** in which case energy cannot be transferred across its boundaries.

A system can possess different types of energy, some of which we have already discussed. These include **kinetic energy**—the energy possessed by a body as a consequence of its motion; **potential energy**—the energy possessed by a body as a consequence of its position in a force field, for example, a mass in a gravity field; and **thermal energy**—the energy possessed by a body as a consequence of its temperature. Other common forms of energy include mechanical energy, electrical energy, and so on.

The transfer of energy between a system and its surroundings can involve the flow of heat and/or work. The meaning of these common terms as they are used in thermodynamics is very specific and will be examined in detail in the next two sections. Here we focus on some of their common features. Both work and heat are energy in motion between a system and its surroundings. They appear only at the boundary of a system and are *not* properties of the system.

The state of a system changes when its energy changes. Thus, work and heat are associated with a change in state. If a system does not change its state there can be no flow of work or heat. When the flow of work or heat is directed from the surroundings to the system, the energy of the system increases. On the other hand, when the flow of work or heat is directed from the system to the surroundings, the energy of the system decreases.

We have so far examined the common features of work and heat. These quantities must also differ in some way since otherwise there would be no reason to make a distinction between them. The difference between work and heat lies in their effects on the surroundings. A flow of work into the surroundings can be completely converted into the potential energy resulting from the lifting of a weight in the surroundings, as illustrated in Figure 10.1. A flow of heat into the surroundings, which occurs when the system is at a higher temperature than the surroundings, can be completely converted into thermal

Energy in Chemical Processes—The First Law of Thermodynamics  10.1  **333**

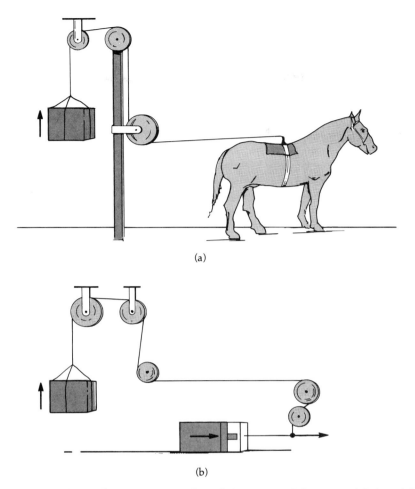

**FIGURE 10.1** The conversion of work into potential energy. (a) A weight is lifted by the work of a horse. (b) The weight is lifted by the work of an expanding gas.

---

energy—the temperature of the surroundings increases and that of the system decreases.

The flow of work and heat between a system and its surroundings can be visualized by means of Figure 10.2. Although in reality a system has a single surroundings, it is useful to represent the surroundings as two reservoirs associated with the flow of either work or heat. The mechanical reservoir can be represented by means of a weight which can be raised or lowered when work flows between the system and the surroundings. The thermal reservoir can be represented by a body of water the temperature of which either rises or falls when heat flows between the system and the surroundings. Figure 10.2a shows a process in which work and heat flow from system to surroundings, resulting in a decrease in the energy of the system. Figure 10.2b shows the opposite type of process: work and heat flow from the surroundings to the system and lead to an increase in the energy of the system

Although Figure 10.2 shows the simultaneous flow of work and heat, a given process need only involve one type of flow. For example, the heat link

334 Energy in Chemical Processes—The First Law of Thermodynamics 10.1

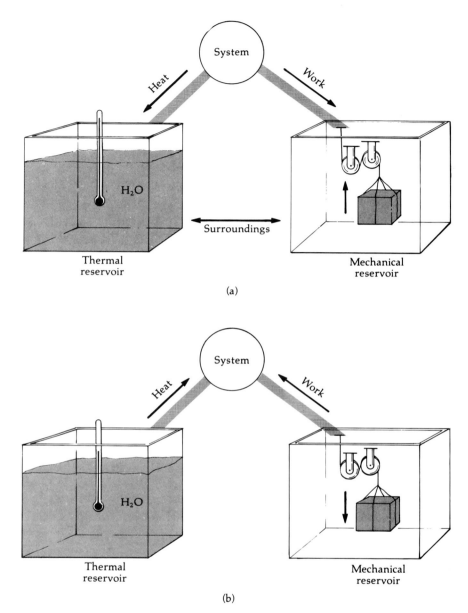

**FIGURE 10.2** The flow of work and heat between a system and its surroundings. The surroundings is represented by mechanical and thermal reservoirs. These reservoirs are connected to the system by work and heat links, respectively. (a) Work and heat flow from system to surroundings, decreasing the energy of the system. (b) Work and heat flow from the surroundings to the system, increasing the energy of the system.

between a system and its surroundings can be severed by constructing the boundary walls so that they prevent the flow of heat. A process in which there is no heat flow is call an **adiabatic** process.

## 10.2 Work

***a. Mechanical Work*** A system can do different types of work on its surroundings. For example, a battery can do electrical work (Section 13.2). We are interested here in mechanical work, examples of which have already been shown in Figure 10.1.

Suppose a constant force $F$ acting on a body displaces it in the direction in which the force acts by a distance $d$ (Figure 10.3a). The work done on the body is defined as

$$w = Fd \tag{10.1}$$

The units in which work is expressed follow from this definition. Recall from Section 3.2 that force is expressed in newtons:

$$F = ma$$
$$1 \text{ N} = (1 \text{ kg})(1 \text{ m s}^{-2})$$

It follows from Equation 10.1 that work is expressed in newton-meters or, equivalently, in joules:

$$w = Fd = mad$$
$$1 \text{ N m} = (1 \text{ kg})(1 \text{ m s}^{-2})(1 \text{ m}) = 1 \text{ kg m}^2 \text{ s}^{-2} = 1 \text{ J}$$

Thus, work is expressed in units of energy.

A familiar example of work is the vertical displacement of a block (Figure 10.3b). A block of mass $m$ is raised to a height $h$ against the gravitational acceleration $g$. The force that must be exerted is equal to $mg$ (Equation 1.2), and the work done on the block is

$$w = mgh \tag{10.2}$$

When the block is raised from $h = 0$ to some arbitrary height, $h$, its potential energy increases from 0 to $mgh$. An amount of work $mgh$ has been performed on a system, the block, thereby increasing its energy by an amount $mgh$.

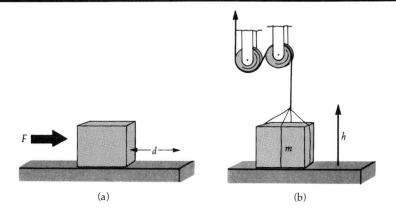

**FIGURE 10.3** Examples of mechanical work. (a) A force $F$ displaces an object by a distance $d$ in the direction in which it is applied. (b) A weight of mass $m$ is raised to a height $h$.

We shall adopt the convention that $w$ is the work done *on* the system. In other words, if $w > 0$, the surroundings have done work on the system and have thereby increased its energy. On the other hand, if $w < 0$, the system has done work on the surroundings, and the energy of the system has decreased by a corresponding amount.

***b. Work Associated with Expansion and Compression of a Gas***  The expansion or compression of a gas is one of the most common chemical processes that involve work. Suppose that a gas is confined to a cylindrical container, one end of which is a movable piston, as illustrated in Figure 10.4. Let the internal pressure of the gas be $P$ and the constant external pressure be $P_{ex}$. If $P = P_{ex}$ the system is at equilibrium and nothing happens. Suppose, though, that $P > P_{ex}$. The gas will then expand and push the piston outward until the internal and external pressures are once again equalized. If the piston is initially at distance $d_1$ and the expansion moves it to distance $d_2$, the work done by the gas against the external pressure is, according to Equation 10.1,

$$w = -F_{ex}(d_2 - d_1) \tag{10.3}$$

where $F_{ex}$ is the force associated with the external pressure. Note that, since $d_2 > d_1$, $w < 0$. This is consistent with the convention that $w < 0$ when the system does the work.

We know from Section 3.2 that force and pressure are related by the equation

$$P_{ex} = F_{ex}/A$$

where $A$ is the cross-sectional area of the cylinder. Substituting into Equation 10.3, we obtain

$$w = -P_{ex}A(d_2 - d_1) \tag{10.4}$$

Since the product of the cross-sectional area and the height of the cylinder is equal to the gas volume ($Ad_2 = V_2$, $Ad_1 = V_1$), Equation 10.4 can be rewritten as

$$w = -P_{ex}(V_2 - V_1) = -P_{ex}\,\Delta V < 0 \tag{10.5}$$

where $\Delta V$ is the change in volume.

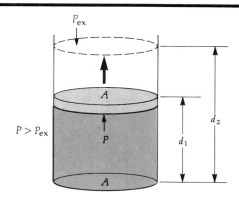

**FIGURE 10.4**  A gas confined to a cylinder of cross-sectional area $A$ expands against a constant external pressure $P_{ex}$. The piston moves from $d_1$ to $d_2$.

Consider, now, the reverse situation, in which the external pressure is larger than the gas pressure, $P_{ex} > P$. Under these conditions, the gas in the cylinder is compressed and the surroundings perform work on the system. If the initial volume is designated $V_1$ and the final volume is designated $V_2$, where $V_2 < V_1$, this process can be represented by Equation 10.5 written in the form

$$w = -P_{ex}(V_2 - V_1) = -P_{ex} \Delta V > 0 \tag{10.6}$$

The two expressions differ only in the sign of $w$. It is well to remember that in an expansion, $\Delta V > 0$, the gas does work on the surroundings, and $w < 0$. On the other hand, in a compression, $\Delta V < 0$, the surroundings do work on the gas, and $w > 0$.

**Example 10.1** An ideal gas maintained at 25°C expands against a constant external pressure of 1.0 atm from an initial volume of 1.0 L to a final volume of 10.0 L.
(a) What is the work done by the system? (b) Assuming that at the end of this process the system is in equilibrium with its surroundings, what was the initial gas pressure?

(a) We use Equation 10.5, noting that in an expansion the system does the work and $w < 0$:

$$w = -P_{ex}\Delta V = -(1.0 \text{ atm})(10.0 \text{ L} - 1.0 \text{ L}) = -9.0 \text{ L atm}$$
$$= -(9.0 \text{ L atm})[101.32 \text{ J (L atm)}^{-1}]$$
$$= -9.1 \times 10^2 \text{ J}$$

(b) If the system is at equilibrium, the internal pressure must be equal to the external pressure. Therefore the final gas pressure $P_2$ is

$$P_2 = 1.0 \text{ atm}$$

Since the temperature remains constant, Boyle's law (Equation 3.6) is applicable and the initial gas pressure $P_1$ is given by

$$P_1 = \frac{P_2 V_2}{V_1} = \frac{(1.0 \text{ atm})(10.0 \text{ L})}{1.0 \text{ L}} = 10 \text{ atm} \quad \blacksquare$$

**c. Expansion Against a Variable External Pressure** The definition of work expressed in Equation 10.5 is of limited usefulness since it is only valid when the external pressure is constant. While this is a common situation in practical applications, thermodynamics frequently deals with expansions or compressions in which the external pressure changes during the process in question. The most important illustration of a process involving variable external pressure is an expansion or compression in which a system and its surroundings are nearly in balance and the external pressure is just marginally different from the variable gas pressure (see Section 10.2d). Let us generalize Equation 10.5 to make it applicable to the case of variable external pressure.

We begin by noting in Figure 10.5a, which is a plot of the external pressure against the gas volume, that the work done by the gas in an expansion against a constant external pressure is equal to the negative of the area of the rectangle defined by $P_{ex}$ and the change in volume, $w = -P_{ex}(V_2 - V_1)$. When

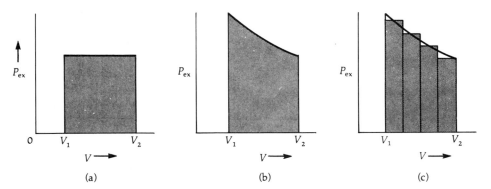

**FIGURE 10.5** Work as the area in a plot of $P_{ex}$ versus $V$. (a) Constant external pressure; (b) $P_{ex} \propto V^{-1}$; (c) approximation to the area in (b) by means of a set of rectangles. The work done by the gas is the negative of the above areas.

the external pressure varies with the volume, the work done is also equal to the negative of the area under the curve of $P_{ex}$ versus $V$. The resulting figure, of course, is no longer a rectangle but has a shape determined by the functional relationship between $P_{ex}$ and $V$. For example, Figure 10.5b shows a plot of $P_{ex}$ versus $V$ in which $P_{ex} \propto V^{-1}$.

The determination of the work performed in an expansion against a changing pressure reduces to calculating the area under a curve such as that depicted in Figure 10.5b. One approach to this problem is to approximate the curve by a number of horizontal segments which can be used to construct rectangles of areas that can easily be evaluated, as shown in Figure 10.5c. The sum of the areas of the individual rectangles is approximately equal to the area under the curve. Clearly, the narrower the rectangles, the more closely their total area agrees with that under the curve.

A more elegant approach to this problem, and one that yields an exact solution instead of an approximate solution, is to apply the methods of integral calculus. The area under a curve is equal to the integral over the independent variable (in this case, $V$). The resulting expression for the work done by the gas is

$$w = -\int_{V_1}^{V_2} P_{ex}\, dV \tag{10.7}$$

where $V_1$ and $V_2$ are the initial and final volumes, respectively, and $dV$ is the differential of $V$, that is, an infinitesimally small increment in volume. The integration is equivalent to a summation of the areas of an infinitely large number of rectangles of sides $P_{ex}$ and $dV$.

In order to perform the integration it is necessary to specify the functional dependence of $P_{ex}$ on $V$. The simplest case is the one already considered, namely that of constant $P_{ex}$. If $P_{ex}$ is independent of $V$, it can be moved outside the integral yielding

$$w = -P_{ex} \int_{V_1}^{V_2} dV = -P_{ex}(V_2 - V_1)$$

(see Appendix 2, Section A2.6). Thus, Equation 10.7 properly reduces to Equation 10.5 for a constant external pressure. In order to evaluate Equation 10.7 for the general case, we must specify how $P_{ex}$ varies with $V$.

**d. Work in Reversible Processes** A **reversible** process is one that can be reversed by an infinitesimal change in the value of an appropriate variable. This definition can be applied readily to the expansion or compression of a gas. The variable whose value must be changed is the external pressure, for its value relative to the gas pressure determines whether the gas undergoes expansion or compression. As noted in Section 10.2b, if $P_{ex} < P$, the gas expands whereas if $P_{ex} > P$, the gas is compressed. It is clear that in order to turn an expansion into a compression, or vice versa, by an infinitesimal change in external pressure, the external pressure can differ from the gas pressure by no more than an infinitesimally small amount $dP$. (An infinitesimal difference between two values of a variable can be expressed as the differential of that variable.) In a reversible expansion $P$ and $P_{ex}$ are thus related by the expression

$$P = P_{ex} + dP \tag{10.8}$$

while in a reversible compression the relationship is

$$P = P_{ex} - dP \tag{10.9}$$

Moreover, these relationships hold during the entire reversible process. Since the gas pressure will change from, say, $P_1$ to $P_2$, the external pressure must change accordingly to maintain this relationship.

A reversible process may be contrasted with an **irreversible** process, one that cannot be reversed by an infinitesimal change in the value of some variable. For example, an irreversible expansion occurs when the external pressure is measurably smaller than the gas pressure, for in this case an infinitesimal increase in the external pressure cannot reverse the expansion. The distinction between these fundamentally different types of processes is illustrated in Figure 10.6.

Reversible processes are closely connected with equilibrium states. A process in which a system remains in equilibrium with its surroundings is necessarily reversible. The special role of equilibrium in chemical processes has been amply illustrated in Chapters 7–9 and is one of the reasons for the importance of reversible processes in thermodynamics. When a confined gas is in true equilibrium with its surroundings, the gas pressure is equal to the external pressure. In an expansion the two pressures are related by Equation 10.8 while in a compression, Equation 10.9 applies. The infinitesimally small difference between $P_{ex}$ and $P$ in a reversible process can be neglected, and $P$ can be substituted for $P_{ex}$ in Equation 10.7 for the work:

$$w = -\int_{V_1}^{V_2} P_{ex}\, dV = -\int_{V_1}^{V_2} P\, dV \tag{10.10}$$

The usefulness of this substitution lies in the fact that the pressure and volume of a gas are related by means of the equation of state which describes the gas behavior (Section 3.4). As we shall see, this makes it possible to evaluate the integral in Equation 10.10 under certain conditions.

The simplest equation of state is that for an ideal gas,

$PV = nRT$

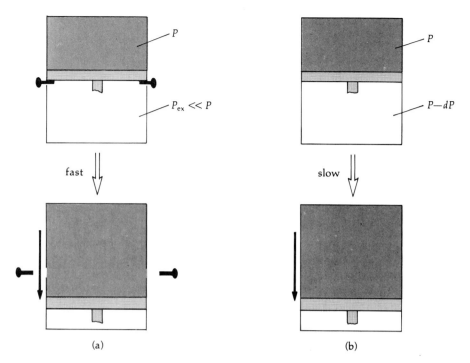

**FIGURE 10.6** The expansion of a gas. (a) The gas expands irreversibly when the retaining pins are removed. (b) Reversible expansion against an external pressure infinitesimally smaller than the gas pressure.

Substituting for $P$ in Equation 10.10, we obtain

$$w = -\int_{V_1}^{V_2} \frac{nRT\,dV}{V} \tag{10.11}$$

This integral cannot as yet be evaluated because the temperature of the gas can vary in some unspecified way as the volume changes. The situation is remedied by requiring $T$ to be constant, i.e., restricting the equation to an isothermal process. This can be achieved in practice by immersing the gas container in a large water bath maintained at constant temperature. The constant quantities can then be moved outside the integral, and the expression becomes

$$w = -nRT \int_{V_1}^{V_2} \frac{dV}{V} \tag{10.12}$$

The integral is equal to $\ln(V_2/V_1)$ and the work performed by an ideal gas in a *reversible isothermal expansion* from initial volume $V_1$ to final volume $V_2$ is

$$w = -nRT \ln(V_2/V_1) \tag{10.13}$$

Note how carefully we have to specify the conditions for which this equation is applicable. This specificity is quite typical of thermodynamics and indicates that care must be exercised in using equations only in those situations where they are applicable.

Further insight into the difference between the work done in a reversible and an irreversible expansion can be obtained from a graphical comparison of

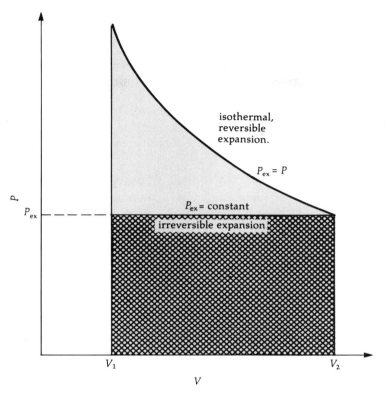

**FIGURE 10.7** The area beneath the isotherm drawn for a temperature $T$ corresponds to the negative of the work done on the surroundings in a reversible isothermal expansion of an ideal gas. The cross-hatched rectangular area corresponds to the negative of the work done against constant external pressure. Observe that the reversible work is greater in magnitude than the irreversible work.

the two. The work done by the system in an expansion is given by the negative of the area under a curve of $P_{ex}$ versus $V$, as already noted in Figure 10.5. For a reversible isothermal expansion of an ideal gas the area in question is that beneath the isotherm connecting the initial and final values of $P$ and $V$, as depicted in Figure 10.7. For comparison, consider the expansion of the gas between the same initial and final volumes against a constant external pressure equal to the final external pressure of the reversible expansion. This gives the rectangular area already depicted in Figure 10.5a.

Figure 10.7 shows that the work performed in a reversible expansion is greater than that performed in an irreversible expansion between the same initial and final states. In fact, the work in a reversible expansion is the *maximum work* that can be performed, for the opposing external pressure has the largest possible value. If the external pressure were any larger, the expansion of the gas would turn into a compression and the surroundings would do work on the gas. This is one of the reasons why reversible processes are so important—they give an indication of the maximum work that is obtainable for a particular set of conditions. To be sure, a truly reversible process is not achievable. The infinitesimally small changes in pressure or other variables that characterize re-

**Example 10.2** An ideal gas at 25°C undergoes a reversible isothermal expansion. The initial gas pressure is 10.0 atm and the final pressure is 1.0 atm while the final volume is 10.0 L. (a) What is the work done by the system? (b) Compare the answer to that obtained in an expansion between the same initial and final states against a constant external pressure of 1.0 atm.

(a) Since the expansion is reversible and isothermal and the gas is ideal, Equation 10.13 applies. We are given $T$ and $V_2$ but must determine $n$ and $V_1$ before Equation 10.13 can be used. These quantities can be found by means of the gas laws. Since $T$ is constant, Boyle's law applies, and so $V_1$ can be determined:

$$V_1 = \frac{P_2 V_2}{P_1} = \frac{(1.0 \text{ atm})(10 \text{ L})}{10.0 \text{ atm}} = 1.0 \text{ L}$$

The number of moles of gas is given by the ideal gas law:

$$n = \frac{PV}{RT} = \frac{(1.0 \text{ atm})(10 \text{ L})}{(0.08206 \text{ L atm mol}^{-1} \text{ K}^{-1})(298 \text{ K})} = 0.409 \text{ mol}$$

Applying Equation 10.13 we obtain

$$w = -nRT \ln(V_2/V_1)$$

$$= -(0.409 \text{ mol})(8.314 \text{ J mol}^{-1} \text{ K}^{-1})(298 \text{ K}) \ln\left(\frac{10 \text{ L}}{1.0 \text{ L}}\right)$$

$$= -2.3 \times 10^3 \text{ J}$$

(b) This second part of the problem is identical to Example 10.1. Note, in particular, that the initial and final pressures, volumes, and temperatures are the same in the two problems, confirming that the initial and final states are indeed the same. The answer obtained in Example 10.1 was

$$w = -9.1 \times 10^2 \text{ J}$$

showing that reversible work is larger in magnitude than irreversible work for processes between the same initial and final states of a system. ■

## 10.3 Heat

***a. The Mechanical Equivalent of Heat*** Like work, heat is a flow of energy between a system and its surroundings. This concept of heat is of relatively recent origin. Because heat appears to flow from a hot body to a colder one, it was once believed that heat was some sort of a weightless fluid surrounding the atoms of substances. It was thought that this fluid drained away from a substance whenever heat was released. This view was formalized in the late eighteenth century in what is historically known as the "caloric theory of heat" by no less a scientist than Lavoisier. The sole legacy of this theory is the term *calorie*, which is used for the common unit of heat. A calorie is the heat needed

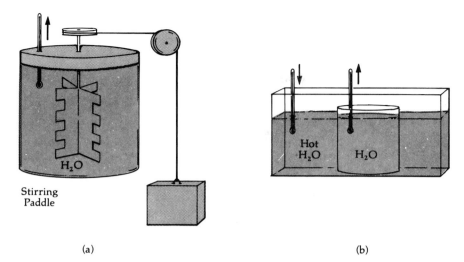

**FIGURE 10.8** Determination of the mechanical equivalent of heat. (a) The work done on the water by the paddle wheel can be calculated from the mass of the weight and the distance it drops (Equation 10.2). The thermal energy of the water increases, as evidenced by the increase in temperature. (b) The same increase in temperature can be achieved by a flow of heat from a hotter body. The mechanical equivalent of heat is the ratio of the amounts of work and heat needed to produce the same increase in temperature.

to raise the temperature of 1 g of water at 15°C by 1°C. Even this unit is being supplanted by the SI unit of energy, the joule.

The modern view of heat stems from a classical experiment by James Joule (1818–1889) on the "mechanical equivalent of heat" (Figure 10.8). Joule measured the increase in the temperature of the water in a container resulting from two different processes, one involving work and the other heat. In the first process (Figure 10.8a), the water is stirred by a paddle driven by a falling weight. The work done by the paddle on the water, which can be computed by Equation 10.2, increases the thermal energy of the water—the temperature increases by a certain number of degrees. In the second process (Figure 10.8b) heat flows from a hot body to the water and, once again, the temperature of the water increases. Joule found that the ratio of the work flow to the heat flow needed to produce a given increase in the water temperature was a constant. This constancy showed that work and heat were in a sense equivalent quantities since they had exactly the same effect on the temperature of a system.

The mechanical equivalent of heat is the ratio of equivalent work and heat flows. The presently accepted value of this ratio is

$$1 \text{ calorie} = 4.184 \text{ J}$$

This value can now be viewed as just another conversion factor between different units, but at the time it was first determined it had the added significance of relating two previously unconnected quantities.

***b. Heat Capacity*** The increase in the temperature of a body resulting from a given flow of heat depends on two factors: the mass of the body and its composition. For a given substance, the rise in temperature is inversely proportional to the mass. For example, if a certain flow of heat causes the temperature of 1 mole of water to increase by 2°C, the same flow of heat will cause the temperature of 2 moles of water to increase by only 1°C. However, for a given amount of a substance, the increase in temperature, $\Delta T$, resulting from a certain flow of heat, $q$, varies from one substance to another.

The proportionality constant between $q$ and $\Delta T$ represents an intrinsic property of a substance called the heat capacity. Because of the dependence of $\Delta T$ on the amount of substance heated, it is necessary to specify a fixed amount of substance. The heat capacity of 1 mole of a substance is called the **molar heat capacity**, designated by $C$. The molar heat capacity is related to $q$ and $\Delta T$ by the expression

$$q = nC\,\Delta T \tag{10.14}$$

where $n$ is the number of moles. The units of $C$ are $J\,mol^{-1}\,K^{-1}$.

Equation 10.14 implicitly assumes that the heat capacity of a substance is independent of its temperature. This assumption is generally not valid. The heat capacity is therefore more properly defined in terms of the infinitesimal heat flow, $dq$, which will increase the temperature of a substance by an infinitesimal amount, $dT$. For one mole of a substance at a specified temperature,

$$dq = C\,dT$$
$$C = dq/dT \tag{10.15}$$

Table 10.1 lists some typical molar heat capacities evaluated at 25°C and atmospheric pressure. A more complete list may be found in Appendix 4. The chief regularity which may be noted is that the molar heat capacities of many

**TABLE 10.1  Molar Heat Capacities at 25°C and Atmospheric Pressure**

| Substance | $C_p$ (J mol$^{-1}$ K$^{-1}$) |
|---|---|
| Al(s) | 24.4 |
| Ar(g) | 20.8 |
| HBr(g) | 29.1 |
| CO$_2$(g) | 37.1 |
| Cu(s) | 24.4 |
| Fe(s) | 25.1 |
| Fe$_2$O$_3$(s) | 103.9 |
| PbO(s) | 45.8 |
| MgCl$_2$(s) | 71.4 |
| Hg(l) | 28.0 |
| Ne(g) | 20.8 |
| N$_2$(g) | 29.1 |
| O$_2$(g) | 29.4 |
| H$_2$O(l) | 75.3 |
| K(s) | 29.6 |
| KF(s) | 49.0 |
| K$_2$CrO$_4$(s) | 146.0 |

solids at 25°C are close to 25 J mol$^{-1}$ K$^{-1}$ per mole of atoms in a formula unit. For example, the molar heat capacity of potassium metal is 29.6 J mol$^{-1}$ K$^{-1}$, that of KF(s) is 49.0 J mol$^{-1}$ K$^{-1}$, and that of K$_2$CrO$_4$(s), with seven atoms per formula unit, is 146.0 J mol$^{-1}$ K$^{-1}$. This regularity is known as the **law of Dulong and Petit** and is considered further in Section 20.2.

As in the case of work, we adopt the convention that $q$ is the heat flowing to the system. If $q > 0$, heat flows from the surroundings to the system and the energy of the system increases; if $q < 0$, heat flows from the system to the surroundings and the energy of the system decreases.

## 10.4 The First Law of Thermodynamics

***a. Statement of the Law*** The **first law of thermodynamics** states that the energy of a system and its surroundings is constant. This law cannot be derived from other principles. It is a postulate based on the universal human experience that energy can neither be created nor destroyed. This concept is so deeply ingrained in our view of the nature of physical reality that any observed violation would be viewed with the deepest suspicion. Indeed, it would be preferable to invent a new form of energy to account for an apparent breakdown of energy conservation than to discard this long held tenet. If nuclear energy had been discovered before Einstein had shown that mass could be converted to energy, scientists might well have postulated that such a transformation was possible rather than accept the fact that energy had been created from nothing.

We have seen that the transfer of energy between a system and its surroundings involves the flow of work and heat. The first law can be formulated in terms of the relation between the **internal energy** of the system, $E$, that is, the energy within the system, and the heat and work flowing to the system from the surroundings. A mathematical statement of the first law is that the change in the internal energy of a system, $\Delta E$, is equal to the sum of the work done on the system, $w$, and the heat flowing into the system, $q$,

$$\Delta E = E_2 - E_1 = q + w \tag{10.16}$$

where $E_2$ is the internal energy of the system in its final state, following the flow of heat and work, and $E_1$ is the internal energy of the system in its initial state. Recall from Sections 10.2 and 10.3 that when energy is transferred to a system from its surroundings, $q + w$ is a positive quantity. Consequently, $\Delta E$ is also positive. The opposite holds true when energy is transferred from the system to the surroundings: $q + w$ is negative, in which case $\Delta E$ is negative. When the system is isolated, $\Delta E$ is zero and so is $q + w$. In a certain sense, the system is analogous to a savings account. The savings correspond to the internal energy. The savings may be increased by deposits and decreased by withdrawals, just as the internal energy can be increased or decreased by the inward or outward flow of work and heat, respectively.

***b. State Functions*** An important distinction in thermodynamics is that between a state function and a path function. A **state function** is a property that depends only on the initial and final states of a system and is independent of the path followed in getting from the one to the other. On the other hand, a **path function** depends on the particular path followed in getting from the initial state to the final state.

**346** Energy in Chemical Processes—The First Law of Thermodynamics  10.4

The following illustration clarifies the meaning of these two concepts. Suppose we wish to travel from Chicago to Miami. The map sketched in Figure 10.9 shows that Chicago is at a latitude of 42° N and a longitude of 88° W, while Miami is at a latitude of 26° N and a longitude of 80° W. The 16° difference in latitude and the 8° difference in longitude are independent of the path we choose to follow in our trip. Clearly, latitude and longitude are state functions. On the other hand, the actual distance traversed depends on the path followed. A direct flight between Chicago and Miami would cover some 1200 miles. A car trip between these two cities might cover a much longer distance,

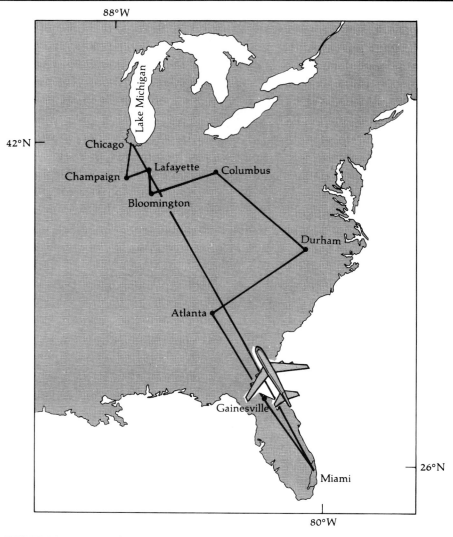

**FIGURE 10.9** A trip between Chicago and Miami, illustrating the difference between a state function and a path function. The latitude and longitude differences are independent of the route followed, indicating that they are state functions. The distance traversed depends on the route followed and is a path function. A direct and a rambling route are shown.

depending on the exact route. The distance travelled is obviously a path function rather than a state function.

In general, properties of a system such as temperature, pressure, volume, and internal energy are state functions. On the other hand, work and heat depend on the path followed and so are path functions, instead.

It is easy to demonstrate that work depends on the specific path followed between initial and final states. Consider the isothermal expansion of a gas from an initial state of pressure $P_a$ and volume $V_a$ to one where these quantities have values $P_b$ and $V_b$. Let us evaluate the work performed in this expansion via the following two different paths, illustrated in Figure 10.10:

Path 1:  The gas expands from $V_a$ to $V_b$ against a constant external pressure $P_a$. When the expansion is over, the external pressure is decreased from $P_a$ to $P_b$ with no change in volume. The work done by the gas is

$$w = -P_a(V_b - V_a)$$

Path 2:  The external pressure is decreased from $P_a$ to $P_b$ while the volume of the gas is maintained at $V_a$. The gas is then expanded from $V_a$ to $V_b$ against constant external pressure $P_b$. The work done by the gas is

$$w = -P_b(V_b - V_a)$$

Clearly, since $P_a > P_b$, more work is performed when the first path is followed.

The amount of heat transferred in a process is also dependent on the particular path followed. This can be readily demonstrated for the two expansions just illustrated. Designating path 1 and path 2 by numerical subscripts and applying the first law to these two processes we obtain

$$\Delta E_1 = q_1 + w_1 \qquad \Delta E_2 = q_2 + w_2$$

Since the initial and final states are identical in the two cases, we have

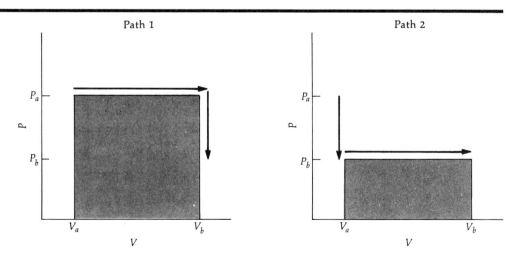

**FIGURE 10.10** Work done by a gas in two different expansions between the same initial and final states. The colored area shows the negative of the work performed in the two cases. The paths are shown by the arrows and are described in the text.

$$\Delta E_1 = \Delta E_2$$

because $E$ is a state function. Therefore,

$$q_1 + w_1 = q_2 + w_2$$

But we have shown that $w_1 \neq w_2$. It follows that $q_1 \neq q_2$, indicating that $q$ depends on the path followed.

We have stated that internal energy is a state function. How do we know that this is so? Examination of a **cyclic process,** that is, a process for which the final state is identical to the inital state, demonstrates the validity of this assertion. Consider the combustion of hydrogen at 25°C to yield water, followed by the electrolysis of water (Section 13.10a) to recover hydrogen and oxygen gas. If, as shown in Figure 10.11, the combustion is designated process 1 and the electrolysis, process 2, we can write for the changes in the internal energy

$$\Delta E_1 = E_B - E_A$$

and

$$\Delta E_2 = E_A - E_B$$

where $E_A$ and $E_B$ represent the internal energy of the hydrogen and oxygen gas and that of the water, respectively. The overall process is obviously cyclic since the end products are hydrogen and oxygen in the same state in which they were present initially. The internal energy change for the overall process is

$$\Delta E = \Delta E_1 + \Delta E_2 = E_B - E_A + E_A - E_B = 0$$

In other words, *in a cyclic process the internal energy of a system does not change.* If this were not the case, we could use a cyclic process to create energy while leaving the system and its surroundings unchanged. Such a process is the basis of many of the so-called perpetual motion machines of the first kind, which are cleverly conceived devices designed to circumvent the law of energy conservation. Unfortunately, but not surprisingly, no one has yet succeeded in building such a device that works.

The difference between a state function and a path function can also be seen by comparing the first law as written for a finite energy change (Equation 10.16) and for an infinitesimal energy change,

$$dE = dq + dw \tag{10.17}$$

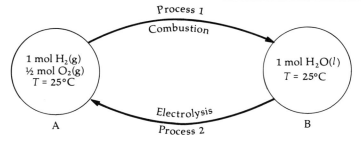

**FIGURE 10.11** A cyclic process: $H_2(g)$ and $O_2(g)$ (state A) are combined to yield water (state B), which upon electrolysis forms $H_2(g)$ and $O_2(g)$. As shown in Section 10.4b, $\Delta E = 0$ for such a process.

An infinitesimal change in a state function gives a finite difference on integration,

$$\int_{State\ 1}^{State\ 2} dE = E_2 - E_1 = \Delta E \tag{10.18}$$

Similar relations can be written for other properties of a system, such as $T$, $P$, and $V$. On the other hand, an infinitesimal change in a path function gives a total quantity rather than a finite difference on integration, for example,

$$\int_{State\ 1}^{State\ 2} dq = q \tag{10.19}$$

Since $q$ and $w$ are *not* properties of the system, it is nonsensical to talk about changes in the heat or work of a system; the symbols $\Delta q$ and $\Delta w$ have no meaning. Differentials such as $dq$ and $dw$ are known as "inexact" differentials.

The first law involves a quantity that is a state function ($\Delta E$) and two quantities ($q$ and $w$) that are not state functions. While $\Delta E$ is independent of path, the manner in which it is apportioned between $q$ and $w$ depends on the particular path.

---

**Example 10.3** One mole of an ideal gas initially occupies a volume of 2.00 L at a pressure of 10.0 atm. The gas is expanded against a constant external pressure of 1.00 atm until its volume is 20.0 L, at which point it is in equilibrium with its surroundings. The gas is then compressed reversibly and isothermally back to a volume of 2.00 L. Calculate $\Delta E$, $q$, and $w$ for this overall process.

The process consists of an expansion followed by a compression. The process appears to be cyclic since the final volume is the same as the initial one. Before we can conclude that this is indeed the case we have to check whether the initial and final pressure and temperature are also equal, since the two states must be identical in *all* respects for the process to be cyclic.

We are given

$P_1 = 10.0$ atm    $V_1 = 2.00$ L

The initial temperature can be obtained by means of the ideal gas law:

$$T_1 = \frac{P_1 V_1}{nR} = \frac{(10.0\ \text{atm})(2.00\ \text{L})}{(1.00\ \text{mol})(0.08206\ \text{L atm mol}^{-1}\ \text{K}^{-1})} = 243.7\ \text{K}$$

After the expansion, the gas is in equilibrium with the surroundings. Therefore, the internal and external pressures are equal. Denoting this state with subscript 2 we have

$P_2 = 1.00$ atm    $V_2 = 20.0$ L

We note that

$$P_2 V_2 = P_1 V_1 = 20.0\ \text{L atm}$$

Since Boyle's law is obeyed, the expansion is isothermal so that

$$T_2 = T_1 = 243.7\ \text{K}$$

The compression is stated to be isothermal. Therefore, since the final volume is 2.00 L, identical to the initial volume, the final gas pressure must also be equal to the initial value, i.e., 10.0 atm, in order for Boyle's law to be satisfied. Since the process is indeed cyclic, we immediately conclude that $\Delta E = 0$.

The work done on the system can be obtained by separate calculations for the expansion and compression steps. In the expansion, we have according to Equation 10.5:

$$w = -P_{ex}(V_2 - V_1)$$
$$= -(1.00 \text{ atm})(20.0 \text{ L} - 2.00 \text{ L})[101.32 \text{ J (L atm)}^{-1}]$$
$$= -1.824 \times 10^3 \text{ J}$$

The work done on the gas in the compression is evaluated for a reversible isothermal process according to Equation 10.13:

$$w = -nRT \ln(V_1/V_2)$$
$$= -(1 \text{ mol})(8.314 \text{ J mol}^{-1} \text{ K}^{-1})(243.7 \text{ K}) \ln(2.00 \text{ L}/20.0 \text{ L})$$
$$= 4.665 \times 10^3 \text{ J}$$

Note that the final volume in the compression step, $V_1$, is identical to the initial volume in the expansion, while the initial volume in the compression, $V_2$, is identical to the final volume in the expansion, which is consistent with the cyclic nature of the process. As required by the convention that $w > 0$ when work is done on the system and that $w < 0$ when work is done by the system, we note that $w < 0$ in the expansion and $w > 0$ in the compression.

The overall work done on the system is the sum of that for the individual steps:

$$w = -1.824 \times 10^3 \text{ J} + 4.665 \times 10^3 \text{ J} = 2.84 \times 10^3 \text{ J}$$

Knowing that $\Delta E = 0$, $q$ can be obtained by means of the first law, Equation 10.16:

$$\Delta E = q + w = 0$$
$$q = -w = -2.84 \times 10^3 \text{ J}$$

The final answers are

$$\Delta E = 0 \qquad q = -2.84 \times 10^3 \text{ J} \qquad w = 2.84 \times 10^3 \text{ J}$$

Note that in a cyclic process the work done on (by) the system is necessarily equal to the heat flowing from (to) the system. From the fact that $w > 0$ and $q < 0$ we conclude that in this particular process work flows to the system and heat flows to the surroundings. Figure 10.12, which shows a sketch of this process on a P–V plot, may be of value in visualizing this example. ∎

## 10.5 Constant Volume and Constant Pressure Processes

Two of the most common types of chemical processes are those performed under conditions in which either the volume or the pressure of the system remains constant. In this section we apply the first law to these two types of processes.

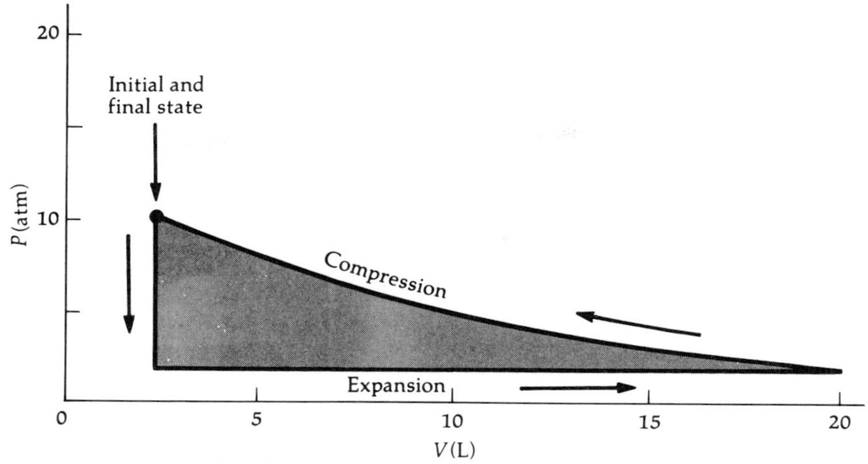

**FIGURE 10.12** The process described in Example 10.3, sketched on a P–V plot. The arrows show the direction of changes in the pressure and volume of the gas. The net work done on the gas is shown by the colored region.

*a. Constant Volume Processes* Chemical reactions involving gases are commonly carried out in closed containers of constant volume because it is necessary to confine these substances. Figure 10.13 pictures a bomb calorimeter, a constant volume device used to measure the heat released in reactions involving gases, such as combustion. It is easily shown that the change in the internal energy of a system is equal to the heat absorbed in a constant volume process. We write the first law,

$$\Delta E = q + w$$

and recall that the work done by the system is, according to Equation 10.5,

$$w = -P_{ex} \Delta V$$

Since $\Delta V = 0$ for a constant volume process, it necessarily follows that $w = 0$. Therefore,

$$\Delta E = q_v \tag{10.20}$$

where the subscript on $q$ emphasizes that this equation is valid only for a process in which the volume does not change.

Equation 10.20 is significant because it reveals that in a constant volume process $q$ becomes equal to $\Delta E$. Therefore, $\Delta E$ values can be determined experimentally by measurement of the heat released or absorbed in constant volume processes. Note that when heat is absorbed by the system, $q_v > 0$. If the process in question is a chemical reaction, the reaction is endothermic. The internal energy of the products must therefore be greater than that of the reactants. The opposite holds true when heat is released by the system, $q_v < 0$, and the reaction is exothermic.

Recall from Section 10.3b that the heat capacity of a substance is a measure of the increase in temperature resulting from a given heat flow. The molar heat capacity was defined by Equation 10.15 as $C = dq/dT$. We have seen that heat is a path function, meaning that $dq$ depends on the path. The heat capacity must consequently also depend on the path. However, for a constant vol-

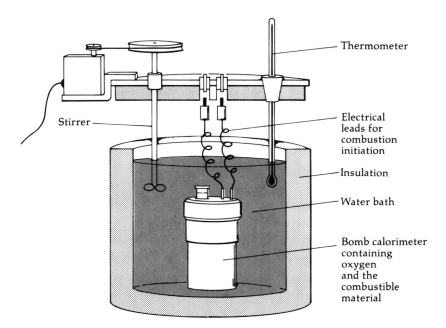

**FIGURE 10.13** A bomb calorimeter for measurement of heats of combustion. The combustible substance and oxygen are introduced into the constant-volume container. The reaction is initiated by a spark caused by an electrical discharge through the oxygen. The energy released is determined from the increase in the temperature of the surrounding water bath.

ume process, Equation 10.20 shows that the heat flow depends only on the initial and final states and is therefore a property of the system. Thus, the heat capacity also becomes a property of the system under these conditions. This property, the heat capacity at constant volume, $C_v$, is defined by the relation

$$C_v = \frac{dq_v}{dT} = \frac{dE}{dT} \quad \text{(constant } V\text{)} \tag{10.21}$$

where the second equality follows from Equation 10.20.

**b. Constant Pressure Processes** Most chemical reactions are run at constant pressure (also known as **isobaric** conditions) rather than at constant volume. A reaction that takes place in solution, for instance, might typically be run in a beaker open to the constant pressure of the atmosphere. Under these conditions the heat flow is not equal to the change in internal energy because there can also be a work flow between a system and its surroundings. Since the quantity of heat flowing in a process is readily measurable, it would be of value to define a property of the system that is equal to the heat flow at constant pressure. Such a function would play an analogous role in isobaric processes as the internal energy does in constant volume processes.

The function that has these properties is the enthalpy $H$. The **enthalpy** of a system is defined by the relation

$$H = E + PV \tag{10.22}$$

Since the enthalpy is the sum of terms that are state functions, it, too, is a state function. In a process that results in a change of a system from state 1 to state 2, the enthalpy change $\Delta H$ is consequently

$$\Delta H = H_2 - H_1 \tag{10.23}$$

We can show that the change in the enthalpy of a system is indeed equal to the heat flow at constant pressure. Applying Equation 10.22 to a process in which the enthalpy changes by an amount $\Delta H$ we obtain

$$\Delta H = \Delta E + \Delta(PV) \tag{10.24}$$
$$= q + w + \Delta(PV)$$
$$\Delta H = q + w + P_{final} V_{final} - P_{initial} V_{initial} \tag{10.25}$$

If we now restrict ourselves to an isobaric process, we have

$$P_{final} = P_{initial} = P_{ex} = P$$

and Equation 10.25 becomes

$$\Delta H = q + w + P \, \Delta V \tag{10.26}$$

where $\Delta V$ is the change in volume. When the pressure is constant, the work done on the system is, according to Equation 10.5,

$$w = -P \, \Delta V$$

Substituting this result into Equation 10.26 we obtain

$$\Delta H = q - P \, \Delta V + P \, \Delta V$$

which reduces to

$$\Delta H = q_P \tag{10.27}$$

where the subscript on $q$ indicates explicitly that the equality holds only at constant pressure.

Because $\Delta H$ is equal to the heat flowing to a system in a process carried out under the most commonly encountered experimental conditions, it is a very useful quantity. Note that since $q > 0$ means that heat flows to the system, $\Delta H$ is positive for endothermic processes and negative for exothermic ones. A calorimeter for the determination of $\Delta H$ values is shown in Figure 10.14.

Just as the molar heat capacity evaluated for a constant volume process is a property of a system, so, too, is the molar heat capacity for a constant pressure process, designated $C_P$. In analogy to Equation 10.21, $C_P$ can be defined by the equation

$$C_P = \frac{dq_P}{dT} = \frac{dH}{dT} \quad \text{(constant } P\text{)} \tag{10.28}$$

Because of the wider use of the enthalpy than of the internal energy, the tabulated values of the heat capacity (e.g., Table 10.1) are usually those of $C_P$.

**c. Relationship between Constant Pressure and Constant Volume Processes** It is instructive to compare the magnitude of the properties of a system that describe its thermal behavior in constant pressure processes with those that describe this behavior in constant volume processes, i.e., $\Delta H$ with $\Delta E$, and $C_P$ with $C_V$.

**FIGURE 10.14** A calorimeter for measurements at constant pressure. The reaction mixture in the central container is open to the atmosphere and is at atmospheric pressure. The enthalpy change in the reaction can be determined from the change in temperature of the water bath. The surrounding insulation minimizes the escape of heat before the measurement is completed.

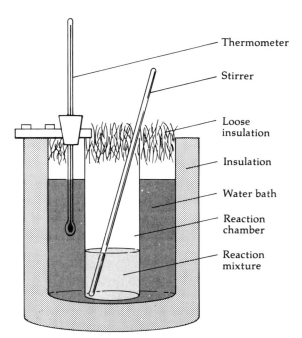

The internal energy and the enthalpy of a system are closely related quantities. When a system is heated at constant volume, all the heat is converted into internal energy. However, when a system is heated at constant pressure, it also expands. In an expansion, a system does work on its surroundings and some of its increased internal energy is thereby transferred to the surroundings. The difference between $\Delta H$ and $\Delta E$ consequently depends on the amount of expansion work that a system can do.

Since solids and liquids are rather incompressible (Section 3.2), chemical reactions, changes of phase, or other processes involving only solids or liquids involve an amount of work that is negligible compared to the change in internal energy. Under these conditions, $\Delta H \simeq \Delta E$ (for processes involving only solids or liquids).

**Example 10.4** The heat liberated when water freezes at 0°C and atmospheric pressure is 6.010 kJ mol$^{-1}$. Calculate $\Delta H$ and $\Delta E$ for the formation of one mole of ice under these conditions. The densities of ice and water at this temperature and pressure are 0.9168 g cm$^{-3}$ and 0.9998 g cm$^{-3}$, respectively.

Since this is a constant pressure process, $\Delta H = q_p$. Therefore,

$\Delta H = -6.010$ kJ mol$^{-1}$

In order to obtain $\Delta E$, we must evaluate the change in volume. The molar volumes of ice and water can be obtained from the respective densities:

$$V_{ice} = \frac{18.02 \text{ g mol}^{-1}}{0.9168 \text{ g cm}^{-3}} = 19.66 \text{ cm}^3 \text{ mol}^{-1}$$

$$V_{H_2O} = \frac{18.02 \text{ g mol}^{-1}}{0.9998 \text{ g cm}^{-3}} = 18.02 \text{ cm}^3 \text{ mol}^{-1}$$

Hence,
$$\Delta V = V_{ice} - V_{H_2O}$$
$$= 19.66 \text{ cm}^3 \text{ mol}^{-1} - 18.02 \text{ cm}^3 \text{ mol}^{-1}$$
$$= 1.64 \text{ cm}^3 \text{ mol}^{-1}$$

The quantity $P \Delta V$ expressed in joules is then equal to
$$P \Delta V = (1.00 \text{ atm})(1.64 \text{ cm}^3 \text{ mol}^{-1})(10^{-3} \text{ L cm}^{-3})[101.3 \text{ J (L atm)}^{-1}]$$
$$= 0.166 \text{ J mol}^{-1}$$

At constant pressure, the relationship between $\Delta E$ and $\Delta H$ expressed in Equation 10.24 can be written as
$$\Delta E = \Delta H - P \Delta V$$
$$= -6.010 \text{ kJ mol}^{-1} - (0.166 \text{ J mol}^{-1})(10^{-3} \text{ kJ J}^{-1})$$
$$= -6.010 \text{ kJ mol}^{-1}$$

In other words, $\Delta E$ and $\Delta H$ are equal to within the given number of significant figures. ∎

In contrast to processes involving only condensed phases, the difference between $\Delta H$ and $\Delta E$ can be substantial when gases are involved because of the large changes in volume that are possible. The difference can be evaluated by means of the ideal gas law. Since
$$PV = nRT$$
it follows that
$$\Delta(PV) = \Delta(nRT)$$
so that
$$\Delta H = \Delta E + \Delta(nRT) \tag{10.29}$$

If we further assume an isothermal process, the only change in the quantity $nRT$ is that in $n$, so that Equation 10.29 becomes
$$\Delta H = \Delta E + RT \Delta n \quad \text{(constant } T) \tag{10.30}$$

Thus, the difference between the enthalpy and internal energy changes depends on the change in the number of moles of gaseous substances.

---

**Example 10.5** The heat liberated when one mole of ammonia is formed from the elements at 25°C in a calorimeter fitted with a piston which can move against the constant pressure of the atmosphere is 46.11 kJ. Determine $\Delta H$ and $\Delta E$. What is the work done in this process?

Since the reaction is performed at constant pressure,
$$\Delta H = q_p = -46.11 \text{ kJ mol}^{-1}$$

The internal energy change can be obtained by means of Equation 10.30 on the assumption that the gases behave ideally,

$$\Delta H = \Delta E + RT\Delta n$$

The value of $\Delta n$ is obtainable from the stoichiometry of the reaction for the formation of 1 mole of $NH_3$

$$\tfrac{1}{2}N_2(g) + \tfrac{3}{2}H_2(g) \rightarrow NH_3(g)$$

which shows that $\Delta n = -1$. Consequently,

$$\Delta E = -46.11 \text{ kJ mol}^{-1} - (8.314 \text{ J mol}^{-1} \text{ K}^{-1})(298 \text{ K})(-1 \text{ mol})(10^{-3} \text{ kJ J}^{-1})$$

$$= -43.63 \text{ kJ mol}^{-1}$$

In contrast to Example 10.4, the difference between $\Delta H$ and $\Delta E$ is significant, reflecting the change in the number of moles of gas and, hence, in gas pressure.

The work done in a constant pressure process is $-P_{ex}\Delta V$. For an isothermal process involving ideal gases, this quantity is equal to $-RT\Delta n$ and thus, according to Equation 10.30, to $\Delta E - \Delta H$. Therefore,

$$w = (-43.63 + 46.11) \text{ kJ} = 2.48 \text{ kJ}$$

Since $w > 0$, work is done on the system by the surrounding atmosphere, i.e., the gases are compressed. This result is consistent with the decrease in the number of moles of gas in the reaction, which at constant temperature and pressure necessarily results in a decrease in volume. ∎

When a certain amount of heat flows to a system at constant pressure, the temperature of the system rises less than when the same amount of heat flows in at constant volume because a system expands and does work on the surroundings when it is heated. For a given $q$, a smaller increase in temperature corresponds to a larger value of the heat capacity. Consequently, the value of $C_p$ of a substance is always larger than that of $C_v$.

The relationship between $C_p$ and $C_v$ can be derived from the definition of the enthalpy, Equation 10.22,

$$H = E + PV$$

The rate of change of the enthalpy with respect to temperature is

$$\frac{dH}{dT} = \frac{dE}{dT} + \frac{d(PV)}{dT}$$

Introducing $C_p$ and $C_v$ by means of Equations 10.28 and 10.21, respectively, we obtain

$$C_p = C_v + \frac{d(PV)}{dT} \tag{10.31}$$

As in the case of $\Delta H$ and $\Delta E$, the difference between $C_p$ and $C_v$ depends on the $PV$ term, in this instance on the magnitude of its rate of change with temperature. Since for solids and liquids $d(PV)/dT$ is generally small, it follows that

$$C_p \simeq C_v \quad \text{(for solids and liquids)} \tag{10.32}$$

The difference between $C_p$ and $C_v$ for gases is much larger. On the basis of the ideal gas approximation, Equation 10.31 can be rewritten as

$$C_p = C_v + \frac{d(RT)}{dT}$$

or

$$C_p - C_v = R\frac{dT}{dT} = R \quad \text{(for ideal gases)} \qquad (10.33)$$

Comparison of the value of $R$, 8.3 J mol$^{-1}$ K$^{-1}$, with the $C_p$ values listed for gases in Table 10.1, shows that the difference between $C_p$ and $C_v$ is a substantial fraction of the heat capacity.

Owing to the absence of intermolecular forces in an ideal gas, the internal energy of such a gas depends only on its temperature (Equation 4.17) and is independent of pressure or volume. Therefore, $\Delta E = 0$ for an isothermal process involving an ideal gas, such as an isothermal expansion. We can also conclude, on the basis of Equation 10.29, which relates the change in the enthalpy and internal energy of an ideal gas, that $\Delta H = 0$ for such a process.

## 10.6 Molecular Interpretation of the Thermodynamic Functions

Thermodynamics deals only with macroscopic processes and is therefore independent of the nature of the microscopic constituents of the systems of interest. The first law, for instance, says nothing about how the internal energy of a system is stored. The same results would be obtained if the constituents of matter were not atoms and molecules but some other entities instead. Nonetheless, we can make use of the vast body of knowledge that chemists and physicists have obtained about the structure of atoms and molecules to develop a more complete understanding of thermodynamics.

The internal energy of a system is determined by the energy of the atoms or molecules of which it is comprised. The relationship between internal energy and the energies of individual molecules is most direct for ideal gases since, because of the absence of intermolecular interactions, the energy of a particular molecule is independent of the energies of the other molecules present. Recall from Section 4.2b that the translational kinetic energy of an ideal gas is $\frac{3}{2}RT$ per mole. If the ideal gas is a monatomic element (e.g., argon), its translational energy normally is its total chemical energy—atoms possess no other forms of chemical energy at ordinary temperatures. This fact makes possible a direct comparison between thermodynamics and the molecular model. If the internal energy of a gas is $\frac{3}{2}RT$ per mole, then, according to Equation 10.21, the molar heat capacity at constant volume is

$$C_v = \frac{dE}{dT} = \frac{d[\frac{3}{2}RT]}{dT} = \frac{3}{2}R$$

Substitution into Equation 10.33 shows that the molar heat capacity at constant pressure has the value

$$C_p = C_v + R = \tfrac{5}{2}R = 20.79 \text{ J mol}^{-1} \text{ K}^{-1}$$

Comparison with the experimental values of $C_p$ for monatomic gases in Table 10.1 shows that both neon and argon have $C_p = 20.8$ J mol$^{-1}$ K$^{-1}$, in excellent agreement with the results of the molecular model. This agreement indicates

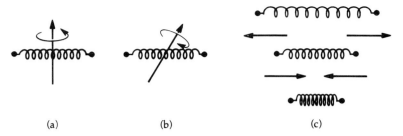

**FIGURE 10.15** (a and b) Rotation and (c) vibration of a diatomic molecule represented as a spring.

that a macroscopic property of a substance, such as its heat capacity, can be understood on the basis of the behavior of its constituent atoms or molecules.

The situation is more complicated for molecular gases. In addition to translational energy, molecules also possess rotational and vibrational energy (Sections 17.2 and 17.3). The rotational energy of a diatomic molecule is associated with the rotation of the molecule about an axis perpendicular to the molecular axis. Figure 10.15 shows the two types of rotational motion that are possible for a diatomic molecule. Vibrational energy is associated with the oscillation of the atoms in a molecule. In a diatomic molecule, the two atoms can oscillate back and forth with respect to each other, as depicted in Figure 10.15c.

Because of the contributions of rotational and vibrational motion, the internal energy of molecular gases exceeds $\frac{3}{2}RT$ per mole. Consequently, $C_p$ must exceed $\frac{5}{2}R$. Table 10.1 shows that the $C_p$ values of $O_2(g)$ and $N_2(g)$ are indeed substantially larger than $\frac{5}{2}R$, as expected from their greater internal energy. The theoretical evaluation of the contribution to the internal energy of a gas by rotational and vibrational motion is beyond the scope of this book. This topic is considered in the study of **statistical mechanics,** the branch of chemistry that deals with the derivation of the properties of macroscopic systems from the properties of molecules.

# Thermochemistry

Thermochemistry is the study of the thermal energy changes that accompany chemical processes. The determination of the heat released or absorbed in chemical processes is of fundamental importance, for it provides the means by which thermodynamic properties such as $\Delta E$, $\Delta H$, and $C_p$ can be determined. Thermochemistry is also a subject of great practical importance. After all, chemical fuels provide over 80% of the energy consumed in the industrialized world.

## 10.7 Thermochemical Equations

A **thermochemical equation** is a balanced equation in which the enthalpy change, or reaction enthalpy, is explicitly stated. Recall from Section 10.5 that the reaction enthalpy is equal to the heat given off or absorbed in a chemical reaction at constant pressure. Thermochemical results are customarily reported for standard conditions of temperature and pressure, $T = 298$ K, $P = 1$ atm. The designation for the standard reaction enthalpy is $\Delta H^\circ_{298}$, where the superscript signifies atmospheric pressure and the temperature (usually 298 K) is given as a subscript.

The following is a typical thermochemical equation, written to show the value of $\Delta H°$ for the formation of 1 mole of water:

$$H_2(g) + \tfrac{1}{2}O_2(g) \rightarrow H_2O(l) \qquad \Delta H°_{298} = -285.83 \text{ kJ}$$

According to the convention established in Section 10.4b, when $\Delta H > 0$, energy is transferred to the system. A reaction for which $\Delta H > 0$ is therefore endothermic. On the other hand, a process for which $\Delta H < 0$ is exothermic. The equation shows that the reaction of hydrogen and oxygen is exothermic.

The energy released or absorbed in a chemical reaction depends on the amount of substance undergoing reaction. Thus, the numerical value of $\Delta H$ depends on the values of the stoichiometric coefficients of the equation. For example, if the preceding reaction is written to show the formation of two moles of water, the thermochemical equation is

$$2H_2(g) + O_2(g) \rightarrow 2H_2O(l) \qquad \Delta H°_{298} = -571.66 \text{ kJ}$$

Furthermore, if the reaction is written in the reverse direction, energy must be absorbed rather than released, and the sign of $\Delta H$ changes:

$$2H_2O(l) \rightarrow 2H_2(g) + O_2(g) \qquad \Delta H°_{298} = 571.66 \text{ kJ}$$

This sign reversal follows from the fact that a chemical reaction followed by its reverse constitutes a cyclic process, for which the overall enthalpy change must be zero (Section 10.4b).

In writing a thermochemical equation, it is essential to specify the phase of each reagent. The energy content of a substance depends on its state of aggregation. As is commonly known, heat must be supplied to a solid in order to melt it, or to a liquid in order to vaporize it, indicating that the energy content of a substance increases as its phase changes from a solid to a liquid to a gas. For example, the following thermochemical changes can be written for the phase transformations of water:

$$H_2O(s) \rightarrow H_2O(l) \qquad \Delta H°_{273} = 6.01 \text{ kJ}$$

$$H_2O(l) \rightarrow H_2O(g) \qquad \Delta H°_{373} = 40.7 \text{ kJ}$$

Table 10.2 gives the enthalpies of fusion (i.e., solid $\rightarrow$ liquid) and vaporization (i.e., liquid $\rightarrow$ vapor) of a number of substances. It is apparent that the enthalpy of a reaction depends on the phases of the reactants and products and that these phases must be specified as part of a thermochemical equation. For example, if the equation for the combustion of hydrogen is written to show gaseous rather than liquid water as the product

$$H_2(g) + \tfrac{1}{2}O_2(g) \rightarrow H_2O(g)$$

then the value of $\Delta H°$ must differ from that given earlier for the formation of $H_2O(l)$. Adding the chemical equations and the corresponding enthalpy changes as follows gives $\Delta H°$ for this reaction:

$$H_2(g) + \tfrac{1}{2}O_2(g) \rightarrow H_2O(l) \qquad \Delta H°_{298} = -285.83 \text{ kJ}$$

$$\underline{H_2O(l) \rightarrow H_2O(g) \qquad \Delta H°_{298} = 44.0 \text{ kJ}\phantom{-000}}$$

$$H_2(g) + \tfrac{1}{2}O_2(g) \rightarrow H_2O(g) \qquad \Delta H°_{298} = -241.8 \text{ kJ}$$

**TABLE 10.2  Enthalpies of Phase Transitions[a]**

| Substance | $\Delta H°_{fusion}$ (kJ mol$^{-1}$) | $\Delta H°_{vaporization}$ (kJ mol$^{-1}$) |
|---|---|---|
| He | 0.021 | 0.084 |
| H$_2$ | 0.117 | 0.904 |
| N$_2$ | 0.720 | 5.58 |
| O$_2$ | 0.444 | 6.82 |
| CH$_4$ | 0.941 | 8.18 |
| CO$_2$ | 25.23[b] | |
| H$_2$O | 6.01 | 40.7 (at 373 K) |
|  |  | 44.0 (at 298 K) |
| NH$_3$ | 5.65 | 23.4 |
| Na | 2.60 | 98.0 |

[a] Unless otherwise stated, these values apply at the normal melting and boiling points of the substances in question.

[b] The solid is directly converted to a vapor (sublimation); the standard enthalpy of sublimation is given.

## 10.8 Hess's Law: The Indirect Determination of Reaction Enthalpies

While the enthalpy change for many reactions can be determined experimentally in a straightforward manner (as illustrated in Figure 10.14), this is not always the case. For example, when more than one reaction can occur between a given pair of reactants, it is difficult to assign the measured enthalpy change to a particular reaction. In order to overcome such difficulties, it is customary to take advantage of the fact that enthalpy is a state function and is independent of path. **Hess's law of constant heat summation,** which states that the reaction enthalpy is independent of path, is an application of the first law to chemical reactions.

The determination of the value of $\Delta H$ for the oxidation of carbon to carbon monoxide provides a good illustration of the use of Hess's law. The enthalpy change for this reaction cannot be measured readily. Under normal conditions, some of the carbon monoxide undergoes further oxidation to carbon dioxide. While this reaction can be minimized by reducing the pressure of O$_2$, the result is that the combustion of carbon then becomes incomplete. The problem can be overcome by use of Hess's law, since the enthalpy change for the direct oxidation of carbon to carbon dioxide as well as that for the oxidation of carbon monoxide to carbon dioxide are readily measurable. The three reactions under consideration are

(1)  C(graphite) + $\frac{1}{2}$O$_2$(g) → CO(g)     $\Delta H°_{298}$ = ?

(2)  CO(g) + $\frac{1}{2}$O$_2$(g) → CO$_2$(g)     $\Delta H°_{298}$ = −283.0 kJ

(3)  C(graphite) + O$_2$(g) → CO$_2$(g)     $\Delta H°_{298}$ = −393.5 kJ

Figure 10.16 shows how these three reactions can be arranged into two distinct paths leading from carbon plus oxygen to carbon dioxide. Applying Hess's law we obtain

$$\Delta H_3 = \Delta H_1 + \Delta H_2$$

**FIGURE 10.16** The formation of $CO_2$ from $C + O_2$ via different paths, illustrating Hess's law.

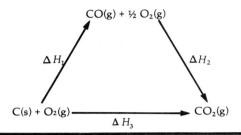

so that

$$\Delta H_1 = \Delta H_3 - \Delta H_2 = [-393.5 - (-283.0)] \text{ kJ} = -110.5 \text{ kJ}$$

Alternatively, we can take advantage of the fact that the enthalpy change for a cyclic process is zero. Starting with carbon plus oxygen we can write for the cycle terminating with these same species

$$\Delta H_1 + \Delta H_2 - \Delta H_3 = 0$$

which, of course, is equivalent to the dual path approach (as can be seen by solving for $\Delta H_3$).

While the graphical approach to Hess's law illustated in Figure 10.16 makes it easy to visualize this procedure, it is customary to perform the calculation algebraically. To do so, the thermochemical equations for the reactions with known enthalpies [i.e., reactions (2) and (3)] are combined algebraically to yield the reaction with unknown enthalpy [reaction (1)] and the $\Delta H$ values are combined appropriately. Inspection of these equations indicates that reaction (1) can be obtained by reversing reaction (2) and adding it to reaction (3):

$$\begin{aligned}
(3)\quad & C(\text{graphite}) + O_2(g) \rightarrow CO_2(g) & \Delta H^\circ_{298} &= -393.5 \text{ kJ} \\
-(2)\quad & CO_2(g) \rightarrow CO(g) + \tfrac{1}{2}O_2(g) & \Delta H^\circ_{298} &= 283.0 \text{ kJ} \\ \hline
(1)\quad & C(\text{graphite}) + \tfrac{1}{2}O_2(g) \rightarrow CO(g) & \Delta H^\circ_{298} &= -110.5 \text{ kJ}
\end{aligned}$$

**Example 10.6** Determine the enthalpy change for the conversion of diamond to graphite from the enthalpies for the complete combustion of these two allotropic forms of carbon:

$$\begin{aligned}
(1)\quad & C(\text{graphite}) + O_2(g) \rightarrow CO_2(g) & \Delta H^\circ_{298} &= -393.51 \text{ kJ} \\
(2)\quad & C(\text{diamond}) + O_2(g) \rightarrow CO_2(g) & \Delta H^\circ_{298} &= -395.41 \text{ kJ}
\end{aligned}$$

Subtraction of reaction (1) from reaction (2) yields the equation for the desired reaction:

$$C(\text{diamond}) \rightarrow C(\text{graphite}) \qquad \Delta H^\circ_{298} = -1.90 \text{ kJ}$$

The negative reaction enthalpy corresponds to a release of energy, indicating that diamond has a higher energy content than graphite at room temperature and atmospheric pressure. The result also illustrates why the allotropic form of a solid element must be specified in writing a thermochemical equation. ■

## 10.9 Standard Enthalpies of Formation

Enthalpy changes are known for many chemical reactions. In order to systematize this information and reduce the list of available reaction enthalpies to manageable proportions, it would be helpful to know the enthalpies of different substances. A list of enthalpies associated with individual substances is necessarily much shorter than a list of possible reaction enthalpies, just as a word dictionary is much shorter than a dictionary containing all possible sentences would be.

Thermochemical measurements provide the values of enthalpy changes but *not* those of the actual enthalpies of substances. The "absolute" values of enthalpies or internal energies cannot be measured—experiments yield only changes in these quantities. How, then, can we assign enthalpies to substances? The procedure followed is to arbitrarily give the value zero to the enthalpy of an element in its **standard state**—the physical state which is most stable at 25°C and a pressure of 1 atm. Consider, for example, the solid element copper. Its molar enthalpy at 25°C and atmospheric pressure has some value which cannot be measured. We are therefore free to assign some arbitrary value to it and the most convenient one is zero. The same procedure can be followed for every other element.

The usefulness of this convention is that the enthalpy of a compound in its standard state is then equal to the enthalpy change for the reaction in which one mole of the compound is made from its constituent elements in their standard states. Consider, for example, the reaction

$$C(\text{graphite}) + O_2(g) \rightarrow CO_2(g) \qquad \Delta H^\circ_{298} = -393.5 \text{ kJ}$$

The indicated reaction enthalpy is equal to the standard enthalpy of 1 mole of $CO_2$ less the standard enthalpies of its constituent elements. Since the latter have been assigned the value zero, the standard molar enthalpy of $CO_2$ is given by the enthalpy change for this reaction. The enthalpy change for the re-

**TABLE 10.3** Standard Molar Enthalpies of Formation at 25°C

| Substance | $\Delta H^\circ_f$ (kJ mol$^{-1}$) |
|---|---|
| $H_2O(l)$ | $-285.83$ |
| $SO_2(g)$ | $-296.83$ |
| $NO(g)$ | $90.25$ |
| $NH_3(g)$ | $-46.11$ |
| $CO(g)$ | $-110.53$ |
| $CO_2(g)$ | $-393.51$ |
| $PbO(s)(\text{yellow})$ | $-217.32$ |
| $PbO(s)(\text{red})$ | $-218.99$ |
| $CaO(s)$ | $-635.09$ |
| $HCl(g)$ | $-92.31$ |
| $AgCl(s)$ | $-127.07$ |
| $NaCl(s)$ | $-411.15$ |
| $HF(g)$ | $-271.1$ |

action in which one mole of a compound is formed in its standard state from its constituent elements in their respective standard states is called the **standard molar enthalpy of formation,** designated $\Delta H_f^\circ$. Table 10.3 lists the standard molar enthalpies of formation of various compounds at 25°C. A more complete list is given in Appendix 4.

**Example 10.7**  Write equations for the chemical reactions whose enthalpy change is equal to the standard enthalpy of formation of (a) CO(g), (b) BrCl(g), and (c) K$_3$[Fe(CN)$_6$](s). Give the values of $\Delta H_f^\circ$.

(a) $C(\text{graphite}) + \tfrac{1}{2}O_2(g) \rightarrow CO(g)$   $\Delta H_f^\circ = -110.5$ kJ mol$^{-1}$

Two points are worthy of note: 1. The equation must be written for the formation of just 1 mole of the product from the appropriate numbers of moles of the elements in their standard states. 2. The standard state of carbon is graphite and not diamond, the other allotropic form of carbon. This can be seen in Appendix 4, which indicates that C(graphite) has $\Delta H_f^\circ = 0$ while C(diamond) has $\Delta H_f^\circ = 1.895$ kJ mol$^{-1}$. The standard enthalpy of formation of an element in its standard state is zero.

(b) $\tfrac{1}{2}Br_2(l) + \tfrac{1}{2}Cl_2(g) \rightarrow BrCl(g)$   $\Delta H_f^\circ = 14.64$ kJ mol$^{-1}$

Note that the standard enthalpy of formation is positive, indicating that the reaction is endothermic.

(c) $3K(s) + Fe(s) + 6C(\text{graphite}) + 3N_2(g) \rightarrow K_3[Fe(CN)_6](s)$

$\Delta H_f^\circ = -249.8$ kJ mol$^{-1}$  ∎

The tabulated standard enthalpies of formation can be used to determine the enthalpy change for any reaction involving the substances in question. The value of $\Delta H^\circ$ for a chemical reaction is equal to the sum of the enthalpies of formation of the products of the reaction less that of the reactants,

$$\Delta H^\circ = \sum_{\text{products}} n_p \Delta H_f^\circ - \sum_{\text{reactants}} n_r \Delta H_f^\circ \qquad (10.34)$$

The standard molar enthalpies of formation for each reactant and product, $\Delta H_f^\circ$, must be multiplied by the respective stoichiometric coefficients ($n_r$ or $n_p$) of these species in the desired thermochemical equation. Equation 10.34 is an application to chemical reactions of Equation 10.23, which states that the enthalpy change in a process is equal to the enthalpy of the final state of the system less that of the initial state. In a chemical reaction, the final state is comprised of the reaction products, and the initial state is comprised of the reactants. The standard enthalpies of formation may be used as measures of the enthalpies of the initial and final states because both reactants and products contain the same numbers of atoms of the same elements, and Hess's law applies (Figure 10.17). Since the number of possible chemical reactions that can occur among the listed substances is much larger than the number of substances, the tabulation of standard enthalpies of formation leads to a useful condensation of the vast body of reaction enthalpies.

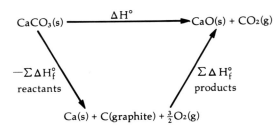

**FIGURE 10.17** The use of standard enthalpies of formation in the calculation of a standard reaction enthalpy. $\Delta H°$ for the reaction $CaCO_3(s) \rightarrow CaO(s) + CO_2(g)$ is given by Equation 10.34 because the decomposition of $CaCO_3(s)$ to the elements followed by the formation of $CaO(s) + CO_2(g)$ from the elements provides an alternate path to the direct reaction.

**Example 10.8** Determine $\Delta H°_{298}$ for the reaction

$$CH_4(g) + 4Cl_2(g) \rightarrow CCl_4(l) + 4HCl(g)$$

by using standard enthalpies of formation.

Using Equation 10.34, we obtain

$$\Delta H°_{298} = \Delta H°_f(CCl_4(l)) + 4\Delta H°_f(HCl(g)) - \Delta H°_f(CH_4(g)) - 4\Delta H°_f(Cl_2(g))$$
$$= [-135.44 + (4)(-92.31) - (-74.81) - (4)(0)] \text{ kJ}$$
$$= -429.87 \text{ kJ}$$

Note that the standard molar enthalpy of formation of each species has been multiplied by its stoichiometic coefficient in the balanced equation. ∎

## 10.10 Combustion Reactions and Natural Fuels

A very important practical application of thermochemistry is the determination of the heat released in combustion reactions, in which a substance is completely oxidized by oxygen. A large fraction of the energy we use is generated by the combustion of natural fuels. Combustion occurs when these substances are burned in air, the reaction products of complete combustion being carbon dioxide and liquid water. The three major naturally occurring fuels are coal, natural gas, and petroleum. These fuels consist primarily of mixtures of hydrocarbons (compounds of carbon and hydrogen). An important class of combustible hydrocarbons are the **alkanes,** which have the general formula $C_nH_{2n+2}$, where $n$ is a positive integer. Natural gas consists of methane ($CH_4$) mixed with some ethane ($C_2H_6$) as well as smaller amounts of propane ($C_3H_8$) and butane ($C_4H_{10}$). Prior to the commercial distribution of natural gas, propane and butane, which are more readily liquefied because of their higher boiling points, are removed and separately sold as liquid petroleum (LP) gas. The main combustion reaction of natural gas is therefore the combustion of methane:

$$CH_4(g) + 2O_2(g) \rightarrow CO_2(g) + 2H_2O(l) \qquad \Delta H°_{298} = -890.3 \text{ kJ}$$

Hydrocarbons with $n \geq 5$ are liquids at ordinary temperatures and are the main components of petroleum, the source of liquid fuels such as gasoline.

Energy in Chemical Processes—The First Law of Thermodynamics   10.10   365

**TABLE 10.4   Standard Enthalpies of Combustion of Some Important Fuels at 25°C**

| Fuel | Formula | Physical state | Enthalpy of combustion (kJ mol$^{-1}$) | Boiling point (°C) |
|---|---|---|---|---|
| Carbon | C | graphite | −393.5 | — |
| Methane | CH$_4$ | gas | −889.3 | −161 |
| Ethane | C$_2$H$_6$ | gas | −1560.1 | −88 |
| Propane | C$_3$H$_8$ | gas | −2220.0 | −42 |
| Butane | C$_4$H$_{10}$ | gas | −2878.5 | 0 |
| Octane | C$_8$H$_{18}$ | liquid | −5470.7 | 125 |
| Methanol | CH$_3$OH | liquid | −726.6 | 65 |
| Ethanol | C$_2$H$_5$OH | liquid | −1366.9 | 79 |
| Hydrogen | H$_2$ | gas | −285.8 | −253 |

In recent years, increasing attention has been paid to the possible large-scale use of fuels such as the alcohols, methanol (CH$_3$OH) and ethanol (C$_2$H$_5$OH). Because they can be obtained from agricultural products that can be continuously grown and harvested, such as corn or sugar cane, such fuels are known as renewable fuels.

Since complete combustion reactions generally occur rapidly and yield a single set of products, their ethalpies are readily measurable. The enthalpies of combustion of a number of important fuels in pure form are summarized in Table 10.4. In addition to their practical value, such enthalpies can also be combined according to Hess's law in order to give enthalpies of formation. The following example illustrates this use of combustion enthalpies.

**Example 10.9**   Calculate the standard enthalpy of formation of ethane using the data in Table 10.4.

The thermochemical equation for the combustion of ethane is

(1)   $C_2H_6(g) + 3\tfrac{1}{2}O_2(g) \rightarrow 2CO_2(g) + 3H_2O(l)$     $\Delta H^\circ_{298} = -1560.1$ kJ

As noted earlier, the combustion products are gaseous CO$_2$ and liquid water. In order to calculate the standard enthalpy of formation of ethane, equations must be written for the combustion of the elements of which ethane is comprised:

(2)   $C(graphite) + O_2(g) \rightarrow CO_2(g)$     $\Delta H^\circ_{298} = -393.5$ kJ

(3)   $H_2(g) + \tfrac{1}{2}O_2(g) \rightarrow H_2O(l)$     $\Delta H^\circ_{298} = -285.8$ kJ

Since the products of all three reactions consist of the same combustion products, suitable algebraic combination of the three equations can be used to eliminate them from the resulting equation for the formation of ethane from the elements. Inspection of the equations shows that since ethane has 2C atoms and 6H atoms, reaction (2) must be multiplied by 2 and reaction (3) by 3 in order to obtain a stoichiometrically correct equation. The desired equation is obtained by adding the reverse of reaction (1) to the sum of reactions (2) and (3), after multiplication by these coefficients:

$2 \times (2)$  $\quad 2C(graphite) + 2O_2(g) \rightarrow 2CO_2(g)$  $\quad \Delta H°_{298} = -787.0$ kJ

$3 \times (3)$  $\quad 3H_2(g) + \frac{3}{2}O_2(g) \rightarrow 3H_2O(l)$  $\quad \Delta H°_{298} = -857.4$ kJ

$-(1)$  $\quad 2CO_2(g) + 3H_2O(l) \rightarrow C_2H_6(g) + 3\frac{1}{2}O_2(g)$  $\quad \Delta H°_{298} = 1560.1$ kJ

$\quad\quad 2C(graphite) + 3H_2(g) \rightarrow C_2H_6(g)$  $\quad \Delta H°_{298} = -84.7$ kJ

The standard enthalpy of formation of ethane at 25°C is $-84.7$ kJ mol$^{-1}$. ∎

## 10.11 Temperature Dependence of Reaction Enthalpies

The enthalpy changes for chemical reactions at 25°C can readily be obtained from the tabulated standard enthalpies of formation. Chemical reactions are, of course, often run at other temperatures and the dependence of the reaction enthalpies on temperature then becomes of interest.

Recall from Section 10.5b that the heat capacity at constant pressure is equal to the rate of change with temperature of the enthalpy (Equation 10.28):

$$dH = C_p \, dT$$

This expression can be integrated over temperature between an initial temperature $T_1$ and a final temperature $T_2$:

$$H_{T_2} - H_{T_1} = \int_{T_1}^{T_2} C_p \, dT \tag{10.35}$$

where $H_{T_2}$ and $H_{T_1}$ are the enthalpies of a substance with heat capacity $C_p$ at temperature $T_2$ and $T_1$, respectively. In general, $C_p$ values do not change rapidly with temperature and may, for the present purpose, be treated as constant. With this assumption, Equation 10.35 can be integrated:

$$H_{T_2} = H_{T_1} + C_p(T_2 - T_1) \tag{10.36}$$

This equation applies to each species participating in a chemical reaction. Since the reaction enthalpy $\Delta H$ is equal to the enthalpy of the products of the reaction less that of the reactants, the reaction enthalpy at temperature $T_2$, $\Delta H_{T_2}$, can be related to the reaction enthalpy at temperature $T_1$, $\Delta H_{T_1}$, by the expression

$$\Delta H_{T_2} = \Delta H_{T_1} + \Delta C_p(T_2 - T_1) \tag{10.37}$$

where $\Delta C_p$ is the difference between the sums of the heat capacities of the products and the reactants. In addition to the data in Table 10.1, values of $C_p°$ are given in Appendix 4 (the superscript on $C_p°$ denotes atmospheric pressure).

---

**Example 10.10**  Using the data in Appendix 4, calculate $\Delta H°_{500}$ for the reaction

$$2NO(g) + O_2(g) \rightarrow 2NO_2(g)$$

Assume that the heat capacities are independent of temperature between 298 K and 500 K.

The standard enthalpies of formation and heat capacities are obtained from Appendix 4:

|  | $\Delta H_f^\circ$ | $C_P^\circ$ |
|---|---|---|
| NO(g) | 90.25 kJ mol$^{-1}$ | 29.84 J mol$^{-1}$ K$^{-1}$ |
| O$_2$(g) | 0 | 29.36 |
| NO$_2$(g) | 33.18 | 37.20 |

We first obtain $\Delta H_{298}^\circ$ from the standard enthalpies of formation:

$$\Delta H_{298}^\circ = 2\Delta H_f^\circ(NO_2) - 2\Delta H_f^\circ(NO) - \Delta H_f^\circ(O_2)$$
$$= (2 \times 33.18 - 2 \times 90.25) \text{ kJ} = -114.14 \text{ kJ}$$

The difference between the heat capacities of products and reactants is

$$\Delta C_P^\circ = 2C_P^\circ(NO_2) - 2C_P^\circ(NO) - C_P^\circ(O_2)$$
$$= (2 \text{ mol})(37.20 \text{ J mol}^{-1} \text{ K}^{-1}) - (2 \text{ mol})(29.84 \text{ J mol}^{-1} \text{ K}^{-1})$$
$$- (1 \text{ mol})(29.36 \text{ J mol}^{-1} \text{ K}^{-1})$$
$$= -14.64 \text{ J K}^{-1}$$

Finally, $\Delta H_{500}^\circ$ is obtained by substitution into Equation 10.37:

$$\Delta H_{500}^\circ = \Delta H_{298}^\circ + \Delta C_P^\circ (500 \text{ K} - 298 \text{ K})$$
$$= -114.14 \text{ kJ} + (-14.64 \text{ J K}^{-1})(202 \text{ K})(10^{-3} \text{ kJ J}^{-1})$$
$$= -117.1 \text{ kJ}$$

We note that the reaction is somewhat more exothermic at the higher temperature. ∎

## Conclusion

The energy of a system and its surroundings must remain constant. The system can interact with its surroundings by the flow of heat or work but the overall energy is conserved. Energy is a state function and is independent of path, but work and heat depend upon the path followed in a particular process. These statements constitute the first law of thermodynamics. We have illustrated these aspects of the first law by considering the expansion or compression of a gas under different conditions: isobaric, reversible isothermal, and cyclic. Isobaric processes are particularly common in chemistry and the enthalpy function and heat capacity at constant pressure are useful in dealing with them.

Thermochemistry, the study of the energy changes accompanying chemical processes, is an important application of the first law. Reaction enthalpies can be calculated from tabulated standard enthalpies of formation as well as by application of Hess's law. The use of tabulated heat capacities permits the correction of reaction enthalpies for changes in temperature.

## Problems

(Use the data in Appendix 4, unless stated otherwise.)

1. Calculate the work that must be expended in raising a 10 kg weight through a height of 10 m (a) on the surface of the Earth and (b) on the surface of the Moon, where $g = 1.6$ m s$^{-2}$.

2. What is the work performed on the surroundings when 2.00 L of an ideal gas at a pressure of 5.00 atm undergoes the following expansions at a constant temperature of 300 K: (a) a reversible expansion to a final pressure of 1.00 atm and (b) an expansion against a constant external pressure of 1.00 atm?

3. It takes 6.0 kJ of work to compress 1 mol of an ideal gas reversibly at $-50°C$. If the initial pressure was 1.0 atm, what is the final pressure of the gas?

4. A 10 g chunk of dry ice (solid carbon dioxide) is allowed to sublime (i.e., evaporate) in a 100 mL container maintained at room temperature (25°C). Calculate the work done when the gas is allowed to expand to a final pressure of 1.0 atm (a) against a constant external pressure of 1.0 atm and (b) reversibly and isothermally.

5. How much heat is lost by 1.00 kg of copper metal when it cools from 500°C to 25°C?

6. A 50.8 g sample of copper is heated to 100°C and dropped into a container with 27.0 mL of water at 25°C. Assuming that no heat escapes from the container, determine the final temperature of the water.

7. A 20.0 kg weight falls through a distance of 4.00 m on Earth, thereby causing 1.00 L of water at 25.00°C to be stirred by a paddle wheel (as in Figure 10.8). What is the final temperature of the water?

8. One mole of an ideal gas at 300 K expands reversibly and isothermally from 10.0 to 30.0 L. Calculate $w$, $q$, $\Delta E$, and $\Delta H$ for this process.

9. A 2.50 mol sample of an ideal gas at a temperature of 370 K and a pressure of 1.30 atm was cooled at constant pressure until its volume was 45.0 L. The heat given off by the gas during this process was 4400 J. Calculate (a) the initial volume, (b) the final temperature, (c) the work done—indicate whether the gas does work or has work done on it, (d) $\Delta E$, (e) $\Delta H$, and (f) $C_p$ for the gas.

10. One mole of an ideal monatomic gas at 0°C and a pressure of 1.00 atm is heated at constant volume to 30°C. Determine $q$, $w$, $\Delta E$, $\Delta H$, and $\Delta P$.

11. The same gas as in Problem 10 is heated at constant pressure to 30°C. Determine $q$, $w$, $\Delta E$, $\Delta H$, and $\Delta V$.

12. What heat flow will raise the temperature of 1 mol of helium by 20 K (a) when it is in a constant volume container at a pressure of 5.0 atm and (b) when it can expand against a constant external pressure of 5.0 atm? How much work does the gas do in each case?

13. A lump of zinc metal weighing 25 g is dropped into a beaker containing dilute hydrochloric acid maintained at a constant temperature of 25°C. How much work does the ensuing reaction do on the surrounding atmosphere ($P = 1$ atm)? Assume that excess HCl(aq) is present.

14. One mole of an ideal monatomic gas at 0°C and a pressure of 10.0 atm is put through the following three-step cyclic process: (1) reversible isothermal expansion to 1.00 atm, (2) cooling at constant pressure until the volume is 2.24 L, and (3) heating at constant volume back to the initial state. (a) For each step of this process determine the initial and final pressure, volume, and temperature, $q$, $w$, $\Delta E$ and $\Delta H$. (b) Determine $q$, $w$, $\Delta E$, and $\Delta H$ for the cyclic process. (c) Sketch the process on a P–V plot.

15.* A gas obeys the equation of state $P(V - b) = RT$, where $b$ is a constant. Derive an expression for the work done by the gas in a reversible isothermal expansion from $V_1$ to $V_2$.

16. Exactly 10 g of ice at 0°C is added to 60.0 g of water at 60°C in an insulated container. What is the final temperature of the water after all the ice has melted?

17. Ice cubes are used to cool 200 mL of water from 25°C to 5°C. Assuming that no heat is lost to the container or the surroundings, what mass of ice will melt?

18. Calculate $q$, $w$, $\Delta E$, and $\Delta H$ for the conversion at a constant pressure of 1 atm of 20.0 g of ice at $-10°C$ to 20.0 g of steam at 130°C. Use the data in Table 10.2; the molar heat capacities of ice, water, and steam are 38, 75, and 36 J mol$^{-1}$ K$^{-1}$, respectively, and may be assumed to be independent of temperature over this range. Treat steam as an ideal gas.

19. Determine the standard enthalpy of formation of ethanol from standard enthalpies of combustion.

20. Determine the enthalpy of combustion of acetaldehyde (CH$_3$CHO) at standard conditions using standard enthalpies of formation.

21. When octane (C$_8$H$_{18}$) is burned in a limited supply of oxygen, the product is CO rather than CO$_2$. Use the data in Table 10.4 and the standard enthalpy of combustion of CO($-283.0$ kJ mol$^{-1}$) to determine (a) the enthalpy change in the incomplete combustion of octane and (b) the change in internal energy.

22. Calculate the standard molar enthalpy of hydration of anhydrous copper(II) sulfate, i.e., the standard reaction enthalpy for

$$Cu(SO_4)(s) + 5H_2O(l) \rightarrow CuSO_4 \cdot 5H_2O(s)$$

23. The reduction of iron(III) oxide by carbon monoxide in the reaction

$$Fe_2O_3(s) + 3CO(g) \rightarrow 2Fe(s) + 3CO_2(g)$$

is one of the steps in the production of iron metal. What is $\Delta H°_{298}$ for this reaction?

24. The enthalpy of combustion at 25°C of benzene, C$_6$H$_6$(l), is $-3273$ kJ mol$^{-1}$ and that of acetylene, C$_2$H$_2$(g), is $-1305$ kJ mol$^{-1}$. (a) Determine the standard enthalpies of formation of benzene and acetylene from these data. (b) Determine the enthalpy change for the reaction

$$3C_2H_2(g) \rightarrow C_6H_6(l)$$

25. At 25°C, the enthalpy of combustion of H$_2$(g) is $-285.8$ kJ mol$^{-1}$ and that of CH$_4$(g) is $-889.3$ kJ mol$^{-1}$. A 10.0 g mixture of H$_2$ and CH$_4$ ignited in a constant pressure calorimeter containing excess oxygen liberates 1020 kJ. What was the composition of the mixture?

26. Using the following two reaction enthalpies:

(1)  $\frac{3}{2}H_2(g) + \frac{1}{2}N_2(g) \rightarrow NH_3(g)$,

$\Delta H = -46.2$ kJ

(2)  $H_2(g) + \frac{1}{2}O_2(g) \rightarrow H_2O(g)$,

$\Delta H = -241.8$ kJ

determine the enthalpy change for the reaction

$$4NH_3(g) + 3O_2(g) \rightarrow 2N_2(g) + 6H_2O(g)$$

at the same temperature.

27. The standard enthalpy of combustion of gaseous ethane (C$_2$H$_6$) is $-1560$ kJ mol$^{-1}$ at 25°C. How much heat is liberated when 1 mol of ethane is burned in a constant volume calorimeter at 25°C? Neglect the volume of the liquid water produced in the reaction.

28. The standard enthalpy of neutralization, i.e., $\Delta H°_{298}$ for the reaction

$$H^+(aq) + OH^-(aq) \rightarrow H_2O(l)$$

is $-55.8$ kJ mol$^{-1}$. (a) Determine the heat liberated when 500 mL of 0.50 $M$ HCl is added to 500 mL of 1.0 $M$ NaOH, where the initial temperature of both solutions is 25°C. (b) Assuming that none of the heat flows out of the solution, what is the maximum temperature of the solution after neutralization? Assume that the heat capacity and density of the solution are equal to those of water.

29. The standard molar enthalpy of formation of sucrose (C$_{12}$H$_{22}$O$_{11}$) is $-2222$ kJ mol$^{-1}$. What is the enthalpy of combustion of a sugar cube (sucrose) weighing 2.0 g? If 25% of this energy can be converted to work by a human body, to what height can a 100 kg man climb on the energy derived from such a sugar cube? Assume the man is on the surface of the Earth.

30. Determine the standard enthalpy of formation of gaseous N$_2$O$_5$ on the basis of the following data:

$N_2(g) + O_2(g) \rightarrow 2NO(g)$     $\Delta H° = 180.5$ kJ

$2NO(g) + O_2(g) \rightarrow 2NO_2(g)$     $\Delta H° = -114.1$ kJ

$4NO_2(g) + O_2(g) \rightarrow 2N_2O_5(g)$     $\Delta H° = -110.2$ kJ

31. When steam is passed through hot coke, water gas, an industrial fuel, is produced by the reaction

$$C(graphite) + H_2O(g) \rightarrow CO(g) + H_2(g)$$

(a) Using values of $\Delta H°_f$, determine the reaction enthalpy at 25°C. (b) Using tabulated values of $C°_p$, determine the reaction enthalpy at 125°C. (c) How much heat is given off when 1 mol of water gas is burned at 25°C and a pressure of 1 atm?

32. Calculate $\Delta H°$ for the synthesis of 1 mol of ammonia from the elements at 600 K. Determine the corresponding change in internal energy.

33. Calcium carbide, used as a source of acetylene, is made at high temperatures in the reaction

$$CaO(s) + 3C(graphite) \rightarrow CaC_2(s) + CO(g)$$

Determine $\Delta H°$ at 1500 K. How does the equilibrium constant for the reaction depend on temperature?

34. The enthalpy of atomization is the enthalpy change for the reaction in which a substance is dissociated into gaseous atoms. Calculate the standard enthalpy of atomization of O$_2$ at (a) 25°C and (b) 600°C.

# 11

# Spontaneity in Chemical Processes—The Second Law of Thermodynamics

In Chapter 10 we examined the energetics of chemical processes. The amount of energy released or absorbed in a particular chemical reaction could be determined from the energies of other reactions on the basis of the first law of thermodynamics. With the aid of tables of standard enthalpies of formation and heat capacities, the energy change accompanying a chemical reaction could be calculated over a broad range of temperatures, a considerable achievement. However, the first law does not address the even more important question of the natural direction of chemical change. Does carbon burn in air or does carbon dioxide decompose to form carbon and oxygen? We know from experience the direction in which this particular reaction occurs, but it would be very useful to be able to predict the direction of change, particularly for less familiar reactions. The second law of thermodynamics provides us with the means of predicting the direction of chemical change. The combination of the first and second laws unifies a number of seemingly diverse aspects of chemistry and shows why thermodynamics is such a far-reaching and powerful discipline.

## Entropy and Spontaneity

### 11.1 Spontaneous Processes

A **spontaneous process** is one that will occur on its own, without any outside interference. Some processes are spontaneous and others are not. Many familiar examples of spontaneous processes can be given (Figure 11.1): ice cubes

## Spontaneity in Chemical Processes—The Second Law of Thermodynamics 11.1

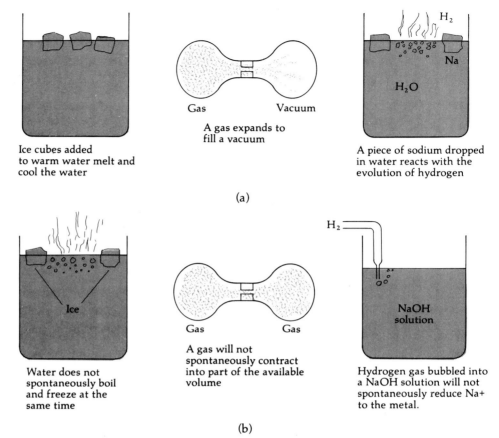

**FIGURE 11.1** (a) Some examples of spontaneous processes. (b) The reverse processes, which are not spontaneous, will not occur by themselves.

added to hot water melt, and the water cools; a gas expands to fill a vacuum; a piece of sodium dropped in water reacts with the evolution of hydrogen. The reverse of these processes is not spontaneous—the reversed processes will not occur without outside interference. One part of a water sample will not heat up while another part freezes; a gas will not contract into part of its available volume; hydrogen bubbled into a solution of sodium hydroxide will not reduce $Na^+$ to the metal. To be sure, these processes can be made to occur, but only when an external agent does some kind of work on the system. For example, a gas can be compressed by means of a pump, and water can be frozen in a freezer and brought to a boil with a heater.

While many spontaneous processes, such as the reaction of sodium with water, occur rapidly, others do not. An aqueous solution of sugar, for example, will not react with oxygen even over a period of years. However, the presence of yeast catalyzes the oxidation of sugar to alcohol in a matter of hours. If some process is known to be spontaneous, it is worthwhile to search for a catalyst that will make it occur rapidly. On the other hand, if a process is not spontaneous, we can be confident that such a search will turn out to be fruitless. Thermodynamics does not deal with the speed at which chemical reactions or other processes occur; this is the subject of kinetics (Chapter 14). A spontaneously occuring process is just as spontaneous if it takes virtually forever to occur as if it occurs in a fraction of a second.

A spontaneous process may be contrasted with a reversible process. Recall from Section 10.2d that a reversible process involves a succession of infinitesimally small changes in which the system remains virtually at equilibrium with its surroundings. The direction of a reversible process can be reversed by an infinitesimal change in the value of the variable which determines the natural direction of the process. In contrast, a spontaneous process is irreversible. The difference between spontaneous and reversible processes is illustrated in Figure 11.2 for several types of comparable cases.

What factors determine the direction of spontaneous change? A change in the total energy of a system and its surroundings cannot be a criterion, for the first law of thermodynamics indicates that the total energy remains constant. But how about the energy of the system alone? It is certainly true that the energy of a system decreases in many spontaneous processes. For example, most spontaneous chemical reactions are exothermic—the energy of the system decreases. However, other spontaneous chemical reactions are endothermic and involve an increase in the energy of the system. The reaction of barium hydroxide octahydrate with ammonium thiocyanate, for instance,

$$Ba(OH)_2 \cdot 8H_2O(s) + 2NH_4SCN(s) \rightarrow 2NH_3(g) + Ba^{2+} + 2SCN^- + 10H_2O$$

occurs spontaneously but is an endothermic process ($\Delta H° = 130$ kJ): a beaker in which the reaction takes place feels cold to the touch, indicating that heat is absorbed from the surroundings. The isothermal expansion of an ideal gas into a vacuum (Figure 11.1) is another example of a spontaneous process in which the energy of the system does not decrease; in fact, we know from Section 10.5 that $\Delta E = 0$ for this process.

It is apparent from these considerations that some additional determinant of spontaneity remains to be identified. This factor can be described only qualitatively at this stage. It is an increase in the disorder of the system on the molecular level. Consider, for example, the expansion into a large evacuated container of a gas confined to a small volume. The molecules of the gas execute the random motion described in Section 4.2, colliding with each other and with the walls of the container, and gradually filling the entire available volume. Suppose, now, that we were able to reverse the motion of the molecules. After some time, the gas would squeeze itself back into the original volume. If we could examine this process in detail we would see that the motion of the molecules involved the same random collisions and paths as in the expansion. And yet, this spontaneous compression never happens. The reason is that there are many paths that the molecules can follow as the gas expands but only a few paths that will result in the molecules occupying a small region of the available space. When all the molecules are allowed to move throughout the entire container, their motion is freer and less constrained than if they are restricted to move in only a small region of the container. Put another way, the probability that the gas molecules occupy the entire volume is much greater than the probability that they occupy only a part of the available volume.

While the drive towards a more disordered and probable state in the preceding example is rather self-evident, subtler changes can be more difficult to interpret in this manner. The second law of thermodynamics provides a more

**FIGURE 11.2** Difference between parallel spontaneous and reversible processes. (a) Expansion of a gas. When $P \gg P_{external}$ the gas expands spontaneously; when $P \approx P_{external}$, a slight increase in $P_{external}$ causes the compression of the gas. (b) Cooling of a system. When $T_{system} \gg T_{bath}$, heat spontaneously flows from the system to the bath; when $T_{system} \approx T_{bath}$, a slight increase in $T_{bath}$ can reverse the direction of heat flow. (c) Change of phase. When $T_{H_2O} \gg 0°C$, ice melts spontaneously; when $T_{H_2O} \approx 0°C$, the flow of a small amount of heat into or out of the system leads to melting of ice or freezing of water, respectively (d) Chemical reactions. The reaction $2Na(s) + 2H_2O(l) \rightarrow 2NaOH(aq) + H_2(g)$ occurs spontaneously when a piece of sodium is dropped into water. The ionization of HOAc, $HOAc(aq) + H_2O(l) \rightleftarrows H_3O^+ + OAc^-$, is a reversible process; an increase in acid concentration decreases the degree of ionization while a decrease in acid concentration increases the degree of ionization.

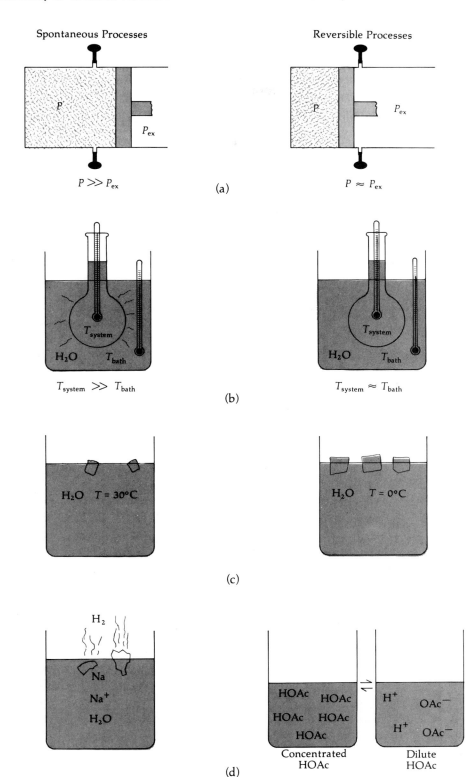

## 11.2 Entropy and the Second Law of Thermodynamics

quantitative criterion by means of which the spontaneity of a process may be determined.

The second law of thermodynamics can be expressed in terms of a property of the system, called the entropy, the change in which measures how the disorder of a system is altered by some process. The **entropy** $S$ of a system is defined as a function that changes in the following way in a process in which a system is taken from state $a$ to state $b$:

$$\Delta S = S_b - S_a = \int_a^b \frac{dq_{\text{rev}}}{T} \tag{11.1}$$

This equation states that the entropy change in a system can be computed by finding a reversible path, evaluating the infinitesimal heat changes occurring along the path, dividing each one by the temperature at which it occurs, and adding the resulting values of $dq/T$. As in the case of the expression for work (Section 10.2c), the summation is actually replaced by an integration. For an infinitesimal process, the change in the entropy of the system, $dS$, is simply

$$dS = dq_{\text{rev}}/T \tag{11.2}$$

Note that the units of entropy are joules per kelvin (J K$^{-1}$). Equation 11.1 asserts that entropy is a state function, with the change in entropy equal to the difference in the entropy of the final and initial states. We will see later that this assertion is indeed valid.

The second law of thermodynamics can be formulated in the following manner in terms of the entropy:

*In a spontaneous process, the total entropy of a system plus its surroundings increases,*

$$\Delta S_{\text{total}} = (\Delta S_{\text{system}} + \Delta S_{\text{surr}}) > 0 \quad \text{(spontaneous process)} \tag{11.3}$$

Furthermore, in a reversible process, in which a system is in equilibrium with its surroundings, the total entropy remains constant:

$$\Delta S_{\text{total}} = (\Delta S_{\text{system}} + \Delta S_{\text{surr}}) = 0 \quad \text{(equilibrium)} \tag{11.4}$$

Finally, if the total entropy were to decrease in some process, the reverse of that process would occur spontaneously, i.e.,

$$\Delta S_{\text{total}} = (\Delta S_{\text{system}} + \Delta S_{\text{surr}}) < 0 \quad \text{(reverse process is spontaneous)}$$

When stated in this manner, the second law provides us with a readily evaluable criterion for both spontaneity and equilibrium. We have to evaluate the entropy change of the system and add it to that of the surroundings. If the result is positive, the process in question will occur spontaneously, while if it is zero the system is already in equilibrium with its surroundings.

In order to demonstrate that the total change in the entropy of a system and its surroundings is indeed a criterion for spontaneity, we must evaluate these quantities for some process that we know to be either spontaneous or reversible and show that Equation 11.3 or 11.4, respectively, is satisfied. A process that received some attention in Chapter 10 is the expansion of an ideal gas. Let us calculate $\Delta S_{\text{total}}$ for a reversible expansion and compare it with the value for an expansion that we know to be spontaneous, e.g., an expansion into a vacuum (Figure 11.1).

First, we evaluate $\Delta S_{total}$ for the reversible isothermal expansion of one mole of an ideal gas from volume $V_1$ to $V_2$. Let us begin by evaluating the entropy change for the system. Since the entropy change is given by Equation 11.1 as

$$\Delta S_{system} = \int \frac{dq_{rev}}{T}$$

we obtain for constant $T$

$$\Delta S_{system} = \frac{1}{T} \int dq_{rev} = \frac{q_{rev}}{T}$$

We know that the internal energy of an ideal gas depends only on the temperature and therefore remains constant in an isothermal process (Section 10.5). Therefore, according to the first law,

$$\Delta E = q + w = 0$$

so that

$$q = -w$$

Recall, now, that the work done by one mole of an ideal gas in a reversible isothermal expansion is given by Equation 10.13:

$$w = -RT \ln(V_2/V_1)$$

Therefore,

$$q_{rev} = RT \ln (V_2/V_1) \tag{11.5}$$

and substituting into the expression for $\Delta S_{system}$, we get

$$\Delta S_{system} = R \ln (V_2/V_1) \tag{11.6}$$

for the change in the entropy of one mole of an ideal gas undergoing an isothermal expansion. This equation shows that entropy is a state function: $\Delta S$ depends only on the initial and final states of the system. Since the volume increases in an expansion, the entropy of a gas must increase in this process. This is a useful result but not the one we are looking for since the second law refers to the change in entropy of the system *plus* surroundings.

Let us now evaluate the entropy change for the surroundings. To do so we must realize that the surroundings is a vast reservoir, containing far more energy than any conceivable system. Consequently, the flow of heat and work between a system and its surroundings changes the properties of the surroundings by only an infinitesimal amount. Since infinitesimal changes are characteristic of reversible processes, we can conclude that *any* process has the same effect on the surroundings as a reversible process. We can therefore write Equation 11.1 for the entropy change of the surroundings as

$$\Delta S_{surr} = \int \frac{dq_{rev}}{T} = \int \frac{dq}{T} = \frac{q}{T} \tag{11.7}$$

where the last equality follows from the fact that the temperature of the surroundings changes only infinitesimally and so remains essentially constant. Now, the heat flowing to the system has been given by Equation 11.5. This heat must come from the surroundings and so we have

$$q = -RT \ln(V_2/V_1)$$

where $q < 0$ means that heat flows from the surroundings. Substituting into Equation 11.7, we obtain

$$\Delta S_{surr} = \frac{-RT}{T} \ln(V_2/V_1) = -R \ln(V_2/V_1) \tag{11.8}$$

Comparison of Equations 11.8 and 11.6 shows that the change in the entropy of the system is equal in magnitude but opposite in sign to that of the surroundings. Therefore, the total entropy change is

$$\Delta S_{total} = \Delta S_{system} + \Delta S_{surr} = R \ln(V_2/V_1) - R \ln(V_2/V_1) = 0$$

Since the total entropy remains constant, the process is reversible, and the system is at all times in equilibrium with its surroundings. If we had not known that this was indeed the case, the constancy of the total entropy would have allowed us to draw this conclusion.

We now repeat this calculation for an expansion into a vacuum. For convenience, we choose the same initial and final states as in the preceding process, i.e., volumes $V_1$ and $V_2$ and constant temperature $T$. Since entropy is a state function, the entropy of the system depends only on the properties of the initial and final states and is independent of path. Since the initial and final states are the same as those chosen for the reversible expansion, the change in the entropy of the system must also be the same. We thus have

$$\Delta S_{system} = R \ln(V_2/V_1)$$

When a gas expands into a vacuum, there is no external pressure resisting the expansion, $P_{ex} = 0$. Therefore, according to Equation 10.7 for the work done by a gas

$$w = -\int P_{ex}\, dV = 0$$

Since the expansion is isothermal, it follows that $\Delta E = 0$. Therefore, $q = -w = 0$. This result shows that there is no flow of heat between the system and the surroundings and immediately allows us to evaluate the change in entropy of the surroundings by means of Equation 11.7:

$$\Delta S_{surr} = q/T = 0$$

Since the entropy of the surroundings does not change, the total entropy change is equal to the change in the entropy of the system:

$$\Delta S_{total} = R \ln(V_2/V_1) > 0$$

Equation 11.3 indicates that since the total entropy increases, the process occurs spontaneously.

---

**Example 11.1** An ideal gas maintained at 25°C expands against a constant external pressure of 1.0 atm to a final volume of 10 L. In its initial state, the gas is at a pressure of 10 atm and occupies a volume of 1.0 L. Determine $\Delta S_{system}$, $\Delta S_{surr}$, and $\Delta S_{total}$.

The entropy change of the system is given by Equation 11.1,

$$\Delta S_{system} = \int \frac{dq_{rev}}{T} = \frac{q_{rev}}{T}$$

where the second equality follows from the constancy of the temperature. To evaluate the entropy change, we must determine the heat absorbed by the gas in a reversible expansion from the initial to the final state. Note that this is *not* the actual heat absorbed in this process, for this process is not reversible. We know from Equation 10.13 that the work done by an ideal gas in a reversible isothermal expansion is

$$w_{rev} = -nRT \ln(V_2/V_1)$$

The number of moles of gas is obtained from the ideal gas law,

$$n = \frac{PV}{RT} = \frac{(1.0 \text{ atm})(10 \text{ L})}{(0.08206 \text{ L atm mol}^{-1} \text{ K}^{-1})(298 \text{ K})} = 0.409 \text{ mol}$$

Therefore,

$$w_{rev} = -(0.409 \text{ mol})(8.314 \text{ J mol}^{-1} \text{ K}^{-1})(298 \text{ K}) \ln(10 \text{ L}/1.0 \text{ L})$$
$$= -2333 \text{ J}$$

Since the process is isothermal and the gas is ideal, $\Delta E = 0$. The first law then gives

$$q_{rev} = -w_{rev} = 2333 \text{ J}$$

The entropy change of the system is

$$\Delta S_{system} = q_{rev}/T = 2333 \text{ J}/298 \text{ K} = 7.8 \text{ J K}^{-1}$$

The entropy change of the surroundings can be evaluated by means of Equation 11.7,

$$\Delta S_{surr} = q/T$$

where $q$ is the *actual* heat flowing into or out of the surroundings. In analogy to the preceding calculation, the actual heat flow can be obtained from the actual work done by the system in the expansion. In an expansion against a constant external pressure, Equation 10.5 gives

$$w = -P_{ex}\Delta V = -(1.0 \text{ atm})[(10 - 1.0)\text{L}][101.3 \text{ J(L atm)}^{-1}]$$
$$= -912 \text{ J}$$

The heat absorbed by the gas in the actual expansion is $q = -w = 912$ J. Since the surroundings is the source of this heat, the heat lost by the surroundings is $q = -912$ J, where according to our convention (Section 10.3), a negative value of $q$ corresponds to an outflow of heat. The entropy change of the surroundings is

$$\Delta S_{surr} = q/T = -912 \text{ J}/298 \text{ K} = -3.1 \text{ J K}^{-1}$$

The total entropy change is, according to Equation 11.3,

$$\Delta S_{\text{total}} = \Delta S_{\text{system}} + \Delta S_{\text{surr}}$$
$$= 7.8 \text{ J K}^{-1} - 3.1 \text{ J K}^{-1} = 4.7 \text{ J K}^{-1}$$

and the final answers are

$$\Delta S_{\text{system}} = 7.8 \text{ J K}^{-1} \qquad \Delta S_{\text{surr}} = -3.1 \text{ J K}^{-1} \qquad \Delta S_{\text{total}} = 4.7 \text{ J K}^{-1}$$

Note that $\Delta S_{\text{total}} > 0$, which is consistent with the fact that the expansion occurs spontaneously. ∎

The same general approach illustrated in Example 11.1 may be applied to other spontaneous processes, e.g., the mixing of hot and cold water or the precipitation of AgCl in the reaction of $AgNO_3$(aq) with $NH_4Cl$(aq). It will be found in all such cases that the entropy of a system plus its surroundings increases. The first two laws of thermodynamics can be succinctly summarized in the statement first made in the nineteenth century by R. Clausius: The energy of the universe is constant; the entropy of the universe always increases.

The formulation of the second law of thermodynamics in terms of the change in total entropy is most directly applicable to chemical processes. However, the second law also can be stated in various other equivalent ways. The following two statements are readily understandable in terms of our daily experience:

1. Heat cannot spontaneously flow from a cold body to a warmer one.
2. Heat cannot be extracted from a heat source and converted entirely into work.

These statements rule out the existence of what are referred to as perpetual motion machines of the second kind, to distinguish them from perpetual motion machines of the first kind, which are based on a violation of the first law. The first statement, for instance, rules out a motorless refrigerator the inside compartment of which is cooled by the spontaneous flow of heat to the warmer surroundings. Similarly, the second statement rules out a ship fueled only by the heat extracted from the ocean it traverses. While refrigerators and ships obviously exist, their operation requires a fuel other than the heat content of the ambient environment.

## 11.3 Molecular Interpretation of Entropy

The concept of entropy is neither familiar from everyday experience nor intuitively obvious. Entropy is most easily understood by examining its meaning on the molecular level. We have already noted that entropy is a measure of the probability, or randomness, of the molecular arrangement. To explore this idea in further detail, it is convenient to pursue a simple analogy to a process involving real molecules. Examination of probabilities in dice tossing provides an understandable context for a discussion of the molecular interpretation of entropy.

Suppose we roll a pair of dice a large number of times and examine the probability of obtaining the various possible combinations between 2 and 12. Although each of the six faces of a die has an equal probability of coming up, certain totals are more probable than others because they can be formed by more combinations. Figure 11.3 shows the possible combinations that can lead to each total as well as the probability of each outcome. The entropy of the sys-

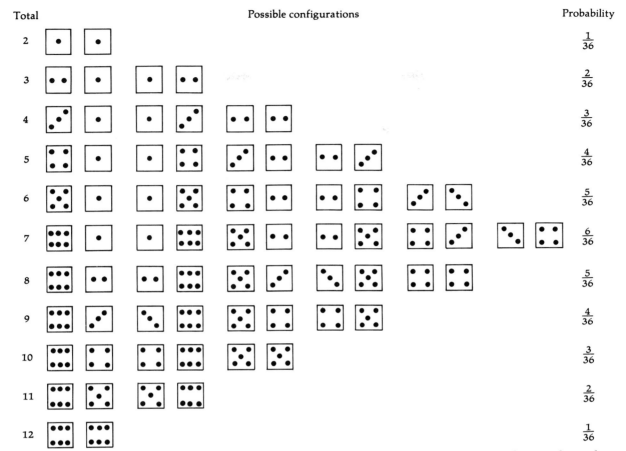

**FIGURE 11.3** Possible outcomes of the toss of a pair of dice, an analogy to the probability of the occurrence of different molecular arrangements. There are 36 different outcomes to the toss of a pair of dice and the probability of each total is equal to the number of outcomes giving that total divided by 36.

tem of dice is simply a measure of the number of configurations corresponding to each point total. For example, the state characterized by a point count of 7 has a higher entropy than the state with a point count of 2.

Pursuing the analogy further, suppose we now consider a single roll of a pair of dice. It is apparent that a process in which the point count changes from 2 to 7 will be six times more probable than one in which the reverse occurs. Since entropy is a measure of probability, the first process involves an increase in entropy while the second involves a decrease in entropy. Suppose, further, that we now increase the number of dice to four. The probability of obtaining a point count of 4 turns out to be $\frac{1}{1296}$ while that of obtaining 14 is $\frac{146}{1296}$. A process in which the point count changes from 4 to 14 is 146 times more likely than the reverse. Thus, a process involving a decrease in entropy becomes increasingly unlikely as the number of dice increases. Imagine, now, that we were to play this sort of game with Avogadro's number of dice. The likelihood of the occur-

**380** Spontaneity in Chemical Processes—The Second Law of Thermodynamics 11.3

rence of a process in which the entropy decreases will now be so small as to be completely negligible.

Let us now turn to a molecular system. Instead of a specific point count, we have a particular set of energy levels. The number of different ways the molecules of the system can arrange themselves in the available levels corresponds to the number of configurations of a particular point count of the dice. As before, the entropy of the system is a measure of the number of possible configurations. The number of distinct ways in which molecules of a given total energy can be arranged increases as the levels of the system become more closely spaced, as shown in Figure 11.4. Consequently, the entropy increases in a process in which the number of levels in a given energy interval increases.

The actual relation between the entropy of a system and the number of molecular configurations, $W$, is given by a simple formula derived by Boltzmann:

$$S = k \ln W \tag{11.9}$$

where $k$ is the Boltzmann constant, that is, the gas constant per molecule

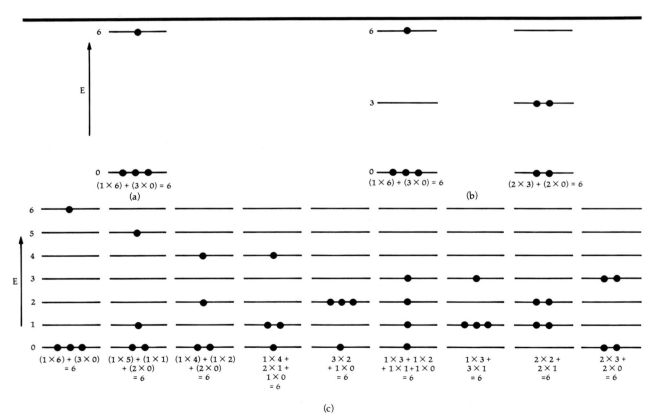

**FIGURE 11.4** A system consisting of four molecules distributed among (a) two, (b) three, and (c) seven energy levels. All arrangements have the same total energy, i.e., 6 (the units are arbitrary). Note that the number of configurations increases from (a) to (b) to (c) and so does the entropy of the system.

(Section 4.2). Note that since a system must have at least one configuration (usually $W \geqslant 1$), $S$ is always greater than or equal to zero.

Equation 11.9 is applied readily to the isothermal expansion of an ideal gas. It is convenient to consider a monatomic gas, for atoms possess only translational levels (Section 10.6). The number of translational levels available to a single atom is proportional to the number of places the atom can be and, hence, to the volume of the container. The particle-in-a-box model (Section 5.9) predicts a similar relationship: the energy difference between the levels of the particle decreases and their number consequently increases as the size of the box increases. When many atoms are present, the number of levels of the system is the number of levels available to the first atom times the number of levels available to the second atom, and so on. Thus, for a system of $N_A$ atoms, we have

$$W \propto V^{N_A}$$

where $V$ is the volume of the container. The change in entropy in an isothermal expansion from $V_1$ to $V_2$ is consequently

$$\Delta S = k \ln W_2 - k \ln W_1$$
$$= k \ln(W_2/W_1)$$
$$= k \ln(V_2/V_1)^{N_A} = kN_A \ln(V_2/V_1)$$
$$= R \ln(V_2/V_1)$$

where we have replaced the product of the Boltzmann constant and Avogadro's number by the gas constant $R$.

Note that this result is identical to Equation 11.6, which was obtained by a thermodynamic analysis, without reference to the microscopic constituents of the system. The molecular interpretation of entropy indicates that a gas expands spontaneously from a small to a large volume because of the enormous increase in the number of configurations. When the volume doubles, the ratio of the number of configurations is

$$W_2/W_1 = 2^{6 \times 10^{23}}$$

This number is so astronomically large that it is fair to say that a gas "always" expands spontaneously into a larger volume. The probability of the reverse process—the spontaneous compression of a gas into half the available volume—is given by the reciprocal of the above number. This number is so infinitesimally small that a gas "never" undergoes a spontaneous compression. Thus, although strictly speaking, we can only talk about the probability of a spontaneous process, the probability is so overwhelmingly large when the system contains a normal number of molecules that the occurrence of a spontaneous process is, in effect, a certainty.

## 11.4 Entropy Changes in Chemical Processes

We have so far considered the entropy change in only one type of process—the expansion of a gas. In this section we evaluate the entropy change in two other processes of importance in chemistry, a change in the temperature of a substance and a change of phase.

***a. Dependence of Entropy on Temperature*** The change in the entropy of a substance when it is heated can be obtained from Equation 11.1,

$$\Delta S = \int \frac{dq_{rev}}{T}$$

Recall from Section 10.5 that two well-defined ways of heating a substance are at constant pressure and at constant volume. At constant pressure, the heat flow is equal to the enthalpy change and, in turn, to the product of $C_p$ and the temperature change (Equation 10.28):

$$(dq_{rev})_P = dH = nC_p\, dT$$

Substituting into Equation 11.1, we obtain

$$\Delta S = \int_{T_1}^{T_2} \frac{nC_p\, dT}{T} \qquad \text{(constant P)} \tag{11.10}$$

The heat capacity of an ideal gas or that of a real monatomic gas is independent of temperature. In these instances, or when the temperature change is sufficiently small to permit the assumption of constant $C_p$, Equation 11.10 can be integrated and the entropy change is

$$\Delta S = nC_p \ln(T_2/T_1) \tag{11.11}$$

Equation 11.11 indicates that the entropy of a substance increases with temperature. This trend can be understood readily within the context of the molecular interpretation of entropy. With increasing temperature, the random motion of the molecules of a gas increases (Section 4.2); the same is true for solids and liquids. Since entropy is a measure of random molecular motion, it must show a concomitant increase.

***b. Entropy and Changes of Phase*** For a given pressure, a pure substance undergoes a change of phase at a specific temperature (Section 12.1). To cite a familiar example, water freezes at 0°C and boils at 100°C when the pressure is 1.0 atm. At the normal freezing or boiling point (i.e., at atmospheric pressure), the two phases are in equilibrium with each other and the phase change can be reversed by a small inflow or outflow of heat. For example, when ice and water are in a container at 0°C, a slight inflow of heat causes some of the ice to melt while a slight outflow of heat results in additional freezing. The process is reversible.

These two characteristics of an equilibrium phase change, namely, constant temperature and pressure, and reversibility, permit a straightforward evaluation of the entropy change. At constant temperature, Equation 11.1 can be integrated:

$$\Delta S = \int \frac{dq_{rev}}{T} = \frac{q_{rev}}{T}$$

As in the derivation of Equation 11.10, $q_{rev}$ can be identified with the enthalpy change for the phase transformation at constant pressure. Consequently, the entropy change is

$$\Delta S = \Delta H/T \tag{11.12}$$

A relation of this type can be written for any phase change. For example, the entropy of vaporization at atmospheric pressure is

$$\Delta S°_{vap} = \Delta H°_{vap}/T_b \qquad (11.13)$$

It is well known that a solid at its melting point must be heated for melting to occur. Similarly, a liquid at its boiling point must be heated for vaporization to occur; the enthalpies of fusion and vaporization are always positive (Section 10.7). It follows from Equation 11.12 that the entropy of a substance increases as its state of aggregation changes from solid to liquid to vapor. This behavior is understandable from the molecular viewpoint. In a crystalline solid, molecules occupy fixed positions; the state is highly ordered. At the other extreme, molecules in a gas are essentially independent and free to move at random; the gaseous state is the most disordered. Liquids occupy an intermediate position in this respect. Since the entropy of a system reflects the disorder and random motion of its constituent molecules, the increase in entropy observed between solid and liquid and between liquid and vapor is understandable.

**Example 11.2** (a) Two moles of water are vaporized at 100°C and atmospheric pressure. Determine $\Delta S_{system}$, $\Delta S_{surr}$, and $\Delta S_{total}$. The standard enthalpy of vaporization of water at 100°C is 40.7 kJ mol$^{-1}$. (b) Calculate $\Delta S_{system}$, $\Delta S_{surr}$, and $\Delta S_{total}$ for the vaporization of two moles of superheated water at 110°C and a pressure of 1.00 atm. Assume that in the temperature range 100–110°C the values of $C°_p$ of water and water vapor are 75.5 J mol$^{-1}$ K$^{-1}$ and 34.4 J mol$^{-1}$ K$^{-1}$, respectively.

(a) The entropy of vaporization is given by Equation 11.13,

$$\Delta S_{system} = \frac{\Delta H_{vap}}{T_b}$$

$$= \frac{(40.7 \text{ kJ mol}^{-1})(2.00 \text{ mol})(10^3 \text{ J kJ}^{-1})}{373 \text{ K}}$$

$$= 218 \text{ J K}^{-1}$$

According to Equation 11.7, the entropy change of the surroundings is equal to the heat absorbed by the surroundings ($q < 0$ if heat flows from the surroundings) divided by the temperature at which the heat transfer occurs. The heat flow from the surroundings is that needed to vaporize two moles of water. Thus,

$$\Delta S_{surr} = \frac{q}{T}$$

$$= \frac{(-40.7 \text{ kJ mol}^{-1})(2.00 \text{ mol})(10^3 \text{ J kJ}^{-1})}{373 \text{ K}}$$

$$= -218 \text{ J K}^{-1}$$

The total entropy change is

$$\Delta S_{total} = \Delta S_{system} + \Delta S_{surr}$$

$$= 218 \text{ J K}^{-1} - 218 \text{ J K}^{-1}$$

$$= 0$$

**FIGURE 11.5** The conversion of superheated water to steam at 110°C (Example 11.2b). Two paths are shown: (a) the spontaneous conversion at 110°C and (b) a reversible path.

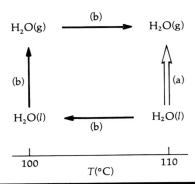

The entropy of the system plus its surroundings does not change because water and water vapor are in equilibrium at the normal boiling point.

(b) The change in the entropy of the system can be determined by devising a reversible path between the initial and final states and using the appropriate relations to determine $\Delta S$ for the specific steps in this path. This must be done because the expressions for $\Delta S$ (e.g., Equation 11.1) apply only to reversible processes. Note that the direct conversion of water to steam at 110°C does not constitute an appropriate path. A change of phase is reversible only at the temperature at which the two phases are in equilibrium. For water at atmospheric pressure, this temperature is 100°C rather than 110°C. Thus, a temperature of 100°C must be used in determining $\Delta S$ for the conversion of water to steam. Keeping this important consideration in mind, we can construct the path shown in Figure 11.5, each step of which can be carried out reversibly:

**Step 1**    $H_2O(l)_{383 K} \rightarrow H_2O(l)_{373 K}$

The entropy change for this step is given by Equation 11.11,

$$\Delta S = nC_p \ln(T_2/T_1)$$
$$= (2.00 \text{ mol})(75.5 \text{ J mol}^{-1} \text{ K}^{-1}) \ln(373 \text{ K}/383 \text{ K})$$
$$= -3.99 \text{ J K}^{-1}$$

**Step 2**    $H_2O(l)_{373 K} \rightarrow H_2O(g)_{373 K}$

The entropy change for this process has already been evaluated in (a) as $\Delta S = 218.2 \text{ J K}^{-1}$.

**Step 3**    $H_2O(g)_{373 K} \rightarrow H_2O(g)_{383 K}$

Once again, we use Equation 11.11,

$$\Delta S = nC_p \ln(T_2/T_1)$$
$$= (2.00 \text{ mol})(34.4 \text{ J mol}^{-1} \text{ K}^{-1}) \ln(383 \text{ K}/373 \text{ K})$$
$$= 1.82 \text{ J K}^{-1}$$

The overall entropy change for the system is the sum of these three contributions:

$$\Delta S_{\text{system}} = (-3.99 + 218.2 + 1.82) \text{ J K}^{-1}$$
$$= 216 \text{ J K}^{-1}$$

The change in the entropy of the surroundings is given by Equation 11.7, $\Delta S_{\text{surr}} = q/T$. In this case $q$ is the actual heat that must flow from the surroundings into the system at 110°C in order to vaporize two moles of water. While this quantity, which is just the enthalpy of vaporization of water at 110°C and atmospheric pressure, is not given, it can be evaluated by means of the same path used to determine the entropy of vaporization at this temperature (Figure 11.5). Using Equation 10.36 for the change in enthalpy with temperature, we have

$$\Delta H_{\text{vap}}(383) = nC_p[\text{H}_2\text{O}(l)](373-383)\text{K} + \Delta H_{\text{vap}}(373)$$
$$+ nC_p[\text{H}_2\text{O}(g)](383-373)\text{K}$$
$$= (2.00 \text{ mol})(75.5 \text{ J K}^{-1})(-10 \text{ K})$$
$$+ (2.00 \text{ mol})(4.07 \times 10^4 \text{ J mol}^{-1})$$
$$+ (2.00 \text{ mol})(34.4 \text{ J K}^{-1})(10 \text{ K})$$
$$= 8.058 \times 10^4 \text{ J}$$

Substituting into Equation 11.7, we obtain

$$\Delta S_{\text{surr}} = (-8.058 \times 10^4 \text{ J})/383 \text{ K} = -210 \text{ J K}^{-1}$$

where $q < 0$ because heat flows *from* the surroundings. The change in total entropy is

$$\Delta S_{\text{total}} = \Delta S_{\text{system}} + \Delta S_{\text{surr}}$$
$$= (216 - 210) \text{ J K}^{-1}$$
$$= 6 \text{ J K}^{-1}$$

The entropy of the system plus its surroundings increases, indicating that the process is spontaneous. This result illustrates the fact that a liquid superheated to above its normal boiling point vaporizes spontaneously. ∎

## 11.5 Third Law Entropies

***a. Entropy Determination*** The expressions for the various entropy changes derived in Section 11.4 may be used to evaluate the molar entropies of substances as a function of temperature. The results are known as **third law entropies** and, when evaluated at 298 K and atmospheric pressure, as **standard entropies,** designated $S^\circ_{298}$.

To see how these entropies can be obtained, consider the following process. Start with a solid at 0 K and heat it to its melting point at constant atmospheric pressure. The entropy change is given by Equation 11.10. When the solid melts, the entropy increases, as given by Equation 11.12. The liquid is now heated to its boiling point and vaporized; the entropy changes are once again given by Equations 11.10 and 11.12, respectively. Finally, the gas is heated to 298 K and the entropy change evaluated by means of Equation 11.10. The various entropy changes may be combined to give the following expression for the standard molar entropy:

$$S^\circ_{298} = S^\circ_0 + \int_0^{T_f} \frac{C^\circ_P(s)\, dT}{T} + \frac{\Delta H^\circ_{fus}}{T_f} + \int_{T_f}^{T_b} \frac{C^\circ_P(l)\, dT}{T}$$
$$+ \frac{\Delta H^\circ_{vap}}{T_b} + \int_{T_b}^{298} \frac{C^\circ_P(g)\, dT}{T} \tag{11.14}$$

This expression assumes that a substance exists as a gas at 25°C and atmospheric pressure. If this is not the case, the superfluous terms must be deleted. For example, only the first integral contributes to the standard entropy of a solid.

The first term in Equation 11.14, $S^\circ_0$, is the entropy at 0 K. At absolute zero, all thermal disorder has been quenched from a substance and, in the case of perfect crystals, i.e., crystals with no imperfections (Section 20.9), the atoms or molecules are in a completely ordered state. It is believed that all elements and compounds which form perfect crystals possess the same entropy at 0 K, and it is convenient to set the value to zero, i.e.,

$$S^\circ_0 = 0 \tag{11.15}$$

This statement constitutes the **third law of thermodynamics.** This postulate is consistent with Boltzmann's expression for the entropy (Equation 11.9), $S = k \ln W$, for only a single configuration is available to a perfectly ordered array of molecules. Thus, $W = 1$ and $S = 0$.

Having disposed of $S^\circ_0$, we can determine the third law entropies by evaluation of Equation 11.14. The various integrals in this expression must be evaluated over a broad range of temperatures, from the melting point to the boiling point of a substance, for example. It is generally not valid to assume that the heat capacity remains constant under these conditions, and Equation 11.11 for the change in entropy on heating therefore cannot be used. Instead, the various integrals are evaluated graphically, as shown in Figure 11.6. Experimental values of $C_P/T$ for a particular substance are plotted versus $T$. The entropy at temperature $T$ is given by the area under the curve evaluated up to $T$. So long as only a single phase is considered, the curves are continuous. Discontinuities occur at phase changes and the change in entropy at these temperatures can be obtained directly by means of Equation 11.12. To illustrate the procedure, Table 11.1 lists the contributions of the separate terms in Equation 11.14 to the third law entropy of $N_2(g)$ at 298 K.

**b. Standard Entropies of Substances**  The procedure described in Section 11.5a has been used to determine the standard entropies $S^\circ_{298}$ of many different substances and aqueous ions. A partial list is given in Table 11.2, and a more complete set of values may be found in Appendix 4. Note that these are "absolute" entropy values. In contrast to the internal energy and enthalpy, the entropy content of a substance can be determined. Thus, the third law entropies are designated $S^\circ$ rather than $\Delta S^\circ$.

The tabulated third law entropies reveal some interesting trends. On average, the entropies of gases are larger than those of liquids which, in turn, are larger than those of solids. This trend is a direct consequence of Equation 11.14 for the third law entropies, which indicates that an additional term contributes to the entropy for each change of phase. Put another way, the molecular disor-

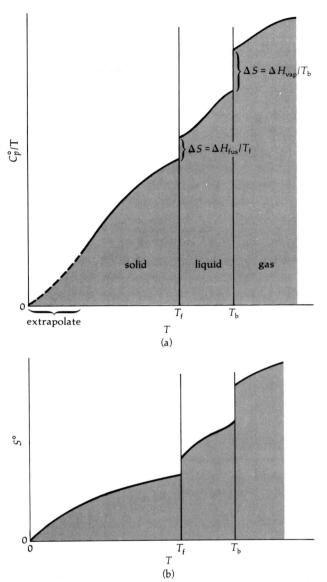

**FIGURE 11.6** Graphical evaluation of Equation 11.14 for the third law entropy of a substance. (a) Plot of $C_P^\circ/T$ versus $T$. The entropy at $T$ is equal to the area under the curve up to $T$. Discontinuities in $C_P^\circ/T$ occur at phase changes and the corresponding entropy change is evaluated by means of Equation 11.12. Experimental values of $C_P^\circ$ at very low temperatures are generally not available; the values of $C_P^\circ/T$ are therefore extrapolated to 0 K as shown. (b) Plot of the third law entropy $S^\circ$ versus $T$. At a given $T$, the value of $S^\circ$ is given by the area under the curve in (a) from $T = 0$ to $T$.

**TABLE 11.1 Evaluation of $S°_{298}$ for $N_2(g)$ by Means of Equation 11.14**

| Step | $\Delta S°$ (J mol$^{-1}$ K$^{-1}$) |
|---|---|
| 0–10 K [extrapolation of higher $T$ data for $N_2(s)$] | 1.92 |
| 10–35.61 K [graphical integration of heat capacity data for $N_2(s)$] | 25.25 |
| 35.61 K (crystal structure transition)[a] | 6.43 |
| 35.61–63.14 K [graphical integration of heat capacity data for $N_2(s)$] | 23.38 |
| 63.14 K (solid $N_2$ melts) | 11.42 |
| 63.14–77.32 K [graphical integration of heat capacity data for $N_2(l)$] | 11.41 |
| 77.32 K (liquid $N_2$ boils) | 72.13 |
| 77.32–298.2 K (ideal gas approximation, Equation 11.11) | 39.20 |
| 77.32–298.2 K [correction for nonideality of $N_2(g)$] | 0.92 |
| | $S°_{298} = 192.06$ |

[a] The crystal structure of $N_2(s)$ changes at 35.61 K. This transition is accompanied by an enthalpy change and $\Delta S$ is evaluated by means of Equation 11.12.

der increases from solids to liquids to gases and the entropy increases accordingly.

The entropy of gases increases both with molecular weight and molecular complexity. The mass effect is seen most clearly for the noble gases, the entropy of which increases from 126 J mol$^{-1}$ K$^{-1}$ for He to 170 J mol$^{-1}$ K$^{-1}$ for xenon, or for the hydrogen halides, whose entropy increases from 174 J mol$^{-1}$ K$^{-1}$ for HF(g) to 207 J mol$^{-1}$ K$^{-1}$ for HI(g). The effect of molecular complexity can be seen in a comparison of molecules having the same mass but different numbers of atoms. For example, the entropy of gaseous Ne and HF, both of which have mass 20, increases in the indicated order.

The increase of entropy with mass and complexity is also displayed by solid substances. For example, the entropies of the metal chlorides of the third period range from 72 J mol$^{-1}$ K$^{-1}$ for NaCl to 111 J mol$^{-1}$ K$^{-1}$ for AlCl$_3$, while

**TABLE 11.2 Standard Entropies of Selected Solids, Liquids, and Gases**

| Solids | $S°_{298}$ (J mol$^{-1}$ K$^{-1}$) | Liquids | $S°_{298}$ (J mol$^{-1}$ K$^{-1}$) | Gases | $S°_{298}$ (J mol$^{-1}$ K$^{-1}$) |
|---|---|---|---|---|---|
| C(graphite) | 5.74 | $H_2O$ | 69.9 | He | 126.2 |
| Cu | 33.2 | $Br_2$ | 152.2 | Ne | 146.3 |
| $CuSO_4$ | 109 | $CH_3OH$ | 126.8 | Ar | 154.8 |
| $CuSO_4 \cdot 5H_2O$ | 300 | $C_2H_5OH$ | 160.7 | $H_2$ | 130.7 |
| Ag | 42.6 | | | HF | 173.7 |
| AgCl | 96.2 | | | $N_2$ | 191.6 |
| AgI | 115.5 | | | $O_2$ | 205.1 |
| NaF | 51.5 | | | $F_2$ | 202.8 |
| NaCl | 72.1 | | | $Cl_2$ | 223.1 |
| NaBr | 86.8 | | | CO | 197.7 |
| NaI | 98.5 | | | $CO_2$ | 213.7 |

the entropies of the sodium halides increase from 52 J mol⁻¹ K⁻¹ for NaF to 99 J mol⁻¹ K⁻¹ for NaI. These various trends reflect the increase in the number of configurations with increasing mass and complexity and the dependence of the entropy on the number of configurations. As in the case of the heat capacity (Section 10.6), the methods of statistical mechanics may be used to evaluate the contributions of translational, rotational, and vibrational molecular motion to the entropies of substances.

***c. Entropy Changes in Chemical Reactions*** The tabulated standard entropies may be used to evaluate the entropy changes in chemical reactions occurring at standard conditions. The procedure is analogous to the use of standard enthalpies of formation to determine reaction enthalpies. Like enthalpy, entropy is a state function. The entropy change in a process is equal to the entropy of the final state less that of the initial state. If the process is a chemical reaction, the entropy change is equal to the entropy of the products less that of the reactants. In analogy to Equation 10.34 for the reaction enthalpy, we write

$$\Delta S°_{298} = \sum n_p S°_{298} - \sum n_r S°_{298} \qquad (11.16)$$

where the standard molar entropies are multiplied by the stoichiometric coefficients appearing in the balanced equation in order to obtain the entropy change in the reaction.

---

**Example 11.3** Calculate $\Delta S°_{298}$ for the reaction

$$2H_2(g) + O_2(g) \rightarrow 2H_2O(l)$$

Applying Equation 11.16 to the standard entropies, we obtain

$$\Delta S°_{298} = 2S°_{298}[H_2O(l)] - 2S°_{298}[H_2(g)] - S°_{298}[O_2(g)]$$

$$= (2 \text{ mol})(69.9 \text{ J mol}^{-1} \text{ K}^{-1}) - (2 \text{ mol})(130.7 \text{ J mol}^{-1} \text{ K}^{-1})$$

$$- (1 \text{ mol})(205.1 \text{ J mol}^{-1} \text{ K}^{-1})$$

$$= -326.7 \text{ J K}^{-1}$$

Although the entropy of the system decreases in the reaction, we cannot draw any conclusions concerning the spontaneity of the reaction from this fact. As noted in Section 11.2, spontaneity is determined by the change in the total entropy of the system *and* the surroundings. ∎

Example 11.3 illustrates a general observation about entropy changes in chemical reactions: a reaction in which the number of moles of gaseous substances increases (decreases) is accompanied by an increase (decrease) in entropy. Thus, in the reaction of interest, $\Delta S < 0$ and $\Delta n_{gas} = -3$. Similar calculations provide the following additional examples of this generalization:

$$C(\text{graphite}) + \tfrac{1}{2}O_2(g) \rightarrow CO(g) \qquad \Delta n_{gas} = \tfrac{1}{2} \qquad \Delta S°_{298} = 89.4 \text{ J K}^{-1}$$

$$N_2O_4(g), \rightarrow 2NO_2(g) \qquad \Delta n_{gas} = 1 \qquad \Delta S°_{298} = 175.8 \text{ J K}^{-1}$$

$$Ca(s) + \tfrac{1}{2}O_2(g) \rightarrow CaO(s) \qquad \Delta n_{gas} = -\tfrac{1}{2} \qquad \Delta S°_{298} = -104.2 \text{ J K}^{-1}$$

The observed correlation is another manifestation of the connection between entropy and molecular disorder already illustrated in Section 11.5b.

The correlation between entropy and disorder can be used to predict the sign of the entropy change for various other types of reactions. For example, the dissolution of solids or liquids in water is generally accompanied by an increase in entropy. The change is particularly noticeable when a substance dissociates upon dissolution, e.g.,

$$KCl(s) \xrightarrow{H_2O} K^+(aq) + Cl^-(aq) \qquad \Delta S^\circ_{298} = 76.4 \text{ J K}^{-1}$$

Even though some water molecules are tied up in hydrating the ions, a system involving separated ions in solution is more disordered than one in which the ions are held in place in a crystal. Thus, the entropy increases on dissolution.

# Free Energy and Chemical Equilibrium

## 11.6 The Free Energy Function

Although an increase in the total entropy is the most general criterion of spontaneity, this criterion is not, in fact, customarily used. The procedure involved in the determination of the total entropy is simply too cumbersome. Recall that the total entropy is obtained as the sum of the separate entropy changes for the system and its surroundings. Moreover, as shown in Examples 11.1 and 11.2, the change in the entropy of the surroundings is derived from the same type of thermodynamic data as that used to obtain the change in the entropy of the system. It is therefore possible to rephrase the criterion for spontaneity solely in terms of properties of the system, without explicit consideration of the surroundings. To do so, we introduce a new state function known as the free energy, or more precisely, as the Gibbs free energy, designated G. We will show that the change in this function is a criterion of spontaneity for the most common type of chemical process, one occurring at constant temperature and pressure. In addition, the free energy is intimately connected with the position of chemical equilibrium. The connection between thermal and equilibrium properties is one of the major accomplishments of thermodynamics.

The **free energy** is defined by the relation

$$G = H - TS \qquad (11.17)$$

Since the free energy is the sum of quantitites that are state functions, it, too, is a state function. An important consequence of this definition is that for an isothermal process the relation

$$\Delta G = \Delta H - T\Delta S \qquad (T \text{ constant}) \qquad (11.18)$$

for the change in the free energy of a system is obeyed.

We wish to show that the change in free energy is a criterion for spontaneity and the direction of spontaneous change, or for the existence of equilibrium in a process carried out at constant temperature and pressure. To do so, we express the change in total entropy in terms of the change in the entropy of the system and the surroundings,

$$\Delta S_{\text{total}} = \Delta S_{\text{system}} + \Delta S_{\text{surr}} \qquad (11.19)$$

The change in the entropy of the surroundings is given by Equation 11.7,

$$\Delta S_{\text{surr}} = q/T$$

For a constant pressure process, $q$ can be identified with the change in enthalpy of the surroundings (Section 10.5) and, therefore, with $-\Delta H$ for the system,

$$\Delta S_{\text{surr}} = -\frac{\Delta H_{\text{system}}}{T} \tag{11.20}$$

Substituting this result into Equation 11.19, we obtain after rearrangement

$$T\Delta S_{\text{total}} = T\Delta S_{\text{system}} - \Delta H_{\text{system}} \tag{11.21}$$

Comparison of Equation 11.21 with Equation 11.18 shows that it is possible to identify $T\Delta S_{\text{total}}$ with $-\Delta G$ for a process carried out at constant temperature and pressure:

$$\Delta G = -T\Delta S_{\text{total}} = \Delta H - T\Delta S \quad (P, T \text{ constant}) \tag{11.22}$$

where the subscript "system" has been dropped from $\Delta G$, $\Delta H$, and $\Delta S$ because we no longer need to refer to the surroundings. Equation 11.22 is important because it indicates that the free energy change of the *system* is equal to the negative of the temperature times the *total* entropy change provided only that the temperature and pressure remain constant. This simple expedient allows us to replace the previously developed criterion of spontaneity for a system plus its surroundings with on that is dependent only on a property of the system. Substituting Equation 11.22 into Equations 11.3 and 11.4 we obtain for a process at constant temperature and pressure:

$$\begin{aligned} &\Delta G < 0 &&\text{the process occurs spontaneously} \\ &\Delta G = 0 &&\text{the process is reversible; the system is at equilibrium} \\ &\Delta G > 0 &&\text{the reverse process occurs spontaneously} \end{aligned} \tag{11.23}$$

Let us illustrate the validity of Equation 11.23 by consideration of a change of phase, a process which we know to occur with no change in temperature or pressure. We calculate $\Delta G$ for the vaporization at atmospheric pressure of two moles of water, first at 100°C and then at 110°C, using the data given in Example 11.2.

At 100°C,

$$\begin{aligned} \Delta G° &= \Delta H° - T\Delta S° \\ &= (40.7 \times 10^3 \text{ J mol}^{-1})(2.00 \text{ mol}) - (373 \text{ K})(218.2 \text{ J K}^{-1}) \\ &= 0 \end{aligned}$$

Since the normal boiling point of water is 100°C, the two phases are in equilibrium at this temperature.

At 110°C,

$$\begin{aligned} \Delta G° &= \Delta H° - T\Delta S° \\ &= 8.058 \times 10^4 \text{ J} - (383 \text{ K})(216.0 \text{ J K}^{-1}) \\ &= -2.15 \times 10^3 \text{ J} \end{aligned}$$

## 11.7 Free Energy Changes in Chemical Reactions

*a. The Makeup of the Free Energy Change* The free energy change at standard conditions can be calculated for any chemical reaction by means of Equation 11.18 and tabulated values of standard entropies and enthalpies. The following example illustrates the procedure.

---

**Example 11.4**  Calculate $\Delta H°$, $\Delta S°$, and $\Delta G°$ for the reaction

$$3H_2(g) + N_2(g) \rightleftarrows 2NH_3(g)$$

Determine whether the reaction occurs spontaneously at standard conditions. Use the following data taken from Appendix 4 (four significant figures):

| Substance | $\Delta H_f°$ (kJ mol$^{-1}$) | $S°$ (J mol$^{-1}$ K$^{-1}$) |
|---|---|---|
| $NH_3(g)$ | −46.11 | 192.5 |
| $H_2(g)$ | 0 | 130.7 |
| $N_2(g)$ | 0 | 191.6 |

The values of $\Delta H°$ and $\Delta S°$ are obtained in the previously described manner (Examples 10.8, 11.3):

$$\Delta H° = 2\Delta H_f°(NH_3) - 3\Delta H_f°(H_2) - \Delta H_f°(N_2)$$
$$= (2 \text{ mol})(-46.11 \text{ kJ mol}^{-1})$$
$$= -92.22 \text{ kJ}$$

$$\Delta S° = 2S°(NH_3) - 3S°(H_2) - S°(N_2)$$
$$= (2 \text{ mol})(192.5 \text{ J mol}^{-1} \text{ K}^{-1}) - (3 \text{ mol})(130.7 \text{ J mol}^{-1} \text{ K}^{-1})$$
$$- (1 \text{ mol})(191.6 \text{ J mol}^{-1} \text{ K}^{-1})$$
$$= -198.7 \text{ J K}^{-1}$$

The value of $\Delta G°$ is obtained from these results by means of Equation 11.18 evaluated at 298 K and atmospheric pressure:

$$\Delta G° = \Delta H° - T\Delta S°$$
$$= -92.22 \text{ kJ} - (298 \text{ K})(-198.7 \text{ J K}^{-1})(10^{-3} \text{ kJ J}^{-1})$$
$$= -33.0 \text{ kJ}$$

Since $\Delta G° < 0$, the reaction occurs spontaneously at standard conditions. ∎

It is apparent from Equation 11.18 and Example 11.4 that there are two distinct contributions to $\Delta G$, that from the enthalpy change of the system, and that from the entropy change of the system. A negative reaction enthalpy favors the spontaneity of the reaction, reflecting the drive of a system to attain a state of lower potential energy. A weight suspended above the ground provides a familiar example of this drive: when the weight is released it falls and its potential energy decreases. In the absence of a change in the entropy of the system, $\Delta G = \Delta H$, and the enthalpy change becomes the sole criterion for spontaneity—the reaction occurs spontaneously if it is exothermic and does not occur spontaneously if it is endothermic. On the other hand, an increase in the entropy of the system also favors spontaneity, reflecting the drive of the system towards greater disorder. In the absence of enthalpy changes, it is the entropy change that determines whether a reaction occurs spontaneously. The free energy represents the combined effect of the drive by the system towards lower enthalpy and higher entropy.

For some reactions, both $\Delta H$ and $\Delta S$ either favor or disfavor spontaneity and their effects reinforce each other. For other reactions, $\Delta H$ and $\Delta S$ oppose each other and the term of larger numerical magnitude determines whether a reaction is spontaneous. In Example 11.4, for instance, the formation of ammonia is favored by a decrease in enthalpy and disfavored by the decrease in entropy associated with the decrease in the number of moles of gas. Since the magnitude of $\Delta H$ is larger than that of $T\Delta S$, the reaction occurs spontaneously.

It is apparent from these considerations that other combinations of enthalpy and entropy changes can also result in a spontaneous reaction. Consider, for instance, the reaction

$$H_2(g) + Br_2(l) \rightarrow 2HBr(g)$$

A calculation of the type performed in Example 11.4 yields the following results:

$$\Delta H° = -72.8 \text{ kJ} \qquad T\Delta S° = 34.1 \text{ kJ} \qquad \Delta G° = -106.9 \text{ kJ}$$

In this instance, both the decrease in enthalpy and the increase in entropy favor the occurrence of the reaction. As a final example, the dissolution of NaCl,

$$NaCl(s) \xrightarrow{H_2O} Na^+(aq) + Cl^-(aq)$$

is an example of a process which occurs spontaneously even though it is endothermic. In this case the entropy term is more important than the enthalpy term and the driving force for the process is the increase in disorder. The results obtained for this reaction are as follows:

$$\Delta H° = 3.44 \text{ kJ} \qquad T\Delta S° = 12.92 \text{ kJ} \qquad \Delta G° = -9.48 \text{ kJ}$$

Figure 11.7 schematically illustrates the three possible combinations of $\Delta H°$ and $T\Delta S°$ that result in a decrease in the free energy, thus causing a process to be spontaneous.

**b. Standard Free Energies of Formation** The evaluation of the free energy change in a chemical reaction in terms of the separate contributions of the enthalpy and entropy changes is useful in pinpointing which of these factors is dominant. However, the procedure is unnecessarily cumbersome when all that may be of interest is the actual value of $\Delta G$. Since $\Delta G$ values are widely used, a more direct method for their determination is desirable. The procedure that has been developed is analogous to that involved in the determination of reaction enthalpies from standard enthalpies of formation. The **standard free energy of formation** of a substance, $\Delta G_f°$, is defined as the free energy change for the reaction in which one mole of the substance is formed at 25°C and one atmosphere from the elements in their standard states. As in the case of the standard enthalpies of formation, it is customary to set $\Delta G_f° = 0$ for the elements in their standard states. This procedure is permissible because only differences in

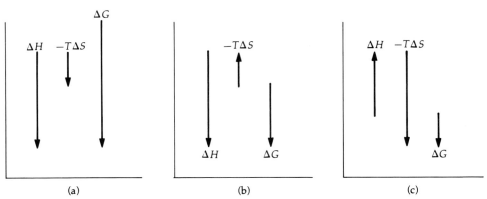

**FIGURE 11.7** Relative contributions of $\Delta H$ and $T\Delta S$ to $\Delta G$ in spontaneous reactions. The length of each arrow is proportional to the magnitude of the quantity it represents; downward arrows represent negative changes. (a) Both $\Delta H$ and $\Delta S$ favor spontaneity; (b) the decrease in $\Delta H$ is the driving force; and (c) the increase in $\Delta S$ is the driving force.

free energy are measurable. In analogy to Equation 10.34, the free energy change for a reaction at standard conditions can then be obtained from the standard free energies of formation of reactants and products by the relation

$$\Delta G°_{298} = \sum n_p \Delta G°_f - \sum n_r \Delta G°_f \tag{11.24}$$

For example, the standard free energy change for the formation of ammonia from the elements (Example 11.4) is

$$\Delta G°_{298} = 2\Delta G°_f(NH_3) - 3\Delta G°_f(H_2) - \Delta G°_f(N_2)$$
$$= (2 \text{ mol})(-16.5 \text{ kJ mol}^{-1}) - 0 - 0$$
$$= -33.0 \text{ kJ}$$

which is in agreement with the value obtained by combining the reaction enthalpies and entropies.

Appendix 4 lists the values of $\Delta G°_f$ for a variety of substances. This list is the final entry in a table comprised of the values of $\Delta H°_f$, $S°$, and $C°_p$. These entries are collectively referred to as the thermodynamic properties of substances at standard conditions and are of use in a variety of calculations, some of which have already been illustrated. The remaining sections of this chapter provide additional illustrations of the use of these tables.

## 11.8 Free Energy and the Equilibrium Constant

Consider a spontaneous process in which there is no heat flow, such as a marble rolling down an incline (Figure 11.8a). It is well known that the marble seeks a position of lowest potential energy; its equilibrium position is at the bottom of the incline. The free energy plays an analogous role in chemical reactions. Starting with either reactants or products, a reaction proceeds spontaneously in the direction of decreasing free energy until equilibrium is attained,

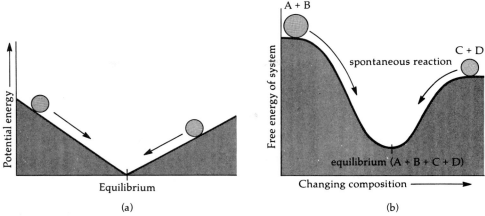

**FIGURE 11.8** Analogy between the drive of a mechanical system towards minimum potential energy and that of a chemical system towards minimum free energy. (a) A marble rolling down an incline and (b) reagents undergoing chemical reaction. The reaction occurs spontaneously until equilibrium is attained.

at which point the free energy of the system is at a minimum (Figure 11.8b). Recall from Section 7.2 that the position of chemical equilibrium can be expressed by a quotient of equilibrium concentrations or partial pressures of products and reactants, as given by the equilibrium constant. These considerations suggest that free energy and the equilibrium constant must be related in some manner.

The relation between the free energy change and the equilibrium constant is most readily derived for a reaction involving ideal gases. The result, however, is of general validity. Since in a reaction involving gases the partial pressure of each substance changes until equilibrium is attained, the dependence of the free energy on pressure must be examined before the relation to the equilibrium constant can be determined.

Recalling the basic definitions given in Equation 11.17 and 10.22, i.e.,

$$G = H - TS \quad \text{and} \quad H = E + PV$$

we can write the combined relation

$$G = E + PV - TS \tag{11.25}$$

The differential of the free energy can then be written as

$$dG = dE + P\,dV + V\,dP - T\,dS - S\,dT \tag{11.26}$$

The first law is written in differential form as

$$dE = dq + dw$$

and for a reversible process in which only pressure–volume work is performed

$$dw = -P\,dV$$

Substituting these last two relations into Equation 11.26 we obtain

$$\begin{aligned} dG &= dq - P\,dV + P\,dV + V\,dP - T\,dS - S\,dT \\ &= dq + V\,dP - T\,dS - S\,dT \end{aligned} \tag{11.27}$$

Since we are considering a reversible process we can, according to Equation 11.2, equate $dq$ with $T\,dS$. Furthermore, if we restrict ourselves to an isothermal process, $T$ is constant and $dT$ must be zero. Under these conditions Equation 11.27 reduces to

$$dG = V\,dP \qquad \text{(constant } T\text{)} \tag{11.28}$$

In order to determine the free energy change associated with a finite change in pressure, Equation 11.28 must be integrated between initial pressure $P_1$ and final pressure $P_2$. The result is

$$\Delta G = G_2 - G_1 = \int_{P_1}^{P_2} V\,dP \tag{11.29}$$

To perform the integration explicitly, $V$ must be expressed in terms of $P$. This can be done readily for an ideal gas, for which $V = RT/P$ per mole. Substituting into Equation 11.29, we obtain

$$G_2 - G_1 = \int_{P_1}^{P_2} \frac{RT\,dP}{P}$$

The integration can be performed provided $T$ is constant, yielding

$$G_2 - G_1 = RT \ln(P_2/P_1) \tag{11.30}$$

This relation gives the change in the free energy of one mole of an ideal gas, the pressure of which is increased isothermally from $P_1$ to $P_2$. Since the free energy is a state function, this result is independent of whether the process is reversible or not.

Let us now specify the initial state to be the standard state, i.e., $P_1 = 1$ atm, and let the final state correspond to some arbitrary pressure $P$. Equation 11.30 can then be written as

$$G = G° + RT \ln(P/1 \text{ atm}) \tag{11.31}$$

or, for $n$ moles,

$$nG = nG° + nRT \ln(P/1 \text{ atm}) \tag{11.32}$$

Now consider a chemical reaction between ideal gases, which we write in the customary general form as

$$aA + bB \rightarrow cC + dD$$

The free energy change in the reaction is equal to the difference between the free energy of the products and that of the reactants:

$$\Delta G = \sum G_{(\text{products})} - \sum G_{(\text{reactants})} \tag{11.33}$$

For the reaction written in the preceding manner we obtain

$$\Delta G = cG_C + dG_D - aG_A - bG_B \tag{11.34}$$

Using Equation 11.32 we can rewrite this relation in terms of the standard free energies and pressures of the participating gases:

$$\Delta G = cG°_C + cRT \ln \frac{P_C}{1 \text{ atm}} + dG°_D + dRT \ln \frac{P_D}{1 \text{ atm}} \tag{11.35}$$

$$-\left[aG_A^\circ + aRT \ln \frac{P_A}{1 \text{ atm}}\right] - \left[bG_B^\circ + bRT \ln \frac{P_B}{1 \text{ atm}}\right] \tag{11.35}$$

Combining terms we obtain

$$\Delta G = \Delta G^\circ + RT(c \ln \frac{P_C}{1 \text{ atm}} + d \ln \frac{P_D}{1 \text{ atm}}$$

$$- a \ln \frac{P_A}{1 \text{ atm}} - b \ln \frac{P_B}{1 \text{ atm}}) \tag{11.36}$$

and further combining the logarithmic terms we get

$$\Delta G = \Delta G^\circ + RT \ln \frac{(P_C/1 \text{ atm})^c (P_D/1 \text{ atm})^d}{(P_A/1 \text{ atm})^a (P_B/1 \text{ atm})^b}$$

$$= \Delta G^\circ + RT \ln Q \tag{11.37}$$

where $Q$ is the reaction quotient (Section 7.3). This relation expresses the free energy change in a chemical reaction in terms of the standard free energy change—the free energy change for the reaction in which reactants at standard conditions ($P = 1$ atm) are converted to products at standard conditions—and the actual partial pressures of the reagents.

Suppose, now, that the partial pressures are those that occur at equilibrium. The reaction quotient $Q$ is then equal to the equilibrium constant $K$. Since the pressure and temperature are constant and the system is at equilibrium, the condition $\Delta G = 0$ must hold true (Equation 11.23). Substitution into Equation 11.37 finally gives the desired relation:

$$\Delta G^\circ = -RT \ln K \tag{11.38}$$

where for reactions involving gases the equilibrium constant $K$ is actually $K_P$.

Equation 11.38 is one of the most important equations of thermodynamics since it ties together two important yet different aspects of chemistry. The standard free energy change of a reaction is obtained from the standard free energies of formation of the participating species. The $\Delta G_f^\circ$ values are in turn based on thermal measurements, e.g., standard enthalpies of formation and heat capacities. On the other hand, the equilibrium constant is based on the partial pressures or concentrations of the participating species at equilibrium. In deriving this equation we have demonstrated that thermal properties and chemical equilibrium, which at first sight appear to be different aspects of chemistry, actually are intimately related.

It should be noted that Equation 11.38 relates the equilibrium constant to the *standard* free energy change $\Delta G^\circ$ rather than to the *actual* free energy change $\Delta G$. The actual free energy of the system changes as the reaction proceeds and $\Delta G$ becomes equal to zero at equilibrium. On the other hand, both $\Delta G^\circ$ and $K$ are constant quantities for a particular reaction at a given temperature. The distinction can be visualized by substituting the expression for $\Delta G^\circ$ from Equation 11.38 into Equation 11.37. After combining the logarithmic terms, we obtain

$$\Delta G = RT \ln(Q/K) \tag{11.39}$$

At the beginning of the reaction, the concentration of the reactants is larger than it is at equilibrium while that of the products is smaller, i.e., $Q < K$. Un-

**FIGURE 11.9** Dependence of $\Delta G$ of a chemical reaction on $Q/K$. When $Q < K$, $\Delta G < 0$ and the reaction occurs spontaneously. When $Q = K$, $\Delta G = 0$, and equilibrium has been reached. When $Q > K$, $\Delta G > 0$, and the reverse reaction occurs spontaneously.

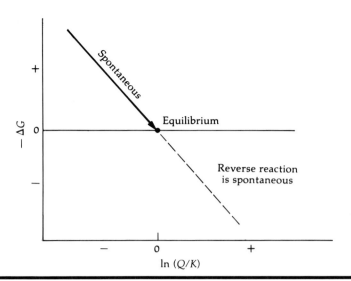

der these circumstances, $\Delta G < 0$, and the reaction occurs spontaneously. As the reaction proceeds, $Q$ increases and $\Delta G$ becomes less negative. Eventually $Q$ becomes as large as $K$, and then $\Delta G$ equals zero. At this point the reaction winds down and equilibrium is attained. Figure 11.9 depicts these changes graphically.

The standard free energy change, $\Delta G°$, is related to the value of the equilibrium constant and, thus, to the relative amounts of reactants and products present at equilibrium. Equation 11.38 indicates that when $\Delta G° < 0$, then $K > 1$. Under these conditions the equilibrium concentration of products is greater than that of reactants and equilibrium favors the formation of products. On the other hand, when $\Delta G° > 0$, then $K < 1$, and the equilibrium mixture consists largely of the reactants. Note that some products can still be formed from the reactants, but their equilibrium concentrations will be smaller than those of the reactants.

Equation 11.38 permits a determination of the equilibrium constant from the standard free energy change and vice versa. For some reactions, the determination of the equilibrium constant may pose experimental difficulties while the value of $\Delta G°$ can be determined readily. The opposite may be true for other reactions. The use of Equation 11.38 offers an alternate route to direct measurement for the determination of these important quantities.

**Example 11.5** Using the standard free energies of formation tabulated in Appendix 4, determine the equilibrium constant of the reaction

$$2NO_2(g) \rightleftharpoons N_2O_4(g)$$

at 25°C.

The standard free energy change for the reaction is obtained by means of Equation 11.24 as

$$\Delta G° = \Delta G°_f(N_2O_4) - 2\Delta G°_f(NO_2)$$

$$= (1 \text{ mol})(97.89 \text{ kJ mol}^{-1}) - (2 \text{ mol})(51.31 \text{ kJ mol}^{-1})$$
$$= -4.73 \text{ kJ}$$

The equilibrium constant is obtained by substitution into Equation 11.38, which can be rewritten as

$$K_p = \exp\left(\frac{-\Delta G°}{RT}\right)$$

$$= \exp\frac{(4.73 \text{ kJ})(10^3 \text{ J kJ}^{-1})}{(\text{mol})(8.314 \text{ J mol}^{-1} \text{ K}^{-1})(298 \text{ K})}$$

$$= 6.75$$

(The (mol) factor in the denominator of the exponential arises in the derivation of Equation 11.38. Equation 11.36 contains the product of RT and the stoichiometric coefficients, the units of which are mol. In the next step of the derivation, Equation 11.37, these coefficients are shifted into the log term. However, since this term is dimensionless, the unit of these coefficients, i.e., mol, remains with the RT term.) ■

**Example 11.6** What is the first acid ionization constant of phosphoric acid in aqueous solution at standard conditions?

The reaction in question and the $\Delta G_f°$ values of the species involved (from Appendix 4) are

$$H_3PO_4(aq) \rightleftharpoons H^+(aq) + H_2PO_4^-(aq)$$

$\Delta G_f°(\text{kJ mol}^{-1})$     −1142.5     0     −1130.3

from which we obtain directly

$$\Delta G° = [-(1130.3) - (-1142.5)] \text{ kJ}$$
$$= 12.2 \text{ kJ}$$

As in Example 11.5, the ionization constant $K_1$ is obtained by the relation

$$K_1 = \exp\left(\frac{-\Delta G°}{RT}\right)$$

$$= \exp\frac{(-12.2 \text{ kJ})(10^3 \text{ J kJ}^{-1})}{(\text{mol})(8.314 \text{ J mol}^{-1} \text{ K}^{-1})(298 \text{ K})}$$

$$= 7.27 \times 10^{-3}$$

This value is in very good agreement with that listed in Table 8.5 or Appendix 3. ■

Equation 11.38, which relates the standard free energy to the equilibrium constant, is as applicable to equilibrium constants expressed in terms of molar concentrations as it is to those expressed in terms of partial pressures. Just as the partial pressures in Equation 11.37 are divided by standard pressure (1 atm) the concentrations appearing in the equilibrium expression are divided by 1 M, the concentration of solutes in the standard state. Furthermore, recall that the actual concentrations or partial pressures must be sufficiently low to permit ideal behavior. Otherwise, these quantities must first be corrected for deviations from ideality in Equations 11.30–11.37 (Section 7.7).

## 11.9 Temperature Dependence of K and ΔG

*a. Variation of the Equilibrium Constant with Temperature* In Section 7.6 we considered the effect of temperature on the position of chemical equilibrium. It was shown that, according to Le Chatelier's principle, the variation of the equilibrium constant with temperature depends on the sign of $\Delta H$: K increases with $T$ if $\Delta H > 0$ and decreases if $\Delta H < 0$. On the basis of the various thermodynamic relations derived in the present chapter, the previous discussion can be recast in a more quantitative form, and the variation of $K$ with $T$ can be determined.

The dependence of the equilibrium constant on temperature is implicit in Equations 11.38 and 11.18, i.e.,

$$\Delta G° = -RT \ln K \quad \text{and} \quad \Delta G° = \Delta H° - T\Delta S°$$

Equating these two expressions and solving for $K$ we obtain

$$\ln K = -(\Delta H°/RT) + (\Delta S°/R) \tag{11.40}$$

To the extent that $\Delta H°$ and $\Delta S°$ are independent of temperature, $\ln K$ varies as $1/T$, increasing with $1/T$ when $\Delta H° < 0$ and decreasing when $\Delta H° > 0$. Moreover, the greater the magnitude of $\Delta H°$, the greater is the variation of the equilibrium constant with temperature. Figure 11.10 displays this dependence graphically for both positive and negative values of $\Delta H°$. The consistency with Le Chatelier's criterion is apparent: $K$ increases with $T$ for an endothermic reaction and decreases with $T$ for an exothermic reaction.

How valid is the assumption of the constancy of $\Delta H°$ and $\Delta S°$? We have already seen (Sections 10.11 and 11.4) that both enthalpy and entropy changes depend on the variation of the heat capacities with temperature:

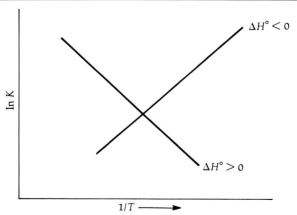

**FIGURE 11.10** Temperature dependence of the equilibrium constant for reactions with $\Delta H > 0$ (colored line) and $\Delta H < 0$ (black line).

### Spontaneity in Chemical Processes—The Second Law of Thermodynamics

$$\Delta H°(T_2) = \Delta H°(T_1) + \int_{T_1}^{T_2} \Delta C_p° \, dT$$

$$\Delta S°(T_2) = \Delta S°(T_1) + \int_{T_1}^{T_2} \frac{\Delta C_p°}{T} \, dT$$

To the extent that the difference between the heat capacities of products and reactants is equal to zero this assumption is strictly correct. While this condition is rarely obeyed exactly, the effect of the heat capacities is generally quite small. Its manifestation will be a slight curvature in the lines in Figure 11.10.

Equation 11.40 can be recast in a form that allows the determination of the equilibrium constant at one temperature from its value at another. Let us evaluate this expression at temperatures $T_2$ and $T_1$:

$$\ln K(T_2) = -(\Delta H°/RT_2) + (\Delta S°/R)$$

$$\ln K(T_1) = -(\Delta H°/RT_1) + (\Delta S°/R)$$

where $\Delta H°$ and $\Delta S°$ are assumed to be independent of temperature. Subtracting the second equation from the first and combining terms we obtain

$$\ln \frac{K(T_2)}{K(T_1)} = \frac{-\Delta H°}{R} \left( \frac{1}{T_2} - \frac{1}{T_1} \right) \tag{11.41}$$

which is known as **van't Hoff's equation** [named after Jacobus van't Hoff (1852–1911)]. This important relation allows the determination of the equilibrium constant at temperature $T_2$ from its value at $T_1$ and the value of $\Delta H°$. Conversely, if $K$ is known at two different temperatures, $\Delta H°$ can be determined without the need to perform calorimetric measurements. This is another manifestation of the unexpected connection between different quantitites predicted by thermodynamics.

---

**Example 11.7** What is the equilibrium constant for the reaction

$$2SO_2(g) + O_2(g) \rightleftarrows 2SO_3(g)$$

at 1000 K? Use the data in Appendix 4.

According to Equation 11.41 we must know the equilibrium constant at one temperature before we can determine it at another. The data in Appendix 4 permit a determination of $K$ at 298 K. First, the standard free energy change is obtained from the tabulated standard free energies of formation (Equation 11.24):

$$\Delta G° = 2\Delta G_f°(SO_3) - 2\Delta G_f°(SO_2) - \Delta G_f°(O_2)$$

$$= [(2)(-371.06) - (2)(-300.194) - 0] \text{ kJ}$$

$$= -141.73 \text{ kJ}$$

The equilibrium constant is obtained from $\Delta G°$ by means of Equation 11.38:

$$\Delta G° = -RT \ln K$$

$$(-141.73 \text{ kJ})(10^3 \text{ J kJ}^{-1}) = -(\text{mol})(8.314 \text{ J mol}^{-1} \text{ K}^{-1})(298 \text{ K}) \ln K$$

$$K(298) = 6.981 \times 10^{24}$$

In order to evaluate $K(1000)$ by means of Equation 11.41, $\Delta H°$ must be known. Using the standard enthalpies of formation tabulated in Appendix 4, $\Delta H°$ can be obtained by the same procedure used to obtain $\Delta G°$:

$$\Delta H° = 2\Delta H_f°(SO_3) - 2\Delta H_f°(SO_2) - \Delta H_f°(O_2)$$
$$= [(2)(-395.72) - (2)(-296.83) - 0]\ kJ$$
$$= -197.78\ kJ$$

Substitution of the results into Equation 11.42,

$$\ln K(T_2) = \ln K(T_1) - \frac{\Delta H°}{R}\left(\frac{1}{T_2} - \frac{1}{T_1}\right)$$

gives

$$\ln K(1000) = \ln(6.981 \times 10^{24}) - \frac{(-197.78\ kJ)(10^3\ J\ kJ^{-1})}{(mol)(8.314\ J\ mol^{-1}\ K^{-1})}\left(\frac{1}{1000\ K} - \frac{1}{298\ K}\right)$$

$$K(1000) = 3.21$$

Note that an increase in temperature of approximately 700 K reduces the equilibrium constant of this exothermic reaction by a factor of $10^{24}$. Clearly, the reaction must be run at low temperature in order to obtain a reasonably high yield of $SO_3$.

The above procedure is not the only way to solve this problem. Another approach involves the direct use of Equation 11.40:

$$\ln K = -(\Delta H°/RT) + (\Delta S°/R)$$

The value of $\Delta S°$ can be obtained from the tabulated standard entropies, assuming that they are independent of temperature:

$$\Delta S° = 2S°(SO_3) - 2S°(SO_2) - S°(O_2)$$
$$= [(2)(256.76) - (2)(248.22) - 205.138]\ J\ K^{-1}$$
$$= -188.06\ J\ K^{-1}$$

Using the value of $\Delta H°$ determined earlier we obtain

$$\ln K(1000) = \frac{-(-197.78\ kJ)(10^3\ J\ kJ^{-1})}{(mol)(8.314\ J\ mol^{-1}\ K^{-1})(1000\ K)} + \frac{(-188.06\ J\ K^{-1})}{(8.314\ J\ mol^{-1}\ K^{-1})(mol)}$$

$$K(1000) = 3.219$$

In yet another method, $\Delta G°$ at 1000 K can be obtained by means of Equation 11.18 evaluated at 1000 K, on the assumption that $\Delta H°$ and $\Delta S°$ are independent of temperature:

$$\Delta G°_{1000} = \Delta H°_{1000} - T\Delta S°_{1000}$$
$$= (-197.78\ kJ)(10^3\ J\ kJ^{-1}) - (1000\ K)(-188.06\ J\ K^{-1})$$
$$= -9720\ J$$

and the equilibrium constant obtained by means of Equation 11.38,

$$\Delta G° = -RT \ln K$$

$-9720 \text{ J} = -(\text{mol})(8.314 \text{ J mol}^{-1} \text{ K}^{-1})(1000 \text{ K}) \ln K$

$K(1000) = 3.219$

The availability of several approaches to the solution of a particular problem constitutes one of the useful, if at times confusing, features of thermodynamics. ∎

**b. Variation of ΔG with Temperature** In Section 11.7a we noted that the spontaneity of a reaction was determined by the signs and relative magnitudes of the enthalpy and entropy changes at a given temperature. We have seen (Section 11.9a) that the values of $\Delta H$ and $\Delta S$ for a particular reaction are fairly insensitive to temperature. It follows from Equation 11.18, $\Delta G = \Delta H - T\Delta S$, that the contribution of the entropy term is directly proportional to the temperature. In the limit as $T$ approaches 0 K, $\Delta G$ approaches $\Delta H$. Consequently, at sufficiently low temperatures the spontaneity of a reaction is determined solely by the sign of $\Delta H$: exothermic reactions occur spontaneously and endothermic reactions do not. On the other hand, at sufficiently high temperatures, $\Delta G$ approaches the value of $-T\Delta S$ and the role of the enthalpy becomes negligible. Reactions for which $\Delta S > 0$ can occur spontaneously at sufficiently high temperatures regardless of the value of $\Delta H$. For instance, all molecules dissociate into their constituent atoms at sufficiently high temperatures, a reaction that can be understood as a consequence of the increase in entropy associated with the increase in the number of particles.

It is instructive to examine graphically the relationships among $\Delta G$, $\Delta H$, and $\Delta S$ as a function of temperature. This can be done conveniently by plotting $-\Delta G/RT$, $\Delta H/RT$, and $\Delta S/R$ versus $T$. These variables are dimensionless and consequently can be compared directly. In addition, when all the reacting species are present at standard pressure or concentration, we have from Equation 11.38

$$-\Delta G/RT = -\Delta G°/RT = \ln K$$

Under these conditions, a plot of $-\Delta G/RT$ shows the variation of $\ln K$ with $T$. Figure 11.11 shows plots of these quantities versus $T$ for the four possible combinations of $\Delta H$ and $\Delta S$:

(a)  $\Delta H > 0$   and   $\Delta S > 0$

(b)  $\Delta H > 0$   and   $\Delta S < 0$

(c)  $\Delta H < 0$   and   $\Delta S > 0$

(d)  $\Delta H < 0$   and   $\Delta S < 0$

For the sake of convenience, the magnitudes of $\Delta H$ and $\Delta S$ have been kept constant throughout. All four diagrams display the asymptotic behavior at high and low temperatures discussed earlier: with increasing $T$, $\Delta H/RT$ approaches zero and $-\Delta G/RT$ approaches $\Delta S/R$ while $\ln K$ approaches $\Delta S°/R$. On the other hand, as $T$ approaches 0 K, $\Delta G/RT$ tends towards $\Delta H/RT$ and $\ln K$ approaches the value of $-\Delta H°/RT$.

Beyond this similarity, the four diagrams differ in systematic ways, reflecting the differences in sign of $\Delta H$ and $\Delta S$. When both $\Delta H$ and $\Delta S$ are positive (Figure 11.11a), the reaction is spontaneous, i.e., $\Delta G < 0$, so long as $\Delta H/RT$ is less than $\Delta S/R$. The diagram shows that this happens at tempera-

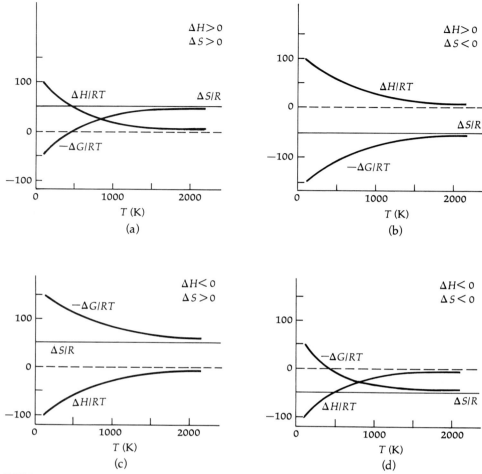

**FIGURE 11.11** Variation of $\Delta H/RT$, $\Delta S/R$, and $-\Delta G/RT$ with temperature for the four possible combinations of positive and negative values of $\Delta H$ and $\Delta S$. A reaction can occur spontaneously whenever $-\Delta G/RT > 0$ (colored curves). If the various quantities are evaluated at conditions of standard pressure or concentration, $\Delta G$ is just $\Delta G°$ and $\Delta G/RT$ becomes equal to ln $K$. For convenience, we have chosen $\Delta H/R = \pm 2.5 \times 10^4$ K and $\Delta S/R = \pm 50$ at all tempeatures in all four diagrams. (a) $\Delta H > 0$ and $\Delta S > 0$; the reaction occurs spontaneously for $T > 500$ K. If $\Delta G$ is evaluated at standard conditions, $K > 1$ when $T > 500$ K. (b) $\Delta H > 0$ and $\Delta S < 0$; the reaction is not spontaneous at any temperature. $K < 1$ at all temperatures but increases with $T$.
(c) $\Delta H < 0$ and $\Delta S > 0$; the reaction is spontaneous at all temperatures. $K > 1$ at all temperatures but decreases with increasing $T$. (d) $\Delta H > 0$ and $\Delta S < 0$; the reaction occurs spontaneously for $T < 500$ K. $K$ decreases with increasing $T$ but remains larger than 1 below 500 K.

---

tures higher than that at which the $\Delta H/RT$ and $\Delta S/R$ curves cross (500 K). At standard conditions of pressure or concentration, this temperature also marks the point at which $K$ changes from $K < 1$ to $K > 1$. An example of a reaction of this type is the decomposition of sulfur trioxide to sulfur dioxide and oxygen.

When $\Delta H > 0$ and $\Delta S < 0$ (Figure 11.11b), the reaction does not occur spontaneously at any temperature; $\Delta G$ is always positive. While $K$ does increase with temperature, its value remains less than unity (ln $K = -\Delta G°/RT < 0$) even in the high-temperature limit. The formation of benzene, $C_6H_6$, from the elements is a case in point. On the other hand, when $\Delta H < 0$ and $\Delta S > 0$ (Figure 11.11c), the reaction occurs spontaneously at all temperatures, but the equilibrium constant decreases with increasing $T$. The decomposition of ozone ($O_3$) to oxygen is an example.

Finally, when $\Delta H < 0$ and $\Delta S < 0$ (Figure 11.11d), the reaction is once again spontaneous, but only so long as $\Delta H/RT < \Delta S/R$. As indicated in the diagram, this condition is only obeyed at relatively low temperatures ($T < 500$ K). The equilibrium constant decreases with $T$ and changes from $K > 1$ to $K < 1$ at this temperature. The formation of $SO_3$ from $SO_2$ and $O_2$ is an example of this type of reaction.

**Example 11.8** Using the data of Example 11.7, determine the temperature regime in which the reaction

$$2SO_2(g) + O_2(g) \rightleftarrows 2SO_3(g)$$

has (a) $\Delta G° < 0$ and (b) $\Delta G° < -50.0$ kJ. Determine the values of $K$ corresponding to these $\Delta G°$ values. Describe qualitatively the composition of the equilibrium mixture.

(a) Example 11.7 gives $\Delta H° = -197.78$ kJ and $\Delta S° = -188.06$ J K$^{-1}$. Since both $\Delta H°$ and $\Delta S°$ are less than 0, $\Delta G°$ will be less than 0 below some maximum temperature (see Figure 11.11d). This temperature can be determined by means of Equation 11.18,

$$\Delta G° = \Delta H° - T\Delta S°$$

$$0 = (-197.78 \text{ kJ})(10^3 \text{ J kJ}^{-1}) - T_{max}(-188.06 \text{ J K}^{-1})$$

$$T_{max} = 1051.69 \text{ K}$$

Since, according to Equation 11.38,

$$\Delta G° = -RT \ln K$$

$$0 = -RT \ln K$$

$$K = 1$$

at this temperature. The equilibrium mixture consists of comparable amounts of reactants and product.

(b) When $\Delta G° = -50.0$ kJ, we have

$$\Delta G° = \Delta H° - T\Delta S°$$

$$(-50.0 \text{ kJ})(10^3 \text{ J kJ}^{-1}) = (-197.78 \text{ kJ})(10^3 \text{ J kJ}^{-1}) - T(-188.06 \text{ J K}^{-1})$$

$$T = 786 \text{ K}$$

At this temperature, the equilibrium constant is

$$(-50.0 \times 10^3 \text{ J}) = -(\text{mol})(8.314 \text{ J mol}^{-1} \text{ K}^{-1})(786 \text{ K}) \ln K$$

$$K = 2.10 \times 10^3$$

The equilibrium mixture consists predominantly of $SO_3$. ∎

## Conclusion

The second law of thermodynamics deals with the occurrence and direction of chemical change. The spontaneity of a chemical process is determined by the change in the total entropy of the system and the surroundings: the process is spontaneous if $\Delta S_{total} > 0$; however, if $\Delta S_{total} = 0$, the system is at equilibrium. Entropy changes in chemical reactions can be evaluated by means of the standard entropies of the reactants and products. These entropies can be obtained by determining the change in the entropy of a substance when it is heated from 0 K to room temperature.

The criterion of spontaneity can be simplified for the most common type of chemical process—one occurring at constant temperature and pressure. A new state function, the free energy, is introduced and shown to be a criterion of spontaneity: a process occurs spontaneously when the free energy of the system decreases; the surroundings need not be considered explicitly. The free energy change in a chemical reaction can be obtained from the enthalpy and entropy changes, or more conveniently, from standard free energies of formation.

The spontaneity of a reaction reflects the combined drive of a system towards lower enthalpy and higher entropy or, from the molecular viewpoint, towards lower potential energy and greater disorder. The free energy decreases as the reaction proceeds towards equilibrium. At equilibrium, the free energies of reactants and products become equal, and no further net change occurs. The equilibrium constant of a reaction can be related to the free energy change at standard conditions of pressure or concentration, and its variation with temperature depends on the sign of the standard reaction enthalpy.

Taken together, Chapters 10 and 11 show that thermodynamics is a subject of far reaching scope. Energy release, spontaneity, and equilibrium constants are treated in a unified way and unsuspected connections between them are exposed. At the same time, the limitations of thermodynamics must be understood. Thermodynamics tells us what *can* happen, not what *will* happen—a chemical reaction can occur spontaneously but too slowly to be observed. Furthermore, by focusing on the macroscopic properties of substances, thermodynamics necessarily ignores the rich complexities of molecular structure and the more detailed understanding that can be obtained from a study of this structure. Thermodynamics might tell us that A and B can react to form C and D but not what it is about the molecules of C and D that makes them more stable than A and B.

## Problems

(Use the data in Appendix 4 unless stated otherwise.)

1. An ideal gas occupies a volume of 5.00 L at atmospheric pressure and 300 K. A constant external pressure of 5.00 atm compresses the gas until equilibrium is attained. Calculate $\Delta S_{system}$, $\Delta S_{surr}$, and $\Delta S_{total}$ for this process assuming that $T$ remains constant.

2. What is the probability that all the molecules of a gas will be found in one half of a container when the number of molecules is (a) 3, (b) 8, and (c) $10^{23}$?

3. Derive an expression for the change in the entropy of $n$ moles of an ideal gas in a constant volume container which is reversibly heated from $T_1$ to $T_2$.

What is the change in entropy if the container at $T_1$ is suddenly placed in a bath at $T_2$?

4. Calculate $\Delta S_{system}$, $\Delta S_{surr}$, and $\Delta S_{total}$ when 3.000 mol of water at atmospheric pressure is cooled from 25.00°C to 10.00°C in each of the following ways: (a) by placing it in a bath at 10.00°C, (b) by placing it in a succession of baths, each 3.00°C cooler than the one before, and (c) by placing it successively in an infinite number of baths, each $dT$ cooler than the one before. Assume that $C_P^\circ$ is independent of temperature in this interval.

5. Determine the change in the entropy of a 1.043 kg iron vessel containing 1.107 kg of water when heated from 296.5 K to 307.2 K at atmospheric pressure. Assume that $C_P^\circ$ is independent of temperature.

6. One mole of water freezes at 0°C and atmospheric pressure, releasing 6.01 kJ of heat. Calculate $\Delta S_{system}$, $\Delta S_{surr}$, and $\Delta S_{total}$. Is this process reversible?

7. One mole of water at atmospheric pressure is supercooled to −10°C, where it freezes. Calculate $\Delta S_{system}$, $\Delta S_{surr}$, and $\Delta S_{total}$. Use the data in Problem 6 and the $C_P^\circ$ values of water and ice, which are 75.3 and 37.7 J mol$^{-1}$ K$^{-1}$, respectively, in this temperature interval. Is this process reversible?

8. Calculate the change in entropy when 50.0 g of ice at 0°C is dropped into 100 g of water at 75°C in an insulated vessel. The enthalpy of fusion of ice is 6.01 kJ mol$^{-1}$, and the heat capacity of water is 75.4 J mol$^{-1}$ K$^{-1}$.

9. One mole of an ideal gas is (1) heated at constant pressure from $T$ to $2T$ and (2) cooled back to $T$ at constant volume. (a) Sketch this process on a plot of $P$ versus $V$. (b) Derive an expression for $\Delta S$ for the overall process. (c) Show that the overall process is equivalent to an isothermal expansion of the gas at $T$ from $V$ to $2V$. (d) Calculate $\Delta S$ for the process in (c) and show that the result is the same as that obtained in (b).

10. Determine the standard entropy $S_{298}^\circ$ of solid silver by graphical integration (see Figure 11.6). This can be done conveniently by plotting the data on graph paper and determining the area under the curve by counting the squares. Use the following heat capacity data:

| $T$ (K) | $C_P^\circ$ (J mol$^{-1}$ K$^{-1}$) | $T$ (K) | $C_P^\circ$ (J mol$^{-1}$ K$^{-1}$) |
|---|---|---|---|
| 15 | 0.669 | 90 | 19.1 |
| 20 | 1.72 | 130 | 22.1 |
| 30 | 4.77 | 170 | 23.6 |
| 40 | 8.41 | 210 | 24.4 |
| 50 | 11.6 | 250 | 24.7 |
| 70 | 16.3 | 298 | 25.4 |

11. Predict the sign of the entropy change in each of the following processes on the basis of the change in molecular disorder:

(a) $H_2(g) + Br_2(l) \rightarrow 2HBr(g)$
(b) $Cl_2(g) \rightarrow 2Cl(g)$
(c) $2Na(s) + Cl_2(g) \rightarrow 2NaCl(s)$
(d) $CaCO_3(s) \rightarrow CaO(s) + CO_2(g)$
(e) $CO_2(s) \rightarrow CO_2(g)$

12. Without consulting thermodynamic data tables, choose the substance with the greater molar entropy from each of the following pairs of substances:

(a) Ne (1 atm), Ne (0.1 atm)
(b) Hg(l) at 300 K, Hg(l) at 400 K
(c) $Cl_2(g)$, $2Cl(g)$
(d) $KCl(s)$, $CaCl_2(s)$
(e) $Cl_2(g)$, $F_2(g)$

13. The standard free energies of formation of $NO_2(g)$ and $CO_2(g)$ are 51.3 kJ mol$^{-1}$ and −394.4 kJ mol$^{-1}$, respectively. Compare these oxides with respect to their tendencies to decompose to the elements at 298 K.

14. Calculate $\Delta H_{298}^\circ$, $\Delta S_{298}^\circ$, and $\Delta G_{298}^\circ$ for each of the following reactions:

(a) $2Ag(s) + Br_2(l) \rightarrow 2AgBr(s)$
(b) $Fe_2O_3(s) + 3C(graphite) \rightarrow 2Fe(s) + 3CO(g)$
(c) $3Cl_2(g) + I_2(s) \rightarrow 2ICl_3(s)$
(d) $CH_4(g) + 2O_2(g) \rightarrow CO_2(g) + 2H_2O(l)$
(e) $2C(graphite) + H_2(g) \rightarrow C_2H_2(g)$

Indicate for each reaction whether it is spontaneous at standard conditions and whether the enthalpy and entropy changes work for or against spontaneity.

15. The normal boiling point of benzene is 80.1°C and the standard enthalpy of vaporization at this temperature is 30.77 kJ mol$^{-1}$. What are the molar entropy of vaporization and the free energy of vaporization of benzene at 80.1°C and standard pressure?

16. One mole of an ideal monatomic gas at 25°C and atmospheric pressure is put through the following reversible processes: (a) cooling to −25°C at constant volume, (b) isothermal compression to 50 atm, and (c) heating at constant pressure to 50°C. For each process, determine $q$, $w$, $\Delta E$, $\Delta H$, and $\Delta S$. For which of these processes can $\Delta G$ be determined by the relation $\Delta G = \Delta H - T\Delta S$?

17. One mole of $N_2$, assumed to be an ideal gas, is compressed isothermally at 300 K from a volume of 100 L to 1.00 L. Calculate $\Delta G$.

18. A substance exists as a gas, assumed to be ideal, at room temperature. A sample of this gas occupies a volume of 2.00 L at 300 K and a pressure of 1.00 atm. The gas is heated to 600 K and allowed to expand while maintaining constant pressure. The value of $S_{300}^\circ$ is 130.0 J mol$^{-1}$ K$^{-1}$ and that of $C_p^\circ$ is 30.0 J mol$^{-1}$ K$^{-1}$. Calculate $\Delta E$, $\Delta H$, $\Delta S$, $\Delta G$, $q$, and $w$ for this process.

19. One mole of water at 20°C is mixed with two moles of water at 50°C at constant atmospheric pressure in a thermally insulated vessel. Calculate (a) the final temperature of the water, (b) $\Delta H$, (c) $\Delta S_{system}$, (d) $\Delta S_{surr}$, (e) $\Delta S_{total}$, and (f) $\Delta G$. The heat capacity of water in this temprature range is 75.3 J mol$^{-1}$ K$^{-1}$, and the standard entropy of water is $S_{298}^\circ = 69.9$ J mol$^{-1}$ K$^{-1}$. Assume that the vessel does not absorb any heat from the water.

20. The following derivation of the relation $\Delta G = 0$ contains several fallacies. For each step in the derivation, point out the fallacy and indicate the specific circumstances for which the equation is valid:

  1. $\Delta S = q/T$
  2. $\Delta H = q$
  3. $\Delta G = \Delta H - T\Delta S$
     $= q - T(q/T) = 0$

  Give an example of a process for which the final result is valid.

21. Thermodynamic relationships are frequently valid only in specific circumstances. For each of the following relationships, indicate whether it is always true, and if not, the specific conditions for which it is applicable:

  (a) $\Delta S = \Delta H/T$
  (b) $C_p = dq/dT$
  (c) $\Delta G = RT \ln(P_2/P_1)$
  (d) $G = E + PV - TS$
  (e) $\Delta S = C_v \ln(T_2/T_1)$

22. Calculate $\Delta G^\circ$ and $K$ for the following reaction at 298 K:

  $$PCl_5(g) \rightleftarrows PCl_3(g) + Cl_2(g)$$

23. A vessel at 25°C contains a mixture of gaseous N$_2$, H$_2$, and NH$_3$ at partial pressures of 75.3 torr, 36.4 torr, and 98.6 torr, respectively. Could additional NH$_3$ be formed at these conditions?

24. For the reaction

  $$CuSO_4 \cdot 5H_2O(s) \rightleftarrows CuSO_4(s) + 5H_2O(g)$$

  at 298 K, determine $\Delta G^\circ$ and $K$. What is the equilibrium vapor pressure?

25. Calculate the solubility product of AgCl at 25°C. Compare with the value given in Appendix 3.

26. Calculate the acid ionization constant of HF at 25°C. Compare with the value given in Appendix 3.

27. The standard free energy of formation of ClF(g) is $-55.94$ kJ mol$^{-1}$. (a) What is the equilibrium constant for the reaction

  $$Cl_2(g) + F_2(g) \rightleftarrows 2ClF(g)$$

  at 25°C? (b) What would be the value of $\Delta G$ for this reaction at 25°C if the partial pressures of Cl$_2$, F$_2$, and ClF are 1.35 atm, 0.87 atm, and 0.35 atm, respectively? Is the reaction more or less favorable under these conditions?

28. Which of the following reactions are spontaneous at 25°C and standard conditions?

  (a) $HCl(g) + NH_3(g) \rightarrow NH_4Cl(s)$
  (b) $2Ag(s) + Cl_2(g) \rightarrow 2AgCl(s)$
  (c) $2HCl(aq) + Mg(s) \rightarrow MgCl_2(aq) + H_2(g)$
  (d) $2NH_4NO_3(s) \rightarrow 2N_2(g) + 4H_2O(g) + O_2(g)$
  (e) $Fe_2O_3(s) + 6HCl(g) \rightarrow 2FeCl_3(s) + 3H_2O(g)$

  For each reaction, which factors favor spontaneity and which disfavor it? Which of the reactions are favored by an increase in temperature?

29. Calculate $\Delta G_{298}^\circ$ for the transformation of one mole of graphite to diamond. Is there a temperature for which the transformation can occur spontaneously at atmospheric pressure?

30. The equilibrium constant of a certain reaction increases by a factor of 10 when the temperature is increased from 300 K to 350 K. What is the reaction enthalpy?

31. Calculate the equilibrium constants for the reactions

  $$2H_2(g) + CO(g) \rightleftarrows CH_3OH(l)$$

  $$3H_2(g) + CO_2(g) \rightleftarrows CH_3OH(l) + H_2O(l)$$

  (a) at 298 K and (b) at 500 K, assuming $\Delta H^\circ$ and $\Delta S^\circ$ are independent of temperature.

32. Is there a temperature at which the equilibrium constants of the two reactions in Problem 31 are equal? If so, what is the temperature and what is the value of $K$ at this temperature?

33. The water gas reaction

  $$CO(g) + H_2O(g) \rightleftarrows CO_2(g) + H_2(g)$$

  may be used for the production of hydrogen gas from low-grade fuels. Assuming that $\Delta H^\circ$ and $\Delta S^\circ$

are independent of temperature, evaluate $\Delta G°$ and $K$ at 450 K. Is the reaction favored by an increase in temperature?

34. Calculate $\Delta G°$ and $K$ at 600 K for the decomposition of $PCl_5(g)$ (see Problem 22). Assume that $\Delta H°$ and $\Delta S°$ are independent of temperature between 298 and 600 K.

35. Calculate the solubility product of AgCl at 80°C. Compare with the value obtained at 25°C (Problem 25). Does the solubility of AgCl increase with temperature?

36. For each of the reactions in Problem 14 found to be spontaneous at standard conditions, determine the temperature range over which it is spontaneous. For each reaction that is not spontaneous, determine the temperature range, if any, in which it is spontaneous. Assume that $\Delta H°$ and $\Delta S°$ are independent of temperature and ignore the possible occurrence of phase changes.

37. The ionization constant of water, $K_w$, decreases from its well-known value at 25°C to $0.67 \times 10^{-14}$ at 20°C. (a) What is the pH of water at 20°C? (b) What is the pH of water at 30°C? Does the result mean that water becomes acidic at 30°C? Explain your answer.

38.* The temperature dependence of experimental heat capacities can be expressed in various analytical forms. A simple representation is

$$C_P° = a + bT + cT^2$$

The values of the coefficients $a$, $b$, and $c$, for $N_2(g)$, $H_2(g)$, and $NH_3(g)$ are as follows:

|         | a      | b                      | c                       |
|---------|--------|------------------------|-------------------------|
| $N_2(g)$  | 26.983 | $5.912 \times 10^{-3}$  | $-3.38 \times 10^{-7}$   |
| $H_2(g)$  | 29.066 | $-8.368 \times 10^{-4}$ | $2.01 \times 10^{-6}$    |
| $NH_3(g)$ | 25.985 | $3.300 \times 10^{-2}$  | $3.05 \times 10^{-6}$    |

Using these data as well as data in Appendix 4 determine the equilibrium constant for the reaction

$$N_2(g) + 3H_2(g) \rightleftarrows 2NH_3(g)$$

at 900 K. Compare your answer with that obtained on the assumption of constant $C_P°$. Use the following procedure: (1) Determine $\Delta H°$ and $\Delta S°$ for the cooling of the reactants from 900 K to 298 K; (2) determine $\Delta H°_{298}$ and $\Delta S°_{298}$ for the reaction; (3) determine $\Delta H°$ and $\Delta S°$ for the heating of the product from 298 K to 900 K. (4) The sum of (1)–(3) gives $\Delta H°_{900}$ and $\Delta S°_{900}$ for the reaction, from which $\Delta G°_{900}$ and the equilibrium constant can be obtained.

# 12

# Phase Equilibria

The phase changes of water—the conversion to ice at 0°C and to steam at 100°C—are among the most familiar physical phenomena known. It is perhaps less well known that the temperatures at which these phase changes occur also depend on pressure. As the external pressure varies, the boiling point of water can range from 0°C to 374°C! Observations of this kind can be understood by application of the laws of thermodynamics developed in Chapters 10 and 11 to phase equilibria of pure substances and of solutions.

## Phase Equilibria of Pure Substances

### 12.1 Changes of Phase

*a. The Molecular Viewpoint* With increasing temperature, most substances change from solids to liquids and eventually to gases. In general, these phase changes occur abruptly at specific temperatures and involve major changes in structure. In a solid, molecules (or atoms or ions) occupy specific sites in a crystal (Section 20.1). They can vibrate about their equilibrium positions but only rarely move to other sites. This rigid structure relaxes upon melting, and the individual molecules are relatively free to move at random throughout the liquid. However, as evidenced by the small difference between the density of a given substance in the solid and liquid phases, the molecules in a liquid are almost as close to each other, on average, as they are in the solid. The attractive forces between molecules decrease and the potential energy of the molecules increases as the distance between them increases (see Figure 4.15). Thus, the forces between molecules in a liquid are still relatively strong and some of the order seen in the solid state remains (Figure 12.1).

There is an important difference in the structural order possessed by the solid and liquid phases. A crystalline solid displays long-range order, that is, the arrangement of the molecules repeats itself indefinitely. On the other hand, a liquid has only short-range order. While the arrangement of the molecules about a given molecule in a liquid may resemble that in a solid, this similarity breaks down beyond a distance of a few molecular diameters. The random motion of the molecules reduces the order with increasing likelihood as the distance from a given molecule increases. A more dynamic view of the similarities and differences between the phases of matter than that provided by Figure 12.1 is shown in Figure 12.2. This figure shows the trajectories of molecules as simulated by a computer calculation based on a detailed model of molecular interactions. Note that the molecules in a solid are confined to essentially fixed sites. The slight motion displayed by these molecules is the result of vibrations about their equilibrium positions. In contrast, molecules in a liquid are free to move away from their initial sites. While remnants of the structure of the solid are discernible, the "noise" introduced by the random motion of the molecules leads to a smearing of the sharp pattern displayed by the solid.

The transition from liquid to vapor results in an even more profound change in bulk structure. As we saw in Section 4.1, the lower densities of gases indicate that the molecules in a gas are much farther apart than they are in a liquid (Figure 12.1). At ordinary pressures, the molecules in a gas are actually so far apart that the intermolecular forces are negligibly small, and the potential energy is highest; the ideal gas law is approximately valid. The molecules in a gas are consequently free to move independently of each other. The result is that the structural order displayed by the condensed phases is absent in the gas phase (Figure 12.2).

The extent to which a particular phase has an ordered structure is determined by the behavior of many molecules. In a change of phase all these molecules must act in concert, and the transformation from solid to liquid or from liquid to gas occurs abruptly at a specific temperature (for a given external

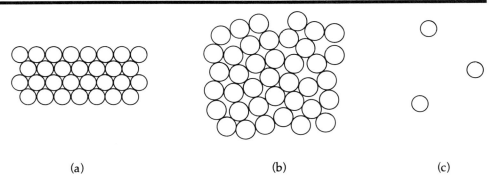

**FIGURE 12.1** Schematic representation of the structure of (a) a crystalline solid, (b) a liquid, and (c) a gas.

412  Phase Equilibria  12.1

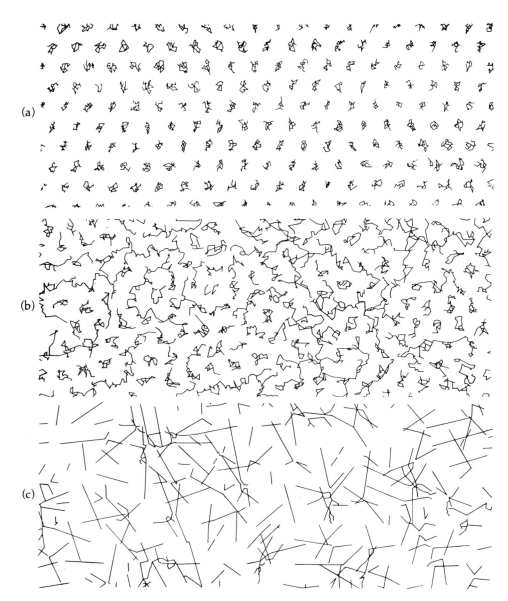

**FIGURE 12.2** Two-dimensional trajectories of molecules in (a) solid, (b) liquid, and (c) gas phases, as simulated by a computer calculation. (Reprinted by permission of J. A. Barber and D. Henderson, *Scientific American*, November 1981, p. 94. Copyright © 1981 by Scientific American, Inc. All rights reserved.)

pressure). At this temperature, the two phases are in equilibrium with each other. However, at other temperatures, a substance is either in one phase or another.

**b. The Thermodynamic Viewpoint** Changes of phase can also be interpreted from the thermodynamic point of view. Transitions from solid to liquid or from liquid to vapor are accompanied by increases in both enthalpy and entropy (Sections 10.7 and 11.4). We have seen that these changes can be understood from the molecular viewpoint. Energy must be supplied in order to increase the average distance between molecules against the action of the intermolecular forces. Since the average intermolecular distance is largest in the gas phase and smallest in the solid phase, the enthalpies of fusion and vaporization are positive. Furthermore, we have seen that entropy is a measure of molecular disorder. Since this disorder also increases from solid to liquid to gas (see Figure 12.2), the entropies of fusion and vaporization must also be positive.

Recall from Section 11.7a that the enthalpy and entropy changes in a chemical reaction sometimes oppose each other and sometimes reinforce each other in determining spontaneity. In a change of phase the enthalpy and entropy changes *always* oppose each other. The enthalpy effect favors the stability of the solid phase, where the molecular potential energy is lowest—the interaction between the molecules is as strong as it can be. On the other hand, the entropy effect favors the stability of the gas phase, where molecular disorder is greatest. The stable phase of a substance at a given temperature represents a compromise between these opposing tendencies. The function that expresses this compromise is the free energy. Since a phase transition occurs at constant temperature and pressure, $\Delta G$ is a criterion of spontaneity. At constant $T$, Equation 11.18 for the free energy change is applicable:

$$\Delta G = \Delta H - T\Delta S$$

If this equation is applied to, say, the melting (fusion) of a solid, we find that so long as $\Delta H_{fus} > T\Delta S_{fus}$, then $\Delta G_{fus} > 0$, and the solid does not melt spontaneously. Rather, the reverse process, i.e., freezing, occurs spontaneously. As shown in Figure 12.3, this happens at relatively low temperatures, and this is where the solid phase is stable. Similar considerations show that at relatively high temperatures $\Delta G_{fus} < 0$, melting occurs spontaneously, and the liquid is the stable phase.

For a given external pressure, two phases are in equilibrium at a unique temperature that marks the boundary between the temperature regimes in which either one or the other is stable. Depending on the phases under consideration, this temperature defines the melting point, the boiling point, or some other transition point. The temperatures at which melting or boiling occur can be related readily to the corresponding enthalpy and entropy changes. Applying Equation 11.18 to vaporization at atmospheric pressure, we have

$$\Delta G°_{vap} = \Delta H°_{vap} - T\Delta S°_{vap}$$

At the boiling point the liquid and vapor phases are in equilibrium and $\Delta G°_{vap} = 0$, giving

$$T_b = \Delta H°_{vap}/\Delta S°_{vap} \tag{12.1}$$

A similar expression relates the melting point to the enthalpy and entropy of

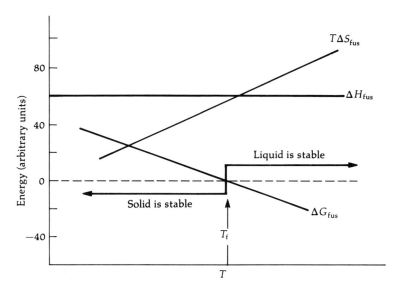

**FIGURE 12.3** Temperature dependence of $\Delta G_{fus}$. The contributions of the enthalpy and entropy of fusion are indicated, with the assumption that $\Delta H_{fus}$ and $\Delta S_{fus}$ are independent of $T$. As long as $\Delta G_{fus} > 0$, the solid phase is stable; when $\Delta G_{fus} < 0$, the liquid phase is stable. The melting point is the temperature at which $\Delta G_{fus} = 0$. The two phases are in equilibrium at this one temperature (for a given pressure).

fusion. Table 12.1 summarizes the nomenclature of enthalpy changes associated with the various phase transformations to be considered in this chapter.

**Example 12.1** To carry out a bromination reaction (addition of bromine to a specific compound) at elevated temperatures an approximate knowledge of the boiling point of $Br_2$ is necessary. Suppose that a table of thermodynamic data (Appendix 4) is the only available information. Estimate the boiling point of $Br_2$.

The boiling point can be estimated by means of Equation 12.1. The values of $\Delta H°_{vap}$ and $\Delta S°_{vap}$ can be obtained from the tabulated standard enthalpies of formation and standard entropies of liquid and vapor:

$$\Delta H°_{vap} = \Delta H°_f[Br_2(g)] - \Delta H°_f[Br_2(l)]$$
$$= (30.9 - 0) \text{ kJ mol}^{-1} = 30.9 \text{ kJ mol}^{-1}$$

**TABLE 12.1 Enthalpy Changes Associated with Phase Transformations**

| Phase transformation | Symbol for enthalpy change | Sign of enthalpy change |
|---|---|---|
| Melting (solid → liquid) | $\Delta H_{fus}$ | + |
| Freezing (liquid → solid) | $\Delta H_{solid}$ | $- \; (= -\Delta H_{fus})$ |
| Boiling (liquid → vapor) | $\Delta H_{vap}$ | + |
| Condensation (vapor → liquid) | $\Delta H_{liq}$ | $- \; (= -\Delta H_{vap})$ |
| Sublimation (solid → vapor) | $\Delta H_{sub}$ | + |
| Condensation (vapor → solid) | $\Delta H_{con}$ | $- \; (= -\Delta H_{sub})$ |

and
$$\Delta S°_{vap} = S°[Br_2(g)] - S°[Br_2(l)]$$
$$= (245.5 - 152.2) \text{ J mol}^{-1} \text{ K}^{-1} = 93.3 \text{ J mol}^{-1} \text{ K}^{-1}$$

The boiling point is

$$T_b = \frac{\Delta H°_{vap}}{\Delta S°_{vap}}$$

$$= \frac{30.9 \text{ kJ mol}^{-1})(10^3 \text{ J kJ}^{-1})}{93.3 \text{ J mol}^{-1} \text{ K}^{-1}}$$

$$= 331 \text{ K} = 58°C$$

This value is very close to the actual boiling point of 58.8°C. The small difference arises because the enthalpy and entropy of vaporization values used are strictly correct only at 25°C. The changes in these quantities over the thirty degree interval between 25°C and the actual boiling point account for the discrepancy. ∎

The enthalpies and entropies associated with various phase changes of diverse substances are given in Table 12.2. A number of regularities may be noted:

First, for a given substance, the enthalpy and entropy of vaporization are, respectively, larger than the enthalpy and entropy of fusion. This is to be expected on the basis of the changes in the structure of the respective phases. More energy is required to overcome the attractive forces between the molecules in a liquid and form a collection of independent gas molecules than to loosen the structure of a solid by melting it. Similarly, a greater increase in molecular disorder occurs upon vaporization than upon melting.

Second, the enthalpies of vaporization of many substances increase in essentially linear fashion with their boiling points; both are a measure of the strength of intermolecular forces. Since the entropy of vaporization is equal to

**TABLE 12.2 Enthalpies and Entropies of Vaporization and Fusion**

| Substance | $T_b$ (°C) | $\Delta H°_{vap}$ (kJ mol$^{-1}$) | $\Delta S°_{vap}$ (J mol K$^{-1}$) | $T_f$ (°C) | $\Delta H°_{fus}$ (kJ mol$^{-1}$) | $\Delta S°_{fus}$ (J mol$^{-1}$ K$^{-1}$) |
|---|---|---|---|---|---|---|
| $N_2$ | −195.8 | 5.577 | 72.13 | −210.0 | 0.720 | 11.4 |
| $O_2$ | −183.0 | 6.820 | 75.60 | −218.8 | 0.444 | 8.16 |
| $CH_4$ | −161.5 | 8.180 | 73.26 | −82.5 | 0.941 | 10.4 |
| HCl | −85.0 | 16.2 | 85.8 | −114.2 | 1.99 | 12.5 |
| $Cl_2$ | −34.1 | 20.41 | 85.35 | −101.0 | 6.406 | 37.2 |
| $NH_3$ | −33.4 | 23.35 | 97.40 | −77.8 | 5.653 | 28.93 |
| $SO_2$ | −10.0 | 24.92 | 94.68 | −75.5 | 7.402 | 37.4 |
| $CCl_4$ | 76.7 | 30.0 | 85.8 | −97.9 | 2.5 | 10 |
| $C_6H_6$ | 80.1 | 30.76 | 87.07 | 5.5 | 10.59 | 35.30 |
| $H_2O$ | 100.0 | 40.67 | 109.0 | 0.0 | 6.009 | 22.00 |
| Hg | 356.6 | 58.12 | 92.30 | −38.9 | 2.33 | 9.92 |
| Zn | 907 | 114.8 | 97.24 | 419.5 | 6.674 | 9.634 |
| NaCl | 1465 | 171 | 99.6 | 808 | 27 | 26 |
| Ag | 2193 | 254.1 | 103.0 | 327.4 | 11.3 | 9.16 |

the ratio of these two quantities, it does not vary much from one substance to another, the average value being approximately 88 J mol$^{-1}$ K$^{-1}$. This nearly constant value is found in spite of the fact that the boiling points cover a temperature range of over 2000 K. This generalization is known as **Trouton's rule.** Its explanation lies in the essentially uniform increase in the entropy associated with translational molecular motion when a substance is vaporized. The molecules of all gases are in random motion, essentially independent of the presence of other molecules, and the entropy of vaporization reflects this uniformity. Trouton's rule can be used to make a crude estimate of the enthalpy of vaporization of a substance from its known boiling point, and vice versa.

Third, the entropies of fusion tend to vary widely. The increase in molecular disorder on melting is less pronounced than it is on vaporization and the actual values of $\Delta S_{fus}$ depend on the detailed structural changes that occur when a particular solid melts.

Our discussion of phase transitions has so far been confined to those involving the solid and liquid, or liquid and vapor states, respectively. While these are the most common phase changes observable at atmospheric pressure, the direct conversion of a solid to a vapor, a phase change known as **sublimation,** is also possible. While this process occurs for all substances, it generally happens at pressures much lower than normal atmospheric pressure and so is not readily observed. A notable exception is carbon dioxide, which sublimes at atmospheric pressure at $-78°C$. Solid carbon dioxide, known as "dry ice", is useful as a portable refrigerant. Since the conversion of a solid to a vapor could be carried out through an intermediate liquid phase as well as by direct sublimation, and because $H$ is a state function, at a given temperature

$$\Delta H_{sub} = \Delta H_{fus} + \Delta H_{vap} \tag{12.2}$$

The transformation of one solid form of a substance to another, e.g., diamond to graphite, is also a change of phase that occurs at a particular temperature and is accompanied by changes in enthalpy and entropy.

## 12.2 Vapor Pressure

A property of a liquid or solid substance that plays an important role in phase equilibria is its vapor pressure. Consider what happens when a liquid is exposed to the atmosphere. The molecules of the liquid are in random motion. Their kinetic energy distribution is similar to that of gas molecules in random motion (as given in Section 4.6 by the Maxwell–Boltzmann equation). At any given moment, a small fraction of the molecules near the surface of the liquid have sufficiently high kinetic energy to overcome the potential energy binding them to the liquid phase. These molecules can escape from the liquid into the atmosphere. As a result, the liquid gradually evaporates. The phenomenon of evaporation is familiar from our daily experience. Small puddles of water dry up and liquid spills evaporate.

Suppose that instead of being exposed to the atmosphere, a liquid is in a closed container. The most energetic molecules can leave the liquid phase, but instead of being free to escape into the atmosphere, will be confined to the enclosed space above the liquid, as illustrated in Figure 12.4. As the number of molecules in the vapor phase increases, collisions become increasingly likely. The result is an increasing probability of molecules in the vapor phase striking the surface of the liquid and reentering the liquid phase.

**FIGURE 12.4** Evaporation of a liquid in an open and a closed vessel.

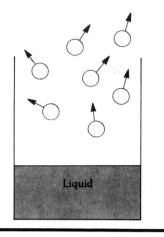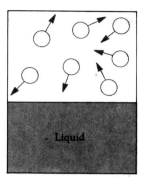

As long as the temperature is kept constant, the rate of evaporation remains unchanged. On the other hand, the rate of condensation increases with time from a value of zero when the liquid is first introduced into the closed container. Thus, a point must eventually be reached when the two rates become equal, as illustrated in Figure 12.5. At this point the liquid and vapor phases come into equilibrium with each other and the number of molecules in each phase becomes constant. The situation is analogous to that depicted in Figure 7.1 for the equilibrium established between reactants and products in a chemical reaction.

The **vapor pressure** of a liquid is the pressure of the vapor in equilibrium with the liquid. The vapor pressure is a characteristic property of a liquid (or solid) and its value depends only on the temperature. In particular, the vapor pressure is independent of the amount of liquid present and of the surface area presented by the liquid to the vapor. It is necessary, however, that there be *some* liquid present, otherwise there can be no equilibrium between the two phases.

Table 12.3 lists the vapor pressures of a number of liquids at room temperature. These values can be determined by the use of an isoteniscope, whose operation is illustrated in Figure 12.6. Note that at room temperature, increasing vapor pressure is correlated with decreasing boiling point. Both properties reflect the strength of the intermolecular forces in the liquid. If the attractive forces are weak, the vapor pressure is high, and the liquid boils at a low temperature.

**FIGURE 12.5** Variation with time of the rates of evaporation and condensation of a liquid in a closed container. Equilibrium between liquid and vapor phases is established when the rates become equal.

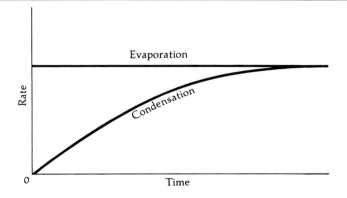

TABLE 12.3 Vapor Pressure of Liquids at 20°C

| Liquid | Vapor pressure (mm Hg) | Boiling point (°C) |
|---|---|---|
| n-Octane, $C_8H_{18}$ | 9.9 | 125.7 |
| Water, $H_2O$ | 17.5 | 100 |
| Ethyl alcohol, $C_2H_5OH$ | 44 | 78.5 |
| Carbon tetrachloride, $CCl_4$ | 91 | 76.8 |
| Acetone, $(CH_3)_2CO$ | 185 | 56.2 |
| Ethyl ether, $(C_2H_5)_2O$ | 442 | 34.5 |

The vapor pressure of a liquid increases with increasing temperature, as illustrated in Figure 12.7 for a number of substances. As the kinetic energy of the molecules of a liquid increases with increasing temperature, more molecules have enough energy to enter the vapor phase and the vapor pressure consequently increases. The plotted curves can be extended all the way up to the critical point. Recall from Section 4.8 that this is the highest temperature at which a substance can exist as a liquid.

The increase of the vapor pressure of a liquid with temperature indicates that at some particular temperature the vapor pressure must become equal to

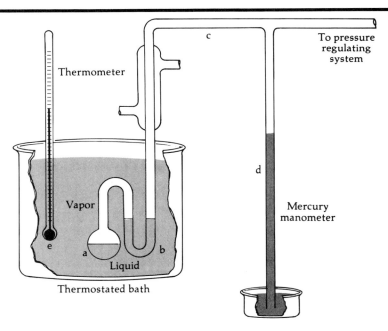

**FIGURE 12.6** Determination of the vapor pressure of a liquid with an isoteniscope. Liquid is added to the bulb (a) and boiled at reduced pressure to expel all the air from the bulb and attached tube. Some liquid is then transferred from the bulb to the attached U-tube (b). The external pressure on (b) is adjusted by admitting air to the attached tube (c) until the heights of the liquid in both arms of the U-tube are equal. Under these conditions, the external pressure is equal to the vapor pressure. The external pressure can be measured with a mercury manometer (d). The isoteniscope is immersed in a water bath (e) maintained at constant temperature.

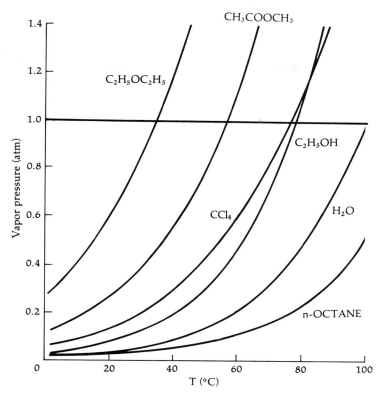

**FIGURE 12.7** The dependence of the vapor pressure of various liquids on temperature. The intersection of these curves with the line drawn at $P = 1$ atm defines the normal boiling point.

exactly 1 atm. By definition, this temperature is the **normal boiling point** of the liquid. When a liquid boils, evaporation can occur from anywhere within the body of the liquid rather than from just the surface. The normal boiling point of a liquid is not the only temperature at which the liquid can boil. A liquid boils whenever its vapor pressure is equal to the external pressure, and the boiling point of a liquid is not unique, as we shall see.

The shape of the curves in Figure 12.7 is predicted by thermodynamics. In order to see that this is so we merely have to recognize that the vapor pressure is, in effect, an equilibrium constant. Since evaporation is an equilibrium process, we can define its equilibrium constant as the ratio of concentrations in the vapor and liquid phases. As the concentration of a pure liquid is constant, its value can be incorporated in the equilibrium constant. Therefore, the equilibrium constant for evaporation is simply equal to the vapor pressure, $K = P$. Recall that the temperature dependence of an equilibrium constant is given by Equation 11.40,

$$\ln K = -(\Delta H°/RT) + (\Delta S°/R)$$

Applying this expression to the vapor pressure we obtain

$$\ln P = -(\Delta H°_{vap}/RT) + (\Delta S°_{vap}/R) \tag{12.3}$$

## 420 Phase Equilibria 12.2

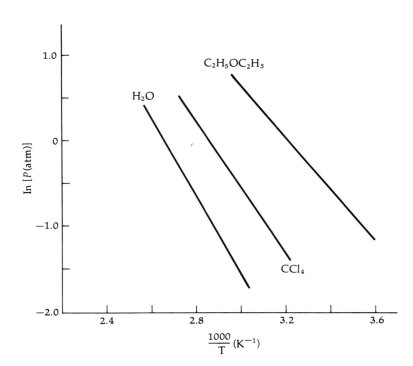

**FIGURE 12.8** Temperature dependence of the vapor pressure in a ln $P$ versus $1/T$ plot.

The data in Figure 12.7 can be replotted as ln $P$ versus $1/T$ to obtain the essentially straight lines shown in Figure 12.8. Since the enthalpy of vaporization is always positive, the lines must have a negative slope in such a plot.

In order to perform calculations involving the temperature dependence of the vapor pressure it is more convenient to use the van't Hoff equation (Equation 11.41). When this relation is applied to vaporization it is known as the **Clausius–Clapeyron equation**:

$$\ln \frac{P_2}{P_1} = \frac{-\Delta H°_{vap}}{R}\left(\frac{1}{T_2} - \frac{1}{T_1}\right) \tag{12.4}$$

Equation 12.4 relates the vapor pressure $P_2$ at temperature $T_2$ to its value $P_1$ at temperature $T_1$.

**Example 12.2** The pressure of the atmosphere on the surface of Mars is 5.3 torr. At what temperature does water boil there?

Since the boiling point is the temperature at which the vapor pressure is equal to the external pressure, we must calculate the temperature $T_2$ at which the vapor pressure $P_2$ is 5.3 torr. In order to use Equation 12.4 we need to know the vapor pressure of water at some other temperature. The most readily available value is the normal boiling point of water, 100°C, at which the vapor pressure is, by definition, 1 atm. Inserting these data as well as the enthalpy of vaporization of water, 40.67 kJ mol$^{-1}$, into Equation 12.4 we obtain

$$\ln\left(\frac{5.3 \text{ torr}}{760 \text{ torr}}\right) = \frac{(-40.67 \text{ kJ mol}^{-1})(10^3 \text{ J kJ}^{-1})}{8.314 \text{ J mol}^{-1} \text{ K}^{-1}} \left(\frac{1}{T_2} - \frac{1}{373 \text{ K}}\right)$$

Solving for $T_2$,

$$T_2 = 2.7 \times 10^2 \text{ K}$$

gives ~0°C as the boiling point of water on Mars. This result assumes only that the enthalpy of vaporization remains constant over this temperature interval. So long as temperatures close to the critical point (where the enthalpy of vaporization rapidly approaches zero) are avoided this assumption is fairly good. ∎

The result of Example 12.2 shows that an astronaut on Mars would have a difficult time in cooking food by boiling it in ice cold water. The same problem, albeit to a much smaller extent, is faced at high elevations on Earth because of the decrease in atmospheric pressure with altitude (Section 3.6). (The use of a pressure cooker, a device that increases the external pressure by confining the steam generated by boiling water, is helpful in cutting down the cooking time under these conditions.)

**Example 12.3** Calculate the vapor pressure of $CCl_4$ at 25°C using (a) the data in Table 12.2 and (b) the data in Appendix 4.

(a) Table 12.2 gives $\Delta H°_{vap}$ and the boiling point of $CCl_4$. The vapor pressure at 25°C can be obtained by means of Equation 12.4:

$$\ln\left(\frac{P_2}{P_1}\right) = -\frac{\Delta H°_{vap}}{R}\left(\frac{1}{T_2} - \frac{1}{T_1}\right)$$

$$\ln\left(\frac{P_2}{760 \text{ torr}}\right) = -\frac{(30.0 \text{ kJ mol}^{-1})(10^3 \text{ J kJ}^{-1})}{8.314 \text{ J mol}^{-1} \text{ K}^{-1}}\left(\frac{1}{298.2 \text{ K}} - \frac{1}{(273.2 + 76.7) \text{ K}}\right)$$

$$P_2 = 127 \text{ torr}$$

(b) Appendix 4 lists values of $\Delta G°_f$ of gaseous and liquid $CCl_4$. These results can be used to obtain $\Delta G°$ for the vaporization at 25°C. The free energy change, in turn, can be used to obtain the equilibrium constant for this process, which we have seen to be equivalent to the vapor pressure. We obtain,

$$\Delta G° = \Delta G°_f[CCl_4(g)] - \Delta G°_f[CCl_4(l)]$$

$$= (-60.59 \text{ kJ mol}^{-1}) - (-65.21 \text{ kJ mol}^{-1}) = 4.62 \text{ kJ mol}^{-1}$$

The equilibrium constant is obtained by means of Equation 11.38,

$$\Delta G° = -RT \ln K$$

$$(4.62 \text{ kJ mol}^{-1})(10^3 \text{ J kJ}^{-1}) = (-8.314 \text{ J mol}^{-1} \text{ K}^{-1})(298.2 \text{ K}) \ln K$$

$$\ln K = -1.86$$

$$P = K = 0.16 \text{ atm}$$

$$= (0.16 \text{ atm})(760 \text{ torr atm}^{-1})$$

$$= 1.2 \times 10^2 \text{ torr}$$

The two answers agree quite well although the second approach yields a somewhat more reliable result. Can you think of why? This example provides another illustration of how a thermodynamic problem can be solved in two distinctly different ways. ∎

## 12.3 Phase Diagrams

***a. The Shape of a Phase Diagram*** We are used to thinking of the physical states of substances as they occur at normal atmospheric pressure. We think of water as a liquid between 0°C and 100°C, and as a solid and vapor below and above these temperatures, respectively. The preceding section has made it clear that pressure is also an important variable in determining the state of aggregation of a substance. A **phase diagram** shows the stability of the various phases of a substance as a function of both pressure and temperature.

Let us explore the nature of a phase diagram with reference to Figure 12.9, the phase diagram of water. The diagram consists of three intersecting lines, known as phase equilibrium lines or phase boundaries, which divide the graph into areas that represent the solid, liquid, and vapor phases. The solid phase is stable at low temperatures and high pressures, the vapor phase at high temperatures and low pressures, and the liquid phase at intermediate temperatures and pressures.

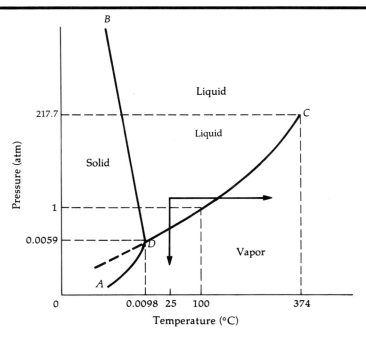

**FIGURE 12.9** The phase diagram of water. The colored curves represent the phase boundaries. The dashed lines, from left to right, mark the temperatures and pressures of the triple point, the normal boiling point, and the critical point. The arrows show two particular paths for the conversion of liquid at $T = 25°$ and $P = 1$ atm to vapor. The dashed extension of line $DC$ represents the formation of supercooled water. Note that the temperature and pressure values are not drawn to scale.

The meaning of the phase boundaries can be understood in several essentially equivalent ways. First, consider line DC, which separates the liquid from the vapor phase. This phase boundary can be interpreted as follows:

1. It is a plot of the $(P, T)$ values at which liquid and vapor are in equilibrium with each other and can, therefore, coexist. Recall that $\Delta G = 0$ for an equilibrium phase change (Section 11.6). The free energy per mole of the liquid and vapor phases of a substance must consequently be equal at the temperatures and pressures defined by line DC. Observe that line DC extends over a wide temperature interval, from $\approx 0$ to 374°C, the critical point of water.
2. It is a plot of the vapor pressure of water as a function of temperature. This interpretation follows from point 1 and the definition of vapor pressure as the pressure of a vapor in equilibrium with the liquid.
3. It is a plot of the boiling point of water as a function of external pressure. This interpretation follows from points 1 and 2 and the definition of the boiling point as the temperature at which the vapor pressure and the external pressure are equal.

Water can be converted to water vapor by a variety of paths. Figure 12.9 shows two distinct ways in which water at 25°C and atmospheric pressure can be vaporized. The horizontal arrow corresponds to heating the liquid at constant pressure and represents the normal process of boiling. The vertical arrow corresponds to a reduction in the external pressure at constant temperature. The boiling point of water is reached when the external pressure has been reduced to a value equal to the vapor pressure of water at 25°C, i.e., 24 torr, or 0.032 atm. It should be apparent that the phase transformation can also be achieved by simultaneous changes in both temperature and pressure.

The other colored lines in the phase diagram have essentially similar meanings as line DC. Line AD is a plot of the $(P, T)$ values for which ice and water vapor are in equilibrium. In analogy to line DC, it can also be understood as being a plot of the vapor pressure of ice as a function of temperature, or of the sublimation point of ice as a function of external pressure. The dashed extension of line DC represents the formation of supercooled water. At temperatures below the freezing point, water is thermodynamically unstable with respect to ice. The molar free energy and the vapor pressure of water are higher than the corresponding quantities for ice, and freezing occurs spontaneously with a decrease in free energy.

Line DB shows the equilibrium between ice and water as a function of temperature and pressure. In this instance, $P$ cannot be regarded as a vapor pressure since the equilibrium does not involve the vapor phase. Rather, $P$ can only be the external pressure on the ice–water system. This pressure can either be that of the atmosphere or that applied by some mechanical device. Note that the freezing point of ice decreases with increasing pressure. This effect can be understood by means of Le Chatelier's principle, which predicts that a two-phase system should respond to an increase in pressure by favoring the formation of the denser phase, i.e., water. Water is unusual in this respect since most solid substances are denser than the corresponding liquids.

While most substances have solid–liquid equilibrium lines of positive slope, these lines are similar to line DB (Figure 12.9) in being much steeper than the corresponding liquid–vapor equilibrium lines. Much larger pressure changes are needed to modify the freezing point than the boiling point of a liquid.

**424** Phase Equilibria 12.3

**FIGURE 12.10** The phase diagram of carbon dioxide.

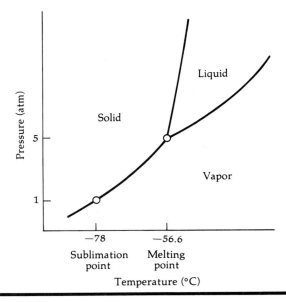

Figure 12.9 shows that all three phase equilibrium lines meet at point D. At this point, called the **triple point,** all three phases coexist simultaneously. The triple point of water occurs at 0.0098°C and a pressure of 4.58 mm Hg (0.0060 atm). The occurrence of a triple point is a characteristic feature of a phase diagram.

The phase diagram of $CO_2$ is displayed in Figure 12.10. Note that the triple point occurs at a pressure of about 5 atm. The triple point defines the lowest external pressure at which the liquid is stable. Therefore, it is apparent that liquid $CO_2$ does not exist at atmospheric pressure, which accounts for the sublimation of dry ice at normal pressures. It may also be seen that, in contrast to the water system, the solid–liquid equilibrium line has a positive slope, indicating that the melting point of solid carbon dioxide increases with the applied pressure.

**Example 12.4** The phase diagram of a certain substance, representing the equilibrium between solid, liquid, and vapor phases, is shown in the accompanying figure.

(a) Indicate whether each of the following statements is true or false:

1. The substance is a liquid at room temperature and atmospheric pressure.
   **False.** The point corresponding to $P = 1$ atm and $T = 25°C$ lies in the bottom region, which represents the vapor.
2. The liquid state cannot be observed at atmospheric pressure.
   **True.** The triple point, which determines the lowest pressure at which the liquid is stable, lies above 1 atm.
3. The melting point increases with an increase in the external pressure.
   **False.** The solid–liquid equilibrium line has negative slope, so the melting point must decrease with increasing pressure.

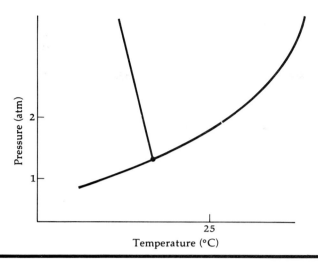

(b) Describe the phase changes that occur when the substance is compressed from an initial pressure of 1 atm to a final pressure of 3 atm at a constant temperature of 25°C.

Initially, the substance is present as a vapor. When compressed to $P \approx 1.7$ atm it becomes liquid. No additional phase changes occur on further compression to 3 atm. ∎

**b. The Phase Rule** When only a single phase of a substance is present, the pressure and temperature can be varied independently over a range of values without causing a change of phase. We say that the system has two **degrees of freedom,** meaning that two variables which describe the state of the system, i.e., $P$ and $T$, can be varied independently.

When two phases of a substance are present in equilibrium only one of the two variables can be varied independently. Considering, for example, line DC, the liquid–vapor equilibrium line in Figure 12.9, we note that the temperature can be varied over a broad range. However, once the temperature is fixed, so is the pressure. Alternatively, the pressure can be varied over a range of values along line DC but, at a given pressure, the temperature is fixed. When two phases are in equilibrium, the system has only one degree of freedom.

Finally, when three phases are simultaneously in equilibrium, both pressure and temperature are fixed and the system has no degrees of freedom. As already noted, the triple point of water occurs at $T = 0.0098°C$ and $P = 4.58$ mm Hg—neither quantity can vary if all three phases are to coexist.

These observations can be summarized by the relationship

$$F = 3 - P \qquad (12.5)$$

where $F$ is the number of degrees of freedom and $P$ is the number of phases present. This relationship, first derived by J. Willard Gibbs (1839–1903), is known as the **phase rule** (see Section 12.5b for a more complete form of the phase rule).

## Phase Equilibria of Solutions

### 12.4 Ideal Solutions

*a. Retrospective View of Solutions* The study of solutions is one of the most important subjects in chemistry, for many chemical processes occur in solution. We have already briefly considered solutions in Chapter 2, treated solutions of acids and bases in Chapter 8, and examined the precipitation of insoluble salts from solution in Chapter 9. The solutions discussed so far have involved liquids or solids, such as acetic acid or sodium chloride, dissolved in water. Solutions of gases in liquids are also common. For example, fish require oxygen dissolved in water for respiration. Gases, liquids, or solids can also dissolve in solids. Many alloys are solid–solid solutions, for example.

Solutions have one common feature which characterizes them as such: they are homogeneous on the macroscopic level and their composition can be varied continuously between some limits. Solutions differ in this respect from compounds, which have a unique composition. We shall restrict ourselves in this chapter to binary solutions, that is, solutions made up of only two components. Recall from Section 2.3a that the component present in the greatest quantity is called the solvent and the other component is the solute.

In considering the phase equilibria of pure substances we noted that pressure and temperature were the important variables. In the case of solutions we must, in addition, specify the composition. The composition of a solution can be specified in a variety of ways, most of which have already been considered. The composition of liquid solutions is commonly expressed in concentration units, such as molarity (Section 2.3b). However, for a consideration of phase equilibria, the most convenient measure of composition is the mole fraction, previously introduced in connection with Dalton's law (Section 3.5). We shall in addition have occasion to use the molality unit. The **molality** $m$ of a solution is defined as the number of moles of solute per kilogram of solvent. For example, a $0.3\ m$ aqueous solution of $Na_2SO_4$ contains 0.3 mol of this salt for each kilogram of water.

*b. Properties of Ideal Solutions* The properties of gases can be explained in terms of a simple model, the ideal gas model. The complications introduced by the interactions between the molecules comprising a real gas proved to be understandable by comparison with the behavior expected for non-interacting ideal gas molecules. The same approach can be used to advantage in an understanding of the properties of solutions.

An **ideal solution** is defined as one in which the forces between molecules are independent of the identity of the molecules. Thus, for an ideal solution composed of molecules of types A and B, the intermolecular forces are equal,

$$F_{AA} = F_{BB} = F_{AB}$$

where $F$ is the force between the molecules identified by the subscripts. Ideal solutions are generally formed by substances of similar molecular structure.

Several consequences immediately follow from this definition. If the forces between the molecules are equal, there can be no difference in energy between the unmixed components A and B and the solution. As a consequence, when the pure components are mixed to form the solution, there is no accompanying evolution or absorption of heat, i.e., $\Delta H_{sol} = 0$, where $\Delta H_{sol}$ is the enthalpy change for the formation of the solution. Similarly, there is no change in volume when the solution is formed, so $\Delta V_{sol} = 0$.

While measurements of $\Delta H$ or $\Delta V$ could be used to determine whether a particular solution is ideal, the behavior of the vapor pressure of a liquid solution constitutes a more useful test of ideality. The vapor pressure of a solution is the sum of the partial pressures, $P_A$ and $P_B$, due to each component.

We have seen that the vapor pressure of a liquid is a measure of the tendency of its molecules to escape to the vapor phase. In an ideal solution, the tendency of a molecule to escape from the liquid is independent of the identity of its neighbors. The partial pressure of each component depends only on its relative abundance in the solution and is proportional to its mole fraction:

$$P_A \propto X_A \qquad P_B \propto X_B$$

When the mole fraction of a component becomes unity, its partial pressure is just the vapor pressure of the pure component. Therefore, the vapor pressures of the pure components are the proportionality constants, giving

$$P_A = X_A P_A^\circ \qquad P_B = X_B P_B^\circ \tag{12.6}$$

where $P_A^\circ$ and $P_B^\circ$ are the vapor pressures of pure liquids A and B, respectively. Equation 12.6 is commonly known as **Raoult's law** and is an operational definition of an ideal solution.

The vapor pressure of a solution can be obtained readily from the partial pressures of the components on the assumption that the vapor is an ideal gas. According to Dalton's law, the vapor pressure $P$ is the sum of the partial pressures,

$$P = P_A + P_B$$

Since the partial pressures of the components of an ideal solution vary linearly with the mole fractions, the vapor pressure must also be a linear function of

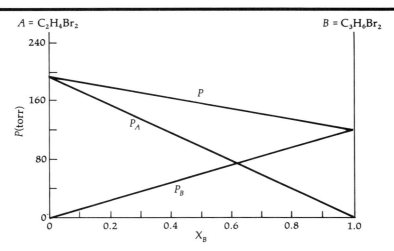

**FIGURE 12.11** Partial pressures and vapor pressure of a solution of $A$, dibromoethane ($C_2H_4Br_2$), and $B$, dibromopropane ($C_3H_6Br_2$) at 85°C.

**428** Phase Equilibria 12.4

mole fraction. This linearity constitutes the most direct proof of ideal behavior, for the vapor pressure of a solution can be measured readily.

Figure 12.11 illustrates Raoult's law for a solution of dibromoethane ($C_2H_4Br_2$) and dibromopropane ($C_3H_6Br_2$) at 85°C. Observe that the partial pressures are a linear function of the mole fraction. The vapor pressure, being a sum of two linear functions, is also linear. Since the vapor pressure of a liquid varies with temperature (Figure 12.7), the temperature has to be specified in such a plot.

Knowing the partial pressures of the components and the vapor pressure of the solution, it is possible to determine the composition of the vapor by means of Dalton's law in the form of Equation 3.18:

$$P_A = X_A^{vap} P \qquad P_B = X_B^{vap} P$$

where the superscripts on the mole fractions are meant to distinguish the composition of the vapor phase from that of the liquid phase.

---

**Example 12.5** Benzene ($C_6H_6$) and toluene ($C_6H_5CH_3$) form an ideal solution. The normal boiling point of benzene is 80°C, and the vapor pressure of toluene at this temperature is 350 torr. (a) Calculate the partial pressures and the vapor pressure at 80°C of a solution whose composition is given by $X_B = 0.200$ (B = benzene, T = toluene). (b) Calculate the composition of the vapor in equilibrium with the solution in (a). (c) What is the composition of a solution of benzene and toluene that will just boil at 80°C under a reduced pressure of 500 torr?

(a) Equation 12.6 yields the desired partial pressures, once it is recognized that the vapor pressure of pure benzene must be 760 torr at 80°C:

$$P_B = X_B P_B^\circ = (0.200)(760 \text{ torr}) = 152 \text{ torr}$$

$$P_T = X_T P_T^\circ = (0.800)(350 \text{ torr}) = 280 \text{ torr}$$

The vapor pressure of the solution is obtained by means of Dalton's law as

$$P = P_B + P_T = 152 \text{ torr} + 280 \text{ torr} = 432 \text{ torr}$$

(b) The composition of the vapor can be determined by means of Dalton's law in the form of Equation 3.18, $P_B = X_B^{vap} P$, which can be solved for $X_B^{vap}$ to obtain

$$X_B^{vap} = P_B/P = 152 \text{ torr}/432 \text{ torr} = 0.352$$

The mole fraction of toluene can be obtained from a similar relation, or more directly, from the fact that the sum of the mole fractions of the two components must be unity. This constitutes a useful check of the correctness of the calculation:

$$X_T^{vap} = P_T/P = 280 \text{ torr}/432 \text{ torr} = 0.648$$

$$X_T^{vap} + X_B^{vap} = 0.648 + 0.352 = 1.000$$

By comparing the composition of the vapor with that of the liquid, we note that the vapor is enriched in the component with the lower boiling point. As we shall presently see, this enrichment forms the basis of a procedure used to separate a solution into its components.

(c) Since the vapor pressure of a liquid is equal to the external pressure at the boiling point, we have

$$P = P_{\text{ext}} = 500 \text{ torr}$$

Using Raoult's law, we relate the vapor pressure to the partial pressures:

$$P = X_B P_B^\circ + X_T P_T^\circ$$

$$500 \text{ torr} = (760 X_B + 350 X_T) \text{ torr}$$

This equation has two unknowns. We know, however, that the sum of the mole fractions is unity, i.e., $X_B + X_T = 1$. Substituting $X_T = 1 - X_B$ into the preceding expression, we get

$$500 \text{ torr} = (760 X_B + 350 - 350 X_B) \text{ torr}$$

The answer is

$$X_B = 0.366 \qquad X_T = 0.634 \quad \blacksquare$$

## 12.5 Phase Diagrams of Ideal Solutions

*a. The Liquid–Vapor Phase Diagram* The liquid–vapor phase diagram of a solution consists of a plot of the temperature and pressure at which liquid and vapor are in equilibrium as a function of the solution composition. Since a graph of three variables in two dimensions is not readily visualized, the phase diagram is usually presented as a plot of just two parameters. This can be done by showing either the variation of the equilibrium pressure with composition at constant temperature or, more customarily, that of the equilibrium temperature with composition at constant pressure. The phase diagram can be determined experimentally by measurement of the composition of liquid and vapor in equilibrium as a function of temperature at some constant pressure. For an ideal solution, calculations such as those illustrated in Example 12.5 can be used to determine the phase diagram.

Figure 12.12 is the phase diagram for $P = 1$ atm of a mixture of benzene and toluene, which form an ideal solution. The diagram consists of two phase equilibrium curves which divide the diagram into three regions. Since the liquid phase is stable at lower temperatures than the vapor phase, the lower region corresponds to the liquid and the upper one to the vapor. The lower phase boundary curve shows the variation of the boiling point and the upper one that of the condensation point, both as a function of composition.

The region between the two phase boundary lines represents the two phases in equilibrium with each other. A horizontal line (called a **tie line**) intersects the phase boundary lines at points which give the composition of the two phases at the temperature in question. For example, line *ab* shows that at a temperature of 103°C, a solution whose composition is given by $X_B = 0.15$ (point *a*) just boils. The vapor in equilibrium with the boiling liquid has composition $X_B^{\text{vap}} = 0.37$ (point *b*) and is thus enriched in the component having the lower boiling point (benzene). Conversely, a vapor consisting of 37 mol% benzene ($X_B = 0.37$) just condenses at 103°C and the composition of the liquid in equilibrium with the condensing vapor is given by $X_B = 0.15$.

**430** Phase Equilibria **12.5**

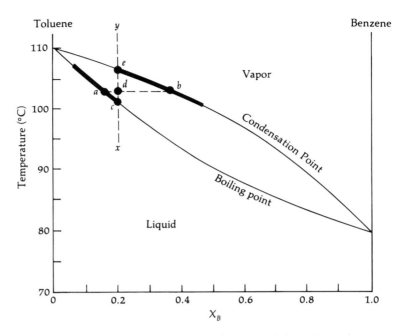

**FIGURE 12.12** Liquid–vapor phase diagram of the toluene–benzene system at atmospheric pressure. The mole fraction $X_B$ of benzene is shown and the mole fraction of toluene is understood to vary from 1 at the left to 0 at the right. The colored portions of the phase boundary lines show the compositions of liquid and vapor present when the solution is heated along line *xy* from point *c* to point *e*. The marked points are discussed in the text.

A more complete understanding of the phase diagram can be obtained by consideration of a process in which the temperature of a solution of given composition is gradually increased. Line *xy* in Figure 12.12 illustrates such a process for a solution with $X_B = 0.20$. As the temperature is increased, point *c* (101°C), at which the liquid commences to boil, is eventually reached. Boiling continues with increasing temperature until point *e* (107°C) is reached, where vaporization is complete. The composition of the liquid and vapor phases in this temperature interval is given by the appropriate segments of the boiling and condensation point curves (drawn in color). Observe that the first vapor that comes off is enriched in benzene ($X_B = 0.46$). As the temperature increases to 107°C, the composition of the vapor necessarily approaches the composition of the initial solution ($X_B = 0.20$). At the same time, the liquid becomes increasingly enriched in toluene, the component with the higher boiling point. Thus, $X_B$ decreases from 0.20 to 0.06. At any specific temperature in this interval, the composition of liquid and vapor is given by the intersection of a horizontal line with the phase boundary lines, as already mentioned. For instance at point *d* (103°C), the composition of the solution is given by point *a* on the boiling-point curve and that of the vapor by point *b* on the condensation point curve. This discussion serves to emphasize the important point that in contrast to a pure substance, a solution does not boil at a fixed temperature (at constant pressure).

**Example 12.6** The liquid–vapor phase diagram of a solution of A and B at atmospheric pressure is shown in the accompanying figure. Answer the following questions about this diagram:

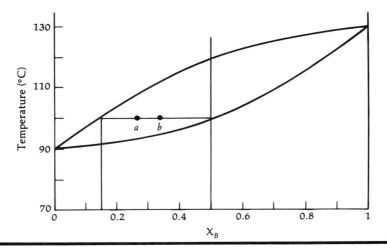

(a) At approximately what temperature does liquid of composition $X_B = 0.5$ first begin to boil?

The temperature at which boiling commences can be determined by drawing a vertical line through $X_B = 0.5$ and noting the intersection with the boiling point curve, i.e., the lower phase boundary. The temperature is $\approx 100°C$.

(b) What is the composition of the vapor in equilibrium with the liquid described in (a)?

The composition of the vapor in equilibrium with liquid of composition $X_B = 0.5$ is obtained by drawing a tie line through the point $X_B = 0.5$, $T \approx 100°C$, and determining its intersection with the condensation curve. At that point, the composition is $X_B = 0.16$. Observe that the vapor is enriched in A, as expected from the lower boiling point of A.

(c) At approximately what temperature is vaporization of the solution with $X_B = 0.5$ complete?

Extending the vertical line through $X_B = 0.5$ up to the condensation curve, we see that the intersection occurs at $T \approx 117°C$. At higher temperatures, a system with this composition exists only as a vapor (when $P = 1$ atm).

(d) Describe the state of the system at the points labelled $a$ and $b$.

Both these points lie on the same horizontal line in the two-phase region. Therefore, the system consists of a mixture of liquid of composition $X_B = 0.5$ and vapor of composition $X_B = 0.16$ in both cases. The difference between points $a$ and $b$ lies in the relative abundance of the two phases. Point (a) is closer to the vapor phase boundary, and the system is therefore richer in vapor at $a$ than at $b$. ■

**b. The Phase Rule for a Two-Component System**  Recall from Section 12.3b that the phase rule provides a simple relationship between the number of coexisting phases, $P$, and the number of degrees of freedom, $F$, i.e., the number of variables describing the system that can be varied independently. The phase rule can be extended to a system consisting of $C$ components. Each additional component introduces an extra degree of freedom, for its abundance can be varied. Thus,

$$F = C - P + 2 \tag{12.7}$$

Notice that for a pure substance, $C = 1$, and Equation 12.7 properly reduces to Equation 12.5, $F = 3 - P$, the phase rule for a one-component system.

The shape of a liquid–vapor phase diagram for a two-component system, such as that depicted in Figure 12.12, can be understood with the aid of the phase rule. Let us specify, first, that the pressure will be fixed at some value, e.g., 1 atm. This uses up one of the degrees of freedom and Equation 12.7 becomes

$$F = C - P + 1 \quad \text{(at constant pressure)} \tag{12.8}$$

Applying this expression to a two-phase region, we have $C = 2$, $P = 2$, and therefore $F = 1$. This single degree of freedom can be either the temperature or the composition of each phase. If the temperature is fixed at some arbitrary value, the composition of each phase is specified. In Example 12.6, for instance, at 100°C, $X_B = 0.50$ for the liquid and $X_B = 0.16$ for the vapor. Conversely, for a given composition of the two phases, the temperature is determined. Since, in fact, the composition of the phases changes while the liquid boils, the temperature must change accordingly. Thus, the liquid boils over a range of temperatures.

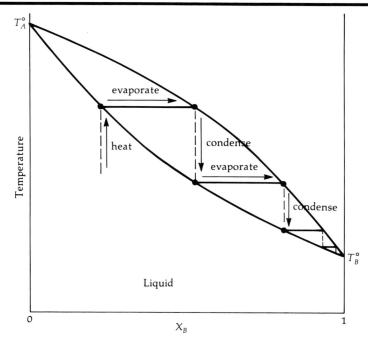

**FIGURE 12.13**  Separation of the components of a solution by fractional distillation.

Equation 12.8 indicates that at either end of the phase diagram, where only a single component is present ($X_B = 0$ or 1), $F = 0$ when two phases are present. Therefore, the temperature cannot vary while the liquid boils. This requires the two phase boundaries to coalesce and results in the distinctive shape of the phase diagram.

*c. Distillation* The separation of a solution into its components is a widely used process in both laboratory and industrial-scale chemistry. The synthesis of organic compounds, for instance, often results in a solution of two or more products which must be separated from each other. Another example is the refining of petroleum, in which a mixture of hydrocarbons must be separated. One procedure by which such separations are carried out can be understood with the aid of a liquid–vapor phase diagram (Figure 12.13). Suppose a solution of a certain composition is heated to its boiling point, as indicated by the dashed line at the left in Figure 12.13. The composition of the vapor which is formed initially is given by the intersection of the uppermost solid horizontal line with the vapor phase boundary. Note that the vapor is enriched in B, the

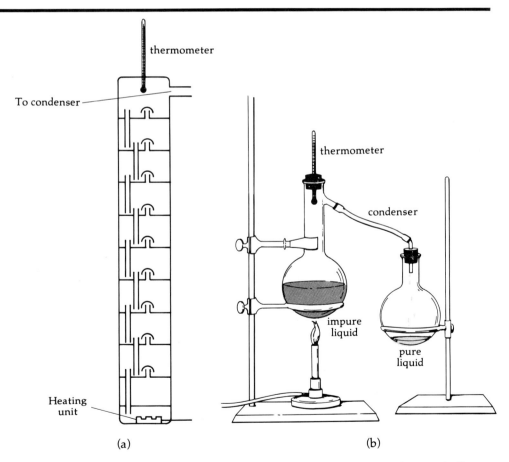

**FIGURE 12.14** (a) A fractional distillation column for the separation of two volatile liquids from each other. (b) A simple distillation setup for the separation of a volatile liquid from non-volatile impurities.

component with the lower boiling point. If this vapor is now condensed and the process repeated, the resulting vapor becomes increasingly enriched in component B. Further repetitions eventually lead to a virtually complete separation of component B, as shown in the figure.

The procedure of separating two volatile components of a solution from each other by successive vaporization and condensation is called **fractional distillation.** Figure 12.14a shows a distillation column that can be used for fractional distillation. The key feature of this column is the series of horizontal plates which protrude from the sides into the interior. As the liquid in the bottom container is boiled, the vapor will rise and condense on the first horizontal plate encountered. As further vapor rises and condenses, the plate is gradually heated. Some of the liquid on this plate will therefore vaporize and the vapor rises to the next horizontal plate, where the process repeats itself. As each successive plate accumulates liquid, the excess, which is depleted in the more volatile component, drips down and recommences the cycle. Eventually, an equilibrium is established in the column. The temperature gradually decreases from bottom to top, and the rising vapor becomes increasingly enriched in the component with the lower boiling point. If the number of horizontal plates is sufficiently large, the vapor drawn off at the top will consist of the virtually pure more volatile component. This vapor can be condensed by cooling.

Fractional distillation may be contrasted with simple distillation, depicted in Figure 12.14b, which is used to separate a liquid from nonvolatile impurities (usually solids). The liquid is simply boiled and the vapor condensed to yield a distillate free of such impurities. The impurities remain behind in the distillation flask.

## 12.6 Non-Ideal Solutions and Their Phase Diagrams

*a. The Nature of Non-Ideal Solutions*  Most solutions exhibit deviations from ideal behavior. This is not surprising since ideal solutions are only formed by substances of very similar molecular structure. Two types of deviations are possible. The more common case occurs when the forces between the unlike molecules are weaker than those between the like ones, i.e.,

$$F_{AB} < F_{AA} \text{ or } F_{BB}.$$

Substances of this type require an input of energy when they form a solution, for the strongly attracting like molecules must be separated to permit them to mix with unlike molecules. Thus, the enthalpy of solution is positive. In addition, the volume of the solution is greater than that of its components since the unlike molecules do not attract each other as strongly as molecules of the same type, and so are farther apart from each other, on average. When the differences in cohesive forces between like and unlike molecules becomes sufficiently great, the two substances may actually be only slightly miscible, like ether and water.

Another consequence of the weaker attractive forces between the molecules in such a solution is a greater tendency of the molecules to escape into the vapor phase than is the case in an ideal solution. The partial pressures and vapor pressure of such a solution are consequently larger than the values predicted by Raoult's law:

$$P_A > X_A P_A^\circ \qquad P_B > X_B P_B^\circ$$

Figure 12.15a displays these results graphically for a carbon tetrachloride (CCl$_4$)–methyl alcohol (CH$_3$OH) solution. A solution of the type represented is said, for rather obvious reasons, to display positive deviations from ideal behavior.

In some solutions, the forces between unlike molecules are stronger than those between like molecules, i.e.,

$$F_{AB} > F_{AA} \text{ or } F_{BB}.$$

This happens when the like molecules are held together by weak van der Waals forces (Section 18.2) while the unlike molecules are held together by hydrogen bonds (Section 18.3), which are stronger. Chloroform (CHCl$_3$) and acetone (CH$_3$COCH$_3$), for example, form a solution of this type, for the hydrogen atom of chloroform can form a hydrogen bond with the oxygen atom of acetone. The vapor pressure diagram for this system is shown in Figure 12.15b. The behavior is just the opposite of that displayed by the CCl$_4$–CH$_3$OH system. The vapor pressure as well as the partial pressures are lower than expected from Raoult's law, the enthalpy of solution is negative, and the volume is smaller than that of the unmixed components. Such a solution is said to exhibit negative deviations from ideal behavior.

**b. Henry's Law** When a non-ideal solution is very dilute, the vapor pressures of its components obey certain limiting laws. As is true for ideal solutions, the vapor pressure of the solvent obeys Raoult's law (Equation 12.6), $P_A = X_A P_A^\circ$, (solvent and solute are designated A and B, respectively). This behavior can be understood as follows: in a very dilute solution, a molecule of the solvent is

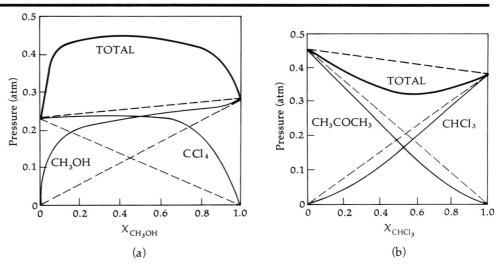

**FIGURE 12.15.** Vapor pressure diagrams for (a) carbon tetrachloride (CCl$_4$)–methyl alcohol (CH$_3$OH) at 35°C, a solution displaying positive deviations from ideal behavior, and (b) acetone (CH$_3$COCH$_3$)–chloroform (CHCl$_3$) at 35°C, a solution displaying negative deviations. The solid curves are the partial pressures of the components and the total vapor pressure of the solution (in color). The corresponding dashed lines represent the values expected from Raoult's law.

surrounded by virtually only other solvent molecules. Therefore, the tendency of solvent molecules to escape to the vapor phase is proportional to their relative abundance in solution. Thus, Raoult's law is obeyed even though the forces between unlike molecules differ from those between like molecules.

The solute molecules, too, are in a uniform environment when a solution is very dilute. Each solute molecule is surrounded practically only by solvent molecules. Consequently, the vapor pressure of the solute is proportional to its mole fraction. However, the proportionality constant is not the vapor pressure of the pure solute, as it would be if the solution were ideal, but a hypothetical vapor pressure, $K_B$, representing the strength of the solute–solvent interaction. The proportionality between the vapor pressure and mole fraction of the solute in a dilute solution is known as **Henry's law**, and is given by the relation

$$P_B = K_B X_B \tag{12.9}$$

Figure 12.16 shows data like that in Figure 12.15b in a plot that focuses on the operation of the above two limiting laws. The vapor pressure of acetone obeys Raoult's law so long as $X_{\text{acetone}} \gtrsim 0.95$ ($X_{\text{CHCl}_3} \lesssim 0.05$). Similarly, the

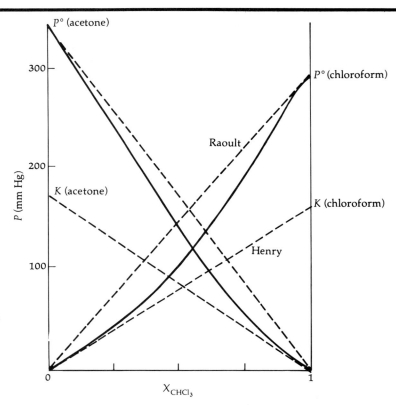

**FIGURE 12.16** The operation of Raoult's and Henry's laws in dilute solutions of acetone and chloroform. When the solvent is acetone ($X_{\text{CHCl}_3} \ll 1$) the vapor pressure of acetone obeys Raoult's law (dashed line) and the vapor pressure of chloroform obeys Henry's law (colored dashed line). The situation is reversed when the solvent is chloroform ($X_{\text{CHCl}_3} \gg 0$). The solid curves represent the actual vapor pressures of the two components. The values of the Henry's law constants $K$ are given by the intersections of the colored dashed lines with the vapor pressure axes.

**TABLE 12.4 Henry's Law Constant for Gases Dissolved in Water at 25°C**

| Gas | K (mm Hg) |
|---|---|
| $CH_4$ | $3.14 \times 10^5$ |
| $CO_2$ | $1.25 \times 10^6$ |
| $O_2$ | $3.30 \times 10^7$ |
| CO | $4.34 \times 10^7$ |
| $H_2$ | $5.34 \times 10^7$ |
| $N_2$ | $6.51 \times 10^7$ |

vapor pressure of chloroform obeys Raoults's law so long as $X_{CHCl_3} \geq 0.94$. On the other hand, the vapor pressure of chloroform obeys Henry's law provided $X_{CHCl_3} \leq 0.21$, while the vapor pressure of acetone obeys Henry's law when $X_{acetone} \leq 0.11$ ($X_{CHCl_3} \geq 0.89$). Note that the values of the constants $K$ of acetone and chloroform, given by the intersections of the Henry's law lines with the vapor pressure axes, are much smaller than the vapor pressures of the pure liquids. This behavior follows from the negative departures from ideality displayed by the acetone–chloroform system.

Since many gases are only slightly soluble in water, their solutions are very dilute and obey Henry's law. Table 12.4 gives the Henry's law constants for several gases dissolved in water at 25°C. These data may be used to determine the solubilities of gases in water.

**Example 12.7** What is the solubility in water at 25°C of atmospheric nitrogen (a) when the total air pressure is 1.00 atm and (b) when the total air pressure is 4.50 atm? Express the answers in terms of mole fraction, molality (Section 12.4a), and concentration.

(a) The mole fraction of nitrogen in air is 0.780 (Section 3.6). Thus the partial pressure of $N_2$ in 1.00 atm of air is, by Dalton's law,

$$P_{N_2} = (0.780)(1.00 \text{ atm})(760 \text{ mm Hg atm}^{-1}) = 593 \text{ mm Hg}$$

The mole fraction of $N_2$ in water is obtained by Henry's law, Equation 12.9,

$$X_{N_2} = P_{N_2}/K_{N_2}$$
$$= (593 \text{ mm Hg})/(6.51 \times 10^7 \text{ mm Hg})$$
$$= 9.11 \times 10^{-6}$$

The molality of a solution is the number of moles of solute per kilogram of solvent. For a very dilute solution, the molality can be obtained by approximating the expression for the mole fraction,

$$X_{N_2} = \frac{n_{N_2}}{n_{N_2} + n_{H_2O}} \approx \frac{n_{N_2}}{n_{H_2O}}$$

This approximation is valid because the number of moles of $H_2O$ is much larger than that of $N_2$. The molality is

$$n_{N_2}/1 \text{ kg H}_2\text{O} = (X_{N_2})(n_{H_2O})/1 \text{ kg H}_2\text{O} = (X_{N_2})(1000 \text{ g}/18.0 \text{ g mol}^{-1})$$
$$= (9.11 \times 10^{-6})(1000 \text{ g}/18.0 \text{ g mol}^{-1}) = 5.06 \times 10^{-4} \text{ m}$$

The solution is sufficiently dilute for its density to be equal to that of water, i.e., 1.0 g cm$^{-3}$. As for the molality, the amount of nitrogen can be neglected relative to that of water, so that 1 L of solution contains 1000 g of water. Consequently the molarity of the solution is equal to its molality.

(b) The mole fraction of $N_2$ scales with the partial pressure of $N_2$, which in turn scales with the total air pressure. Thus,

$$X_{N_2} = \frac{(4.50 \text{ atm})}{(1.00 \text{ atm})}(9.11 \times 10^{-6}) = 4.10 \times 10^{-5}$$

the molality is

$$\frac{4.50 \text{ atm}}{1.00 \text{ atm}} (5.06 \times 10^{-4} \, m) = 2.28 \times 10^{-3} \, m$$

and the concentration is $2.28 \times 10^{-3} \, M$. ∎

Example 12.7 illustrates the increase in the solubility of a gas with increasing partial pressure. There are several common manifestations of this effect. The conditions prevalent in deep sea diving are approximately those in Example 12.7b. A diver experiences a water pressure of several atmospheres and the air for breathing must be compressed to several atmospheres in order to permit exhalation. Nitrogen and oxygen are more soluble in blood under these conditions. While oxygen is used up by metabolism, nitrogen stays in solution. When the diver ascends to sea level and atmospheric pressure, the solubility of nitrogen decreases and bubbles of $N_2$ gas form in the blood. If the ascent is rapid, these bubbles can accumulate and cause a painful and dangerous condition known as "the bends," so called because the afflicted diver commonly bends over in pain. The condition can be treated by placing the diver in a hyperbaric (i.e., high-pressure) chamber and decompressing slowly.

The storage of carbonated beverages provides another illustration of Henry's law. These beverages contain dissolved carbon dioxide gas under pressures exceeding 1 atm. When the container is opened, the pressure on the liquid is reduced and the $CO_2$ bubbles out of solution, producing the characteristic fizz of carbonated beverages such as soda.

*c. Liquid–Vapor Phase Diagrams*  The liquid–vapor phase diagrams of non-ideal solutions reflect the variation of the total vapor pressure with composition. We have noted that the vapor pressure of a liquid is inversely correlated with the boiling point—a liquid with high vapor pressure at a particular temperature has a relatively low boiling point and vice versa. This relationship also applies to non-ideal solutions: if a solution has a minimum (maximum) vapor pressure at some composition, it will generally have a maximum (minimum) in the boiling point curve at roughly the same composition.

Figure 12.17a is the temperature–composition phase diagram for the acetone–chloroform system at atmospheric pressure. As shown in Figure 12.15b, the vapor pressure of the solution has a minimum for $X_{CHCl_3} \approx 0.6$. As expected from the preceding considerations, the phase diagram shows the occurrence of a maximum in the boiling point curve for $X_{CHCl_3} \approx 0.6$. A solution having this composition has a higher boiling point than either of the pure liquids making up the solution. When a solution of this composition is heated, as indicated by line *ab* in the figure, the solution boils at a constant temperature until all the liquid has been vaporized. A solution exhibiting this property is known as a constant-boiling mixture, or **azeotrope.** The same term is applied to a solution whose composition corresponds to a minimum in a boiling point curve, as displayed for the ethanol–benzene system in Figure 12.17b.

The constancy of the boiling point of an azeotropic solution can be understood by application of the phase rule. Since the azeotrope consists of two components and two phases present at constant pressure, it follows from Equation 12.8 that $F = 1$. As an azeotropic solution has a specific composition, the one degree of freedom must be used to specify this composition. Consequently, the

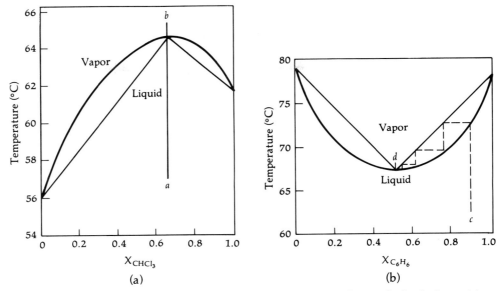

**FIGURE 12.17** Temperature–composition phase diagrams of non-ideal solutions. (a) The acetone (CH$_3$COCH$_3$)–chloroform (CHCl$_3$) system at atmospheric pressure. The vapor pressure of this solution at 35°C is shown for different compositions in Figure 12.15b. (b) The benzene (C$_6$H$_6$)–ethanol (C$_2$H$_5$OH) system at atmospheric pressure. Lines *ab* and *cd* are discussed in the text.

---

temperature is not free to vary and the solution has a definite boiling point. While an azeotrope mimics the behavior of a pure substance in this respect, it can in no sense be regarded as a compound. This is most readily demonstrated by examining the phase diagram for a different external pressure. The composition of the azeotrope changes with pressure, indicating that it is not a compound.

A solution with an azeotropic point cannot be separated into its components by fractional distillation. Consider a solution with a minimum boiling point, such as the benzene–ethanol system. As shown by line *cd* in Figure 12.17b, the composition of the vapor resulting from distillation of such a solution approaches that of the azeotrope instead of that of the lower boiling component. On the other hand, so long as the starting solution is not the azeotrope, a solution with a maximum boiling point does yield the vapor of either pure component upon fractional distillation. However, the remaining liquid approaches the azeotropic composition. In neither case is a complete separation into the pure liquids possible.

## 12.7 Solid–Liquid Phase Equilibria

*a. Phase Equilibrium Diagrams*   If the temperature is lowered to a sufficient extent, a solution eventually freezes. As in the case of a liquid–vapor phase diagram, the variables of importance in solid–liquid phase diagrams are temperature, pressure, and composition. In order to present the results for a particular system as a two-dimensional plot, either temperature or pressure must be kept constant. Since, as already noted in Section 12.3, the effect of pressure changes on the freezing point of a substance is relatively small, the results are usually presented as a plot of temperature versus composition at constant pressure.

Solid–liquid phase diagrams display a variety of shapes depending primarily on the nature of the solid phase. The simplest type of diagram is obtained when the substances that freeze out of solution upon cooling are just the pure components. An illustration of such a phase diagram is provided by the benzene–naphthalene system, as displayed in Figure 12.18.

At temperatures higher than those given by line $AEB$ in Figure 12.18, the ends of which correspond to the freezing points of the pure components, benzene and naphthalene exist as a liquid solution. Line $AEB$, which is the freezing point curve, shows that when a small amount of naphthalene is added to benzene, the freezing point of the resulting solution is lower than that of pure benzene. The same holds true when a small amount of benzene is added to naphthalene. This lowering of the freezing point can be understood on the basis of thermodynamics and is treated in detail in Section 12.8. For now, we content ourselves with noting that this decrease in freezing point gives the diagram its characteristic shape.

To understand the phase diagram in detail, it is helpful to trace the changes that occur when a solution of a given composition and temperature is cooled. Consider, for instance, what happens when the solution at point $b$ in Figure 12.18 is cooled. When point $c$ is reached, the temperature has dropped sufficiently to permit freezing to occur. The substance freezing out is pure naphthalene. The region defined by $BED$ consists of solid naphthalene in equilibrium with the solution. As the temperature drops further, more naphthalene freezes, and the remaining solution gradually becomes enriched in benzene. The composition of the solution at any point within the two-phase region is given by the intersection of a tie line through the point in question with phase boundary line $BE$. For instance, at point $x$, the composition of the solution in

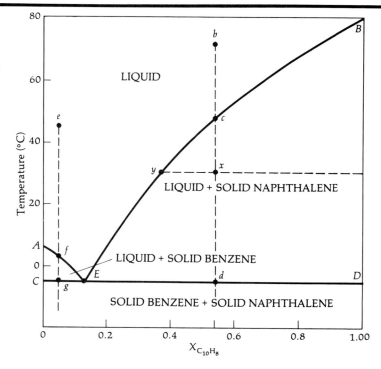

FIGURE 12.18 Freezing point diagram for the benzene ($C_6H_6$)–naphthalene ($C_{10}H_8$) system at a pressure of 1 atm. The solid curves represent the solid–liquid phase boundaries. The dashed lines are discussed in the text.

equilibrium with solid naphthalene is given by point *y*; the solution consists of approximately 35 mol% naphthalene. As the system is cooled further, point *d* is eventually reached. At this temperature, the composition of the remaining liquid is given by point *E*. Note that a solution freezes over a range of temperatures rather than at a fixed temperature, in analogy to vaporization.

Let us now perform the same analysis for a liquid of temperature and composition given by point *e*. As this liquid is cooled, freezing is first observed at point *f*. The substance freezing out of solution in this region of the phase diagram is pure benzene. Region *ACE* defines a two-phase region consisting of solid benzene in equilibrium with the solution. As the temperature decreases further, additional benzene freezes out and the solution becomes increasingly enriched in naphthalene. When the temperature has fallen to $-7°C$ (point *g*) the composition of the solution is given by point *E*.

Note that as point *E* is approached from the right, the solution is in equilibrium with solid naphthalene, while as it is approached from the left, the solution is in equilibrium with solid benzene. Consequently, at point *E* the solution must be in equilibrium with both solids. Point *E* is called the **eutectic point** and corresponds to the lowest temperature at which a liquid solution is stable. When the solution attains this composition both solids freeze out with no change in temperature or composition. A solution of eutectic composition behaves like a pure compound in this respect. However, careful examination of the solid forming at the eutectic point shows that two different types of crystals, corresponding to the two pure components, are present.

The constancy of the temperature at the eutectic point can be understood by means of the phase rule. Since the system consists of two components and three phases at this point, the number of degrees of freedom is given by

$$F = C - P + 2 = 1$$

The single degree of freedom is used to specify the pressure and therefore neither the temperature nor the composition can vary.

Once the remaining solution has solidified at the eutectic point, the system consists of solid benzene and naphthalene. These two solids are present at temperatures below the eutectic point, i.e., in the region of the phase diagram below line *CD*.

Solid–liquid phase diagrams frequently are more complex than the diagram shown in Figure 12.18. One common complication is the formation of compounds in the solid state. An aqueous solution of sulfuric acid, for example, freezes at different temperatures in the form of hydrates, such as $H_2SO_4 \cdot H_2O$, $H_2SO_4 \cdot 2H_2O$, and $H_2SO_4 \cdot 4H_2O$. The phase diagram, shown in Figure 12.19, at first sight appears to be exceedingly complicated. In actuality, it just consists of a number of simple diagrams of the type shown in Figure 12.18 placed side by side. For example, the region corresponding to an $H_2SO_4$ mole fraction of 0–0.2 consists of a simple diagram with one eutectic point in which the substances freezing out of the solution are $H_2O$ and $H_2SO_4 \cdot 4H_2O$. This diagram is followed by one in which the substances which freeze out are $H_2SO_4 \cdot 4H_2O$ and $H_2SO_4 \cdot 2H_2O$, and so forth (see Example 12.8).

It is also possible for two solids to form a solid solution, either over a limited range of compositions or for all compositions. Many alloys are just solid solutions of two metals. For example, brass consists of a solution of zinc in copper in a 1 : 2 mole fraction ratio. The more complicated phase diagrams of such systems are of importance in such varied fields as metallurgy and geochemistry.

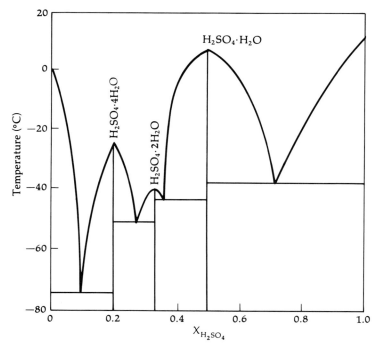

**FIGURE 12.19.** Freezing point diagram for the H$_2$SO$_4$–H$_2$O system at atmospheric pressure. The diagram shows the formation of three different solid hydrates.

---

**b. Thermal Analysis**  A phase diagram like Figure 12.18 is constructed by a technique called **thermal analysis**. The technique involves the determination of the rate at which a solution cools when it has been heated above its melting point. A plot of temperature versus time is known as a **cooling curve** and represents the changes along vertical lines in a phase diagram, such as line *bd* in Figure 12.18.

Figure 12.20 shows two typical cooling curves for the benzene–naphthalene system. Figure 12.20a corresponds to cooling along line *bd* in Figure 12.18. So long as only liquid is present, the temperature decreases uniformly with time. When the temperature drops to point *c*, naphthalene begins to freeze out. Since $\Delta H_{\text{solid}} < 0$, heat is released as long as freezing continues. The temperature therefore decreases less rapidly than before. The change in the slope of the cooling curve defines the occurrence of point *c* on the phase diagram. The small jog observed at this point reflects the supercooling of the solution below its normal freezing point and can be minimized by vigorous stirring. Eventually, the temperature drops to the eutectic value and remains constant while the remaining solution freezes. Once the solution is completely solidified, the temperature resumes its uniform decrease with time.

The phase diagram in Figure 12.18 can be constructed on the basis of the temperature at which freezing is first observed for a number of solutions of different composition. The cooling curve shown in Figure 12.20a is typical of the type of curve obtained for a solution of any composition except that of the eutectic mix or the pure components. At these three compositions the liquid has a

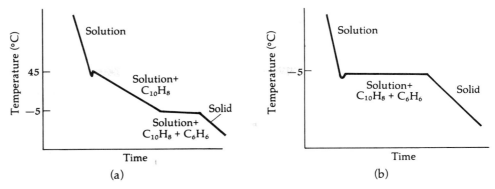

**FIGURE 12.20** Typical cooling curves used to construct the solid–liquid phase diagram in Figure 12.18. (a) A cooling curve along line *bd* in Figure 12.18 and (b) a cooling curve for a solution having the eutectic composition.

unique freezing point, and the cooling curve is like that depicted in Figure 12.20b.

Example 12.8  (a) Describe the changes that occur in the $H_2SO_4$–$H_2O$ phase diagram (Figure 12.19) when samples of the following compositions are heated from $-80°C$ to $20°C$:

(1) $X_{H_2SO_4} = 0.60$.

At $-80°C$, the sample consists of a mixture of solid $H_2SO_4 \cdot H_2O$ and solid $H_2SO_4$. When the solid is warmed to $\approx -38°C$ it begins to melt and continues to do so until a temperature of $\approx -3°C$ is reached. In this temperature interval the sample consists of solid $H_2SO_4 \cdot H_2O$ and solution (the composition of the solution at a particular temperature can be read off the phase boundary extending between $X_{H_2SO_4} = 0.50$ and $\approx 0.72$). Above $\approx -3°C$, the sample consists of an $H_2SO_4$–$H_2O$ solution with the given composition.

(2) $X_{H_2SO_4} = 0.20$.

At $-80°C$, the sample consists of solid $H_2SO_4 \cdot 4H_2O$. This compound melts at $\approx -26°C$. At higher temperatures the sample consists of an $H_2SO_4$–$H_2O$ solution with the given composition.

(b) Which solid species has the highest melting point?

$H_2SO_4$. It melts at $\approx 10°C$, while $H_2SO_4 \cdot H_2O$ melts at $\approx 8°C$.

(c) What is the lowest temperature at which the liquid phase is stable? What is the composition of this liquid?

$T \approx -73°C$. The liquid that freezes at this temperature is the eutectic with $X_{H_2SO_4} \approx 0.10$.

(d) For how many compositions does the cooling curve look like Figure 12.20b?

Nine—the four eutectics, the three compounds, and the two components ($H_2SO_4$, $H_2O$). ∎

## 12.8 Colligative Properties

The addition of a soluble solid to a pure liquid depresses the freezing point of the liquid. A familiar example is the lowering of the freezing point of water by the addition of salt, a procedure that is commonly used to melt the ice on slippery roads in winter. Solid–liquid phase diagrams provide further evidence to support this assertion. For example, the addition of solid naphthalene to liquid benzene (Figure 12.18) lowers the freezing point of benzene and vice versa.

The lowering of the freezing point is just one of several related phenomena, known as **colligative properties,** which occur when a small amount of a non-volatile solute, that is, a solute with negligibly small vapor pressure, is added to a liquid. The other colligative properties are a decrease in the vapor pressure of the liquid, an increase in its boiling point, and a phenomenon known as osmotic pressure. The magnitude of the observed changes depends on the solute concentration but not on its identity. These phenomena have a common origin. When two phases of a pure substance are in equilibrium, $\Delta G = 0$, and the free energy per mole is the same in both phases (Section 11.6). The addition of a solute changes the free energy of the solution but not that of the solid or vapor phases. These phases are unaffected because the solute is not present in them. Since the molar free energies of the two phases are now unequal, the system is no longer in equilibrium. In order to restore equilibrium, the pressure or temperature of the system must be changed, giving rise to the observed effects.

***a. Vapor Pressure of a Solution of a Non-volatile Solute***  Consider a dilute solution in which we designate the solvent as component A and the solute as B. Since the solute is assumed to be non-volatile, the vapor pressure of the solution is equal to the partial pressure of the solvent, $P_A$. If the solution is sufficiently dilute, the solvent obeys Raoult's law (Section 12.6b). Expressing these statements in the form of an equation, we have

$$P = P_A = X_A P_A^\circ$$

If we designate the difference between the vapor pressure of the solution and the solvent as $\Delta P$, i.e., $\Delta P = P - P_A^\circ$, we obtain after replacing $X_A$ by $(1 - X_B)$ and rearrangement

$$\Delta P = - X_B P_A^\circ \tag{12.10}$$

Since $X_B$ and $P_A^\circ$ are positive quantities, $\Delta P$ must be negative, showing that the vapor pressure of the solution is lower than that of the solvent. Furthermore, the magnitude of the vapor pressure lowering is proportional to the amount of solute present but is independent of the identity of the solute, in accord with the definition of a colligative property. Two different non-volatile solutes will reduce the vapor pressure of a pure liquid by an identical amount provided that their mole fractions in the solution are equal.

***b. Boiling Point Elevation and Freezing Point Depression***  The increase in the boiling point and decrease in the freezing point of a dilute solution of a non-volatile solute are consequences of the reduction in the vapor pressure of the solution. This statement can be understood with reference to Figure 12.21, which compares the phase diagram of water with that of a dilute solution of

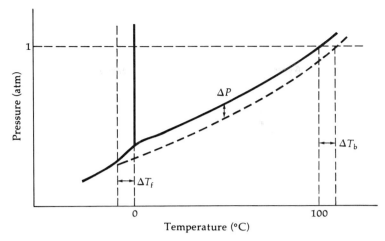

**FIGURE 12.21** Effect of the reduction in the vapor pressure of water on the boiling and freezing points due to the addition of a nonvolatile solute. The colored dashed line corresponds to a reduction in the vapor pressure, $\Delta P$, resulting, in turn, in changes in the boiling and freezing points, designated $\Delta T_b$ and $\Delta T_f$, respectively.

some non-volatile solid, such as NaCl. Consider, first, the boiling point of the solution. Owing to the reduction in the vapor pressure of the solution, a value of 1 atm is attained only when the temperature has been increased by an amount $\Delta T_b$ above the normal boiling point of water. The boiling point of the solution therefore exceeds that of water by this amount.

The decrease in the freezing point of a solution can be understood in terms of the effect of the solute on the triple point. The triple point of the solution is given by the intersection of the vapor pressure curve of the solution with the vapor pressure curve of ice, in complete analogy to the triple point of water. Since the vapor pressure curve of the solution lies below that of water, the triple point of the solution lies at a lower temperature than that of water. Recall from Section 12.3a that the liquid–solid equilibrium curve begins at the triple point and extends in the direction of increasing pressure. The normal freezing point of the liquid is given by the intersection of this curve with a horizontal line at $P = 1$ atm. As shown in Figure 12.21, the liquid–solid equilibrium curves of both water and the solution are nearly vertical—a change in pressure has only a small effect on the freezing point. Consequently, the normal freezing point of the solution is lower than that of water, the difference being approximately equal to that in the triple points.

The magnitude of the boiling point elevation can be derived on the basis of thermodynamics. Consider an ideal solution, for which the mole fraction of the solvent is $X_A$, in equilibrium with vapor. If the solute is non-volatile, the vapor consists of solvent molecules. At the normal boiling point, the vapor pressure is 1 atm, and the equilibrium can be represented as

liquid $(X_A) \rightleftarrows$ vapor (1 atm)

The equilibrium constant for this process can be approximated as

$$K = [\text{vapor}]/[\text{liquid}] = 1/X_A \tag{12.11}$$

The variation of the equilibrium constant with temperature is given by van't Hoff's equation (11.41):

$$\ln \frac{K_2}{K_1} = -\frac{\Delta H°}{R}\left(\frac{1}{T_2} - \frac{1}{T_1}\right)$$

To apply this equation to the present case, let $T_1$ be the boiling point of the pure solvent, designated $T_0$, and $T_2$ be the boiling point of the solution, designated $T$. The enthalpy change is that for vaporization and, since the solution is dilute, $\Delta H°$ can be approximated by $\Delta H°_{vap}$ for the solvent. We know (Section 11.6) that at the normal boiling point of the solvent the free energy change for the phase change is zero, $\Delta G° = 0$. Consequently, $K_1 = 1$. Substituting this value, and that for $K_2$ from Equation 12.11, into the preceding expression we obtain

$$\ln \frac{1}{X_A} = -\frac{\Delta H°_{vap}}{R}\left(\frac{1}{T} - \frac{1}{T_0}\right)$$

$$= \frac{\Delta H°_{vap}\, \Delta T_b}{RTT_0} \qquad (12.12)$$

where $\Delta T_b = T - T_0$ is the boiling point elevation. This expression can be simplified in two ways. First, since the solution is dilute, the change in boiling point will be small, $T \approx T_0$, and $TT_0$ can therefore be replaced by $T_0^2$. Second, the mole fraction of the solvent can be expressed in terms of that of the solute as $X_A = 1 - X_B$. The left side of Equation 12.12 becomes

$$\ln \frac{1}{X_A} = \ln \frac{1}{1 - X_B} = -\ln(1 - X_B) \approx X_B \qquad (12.13)$$

where the last result is an approximation that becomes increasingly valid as $X_B$ approaches zero, i.e., for dilute solutions (see Appendix 2, Section A2.4). Substituting the results of these approximations into Equation 12.12, we get

$$X_B = \Delta H°_{vap}\, \Delta T_b / RT_0^2$$

which upon rearrangement yields an expression for the boiling point elevation:

$$\Delta T_b = RT_0^2 X_B / \Delta H°_{vap} \qquad (12.14)$$

For a dilute solution, it is more practical to express the concentration in terms of the molality (Section 12.4a) instead of the mole fraction. If $n_A$ and $n_B$ are the number of moles of solvent and solute, we have

$$X_B = \frac{n_B}{n_A + n_B} \approx \frac{n_B}{n_A} = \frac{m_B \overline{M}_B}{m_A \overline{M}_A} \qquad (12.15)$$

where $m_A$ and $m_B$ are the masses, and $\overline{M}_A$ and $\overline{M}_B$ are the molecular weights of solvent and solute, respectively. Recall that the molality of a solution is equal to the number of moles of solute per kilogram of solvent, i.e.,

$$m = (m_B / \overline{M}_B)/1000$$

where we replace 1 kg by 1000 g because the molecular weight of the solvent is expressed in g mol$^{-1}$ rather than kg mol$^{-1}$. We finally obtain upon substitution into Equation 12.14

## Phase Equilibria 12.8

**TABLE 12.5  Molal Boiling Point Elevation and Freezing Point Depression Constants**

| Solvent | $T_b$ (°C) | $K_b$ (K/mol kg$^{-1}$) | $T_f$ (°C) | $K_f$ (K/mol kg$^{-1}$) |
|---|---|---|---|---|
| Water, $H_2O$ | 100 | 0.52 | 0 | 1.86 |
| Ethanol, $C_2H_5OH$ | 78.5 | 1.2 | −117.2 | 2.0 |
| Ether, $(C_2H_5)_2O$ | 34.5 | 2.2 | −116.2 | 1.8 |
| Benzene, $C_6H_6$ | 80.1 | 2.5 | 5.5 | 5.1 |
| Carbon disulfide, $CS_2$ | 76.0 | 2.3 | −111.5 | 3.8 |
| Naphthalene, $C_{10}H_8$ | 218 | 5.7 | 80.5 | 32 |

$$\Delta T_b = RT_0^2 \overline{M}_A m / 1000 \, \Delta H^\circ_{vap} \tag{12.16}$$

This expression shows that the boiling point elevation of a dilute solution is proportional to its molality, the proportionality constant being made up of a number of quantities that are properties of the solvent, including its standard enthalpy of vaporization, molecular weight, and boiling point temperature. All these constants are usually combined into a single constant called the **boiling point elevation constant** $K_b$, and Equation 12.16 is commonly used in the form

$$\Delta T_b = K_b m \tag{12.17}$$

The depression of the freezing point of a dilute solution can be obtained by means of a parallel thermodynamic analysis. The result is

$$\Delta T_f = -K_f m \tag{12.18}$$

where the freezing point depression constant has an analogous definition as that given in Equation 12.16 for the boiling point elevation constant, i.e.,

$$K_f = RT_0^2 \overline{M}_A / 1000 \, \Delta H^\circ_{fus} \tag{12.19}$$

Table 12.5 lists the boiling point elevation and freezing point depression constants for a number of solvents. Note that the units of $K_f$ or $K_b$ are those of $T/m$, i.e., K/(mol kg$^{-1}$). These data can be used to determine the molecular weights of soluble substances. A weighed amount of material of unknown molecular weight is dissolved in a given amount of solvent and the change in either the boiling or freezing point is measured. The molality is determined by means of either Equation 12.17 or 12.18, respectively, and the molecular weight of the solute is then calculated.

---

**Example 12.9**  A solution of 16.1 g of sucrose in 200 g of water freezes at −0.470°C. What is the molecular weight of sucrose?

Using Equation 12.18 and the data in Table 12.5 we have

$$\Delta T_f = -K_f m$$

$$(-0.470°C)(1 \text{ K } 1°C^{-1}) = -(1.86 \text{ K kg mol}^{-1}) \, m$$

$$m = 0.253 \text{ mol kg}^{-1}$$

The number of moles of solute is $(16.1/\overline{M})$, where $\overline{M}$ is the molecular weight of sucrose. Since the molality is the number of moles per kilogram of solvent, we obtain

$$\frac{(16.1 \text{ g})/\overline{M}}{0.200 \text{ kg}} = 0.253 \text{ mol kg}^{-1}$$

$$\overline{M} = 318 \text{ g mol}^{-1} \quad \blacksquare$$

A study of the colligative properties of solutions of strong electrolytes reveals that the effects on the freezing or boiling points are systematically larger than expected from the nominal molality. For example, the freezing point depression of a dilute NaCl solution is nearly twice as large as expected. The reason for this difference is that colligative effects depend on the total number of solute particles and therefore, for the same molality, are larger for substances that undergo dissociation than for those that do not. The freezing point depression of an electrolytic solution is given by the expression

$$\Delta T_f = -\nu K_f m \tag{12.20}$$

and the boiling point elevation is similarly given by

$$\Delta T_b = \nu K_b m \tag{12.21}$$

where the quantity $\nu$ is called the **van't Hoff factor**. For an electrolytic solution that is sufficiently dilute to be ideal, $\nu$ is equal to the number of ions per formula unit. For example, for a 1 : 1 electrolyte such as KCl, $\nu = 2$ whereas for 2 : 1 electrolyte such as $K_2CrO_4$, $\nu = 3$. Owing to the strong interaction between oppositely charged ions (Section 18.1), even dilute solutions of electrolytes show departures from ideality and the values of $\nu$ are somewhat lower than the expected integral values.

***c. Osmotic Pressure*** The determination of molecular weights by the measurement of $\Delta T_f$ or $\Delta T_b$ is limited to substances of relatively low molecular weight. The maximum solubility of massive molecules is generally sufficiently low that their solutions are of very low molality. The resulting changes in the freezing or boiling points are therefore too small to be measured with the necessary accuracy. The molecular weights of such macromolecules can instead be determined by measurement of the osmotic pressure of their solutions.

**Osmosis** is the selective passage of solvent molecules through a semipermeable membrane, that is, a membrane through which molecules of the solute cannot pass. Consider a solution of some macromolecular substance in water, e.g., an aqueous solution of a protein, in the container depicted in Figure 12.22. The left compartment, containing only water, is separated from the right compartment, containing the solution, by a semipermeable membrane made, for instance, of cellulose acetate. Molecules of water can pass through the membrane in either direction but the protein molecules are too big to pass through. Osmosis leads to a net flow of water from the solvent to the solution compartment. As a result, the level of the liquid in the solution compartment rises. Just as the mercury in a barometer (Figure 3.2), the additional liquid exerts a pressure on the solution, $P = \rho g h$ (Equation 3.5). This pressure opposes the flow of solvent through the membrane. Eventually, equilibrium is reached and there is no further net flow. The **osmotic pressure** $\pi$ is the equilibrium

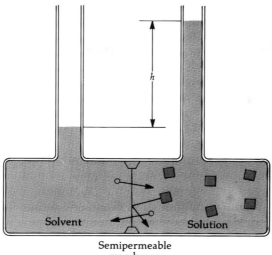

**FIGURE 12.22** Apparatus for measuring osmotic pressure. The left compartment contains pure solvent and the right compartment contains a solution of a substance that cannot pass through the semipermeable membrane separating the two compartments. Initially, the liquid level is at the same height in both compartments. At equilibrium, the liquid level in the solution has risen to a height $h$. The pressure exerted by this excess liquid is the osmotic pressure.

pressure on the solution resulting from the additional solvent present. Alternatively, the flow of solvent into the solution can just be prevented by applying an external pressure to the solution, the magnitude of which is equal to the osmotic pressure.

The free energy of one mole of solvent in the solution is less than it is in the pure solvent. The difference arises from the greater entropy of the solution which, in turn, arises from the greater molecular disorder associated with the presence of two different kinds of molecules (solvent and solute). The increase in the pressure on the solution increases the free energy (Equation 11.30) and tends to counteract the effect of osmosis. At equilibrium, the free energy of one mole of solvent must be the same on both sides of the membrane.

The osmotic pressure can be related to the concentration of the solution by means of a thermodynamic analysis in which the change in the free energy of the solution due to its dilution is equated to that associated with the increase in pressure. The resulting relation between $\pi$ and the molar concentration $c$ is similar to the ideal gas formula:

$$\pi V = nRT$$
$$\pi = cRT \tag{12.22}$$

where $c = n/V$. Note that since the osmotic pressure depends on the concentration of solute but not on its identity, it is a colligative property.

**Example 12.10** An aqueous solution containing 10.0 g L$^{-1}$ of hemoglobin is separated from pure water by a semipermeable membrane. At equilibrium, the height of liquid in the solution exceeds that in the surrounding water by 3.89 cm. The temperature is maintained at 25°C. What is the molecular weight of hemoglobin? Assume that the density of the solution is the same as that of water.

The given height of liquid must first be converted to the osmotic pressure in atmospheres. Using Equation 3.5,

$$\pi = P = \rho g h$$
$$= (1.00 \text{ g cm}^{-3})(10^{-3} \text{ kg g}^{-1})(10^6 \text{ cm}^3 \text{ m}^{-3})(9.807 \text{ m s}^{-2})(3.89 \text{ cm})$$
$$\times (10^{-2} \text{ m cm}^{-1})(1 \text{ N kg}^{-1} \text{ m}^{-1} \text{ s}^2)(1 \text{ atm}/1.013 \times 10^5 \text{ N m}^{-2})$$
$$= 3.766 \times 10^{-3} \text{ atm}$$

Knowing the osmotic pressure, the concentration of the solution is next determined by means of Equation 12.22:

$$\pi = cRT$$
$$c = \frac{3.766 \times 10^{-3} \text{ atm}}{(0.08206 \text{ L atm mol}^{-1} \text{ K}^{-1})(298 \text{ K})}$$
$$= 1.540 \times 10^{-4} \text{ mol L}^{-1}$$

Since the mass corresponding to this concentration is 10.0 g, the molecular weight is

$$\overline{M} = \frac{m/V}{c} = \frac{10.0 \text{ g L}^{-1}}{1.540 \times 10^{-4} \text{ mol L}^{-1}} = 6.49 \times 10^4 \text{ g mol}^{-1}$$

Note that even though the solution is very dilute, the osmotic pressure corresponds to a height difference of several centimeters, and is readily measured. By contrast, such a dilute solution would lower the freezing point of water by only about $3 \times 10^{-4}$ °C, rendering impossible a meaningful determination of the molecular weight by measurement of $\Delta T_f$. ∎

Osmotic pressure effects are of importance in many fields. They are particularly significant in biological systems, in which the concentrations of solutions are controlled by osmosis. In a totally different application, **reverse osmosis,** in which a pressure higher than the osmotic pressure is applied to a solution, is used in the desalination of sea water. Excess pressure is applied to a salt solution and the direction of the osmotic flow is thereby reversed. Pure water then passes through a membrane into the solvent compartment leaving a concentrated brine behind. While other processes considered in this chapter, such as distillation and freezing, can also be used for desalination, they are economically unattractive because of the relatively high energy required to effect the phase changes in question. Since osmosis does not involve a change of phase it is inherently more economical. Sea water, which is appromimately 0.7 $M$ in NaCl, has an osmotic pressure of about 30 atm. Since substantially higher pressures would be required to reverse the natural flow, the process requires membranes permeable to water, but not to salt, that can withstand such pressures over long periods of time. In an extension of this process, the flow of water is directed at a turbine generator in order to produce electricity as well as potable water. Considerable research and development is still needed in order to make these processes economical on a large scale.

## 12.9 Colloids

A solution in which very small particles that do not settle out are suspended is called a **colloid**, or colloidal solution. The suspended particles may be solid or liquid aggregates of an insoluble solute or they may be individual macromolecules. The distinction between a colloidal solution and a true solution can be made on the basis of particle size but is not sharp. Particles with dimensions of $10-10^4$ nm frequently form colloidal solutions.

Colloidal particles in suspension undergo **brownian motion**. The incessant zigzag motion, which may be seen under a microscope, is the result of collisions between the suspended particle and solvent molecules. Although a particle collides with a large number of solvent molecules at a given time, fluctuations in the number of molecules approaching the particle from a given direction result in a momentum imbalance which causes the particle to change direction abruptly (Figure 12.23). The observation of brownian motion provided early experimental evidence for the reality of atoms and molecules.

Although colloidal particles are frequently too small to be observable even under a microscope, they can be detected by their effect on light. When a light beam is incident on a colloidal solution, the suspended particles scatter some of the light sideways, thereby rendering visible the passage of the beam through the solution. This phenomenon is known as the **Tyndall effect** and can be used to distiguish between a colloidal and a true solution. The effect is observed when a beam of light passes through soapy water or smoke-filled air.

The molecular weight of colloidal-sized molecules can be determined by osmotic pressure measurements. The inability of macromolecules to pass through a semiporous membrane can also be used to remove small molecules and ions from a colloidal suspension. The dissolved molecules and ions can freely pass through such a membrane into the solvent, leaving the macromolecules behind. This purification method is called **dialysis** and is the means by which natural and artifical kidneys remove small molecules and salts from the bloodstream.

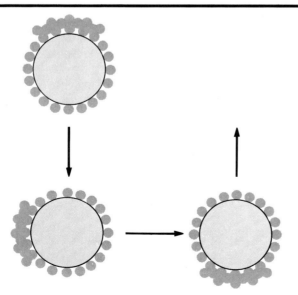

**FIGURE 12.23** Brownian motion. A colloidal particle is simultaneously struck by many solvent molecules. Owing to statistical fluctuations, more molecules may strike the particle on one particular side and the particle recoils in the opposite direction.

## Conclusion

In this chapter, the laws of thermodynamics have been applied to phase equilibria. The phase diagram of a pure substance is determined by the temperature and pressure, and consists of phase equilibrium curves which intersect at the triple point, the point at which three phases coexist. The liquid–vapor and solid–vapor equilibrium curves show the temperature dependence of the vapor pressure, and can be related to the temperature dependence of the equilibrium constant.

The phase diagram of a solution is determined by its composition, in addition to the temperature and pressure. Ideal solutions have the simplest liquid–vapor phase diagrams, and their shape can be understood with the aid of the phase rule. The more complicated phase diagrams of non-ideal solutions can be understood in terms of deviations from ideal behavior. The solid–liquid phase diagrams of solutions are characterized by the occurrence of an eutectic point. This is the lowest temperature at which a solution of two components can exist as a liquid, a temperature that can be substantially lower than the freezing points of the individual components.

The dissolution of a non-volatile solute in a liquid lowers its vapor pressure and freezing point, raises its boiling point, and, under suitable experimental conditions, generates osmotic pressure. These colligative properties depend on the concentration of the solute but are independent of its identity. Thermodynamics shows that colligative measurements can be used to deduce the molecular weight of the solute. These techniques are widely used to characterize newly synthesized substances.

## Problems

1. The enthalpy of vaporization of NO is 13.78 kJ mol$^{-1}$ and the normal boiling point is 121.4 K. What are the entropy and free energy of vaporization at this temperature?

2. For the graphite → diamond phase transition, $\Delta H = 1.896$ kJ mol$^{-1}$ and $\Delta S = -3.255$ J mol$^{-1}$ K$^{-1}$. What is $\Delta G$ at 298 K? Does a rise in temperature favor the spontaneous direction?

3. Show that, according to Trouton's rule and on the assumption that the enthalpy of vaporization is independent of temperature, all liquids should have the same hypothetical vapor pressure at infinite temperature. What is its numerical value?

4. Use the data in Appendix 4 to calculate the vapor pressure of chloroform (CHCl$_3$) (a) at 25°C and (b) at 50°C.

5. Mount Everest is 8900 m high. What is the boiling point of water on the summit? Use Equation 3.19 to determine the variation of atmospheric pressure with altitude. Assume an average temperature of 0°C.

6. An open container filled with water stands in a room measuring 7 × 4 × 3 m. The temperature is maintained at 22°C; the vapor pressure of water at this temperature is 19.8 torr. Assuming that air cannot flow in or out of the room, determine the mass of water vapor present in the air at equilibrium.

7. Vapor pressures are frequently tabulated in terms of the constants $a$ and $b$ in the equation

$$\log P = -(0.05223a/T) + b$$

where $P$ is expressed in torr and $T$ in K. (a) Express $\Delta H_{vap}$ in terms of $a$ and $b$. (b) In the temperature range 525–1325°C, the values for liquid lead are $a = 1.885 \times 10^5$ K and $b = 7.827$. Determine the vapor pressure and molar enthalpy of vaporization of lead at 1000°C.

8. Nitrogen gas is bubbled through liquid benzene (C$_6$H$_6$) at 20.0°C and 50.0 L of gas is collected. The gas is passed through a cold trap which removes all the benzene from the nitrogen and 12.4 g of benzene is collected. What is the vapor pressure of benzene at 20°C?

9. A sealed vessel with a volume of 12.0 L contains 1.0 g of water. At what temperature will half the water be in the vapor phase? (The vapor pressure of water is tabulated in the Handbook of Chemistry and Physics.)

10. The vapor pressure of liquid ammonia at −50.0°C and at −40.0°C is 306.6 torr and 538.3 torr, respectively. Determine $\Delta H_{vap}$ and the normal boiling point of ammonia.

11. The vapor pressure of ice is 4.579 torr at 0°C and 0.776 torr at −20.0°C. At what temperature does ice sublime when the external pressure is reduced to 3.000 torr? What is $\Delta H_{sub}$ of ice?

12. (a) Use the data in Appendix 4 to determine the free energy of vaporization and the vapor pressure of benzene ($C_6H_6$) at 25°C. (b) Assuming that the enthalpy and entropy of vaporization are independent of temperature, determine the normal boiling point of benzene.

13. The relative humidity of air may be defined as the ratio of the actual partial pressure of water vapor in air to the vapor pressure of water at a given temperature. Is the partial pressure of water vapor higher when the temperature is 20°C and the relative humidity is 90% or when the temperature is 35°C and the relative humidity is 40%? (The vapor pressure of water may be found in the Handbook of Chemistry and Physics.)

14. A mixture of nitrogen and water vapor at 70°C and 736 torr is cooled to 22°C. It is observed that some water condenses and the pressure of the remaining gas drops to 452 torr. Assuming that the gas volume remains unchanged, determine the partial pressure of nitrogen in the gas at 22°C and at 70°C. (The vapor pressure of water at 22°C is 19.8 torr.)

15. The weight loss of a sample of water at 30°C observed when 5.00 L of dry air was slowly bubbled through was 0.152 g. Calculate the vapor pressure of water at 30°C on the assumption that 5.00 L of saturated water vapor were formed.

16. Acetone boils at 56°C and its enthalpy of vaporization is 30.21 kJ mol$^{-1}$. What is the vapor pressure of acetone at 20°C?

17. On the basis of Figure 12.9, describe the changes that occur when a sample of steam at 120°C is cooled to −10°C at a constant pressure of (a) 1 atm and (b) 0.005 atm.

18. A sample of carbon dioxide at −35°C and 1 atm is subjected to the following cycle: (1) isobaric cooling to −100°C, (2) isothermal compression to 10 atm, (3) isobaric heating to −35°C, and (4) isothermal decompression to 1 atm. On the basis of Figure 12.10, describe the phase changes, if any, that occur in each step of the cycle.

19. When the vapor pressure of anthracene ($C_{14}H_{10}$) is represented by the empirical formula

$$\log P = -(0.05223a/T) + b$$

where $P$ is in torr and $T$ in K, the values for the solid are $a = 7.039 \times 10^4$ K and $b = 8.706$, while those for the liquid are $a = 5.922 \times 10^4$ K and $b = 7.910$. (a) Which phase is more stable at 200°C? (b) At what temperature and pressure does the triple point of anthracene occur?

20. An aqueous solution of ethanol ($C_2H_5OH$) was prepared by adding 34.50 g of ethanol to enough water to make 500.0 mL of solution, the density of which is 0.9864 g cm$^{-3}$. What are the molarity, mole fraction, and molality of this solution?

21. At 135°C, the vapor pressure of chlorobenzene ($C_6H_5Cl$) is 863 torr and that of bromobenzene ($C_6H_5Br$) is 453 torr. What is the composition of a mixture of these two liquids with a vapor pressure of 750 torr at 135°C? What is the composition of the vapor? Assume the solution is ideal.

22. At 90°C the vapor pressure of o-xylene is 150 torr and that of toluene is 400 torr. Determine the composition of the liquid mixture that will boil at 90°C at a reduced pressure of 0.50 atm. What is the composition of the resulting vapor? Assume the solution is ideal.

23. The partial pressure of water over an aqueous solution of ammonia which is 16.0% by weight of $NH_3$ is 44.7 torr at 40°C. Is the solution ideal? If not, does it show positive or negative deviations from Raoult's law? The vapor pressure of water at 40°C is 55.32 torr.

24. Liquids A and B form an ideal solution. The vapor pressures of pure A and B at 80°C are 400 torr and 200 torr, respectively. The vapor above an equimolar solution of A and B is collected and condensed. (a) What is the composition of the resulting liquid? (b) This liquid is reheated to 80°C and the vapor in equilibrium is again collected and condensed. What is the composition of the condensate?

25. The following data on the vapor pressure of solutions of silicon tetrachloride ($SiCl_4$) and carbon tetrachloride ($CCl_4$) at 25°C have been reported:

| $X_{SiCl_4}$ | $X_{SiCl_4}^{vapor}$ | Total vapor pressure (torr) |
| --- | --- | --- |
| 0 | 0 | 114.9 |
| 0.266 | 0.436 | 153.0 |
| 0.472 | 0.648 | 179.1 |
| 0.632 | 0.773 | 198.5 |
| 1.00 | 1.00 | 238.3 |

(a) Make a graph of the vapor pressures of the components and the total vapor pressure versus $X_{SiCl_4}$. (b) Draw the comparable lines expected for an ideal solution. (c) Does this solution display positive or negative deviations from ideal behavior? (d) Without looking up the boiling points, indicate which of the pure liquids has the lower boiling point.

26. What is the carbon dioxide content of soda water kept under a $CO_2$ pressure of 5.0 atm at 25°C? Express the answer in terms of mole fraction and molality.

27. What are the molalities of oxygen and nitrogen in water exposed to air at a pressure of 1.00 atm and 25°C? Air consists of 78 mol% nitrogen and 21 mol% oxygen.

28. Copper and aluminum are miscible as liquids but do not dissolve each other as solids. The melting points of the following mixtures were determined:

| % Cu by mass | 0 | 20 | 40 | 60 | 80 | 100 |
|---|---|---|---|---|---|---|
| $T_{mp}$ (°C) | 660 | 597 | 538 | 606 | 925 | 1083 |

(a) Convert the % Cu by mass to mol% Cu. (b) Draw the phase diagram. (c) Use the diagram to determine the physical state and estimate the composition of every phase present at equilibrium in an equimolar mixture of the two metals at $T = 600°C$.

29. Magnesium and copper are miscible in the liquid phase and immiscible in the solid. However, they form two solid intermetallic compounds: $MgCu_2$ and $Mg_2Cu$. The following freezing points are determined: $T_f(Cu) = 1085°C$, $T_f(Mg) = 648°C$, $T_f(MgCu_2) = 800°C$, $T_f(Mg_2Cu) = 580°C$. The system has three eutectic points: $X_{Mg} = 0.225$, $T_f = 690°C$; $X_{Mg} = 0.563$, $T_f = 560°C$; $X_{Mg} = 0.829$, $T_f = 380°C$. Sketch the phase diagram.

30. A sample of a MgCu alloy with $X_{Mg} = 0.40$ is heated to 900°C and then allowed to cool slowly to room temperature. On the basis of the phase diagram sketched in Problem 29, describe the phase changes that occur and state the temperature at which they occur. Draw a cooling curve for this process.

31. The solid–liquid phase diagram for the $FeCl_3 + H_2O$ system is sketched on p. 455. Answer the following questions about this diagram: (a) What are the formulas of the hydrates of $FeCl_3$? (b) For how many compositions does the cooling curve look like Figure 12.20a? (c) For how many compositions does the cooling curve look like Figure 12.20b? (d) Describe the phases present in region a. (e) Describe region b. (f) Describe region d. (g) Describe the phases present at point e. (h) What pure substance has the highest melting point? (i) What pure substance has the lowest melting point? (j) What is the composition of the eutectic with the lowest melting point?

32. The phase diagram of substances A and B is sketched on p. 455 for a sufficiently broad range of temperatures that solid, liquid, and vapor phases are present. (a) What substances are present in each of the regions labelled 1–7 and what are their phases? (b) Describe point a. (c) What are the compositions of the distillate and the residue obtained in fractional distillation of a solution of composition given by point b? (d) For how many different compositions does the liquid have either a constant boiling point or a constant freezing point? (e) How does the maximum solubility of B in liquid A at a temperature just above the melting point of A compare with the maximum solubility of A in liquid B just above the melting point of B?

33. At a temperature for which the vapor pressure of water is 28.2 torr, determine the vapor pressure lowering (a) when 5.00 g of sucrose ($C_{12}H_{22}O_{11}$) is dissolved in 100 g of water and (b) when 5.00 g of NaCl is dissolved in 100 g of water.

34. A solution containing 31.5 g of a non-volatile compound in 140 g of water boils at 373.18 K at a pressure at which pure water boils at 372.91 K. Assuming that the compound does not dissociate, what is its molecular weight?

35. Estimate the mass of $CaCl_2$ required to melt completely 3.0 kg of ice at a temperature of −4°C. (Only an estimate is possible because the resulting solution is rather concentrated.)

36. Exactly one gram of a certain polymer (a large molecule made up of identical repeating units) is dissolved in 38.5 g of naphthalene. The freezing point of this solution is 0.41°C lower than the normal freezing point of naphthalene. What is the molecular weight of the polymer?

37. Estimate the osmotic pressure of a solution of 0.050 g of sucrose ($C_{12}H_{22}O_{11}$) in 100 g of water at 25°C. How high a column of solution corresponds to this pressure? The density of the solution is the same as that of water, 1.00 g cm$^{-3}$.

38. At which temperature will 200 mL of water sweetened with 2 lumps of sugar (5.00 g of sucrose, $C_{12}H_{22}O_{11}$) freeze?

39. A solution of 0.800 g of naphthalene ($C_{10}H_8$) in 100 g of benzene ($C_6H_6$) freezes at 5.11°C. The normal freezing and boiling points of benzene are 5.42°C and 80.0°C, respectively, and the enthalpy of vaporization is 30.76 kJ mol$^{-1}$. On the basis of these data determine (a) the enthalpy of fusion of benzene, (b) the vapor pressure of the above solution at 80.0°C, and (c) the boiling point of the solution at atmospheric pressure.

40. A solution is prepared by dissolving 3.10 g of an unknown non-volatile solute in 100 g of benzene. The freezing point of this solution is 1.95°C lower than the normal freezing point of benzene. (a) What is the molecular weight of the solute? (b) What is the boiling point of the solution?

41. A solution of a polymer is prepared by dissolving 0.605 g in 100 mL of water. The osmotic pressure of the solution at 25°C is 3.05 torr. What is the molecular weight of the polymer? What is its mole fraction in the solution?

42. Determine the approximate osmotic pressure at 20°C of an aqueous solution of a non-volatile solute that freezes at −0.043°C.

43. Exactly 1 g each of NaCl, NaBr, and NaI are added to 250 g of water. What is the boiling point of the resultant solution at atmospheric pressure?

44. A solution of 15.0 g of sucrose in 800 g of ethanol boils 0.107°C higher than pure ethanol, while a solution of 10.0 g of an unknown substance in 800 g of ethanol boils 0.156°C higher. From these data determine the molecular weight of the unknown and the value of $K_b$ of ethanol.

45. Determine the boiling and freezing points of an 0.0600 $m$ aqueous solution of $Na_3PO_4$ at atmospheric pressure. What is the vapor pressure and the osmotic pressure of this solution at 30°C? The vapor pressure of water at this temperature is 31.8 torr. Neglect the hydrolysis of $PO_4^{3-}$. Assume that the density of the solution is 1.00 g mL$^{-1}$.

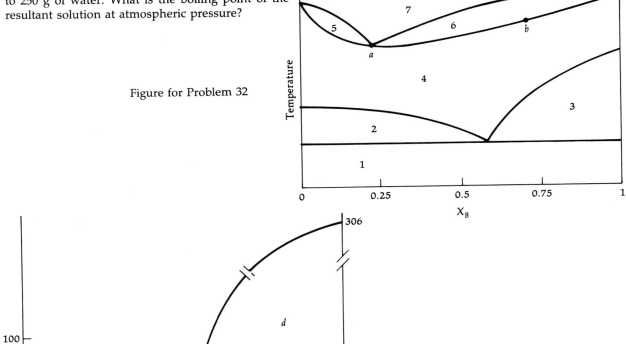

Figure for Problem 32

Figure for Problem 31

# 13
# Electrochemistry

Oxidation–reduction reactions often involve a transfer of electrons between the reacting species (Section 2.6). One of the remarkable features of this type of reaction is that it can occur even when the reactants are physically separated, provided only that they are incorporated in a closed electrical circuit. The advantage of this arrangement is that a spontaneous redox reaction can be harnessed to perform electrical work. In few other areas of chemistry do practical applications follow so directly from the fundamental aspects of the subject. The generation of electricity by means of chemical batteries and the production of the most active metals by electrolytic reduction are among the most familiar of these applications. Electrochemical measurements also provide data from which equilibrium constants for many chemical reactions can be obtained by means of a thermodynamic analysis. The results are frequently far more accurate than those obtainable by means of other techniques.

## Fundamentals of Electrochemistry

### 13.1 Galvanic Cells

*a. The Operation of a Cell* A **galvanic cell** (also called a voltaic cell, names honoring Luigi Galvani 1739–1798 and Alessandro Volta 1745–1827, two pioneers in the study of electrochemistry) is a device for the generation of electricity through the occurrence of a redox reaction. Figure 13.1 shows the compo-

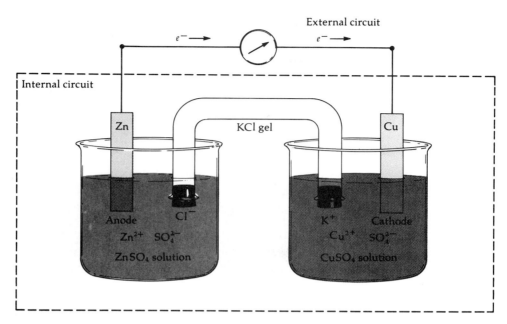

**FIGURE 13.1** Operation of the Daniell cell, a galvanic cell. The cell is described in the text.

nents of a typical cell and illustrates its operation. This particular cell is based on the reaction

$$Zn(s) + Cu^{2+} \rightarrow Zn^{2+} + Cu(s) \tag{13.1}$$

and is called the Daniell cell. However, the principles of cell operation are independent of the specific reaction chosen.

The cell is divided into two compartments each of which is filled with a solution of an electrolyte. A metal strip called an **electrode** is partially submerged in each solution. The two electrodes are connected by means of an external electrical circuit, which may be as simple as a metallic wire, but which may also include a meter to monitor the operation of the cell or some device to be operated by the electricity generated, e.g., a light bulb. The solutions in the two compartments are connected by a **salt bridge,** filled with an electrolyte.

The overall redox reaction can be divided into two half-reactions (Section 2.8) each of which occurs at one of the electrodes. The electrode at which oxidation takes place is called the **anode** and the electrode at which reduction occurs is called the **cathode.** It is customary to place the anode of a galvanic cell on the left side of a cell diagram, and this is the convention followed in Figure 13.1. The electrodes serve as the interfaces between the solutions and the external circuit. At the anode, oxidation liberates electrons which are released to the external circuit. These electrons flow to the cathode, where their acceptance by some species results in reduction.

The oxidation half-reaction in the Daniell cell is

$$Zn(s) \rightarrow Zn^{2+} + 2e^-$$

The anode consists of a strip of zinc which is immersed in a solution containing $Zn^{2+}$ ions and a suitable anion to balance the charge, e.g., $SO_4^{2-}$ (the anion must not react with $Zn^{2+}$). The reduction half-reaction is

$$Cu^{2+} + 2e^- \rightarrow Cu(s)$$

The cathode consists of a strip of copper immersed in a $CuSO_4$ solution. As the reaction proceeds, the concentration of $Zn^{2+}$ in the anode compartment increases while that of $Cu^{2+}$ in the cathode compartment decreases. To balance the net positive charge building up at the anode and the net negative charge building up at the cathode, ions flow from the salt bridge into the electrode compartments. The salt bridge is filled with some electrolyte the ions of which do not react at the electrodes, such as $KCl$ or $NH_4NO_3$. This electrolyte is incorporated into a semi-rigid gel, whose purpose is to prevent the mixing of the contents of the two compartments while still permitting the migration of individual ions. Since positive charge builds up at the anode, negative ions (i.e., anions) migrate from the salt bridge to this electrode. Similarly, since negative charge builds up at the cathode, positive ions (i.e., cations) migrate towards this electrode. Without this connection between the two solutions, the buildup of charge in the two compartments (called **polarization**) would quickly prevent the cell from operating.

When all the components of the Daniell cell are assembled in the manner depicted in Figure 13.1, electrons flow through the external circuit from anode to cathode. At the same time the redox reaction (Equation 13.1) occurs in the cell and ions migrate towards the electrodes to balance the charge. The zinc anode is gradually consumed and the concentration of $Zn^{2+}$ increases, while the copper cathode accretes additional copper and the concentration of $Cu^{2+}$ decreases.

The conditions under which a redox reaction occurs in a galvanic cell may be contrasted with the normal reaction conditions. The reduction of copper(II) ions by zinc, for example, can be effected by dipping a strip of zinc metal in a copper sulfate solution (Figure 13.2). The zinc strip is gradually eroded and at the same time a dark layer of copper deposits on its surface. Electron transfer occurs directly between Zn atoms and $Cu^{2+}$ ions in contact. By contrast, the physical separation of the reactants in a galvanic cell forces the flow of electrons and, as we will see, makes it possible to use this flow to perform electrical work.

The cell diagram in Figure 13.1 can be expressed in condensed form by listing in sequence the various phases in contact, beginning with the anode on the left and ending with the cathode on the right. The concentrations or pressures of all participating ions or gases are given in parentheses. A single solid line shows the presence of a phase boundary, and a double solid line represents a salt bridge between the two solutions listed on either side. The line notation for the Daniell cell, with $Cu^{2+}$ and $Zn^{2+}$ present in 1 M concentration, is

$$Zn/Zn^{2+}(1\ M)//Cu^{2+}(1\ M)/Cu$$

indicating that a zinc anode is in contact with a solution containing $1\ M\ Zn^{2+}$ and that this solution is separated by a salt bridge from a $1\ M\ Cu^{2+}$ solution in contact with a copper cathode. Since, by convention, the anode is on the left, the chemical reaction occurring in the cell must involve the oxidation of Zn and the reduction of $Cu^{2+}$.

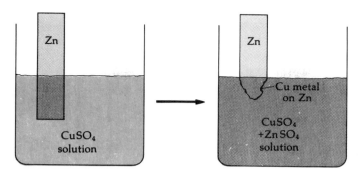

**FIGURE 13.2** The direct redox reaction between Zn and $Cu^{2+}$. Contrast the conditions with those in a galvanic cell based on the same reaction (Figure 13.1).

---

Example 13.1   Iron can be oxidized by nickel(II) forming iron(II) and nickel metal. Design and sketch a galvanic cell based on this reaction. Summarize the cell arrangement in line notation.

It is helpful to first write the half-reactions for the processes occurring at the electrodes:

|   |   |
|---|---|
| oxidation at the anode: | $Fe(s) \rightarrow Fe^{2+} + 2e^-$ |
| reduction at the cathode: | $Ni^{2+} + 2e^- \rightarrow Ni(s)$ |
| overall reaction: | $Fe(s) + Ni^{2+} \rightarrow Fe^{2+} + Ni(s)$ |

The anode compartment can be made by immersing a strip of iron in a solution containing $Fe^{2+}$ ions. Similarly, the cathode compartment can be made by immersing a strip of nickel in a solution containing $Ni^{2+}$. The two solutions must be connected by a salt bridge. The two electrodes are connected by a wire and electrons flow from the iron to the nickel electrode. The cell is sketched in the accompanying figure.

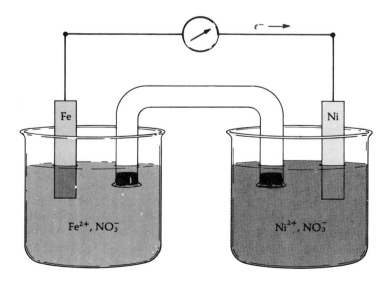

The line notation is

$$\text{Fe}/\text{Fe}^{2+}(c) \,//\, \text{Ni}^{2+}(c)/\text{Ni}$$

where $c$ signifies the concentration of $\text{Fe}^{2+}$ and $\text{Ni}^{2+}$. ∎

**b. The Cell EMF** An **electric current** is a flow of charge. The amount of charge $Q$ moving past a given point in a time $t$ is

$$Q = It \tag{13.2}$$

1 coulomb (C) = 1 ampere (A) × 1 second (s)

where the current $I$ is expressed in amperes. Thus, a current of 1 A is the flow of one coulomb per second.

The flow of electrons from the anode to the cathode of a galvanic cell constitutes an electric current. It is convenient to express the charge flowing through the external circuit of such a cell in terms of the electronic charge, $1.602 \times 10^{-19}$ C. The charge on one mole of electrons is called a **faraday** ($F$). The faraday is a very useful unit of charge, for if $n$ electrons are transferred in the oxidation or reduction of *one atom* or ion, then the oxidation or reduction of *one mole* of a substance involves $n$ faradays. For example, 2 electrons are transferred for each Zn atom oxidized to $\text{Zn}^{2+}$ in the Daniell cell. Consequently, the oxidation of one mole of zinc (or the reduction of one mole of copper(II) ions) requires two faradays of charge. The magnitude of 1 faraday expressed in coulombs is

$$F = (1.602189 \times 10^{-19}\,\text{C}/e^-)(6.022045 \times 10^{23}\,e^-/\text{mol}) = 96{,}485\,\text{C mol}^{-1}$$

An electric current flows between two points because there is a difference in electrical potential energy between them. This flow is analogous to that of water down a waterfall, which occurs because of the difference in gravitational potential energy between the top and bottom of the fall. The electrical potential energy difference corresponds to the work that can be performed when a charge moves between the two points in question. The work per unit charge is called the **electrical potential difference** and is expressed in joules per coulomb. This unit is called the volt, V,

$$w = -QV \tag{13.3}$$

1 joule = 1 coulomb × 1 volt

where, adopting the convention of Section 10.2, $w < 0$ means that the cell (i.e., the system) does the work.

The difference in electrical potential between the anode and the cathode of a galvanic cell is called the cell **EMF** (electromotive force), or **cell potential,** and is designated by $\epsilon$. It is customary to report EMF values at standard conditions: 1 $M$ concentration of all ionic species and 1 atm pressure of any gases involved in the cell reaction. The standard EMF of a cell is symbolized by $\epsilon°$ and is a characteristic property of the cell reaction, dependent only on the temperature. For example, the Daniell cell has $\epsilon° = 1.10$ V at 25°C.

It is important to distinguish between the standard EMF of a cell and the *actual* EMF. The standard EMF of a cell is a constant at a given temperature, just like the standard free energy of the cell reaction, to which it is intimately related (Section 13.2). By contrast, the actual EMF changes as the concentrations or partial pressures of the various species present change. The magnitude of $\epsilon$ is related to that of $\Delta G$, as we will see. Just as $\Delta G$ approaches zero as reactants and products approach a state of chemical equilibrium, the cell EMF also decreases to zero—the cell runs down as chemical equilibrium is approached.

**Example 13.2** An electronic calculator can run for 5.0 h on a 3.6 V battery, the average operating current being 150 mA. (The battery consists of several galvanic cells the EMF of which add up to 3.6 V.) (a) How much energy does the battery provide? (b) How many faradays are delivered by the battery?

(a) The energy provided by the battery is equal to the electrical work that it can perform. According to Equation 13.3, Energy $= -w = QV$. The current can be introduced into this expression by means of Equation 13.2, $Q = It$, giving

$$\text{Energy} = ItV$$
$$= (150 \text{ mA})(10^{-3} \text{ A mA}^{-1})(5.0 \text{ h})(3600 \text{ s h}^{-1})(3.6 \text{ V})[1 \text{ J(A s V)}^{-1}]$$
$$= 9.7 \times 10^3 \text{ J}$$

(b) The total charge is

$$Q = It$$
$$= \frac{(150 \text{ mA})(10^{-3} \text{ A mA}^{-1})(5.0 \text{ h})(3600 \text{ s h}^{-1})[1 \text{ C (A s)}^{-1}]}{(96{,}485 \text{ C } F^{-1})}$$
$$= 0.028 \, F \quad \blacksquare$$

**c. Types of Electrodes** We have so far examined galvanic cells in which the electrodes are the reduced states of a redox couple. (A **redox couple** consists of the oxidized and reduced forms of some species undergoing oxidation or reduction. It is customary to symbolize a redox couple as ox/red, e.g., $Fe^{2+}/Fe$.) This type of arrangement is most suitable when the reduced state is a metallic element. For example, the electrodes in the Daniell cell are metallic zinc and copper, the reduced forms of the $Zn^{2+}/Zn$ and $Cu^{2+}/Cu$ couples (Figure 13.3a). Many redox reactions, however, do not involve oxidation of a metal or reduction to the metallic state. In these cases, other types of electrodes are commonly used.

When both the oxidized and reduced states of a species are ions, an inert electrode is used. Such an electrode is made of a chemically inert, conducting material, such as platinum metal or graphite. The sole function of this electrode is to provide a surface at which ions in solution can donate or accept electrons from the external circuit. Some examples of redox couples requiring the use of an inert electrode are $Fe^{3+}/Fe^{2+}$ and $Ce^{4+}/Ce^{3+}$. Figure 13.3b illustrates this type of electrode.

When one of the reactants or products of a redox reaction is a gas, the use of a gas electrode (Figure 13.3c) is indicated. A gas electrode is designed so that electron transfer between gaseous molecules and the external circuit can occur. Typically, the electrode is made of platinum, the surface area of which has been increased and thereby rendered more adsorptive (i.e., a large number of

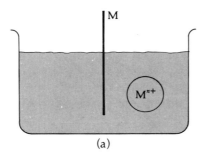
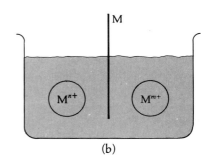
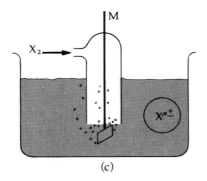
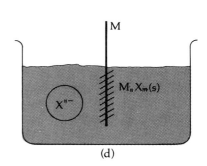

**FIGURE 13.3** Different types of electrodes for galvanic cells. (a) Reduced metal electrode; the solution contains the oxidized form of the metal, e.g., $M^{2+}$. (b) Inert electrode; the solution contains both the reduced and oxidized states, e.g., $M^{2+}$ and $M^{3+}$. (c) Gas electrode; the gas brought to the electrode can be either the reduced or the oxidized member of a redox couple; the other member of the couple is in solution. (d) Metal/insoluble metal salt electrode; the metal is coated with an insoluble salt of the metal ion, e.g., Ag + AgCl. The concentration of $Cl^-$ in the solution fixes the $Ag^+$ concentration.

gas molecules can adhere to the surface) by the deposition of finely divided platinum. In addition, the electrode must be connected to a gas delivery system which provides gas at the desired pressure from an external reservoir. The hydrogen electrode, based on the $H^+/H_2$ redox couple, and the chlorine electrode, based on the $Cl_2/Cl^-$ couple, are typical gas electrodes. In each case, the electrode is immersed in a solution containing the ionic form of the couple.

A metal/insoluble metal salt electrode (Figure 13.3d) has somewhat different properties than the electrodes described so far. This electrode consists of a metal which is the reduced form of the redox couple, coated with an insoluble salt of the metal ion, for example, Ag coated with AgCl. The electrode is immersed in a solution containing the anion of the salt. The concentration of $Ag^+$ ions in the solution is fixed by the concentration of $Cl^-$ and the value of the solubility product of AgCl,

$$[Ag^+] = K_{sp(AgCl)}/[Cl^-]$$

The ability to maintain a constant Ag⁺ concentration makes an electrode of this type useful as a reference electrode, able to supply a constant known voltage. In addition, this electrode can be used for the determination of the solubility product of the insoluble salt in question (Section 13.6).

A galvanic cell can be made from a combination of any two of the various electrodes described in this Section (Figure 13.3) provided that the overall redox reaction occurs spontaneously. As noted for the Daniell cell, the two half-cells must generally be kept physically separate in order to prevent the direct reaction from occurring, and must be connected internally by a salt bridge.

**Example 13.3**    What chemical reaction occurs in the following cell?

$$Pt/H_2(g)(P = 1 \text{ atm})/H^+(1 \text{ M})//Ag^+(1 \text{ M})/Ag(s)$$

Make a sketch of this cell.

By convention, the anode is described first in the shorthand notation. The half-reaction occurring at the anode must therefore be the oxidation of $H_2$ to $H^+$. At the cathode, $Ag^+$ is reduced to Ag. The half-reactions are

| | |
|---|---|
| oxidation at anode: | $H_2(g) \rightarrow 2H^+ + 2e^-$ |
| reduction at cathode: | $2 \times [Ag^+ + e^- \rightarrow Ag(s)]$ |
| overall reaction: | $H_2(g) + 2Ag^+ \rightarrow 2H^+ + 2Ag(s)$ |

Note that the stoichiometric coefficients of the reduction half-reaction have been multiplied by two in order to cancel the electrons in the overall reaction.

The cell can be sketched by combining half-cells such as those shown in Figure 13.3 for a gas electrode and a reduced metal electrode. The two half-cells are combined by means of a salt bridge containing a non-reactive electrolyte, e.g., $KNO_3$. In the external circuit, electrons flow from the hydrogen anode to the silver cathode. The cell is sketched in the accompanying figure. ∎

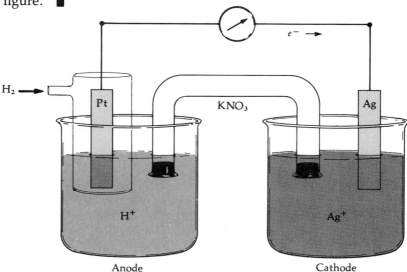

## 13.2 Cell EMF and the Reaction Free Energy

The EMF of a particular galvanic cell can be related to the free energy change of the cell reaction. The relationship can be obtained by a modification of the derivation for the dependence of the free energy on pressure, $dG = V\,dP$ (Equation 11.28), which applies to a reversible process at constant temperature in which the only work that can be performed is that resulting from expansion. Here we wish to examine the case where the system also can perform electrical work.

Let us recapitulate briefly the argument presented in Section 11.8. We started with the relation between the free energy and the internal energy and entropy (Equation 11.25)

$$G = E + PV - TS$$

From this we obtained the relation (Equation 11.26)

$$dG = dE + P\,dV + V\,dP - T\,dS - S\,dT$$

for a process in which the free energy changes by an infinitesimal amount. If this process is carried out at constant temperature and pressure, $dT = 0$ and $dP = 0$, so that

$$dG = dE + P\,dV - T\,dS \tag{13.4}$$

According to the first law of thermodynamics,

$$dE = dq + dw$$

Substituting into Equation 13.4, we obtain

$$dG = dq + dw + P\,dV - T\,dS \tag{13.5}$$

Furthermore, if the process is carried out reversibly, the second law of thermodynamics permits us to equate $dq$ and $T\,dS$. Under these conditions, Equation 13.5 simplifies to

$$dG = dw + P\,dV \tag{13.6}$$

So long as only reversible expansion work is possible,

$$dw = -P\,dV$$

In this case Equation 13.6 reduces to $dG = 0$. This is the familiar result that the free energy change is zero in a reversible process at constant temperature and pressure in which only expansion work is possible. If we relax this last restriction and allow electrical as well as expansion work, i.e.,

$$dw = dw_e - P\,dV \tag{13.7}$$

where $w_e$ is the electrical work, Equation 13.6 becomes

$$dG = dw_e - P\,dV + P\,dV$$

$$dG = dw_e \tag{13.8}$$

or, for a finite change,

$$\Delta G = w_e \tag{13.9}$$

In Section 13.1b we identified the electrical work done by a cell with the product of the total charge delivered by the cell and the cell potential. If the charge is expressed in units of faradays, Equation 13.3 for the electrical work can be written as

$$w_e = -nF\epsilon \tag{13.10}$$

where $n$ is the number of moles of electrons flowing through the circuit. (The voltage V is just the EMF $\epsilon$.) Substitution of this result into Equation 13.9 gives the desired relation between the free energy change and the cell EMF

$$\Delta G = -nF\epsilon \tag{13.11}$$

Two additional important relationships follow immediately. If a cell is run under standard conditions, $\epsilon$ becomes $\epsilon°$ and $\Delta G$ becomes $\Delta G°$. Equation 13.11 can then be written as

$$\Delta G° = -nF\epsilon° \tag{13.12}$$

In Section 11.8 we related the standard free energy change to the equilibrium constant (Equation 11.38), i.e., $\Delta G° = -RT \ln K$. Combining this expression with Equation 13.12 gives a relationship between the standard cell EMF and the equilibrium constant for the cell reaction:

$$\epsilon° = (RT/nF) \ln K \tag{13.13}$$

If the cell is run at 25°C, all the constants can be evaluated numerically, giving

$$\epsilon° = \frac{(8.314 \text{ J mol}^{-1} \text{ K}^{-1})(298.15 \text{ K})(2.303 \log K)}{(96{,}485 \text{ J V}^{-1} \text{ mol}^{-1})n}$$

where it is customary to express the logarithm of $K$ to the base 10. Combining terms, we get

$$\epsilon° = (0.0592/n) \log K \tag{13.14}$$

The relationships between the EMF and the free energy change or the equilibrium constant, together with the previously discussed properties of these quantitites, make a more precise understanding of the EMF possible. Since a spontaneous chemical reaction (at constant $T$ and $P$) occurs when $\Delta G < 0$, it follows from Equation 13.11 that the EMF of a cell based on such a reaction must be positive. Thus, a cell with $\epsilon > 0$ will generate an electric current and can be used to perform electrical work. On the other hand, a reaction for which $\Delta G > 0$ cannot occur spontaneously. Rather, the reverse reaction is spontaneous. It follows that a reaction for which $\epsilon < 0$ does not occur spontaneously and cannot be used to power a galvanic cell. Since reversing a reaction changes the sign of $\Delta G$ without changing its magnitude, it also follows that if a reaction has an EMF $\epsilon$, the reverse reaction will have an EMF of $-\epsilon$.

The free energy of a reaction depends on the way the reaction is written, i.e., on the stoichiometric coefficients of reactants and products (Equation 11.24), and is consequently an extensive property. We have seen that the cell EMF is proportional to $\Delta G/n$. Now the number of electrons transferred in a redox reaction, $n$, depends on the way the reaction is written in the same manner as $\Delta G$. It follows that $\Delta G/n$ is independent of the way the reaction is written. The EMF of a cell is therefore an intensive property. At a particular temperature, a given cell will generate the same EMF regardless of the amounts of substances present; only the concentrations of species in solution and the partial pressures of gases matter.

Many redox reactions go practically to completion and their equilibrium constants consequently are very large. Such large values of $K$ are difficult to measure, for the equilibrium concentrations of the reactants are necessarily extremely low and may, in fact, be below the limit of detectability. The relationship between the equilibrium constant and the standard cell EMF makes it possible to infer equilibrium constants from electrochemical data.

---

**Example 13.4**  The standard EMF of a cell based on the reaction

$$2Co^{3+} + 2I^- \rightarrow 2Co^{2+} + I_2(s)$$

is 1.272 V at 25°C. (a) What is the equilibrium constant? (b) What is the maximum amount of electrical work done by this cell at standard conditions in a process in which 100 g of $I_2$ is deposited?

(a) Equation 13.14, $\epsilon° = (0.0592/n) \log K$, indicates that the number of electrons, $n$, transferred in the balanced equation must be determined. It is easy to see from the given equation for the cell reaction that $n = 2$. If the value of $n$ is not obvious by inspection, and there are more complicated equations where this will surely be the case, breaking down the equation into those for the half-reactions yields the value of $n$ explicitly:

$$2I^- \rightarrow I_2(s) + 2e^-$$
$$2Co^{3+} + 2e^- \rightarrow 2Co^{2+}$$

Substitution into Equation 13.14 gives

$$1.272 \; V = (0.0592 \; V/2) \log K$$
$$\log K = 43.0$$
$$K = 1 \times 10^{43}$$

(b) The maximum electrical work, which is obtained if the cell operates without losses, is given by the product of the total charge and the EMF (Equation 13.3),

$$w_e = -Q\epsilon = -Q\epsilon°$$

The total charge passing through the circuit can be obtained from the mass of $I_2$ produced. Expressed in terms of the number of moles of $I_2$, we have

$$n_{I_2} = 100 \; g/253.8 \; g \; mol^{-1} = 0.3940 \; mol$$

Since two electrons are transferred for each molecule of $I_2$ formed, the total charge is

$$Q = (2)(0.3940 \; mol)(96,485 \; C \; mol^{-1}) = 76,030 \; C$$

The electrical work is

$$w_e = -Q\epsilon°$$
$$= -(76,030 \; C)(1.272 \; V)[1 \; J \; (CV)^{-1}](10^{-3} \; kJ \; J^{-1})$$
$$= -96.7 \; kJ \quad \blacksquare$$

## 13.3 The Nernst Equation

The EMF of a cell depends on the concentrations or partial pressures of the reactants and products of the cell reaction. The form of this dependence follows from the relationship between the EMF and the free energy change and from that between the free energy change and the concentrations or partial pressures of reactants and products (Section 11.8). The dependence of $\Delta G$ on the composition of the reaction mixture is given by Equation 11.37,

$$\Delta G = \Delta G° + RT \ln Q$$

where $Q$ is the reaction quotient. Since $\Delta G$ is related to $\epsilon$ and $\Delta G°$ to $\epsilon°$ (Equations 13.11 and 13.12), we obtain upon substitution

$$-nF\epsilon = -nF\epsilon° + RT \ln Q$$

which can be rewritten as

$$\epsilon = \epsilon° - (RT/nF) \ln Q \tag{13.15}$$

Evaluation of the numerical constants at 25°C (see Equation 13.14) gives

$$\epsilon = \epsilon° - (0.0592/n) \log Q \tag{13.16}$$

Equations 13.15 or 13.16 are known as the **Nernst equation.**

The Nernst equation shows how the cell potential depends on the concentrations of reactants and products. When reactants and products are present at standard concentration, the cell EMF is just equal to the standard cell EMF. Whenever $Q < 1$, that is, when the reactants are present in higher concentration than the products, $\epsilon > \epsilon°$. As a spontaneous cell reaction takes place, the concentrations of the products increase while those of the reactants decrease. Consequently, $\epsilon$ gradually decreases. When equilibrium is finally reached, $Q$ becomes equal to $K$. Since $\epsilon°$ and $K$ are related by means of Equation 13.13, the value of $\epsilon$ at equilibrium can be obtained by substitution of this relationship into the Nernst equation,

$$\epsilon = \epsilon° - \left(\frac{RT}{nF}\right)\left(\frac{nF\epsilon°}{RT}\right) = \epsilon° - \epsilon° = 0 \tag{13.17}$$

Thus, the cell potential drops to zero when chemical equilibrium is attained. As is the case for the equilibrium constant, the Nernst equation is strictly valid only for dilute solutions. When a solution is concentrated, the concentrations must be corrected for departures from ideal behavior.

---

**Example 13.5** Calculate the EMF of the cell

$$Cu/Cu^{2+}(0.200\ M)//Ag^+(0.0500\ M)/Ag$$

at 25°C. Use data in Appendix 4.

The half-reactions and overall reaction are

$$Cu \rightarrow Cu^{2+} + 2e^-$$
$$\underline{2(Ag^+ + e^- \rightarrow Ag)}$$
$$Cu + 2Ag^+ \rightarrow Cu^{2+} + 2Ag$$

In order to use the Nernst equation, the value of $\epsilon°$ must be known. While the value is not given, it can be determined from the tabulated standard free energies of formation. We first obtain the standard free energy change in the reaction,

$$\Delta G° = \sum_{\text{products}} \Delta G_f° - \sum_{\text{reactants}} \Delta G_f°$$

$$= [65.49 - (2 \times 77.11)] \text{ kJ} = -88.73 \text{ kJ}$$

The value of $\epsilon°$ is next obtained by means of Equation 13.12, $\Delta G° = -nF\epsilon°$. Since two electrons are transferred in the reaction for which $\Delta G°$ was calculated, we obtain

$$\epsilon° = \frac{(-88.73 \text{ kJ})(10^3 \text{ J kJ}^{-1})[1\text{V}/(\text{J C}^{-1})]}{-(2 \text{ mol})(96{,}485 \text{ C mol}^{-1})}$$

$$= 0.4598 \text{ V}$$

Note that if the reaction had been written as a one-electron transfer process,

$$\tfrac{1}{2}\text{Cu} + \text{Ag}^+ \rightarrow \tfrac{1}{2}\text{Cu}^{2+} + \text{Ag}$$

the same value of $\epsilon°$ would have been obtained since $\Delta G°$ would have been smaller by a factor of two, as would $n$. As already noted, the value of $\varepsilon°$ is independent of factors multiplying the stoichiometric coefficients in the balanced equation.

The value of $\epsilon$ is finally obtained by means of Equation 13.16, which for the reaction of interest can be written as

$$\epsilon = \epsilon° - \frac{0.0592}{n} \log \frac{[\text{Cu}^{2+}]}{[\text{Ag}^+]^2}$$

The concentrations of solid silver and copper are constant and need not be stated explicitly. Substituting the appropriate numerical values, we obtain

$$\epsilon = 0.4598 \text{ V} - \left(\frac{0.0592 \text{ V}}{2}\right) \log \left[\frac{(0.200)}{(0.0500)^2}\right]$$

$$= 0.403 \text{ V}$$

Note that the potential is smaller than the value obtained at standard conditions. This decrease is consistent with Le Chatelier's principle. Since the concentration of product relative to that of reactant is higher than it is at standard conditions, the equilibrium point must shift to the left, resulting in a less negative value of $\Delta G$ and thus in a smaller value of $\epsilon$. ∎

## 13.4 Concentration Cells

When a concentrated and a dilute solution of an ion or some other species are mixed together, a solution of intermediate concentration is formed. The main driving force for the mixing process is the increase in entropy. The solute ions or molecules are in a more random arrangement in the mixed than in the unmixed configuration and the entropy consequently increases on mixing. This drive towards a uniform concentration can be made the basis of a galvanic cell known as a concentration cell.

A **concentration cell** is a galvanic cell in which a given half-reaction and its reverse make up the cell reaction. The concentration of the species undergoing the redox reaction must differ in the two half-cells. Consider, for example, the concentration cell

$$Cu/Cu^{2+}(0.01\ M)//Cu^{2+}(0.1\ M)/Cu$$

whose operation is illustrated in Figure 13.4. The two half-reactions and the overall cell reaction are

anode:  $Cu \rightarrow Cu^{2+}\ (0.01\ M) + 2e^-$

cathode:  $Cu^{2+}(0.1\ M) + 2e^- \rightarrow Cu$

overall:  $Cu^{2+}(0.1\ M) \rightarrow Cu^{2+}\ (0.01\ M)$

The net result of this process is that the concentration of the concentrated $Cu^{2+}$ solution decreases and that of the dilute $Cu^{2+}$ solution increases, while a current flows through the external circuit from anode to cathode.

The EMF of a concentration cell can be obtained by means of the Nernst equation. Applying this equation to the cell depicted in Figure 13.4, we have

$$\epsilon = \epsilon^\circ - \frac{0.0592}{n} \log\left(\frac{[Cu^{2+}]_{anode}}{[Cu^{2+}]_{cathode}}\right)$$

where, as indicated by the cell reaction, the more dilute solution (i.e., the anode solution) is the "product" and the more concentrated solution (i.e., the cathode solution) is the "reactant". The standard cell potential must be zero, for at standard conditions the concentration of $Cu^{2+}$ is 1 $M$ in both compartments and no reaction occurs. The Nernst equation then reduces to

$$\epsilon = -\frac{0.0592}{2} \log\left(\frac{0.01}{0.1}\right)$$

$$= 0.03\ V$$

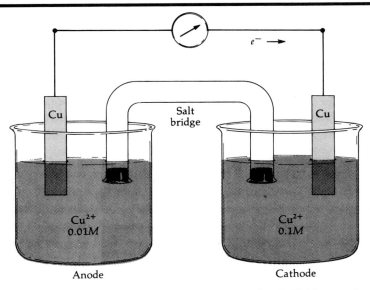

**FIGURE 13.4** A concentration cell based on the $Cu^{2+}/Cu$ couple.

A factor of 10 difference in concentration leads to a cell potential of $(0.0592/n)$ V.

Concentration cells are of importance in biological processes. For example, differences in the concentration of $Na^+$ and $K^+$ ions at the exterior and interior surfaces of nerve cells are involved in the generation of an electric current which travels along the nerve. Periodic concentration differences in these ions accompany the contraction of the heart and give rise to a current, the detection of which forms the basis of the electrocardiogram. This record of the variation of the current with time is of great value in the diagnosis of heart diseases.

## 13.5 Standard Electrode Potentials

*a. Determination of Standard Electrode Potentials* Standard cell potentials are known for a large number of reactions. As in the case of $\Delta G°$ and $\Delta H°$ values, it is desirable to tabulate this information in a compact yet readily usable form. Since a redox reaction is made up of two half-reactions, the standard reaction potential can be divided into two standard half-reaction potentials, commonly known as **standard electrode potentials.** These values can, in turn, be combined to yield standard cell potentials for all the possible combinations of half-reactions.

Since oxidation and reduction always occur together, it is impossible to measure a standard electrode potential directly. Only cell potentials are measurable. Recall, however, that the cell potential is equal to the difference between the potentials of its two electrodes. Consequently, a particular electrode can be picked as a reference electrode and arbitrarily assigned a potential of zero. The values of all other standard electrodes are expressed relative to the value of this reference electrode. The procedure works because only differences between electrode potentials can be measured and such differences are independent of the choice of scale zero.

The electrode chosen to mark this arbitrary reference point is the standard hydrogen electrode, that is, the electrode at which the half-reaction

$$2H^+(1\ M) + 2e^- \rightarrow H_2\ (1\ atm)$$

takes place. For this electrode, $\epsilon° = 0$ at 298 K. The value of any other standard electrode potential can then be determined by combining it with the standard hydrogen electrode, measuring the resulting cell potential, and assigning the measured value to the electrode in question.

To illustrate this procedure, let us consider some specific examples. Suppose the standard electrode potential for the reduction of $Cu^{2+}$ to the metal is to be measured, i.e.,

$$Cu^{2+} + 2e^- \rightarrow Cu \qquad \epsilon° = ?$$

This can be done by combining a $Cu^{2+}/Cu$ half-cell with a $H^+/H_2$ half-cell, both at standard conditions, measuring the cell potential, and observing the direction in which the current flows. It is found that the cell potential is 0.34 V and that the current flows through the external circuit from the hydrogen electrode to the copper electrode (Figure 13.5a). Consequently, the hydrogen electrode acts as the anode and the copper electrode as the cathode. The overall cell reaction is

$$H_2(g) + Cu^{2+} \rightarrow 2H^+ + Cu(s)$$

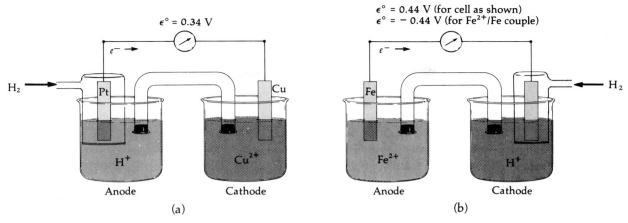

**FIGURE 13.5** Determination of standard electrode potentials by use of a hydrogen reference electrode. (a) Electrons flow from the hydrogen to the copper electrode; the standard electrode potential of the Cu²⁺/Cu couple is positive. (b) Electrons flow from the iron to the hydrogen electrode; the standard electrode potential of the Fe²⁺/Fe couple is negative.

This reaction can be divided into the half-reactions

$$H_2(g) \rightarrow 2H^+ + 2e^- \qquad \epsilon° = 0$$
$$\underline{Cu^{2+} + 2e^- \rightarrow Cu(s) \qquad \epsilon° = ?}$$
$$H_2(g) + Cu^{2+} \rightarrow 2H^+ + Cu(s) \qquad \epsilon° = 0.34 \text{ V}$$

from which we immediately infer that the standard electrode potential for the Cu²⁺/Cu half-reaction is $\epsilon° = 0.34$ V. The positive value of $\epsilon°$ means that the reaction is spontaneous as written, i.e., Cu²⁺ is more readily reduced to Cu than H⁺ is to H₂.

Since the Cu²⁺/Cu half-reaction involves reduction, its standard electrode potential is called a standard reduction potential $\epsilon°_{red}$. The value of $\epsilon°$ for the reverse half-reaction,

$$Cu \rightarrow Cu^{2+} + 2e^-$$

known as a standard oxidation potential $\epsilon°_{ox}$, can be obtained from $\epsilon°_{red}$ by the relation

$$\epsilon°_{ox} = -\epsilon°_{red} \qquad (13.18)$$

The validity of this relation follows from Section 13.2, where we saw that when a reaction is reversed, the sign of its EMF changes while the magnitude remains unchanged. Since it is usually clear whether a particular half-reaction involves oxidation or reduction, the subscripts "ox" and "red" on $\epsilon°$ values will generally be omitted.

As a second example, consider the half-reaction

$$Fe^{2+} + 2e^- \rightarrow Fe(s) \qquad \epsilon° = ?$$

Once again, the standard electrode potential for the reduction of Fe²⁺ can be determined by combining a Fe²⁺/Fe half-cell with a H⁺/H₂ half-cell and measuring the cell potential and direction of current flow. It is observed that $\epsilon° = 0.44$ V and that electrons flow from the iron electrode to the hydrogen electrode. Evidently, the iron electrode acts as the anode and the hydrogen electrode as the cathode (Figure 13.5b). The cell reaction is

$$Fe(s) + 2H^+ \rightarrow Fe^{2+} + H_2(g)$$

indicating that the standard electrode potential for the half-reaction

$$Fe \rightarrow Fe^{2+} + 2e^-$$

is $\epsilon° = 0.44$ V. Therefore, for the half-reaction

$$Fe^{2+} + 2e^- \rightarrow Fe(s) \qquad \epsilon° = -0.44 \text{ V}$$

The negative sign of $\epsilon°$ means that $Fe^{2+}$ is not as readily reduced as $H^+$.

**b. The EMF Series** It is customary to tabulate the standard electrode potentials written as standard reduction potentials. Table 13.1 gives a list of standard reduction potentials arranged in order of decreasing $\epsilon°$; a more complete list is given in Appendix 5. Such a list is also known as an **EMF series.** Since some substances are present as different species in acidic and basic solutions, the standard reduction potentials are different in these two media and are listed separately. Half-reactions with $\epsilon°$ independent of pH, e.g., $K^+ + e^- \rightarrow K$, are included in the table for acidic solutions.

The list of standard reduction potentials in acidic solution is headed by the half-reation

$$F_2 + 2e^- \rightarrow 2F^- \qquad \epsilon° = 2.87 \text{ V}$$

indicating that of all the listed substances, fluorine displays the greatest tendency to be present in the reduced state. In other words, fluorine is the strongest oxidizing agent listed. An obvious corollary of this statement is that the reverse reaction,

$$2F^- \rightarrow F_2 + 2e^- \qquad \epsilon° = -2.87 \text{ V}$$

has the least tendency of all oxidation half-reactions to occur.

The bottom entry in the table is for the half-reaction

$$Li^+ + e^- \rightarrow Li \qquad \epsilon° = -3.05 \text{ V}$$

Lithium displays the smallest tendency of all the listed substances to be present in the reduced form, or conversely, the greatest tendency to be in the oxidized form. Lithium is therefore the strongest reducing agent listed.

The arrangement of standard reduction potentials in order of decreasing numerical value has the virtue of ordering the oxidized species in order of decreasing strength as oxidizing agents. Any entry in the list should be able to oxidize the reduced forms of all species listed below it. For example, $F_2$ can oxidize $H_2O$ to $H_2O_2$, $Mn^{2+}$ to $MnO_4^-$, $Cl^-$ to $Cl_2$, and so on, whereas $MnO_4^-$ can oxidize $Cl^-$ to $Cl_2$ but not $H_2O$ to $H_2O_2$. Conversely, the reduced species in the list are arranged in order of increasing strength as reducing agents. For example, metallic lithium is the strongest reducing agent and is able to reduce $K^+$ to K, $Ba^{2+}$ to Ba, and so on. Figure 13.6 summarizes the ordering scheme graphically. It must be kept in mind when considering the possible occurrence of specific redox reactions that the values of $\epsilon°$, like those of $\Delta G°$ or $K$, convey no

**TABLE 13.1 Standard Reduction Potentials at 298 K**

### Acidic Solution

| Half-reaction | $\epsilon°$ (V) |
|---|---|
| $F_2 + 2e^- \rightarrow 2F^-$ | 2.87 |
| $H_2O_2 + 2H^+ + 2e^- \rightarrow 2H_2O$ | 1.78 |
| $MnO_4^- + 8H^+ + 5e^- \rightarrow Mn^{2+} + 4H_2O$ | 1.51 |
| $Cl_2 + 2e^- \rightarrow 2Cl^-$ | 1.36 |
| $Cr_2O_7^{2-} + 14H^+ + 6e^- \rightarrow 2Cr^{3+} + 7H_2O$ | 1.33 |
| $O_2 + 4H^+ + 4e^- \rightarrow 2H_2O$ | 1.23 |
| $Br_2(l) + 2e^- \rightarrow 2Br^-$ | 1.07 |
| $NO_3^- + 4H^+ + 3e^- \rightarrow NO + 2H_2O$ | 0.96 |
| $Ag^+ + e^- \rightarrow Ag$ | 0.80 |
| $Fe^{3+} + e^- \rightarrow Fe^{2+}$ | 0.77 |
| $I_2 + 2e^- \rightarrow 2I^-$ | 0.54 |
| $Cu^+ + e^- \rightarrow Cu$ | 0.52 |
| $Cu^{2+} + 2e^- \rightarrow Cu$ | 0.34 |
| $AgCl + e^- \rightarrow Ag + Cl^-$ | 0.22 |
| $2H^+ + 2e^- \rightarrow H_2$ | 0.00 |
| $Pb^{2+} + 2e^- \rightarrow Pb$ | −0.13 |
| $AgI + e^- \rightarrow Ag + I^-$ | −0.15 |
| $PbSO_4 + 2e^- \rightarrow Pb + SO_4^{2-}$ | −0.36 |
| $Fe^{2+} + 2e^- \rightarrow Fe$ | −0.44 |
| $Zn^{2+} + 2e^- \rightarrow Zn$ | −0.76 |
| $Al^{3+} + 3e^- \rightarrow Al$ | −1.66 |
| $Mg^{2+} + 2e^- \rightarrow Mg$ | −2.36 |
| $Na^+ + e^- \rightarrow Na$ | −2.71 |
| $Ba^{2+} + 2e^- \rightarrow Ba$ | −2.91 |
| $K^+ + e^- \rightarrow K$ | −2.93 |
| $Li^+ + e^- \rightarrow Li$ | −3.05 |

### Basic Solution

| Half-reaction | $\epsilon°$ (V) |
|---|---|
| $MnO_4^- + 2H_2O + 3e^- \rightarrow MnO_2 + 4OH^-$ | 0.59 |
| $O_2 + 2H_2O + 4e^- \rightarrow 4OH^-$ | 0.40 |
| $S + 2e^- \rightarrow S^{2-}$ | −0.45 |
| $Cd(OH)_2 + 2e^- \rightarrow Cd + 2OH^-$ | −0.81 |
| $Mn(OH)_2 + 2e^- \rightarrow Mn + 2OH^-$ | −1.55 |
| $Ca(OH)_2 + 2e^- \rightarrow Ca + 2OH^-$ | −3.02 |

information about the rates at which such reactions occur. Furthermore, the values of $\epsilon°$ are valid only at standard conditions; the Nernst equation must be used to correct these values for changes in concentrations or partial pressures. In spite of these qualifications, the EMF series is a most useful summary of the chemical behavior of the elements and their ions in aqueous solution.

Any two half-reactions may be combined to give a redox reaction, the standard EMF of which is obtained by combining their standard electrode potentials. The procedure is straightforward: reverse the half-reaction with the

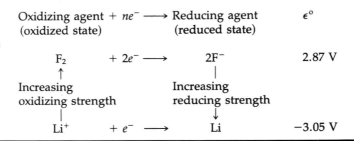

**FIGURE 13.6** Ordering of the entries in Table 13.1 according to their oxidizing and reducing strength. The arrows point in the direction of increasing strength.

more negative (or less positive) EMF and add it to the other half-reaction, multiplying the stoichiometric coefficients by whatever factors are needed to ensure the cancellation of electrons. The standard cell EMF is obtained by reversing the sign of the standard electrode potential of the half-reaction which was reversed and adding the result to the standard electrode potential of the other half-reaction. Thus, if two half-reactions have standard reduction potentials $\epsilon_1^\circ$ and $\epsilon_2^\circ$, then the standard reaction EMF $\epsilon^\circ$ is either

$$\epsilon^\circ = \epsilon_1^\circ - \epsilon_2^\circ \qquad (\epsilon_1^\circ - \epsilon_2^\circ > 0) \tag{13.19}$$

or

$$\epsilon^\circ = \epsilon_2^\circ - \epsilon_1^\circ \qquad (\epsilon_2^\circ - \epsilon_1^\circ > 0) \tag{13.20}$$

This procedure ensures that the standard cell potential will be positive and that the corresponding reaction will occur as written.

**Example 13.6** (a) What is the standard cell potential for the reaction of copper with nitric acid? (b) Which reaction has a stronger tendency to occur at standard conditions: the oxidation of Cu by $H^+$ or the reduction of $Cu^{2+}$ by $H_2$?

(a) Table 13.1 lists the following pertinent half-reactions and standard reduction potentials:

$$NO_3^- + 4H^+ + 3e^- \rightarrow NO + 2H_2O \qquad \epsilon^\circ = 0.96 \text{ V}$$

$$Cu^{2+} + 2e^- \rightarrow Cu \qquad \epsilon^\circ = 0.34 \text{ V}$$

To obtain the desired reaction, the second half-reaction and its electrode potential must be reversed,

$$Cu \rightarrow Cu^{2+} + 2e^- \qquad \epsilon^\circ = -0.34 \text{ V}$$

To cancel the electrons, we must multiply the stoichiometric coefficients of the first half-reaction by 2 and those of the second half-reaction by 3. The resulting half-reactions are added, as are the corresponding values of $\epsilon^\circ$:

$$3Cu + 2NO_3^- + 8H^+ \rightarrow 3Cu^{2+} + 2NO + 4H_2O \qquad \epsilon^\circ = 0.62 \text{ V}$$

The positive value of $\epsilon^\circ$ indicates that copper should dissolve in 1 $M$ nitric acid. Note that the half-reaction $\epsilon^\circ$ values are *not* multiplied by the factors used to balance the numbers of electrons; electrode potentials are intensive properties.

(b) The pertinent half-reactions are

$$Cu^{2+} + 2e^- \rightarrow Cu \qquad \epsilon° = 0.34 \text{ V}$$

$$2H^+ + 2e^- \rightarrow H_2 \qquad \epsilon° = 0$$

To obtain a positive cell potential, the second half-reaction must be reversed. Addition of the half-reactions and $\epsilon°$ values gives

$$Cu^{2+} + H_2 \rightarrow Cu + 2H^+ \qquad \epsilon° = 0.34 \text{ V}$$

This result indicates that copper cannot dissolve in 1 $M$ HCl at 25°C, as may readily be confirmed in the laboratory. ∎

*c. Standard Electrode Potentials of Elements with Three or More Oxidation States* Many elements display a number of different oxidation states. Each species excepting the one in the lowest oxidation state can be characterized by standard potentials for reduction to all of the species of lower oxidation state. How are these various half-reaction potentials related?

Consider, for example, the entries in Table 13.1 for iron species in acidic solution

$$Fe^{3+} + e^- \rightarrow Fe^{2+} \qquad \epsilon° = 0.77 \text{ V}$$

$$Fe^{2+} + 2e^- \rightarrow Fe \qquad \epsilon° = -0.44 \text{ V}$$

While no entry is listed for the direct reduction of $Fe^{3+}$ to Fe, the standard electrode potential for this half-reaction should be obtainable from the entries for the stepwise process. It might appear, at first sight, that the desired $\epsilon°$ should just be the sum of the listed values since the desired half-reaction is obtained by adding the given half-reactions:

$$Fe^{3+} + 3e^- \rightarrow Fe$$

While this procedure is just the one followed in combining two standard electrode potentials to obtain the standard cell potential, it is not valid in the special case in which two half-reactions are combined to yield a third half-reaction. Strictly speaking, the quantitites that are actually additive are the free energy changes for the processes in question. So long as the same number of electrons is transferred in each half-reaction and in the overall reaction, as is necessarily the case when two half-reactions are combined into a balanced redox reaction, the same answer is obtained whether the free energies or the half-reaction potentials are combined. As may be inferred from Equation 13.12, $\Delta G° = -nF\epsilon°$, the potential is a measure of the free energy change per electron. When two half-reactions are combined into a third half-reaction, the values of $n$ of the three half-reactions necessarily cannot all be the same. Consequently, the standard electrode potentials must be converted to standard free energies prior to combination.

To demonstrate how a desired half-reaction potential is obtained from two others, we calculate $\epsilon°$ for reduction of $Fe^{3+}$ to Fe. The half-reactions and corresponding $\Delta G°$ values are

$$Fe^{3+} + e^- \rightarrow Fe^{2+} \qquad \Delta G° = -0.77 \text{ F V C}$$

$$Fe^{2+} + 2e^- \rightarrow Fe \qquad \Delta G° = -(2)(-0.44 \text{ F}) \text{ V C} = 0.88 \text{ F V C}$$

$$Fe^{3+} + 3e^- \rightarrow Fe \qquad \Delta G° = 0.11 \text{ F V C}$$

**476** Electrochemistry 13.5

where the unit F V C arises as follows:

$$\Delta G° = -nF\epsilon°$$
$$= (\text{mol})(F \text{ C mol}^{-1})(V) = F \text{ V C}$$

The desired standard electrode potential is obtained by once again applying Equation 13.12 to the appropriate $\Delta G°$ value,

$$\epsilon° = \frac{-\Delta G°}{nF} = \frac{-0.11 \text{ F V C}}{(3 \text{ mol})(F \text{ C mol}^{-1})} = -0.037 \text{ V}$$

Note that it is not necessary to explicitly calculate $\Delta G°$, since the value of $F$ cancels out in the conversion between $\Delta G°$ and $\epsilon°$. The above procedure can be summarized by the relation

$$\epsilon_3° = \frac{n_1 \epsilon_1° + n_2 \epsilon_2°}{n_3} \tag{13.21}$$

where $n_1$, $n_2$, and $n_3$ are the number of electrons transferred in half-reactions with standard electrode potentials $\epsilon_1°$, $\epsilon_2°$ and $\epsilon_3°$, respectively.

---

**Example 13.7** Is $Cu^+$ stable to disproportionation in aqueous solution at standard conditions?

In a disproportionation reaction (Section 2.9), a given species undergoes both oxidation and reduction. The disproportionation of $Cu^+$ involves the reaction

$$2 \text{ Cu}^+ \rightarrow \text{Cu} + \text{Cu}^{2+}$$

which is made up of the half-reactions

(1) $\quad \text{Cu}^+ \rightarrow \text{Cu}^{2+} + e^- \qquad \epsilon° = ?$

(2) $\quad \text{Cu}^+ + e^- \rightarrow \text{Cu} \qquad \epsilon° = 0.52 \text{ V}$

While the $\epsilon°$ value for the first half-reaction is not listed in Table 13.1, a value is given for the half-reaction

$$\text{Cu}^{2+} + 2e^- \rightarrow \text{Cu} \qquad \epsilon° = 0.34 \text{ V}$$

When the reverse of this half-reaction is combined with (2), the result is (1):

$$\text{Cu} \rightarrow \text{Cu}^{2+} + 2e^- \qquad \epsilon° = -0.34 \text{ V}$$
$$\underline{\text{Cu}^+ + e^- \rightarrow \text{Cu} \qquad \epsilon° = 0.52 \text{V}}$$
$$\text{Cu}^+ \rightarrow \text{Cu}^{2+} + e^- \qquad \epsilon° = ?$$

Applying Equation 13.21 we obtain

$$\epsilon° = \frac{2 \times -0.34 \text{ V} + 0.52 \text{ V}}{1} = -0.16 \text{ V}$$

Note that this value is quite different from that obtained by direct addition of the $\epsilon°$ values.

In order to finish the problem, the $\epsilon°$ values for half-reactions (1) and (2) must be added to obtain $\epsilon°$ for the disproportionation reaction. We obtain

$$\epsilon° = -0.16 \text{ V} + 0.52 \text{ V} = 0.36 \text{ V}$$

Since $\epsilon°$ is positive, the disproportionation of $Cu^+$ occurs spontaneously and this species is unstable in aqueous solution at standard conditions. ■

## 13.6 Determination of Equilibrium Constants from Standard Reduction Potentials

Standard reduction potentials may be used to calculate equilibrium constants for redox reactions. The procedure is straightforward: the appropriate two half-reactions and the corresponding standard electrode potentials are first combined to give the overall reaction and its EMF. The equilibrium constant is then obtained by means of Equation 13.13 or 13.14.

**Example 13.8** What is the equilibrium constant at 25°C for the reaction of zinc metal with $Al^{3+}$ ions? What is the equilibrium concentration of $Zn^{2+}$ ions when a piece of zinc is placed in a 1 M $Al^{3+}$ solution?

The half-reactions and their standard reduction potentials are

$$Zn^{2+} + 2e^- \rightarrow Zn \qquad \epsilon° = -0.76 \text{ V}$$
$$Al^{3+} + 3e^- \rightarrow Al \qquad \epsilon° = -1.66 \text{ V}$$

To obtain the reaction of interest, the first half-reaction must be reversed and multiplied by 3 while the second one must be multiplied by 2

$$3Zn \rightarrow 3Zn^{2+} + 6e^- \qquad \epsilon° = 0.76 \text{ V}$$
$$\underline{2Al^{3+} + 6e^- \rightarrow 2Al \qquad \epsilon° = -1.66 \text{ V}}$$
$$3Zn + 2Al^{3+} \rightarrow 3Zn^{2+} + 2Al \qquad \epsilon° = -0.90 \text{ V}$$

The equilibrium constant is obtained by means of Equation 13.14,

$$\epsilon° = \left(\frac{0.0592}{n}\right) \log K$$

$$-0.90 \text{ V} = \left(\frac{0.0592 \text{ V}}{6}\right) \log K$$

$$\log K = -91$$

$$K = 10^{-91}$$

The equilibrium concentration of $Zn^{2+}$ is obtained in the manner described in detail in Section 7.4:

$$K = \frac{[Zn^{2+}]^3}{[Al^{3+}]^2}$$

$$10^{-91} = \frac{[Zn^{2+}]^3}{(1)^2}$$

$$[Zn^{2+}] = 10^{-31} \text{ mol L}^{-1}$$

The equilibrium point for this reaction lies so far to the left that not a single $Zn^{2+}$ ion is likely to be present in a liter of the $Al^{3+}$ solution! ∎

Standard electrode potentials may also be used to evaluate solubility products. While this assertion may at first seem surprising since dissolution of a sparingly soluble salt generally does not involve a redox reaction, the use of a metal/insoluble metal salt electrode (Section 13.1c) makes it possible to make such a determination. Consider, for example, the dissolution of AgCl(s). This process can be viewed as the net result of the following redox reaction:

$$Ag \rightarrow Ag^+ + e^-$$
$$\underline{AgCl(s) + e^- \rightarrow Ag + Cl^-}$$
$$AgCl(s) \rightarrow Ag^+ + Cl^-$$

The electrochemical determination of solubility products has proved to be useful because the conventional determination of $K_{sp}$ by measurement of the equilibrium ionic concentrations is often not possible, for the ionic concentrations may be below the limit of detectability.

---

**Example 13.9** Determine the solubility product of AgI at 25°C on the basis of data in Table 13.1.

The half-reactions of interest are

$$Ag^+ + e^- \rightarrow Ag \qquad \epsilon° = 0.80 \text{ V}$$
$$AgI(s) + e^- \rightarrow Ag + I^- \qquad \epsilon° = -0.15 \text{ V}$$

Reversing the first half-reaction and adding the two half-reactions and corresponding potentials we obtain

$$AgI(s) \rightarrow Ag^+ + I^- \qquad \epsilon° = -0.95 \text{ V}$$

The value of $K_{sp}$ is obtained by means of Equation 13.14,

$$\epsilon° = (0.0592/n) \log K$$
$$-0.95 \text{ V} = (0.0592 \text{ V}) \log K$$
$$\log K = -16$$
$$K_{sp} = K = 10^{-16}$$

This result is in good agreement with the value listed in Table 9.2. ∎

It is of interest to note, in Example 13.9, that the standard electrode potential for the reduction of AgI is much smaller than that for the reduction of $Ag^+$. The difference between the two values can be understood by means of the Nernst equation. The reduction potential for the half-reaction $Ag^+ + e^- \rightarrow Ag$ can be expressed as a function of the $Ag^+$ concentration by the relation

$$\epsilon = \epsilon° - \frac{0.0592}{n} \log \frac{1}{[Ag^+]}$$
$$= 0.80 \text{ V} - 0.0592 \text{ V} \log \frac{1}{[Ag^+]}$$

The value of $\epsilon$ therefore decreases with decreasing [Ag$^+$]. For instance, when the concentration of Ag$^+$ is $10^{-4}$ mol L$^{-1}$, the reduction potential is only 0.56 V. This result is consistent with Le Cateliers's principle since a decrease in the concentration of a reactant shifts the equilibrium point to the left. Consider, now, the case when Ag$^+$ is in equilibrium with solid AgI. The concentration of Ag$^+$ must then satisfy the solubility product relation

$$[\text{Ag}^+] = \frac{K_{sp}}{[\text{I}^-]}$$

Under these circumstances the half-reaction potential becomes

$$\epsilon = 0.80 \text{ V} - 0.0592 \text{ V} \log \frac{[\text{I}^-]}{K_{sp}}$$

Substituting the value $K_{sp} = 10^{-16}$ yields

$$\epsilon = -0.15 \text{ V} - 0.0592 \text{ V} \log[\text{I}^-]$$

When [I$^-$] = 1 M, the result is just the standard electrode potential for the reduction of AgI.

In general, the more stable the species in which the silver ion is bound, the lower will be the reduction potential of silver. Table 13.2 summarizes the standard reduction potentials of a number of such species. Note, for example, that $\epsilon°$ for the reduction of AgI is lower than that for the reduction of AgCl, reflecting the fact that AgI is the less soluble salt. We have seen in Section 9.6 that complex ions also reduce the concentration of the free ion. Table 13.2 includes entries for some complex ions of silver. The $\epsilon°$ values for these species are related to the complex formation constants, which can be determined electrochemically in the same manner as solubility product constants. Observe that the order of the entries in Table 13.2 generally parallels the sequence of processes which happen as various reagents are sequentially added to a solution of Ag$^+$ (Figure 9.6). The similarity is not accidental. The occurrence of precipitation or complex ion formation is determined by the values of the solubility product or complex formation constants, and, as we have seen, these constants are closely related to the standard potentials.

Data such as that given in Table 13.2 indicate that all the processes which can occur when two reagents are mixed must be considered when standard electrode potentials are used to predict the outcome of such mixing. Consider,

TABLE 13.2 Standard Reduction Potentials of Species Containing Ag$^+$

| Half-reaction | $\epsilon°$ (V) |
|---|---|
| Ag$^+$ + $e^-$ → Ag(s) | 0.80 |
| Ag(NH$_3$)$_2^+$ + $e^-$ → Ag(s) + 2NH$_3$(aq) | 0.37 |
| AgCl(s) + $e^-$ → Ag(s) + Cl$^-$ | 0.22 |
| AgBr(s) + $e^-$ → Ag(s) + Br$^-$ | 0.07 |
| AgI(s) + $e^-$ → Ag(s) + I$^-$ | −0.15 |
| Ag(CN)$_2^-$ + $e^-$ → Ag(s) + 2CN$^-$ | −0.31 |
| Ag$_2$S(s) + $e^-$ → 2Ag(s) + S$^{2-}$ | −0.69 |

for example, the reaction

$$Ag^+ + I^- \rightarrow Ag(s) + \tfrac{1}{2}I_2(s)$$

The data in Table 13.1 indicate that this reaction should occur spontaneously under standard conditions since $\epsilon° = 0.26$ V. Yet all that happens when 1 $M$ $Ag^+$ and $I^-$ are mixed is the precipitation of AgI, a process for which $\epsilon° = 0.95$ V (Example 13.9). Evidently, the reduction in the $Ag^+$ and $I^-$ concentrations resulting from the precipitation of AgI is sufficiently large to prevent the redox reaction from occurring.

# Applications of Electrochemistry

## 13.7 Batteries

Since galvanic cells are devices that produce a spontaneous flow of electrons, it is not surprising that they are widely used as portable sources of electric current, or **batteries,** as they are commonly known. The large variety of batteries that have become available in recent years attests to the growing importance of electrochemical energy generation. While any two half-cells can in principle be used in a battery, a number of practical constraints severely limit the actual choices. For example, the chemical substances comprising the battery must be available at a reasonable cost, the battery must be mechanically stable and safe, and the total energy available per unit weight (energy density) must be reasonably high and available over the desired temperature range.

It is convenient to divide batteries into three broad categories: primary cells, storage (secondary) cells, and fuel cells. **Primary cells** contain a limited amount of chemical fuel. Once this material has been consumed by the redox reaction, the battery is no longer of use. **Storage cells** also contain a limited amount of chemical fuel but are designed to take advantage of the reversibility of the cell reaction. A storage cell can be recharged periodically by use of an external power source (Section 13.10). **Fuel cells** differ from both primary and storage cells in that the chemical substances which power the cell are continuously replenished, thereby permitting continuous operation over long periods of time.

*a. Primary Cells* The oldest of the widely used primary cells, now gradually falling into disuse, is the Leclanché cell. This "flashlight" battery (Figure 13.7a) consists of a zinc can, which serves as the anode, and a carbon rod immersed in a paste consisting of manganese dioxide, aqueous ammonium chloride and zinc chloride, and graphite. The carbon rod and manganese dioxide serve as the cathode and the other ingredients constitute the electrolyte. The cell reaction is made up of the following half-reactions

anode: $\quad Zn(s) \rightarrow Zn^{2+} + 2e^-$

cathode: $\quad 2MnO_2(s) + 2NH_4^+ + 2e^- \rightarrow Mn_2O_3(s) + 2NH_3(aq) + H_2O$

cell: $\quad Zn(s) + 2MnO_2(s) + 2NH_4^+ \rightarrow Zn(NH_3)_2^{2+} + Mn_2O_3(s) + H_2O$

Note that the electrolyte is a paste rather than a liquid. Batteries of this type are

Electrochemistry 13.7 **481**

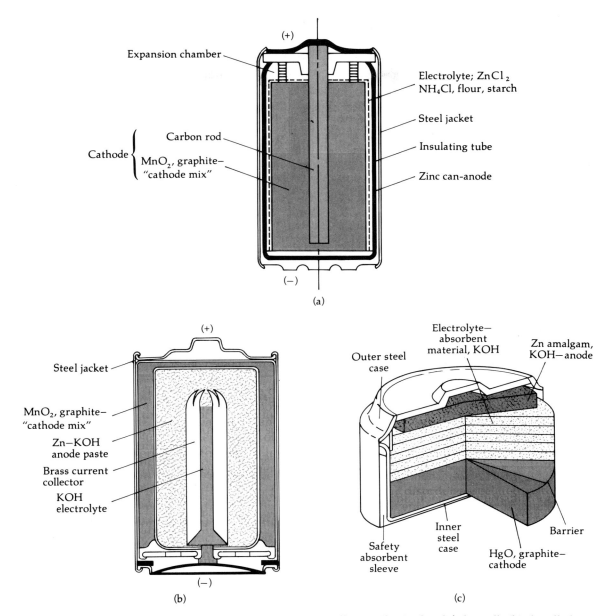

**FIGURE 13.7** Some common primary cells. (a) The Leclanché dry cell, (b) the alkaline managanese(IV) oxide–zinc cell, and (c) the mercury "button" cell.

commonly known as dry cells and are particularly useful in portable devices, where they can be inverted, jarred, and otherwise abused. The nominal voltage of the Leclanché cell is 1.5 V, but the actual value drops on usage because of the accumulation of byproducts which hinder the cathode reaction.

A useful variant of the Leclanché cell is the alkaline manganese(IV) oxide–zinc cell (Figure 13.7b). The main difference from the Leclanché cell is that the electrolyte is replaced by KOH. This results in a cell having a greater energy density and thus a longer operating life. This cell is commonly used to power cameras, toys, and other small appliances. The half-reactions and the overall reaction of the cell are

$$\text{anode:} \quad \text{Zn(s)} + 2\text{OH}^- \rightarrow \text{Zn(OH)}_2(s) + 2e^-$$

$$\text{cathode:} \quad \underline{2\text{MnO}_2(s) + 2\text{H}_2\text{O} + 2e^- \rightarrow 2\text{MnOOH}(s) + 2\text{OH}^-}$$

$$\text{cell:} \quad \text{Zn(s)} + 2\text{MnO}_2(s) + 2\text{H}_2\text{O} \rightarrow \text{Zn(OH)}_2(s) + 2\text{MnOOH}(s)$$

and the cell potential is 1.54 V.

The mercury cell is another commonly used primary cell (Figure 13.7c). It has the advantage that its potential, 1.35 V, does not change as the cell is discharged and is quite constant from one cell to another, as explained below. This feature makes it of use in scientific and medical equipment, which frequently requires a constant voltage source. The mercury cell contains a zinc amalgam anode and a mercury(II) oxide cathode. (An amalgam is an alloy of a metal with mercury). The electrolyte consists of an absorbent material containing sodium or potassium hydroxide. These materials are in the form of compacted powders, thereby making possible the construction of a compact "button" type cell. The reactions occurring in the cell are

$$\text{anode:} \quad \text{Zn(amalgam)} + 2\text{OH}^- \rightarrow \text{ZnO(s)} + \text{H}_2\text{O} + 2e^-$$

$$\text{cathode:} \quad \underline{\text{HgO(s)} + \text{H}_2\text{O} + 2e^- \rightarrow \text{Hg(l)} + 2\text{OH}^-}$$

$$\text{cell:} \quad \text{Zn(amalgam)} + \text{HgO(s)} \rightarrow \text{ZnO(s)} + \text{Hg(l)}$$

Note that the only species whose concentration changes during cell operation is the zinc in the amalgam. It has been found that the change in this concentration has a negligible effect on the cell potential. It follows, according to the Nernst equation, that $\varepsilon \approx \varepsilon^\circ$—the cell potential is constant.

**b. Storage Cells** While galvanic cell reactions can in principle be reversed, practical difficulties preclude the recharging of most types of batteries. The changes in electrodes and electrolyte which occur as the cell is discharged are not readily reversed. Nonetheless, several types of storage cells are in wide use and others are under active development.

The best known of these cells is the lead storage cell, widely used as the automobile battery. Figure 13.8 shows a schematic diagram of this cell. The half-reactions occurring at the electrodes are

$$\text{anode:} \quad \text{Pb(s)} + \text{HSO}_4^- \underset{\text{charge}}{\overset{\text{discharge}}{\rightleftarrows}} \text{PbSO}_4(s) + \text{H}^+ + 2e^-$$

$$\text{cathode:} \quad \underline{\text{PbO}_2 + 3\text{H}^+ + \text{HSO}_4^- + 2e^- \underset{\text{charge}}{\overset{\text{discharge}}{\rightleftarrows}} \text{PbSO}_4(s) + 2\text{H}_2\text{O}}$$

$$\text{cell:} \quad \text{Pb(s)} + \text{PbO}_2(s) + 2\text{H}_2\text{SO}_4(aq) \underset{\text{charge}}{\overset{\text{discharge}}{\rightleftarrows}} 2\text{PbSO}_4(s) + 2\text{H}_2\text{O}$$

The cell potential is approximately 2 V, and six cells connected in series make up the common 12 V battery.

A fully charged lead storage cell consists of a spongy Pb anode and a solid $\text{PbO}_2$ cathode immersed in 6 M $\text{H}_2\text{SO}_4$. As the cell is discharged, $\text{PbSO}_4$ is formed at both electrodes and the sulfuric acid electrolyte is gradually diluted. Since $\text{H}_2\text{SO}_4$ is denser than $\text{H}_2\text{O}$, the density of the electrolyte provides a use-

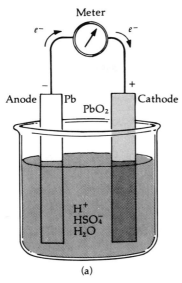 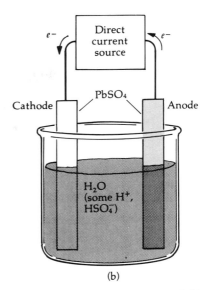

**FIGURE 13.8** The lead storage cell operating (a) in the discharge mode and (b) in the charge mode. When operating in the discharge mode, the anode is the negative terminal of the battery and the cathode is the positive terminal.

ful measure of the extent of discharge. The cell is reversible because the insoluble PbSO$_4$ product clings to the electrodes instead of falling off or diffusing into the electrolyte. The battery can be recharged by operating it in the reverse or electrolytic mode (see Section 13.10) by means of an external direct-current electrical energy source. Under these conditions, the cell reaction is reversed and PbSO$_4$ is converted back to Pb and PbO$_2$. As is well known from practical experience with lead storage batteries, the recycling process cannot be repeated indefinitely. The condition of the electrodes gradually deteriorates and eventually an internal short circuit causes a cell to fail.

While the lead storage battery is well suited for use in connection with the startup of motors, its substantial weight and corrosive liquid electrolyte make it unsuitable for the growing number of small electronic devices that are powered by rechargeable batteries. The nickel–cadmium cell is commonly used for this purpose. This cell operates on the basis of the following reaction

anode:    $Cd(s) + 2OH^- \rightarrow Cd(OH)_2(s) + 2e^-$

cathode:  $2NiOOH(s) + 2H_2O + 2e^- \rightarrow Ni(OH)_2(s) + 2OH^-$

cell:     $Cd(s) + 2NiOOH(s) + 2H_2O \underset{\text{charge}}{\overset{\text{discharge}}{\rightleftarrows}} Cd(OH)_2(s) + 2Ni(OH)_2(s)$

The electrolyte is KOH and the cell EMF is 1.3 V.

The potential scarcity of gasoline has spurred efforts to develop a battery to supply the power for electric automobiles. Such vehicles would have the added advantages of being much quieter and less polluting than conventionally powered automobiles. Because of its large weight and consequent low energy density, the lead storage cell is unsuitable for this purpose. While specialized electrical vehicles powered by lead batteries exist, their speed and range are as yet far too low to make them suitable for general use.

A number of different batteries that may eventually be practical for use in electric automobiles are under active development. The sodium–sulfur battery has attracted considerable interest because its energy density is some six times larger than that of the lead battery, its recharging time is considerably shorter, and it appears to have a potentially longer useful lifetime. The basic cell reaction can be represented as

$$\text{anode:} \quad 2\text{Na}(l) \to 2\text{Na}^+ + 2e^-$$
$$\text{cathode:} \quad \underline{S(l) + 2e^- \to S^{2-}}$$
$$\text{cell:} \quad 2\text{Na}(l) + S(l) \to 2\text{Na}^+ + S^{2-}$$

although the actual process is more complicated, involving the formation of complex sulfides. The design of this cell is rather complicated. The anode consists of molten sodium, and the cathode of a mixture of carbon, sulfur, and $Na_2S$. The electrolyte consists of a solid ceramic material, $\beta$-alumina ($Na_2O \cdot 11Al_2O_3$), which readily allows $Na^+$ ions to diffuse through its lattice. The chief drawback of this cell is that it requires a temperature of 300–350°C for its operation and contains sodium, a rather corrosive metal, particularly in the molten state.

***c. Fuel Cells*** A fuel cell differs from the other cells described in the preceding sections in that fuel is continuously supplied during operation. A fuel cell therefore never needs recharging and continues to operate as long as fuel is supplied. The fuel used in these cells is frequently the same as that used in a conventional combustion engine. In such an engine, some fuel is burned and the resulting thermal energy is converted to electrical energy. This two-stage process is substantially less efficient than the direct electrochemical oxidation of fuels, thus providing a strong economic incentive for the development of fuel cells.

If fuel cells are to have a significant practical impact, they must utilize some readily available fuel, such as methane or hydrogen. These fuels consist of substances in a reduced state and the cell reaction involves their oxidation by some readily available oxidant, such as oxygen. The reaction products are generally non-polluting substances, such as water and carbon dioxide.

Fuel cells based on the oxidation of hydrogen by oxygen have undoubtedly received the greatest attention. These cells have been used successfully in manned satellites both for the generation of electricity and the production of water. The virtually inexhaustible supply of hydrogen available from the decomposition of water raises the possibility that this substance will succeed fossil fuels as a major ingredient of our energy economy.

Figure 13.9 shows a schematic diagram of a $H_2$–$O_2$ fuel cell. Hydrogen is introduced at the anode and oxygen at the cathode; the electrolyte consists of KOH solution. The cell operation involves the following half-reactions

$$\text{anode:} \quad 2H_2(g) + 4OH^- \to 4H_2O + 4e^-$$
$$\text{cathode:} \quad \underline{O_2(g) + 2H_2O + 4e^- \to 4OH^-}$$
$$\text{cell:} \quad 2H_2(g) + O_2(g) \to 2H_2O(l)$$

The successful operation of this cell depends critically on the use of electrodes which catalyze the reaction and thereby permit the conversion of a significant

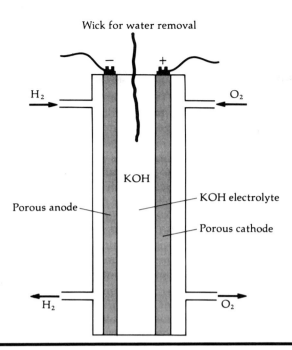

**FIGURE 13.9** A hydrogen–oxygen fuel cell.

fraction of the fuel as it flows through the cell. Porous nickel and nickel oxide have been used but there is room for much improvement.

## 13.8 Electrochemical Determination of pH

A number of electrodes whose EMF is a measure of the concentration of a specific ion have been developed. These ion-selective electrodes are widely used in analytical chemistry for the determination of the concentrations of various ions in clinical samples, drinking water samples, and so on.

One of the most widespread applications of ion-selective electrodes is in the determination of the concentration of hydrogen ions in solution, that is, in the measurement of pH. Consider the operation of the following cell:

Pt/H$_2$ (1 atm)/solution of unknown pH//KCl(saturated solution)
/Hg$_2$Cl$_2$(s)/Hg(l)

The electrode half-reactions are

anode: $\frac{1}{2}$H$_2$(g) $\rightarrow$ H$^+$ + $e^-$

cathode: $\frac{1}{2}$Hg$_2$Cl$_2$(s) + $e^-$ $\rightarrow$ Hg(l) + Cl$^-$

cell: $\frac{1}{2}$H$_2$(g) + $\frac{1}{2}$Hg$_2$Cl$_2$(s) $\rightarrow$ H$^+$ + Hg(l) + Cl$^-$

The anode consists of a hydrogen electrode immersed in a test solution of unknown pH. The cathode is known as a **calomel electrode** and the half-cell reaction is the reduction of insoluble Hg$_2$Cl$_2$ to the metal. The calomel electrode is a metal/insoluble metal salt electrode. Recall from Section 13.1c that the concentrations of all species involved in the half-reaction occurring at such an elec-

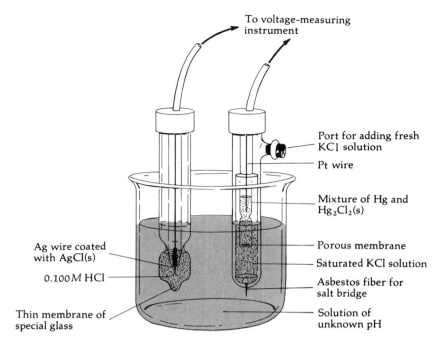

**FIGURE 13.10** The electrodes of a pH meter. The glass electrode (left) consists of a thin glass membrane enclosing an HCl solution. An electrical potential is generated across this membrane when the pH in the internal and external solutions differs. The inclusion of the Ag/AgCl(s) electrode improves the reproducibility of the EMF developed across the glass membrane. The reference electrode, the potential of which is independent of the solution pH, is a calomel electrode, which is described in the text. The two electrodes are connected to a voltmeter, calibrated in pH units. In practice, the pH meter must be calibrated before use by placing the electrodes in a solution of known pH.

trode remain constant. Consequently, the only variable concentration in the overall reaction is that of H$^+$ in the test solution. Its value, and therefore the pH of the solution, can be obtained from the observed cell potential by means of the Nernst equation:

$$\epsilon = \epsilon° - 0.0592 \log[H^+]$$

In practice, the measurement is usually performed with a pH meter (Figure 13.10). Because of its inconvenience, the hydrogen gas electrode is replaced in this instrument by the so-called glass electrode, which serves as the sensing probe of the pH meter. This somewhat complicated electrode is specific in sensing only H$^+$ ions and serves exactly the same function as the hydrogen gas electrode.

## 13.9 Corrosion

Corrosion can be defined as the process by which a metallic object is gradually destroyed by the environment. The rusting of iron, which is undoubtedly the most ubiquitous example of corrosion, results in worldwide annual losses in the billion dollar range. Most corrosion is the result of electrochemical processes. It should come as no surprise, then, that an understanding of the electrochemistry of corrosion has provided the basis for the development of procedures to minimize the extent of the damage.

The electrochemistry of the rusting of iron is a complex process. It depends on the fact that certain parts of a piece of iron are more susceptible to oxidation than others. This difference in susceptibility is due to the non-uniformity of the iron lattice, which actually consists of randomly oriented microcrystals. Atoms located at the edges of these crystals, or those in the vicinity of impurity atoms, are more readily oxidized than those in a more uniform microenvironment. The first step in the corrosion process consists of the following redox reaction

anodic oxidation: $\quad\quad\quad\quad\quad\quad Fe(s) \rightarrow Fe^{2+} + 2e^- \quad\quad \epsilon° = 0.44$ V

cathodic reduction: $\quad \frac{1}{2}O_2(g) + H_2O + 2e^- \rightarrow 2OH^- \quad\quad \epsilon° = 0.40$ V

or $\quad\quad\quad\quad\quad\quad\quad 2H^+ + 2e^- \rightarrow H_2(g) \quad\quad\quad\quad \epsilon° = 0.00$ V

Note that the reaction depends on the presence of both air and moisture and is further aided if the moisture is acidic. These conditions are readily met in most practical situations.

While these half-reactions are identical to those that might occur in a galvanic cell, the actual conditions are quite different, of course. There is no distinct difference between external and internal circuits; the electrons simply flow through the conducting metal. Furthermore, instead of electrodes, there are anodic and cathodic regions of microscopic size and separation. Small amounts of electrolytes dissolved in the moisture fill the role played by the salt bridge. Although the standard electrode potentials are listed here, the conditions are obviously not standard. Nonetheless, the cell potential is distinctly positive for the overall process.

Once the $Fe^{2+}$ ions are formed, they undergo a series of further reactions involving precipitation as the oxide or hydroxide and eventual oxidation by air to hydrated $Fe_2O_3$, which is the substance constituting rust. The oxide does not adhere particularly well to the metal and tends to flake off. This exposes a fresh layer of iron metal to air and moisture permitting the process to repeat itself. The end result is the destruction of the corroded object and the return of the iron atoms to the oxidized state in which they are normally found on Earth. The overall process is depicted schematically in Figure 13.11.

The electrochemical nature of corrosion suggests some methods for inhibiting or even preventing this process from occurring. Since corrosion requires air and moisture, it is helpful to keep these substances away from the surface of the metal. This can be done by applying a protective coating of paint. An even more effective method is to connect the iron to a more readily oxidized metal. A steel pipe, such as an oil pipe for instance, is connected electrically at periodic intervals to pieces of magnesium or some other metal more active than iron. Since magnesium is more readily oxidized than iron, it will act as the anode in the half-reaction

$$Mg \rightarrow Mg^{2+} + 2e^-$$

The electrons flow to the steel pipe, which acts as a cathode where $O_2$ or $H^+$ are reduced. The magnesium metal is gradually consumed in this process and must be replaced periodically. It is therefore known as a "sacrificial" anode. Since the magnesium causes the iron to act as a cathode, it is said to afford ca-

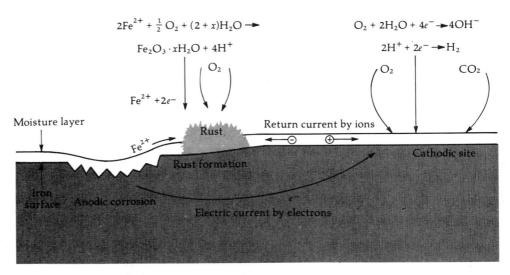

**FIGURE 13.11** Electrochemical corrosion of iron. Iron atoms located at an anodic corrosion site are oxidized to $Fe^{2+}$ ions. Migration of these ions to sites where they are exposed to the atmosphere leads to rust formation. Electrons flow to a cathodic region where reduction of $O_2$ or $H^+$ (acid moisture) occurs.

thodic protection to the pipe. The same principle is used when an iron object is plated with a coating of a more active metal. For example, galvanized iron has a thin coating of zinc. The zinc acts as the anode and cathodically protects the iron from rusting. Since zinc is in turn protected against excessive oxidation by an adherent layer of zinc oxide, it does not have to be replaced. This technique offers better protection than paint because the iron object does not have to be completely covered to be resistant to attack. In contrast, a painted iron object will begin to corrode as soon as even the smallest chip of paint has broken off. For expensive equipment subject to corrosion, e.g., filter presses used in pulp and paper mills, cathodic protection is provided by the use of an external power supply inserted between the steel press and an inert electrode. The external potential is applied so as to make the iron in the press the cathode of a cell, thereby inhibiting its oxidation.

The corrosion of iron may be compared with that of aluminum. The standard reduction potential of $Al^{3+}$ is considerably more negative than that of $Fe^{2+}$ indicating that aluminum should be much more reactive than iron. Yet aluminum is quite resistant to corrosion. This difference arises because aluminum exposed to the atmosphere forms a surface layer of $Al_2O_3$ which tightly adheres to the metal. This protective layer prevents oxygen from penetrating to the metal and effectively prevents corrosion. Many common aluminum products, e.g., kitchenware, are made of "anodized" aluminum. This material consists of aluminum which has been electrochemically protected by anodic oxidation. The resulting oxide layer is thicker and more uniform than that formed by exposure to the atmosphere and thus affords extra protection against corrosion.

## 13.10 Electrolysis and Its Applications

*a. Electrolytic Cells* We have seen that in a galvanic cell a spontaneous chemical reaction can be used to do electrical work. The opposite process, in which electrical work is used to reverse the direction of a spontaneous chemical reac-

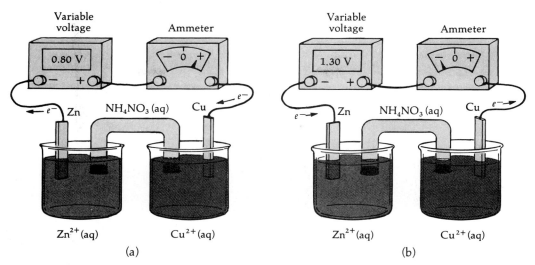

**FIGURE 13.12** Relationship between galvanic and electrolytic cells as demonstrated for the Daniell cell with an external variable voltage opposing the cell potential. (a) The external potential is smaller than the cell potential and electrons flow from the zinc to the copper electrode. (b) The external potential exceeds the cell potential and the direction of flow is reversed.

tion, is also possible. A cell in which a non-spontaneous redox reaction is made to occur by the action of an external potential is called an **electrolytic cell.** The process occurring in such a cell is known as **electrolysis.**

The relationship between an electrolytic cell and a galvanic cell is illustrated in Figure 13.12, which shows a Daniell cell at standard conditions. A variable voltage source has been inserted in the external circuit so as to oppose the cell potential, thereby impeding the normal flow of electrons. As long as the external voltage is less than 1.10 V, the Daniell cell continues to operate. When the external potential is just equal to 1.10 V, the flow of electrons stops. Increasing the external voltage further, reverses the direction of flow of the electrons and causes the reverse reaction to occur, i.e.,

$$Zn^{2+} + Cu \rightarrow Zn + Cu^{2+} \quad \text{if opposing external potential} > 1.10 \text{ V}$$

The cell is now operating in the electrolytic mode.

The recharging of a storage cell, such as the lead storage cell, provides a practical illustration of the relationship between galvanic and electrolytic cells. When operated in the galvanic, or discharge mode, the spontaneous reaction between lead and lead dioxide is used to generate electrical energy. A discharged cell can be recharged by operating it in the electrolytic mode—external electrical energy is used to convert lead sulfate back to lead and lead dioxide.

Electrolytic and galvanic cells share many common features and are collectively known as **electrochemical cells.** Both types of cells consist of essentially the same components: electrodes, electrolyte, external circuit. Oxidation occurs at the anode and reduction at the cathode in both types of cells. However, electrolytic and galvanic cells differ in a fundamental way: in a galvanic cell, the free energy of a spontaneous redox reaction can be used to perform electrical work; in an electrolytic cell, an external source provides the energy required to reverse the direction of a spontaneous reaction. This feature makes electrolysis useful for the production of unstable or highly reactive substances, as we shall see.

An apparatus for demonstrating a familiar electrolysis reaction, the electrolysis of water, is shown in Figure 13.13. An external source of dc (direct) current, such as a battery, provides a flow of electrons to the electrolytic cell. Electrons flow from the negative terminal of the battery towards the cathode and from the anode towards the positive terminal of the battery. The cell is filled with water to which an electrolyte such as $Na_2SO_4$ has been added to make the solution conducting. The particular half-reactions occurring in the electrolysis of water are

| anode: | $2H_2O \rightarrow O_2(g) + 4H^+ + 4e^-$ |
| cathode: | $4H_2O + 4e^- \rightarrow 2H_2(g) + 4OH^-$ |
| overall: | $6H_2O \rightarrow 2H_2(g) + O_2(g) + 4H^+ + 4OH^-$ |
| net: | $2H_2O \rightarrow 2H_2(g) + O_2(g)$ |

While the $H^+$ and $OH^-$ ions eventually recombine to form $H_2O$, as indicated in the equation for the net reaction, the solution in the vicinity of the electrodes remains distinctly acidic or basic while the current flows.

The quantitative aspects of electrolysis were first investigated in the nineteenth century by Michael Faraday (1791–1867). In modern terms, Faraday's

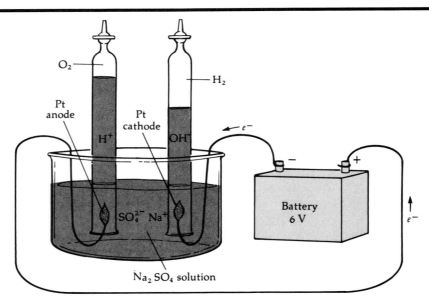

**FIGURE 13.13** The electrolysis of water. A 6 V battery provides sufficiently energetic electrons to reverse the spontaneous reaction [i.e., $2H_2(g) + O_2(g) \rightarrow 2H_2O(l)$]. The cell is filled with water containing a non-reactive electrolyte such as $Na_2SO_4$, and inert platinum electrodes can be used. Since the reagents in the solution do not react with each other, both electrodes can be immersed in the same solution. Note that the volume of $H_2(g)$ is twice as large as that of $O_2(g)$ as required by the reaction stoichiometry. A pH indicator shows that the solution in the anode compartment becomes acidic while that in the cathode compartment becomes basic, consistent with the half-reactions occurring at the electrodes.

laws state that the amount of substance produced in electrolysis is proportional to the quantity of electricity passing through the cell, 96,485 C liberating one equivalent of material (i.e., the amount of a substance that can combine with one mole of electrons). While Faraday's work preceded the discovery of the electron, it indicated that electrolysis caused the charge on an ion to change in multiples of a fundamental unit of charge, with 96,485 C corresponding to one mole of these units.

***b. Preparation of Active Metals*** Electrolysis can be used to form species in oxidation states in which they are not commonly found in nature. The production of the most active metals in the elemental state is an important application of electrolysis. Table 13.1 shows that the standard reduction potentials of the alkali elements are so negative that it is difficult to find substances that will reduce their ions to the metals. Moreover, such reduction cannot take place in aqueous solution since the alkali metals spontaneously react with water. However, these ions can be reduced with ease in electrolytic cells since the requisite potentials are readily obtainable. Thus, electrolytic reduction is used in the production of the alkali metals from their naturally occurring salts.

Figure 13.14 is a sketch of a Downs cell, used in the electrolytic production of sodium. The electrolyte consists of molten sodium chloride. Although the solid salt does not conduct electricity, the molten salt is a good conductor because the ions are free to move in the liquid state (Section 12.1). The follow-

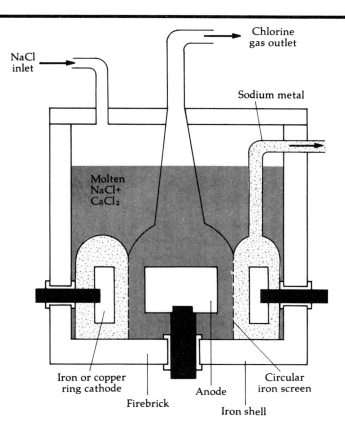

**FIGURE 13.14** The Downs cell for the production of sodium metal in the electrolysis of molten sodium chloride. The addition of $CaCl_2$ reduces the melting point of NaCl from 808°C, which is close to the boiling point of sodium, to about 600°C. The heat generated by the passage of the electric current is sufficient to keep the salt mixture molten.

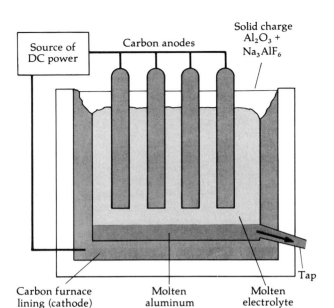

**FIGURE 13.15** Electrolytic cell for production of aluminum by the Hall process. The electrolyte consists of molten cryolite ($Na_3AlF_6$) at a temperature of 900–1000°C. Aluminum oxide, the substance to be reduced, readily dissolves in cryolite under these conditions. The liquid metal collects at the carbon cathode, which lines the bottom of the container. Oxygen is formed at the carbon anode.

ing reactions occur in the Downs cell:

cathode: $Na^+ + e^- \rightarrow Na(l)$

anode: $Cl^- \rightarrow \frac{1}{2}Cl_2(g) + e^-$

An external potential of 5 or 6 V is sufficient to cause the reaction to occur. The products must be kept apart in order to prevent the occurrence of the spontaneous reaction between sodium metal and chlorine gas to form sodium chloride.

Aluminum is another metal made by electrolytic reduction. The most important source of aluminum is the mineral bauxite, which contains a high concentration of hydrated aluminum oxide, $Al_2O_3 \cdot xH_2O$. After anhydrous $Al_2O_3$ is separated from bauxite by chemical means, the metal is obtained by an electrolytic method known in this country as the Hall process (Figure 13.15). Because of the widespread use of aluminum, large amounts of electricity are required for aluminum production.

**Example 13.10** How much aluminum metal is deposited in an electrolytic cell through which a current of $1.00 \times 10^5$ A is passed for exactly 1 hour at a potential of 5.00 V? Neglecting losses, how much energy is expended in this process?

This problem reduces to an ordinary stoichiometry problem once the number of electrons passing through the circuit is obtained from the given electrical data. The total amount of charge is given by Equation 13.2,

$$Q = It$$
$$= (1.00 \times 10^5 \text{ A})(1.00 \text{ h})(3600 \text{ s h}^{-1})(1 \text{ C/A s})$$
$$= 3.60 \times 10^8 \text{ C}$$

Since the charge of one mole of electrons is one faraday, the number of moles of electrons corresponding to the total charge is

$$n = \frac{Q}{F} = \frac{3.60 \times 10^8 \text{ C}}{96{,}485 \text{ C mol}^{-1}} = 3731 \text{ mol}$$

The half-reaction for the reduction of $Al^{3+}$ to Al is

$$Al^{3+} + 3e^- \rightarrow Al$$

Consequently, the mass of aluminum produced is

$$m_{Al} = (3731 \text{ mol } e^-)[(1 \text{ mol Al})/(3 \text{ mol } e^-)](26.98 \text{ g mol}^{-1})(10^{-3} \text{ kg g}^{-1})$$
$$= 33.6 \text{ kg}$$

The energy expended is equal to the electrical work done *on* the system,

$$E = w = QV$$
$$= (3.60 \times 10^8 \text{ C})(5.00 \text{ V})(1 \text{ J/C V})$$
$$= 1.80 \times 10^9 \text{ J}$$

Electrical energy is frequently expressed in kilowatt hours (kW h), where

$$1 \text{ J} = (1 \text{ W})(1 \text{ s})$$
$$(\text{joule}) = (\text{watt})(\text{second})$$

Making this conversion, we obtain

$$E = \frac{(1.80 \times 10^9 \text{ J})(1 \text{ W s/J})(10^{-3} \text{ kW W}^{-1})}{3.60 \times 10^3 \text{ s h}^{-1}}$$
$$= 500 \text{ kW h} \quad \blacksquare$$

***c. Electrolysis of Aqueous Solutions*** When aqueous solutions of salts are electrolyzed, the products can be either those of the electrolysis of the salt or those of the electrolysis of water. Consider, for example, the electrolysis of an aqueous solution of sodium chloride. The possible half-reactions and their standard potentials are

| | | | |
|---|---|---|---|
| anode: | $2Cl^- \rightarrow Cl_2(g) + 2e^-$ | | $\epsilon^\circ = -1.36$ V |
| | $H_2O \rightarrow \frac{1}{2}O_2(g) + 2H^+ + 2e^-$ | | $\epsilon^\circ = -1.23$ V |
| cathode: | $Na^+ + e^- \rightarrow Na$ | | $\epsilon^\circ = -2.71$ V |
| | $H_2O + e^- \rightarrow \frac{1}{2}H_2(g) + OH^-$ | | $\epsilon^\circ = -0.83$ V |

On the basis of these standard potentials, the electrolysis should liberate oxygen gas at the anode and hydrogen gas at the cathode. In actuality, the electrolysis products are $Cl_2(g)$ at the anode and $H_2(g)$ at the cathode. As may be

**TABLE 13.3  Electrolysis of Some Common Aqueous Solutions and Melts**

| Electrolyte | Anode product | Cathode product |
|---|---|---|
| HCl solution | $Cl_2$ | $H_2$ |
| $H_2SO_4$ solution | $O_2$ | $H_2$ |
| $Na_2SO_4$ solution | $O_2$ | $H_2$ |
| $CuSO_4$ solution | $O_2$ | Cu |
| NaCl solution | $Cl_2$ | $H_2$ |
| NaCl melt | $Cl_2$ | Na |
| $AgNO_3$ solution | $O_2$ | Ag |
| $AlCl_3$ solution | $Cl_2$ | $H_2$ |
| $AlCl_3$ melt | $Cl_2$ | Al |
| KI solution | $I_2$ | $H_2$ |

inferred from the corresponding half-reactions, the solution becomes enriched in sodium hydroxide—the electrolysis of concentrated NaCl solution is a common procedure for the production of NaOH.

The electrolysis of sodium chloride solution illustrates the fact that thermodynamics, i.e., $\epsilon°$ values, does not offer a completely reliable guide for the prediction of electrolysis products. An additional factor that must be considered is the rate at which a particular half-reaction occurs on the electrode surface. When two half-reactions are possible, the one with the more favorable half-cell potential does not necessarily occur if it is sufficiently slower than the other half-reaction. This behavior can be described quantitatively in terms of the **overpotential**—the difference between the actual external potential required for a particular electrolysis reaction and the thermodynamic cell potential. Table 13.3 lists the actual products of the electrolysis of some typical solutions and melts.

*d. Electrolytic Refining and Electroplating of Metals*  Electrolytic refining is a process used in the purification of certain metals produced by chemical reduction of their ores. Copper, for example, occurs naturally in the form of various compounds of copper(II). Copper ores are chemically reduced by carbon, yielding a product with impurities at the percent level. Figure 13.16 is a schematic diagram of a cell used to increase the purity of the metal to more than 99.9%. An impure slab of copper is made the anode and a thin sheet of pure copper serves as the cathode. The electrodes are immersed in copper sulfate solution. As the electrolysis proceeds, the copper anode is gradually consumed while pure copper plates out on the cathode. Under suitable conditions, metals more active than copper, such as iron and zinc, are also oxidized at the anode but are not reduced at the cathode, remaining instead in solution. On the other hand, impurities of less active metals, such as silver and gold, are not oxidized but collect below the anode as part of a muddy deposit. The recovery of these valuable impurities can often pay for the entire cost of purification.

A related application of electrolytic cells is the electroplating of noble metals. Gold and silver are often deposited in the form of a thin layer on jewelry or tableware made of a base metal. The object to be plated is made the cathode and the noble metal constitutes the anode. The electrolyte must contain the ion of the noble metal. Passage of current plates the noble metal onto the cathode, while the anode continuously replenishes the solution.

**FIGURE 13.16** Electrolytic purification of copper metal.

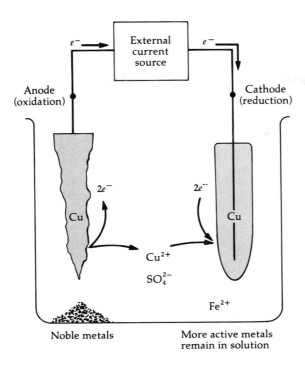

## Conclusion

Galvanic cells are devices in which the electrons transferred in a spontaneously occurring redox reaction are incorporated into an electric current which can be used to perform electrical work. Cell potentials provide a quantitative measure of the tendency of reactions to proceed as written and are intimately connected with the free energy change and the position of chemical equilibrium. The use of standard electrode potentials makes it possible to rank individual reagents in the EMF series according to their ability to act as oxidizing or reducing agents. The applications of electrochemistry are numerous and diverse. They include the development of batteries, methods for the production and purification of various metals, and techniques for the prevention of electrochemical corrosion.

## Problems

1. Sketch cells in which the following reactions can take place
   - (a) $Ni(s) + 2Ag^+ \rightarrow Ni^{2+} + 2Ag(s)$
   - (b) $H_2(g) + 2Ag^+ \rightarrow 2H^+ + 2Ag(s)$
   - (c) $6Fe^{2+} + Cr_2O_7^{2-} + 14H^+ \rightarrow 6Fe^{3+} + 2Cr^{3+} + 7H_2O$
   - (d) $Ag^+ + I^- \rightarrow AgI(s)$

   Summarize each diagram by means of the cell line notation.

2. Write the half-reactions and cell reaction which occur in the following cells
   - (a) $Pt/Fe^{2+}, Fe^{3+}//Sn^{4+}, Sn^{2+}/Pt$
   - (b) $Ag/Ag^+//Mn^{2+}, H^+/MnO_2(s)/Pt$
   - (c) $Pt/H_2(g)/H^+//Cl^-/AgCl(s)/Ag$

(d) $Zn/Zn(OH)_4^{2-}$, $OH^-//MnO_4^-$, $OH^-/MnO_2/Pt$

3. Devise cells in which the following reactions occur. Express your answers using the cell line notation.
   (a) $H_2(g) + \frac{1}{2}O_2(g) \rightarrow H_2O(l)$
   (b) $Zn(s) + Ni(NO_3)_2(aq) \rightarrow Zn(NO_3)_2(aq) + Ni(s)$
   (c) $AgCl(s) + \frac{1}{2}H_2(g) \rightarrow HCl(aq) + Ag(s)$
   (d) $3Ag(s) + MnO_4^- + 4H^+ \rightarrow 3Ag^+ + MnO_2(s) + 2H_2O$

4. The EMF for the cell
   $$Cd/Cd(NO_3)_2(0.1\ M)//AgNO_3(0.01\ M)/Ag$$
   is 1.12 V at 25°C. (a) Write the half-reactions and the cell reaction. (b) What is the standard cell EMF?

5. How do each of the following changes affect the EMF of the Daniell cell? (a) Excess ammonia is added to the cathode compartment. (b) $H_2S$ is bubbled into the anode compartment. (c) $ZnCl_2(s)$ is added to the anode compartment. (d) $NaCN(s)$ is added to the cathode compartment.

6. For the cell
   $$Cd(s)/Cd^{2+}//I^-/I_2(s)/Pt$$
   (a) write the electrode and cell reactions. (b) Using the data in Appendix 5, determine $\epsilon°$, $\Delta G°$, and $K$ per mole of cadmium. (c) What are the effects on $\epsilon°$, $\Delta G°$, and $K$ of (1) doubling the amount of material undergoing the cell reaction and (2) reversing the cell reaction?

7. What is the pressure of hydrogen necessary to maintain equilibrium with respect to the following reaction at 25°C if the concentration of $Sn^{2+}$ is 0.020 mol $L^{-1}$ and the solution is buffered at pH 1.50:
   $$Sn(s) + 2H^+(aq) \rightarrow Sn^{2+} + H_2(g)$$
   Use the data in Appendix 5.

8. Calculate $\epsilon°$, $\epsilon$, and $K$ for the cell
   $$Cu/Cu^{2+}(0.15\ M)//Ag^+(0.050M)/Ag$$
   at 25°C. Use data in Appendix 5.

9. What is the EMF of a cell consisting of a cadmium electrode in a 0.250 M solution of $Cd^{2+}$ and a nickel electrode in a 0.100 M solution of $Ni^{2+}$? Which electrode is the anode?

10. The EMF of a cell based on the reaction $Fe(s) + 2H^+ \rightarrow Fe^{2+} + H_2(g)$ is 0.38 V. If the pressure of $H_2$ is 1 atm and the pH of the hydrogen electrode solution is 1.0, what is the concentration of $Fe^{2+}$?

11. Which of the following reactions tend to occur to a significant extent in aqueous solution? Assume that the initial concentrations are 1 M. Calculate the equilibrium constant of each reaction.
    (a) $Pb(s) + Ni^{2+} \rightarrow Pb^{2+} + Ni(s)$
    (b) $2Fe^{2+} + Br_2(l) \rightarrow 2Fe^{3+} + 2Br^-$
    (c) $3Hg_2^{2+} + Cr_2O_7^{2-} + 14H^+ \rightarrow 6Hg^{2+} + 2Cr^{3+} + 7H_2O$

12. The EMF of the cell
    $$Pb/Pb^{2+}(0.10\ M)//VO_2^+(0.25\ M),$$
    $$V^{3+}(0.010\ M),\ H^+(0.25\ M)/Pt$$
    is 0.49 V. (a) Write equations for the half-reactions and the cell reaction. (b) What is the value of $\epsilon°$ for the $VO_2^+/V^{3+}$ couple? (c) What is the equilibrium constant of the cell reaction?

13. An electrochemical cell consists of a cadmium electrode immersed in 0.20 M $Cd(NO_3)_2$ and a silver electrode immersed in 0.10 M $AgNO_3$. (a) Determine the initial cell potential. (b) Write the overall reaction that occurs when the cell operates in the galvanic mode. (c) Identify the anode and the cathode and state in which direction the electrons flow in the external circuit. (d) If a steady current of 0.35 A is supplied by the cell for 2.5 h, what weight of cadmium and of silver will be deposited or dissolved? (e) What will be the approximate final concentrations of $Ag^+$ and $Cd^{2+}$ if the cell is operated until equilibrium is reached? Assume that the volumes of the solutions in the two half-cells are equal.

14. Indicate which of the following oxidizing agents become stronger, which become weaker, and which remain unchanged with increasing $H^+$ concentration. (a) $F_2$, (b) $H_2O_2$, and (c) $Cr_2O_7^{2-}$.

15. The magnitudes of the standard electrode potentials of two metals A and B were determined:
    (1) $A^{2+} + 2e^- \rightarrow A$  $|\epsilon°| = 0.40$ V
    (2) $B^{2+} + 2e^- \rightarrow B$  $|\epsilon°| = 0.70$ V

    Note that the signs of the $\epsilon°$ values are unknown. When half-cells (1) and (2) are connected, electrons flow from A to B in the external circuit. When half-cell (1) is connected to a hydrogen gas electrode half-cell at standard conditions, electrons flow from A to the hydrogen electrode. (a) What are the signs of the $\epsilon°$ values of half-reactions (1) and (2)? (b) What reaction occurs in the above cell and what is its standard EMF?

16. A cell is made up of the following two couples: $Ni^{2+}(0.10\ M)/Ni$ and $Sn^{2+}(1.0\ M)/Sn$. (a) What reaction will occur spontaneously? (b) Which of the two metals is the anode and which the cathode? (c) What is the value of $\epsilon°$ for the cell? (d) What is

the value of $\epsilon$ for the given concentrations? (e) For what concentration of $Ni^{2+}$ will there be no flow of current, i.e., $\epsilon = 0$, assuming $[Sn^{2+}] = 1.0\ M$?

17. Is hydrogen peroxide stable to disproportionation in acidic solution? Answer this question by evaluating $\epsilon°$ for the reaction

$$2H_2O_2 \rightarrow O_2 + 2H_2O$$

by using Appendix 5.

18. Arrange the oxidant members of the following redox couples in order of increasing strength; arrange the reductant memebers of the same couples in order of increasing strength: $MnO_4^-/Mn^{2+}$ (acid), $MnO_4^-/MnO_4^{2-}$, $Br_2/Br^-$, $Mn^{2+}/Mn$, $Fe^{3+}/Fe^{2+}$, $Mg^{2+}/Mg$, $Ni^{2+}/Ni$.

19. A galvanic cell is made from the following two couples: $Hg_2^{2+}/Hg$ and $Cu^{2+}/Cu$. (a) Describe the cell in line notation. (b) Write the overall reaction that occurs spontaneously. (c) Determine the standard cell potential.

20. Using data in Appendix 5, calculate the value of $K_{sp}$ for AgCl at 25°C.

21. Using the data in Appendix 5 calculate the equilibrium constant for the reaction

$$Zn^{2+} + 4OH^- \rightarrow Zn(OH)_4^{2-}$$

22. Determine the standard electrode potential for (a) $MnO_4^-/Mn(OH)_2$ in basic solution and (b) $MnO_4^-/Mn$ in basic solution.

23. Using data in Appendix 5 determine $K_{sp}$ for $Mg(OH)_2$.

24. Calculate the standard electrode potential for the $Hg^{2+}/Hg$ couple.

25. Two $H^+/H_2$ half-cells are connected. The pH in one of the half-cells is 1.5 and the pH in the other half-cell is unknown. The measured cell potential is 0.11 V, with the electrons flowing in the external circuit towards the half-cell of unknown pH. What is the pH of this half-cell?

26. The solubility product constant of $FeCO_3$ at 25°C is $2.1 \times 10^{-11}$. Using this information along with Appendix 5 calculate the standard electrode potential for the half-reaction $FeCO_3(s) + 2e^- \rightarrow Fe(s) + CO_3^{2-}$.

27. The potential of the cell

$$Cu(s)/CuCl(s)/KCl(0.200\ M)/Cl_2(1\ atm)/Pt$$

is 1.23 V. (a) Write the half-reactions and overall reaction which take place. (b) Determine $K_{sp}$ for CuCl. (c) What is the concentration of $Cu^+$ at the anode?

28. For the cell

$$Al(s)/Al^{3+}//Fe^{2+},\ Fe^{3+}/Pt$$

(a) write the cell reaction, (b) determine the value of $\epsilon°$ from data in Appendix 5, and (c) find $\Delta G°$ and K.

29. For the galvanic cell

$$Ag/Ag^+(0.050\ M)//Ag^+(1.5\ M)/Ag$$

(a) write the reactions occurring at the anode, cathode, and overall; and (b) determine the values of $\epsilon°$ and $\epsilon$.

30. A copper cathode is immersed in 1.00 M $CuNO_3$. This half-cell is connected to a hydrogen half-cell in which the pressure of $H_2$ is 1 atm and the $H^+$ concentration is unknown. If the cell potential is 0.50 V, what is the pH in the hydrogen half-cell?

31. Calculate the standard free energies of formation of $Cl^-$, $Mg^{2+}$, and $Al^{3+}$ in aqueous solution at 25°C from the tabulated standard electrode potentials.

32. How many faradays of charge will a lead storage cell supply when the concentration of the sulfuric acid electrolyte decreases from 34.5 wt.% ($\rho = 1.255$ g cm$^{-3}$) to 28.5 wt.% ($\rho = 1.206$ g cm$^{-3}$), where the electrolyte volume is 750 mL? How long does it take to recharge the battery back to its initial state with a current of 10.0 A?

33. In the electrolysis of a solution containing $Zn^{2+}$, 14.72 g of zinc is deposited when a current of 0.530 A is used. How long did the electrolysis take?

34. Three electrolytic cells containing solutions of $AgNO_3$, $Pb(NO_3)_2$, and $Au(NO_3)_3$, respectively, were connected in series. A current of 2.50 A was passed through the cells until 2.85 g of silver was deposited in the first cell. What weights of lead and gold were deposited? How long did the current flow?

35. An aqueous solution of gold(III) is electrolyzed and gold metal is deposited at the cathode while oxygen gas evolves at the anode. (a) What volume of oxygen at STP is liberated when 450 mg of gold is deposited? (b) What current was used if the electrolysis lasted 15.0 minutes? Assume that the solubility of $O_2$ in the solution is negligibly small.

36. The electrolysis of an aqueous solution of calcium iodide results in the formation of hydrogen (from the decomposition of water) and iodine at the electrodes. When 800 mL of a 0.150 M solution of $CaI_2$ was electrolyzed, the volume of hydrogen gas collected over water at 20°C and a pressure of 752 torr was 890 mL. The vapor pressure of water at 20°C is 18 torr. (a) How much charge, expressed in fara-

days, was delivered to the cell? (b) If a current of 6.00 A was used, how long did the electrolysis take? (c) What were the final concentrations of $Ca^{2+}$, $I^-$, and $OH^-$ in the solution? Assume that the volume remains unchanged and that there is no reaction between these ions.

37. The products of the electrolysis of a silver nitrate solution are listed in Table 13.3. (a) Write the half-reactions which occur at the anode and cathode. (b) What mass of silver can be deposited in exactly 1 hour of operation with a current of 3.00 A? (c) If the external potential is 6.00 V, how much electrical energy is required? Express your answer in joules and in kilowatt hours.

# 14

# Kinetics—The Rates and Mechanisms of Chemical Reactions

Two central aspects are part of the study of chemical reactions. The first deals with the spontaneity of chemical reactions and with the position of chemical equilibrium. This is the province of thermodynamics and has been explored in the last four chapters. Thermodynamics relates the equilibrium constant of a reaction to the standard free energy change and, in turn, to the thermal properties of reactants and products that determine the reaction enthalpy and entropy.

The second principal aspect of chemical reactions is the rate at which they occur. Some reactions occur nearly instantaneously, on a time scale of nanoseconds or less; others occur so slowly that the change is imperceptible. For example, when heated, magnesium rapidly burns in air to form the oxide, while iron, on the other hand, slowly rusts when exposed to air. The rates of chemical reactions do not depend on the values of the equilibrium constants and cannot be predicted by thermodynamics. The reason is that the rate of a reaction depends on the path followed by molecules of reactants in their conversion to molecules of products. Since thermodynamics deals primarily with state functions, the values of which are independent of path, it cannot be used to predict the behavior of path-dependent properties.

The branch of chemistry that deals with the rates of chemical reactions and with the examination of how they proceed on the molecular level is **kinetics.** This subject is important for both fundamental and practical reasons. On the fundamental level, we can learn exactly how a reaction occurs. For example, the reaction between $H_2$ and $I_2$ to form HI is believed to involve the splitting of an $I_2$ molecule into I atoms, followed by the attack on a $H_2$ molecule by an I atom, and so forth. The behavior of these species at the moment of reaction and the dependence of this behavior on properties such as energy and orientation, provides a detailed picture of the way in which chemical reactions occur.

On the more practical level, kinetics is important because it allows us to predict how quickly a particular reaction will occur. Moreover, the rate will, in general, be dependent on such factors as the temperature, concentrations, pressures, and the presence of a catalyst. Understanding how these factors affect the rate of a reaction frequently makes it possible to optimize the conditions so that the maximum yield of a desired product is obtained in the shortest possible time. The economic impact on industrial scale chemistry can be very great.

# Empirical Rate Laws

## 14.1 An Overview of Kinetics

Kinetics offers an excellent illustration of how chemistry progresses from experimental observations, to phenomenological models, to basic theories. In turn, the validity of these theories can be confirmed on the basis of their experimental predictions. This continuous interplay between experiment and theory provides the means by which scientific understanding of the phenomena of interest can be increased and refined.

Experiments in kinetics measure the rates at which chemical reactions occur, as given by the change in concentration of a reactant or product per unit time. This information is reduced to the form of a **rate law,** a statement of the dependence of the rate of a reaction on the concentrations of the species participating in the reaction.

Once the rate law has been established, the next goal of kinetics is the determination of the reaction mechanism. The **reaction mechanism** is a statement of the sequence of elementary reactions by which the reactants are converted to products. An **elementary reaction** is one that occurs in a single step—the equation for such a reaction represents the actual species undergoing that particular reaction. For example, the mechanism for the overall reaction represented by the chemical equation

$$2A + B \rightarrow C + D$$

might consist of the following two elementary reactions:

(1) $\quad A + B \rightarrow E$

(2) $\quad \underline{E + A \rightarrow C + D}$

$\quad\quad\quad 2A + B \rightarrow C + D$

In reaction (1) molecule A combines with molecule B to form molecule E. This species is a reaction **intermediate**—it does not appear in the equation for the overall reaction. Thus, in reaction (2), E reacts with A to form C and D. Note that the mechanism cannot be inferred from the chemical equation for the reaction since the intermediate E, which plays a vital role in the second step, does not appear in the overall equation.

The determination of a reaction mechanism is a difficult task. In many instances a reaction proceeds via some unstable intermediate which cannot be isolated or detected because of its low concentration. Specific intermediates are frequently postulated on the basis of chemical intuition and by drawing on analogies to similar reactions. The rate law provides an important constraint on possible mechanisms since, as we shall see, not all of the likely mechanisms are necessarily consistent with a particular rate law.

Since elementary reactions procede in a single step, they are most amenable to theoretical treatment. The theory of molecular collisions in gases (Section 4.5) can be used to determine the rate at which molecules collide. Reaction rates are usually much slower than collision rates, but increase exponentially with temperature, as we shall see. This increase parallels that in the number of very energetic molecules in a gas (Section 4.6), indicating that only a fraction of the colliding molecules have enough energy to react. This result suggests that there is some sort of "barrier" that inhibits the occurrence of a reaction. The theory of elementary reactions focuses on the details of how the reacting molecules approach each other, how existing bonds break, and how new bonds are formed as the reactant molecules pass over the barrier and become product molecules.

## 14.2 The Rate Law

*a. The Rate of a Reaction*   Consider a reaction such as

$$2NO(g) + O_2(g) \rightarrow 2NO_2(g)$$

Figure 14.1 shows the variation of the concentrations of reactants and product

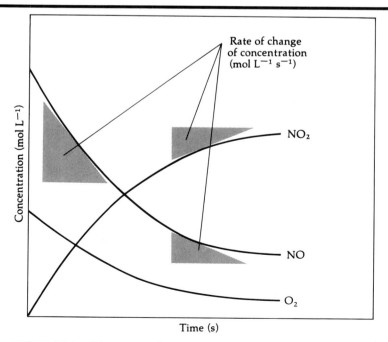

**FIGURE 14.1** Changes with time in the concentrations of the reactants and products of the reaction $2NO + O_2 \rightarrow 2NO_2$. The rates of change of the concentrations are given by the slopes of the tangents to the concentration curves.

with time. The concentrations of the reactants decrease as equilibrium is approached. In any given time interval the concentration of NO decreases by twice as much as that of $O_2$ because of the reaction stoichiometry. In contrast, the concentration of the $NO_2$ product increases with time towards its equilibrium value. Once again, because of the reaction stoichiometry, the change in its concentration in a given time interval is equal in magnitude but opposite in sign to that of NO.

The rates at which the concentrations change are represented graphically by the slopes of tangents to the concentration curves in Figure 14.1. As illustrated for NO, the slopes of these lines decrease with time, indicating that the rates of change of the concentrations gradually decrease—the reaction occurs rapidly at first and then, as equilibrium is approached, more slowly.

The rate of change of a concentration can be represented mathematically by its derivative with respect to time, $d[NO]/dt$, $d[NO_2]/dt$, and so on. These derivatives are related to each other by the reaction stoichiometry. We have noted that in a given time interval the NO concentration decreases by twice as much as the $O_2$ concentration. This relationship holds true even for an infinitesimally short time interval and consequently applies to these derivatives,

$$-\frac{1}{2}\frac{d[NO]}{dt} = -\frac{d[O_2]}{dt} = \frac{1}{2}\frac{d[NO_2]}{dt}$$

Note that the derivative of the concentration of $NO_2$, the reaction product, has the opposite sign of those of the other derivatives, which refer to reactants. Since the concentrations of reactants decrease with time while that of product increases, the rate of change of the former is negative while that of the latter is positive.

The rate of the reaction can be expressed in terms of the rate at which the concentration of any of the reactants or products changes. It is customary to equate the **reaction rate** to the rate of change of any of the product concentrations or to the negative of that of any of the reactant concentrations, each derivative being divided by the stoichiometric coefficient of the species in question. This procedure ensures that the same positive value is obtained for the rate of a given reaction regardless of which species is used to determine this rate. Thus the rate $R$ of the preceding reaction can be expressed as

$$R = -\frac{d[O_2]}{dt} = -\frac{1}{2}\frac{d[NO]}{dt} = \frac{1}{2}\frac{d[NO_2]}{dt}$$

Note that the units of $R$ are those of concentration/time, mol $L^{-1}$ $s^{-1}$.

Let us now consider the general reaction

$$aA + bB \rightarrow cC + dD$$

Drawing on the preceding example we can write the following relation between the reaction rate and the derivatives of the concentrations with respect to time:

$$R = -\frac{1}{a}\frac{d[A]}{dt} = -\frac{1}{b}\frac{d[B]}{dt} = \frac{1}{c}\frac{d[C]}{dt} = \frac{1}{d}\frac{d[D]}{dt} \tag{14.1}$$

Strictly speaking, the reaction rate defined in Equation 14.1 is the *net* reaction rate, that is, the rate of the forward reaction less that of the reverse reaction. We know from Section 7.1 that the two rates approach each other as the equilibrium point is approached. However, when the reaction is first initiated, the concentrations of the products are small compared to those of the reactants and the rate of the reverse reaction can be neglected. This is also the case when the equilibrium point lies so far to the right that the reverse reaction is unimportant. In this section we shall neglect the effect of the reverse reaction on the reaction rate.

**b. Types of Rate Laws** The rate of a particular reaction depends on a number of variables. These generally include the concentrations of at least some of the reactants, the temperature, and, for reactions in solution, the identity of the solvent. The first goal of kinetics is to determine the dependence of the rate of a reaction on the concentrations. This relation is known as the **rate law**. For the general reaction

$$a\text{A} + b\text{B} \rightarrow \text{products} \tag{14.2}$$

the rate law can often be written in the form

$$R = k[\text{A}]^n[\text{B}]^m \tag{14.3}$$

where $n$ and $m$ are either small integers (e.g., 1, 2) or small fractions (e.g., $\frac{1}{2}$, $\frac{3}{2}$). We have deliberately used different symbols for the exponents and the stoichiometric coefficients to underscore the fact that these two sets of numbers need bear no relation to each other.

The exponents in a rate law having the form of Equation 14.3 are called the **reaction orders**. Reaction 14.2 is said to have order $n$ with respect to reactant A and order $m$ with respect to B. The sum of $n$ and $m$ gives the overall reaction order. The proportionality constant $k$ between the rate and the concentrations is called the **rate constant**. The rate constant depends on the particular reaction, on the temperature, and on the identity and concentration of solvent and/or catalyst, if present. It does *not* depend on the concentrations of the reacting species. From the form of Equation 14.3, it can be seen that the units of $k$ must be $(\text{mol L}^{-1})^{(1-n-m)}\,\text{s}^{-1}$. Note that the rate constant is numerically equal to the reaction rate when the reactants are present at unit concentration.

Let us illustrate these concepts with some specific examples:

The decomposition of dinitrogen pentoxide,

$$\text{N}_2\text{O}_5(g) \rightarrow 2\text{NO}_2(g) + \tfrac{1}{2}\text{O}_2(g)$$

is experimentally found to obey the rate law

$$R = -\frac{d[\text{N}_2\text{O}_5]}{dt} = k[\text{N}_2\text{O}_5] \tag{14.4}$$

The reaction is first order, and the units of $k$ are $\text{s}^{-1}$.

The reaction between hydrogen and iodine,

$$\text{H}_2(g) + \text{I}_2(g) \rightarrow 2\text{HI}(g)$$

follows the rate law

$$R = k[\text{H}_2][\text{I}_2] \tag{14.5}$$

The reaction is first order with respect to hydrogen and iodine and second order overall. The units of $k$ are $(\text{mol L}^{-1})^{-1}\,\text{s}^{-1}$.

The reaction of bromate and bromide ions in acidic solution,

$$BrO_3^- + 5Br^- + 6H^+ \rightarrow 3Br_2(aq) + 3H_2O$$

obeys the rate law

$$R = -\frac{d[BrO_3^-]}{dt} = k[BrO_3^-][Br^-][H^+]^2 \qquad (14.6)$$

The reaction is first order with respect to $BrO_3^-$ and $Br^-$, second order with respect to $H^+$, and fourth order overall. The units of $k$ are $(mol\ L^{-1})^{-3}\ s^{-1}$.

The gas-phase decomposition of acetaldehyde at elevated temperatures,

$$CH_3CHO(g) \rightarrow CH_4(g) + CO(g)$$

follows the rate law

$$R = k[CH_3CHO]^{3/2} \qquad (14.7)$$

and is of order $\frac{3}{2}$. The units of $k$ are $(mol\ L^{-1})^{-1/2}\ s^{-1}$.

In the presence of a tungsten catalyst, the decomposition of ammonia to the elements

$$2NH_3(g) \xrightarrow{W} N_2(g) + 3H_2(g)$$

obeys the rate law

$$R = k \qquad (14.8)$$

The rate is independent of concentration and the reaction is zeroth order. The units of $k$ are $mol\ L^{-1}\ s^{-1}$. The rate of a zeroth order reaction is determined by factors other than the concentrations of the reactants or products (e.g., by the catalyst concentration).

Observe that in some of these examples the exponents of the concentrations are equal to the stoichiometric coefficients; in others they are not. *The particular form of the rate law depends on the reaction mechanism and must be determined experimentally; it cannot be inferred from the balanced equation.*

Not all reactions have rate laws that can be put in the simple form of Equation 14.3. A classical example of a reaction with a very complicated rate law is that between hydrogen and bromine,

$$H_2(g) + Br_2(g) \rightarrow 2HBr(g)$$

The rate of this reaction depends on the concentrations in the following manner:

$$R = \frac{k[H_2][Br_2]^{1/2}}{1 + k'[HBr][Br_2]^{-1}} \qquad (14.9)$$

When the rate law deviates from the form of Equation 14.3 in such a drastic manner, it is not possible to assign an overall order to the reaction. It is worth noting that two chemically similar reactions, that between hydrogen and either bromine (Equation 14.9) or iodine (Equation 14.5), have such vastly different rate laws. Although the overall reactions are chemically similar, they apparently procede via very different mechanisms.

---

**Example 14.1** The reaction

$$2HCrO_4^- + 3HSO_3^- + 5H^+ \rightarrow 2Cr^{3+} + 3SO_4^{2-} + 5H_2O$$

follows the rate law

$$R = k[\text{HCrO}_4^-][\text{HSO}_3^-]^2[\text{H}^+]$$

What happens to the rate if, in separate experiments, (a) the concentration of $\text{HSO}_3^-$ is doubled; (b) the pH is decreased by one unit; (c) the solution is diluted to twice its volume; and (d) the solution is diluted to twice its volume but the pH is kept constant by means of a buffer.

(a) Since $R \propto [\text{HSO}_3^-]^2$, doubling the concentration of hydrogen sulfite ion increases the rate by a factor of four.

(b) A decrease of one pH unit corresponds to a factor-of-ten increase in $[\text{H}^+]$; therefore the rate increases by a factor of ten.

(c) Doubling the volume corresponds to halving the concentrations. Since the reaction is fourth order overall, the rate is changed by a factor of $(\frac{1}{2})^4$, that is, it is reduced by a factor of sixteen.

(d) The $\text{HCrO}_4^-$ and $\text{HSO}_3^-$ concentrations are reduced by a factor of two but the $\text{H}^+$ concentration remains unchanged. The reaction rate is in this instance reduced by a factor of eight. ■

*c. The Initial Rate Method for the Determination of the Rate Law*  The particular form of the rate law must be determined experimentally for a given reaction. Example 14.1 suggests one possible way of performing such a determination: a series of mixtures of reactants A and B is prepared in which the concentration of B is kept constant while that of A is varied. For each mixture, the initial rate of the reaction (i.e., the rate measured when the reactants are first mixed) is measured by determining the change in the concentration of one of the reactants or products occuring in a short time interval after the reactants are mixed. The dependence of the initial rate on the initial concentration of A determines the reaction order with respect to A as well as the rate constant. Similarly, if a series of mixtures of A and B is prepared in which the concentration of B is varied while that of A is kept constant, the variation of the initial rate with the initial concentration of B determines the order with respect to B and also yields the value of the rate constant. This method of determining the rate law is known as the **initial rate method**. As already noted in Figure 14.1, the initial rate of a reaction is the largest rate for a particular experiment and can be measured most accurately.

To see how the method works, we suppose that the rate law can be written in the form of Equation 14.3, $R = k[\text{A}]^n[\text{B}]^m$. Let the initial concentrations be $[\text{A}]_0$ and $[\text{B}]_0$. The initial reaction rate then is

$$R_{\text{in}} = k[\text{A}]_0^n[\text{B}]_0^m \tag{14.10}$$

Taking the logarithm of both sides, we get

$$\log R_{\text{in}} = \log k + n \log[\text{A}]_0 + m \log[\text{B}]_0 \tag{14.11}$$

When the initial concentration of B remains constant while that of A is varied, a plot of $\log R_{\text{in}}$ versus $\log[\text{A}]_0$ gives a straight line with slope $n$ and an intercept which yields $k$. A similar plot versus $\log[\text{B}]_0$ for constant $[\text{A}]_0$ determines the order with respect to B. The determination by this method of the rate law for the combination of iodine atoms in the presence of argon is illustrated in

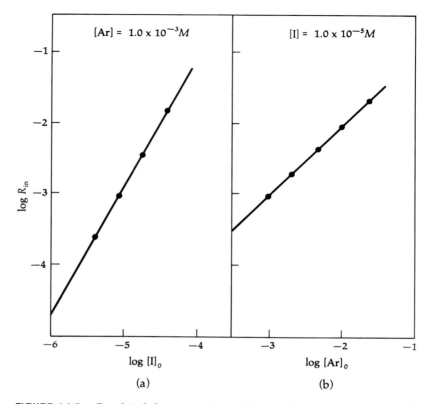

**FIGURE 14.2** Graphical determination of the rate law for the reaction $2I(g) + Ar(g) \rightarrow I_2(g) + Ar(g)$ by the initial rate method. (a) Dependence of the reaction rate on the initial concentration of $I(g)$ for a constant concentration of Ar. (b) Dependence of the reaction rate on the initial concentration of $Ar(g)$ for a constant concentration of $I(g)$. The slope of the line in (a) is 2 and that of the line in (b) is 1; therefore, the rate law is $R = k[I]^2[Ar]$. The value of $k$ is obtained from the intercept of either line.

Figure 14.2. Equation 14.10 can also be solved algebraically for the values of $n$ and $k$ and those of $m$ and $k$, as illustrated in the following example.

**Example 14.2** Hypophosphite ion ($H_2PO_2^-$) decomposes in alkaline solution as a result of the following reaction:

$$H_2PO_2^- + OH^- \rightarrow HPO_3^{2-} + H_2(g)$$

The kinetics of this reaction were studied at 100°C by use of the initial rate method. The following data were obtained on the concentration of hydrogen phosphite ion ($HPO_3^{2-}$) formed during 30 s following mixing of the reactants:

| Initial concentration | | |
|---|---|---|
| $[H_2PO_2^-]_0$ (mol L$^{-1}$) | $[OH^-]_0$ (mol L$^{-1}$) | $[HPO_3^{2-}]$ (mol L$^{-1}$) |
| 0.10 | 1.0 | $1.6 \times 10^{-5}$ |
| 0.30 | 1.0 | $4.8 \times 10^{-5}$ |
| 0.10 | 2.0 | $6.4 \times 10^{-5}$ |

Determine the rate law. What fraction of the hypophosphite ion present reacts in the first 30 s in a 1.0 M OH⁻ solution?

Because of the small number of experiments, it is more convenient to obtain an algebraic rather than a graphical solution. In the first two experiments, the OH⁻ concentration is kept constant; these data can therefore be used to establish the order with respect to $H_2PO_2^-$. We see that a threefold increase in $[H_2PO_2^-]$ (from 0.10 to 0.30 mol L⁻¹) changes the product concentration by a factor of

$$\frac{4.8 \times 10^{-5} \text{ mol L}^{-1}}{1.6 \times 10^{-5} \text{ mol L}^{-1}} = 3.0$$

The reaction rate is therefore proportional to $[H_2PO_2^-]$ and the reaction is first order in this reactant.

The first and third experiments similarly allow the order with respect to OH⁻ to be determined. Doubling the OH⁻ concentration (from 1.0 to 2.0 M) changes the concentration of product formed by a factor of

$$\frac{6.4 \times 10^{-5} \text{ mol L}^{-1}}{1.6 \times 10^{-5} \text{ mol L}^{-1}} = 4.0$$

The reaction rate is therefore proportional to $[OH^-]^2$. The rate law for this reaction is

$$R = k[H_2PO_2^-][OH^-]^2$$

Equivalently, we can use Equation 14.11 explicitly to obtain

$$m = \frac{\log (R_2/R_1)}{\log (C_{H_2PO_2^-,2}/C_{H_2PO_2^-,1})} = \frac{\log (4.8 \times 10^{-5}/1.6 \times 10^{-5})}{\log (0.30/0.10)} = 1$$

$$n = \frac{\log (R_3/R_1)}{\log (C_{OH^-,3}/C_{OH^-,1})} = \frac{\log (6.4 \times 10^{-5}/1.6 \times 10^{-5})}{\log (2.0/1.0)} = 2$$

where $m$ and $n$ are the orders with respect to $H_2PO_2^-$ and OH⁻, respectively.

The value of the rate constant can be determined by substituting the data for $HPO_3^{2-}$ formation into the rate law. (According to Equation 14.1, $R$ is proportional to the rate of product formation.) Using the results of the first experiment we obtain for the reaction rate

$$R = \frac{d[HPO_3^{2-}]}{dt} \approx \frac{\Delta[HPO_3^{2-}]}{\Delta t} = \frac{1.6 \times 10^{-5} \text{ mol L}^{-1}}{30 \text{ s}} = 5.33 \times 10^{-7} \text{ mol L}^{-1} \text{ s}^{-1}$$

Substitution of this result into the rate law yields the value of $k$,

$$5.33 \times 10^{-7} \text{ mol L}^{-1} \text{ s}^{-1} = k(0.10 \text{ mol L}^{-1})(1.0 \text{ mol L}^{-1})^2$$

$$k = 5.3 \times 10^{-6} \text{ (mol L}^{-1})^{-2} \text{ s}^{-1}$$

Because of the reaction stoichiometry, the decrease in the $H_2PO_2^-$ concentration is equal to the concentration of the $HPO_3^{2-}$ product formed in a given time interval. Using the first set of data we can compute the fraction of hypophosphite that reacts in the first 30 s:

$$\frac{1.6 \times 10^{-5} \text{ mol L}^{-1}}{0.10 \text{ mol L}^{-1}} = 1.6 \times 10^{-4} = 0.016\%$$

This is such a small fraction of the total amount of material present that it is indeed justifiable to replace $d[HPO_3^{2-}]/dt$ by $\Delta[HPO_3^{2-}]/\Delta t$, the change in concentration in the 30-s time interval. ∎

## 14.3 Integrated Rate Laws

The rate law expresses the dependence of the rate of a reaction on the concentrations of the reactants. Based on this dependence, it is possible to infer how the concentrations themselves must change with time. Let us reexamine Figure 14.1 from this point of view. The smooth curves in this figure show how the concentrations of the reactants and product vary with time. We have noted that the rate of change of a concentration is given by the slope of a tangent to the curve, and thus, by the derivative of the concentration with respect to time. We are now interested in the reverse process: how to infer the functional form of the dependence of concentration on time from its derivative. This can be done by integration of the rate expression. Comparison of the experimentally determined changes with time in the concentrations to the integrated forms of the rate expressions constitutes a common method for the determination of the rate law—it is generally easier to measure concentrations of reactants or products than the rates at which they change.

*a. First-Order Kinetics* The integration of the rate law is particularly straightforward for reactions obeying a first-order rate law,

$$-\frac{d[A]}{dt} = k[A] \tag{14.12}$$

where A is a reactant. This equation can be rearranged to the form

$$\frac{d[A]}{[A]} = -k\,dt \tag{14.13}$$

where each side now contains terms which depend on only a single variable, either the concentration of A or time. Both sides of the equation can consequently be integrated. The integration limits are the initial concentration of A at $t = 0$, designated $[A]_0$, and that at an arbitrary time $t$, designated $[A]$. The integrals are

$$\int_{[A]_0}^{[A]} \frac{d[A]}{[A]} = \int_0^t -k\,dt \tag{14.14}$$

and integration gives

$$\ln[A] - \ln[A]_0 = -kt \tag{14.15}$$

Combination of the logarithmic terms gives

$$\ln \frac{[A]}{[A]_0} = -kt \tag{14.16}$$

or, equivalently,

$$\frac{[A]}{[A]_0} = e^{-kt} \tag{14.17}$$

Solving for the concentration of A, we obtain

$$[A] = [A]_0 e^{-kt} \tag{14.18}$$

The exponential decrease in the reactant concentration (or the linear decrease in the logarithm of the reactant concentration) is a characteristic feature of a first-order reaction.

An alternative measure of the rate of a reaction is its **half-life,** the time required for the concentration of a reactant to decrease to half its initial value. The half-life, denoted by $t_{1/2}$, is related to the initial concentration of a reactant in a way that depends on the reaction order. The relation is particularly simple for a first-order reaction and can be derived by means of Equation 14.16,

$$\ln \frac{[A]}{[A]_0} = -kt$$

When the elapsed time is equal to $t_{1/2}$, the concentration of A has decreased to $0.5[A]_0$. Making this substitution, we obtain

$$\ln \frac{0.5[A]_0}{[A]_0} = \ln 0.5 = -kt_{1/2} \tag{14.19}$$

Since $\ln 0.5 = -\ln 2$, this relation can be written as

$$t_{1/2} = \ln 2/k = 0.693/k \tag{14.20}$$

The half-life of a first-order reaction is inversely proportional to the rate constant and is independent of concentration.

Of the many known reactions obeying first-order kinetics, we select the thermal decomposition of dinitrogen pentoxide at 25°C,

$$N_2O_5(g) \rightarrow 2NO_2(g) + \tfrac{1}{2}O_2(g)$$

to illustrate these concepts. Figure 14.3 shows plots of $[N_2O_5]$ and $\ln[N_2O_5]$ as a function of time. The linear decrease of the latter confirms the occurrence of first-order kinetics. The value of the rate constant, $k = 3.4 \times 10^{-5}$ s$^{-1}$, is obtained from the slope of the straight line. The half-life of the reaction, $t_{1/2} = 5.7$ h, is obtained by means of Equation 14.20 or directly from the plot of $[N_2O_5]$ as the time required for the concentration to decrease by one-half.

---

**Example 14.3** The reaction

$$(CH_3)_3CBr(l) + H_2O \rightarrow (CH_3)_3COH(l) + HBr(aq)$$

procedes sufficiently slowly that its rate can be measured by determining the HBr content of successive samples by titration. The following table summarizes some of the data obtained at 25°C expressed as the concentration of $(CH_3)_3CBr$ as a function of time:

| Time (h) | [(CH$_3$)$_3$CBr] (mol L$^{-1}$) |
|---|---|
| 0 | 0.104 |
| 3.15 | 0.0896 |
| 10.0 | 0.0648 |

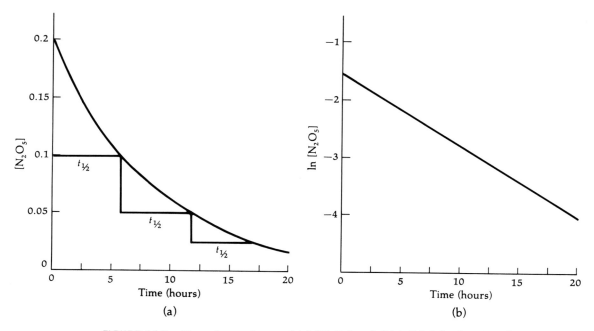

**FIGURE 14.3** Time dependence of (a) [N$_2$O$_5$] and (b) ln[N$_2$O$_5$], showing the occurrence of first-order kinetics in the reaction N$_2$O$_5$(g) → 2NO$_2$(g) + ½O$_2$(g). The time required for the concentration to decrease by a factor of two, i.e., the half-life, is marked on the graph in (a).

---

(a) Does the reaction follow first-order kinetics? (b) If so, determine the value of the rate constant and the half-life.

(a) In order to determine whether the reaction follows first-order kinetics a logarithmic plot of the data such as that shown in Figure 14.3 can be prepared. Alternatively, since there are only three data points, the question can be answered algebraically. One way to procede is to assume a first-order rate law and use the first two entries to calculate the rate constant. This value can then be used to calculate the concentration expected at the time corresponding to the third entry. If the result agrees with the data, then the reaction is first order. Applying Equation 14.16 to the first two entries we obtain

$$\ln \frac{[A]}{[A]_0} = -kt$$

$$\ln \frac{(0.0896 \text{ mol L}^{-1})}{(0.104 \text{ mol L}^{-1})} = -k(3.15 \text{ h})$$

$$k = 0.0473 \text{ h}^{-1}$$

Using this rate constant and the data from the first entry to calculate the concentration expected after 10.0 h, we obtain by means of Equation 14.17

$$[A] = [A]_0 e^{-kt}$$

$$[(CH_3)_3CBr] = (0.104 \text{ mol L}^{-1})[\exp(-0.0473 \text{ h}^{-1})(10.0 \text{ h})]$$

$$= 0.0648 \text{ mol L}^{-1}$$

which is in agreement with the data. It should be pointed out that, as a consequence of experimental uncertainties, the agreement is seldom as perfect as in this example. It then becomes necessary to perform additional measurements to ensure that the data are consistent with the rate law.

(b) Once the value of the rate constant is known ($k = 0.047$ h$^{-1}$), the half-life can be calculated from Equation 14.20:

$$t_{1/2} = \ln 2/k = (\ln 2)/0.047 \text{ h}^{-1} = 15 \text{ h} \quad \blacksquare$$

Radioactive decay (Section 23.8) is a process that is not a chemical reaction but which nonetheless is an excellent example of first-order kinetics. The rate at which a particular radioactive isotope decays is proportional to the number $N$ of atoms of that isotope. The decay rate, $-dN/dt$, is therefore given by the relation

$$-dN/dt = \lambda N \tag{14.21}$$

where the decay constant $\lambda$ is analogous to the rate constant $k$. Equation 14.21 has the same form as Equation 14.12 and can be integrated in the same manner. It is customary to characterize a radioisotope by its half-life, e.g., 20.3 m $^{11}$C, 15.0 h $^{24}$Na, and 44.6 d $^{59}$Fe.

**b. Second-Order Kinetics** The integration of the rate law of a second-order reaction with just one reactant is straightforward. The rate law can be written as

$$-d[A]/dt = k[A]^2 \tag{14.22}$$

This expression can be rearranged so that only a single variable appears on each side,

$$\frac{d[A]}{[A]^2} = -k \, dt \tag{14.23}$$

and then integrated between the limits $[A] = [A]_0$ at $t = 0$ and $[A]$ at an arbitrary time $t$:

$$\int_{[A]_0}^{[A]} \frac{d[A]}{[A]^2} = \int_0^t -k \, dt \tag{14.24}$$

The result is

$$\frac{1}{[A]} - \frac{1}{[A]_0} = kt \tag{14.25}$$

showing that the reciprocal of the reactant concentration increases linearly with time. Comparison of Equation 14.25 with Equation 14.18, the corresponding expression for first-order kinetics, shows that the reactant concentration has a different time dependence in the two cases. Thus, measurements of the variation of concentration with time can be used to distinguish between the two rate laws.

The half-life of a reaction with a rate law given by Equation 14.22 can be obtained by means of Equation 14.25. When the elapsed time is equal to the half-life $t_{1/2}$, the reactant concentration has decreased from $[A]_0$ to $0.5[A]_0$. Substituting these values into Equation 14.25, we obtain

$$\frac{1}{0.5[A]_0} - \frac{1}{[A]_0} = kt_{1/2} \tag{14.26}$$

**FIGURE 14.4** Plot of 1/[NO₂] versus time showing the occurrence of second-order kinetics in the decomposition of NO₂ at 300°C. The rate constant, $k = 0.54$ (mol L$^{-1}$)$^{-1}$ s$^{-1}$ is equal to the slope of the line.

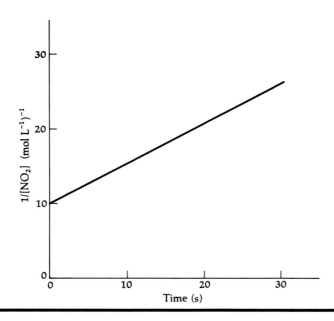

which, on rearrangement, gives

$$t_{1/2} = \frac{1}{k[A]_0} \tag{14.27}$$

In contrast to the half-life of a first-order reaction, which is independent of concentration, the half-life of a second-order reaction depends on the initial concentration. The higher the initial concentration, the shorter is the half-life and the more rapidly the reaction occurs. This different behavior of the half-lives provides yet another way of distinguishing between first-order and second-order kinetics.

The decomposition of nitrogen dioxide at elevated temperatures by the reaction

$$2NO_2(g) \rightarrow 2NO(g) + O_2(g)$$

illustrates the occurrence of second-order kinetics. Figure 14.4 shows a plot of the reciprocal of the NO₂ concentration versus time. The linear variation shows that the reaction is second order in NO₂. The rate constant is equal to the slope of the line.

**Example 14.4** The kinetics of the reaction

$$2NOBr(g) \rightarrow 2NO(g) + Br_2(g)$$

was studied at 20°C in a solvent in which NOBr and Br₂ are soluble but NO is not and escapes into an evacuated 1.00 L vessel where its pressure can be measured. The following data were obtained for 0.500 L of a solution whose initial NOBr concentration was 0.0200 $M$:

| Time (min) | $P_{NO}$ (mm Hg) |
|---|---|
| 0 | 0 |
| 2.00 | 89.5 |
| 5.00 | 129 |

Determine the rate law and the value of the rate constant.

Before the integrated rate laws can be tested, the concentration of NOBr must be obtained from the NO pressure. This can be done by first determining the number of moles of NO by means of the ideal gas law:

$$n_{NO} = \frac{P_{NO}V}{RT}$$

For the 2.00 min entry we have

$$n_{NO} = \left(\frac{89.5 \text{ mm Hg}}{760 \text{ mm Hg atm}^{-1}}\right)\frac{1.00 \text{ L}}{(0.08206 \text{ L atm mol}^{-1} \text{ K}^{-1})(293 \text{ K})}$$

$$= 4.898 \times 10^{-3} \text{ mol}$$

Similarly, 5.00 min after the start we obtain

$$n_{NO} = \left(\frac{129 \text{ mm Hg}}{760 \text{ mm Hg atm}^{-1}}\right)\frac{1.00 \text{ L}}{(0.08206 \text{ L atm mol}^{-1} \text{ K}^{-1})(293 \text{ K})}$$

$$= 7.060 \times 10^{-3} \text{ mol}$$

The concentration of unreacted NOBr can be obtained from these results since the number of moles of NOBr that have decomposed is equal to the number of moles of NO formed. The number of moles of NOBr present initially is first obtained from the given NOBr concentration and the volume as

$$n_{NOBr} = (0.0200 \text{ mol L}^{-1})(0.500 \text{ L}) = 0.0100 \text{ mol}$$

The concentrations of NOBr at the various times are

| Time (min) | [NOBr] (mol L$^{-1}$) |
|---|---|
| 0 | 0.0200 |
| 2.00 | $\dfrac{(0.0100 - 4.898 \times 10^{-3}) \text{ mol}}{0.500 \text{ L}} = 0.010$ |
| 5.00 | $\dfrac{(0.0100 - 7.060 \times 10^{-3}) \text{ mol}}{0.500 \text{ L}} = 0.0059$ |

We are now in a position to test the integrated rate laws. A quick look at the data shows that they are not consistent with first-order kinetics. The NOBr concentration decreases by just a factor of two in the first two minutes. The half-life is therefore two minutes and for first-order kinetics, the concentration would have to drop by another factor of two to 0.005 mol L$^{-1}$ in the next two minutes. Since the concentration decreases by less than this factor in the next three minutes, the half-life is not constant. The reaction consequently does not obey a first-order rate law.

We therefore turn to second-order kinetics. Following the procedure outlined in Example 14.3, we first evaluate the rate constant by substituting the first two entries into Equation 14.25:

$$\frac{1}{[\text{NOBr}]} = \frac{1}{[\text{NOBr}]_0} + kt$$

$$\frac{1}{0.010 \text{ mol L}^{-1}} = \frac{1}{0.0200 \text{ mol L}^{-1}} + k(2.00 \text{ min})$$

$$k = 25 (\text{mol L}^{-1})^{-1} \text{ min}^{-1}$$

To check whether the third datum is consistant with this result, we can repeat this calculation using the first and third values:

$$\frac{1}{0.0059 \text{ mol L}^{-1}} = \frac{1}{0.0200 \text{ mol L}^{-1}} + k(5.00 \text{ min})$$

$$k = 24 \ (\text{mol L}^{-1})^{-1} \text{ min}^{-1}$$

Since the two values of $k$ agree within the round-off error, the reaction follows second-order kinetics. ■

Example 14.4 shows that it is not necessary to measure concentrations directly in order to study kinetics. Any quantity proportional to the concentration, such as the pressure of a gas or the absorption of light by a particular species, can be used.

Second-order reactions frequently are first order with respect to each of two different reactants,

$$-d[\text{A}]/dt = k[\text{A}][\text{B}] \tag{14.28}$$

The reaction between $H_2$ and $I_2$ (Equation 14.5) is an example of such a reaction. In the special case that the concentrations of the two reactants are equal, Equation 14.28 reduces to Equation 14.22, $-d[\text{A}]/dt = k[\text{A}]^2$, and the integrated rate law given by Equation 14.25 applies. The integration of Equation 14.28 for the general case in which the concentrations of A and B are unequal is more complicated and is beyond the scope of this book.

*c. The Isolation Method*  For reactions that are more complex than second order, the integrated rate laws are usually too complicated to be very useful. An approach that is helpful in such cases is known as the **isolation method.** In this method, a large excess of one of the reactants is added. The concentration of this reactant hardly changes during the reaction and is assumed to be constant.

Let us illustrate the use of this method to simplify the second-order rate law given by Equation 14.28. Suppose reactant B is added in large excess. To a good approximation, the concentration of B remains equal to its initial value. The rate law can then be written as

$$-d[\text{A}]/dt = k[\text{A}][\text{B}]_0 = k'[\text{A}] \tag{14.29}$$

where the constant initial concentration of B has been incorporated in the rate constant $k'$. This is a **pseudo-first-order rate law** and its integrated form has already been obtained (Equation 14.18). The isolation method can be applied, in turn, to each reactant and is of value in unraveling complicated rate laws.

In dilute solutions, reactions whose rates depend on the solvent concentration commonly obey pseudo-first- (or second-) order kinetics. A common case involves hydrolysis reactions of the type

$$A + H_2O \rightarrow \text{products}$$

Such reactions frequently obey a second-order rate law of the type

$$R = -d[A]/dt = k[A][H_2O] \tag{14.30}$$

Since the concentration of water in dilute aqueous solution ($\approx 55.5$ $M$) is so much larger than that of solute A, it remains virtually constant during the reaction. Thus, the observed rate law is

$$R = k'[A] \tag{14.31}$$

and is pseudo-first order. Since the $H_2O$ concentration can be varied only within narrow limits, its effect on the reaction rate can be difficult to determine.

---

**Example 14.5** The rate of the reaction

$$2NO(g) + Cl_2(g) \rightarrow 2NOCl(g)$$

is studied by the isolation method at a certain temperature. In one experiment, the initial concentration of NO is 0.100 mol $L^{-1}$ and that of $Cl_2$ is $2.00 \times 10^{-3}$ mol $L^{-1}$. The concentration of $Cl_2$ varies with time as follows:

| Time (h) | [$Cl_2$] (mol $L^{-1}$) |
|---|---|
| 0 | $2.00 \times 10^{-3}$ |
| 1.00 | $1.17 \times 10^{-3}$ |
| 2.50 | $5.18 \times 10^{-4}$ |

In a second experiment, the initial concentration of NO is $1.50 \times 10^{-3}$ mol $L^{-1}$ and that of $Cl_2$ is 0.150 mol $L^{-1}$. The concentration of NO varies with time as follows:

| Time (h) | [NO] (mol $L^{-1}$) |
|---|---|
| 0 | $1.50 \times 10^{-3}$ |
| 30.0 | $1.10 \times 10^{-3}$ |
| 100 | $6.79 \times 10^{-4}$ |

Determine the form of the rate law and the value of the rate constant.

The first experiment can be used to establish the order with respect to $Cl_2$. Let us assume that the reaction is first order with respect to $Cl_2$ and determine the rate constant from the first two sets of data by means of Equation 14.18,

$$[Cl_2] = [Cl_2]_0 e^{-k't}$$

$$1.17 \times 10^{-3} \text{ mol } L^{-1} = (2.00 \times 10^{-3} \text{ mol } L^{-1}) \exp[-k'(1.00 \text{ h})(3600 \text{ s h}^{-1})]$$

$$(-3600 \text{ s})k' = -0.54$$

$$k' = 1.5 \times 10^{-4} \text{ s}^{-1}$$

Applying the same procedure to the first and third sets of data we obtain

$$5.18 \times 10^{-4} \text{ mol L}^{-1} = (2.00 \times 10^{-3} \text{ mol L}^{-1}) \exp[-k'(2.50 \text{ h})(3600 \text{ s h}^{-1})]$$

$$(-9000 \text{ s})k' = -1.4$$

$$k' = 1.6 \times 10^{-4} \text{ s}^{-1}$$

The slight difference in $k'$ may arise from round off, or else the reaction may be second order. To check the latter possibility we assume second-order kinetics and repeat the procedure using Equation 14.25:

$$\frac{1}{[Cl_2]} = \frac{1}{[Cl_2]_0} + k't$$

From the first two sets of data we obtain

$$\frac{1}{1.17 \times 10^{-3} \text{ mol L}^{-1}} = \frac{1}{2.00 \times 10^{-3} \text{ mol L}^{-1}} + k'(3600 \text{ s})$$

$$k' = 9.85 \times 10^{-2} \text{ (mol L}^{-1})^{-1} \text{ s}^{-1}$$

and from the first and third sets we get

$$\frac{1}{5.18 \times 10^{-4} \text{ mol L}^{-1}} = \frac{1}{2.00 \times 10^{-3} \text{ mol L}^{-1}} + k'(2.5 \text{ h})(3600 \text{ s h}^{-1})$$

$$k' = 1.59 \times 10^{-1} \text{ (mol L}^{-1})^{-1} \text{ s}^{-1}$$

Since the two values of $k'$ differ by nearly a factor of two, the reaction is not second order with respect to $Cl_2$ but first order.

The second experiment can be used to establish the order with respect to NO. Let us assume that the reaction is second order with respect to NO and, as before, use Equation 14.25. We get from the first two sets of values

$$\frac{1}{1.10 \times 10^{-3} \text{ mol L}^{-1}} = \frac{1}{1.50 \times 10^{-3} \text{ mol L}^{-1}} + k''(30.0 \text{ h})(3600 \text{ s h}^{-1})$$

$$k'' = 2.24 \times 10^{-3} \text{ (mol L}^{-1})^{-1} \text{ s}^{-1}$$

and from the first and third sets

$$\frac{1}{6.79 \times 10^{-4} \text{ mol L}^{-1}} = \frac{1}{1.50 \times 10^{-3} \text{ mol L}^{-1}} + k''(100 \text{ h})(3600 \text{ s h}^{-1})$$

$$k'' = 2.24 \times 10^{-3} \text{ (mol L}^{-1})^{-1} \text{ s}^{-1}$$

The agreement shows that the reaction is second order with respect to NO. The overall rate law therefore is

$$R = k[NO]^2[Cl]$$

The value of $k$ can be obtained from either experiment by recognizing that the calculated rate constants $k'$ or $k''$ are apparent rate constants which include the value of the constant concentration of the reactant present in excess. The second experiment yields the most precise answer. Using Equation 14.29, we obtain

$$R = k[NO]^2[Cl_2]_0 = k''[NO]^2$$

Solving for $k$, we get

$$k = \frac{k''}{[Cl_2]_0} = \frac{2.24 \times 10^{-3} (\text{mol L}^{-1})^{-1} \text{ s}^{-1}}{0.150 \text{ mol L}^{-1}}$$

$$k = 1.49 \times 10^{-2} \text{ (mol L}^{-1})^{-2} \text{ s}^{-1}$$

The first experiment yields essentially the same answer with lower precision (owing to the loss of a significant figure):

$$k = \frac{k'}{[NO]_0^2} = \frac{1.6 \times 10^{-4} \text{ s}^{-1}}{(0.100 \text{ mol L}^{-1})^2} = 1.6 \times 10^{-2} \text{ (mol L}^{-1})^{-2} \text{ s}^{-1} \quad \blacksquare$$

The combined use of the various techniques described so far—the initial rate method, the isolation method, and the application of the integrated rate laws—generally permits the correct determination of the rate law of a chemical reaction. Table 14.1 summarizes some of the results for various rate laws.

## Reaction Mechanisms

### 14.4 Elementary Reactions

*a. Molecularity and Reaction Order* Most chemical reactions are complex and can be broken down into a sequence of **elementary reactions,** that is, reactions that each occur in a single step. The particular sequence of steps by which reactant molecules are converted to product molecules constitutes the **mechanism** of the overall reaction in question. The overall reaction is the sum of the elementary reactions comprising the mechanism.

To illustrate the distinction between an overall reaction and its constituent elementary reactions, consider the formation of HI from the elements,

$$H_2(g) + I_2(g) \rightarrow 2HI(g)$$

which obeys the second order rate law

$$\frac{-d[I_2]}{dt} = k[H_2][I_2]$$

This reaction is thought to involve the following sequence of steps:

(1) $\quad I_2 \rightleftarrows 2I$

(2) $\quad H_2 + I \rightleftarrows H_2I$

(3) $\quad H_2I + I \rightarrow 2HI$

$\quad\quad\quad H_2 + I_2 \rightarrow 2HI$

**TABLE 14.1 A Summary of Rate Laws**

| Order | Reaction | Rate law | Units of k | Integrated rate law | | Half-life |
|---|---|---|---|---|---|---|
| 0 | $aA \rightarrow$ products | $R = k$ | mol L$^{-1}$ s$^{-1}$ | $[A] = [A]_0 - kt$ | $(kt \leq [A]_0)$ | $[A]_0/(2k)$ |
| 1 | $aA \rightarrow$ products | $R = k[A]$ | s$^{-1}$ | $[A] = [A]_0 e^{-kt}$ | | $(\ln 2)/k$ |
| 2 | $aA \rightarrow$ products | $R = k[A]^2$ | (mol L$^{-1}$)$^{-1}$ s$^{-1}$ | $\frac{1}{[A]} = \frac{1}{[A]_0} + kt$ | | $1/(k[A]_0)$ |
| 2 | $aA + bB \rightarrow$ products | $R = k[A][B]$ | (mol L$^{-1}$)$^{-1}$ s$^{-1}$ | | | |
| 3 | $aA + bB \rightarrow$ products | $R = k[A][B]^2$ | (mol L$^{-1}$)$^{-2}$ s$^{-1}$ | | | |

Each of these steps is an elementary reaction and describes the way the individual atoms or molecules react. The overall reaction is the sum of the three steps, but the occurrence of these steps cannot be inferred from it. The impossibility of deducing a mechanism from the stoichiometry of an overall reaction is underscored by the fact that, until recently, it was believed that the reaction between $H_2$ and $I_2$ was an elementary reaction in which the two molecules exchanged hydrogen and iodine atoms.

In order to determine the mechanism of a reaction it is necessary to determine the presence and identity of **intermediate species,** that is, species that are produced and consumed in the course of a reaction but which do not appear in its overall stoichiometry, and to infer from known chemical behavior how these species might react. The mechanism for HI formation, for example, requires the presence of I and $H_2I$ intermediates. In many instances intermediates are difficult to detect because they are short-lived, thus making the determination of the mechanism a challenging task.

An elementary reaction can be characterized by the number of reactant molecules it involves, commonly known as the **molecularity.** A **unimolecular reaction** involves the participation of just a single molecule. The first step in the HI mechanism,

$$I_2 \rightarrow 2I$$

is an example of unimolecular decomposition. This type of process frequently occurs when a molecule has been excited in an earlier collision and then decomposes or undergoes rearrangement.

A **bimolecular reaction** is a reaction between two molecules. The two molecules can associate to form a third species, as in the second step of the HI formation mechanism,

$$H_2 + I \rightarrow H_2I$$

they can undergo transfer, as in the third step,

$$H_2I + I \rightarrow HI + HI$$

and so on.

The vast majority of elementary reactions are unimolecular or bimolecular. The probability that more than two molecules will simultaneously collide and undergo reaction is small. **Termolecular reactions,** those involving three molecules, are rare, and elementary processes involving more than three molecules are unknown.

It is important to understand the distinction between molecularity and reaction order. The molecularity refers to the number of molecules participating in an elementary reaction; the reaction order refers to the observed dependence of the rate of a reaction on reactant concentrations. *For an elementary reaction, the order is necessarily equal to the molecularity.* For example, a bimolecular reaction,

$$A + B \rightarrow products$$

is necessarily second order and obeys the rate law

$$-d[A]/dt = k[A][B]$$

Since the reaction requires molecules A and B to come together, its rate must be proportional to the product of their concentrations. On the other hand, the order of a multistep reaction must be obtained experimentally. The concept of molecularity cannot be applied to such a reaction.

**b. Elementary Reactions Attaining Equilibrium** Some elementary reactions are reversible and chemical equilibrium between reactants and products is established. For example, the one-step reaction between carbon monoxide and nitrogen dioxide can be characterized by the following equilibrium

$$CO(g) + NO_2(g) \underset{k_{-1}}{\overset{k_1}{\rightleftharpoons}} CO_2(g) + NO(g)$$

where $k_1$ is the rate constant for the forward reaction and $k_{-1}$ is the rate constant for the reverse reaction. For an elementary reaction, the values of these rate constants can be related to the equilibrium constant for the reaction.

Consider the general elementary reaction

$$A + B \underset{k_{-1}}{\overset{k_1}{\rightleftharpoons}} C + D \tag{14.32}$$

The rate of the forward reaction is

$$R_f = k_1[A][B] \tag{14.33}$$

while that of the reverse reaction is

$$R_b = k_{-1}[C][D] \tag{14.34}$$

These rate laws follow from the molecularity of the reaction. We know from Section 7.1 that the rates of the forward and reverse reactions become equal at equilibrium. Applying this condition to Equations 14.33 and 14.34, we obtain

$$k_1[A]_{eq}[B]_{eq} = k_{-1}[C]_{eq}[D]_{eq} \tag{14.35}$$

where the subscript eq denotes that the concentrations are evaluated at equilibrium. Since the equilibrium constant of the reaction is

$$K = \frac{[C]_{eq}[D]_{eq}}{[A]_{eq}[B]_{eq}} \tag{14.36}$$

the rate constants and the equilibrium constant must be related. Combining Equation 14.35 and 14.36, we obtain

$$K = k_1/k_{-1} \tag{14.37}$$

The equilibrium constant is equal to the ratio of the rate constants of the forward and reverse reactions. This important relationship permits the equilibrium constant to be obtained from the rate constants or, alternatively, permits one of the rate constants to be obtained from the equilibrium constant and the other rate constant. We emphasize that this relationship is always valid only for an elementary reaction—it does not necessarily hold true for a reaction that occurs in more than one step.

## 14.5 Understanding Reaction Mechanisms

Although the determination of a reaction mechanism is a difficult task, several general principles are useful aids. Moreover, the mechanisms of many reactions are basically similar and can be classified into a few broad categories. The mechanism of a specific reaction can be viewed as a variant of one of these basic types.

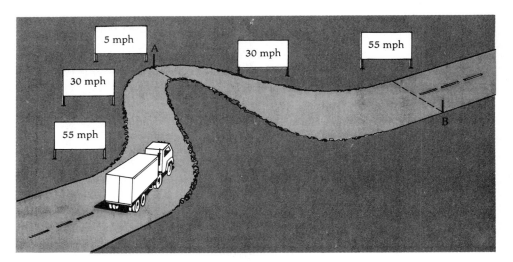

**FIGURE 14.5** Illustration of the concept of a rate-determining step in terms of a freeway with a detour. The rate at which vehicles move past point B is limited by the rate at which they can move past point A in the detour. The traversal of the detour is the rate-determining step.

***a. The Rate-Determining Step***  In the sequence of elementary reactions that comprise a mechanism it frequently happens that one step is much slower than all the others. The overall reaction cannot occur faster than this slowest reaction, which is therefore called the **rate-determining step.** This step is important because it determines the rate of the overall reaction and the form of the rate law. The analogy illustrated in Figure 14.5 may help to clarify this concept.

Consider the following two-step mechanism for the reaction

$$A + 2B \rightarrow 2C$$

involving the intermediate I:

$$(1) \quad A + B \xrightarrow{k_1} I + C \quad \text{slow}$$
$$(2) \quad \underline{I + B \xrightarrow{k_2} C} \quad \text{fast}$$
$$A + 2B \xrightarrow{k} 2C$$

The first reaction is much slower than the second and is the rate-determining step. Since the rate of a reaction depends on the magnitude of the rate constant, $k_1$ must be much smaller than $k_2$.

The rate law for the overall reaction can be deduced readily from the simple mechanism postulated for the reaction. The rate of the reaction can be expressed in terms of the rate of change of the concentration of reactant A as

$$R = -d[A]/dt = k_1[A][B] \tag{14.38}$$

Since the second step does not affect the reaction rate, the experimentally determined rate law should be identical to Equation 14.38, where the empirical rate constant $k$ can be identified with the rate constant of the first elementary reaction, $k_1$.

**Example 14.6** The formation of nitryl fluoride in the reaction

$$2NO_2(g) + F_2(g) \rightarrow 2NO_2F(g)$$

obeys the second-order rate law

$$R = k[NO_2][F_2]$$

Devise a two-step mechanism that is consistent with the rate law in which the rate-determining step involves the formation of atomic fluorine.

The following mechanism involves the formation of F in the first step and is consistent with the reaction stoichiometry:

(1) $NO_2 + F_2 \xrightarrow{k_1} NO_2F + F$     slow
(2) $\underline{NO_2 + F \xrightarrow{k_2} NO_2F}$     fast

$$2NO_2 + F_2 \xrightarrow{k} 2NO_2F$$

The rate of the reaction is determined by the slow first step and is equal to $-d[F_2]/dt$,

$$R = -d[F_2]/dt = k_1[NO_2][F_2]$$

This is identical to the empirical rate law provided the experimentally determined rate constant $k$ is identified with $k_1$.

It must be emphasized that while this mechanism is consistent with the facts, such consistency is *not* proof that the mechanism is correct. A proof of the correctness of the mechanism would require the detection of the F atoms formed as intermediates and an indication that their concentration varies with time in a manner that is consistent with the assumed mechanism. ∎

**b. The Steady-State Approximation** The formation of nitryl fluoride (Example 14.6) has a very simple two-step mechanism. Although the mechanism does involve the formation of an intermediate, the concentration of the intermediate does not affect the speed of the rate-determining step. Frequently, the rate-determining step depends on the concentration of an intermediate. Since the experimental rate law expresses the dependence of the reaction rate on the concentrations of observable reactants or products and not on that of some intermediate that may actually be undetectable, determining the mechanism of such reactions is more complicated.

A frequently encountered mechanism has as its first step a reversible reaction in which at least one intermediate is formed. The second step can be the decay of the intermediate, a reaction of the intermediate with a reactant, or even a reaction between intermediates. The mechanism

(1) $A + B \underset{k_{-1}}{\overset{k_1}{\rightleftarrows}} I$
(2) $\underline{I \xrightarrow{k_2} C}$

$$A + B \xrightarrow{k} C$$

where I is an intermediate, is one of many variations of a common mechanism for the reaction

$$A + B \rightarrow C \tag{14.39}$$

Let us derive the rate law for the reaction from the postulated mechanism. The rate of the reaction can be expressed in terms of the rate of decrease of the concentration of a reactant or in terms of the rate of increase of the concentration of a product (Equation 14.1). Since the reactants are also the products of the reverse reaction in step (1), the rate of change of their concentrations is complex. It is consequently simpler to express the reaction rate in terms of the rate of product formation:

$$R = d[C]/dt = k_2[I] \tag{14.40}$$

where the second equality follows from the unimolecularity of step (2).

To solve Equation 14.40, we must determine the concentration of the intermediate I. This can be done by first determining the rate of change of the concentration of I. The value of $d[I]/dt$ can be obtained by noting that I is produced in the forward reaction of step (1) and is consumed in the reverse reaction of step (1) as well as in step (2). We can consequently write

$$\frac{d[I]}{dt} = k_1(A)(B) - k_{-1}(I) - k_2(I) \tag{14.41}$$

The positive term of this equation, $k_1(A)(B)$, gives the rate for the production of I in the forward reaction of step (1), while the negative terms give the rate for the consumption of I. Since I is both a reactant and a product, both positive and negative terms contribute to the rate of change of its concentration.

In principle, Equation 14.41 can be solved for the concentration of I in terms of known quantities in the same manner as the differential equations for first-order or second-order kinetics. In practice, the solution of such an equation is complex. Fortunately, a simplifying approximation can be made. Let us think of how the concentration of the intermediate I changes with time. Initially, I is not present in the mixture of reactants. Its concentration consequently increases as a result of step (1). However, once formed, I reacts via step (2) as well as via the reverse reaction in step (1) and, since it also continues to be made in step (1), its concentration approaches a constant value. Since I is not a product, its concentration must eventually decrease to zero by the end of the reaction. Figure 14.6 illustrates these changes in concentration. It is customary to assume that the concentration of the intermediate remains constant during the reaction. It follows that the rate of change of its concentration is zero:

$$\frac{d[I]}{dt} = 0 \tag{14.42}$$

This assumption is **known as the steady-state approximation.**

The use of this simplifying approximation makes it possible to obtain the rate law for the reaction represented by Equation 14.39. First, Equation 14.41 is solved for [I]:

$$\frac{d[I]}{dt} = 0 = k_1[A][B] - k_{-1}[I] - k_2[I]$$

$$[I] = \frac{k_1[A][B]}{k_{-1} + k_2} \tag{14.43}$$

Next, this result is introduced in the rate law, Equation 14.40, giving

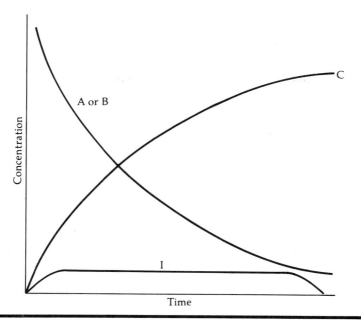

**FIGURE 14.6** Time dependence of concentrations of species involved in the reaction A + B → C. Note that the concentration of the intermediate I increases to a certain small value, remains constant during most of the reaction, and then decreases to zero. The constancy of [I] is the basis of the steady-state approximation.

$$R = \frac{k_1 k_2}{k_{-1} + k_2} [A][B] = k[A][B] \tag{14.44}$$

This result shows that the mechanism predicts a second-order rate law, with the empirical rate constant $k$ corresponding to the indicated combination of elementary reaction rate constants.

Depending on the relative rates of the elementary reactions, additional simplifications of Equation 14.44 may be possible. If step (2) is much slower than step (1) (that is, if $k_2 \ll k_{-1}$), the rate law reduces to

$$R = Kk_2[A][B] \tag{14.45}$$

where $K = k_1/k_{-1}$ is the equilibrium constant for the reversible reaction in step (1) (see Equation 14.37). The reaction is still second order but the empirical rate constant now consists of the product of the rate constant for the slow step and the equilibrium constant for the fast step.

If the first step of the mechanism consists of a rapid equilibrium, it is frequently possible to derive the rate law without recourse to the steady-state approximation. Instead, the equilibrium expression can be used directly to relate the concentration of the intermediate to those of the reactants. For the reaction of present interest, the concentration of intermediate I is obtained from the equilibrium constant expression for step (1) as

$$[I] = K[A][B] \tag{14.46}$$

Insertion of this result into Equation 14.40 for the reaction rate immediately gives Equation 14.45, the rate law for the case in which the equilibrium first step occurs much faster than the second step.

On the other hand, if $k_{-1} \ll k_2$, that is, if the decomposition of I to give A + B is negligible compared to that to give C, Equation 14.44 reduces to

$$R = k_1[A][B] \tag{14.47}$$

## Example 14.7

The reaction

$$2NO(g) + H_2(g) \rightarrow N_2O(g) + H_2O(g)$$

has been found to be second order in NO, first order in $H_2$, and third order overall. The following mechanism has been proposed:

(1) $\qquad 2NO \underset{k_{-1}}{\overset{k_1}{\rightleftarrows}} N_2O_2$

(2) $\quad N_2O_2 + H_2 \overset{k_2}{\longrightarrow} N_2O + H_2O$

$\overline{\phantom{XX} 2NO + H_2 \overset{k}{\longrightarrow} N_2O + H_2O \phantom{XX}}$

Show that this mechanism leads to the experimentally determined rate law. Relate the empirical rate constant $k$ to the rate constants of the elementary reactions.

The rate of the reaction is equal to the rate at which $N_2O$ is formed and, on the basis of step (2), can be written as

$$R = d[N_2O]/dt = k_2[N_2O_2][H_2]$$

The rate of change of the concentration of the $N_2O_2$ intermediate is

$$d[N_2O_2]/dt = \underset{\substack{\text{Step (1)}\\\text{(forward)}}}{k_1[NO]^2} - \underset{\substack{\text{Step (1)}\\\text{(reverse)}}}{k_{-1}[N_2O_2]} - \underset{\text{Step (2)}}{k_2[N_2O_2][H_2]}$$

Applying the steady-state approximation, we set

$$d[N_2O_2]/dt = 0$$

and solving for $[N_2O_2]$, we obtain

$$[N_2O_2] = \frac{k_1[NO]^2}{k_{-1} + k_2[H_2]}$$

Substitution of this result into the expression for the reaction rate $R$ gives

$$R = \frac{k_1 k_2 [NO]^2 [H_2]}{k_{-1} + k_2[H_2]}$$

Except for the presence of the $[H_2]$ term in the denominator, this expression looks like the empirical third-order rate law. To obtain the desired empirical rate law, it must be assumed that the second step of the mechanism is the rate-determining step. Therefore, unless the concentration of $H_2$ is unusually high, the approximation

$$k_{-1} + k_2[H_2] \approx k_{-1}$$

will be valid. Under these circumstances we obtain

$$R = \frac{k_1 k_2}{k_{-1}} [NO]^2[H_2]$$

which is second order in NO, first order in H$_2$, and third order overall. The empirical rate constant corresponds to the indicated combination of rate constants,

$$k = \frac{k_1 k_2}{k_{-1}} = K k_2$$

where $K$ is the equilibrium constant for the reversible reaction in step (1). ∎

It is worth emphasizing that while the proposed mechanism accounts for the observed rate law in Example 14.7, this is not proof that the reaction actually procedes by this mechanism. The following example shows that an alternate mechanism leads to the same empirical rate law.

**Example 14.8** Show that the following mechanism leads to the rate law $R = k[NO]^2[H_2]$ for the reaction

$$2NO(g) + H_2(g) \rightarrow N_2O(g) + H_2O(g)$$

(1) $NO + H_2 \underset{k_{-1}}{\overset{k_1}{\rightleftarrows}} NOH_2$      fast

(2) $NOH_2 + NO \xrightarrow{k_2} N_2O + H_2O$      slow

The rate of the overall reaction is obtained from the second step as

$$R = d[N_2O]/dt = k_2[NOH_2][NO]$$

Since the first equilibrium step is known to be fast, we can use the equilibrium expression to obtain the concentration of the intermediate:

$$K = \frac{k_1}{k_{-1}} = \frac{[NOH_2]}{[NO][H_2]}$$

$$[NOH_2] = K[NO][H_2]$$

Alternatively, [NOH$_2$] can be obtained by means of the steady-state approximation. However, the use of this approximation involves more algebraic manipulation (see Example 14.7). Substituting for [NOH$_2$] into the expression for $R$, we obtain

$$R = k_2 K[NO]^2[H_2]$$

This expression is identical to the empirical rate law once the quantity $k_2 K$ is identified with the experimental rate constant $k$. ∎

Examples 14.7 and 14.8 show that an empirical rate law can be consistent with two different mechanisms involving different intermediates. One way of distinguishing between the two mechanisms is to determine whether NOH$_2$ or N$_2$O$_2$ is formed as intermediate. Special techniques are required to detect the very low concentrations of such intermediates in the brief time in which the reaction occurs.

## 14.6 Chain Reactions

A **chain reaction** involves the formation of an intermediate which is alternately consumed and regenerated in a repeating series of reactions. Each cycle results in the formation of the reaction products and constitutes a link in a chain which can involve the conversion of a large number of reactant molecules to products.

The formation of hydrogen bromide in the reaction

$$H_2(g) + Br_2(g) \rightarrow 2HBr(g)$$

is a classical example of a chain reaction. The following steps are believed to be involved in the mechanism:

(1) $Br_2 + M \xrightarrow{k_1} 2Br + M$ (chain initiation)

(2) $Br + H_2 \underset{k_4}{\overset{k_3}{\rightleftarrows}} HBr + H$ (chain propagation)

(3) $H + Br_2 \xrightarrow{k_5} HBr + Br$ (chain propagation)

(4) $2Br + M \xrightarrow{k_2} Br_2 + M$ (chain termination)

The first reaction initiates the chain and involves a collision of a bromine molecule with some other molecule M that can provide enough energy to dissociate $Br_2$. Steps (2) and (3) constitute the chain reaction, where the combination of these reactions corresponds to the overall reaction. Note, however, that step (2) is reversible, the reverse reaction inhibiting the advancement of the overall reaction. Contrary to the mechanisms considered in the preceding section, there is no net consumption of the intermediates in steps (2) and (3) and this two-step cycle can therefore occur repeatedly. Eventually, however, the collision of two bromine atoms in the presence of some molecule M, whose function is to remove their excess energy, results in their recombination. The chain reaction thus ends at step (4).

The rate law expected for this mechanism can be derived in the manner described in Section 14.5b. The concentrations of H and Br are obtained by means of the steady-state approximation and the results used to obtain the rate of the reaction. The result is

$$R = \frac{1}{2}\frac{d[HBr]}{dt} = \frac{k_3(k_1/k_2)^{1/2}[H_2][Br_2]^{1/2}}{1 + (k_4[HBr]/k_5[Br_2])} \qquad (14.48)$$

which has the exact form of the experimental rate law given earlier (Equation 14.9),

$$R = -\frac{d[Br_2]}{dt} = \frac{1}{2}\frac{d[HBr]}{dt} = \frac{k[H_2][Br_2]^{1/2}}{1 + k'[HBr][Br_2]^{-1}}$$

provided the empirical rate constants $k$ and $k'$ are identified with the appropriate combinations of elementary reaction rate constants. Note that the mechanism accounts for the unusual reduction in rate associated with a high HBr concentration. Hydrogen bromide inhibits the overall reaction as a consequence of its reaction with atomic hydrogen in step (2).

A characteristic feature of a chain reaction is the presence of free-radical intermediates (Section 3.8a). Recall that a free radical is a species containing an unpaired electron. Free radicals usually are reactive molecular fragments, such as ·$CH_3$, and are of particular importance in organic chemistry. The free radicals contributing to the formation of hydrogen bromide are the atomic intermediates, H· and Br·, each of which possesses an unpaired electron.

Chain reaction mechanisms occur in a large variety of processes, including many of importance in biochemistry, atmospheric chemistry (Section 3.8a), and polymer chemistry. A particularly interesting case is that of explosive reactions. Explosions can occur in chain reactions involving steps that increase the number of intermediates. Since each intermediate propagates the reaction, the process can occur at an ever-increasing rate and an explosion ensues.

The formation of water when the reaction of hydrogen and oxygen is initiated by a spark is a good example of a reaction that under proper conditions is an explosive chain reaction. The overall equation

$$2H_2(g) + O_2(g) \rightarrow 2H_2O(l)$$

is deceptively simple but the mechanism is very complex and has not been completely worked out. However, the following steps are known to be important:

(1) $H_2 + O_2 \rightarrow HO_2 + H$ (chain initiation)
(2) $H_2 + HO_2 \rightarrow HO + H_2O$ (chain propagation)
(3) $HO + H_2 \rightarrow H + H_2O$ (chain propagation)
(4) $H + O_2 \rightarrow HO + O$ (branching)
(5) $O + H_2 \rightarrow HO + H$ (branching)

The first reaction initiates the chain and the second and third reactions form the $H_2O$ product and propagate the chain. The last two reactions involve branching, a characteristic feature of an explosive chain reaction. In each of these steps, an intermediate reacts with a molecule of one of the starting materials to form two new intermediates. These additional intermediates can also propagate the chain and the reaction rate increases ever more rapidly, resulting in the well-known explosive formation of water.

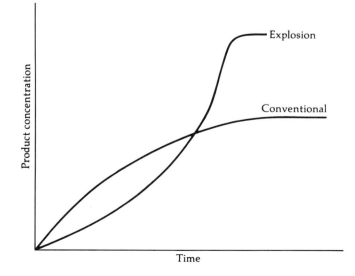

FIGURE 14.7 Distinction between an explosive and a conventional reaction. In an explosion, the product concentration increases ever more rapidly—the rate of the reaction increases with time. In a conventional reaction, the rate gradually decreases with time.

In addition to occurring via branching chain reactions, explosions can also occur in exothermic reactions which have rates that increase sharply with increasing temperature. If conditions are such that the evolved heat is not readily dissipated, the resulting increase in temperature can accelerate the rate at an increasingly fast pace, and an explosion occurs. The key distinction between either type of explosive reaction and a conventional reaction lies in the opposite dependence of the reaction rate on time, as illustrated in Figure 14.7.

# The Rates of Elementary Reactions

The approach we have used so far to examine various aspects of chemical kinetics has been largely empirical. We are now ready to adopt a more fundamental viewpoint and see whether the rate constant of a reaction can be predicted on the basis of the interaction between the reacting molecules.

The discussion of reaction mechanisms has made it clear that most reactions are complex; their rate constants in general can be understood as combinations of the rate constants of some of the elementary processes that comprise the mechanism. In order to focus on the essential factors determining the magnitude of a rate constant we must examine an elementary reaction, for which this complication does not occur. The simplest case involves a bimolecular reaction in the gas phase, and it is on this process that we will focus at the outset.

## 14.7 The Rate of Molecular Collisions

It is evident that in order to react with each other two molecules must first come into close contact. The maximum rate of a reaction must therefore be roughly equal to the frequency of molecular collisions. It is well known, however, that under similar conditions different bimolecular reactions can proceed at very different rates. Evidently, only a fraction of molecular collisions usually result in a chemical reaction. The collision frequency therefore represents an upper limit to the reaction rate rather than its actual value. In spite of this limitation, knowledge about the rate of collisions is useful because of the insight it provides about at least some of the basic factors governing reaction rates.

In Section 4.5 we developed an expression for the frequency of collisions experienced by a molecule in the gas phase, Equation 4.24,

$$Z_1 = (\sqrt{2}\pi d^2 \bar{v} N^*) \text{ s}^{-1}$$

where $d$ is the molecular diameter, $\bar{v}$ the mean speed, and $N^*$ the number of molecules per unit volume. This expression can be adapted readily to yield the total rate of collisions between molecules of types A and B, which we shall designate $Z_{AB}$. Let the concentration of molecules of type B be $N_B^*$ m$^{-3}$. Equation 4.24 indicates that the number of collisions made by one A molecule with B molecules in one second is $(\sqrt{2}\pi d^2 \bar{v} N_B^*)$ s$^{-1}$. If the concentration of A is $N_A^*$ m$^{-3}$ the total collision rate between the two types of molecules is

$$Z_{AB} = (\sqrt{2}\pi d^2 \bar{v} N_A^* N_B^*) \text{ m}^{-3} \text{ s}^{-1} \tag{14.49}$$

Let us evaluate the collision rate for typical conditions. If both gases are present at 0°C and a partial pressure of 1 atm, the number of molecules per unit volume is

$$N_A^* = N_B^* = \left(\frac{1}{22.4} \text{ mol L}^{-1}\right)(6.022 \times 10^{23} \text{ mol}^{-1})(10^3 \text{ L m}^{-3})$$

$$= 2.69 \times 10^{25} \text{ m}^{-3}$$

Typical values of $d$ and $\bar{v}$ are $3 \times 10^{-10}$ m and 500 m s$^{-1}$, respectively. Combining these quantities we obtain

$$Z_{AB} = (\sqrt{2}\pi)(3 \times 10^{-10} \text{ m})^2(500 \text{ m s}^{-1})(2.69 \times 10^{25} \text{ m}^{-3})^2$$

$$= 1 \times 10^{35} \text{ m}^{-3} \text{ s}^{-1}$$

or, converting to units customarily used to express reaction rates,

$$Z_{AB} = (1 \times 10^{35} \text{ m}^{-3} \text{ s}^{-1})\left(\frac{1}{6.022 \times 10^{23} \text{ mol}^{-1}}\right)(10^{-3} \text{ m}^3 \text{ L}^{-1})$$

$$= 2 \times 10^8 \text{ mol L}^{-1} \text{ s}^{-1}$$

This is an enormously fast rate. If a collision between two molecules always led to a reaction, over $10^8$ mol L$^{-1}$ of gas would react in one second. Put another way, at ordinary pressures a bimolecular gas phase reaction would occur in less than $10^{-9}$ s. While some reactions occur at a rate that approaches this value, most do not.

To make the comparison between the collision rate and the rate of bimolecular reactions more quantitative, it is helpful to evaluate a rate constant $k_{col}$ for the collision process. This rate constant can be compared with experimental rate constants of reactions known to involve an elementary bimolecular reaction.

The collision rate constant can be obtained by writing the collision rate as a bimolecular second-order rate law:

$$Z_{AB} = k_{col}[A][B] \tag{14.50}$$

To evaluate this expression, let us assume that A and B are present at STP conditions, so that their molar volume is 22.4 L. Using the value of $Z_{AB}$ calculated earlier for these conditions, we obtain

$$2 \times 10^8 \text{ mol L}^{-1} \text{ s}^{-1} = k_{col}\left(\frac{1 \text{ mol}}{22.4 \text{ L}}\right)^2$$

from which we get

$$k_{col} = 1 \times 10^{11} (\text{mol L}^{-1})^{-1} \text{ s}^{-1}$$

The collision rate constant has been evaluated for $T = 273$ K, and collision theory indicates how $k_{col}$ must vary with temperature. The dependence of $Z_{AB}$ on temperature comes about from that of the mean molecular speed $\bar{v}$. According to the kinetic theory of gases, the mean speed varies as $T^{1/2}$ (Equation 4.19). Since Equations 14.49 and 14.50 show that $k_{col}$ is proportional to $\bar{v}$, the collision rate constant must increase as the square root of the absolute temperature.

Table 14.2 lists some values of bimolecular rate constants determined at 273 K. The most striking feature is the large variability among the listed

**TABLE 14.2 Rate Constants of Some Bimolecular Gas-phase Reactions at 273 K**

| Reaction | $k$ $(mol\ L^{-1})^{-1}\ s^{-1}$ |
|---|---|
| $H + Br_2 \rightarrow HBr + Br$ | $2.1 \times 10^{10}$ |
| $NO + O_3 \rightarrow NO_2 + O_2$ | $1.5 \times 10^{8}$ |
| $NO + NO_2Cl \rightarrow NOCl + NO_2$ | $4.6 \times 10^{3}$ |
| $CH_3 + CHCl_3 \rightarrow CH_4 + CCl_3$ | $4.8 \times 10^{2}$ |
| $CH_3 + H_2 \rightarrow CH_4 + H$ | $2.9$ |
| $Br + H_2 \rightarrow HBr + H$ | $2.8 \times 10^{-4}$ |
| $NO + Cl_2 \rightarrow NOCl + Cl$ | $3.4 \times 10^{-7}$ |
| $NOCl + NOCl \rightarrow 2NO + Cl_2$ | $1.2 \times 10^{-10}$ |
| $NO_2 + NO_2 \rightarrow 2NO + O_2$ | $2.3 \times 10^{-12}$ |

values. The rate constants range from that for the reaction between H and $Br_2$, which is nearly comparable to that of $k_{col}$, to that for the reaction between $NO_2$ molecules, which is smaller than $k_{col}$ by a factor of $10^{23}$. Evidently, only a very small but highly variable fraction of the collisions generally result in a chemical reaction. Most of the time, the molecules bounce off each other with no change in identity. What factors could be responsible for this very large and variable difference? It turns out that the temperature dependence of the rates of chemical reactions provides some important insights to this problem and we turn to an examination of this dependence.

## 14.8 Temperature Dependence of Reaction Rates and the Magnitude of Rate Constants

*a. The Arrhenius Equation* The rates of chemical reactions generally display a marked increase with temperature. Typically, a ten degree increase in temperature results in a doubling of the reaction rate. This behavior suggests an exponential variation of the rate constant with temperature and such a variation is, in fact, observed. The following equation proposed by Arrhenius gives a reasonably good fit to the data obtained for most reactions:

$$k = Ae^{-E_a/RT} \tag{14.51}$$

where $A$ is a constant known as the **frequency factor,** the units of which are the same as those of the rate constant. The exponential consists of the ratio of an energy $E_a$, called the **activation energy,** to the thermal energy $RT$. The magnitude of the activation energy determines the variation of the rate constant with temperature.

The two parameters of the Arrhenius equation, $A$ and $E_a$, can be obtained from the measured temperature dependence of the rate constant of a reaction. This is most conveniently done by plotting $\ln k$ (or $\log k$) versus $1/T$. Taking the natural logarithm of both sides of Equation 14.51 gives

$$\ln k = \ln A - (E_a/RT) \tag{14.52}$$

indicating that such a plot should be linear. Figure 14.8 shows the data for a typical reaction plotted in this fashion. The linearity of the plot indicates that the Arrhenius equation provides a valid representation of the temperature dependence of the rate constant. Once the applicability of the Arrhenius equation

**FIGURE 14.8** Arrhenius plot of the temperature dependence of the rate constant of the reaction $H_2(g) + I_2(g) \rightarrow 2HI(g)$. When $\log k$ is plotted versus $1/T$, the slope of the line is equal to $-E_a/2.303R$, and the intercept is equal to $\log A$.

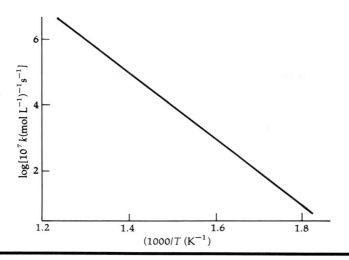

to a particular reaction has been established, Equation 14.52 can be used to determine the activation energy as well as to relate the rate constant at one temperature to that at another. Evaluating Equation 14.52 at two temperatures, $T_2$ and $T_1$, for which the rate constants are $k_2$ and $k_1$, respectively, and subtracting the equation for $T_1$ from that for $T_2$, we obtain

$$\ln \frac{k_2}{k_1} = -\frac{E_a}{R}\left(\frac{1}{T_2} - \frac{1}{T_1}\right) \tag{14.53}$$

**Example 14.9** The rate constant of the reaction

$$NO_2(g) + O_3(g) \rightarrow NO_3(g) + O_2(g)$$

at 50°C is $2.10 \times 10^5$ (mol L$^{-1}$)$^{-1}$ s$^{-1}$. The reaction occurs 1.90 times faster at 70°C. (a) What is the activation energy? (b) What is the rate constant of the reaction at 100°C?

(a) Letting $T_2 = 343$ K and $T_1 = 323$ K, we obtain on substitution into Equation 14.53

$$\ln 1.90 = -\left(\frac{E_a}{8.314 \text{ J mol}^{-1} \text{ K}^{-1}}\right)\left(\frac{1}{343 \text{ K}} - \frac{1}{323 \text{ K}}\right)$$

$$E_a = 3.0 \times 10^4 \text{ J mol}^{-1} = 30 \text{ kJ mol}^{-1}$$

(b) The rate constant at 100°C is obtained by substituting this value of $E_a$ into Equation 14.53 and solving for $k_2$ at $T = 373$ K:

$$\ln\left(\frac{k_2}{2.10 \times 10^5 \text{ (mol L}^{-1})^{-1} \text{ s}^{-1}}\right) = -\frac{(30 \text{ kJ mol}^{-1})(10^3 \text{ J kJ}^{-1})}{8.314 \text{ J mol}^{-1} \text{ K}^{-1}}\left(\frac{1}{373 \text{ K}} - \frac{1}{323 \text{ K}}\right)$$

$$k_2 = 9 \times 10^5 \text{ (mol L}^{-1})^{-1} \text{ s}^{-1} \quad \blacksquare$$

**b. The Basis of the Arrhenius Equation** We have noted that the rates of molecular collisions in general far exceed the rates of bimolecular reactions. Furthermore, the collision rate constant increases slowly as the square root of the temperature, in marked contrast to the rapid exponential increase of the reaction rate constant. What additional factors must be considered in order to account for the magnitudes and temperature dependence of the rate constants? The kinetic theory of gases (Chapter 4) provides the answer.

Recall that, as a result of the large frequency of molecular collisions, the molecules in a gas display a characteristic distribution of speeds, the Maxwell–Boltzmann distribution (Section 4.6). Figure 14.9 shows the predicted kinetic energy distribution at two different temperatures. The area under either curve extending upwards from some energy $E$ defines the fraction of the molecules with kinetic energy greater than or equal to $E$. If $E$ is sufficiently large to lie in the tail of the distribution then, as shown in the figure, the number of molecules in this interval increases markedly with temperature. The collision energy, that is, the total energy available when two molecules collide, is closely related to the energies of the two molecules and is equal to the sum of their energies in the special case of a head-on collision. Consequently, the probability of large collision energies must also increase markedly with the temperature, suggesting that two molecules can only react if they collide with enough energy. The activation energy $E_a$ can, therefore, be understood as a "barrier" which must be exceeded in a collision in order for a reaction to occur. If the collision energy is less than $E_a$, the molecules merely bounce off each other without reacting. This hypothesis explains the two problematic aspects of the rate constant: first, the reaction rate must be much smaller than the collision rate because only a small fraction of the molecules have high enough energy. Second, the rate increases sharply with temperature owing to the corresponding increase in the fraction of energetic molecules.

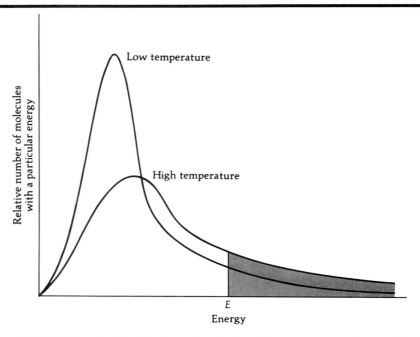

**FIGURE 14.9** The distribution of molecular kinetic energies at low and high temperatures. The shaded area corresponds to the fraction of molecules whose energy is greater than or equal to $E$.

The collision theory is readily modified to take into account the effect of the activation energy. The reaction rate is set equal to the collision rate multiplied by the fraction of collisions with energy at least as large as $E_a$. This fraction can be obtained by integration of the energy distribution and the result is $\exp(-E_a/RT)$. Since the rate constants are proportional to the rates, we can relate the reaction rate constant $k$ to the collision rate constant $k_{col}$ by the expression

$$k = k_{col} e^{-E_a/RT} \tag{14.54}$$

Note that this expression has exactly the same form as the Arrhenius equation, $k = Ae^{-E_a/RT}$, indicating that the frequency factor $A$ can be identified with the collision rate constant. While $k_{col}$ increases slowly with $T$, this variation is much slower than that of the exponential factor and can be neglected. Consequently, $\log k$ varies approximately as $1/T$ (Figure 14.8).

Collision theory modified in this manner leads to fairly good estimates of the rate constants of many bimolecular reactions. A rate constant can frequently be many orders of magnitude smaller than the collision constant because of the effect of the activation energy factor. Table 14.3 lists the activation energies of the bimolecular reactions in Table 14.2 and the values of the exponential factors at two different temperatures. It is seen that relatively modest variations in activation energy result in large changes in the exponential factor and thus account for the great variability of the experimental rate constants. Furthermore, since the exponential involves a ratio of the activation and thermal energies, a moderate increase in temperature can have a major effect on the rate of a given reaction, as already noted.

*c. Orientation Effects*  While the inclusion of the activation energy factor in the expression for the calculated rate constant (Equation 14.54) constitutes a marked improvement over collision theory (Equation 14.49), sizeable discrepancies remain in many cases. Consider, for example, the bimolecular reaction

$$CH_3 + CHCl_3 \rightarrow CH_4 + CCl_3$$

which involves moderately complex reactants. Based on the experimental activation energy of 24 kJ mol$^{-1}$ and a collision theory estimate of the frequency factor, the calculated rate constant at 0°C is about $6 \times 10^6$ (mol L$^{-1}$)$^{-1}$ s$^{-1}$. This

**TABLE 14.3** Activation Energies and $e^{-E_a/RT}$ Factors for the Bimolecular Reactions of Table 14.2

| Reaction | $E_a$ (kJ mol$^{-1}$) | $e^{-E_a/RT}$ $T = 273$ K | $T = 473$ K |
|---|---|---|---|
| H + Br$_2$ → HBr + Br | 3.8 | $1.9 \times 10^{-1}$ | $3.8 \times 10^{-1}$ |
| NO + O$_3$ → NO$_2$ + O$_2$ | 9.6 | $1.4 \times 10^{-2}$ | $8.7 \times 10^{-2}$ |
| NO + NO$_2$Cl → NOCl + NO$_2$ | 28 | $4.4 \times 10^{-6}$ | $8.1 \times 10^{-4}$ |
| CH$_3$ + CHCl$_3$ → CH$_4$ + CCl$_3$ | 24 | $2.6 \times 10^{-5}$ | $2.2 \times 10^{-3}$ |
| CH$_3$ + H$_2$ → CH$_4$ + H | 42 | $9.3 \times 10^{-9}$ | $2.3 \times 10^{-5}$ |
| Br + H$_2$ → HBr + H | 74 | $7.0 \times 10^{-15}$ | $6.8 \times 10^{-9}$ |
| NO + Cl$_2$ → NOCl + Cl | 82 | $2.1 \times 10^{-16}$ | $8.8 \times 10^{-10}$ |
| NOCl + NOCl → 2NO + Cl$_2$ | 108 | $2.2 \times 10^{-21}$ | $1.2 \times 10^{-12}$ |
| NO$_2$ + NO$_2$ → 2NO + O$_2$ | 111 | $5.9 \times 10^{-22}$ | $5.6 \times 10^{-13}$ |

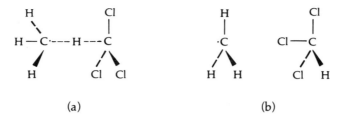

**FIGURE 14.10** Two possible relative orientations of a CH₃ radical and a CHCl₃ molecule in the reaction to form CH₄ and CCl₃: (a) the geometry favors the reaction—the dashed lines represent the new and the old C—H bonds in the act of being formed and broken, respectively; (b) the geometry disfavors the reaction—the carbon and hydrogen atoms are on opposite sides of the approaching molecules.

value is approximately $10^4$ times larger than the measured rate constant at this temperature (Table 14.2). Evidently, the inclusion of the activation energy term does not sufficiently reduce the collision rate constant to result in agreement with experiment. Something is still missing from the theory.

The missing factor has to do with the relative orientation of the reacting molecules at the moment in which old bonds are broken and new ones are formed. Two molecules must approach each other not only with enough energy but with a relative orientation that brings the atoms that are actually involved in the reaction into close proximity. Figure 14.10 shows two possible orientations of a CH₃ radical relative to a CHCl₃ molecule at the moment the reaction occurs. In one case, the carbon atom of the methyl radical is properly oriented with respect to the hydrogen of CHCl₃ to permit the formation of a C—H bond. In the other case, however, the two atoms are in an unfavorable relative position, and the reaction is unlikely to occur.

The factor by which a calculated rate constant must be reduced owing to orientation effects is known as the **steric factor.** As just illustrated, the effect can be important, particularly for complex molecules. The calculation of steric factors is one of the more difficult aspects of chemical kinetics.

## 14.9 The Activated Complex

We are now ready to examine the reasons for the presence of an activation energy in chemical reactions. It must be made clear at the outset that the magnitude of the activation energy is not related to the overall energy released or absorbed in the reaction. A particular reaction can be either exothermic or endothermic and, at the same time, have either a small or a large activation energy. The activation energy is not related to the properties of the reactants and products but, rather, to the intimate details of the conversion of the one to the other. Put rather simply (Figure 14.11), when two molecules approach each other closely enough for their electron clouds to overlap, they feel a repulsive potential (see Figure 4.15). The repulsion causes them to slow down as their kinetic energy is converted to potential energy. If the relative velocity of the molecules prior to collision is low, the molecules fly apart before there is significant overlap of their electron clouds. However, when the molecules approach each other with a high relative velocity, their electron clouds can interpenetrate. Under these conditions chemical bonds can be broken and formed.

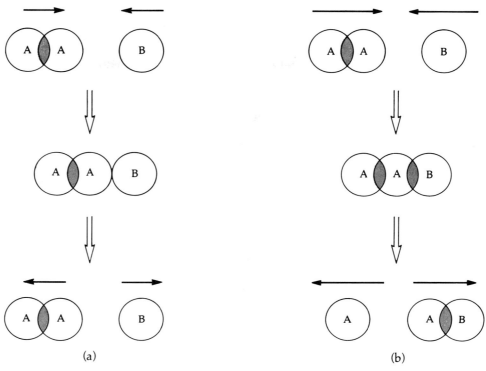

**FIGURE 14.11** Possible outcomes of the collision between two molecules. (a) The molecules collide with low relative velocity and bounce off before their electron clouds interpenetrate significantly. (b) The molecules collide with high relative velocity. The A—A bond is broken, and an A—B bond is formed. The reaction is $A_2 + B \rightarrow A + AB$.

In other words, a chemical reaction can occur. In actuality, this process is rather complicated, involving changes in molecular shapes, redistribution of energy among different bonds, and ultimately, the breaking of old bonds and the formation of new ones.

Let us examine the changes in potential energy that accompany a molecular collision such as that illustrated in Figure 14.11. To be specific, consider the following simple reaction, involving the participation of only three atoms,

$$H + F_2 \rightarrow HF + F$$

The simplest situation is that in which the nuclei of the three atoms are colinear, that is, they lie on a single straight line throughout the reaction. It is this case, illustrated in Figure 14.12, that we will examine in detail. We plot the potential energy of the system as a function of a "reaction coordinate" that represents the gradual change of the three atoms from reactants to products, and is a measure of the extent to which the reaction has proceeded. Initially, as the hydrogen atom approaches the fluorine molecule from afar, the energy of the system is that of the isolated atom and molecule and is constant (reaction coordinate = 0). As the two species come sufficiently close that their orbitals start to overlap, the F—F bond begins to stretch and simultaneously an incipient

**536** Kinetics—The Rates and Mechanisms of Chemical Reactions  14.9

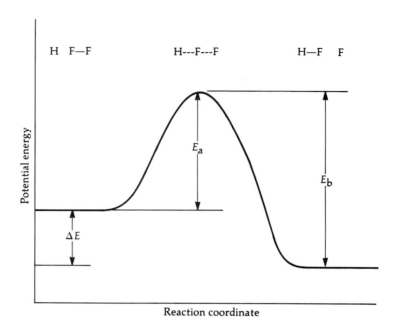

**FIGURE 14.12** Potential energy versus reaction coordinate for the linear system of atoms in the reaction H + F$_2$ → HF + F. The configuration of the atoms at various stages of the reaction is shown at the top. The activation energy $E_a$ is equal to the energy difference between the activated complex and the reactants. The energy difference between products and reactants corresponds to the energy change $\Delta E$ in the overall reaction. The activation energy for the backward reaction, $E_b$, is shown.

bond between the hydrogen atom and one of the fluorine atoms begins to form. This configuration is less stable than that of the separated species, and the potential energy consequently rises. As the hydrogen atom comes closer, the F—F bond stretches even further, and the H—F bond becomes stronger. The potential energy rises to a maximum value corresponding to a state in which one of the fluorine atoms is partially bonded to both the hydrogen atom and the other fluorine atom. This state is known as the **activated complex** or **transition state** and its energy relative to that of the separated reactants is equal to the activation energy. The transition state is not a stable state and falls apart rapidly as the H—F bond becomes stronger while the F—F bond breaks. The new species then move apart and the potential energy of the system decreases to that of the products. The shape of the potential energy curve is that of a barrier which the reactants must surmount in order to react.

Figure 14.12 shows that the products of the reaction between H and F$_2$ lie at a lower energy than the reactants. This energy change corresponds to $\Delta E$ for the overall reaction and, as noted in Section 10.5, is closely related to the enthalpy change $\Delta H$. In this particular case, the reaction is exothermic and $\Delta E$ is negative. The reverse reaction would, of course, be endothermic. For this reaction to occur, the product molecules must follow the path shown in Figure 14.12 in the reverse direction. Since both forward and reverse reactions go through the same activated complex, the activation energy for the reverse reaction, $E_b$, is related to that for the forward reaction and to the overall energy change as

$$\Delta E = E_a - E_b \tag{14.55}$$

The preceding description of the interaction between a hydrogen atom and a fluorine molecule must be recognized as an oversimplification. In any real reaction, the three atoms are not likely to be colinear but instead will approach each other at all angles. Each direction of approach generally results in a slightly different potential energy curve with a somewhat different value of the activation energy.

Potential energy curves of the type shown in Figure 14.12 are derived from more complete diagrams of the potential energy as a function of all the relevant internuclear distances in the reacting system. Consider, for example, the reaction

$$H_2 + Br \rightarrow H + HBr$$

in which, for simplicity, the three atoms are assumed to be colinear at all times. The two relevant internuclear distances are $R_{H-H}$, the distance between the two hydrogen atoms, and $R_{H-Br}$, the distance between hydrogen and bromine. The potential energy of the three atoms corresponds to a three-dimensional surface in a plot of energy versus $R_{H-H}$ and $R_{H-Br}$. Instead of attempting to represent such a three-dimensional plot in two dimensions, it is customary to make cuts through the potential energy surface at various energy values. In a two-dimensional representation, each cut is shown as a curve of constant potential energy, called a contour line. The contour diagram of the potential energy consists of the various contour lines in a plot of $R_{H-H}$ versus $R_{H-Br}$.

Figure 14.13a is a contour diagram of the potential energy in the reaction between $H_2$ and Br. Point (1) corresponds to the initial state, a bromine atom at a large distance (>2.5 Å) from a hydrogen molecule, the atoms of which are at the normal internuclear distance of 0.74 Å. Point (2) corresponds to the final state, a hydrogen atom at a large distance (>2.0 Å) from an HBr molecule, for which $R_{H-Br} = 1.41$ Å. The reaction can be represented by a path between points (1) and (2). The dashed curve labelled A represents one such path. Observe that this curve crosses the potential energy contours of 20, 40, 60, 85, 105, 125, and 150 kJ mol$^{-1}$. Figure 14.13b gives a plot of the potential energy along this reaction path. Clearly, many such curves can be drawn. The path that is energetically most favored is the one that involves the lowest maximum potential energy. This path is shown by curve B in Figure 14.13a (drawn in color), while Figure 14.13b shows how the potential energy varies along this path. The maximum potential energy is reached at point (3), and this point corresponds to the activated complex.

Figure 14.13 provides a more complete understanding of the activated complex than can be obtained from Figure 14.12. All paths between reactants and products involve a maximum in the potential energy, as illustrated in Figure 14.13b. The path, involving the lowest maximum potential energy (curve B) is favored energetically and its maximum energy relative to the reactants corresponds to the activation energy. Figure 14.14 shows an analogy to the passage through the activated complex that may be of help in visualizing this process.

Calculated potential energy surfaces, such as that depicted in Figure 14.13a, must be checked by experiment. A powerful modern experimental technique is the study of chemical reactions by means of molecular beams (Figure 14.15). The two reactants are formed into separate beams, and the reac-

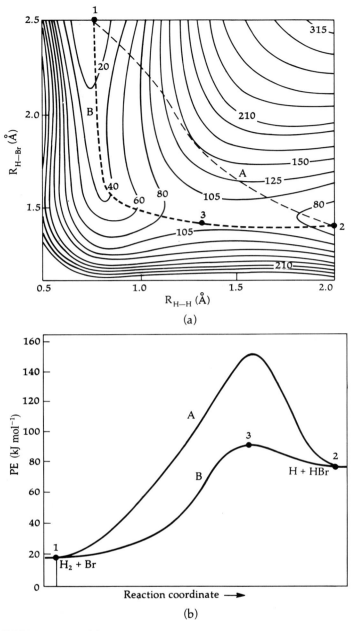

**FIGURE 14.13** (a) Contour diagram of the potential energy surface in the reaction $H_2 + Br \rightarrow H + HBr$ as a function of H—H and H—Br internuclear distances (labelled $R_{H-H}$ and $R_{H-Br}$). All three atoms are assumed to be colinear during the entire reaction. The solid curves represent contours of equal potential energy the values of which are given in kJ mol$^{-1}$. Point (1) is the initial state—a bromine atom far from an $H_2$ molecule; point (2) is the final state—a hydrogen atom far from an HBr molecule. The dashed curves show two possible paths between (1) and (2). Path A crosses some high-energy contours (e.g., 125 kJ mol$^{-1}$) and is energetically unfavored. Path B (in color) represents the route requiring the crossing of the lowest possible potential energy contours between points (1) and (2). Point (3) (at 95 kJ mol$^{-1}$) lies at the highest energy along this path and is the activated complex. (b) Variation of the potential energy along the reaction coordinate for paths A and B. Points (1), (2), and (3), have the same meaning as in (a). Observe that the products lie at a higher energy than the reactants—the reaction is endothermic.

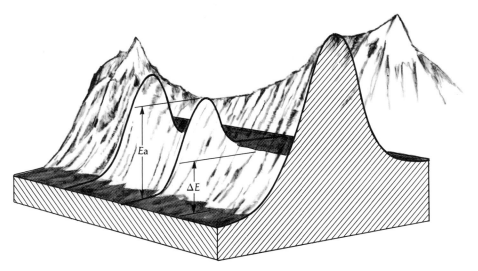

**FIGURE 14.14** Traversal of a mountain range separating two valleys—an analogy to the passage of reactants through the activated complex. The difference in the elevation of the valley floors corresponds to the energy difference $\Delta E$ between reactants and products. There are many paths that can be followed in order to reach the second valley, and a few are shown. Each path involves a climb to a different maximum elevation. The energetically most favored path (in color) involves the traversal of the lowest mountain pass. This pass corresponds to the activated complex, and its height relative to the first valley corresponds to the activation energy $E_a$.

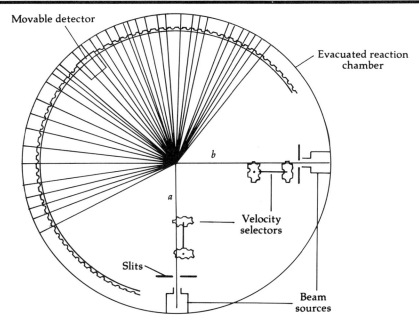

**FIGURE 14.15** Diagram of a crossed molecular beam apparatus. The beam sources are containers from which molecules of each species effuse with a distribution of speeds. A velocity selector (see Figure 4.11) permits only molecules of a specific energy to pass. The two beams (labelled *a* and *b*) are collimated by means of slits so that they intersect in a well-defined region. Reaction products made in this intersection region are scattered at various angles where they can be detected by a movable detector such as a mass spectrometer. The reaction chamber is highly evacuated in order to minimize the likelihood of collisions with air molecules.

539

tion occurs in the region in which the beams cross. Each beam consists of molecules of a specific energy, and the reaction products can be detected as a function of angle relative to either beam. The location of the activated complex in the potential energy surface, as well as the relative orientation of the reacting atoms in the activated complex, can be inferred from the dependence of the yield of product molecules on angle and on the energies of the reacting molecules.

## 14.10 Rates and Mechanisms of Chemical Reactions in Solution

The presence of a solvent substantially alters the reaction path followed by two molecules or ions from that described for gas phase reactions. Since molecules in a condensed phase are in close proximity to each other, their motion is very different from what it is in a gas. A solute molecule is surrounded by a number of solvent molecules in what is known as a **solvent cage,** schematically shown in Figure 14.16a. A molecule confined to such a cage will typically bounce a number of times against the nearby solvent molecules that make up the walls of the cage before diffusing away to a new location. In order for two different solute molecules to react with each other they must first diffuse into the same solvent cage, as illustrated in Figure 14.16b. Such a process is called an **encounter** and constitutes the key first step in a reaction occurring in solution.

The frequency of encounters is determined by the rate at which solute molecules diffuse through the solvent as well as by their concentration. Diffusive motion in a liquid requires that the solute molecules squeeze through the constantly changing openings between the densely packed, ever-moving solvent molecules. It is therefore reasonable to suppose that the motion of molecules in solution depends on properties of both the solvent and the solute. If the solvent has high viscosity, that is, if it is comprised of large, non-spherical, strongly interacting molecules, the passage of solute molecules will be impeded. Non-polar liquids of low molecular weight (e.g., heptane) have low viscosity and permit more rapid diffusion than polar solvents such as water. The effective size of the solute molecules is also of importance, smaller molecules being able to diffuse more rapidly than large ones. However, if the solute molecules are strongly solvated (i.e., attached to molecules of solvent), their effective size will be larger than that of bare molecules and their diffusion rate will be correspondingly slower. Additional factors come into play if the two re-

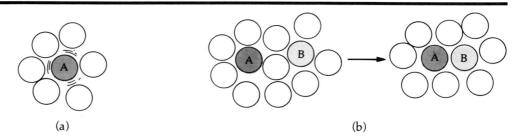

**FIGURE 14.16** Molecular collisions in solution. (a) Solute molecule A confined to a solvent cage. (b) Solute molecules of type A and B diffuse into the same cage and encounter each other.

acting species are ionic. For example, the long-range attraction between oppositely charged ions causes them to diffuse towards each other more rapidly than would otherwise be the case.

If two solute molecules were to react as soon as they came into contact in the same solvent cage, their reaction rate would be limited only by their diffusion rate. The rate constant of such **diffusion-controlled reactions** is typically $10^9$ (mol L$^{-1}$)$^{-1}$ s$^{-1}$ or greater. Reactions of this type occur on a time scale of nanoseconds or less, far more rapidly than other reactions in solution. Reactions are diffusion-controlled when they have little or no activation energy. This behavior is characteristically displayed by reactions in which two atoms with unpaired electrons or two free radicals combine. For example, the recombination of iodine atoms dissolved in n-hexane,

$$I\cdot + \cdot I \rightarrow I{-}I$$

has a rate constant of $1.8 \times 10^{10}$ (mol L$^{-1}$)$^{-1}$ s$^{-1}$ at 50°C and is diffusion-controlled.

Most reactions in solution are not limited by the rate of diffusion but occur more slowly. In order to understand the mechanism of such slower reactions we must examine what happens in an encounter in more detail. As already indicated in Figure 14.16, two reactant molecules move into the same solvent cage at a rate determined by their diffusion rates and their concentrations. The duration of such an encounter (typically $10^{-11}$ to $10^{-10}$ s) is relatively long compared to the time it takes the closely confined molecules to bump into each other. Consequently, two molecules may collide as many as a few hundred times before they diffuse away from each other. The rate of molecular collisions in solution is therefore very non-uniform: active periods of very frequent collisions that characterize an encounter alternate with quiet periods when the reacting molecules are in separate solvent cages. Figure 14.17 schematically illustrates this behavior and contrasts it with that in the gas phase. The rate of molecular collisions in a gas is determined by the random motion of the molecules and is essentially uniform. It turns out that the rates of collisions in the two phases are roughly comparable; however, their distribution in time is totally different.

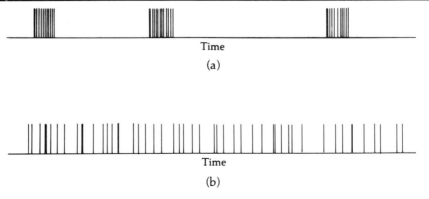

**FIGURE 14.17** (a) Distribution in time of the collisions between molecules of type A and B in solution. Each vertical line represents a collision and each bunch of collisions marks the occurrence of an encounter. (b) A similar distribution for collisions in a gas.

## 542 Kinetics—The Rates and Mechanisms of Chemical Reactions 14.10

**TABLE 14.4 Experimental Values of Arrhenius Equation Parameters for Second-order Reactions in Solution**

| Reactants | Solvent | $A$ $(mol\ L^{-1})^{-1}\ s^{-1}$ | $E_a$ $(kJ\ mol^{-1})$ |
|---|---|---|---|
| $CH_3Br + I^-$ | $H_2O$ | $1.68 \times 10^{10}$ | 76.4 |
|  | $CH_3OH$ | $2.6 \times 10^{10}$ | 76.4 |
| $CH_3Cl + CH_3O^-$ | $CH_3OH$ | $1.76 \times 10^{12}$ | 100 |
| $C_2H_5Br + (C_2H_5)_2S$ | $C_6H_5OH$ | $1.40 \times 10^{11}$ | 107 |

The outcome of an encounter depends on the activation energy of the reaction in question. If the activation energy is small or non-existent, as in the case of diffusion-controlled reactions, the reaction occurs in the course of a single encounter. However if the activation energy is sufficiently large, many encounters are needed before a reaction can occur. Such reactions are said to be **activation controlled**. The difference in the distribution in time of the collision frequencies in a gas and a liquid illustrated in Figure 14.17 becomes unimportant in such circumstances and the modified collision theory developed in Section 14.8 is applicable. Table 14.4 gives the parameters in the Arrhenius equation (Equation 14.51) for some activation controlled reactions.

---

**Example 14.10**  Assuming 200 collisions per encounter, estimate the average number of collisions and encounters required for the reaction between a methyl bromide ($CH_3Br$) molecule and an iodide ion in methanol at 25°C (see data in Table 14.4).

The frequency factor $A$ in the Arrhenius equation, $k = Ae^{-E_a/RT}$, can be interpreted as the collision rate constant $k_{col}$ (see Equation 14.54), on the assumption that the steric factor is unity. Since the rate constant is proportional to the reaction rate, the ratio $k/k_{col} = e^{-E_a/RT}$ is equal to the ratio of reaction and collision rates. The reciprocal of this ratio is therefore equal to the ratio of the collision to the reaction rates and is the desired number of collisions per reaction:

$$\frac{\text{Rate of collisions}}{\text{Rate of reactions}} = \frac{\text{no. of collisions/s}}{\text{no. of reactions/s}}$$

$$= \text{no. of collisions per reaction} = e^{E_a/RT}$$

$$= \exp\frac{(76.4\ kJ\ mol^{-1})(10^3\ J\ kJ^{-1})}{(8.314\ J\ mol^{-1}\ K^{-1})(298\ K)}$$

$$= 2 \times 10^{13}$$

If there are 200 collisions per encounter, the average number of encounters per reaction is

$$(2 \times 10^{13})/200 = 1 \times 10^{11}$$

This large value may be contrasted with a value of one encounter per reaction for a diffusion controlled reaction. ∎

The properties of the activated complex must be of great importance for all but diffusion-controlled reactions, for which there is essentially no activation barrier. Recall from Section 14.9 that the properties of the activated complex in gas phase reactions depend on the relative potential energy and orientation of the reacting molecules. For reactions in solution, the solvent must also be of importance since the activated complex is confined to a cage where it is in intimate contact with solvent molecules. In general, if a particular solvent can stabilize the activated complex of a reaction, the reaction in question will occur more rapidly in that solvent. For example, activated complexes which are ionic are more stable in a polar solvent such as water, in which they can be stabilized by solvation, than in a non-polar solvent such as heptane. The conversion of $(CH_3)_3CCl$ to $(CH_3)_3COH$ in the presence of base, for example, occurs more rapidly in water than in organic solvents, presumably because the activated complex involves the formation of the $(CH_3)_3C^+$ ion.

## 14.11 Catalysis

The rate of a chemical reaction can frequently be speeded up by the presence of a **catalyst,** a substance that is not permanently changed by the reaction and therefore does not appear in the overall chemical equation. The effect of a catalyst can be dramatic. The reaction between hydrogen and oxygen, for instance, occurs so slowly in the absence of an initiating spark as to be undetectable. However, the addition of finely divided platinum, commonly known as platinum black, causes the reaction to occur with explosive speed. Catalysis is of great importance in the industrial production of chemicals. The reactions which form the basis of the manufacture of such important substances as ammonia and sulfuric acid only become economical as a result of catalysis. To cite a totally different example, many of the complex reactions occuring in living systems happen as a result of the action of highly specific catalysts known as **enzymes.**

A catalyst works by providing a different mechanism for the reaction it catalyzes. The activation energy of the rate-determining step of the catalyzed reaction is lower than that of rate-determining step of the uncatalyzed reaction. Recall from Section 14.8 that the presence of the activation barrier prevents most collisions between two reactant molecules from leading to a reaction. Only when the molecules have enough energy can they surmount the barrier and react. Thus, if the activation barrier is lowered, a larger fraction of the collisions lead to a reaction, and the rate increases.

To cite a specific example, the decomposition of hydrogen peroxide,

$$2H_2O_2(aq) \rightarrow 2H_2O + O_2(g)$$

has an activation energy of 76 kJ mol$^{-1}$ and occurs rather slowly at room temperature. The addition of some iodide ion provides a different path for the reaction, with an activation energy of only 57 kJ mol$^{-1}$. The increase in the rate of the catalyzed reaction can be estimated by means of the Arrhenius equation. We write Equation 14.51 as a ratio of the rate constants of the catalyzed reaction, $k_{cat}$, and the uncatalyzed reaction, $k$:

$$\frac{k_{cat}}{k} = \frac{Ae^{-E_a(cat)/RT}}{Ae^{-E_a/RT}} = e^{-[(E_a(cat)-E_a)/RT]} \tag{14.56}$$

Substituting the preceding values of the activation energies, we obtain at $T = 298$ K

$$\frac{k_{cat}}{k} = \exp -\left[\frac{(57 - 76) \text{ kJ mol}^{-1} (10^3 \text{ J kJ}^{-1})}{(8.314 \text{ J mol}^{-1} \text{ K}^{-1})(298 \text{ K})}\right]$$

$$= 2.1 \times 10^3$$

indicating that the catalyzed reaction occurs some two thousand times faster than the uncatalyzed reaction.

Figure 14.18 gives a graphical comparison of the energy changes occurring in a catalyzed and an uncatalyzed reaction. Catalysis does not change the position of chemical equilibrium. Thus, a reduction in the activation energy of a reaction must be accompanied by a concomitant reduction in the activation energy of the reverse reaction. If this were not the case, the ratio of the rate constants of the forward and backward reactions would be different in the presence of a catalyst, and so would the equilibrium constant (see Equation 14.37). Figure 14.18 also shows that the overall energy change in the reaction is unaffected by the addition of a catalyst.

Although the mechanisms of catalysis are highly variable, it is nonetheless convenient to group them into two broad categories: **homogeneous catalysis,** in which the catalytic process occurs in a single phase, such as a solution, and **heterogeneous catalysis,** in which the reaction occurs at a phase boundary. A typical example of the latter is a reaction between gaseous molecules on the surface of a metal catalyst.

A large variety of reactions in aqueous solution can be speeded up by the addition of catalysts. It is instructive to examine the mechanism of one such reaction in detail in order to see precisely how a catalyst works.

The rates of many redox reactions can be accelerated by the addition of small amounts of transition metal ions. These metals typically have several oxidation states that can be formed under the conditions of the reaction and their alternate oxidation and reduction catalyzes the reaction of interest. Consider,

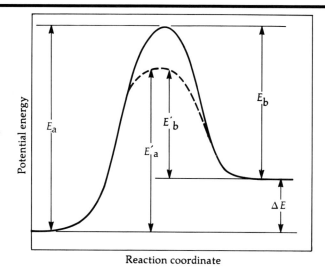

**FIGURE 14.18** Effect of a catalyst on the potential energy changes occurring in a reaction. The activation energy is lowered from $E_a$ to $E'_a$. The activation energy of the reverse reaction is lowered to exactly the same extent from $E_b$ to $E'_b$. The overall energy change $\Delta E$, as well as the equilibrium constant, remain unchanged.

for example, the oxidation of thallium(I) by cerium(IV),

$$Tl^+ + 2Ce^{4+} \rightarrow Tl^{3+} + 2Ce^{3+}$$

The uncatalyzed reaction occurs slowly because it requires the simultaneous interaction of three ions. The addition of a small amount of manganese(II) makes the following three-step mechanism possible:

(1) $Ce^{4+} + Mn^{2+} \rightarrow Ce^{3+} + Mn^{3+}$

(2) $Ce^{4+} + Mn^{3+} \rightarrow Ce^{3+} + Mn^{4+}$

(3) $\underline{Tl^+ + Mn^{4+} \rightarrow Tl^{3+} + Mn^{2+}}$

$\phantom{(3)}\ Tl^+ + 2Ce^{4+} \rightarrow Tl^{3+} + 2Ce^{3+}$

The catalyzed reaction obeys the rate law

$$R = k[Ce^{4+}][Mn^{2+}]$$

indicating that the first step is rate-determining. In this particular instance, the addition of an $Mn^{2+}$ catalyst makes possible a reaction path in which three relatively fast bimolecular reactions replace one very slow termolecular reaction.

Many other examples of homogeneous catalysis can be cited. A particularly important category is the catalysis of organic reactions by acids or bases (Section 22.4). The enzyme catalysis of biochemical reactions in living organisms is another example of homogeneous catalysis. Specific enzymes are particularly adept at catalyzing particular reactions. Typically, the catalyzed reaction occurs many orders of magnitude more rapidly than the uncatalyzed reaction. Many of the reactions required to maintain life would not occur in the absence of enzyme catalysis.

A number of metals catalyze gas phase reactions by providing a surface to which the reacting species can attach themselves and, in so doing, be converted into a more reactive form. This chemical attachment is called **chemisorption,** and the reactions in which it occurs are examples of heterogeneous catalysis. To be effective, the catalyst must have a large surface area and must thus be present in finely divided form. As in the case of homogeneous catalysis, a variety of different catalytic mechanisms are possible.

A common mechanism involves the dissociation of the chemisorbed species. The resulting molecular fragments can wander across the surface until they encounter another fragment or until they react with a gaseous molecule incident on the surface.

The hydrogenation of ethylene in the presence of platinum or nickel is a good example of such a process. The uncatalyzed reaction,

$$C_2H_4(g) + H_2(g) \rightarrow C_2H_6(g)$$

occurs at immeasurably slow speed at moderate temperatures. The catalytic surface promotes the dissociation of chemisorbed hydrogen, a process that can schematically be represented by the equation

$$H_2 + 2M \rightarrow \underset{\underset{|}{M}}{H}\text{---}\underset{\underset{|}{M}}{H} \rightarrow \underset{\underset{|}{M}}{H} + \underset{\underset{|}{M}}{H}$$

where M is a surface metal atom. At the same time, ethylene molecules can attach to the surface by forming bonds from the two carbon atoms:

$$H_2C=CH_2 + 2M \rightarrow \begin{array}{c} H_2C-CH_2 \\ | \quad \; | \\ M \quad M \\ | \quad \; | \end{array}$$

Collisions between these chemisorbed species will occur in due course and eventually result in the successive addition of two hydrogen atoms:

$$\begin{array}{c} H_2C-CH_2 \\ | \quad \; | \\ M \quad M \\ | \quad \; | \end{array} + \begin{array}{c} H \\ | \\ M \\ | \end{array} \rightarrow \begin{array}{c} H_3C-CH_2 \\ | \\ M \\ | \end{array} + 2M$$

followed by

$$\begin{array}{c} H_3C-CH_2 \\ | \\ M \\ | \end{array} + \begin{array}{c} H \\ | \\ M \\ | \end{array} \rightarrow H_3C-CH_3 + 2M$$

The activation energy barrier for the addition of atomic hydrogen is substantially lower than that for the addition of molecular hydrogen and the reaction is consequently speeded up.

This is just one example of a large variety of important gas phase reactions catalyzed by various metals, metal oxides, and other substances. Additional examples include the formation of ammonia from the elements in the presence of various metal oxides, the catalytic oxidation of various nitrogen oxides formed in automobile engine operation (Section 3.8a), and the large variety of reactions used in petroleum refining.

## Conclusion

A chemical reaction can be characterized by a rate law—a statement of the dependence of the reaction rate on the concentrations of the reacting species. The proportionality constant in this relation, which is a measure of the reaction rate, is the rate constant. The rates of chemical reactions range from imperceptibly slow to near instantaneous, reflecting the nature of the path followed by two reacting molecules in their conversion to products. This path can be changed by the addition of suitable catalysts and in this way the rate of a reaction can be increased, frequently by many orders of magnitude.

The sequence of elementary reactions that comprise the reaction path is known as the reaction mechanism. While the rate law imposes constraints on the nature of the mechanism, it does not uniquely determine it. The slowest elementary reaction in a reaction mechanism generally determines the overall reaction rate. The maximum rate at which an elementary reaction can occur is determined by the rate of molecular collisions. Most reactions occur considerably slower because only molecules with enough energy to overcome the potential energy barrier that generally is found in the reaction path can react. The height of this barrier, commonly known as the activation energy, can be deduced from the exponential temperature dependence of the reaction rate.

## Problems

1. The reaction $2MnO_4^- + 10I^- + 16H^+ \rightarrow 2Mn^{2+} + 5I_2 + 8H_2O$ occurs at a rate R. Write expressions relating R to the rate of change of the concentration of each of the reactants and products.

2. In acidic solution $Cr_2O_7^{2-}$ oxidizes $HNO_2$ to $NO_3^-$ and is reduced to $Cr^{3+}$. During a certain short time interval, the concentration of $Cr_2O_7^{2-}$ decreases at a rate of $1.5 \times 10^{-4}$ mol $L^{-1}$ $s^{-1}$. Find the rates of change of the concentration of $HNO_2$ and of the reaction products during this time interval.

3. The rate of the reaction $2Br^- + H_2O_2 + 2H^+ \rightarrow Br_2 + 2H_2O$ is first order in each of the reactants. Write the rate law. What are the units of the rate constant?

4. The reaction $2NO + O_2 \rightarrow 2NO_2$ is first order in the concentration of oxygen and second order in that of nitric oxide. Write the rate law in terms of the rate of change of the concentration of each of the reactants and product.

5. The thermal decomposition of dinitrogen pentoxide, $N_2O_5(g) \rightarrow 2NO_2(g) + \frac{1}{2}O_2(g)$ obeys first-order kinetics. The rate constant at 25°C is $3.4 \times 10^{-5}$ $s^{-1}$. At what rate does $N_2O_5$ decompose at this temperature if its concentration is 0.100 mol $L^{-1}$?

6. The kinetics of the reaction $A + B \rightarrow C$ was studied by the initial rate method. The rate of change of the initial concentration of A, $(d[A]/dt)_0$, was determined by drawing a tangent through the curve of [A] versus $t$ at $t = 0$. The following results were obtained in separate experiments:

| $[A]_0$ (mol $L^{-1}$) | $[B]_0$ (mol $L^{-1}$) | $(d[A]/dt)_0$ (mol $L^{-1}$ $s^{-1}$) |
|---|---|---|
| 0.50 | 1.0 | $-1.9 \times 10^{-3}$ |
| 0.25 | 0.50 | $-2.4 \times 10^{-4}$ |
| 0.25 | 0.25 | $-1.2 \times 10^{-4}$ |
| 0.10 | 0.10 | $-7.4 \times 10^{-6}$ |
| 0.50 | 0.10 | $-1.9 \times 10^{-4}$ |

Assuming that the rate law has the form $-d[A]/dt = k[A]^m[B]^n$, determine the values of $k$, $m$, and $n$. What is the overall reaction order?

7. The following data on the rate of the reaction $2NO(g) + Cl_2(g) \rightarrow 2NOCl(g)$ were obtained at a certain temperature:

| [NO] (mol $L^{-1}$) | $[Cl_2]$ (mol $L^{-1}$) | $-d[Cl_2]/dt$ (mol $L^{-1}$ $s^{-1}$) |
|---|---|---|
| 0.200 | 0.200 | $8.00 \times 10^{-2}$ |
| 1.00 | 0.200 | 2.00 |
| 0.200 | 1.00 | 0.400 |

(a) What is the order of the reaction with respect to each reactant, and what is the overall order? (b) What is the rate constant? (c) If 1.00 mol of NO and 2.00 mol of $Cl_2$ are mixed in a 5.00 L vessel, how much $Cl_2$ reacts during the first $1.00 \times 10^{-6}$ s?

8. Under certain conditions, the reaction $H_2 + Br_2 \rightarrow 2HBr$ obeys the rate law $R = k[H_2]^l[Br_2]^m[HBr]^n$. The rate was found to depend on the concentrations in the following manner.

| $[H_2]$ (mol $L^{-1}$) | $[Br_2]$ (mol $L^{-1}$) | [HBr] (mol $L^{-1}$) | Rate (mol $L^{-1}$ $s^{-1}$) |
|---|---|---|---|
| 0.10 | 0.10 | 1.0 | R |
| 0.10 | 0.40 | 1.0 | 8R |
| 0.20 | 0.10 | 1.0 | 2R |
| 0.10 | 0.20 | 1.5 | 1.89R |

Determine the values of the exponents $l$, $m$, and $n$.

9. The rate law for the hydration of carbon dioxide is $-d[CO_2]/dt = k_1[CO_2] + k_2[CO_2][OH^-]$, where $k_1 = 4.2 \times 10^{-2}$ $s^{-1}$ and $k_2 = 9.3 \times 10^3$ (mol $L^{-1})^{-1}$ $s^{-1}$ at 27°C. The presence of two terms in the rate law indicates that the reaction proceeds via two pathways. At what pH will the reaction proceed at an equal rate via the two pathways?

10. Reactant A decomposes according to the reaction

A → B + C. In an experiment in which $[A]_0 = 0.840$ mol L$^{-1}$, the concentration of B was measured as a function of time with the following results:

| $t$ (s) | [B] (mol L$^{-1}$) | $t$ (s) | [B] (mol L$^{-1}$) |
|---|---|---|---|
| 0 | 0.0 | | |
| 10.0 | 0.202 | 50.0 | 0.526 |
| 20.0 | 0.330 | 60.0 | 0.562 |
| 30.0 | 0.412 | 70.0 | 0.588 |
| 40.0 | 0.478 | 80.0 | 0.612 |

By plotting the data in the form of ln[A] versus $t$ and 1/[A] versus $t$ determine whether the reaction is first or second order in A, or whether some other rate law applies. If a fit to first- or second-order kinetics is obtained, determine the value of the rate constant.

11. The decomposition of a certain substance follows first-order kinetics. If the half-life of the decomposition is 1.05 h, how long does it take for 99.0% of the substance to decompose?

12. A certain substance decomposes according to a second-order rate law. When the initial concentration at 25°C is 0.200 $M$, it is observed that the concentration decreases to 0.150 $M$ in 2.00 h. If the initial concentration is 0.400 $M$, what will the concentration be after 4.50 h?

13. A certain reaction obeys first-order kinetics. It is found that 30% of the reactant initially present reacts in 30 min. What are the values of the half-life and the rate constant?

14. A plot of the reciprocal of the partial pressure of a certain reactant versus time is linear. The slope is $4.5 \times 10^{-3}$ atm$^{-1}$ s$^{-1}$, and the intercept is 200 atm$^{-1}$. What is the half-life for this reaction?

15. A certain compound, A, decomposes in aqueous solution, and its concentration varies with time in the following manner:

| $t$ (min) | [A] mol L$^{-1}$ |
|---|---|
| 0 | 1.00 |
| 1.00 | 0.871 |
| 6.50 | 0.406 |
| 10.0 | 0.250 |

Determine the rate law and evaluate the rate constant. Predict the remaining concentration of A after 30 min.

16. The composition of the reaction mixture in the reaction 2A(g) → B(g) is monitored by measurement of the total gas pressure. The following data are obtained:

| $t$ (s) | $P$ (torr) |
|---|---|
| 0 | 800 |
| 10 | 622 |
| 20 | 554 |
| 30 | 518 |
| 40 | 495 |

Determine the order of the reaction and the value of the rate constant. At what time will the reaction be 99.9% complete?

17. The composition of the reaction mixture in the reaction A(g) → 2B(g) is determined by measurement of the total pressure. The following data are obtained:

| $t$ (s) | $P$ (torr) |
|---|---|
| 0 | 400 |
| 10 | 504 |
| 20 | 580 |
| 30 | 637 |
| 40 | 680 |

(a) Determine the order of the reaction and the value of the rate constant. (b) At what time will the reaction be 99.9% complete? (c) Can the total pressure be used to determine the kinetics of any gas phase reaction? Explain.

18. The decomposition of $N_2O_5$ dissolved in $CCl_4$ is studied. The rate constant at a certain temperature is determined as $5.6 \times 10^{-4}$ s$^{-1}$. If the initial concentration of $N_2O_5$ is 0.360 mol L$^{-1}$, how long will it take for the concentration to decrease to 0.240 mol L$^{-1}$?

19. The decomposition of oxalic acid at elevated temperatures involves the reaction $H_2C_2O_4(g) \rightarrow$ HCOOH(g) + $CO_2$(g). Find the rate law of the reaction from the following data, which give the total pressure reached after 5.56 h from the indicated starting pressures of oxalic acid. For each set of data, determine the reaction half-life.

| $P_{H_2C_2O_4}$(initial) (torr) | $P_{tot}$ (torr) |
|---|---|
| 5.00 | 7.19 |
| 7.00 | 10.0 |
| 8.44 | 12.2 |

20. The rate of decomposition of $N_2O_3$ to $NO_2$ and NO is studied. In a certain experiment, the following data on the variation of [$NO_2$] with time are obtained:

| Time (s) | [$NO_2$] (mol L$^{-1}$) | Time (s) | [$NO_2$] (mol L$^{-1}$) |
|---|---|---|---|
| 0 | 0 | 2800 | 0.502 |
| 500 | 0.135 | $5.0 \times 10^4$ | 0.910 |
| 1500 | 0.347 | $7.0 \times 10^4$ | 0.910 |

Determine the rate law and the value of the rate constant. Assume that the reaction goes to completion.

21. The kinetics of the reaction $CH_3CHO(g) \rightarrow CH_4(g) + CO(g)$ is studied. The following data on the variation of $[CH_3CHO]$ with time are obtained in a particular experiment:

| Time (s) | $[CH_3CHO]$ (mol L$^{-1}$) |
|---|---|
| 0 | 0.250 |
| 1000 | 0.118 |
| 2000 | 0.0770 |
| 3000 | 0.0572 |
| 4000 | 0.0455 |

(a) Determine the order of the reaction in $CH_3CHO$ and the value of the rate constant. (b) At what time will 99.99% of the reactant have decomposed?

22. In the reaction $A \rightarrow B + C$, it is observed that the concentration of A decreases from 0.40 mol L$^{-1}$ to 0.20 mol L$^{-1}$ in 100 s. What will the concentration of A be after another 100 s if the reaction is (a) first order in A, (b) second order in A, (c) zeroth order in A?

23. Two substances, A and B, each decompose by first-order kinetics and their half-lives are 40.0 min and 15.0 min, respectively. A mixture of A and B initially contains equal concentrations of the two. At what time will the concentration of A be 3.00 times larger than that of B?

24. The reaction $2A \rightarrow B$ obeys the rate law $dB/dt = kA^2$. In an experiment in which the initial concentration of A is 0.165 mol L$^{-1}$, the concentration of B after 400 s has increased from 0 to 0.0714 mol L$^{-1}$. (a) What is the value of the rate constant? (b) What is the concentration of A after 850 s?

25.* Derive (a) the integrated rate law and (b) an expression for the half-life of a reaction that obeys the rate law $-d[A]/dt = k[A]^3$.

26.* Derive (a) the integrated rate law and (b) an expression for the half-life of a reaction that follows zeroth-order kinetics. (c) A zeroth-order reaction is unusual because there is a maximum time beyond which the integrated rate law cannot be valid. What is this time expressed in terms of the initial concentration of the reactant $[A_0]$? To how many half-lives does this maximum time correspond?

27. The rate law of the reaction $2NO(g) + H_2(g) \rightarrow N_2O(g) + H_2O(g)$ is investigated at a certain temperature by means of the isolation method. The following two experiments are performed: (1) 2.0 mol L$^{-1}$ of NO is mixed with 0.010 mol L$^{-1}$ of $H_2$, and the time dependence of $[H_2]$ is determined with the following results:

| t (s) | $[H_2]$ (mol L$^{-1}$) |
|---|---|
| 0 | $1.0 \times 10^{-2}$ |
| 10 | $6.2 \times 10^{-3}$ |
| 20 | $3.8 \times 10^{-3}$ |
| 30 | $2.4 \times 10^{-3}$ |

(2) 2.0 mol L$^{-1}$ of $H_2$ is mixed with 0.010 mol L$^{-1}$ of NO, and the time dependence of [NO] is determined with the following results:

| t (s) | [NO] (mol L$^{-1}$) |
|---|---|
| 0 | $1.0 \times 10^{-2}$ |
| 1000 | $8.1 \times 10^{-3}$ |
| 2000 | $6.8 \times 10^{-3}$ |
| 3000 | $5.8 \times 10^{-3}$ |

Determine the rate law of the reaction and the value of the rate constant.

28. The following reactions are elementary. Write expressions for their rate laws.
    (a) $NO + NO_2Cl \rightarrow NOCl + NO_2$
    (b) $2NO_2 \rightarrow N_2O_4$
    (c) $C_3H_7I \rightarrow C_3H_6 + HI$
    (d) $2NO + O_2 \rightarrow 2NO_2$

29. The mechanism of the reaction $2NO(g) + H_2(g) \rightarrow N_2O(g) + H_2O(g)$ is considered in Example 14.7. For what conditions will this mechanism lead to the rate law $R = k[NO]^2$ rather than to the observed rate law, $R = k[NO]^2[H_2]$?

30. On the basis of the mechanism for the reaction $H_2(g) + Br_2(g) \rightarrow 2HBr(g)$, given in Section 14.6, derive the experimental rate law (Equation 14.9).

31. The gas-phase reaction between molecular hydrogen and atomic iodine, $H_2(g) + 2I(g) \rightarrow 2HI(g)$, obeys the rate law $R = k[H_2][I]^2$. Devise a mechanism consistent with this rate law involving only unimolecular or bimolecular elementary reactions.

32. Ozone ($O_3$) is generated by the action of sunlight in the upper atmosphere. The following is a somewhat simplified version of the mechanism believed to be involved:

$$NO_2 \xrightarrow{h\nu}_{k_1} NO + O$$
$$O + O_2 \xrightarrow{k_2} O_3$$
$$NO + O_3 \xrightarrow{k_3} NO_2 + O_2$$

Show that the steady state concentration of ozone is $[O_3] = k_1[NO_2]/k_3[NO]$ provided that the sunlight intensity is constant.

33. The following mechanism is proposed for the decomposition of $NO_2Cl$:

(1) $NO_2Cl \underset{k_{-1}}{\overset{k_1}{\rightleftharpoons}} NO_2 + Cl$

(2) $NO_2Cl + Cl \overset{k_2}{\longrightarrow} NO_2 + Cl_2$

(a) What is the overall reaction? (b) Derive the rate law by means of the steady state approximation.

34. In aqueous solution, $HNO_2$ disproportionates by the reaction $3HNO_2(aq) \rightarrow 2NO(g) + H^+ + NO_3^- + H_2O$. The mechanism of this reaction is believed to be

(1) $2HNO_2 \overset{K_1}{\rightleftharpoons} NO + NO_2 + H_2O$    fast equilibrium

(2) $2NO_2 \overset{K_2}{\rightleftharpoons} N_2O_4$    fast equilibrium

(3) $N_2O_4 + H_2O \overset{k}{\longrightarrow} HNO_2 + H^+ + NO_3^-$    slow

Find the rate law, expressing the rate constant in terms of the rate constant $k$ and the equilibrium constants $K_1$ and $K_2$.

35. The following mechanism has been proposed for the decomposition of $O_3(g)$ to $O_2(g)$:

(1) $O_3 \underset{k_{-1}}{\overset{k_1}{\rightleftharpoons}} O_2 + O$    $K = k_1/k_{-1}$

(2) $O + O_3 \overset{k_2}{\longrightarrow} 2O_2$

(a) Derive a general expression for the rate law. (b) To what simple expression does this rate law reduce if the first step is rate determining? (c) To what simple expression does this rate law reduce if the second step is rate determining? (d) How could you distinguish between the two?

36. The equilibrium constant for the reaction $NO_2Cl(g) + NO(g) \rightarrow NO_2(g) + NOCl(g)$ at 500 K is 640. The reaction is elementary and the rate constant for the forward reaction is $7.50 \times 10^5$ (mol $L^{-1})^{-1}$ $s^{-1}$ at this temperature. What is the rate constant for the reverse reaction?

37. The reaction $2NO_2(g) \rightarrow 2NO(g) + O_2(g)$ is an elementary bimolecular reaction with a rate constant at 25°C of $3.10 \times 10^{-3}$ (mol $L^{-1})^{-1}$ $s^{-1}$. On the basis of this value and by use of the data in Appendix 4, determine the rate constant of the reaction $2NO(g) + O_2(g) \rightarrow 2NO_2(g)$ at 25°C.

38. A certain reaction occurs twice as fast at 308 K as at 298 K. What is the activation energy?

39. The rate constant of the reaction $H_2 + I_2 \rightarrow 2HI$ is determined at 575 K and at 700 K and the results are $k = 1.22 \times 10^{-6}$ (mol $L^{-1})^{-1}$ $s^{-1}$ and $k = 1.16 \times 10^{-3}$ (mol $L^{-1})^{-1}$ $s^{-1}$, respectively. What are the values of the activation energy and the frequency factor?

40. Two bimolecular reactions have identical values of the frequency factor $A$ in the Arrhenius equation. However, the activation energy of the first reaction is 15 kJ mol$^{-1}$ greater than that of the second reaction. Determine the ratio of their rate constants at (a) 15°C, and (b) 200°C.

41. The decomposition of ethyl chloride by the reaction $CH_3CH_2Cl(g) \rightarrow C_2H_4(g) + HCl(g)$ obeys first-order kinetics with $k = 2.021 \times 10^{-30}$ s$^{-1}$ at 300 K. If the activation energy of the reaction is 254.4 kJ mol$^{-1}$, how long will it take for the pressure of ethyl chloride to drop from 1.00 atm to 0.300 atm at 600 K because of this reaction?

42. Substances A and B simultaneously undergo the following two second-order reactions:

$A + B \overset{k_1}{\longrightarrow} C + D$

$A + B \overset{k_2}{\longrightarrow} E$

The ratio of the rate constants at 50°C is $k_1/k_2 = 10.0$. Assuming that the frequency factor is the same for both reactions, what is the ratio of rate constants at 500°C?

43. The activation energy of the bimolecular reaction $NO_2(g) + CO(g) \rightarrow NO(g) + CO_2(g)$ is 132 kJ mol$^{-1}$. When both reactants are present at a concentration of 0.250 mol L$^{-1}$ and a temperature of 500 K, the reaction occurs at a rate of $1.26 \times 10^{-7}$ (mol L$^{-1}$ s$^{-1}$). What is the rate constant of this reaction at 600 K?

44. A certain reactant can undergo either of two first-order reactions,

$A \overset{k_1}{\longrightarrow} B$    $A \overset{k_2}{\longrightarrow} 2C$

The activation energies of the two reactions are 56.8 kJ mol$^{-1}$ for $k_1$ and 77.4 kJ mol$^{-1}$ for $k_2$. The values of the rate constants are equal at 350 K. (a) At what temperature is $k_2/k_1 = 2.00$? (b) At what temperature is $k_1/k_2 = 2.00$?

45. The mechanism for the reaction $2NO(g) + O_2 \rightarrow 2NO_2(g)$ is believed to involve the following steps:

(1) $2NO \underset{k_{-1}}{\overset{k_1}{\rightleftharpoons}} N_2O_2$    fast

(2) $N_2O_2 + O_2 \overset{k_2}{\longrightarrow} 2NO_2$    slow

(a) Derive a third-order rate law consistent with this mechanism.
(b) The rate of the reaction *decreases* with increasing temperature. Explain this unusual behavior on the basis of the postulated mechanism, the derived rate law, and the fact that the reaction is exothermic.

46. The diameter of the methyl radical, $\cdot CH_3$, is about 3.80 Å. What is the maximum possible rate constant

for the second-order gas-phase recombination reaction, $2 \cdot CH_3 \rightarrow C_2H_6$ at 40°C and atmospheric pressure. (Because there is only a single reactant, the quantity $N_A^* N_B^*$ in Equation 14.49 must be replaced by $N^{*2}/2$. Division by 2 ensures that each collision between identical molecules will be counted only once.)

47. The figure below shows a plot of the potential energy of three colinear hydrogen atoms in the reaction $H + H_2 \rightarrow H_2 + H$. Three paths between reactants and products are shown. (a) Sketch the variation of the potential energy with reaction coordinate for each path. (b) What is the activation energy for each path? On the basis of these values determine which path is most likely to be followed. (c) What are the internuclear distances in the activated complex? (d) What is the overall energy change in the reaction, $\Delta E$?

48. The activation energy of a certain bimolecular reaction is 50.0 kJ mol$^{-1}$. The addition of a catalyst reduces the activation energy to 25.0 kJ mol$^{-1}$. How much faster does the catalyzed reaction occur at (a) 25°C and (b) 250°C?

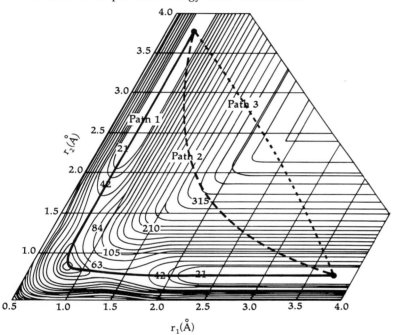

# 15

Sections 15.1 – 15.9

# Lewis Structures and Chemical Bonding

The existence of chemical compounds indicates that atoms of the same or different elements can attract each other with sufficient strength to form the distinct aggregates that we know as molecules and polyatomic ions. The nature of the interaction responsible for the stability of ionic substances has been understood since the early days of the atomic theory of matter. Charles Coulomb (1736–1806) proposed in the late eighteenth century the law governing the interaction of charged particles that bears his name. Nineteenth century chemists fruitfully applied this electrical view of the nature of matter to explain the properties of ionic substances. However, this same approach proved unsuccessful when applied to the very large class of substances that we now recognize as being comprised of molecules. A proper understanding of how neutral atoms combine to form molecules had to await the development of quantum mechanics in the first quarter of the twentieth century. The application of quantum theory to molecules has proved to be one of the great success stories of modern chemistry.

Chemical bonding is examined in this chapter and in the next two chapters. We begin with a simple pictorial representation of chemical bonds by means of Lewis structures, first proposed by Gilbert N. Lewis (1875–1946). Although the quantum theories of bonding provide a far more detailed view of bond properties, Lewis structures continue to be widely used because of their simplicity.

# Lewis Structures

## 15.1 An Overview of Chemical Bonding

The concept of the **chemical bond** as the interaction by which two or more atoms are held together to form a species with distinctive properties developed in parallel with experimental knowledge of the structure of these species. A number of different interactions of importance on the molecular level are known and will be examined in the course of the next few chapters. We begin by focusing on the covalent bond, the interaction by which we explain the stability of individual molecules. A **covalent bond** between two atoms consists of a pair of electrons shared between them. The electrons may be equally shared, as is necessarily the case in a **homonuclear diatomic molecule,** that is, one consisting of two identical atoms, such as $H_2$ or $O_2$. The electrons may also be unequally shared, for example, in the case of a **heteronuclear diatomic molecule,** i.e., one consisting of two different atoms, such as HCl or CO. Recall that the various elements differ in **electronegativity** (Section 6.7), that is, in the tendency of their bonded atoms to attract electrons. When two atoms of different electronegativities form a covalent bond, the shared electrons are held more closely by the more electronegative atom. This atom acquires a partial negative charge and the less electronegative atom consequently acquires a partial positive charge. Covalent bonds of this type are known as **polar covalent bonds.** The **ionic bond,** the interaction between two oppositely charged ions, can be regarded as a limiting case of polar covalent bonding—one or more electrons are transferred from the less electronegative atom to the more electronegative atom. Chemical compounds in which ionic bonding occurs are usually solids, and there is a simultaneous interaction between all the ions in a particular crystal (Section 20.10b).

As in other areas of chemistry, the theories of chemical bonding must be checked against experiment. Chemical bonds have a number of measurable properties. Two of the most useful properties are the bond energy and the bond length. The **bond energy** (Section 17.1) is the energy required to break a bond between two atoms or, equivalently, the energy released on bond formation, and is a measure of the strength of the bond. The formation of a chemical bond *always* releases energy—atoms form bonds because in so doing they attain a state of lower energy. The **bond length** (Section 17.2) is the distance between the nuclei of two bonded atoms. In addition to constituting a body of data that can be used to check the predictive ability of a particular theory, bond energies and lengths are a rich source of information about regularities and trends that can be correlated with chemical properties.

The experimentally determinable properties of chemical bonds are of value in elucidating the nature of the bonds. In turn, the properties and arrangement of the bonds in a particular molecule help to determine the chemical and physical properties that characterize the substance in question. In Chapter 17 we will discuss further the measurable properties of chemical bonds.

The relation between bonding and properties is clearly seen in a comparison of two **structural isomers,** molecules comprised of identical atoms bonded in different ways. A good example is provided by two compounds that both have the composition $C_2H_6O$. Ethyl alcohol (ethanol), for which the molecular formula is more commonly written as C₂H₅OH, and dimethyl ether, usually written as (CH₃)₂O, are structural isomers:

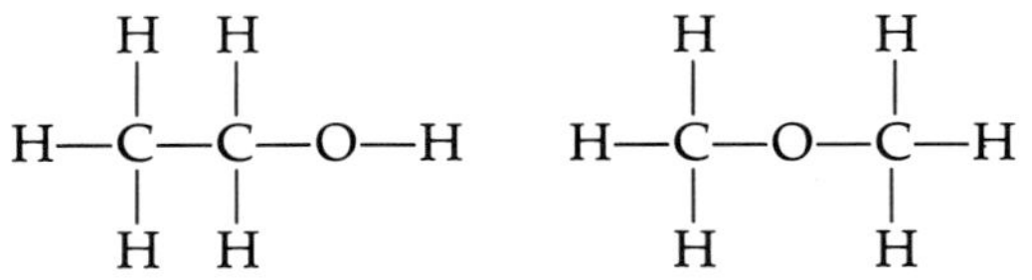

```
    H   H                     H       H
    |   |                     |       |
H—C—C—O—H              H—C—O—C—H
    |   |                     |       |
    H   H                     H       H
  ethyl alcohol           dimethyl ether
```

The arrangement of the bonds in the two molecules differs as shown. The properties of these compounds are quite different. For example, the ether boils at −25°C and does not react with nitric acid, while the alcohol boils at 78°C and reacts with nitric acid to form the explosive substance ethyl nitrate.

## 15.2 Simple Lewis Structures

*a. Bonded Atoms with Complete Valence Shells* The Lewis symbol for an atom depicts the nucleus and inner shell electrons by the chemical symbol of the element and the valence shell electrons by dots surrounding the chemical symbol. The elements in the first two periods are represented in the following manner:

H·                                                                He:

Li·    Be:    ·B:    ·C̈:    ·N̈:    ·Ö:    ·F̈:    :N̈e:

The heavier representative elements have the same number of valence electrons as the corresponding elements in the second period. The number of valence electrons in transition element atoms may be obtained from Table 6.3. The placement of the dots can be understood in terms of the atomic orbitals of the valence shell (Section 6.3). A full orbital, containing a pair of electrons with opposite spin, is represented as a pair of dots, while a half-filled orbital is shown as a single dot. Note that the elements in the first period can be assigned at most one pair of electrons, corresponding to a filled 1s orbital. Similarly, the elements in the second period can be assigned up to four pairs of electrons—an electron octet—corresponding to the filling of the 2s and the three 2p orbitals. The placement of the dots reflects the quantum mechanical principles embodied in the Pauli exclusion principle and in Hund's rule (Section 6.2). For example, the Lewis symbol for the nitrogen atom has one pair of dots for the electron pair filling the 2s orbital and three single dots, each representing an unpaired electron in one of the 2p orbitals.

The **Lewis structure** of a molecule represents covalent bonds by lines connecting the atomic symbols and valence electrons not involved in bonding as dots. (Some chemists represent covalent bonds by a pair of dots.) For example, the Lewis structure of HF is

H—F̈:

where the Lewis symbols of the two constituent atoms, H· and ·F̈:, show that each atom contributes one of the bonding electrons. Observe that the fluorine atom is surrounded by eight electrons, two of which are involved in bonding and six of which are not. Paired valence shell electrons not involved in bonding are known as **lone pairs** (or nonbonded electron pairs).

Figure 15.1 shows some typical Lewis structures. As illustrated by the structure of the hydroxide ion, Lewis structures can be written for ions as well

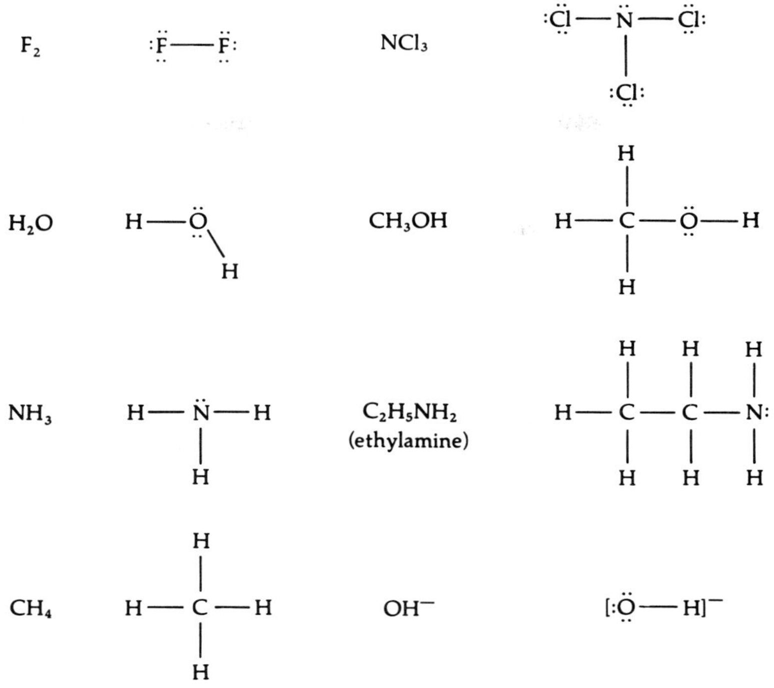

| | | | |
|---|---|---|---|
| F$_2$ | :F̈—F̈: | NCl$_3$ | :C̈l—N̈—C̈l:<br>         \|<br>        :C̈l: |
| H$_2$O | H—Ö<br>       \\<br>        H | CH$_3$OH | H—C—Ö—H (with H above and below C) |
| NH$_3$ | H—N̈—H<br>    \|<br>    H | C$_2$H$_5$NH$_2$<br>(ethylamine) | H—C—C—N̈: (with H's above and below) |
| CH$_4$ | H—C—H (with H above and below) | OH$^-$ | [:Ö—H]$^-$ |

**FIGURE 15.1** Examples of Lewis structures of second-period elements in which the octet rule is obeyed. The relative positions of the atoms in a molecule convey an approximate two-dimensional picture of the shape of the molecule. For example, a molecule of water is bent, not linear.

as for molecules. The number of valence electrons must be appropriately adjusted to reflect the gain or loss of negative charge. The hydroxide ion, for example, has eight valence electrons, six from the oxygen atom, one from the hydrogen, and one that gives rise to the negative ionic charge. Note that the stucture is drawn within brackets because the charge is ascribed to the ion as a whole rather than to a single atom.

In the Lewis structures depicted in Figure 15.1, each of the atoms excepting hydrogen is surrounded by eight electrons, counting both the bonded and lone pair electrons. The tendency of many bonded atoms to have valence shells completely filled by eight electrons is called the **octet rule.** This tendency reflects the unusual stability of a closed valence shell, as illustrated, for example, by the stability of the noble gases. The octet rule is not without exceptions. Elements with very few valence electrons, such as beryllium and boron, form covalent compounds in which they are surrounded by fewer than eight electrons. On the other hand, many elements in the third and higher periods form covalent compounds in which they are surrounded by more than eight electrons. In spite of these exceptions, which are discussed in the next two subsections, the octet rule is obeyed in a sufficiently large number of molecules to be a useful guide in drawing Lewis structures. Hydrogen, of course, is a special case because the filled K shell contains only two electrons. Thus, a bonded hydrogen atom must be surrounded by two electrons.

**556** Lewis Structures and Chemical Bonding  15.2

In the compounds included in Figure 15.1, each atom contributes one electron to each of its bonds. Consequently, the number of bonds made by an atom of a particular element is usually limited by the number of unpaired electrons available to be shared. Fluorine and hydrogen therefore generally form only one bond in their compounds, oxygen generally is limited to two bonds, nitrogen to three bonds, and carbon to four. (The Lewis symbol for carbon suggests that carbon should form only two bonds; however, four bonds are formed as discussed in Section 16.1b.)

Before a Lewis structure of a molecule can be written, it is necessary to know how the atoms in the molecule are connected. For example, in order to write the structure of HCN, we must know that the carbon atom is connected to the other two atoms. This information can be obtained experimentally and can also frequently be deduced by analogy to the structure of known, related compounds. The following are some general guidelines: Many compounds can be characterized by the formula $AB_n$, for example, $CCl_4$, $NF_3$, and $H_2S$. In these molecules the single atom of type A, called the **central atom,** is usually attached to all the atoms of type B. Molecules or ions with two central atoms, e.g., $C_2H_6$, $N_2O_4$, and $Cr_2O_7^{2-}$, are usually symmetrical. Oxygen atoms do not bond to other oxygen atoms except in peroxides (e.g., $H_2O_2$) and in superoxides (e.g., $KO_2$).

Once it is known how the atoms of a molecule are connected, it is possible to write its Lewis structure on the basis of the following rules:

1. Connect all adjacent atoms in the molecule by electron pair bonds. The resulting structure is called a **structural formula.**
2. Determine the total number of valence shell electrons contributed by all of the atoms in the molecule.
3. Add any electrons not accounted for by Rule 1 to the structure. These electrons will either be present as lone pairs or as additional bonds between one or more pairs of bonded atoms (Section 15.4).

---

**Example 15.1**   Draw Lewis structures for $H_3PO_4$ and $H_2PO_4^-$ in which the octet rule is obeyed by oxygen and phosphorus. The phosphorus atom is attached to each of the oxygen atoms and each of the hydrogen atoms is attached to a different oxygen atom.

The structural formula of $H_3PO_4$ consistent with the given information is

$$\begin{array}{c} \text{O} \\ | \\ \text{H—O—P—O—H} \\ | \\ \text{O} \\ | \\ \text{H} \end{array}$$

where it is immaterial to which oxygen atoms the hydrogen atoms are attached except that only one hydrogen is attached to a given oxygen. To complete the structure, the lone pairs must be added. We determine first the total number of valence electrons:

| Atom | Number of valence electrons |
|---|---|
| P | 5 |
| 4 O | 4 × 6 = 24 |
| 3 H | 3 × 1 = 3 |
| | 32 |

Since 7 bonds are needed to connect all the atoms, 14 electrons are accounted for. The number of lone pairs is (32 − 14)/2 = 9. These are arranged about the oxygen atoms to give the following Lewis structure:

$$\begin{array}{c} :\ddot{O}: \\ | \\ H-\ddot{O}-P-\ddot{O}-H \\ | \\ :\ddot{O}: \\ | \\ H \end{array}$$

Observe that phosphorus and each of the oxygen atoms are surrounded by eight electrons, while each hydrogen atom is surrounded by two electrons.

The dihydrogen phosphate ion ($H_2PO_4^-$) can be made from phosphoric acid by removal of a proton; the number of electrons remains unchanged:

$$\left[\begin{array}{c} :\ddot{O}: \\ | \\ H-\ddot{O}-P-\ddot{O}-H \\ | \\ :\ddot{O}: \end{array}\right]^-$$

It is immaterial which $H^+$ ion is removed from the molecule. ∎

**b. Bonded Atoms with Incomplete Valence Shells** The elements beryllium and boron in the second period, and aluminum in the third period, form a number of covalently bonded compounds. Beryllium, being in Group IIA, has two valence shell electrons, and boron and aluminum, which are in Group IIIA, have three valence shell electrons. Even when all the valence electrons in atoms of these elements are involved in bonding, the octet rule is generally not obeyed. The Lewis structure of boron trifluoride is illustrative:

$$\begin{array}{c} :\ddot{F}: \\ | \\ B \\ / \quad \backslash \\ :\ddot{F}: \quad :\ddot{F}: \end{array}$$

While each fluorine is surrounded by eight electrons, boron is surrounded by only six electrons. Boron trifluoride is an example of an **electron deficient molecule,** that is, one in which an atom retains an incomplete valence shell after bonding.

Atoms in molecules in which the total number of electrons is an odd number also violate the octet rule. For example, chloride dioxide ($ClO_2$) is a molecule with an odd number of electrons since the chlorine atom has 7 valence electrons and each oxygen atom has six. A reasonable Lewis structure for this molecule is

$$:\ddot{O}-\dot{C}l-\ddot{O}:$$

**FIGURE 15.2** A magnetic balance used for the measurement of paramagnetism.

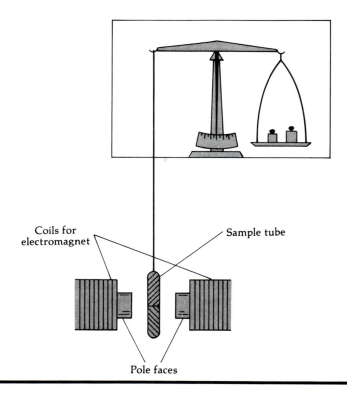

in which the less electronegative chlorine atom is surrounded by only 7 electrons.

Molecules with unpaired electrons are attracted by a magnetic field and are said to be **paramagnetic.** This property is the result of the net spin of the unpaired electron. Paramagnetism can be measured by means of a magnetic balance, as illustrated in Figure 15.2. The apparent weight of a paramagnetic substance changes when the magnetic field is turned on because of the attractive force exerted by the field. When a substance all of whose electrons are paired is introduced into a magnetic field, it is weakly repelled by the field. Such substances are said to be **diamagnetic.**

*c. Bonded Atoms with Expanded Valence Shells* Elements in the third and higher periods form compounds in which the octet rule is obeyed and also compounds in which the atom is surrounded by more than eight electrons. Figure 15.3 shows some structures of each type. Valence shell expansion becomes possible for elements in the third and higher periods because of the availability of d orbitals. As already indicated in Figure 6.7 and Figure 6.9, the $nd$ orbitals of these elements do not lie much higher in energy than the $np$ orbitals and so are available for bonding. By contrast, second-period elements do not have d orbitals, and, therefore, their atoms *never* form more than four bonds.

Among the third-period elements, valence shell expansion occurs most commonly for the elements phosphorus, sulfur, and chlorine in Groups VA–VIIA, respectively, in their compounds with small, highly electronegative atoms, such as fluorine, chlorine, and oxygen. Some typical examples, shown in Figure 15.3b, are phosphorus pentachloride ($PCl_5$), sulfur hexafluoride ($SF_6$),

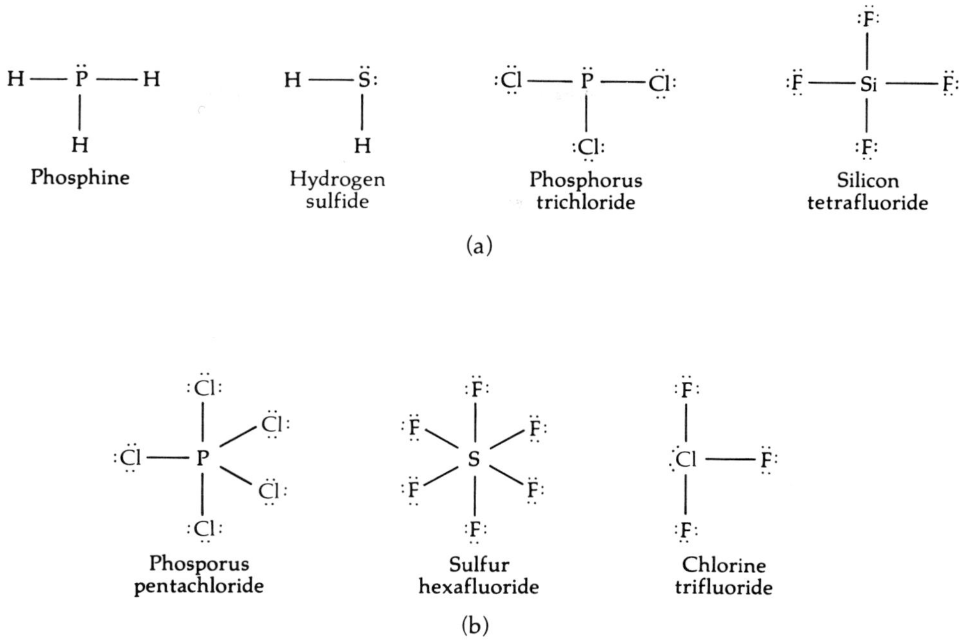

**FIGURE 15.3** (a) Compounds of third-period elements in which the octet rule is obeyed. (b) Compounds of third-period elements in which valence shell expansion occurs.

and chlorine trifluoride (ClF$_3$), in which some of the expanded valence shell electrons are lone pairs.

Let us examine the bonding in one of these molecules, say PCl$_5$, in more detail. The electronic configuration of phosphorus is [Ne]$3s^2 3p^3$; the valence shell contains 5 electrons. Bonding with 5 chlorine atoms surrounds the phosphorus atom with 10 electrons indicating the participation of one 3d orbital. Thus, the electronic configuration of phosphorus in PCl$_5$ can be represented as

where each of the five electron pairs corresponds to one of the covalent bonds. Note that four of the 3d orbitals remain vacant. In principle, more electrons could be accommodated. However, since all five valence electrons of phosphorus are already involved in bonding, additional bonds in which each atom contributes one electron are not formed (however, see Section 15.3).

In addition to the limit on the number of bonds imposed by the number of valence shell electrons, there is yet another limit on this number. As the number of bonding electrons surrounding a given atom increases, electron–electron repulsions also increase. Up to a point, this increase is balanced by the energy released in bond formation. However, when the number of bonding electrons reaches 12, the repulsive force begins to predominate and further increases are rarely found. There are only a few compounds in which more than six atoms are bonded to a single atom. In principle, halogen atoms, which have seven valence shell electrons, could form molecules of the type $XF_7$. In fact, the only known heptafluoride is $IF_7$. Evidently, the iodine atom is sufficiently large that seven electron pairs can be accommodated without excessive repulsion. The smaller size of the bromine and chlorine atoms precludes the existence of the corresponding molecules. The heaviest fluorides of these elements are $BrF_5$ and $ClF_3$. As we shall see in Section 15.9, electron pair repulsion plays an important role in determining the shape of a molecule.

***d. Ionic Compounds*** In addition to their usefulness in the description of covalently bonded species, Lewis structures can be used to describe the structure of individual ion pairs, which can be formed at elevated temperatures in the gaseous state (Section 18.1). For example, the formation of an NaCl ion pair from the atoms can be represented by means of the following equation written in terms of Lewis structures:

$$Na\cdot \; + \; \cdot\ddot{\underset{..}{Cl}}: \; \rightarrow \; [Na]^+[:\ddot{\underset{..}{Cl}}:]^-$$

Observe that the octet rule is obeyed in NaCl—the sodium ion has the $[He]2s^22p^6$ electronic configuration, and the chloride ion has the $[Ne]3s^23p^6$ configuration.

While covalently bonded atoms can attain a closed-shell configuration by electron sharing, this configuration is achieved by ions as a result of electron transfer. Note that the ionic bond is not shown explicitly in the representation of a sodium chloride ion pair. The ionic bond consists of the attraction between the oppositely charged ions and is represented by the positive and negative charges on the two ions.

## 15.3 Coordinate Covalent Bonds

***a. The Nature of Coordinate Covalent Bonds*** Electron deficient molecules contain a central atom surrounded by fewer than 8 electrons. These molecules can acquire the additional electrons needed to complete the octet by combining with molecules containing an atom capable of donating electrons. For example, boron trifluoride reacts with ammonia to form an addition compound, or adduct. The formation of $BF_3NH_3$ can be understood on the basis of the Lewis structures of the two molecules:

$$\begin{array}{c}:\ddot{F}:\\|\\:\ddot{F}-B\\|\\:\ddot{F}:\end{array} \; + \; \begin{array}{c}H\\|\\:N-H\\|\\H\end{array} \; \rightarrow \; \begin{array}{c}:\ddot{F}:\;\;H\\|\;\;\;\;|\\:\ddot{F}-B-N-H\\|\;\;\;\;|\\:\ddot{F}:\;\;H\end{array}$$

Note that the nitrogen atom has a lone pair of electrons which it can donate to the boron atom, thereby forming a compound in which each atom in the molecule has a completed valence shell. A bond in which both electrons originate in one atom is called a **coordinate covalent** or **dative** bond. While this term emphasizes the source of the electrons in such a bond, the bond itself is no different than any other covalent bond. Once the molecule is formed, it is immaterial whether the electrons in a particular bond originated in one or two atoms.

Coordinate covalent bonding permits an atom to form more bonds than correspond to the number of unpaired electrons it possesses. The formation of the ammonium ion is a case in point. The nitrogen atom possesses five electrons, three of which are unpaired. It can attain a closed valence shell by forming three covalent bonds, as in ammonia or in nitrogen trichloride (Figure 15.1). The formation of the ammonium ion, in which nitrogen forms four bonds, is possible because the nitrogen atom donates both electrons to the fourth N—H bond:

$$\text{H}-\overset{\overset{\text{H}}{|}}{\underset{\underset{\text{H}}{|}}{\text{N}}}: \; + \; \text{H}^+ \; \rightarrow \; \left[ \text{H}-\overset{\overset{\text{H}}{|}}{\underset{\underset{\text{H}}{|}}{\text{N}}}-\text{H} \right]^+$$

The fourth N—H bond is equivalent to the three original N—H bonds: all four N—H bonds have the same bond length and bond energy. The hydronium ion, in which oxygen is bonded to three hydrogen atoms, provides another illustration of this effect:

$$\text{H}-\underset{\underset{\text{H}}{|}}{\ddot{\text{O}}}: \; + \; \text{H}^+ \; \rightarrow \; \left[ \text{H}-\underset{\underset{\text{H}}{|}}{\ddot{\text{O}}}-\text{H} \right]^+$$

The number of coordinate covalent bonds in a given molecule is generally limited to one (however, see Section 21.4a). Thus, the $H_3O^+$ ion does not combine with yet an additional $H^+$ to form $H_4O^{2+}$ even though the oxygen atom still has a lone electron pair.

Atoms with expanded valence shells can also form coordinate covalent bonds. For example, $PCl_5$ can form the $PCl_6^-$ anion, in which the sixth chlorine donates both electrons to the bond:

$$\underset{\text{Cl}}{\overset{\text{Cl}}{\diagdown}}\overset{\overset{\text{Cl}}{|}}{\underset{\underset{\text{Cl}}{|}}{\text{P}}}-\text{Cl} \; + \; :\ddot{\text{Cl}}:^- \; \rightarrow \; \left[ \underset{\text{Cl}}{\overset{\text{Cl}}{\diagdown}}\overset{\overset{\text{Cl}}{|}}{\underset{\underset{\text{Cl}}{|}}{\text{P}}}\underset{\text{Cl}}{\diagup} \right]^-$$

---

**Example 15.2** Which of the following species have (a) an odd number of electrons, (b) an atom with an expanded valence shell, and (c) a coordinate covalent bond: $AsF_5$, $(CH_3)_3NO$, $NO$, $XeF_2$? Write an appropriate Lewis structure for each molecule.

(a) NO has 11 valence electrons (6 from the oxygen atom and 5 from the nitrogen atom). The more electronegative oxygen atom is given a complete octet:

$$\ddot{\text{N}}-\ddot{\ddot{\text{O}}}:$$

(b) $AsF_5$. Arsenic has 5 valence electrons and in bonding to 5 fluorine atoms becomes surrounded by 10 electrons:

$$:\ddot{\text{F}}-\underset{\underset{:\ddot{\text{F}}:}{|}}{\overset{\overset{:\ddot{\text{F}}:}{|}}{\text{As}}}\underset{:\ddot{\text{F}}:}{\diagdown}^{\diagup :\ddot{\text{F}}:}$$

XeF$_2$. Xenon has 8 valence electrons. Two electrons are used to form bonds with the fluorine atoms, thereby surrounding Xe by 10 electrons:

$$:\ddot{\ddot{F}}-\dot{X}\dot{e}\dot{.}-\ddot{\ddot{F}}:$$

(c) (CH$_3$)$_3$NO. The occurrence of coordinate covalent bonding can be seen by first writing the structure of (CH$_3$)$_3$N and then bonding an oxygen atom to the nitrogen atom:

$$\begin{array}{c} H \\ | \\ H-C-H \\ H\ \ | \\ |\ \ \ \ \\ H-C-N: \\ |\ \ \ \ \\ H\ \ | \\ H-C-H \\ | \\ H \end{array} \ +\ :\ddot{O}: \ \rightarrow\ \begin{array}{c} H \\ | \\ H-C-H \\ H\ \ | \ \ \ \ \ \ \ \ \\ |\ \ \ \ \ \ \ \ \ \ \ \ \\ H-C-N-\ddot{\ddot{O}}: \\ |\ \ \ \ \ \ \ \ \ \ \ \ \\ H\ \ | \ \ \ \ \ \ \ \ \\ H-C-H \\ | \\ H \end{array}$$

The nitrogen atom donates an electron pair to the oxygen atom. ∎

***b. Lewis Acids and Bases and Coordinate Covalent Bonds*** Recall that a Lewis acid was defined as a species acting as an electron-pair acceptor, while a Lewis base was defined as a species acting as an electron-pair donor (Section 8.3). A reaction involving the formation of a coordinate covalent bond can consequently be understood as a Lewis acid–base reaction. In the preceding illustrations, the electron-deficient species, such as BF$_3$, act as Lewis acids, while the species with lone electron pairs, such as NH$_3$ or H$_2$O, act as Lewis bases. The generalization of the acid–base concept provided by the Lewis model is of importance in organic chemistry. Many organic reactions are catalyzed by substances classified as acid catalysts, which are usually Lewis acids, such as BF$_3$.

The formation of complex ions (Section 9.7) can also be interpreted as Lewis acid–base reactions. The formation of the diammine silver complex, for example,

$$\text{Ag}^+ + 2\ :\!\!\begin{array}{c} H \\ | \\ N-H \\ | \\ H \end{array} \ \rightarrow\ \left[ \begin{array}{ccc} H & & H \\ | & & | \\ H-N-\text{Ag}-N-H \\ | & & | \\ H & & H \end{array} \right]^+$$

can be understood as a reaction between the acid Ag$^+$ and the base NH$_3$. More precisely, the reaction actually involves the replacement of H$_2$O ligands by NH$_3$,

$$\text{Ag(H}_2\text{O)}_2^+ + 2\text{NH}_3 \ \rightarrow\ \text{Ag(NH}_3\text{)}_2^+ + 2\text{H}_2\text{O}$$

and can be viewed as the displacement of a weak Lewis base (H$_2$O) by a stronger Lewis base (NH$_3$).

## 15.4 Multiple Bonds

In order to attain the stability conferred by a closed valence shell, covalently bonded molecules frequently require multiple bonds between two atoms. A **multiple bond** is a bond involving more than one pair of electrons; a double

| Structure | H—C—C—H with H's (ethane structure) | H₂C=CH₂ structure | H—C≡C—H |
|---|---|---|---|
| Compound name | Ethane ($C_2H_6$) | Ethylene ($C_2H_4$) | Acetylene ($C_2H_2$) |
| Carbon-carbon bond energy (kJ mol$^{-1}$) | 347 | 611 | 828 |
| Carbon-carbon bond length (Å) | 1.54 | 1.35 | 1.21 |

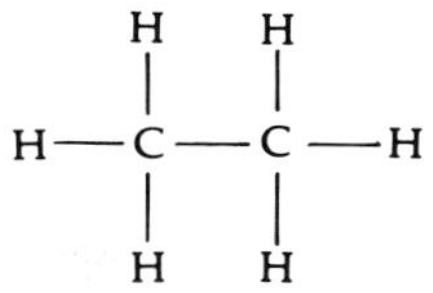

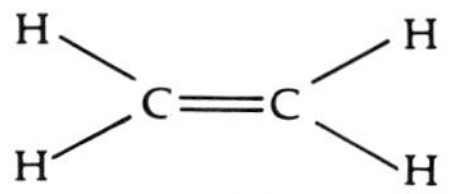

**FIGURE 15.4** Correlation between bond order and bond length and energy for some simple hydrocarbons.

bond involves the sharing of two electron pairs and a triple bond involves the sharing of three electron pairs. It is customary to refer to the bond order when dealing with multiple bonds. The **bond order** of a bond is given by the number of electron pairs shared between the two bonded atoms. The bond order of a single, double, or triple bond is one, two, or three, respectively.

The Lewis structures of carbon dioxide and nitrogen illustrate multiple bond formation:

$$\ddot{O}=C=\ddot{O} \qquad :N\equiv N:$$

Observe that each of the atoms in these molecules obeys the octet rule. Multiple bond formation is most common for carbon, nitrogen, and oxygen, but also occurs for some of the heavier elements of Groups IVA–VIA.

The experimental evidence for the occurrence of multiple bonds may be found in a comparison of the lengths and energies of single and multiple bonds between a given pair of atoms. The electron density in the region between the nuclei of two atoms increases with the number of bonds between the atoms. Now the nuclei of the two atoms repel each other because of their positive charge. However, each nucleus is attracted to the negative charge cloud of the electrons in the bond, and this attraction increases with the electron density. Consequently, we can expect the length of the bond between two atoms to decrease with increasing bond order and the bond energy to vary in the opposite way: it must increase with the bond order. These expected trends are illustrated by the hydrocarbons ethane ($C_2H_6$), ethylene ($C_2H_4$), and acetylene ($C_2H_2$). Figure 15.4 shows the Lewis structures of these molecules and lists the carbon–carbon bond lengths and energies. We see that the carbon–carbon bond length in acetylene is some 20% shorter than it is in ethane, while the bond energy is nearly three times greater.

**Example 15.3** Draw a Lewis structure for carbon monoxide in which the octet rule is obeyed by each atom.

We begin by writing the formula in a manner which shows how the atoms are attached:

C—O

Next, we determine the total number of valence electrons: 4 from carbon and 6 from oxygen, a total of 10 electrons. Finally, these electrons are added to the structure in a manner consistent with the octet rule. It is impossible to place the eight remaining electrons as lone pairs and satisfy the octet rule for both atoms. A triple bond allows each atom to share in an electron octet:

:C≡O: ■

Comparison of the structure of CO with that of $N_2$ shows that they are very similar: in each case the two atoms are attached by triple bonds and, in addition, each atom possesses a lone pair. These two molecules are isoelectronic (Section 6.5). Recall that isoelectronic atoms or simple ions have the same number of electrons. When applying this concept to molecules, it is useful to add the restriction that isoelectronic molecules must contain the same number of atoms. With this additional stipulation, it is possible to state that, in general, isoelectronic molecules or ions have similar Lewis structures.

The Lewis structure of molecular oxygen poses an interesting problem. At first sight, a satisfactory structure, in which each oxygen atom is surrounded by eight electrons, can be written as

Ö=Ö

Since this Lewis structure indicates that all the electrons are paired, oxygen should be diamagnetic. However, experiment shows that molecular oxygen is paramagnetic and must therefore contain unpaired electrons. While Lewis structures containing unpaired electrons can, of course, be written, it is fair to conclude that the Lewis approach does not provide a satisfactory representation of $O_2$ (see Section 16.6a).

## 15.5 The Distribution of Electronic Charge

*a. Formal Charge* The **formal charge** of an atom is defined as the number of valence electrons in the isolated atom minus the number of valence electrons assigned to it in a molecule. The number of electrons assigned to an atom in a molecule is equal to the number of its nonbonded electrons plus half the number of its bonded electrons. Thus, the formal charge of an atom is given by the relation

formal charge = (number of valence electrons)

− (number of nonbonded valence electrons)     (15.1)

$-\frac{1}{2}$(number of bonded valence electrons)

When more than one Lewis structure can be written for a molecule, the formal charges can often be used to pick out the most realistic structure, as we shall see.

To illustrate the evaluation of formal charges, let us examine the following two possible structures of sulfuric acid:

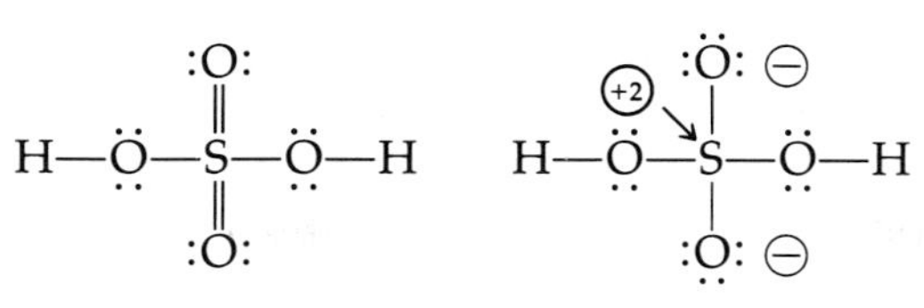

## 15.5 Lewis Structures and Chemical Bonding

$$\text{(a)} \quad H-\overset{..}{\underset{..}{O}}-\overset{\overset{:\overset{..}{O}:}{\|}}{\underset{\underset{:\overset{..}{O}:}{\|}}{S}}-\overset{..}{\underset{..}{O}}-H \qquad \text{(b)} \quad H-\overset{..}{\underset{..}{O}}-\overset{\overset{:\overset{..}{O}:\ominus}{\nwarrow}}{\underset{\underset{:\overset{..}{O}:\ominus}{\downarrow}}{S^{+2}}}-\overset{..}{\underset{..}{O}}-H$$

In structure (a) the sulfur atom is surrounded by 12 shared electrons, half of which are assigned to it. Since sulfur has 6 valence electrons, the formal charge is zero. Similarly, each oxygen atom is surrounded by two lone electron pairs and is attached by two bonds. The number of electrons assigned to each oxygen is six (four from the lone pairs and two from the bonded pairs). The formal charge of each oxygen atom is zero.

Now consider structure (b). The sulfur atom is surrounded by four electron pair bonds and is therefore assigned four electrons. The formal charge on sulfur is $6 - 4 = 2$, and is indicated by the +2 symbol written next to the S in structure (b). Similarly, two of the oxygen atoms in this structure are surrounded by three lone pairs and one bond, and consequently have a formal charge of $6 - 7 = -1$, as indicated. The other oxygen atoms have six assigned electrons; their formal charge is zero. Note that the sum of the formal charges—the net formal charge—is zero for either structure. The net formal charge of a neutral molecule must be zero, while that of an ion must be equal to the ionic charge.

Formal charges do *not* represent the actual charges in a molecule. The reason is that the electrons in a covalent bond are usually not shared equally between the bonded atoms because different elements differ in electronegativity. Nonetheless, formal charges give an indication of the charge distribution and can be used as a guide to the formulation of reasonable Lewis structures. A reasonable Lewis structure generally does not have large formal charges on any of its atoms. Furthermore, when two atoms in a Lewis structure carry a formal charge, it is most likely that the negative charge will reside on the more electronegative atom, which has a greater tendency to attract electrons. (Molecules with coordinate covalent bonds are an exception, since the donor atom acquires a positive formal charge.) Finally, like charges should not reside on adjacent atoms because the repulsion between like charges reduces the stability of the molecule.

The application of these guidelines to the structures of sulfuric acid indicates that structure (a), which has no formal charges, is more realistic than (b), in which the sulfur atom has a formal charge of +2. Experimentally determined bond lengths (see Section 17.2) support this conclusion since the S—O bond in sulfuric acid is shorter than expected for a typical single bond between sulfur and oxygen atoms.

---

**Example 15.4** Two Lewis structures are given for each of the following species. In each instance, pick the more realistic structure and give the reason for the choice.

(a) 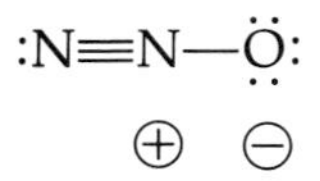 $\quad :N\equiv N-\overset{..}{\underset{..}{O}}: \quad$ or $\quad :\overset{..}{N}-N\equiv O:$
$\qquad\qquad\;\;\oplus \;\;\, \ominus \qquad\qquad\; \ominus_2 \;\; \oplus \;\; \oplus$

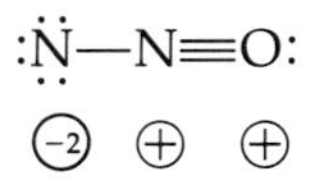

The first structure for dinitrogen oxide is more realistic. The second structure contains a large negative formal charge on a nitrogen atom as well as a positive formal charge on the more electronegative oxygen atom.

(b)  $[\text{H}-\text{C}=\ddot{\text{O}}]^+$  or  $[\text{H}-\text{C}\equiv\text{O}:]^+$

The second structure is more realistic because both the carbon and oxygen atoms obey the octet rule whereas the carbon atom is surrounded by only 6 electrons in the first structure. The octet rule for elements such as C, N, O, and F is more important than any of the formal charge guidelines.

(c)  
$$\text{H}-\overset{\overset{\displaystyle :\text{O}:}{\|}}{\text{C}}-\ddot{\text{O}}-\text{H} \quad \text{or} \quad \text{H}-\overset{\overset{\displaystyle :\ddot{\text{O}}: \ominus}{|}}{\text{C}}=\overset{\oplus}{\ddot{\text{O}}}-\text{H}$$

The first structure of formic acid is more realistic because it contains no formal charges. ■

***b. Oxidation Number***  The concept of formal charge is closely related to that of oxidation number. Recall that oxidation numbers were introduced in Section 2.7 as an aid in balancing equations for redox reactions. Oxidation numbers are also exceedingly useful in a systematization of the redox chemistry of the elements, as we shall see (Chapter 19). A more meaningful but equivalent method for the assignment of oxidation numbers than the set of rules in Section 2.7 can be given by analogy to formal charge.

As in the case of formal charge, the oxidation number of an atom is obtained as the difference between the number of valence-shell electrons in the isolated atom and those assigned to it when bonded. Oxidation numbers differ from formal charges in the way electrons are assigned: nonbonded electrons are all assigned to the atom on which they reside, the same as for formal charge; however, bonded electrons are not divided between the attached atoms but, rather, are assigned to the more electronegative atom. Thus, the difference between formal charge and oxidation number lies in the way the electrons constituting a covalent bond are assigned to atoms. In assigning formal charges it is assumed that there is equal electron sharing; in assigning oxidation numbers it is assumed that there is no electron sharing. Formal charge and oxidation number represent possible limits to the distribution of charge in a molecule. In analogy to Equation 15.1, we can write the following expression for the oxidation number of atom A bonded to atom B:

oxidation no. = (no. of valence electrons)  (15.2)

− (no. of nonbonded valence electrons)

$$\begin{cases} - \text{(no. of bonded electrons)}, & \text{if } \chi_A > \chi_B \\ - \tfrac{1}{2}\text{(no. of bonded electrons)}, & \text{if } \chi_A = \chi_B \\ - 0, & \text{if } \chi_A < \chi_B \end{cases}$$

where $\chi_A$ and $\chi_B$ are the electronegativities of A and B. Note that the first two terms of Equation 15.2 are common to all three forms of the equation while the third term depends on the values of the electronegativities.

The structure of phosphoryl chloride (POCl₃) may be used to illustrate the difference between formal charge and oxidation number. The Lewis structure of this molecule is

in which the 32 valence electrons of the five atoms in the molecule are arranged so that the octet rule is obeyed by each atom. The formal charge on phosphorus is $5 - 4 = 1$ and that on oxygen is $6 - 7 = -1$, as indicated. In determining the oxidation numbers, we note that both oxygen and chlorine are more electronegative than phosphorus (see Figure 6.20). Thus, the electrons in the P—O and P—Cl bonds are assigned to either oxygen or chlorine. The oxidation number of oxygen is $6 - 8 = -2$, that of chlorine is $7 - 8 = -1$, and that of phosphorus is $5 - 0 = 5$.

## Example 15.5

Draw a Lewis structure for chloric acid, HClO₃, in which the octet rule is obeyed by each atom. Determine the formal charges and oxidation numbers. Note that oxygen is more electronegative than chlorine.

Attaching each of the oxygen atoms to the chlorine atom, we can write the structural formula

$$\begin{array}{c} O \\ | \\ H-O-Cl-O \end{array}$$

The total number of valence electrons is $7 + 3 \times 6 + 1 = 26$. In addition to the 4 bonds joining the various atoms, 9 lone pairs can be added, thereby satisfying the octet rule:

$$\begin{array}{c} :\ddot{O}: \\ | \\ H-\ddot{O}-\ddot{Cl}-\ddot{O}: \end{array}$$

The formal charge on Cl is $7 - 5 = 2$ and that on the end oxygen atoms is $6 - 7 = -1$. The oxidation number of chlorine is $7 - 2 = 5$ since all the bonded electrons are assigned to oxygen. The oxidation number of each oxygen is $6 - 8 = -2$. ∎

## 15.6 Resonance

It is often possible to write more than one reasonable Lewis structure for a particular covalently bonded species. It is then necessary to determine which structure is consistent with experiment. In many instances, the properties of a molecule are not consistent with any single Lewis structure but, rather, with an intermediate arrangement of electrons that appears to be a blend of the individual structures.

The structure of the nitrate ion is a case in point. In each of the following three Lewis structures the 24 valence electrons are distributed in a way that satisfies the octet rule:

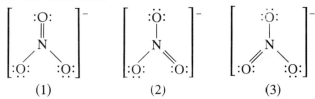

The three structures differ only in the location of the double bond. If any single one of these structures were correct, the nitrate ion would contain one N=O double bond, the length of which is known to be 1.19 Å, and two N—O single bonds, 1.36 Å in length. However, experiment shows that there is only one type of bond between nitrogen and oxygen in $NO_3^-$. The length of this bond is 1.23 Å, that is, between the single and double bond lengths. Apparently none of the preceding structures is a correct representation of the nitrate ion. Rather, the actual structure can be represented as a blend or hybrid of the three individual structures.

To describe such an intermediate configuration, the concept of **resonance** has been introduced. The individual representations of the structure are said to be in resonance, and the actual structure is their **resonance hybrid.** It is difficult to write a Lewis structure of a resonance hybrid. A common approach is to write the contributing structures separated by double-headed arrows (↔). Thus, for the nitrate ion, the structure of the resonance hybrid may be written as

The contributing resonance structures are *equivalent*—they have the same number of covalent bonds.

The double-headed arrow between structures in resonance should not be confused with the double arrow (⇌) used to represent equilibrium. The experimental evidence indicates that *the individual structures do not exist as such.* Consequently, the resonance hybrid does *not* consist of an equal mixture of the individual structures nor of the individual structures in equilibrium. The correct description is that the three oxygen atoms are attached to the nitrogen by a combination of three single bonds and two extra electrons which are smeared out, or **delocalized,** over all three bonds. The bond order of the equivalent bonds in a resonance hybrid is given by half the number of electrons shared between two atoms,

$$\text{Bond order (equivalent bonds)} = \frac{1}{2}\left(\frac{\text{total no. of electrons in equivalent bonds}}{\text{total no. of equivalent bonds}}\right) \quad (15.3)$$

For example, there are 8 electrons (two in each of four covalent bonds) shared among three different pairs of bonded atoms in $NO_3^-$, giving an average of $\frac{8}{3}$ electrons per pair of atoms. The bond order is $(\frac{1}{2})(\frac{8}{3}) = 1.33$.

The concept of resonance is particularly important in organic chemistry, where it is invoked to explain the structure and chemical behavior of the very important aromatic compounds (Section 22.1e). The prototype of these compounds is benzene (C₆H₆). Structural evidence indicates that the benzene molecule consists of six coplanar carbon atoms joined in a regular hexagon with a hydrogen atom attached to each carbon. The bond lengths indicate that the carbon atoms are joined by bonds that are intermediate between single and double bonds. The following two structures contribute to the resonance hybrid:

[benzene resonance structures]

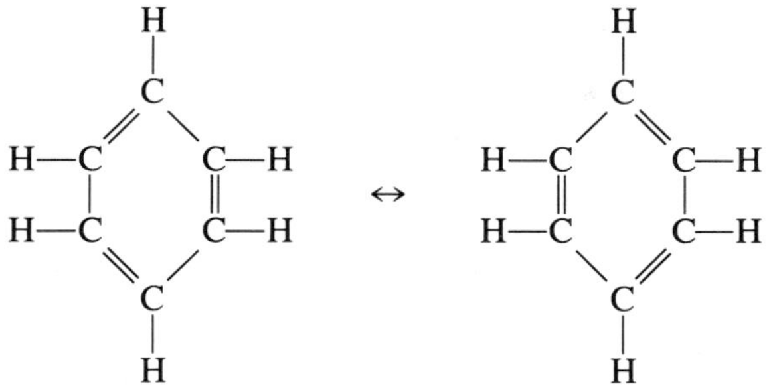

Note that the two structures differ only in the locations of the three double bonds. The six electrons in these bonds are delocalized over the entire carbon ring.

## Example 15.6

Experiment shows that the bond between sulfur and oxygen in $SO_4^{2-}$ has order 1.5. Draw equivalent resonance structures consistent with this fact. What are the formal charges?

The bond order indicates that there should be, on average, three electrons between the sulfur atom and each oxygen atom, equivalent to an equal number of single and double bonds. The following structure disposes of the 32 valence electrons in a manner that is consistent with the bond order and also provides each oxygen atom with an electron octet:

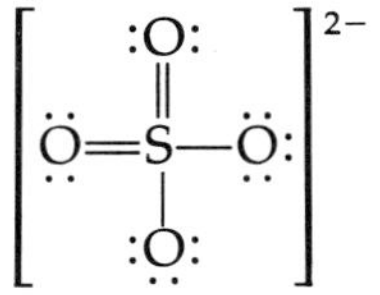

[Lewis structure of SO₄²⁻]

This is just one of several equivalent structures differing only in the location of the single and double bonds. The following structures contribute to the resonance hybrid:

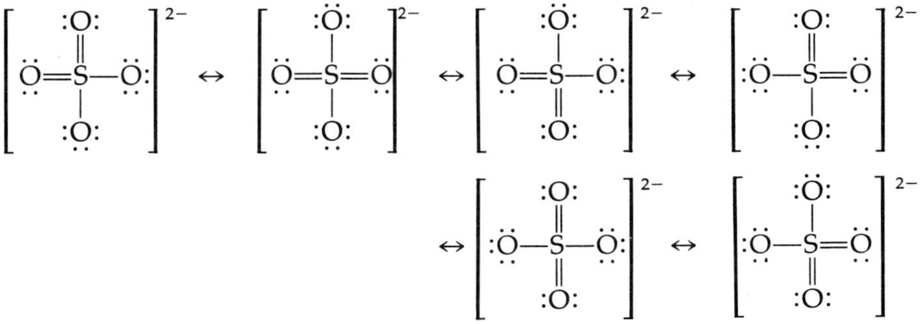

[six resonance structures of SO₄²⁻]

The sulfur atom and each double-bonded oxygen atom have a formal charge of 6 − 6 = 0 while each of the single-bonded oxygen atoms has a formal charge of $6 - 7 = -1$. Since the four oxygen atoms in the resonance hybrid are actually bonded in the same way, their formal charges can also be said to be equal to the average of the formal charges on each of the two single-bonded and two double-bonded atoms, that is, $-0.5$. Note that the sum of the formal charges is equal to the ionic charge. ∎

We have so far considered resonance between equivalent structures that differ only in the location of one or more double bonds. Many molecules can be represented by Lewis structures that show more substantial differences in electron distribution and are consequently not equivalent. The sulfate ion (Example 15.6) may once again be used to illustrate this point. We can write a structure involving only single bonds in which the sulfur atom has a formal charge of $+2$ and each oxygen atom one of $-1$,

$$\left[ \ominus \; :\ddot{\underset{..}{\text{O}}}-\overset{\overset{:\ddot{\underset{..}{\text{O}}}:\; \ominus}{|}}{\underset{\underset{\ominus}{:\ddot{\underset{..}{\text{O}}}:}}{\text{S}}}^{(+2)}\ddot{\underset{..}{\text{O}}}: \; \ominus \right]^{2-}$$

a structure involving only double bonds in which sulfur has a formal charge of $-2$,

$$\left[ \ddot{\underset{..}{\text{O}}}=\overset{\overset{:\text{O}:}{\|}}{\underset{\underset{:\text{O}:}{\|}}{\text{S}}}^{(-2)}\ddot{\underset{..}{\text{O}}} \right]^{2-}$$

and various intermediate structures involving one double bond,

$$\left[ \ominus \; :\ddot{\underset{..}{\text{O}}}-\overset{\overset{:\text{O}:}{\|}}{\underset{\underset{\ominus}{:\ddot{\underset{..}{\text{O}}}:}}{\text{S}}}^{\oplus}\ddot{\underset{..}{\text{O}}}: \; \ominus \right]^{2-}$$  (four equivalent structures)

two double bonds,

$$\left[ \ominus \; :\ddot{\underset{..}{\text{O}}}-\overset{\overset{:\text{O}:}{\|}}{\underset{\underset{:\text{O}:}{\|}}{\text{S}}}-\ddot{\underset{..}{\text{O}}}: \; \ominus \right]^{2-}$$  (six equivalent structures; see Example 15.6)

and three double bonds,

$$\left[ \ddot{\underset{..}{\text{O}}}=\overset{\overset{:\text{O}:}{\|}}{\underset{\underset{:\ddot{\underset{..}{\text{O}}}: \; \ominus}{|}}{\text{S}}}^{\ominus}\ddot{\underset{..}{\text{O}}} \right]^{2-}$$  (four equivalent structures)

Although the structure containing two double bonds has properties that most closely agree with experiment, all of these structures may contribute, although the first two are unrealistic because of their large formal charges. However, in contrast to the case of resonance between equivalent structures, non-equivalent structures do *not* make equal contributions to the resonance hybrid.

While the possibility of non-equivalent contributors to a resonance hybrid greatly expands the number of possible structures, two restrictions do limit the number of possibilities: first, each contributing structure must have the same arrangement of atoms. Resonance is possible only between structures differing in electron distribution, not between structures differing in the position of their constituent atoms. For example, the cyanate ion (NCO$^-$) and the isocyanate ion (CNO$^-$) are different structural isomers. There can be no resonance between them. Second, each contributing structure must have the same number of unpaired electrons.

# Molecular Geometry

## 15.7 The Shapes of Molecules

A characteristic feature of covalent bonds is that they are highly directional and make specific angles with respect to each other. A **bond angle** is the angle between two imaginary lines joining the nucleus of a central atom with the nuclei of two atoms bonded to it. For example, the bond angle in the H$_2$O molecule is the angle between the lines connecting the nucleus of the oxygen atom with those of the hydrogen atoms (Figure 15.5). Since the positions of atoms can be determined experimentally, bond angles are measurable quantities. The shape of a molecule is determined by the lengths of the various bonds between its constituent atoms and by the corresponding bond angles.

In dealing with diatomic molecules we need not be concerned with the concept of shape since such a molecule is necessarily linear. The concept of shape first arises for triatomic molecules. A triatomic molecule must be planar since three points, corresponding to the three atomic nuclei, define a plane. Molecules of this type are either linear, with a bond angle of exactly 180°, or bent, with a bond angle of some smaller value. For example, CO$_2$ is a linear molecule while H$_2$O is bent, with a bond angle of 104.5°.

Molecules comprised of more than three atoms can have a three-dimensional structure. Although many shapes are possible, it turns out that just a few of them suffice to describe the structure of most simple molecules. Moreover, these simple structures also appear as subunits of more complex molecules. Figure 15.6 shows the five most important shapes of molecules with more than three atoms. Let us examine their geometry in detail.

**Trigonal planar** molecules, such as BF$_3$, are planar and have bond angles of 120°. The BF$_3$ structure can be described as an equilateral triangle defined by the three fluorine atoms, with the boron at the center of the triangle. The **trigonal pyramidal** structure, exhibited by molecules such as NH$_3$, differs from the BF$_3$ structure in being non-planar. The three hydrogen atoms from an equilateral triangle and the nitrogen atom is equidistant from them but lies out of plane. The bond angles, while equal, are significantly smaller than those in the corresponding planar structure.

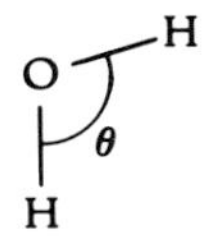

The **tetrahedral** structure, exemplified by methane (CH$_4$), is the fundamental shape used in building up the complicated molecules of organic chemistry and so is of special importance. In this structure the four hydrogen atoms

**FIGURE 15.5** The meaning of bond angle, as illustrated for H$_2$O.

572 Lewis Structures and Chemical Bonding 15.7

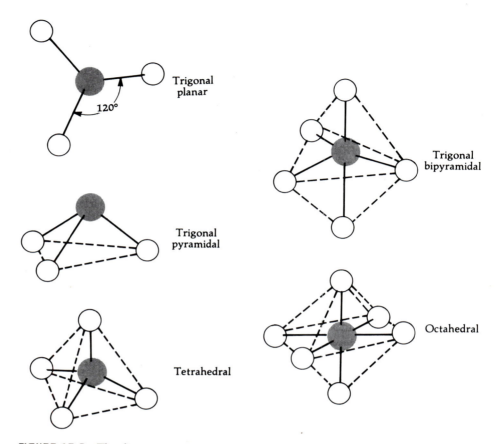

**FIGURE 15.6** The five common shapes of molecules containing four or more atoms.

are at the corners of a regular tetrahedron and the carbon atom is at its center. The C—H bonds are equal in length and the H—C—H bond angles are 109.47°, a value commonly known as the **tetrahedral angle.** Although the tetrahedral structure may, at first sight, appear to be somewhat complicated, it can be readily understood with reference to a cube. If the hydrogen atoms are placed at alternate corners of the cube and the C atom at its center, the tetrahedral structure depicted in Figure 15.6 is obtained. Figure 15.7 illustrates the procedure graphically. A tetrahedron can be obtained by connecting the hydrogen atoms, a procedure that is equivalent to drawing the appropriate face diagonals of the cube. Use of the simple geometric properties of the cube permits a determination of the tetrahedral geometry.

The **trigonal bipyramidal** shape, displayed by molecules such as $PCl_5$, differs from the preceding structures in that not all of the P—Cl bonds and angles are identical. Three of the chlorine atoms are arranged in an equilateral triangle, the center of which is occupied by the phosphorus atom. The other two chlorine atoms are colinear with the phosphorus in a line perpendicular to the plane of the triangle. In order to distinguish between the two types of chlorine atoms the designations **equatorial** for the in-plane atoms and **axial** for the colinear atoms are commonly used. The distinction is important because the equatorial bond angle is 120° while the angle between an axial and an equato-

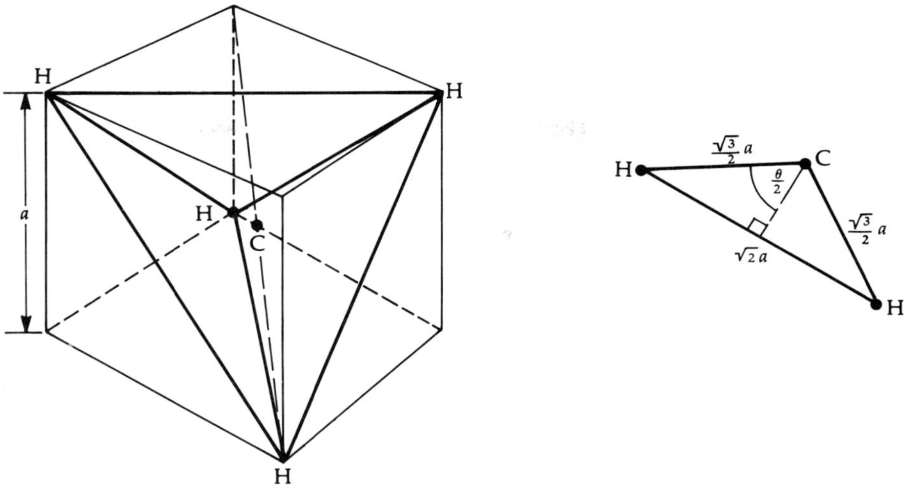

**FIGURE 15.7** Understanding tetrahedral geometry in CH₄ by means of a cube. On the right, the distance between H—H and C—H atoms is expressed in terms of the length of a side of the cube, $a$.

rial bond is 90°. Each equatorial chlorine atom has two nearest-neighbor chlorine atoms at 120° while each axial chlorine atom has three nearest-neighbor chlorine atoms at 90°. This difference leads to an equatorial P—Cl bond that is shorter than the axial P—Cl bond, for this arrangement minimizes the potential energy of the molecule. This effect is usually found in molecules with the general formula $AB_5$.

The final important structure is the **octahedral** structure, displayed by molecules such as $SF_6$. In this molecule, the sulfur atom is at the center of a square whose corners consist of four fluorine atoms. The remaining two fluorines are colinear with the sulfur and their bonds make a right angle with the plane of the square. In contrast to the trigonal bipyramidal structure, all the bond lengths are equal and all the bond angles are 90°. Thus, all six fluorine atoms occupy equivalent sites. The designation of this structure comes from the fact that the solid figure formed by joining all the fluorine atoms is an octahedron, a structure whose eight sides consist of equilateral triangles.

## 15.8 Dipole Moments

We have remarked in Section 15.7 that the $CO_2$ molecule is linear while $H_2O$ is bent. How do we know this? One of the measurable properties of a molecule that provides information about its shape is the dipole moment. The dipole moment is a measure of the way the electronic charge is distributed in a molecule. We have seen that, due to differences in electronegativity, many covalent bonds are polar—the bonding electrons are more strongly attracted to the more electronegative atom. The resulting charge asymmetry can give rise to a measurable dipole moment.

*a. Diatomic Molecules* Consider a diatomic molecule in which the atoms have charges of $\pm\delta$. (If $\delta = e$, the charge of an electron, the molecule actually is an ion pair, and if $\delta < e$, the molecule is polar.) We assume that the atoms can be

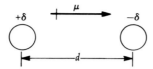

**FIGURE 15.8** The dipole moment of a diatomic molecule. The atoms are treated as small, hard spheres with charge $\pm\delta$, and are separated by a distance $d$. The arrow directed from the positive to the negative charge is a common representation of the dipole moment.

---

represented by hard charged spheres separated by a distance $d$ (Figure 15.8). The **dipole moment** $\mu$ is given by the relation

$$\mu = \delta d \tag{15.4}$$

that is, by the product of the magnitude of the charges and the distance between them. Since charge is expressed in coulombs and distance in meters, the units of $\mu$ are C m (coulomb meters).

An indication of the magnitude of molecular dipole moments can be obtained by assuming that two singly charged ions are separated by a distance of 1 Å. Inserting the appropriate numerical values into Equation 15.4 we obtain

$$\mu = (1.602 \times 10^{-19} \text{ C})(1.0 \times 10^{-10} \text{ m}) = 16 \times 10^{-30} \text{ C m}$$

Molecular dipole moments are commonly expressed in units of debyes [named after Peter Debye (1884–1966), a pioneer in the study of polar molecules], where

$$1 \text{ D} = 3.338 \times 10^{-30} \text{ C m}$$

---

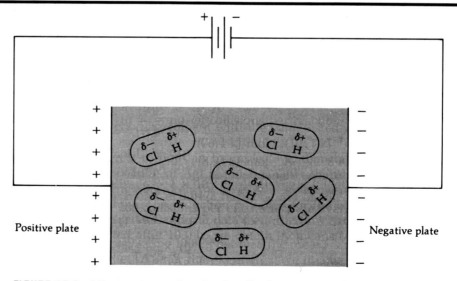

**FIGURE 15.9** Measurement of molecular dipole moments of gases by the alignment of the molecules in an electric field.

Thus, the dipole moment resulting from the separation of two opposite charges of electronic magnitude by a distance of 1 Å is

$$\mu = \frac{(16 \times 10^{-30} \text{ C m})}{(3.338 \times 10^{-30} \text{ C m D}^{-1})} = 4.8 \text{ D}$$

A dipole moment is characterized by direction as well as magnitude. It is customary to represent the direction of a dipole moment by an arrow pointing from the positive charge towards the negative charge and its relative magnitude by the length of the arrow, as shown in Figure 15.8.

Dipole moments can be determined experimentally by the method illustrated in Figure 15.9. An electric circuit containing two parallel plates is set up and the electric field is measured when the region between the plates is evacuated. Next, the polar substance of interest is introduced. Because the molecules have a dipole moment, they align along the electric field direction and thereby change the magnitude of the field. The dipole moment can be determined from the observed change in the field. Some typical values are listed in Table 15.1. The values are in the range of 0–10 D. Note that homonuclear diatomic molecules necessarily have zero dipole moment—there can be no permanent charge asymmetry between identical atoms.

*b. Polyatomic Molecules*  The dipole moments of diatomic molecules reflect the polarity of the one bond such molecules possess. Polyatomic molecules

**TABLE 15.1  Dipole Moments of Diatomic and Polyatomic Gaseous Molecules**

| Diatomic Molecules | | Polyatomic Molecules | | |
|---|---|---|---|---|
| Molecule | Dipole Moment (D) | Molecule | Geometry | Dipole Moment (D) |
| $H_2$, $N_2$, $O_2$, $F_2$, $Cl_2$, $I_2$ | 0 | $H_2O$ | Bent | 1.84 |
| | | $H_2S$ | Bent | 0.89 |
| HF | 1.94 | $SO_2$ | Bent | 1.62 |
| HCl | 1.08 | $CO_2$, $CS_2$ | Linear | 0 |
| HBr | 0.78 | $N_2O$ | Linear | 0.166 |
| HI | 0.38 | $BCl_3$, $BF_3$ | Trigonal planar | 0 |
| BrCl | 0.57 | $NH_3$ | Pyramidal | 1.45 |
| NO | 0.16 | $PH_3$ | Pyramidal | 0.55 |
| CO | 0.12 | $PF_3$ | Pyramidal | 1.02 |
| LiF | 6.33 | $CH_4$, $SiH_4$, $CCl_4$, $SnCl_4$ | Tetrahedral | 0 |
| LiCl | 7.13 | | | |
| LiBr | 7.27 | $CHCl_3$ | Tetrahedral | 1.02 |
| NaCl | 9.00 | $SO_2Cl_2$ | Tetrahedral | 1.81 |
| KF | 8.60 | $PCl_5$ | Trigonal bipyramidal | 0 |
| KCl | 10.27 | | | |
| KBr | 10.41 | $SF_6$ | Octahedral | 0 |
| CsCl | 10.42 | | | |

have two or more bonds, and these bonds may differ in polarity. It is therefore necessary to distinguish between the dipole moment of a molecule and the dipole moment associated with a specific bond, called the **bond moment.** The dipole moment of a molecule depends on both the magnitude and direction of its bond moments.

That molecular geometry affects the dipole moment of a molecule can be seen by considering the carbon dioxide molecule, $CO_2$. The carbon-oxygen bond is polar, as shown by the difference in the eletronegativities of these elements as well as by the dipole moment of carbon monoxide, CO. Yet the dipole moment of carbon dioxide is zero. The reason for this somewhat surprising result is that the $CO_2$ molecule is linear. Figure 15.10a shows the structure of $CO_2$ with the bond moments represented by arrows. Note that the bond moments point in opposite directions. Since the two carbon–oxygen bonds are identical, the two bond moments are equal in magnitude and exactly cancel each other. Thus, the fact that $CO_2$ has $\mu = 0$ shows that the molecule is linear and symmetric.

The exact cancellation of the bond moments occurs for molecules with a symmetric structure. As may be noted in Table 15.1, this is the case for molecules that are linear, trigonal planar, tetrahedral, trigonal bipyramidal, or octahedral, provided all the atoms linked to the central atom are identical. Molecules with one of these structures, such as $BF_3$, $CH_4$, $PCl_5$, or $SF_6$, have zero dipole moments.

In order for a polyatomic molecule to have a dipole moment, it must be asymmetric. Among the molecules fulfilling this criterion are bent molecules, such as $H_2O$, and pyramidal molecules, such as $NH_3$. The relationship between the bond moments and the dipole moment of $H_2O$ is illustrated in Figure 15.10b. Note that since the negative charge resides on the oxygen atom, the positive charge must be at a resultant position between the two hydrogen atoms.

Table 15.2 lists some typical bond moments. To a good approximation, the same bond moment is obtained for a given bond, e.g., an N—C single bond, for all molecules containing this bond; bond moments are largely independent of the molecular environment. Therefore, bond moments can be combined to give an estimate of the dipole moment of the molecule comprised of the bonds in question. This combination does not involve ordinary addition because the bond moments generally point in different directions. A quantity characterized by both magnitude and direction is a **vector** quantity. Vector quantities can be combined by vector addition, as illustrated for $H_2O$ in Exam-

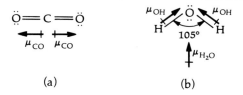

**FIGURE 15.10** Relationship between bond moments and the dipole moment of a polyatomic molecule. (a) $CO_2$, a symmetric molecule and (b) $H_2O$, an asymmetric bent molecule. See Example 15.7 for details of the vector addition of the bond moments of $H_2O$.

## Lewis Structures and Chemical Bonding 15.8

**TABLE 15.2  Some Typical Bond Moments**

| Bond[a] | Moment (D) | Bond[a] | Moment (D) |
|---|---|---|---|
| C—H | 0.4 | Cl—C | 1.46 |
| N—H | 1.31 | Br—C | 1.38 |
| H—P | 0.36 | C—I | 1.19 |
| O—H | 1.51 | O—N | 0.3 |
| S—H | 0.68 | O=N | 2.0 |
| F—H | 1.94 | F—N | 0.17 |
| Cl—H | 1.08 | O=P | 2.7 |
| Br—H | 0.78 | S=P | 3.1 |
| I—H | 0.38 | Cl—P | 0.81 |
| N—C | 0.22 | Br—P | 0.36 |
| N=C | 0.9 | I—P | 0 |
| N≡C | 3.5 | O=S | 2.8 |
| O—C | 0.74 | Cl—S | 0.7 |
| O=C | 2.3 | O—Cl | 0.7 |
| C—S | 0.9 | F—Cl | 0.88 |
| F—C | 1.41 | | |

[a]The more electronegative atom is given first.

ple 15.7. Note that the vectors representing the bond moments are resolved into their components along the $x$ and $y$ axes (or along the $x$, $y$, and $z$ axes for non-planar molecules), and these components are then added algebraically in order to obtain the dipole moment of the molecule.

**Example 15.7**  The bond angle in $H_2O$ is 104.5°. Use the O—H bond moment given in Table 15.2 to determine the dipole moment of $H_2O$. Compare with the value given in Table 15.1.

A suitable coordinate system for the evaluation of the dipole moment involves one axis (we call it the $x$ axis) drawn along a line joining the two hydrogen atoms and a second axis (the $y$ axis) perpendicular to the $x$ axis and passing through the oxygen atom. The following diagram shows the molecule drawn on this coordinate system in more detail:

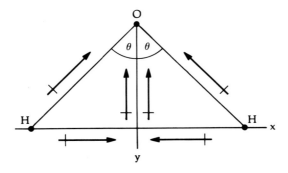

Note that the y axis bisects the HOH bond angle forming two right triangles each of which can be characterized by an angle $\theta$, where $\theta = 52.25°$ is one-half the bond angle. The bond moments point along the hypotenuse of each triangle and can be resolved into $x$ and $y$ components. The $x$ components point in opposite directions, as shown, and cancel. The $y$ components point in the same direction, towards the oxygen atom, and their sum is equal to the dipole moment of the molecule.

Let the O—H bond moment be designated $m$ and the dipole moment of the molecule be $\mu$. The $y$ component of $m$, $m_y$, is given by the trigonometric relation

$$\cos \theta = m_y/m$$

Since both bond moments are identical, the dipole moment is

$$\mu = 2m_y = 2m \cos \theta$$

Substituting the appropriate numerical values, we obtain

$$\mu = (2)(1.51 \text{ D})[\cos(52.25°)] = 1.84 \text{ D}$$

This value is in agreement with the tabulated value for $H_2O$. ∎

## 15.9 The VSEPR Model

The **valence shell electron pair repulsion** (**VSEPR**) model explains the shapes of molecules on the basis of the electrostatic repulsion between the electron pairs surrounding the central atom of the molecule. Because of this repulsion, the electron pairs about a central atom are arranged in a manner that maximizes the distance between them. Atoms covalently bonded to a central atom are therefore also located at positions that tend to maximize the distance between them.

The VSEPR model is easily applied to structures in which all the valence electrons of the central atom participate in bonding. So long as all the bonds made by the central atom are of equal length, the structure resulting from maximum separation between these bonds can be drawn by placing the attached atoms on the surface of a sphere the center of which is defined by the central atom. The shape of a molecule then corresponds to the geometric arrangement which maximizes the distance between the appropriate number of points inscribed on the surface of a sphere, as shown in Figure 15.11.

The structure of molecules described by the general formula $AB_2$, in which all the valence electrons surrounding A are bonded, is linear, for this structure maximizes the distance between the bonding electron pairs. Similarly, molecules of the type $AB_3$ are planar and the BAB bond angle is 120°. Molecules of type $AB_4$ have tetrahedral geometry, those of type $AB_5$ are trigonal bipyramidal, and those of type $AB_6$ are octahedral. Some examples of each type are listed in Figure 15.11. Note that most of the common shapes of molecules depicted in Figure 15.6 are included.

Two additional comments concerning the VSEPR structures in Figure 15.11 must be made. First, the shape of a molecule is determined primarily by the number of atoms attached to the central atom, not by the number of bonds. Consequently, molecules have the same general shape regardless of whether the central atom is attached to its neighbors by single bonds or by multiple bonds, provided only that all the electrons on the central atom are bonded. For

| Number of points | Geometrical description | General formula | Angles | Examples |
|---|---|---|---|---|
| 2 | 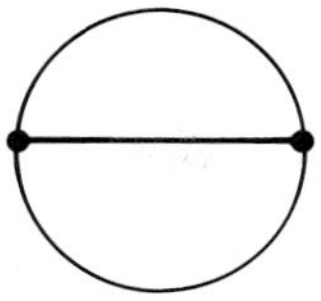 Linear | $AB_2$ | 180° | $BeCl_2(g)$, $CO_2$, $CS_2$ |
| 3 | 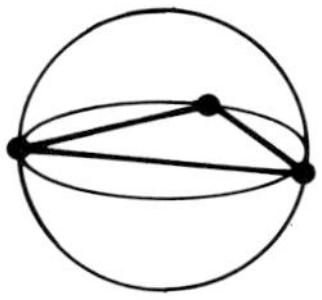 Trigonal planar (Equilateral triangle) | $AB_3$ | 120° | $BX_3$, $CO_3^{2-}$, $SO_3$ |
| 4 | 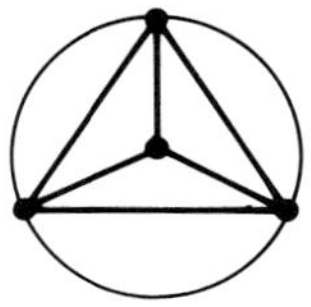 Tetrahedron | $AB_4$ | 109.5° | $CH_4$, $CX_4$, $NH_4^+$, $SiX_4$, $BX_4^-$, $ClO_4^-$, $PO_4^{3-}$ |
| 5 | 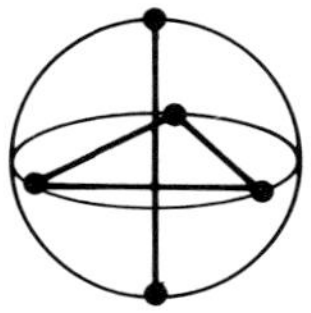 Trigonal bipyramid | $AB_5$ | 90°, 120° | $PF_5$, $PCl_5(g)$, $AsF_5$ |
| 6 | 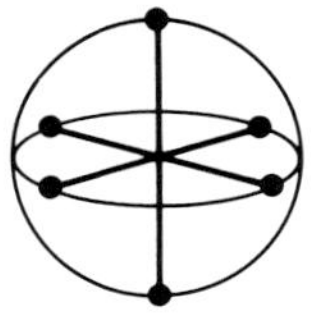 Octahedron | $AB_6$ | 90° | $SF_6$, $SeF_6$, $PF_6^-$, $SiF_6^{2-}$ |

**FIGURE 15.11** Positions of points on the surface of a sphere maximizing the distances between them. The points correspond to the atoms attached to a central atom located at the center of the sphere (not shown). The central atom is surrounded by from two to six atoms. Some examples of molecules with the indicated shapes are listed.

example, both beryllium chloride (at elevated temperatures) and carbon disulfide ($CS_2$) are linear in spite of the different types of bonds made by the central atom:

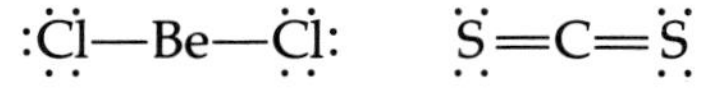

$$:\ddot{Cl}-Be-\ddot{Cl}: \qquad \ddot{S}=C=\ddot{S}$$

Second, the replacement of one of the outer atoms in a molecule by an atom of a different element introduces some distortion into the basic structure. For example, both fluoromethane ($CH_3F$) and chloromethane ($CH_3Cl$) are approximately tetrahedral. However, the HCH bond angles are 110.0° and 110.3°, respectively, compared to the tetrahedral angle (109.47°) in methane.

Let us now apply the VSEPR model to the slightly more difficult case of species with a central atom possessing lone electron pairs. The simplest example is a central atom that has four electron pairs, one of which is a lone pair. The ammonia molecule,

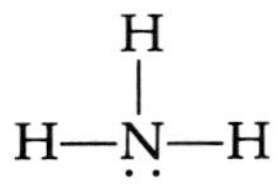

$$H-\underset{..}{N}-H \quad \text{with H above N}$$

**580** Lewis Structures and Chemical Bonding 15.9

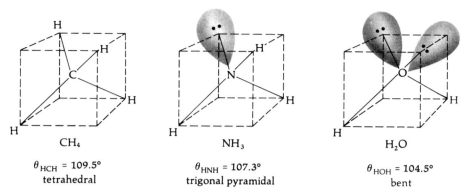

$\theta_{HCH} = 109.5°$  $\theta_{HNH} = 107.3°$  $\theta_{HOH} = 104.5°$
tetrahedral  trigonal pyramidal  bent

**FIGURE 15.12** The structures of methane (CH$_4$), ammonia (NH$_3$), and water (H$_2$O). The structures of NH$_3$ and H$_2$O are analogous to those resulting from the removal of one and two protons from CH$_4$, respectively. The ellipses indicate the presence of lone electron pairs.

---

is a good example. It is reasonable to assume that, to a first approximation, the four electron pairs will be directed towards the corners of a tetrahedron, in analogy to a molecule with four bonded electron pairs. However, the structure of ammonia cannot be tetrahedral, for the structure of a molecule is determined by the position of its constituent atoms and not by that of lone pair electrons. Consequently, the ammonia molecule can be characterized as a tetrahedron with a missing corner. As indicated in Figure 15.12, this shape is that of a trigonal pyramid, a common molecular shape already characterized in Figure 15.6.

There is one other significant difference between the structure of ammonia and that of a tetrahedral molecule such as methane. The HNH bond angle, 107.3°, is somewhat smaller than the tetrahedral value. This difference suggests that the lone electron pair on the nitrogen atom repels the bonded electron pairs more strongly than the latter repel each other. Because of this difference in repulsion, the potential energy of the molecule is reduced when the bonded electron pairs are closer to each other than they are in the normal tetrahedral structure. The lone pair essentially pushes the bonded electron pairs towards each other and the bond angle decreases. This effect results from the tendency of bonded electron pairs to be localized to the region between the two atoms that they attach. Lone pair electrons, however, are not confined in this manner and electron–electron repulsion causes them to spread out and occupy the available space.

In the H$_2$O molecule, as shown in Figure 15.12, two of the four electron pairs surrounding the central atom are nonbonded. The presence of two lone pairs should exert a stronger repulsive effect on the bonded electrons than that of one lone pair. As expected, the distortion of the tetrahedral geometry is greater than that observed for ammonia and the bond angle is reduced to 104.5°.

Figure 15.13 summarizes a variety of different molecular structures that result from the combination of various numbers of bonded and lone electron pairs about the central atom. All these structures can be understood within the context of the VSEPR model.

Let us consider in more detail the structures of some molecules in which

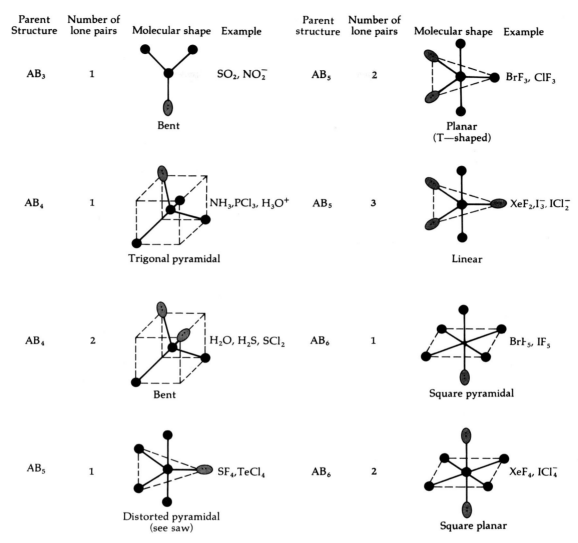

**FIGURE 15.13** The shapes of molecules with central atoms possessing one or more lone electron pairs.

the central atom has an expanded valence shell. Our earlier discussion of this concept (Section 15.2c) indicated that the elements in Group VA–VIIA occupying the third and higher periods frequently display such an expansion.

The structure of molecules whose central atom is surrounded by five bonded electron pairs is that of a trigonal bipyramid. The replacement of one or more of these pairs by lone pairs leads to shapes derived from this structure. For example, in sulfur tetrafluoride (SF$_4$) the sulfur atom has one lone pair and four bonded pairs. There are two choices for the placement of the lone pair: an equatorial or an axial site. According to the model, the lone pair should occupy the site in which it is as far from the other electron pairs as possible. The stronger repulsion caused by a lone pair indicates that this placement will re-

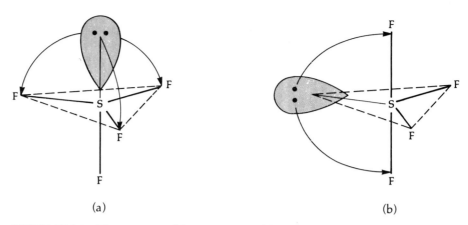

**FIGURE 15.14** The two possible structures of SF$_4$. (a) Lone pair in an axial site; (b) lone pair in an equatorial site. The correct structure is (b) because the lone pair is farther from the bonded electrons.

sult in a state of lower energy. Figure 15.14 illustrates the two choices. Note that the axial site is adjacent to all three equatorial sites at an angle of 90°. On the other hand, the equatorial site is adjacent to only two axial sites at 90°. The other two equatorial sites are not as close since the angle between them is 120°. The lone pair therefore occupies an equatorial site. Figure 15.14b shows that the result is a distorted pyramid, resembling a seesaw in shape. Furthermore, recall that in trigonal bipyramidal molecules of type AB$_5$ the axial bonds are generally longer than the equatorial bonds. This distinction is maintained in lone-pair structures derived from this parent structure. Thus, in sulfur tetrafluoride two of the S—F bonds are 1.646 Å long and the other two S—F bonds are 1.545 Å long.

Additional lone pairs in molecules derived from the trigonal bipyramidal structure also occupy equatorial sites. Chlorine trifluoride (ClF$_3$) is an example of such a molecule. Figure 15.13 shows that the resulting structure is planar and T-shaped. One additional remark must be made about both the sulfur tetrafluoride and chlorine trifluoride structures. As in the case of ammonia or water, the presence of a lone pair decreases the angles between the other bonds from the values found in the parent structure. Figure 15.15 shows these distortions in more detail. In SF$_4$, the FSF equatorial bond angle is reduced from 120° to 101° while the angle between equatorial and axial F—S bonds is reduced from 90° to 86.5°. Similarly, in ClF$_3$ the FClF angle is reduced from 90° to 87.5°.

The final molecular structure to be examined is the octahedral structure. As noted in Section 15.7, all six sites in the structure are equivalent. A single lone pair can therefore occupy any site. It is customary to draw the structure with the lone pair occupying an axial site, as shown in Figure 15.13. This representation makes it easiest to see that molecules with one lone pair, such as iodine pentafluoride (IF$_5$), have a square pyramidal shape. Once again, the presence of the lone pair introduces some distortion, and the angle between the axial and equatorial I—F bonds is only 84°. The presence of two lone electron pairs, as in xenon tetrafluoride (XeF$_4$), results in a symmetric square planar

**FIGURE 15.15** Effect of lone electron pairs on the bond angles in SF₄ and ClF₃.

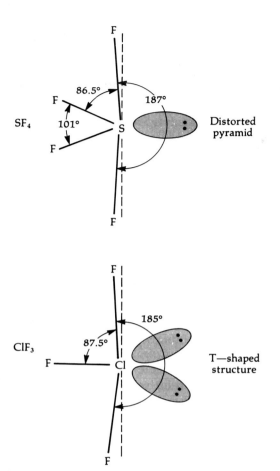

structure. There is no distortion because the lone pairs are exactly opposite to each other in the two axial positions.

The application of the VSEPR model may be summarized by the following rules:

1. Draw the Lewis structure of the species in question.
2. Examine the valence electrons surrounding the central atom. If all of these electrons are bonded, count the number of atoms to which the central atom is attached. Each number corresponds to a different molecular shape, as illustrated in Figure 15.11.
3. If the central atom is surrounded by one or more lone pairs, first determine the parent structure. This is the structure that would be obtained if all the lone pairs were bonded. The actual structure is derived from the parent structure by removing an atom for each lone pair. If only one lone pair is present, the atom is removed from the site which is farthest from the largest number of other sites. If more than one lone pair is present, atoms are removed from sites which are farthest from each other except for trigonal bipyramidal structures, where lone pairs are removed from equatorial sites. (The total repulsion between electron pairs thereby is minimized.)

# 584 Lewis Structures and Chemical Bonding 15.9

**Example 15.8** Determine the shape of (a) $ICl_2^-$, (b) $O_3$, and (c) $ClF_5$.

(a) $ICl_2^-$. We begin by writing the Lewis structure. Since each halogen atom has 7 valence electrons, the structure must show 22 electrons. The most plausible structure is

$$[:\ddot{\underset{..}{Cl}}-\cdot\underset{..}{\overset{..}{I}}\cdot-\underset{..}{\overset{..}{Cl}}:]^-$$

in which each of the chlorine atoms is surrounded by 8 electrons while the central iodine atom is surrounded by 10 electrons.

A molecule with a central atom surrounded by five electron pairs has a structure based on a trigonal bipyramid (Figure 15.11). The actual shape is determined by the location of the three lone electron pairs. Since these pairs occupy more space than bonded electron pairs, the overall electron pair repulsion in the molecule is minimized when the lone pairs are as far from each other as possible. This is accomplished by placing all three lone pairs in equatorial positions, where the angle between them is 120°. As shown below, the placement of one or two lone pairs in axial positions would result in 90° angles between lone pairs in axial and equatorial positions. The closer lone pair–lone pair proximity makes this configuration less stable. The result is that, since the bonded electron pairs occupy the axial sites, $ICl_2^-$ is linear [structure (a)].

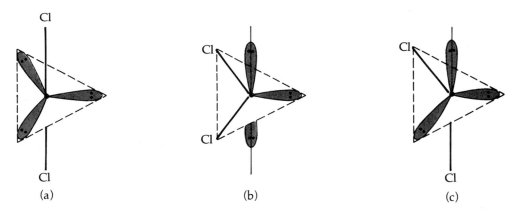

(b) $O_3$. The Lewis structure must include 18 valence electrons. The most reasonable structures are

$$:\underset{..}{\overset{..}{O}}\diagup\overset{\overset{..}{O}}{\diagdown\!\!\!=}\underset{..}{O}: \quad \leftrightarrow \quad :\underset{..}{O}\overset{\overset{..}{O}}{=\!\!\!\diagup}\diagdown\underset{..}{\overset{..}{O}}:$$

since each oxygen is surrounded by 8 electrons. The central oxygen atom is attached to two atoms and has one lone pair. The shape of the molecule is derived from the trigonal planar parent structure (Figure 15.11) and is bent. The bond angle should be less than 120° because of the greater repulsion exerted by the lone pair and, in fact, is 117°.

(c) ClF$_5$. The molecule has $6 \times 7 = 42$ valence electrons. The Lewis structure is

in which each fluorine atom obeys the octet rule and the central chlorine atom has an expanded valence shell containing 6 electron pairs. The parent structure is the octahedron (Figure 15.11). Since all the electron pair sites in an octahedral structure are identical, the lone pair can occupy any one of the corners. The resulting structure is square pyramidal (Figure 15.13). ■

## Conclusion

A covalent bond can be described as an electron pair shared between two atoms; the formation of a covalent bond lowers the energy of the system. This rather simple concept has far reaching consequences, for it is how we explain the existence of molecules.

Covalent bonding can be understood in a qualitative but, nonetheless, very informative way by means of Lewis structures. This pictorial model of molecular structure shows the location of all the valence electrons in a molecule, either as electron pair bonds or as lone pairs. By making use of concepts such as the octet rule, coordinate covalent bonding, valence shell expansion, multiple bonding, formal charges, and resonance, the model depicts covalent bonding in a manner useful to a qualitative description of many chemical phenomena. When combined with the concept of electron pair repulsion in the VSEPR model, Lewis structures also satisfactorily account for the shapes of many molecules, as determined, for example, by measurement of their dipole moments.

## Problems

1. Draw a Lewis structure for each of the following molecules: BeH$_2$, SF$_2$, AsF$_3$, and H$_2$O$_2$.

2. Draw a Lewis structure for each of the following ions: ClO$_2^-$, ClO$_3^-$, ClO$_4^-$, and CO$_3^{2-}$.

3. Write Lewis structures for the following ionic substances: lithium fluoride, ammonium nitrate, and potassium perchlorate.

4. Write Lewis structures for the following species and identify those cases in which the octet rule is not obeyed: XeF$_2$, NO, Cl$_2$, and COCl$_2$. Which of these species have unpaired electrons?

5. Draw the Lewis structures of the following species containing nitrogen and hydrogen: NH, NH$_2$, and H$_2$NNH$_2$.

6. Which of the following molecules are expected to be paramagnetic: N$_2$O, NO, NO$_2$, and N$_2$O$_4$? Write Lewis structures for all of these molecules.

7. Aluminum hydroxide is amphoteric and reacts with acids and bases as follows:

$$Al(OH)_3(s) + 3H^+(aq) \rightarrow Al^{3+}(aq) + 3H_2O$$

$$Al(OH)_3(s) + OH^-(aq) \rightarrow Al(OH)_4^-(aq)$$

Interpret these reactions in terms of the Lewis theory of acids and bases.

8. Aluminum chloride (AlCl₃) dimerizes (i.e., attaches to another like molecule) in the gas phase to form Al₂Cl₆. (a) Write the Lewis structure of AlCl₃ and determine the formal charges. (b) Write the Lewis structure of Al₂Cl₆ and determine the formal charges. (The molecule does *not* contain an Al—Al bond.) (c) Which of the molecules is electron deficient? (d) Which of the molecules has coordinate covalent bonds? How many?

9. Phosphorus can form coordinate covalent bonds in which it either donates an electron pair or accepts an electron pair into a d orbital. The formation of adducts in the reactions (a) PCl₃ + N(CH₃)₃ and (b) PCl₃ + BBr₃ exemplifies these two cases. Determine in which of these reactions phosphorus acts as a donor and in which it acts as an acceptor by writing Lewis structures of the reactants and adducts. What are the formal charges in the adducts?

10. Which of the following species are isoelectronic: BF₃, CO₃²⁻, SO₃²⁻, SO₃, NH₃, and NO₃⁻? Write Lewis structures for all the given species.

11. Determine the formal charges and oxidation numbers in the ClO₂⁻ ion. Draw at least two different reasonable Lewis structures.

12. In nitromethane (CH₃NO₂) the nitrogen–oxygen bond lengths are found to be equal. Draw Lewis structures consistent with this observation.

13. Draw the Lewis structure of the bicarbonate ion, HOCO₂⁻. Rank the three carbon–oxygen bonds in order of increasing length.

14. Draw Lewis structures for the following species, showing important resonance structures where appropriate: H₃PO₂ (the phosphorus atom is bonded to an oxygen atom, two hydrogen atoms and an OH group), PH₄⁺, N₂O (the two nitrogen atoms are linked), and C₅H₅N (containing a six membered ring consisting of the carbon and nitrogen atoms).

15. Iodine forms several oxoanions, such as IO₃⁻, IO₄⁻, and IO₆⁵⁻. (a) Write Lewis structures for these species. (b) What is the bond order in each oxoanion?

16. The structural formula of diazomethane (CH₂N₂) is

$$H-C-N-N$$
$$\phantom{H-}|$$
$$\phantom{H-C-N-}H$$

Draw Lewis structures for this molecule in which the two nitrogen atoms are attached by (a) a single bond (b) a double bond, and (c) a triple bond. Determine formal charges and on this basis decide which of these structures, if any, are unrealistic.

17. The two carbon–oxygen bonds in acetic acid (CH₃COOH) differ in length but the two carbon–oxygen bonds in the acetate ion (CH₃COO⁻) are equal. Discuss this difference in terms of the various Lewis structures.

18. Write three Lewis structures in which the octet rule is obeyed by each atom for (a) the cyanate ion (NCO⁻) and (b) the isocyanate ion (CNO⁻). On the basis of the formal charges, which are the most important structures of each ion?

19. Write non-equivalent resonance structures for the phosphate ion (PO₄³⁻). State how many equivalent resonance forms there are for each of these structures. Which structures are most realistic? Determine the oxidation number of phosphorus in each of the structures and explain why it remains constant.

20. Draw the important resonance structures for the allyl radical (H₂CCHCH₂).

21. The skeleton structure of *N,N*-dimethylacetamide is

$$\begin{array}{c}
\phantom{H-C-C-N}H \\
\phantom{H-C-C-N}| \\
H\phantom{-}O\phantom{-}H-C-H \\
|\phantom{-}|\phantom{-}\diagdown\phantom{H} \\
H-C-C-N\phantom{-C-H} \\
|\phantom{-C-C-}\diagdown\phantom{H} \\
H\phantom{-C-C-N-}C\diagdown \\
\phantom{H-C-C-N-H}H\phantom{-}H
\end{array}$$

Draw two resonance structures and pick the one that is most important.

22. Consider the following Lewis structures for the thiocyanate ion (SCN⁻):

[:S̈=C=N:]⁻    [S̈=C=N̈]⁻    [:S≡C—N̈:]⁻
(a)            (b)            (c)

[:S̈—C≡N:]⁻
(d)

Knowing that nitrogen is more electronegative than sulfur, pick the most reasonable Lewis structure. Indicate the drawbacks of each of the other structures.

23. By referring to Figure 15.7, derive the value of the tetrahedral angle.

24. The bond angle in H₂S is 92.2°, the H—S bond length is 1.335 Å and the H—S bond moment is 0.68 D. Evaluate the dipole moment of H₂S.

25. Which of the following species should have a dipole moment of zero: CO, BF₃, BrF₃, SO₂, and N₂O? Explain.

26. Three different isomers of dichlorobenzene are known. The structure of these planar molecules is as follows:

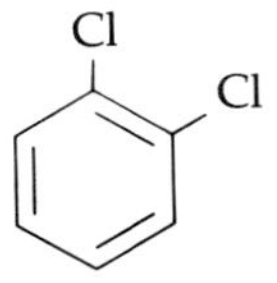

o-dichlorobenzene

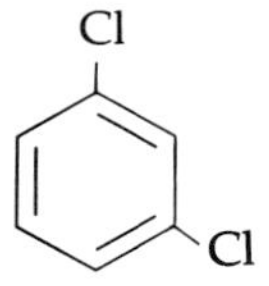

m-dichlorobenzene

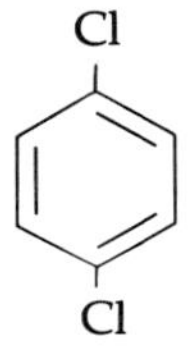

p-dichlorobenzene

(The six carbon atoms in the ring and the four hydrogen atoms attached to the carbons are not shown for simplicity.) The C—Cl bond moment in these molecules is given by the experimental dipole moment of chlorobenzene and is 1.57 D. The angles between the two chlorine atoms and the center of the ring are 60°, 120°, and 180°, respectively, as shown. Make estimates of the dipole moments of the three dichlorobenzenes.

27. Predict the shape of the following molecules on the basis of the VSEPR model: $SO_3$, $SF_2$, and $SbCl_3$.

28. Predict the shape of the following ions by means of the VSEPR model: $ClO_3^-$, $ClO_4^-$, $CO_3^{2-}$, and $SO_3^{2-}$.

29. Use the VSEPR model to predict the shapes of the following species: $Cl_2CO$, $IO_4^-$, and $ClO_2$.

30. Write the Lewis structure of the formate ion (HCOO$^-$), assign formal charges, and use the VSEPR model to determine the geometry.

31. Use the VSEPR model to predict the shapes of the following molecules: $SO_2$, $PbCl_4$, and $SbH_3$.

32. Compare the Lewis structures and VSEPR shapes of $SO_3$ and $SO_3^{2-}$. For each species, write the main resonance forms and the formal charges. Which species is expected to have a longer sulfur–oxygen bond length? Explain.

33. Draw a Lewis structure for $XeO_3$. Determine the formal charges and the VSEPR structure.

34. Use the VSEPR model to determine the geometry of (a) $TeF_5^-$ and (b) $ICl_4^-$.

# 16    SECTIONS All

# Theories of Chemical Bonding

Chemical bonding can be described on different levels of sophistication. Lewis structures provide a simple pictorial description of the manner in which atoms form bonds. However, an accurate description of covalent bonding requires the application of quantum mechanics. The Schrödinger wave equation (Section 5.8) appropriate to a particular molecule is solved to obtain the wave function, the energy of the molecule in its ground and excited states, and the mean internuclear distances, much as in the case of the hydrogen atom (Section 5.12). The calculated energy is closely related to the bond energies, and the calculated internuclear distances correspond to the bond lengths. Since bond energies and bond lengths can be determined by experiment (Sections 17.1 and 17.2), the calculations can be tested and refined by comparison of their predictions with the data.

    The Schrödinger equation for systems containing more than one atom is so complicated that an exact solution has not been obtained except for the simplest molecules. Instead, different types of approximate methods have been developed. The two most widely used approaches are the valence bond and the molecular orbital theories. Both theories are based on the principles of quantum mechanics and picture covalent bonding as resulting from the overlap of the atomic orbitals of neighboring atoms. However, they differ in their approach and in their approximations. Nonetheless, valence bond theory and molecular orbital theory should not be regarded as competing theories but, rather, as complementary descriptions. One approach may be more useful than the other in a particular application, but both are needed for an understanding of the many facets of covalent bonding. Before reading this chapter, it would be beneficial to review the principles of quantum mechanics developed in Chapters 5 and 6 (Sections 5.8–5.12, and 6.1–6.3).

# Valence Bond Theory

## 16.1 The Hydrogen Molecule

According to **valence bond theory**, covalent bonding between two atoms occurs as a consequence of the overlap of two of their respective atomic orbitals. For bonding to occur, the two orbitals must have comparable energies, they must be directed toward each other, and they must each contain an electron of opposite spin. A covalent bond is formed when the two electrons are shared between the two orbitals. In this sense, valence bond theory is a quantum mechanical extension of the Lewis electron pair model.

It is helpful to begin by applying valence bond theory to a very simple molecule, the $H_2$ molecule, for then the essential features can be seen most clearly. The formation of a hydrogen molecule from two hydrogen atoms is depicted in Figure 16.1. When the two atoms are far apart, say more than 10 Å, there is essentially no interaction between them (Figure 16.1a). As the two atoms approach each other, their 1s orbitals begin to overlap (Figure 16.1b). The electron density in the region between the two nuclei begins to increase, primarily because of the attraction of the nucleus of one atom for the electron of the other. At a sufficiently close distance of approach, good overlap of the 1s orbitals occurs, and a bond forms (Figure 16.1c). The two atoms do not move significantly closer to each other because the repulsion between the two protons and that between the electrons become important at short distances.

The process depicted in Figure 16.1 can be described more quantitatively in terms of the potential energy of the H—H system as a function of internuclear separation, depicted in Figure 16.2. The colored curve in this figure is consistent with the experimental data. As already noted, there is no interaction between the two atoms at large internuclear distances, and the potential energy therefore is zero. As the internuclear distance decreases, the potential energy at first becomes increasingly negative, reflecting the net attractive force between the two atoms. However, at sufficiently short distances, the force between the two atoms changes from attractive to repulsive, and the potential energy rapidly increases. The combination of an attractive and a repulsive potential necessarily results in the observed minimum at an intermediate distance, which for $H_2$ occurs at 0.74 Å. This distance is equal to the H—H bond length, while the depth of the potential "well" at this distance, 458 kJ mol$^{-1}$, is closely related to the H—H bond energy.

Let us now examine the predictions of the valence bond theory. When the two hydrogen atoms (designated A and B) are far apart, they can be described by wave functions $\psi_A(1)$ and $\psi_B(2)$, where electron 1 is on atom A and electron 2 is on atom B. These are just the 1s wave functions in Table 5.6. The overall wave function of the system of two hydrogen atoms is the product of the individual wave functions,

$$\psi = \psi_A(1)\psi_B(2) \tag{16.1}$$

590   Theories of Chemical Bonding   16.1

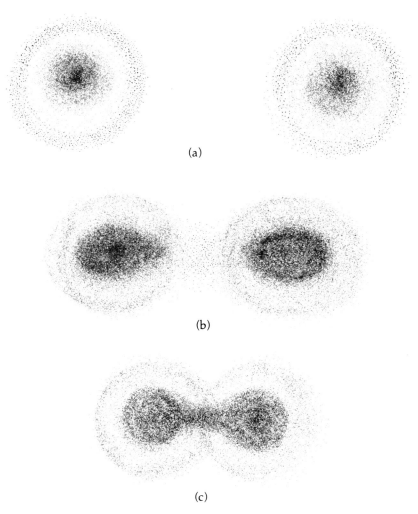

**FIGURE 16.1**   The interaction of two approaching hydrogen atoms to form an $H_2$ molecule. (a) The two atoms are far apart and there is no interaction between them; (b) the 1s atomic orbitals begin to interact; (c) the 1s orbitals overlap and form a covalent bond.

When the two atoms are brought together, the wave function will change owing to the interaction. Nonetheless, the wave function in Equation 16.1 is a useful starting point. The variation of the energy with the internuclear distance corresponding to this wave function is given by curve $a$ in Figure 16.2. Note that this curve has a shallow minimum at a distance of about 0.9 Å. While this value is encouragingly close to the experimental bond length, the depth of the well is only about 24 kJ mol$^{-1}$, nearly 20 times smaller than the experimental value.

It is apparent that some major factor has not been included in Equation 16.1. We have neglected the fact that electrons are indistinguishable and it is consequently just as likely that electron (2) is on atom A and electron (1) on B as that electron (2) is on atom B and electron (1) on A. The wave function is modified readily for this effect by addition of a term in which the positions of

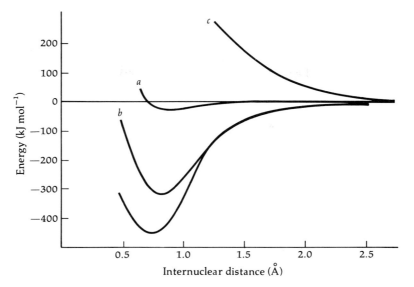

**FIGURE 16.2** Potential energy curve for the interaction of two hydrogen atoms to form an H$_2$ molecule. Curves (a) and (b) represent the results of successive improvements in the valence bond wave function of the system. The colored curve is based on experimental data and also represents the most accurate version of valence bond theory. Curve (c) is the potential energy of two hydrogen atoms whose electrons have like spin.

the electrons are exchanged:

$$\psi = \psi_A(1)\psi_B(2) + \psi_A(2)\psi_B(1) \tag{16.2}$$

This modification leads to energy curve $b$ in Figure 16.2, and provides a remarkable improvement: the depth of the well is 303 kJ mol$^{-1}$ and the bond length is approximately 0.87 Å. The lowering of the energy can be understood by analogy to the quantum mechanics of the particle in a box (Section 5.9). Recall that the energy of a particle is inversely proportional to the square of the length of the box in which it is confined (Equation 5.39). By allowing each electron to reside on both atoms, the "box" has been effectively enlarged, and the energy is therefore lower.

Although the inclusion of electron exchange constitutes a vast improvement, the wave function of Equation 16.2 is still too restrictive. We must consider the additional possibility that both electrons will reside on a single atom part of the time, thereby forming H$^+$H$^-$ ion pairs. While the probability of such ionic structures is substantially lower than that of the covalent structure due to the repulsion between two electrons attached to a single proton, such structures cannot be excluded. The actual structure of H$_2$ is likely to be a resonance hybrid of the covalent and ionic structures,

$$\text{H—H} \leftrightarrow \underbrace{\text{H}^+\text{H}^- \leftrightarrow \text{H}^-\text{H}^+}_{}$$
$$\text{covalent} \qquad \text{ionic}$$

and the corresponding wave function can be written as

$$\psi = \psi_A(1)\psi_B(2) + \psi_A(2)\psi_B(1) + \lambda[\psi_A(1)\psi_A(2) + \psi_B(1)\psi_B(2)] \tag{16.3}$$

where the coefficient λ, which is substantially smaller than unity, is introduced because the ionic structure is less important than the covalent structure. The inclusion of these ionic terms in the wave function further improves the agreement of the theory with experiment. When some additional refinements are made, e.g., electron screening (Section 6.2), the results are in excellent agreement with the data.

What happens when the electrons of the two hydrogen atoms have like spin? According to the Pauli exclusion principle, two electrons of like spin cannot occupy the same orbital. Thus, bond formation cannot occur and the potential energy of the system increases continuously with decreasing internuclear separation (curve c, Figure 16.2).

## 16.2 Hybridization of Atomic Orbitals

The formation of a hydrogen molecule requires the overlap of just one type of orbital of each atom—a 1s orbital. For many molecules, overlap of atomic orbitals of different types is required. When this is the case, valence bond theory must be developed further if it is to agree with experiment.

Consider the formation of methane ($CH_4$) from the constituent elements. The electronic configuration of the valence shell of carbon can be written as

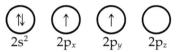

where, in conformity with Hund's rule, the p electrons are placed in separate orbitals. At first sight, it appears that carbon has only two unpaired electrons available for sharing. If this were indeed the case, the most stable molecule formed with hydrogen would be $CH_2$ which, in fact, exists only as an unstable intermediate. What occurs instead is that the 2s electron pair is broken and one of the electrons is promoted to the empty 2p orbital:

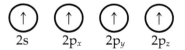

The promotion of an electron from the 2s to the 2p orbital requires some 400 kJ mol$^{-1}$. However, the formation of the two extra C—H bonds made possible by this promotion releases a sufficiently large amount of energy to make $CH_4$ some 900 kJ mol$^{-1}$ more stable than $CH_2$ + 2H.

One puzzle still remains. According to this picture, a methane molecule should contain three identical bonds and one bond that is different. The bonds formed by electrons in the three p orbitals of carbon are clearly identical but that formed by the electron in the s orbital of carbon should be different. Recall that both the radial and angular wave functions of s and p electrons are different (Section 5.12), and consequently the electron density distribution in the region between the carbon and hydrogen nuclei should also be different.

The actual structure of $CH_4$ does not reflect these expectations. All four C—H bonds have the same length and all the bond angles are tetrahedral. To account for these observations, Linus Pauling (b. 1901) introduced the concept of hybridization. **Hybridization** involves the mixing of different orbitals of an atom to form new hybrid orbitals. For example, the valence orbitals of carbon can be described as four energetically and spatially equivalent sp$^3$ hybrid orbitals formed from the 2s orbital and the three 2p orbitals. Note that hybridization does not change the number of orbitals: four pure orbitals (one 2s and three 2p) are transformed into four hybrid sp³ orbitals. According to valence bond theory, bonding in methane can be described in terms of the overlap of four sp³ hybrid orbitals of carbon with the s orbitals of four hydrogen atoms. This overlap results in the formation of four identical C—H bonds, as is observed experimentally.

To examine hybridization in more detail, let us focus on the simpler case of sp hybridization, in which one s orbital and one p orbital of an atom mix to form two hybrid sp orbitals. The bonding in molecular $BeCl_2$, which exists at high temperatures in the gas phase, can be described by sp hybridization. Let the s and p wave functions be $\psi_s$ and $\psi_p$, respectively, and the hybrid wave functions be $\psi_{sp}(1)$ and $\psi_{sp}(2)$. The formation of the hybrid wave functions is described by the following relations:

$$\psi_{sp}(1) = \psi_s + \psi_p$$
$$\psi_{sp}(2) = \psi_s - \psi_p \qquad (16.4)$$

Note that the hybrid orbital wave functions are given by the sum or difference of the pure orbital wave functions. The combination of two wave functions by addition or subtraction is analogous to the superposition of two light waves (Section 5.2). As in the case of light waves, addition of the two wave functions results in constructive interference and subtraction results in destructive interference.

The hybridization described by Equation 16.4 is shown graphically in Figure 16.3. The figure shows the addition or subtraction of the s and p wave

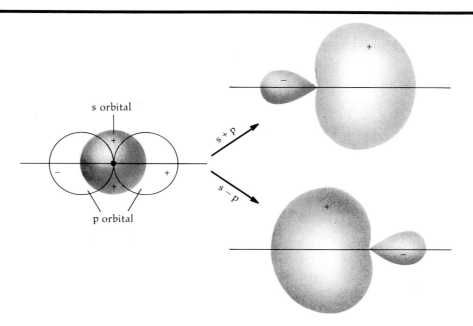

**FIGURE 16.3** Formation of sp hybrid orbitals by the addition and subtraction of s and p wave functions. The s and p wave functions are overlapped on the left, and the hybrid wave functions are shown on the right. The signs refer to the signs of the wave functions.

functions (see Figure 5.30). Observe that reinforcement, i.e., constructive interference, occurs in the region where the wave functions have the same sign and cancellation, i.e., destructive interference, occurs in the region where the wave functions have the opposite sign. The result is the formation of orbitals with asymmetric lobes pointing in opposite directions. The electron density is concentrated in the large lobe and this is the bonding lobe. Since the large lobes point in opposite directions, the two Be—Cl bonds should likewise be oppositely directed and the molecule should therefore be linear, as is indeed observed experimentally. Observe that the hybrid orbitals are strongly dependent on direction. This directionality allows for good overlap of the orbitals, and this overlap is important for bond formation.

The hybridizaton of atomic orbitals is a concept of considerable generality. Provided that the energy differences between them are not too large, orbitals of different types can be combined to form hybrid orbitals. In practice, this usually means that orbitals can be described as hybrid orbitals if they belong to the same shell (Section 6.2). For example a 2s orbital can hybridize with a 2p orbital but not with a 3p orbital. Since the electron in the hydrogen atom occupies a 1s orbital, and all other orbitals lie at significantly higher energy, hydrogen does not form hybrid orbitals.

In addition to sp and $sp^3$ hybrids, some commonly encountered hybrid orbitals are $sp^2$, $sp^3d$, and $sp^3d^2$. The formation of $sp^2$ hybrids is common for elements with three valence electrons, such as boron, with a $1s^22s^22p^1$ electronic configuration. Hybridization can be described as the formation from two s orbitals and one p orbital of three coplanar $sp^2$ hybrid orbitals directed at 120° to each other. These hybrid orbitals form the bonds in molecules such as $BF_3$.

The participation of d orbitals in hybridization occurs when the central atom of a molecule has an expanded valence shell. As already noted (Section 15.2c), valence shell expansion is common for atoms with low-lying d orbitals. When only one d orbital is involved, the combination of this orbital with one s orbital and three p orbitals results in the formation of five $sp^3d$ hybrid orbitals, as illustrated in Figure 16.4. Contrary to the other cases considered so far, the

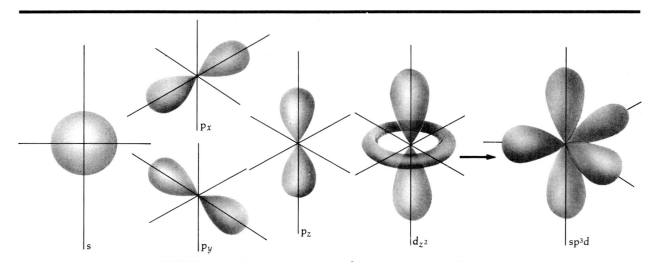

**FIGURE 16.4** The formation of $sp^3d$ hybrid orbitals. The angular parts of the parent and hybrid wave functions are depicted.

TABLE 16.1  Hybrid Orbitals and Their Orientation

| Hybrid orbital | Orientation | Bond angles |
|---|---|---|
| sp | Linear | 180° |
| sp² | Planar | 120° |
| sp³ | Tetrahedral | 109.5° |
| sp²d or dsp² | Square planar | 90° |
| sp³d or dsp³ | Trigonal bipyramidal | 90° and 120° |
| sp³d² or d²sp³ | Octahedral | 90° |

five hybrid orbitals are not identical. Rather, they divide into two groups of equivalent orbitals, three lying in the equatorial plane at 120° to each other and two directed axially at 180° to each other. When two d orbitals are involved, the central atom forms six hybrid sp³d² orbitals, as in $SF_6$. These orbitals point to the corners of a regular octahedron. In some transition metals, hybridization involves the $ns$, $np$, and $(n-1)d$ orbitals. The resulting hybrid orbitals are called dsp³ and d²sp³ orbitals. Their directional properties are similar to those of sp³d and sp³d² orbitals, respectively. Table 16.1 summarizes the hybrid orbitals of importance in chemical bonding and lists the angles between the resulting bonds. Note that the shapes are identical to those explained by the more intuitive but less rigorous VSEPR model (Section 15.9).

## Example 16.1

Determine the hybrid orbitals and bond angles formed by the central atom in the following molecules or ions: (a) $GeCl_4$, (b) $AsF_6^-$, (c) $BrF_3$, and (d) $AgCl_2^-$.

(a) $GeCl_4$. We must first identify the orbitals that might be used by the central atom for bonding and write a reasonable Lewis structure. The electronic configuration of Ge is $[Ar]3d^{10}4s^24p^2$. The valence orbitals are the 4s, the three 4p, and the five 4d orbitals, and these are the orbitals available for bonding. The Lewis structure is

$$\begin{array}{c} \ddot{\text{Cl}}: \\ | \\ :\ddot{\text{Cl}}-\text{Ge}-\ddot{\text{Cl}}: \\ | \\ :\ddot{\text{Cl}}: \end{array}$$

The central atom is surrounded by four valence electron pairs which can be accommodated by the 4s and 4p orbitals. The hybridization is sp³ and the bond angles are 109.5°.

(b) $AsF_6^-$. The configuration of arsenic is $[Ar]3d^{10}4s^24p^3$. The available valence orbitals are the 4s, the three 4p, and the five 4d orbitals. The Lewis structure is

$$\left[ \begin{array}{c} :\ddot{\text{F}}: \\ :\ddot{\text{F}}\diagdown | \diagup \ddot{\text{F}}: \\ \text{As} \\ \diagup | \diagdown \\ :\ddot{\text{F}}: \; :\ddot{\text{F}}: \; :\ddot{\text{F}}: \end{array} \right]^-$$

The As atom is surrounded by six electron pairs, requiring the use of the 4s, the three 4p, and two 4d orbitals. The hybridization is sp³d² and the bond angles are 90°.

(c) $BrF_3$. The configuration of Br is $[Ar]3d^{10}4s^24p^5$. The valence orbitals are the 4s, the three 4p, and the five 4d orbitals. The Lewis structure of $BrF_3$ is

$$:\!\ddot{F}\!-\!\ddot{B}r\!-\!\ddot{F}\!:$$
$$\qquad\;\;|$$
$$\quad\;\;:\!\ddot{F}\!:$$

indicating that the 4s orbital, the three 4p orbitals, and one 4d orbital must be used to accommodate the 10 electrons surrounding the bromine. Thus, the hybridization is sp³d. Note that two of the hybrid orbitals contain lone pairs. It follows that although the hybridized orbitals are arranged in the shape of a trigonal bipyramid, the molecule itself is T-shaped because the lone pairs occupy equatorial sites (Section 15.9). Consequently, the bond angles are close to 90° (see Figure 15.15).

(d) $AgCl_2^-$. The available orbitals of silver, $[Kr]4d^{10}5s^1$, are the 5s, the three 5p, and the five 5d orbitals. The Lewis structure of $AgCl_2^-$ is

$$[:\!\ddot{C}l\!-\!Ag\!-\!\ddot{C}l\!:]^-$$

where an extra electron has been added to account for the negative ionic charge. Since the Ag atom is surrounded by two valence electron pairs, sp hybridization is indicated and the bond angle is 180°. ∎

## 16.3 Ionic Character of Covalent Bonds

The bond between a hydrogen and a chlorine atom in an HCl molecule can be described, according to valence bond theory, as a covalent bond resulting from the overlap of the 1s orbital of hydrogen with a hybrid sp³ orbital of chlorine. Owing to the substantial difference between the electronegativities of the hydrogen and chlorine atoms, this bond is rather polar—the electron cloud is concentrated in the vicinity of the chlorine atom.

The dipole moment of a molecule provides some experimental information on bond polarity. Equation 15.4 for the dipole moment can be recast in the form

$$\delta = \mu/d \qquad (16.5)$$

thereby relating the magnitude of the fractional charge on each atom, $\delta$, to the experimentally determined dipole moment $\mu$, and bond length $d$. The ratio $\delta/e$, expressed as a percentage, is called the **ionic character** of a bond ($e$ is the electronic charge). The ionic character of a bond can range from 0 to 100%. A completely nonpolar bond, in which there is no charge separation, has 0% ionic character, while a completely ionic bond, formed by the transfer of an electron from one atom to another ($\delta = e$), has 100% ionic character.

Let us illustrate the use of Equation 16.5 by evaluating the ionic character of HCl. Using the experimental dipole moment tabulated in Table 15.1 and the bond length in Table 17.1, we obtain

$$\frac{\delta}{e} = \frac{\mu}{de} = \frac{(1.08\text{ D})(3.338\times 10^{-30}\text{ C m D}^{-1})}{(1.27\times 10^{-10}\text{ m})(1.602\times 10^{-19}\text{ C})} = 0.177 = 17.7\%$$

This result means that the chlorine atom exerts a stronger pull on the bonding electrons in HCl than the hydrogen atom and acquires a negative charge of 0.177e. An equivalent viewpoint is that the HCl molecule can be described as a resonance hybrid of a nonpolar covalent structure (82.3%) and an ion pair structure (17.7%):

$$\text{H}-\ddot{\text{Cl}}: \leftrightarrow [\text{H}]^+[:\ddot{\text{Cl}}:]^-$$
$$\quad 82.3\% \qquad\qquad 17.7\%$$

In valence bond language, the wave function describing HCl consists of a function similar to Equation 16.2, according to which the electrons are equally shared between the two atoms, plus a function for which both electrons reside on the Cl atom. This second function has a small fractional coefficient $\lambda$, consistent with the much greater contribution of the covalent structure:

$$\psi_{HCl} = \psi_H(1)\psi_{Cl}(2) + \psi_H(2)\psi_{Cl}(1) + \lambda\psi_{Cl}(1)\psi_{Cl}(2) \tag{16.6}$$

The ionic character of a diatomic molecule determined in this manner must be correlated with the difference in the electronegativities of its two atoms (Section 6.7). When this difference is small, the two atoms exert a nearly equal attraction for the bonding electrons; the ionic character must be correspondingly small. However, a large difference in electronegativity must lead to a bond that is largely ionic in character. Figure 16.5 summarizes the electronegativities of the representative elements and Figure 16.6 shows the correlation between ionic character and difference in electronegativities. It is seen that the two quantities are well, though not perfectly, correlated.

The solid curve in Figure 16.6 represents an empirical fit to the data points given by the relation

$$\text{Ionic character} = 100\left[1 - \exp\left(-\left(\frac{\chi_A - \chi_B}{2}\right)^2\right)\right] \tag{16.7}$$

where $\chi_A$ and $\chi_B$ are the electronegativities of the bonded atoms. This relation permits an estimate of bond moments from electronegativities for cases in which data from diatomic molecules are not available.

| H    |      |      |      |      |      |      |
|------|------|------|------|------|------|------|
| 2.1  |      |      |      |      |      |      |
| Li   | Be   | B    | C    | N    | O    | F    |
| 1.0  | 1.5  | 2    | 2.5  | 3.0  | 3.5  | 4.0  |
| Na   | Mg   | Al   | Si   | P    | S    | Cl   |
| 0.9  | 1.2  | 1.5  | 1.8  | 2.1  | 2.5  | 3.0  |
| K    | Ca   | Ga   | Ge   | As   | Se   | Br   |
| 0.8  | 1.0  | 1.6  | 1.8  | 2.0  | 2.4  | 2.8  |
| Rb   | Sr   | In   | Sn   | Sb   | Te   | I    |
| 0.8  | 1.0  | 1.7  | 1.8  | 1.9  | 2.1  | 2.5  |
| Cs   | Ba   | Tl   | Pb   | Bi   | Po   | At   |
| 0.7  | 0.9  | 1.8  | 1.8  | 1.9  | 2.0  | 2.2  |

**FIGURE 16.5** Electronegativities of the representative elements. The dashed line separates the nonmetals from the other elements.

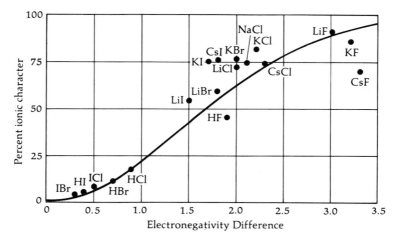

**FIGURE 16.6** Correlation between ionic character and difference in electronegativity, $|\chi_A - \chi_B|$, between atoms A and B in diatomic molecules. The points show the data for specific diatomic molecules. The solid curve represents the empirical relation given by Equation 16.7.

---

**Example 16.2** Use the correlation given by Equation 16.7 to estimate the dipole moment of BrF. The bond length is 1.76 Å.

The electronegativities of bromine and fluorine are $\chi_{Br} = 2.8$ and $\chi_F = 4.0$ (Figure 16.5). Inserting these values into Equation 16.7 we obtain

$$\text{Ionic character} = 100\left[1 - \exp\left(-\left(\frac{\chi_F - \chi_{Br}}{2}\right)^2\right)\right]$$

$$= 100\left[1 - \exp\left(-\left(\frac{1.2}{2}\right)^2\right)\right]$$

$$= 30\%$$

The charge on the atoms is therefore $\delta = 0.30e$. The dipole moment is given by Equation 15.4,

$$\mu = \delta d = \frac{(0.30)(1.602 \times 10^{-19} \text{ C})(1.76 \times 10^{-10} \text{ m})}{(3.338 \times 10^{-30} \text{ C m D}^{-1})} = 2.5 \text{ D} \quad \blacksquare$$

The data in Figure 16.6 indicate that even when the electronegativity difference between two atoms is very large, the ionic character of the bond between them is somewhat less than 100%. For example, the ionic character of LiF is only 87% even though lithium and fluorine differ in electronegativity by 3.0 units. This discrepancy arises because of an effect known as polarizability. The **polarizability** of an atom or ion is the ease with which its electron cloud can be distorted by an electric field. Each ion in an ion pair has an associated electric field which polarizes the electron cloud of the other ion in the manner illustrated in Figure 16.7. Evidently, real atoms cannot be represented as hard spheres, and their dipole moments are somewhat lower than expected from the hard-sphere approximation (Equation 15.4). In general, the polarizability of anions increases with their size and ionic charge, and is largest in the presence

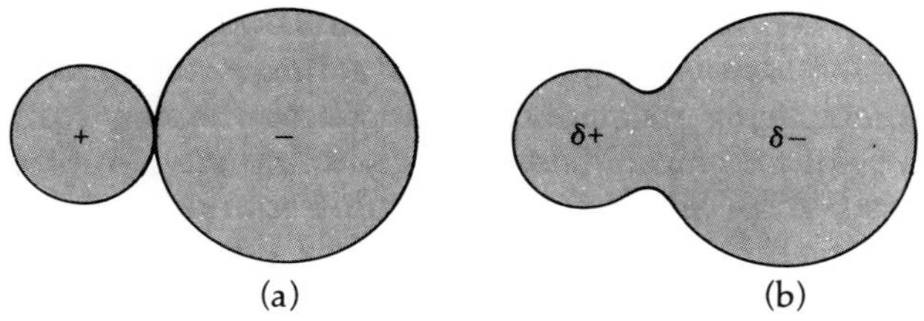

**FIGURE 16.7** A gaseous ion pair consisting of (a) hard sphere ions and (b) polarizable ions.

of a small, highly charged cation. For example, polarizability effects are more important in LiI, which consists of a large anion and a small cation, than in KCl, for which the ions are more comparable in size.

# Molecular Orbital Theory

## 16.4 Molecular Orbitals

Molecular orbital theory offers an alternative treatment of chemical bonding. The essence of the approach is that a molecule is viewed as a single entity rather than as a collection of bonded atoms. If two nuclei of a diatomic molecule are positioned at their equilibrium distance and electrons are added, they will go into **molecular orbitals** (MOs). These orbitals have many similarities to atomic orbitals (Sections 5.12 and 6.2). Just as atoms have s, p, d, ... orbitals which can be characterized by their energies and a set of quantum numbers, molecules have $\sigma$, $\pi$, ... orbitals with specific energies and associated quantum numbers. The filling of these molecular orbitals is similar in many respects to the filling of atomic orbitals in the elements (Section 6.3): orbitals are filled in order of increasing energy, and the Pauli exclusion principle and Hund's rule are obeyed.

As in the case of atoms heavier than hydrogen, the Schrödinger equation for molecules cannot, in general, be solved exactly. Molecular orbitals must consequently be obtained by approximate methods. A widely used approximation, known as the **LCAO-MO** method, describes a molecular orbital as a linear combination, i.e., a sum or difference, of atomic orbitals.

To illustrate the molecular orbital approach, let us consider the simplest of all molecular systems, the hydrogen molecule–ion, $H_2^+$. This species, which has been observed in electrical discharges through hydrogen gas, consists of two protons held together by a single electron.

When the electron is close to one of the protons, designated A, the effect of the second proton is relatively small, and the molecular wave function is nearly equal to the 1s atomic wave function of hydrogen atom A, $\psi_A(s)$. Similarly, when the electron is close to proton B, the molecular wave function is nearly equal to $\psi_B(s)$. The simultaneous effect of the two protons can be approximated by combining the two atomic wave functions to obtain two molecular wave functions:

$$\psi(\sigma) = \psi_A(s) + \psi_B(s)$$
$$\psi(\sigma^*) = \psi_A(s) - \psi_B(s)$$

(16.8)

Observe the similarity between these relations and Equation 16.4 for the formation of hybrid atomic orbitals. In both cases, the sum and difference of two orbitals leads to the formation of two new orbitals. Note, however, that in hybridization, two atomic orbitals are combined to form two new *atomic* orbitals whereas in the LCAO-MO approximation two atomic orbitals are combined to form two *molecular* orbitals.

The distinction between the MO designated $\sigma$ (sigma) and that designated $\sigma^*$ (sigma star) is of fundamental importance and can be seen graphically. Figure 16.8a shows the 1s wave functions of two non-interacting hydrogen atoms. The protons are at their average internuclear distance in an $H_2^+$ molecule–ion. Figure 16.8b shows the wave function $\psi(\sigma)$, obtained as the sum of the two 1s wave functions, while Figure 16.8c shows $\psi(\sigma^*)$, the wave function obtained from the difference of the 1s wave functions. The difference in shape of these molecular wave functions is striking. The function $\psi(\sigma)$ has a large value in the region between the two protons—the two waves reinforce each other. By contrast, when forming $\psi(\sigma^*)$ the atomic wave functions interfere destructively and tend to cancel in the internuclear region. Note that $\psi(\sigma^*)$ has a **node** halfway between the two protons, that is, it goes through zero.

The formation of a chemical bond is possible if the electron density in the internuclear region is high; bonding cannot occur if the electron density in this region is low. The electron density, or more formally, the electron probability distribution, is given by the square of the wave function (Section 5.8). Thus, the electron density for the two MOs formed from the 1s atomic wave functions is given by the relations

$$\psi^2(\sigma) = [\psi_A(s) + \psi_B(s)]^2 = \psi_A^2(s) + \psi_B^2(s) + 2\psi_A(s)\psi_B(s)$$
$$\psi^2(\sigma^*) = [\psi_A(s) - \psi_B(s)]^2 = \psi_A^2(s) + \psi_B^2(s) - 2\psi_A(s)\psi_B(s)$$
(16.9)

(To obtain the actual probability, the wave function must be multiplied by a normalization constant, as discussed in Section 5.9). The two density distributions differ in the sign of the $\psi_A(s)\psi_B(s)$ term. This term is positive in $\psi^2(\sigma)$ and increases the electron density in the internuclear distance. On the other hand, this term is negative in $\psi^2(\sigma^*)$ and decreases the electron density between the two protons.

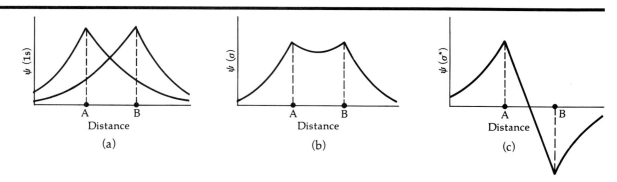

**FIGURE 16.8** Atomic and molecular wave functions in hydrogen. (a) Atomic wave functions (1s) of two hydrogen atoms located at the average internuclear distance in $H_2^+$. Points A and B mark the position of the two nuclei. (b) Molecular wave function $\psi(\sigma)$ and (c) molecular wave function $\psi(\sigma^*)$. The dashed lines show the values of the wave functions at the position of the nuclei.

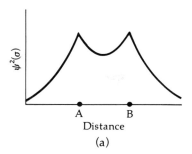 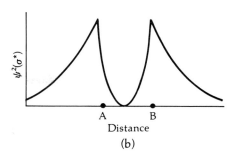

**FIGURE 16.9** Electron probability distribution in $H_2^+$. (a) The $\sigma$ bonding orbital and (b) the $\sigma^*$ antibonding orbital. The quantity plotted is the square of the molecular wave function, $\psi^2(\sigma)$ or $\psi^2(\sigma^*)$. Points A and B mark the position of the two nuclei.

A graphical depiction of the difference in electron probability distributions is given in Figure 16.9. When an electron occupies the $\sigma$ orbital the electron density in the internuclear region is high. The two protons are consequently shielded from each other and the energy of the $H_2^+$ system is lower than that of the separate $H + H^+$ species. Since bonding can occur under these circumstances, the $\sigma$ orbital is called a **bonding orbital.**

By contrast, when the electron occupies the $\sigma^*$ orbital the electron density in the internuclear region is low—the density actually drops to zero in the nodal plane. The two protons are therefore partially exposed to each other, and their mutual repulsion increases the energy of the system relative to that of $H + H^+$. Bonding cannot occur under these circumstances, and the $\sigma^*$ orbital is called an **antibonding orbital.** Figure 16.10 shows another representation of the electron distribution in the $\sigma$ and $\sigma^*$ orbitals. The curves of the electron density contours vividly show the striking difference between the two types of orbitals.

As is the case for atomic orbitals, molecular orbitals have characteristic shapes. Figure 16.10 shows the shape of the $\sigma$ and $\sigma^*$ orbitals. Both orbitals are

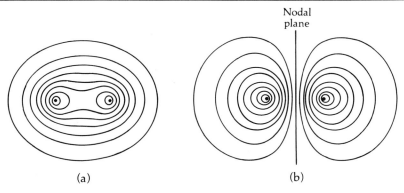

**FIGURE 16.10** Electron density contours in $H_2^+$. (a) The $\sigma$ bonding orbital and (b) the $\sigma^*$ antibonding orbital. The points show the positions of the two nuclei. Each curve represents a region of constant electron density. The density is highest in the region near the nuclei and decreases in the outward direction. The vertical line in (b) represents the nodal plane extending in and out of the plane of the paper. The electron density is zero in this plane.

symmetric about the internuclear axis, that is, the electron density at a given distance from the internuclear axis is independent of direction; the orbitals are circular in cross section. Other types of molecular orbitals have different shapes, as we shall see. Regardless of shape and designation, there are two varieties of every molecular orbital—a bonding and an antibonding orbital corresponding, respectively, to the sum and difference of the atomic orbitals involved. As in the case of the $\sigma$ and $\sigma^*$ orbitals, the unstarred Greek letter designates the bonding orbital and the starred Greek letter designates the antibonding orbital ($\pi$, $\pi^*$, and so on).

The molecular potential energy curve for $H_2^+$ as a function of internuclear distance is shown in Figure 16.11a. A similar curve is obtained for $H_2$. Observe that the curve corresponding to the bonding orbital closely resembles that ob-

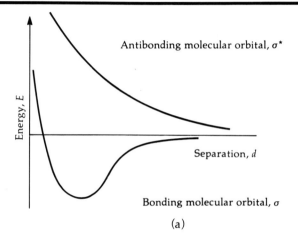

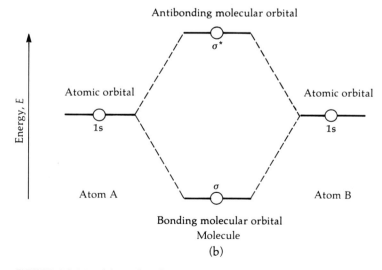

**FIGURE 16.11** (a) Molecular potential energy curve of $H_2^+$ (the electron occupies the bonding orbital). The potential energy curve for the $\sigma^*$ antibonding orbital also is shown. (b) Energies of the $\sigma_{1s}$ and $\sigma_{1s}^*$ molecular orbitals relative to those of the 1s atomic orbitals.

tained by means of the valence bond theory (Figure 16.2). When some refinements are added to our simplified development, it is found that molecular orbital theory can fit the bond length and bond energy of H$_2$ as accurately as valence bond theory.

Figure 16.11a also shows the potential energy curve corresponding to the $\sigma^*$ antibonding orbital. Observe that the energy is higher than that for the $\sigma$ orbital at all internuclear distances. Since the electron density in the internuclear region is low, the repulsion between the protons increases continuously with decreasing internuclear distance. Thus, the curve has no minimum, and there is no internuclear distance at which a bond can be formed.

Figure 16.11b shows a molecular orbital energy level diagram based on Figure 16.11a. The $\sigma$ orbital lies at a lower energy than the 1s atomic orbitals, while the $\sigma^*$ orbital lies at a higher energy. A particular orbital can accomodate two spin-paired electrons. Additional electrons must occupy higher energy orbitals. The filling of molecular orbitals is analogous in this respect to the filling of atomic orbitals, and the procedure used to determine the electronic configuration of the elements may be applied to the case of homonuclear diatomic molecules, as we shall see.

Let us examine how this orbital filling scheme works in the case of H$_2$. The two spin-paired electrons populate the $\sigma$ orbital, which is the lowest energy orbital. The electronic configuration is said to be $(\sigma_{1s})^2$, where the subscript indicates the atomic orbitals forming the molecular orbital in question and the superscript gives the number of electrons in the orbital. What happens if two H atoms with parallel spins attempt to form a molecule? The exclusion principle indicates that while one of the electrons can occupy the $\sigma$ orbital, the other must be placed in the $\sigma^*$ orbital. The electronic configuration of this state can be written as $(\sigma_{1s})^1(\sigma_{1s}^*)^1$. If this molecule were bound, it would be an excited electronic state of the H$_2$ molecule, that is, a state of higher internal energy. However, because of the antibonding character of the $\sigma^*$ orbital, this state is unstable with respect to two hydrogen atoms and spontaneously dissociates.

The stability of a molecule is related to the relative number of electrons in bonding and in antibonding orbitals. To quantify this relation it is customary to use a somewhat more general definition of bond order than that given in Section 15.4. The **bond order** is defined as one-half the difference between the number of electrons in bonding orbitals and the number of electrons in antibonding orbitals:

$$\text{Bond order} = \tfrac{1}{2}(\text{number of bonding } e^- - \text{number of antibonding } e^-)$$

(16.10)

In order to be stable, a molecule must have a positive bond order. Thus, H$_2$ in its ground state has a bond order of one and is stable; H$_2$ in its excited state has bond order zero and is unstable. As already noted, the bond order gives a qualitative measure of the strength of a given type of bond: the greater the bond order, the greater the bond energy and the shorter the bond length.

---

Example 16.3   Write the electronic configurations of (a) H$_2^+$, (b) H$_2^-$, (c) H$_2^{2-}$, and (d) He$_2$. What is the bond order of each of these species? How do the bond energy and bond length compare with those in H$_2$?

(a) $H_2^+$. The single electron must go into the $\sigma_{1s}$ orbital, and the electronic configuration is $(\sigma_{1s})^1$. The bond order, as given by Equation 16.10, is $(1 - 0)/2 = \frac{1}{2}$. Since the bond order is lower than that in $H_2$, the bond energy should be lower, and the bond length longer than they are in $H_2$. (The experimental values for $H_2^+$ are 255 kJ mol$^{-1}$ and 1.06 Å, compared to 436 kJ mol$^{-1}$ and 0.74 Å for $H_2$.)

(b) $H_2^-$. This molecule–ion has three electrons. The first two can occupy the $\sigma_{1s}$ bonding orbital, but the third electron must go into the $\sigma_{1s}^*$ antibonding orbital. The electronic configuration is $(\sigma_{1s})^2 (\sigma_{1s}^*)^1$. The bond order is $(2 - 1)/2 = \frac{1}{2}$, indicating that the bond energy should be lower and the bond length longer than in $H_2$.

(c) $H_2^{2-}$. With four electrons, the electronic configuration is $(\sigma_{1s})^2 (\sigma_{1s}^*)^2$. The bond order is $(2 - 2)/2 = 0$. A species with bond order zero dissociates spontaneously. Thus, the concepts of bond energy and bond length have no meaning.

(d) $He_2$. The molecule has four electrons and its electronic configuration consequently is identical to that of $H_2^{2-}$. The bond order is zero and the molecule is unstable. (The very weak van der Waals forces can hold two helium atoms together briefly. This is not a stable $He_2$ molecule.) ∎

## 16.5 Sigma and Pi Orbitals

The stability of $H_2$ and the instability of $He_2$ can be understood in terms of the filling of the $\sigma_{1s}$ and $\sigma_{1s}^*$ orbitals. Additional MOs are needed in order to extend the molecular orbital approach to homonuclear diatomic molecules of the second period. Since several different atomic orbitals, i.e., 1s, 2s, and 2p, can be occupied in the isolated atoms of these elements, we must at the outset address the question of which of these orbitals can combine with each other to form molecular orbitals. This question can be simplified immediately by noting that for elements in the second period the 1s electrons are no longer part of the valence shell. These electrons are pulled in close to the nucleus and do not give rise to appreciable orbital overlap. Inner-shell electrons can therefore be regarded as nonbonding, and bonding of second period elements must involve the molecular orbitals that can be formed from the 2s and 2p atomic orbitals.

In order for atomic orbitals to interact and form a molecular orbital they must, first of all, have comparable energies. Figure 16.12 shows the relative en-

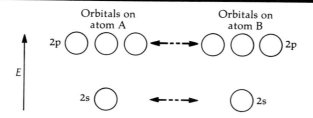

**FIGURE 16.12** Schematic diagram of 2s and 2p orbitals showing their relative energies. The dashed double arrows identify the orbitals that can interact in the formation of a homonuclear diatomic molecule.

ergies of the 2s and 2p orbitals, as adapted from Figure 6.7 or 6.9. It is apparent that molecular orbitals can form from the overlap of 2s orbitals and from the overlap of 2p orbitals. Let us consider these two cases in turn.

The interaction between two 2s atomic orbitals is similar to that between two 1s orbitals and gives rise to a $\sigma_{2s}$ and a $\sigma_{2s}^*$ molecular orbital. As in the case of the $\sigma_{1s}$ and $\sigma_{1s}^*$ orbitals, the bonding $\sigma_{2s}$ orbital lies at a lower energy than the 2s atomic orbitals while the antibonding $\sigma_{2s}^*$ orbital lies at a higher energy than the 2s orbitals.

The interaction between the 2p atomic orbitals involves some new features. Recall from Section 5.12 that the three p orbitals, designated $p_x$, $p_y$, and $p_z$, differ in orientation, being directed along the x, y, and z axes, respectively (Figure 5.30). We are free, for the sake of simplicity, to identify the internuclear axis of a diatomic molecule with one of these three axes. It is customary to designate this axis the z axis. The $p_z$ orbitals of the two atoms must consequently point along the internuclear axis while the $p_x$ and $p_y$ orbitals point along axes that are perpendicular to the internuclear axis (Figure 16.13).

We must now decide which of the p orbitals of two atoms can interact to form a molecular orbital. The preceding requirement that the atomic orbitals have comparable energy is of no help in this respect since the energies of the three p orbitals are the same. An additional requirement comes into play: in order to interact, two atomic orbitals must overlap to a significant extent. Owing to the directional properties of the p orbitals, it is possible for two orbitals of

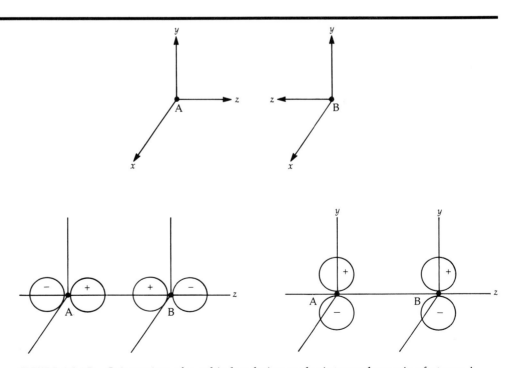

**FIGURE 16.13** Orientation of p orbitals relative to the internuclear axis of atoms A and B. The $p_z$ orbitals are shown on the left and the $p_y$ orbitals on the right. The $p_x$ orbitals (not shown) are oriented parallel to the x axis. The signs in the diagram indicate the signs of the wave functions.

the same designation, e.g., $p_z$, to overlap and form a molecular orbital. However two orbitals of different designation, e.g., $p_x$ and $p_z$, are mutually perpendicular and there is no net overlap between them. This point is illustrated in Figure 16.14, which shows that the regions in which the $p_x$ and $p_y$ orbitals overlap with the same signs of the wave functions are equal in area to the regions in which they overlap with opposite signs of the wave functions. Thus, the contributions of constructive and destructive interference in the wave functions cancel exactly—the increase in electron density associated with the reinforcement of the wave functions is exactly balanced by the decrease in electron density associated with the cancellation of the wave functions. In examining the formation of molecular orbitals from p atomic orbitals we need therefore only be concerned with the interaction of two p orbitals with the same orientation.

The combination of two $p_z$ orbitals is shown schematically in Figure 16.15b. The two orbitals overlap strongly. The combination corresponding to the sum of the wave functions results in constructive interference and leads to an increase in electron density in the internuclear region—a bonding molecular orbital is formed. On the other hand, the combination corresponding to the difference of the wave functions results in destructive interference and leads to a decrease in electron density in the internuclear region—an antibonding molecular orbital is formed. Both molecular orbitals are symmetric about the internuclear axis and are therefore $\sigma$ orbitals. They are designated $\sigma_{2p}$ and $\sigma_{2p}^*$ in order to distinguish them from the $\sigma_{2s}$ and $\sigma_{2s}^*$ orbitals (which are illustrated in Figure 16.15a).

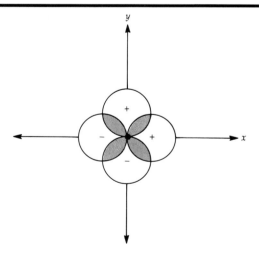

**FIGURE 16.14** Examination of the overlap between mutually perpendicular p orbitals (e.g., $p_x$ and $p_y$). The colored areas show overlap between regions in which the atomic wave functions have like sign (+ and +, − and −) and correspond to enhanced electron density. The gray areas show overlap between regions in which the atomic wave functions have opposite sign and correspond to reduced electron density. The two regions cancel exactly, showing that there is no net overlap. The nuclei of the two atoms (heavy dot) lie along the z axis (in and out of the plane of the paper).

Theories of Chemical Bonding   16.5   607

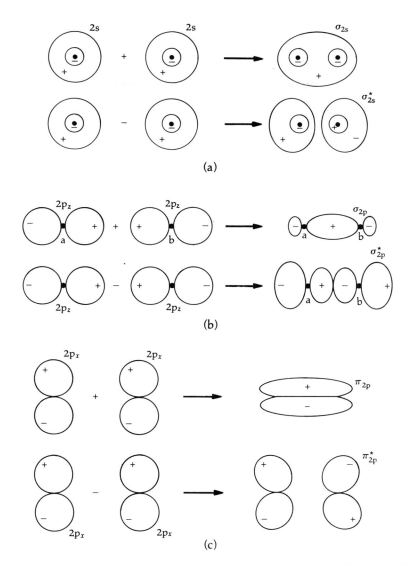

**FIGURE 16.15** The formation of molecular orbitals by the linear combination of atomic orbitals. (a) $\sigma_{2s}$ and $\sigma_{2s}^*$ orbitals formed from the sum and difference of 2s atomic orbitals, respectively. (b) $\sigma_{2p}$ and $\sigma_{2p}^*$ orbitals formed from the sum and difference of $2p_z$ atomic orbitals, respectively (note that the z axis is the internuclear axis). (c) $\pi_{2p}$ and $\pi_{2p}^*$ orbitals formed from the sum and difference of either $2p_x$ or $2p_y$ atomic orbitals, respectively. The heavy dots show the positions of the two atomic nuclei. The + and − signs on the orbitals refer to the signs of the wave functions (see Figure 5.30).

The combination of two $2p_x$ or two $2p_y$ orbitals results in the formation of a different type of molecular orbital called a $\pi$ orbital. The formation of a $\pi$ orbital is illustrated in Figure 16.15c. Since the atomic orbitals have their maximum amplitude away from the internuclear axis, their constructive interference leads to enhanced electron density in the general region between the two nuclei but off the internuclear axis. The $\pi$ orbital has a nodal plane through the

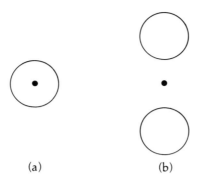

**FIGURE 16.16** Cross-sectional view of (a) $\sigma$ or $\sigma^*$ orbitals and (b) $\pi$ or $\pi^*$ orbitals. The internuclear axis runs through the heavy dot in and out of the plane of the paper.

internuclear axis and the region of increased electron density is on either side of this plane. Due to this increased electron density, the $\pi$ orbital is a bonding orbital. On the other hand, the destructive interference of two $2p_x$ or two $2p_y$ orbitals reduces the electron density in the internuclear region and leads to the formation of an antibonding $\pi^*$ orbital.

It should be noted that there are two $\pi_{2p}$ orbitals. They are formed from the combination of either the $2p_x$ or $2p_y$ atomic orbitals and are completely equivalent: the electron density is concentrated on either side of the internuclear axis (to the "left" and "right" or "above" and "below") and the orbitals are degenerate in energy (i.e., they lie at the same energy, Section 5.11). Similarly, there are two $\pi_{2p}^*$ orbitals which are also degenerate in energy and lie at a higher energy than the $\pi_{2p}$ orbitals.

To recapitulate, the combination of s and p atomic orbitals results in the formation of either $\sigma$ or $\pi$ molecular orbitals (and their corresponding antibonding orbitals). These molecular orbitals differ in the distribution of electron density: the $\sigma$ orbital concentrates the electron density along the internuclear axis and is axially symmetric. The $\pi$ orbital concentrates the electron density on either side of the internuclear axis and is not axially symmetric—the electron density is not equal in all directions about the internuclear axis (Figure 16.16). Electrons occupying $\sigma$ orbitals form **sigma bonds** while electrons occupying $\pi$ orbitals form **pi bonds.** The experimental evidence indicates that a sigma bond is generally stronger than a pi bond. This result can be explained in terms of the enhanced electron density along the internuclear axis in a $\sigma$ orbital.

## 16.6 Homonuclear Diatomic Molecules

*a. Electronic Configuration* The electronic configurations of homonuclear diatomic molecules of the elements in the second period can be determined by feeding electrons into the molecular orbitals formed by the combination of the 2s and 2p atomic orbitals. As in the case of atoms (see Section 6.2), the orbitals are filled in order of increasing energy following the restrictions imposed by the Pauli exclusion principle and Hund's rule. The orbital order can be determined by a combination of quantum mechanical calculations, measurements of electronic energy levels using photoelectron spectroscopy (Section 6.5c), and

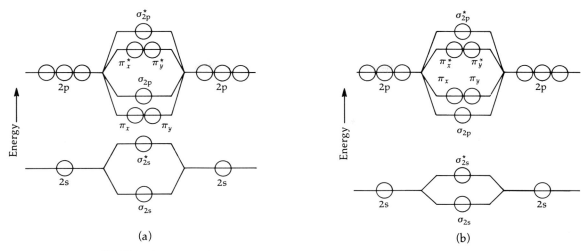

**FIGURE 16.17** Molecular orbital energy level diagram for homonuclear diatomic molecules formed by second-period elements. (a) Diagram for $Li_2$ through $N_2$ and their ions and (b) diagram for $O_2$ through $Ne_2$ and their ions.

measurements of the magnetic properties of molecules. The experimental evidence indicates that the filling order of the molecular orbitals is not the same for all the second-period elements. The molecular orbital level scheme shown in Figure 16.17a applies to the diatomic molecules of the elements lithium through nitrogen as well as to their positive and negative ions. The level scheme shown in Figure 16.17b applies to oxygen through neon and their ions. In actuality, the only observable difference between the two filling orders occurs for $B_2$ and $C_2$, as we shall see.

Both diagrams indicate that the $\sigma_{2s}$ and $\sigma_{2s}^*$ orbitals are the lowest-energy molecular orbitals, chiefly because the 2s atomic orbitals lie below the 2p atomic orbitals, as shown. The two diagrams differ in the order of the $\sigma_{2p}$ and $\pi_{2p}$ orbitals. For the lighter molecules, the $\pi_{2p}$ orbitals lie below the $\sigma_{2p}$ orbitals, whereas the reverse is true for the heavier molecules. This second sequence ($\sigma_{2p}$ before $\pi_{2p}$) is easy to understand: a $\sigma_{2p}$ orbital concentrates the electron density along the internuclear axis whereas a $\pi_{2p}$ orbital does not. Consequently the $\sigma_{2p}$ orbital is more stable and lies at a lower energy. This sequence must be reversed for the lighter diatomic molecules if the magnetic properties of $B_2$ and $C_2$ are to be predicted correctly. The change in sequence can be attributed to the small energy difference between the 2s and 2p orbitals in the first part of the second period. Under these circumstances, molecular orbitals may be formed by the simultaneous interaction of the 2s and 2p orbitals, rather than by their separate interaction. Quantum mechanical calculations indicate that when such complex interactions occur, alterations in the expected sequence of molecular orbitals are possible.

The electronic configurations of the second-period homonuclear diatomic molecules, $Li_2$ through $Ne_2$, are obtained by feeding electrons into the molecular orbitals following the appropriate sequence in Figure 16.17. Table 16.2 summarizes the results, including some of the important bond properties. The results for $H_2$ and $He_2$ are included for completeness.

TABLE 16.2  Electronic Configuration and Bond Properties of Homonuclear Diatomic Molecules

| Molecule | Electronic configuration | Bond order | Number of unpaired electrons | Bond energy (kJ mol$^{-1}$) | Bond length (Å) |
|---|---|---|---|---|---|
| H$_2$ | $(\sigma_s)^2$ | 1 | 0 | 436 | 0.741 |
| He$_2$ | $(\sigma_s)^2(\sigma_s^*)^2$ | 0 | 0 | —[a] | —[a] |
| Li$_2$ | KK$(\sigma_s)^2$ | 1 | 0 | 101 | 2.673 |
| Be$_2$ | KK$(\sigma_s)^2(\sigma_s^*)^2$ | 0 | 0 | —[a] | —[a] |
| B$_2$ | KK$(\sigma_s)^2(\sigma_s^*)^2(\pi_x)^1(\pi_y)^1$ | 1 | 2 | 291 | 1.59 |
| C$_2$ | KK$(\sigma_s)^2(\sigma_s^*)^2(\pi_x)^2(\pi_y)^2$ | 2 | 0 | 599 | 1.243 |
| N$_2$ | KK$(\sigma_s)^2(\sigma_s^*)^2(\pi_x)^2(\pi_y)^2(\sigma_p)^2$ | 3 | 0 | 945 | 1.098 |
| O$_2$ | KK$(\sigma_s)^2(\sigma_s^*)^2(\sigma_p)^2(\pi_x)^2(\pi_y)^2(\pi_x^*)^1(\pi_y^*)^1$ | 2 | 2 | 494 | 1.207 |
| F$_2$ | KK$(\sigma_s)^2(\sigma_s^*)^2(\sigma_p)^2(\pi_x)^2(\pi_y)^2(\pi_x^*)^2(\pi_y^*)^2$ | 1 | 0 | 158 | 1.41 |
| Ne$_2$ | KK$(\sigma_s)^2(\sigma_s^*)^2(\sigma_p)^2(\pi_x)^2(\pi_y)^2(\pi_x^*)^2(\pi_y^*)^2(\sigma_p^*)^2$ | 0 | 0 | —[a] | —[a] |

[a] The molecule is not stable.

The first diatomic molecule of interest is Li$_2$. A lithium atom has one electron outside the K shell so that two electrons are available for bonding in Li$_2$. These electrons fill the $\sigma_{2s}$ orbital, and the configuration can be written as KK$(\sigma_s)^2$, where each K denotes the filled atomic K shell of one of the two bonded atoms (Section 6.2) and the notation has been simplified in an obvious manner. The bond order is one, and the molecule is held together by a single sigma bond. Each succeeding diatomic molecule has two additional electrons. The configuration of Be$_2$ consequently is KK$(\sigma_s)^2(\sigma_s^*)^2$. Note that this molecule has no net bonding electrons and the bond order is zero, meaning that Be$_2$ should not be stable. The experimental facts are consistent with these configurations: lithium is diatomic in the gas phase, and beryllium is monatomic.

The next two molecules, B$_2$ and C$_2$, are not commonly observed chemical species. However, their properties are important in establishing the sequence of molecular orbital energies. Figure 16.17a indicates that the two additional electrons in B$_2$ must occupy the $\pi$ orbitals. According to Hund's rule, it is energetically favorable to place one electron in each of these orbitals so that the configuration of B$_2$ is KK$(\sigma_s)^2(\sigma_s^*)^2(\pi_x)^1(\pi_y)^1$. Since the two electrons are unpaired and have parallel spin, B$_2$ must be paramagnetic. In fact, it was the observation of paramagnetism in B$_2$ that indicated that the $\pi_{2p}$ orbitals must lie at a lower energy than the $\sigma_{2p}$ orbital. If this were not the case, the two electrons in question would fill the $\sigma_{2p}$ orbital and B$_2$ would be diamagnetic.

The magnetic properties of C$_2$ lend further support to the orbital energy scheme in Figure 16.17a. The two additional electrons fill the two $\pi$ orbitals, and the molecule should consequently be diamagnetic, as is observed experimentally. Note that according to Figure 16.17b, C$_2$ is predicted to be paramagnetic instead.

The Lewis structure of C$_2$ that corresponds to the molecular orbital description is

:C=C:

The double bond is unusual in that it consists of two pi bonds formed by the electrons in the filled $\pi_x$ and $\pi_y$ orbitals. Double bonds normally consist of a sigma and a pi bond (Section 16.8). The lone electron pairs in C₂ can be identified with the bonding–antibonding $\sigma_{2s}$–$\sigma_{2s}^*$ orbital pair. These orbitals lie at a much lower energy than the $\pi$ and $\pi^*$ orbitals and can be regarded as two nonbonding pairs of 2s electrons. Although C₂ is stable in the gas phase at high temperatures, it is an electron-deficient molecule in which neither carbon is surrounded by an octet of electrons. At ordinary temperatures, a carbon atom is bonded to additional carbon atoms to form the structures of graphite or diamond (Section 20.11).

The next molecule in the sequence is N₂. The two additional electrons fill the $\sigma_{2p}$ orbital, and the resulting bond order is three. The two atoms are held together by a triple bond comprised of one sigma bond, corresponding to the filled $\sigma_p$ orbital, and two pi bonds, corresponding to the filled $\pi_x$ and $\pi_y$ orbitals. As in the case of C₂, electrons in the filled $\sigma_{2s}$ and $\sigma_{2s}^*$ orbitals can be regarded as nonbonding s electrons and assigned as lone pairs to the individual atoms, consistent with the familiar Lewis structure of N₂:

:N≡N:

Figure 16.18 shows the electron density distribution in N₂. The extraordinary stability of molecular nitrogen is due to the high electron density in the internuclear region.

The explanation of the structure of O₂ undoubtedly constitutes one of the great successes of molecular orbital theory. Recall from Section 15.4 that the paramagnetism of molecular oxygen indicates the presence of two unpaired electrons with parallel spin. This result shows that the expected Lewis structure of O₂,

Ö=Ö

must be incorrect. The molecular orbital electronic configuration of O₂ is $KK(\sigma_s)^2(\sigma_s^*)^2(\sigma_p)^2(\pi_x)^2(\pi_y)^2(\pi_x^*)^1(\pi_y^*)^1$. Since the two $\pi^*$ orbitals are degenerate in energy, Hund's rule demands that they each contain one electron, with both electrons having parallel spin. The paramagnetism of O₂ thereby is explained

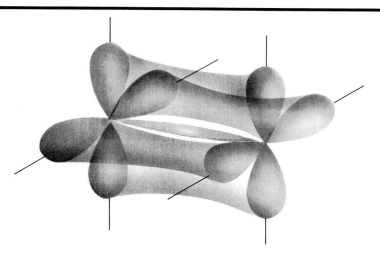

**FIGURE 16.18** Electron density distribution in N₂.

in a natural way. Note that the population of the $\pi^*$ orbitals by unpaired electrons is predicted by either of the orbital energy sequences shown in Figure 16.17 and is therefore independent of the details of the filling sequence.

The $O_2$ molecule has four net bonding electrons, two in the $\sigma_p$ orbital and two in the $\pi$ orbitals. Consequently the bond order is two. While the Lewis structure of $O_2$ also indicates a bond order of two, the molecular orbital approach is more general since the bond order is determined by the number of bonding and antibonding electrons, and electron pairing is not required.

The sequence of homonuclear diatomic molecules continues with $F_2$. The additional two electrons fill the $\pi^*$ orbitals and the molecule is diamagnetic, in accord with the molecular orbital electron configuration. The resulting bond order is one, and the molecule is held together by a sigma bond formed by the electrons in the $\sigma_{2p}$ orbital. The second period terminates with neon. All the orbitals displayed in Figure 16.17 are filled at this point. Consequently $Ne_2$ has bond order zero and is not stable, although van der Waals forces do result in its transitory existence.

***b. Bond Properties*** Table 16.2 lists the bond energies and bond lengths of the homonuclear diatomic molecules examined in Section 16.6a. The increase in bond energy with increasing bond order as well as the corresponding decrease in bond length are evident. Figure 16.19, which is a plot of the bond parameters versus bond order, shows that the correlation extends to the ions derived from these molecules. Evidently, the bond order is one of the important factors determining the strength of a covalent bond. However, the scatter in the data indicates that it cannot be the only factor. Recall that the magnitude of the nuclear charge affects the strength with which an atom retains its valence electrons (Section 6.5). The effect of nuclear charge on bond strength can be seen in a comparison of the peroxide ion ($O_2^{2-}$) with $F_2$. These two species are

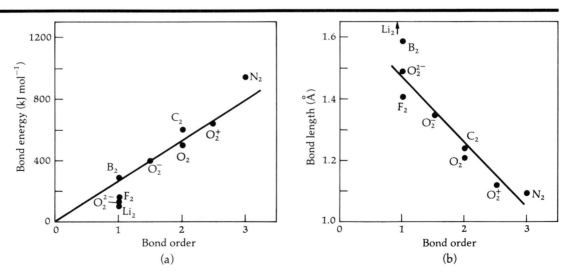

**FIGURE 16.19** Dependence of (a) bond energies and (b) bond lengths of second-period homonuclear diatomic molecules on the bond order predicted by molecular orbital theory. Results for some molecular ions are included.

isoelectronic since they both contain 14 valence electrons. The greater nuclear charge of fluorine causes the electron cloud around the atom to shrink (Section 6.8) and results in a stronger and shorter bond.

**Example 16.4** The following molecular ions of oxygen are known: $O_2^+$ [dioxygenyl cation, as in $O_2(PtF_6)$], $O_2^-$ (superoxide ion, as in $KO_2$), and $O_2^{2-}$ (peroxide ion, as in $Na_2O_2$). Write the electronic configurations of these species, determine the bond orders, and arrange the species in order of increasing bond energy and length. Indicate which of these ions are paramagnetic. Include $O_2$ in this analysis.

Starting with the configuration of $O_2$, $KK(\sigma_s)^2(\sigma_s^*)^2(\sigma_p)^2(\pi_x)^2(\pi_y)^2(\pi_x^*)^1(\pi_y^*)^1$, we obtain the following results by adding or removing electrons, where we only show the occupation of the $\pi^*$ orbitals since that of the other orbitals remains unchanged.

| Species | Configuration | Bond order | Unpaired electrons | Bond energy (kJ mol$^{-1}$) | Bond length (Å) |
|---|---|---|---|---|---|
| $O_2^+$ | $(\pi_x^*)^1$ | 2.5 | 1 | 643 | 1.12 |
| $O_2$ | $(\pi_x^*)^1(\pi_y^*)^1$ | 2 | 2 | 498 | 1.21 |
| $O_2^-$ | $(\pi_x^*)^2(\pi_y^*)^1$ | 1.5 | 1 | 395 | 1.35 |
| $O_2^{2-}$ | $(\pi_x^*)^2(\pi_y^*)^2$ | 1 | 0 | 126 | 1.49 |

Since the bond order decreases from $O_2^+$ to $O_2^{2-}$, the bond energies should decrease while the bond lengths increase in the same sequence. The experimental values, shown in the last two columns, confirm these predictions. All but $O_2^{2-}$ have unpaired electrons and are consequently paramagnetic. ∎

## 16.7 Heteronuclear Diatomic Molecules

The molecular orbital analysis described for homonuclear diatomic molecules may be applied, with some modifications, to heteronuclear diatomic molecules. Since we are already familiar with their orbital structures, let us consider some heteronuclear molecules formed by the elements in the first two periods. The chief difference, depicted in Figure 16.20, is that the atomic orbitals no longer

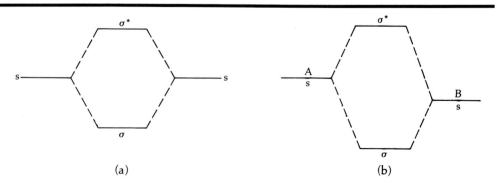

**FIGURE 16.20** Molecular orbital energies in (a) a homonuclear diatomic molecule and (b) a heteronuclear diatomic molecule. Atom B is more electronegative than atom A.

lie at the same energy, as they necessarily do in the case of two atoms of the same element. The orbitals of the more electronegative of the two atoms lie at a lower energy and exert a greater attraction for the bonding electrons. Instead of being an equal combination of the two atomic orbitals, the bonding molecular orbital will more closely resemble the atomic orbital of the more electronegative element; the opposite is true for the antibonding orbital. So long as the energies of the two atomic orbital levels differ only slightly, the molecular orbital will be made up of the same combination of atomic orbitals as if the molecule were homonuclear. The only effect in this case is that the electron density is distorted towards the more electronegative atom. This is the situation when the combining elements differ only slightly in electronegativity. However, if the difference is substantial, the combining atomic orbitals may actually differ from those forming the homonuclear molecular orbitals. In addition to exhibiting a distorted electron density, such a molecule will also have a different electronic configuration.

Carbon monoxide is a good example of a molecule whose structure can be described in terms of the same molecular orbitals formed by the homonuclear diatomic molecules. The molecular orbital energy diagram is shown in Figure 16.21. Note that the levels of atomic oxygen lie lower than those of carbon, reflecting the greater electronegativity of oxygen. The actual placement of the atomic orbitals is determined by the values of the first ionization energy $I_1$ of the elements (Section 6.5). Recall that it is customary to report the values of the energies of atomic levels relative to a value of zero for the singly ionized atom

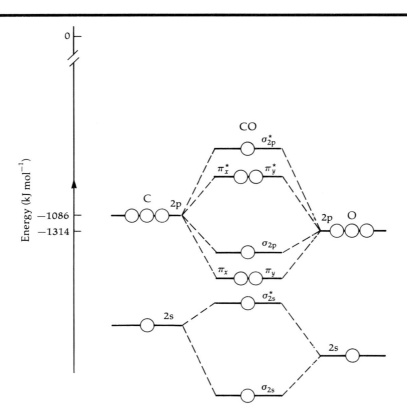

**FIGURE 16.21**
Molecular orbital energy level diagram of carbon monoxide.

(Section 5.5). The energy of the highest occupied atomic orbital consequently lies at an energy $I_1$ below zero. Since the first ionization energy of oxygen is 1314 kJ mol$^{-1}$ while that of carbon is 1086 kJ mol$^{-1}$ (Table 6.4), the 2p valence orbitals of oxygen lie 228 kJ mol$^{-1}$ below those of carbon. This difference is not large enough to prevent the two sets of orbitals from interacting. Thus, the molecular orbitals are formed from the same combination of atomic orbitals as in the homonuclear case, and the electronic configuration of CO is identical to that of N$_2$, the two molecules being isoelectronic. The difference is that the bonding molecular orbitals of CO lie closer in energy to the atomic orbitals of oxygen. Consequently, the electron density distribution in CO is asymmetric, as shown in Figure 16.22.

In addition to a graphical depiction, the asymmetry of the molecular orbitals of CO can be represented mathematically. Recall that in the LCAO-MO approach, the molecular wave function of a homonuclear diatomic is given by the sum or difference of the atomic wave functions, (e.g., Equation 16.8). To represent the molecular orbitals of a heteronuclear diatomic, the atomic wave functions must be multiplied by unequal coefficients. For example, the $\sigma_{2p}$ and $\sigma_{2p}^*$ wave functions of CO are given by the relations

$$\psi(\sigma_{2p}) = C_C \psi_C(2p_z) + C_O \psi_O(2p_z)$$
$$\psi(\sigma_{2p}^*) = C'_C \psi_C(2p_z) - C'_O \psi_O(2p_z) \tag{16.11}$$

where the $C$'s are constants such that $C_O > C_C$ and $C'_C > C'_O$, and $\psi_O$ and $\psi_C$ are the atomic wave functions of oxygen and carbon, respectively. Note that the oxygen p$_z$ orbital contributes more than the carbon p$_z$ orbital to the bonding molecular orbital (since $C_O > C_C$); the reverse is true for the antibonding molecular orbital.

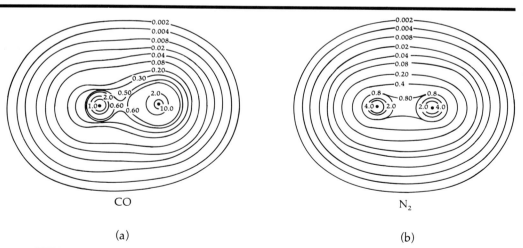

(a)          (b)

**FIGURE 16.22** Electron density contours in (a) CO and (b) N$_2$. The positions of the nuclei are shown by the dots, the oxygen nucleus in CO being at the right. The electron density is constant along each contour and decreases by about a factor of two for each succeeding contour. It is highest in the region near the nuclei (the relative values of the density are given by the numbers on the contours). Note the difference between the asymmetric electron density distribution in CO and the symmetric distribution in N$_2$. [From R. F. W. Bader and A. D. Bandrauk, *J. Chem, Phys,* **49,** 1653 (1968).]

**616** Theories of Chemical Bonding   16.7

**FIGURE 16.23**
Relative energies of atomic and molecular orbitals in HF.

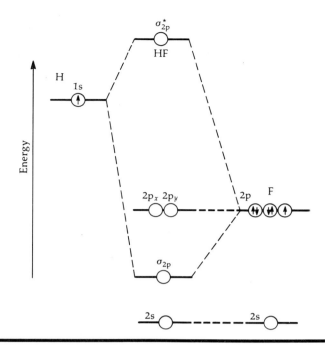

When the difference in electronegativity of two elements becomes sufficiently large, the energies of the equivalent atomic orbitals of the two atoms differ widely. Consequently, the orbitals are too different in spatial distribution (see Figure 5.32) to permit overlap. In this instance, a molecular orbital is formed from two atomic orbitals lying closer in energy. The HF molecule provides a good illustration (Figure 16.23). The energy of the fluorine 2p orbitals is nearly 400 kJ mol$^{-1}$ lower than the energy of the 1s orbital in hydrogen, reflecting the difference in the first ionization energies of the two elements. The 2s orbital in fluorine lies even lower in energy—so low that it essentially does not interact with the 1s orbital in hydrogen. Consequently, the formation of a molecular orbital in HF can result only from the overlap of the 1s orbital in hydrogen with one of the 2p orbitals in fluorine.

The overlap of an s orbital of an atom with the various p orbitals of another atom may be examined in the same way as the overlap of two p orbitals. Since the $p_z$ orbital is directed along the internuclear axis it can overlap with an s orbital to form a $\sigma$ orbital (Figure 16.24a). On the other hand, the $p_x$ and $p_y$ orbitals have a node along the internuclear axis and there is no net overlap between them and an s orbital (Figure 16.24b). Note that the region in which the two atomic wave functions have the same sign and reinforce each other is just equal to the region in which they have the opposite signs and cancel.

We can conclude that the HF molecule is held together by a sigma bond formed by the overlap of the 1s orbital of hydrogen with the $2p_z$ orbital of fluorine. The $p_x$ and $p_y$ orbitals are strictly nonbonding since they have the wrong symmetry to interact with the 1s orbital. The 2s orbital of fluorine is also nonbonding but for a different reason: its energy is too low to interact significantly with the 1s orbital of hydrogen. The molecular orbital analysis of

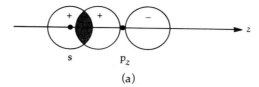

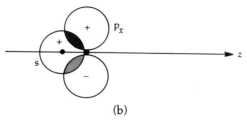

**FIGURE 16.24** Combination of s and p atomic orbitals. (a) The combination of an s and a $p_z$ orbital results in significant net overlap and forms a $\sigma$ molecular orbital. (b) The combination of an s and a $p_x$ (or $p_y$) orbital leads to zero net overlap; a molecular orbital cannot be formed.

---

HF is consistent with the Lewis structure of the molecule,

H—F̈:

where the three lone pairs on the fluorine correspond to the nonbonding 2s, $2p_x$, and $2p_y$ orbitals. Since the $\sigma$ orbital lies closer in energy to the $2p_z$ orbital in fluorine than to the 1s orbital in hydrogen, the bond in HF must be polar—the electron density is greater in the vicinity of the fluorine atom. The rather large dipole moment of HF (Section 15.8) confirms the polarity of the bond and shows that the H—F bond has a substantial ionic character.

## 16.8 Bonding in Simple Hydrocarbons

*a. Localized Bonding* The molecular orbitals of diatomic molecules are formed from the combination of pure atomic orbitals—s with s, s with $p_z$, and so on. Recall from Section 16.2 that the use of hybrid atomic orbitals frequently makes possible a more satisfactory description of covalent bonding than does the use of pure atomic orbitals. The hydrocarbons ethane, ethylene, and acetylene (see Figure 15.4) may be used to illustrate the formation of molecular orbitals from hybrid atomic orbitals. This procedure involves a combination of the valence bond and molecular orbital approaches: first, hybrid atomic orbitals are formed as described by valence bond theory; second, molecular orbitals are formed by the overlap of the hybrid atomic orbitals using the LCAO-MO method. While this is not the only possible description of covalent bonding in these molecules, it is widely used and ties together many of the bonding concepts developed so far.

When a carbon atom forms bonds to four other atoms, its atomic orbitals can be described as $sp^3$ hybrids pointing towards the corners of a tetrahedron

618  Theories of Chemical Bonding  16.8

(Table 16.1). In ethane (C$_2$H$_6$) each carbon atom is bonded to three hydrogen atoms as well as to the other carbon atom,

$$\begin{array}{c} \text{H} \ \ \text{H} \\ | \ \ \ | \\ \text{H}-\text{C}-\text{C}-\text{H} \\ | \ \ \ | \\ \text{H} \ \ \text{H} \end{array}$$

and so fulfills this criterion. The molecular orbitals are formed from the combination of these atomic orbitals as follows: three sp$^3$ orbitals of each carbon atom combine with the s orbitals of three hydrogen atoms; the fourth sp$^3$ orbitals of the carbon atoms combine with each other. In both cases, the resulting molecular orbitals are symmetric about the internuclear axis and are consequently $\sigma$ orbitals. The ethane molecule can be described as having 7 $\sigma$ orbitals filled with the 14 valence electrons, or equivalently, as having 7 sigma bonds. The $\sigma^*$ orbitals lie at a higher energy and, since they are empty, do not figure in the description of the molecule. Figure 16.25 shows the electron density distribution and the molecular structure of ethane. Observe that they are consistent with the preceding description—the electron density is axially symmetric and the bond angles are tetrahedral.

The structure of ethylene (C$_2$H$_4$) is shown in Figure 16.26a. Each carbon atom is attached to three other atoms and the bond angles are close to 120°. The sum of the three bond angles about each carbon atom is 360°, indicating that the molecule is planar. These facts suggest that the s orbital and two of the p orbitals of carbon form hybrid sp$^2$ orbitals, leaving one p orbital unhybridized. The distribution of the valence electrons of carbon in these orbitals can be represented schematically as follows:

$$\underset{2s}{\textcircled{\uparrow\downarrow}} + \underset{2p}{\textcircled{\uparrow}\textcircled{\uparrow}\textcircled{\phantom{\uparrow}}} \rightarrow \underset{sp^2}{\textcircled{\uparrow}\textcircled{\uparrow}\textcircled{\uparrow}} + \underset{p}{\textcircled{\uparrow}}$$

Two of the sp$^2$ orbitals of each carbon atom combine with the s orbitals of the two hydrogen atoms to form $\sigma$ molecular orbitals. The remaining sp$^2$ orbitals of

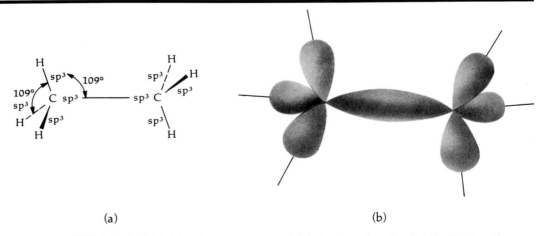

(a)                                    (b)

**FIGURE 16.25** (a) Molecular structure and (b) electron density distribution in ethane.

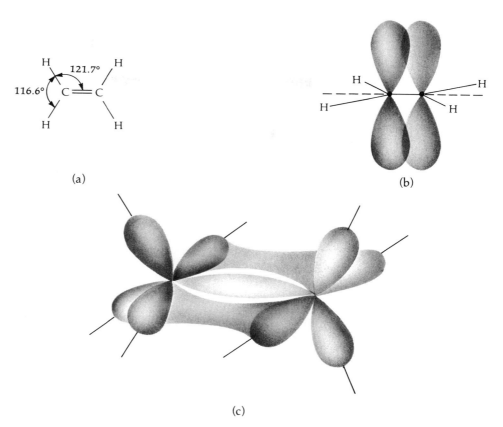

**FIGURE 16.26** Ethylene. (a) Molecular structure, (b) orientation of the p orbitals of carbon relative to the sigma framework of the molecule, and (c) electron density distribution.

the two carbons are directed at each other and overlap to form a $\sigma$ molecular orbital. The unhybridized p orbitals of each carbon atom are perpendicular to the sigma framework of the molecule (Figure 16.26b) and, since the electron density in these orbitals is concentrated off the carbon–carbon axis, they overlap to form a $\pi$ orbital (Figure 16.26c). The ethylene molecule can therefore be described as having five $\sigma$ orbitals and one $\pi$ orbital, which together accommodate the 12 valence electrons of the two carbon and four hydrogen atoms. The molecule also has five $\sigma^*$ orbitals and one $\pi^*$ orbital; however, these orbitals lie at higher energies and do not contain any electrons.

The ethylene molecule contains five sigma bonds and one pi bond, corresponding to the filled molecular orbitals. In particular, the double bond between the two carbon atoms consists of one sigma bond and one pi bond, a very common occurrence in molecules with double bonds. These bonds are clearly different; they differ in their symmetry and electron density distributions as well as in their strength, the sigma bond being stronger than the pi bond. These differences manifest themselves in the chemical reactivity of molecules such as ethylene (Section 22.1d). The pi bond is readily attacked by a variety of reagents in reactions such as

$$\begin{array}{c} H \\ \phantom{H} \\ H \end{array} C=C \begin{array}{c} H \\ \phantom{H} \\ H \end{array} + Cl_2 \rightarrow H-\underset{\underset{H}{|}}{\overset{\overset{Cl}{|}}{C}}-\underset{\underset{H}{|}}{\overset{\overset{Cl}{|}}{C}}-H$$

The addition of a chlorine molecule replaces the pi bond with two additional sigma bonds, and the two carbon atoms remain attached by a sigma bond. The reactivity of ethylene may be contrasted with the relative inertness of ethane which, since it contains only sigma bonds, cannot undergo such addition reactions.

Bonding in acetylene (H—C≡C—H) can be understood by a simple extension of the analysis of bonding in ethylene. The acetylene molecule is linear, indicating that each carbon atom can be characterized by two sp hybrid orbitals. Two of these orbitals, one from each carbon, overlap to form a $\sigma$ molecular orbital, and the other two overlap with the s orbitals of the two hydrogen atoms to form two additional $\sigma$ orbitals. The two remaining p orbitals of each carbon are perpendicular to the axis of the molecule as well as to each other and overlap to form two $\pi$ orbitals, as shown in Figure 16.27. The 10 valence electrons occupy the three $\sigma$ orbitals and the two $\pi$ orbitals. The triple bond in acetylene can therefore be described as a combination of one sigma bond and two pi bonds.

Ethane, ethylene, and acetylene are held together by **localized bonds**—the molecular orbitals are formed by the overlap of the orbitals of two adjacent atoms and the electron density is concentrated in the region between their nuclei. Molecules can also be held together by "delocalized" bonds, which we shall now examine.

***b. Delocalized Bonding*** Molecules in which there is an alternation between single and double bonds, known as **conjugated** molecules, involve delocalized molecular orbitals and bonds. As the name implies, **delocalized orbitals** are spread out over part or all of the molecule, instead of being concentrated on a pair of atoms. Delocalized orbitals are found in molecules that are described by

**FIGURE 16.27**
Electron density distribution in acetylene.

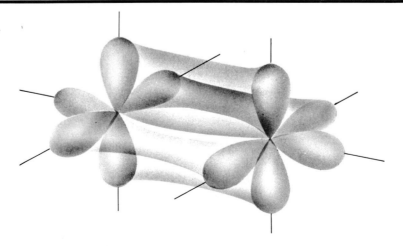

**FIGURE 16.28** The sigma framework of benzene, showing the orientation of the p orbitals.

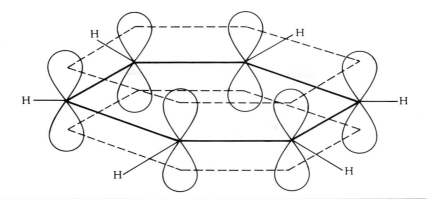

resonance hybrids in the Lewis description (Section 15.6). The classical example of delocalized bonding in hydrocarbons is provided by benzene.

Recall from Section 15.6 that the benzene molecule, $C_6H_6$, is planar and has 120° bond angles. This geometry suggests that, as in the case of ethylene, three of the orbitals of each carbon form $sp^2$ hybrid orbitals. Each carbon atom forms three sigma bonds, one by overlap with a 1s orbital of hydrogen, and two by overlap with $sp^2$ orbitals on the adjacent carbon atoms. Figure 16.28 shows the hexagonal sigma framework of the molecule.

In further analogy to ethylene, each carbon atom has an unhybridized, half-filled p orbital. These p orbitals are directed perpendicular to the plane of the molecule and are parallel to each other, as shown in Figure 16.28. Owing to their proximity, adjacent p orbitals can overlap and thereby form $\pi$ molecular orbitals. It is here that we see a significant difference between benzene, with six adjacent p orbitals, and ethylene, with only two adjacent p orbitals. Once more than two adjacent p orbitals are present, the overlap cannot be restricted to just a pair of orbitals. Each p orbital overlaps equally with the p orbitals on either side of it. Since the six p orbitals form a ring, the interaction between them cannot be localized. Rather, all six p orbitals interact to form six delocalized $\pi$ orbitals extending over the entire ring. Figure 16.29a shows the $\pi$ elec-

**FIGURE 16.29** The $\pi$ orbitals of benzene. (a) $\pi$-electron density and (b) energies of the $\pi$ orbitals. The arrows represent the spin-paired electrons filling the lowest three orbitals.

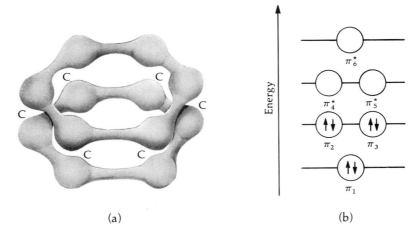

(a)                    (b)

tron density in benzene: the $\pi$ electron cloud extends over the ring in planes above and below the sigma skeleton.

The mathematical description of delocalized $\pi$ orbitals is necessarily more complicated than that of orbitals localized on just two atoms. Recall that in the LCAO-MO procedure, the molecular wave function formed from two atomic wave functions is given by their sum or difference, each atomic wave function being multiplied by a coefficient which indicates the contribution of the atomic orbital in question to the molecular orbital (see Equation 16.11, for example). The linear combination of six atomic wave functions results in the formation of six molecular orbitals. Figure 16.29b shows the energy levels of these $\pi$ orbitals. The orbitals change from strongly bonding to strongly antibonding with increasing energy. Note that two of the $\pi$ orbitals and two of the $\pi^*$ orbitals are degenerate in energy. Each carbon atom contributes one electron to the $\pi$ orbitals and these orbitals are filled in order of increasing energy, two spin-paired electrons to an orbital, as shown. The $\pi$ electron configuration of benzene consequently is $(\pi_1)^2(\pi_2, \pi_3)^4$.

The delocalization of the $\pi$ electrons over the entire molecule reduces the energy of the molecule and enhances its stability. This stabilization is nicely explained by the particle-in-a-box model (Section 5.9). Delocalization increases the size of the "box" to which the electrons are confined and reduces the energy of the system.

The structure of molecules such as benzene can be examined by the use of resonance structures, as in Section 15.6, or by means of molecular orbitals, as done here. It must be emphasized that the two approaches are complementary rather than contradictory. For example, the molecular orbital approach makes it possible to calculate the energies of the various orbitals and to interpret electronic spectra in terms of transitions between these orbitals (Section 17.4). The use of resonance structures provides a more pictorial description of the electron distribution which is useful in qualitative correlations with chemical behavior. Both descriptions are needed for a complete understanding of the chemistry of conjugated molecules.

---

**Example 16.5** The benzene radical anion ($C_6H_6^-$) results from the addition of an electron to benzene. Which molecular orbital is occupied by this electron? Write the $\pi$-electron configuration of this species. On the basis of the $\pi$ bond order compare the stability of $C_6H_6^-$ and $C_6H_6$. Write the Lewis structure of this species.

The orbital energy diagram in Figure 16.29b indicates that an additional electron in benzene must occupy one of the degenerate $\pi_4^*$ or $\pi_5^*$ orbitals. The $\pi$-electronic configuration is $(\pi_1)^2(\pi_2, \pi_3)^4(\pi_4^*)^1$. Since the extra electron is placed in an antibonding orbital, the net number of $\pi$-bonding electrons is five, compared to six for benzene. These electrons are distributed over six carbon–carbon bonds. Each bond thus has $\pi$-order of $(\frac{1}{2})(\frac{5}{6}) = \frac{5}{12}$ compared to $(\frac{1}{2})(\frac{6}{6}) = \frac{1}{2}$ for benzene. Consequently $C_6H_6^-$ is less stable than $C_6H_6$.

To write the Lewis structure of $C_6H_6^-$, we begin by writing one of the resonance structures of benzene:

We wish to add an electron to this structure. However, since each carbon is already surrounded by eight electrons and each hydrogen by two electrons, this cannot be done. The electron distribution must be represented by a different structure than that of benzene. The following Lewis structures, containing only two double bonds, contain the requisite number of 31 valence shell electrons:

Only two of the various possible resonance forms are shown. ∎

## 16.9 Three-Center Bonds in Boron Hydrides

Boron forms a series of volatile compounds with hydrogen called boranes, which are held together by an unusual type of bond. While the simplest of these compounds might have been expected to have the formula $BH_3$, this substance is unstable and difficult to isolate. The simplest stable compound is diborane ($B_2H_6$), which can be prepared by the reaction of gaseous boron trifluoride with alkali metal hydrides in liquid polyethers:

$$8BF_3 + 6NaH \xrightarrow{\text{polyethers}} B_2H_6 + 6NaBF_4$$

Before considering bonding in diborane it is helpful to examine the structure of this molecule (Figure 16.30). Each boron atom is surrounded by four hydrogen atoms arranged in a nearly tetrahedral geometry. The boron atoms and the four end hydrogen atoms all lie in a plane. The two remaining hydrogen atoms are called the bridging atoms since they each form an out-of-plane "bridge" between the boron atoms.

An indication that bonding in $B_2H_6$ must have some unusual features can be seen by distributing the valence shell electrons among the various bonds. Since each boron has 3 valence electrons and each hydrogen has 1, there are 12 electrons available in total. Figure 16.30 indicates that there are eight bonds in the molecule, so that there are not enough electrons to form ordinary electron pair bonds. Thus, diborane is an electron deficient molecule. It is reasonable to

**FIGURE 16.30** The structure of diborane.

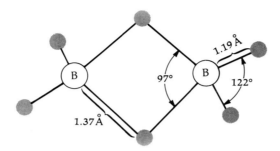

assume that the four B—H bonds with the end hydrogen atoms are ordinary electron pair bonds. These bonds account for 8 electrons, leaving a pair of electrons for each B—H—B unit. A bond in which an electron pair serves to attach three atoms is called a **three-center bond**.

Let us first see whether a resonable Lewis structure can be written for diborane. Although the presence of three-center bonds would appear to render this approach impossible, the use of resonance structures can, at least partially, overcome the difficulty. The following resonance structures appear to be most reasonable:

$$\begin{bmatrix} H & H & H \\ \diagdown & \diagup & \diagdown \\ B & B & \\ \diagup & \diagdown & \diagup \\ H & H & H \end{bmatrix} \leftrightarrow \begin{bmatrix} H & H & H \\ \diagdown & & \diagup \\ B & B & \\ \diagup & \diagdown & \diagdown \\ H & H & H \end{bmatrix} \leftrightarrow \begin{bmatrix} H & H & H \\ \diagdown & & \diagup \\ \oplus B & B \ominus & \\ \diagup & & \diagdown \\ H & H & H \end{bmatrix}$$

$$\leftrightarrow \begin{bmatrix} H & H & H \\ \diagdown & & \diagup \\ \ominus B & B \oplus & \\ \diagup & & \diagdown \\ H & H & H \end{bmatrix}$$

Note that each individual representation divides the molecule into nonbonded halves. However the overall hybrid correctly shows that a single pair of electrons resides on each B—H—B bridge. Nonetheless, the use of "no-bond" structures is not completely satisfactory.

The molecular orbital approach involves the combination of appropriate atomic orbitals. The four valence orbitals of boron form hybrids that, given the observed bond angles, can be described as having both $sp^3$ and $sp^2$ character. The exact form of the hybridization is not important since in either case two of the orbitals of each boron atom form $\sigma$ orbitals by overlap with the 1s orbitals of the end hydrogen atoms. This leaves a hybrid orbital on each boron to combine with the 1s orbital of the bridging hydrogen atom in each B—H—B linkage. The procedure is analogous to that described for benzene except that the overlapping atomic orbitals have different character (they are not p orbitals). The combination of three atomic orbitals leads to the formation of three molecular orbitals. The molecular orbital of lowest energy is a bonding orbital and is filled with the two available valence electrons. Since this orbital is formed from three atomic orbitals, the electron density is delocalized over each of the B—H—B linkages. Figure 16.31 is a schematic representation of the molecular orbitals in diborane.

Diborane is the simplest of a series of boranes. These compounds can be represented by the general formulas $B_nH_{n+4}$ and $B_nH_{n+6}$, e.g., $B_4H_{10}$, $B_5H_9$, $B_{10}H_{14}$, and so on. In addition to the types of bonds considered so far, these molecules also contain B—B—B three-center bonds. The elucidation of bond-

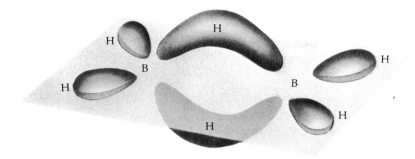

**FIGURE 16.31** The molecular orbitals of diborane.

ing in boranes is chiefly due to William N. Lipscomb (b. 1919), who was awarded the 1976 Nobel prize in chemistry for this work.

## Conclusion

A quantitative understanding of covalent bonding requires the use of quantum mechanics. Molecular Schrödinger equations are rather complicated and various approximate methods have been devised in order to make molecular quantum mechanics tractable. The valence bond theory focuses on the formation of bonds as a result of the overlap between atomic orbitals. These orbitals frequently are hybrids of the orbitals of isolated atoms and their unique directional properties explain the shapes of different molecules. The molecular orbital theory focuses on the molecule as a single entity described by molecular orbitals. These orbitals can be filled with electrons, much in the same way as atomic orbitals, and the properties of the molecular orbitals determine many of the bond properties. We have focused on the application of bonding theories to homonuclear and heteronuclear diatomic molecules, to hydrocarbons with localized and delocalized bonds, and to the boranes. The approaches used in these applications can be extended readily to other types of molecules.

## Problems

1. What is the hybridization of the central atom in each of the following: $AlCl_3$, $ClO_4^-$, $NCl_3$, and $NH_4^+$?

2. What hybrid orbitals are involved in the bonding of carbon in the following species: CO, $CO_2$, and $CO_3^{2-}$?

3. What is the hybridization of the nitrogen atom in each of the following oxides: $NO_2$, $N_2O_3$, and $N_2O_5$?

4. Bonding in HF involves the overlap of the 1s orbital of hydrogen and one of the 2p orbitals of fluorine. Let $\psi(1s)$ be the 1s wave function of hydrogen and $\psi(2p)$ be the 2p wave function of fluorine. Write the valence bond wave function of HF assuming that (a) the bond is purely covalent; (b) the bond is purely ionic, with the electron transferred from H to F; and (c) the bond is partly covalent and partly ionic.

5. The dipole moment of gaseous CsF is 7.88 D, and the bond length is 2.55 Å. What is the ionic character of the Cs—F bond?

6. On the basis of electronegativities, arrange the bonds in the following substances in order of decreasing ionic character: HCl, ClF, $CCl_4$, KCl, and $Cl_2$.

7. Arrange the following molecules in order of increasing dipole moment: $CH_3F$, $CH_3Cl$, $CH_4$, $CCl_4$, and $CH_3Br$. Which of these molecules have zero dipole moment?

8. On the basis of molecular structure and electronega-

tivities, compare the dipole moments of the following pairs of molecules. In each case, pick the molecule with the larger dipole moment or indicate that the dipole moments are equal. (a) HCl, HBr; (b) $CCl_4$, $CBr_4$; (c) $BF_3$, $NF_3$; (d) $PCl_3$, $NCl_3$; and (e) $H_2O$, $H_2S$.

9. Arrange the bonds formed between atoms of the following elements in order of increasing ionic character: N and H, C and O, C and H, H and H, and Ca and O.

10. The dipole moment of CO is 0.12 D, and the bond length is 1.13 Å. (a) Determine the ionic character of the bond. (b) Determine the ionic character on the basis of electronegativities (use Equation 16.7). Explain the discrepancy by examining the formal charges based on the Lewis structure of CO in light of the electronegativities of carbon and oxygen.

11. The bond angles in $F_2O$ and $H_2O$ are nearly equal. Which molecule is likely to have a larger dipole moment? Explain.

12. The dipole moment of $SO_2$ is 1.61 D, and the SO bond length is 1.45 Å. (a) Estimate the ionic character of the SO bond on the basis of the electronegativities of sulfur and oxygen. (b) Determine the SO bond moment. (c) Use the bond moment and the dipole moment of $SO_2$ to estimate the value of the bond angle in $SO_2$.

13. Write the electronic configuration of (a) $He_2$ and (b) $He_2^+$. What is the bond order? Is either one of these species likely to be sufficiently stable to be observable?

14. The valence bond wave function for the $H_2$ molecule is given by Equation 16.3. The molecular orbital wave function for $H_2$ is given by the expression $\psi = [\psi_A(1) + \psi_B(1)][\psi_A(2) + \psi_B(2)]$, where the notation of Equation 16.3 is used. Show how these two wave functions are related.

15. In the LCAO-MO approximation, the molecular wave function of $H_2^+$ is obtained as the linear combination of two 1s atomic hydrogen wave functions, listed in Table 5.6. By plotting the sum and difference of the 1s wave functions along the internuclear axis, confirm the shape of the molecular wave functions shown in Figure 16.8. To simplify the procedure, approximate the atomic wave function by $\psi(1s) = \exp(-r/a_0)$, where $a_0 = 0.53$ Å. Furthermore, neglect the normalization constant, i.e., set $\psi(\sigma) = \psi_A(1s) + \psi_B(1s)$, that is, $\psi(\sigma) = \exp(-r_A/a_0) + \exp(-r_B/a_0)$ and set $\psi(\sigma^*) = \psi_A(1s) - \psi_B(1s)$. The internuclear distance in $H_2^+$ is 1.06 Å.

16. Use the wave function $\psi(\sigma)$ given in Problem 15 to obtain an unnormalized expression for the electron probability distribution in $H_2^+$. Use this expression to calculate the ratio of the probability of finding the electron on the internuclear axis midway between the two protons to (a) the probability of finding the electron on the internuclear axis 0.265 Å from one proton and 0.795 Å from the other proton, (b) the probability of finding the electron on an extension of the internuclear axis 0.53 Å from one proton and 1.59 Å from the other proton, (c) the probability of finding the electron at a point located 0.265 Å from proton A along the axis towards proton B and then 0.530 Å off the axis in the perpendicular direction.

17. Write the electronic configuration of $N_2^+$. How do you expect the bond length and bond energy to compare with those in $N_2$?

18. For each pair of species, pick the one with the greater bond energy: (a) $B_2$, $B_2^+$, (b) $O_2^+$, $O_2^-$, (c) $Be_2$, $Be_2^+$, (d) $F_2$, $F_2^+$, and (e) $F_2$, $F_2^-$.

19. The $C_2^{2-}$ ion is found in compounds such as calcium carbide, $CaC_2$. What is the electronic configuration of $C_2^{2-}$? How do you expect the bond length and bond energy to compare with those in $C_2$?

20. Assuming that the order in which the molecular orbitals of third-period homonuclear diatomic molecules are filled is the same as it is in the second period, determine (a) the electronic configuration of $P_2$ and $S_2$ and (b) the bond order of each molecule. (c) Which molecule do you expect to have the greater bond energy and which the greater bond length?

21. Write the electronic configuration of $Li_2^+$. What is the bond order? Is this ion expected to be paramagnetic? Explain.

22. The ionization energy of molecular oxygen, i.e., the energy required for the reaction $O_2 \rightarrow O_2^+ + e^-$ is 1164 kJ $mol^{-1}$; this value is *smaller* than the ionization energy of atomic oxygen (1314 kJ $mol^{-1}$). On the other hand, the ionization energy of molecular nitrogen (1503 kJ $mol^{-1}$) is *larger* than the ionization energy of atomic nitrogen (1402 kJ $mol^{-1}$). Explain this contrasting behavior in terms of the molecular orbital electronic configurations.

23. Construct a molecular orbital energy diagram for NO similar to that shown for CO in Figure 16.21. On the basis of this diagram, write the electronic configuration of NO.

24. Write the electronic configuration of the cyanide ion, $CN^-$. Name a molecule that is isoelectronic with $CN^-$.

25. On the basis of the electronic configuration of NO (Problem 23), determine the electronic configurations of $NO^+$ and $NO^-$. (b) Arrange these three spe-

cies in order of increasing bond order. (c) Which of these species is isoelectronic with CO?

26. Write the electronic configurations of NF, NF⁺, and NF⁻. Arrange these species in order of increasing (a) bond order, (b) bond energy, and (c) bond length. Which of these species are paramagnetic? How many unpaired electrons does each paramagnetic species have?

27. Write the electronic configuration of (a) PO and (b) SO. (c) For each of these molecules, determine the bond order and the number of unpaired electrons. (d) Which molecule do you expect to have the greater bond energy and which the greater bond length?

28. On the basis of the electronic configurations of $C_2$, $N_2$, $O_2$, $F_2$, CN, and NO, determine which of these species would be stabilized by (a) the addition of an electron to form the corresponding anion, (b) the ionization of an electron to form the corresponding cation.

29. What are the electronic configurations of OF, OF⁺, and OF⁻? Compare these three species with respect to bond order, bond energy, bond length, and paramagnetism.

30. Identify the atomic orbitals associated with the formation of sigma bonds and pi bonds in the following molecules: (a) $NH_3$, (b) $O_3$, (c) $Na_2$, (d) $H_2O$, and (e) $CO_2$.

31. Nitrogen forms a number of compounds with hydrogen including diimide, ($N_2H_2$) and hydrazine ($N_2H_4$). (a) Write the Lewis structure of each of these molecules. (b) Indicate the hybridization of the orbitals of the nitrogen atoms. (c) Deduce which molecular orbitals are formed and state the number of sigma and pi bonds.

32. The structure of cyclopentadiene ($C_5H_6$) is

(a) Describe the hybridization of each of the carbon atoms. (b) How many $\pi$ electrons does the molecule have? (c) Are these $\pi$ electrons delocalized over the entire ring? Explain.

33. The structure of the cyclopentadienyl anion ($C_5H_5^-$) is

(a) Describe the hybridization of each of the carbon atoms. (b) How many $\pi$ electrons does the anion have? (c) Are these $\pi$ electrons delocalized over the entire ring? Contrast the cyclopentadienyl anion with cyclopentadiene (Problem 32) in this respect.

34. Pyridine ($C_5H_5N$) is structurally related to benzene:

(a) Describe the hybridization of each of the carbon atoms and that of the nitrogen atom. (b) How many $\pi$ electrons does pyridine have? (c) Are these electrons delocalized over the entire ring? Explain.

35. The structural formula of formaldehyde is

(a) What is the hybridization of the atomic orbitals of carbon? (b) Categorize each bond in the molecule as a sigma or a pi bond, indicating which atomic orbitals are used in its formation. (c) What is the expected shape of the molecule?

36. What is the $\pi$-electronic configuration of $C_6H_6^+$? What is the $\pi$-bond order? On the basis of this bond order, compare the stability of $C_6H_6^+$ with that of $C_6H_6$. Draw the Lewis structure of this ion.

37. The structural formula of acetic acid is

(a) What is the hybridization of each of the carbon atoms? (b) What is the geometric arrangement of the bonds about each carbon? (c) Which of the carbon-oxygen bonds is longer?

# 17

# Molecular Structure and Chemical Bonding

Advances in chemistry depend on the parallel development of new experimental techniques and new theories. A body of new experimental data stimulates the development of models and theories to explain the observations. Without theoretical guidance, data can remain an unrelated collection of facts. The relationship between theory and experiment is particularly intimate in the area of molecular structure. The quantum theories of chemical bonding considered in Chapter 16 are quite abstract. Such theories must be confronted with a varied body of data before their validity can be fully established. The various experimentally determined molecular properties examined in this chapter constitute such a body of data. The determination of these properties has required a great deal of ingenuity and has been a task that has challenged and occupied generations of chemists.

## 17.1 Bond Energies

***a. Diatomic Molecules*** The concept of bond energy can be applied straightforwardly to a diatomic molecule. The **bond energy** $D$ of a diatomic molecule is equal to the enthalpy change for the reaction in which the gaseous molecule is dissociated into gaseous atoms. (The difference between energy and enthalpy changes in this type of reaction is very small and can be ignored.) For this reason, the bond energy of a diatomic molecule is also called the **dissociation energy**. The bond energy of HCl, for example, is given by $\Delta H°$ for the reaction

$$HCl(g) \rightarrow H(g) + Cl(g) \quad D = \Delta H° = 432.0 \text{ kJ}$$

Bond energies are positive quantities, for it takes energy to separate a molecule into its constituent atoms.

Bond energies can be measured spectroscopically and can also be inferred from standard enthalpies of formation. Recall that the value of $\Delta H°$ for a chemical reaction is equal to the sum of the enthalpies of formation of the products of a reaction less that of the reactants (Equation 10.34). Applying this relation to the dissociation of a diatomic molecule, we see that the bond energy is equal to the standard enthalpies of formation of the gaseous atoms less that of the gaseous molecule.

**Example 17.1** Calculate the bond energy of HCl by means of the data in Appendix 4.

Application of Equation 10.34 to the dissociation of HCl gives

$$D = \Delta H° = \Delta H_f°(H(g)) + \Delta H_f°(Cl(g)) - \Delta H_f°(HCl(g))$$
$$= [217.965 + 121.679 - (-92.307)] \text{ kJ}$$
$$= 431.95 \text{ kJ} \quad \blacksquare$$

Table 17.1 lists the bond energies of a number of diatomic molecules. Note that there is a sizeable variation in bond energy—some covalent bonds are stronger than others. Several regularities may be noted. The bond energies of the hydrogen halides, for example, decrease with increasing halogen atomic number. Since the atomic radius of a halogen atom increases with its atomic number (Section 6.8), the bond lengths of the hydrogen halides must show a similar increase. Thus, the data show an inverse correlation between bond energy and bond length. A second regularity, already discussed in Section 16.6b, is displayed by the homonuclear diatomic molecules of the second-period elements, nitrogen through fluorine, for which the bond order decreases from three to one. Observe that the bond energy decreases between $N_2$ and $F_2$, illustrating the correlation between bond strength and bond order. The large bond energy of carbon monoxide can be understood in the same way. This molecule is isoelectronic with $N_2$ and has the same bond order as well as comparable bond energy.

**TABLE 17.1 Bond Energies and Bond Lengths of Some Gaseous Diatomic Molecules**

| Molecule | Bond Energy (kJ mol$^{-1}$) | Bond Length (Å) |
| --- | --- | --- |
| $I_2$ | 151.2 | 2.67 |
| $F_2$ | 158.0 | 1.41 |
| BrCl | 218.9 | 2.14 |
| HI | 298.3 | 1.61 |
| HBr | 366.3 | 1.41 |
| HCl | 432.0 | 1.27 |
| $H_2$ | 435.9 | 0.74 |
| $O_2$ | 498.3 | 1.21 |
| HF | 568.1 | 0.92 |
| NO | 631.6 | 1.15 |
| $N_2$ | 945.4 | 1.10 |
| CO | 1076.4 | 1.13 |

**b. Polyatomic Molecules** The concept of bond energy is readily extended to polyatomic molecules of type $AB_n$, that is, to molecules in which all the bonds are identical. A good example is methane, $CH_4$, which is held together by four identical C—H bonds (Section 16.2). The C—H bond energy consequently is $\frac{1}{4}$ of the energy required to dissociate the molecule into its constituent atoms. The dissociation reaction is

$$CH_4(g) \rightarrow C(g) + 4H(g) \qquad \Delta H° = 1640 \text{ kJ} \qquad (17.1)$$

and the C—H bond energy is 410 kJ mol$^{-1}$. The enthalpy change in a reaction such as (17.1), in which a substance is dissociated into gaseous atoms is sometimes called an **enthalpy of atomization**.

The meaning of "bond energy" is somewhat ambiguous for polyatomic molecules containing more than one kind of bond. Consider the bond energies in ethane ($C_2H_6$), for example. The molecule contains six C—H bonds and one C—C bond. The enthalpy of atomization of ethane is

$$C_2H_6(g) \rightarrow 2C(g) + 6H(g) \qquad \Delta H° = 2800 \text{ kJ}$$

It obviously makes no sense to divide this value by the number of bonds to obtain an average bond energy—an average of the energies of one C—C bond and six C—H bonds has no physical meaning. However, if we assume that the C—H bond energy derived from the atomizaton of methane applies to the C—H bonds in ethane, the C—C bond energy can be obtained directly:

$$D_{C-C} = 2800 \text{ kJ mol}^{-1} - (6)(410 \text{ kJ mol}^{-1}) = 340 \text{ kJ mol}^{-1}$$

This assumption is reasonable since the C—H bond in both molecules is a sigma bond (Section 16.8).

The procedure used to obtain the C—C bond energy in ethane can be extended to additional molecules, and in this fashion it is possible to construct a table of consistent single-bond energies. This consistency shows that the energy of the bond between a given pair of atoms is fairly independent of the molecular environment and is nearly the same in most molecules. While bond energies obtained in this way are only approximately correct, the approximation is good enough to be useful, as we shall see. Since the bond energies derived from different molecules do differ from each other slightly, it is customary to use an average value.

Table 17.2 lists some representative values of average bond energies. Note that the value of the C—H bond energy, 414 kJ mol$^{-1}$, is slightly different from the value derived from the data for methane; it represents an average value for many molecules. The energies of double and triple bonds can be obtained in the same manner, and the results are included in the table. The variation of bond energy with bond order for bonds between a given pair of atoms is quite striking and is depicted in Figure 17.1. The increase in bond strength with bond order supports the theoretical analysis of covalent bonding, which showed that the strength of a bond depends on the electron density in the internuclear region.

Experiments on molecular dissociation can be used to obtain even more detailed data on bond energies than those summarized in Table 17.2. Recall the assertion that a sigma bond between two atoms is stronger than a pi bond between them (Section 16.5). This statement can be confirmed for the sigma and pi bonds comprising the double bond between two carbon atoms in molecules such as ethylene ($C_2H_4$), as shown in the following example.

Molecular Structure and Chemical Bonding  17.1  **631**

**TABLE 17.2  Average Bond Energies in Kilojoules per Mole**

|      | H   | C   | N   | O    | F   | S   | Cl  | Br  | I   |
|------|-----|-----|-----|------|-----|-----|-----|-----|-----|
| H—   | 436 | 413 | 391 | 463  | 563 | 339 | 432 | 366 | 299 |
| C—   |     | 348 | 292 | 351  | 441 | 259 | 328 | 276 | 240 |
| C=   |     | 615 | 515 | 800  | —   | 582 | —   | —   | —   |
| C≡   |     | 812 | 891 | 1075 | —   | —   | —   | —   | —   |
| N—   |     |     | 161 | 175  | 270 | —   | 200 | 163 | 151 |
| N=   |     |     | 473 | 590  | —   | —   | —   | —   | —   |
| N≡   |     |     | 945 | —    | —   | —   | —   | —   | —   |
| O—   |     |     |     | 139  | 212 | 423 | 210 | —   | —   |
| O=   |     |     |     | 495  | —   | 523 | —   | —   | —   |
| F—   |     |     |     |      | 158 | 343 | 251 | 249 | 281 |
| S—   |     |     |     |      |     | 266 | 277 | 239 | —   |
| Cl—  |     |     |     |      |     |     | 243 | 218 | 210 |
| Br—  |     |     |     |      |     |     |     | 193 | 178 |
| I—   |     |     |     |      |     |     |     |     | 151 |

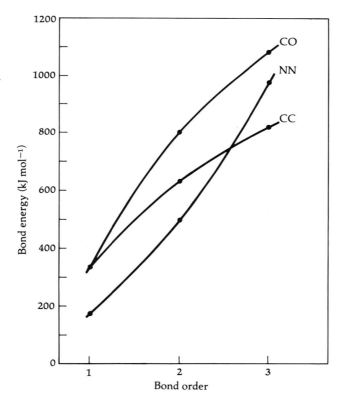

**FIGURE 17.1** Variation of bond energy with bond order for carbon–carbon, nitrogen–nitrogen, and carbon–oxygen bonds.

**Example 17.2**  The energy required to dissociate $C_2H_4$ into two $CH_2$ fragments is 598 kJ mol$^{-1}$ while that required to dissociate $C_2H_6$ into two $CH_3$ fragments is 347 kJ mol$^{-1}$. Use these data to estimate the difference in the energy of the sigma and the pi carbon–carbon bonds in $C_2H_4$. What assumptions must be made?

# 632 Molecular Structure and Chemical Bonding 17.1

The structural formulas of these molecules are

$$\begin{array}{c} H \\ \diagdown \\ H \end{array} C=C \begin{array}{c} H \\ \diagup \\ H \end{array} \quad \text{and} \quad \begin{array}{c} H\ H \\ | \ | \\ H-C-C-H \\ | \ | \\ H\ H \end{array}$$

The energy required to break ethylene into two $CH_2$ fragments (598 kJ mol$^{-1}$) is the energy required to break both the sigma and the pi bond. The energy needed to dissociate ethane into two $CH_3$ fragments (347 kJ mol$^{-1}$) is the energy required to break a sigma bond. On the assumption that the energy of a carbon–carbon sigma bond has the same value in ethylene and in ethane, the pi bond energy is equal to the difference between these values:

$$D_\pi = (D_\sigma + D_\pi) - D_\sigma$$
$$D_\pi = (598 - 347) \text{ kJ mol}^{-1} = 251 \text{ kJ mol}^{-1}$$

The difference in bond energy is

$$D_\sigma - D_\pi = (347 - 251) \text{ kJ mol}^{-1} = 96 \text{ kJ mol}^{-1}$$

confirming that a sigma bond is stronger than a pi bond. ∎

**c. Bond Energies and Reaction Enthalpies** Bond energies can be used to obtain approximate values of reaction enthalpies for reactions in which all reactants and products are gases. Energy must be supplied to break the bonds in the reactant molecules; in turn, energy is released when the bonds of the product molecules are formed. The reaction enthalpy corresponds to the difference between the two:

$$\Delta H° = \sum D_{\text{reactants}} - \sum D_{\text{products}} \qquad (17.2)$$

This equation is an application of Hess's law (Section 10.8): the reaction enthalpy is obtained by means of an alternate path in which the reactants are first broken up into their constituent atoms and these atoms are then reassembled to form the products. Note that if the bond energy of the products is greater than that of the reactants the reaction is exothermic and $\Delta H < 0$, while if the opposite holds true the reaction is endothermic and $\Delta H > 0$.

---

**Example 17.3** Calculate the enthalpy of combustion of ethanol using bond energies. Assume that the reactants and products are in the gas phase.

The reaction is

$$C_2H_5OH(g) + 3O_2(g) \rightarrow 2CO_2(g) + 3H_2O(g)$$

To determine the number and order of all the bonds that are broken and formed, it is helpful to write Lewis structures of all the species involved in the reaction:

$$\begin{array}{c} H\ H \\ | \ | \\ H-C-C-\ddot{\underset{..}{O}}-H \\ | \ | \\ H\ H \end{array} \quad \ddot{\underset{..}{O}}=\ddot{\underset{..}{O}} \quad \ddot{\underset{..}{O}}=C=\ddot{\underset{..}{O}} \quad H-\ddot{\underset{..}{O}}-H$$

Using these structures, the balanced equation, and the tabulated bond energies, we can evaluate Equation 17.2 in tabular form:

| Reactant bonds | Bond energy (kJ mol$^{-1}$) | Product bonds | Bond energy (kJ mol$^{-1}$) |
|---|---|---|---|
| 1 C—C | 348 | 4 C=O | 4 × 800 |
| 5 C—H | 5 × 413 | 6 O—H | 6 × 463 |
| 1 C—O | 351 | | |
| 1 O—H | 463 | | |
| 3 O=O | 3 × 495 | | |

The total bond energy of the reactants is 4712 kJ and that of the products is 5978 kJ. Thus, $\Delta H° = (4712 - 5978)$ kJ $= -1266$ kJ. The experimental enthalpy of combustion is $-1275$ kJ, indicating that for this particular reaction the bond energy method agrees with the experimental value to within 1%. ∎

Although Equation 17.2 is only applicable to gaseous reactants and products, it can readily be extended to substances in condensed phases by the additional use of enthalpies of vaporization and fusion.

Example 17.3 is typical with respect to the accuracy obtainable by the use of bond energies in thermochemical calculations. In some instances, however, the bond energy method gives results that are in marked disagreement with thermochemical data. Such cases are interesting because they indicate that the molecules in question have some unusual structural features. For example, it is generally found that molecules with a structure best described as a resonance hybrid (Section 15.6) have a lower energy than that calculated on the basis of any individual contributing structure. The difference between the two energies is called the **resonance energy** and is a measure of the enhanced stability of the resonance hybrid relative to its contributing structures. A good illustration of this effect is provided by benzene. Recall that the Lewis structure of the benzene ring consists of alternate single and double carbon–carbon bonds. We know that in actuality these bonds have order 1.5 and that the six $\pi$ electrons are delocalized over the entire ring. Delocalization increases the stability of the molecule and decreases its energy (Section 16.8). Resonance energies calculated on the basis of experimental data serve as an independent confirmation of the occurrence of resonance in certain molecules.

To illustrate the concept of resonance energy, let us evaluate the resonance energy of gaseous benzene. We know from Section 10.9 that the standard enthalpy of formation of a substance is a measure of its energy relative to that of its constituent elements in their standard states. The resonance energy, therefore, can be determined by evaluation of the standard enthalpy of formation of either of the contributing structures of benzene by means of the bond energy method and comparison of this result with the experimental standard enthalpy of formation.

The standard enthalpy of formation of gaseous benzene corresponds to the enthalpy change in the reaction

$$6C(\text{graphite}) + 3H_2(g) \rightarrow C_6H_6(g) \tag{17.3}$$

In order to evaluate $\Delta H°$ for this reaction by the bond energy method it is convenient to express the reaction as the sum of the following three processes:

$$6C(\text{graphite}) \rightarrow 6C(g) \tag{17.4}$$

$$3H_2(g) \rightarrow 6H(g) \tag{17.5}$$

$$6C(g) + 6H(g) \rightarrow C_6H_6(g) \tag{17.6}$$

This procedure isolates the formation of a gaseous molecule from the gaseous atoms (Equation 17.6), the enthalpy of which can be calculated by the bond energy method, from the other processes contributing to the standard enthalpy of formation. The standard enthalpies of reactions (17.4) and (17.5) can be obtained from the thermodynamic data tables in Appendix 4, and the results are $\Delta H° = 4300.1$ kJ and $\Delta H° = 1307.8$ kJ, respectively. The enthalpy change for reaction (17.6) can be obtained by use of bond energies. Recall that the structures contributing to the benzene resonance hybrid are of the form

The sum of the bond energies can be obtained by use of the data in Table 17.2 in the manner illustrated in Example 17.3:

| Bond | Bond energy (kJ mol$^{-1}$) |
|---|---|
| 6 C—H | 6 × 413 |
| 3 C—C | 3 × 348 |
| 3 C=C | 3 × 615 |
| | 5367 kJ mol$^{-1}$ |

The enthalpy change in reaction (17.6) must be $-5367$ kJ, for the sum of the bond energies is the energy that must be supplied to atomize the molecule—the reverse of reaction (17.6). The calculated standard enthalpy of formation of gaseous benzene is the sum of the enthalpy changes in reactions (17.4)–(17.6):

$$\Delta H_f° = (4300.1 + 1307.8 - 5367) \text{ kJ} = 241 \text{ kJ}$$

The thermochemical value of the standard enthalpy of formation of gaseous benzene is 82.9 kJ mol$^{-1}$. This value is substantially smaller than that calculated above, indicating that benzene is more stable than would be expected for a molecule of $C_6H_6$ containing three C—C bonds and three C=C bonds. The difference between the two values (158 kJ mol$^{-1}$) corresponds to the resonance energy of benzene.

## 17.2 Rotational Spectra and Bond Lengths

***a. Molecular Spectroscopy*** Molecules have quantized energy levels and transitions between these levels can be studied by the techniques of molecular spectroscopy. When a molecule in an excited state makes a transition to a lower energy state (Figure 17.2a), electromagnetic radiation is emitted. The spectrum of radiation emitted by a sample containing excited molecules is called an **emission spectrum** (Figure 17.2b). When radiation is incident on a

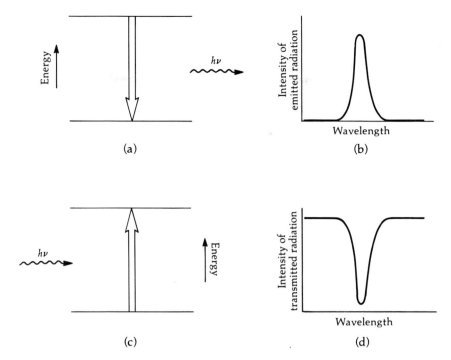

**FIGURE 17.2** Molecular spectroscopy. (a) Emission of radiation in a transition between two levels. (b) Emission spectrum for the transition in (a). (c) Absorption of radiation in a transition between two levels. (d) Absorption spectrum for the transition in (c). Actual emission and absorption spectra have many peaks.

molecular sample, photons can be absorbed by the molecules and transitions to higher-energy states take place (Figure 17.2c). The radiation transmitted by the sample is depleted in the wavelengths of the absorbed radiation, and the resulting spectrum is called an **absorption spectrum** (Figure 17.2d).

Absorption and emission spectroscopy are two of the important types of molecular spectroscopy. As in the case of atomic spectroscopy (Section 5.5), the wavelength of the absorbed or emitted radiation is related to the energy difference between the initial and final states of the molecule by the Bohr formula (Equation 5.18),

$$\Delta E = h\nu = hc/\lambda = hc\bar{\nu}$$

where the observed radiation can be characterized by its frequency $\nu$, its wavelength $\lambda$, or its wave number $\bar{\nu}$ (Section 5.2).

The transitions observed in atomic spectroscopy occur when an electron jumps from one level to another. Molecular transitions are more complex. In addition to electronic energy levels, which correspond to electron jumps, molecules also possess vibrational and rotational energy levels. These levels are associated with the vibrational motion of the atoms in the molecule and with rotational motion of the molecule, respectively (Section 10.6). A characteristic feature of these different types of energy levels is that their spacing differs widely: rotational energy levels are closely spaced, vibrational energy levels are more widely spaced, and electronic energy levels are most widely spaced. Figure 17.3 is a schematic diagram of these molecular energy levels. When a

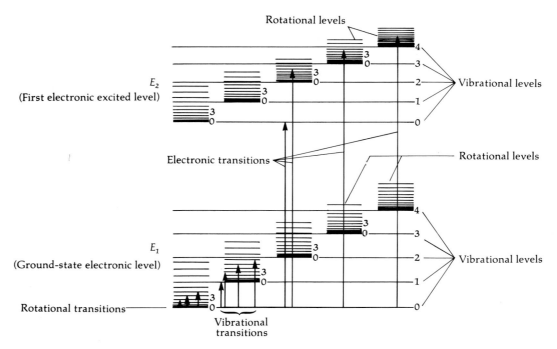

**FIGURE 17.3** Schematic molecular energy level diagram. Each vibrational level has a set of rotational levels and each electronic level has a set of vibrational levels. The black arrows show electronic, vibrational, and rotational transitions (in order of decreasing length). The colored arrows show transitions in which two or more of these excitations occur simultaneously, e.g., vibrational–rotational transitions. The levels are identified by the values of the vibrational ($v$) and rotational ($J$) quantum numbers.

molecule is excited to one of its higher vibrational levels it can simultaneously acquire rotational energy. Consequently, each vibrational energy level has a set of rotational energy levels associated with it, as shown. Similarly, when a molecule is excited electronically, it can at the same time undergo vibrational and rotational excitations. The occurrence of these simultaneous excitations accounts for the characteristic "band" structure of the energy level diagram and of the corresponding emission and absorption spectra.

Molecular spectra reflect the transitions that occur between molecular levels. The inverse relation between the energy spacing and the wavelength of the radiation emitted or absorbed in a transition indicates that transitions between rotational levels should be observed at long wavelengths, those between vibrational levels should be observed at intermediate wavelengths, and those between electronic levels should be observed at short wavelengths. For most molecules, rotational transitions are observed in the microwave or far-infrared region of the electromagnetic spectrum (Section 5.3), vibrational transitions are observed in the infrared, and electronic transitions are observed in the visible or ultraviolet (Figure 17.4). Molecular spectra are a rich source of information about molecular structure. Rotational spectra permit a determination of bond lengths, vibrational spectra give information on the stiffness of bonds and on molecular geometry, and electronic spectra yield data on electronically excited molecules.

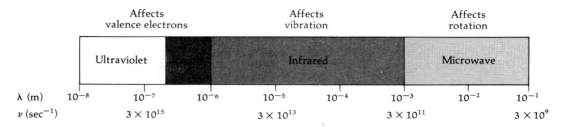

**FIGURE 17.4** Relation between the various regions of the electromagnetic spectrum and the type of molecular energy level affected.

**b. Rotational Spectra of Diatomic Molecules** The rotational motion of a diatomic molecule can be understood by analogy to that of a macroscopic object, such as a dumbell. Figure 17.5 illustrates the rotation of such a body. The motion can be expressed in terms of the **moment of inertia** of the body, $I$, and its angular momentum $p_{ang}$. These quantities play an analogous role in rotational motion to that of mass and linear momentum, respectively, in linear motion. Just as the kinetic energy of a body is related to its linear momentum $p$ and mass $m$ by the formula

$$E_{kin} = \tfrac{1}{2}mv^2 = m^2v^2/2m = p^2/2m \tag{17.7}$$

the rotational energy of a body is given by the analogous relation

$$E_{rot} = p_{ang}^2/2I \tag{17.8}$$

The moment of inertia depends on the masses of the two parts of the dumbell and on the distance between their centers:

$$I = \frac{m_1 m_2}{m_1 + m_2} d^2 \tag{17.9}$$

and is expressed in units of kg m². The ratio of the product and sum of the two masses is called the **reduced mass** $\mu$, so that in terms of this quantity

$$I = \mu d^2 \qquad \mu = \frac{m_1 m_2}{m_1 + m_2} \tag{17.10}$$

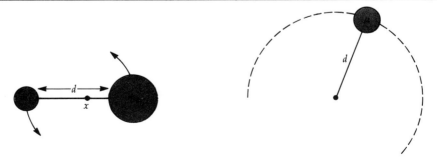

**FIGURE 17.5** The rotational motion of a dumbell—a model for a diatomic molecule. The rotation of an object of mass $\mu$ about the origin (right) yields the same rotational energy as that of the dumbell about its center of mass, point $x$ (left).

As shown in Figure 17.5, the significance of the reduced mass is that the rotation of a dumbell about its center of mass is equivalent to that of a single body of mass $\mu$ located at a distance $d$ from the origin.

The rotational energy and angular momentum of a macroscopic body can take on any value. However, we know from Chapter 5 that a particle of molecular dimensions can only have quantized energy and angular momentum. The preceding relations can be applied to the rotation of a molecule by replacing the classical angular momentum by its quantum mechanical equivalent (Section 5.11),

$$p_{ang} = \hbar\sqrt{J(J+1)} \quad J = 0, 1, 2, \ldots \tag{17.11}$$

where $J$ is the rotational quantum number and $\hbar$ is Planck's constant $h$, divided by $2\pi$. The rotational energy of a diatomic molecule is obtained by substitution of this result into Equation 17.8:

$$E_{rot} = \frac{\hbar^2 J(J+1)}{2I} \quad J = 0, 1, 2, \ldots \tag{17.12}$$

The allowed rotational energy levels of a molecule are shown schematically in Figure 17.6a. Note that the spacing between the levels widens with increasing value of the rotational quantum number $J$, reflecting the variation of the energy with $J(J+1)$.

When a photon of the appropriate frequency is incident upon a molecule in some rotational energy level, the molecule can make a transition to an adjacent level provided it has a dipole moment. (The dipole moment provides the means by which the molecule can interact with the photon and undergo a rotational transition.) The frequency can be related to the energy difference by Bohr's formula. For a transition between levels with rotational quantum numbers $(J-1)$ and $J$ we have

$$E_J - E_{J-1} = h\nu = \frac{\hbar^2}{2I}[J(J+1) - (J-1)J]$$

$$= \frac{\hbar^2}{2I}(2J) \tag{17.13}$$

from which we obtain

$$\nu = hJ/4\pi^2 I \tag{17.14}$$

Since the value of $J$ corresponding to a given transition can be inferred from the spectrum, the absorption frequency yields the value of the moment of inertia and, hence, that of the bond length ($d$ in Equation 17.9). Because photons absorbed in rotational transitions lie in the microwave or far-infrared regions of the electromagnetic spectrum, the experimental study of these transitions is called **microwave spectroscopy**. Figure 17.6b shows the spectrum expected on the basis of the transitions shown in Figure 17.6a. The spectrum consists of a series of equally spaced lines provided that the plotted variable is either the frequency or some quantity proportional to it, such as the wave number. This result follows from Equation 17.14 since the difference between the frequencies at which molecules in adjacent rotational levels absorb radiation,

$$\Delta\nu = \nu_J - \nu_{J-1} = (h/4\pi^2 I)[J - (J-1)] = h/4\pi^2 I \tag{17.15}$$

is a constant.

**FIGURE 17.6** (a) Rotational energy levels of a diatomic molecule showing transitions between adjacent levels. (b) Rotational spectrum expected from the transitions in (a). The lines are labelled by the quantum numbers of the final state.

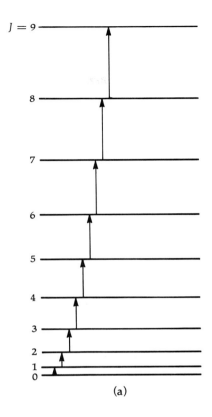

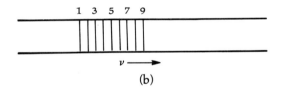

---

**Example 17.4** The microwave absorption spectrum of carbon monoxide consists of a series of lines with a spacing of $1.153 \times 10^5$ MHz. What is the CO bond length?

The moment of inertia of CO is first obtained from the frequency spacing by means of Equation 17.15,

$$\Delta \nu = h/4\pi^2 I$$

Solving for $I$, we obtain

$$I = \frac{h}{4\pi^2 \Delta \nu}$$

$$= \frac{(6.626 \times 10^{-34} \text{ J s})(1 \text{ kg m}^2 \text{ s}^{-2}/\text{J})}{(4\pi^2)(1.153 \times 10^5 \text{ MHz})(10^6 \text{ Hz/MHz})(1\text{s}^{-1}/\text{Hz})}$$

$$= 1.4557 \times 10^{-46} \text{ kg m}^2$$

Knowing the moment of inertia, we can obtain the bond length by means of Equation 17.9

$$I = \frac{m_1 m_2 d^2}{m_1 + m_2}$$

$$1.4557 \times 10^{-46} \text{ kg m}^2 = \frac{(12.01 \text{ g mol}^{-1})(16.00 \text{ g mol}^{-1})(10^{-3} \text{ kg g}^{-1}) d^2}{(12.01 \text{ g mol}^{-1} + 16.00 \text{ g mol}^{-1})(6.022 \times 10^{23} \text{ mol}^{-1})}$$

$$d = 1.130 \times 10^{-10} \text{ m} = 1.130 \text{ Å} \quad \blacksquare$$

The analysis of rotational spectra of polyatomic molecules involves the same general approach described for diatomic molecules. However, polyatomic molecules can rotate about as many as three different axes and can be characterized by up to three different moments of inertia. Given these additional variables, it is not suprising that the analysis of rotational spectra of polyatomic molecules can become fairly complicated.

**c. Bond Lengths** Bond lengths—the internuclear distances between bonded atoms—can be obtained in the manner described in Section 17.2b. The bond lengths of some diatomic molecules are summarized in Table 17.1. As was found to be true for bond energies, bond lengths depend on the identity of the bonded atoms and on bond order but are nearly independent of the molecular environment. Thus, average bond lengths analogous to the average bond energies can be obtained. Table 17.3 lists some representative values. Due to the vibration of atoms in a molecule (Section 17.3), the internuclear distances between bonded atoms are continuously changing. The tabulated values therefore represent average internuclear distances.

To the extent that the bond length between a given pair of atoms is the same in all molecules, it can be apportioned between the two atoms to obtain their **covalent radii**. This is done most conveniently by means of data for homonuclear diatomic molecules, for the covalent radius is just one-half the bond length of such a molecule. The covalent radii of any two atoms can, in turn, be added to obtain a moderately accurate estimate of the bond length between them. In view of the dependence of bond length on bond order (Section 16.6), the addition of covalent radii must be restricted to bonds of the same order as those from which they were derived.

**TABLE 17.3 Average Bond Lengths in Angstroms**

|      | H    | C    | N    | O    | F    | Cl   | Br   | I    |
|------|------|------|------|------|------|------|------|------|
| H—   | 0.74 | 1.09 | 1.01 | 0.97 | 0.92 | 1.27 | 1.41 | 1.61 |
| C—   |      | 1.54 | 1.48 | 1.43 | 1.38 | 1.77 | 1.94 | 2.14 |
| C=   |      | 1.34 | 1.28 | 1.21 | —    | —    | —    | —    |
| C≡   |      | 1.20 | 1.16 | —    | —    | —    | —    | —    |
| N—   |      |      | 1.45 | —    | —    | —    | —    | —    |
| N=   |      |      | 1.25 | —    | —    | —    | —    | —    |
| N≡   |      |      | 1.09 | —    | —    | —    | —    | —    |
| O—   |      |      |      | 1.45 | —    | —    | —    | —    |
| O=   |      |      |      | 1.21 | —    | —    | —    | —    |
| F—   |      |      |      |      | 1.42 | —    | —    | —    |
| Cl—  |      |      |      |      |      | 1.99 | —    | —    |
| Br—  |      |      |      |      |      |      | 2.28 | —    |
| I—   |      |      |      |      |      |      |      | 2.67 |

To illustrate this procedure, let us estimate the Br—F bond length by means of the data in Table 17.3. This bond length is given by the sum of the covalent radii of bromine and fluorine which, in turn, are just one-half the Br—Br and F—F bond lengths, respectively. Designating the three bond lengths in question $d_{Br-F}$, $d_{Br-Br}$, and $d_{F-F}$, we can write the following relation between them:

$$d_{Br-F} \approx \tfrac{1}{2}(d_{Br-Br} + d_{F-F}) \tag{17.16}$$

$$\approx \tfrac{1}{2}(2.28 \text{ Å} + 1.42 \text{ Å}) = 1.85 \text{ Å}$$

This value differs by about 5% from the experimental Br—F bond length (1.76 Å), a result of typical accuracy. The atomic radii of the nonmetallic elements (Section 6.8b) are actually covalent radii obtained from bond lengths.

## 17.3 Vibrational Spectra of Diatomic Molecules and Bond Force Constants

The vibrational motion in a diatomic molecule is illustrated in Figure 17.7a. The two atoms continuously oscillate back and forth, and the internuclear distance varies in a periodic manner (Figure 17.7b). The classical analogue of this type of

**FIGURE 17.7** (a) Vibrational motion of a diatomic molecule. (b) Time dependence of the internuclear distance between the atoms of a vibrating molecule relative to the equilibrium distance $d_e$. (c) Classical analogue of molecular vibrations: oscillatory motion of a body attached by a spring to an immovable wall.

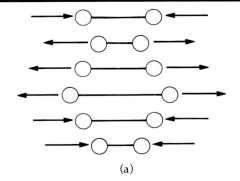

(a)

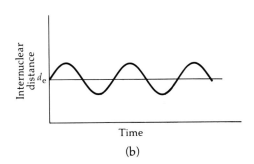

(b)

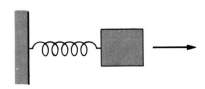

(c)

motion is that of a body attached by means of a spring to an immovable wall (Figure 17.7c). Suppose that the body is pulled away from the wall and then released. It will then move back and forth, with its displacement from the equilibrium position being similar to that depicted in Figure 17.7b. This type of periodic motion is known as **simple harmonic motion** and obeys Hooke's law. Let the body in Figure 17.7c be displaced away from the wall by a distance $x$. **Hooke's law** states that the restoring force with which the spring pulls the body back towards its equilibrium position is given by the relation

$$F = -kx \qquad (17.17)$$

where the constant $k$ is called the **force constant** and is a measure of the stiffness of the spring. The larger the value of $k$, the stiffer the spring, and the greater the magnitude of the restoring force. The units of $k$ are those of force divided by distance, newtons per meter. The negative sign in Hooke's law indicates that the restoring force and the displacement are oppositely directed.

Simple harmonic motion can be described in terms of the **vibrational frequency**, the number of vibrational cycles completed by the body in one second. The vibrational frequency $\nu_0$ can be expressed in terms of the force constant and the mass of the body, $m$, by application of Newton's second law ($F = ma$). The result is

$$\nu_0 = (1/2\pi)(k/m)^{1/2} \qquad (17.18)$$

The vibrational frequency increases as the square root of the force constant and decreases as the square root of the mass.

A spring is an energy-storing device. Thus, a stretched spring can do work as it retracts and a compressed spring can do work as it expands. The potential energy of the system depends on the magnitude of the displacement and can be derived from Hooke's law as

$$U = \tfrac{1}{2}kx^2 \qquad (17.19)$$

The kinetic energy of the body varies continuously between a value of zero at the limits of the displacement and a value of $U$ at the equilibrium position ($x = 0$), where the potential energy is zero. The sum of the potential and kinetic energies remains constant—the total energy is conserved.

The equations describing simple harmonic motion can be applied in a straightforward manner to the vibration of a diatomic molecule. The harmonic oscillator model is applicable because the molecular potential energy has a parabolic dependence on the internuclear distance in the vicinity of the equilibrium position of the atoms. Thus Equation 17.19 is applicable. This is demonstrated in Figure 17.8, which shows the potential energy of $H_2$ (Figure 16.2) with a parabola superimposed on it. So long as the two atoms are not far from their average positions, the two curves practically coincide.

The vibration of a diatomic molecule can be characterized by its bond force constant $k$, which is a measure of the stiffness of the bond. In analogy to the classical motion of a spring, a large force constant corresponds to a stiff bond, for which the nuclei do not move far from their average positions. The frequency of molecular vibrations is given by Equation 17.18 modified by the replacement of the mass of the body by the reduced mass of the molecule (Equation 17.10):

$$\nu_0 = (1/2\pi)(k/\mu)^{1/2} \qquad (17.20)$$

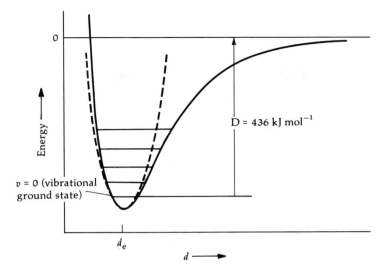

**FIGURE 17.8** Potential energy of an H₂ molecule showing the location of the vibrational energy levels. The dashed curve is a parabolic approximation to the potential energy known as the harmonic oscillator model. Observe that the dissociation energy is measured from the lowest vibrational energy level (zero-point energy).

In contrast to classical vibrational motion, the vibrational energy of a molecule cannot vary continuously but must, instead, be quantized. The vibrational energy, obtained by solving the Schrödinger equation for a **harmonic oscillator** [i.e., a system whose potential energy obeys Equation (17.19)], is

$$E = (v + \tfrac{1}{2})h\nu_0 \qquad v = 0, 1, 2, \ldots \tag{17.21}$$

where $v$ is the vibrational quantum number. The lowest vibrational energy levels are shown in the potential energy diagram of Figure 17.8. As expected from Equation 17.21, the energy levels are equally spaced. The energy of the vibrational ground state, that is, the state with $v = 0$, is $\tfrac{1}{2}h\nu_0$ (as measured from the minimum in the potential energy curve). This **zero-point energy** is the irreducible minimum energy of a molecule, even in the absence of thermal motion. A molecule cannot be completely at rest, for this would violate the uncertainty principle (Section 5.7). For example, it is the zero-point energy of helium that prevents liquid helium from freezing at atmospheric pressure no matter how low the temperature.

As for rotational energy levels, information on vibrational energy levels is obtained from experimental data on transitions between levels. When radiation of suitable wavelength is incident on a diatomic molecule with a dipole moment, absorption can take place, and the molecule is excited from one vibrational level to the adjacent level. Applying the Bohr formula to the energy difference between levels characterized by vibrational quantum numbers $(v + 1)$ and $v$ we obtain

$$\nu = \frac{E_{(v+1)} - E_{(v)}}{h} = \frac{(v + \tfrac{3}{2})h\nu_0 - (v + \tfrac{1}{2})h\nu_0}{h}$$

$$\nu = \nu_0 = \frac{1}{2\pi}\left(\frac{k}{\mu}\right)^{1/2} \tag{17.22}$$

where the second equality follows from Equation 17.20. Equation 17.22 indicates that the vibrational spectrum consists of a single line, the frequency of which depends on the bond force constant of the molecule and on its reduced mass. Because vibrational transitions occur in the infrared, their study involves **infrared spectroscopy**.

---

**Example 17.5** Hydrogen fluoride strongly absorbs at 4141 cm$^{-1}$ (2.415 $\mu$m photons). (a) What is the H—F bond force constant? (b) What is the HF zero-point energy?

(a) The force constant is related to the radiation frequency by Equation 17.22:

$$\nu = \frac{c}{\lambda} = \frac{1}{2\pi}\left(\frac{k}{\mu}\right)^{1/2} = \frac{1}{2\pi}k^{1/2}\left(\frac{m_H + m_F}{m_H m_F}\right)^{1/2}$$

where the reduced mass of the molecule has been expressed in terms of the masses of its constituent atoms. Numerical substitution gives

$$\frac{2.998 \times 10^8 \text{ m s}^{-1}}{2.415 \times 10^{-6} \text{ m}} = \frac{1}{2\pi}\left[\frac{(1.008 \text{ g mol}^{-1} + 19.00 \text{ g mol}^{-1})(6.022 \times 10^{23} \text{ mol}^{-1})}{(1.008 \text{ g mol}^{-1})(19.00 \text{ g mol}^{-1})(10^{-3} \text{ kg g}^{-1})}\right]^{1/2} k^{1/2}$$

$$k = 967.1 \text{ kg s}^{-2} = 967.1 \text{ N m}^{-1}$$

(b) The zero-point energy is $\frac{1}{2}h\nu_0$, where the vibrational frequency is equal to the frequency of the absorbed radiation. Thus,

$$E_0 = \frac{1}{2}h\nu$$

$$= (0.5)(6.626 \times 10^{-34} \text{ J s})\frac{(2.998 \times 10^8 \text{ m s}^{-1})}{(2.415 \times 10^{-6} \text{ m})}(6.022 \times 10^{23} \text{ mol}^{-1})(10^{-3} \text{ kJ J}^{-1})$$

$$= 24.77 \text{ kJ mol}^{-1}$$

The value amounts to less than 5% of the HF bond energy (568.1 kJ mol$^{-1}$). ∎

Vibrational spectra are generally more complicated than indicated here, even for diatomic molecules. Because the potential energy of a molecule is not strictly parabolic except close to the average internuclear distance (Figure 17.8), vibrational levels are not, in fact, completely evenly spaced. The uneven spacing leads to additional lines in the spectrum; however, these are weak compared to the fundamental absorption line.

The number of different vibrational modes of a molecule increases with the number of atoms. The analysis of the infrared spectra of polyatomic molecules consequently becomes quite complicated. Nonetheless, force constants have been deduced for specific bonds. Typical results for a number of molecules are given in Table 17.4. It can be noted that the force constant of a bond between a given pair of atoms increases with the bond order. Evidently, a double bond between, say, two carbon atoms is stiffer than a single bond, and a triple bond is stiffer than a double bond. This trend is in line with that of the bond energies and indicates that the bond force constant is indicative of the bond strength. A similar correlation may be seen in the bonds in $NH_3$, $H_2O$, and HF, for which the force constant increases with the bond energy (Table 17.3).

**TABLE 17.4** Representative Bond Force Constants

| Molecule | Bond | Force constant (N m$^{-1}$) |
|---|---|---|
| $C_2H_6$ | —C—C— | 450 |
|  | —C—H | 490 |
| $C_2H_4$ | —C=C— | 930 |
|  | =C—H | 510 |
| $C_2H_2$ | —C≡C— | 1590 |
|  | ≡C—H | 600 |
| $NH_3$ | N—H | 710 |
| $H_2O$ | O—H | 840 |
| HF | F—H | 970 |
| $CH_3CHO$ | C=O | 1210 |
| CO | C≡O | 1903 |

Infrared (IR) spectroscopy is widely used in chemistry. Absorption spectra can be obtained by use of an IR spectrophotometer (Figure 17.9). The spectra generally consist of a large number of absorption bands. Even the spectrum of a relatively simple molecule, such as dichloromethane ($CH_2Cl_2$), has a substantial number of absorption bands (Figure 17.10). Some of the features of such a spectrum can readily be understood because many of the absorption bands correspond to the stretching of specific bonds, nearly independent of the

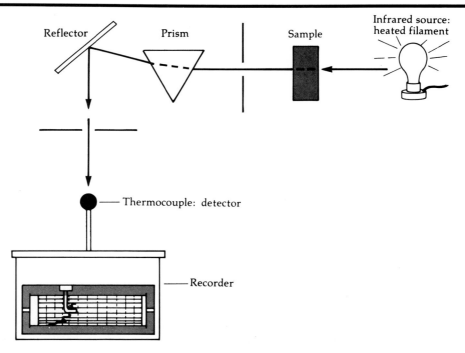

**FIGURE 17.9** Diagram of an infrared spectrophotometer. A heated filament emits a broad range of IR radiation. The sample absorbs at specific wavelengths. The transmitted radiation is dispersed into its component wavelengths by a crystal prism and is detected by a thermocouple. The spectrum can be plotted on a strip chart recorder.

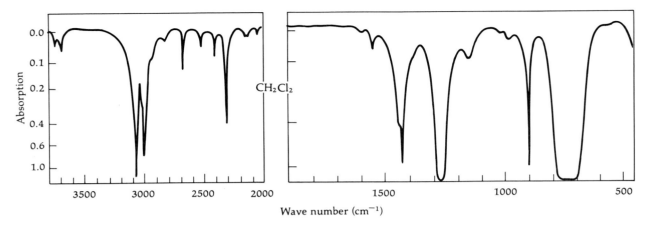

**FIGURE 17.10** Infrared absorption spectrum of dichloromethane, CH$_2$Cl$_2$. Note the change in scale at 2000 cm$^{-1}$.

molecular environment. This behavior is analogous to that of the bond energies and bond lengths, the values of which are also nearly independent of the molecular environment. Table 17.5 lists the wave numbers corresponding to the stretching of some specific bonds. The interpretation of the IR spectrum of dichloromethane in terms of the data in this table indicates that the bands at

**TABLE 17.5** Location of IR Absorption Bands Associated with Bond Stretching

| Bond | $\bar{\nu}$ (cm$^{-1}$) |
|---|---|
| —C—H | 2880–3030 |
| =C—H | 3000–3120 |
| ≡C—H | 3300 |
| C=C | 1600–1680 |
| —C≡C— | 2200–2260 |
| —C≡N | 2150–2250 |
| C=O | 1660–1870 |
| —C—Cl | 650–750 |
| —O—H | 3500–3700 |
| N—H | 3300–3500 |

≈3000 cm$^{-1}$ and at ≈700 cm$^{-1}$ can be assigned to the C—H and C—Cl bonds, respectively.

Two important uses of IR spectroscopy are in the elucidation of molecular structure and in analytical chemistry. An example of a structural problem involves the bonding of the thiocyanate ion (NCS$^-$) to compounds containing phosphorus. The thiocyanate ion can attach to the phosphorus at either end by forming an N—P bond or an S—P bond. The presence of a strong absorption band at ≈530 cm$^{-1}$ indicates the presence of an N—P bond. A typical analytical application involves the identification of substances of unknown composition. Infrared spectra are usually so complex that they constitute a unique fingerprint of a substance. If such a complex IR spectrum matches that of a known substance, the unknown is almost certain to be a sample of the known substance.

## 17.4 Electronic Spectra

*a. Transitions Involving Pi Electrons* Electronic transitions involve larger energy changes than rotational or vibrational transitions (Figure 17.3). The radiation emitted by the molecule upon deexcitation is correspondingly more energetic and generally lies in the visible or ultraviolet. Thus, the study of electronic transitions in molecules involves **visible** and **ultraviolet spectroscopy.**

The molecular orbital model provides a convenient framework for a discussion of electronic spectra. We have seen that the valence electrons in a molecule can be assigned to specific molecular orbitals (Section 16.6). Electronic transitions frequently involve the promotion of a single electron from the highest occupied molecular orbital to an empty higher-energy molecular orbital. In these instances, there is a close correspondence between the molecular orbital energy diagram and the observed spectrum. Electronic spectra can therefore be used to test molecular orbital calculations.

The electronic spectrum of ethylene (C$_2$H$_4$) illustrates this connection. Because they are less tightly held than electrons in $\sigma$ orbitals, it is the $\pi$ electrons forming the double bond that are excited most readily (see Example 17.2). Figure 17.11a shows the $\pi$-orbital energy level diagram of ethylene. When the molecule is in its electronic ground state both $\pi$ electrons occupy the $\pi$ orbital. The absorption of 175 nm radiation excites one of the electrons from the $\pi$ orbital to the $\pi^*$ orbital, in a "$\pi$ to $\pi^*$" transition. The wavelength of this radiation is very short—it lies in the far UV—indicating that the energy difference between the $\pi$ and $\pi^*$ orbitals in ethylene is substantial.

The energy of a $\pi \rightarrow \pi^*$ transition associated with a carbon–carbon double bond is substantially reduced when the double bond is part of a conjugated chain, that is, when the molecule has alternating single and double bonds (Section 16.8b). Figures 17.11b and 17.11c show the $\pi$-orbital energy level diagrams and structural formulas of butadiene (C$_4$H$_6$) and hexatriene (C$_6$H$_8$), two conjugated molecules containing two and three C=C double bonds, respectively. The molecular orbital analysis of these molecules is similar to that for benzene (Section 16.8b). Each carbon atom is attached by sigma bonds to its adjacent carbon atoms. In addition, each carbon atom has an unhybridized half-filled p orbital; the overlap of these orbitals forms the molecular $\pi$ orbitals. The number of $\pi$ orbitals is equal to the number of carbon atoms and half of the orbitals are filled with electrons when these molecules are in the ground state.

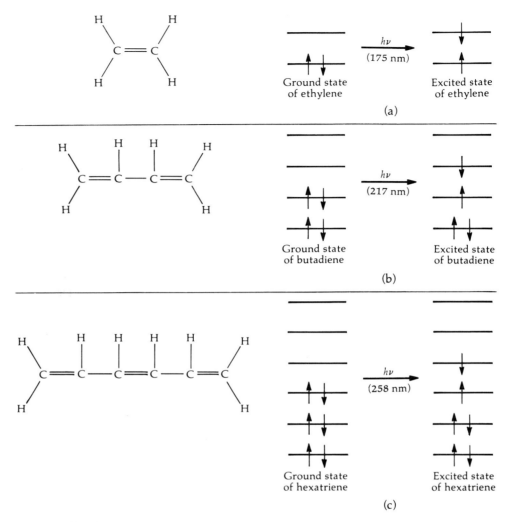

**FIGURE 17.11** Electronic transitions associated with the $\pi$ electrons in carbon–carbon double bonds ($\pi \rightarrow \pi^*$ transitions). (a) Ethylene, (b) butadiene, and (c) hexatriene. The number of electrons in each orbital is shown both for the ground state and the first excited state. The wavelength of the radiation absorbed in a transition between these two states is indicated. On the left, the structural formulas are given.

The levels associated with the $\pi$ orbitals lie increasingly close in energy as the number of double bonds increases. The evidence for this trend comes from the wavelengths of the radiation absorbed in the $\pi \rightarrow \pi^*$ transition. Note that the wavelengths increase from 175 nm for ethylene, to 217 nm for butadiene, to 258 nm for hexatriene; since $\Delta E = hc/\lambda$, the energy difference between levels must show a corresponding decrease.

The increase in the wavelength of the radiation absorbed in the $\pi \rightarrow \pi^*$ transition with the number of carbon atoms in conjugated hydrocarbons can be understood by means of the particle-in-a-box model (Section 5.9). As for benzene, the $\pi$ electrons are delocalized over the entire carbon skeleton of the molecule; the entire molecule can be treated as a box in which the $\pi$ electrons can move back and forth. The particle-in-a-box model is called the **free electron model** in this particular context. Recall that the energy levels of a particle in a one-dimensional box are given by Equation 5.39,

$$E_n = n^2h^2/8mL^2$$

where $L$ is the length of the box, $m$ is the mass of the particle, and $n$ is an integer quantum number.

To apply this equation to conjugated hydrocarbons, we must relate the quantum numbers of the levels involved in the $\pi \rightarrow \pi^*$ transition to the number of carbon atoms in the molecule. As shown in Figure 17.11, the number of levels filled with electrons when a molecule is in its ground state is $N/2$, where $N$ is the number of carbon atoms. This number is equal to the quantum number of the highest occupied level. For example, the highest occupied level in butadiene ($N = 4$) has $n = N/2 = 2$.

In the $\pi \rightarrow \pi^*$ transition, an electron is promoted from the highest occupied level to the next higher level, with quantum number $n' = (N/2) + 1$. According to the model, the energy difference between these levels is equal to

$$\Delta E = \frac{(n')^2 h^2}{8mL^2} - \frac{n^2 h^2}{8mL^2} = \frac{[(N/2) + 1]^2 h^2}{8mL^2} - \frac{(N/2)^2 h^2}{8mL^2}$$

$$\Delta E = \frac{h^2}{8mL^2} \left[ \left(\frac{N}{2} + 1\right)^2 - \left(\frac{N}{2}\right)^2 \right] = \frac{h^2}{8mL^2}(N + 1) \tag{17.23}$$

The wavelength of the absorbed radiation can be obtained from this equation by expressing $\lambda$ in terms of $\Delta E$ by means of the Bohr formula,

$$\lambda = \frac{hc}{\Delta E} = \frac{8mL^2 hc}{h^2(N + 1)} = \frac{8mL^2 c}{h(N + 1)} \tag{17.24}$$

The length of the box in which the electrons are free to move is approximately proportional to the number of carbon atoms in the molecule. This means that $L^2$ is roughly proportional to $N^2$, indicating that the wavelength varies as $N^2/(N + 1)$ and therefore increases with the number of carbon atoms. The predicted increase of $\lambda$ with $N$ is in conformity with experimental observations.

---

**Example 17.6** Octatetraene ($C_8H_{10}$) contains four conjugated double bonds. Calculate the wavelength of the $\pi \rightarrow \pi^*$ transition in this molecule. The length of the $\pi$ bond system is 0.85 nm.

The wavelength is obtained by means of Equation 17.24:

$$\lambda = \frac{8mL^2 c}{h(N + 1)}$$

$$= \frac{(8)(9.110 \times 10^{-31} \text{ kg})(0.85 \text{ nm} \times 10^{-9} \text{ m nm}^{-1})^2 (2.998 \times 10^8 \text{ m s}^{-1})}{(6.626 \times 10^{-34} \text{ J s})(9)(1 \text{ kg m}^2 \text{ s}^{-2} \text{ J}^{-1})}$$

$$= 2.6 \times 10^{-7} \text{ m} = 2.6 \times 10^2 \text{ nm}$$

Note that the $(N + 1)$ term has been set equal to 9 since the molecule has 8 carbon atoms. The observed $\pi \rightarrow \pi^*$ absorption in octatetraene occurs at a wavelength of 286 nm, in reasonably good agreement with the model. ∎

The increase in wavelength of the $\pi \to \pi^*$ transition in conjugated hydrocarbons with the number of double bonds indicates that the absorption bands must eventually fall in the visible region. This occurs when the number of carbon atoms reaches about twenty. Long conjugated chains incorporated in proteins (Section 22.6) are responsible for the perception of light by the human eye.

**b. Deexcitation of Electronic States** A number of different processes can occur following the formation of an electronically excited species by the absorption of electromagnetic radiation. Excited molecules can undergo chemical reactions. Since they possess more energy than ordinary molecules, excited molecules frequently react in different ways than ordinary molecules. The study of the reactions of electronically excited species is the subject of **photochemistry**. Many of the reactions occurring in the atmosphere (Sections 3.7 and 3.8) are photochemical, being induced by the absorption of sunlight.

Electronically excited molecules can also undergo **radiative decay,** that is, give up their excess energy by photon emission. In many instances the electronically excited molecules exist for only a brief moment, typically $10^{-9}$–$10^{-7}$ s, before they decay. The radiation emitted by such short-lived states is called **fluorescence**. The radiative decay of excited butadiene and the other molecules considered in Section 17.4a involves this process.

When a molecule absorbs enough energy for an electronic excitation, it generally also undergoes vibrational and rotational excitations. These excitations populate the closely spaced vibrational and rotational levels which are built on the excited electronic level (Figure 17.3). When a molecule in such an excited state decays back to the ground state, part of its energy is given off in the form of visible or ultraviolet radiation and part in the form of infrared radiation (Figure 17.12). The latter is associated with transitions between vibrational energy levels. As a consequence of these vibrational transitions, less energy is available for the electronic transition to the ground state than was absorbed in the electronic excitation. Since the wavelength of the emitted or absorbed radiation is inversely proportional to the transition energy, the radiation emitted in fluorescent decay is shifted to longer wavelengths relative to that of the radiation absorbed in the excitation. Fluorescent dyes are a practical application of this wavelength shift. These substances absorb radiation in the ultraviolet and fluoresce in the visible. Their vivid colors are particularly striking in a darkened room illuminated by UV lamps.

---

**Example 17.7** A certain substance absorbs 350 nm radiation and fluoresces at 465 nm. Determine (a) the energy change on absorption, (b) the energy given up in fluorescence, and (c) the energy dissipated in vibrational deexcitation. Express the answers in kJ mol$^{-1}$.

(a) The change in the energy of a species is related to the wavelength of the absorbed or emitted radiation by the Bohr formula, $\Delta E = h\nu = hc/\lambda$. Thus, the energy change on absorption is

$$\Delta E = \frac{(6.626 \times 10^{-34} \text{ J s})(2.998 \times 10^8 \text{ m s}^{-1})(6.022 \times 10^{23} \text{ mol}^{-1})(10^{-3} \text{ kJ J}^{-1})}{(350 \text{ nm})(10^{-9} \text{ m nm}^{-1})}$$

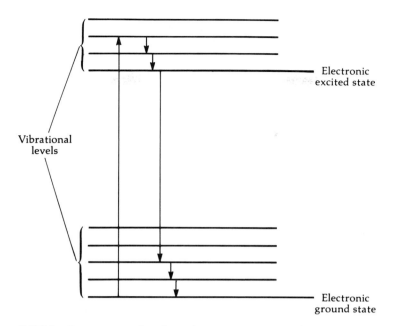

**FIGURE 17.12** Example of an electronic transition from the ground state to the first excited electronic state and back. Each electronic state has vibrational levels built on it. The electronic excitation (upward arrow) populates a vibrational excited state. The decay of this state involves both an electronic transition (black downward arrow) and vibrational transitions (colored arrows). Rotational transitions account for a negligibly small energy change and are not shown.

$\Delta E = 342$ kJ mol$^{-1}$

(b) The energy given up in fluorescence is

$$\Delta E = \frac{(6.626 \times 10^{-34} \text{ J s})(2.998 \times 10^8 \text{ m s}^{-1})(6.022 \times 10^{23} \text{ mol}^{-1})(10^{-3} \text{ kJ J}^{-1})}{(465 \text{ nm})(10^{-9} \text{ m nm}^{-1})}$$

$\Delta E = 257$ kJ mol$^{-1}$

(c) The energy dissipated in vibrational transitions is equal to the difference between the energy gained on absorption and that lost on fluorescence:

$\Delta E = 342$ kJ mol$^{-1}$ $-$ 257 kJ mol$^{-1}$ $=$ 85 kJ mol$^{-1}$ ∎

While most electronically excited molecules deexcite rapidly, some electronically excited molecules deexcite slowly. In contrast to fluorescence, the light emitted in the decay of the excited state persists after the source of the exciting radiation has been switched off and is visible as a distinct afterglow which can last for seconds or even longer. This slowly fading radiation is called **phosphorescence.**

The difference between fluorescence and phosphorescence arises from differences in the way the molecular orbitals of the species exhibiting these decay modes are filled with electrons (Figure 17.13). Consider the possible

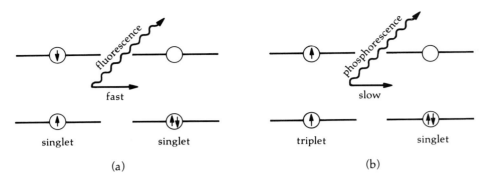

**FIGURE 17.13** The difference between (a) a singlet–singlet and (b) a triplet–singlet electronic transition.

changes in the net electron spin of a molecule upon electronic excitation. If all the electron spins are paired, the net electronic spin $S$ is zero. Such a state is called a **singlet state**. This designation is based on the fact that a state with angular momentum $S$ has $2S + 1$ substates (Section 5.11). When $S = 0$, $2S + 1 = 1$, hence the designation "singlet." Figure 17.13a shows a transition between two singlet states. Such a transition occurs rapidly and the emitted radiation is classified as fluorescence.

If two electrons in a molecule have parallel spins, the net spin of the molecule is $S = 1$. A state characterized by $S = 1$ is called a **triplet state** because $2S + 1 = 3$. Figure 17.13b shows a transition between a triplet state and a singlet state. Such transitions occur relatively slowly because a change in spin is a very unfavored process. An excited molecule undergoing a triplet to singlet transition (or vice versa) requires a relatively long time to release its energy. The radiation emitted in such a transition is consequently classified as phosphorescence.

Numerous examples of fluorescent decay can be given. Figure 17.11 shows that the electronic transitions in ethylene and the other depicted molecules can be classified as singlet → singlet transitions. Consequently, the excited molecules decay by emission of fluorescent radiation.

Phosphorescent decay occurs less commonly than fluorescent decay. Electronic transitions in $O_2$ provide an illustration of this phenomenon. Figure 17.14, which is adapted from Figure 16.17, shows the filling of the highest lying molecular orbitals of $O_2$ in the ground state and in two excited states. Since normal $O_2$ contains two unpaired electrons with parallel spin, the ground state can be classified as a triplet state. Oxygen has two singlet excited states, in which the two unpaired electrons are spin-paired. The radiative decay of either singlet state to the triplet ground state involves phosphorescence.

While the customary method of inducing electronic excitations in molecules is by exposure to ultraviolet or visible light, it is possible in certain instances to form excited molecules in chemical reactions. For example, singlet oxygen $O_2^*$, is formed by the action of hypochlorite on hydrogen peroxide, as in the reaction

$$OCl^-(aq) + H_2O_2(aq) \rightarrow H_2O + Cl^- + O_2^*(aq)$$

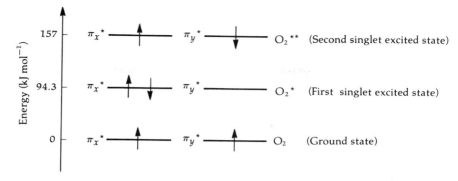

**FIGURE 17.14** The ground triplet state and excited singlet states of $O_2$. Only the highest occupied molecular orbitals are shown (see Figure 16.17 for the complete molecular orbital energy diagram of ground-state oxygen). The relative energies of the three states are given.

When two excited oxygen molecules collide, they can both revert to the ground state with the emission of electromagnetic radiation. The weak red light that is observed is called **chemiluminescence**. The name indicates that the radiation is emitted as the result of a chemical reaction.

## 17.5 The Laser

The laser is a device with increasingly widespread applications in chemistry and other basic sciences, as well as in technology. The term LASER is an acronym for Light Amplification by the Stimulated Emission of Radiation. As indicated by this phrase, laser action involves the amplified emission of light. Laser light differs from ordinary light in being **coherent**: its waves all have the same phase (Section 5.2), the same wavelength, and they travel in the same direction. Laser light can be obtained in the form of extremely short pulses, lasting for only a few picoseconds or less. These properties make the laser a powerful tool for probing the behavior of molecules on the time scale in which many molecular processes actually take place.

The laser can be understood on the basis of what we have learned about excited states of molecules. Figure 17.15 summarizes some of the processes that are of importance to an understanding of laser action. To simplify the discussion we focus on just two states of a molecule, the electronic ground state and an electronic excited state. Figure 17.15a represents the molecule in its ground state. A photon whose energy is just equal to the energy difference $\Delta E$ between the two levels strikes the molecule. The photon is absorbed, an electronic transition occurs, and the molecule is raised to its excited state. A specific example of this process is the $\pi \rightarrow \pi^*$ transition in ethylene when 175 nm radiation is incident, as already shown in Figure 17.11a. In Figure 17.15b the molecule returns to the ground state by the emission of a photon. Once again, the energy of the photon is equal to the difference between the energies of the two levels. Since this process occurs without outside stimulus, it is called **spontaneous emission**. Fluorescence and phosphorescence are examples of spontaneous emission.

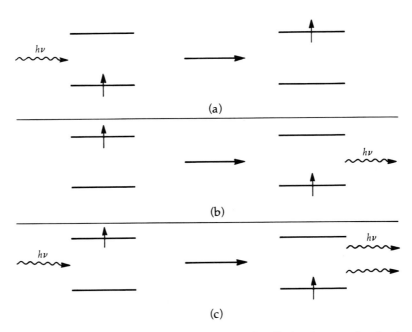

**FIGURE 17.15** Absorption and emission of radiation by a molecule characterized by two levels. (a) Absorption, (b) spontaneous emission, and (c) stimulated emission.

Figure 17.15c shows the process which is responsible for laser action: **stimulated emission.** The molecule is initially in the excited state, for instance as a result of the absorption process illustrated in Figure 17.15a. A photon with energy $\Delta E$ is incident upon the excited molecule. The interaction stimulates the molecule to give up its excess energy in the form of a second photon of energy $\Delta E$ and the electron returns to the ground state. This process demonstrates the key features of laser action. First, amplification: one photon is incident but two photons are emitted. Second, coherence: the two emitted photons are in phase and travel in the same direction.

Under normal conditions, excited molecules decay by spontaneous rather than stimulated emission. The observed light is incoherent: photons are emitted out of phase, in random directions, over a range of time, with no amplification. In order for stimulated emission to occur there must be more molecules in the upper of the two energy levels. This is an unusual way for the molecules to be distributed. Under normal equilibrium conditions the number of molecules in the lower energy level is much larger than that in the upper level. In order for laser action to be possible, the population of the two levels must be inverted so that more molecules reside in the upper level. (In practice, both levels usually are excited states of the molecule.) Population inversion can be achieved by depositing energy in the system, for example, by the use of an intense light source or a chemical reaction. This process is known as pumping. Once this inversion has been achieved, stimulated emission of laser light is possible.

The sequence of steps is shown in Figure 17.16. The normal equilibrium population distribution displayed in step (1) is inverted by pumping in (2).

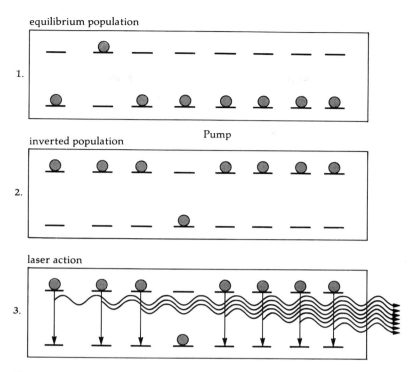

**FIGURE 17.16** Steps in a sequence leading to laser action.

Laser action is demonstrated in (3): spontaneous emission by one of the many excited molecules stimulates emission from the other excited molecules. The result is an amplified and coherent pulse of laser light.

The preceding description has focused on the essential features of the process, neglecting consideration of the various complications that are an inevitable part of such a complex phenomenon. In spite of these complexities, it is surprising that the laser was not invented until some 30 years ago. In 1954 Charles Townes (b. 1915) discovered stimulated emission in the microwave region resulting from the deexcitation of molecular rotational levels and, hence, called it maser action. The first laser using visible light was developed by Theodore Maiman (b. 1927) in 1960. That such a major development in a field as mature as molecular spectroscopy could be achieved so recently is an indication that chemistry still provides ample opportunities for truly creative discoveries.

# Conclusion

The theoretical models of covalent bonding must be confronted with experimental data on bond properties if their validity is to be established. In this chapter we have examined bond energies, bond lengths, bond force constants, and the decay of excited molecules. These bond properties have been obtained

## Problems

1. Using bond energies calculate $\Delta H_f^\circ$ of $H_2O(g)$. Compare your result with the value in Appendix 4.
2. Use average bond energies to estimate reaction enthalpies for the following reactions. Assume that all species are in the gas phase.
   - (a) $H_2CO + H_2 \rightarrow H_3COH$
   - (b) $C_2H_4 + Cl_2 \rightarrow C_2H_4Cl_2$

$$(H-\underset{\underset{H}{|}}{\overset{\overset{Cl}{|}}{C}}-\underset{\underset{H}{|}}{\overset{\overset{Cl}{|}}{C}}-H)$$

   - (c) $C_2H_4 + H_2O \rightarrow CH_3CH_2OH$
   - (d) $C_2H_4 + HBr \rightarrow CH_3CH_2Br$
3. On the basis of standard enthalpies of formation (Appendix 4), determine the average SO bond energy in (a) $SO_2$ and (b) $SO_3$.
4. Use average bond energies to estimate reaction enthalpies for the following gas phase reactions:
   - (a) $2H_2 + N_2 \rightarrow N_2H_4$
   - (b) $H_2 + Br_2 \rightarrow 2HBr$
   - (c) $3I_2 + N_2 \rightarrow 2NI_3$
   - (d) $CH_4 + Cl_2 \rightarrow CH_3Cl + HCl$
5. Use average bond energies and thermochemical data to calculate the standard enthalpy of formation of acetaldehyde vapor, $CH_3CHO(g)$.
6. Calculate the standard enthalpy of the following reaction by the bond energy method: $CO(g) + \frac{1}{2}O_2(g) \rightarrow CO_2(g)$. Compare the result with that obtained thermochemically.
7. Phosphorus vapor contains $P_4$ molecules in which the attached P atoms are at the corners of a tetrahedron. Determine the P—P bond energy by considering the reaction $P_4(g) + 6Cl_2(g) \rightarrow 4PCl_3(g)$. Determine $\Delta H_{298}^\circ$ for this reaction using thermochemical data (Appendix 4). Use a value of 318 kJ mol$^{-1}$ for the P—Cl bond energy.
8. Use the bond energy method to estimate the enthalpies of combustion of the following fuels: (a) methane, (b) acetylene, (c) methanol, and (d) hydrogen. Assume that all reactants and products are gaseous. Which fuel delivers the most heat per gram of fuel?
9. Calculate the resonance energy of $NO_2$ on the basis of tabulated bond energies (Table 17.2) and thermochemical data (Appendix 4).
10. Cyclic hydrocarbons consist of a ring of carbon atoms, with two hydrogen atoms bonded to each carbon. The simplest of these compounds is cyclopropane ($C_3H_6$), which has the structure

    Use the bond energy method to calculate the standard enthalpy of formation of (a) cylopropane, $C_3H_6(g)$ and (b) cyclopentane, $C_5H_{10}(g)$. Compare your results with the values tabulated in Appendix 4. Comment on the agreement or disagreement between the two comparisons in terms of the structure of the molecules.
11. What is the reduced mass (in grams) of the following molecules (a) $^1H^1H$, (b) $^1H^2H$, (c) $^1H^{19}F$, and (d) $^1H^{35}Cl$? Assume that the mass of each atom (in u) is given by its mass number.
12. The rotational absorption spectrum of HI consists of a series of lines separated by 13.10 cm$^{-1}$. What is the bond length?
13. The microwave absorption spectrum of HCl (actually $^1H^{35}Cl$) shows lines at 104.1, 124.7, 145.3, and 165.9 cm$^{-1}$. (a) What are the wavelengths of these lines? (b) What is the moment of inertia of the molecule? (c) What is the bond length? Use the mass numbers for the isotopic masses.
14. Assuming that the bond length of DCl ($^2H^{35}Cl$) is the same as that of HCl (Problem 13) predict the wave number spacing of the DCl microwave ab-

sorption lines. Use the mass numbers for the isotopic masses.

15. A microwave absorption spectrometer was mounted on a spacecraft in order to search for the presence of NO in the atmosphere of Venus. Given a bond length of 1.17 Å, at what wavelengths will the three lowest energy lines be observed?

16. The microwave spectrum of KCl vapor, which can be measured at high temperatures, shows an absorption at 7687.94 MHz that can be identified with the $J = 0 \to J = 1$ transition in the $^{39}K^{35}Cl$ molecule. Determine (a) the moment of inertia of the molecule and (b) the bond length. Use the mass numbers for the isotopic masses.

17. Arrange the following molecules in order of increasing length of the CN bond: $CH_3NH_2$, HCN, and $CH_2NH$. Explain your answer.

18. Use the atomic radii in Table 6.5 to estimate bond lengths in the following molecules: $H_2$, CO, $N_2$, and $F_2$. Compare with the experimental values in Table 17.1. Explain any sizeable discrepancies.

19. The fundamental vibrational frequency of HBr ($^1H^{81}Br$) is at 2650 cm$^{-1}$. Find the bond force constant. What is the zero-point energy of HBr? Use the mass numbers for the isotopic masses.

20. Using the data in Problem 19, predict the fundamental absorption frequency of deuterium bromide ($^2H^{81}Br$). Use the mass numbers for the isotopic masses.

21. The potential energy of $H_2$ at the average internuclear separation of the hydrogen atoms is $-457.9$ kJ mol$^{-1}$, and the zero-point energy is 25.9 kJ mol$^{-1}$. (a) What is the dissociation energy of $H_2$? (b) What is the value of the H—H bond force constant? (c) The potential energy and bond force constant of $D_2$ ($^2H_2$) are the same as the corresponding quantities for $H_2$. What is the dissociation energy of $D_2$? Use the mass numbers for the isotopic masses.

22. The force constant of the Cl—Cl bond is 320 N m$^{-1}$. (a) What is the separation (in cm$^{-1}$) of the vibrational energy levels of $^{35}Cl_2$? (b) What are the zero-point energies of $^{35}Cl_2$ and $^{37}Cl_2$? (c) The dissociation energy of $^{35}Cl_2$ is 239 kJ mol$^{-1}$. What is the dissociation energy of $^{37}Cl_2$? The potential energy curve of the molecule is independent of mass. Use the mass numbers for the isotopic masses.

23. The energy difference between adjacent vibrational levels in $O_2$ corresponds to a wave number of 1580.4 cm$^{-1}$. Calculate the energy difference in kJ mol$^{-1}$ and compare the value with (a) the dissociation energy of $O_2$ and (b) the average translational kinetic energy of $O_2$ at 25°C.

24. The infrared absorption spectrum of CO has an absorption band at 2170 cm$^{-1}$. What is the force constant of the CO bond?

25. A certain dye has an electronic absorption band at 450 nm. The molecule has a conjugated series of double bonds. Calculate the length of the pi bond system on the assumption that the transition is from the $n = 1$ to the $n = 2$ orbital for an electron free to move in one dimension.

26. Use the free electron model to calculate the wavelength of the $\pi \to \pi^*$ transition in benzene. The carbon–carbon bond length is 1.39 Å and the $\pi$ electrons are free to move linearly over all six bonds.

27. The structure of polyenes, which are straight-chain conjugated hydrocarbons, can be represented by the general formula

$$H\text{—}[CH\text{=}CH\text{—}]_n\text{—}H$$

For example, butadiene corresponds to $n = 2$. The electronic spectrum of polyenes can be described by considering the transitions of the $2n$ $\pi$ electrons in a one-dimensional box of length $nL$, where $L = 2.81$ Å is the length of the CH=CH— unit. What is the minimum value of $n$ for which a polyene absorbs in the visible ($\lambda > 400$ nm)? What is the formula of this polyene?

28. An electronic transition in CO is responsible for an absorption band in the far UV. This band can be resolved into narrow lines at wave numbers of 64,703, 66,231, 67,675 cm$^{-1}$, and so on. These lines can be interpreted as transitions from the $v = 0$ vibrational state of the electronic ground state to the $v = 0, 1, 2, \ldots$ vibrational levels of the excited electronic state. (a) What is the energy difference between $v = 0$ and $v = 1$ vibrational levels of the excited electronic state? (b) What is the bond force constant associated with this transition? (c) Compare this value with that for the molecule in its electronic ground state (Table 17.4). Comment on the difference.

# 18

Sections 18.1–18.4

# Intermolecular Interactions

Atoms are held together in molecules by covalent bonds. But what are the forces between molecules in condensed phases—liquids and solids? We have already seen the effect of the interactions between molecules in the deviations from ideal gas behavior displayed by real gases at high pressures and low temperatures (Section 4.7). These interactions are more fully displayed in the liquid and solid states. Many of the properties of condensed states of matter, such as the boiling points of liquids, are in large measure determined by the strength of intermolecular interactions. These interactions are generally much weaker than covalent bonds but nonetheless display great variability in strength. The practical consequences of the strength of the intermolecular forces present in a particular substance can be far reaching. For example, liquid $H_2O$ is held together by relatively strong hydrogen bonds and has a normal boiling point of 100°C. Replacement of the oxygen atom in water by a sulfur atom gives $H_2S$. Liquid $H_2S$ is held together by weaker forces than liquid $H_2O$, and its normal boiling point is only −60.7°C. If $H_2O$ resembled $H_2S$ in its boiling point, the Earth would obviously be a far different place than it is.

## Interactions between Molecules or Ions

### 18.1 The Ion Pair Interaction

In Section 17.1 we examined the strength of covalent bonds, and in the following sections we examine the strength of intermolecular interactions. First, however, for purposes of comparison, we look at the strength of the attraction between a pair of oppositely charged ions, essentially the strength of a single, isolated ionic bond. The interaction between a single pair of ions is most readily seen in the gas phase, for in an ionic solid, a large number of ions interact with each other (Section 20.10b). At elevated temperatures, ionic compounds vaporize and form ion pairs, such as Na⁺Cl⁻. The dipole moments of the gaseous alkali halides are large—about a factor of ten larger than those of the hydrogen halides (Table 15.1)—and provide the evidence for the formation of ion pairs. The formation of gaseous ion pairs is also observed for other compounds formed by elements with large differences in electronegativity, such as the alkaline earth halides and the alkali and alkaline earth oxides.

An ion pair consists of the two ions held closely together by an ionic bond. The formation of an ion pair from the separated atoms can be represented by a reaction such as

$$\text{Na}(g) + \text{Cl}(g) \rightarrow \text{Na}^+\text{Cl}^-(g) \tag{18.1}$$

The energy of the Na⁺Cl⁻ bond is equal to $\Delta H°$ for the reverse reaction—the dissociation of the gaseous ion pair into the gaseous atoms (Section 17.1a). To evaluate the bond energy, reaction 18.1 must be broken down into reactions for which the enthalpy changes can be evaluated more readily. Reaction 18.1 can be written as the sum of the following two processes:

$$\text{Na}(g) + \text{Cl}(g) \rightarrow \text{Na}^+(g) + \text{Cl}^-(g) \tag{18.2}$$

and

$$\text{Na}^+(g) + \text{Cl}^-(g) \rightarrow \text{Na}^+\text{Cl}^-(g) \tag{18.3}$$

Reaction 18.2 involves the formation of gaseous ions from gaseous atoms, while in reaction 18.3 the gaseous ions come together to form the ion pair.

The enthalpy change in reaction 18.2 can be evaluated by noting that this reaction can be written as the sum of the following two processes:

| | | |
|---|---|---|
| $\text{Na}(g) \rightarrow \text{Na}^+(g) + e^-$ | $\Delta H° = I_1$ | (18.4) |
| $\text{Cl}(g) + e^- \rightarrow \text{Cl}^-(g)$ | $\Delta H° = -\varepsilon$ | (18.5) |
| $\text{Na}(g) + \text{Cl}(g) \rightarrow \text{Na}^+(g) + \text{Cl}^-(g)$ | $\Delta H° = I_1 - \varepsilon$ | |

where $I_1$ is the first ionization energy and $\varepsilon$ is the electron affinity. (The difference between enthalpy and energy may be ignored in the present context.) Using the values in Table 6.4, we find that $\Delta H°$ for reaction 18.2 is $\Delta H° = (495.6 - 348.8)$ kJ mol⁻¹ = 146.8 kJ mol⁻¹. It is worth noting that the standard free energy change $\Delta G°$ for reaction 18.2 is essentially equal to $\Delta H°$ because the entropy difference between the gaseous atoms and ions is very small. Since $\Delta H°$ is positive, so is $\Delta G°$, and the reaction does not occur spontaneously at standard conditions. Similar positive enthalpies and free energies are found for other reactions of this type indicating that they too do not occur spontaneously. Evidently, the formation at standard conditions of gaseous ions from gaseous atoms of elements with a large difference in electronegativity does not occur spontaneously even though these ions have a filled valence shell.

If reaction 18.1 is to have $\Delta H° < 0$, as required by the stability of Na⁺Cl⁻, it follows that reaction 18.3 must be strongly exothermic. In this reaction, two oppositely charged ions come together to form an ion pair. The force responsible for the attraction between oppositely charged ions is the Coulomb force. The potential energy associated with this force is given by Equation 5.46. For the case of two ions with charges of magnitude $Z_+$ and $Z_-$, this equation can be written as

$$U = -\frac{Z_+ Z_-}{\alpha d} \tag{18.6}$$

where $\alpha$ is a proportionality constant and $d$ is the distance between the ions. If the ions are singly charged, this equation reduces to $U = -e^2/\alpha d$ (Equation 5.46), where $e$ is the electronic charge. Figure 18.1 shows a plot of the potential energy of an alkali halide ion pair. Starting at the right from a value of zero for the separated ions, the potential becomes increasingly negative as the ions approach each other, reflecting the increasing strength of the Coulomb force. The decrease in energy follows the $1/d$ law until the two ions are sufficiently close for their electron clouds to overlap. At this point the potential changes from attractive to repulsive because the dominant interaction at short distances is the repulsion between the inner electron shells as well as that between the positively charged nuclei. The repulsive energy can be represented by the empirical function $b/d^n$, where the exponent $n$ can range from approximately 8 to 12 and $b$ is a proportionality constant. The total potential energy is equal to the sum of the attractive and repulsive terms and is given by

$$U = -\frac{Z_+ Z_-}{\alpha d} + \frac{b}{d^n} \tag{18.7}$$

The value of $b$ can be determined by noting that the combination of an attractive potential at long distances and a repulsive potential at short distances results in a potential energy curve with a minimum at some particular internuclear distance, as shown in Figure 18.1. At this distance, designated $d_0$, the derivative of $U$ with respect to $d$ must be zero:

$$\frac{dU}{dd} = 0 = \frac{Z_+ Z_-}{\alpha d_0^2} - \frac{nb}{d_0^{n+1}}$$

Solving for $b$, we obtain

$$b = \frac{Z_+ Z_-}{\alpha n} (d_0)^{n-1}$$

and substituting this result into Equation 18.7 we get $U_0$, the potential energy evaluated at the most stable internuclear separation:

$$U_0 = -\frac{Z_+ Z_-}{\alpha d_0} + \frac{Z_+ Z_-}{\alpha n d_0}$$

$$U_0 = -\frac{Z_+ Z_-}{\alpha d_0}\left(1 - \frac{1}{n}\right) \tag{18.8}$$

The distance $d_0$ is the bond length and the potential energy $U_0$ corresponds to the enthalpy change of reaction 18.3. Since $n$ is in the range of 8 to 12, the energy associated with ion–ion repulsion reduces the Coulomb energy by about 10%.

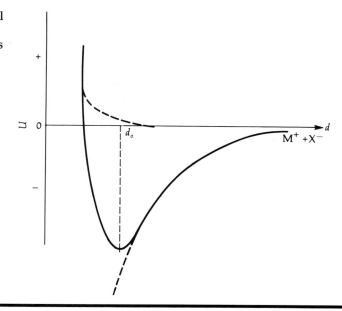

**FIGURE 18.1** Potential energy $U$ of a gaseous alkali halide ion pair as a function of internuclear distance $d$. The dashed curves show the separate contributions of the attractive and repulsive terms and the solid curve is their sum.

The NaCl bond energy can be evaluated by combining the enthalpy changes in reactions 18.3–18.5. The process is illustrated in Figure 18.2. Note that the direct reaction, that is, reaction 18.1, is expressed as a sum of three other reactions. This procedure is an application of Hess's law (Section 10.8), which states that the enthalpy change in a reaction is independent of path. The reactions involved in the preceding calculation form a closed cycle. Calculations based on such reaction cycles are frequently used to determine the stability of various species. Max Born (1882–1970) and Fritz Haber (1868–1934) performed the first such calculation in order to determine the stabilities of ionic crystals (Section 20.10c), a calculation now known as a **Born–Haber cycle**.

Figure 18.2 indicates that the bond energy of an alkali halide ion pair is given by the expression

$$D = -I_1 + \varepsilon - U_0 \tag{18.9}$$

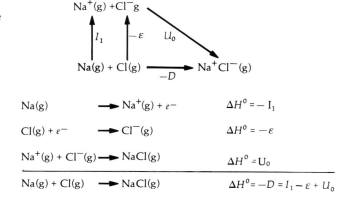

**FIGURE 18.2** Born–Haber-type cycle for the evaluation of the bond energy in gaseous NaCl.

To evaluate this expression for NaCl(g), we need only evaluate the potential energy $U_0$; the other quantities have already been evaluated. Using the NaCl(g) bond length of 2.36 Å and a value of 10 for the exponent $n$ we obtain by means of Equation 18.8,

$$U_0 = -\frac{Z_+ Z_-}{\alpha d_0}\left(1 - \frac{1}{n}\right)$$

$$= -\frac{(1.602 \times 10^{-19}\ \text{C})^2 (6.022 \times 10^{23}\ \text{mol}^{-1})(1 - 1/10)}{(1.113 \times 10^{-10}\ \text{C}^2\ \text{N}^{-1}\ \text{m}^{-2})(1\ \text{N m/J})(10^3\ \text{J kJ}^{-1})(2.36 \times 10^{-10}\ \text{m})}$$

$$= -529.5\ \text{kJ mol}^{-1}$$

where the magnitudes of the ionic charges $Z_+$ and $Z_-$ are set equal to the electronic charge. Combining this result with the values of $I_1$ and $\varepsilon$, we obtain

$$D = (-495.6 + 348.8 + 529.5)\ \text{kJ mol}^{-1} = 383\ \text{kJ mol}^{-1}$$

This calculated value is a few percent smaller than the experimental bond energy of NaCl listed in Table 18.1 (412 kJ mol$^{-1}$). The discrepancy reflects the approximate form of Equation 18.7 for the energy of repulsion, the effect of ion polarizabilities (Section 16.3), and other factors.

The ion pair bond energies listed in Table 18.1 are approximately 400–500 kJ mol$^{-1}$. These values may be compared with the covalent bond energies of diatomic molecules (Table 17.1). While the covalent bond energies are more variable, the main conclusion to be drawn from this comparison is that these two very different types of chemical bonds have comparable strengths.

Although, as shown in Equation 18.9, the bond energy of an ion pair is determined by the combined values of the Coulomb energy, the repulsion energy, the ionization energy, and the electron affinity, the last two of these quantitites tend to cancel each other to a significant extent, leaving the Coulomb energy as the main net contributor to the bond energy. Since the Coulomb energy is inversely proportional to the distance between the nuclei of the two ions, there should therefore be an inverse correlation between bond energy and bond length. The correlation should be particularly noticeable for ion pairs containing one common ion, such as the potassium halides, since the number of variables is reduced thereby. The data in Table 18.1 do indeed bear out this expectation, as may be noticed, for instance, in the inverse variation of the bond energies and bond lengths of the potassium or lithium halides. Since

**TABLE 18.1   Bond Energies and Bond Lengths in Gaseous Alkali Halide Ion Pairs**

| Ion pair | Bond energy (kJ mol$^{-1}$) | Bond length (Å) |
|---|---|---|
| LiF | 568 | 1.51 |
| KF | 497 | 2.17 |
| LiCl | 474 | 1.95 |
| CsCl | 426 | 2.91 |
| LiBr | 423 | 2.17 |
| KCl | 422 | 2.67 |
| NaCl | 412 | 2.36 |
| KBr | 382 | 2.82 |

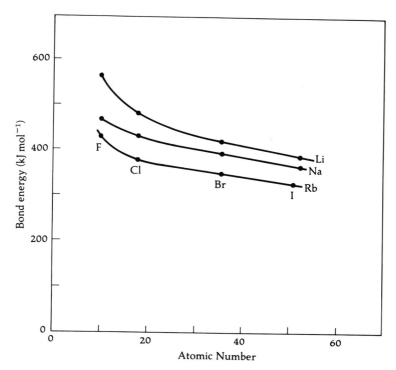

**FIGURE 18.3** Variation of the bond energy of gaseous alkali halides with halogen atomic number.

the bond lengths are approximately equal to the sums of the ionic radii, which increase with atomic number within a group, the bond energies of the halides of a given alkali metal should decrease with increasing halogen atomic number. Figure 18.3 shows that this correlation is indeed observed.

**Example 18.1** Use a Born–Haber-type cycle applied to the formation of gaseous KCl from the gaseous atomic elements to determine the KCl bond length. Compare with the value in Table 18.1.

The following cycle, similar to that in Figure 18.2, may be used:

$$K^+(g) + Cl^-(g)$$
$$\uparrow I_1 \quad \uparrow -\varepsilon \quad \searrow U_0$$
$$K(g) + Cl(g) \longrightarrow KCl(g)$$
$$\Delta H° = -D$$

The Coulomb energy corrected for repulsion, $U_0$, is equal to

$$U_0 = \varepsilon - I_1 - D$$

(also see Equation 18.9). The quantities on the right side of the equation have been tabulated:

$$I_1(K) = 418.8 \text{ kJ mol}^{-1} \quad \text{(Table 6.4)}$$
$$\varepsilon(Cl) = 348.8 \text{ kJ mol}^{-1} \quad \text{(Table 6.4)}$$
$$D(KCl) = 422 \text{ kJ mol}^{-1} \quad \text{(Table 18.1)}$$

Therefore,
$$U_0 = (348.8 - 418.8 - 422) \text{ kJ mol}^{-1} = -492 \text{ kJ mol}^{-1}$$

Knowing $U_0$, the bond length $d_0$ may be obtained by means of Equation 18.8,
$$U_0 = -\frac{Z_+ Z_-}{\alpha d_0}\left(1 - \frac{1}{n}\right)$$

$$d_0 = -\frac{(1.602 \times 10^{-19} \text{ C})^2(1 - \tfrac{1}{10})(6.022 \times 10^{23} \text{ mol}^{-1})}{(1.113 \times 10^{-10} \text{ C}^2 \text{ N}^{-1} \text{ m}^{-2})(1 \text{ N m/J})(10^3 \text{ J kJ}^{-1})(-492 \text{ kJ mol}^{-1})}$$

$$= 2.54 \times 10^{-10} \text{ m} = 2.54 \text{ Å}$$

The result agrees fairly well with the experimental value of 2.67 Å. ∎

## 18.2 Van der Waals Forces

The term **van der Waals forces** is applied to several distinct types of relatively weak forces between molecules. The exact nature of these forces depends on whether the molecules are polar. However, they all share the common feature of being responsible for the deviations from ideal gas behavior observed at high pressures and low temperatures, and for the eventual condensation of molecular gases at sufficiently low temperatures.

***a. Interaction between Instantaneous and Induced Dipoles*** Nonpolar molecules, such as the homonuclear diatomic molecules, interact in a rather subtle way with each other. Molecules of this type do not have a dipole moment and, consequently, there can be no permanent separation of charge between the two ends of each molecule. Nonetheless, the motion of the electrons inevitably leads to fluctuations in the charge distribution, such as those depicted in Figure 18.4. At a given instant, the electron density is greater on one side of a molecule than on the other. In effect, the molecule becomes an instantaneous dipole, in which one end acquires a negative charge and the other a positive charge. This charge asymmetry induces a polarization in the electron cloud of a nearby molecule. Thus, the negatively charged side of one molecule causes the closest side of a neighboring molecule to acquire a positive charge. The two molecules acquire temporary dipole moments and attract each other electrostatically because of the proximity of the oppositely charged ends.

The interaction depicted in Figure 18.4 can last only for the briefest of moments because the electron density distribution continually changes. However, the resulting fluctuations in the charge distributions of neighboring molecules are, to a certain extent, synchronized with each other. That is, if the charge asymmetry in molecule A reverses itself so that the side which was negative now becomes positive, a similar change tends to occur in molecule B. The net effect of these charge fluctuations is that neighboring molecules experience a small mutual attraction. A similar interaction occurs between nonbonded atoms, such as those of the noble gases.

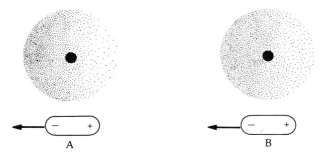

**FIGURE 18.4** Interaction between nonpolar molecules. A fluctuation in the electronic charge distribution of molecule A induces charge polarization in molecule B. The two dipoles attract each other electrostatically.

This interaction between induced dipoles is also known as the **London dispersion force,** after F. London, who performed the first calculation of its magnitude. The interaction is found to be very weak when compared to that between covalently bonded atoms and decreases sharply with increasing interatomic distance $d$. The potential energy has the form

$$U = -A/d^6 \tag{18.10}$$

where $A$ is a proportionality constant. The inverse dependence of the potential energy on such a high power of the distance shows that it rapidly approaches zero with increasing $d$. Since a molecule that is more readily polarizable will more readily acquire an induced dipole moment, the magnitude of the attractive London interaction varies with the polarizability. The **polarizability** (Section 16.3) is a measure of the distortion of the electron cloud of an atom induced by the electric field of a nearby atom. Atomic electrons can be shifted most easily when they are far from the nucleus and when they are loosely held. On average, electrons are relatively far from the nucleus of a large atom, with large atomic number or mass. They are loosely held in an atom with low ionization energy. Thus, the polarizability of an atom varies approximately as the ratio of the atomic number to the ionization energy, $Z/I$. The shape of a molecule can also affect its polarizability. Spherical molecules make contact over a smaller region than elongated molecules and consequently are less polarizable.

*b. Dipole–Dipole Interactions* When two polar molecules are in proximity, there is an interaction between their two dipoles. This interaction can be either attractive or repulsive depending on the relative orientation of the two molecules, as shown in Figure 18.5. When the two molecules approach so that their oppositely charged ends are close (Figure 18.5a), they attract each other. When the two molecules approach so that their like charge ends are in proximity (Figure 18.5b), they repel each other. If both arrangements were equally probable, the net interaction would average to zero. Since the attractive orientations are energetically more favorable, they tend to predominate. Consequently, the net effect of dipole–dipole interactions is attractive. The dipole–dipole interaction acts only over short distances, and the net attraction shows a similar dependence on distance as the London interaction. As might be expected, the in-

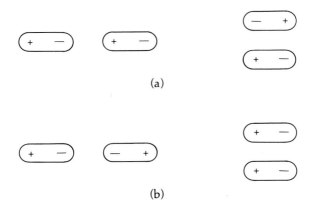

**FIGURE 18.5** Interaction between polar molecules. (a) Attractive interaction—the two dipoles are either in a head to tail orientation (left) or in an antiparallel orientation (right). (b) Repulsive interaction—the two dipoles are either in a head to head orientation (left) or in a parallel orientation (right).

teraction becomes increasingly strong with increasing value of the dipole moment.

Polar molecules also interact by a **dipole-induced dipole interaction.** As in the case of nonpolar molecules, the charge asymmetry in one molecule polarizes and consequently attracts nearby molecules. This interaction depends on the polarizabilities of the molecules, while the dipole–dipole interaction depends on their dipole moments. However, since both interactions show the same dependence on distance, they can be lumped together and treated as a single interaction, with the potential energy given by Equation 18.10.

*c. Repulsive Interactions* The interactions considered so far are attractive and, in the absence of other effects, would cause the molecules to squeeze ever more tightly together. However, when molecules approach too closely, the electrons in the filled orbitals of the inner atomic shells begin to overlap and strongly repel each other. The electrostatic repulsion between particles of like charge is enhanced by the operation of the Pauli exclusion principle: electrons with the same spin cannot occupy the same orbital (i.e., the same region of space) and strongly repel each other. The repulsion decreases rapidly with increasing distance, and the potential energy can be represented by a function such as

$$U = B/d^{12} \tag{18.11}$$

where $B$ is a proportionality constant and the positive sign of $U$ indicates that the potential is repulsive (an exponential function such as $ae^{-bd}$, where $a$ and $b$ are constants, also can be used). Note that the repulsive interaction drops off far more steeply with increasing intermolecular distance than the attractive interaction. Therefore, at moderate distances the attractive forces are stronger than the repulsive forces, and there is a net attraction between molecules.

*d. The Lennard–Jones Potential* The overall potential energy between two molecules is given by the sum of the attractive and repulsive terms,

$$U = -\frac{A}{d^6} + \frac{B}{d^{12}} \tag{18.12}$$

This potential energy function is known as the **Lennard–Jones potential** or, owing to its characteristic dependence on distance, as the **(6, 12) potential**. The parameters $A$ and $B$ can be obtained from the observed deviations from ideal gas behavior displayed by real gases.

Equation 18.12 can be rewritten in a form in which the parameters $A$ and $B$ are replaced by two parameters, $\phi$ and $\sigma$, which have a more physical interpretation:

$$U = 4\phi\left[-\left(\frac{\sigma}{d}\right)^6 + \left(\frac{\sigma}{d}\right)^{12}\right] \tag{18.13}$$

The meaning of these parameters can be understood with reference to Figure 18.6, which shows a plot of the Lennard–Jones potential for argon. As in the case of the potential energy between two ions, Figure 18.1, the portion of the curve to the left of the minimum corresponds to the repulsive interaction while that to the right of the minimum corresponds to the attractive interaction. The parameter $\phi$ is the magnitude of the minimum value of the potential energy—the depth of the potential "well"—and corresponds to the van der Waals interaction energy. The parameter $\sigma$ is the separation distance at which the potential energy changes from negative to positive and corresponds to the distance of closest approach.

Table 18.2 lists the Lennard–Jones potential parameters for a number of substances. The intermolecular interaction energies $\phi$ range from a fraction of a kilojoule per mole to a few kilojoules per mole. The bond energies of those substances which exist as diatomic molecules are listed for comparison. Note

**FIGURE 18.6**
Lennard–Jones potential for argon. The parameters $\phi$ (the depth of the potential "well") and $\sigma$ (the internuclear distance at which the potential energy goes through zero) are shown.

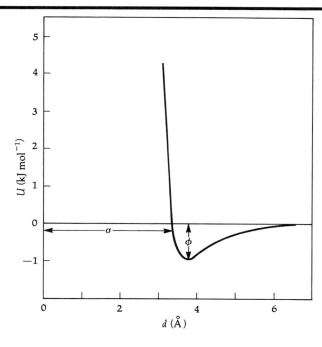

## TABLE 18.2  Lennard–Jones Potential Parameters

| Substance | $\phi$ (kJ mol$^{-1}$) | Bond energy[a] (kJ mol$^{-1}$) | $a$[b] (L$^2$ atm mol$^{-2}$) | $\sigma$ (Å) | Bond length[a] (Å) | $b$[b] (L mol$^{-1}$) |
|---|---|---|---|---|---|---|
| Ne | 0.290 | — | 0.0341 | 2.78 | — | 0.0171 |
| Ar | 0.996 | — | 1.35 | 3.40 | — | 0.0322 |
| Kr | 1.42 | — | 2.32 | 3.60 | — | 0.0398 |
| Xe | 1.84 | — | 4.19 | 4.10 | — | 0.0511 |
| H$_2$ | 0.316 | 432.0 | 0.244 | 2.92 | 0.74 | 0.0266 |
| N$_2$ | 0.791 | 941.7 | 1.39 | 3.70 | 1.10 | 0.0391 |
| O$_2$ | 0.981 | 493.5 | 1.36 | 3.46 | 1.21 | 0.0318 |
| F$_2$ | 0.931 | 139 | — | 3.65 | 1.42 | — |
| Cl$_2$ | 2.14 | 243 | — | 4.40 | 1.99 | — |
| Br$_2$ | 4.32 | 192 | — | 4.27 | 2.28 | — |
| I$_2$ | 4.57 | 149 | — | 4.98 | 2.67 | — |
| CO$_2$ | 1.57 | — | — | 4.49 | — | — |
| CH$_4$ | 1.23 | — | — | 3.82 | — | — |
| CF$_4$ | 1.26 | — | — | 4.70 | — | — |

[a] In the diatomic molecule.
[b] van der Waals equation parameters.

that the bond energies are some two to three orders of magnitude larger, showing that intermolecular interactions are much weaker than covalent bonds.

The values of the parameter $\sigma$ correspond to the sum of the **van der Waals radii** of the adjacent atoms—the radii of nonbonded atoms in contact. They tend to be larger than the covalent radii of bonded atoms, the sum of which is equal to the bond length. This difference, illustrated in Figure 18.7a, comes about because the electron clouds of two atoms overlap significantly when the atoms form a bond. However, if the atoms do not form a bond, there is very little overlap—the atoms are farther apart. Table 18.2 shows that the values of $\sigma$ are indeed larger than those of the bond length. The combined effect of the differences in interatomic distance and interaction energy between bonded and nonbonded atoms in contact is illustrated in Figure 18.7b.

Recall that the van der Waals equation of state (Section 4.9) describes the effects of intermolecular forces on gas behavior. Since the Lennard–Jones potential refers to this very same interaction, the parameters of the van der Waals equation should reflect those of the Lennard–Jones potential. Thus, the parameter $a$ in the van der Waals equation, which is a measure of the strength of the attractive forces, should reflect the value of the interaction energy $\phi$. Similarly, the parameter $b$, which is a measure of the molar volume, should reflect the value of the molecular size parameter $\sigma$. The values of the van der Waals parameters (Table 4.3) are included in Table 18.2. Observe that the trends in the values of $a$ and $b$ closely parallel those in the values of $\phi$ and $\sigma$, respectively.

The Lennard–Jones parameters may be examined for trends. Table 18.2 shows that the values of the interaction energy $\phi$ increase with mass for similar molecules. For example, the values of $\phi$ for the halogens increase from F$_2$ to I$_2$ and those for the noble gases increase from Ne to Xe. These trends reflect the increase in polarizability with size, and consequently with mass, for atoms or molecules of the same type. The polarizability also depends on how tightly the

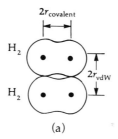

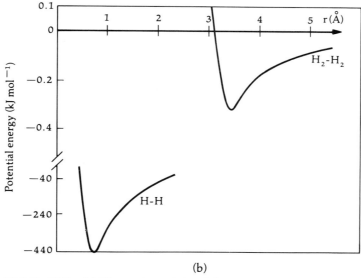

**FIGURE 18.7** (a) Illustration of the difference between van der Waals radii and covalent radii in the H$_2$ molecule. (b) Potential energy of two H$_2$ molecules (van der Waals interaction) and of two hydrogen atoms (covalent bonding). Note the change in energy scale.

electrons are held, an effect shown by a comparison of the values of $\phi$ for CH$_4$ and CF$_4$. Even though the molecular weight of CF$_4$ is much larger than that of CH$_4$, the values of $\phi$ are comparable because the fluorine atoms hold on to their electrons very tightly and are therefore not readily polarized.

The melting and subsequent vaporization of a molecular solid occur when enough thermal energy is supplied to the solid or liquid to overcome the potential energy holding the molecules together. Since the van der Waals force is much weaker than the electrostatic attraction between oppositely charged ions, molecular solids have much lower melting and boiling points than ionic solids. This difference has already been illustrated for the melting points of the fluorides of the elements in the second and third periods (see Figure 6.26). Thus, the ionic fluorides, such as NaF or MgF$_2$, melt some 1400°C higher than the molecular fluorides, such as PF$_3$ or ClF. A similar, albeit much less dramatic trend might be expected for the molecular substances listed in Table 18.2, since these substances also differ in interaction energy. Figure 18.8 shows the melting and boiling points of these substances as a function of the intermolecular

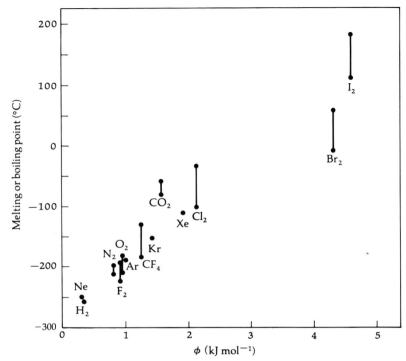

**FIGURE 18.8** Dependence of melting and boiling points of molecular substances on the intermolecular potential energy. The data are from Table 18.2. The two points joined by a vertical line represent the melting and boiling points of the indicated substances. Only single points are shown for substances which have melting and boiling points within 10°C of each other.

potential energy $\phi$. The melting and boiling points increase with $\phi$ in a fairly regular manner. To be sure, such a correlation is far from perfect. Other factors, such as changes in entropy, also affect melting and boiling points. Nonetheless, a correlation extending over temperatures ranging from approximately −250°C to 150°C may be noted.

## 18.3 The Hydrogen Bond

The covalent bonds formed between hydrogen and the most electronegative elements, such as fluorine, oxygen, and nitrogen, are highly polar, as shown by the large dipole moments of diatomic molecules containing these bonds. Molecules containing a hydrogen atom bonded to an atom of a very electronegative element can interact with each other by a strong intermolecular interaction called a **hydrogen bond**.

Figure 18.9 shows some examples of hydrogen bonding between identical molecules. All these species are characterized by the presence of an A—H---A unit, where A is a very electronegative atom. The solid line joining one A atom to the H atom represents an ordinary covalent bond. We have seen that these bonds are relatively strong and short. For example, an O—H bond has an average bond energy of 463 kJ mol$^{-1}$ and an average bond length of 0.97 Å. The dashed line joining the hydrogen atom to a second A atom represents a hydrogen bond, which is both weaker and longer than a covalent

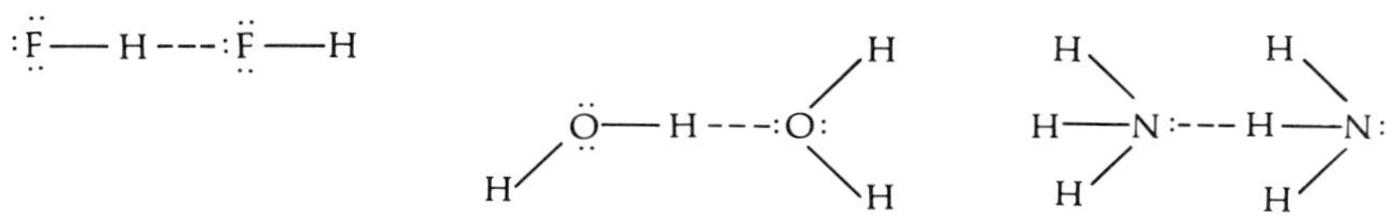

$:\ddot{F}\!-\!H\cdots:\ddot{F}\!-\!H$

$\underset{H}{\overset{\phantom{H}}{\ddot{O}}}\!-\!H\cdots:\ddot{O}\underset{\phantom{H}}{\overset{H}{\phantom{O}}}\!\!\!\diagdown H$

$\underset{H}{\overset{H}{\diagdown}}N\!:\cdots H\!-\!\underset{H}{\overset{H}{\diagdown}}N\!:$

**FIGURE 18.9** Examples of hydrogen bonding in identical molecules. Each pair of identical molecules has an A—H---A structure where A is a very electronegative atom (F, O, or N in the examples). The H---A bond is a hydrogen bond and the A—H bond is an ordinary covalent bond.

bond between the same two types of atoms. An O---H hydrogen bond, for example, has an energy of some 8 to 34 kJ mol$^{-1}$, depending on the identity of the molecule, and a bond length of approximately 1.70 Å.

A hydrogen bond can also be distinguished from the van der Waals interaction examined in Section 18.2 on the basis of its strength and length. Dipole–dipole interactions are usually no larger than a few kilojoules per mole, and the internuclear distances are given by the sum of the van der Waals radii of the atoms in contact. For example, the sum of the van der Waals radii of oxygen and hydrogen is about 2.6 Å, which is significantly larger than the O---H hydrogen bond length. These comparisons show that the hydrogen bond is intermediate in length and strength between a true covalent chemical bond and an ordinary intermolecular interaction.

To a large extent, the hydrogen bond can be described as an especially strong type of dipole–dipole interaction. Hydrogen is unique among the elements capable of forming covalent bonds in that it has no inner electron shells. This permits the positively charged hydrogen atom of one polar molecule to approach a negatively charged atom of a neighboring polar molecule very closely. Since the electrostatic dipole–dipole energy is a strong inverse function of the distance, the hydrogen bond is much stronger than other dipole–dipole interactions. Hydrogen bonding tends to concentrate the positive charge on the hydrogen atom and the negative charge on the electronegative atoms. Consequently, two hydrogen-bonded molecules usually form a linear system (Figure 18.9), for this tends to minimize the electrostatic repulsion between the negative ends.

The structure of certain acid fluoride salts shows, however, that this electrostatic description of the hydrogen bond must be incomplete. Hydrogen fluoride reacts with alkali fluorides to form stable hydrogen fluoride salts, for example,

$$\text{NaF(s)} + \text{HF(g)} \rightarrow \text{NaHF}_2\text{(s)}$$

This salt consists of Na$^+$ and HF$_2^-$ ions. If the electrostatic description of hydrogen bonding were completely adequate, the structure of the HF$_2^-$ ion would be

$$[:\ddot{F}\!-\!H\cdots:\ddot{F}:]^-$$

in which one of the HF bonds is a short covalent bond and the other one is a longer hydrogen bond. The fact is that both HF bonds have the same length. This result suggests that the hydrogen bond may be partly covalent, and therefore amenable to a molecular orbital description. The approach is analogous to that used to explain bonding in diborane (Section 16.9). The 1s orbital of hydrogen and one of the p orbitals of each of two adjacent fluorine atoms combine to form three molecular orbitals, ranging from a strongly bonding $\sigma$ orbital to a

**FIGURE 18.10**
Molecular orbital description of hydrogen bonding in $HF_2^-$. The four bonding electrons fill the lowest two $\sigma$ orbitals.

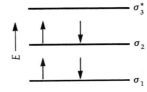

strongly antibonding $\sigma^*$ orbital (Figure 18.10). The lowest two orbitals are filled with four electrons, two from the H—F covalent bond plus the lone pair residing on the second fluorine atom and pointing towards the hydrogen atom. The F—H---F structure can, in this view, be described as a four-electron, three-center bond. While this is not the same as the two-electron, three-center bond in the B—H—B bridge in diborane, it is similar in that the electrons are delocalized over three atoms. It is generally believed that the electrostatic dipole–dipole interaction accounts for most of the strength of the hydrogen bond but that a small degree of covalent bonding also contributes to its stability.

We have focused so far on hydrogen bonding between like molecules. A hydrogen bond can also be formed between unlike molecules. If A and B are electronegative atoms belonging to different molecules, the hydrogen bond between them can be written as —A—H---B— or as —A---H—B—. Hydrogen bonding also can occur between ions and hydrogen atoms attached to electronegative atoms, as in $[M\text{---}H\text{---}A]^+$ or $[X\text{---}H\text{---}A]^-$. The $HF_2^-$ ion is an example of a hydrogen-bonded anion.

**Example 18.2** Which of the following pairs of species are expected to form hydrogen bonds? Write structures of those species that form hydrogen bonds:
(a) $AlCl_3 + AlCl_3$, (b) $H_3C\text{—}O\text{—}CH_3$ (dimethyl ether) + $H_2O$, and (c) HI + HI.

(a) Aluminum chloride does not contain hydrogen. Therefore, hydrogen bonding cannot occur.

(b) A hydrogen bond can be formed between the hydrogen atom in water and the oxygen atom in dimethyl ether:

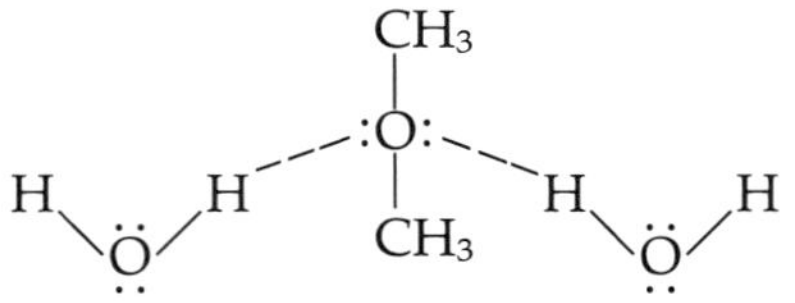

Note that the oxygen atom in dimethyl ether, with its two lone pairs, can form hydrogen bonds with two water molecules.

(c) Iodine is only slightly more electronegative than hydrogen. Consequently, hydrogen iodide is not sufficiently polar to form hydrogen bonds. ∎

The various intermolecular interactions examined in this chapter are summarized and compared with intramolecular interactions, that is, with chemical bonds, in Table 18.3. The van der Waals interactions range from the weak in-

TABLE 18.3  The Strength of Intermolecular and Intramolecular Interactions

| Interaction | Example | Typical interaction energy (kJ mol$^{-1}$) |
| --- | --- | --- |
| Van der Waals | | |
|   Instantaneous dipole–induced dipole (London dispersion force) | $N_2$—$N_2$, $O_2$—$O_2$, Ar—Ar | 0.1–5 |
|   Dipole–dipole (including dipole–induced dipole) | BrF—BrF, $NF_3$—$NF_3$ | 5–20 |
| Hydrogen bond | $H_2O$—$H_2O$, HF—HF | 5–50 |
| Ion-pair bond | $K^+Cl^-$, $Na^+F^-$ | 400–500 (for MX ion pairs) |
| Covalent bond | H—H, H—Cl, O—O | 150–900 |

stantaneous dipole–induced dipole interaction (London dispersion force) to the generally stronger dipole–dipole interaction. The hydrogen bond is usually still stronger, but the "true" chemical bonds—ionic and covalent—are stronger by an order of magnitude.

## 18.4 Experimental Consequences of Hydrogen Bonding

Hydrogen bonding has a significant effect on the physical and chemical properties of the substances in which it occurs. Some of these effects are examined in this section.

*a. Properties of the Hydrides of the Nonmetals*  The hydrides of the elements in Groups IVA–VIIA are covalently bonded compounds with a range of bond polarities. Depending on the polarity, the condensed hydrides exhibit in varying degrees the effects of London dispersion forces, dipole–dipole interactions, and hydrogen bonding. The melting and boiling points of the hydrides in a given group show the varying interplay between these interactions (Figure 18.11).

Starting with the second element in each group, the hydrides display a nearly uniform increase in melting and boiling points in going down the group. The polarity of most of these compounds is relatively low, and the observed variation reflects the increase in polarizability with size for molecules of similar type. The observed trend is similar to that displayed by the noble gases or by the halogens (Figure 18.8).

The striking feature in Figure 18.11 is the upturn in the melting and boiling points displayed by the hydrides of the first elements in Group VA–VIIA, $NH_3$, $H_2O$, and HF. Since these elements have large electronegativities, their hydrides are most likely to form hydrogen bonds in the condensed phase. The anomalously high melting and boiling points of these hydrides can consequently be attributed primarily to hydrogen bonding. In support of this view we note that methane, $CH_4$, which does not form hydrogen bonds because of the comparable electronegativities of carbon and hydrogen, does not exhibit an upturn in melting or boiling points. Figure 18.11b shows that, were it not for hydrogen bonding, water would boil at temperatures far below those normally found on the surface of the Earth. In a very real sense, the occurrence of life as we know it can be attributed to the effect of hydrogen bonding. In addition,

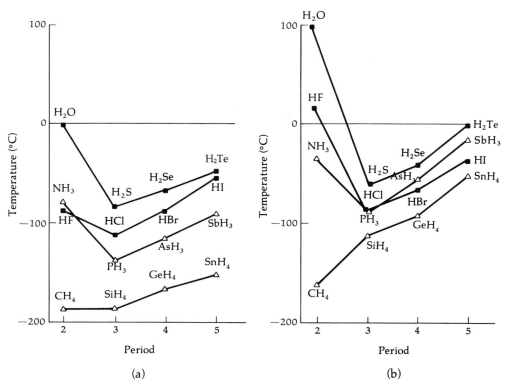

**FIGURE 18.11** (a) Melting and (b) boiling points of the hydrides of the group IVA–VIIA elements.

hydrogen bonding has a more direct bearing on living matter by virtue of its role in the structure of proteins and DNA (Section 22.6).

When a liquid is vaporized, the distance between its molecules generally becomes sufficiently large that the intermolecular interactions can be neglected. The enthalpy of vaporization consequently is a measure of the strength of these interactions. Figure 18.12 shows the variation of the enthalpies of vaporization of the hydrides of the elements in Groups IVA–VIIA. As expected from Trouton's rule (Section 12.1b), the trends are similar to those of the boiling points: the values of $\Delta H_{vap}$ of HF, $H_2O$, and $NH_3$ are larger than expected from the values for the heavier hydrides, confirming the occurrence of hydrogen bonding.

A liquid in which molecules are connected by hydrogen bonds has a more ordered structure than one in which there is no hydrogen bonding. For example, hydrogen bonding restricts the ability of molecules to rotate. When hydrogen bonds are broken upon vaporization, the molecules can rotate freely. The molecular disorder associated with vaporization is consequently greater for a hydrogen-bonded substance. In other words, the entropy of vaporization should be larger than expected. Trouton's rule states that the entropies of vaporization of most substances are in the vicinity of 88 J $mol^{-1}$ $K^{-1}$. However, the value for water is significantly larger, 109 J $mol^{-1}$ $K^{-1}$, showing that the breaking of hydrogen bonds affects the entropy of vaporization as well as the enthalpy of vaporization.

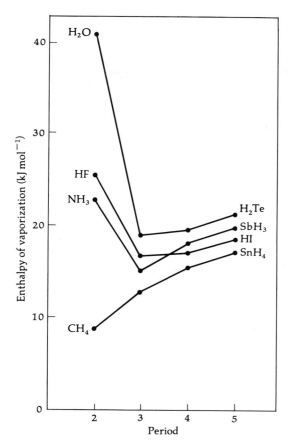

**FIGURE 18.12** Enthalpies of vaporization of the hydrides of group IVA–VIIA elements.

---

**Example 18.3**  On the basis of the intermolecular interactions expected to be of importance, pick the substance with the higher boiling point from each of the following pairs: (a) $Cl_2$ or $I_2$, (b) HCN or $C_2H_6$, (c) $AlF_3$ or $SiF_4$, and (d) $CH_4$ or $CH_3OH$.

(a) London dispersion forces are important in both substances. Iodine is more polarizable (it is more massive and has the same valence shell electronic configuration) and therefore has the higher boiling point.

(b) Only London dispersion forces are of importance in $C_2H_6$ while dipole–dipole interactions and hydrogen bonding are present in HCN. Consequently, HCN has the higher boiling point.

(c) $AlF_3$ is primarily ionic whereas $SiF_4$ is a molecular compound held together by London dispersion forces. Thus, $AlF_3$ has the higher boiling point.

(d) $CH_3OH$ molecules are held together by hydrogen bonds as well as by dipole–dipole and London dispersion forces. Therefore $CH_3OH$ has a higher boiling point than $CH_4$, the molecules of which are held together only by London dispersion forces.  ■

***b. Azeotropic Solutions*** Many binary liquid solutions deviate from Raoult's law (Section 12.4). Negative deviations, in which the total vapor pressure is smaller than the predicted value, occur when the attraction between unlike molecules is greater than that between like molecules; positive deviations occur when the opposite is true (Section 12.6). When either type of deviation becomes sufficiently large, an azeotropic solution involving either a maximum boiling point (for negative deviations) or a minimum boiling point is formed (see Figure 12.17).

The formation of maximum boiling mixtures is frequently attributable to hydrogen bonding. This effect occurs when two liquids form strong hydrogen bonds with each other. Typically, one substance contains a hydrogen atom which has been rendered electron deficient as a result of the stronger attraction for electrons exerted by the rest of the molecule. For example, highly halogenated hydrocarbons have the electron density concentrated in the vicinity of the halogen atoms, thereby imparting an abnormally large positive charge to the remaining hydrogen atoms. These atoms can therefore form hydrogen bonds with molecules containing electronegative atoms. When the composition of a solution is such that the largest number of hydrogen bonds can form, the intermolecular forces in the solution are strongest. Thus, the solution has a maximum boiling point at this composition. Figure 18.13 shows some examples of hydrogen-bonded substances forming azeotropic solutions of this type.

**FIGURE 18.13**
Examples of hydrogen bonding between unlike molecules, the solutions of which have an azeotropic point.

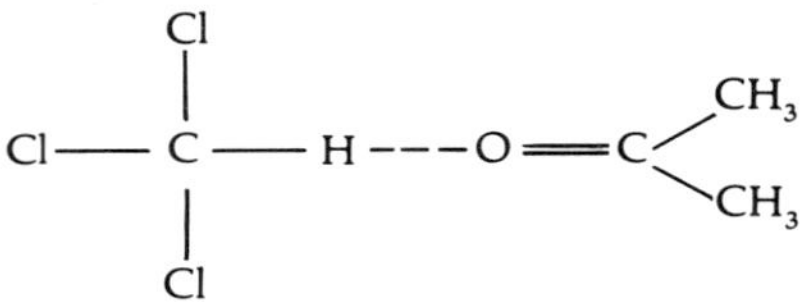

chloroform-acetone

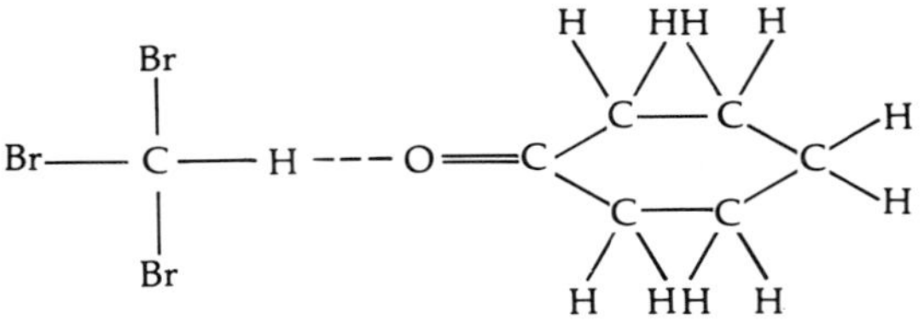

bromoform-cyclohexanone

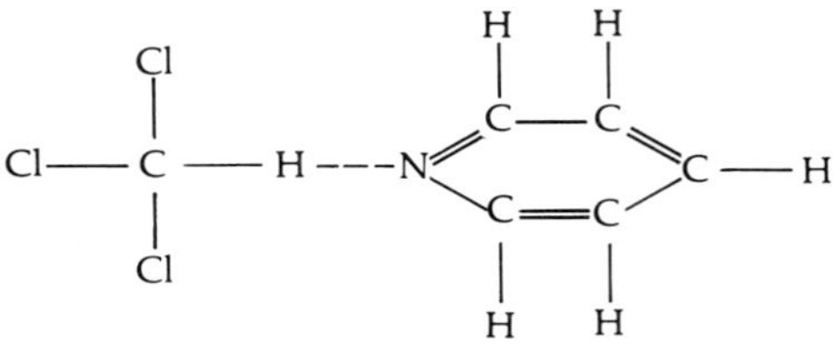

chloroform-pyridine

The second type of azeotropic solution, in which very positive deviations from Raoult's law result in a minimum boiling point, can frequently also be attributed to the occurrence of hydrogen bonding. If one of the liquids constituting the solution forms hydrogen bonds while the other one does not, the forces between the like molecules will generally be much stronger than those between the unlike ones, thereby leading to the formation of an azeotropic solution. For example, when hydrocarbons, which do not form hydrogen bonds, are mixed with a hydrogen-bonded alcohol such as $CH_3OH$, a minimum boiling mixture frequently results. If the deviation from Raoult's law is extreme, separation of the system into two immiscible phases will result, as in the case of water and hydrocarbons.

## Water

Water is so common a liquid that we seldom dwell on its importance, although it is one of the most important substances for the support of life. The evidence to date indicates, in fact, that its presence on a planet is required for the occurrence of life. Many of the unusual features of water, such as its unparalleled ability to act as a solvent, can be traced to the polarity of the molecule and the consequent formation of strong hydrogen bonds.

### 18.5 Hydrogen Bonding in Ice and Water

The structure of ice is shown in Figure 18.14. Each oxygen atom forms covalent bonds with two hydrogen atoms, 0.99 Å in length, and hydrogen bonds with two other hydrogen atoms at a distance of 1.77 Å. The four bonds form tetrahedral angles. The result is that each water molecule is tetrahedrally surrounded by four other water molecules. The resulting structure is very open and has the appearance of layers of irregular hexagonal rings, which are macroscopically expressed in the hexagonal pattern of snowflakes.

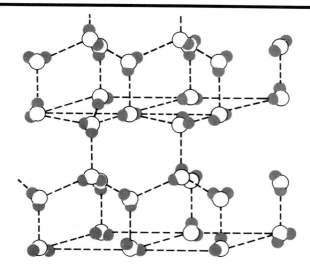

**FIGURE 18.14** The structure of ice.

When ice melts, some of the hydrogen bonds are broken and the structure collapses into a more compact arrangement. The density of water at 0°C is consequently greater than that of ice. This behavior is exceptional and accounts for the unusual shape of the phase diagram of water (Section 12.3).

As the temperature of water increases above 0°C, additional hydrogen bonds are broken, and the density continues to increase. This effect is counteracted by the thermal expansion of the liquid with increasing temperature. This expansion is associated with the increase in the mean kinetic energy of the molecules which, in essence, causes them to occupy more space. The result of these opposing factors is that the density of water attains a maximum value (1.0000 g cm$^{-3}$) at 3.98°C and decreases thereafter, as shown in Figure 18.15.

The dependence of the density of water on temperature has important consequences for the response of large bodies of water to changes in ambient temperature. At the onset of cold weather, the temperature of surface water decreases to the maximum density point, 4°C, causing this layer to sink through the warmer and less dense water below it. This process promotes vertical circulation in lakes and oceans. When the temperature of the lower layers also decreases to 4°C, the coldest layer of water remains on the surface and eventually freezes. The resulting ice, being less dense, floats on the surface and helps to insulate lower layers from extremely low temperatures. This behavior ensures that moderately deep bodies of water will not freeze solid and permits the survival of aquatic life forms in even the harshest climates.

The fraction of water molecules that are hydrogen bonded ranges from essentially 100% in ice to virtually 0% in the vapor phase. A rough estimate of

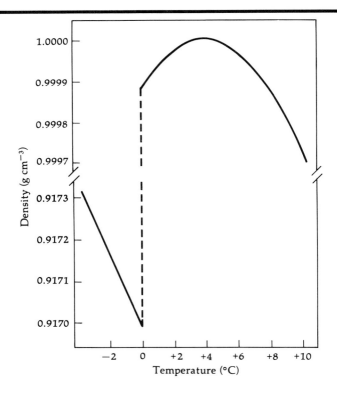

**FIGURE 18.15** Temperature dependence of the densities of ice and water.

the fraction of the hydrogen bonds that are broken at each stage in the progression

$$\text{ice } (0°C) \rightarrow \text{water } (0°C) \rightarrow \text{water } (100°C) \rightarrow \text{steam } (100°C)$$

can be obtained from the enthalpy changes of the various phase transformations. We assume that the enthalpy change associated with a particular phase change is expended in breaking hydrogen bonds. Since the other intermolecular interactions in water, such as London dispersion and dipole–dipole interactions, are weak compared to hydrogen bonds, this assumption is adequate for a rough estimate. As a measure of its validity, we can evaluate the hydrogen bond energy in ice by means of the molar enthalpy of sublimation of ice. This enthalpy change corresponds to the energy needed to break all the hydrogen bonds in 1 mole of ice. Each $H_2O$ molecule is hydrogen bonded to four other $H_2O$ molecules. Since each hydrogen bond is shared by two molecules, there must be two hydrogen bonds per molecule. The molar enthalpy of sublimation is 51 kJ mol$^{-1}$ indicating that the hydrogen bond energy in water is about 25 kJ mol$^{-1}$. This value is in accord with other estimates of the hydrogen bond energy in $H_2O$ and is consistent with the range of values in Table 18.3.

If the enthalpy of sublimation is taken as a measure of the energy required to break *all* the hydrogen bonds in a mole of ice, the enthalpy of fusion similarly is a measure of the energy of the hydrogen bonds broken when ice melts. The ratio of the two enthalpies consequently corresponds to the fraction of the hydrogen bonds broken on melting:

$$\frac{\Delta H_{fus}}{\Delta H_{sub}} = \frac{6.0 \text{ kJ mol}^{-1}}{51 \text{ kJ mol}^{-1}} = 0.12$$

Similarly, the ratio of the enthalpy of vaporization to that of sublimation provides an estimate of the fraction of the hydrogen bonds broken on vaporization:

$$\frac{\Delta H_{vap}}{\Delta H_{sub}} = \frac{40.6 \text{ kJ mol}^{-1}}{51 \text{ kJ mol}^{-1}} = 0.80$$

If 12% of the hydrogen bonds are broken on melting and 80% are broken on vaporization, the remaining 8% must be broken as water is heated from 0°C to 100°C. While these numerical estimates should not be taken too seriously, they do correctly indicate that hydrogen bonding is important in both water and ice. The difference is that ice consists of large hydrogen-bonded aggregates whereas water consists of an equilibrium mixture of smaller aggregates and individual molecules (Figure 18.16).

The enthalpies of fusion and vaporization of $H_2O$ are unusually large, reflecting the energy that must be expended in breaking hydrogen bonds. The large heat capacity of water, 75 J mol$^{-1}$ K$^{-1}$, similarly reflects the energy of the hydrogen bonds broken when water is heated to its boiling point. As a consequence of these properties, the melting of ice and the warming of water are processes that require a relatively large inflow of heat. Conversely, a large amount of heat flows to the surroundings when water cools and freezes. The ability to store great quantities of thermal energy enables large bodies of water to minimize ambient temperature variations. This explains why coastal regions of the Earth have a more moderate climate than inland regions.

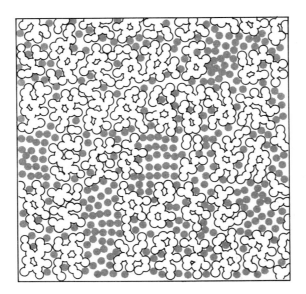

**FIGURE 18.16** The structure of water at room temperature and atmospheric pressure. Clusters of hydrogen bonded molecules are outlined and individual molecules are shaded. The particular arrangement in the figure lasts for about $10^{-11}$ s, and is succeeded by a different arrangement which, however, is similar in the fraction of $H_2O$ molecules that are hydrogen bonded and in the average cluster size. [Source: R. A. Courant, B. J. Ray, and R. A. Horne, *Zh. Struk. Khim.* **4**, 581 (1972).]

## 18.6 Solvent Properties of Water

Water dissolves a large variety of substances, ranging from ionic salts such as sodium chloride, to gases such as ammonia, to various organic compounds such as methanol and ethanol. The unparalleled ability of water to act as a solvent is due to the polarity of the $H_2O$ molecule and its ability to form hydrogen bonds with a variety of other molecules.

*a. The Hydration of Ions* Ions are held in a solid by their mutual electrostatic interaction. The energy that must be supplied to a solid to break the cohesive forces is called the **crystal lattice energy** (Section 20.10b). Since most ionic solids dissolve exothermically, an amount of energy exceeding the crystal lattice energy must generally be released in the dissolution process. This energy comes from the interaction between the ions and water molecules and is called the **hydration energy**.

Ions are **hydrated**, or **solvated** in aqueous solution, meaning that they are attached to several molecules of water. This number changes as the ions move through the solution; the average number of water molecules held by an ion is known as its **hydration number**. The hydration numbers of cations are generally the same as their **coordination numbers**, the numbers of ligands attached to cations in complex ions (Sections 9.5 and 21.4). For example, copper(II) forms the complex ion $[Cu(NH_3)_4]^{2+}$ in which its coordination number is four. Similarly, copper(II) is present in aqueous solution as $[Cu(H_2O)_4]^{2+}$, and its hydration number is four. Figure 18.17 shows the structures of some hydrated cations.

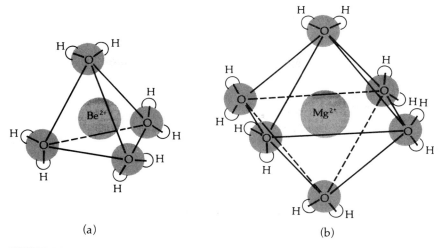

**FIGURE 18.17** The hydration of metal cations. (a) Tetrahedrally coordinated beryllium(II), Be(H$_2$O)$_4^{2+}$, and (b) octahedrally coordinated magnesium(II), Mg(H$_2$O)$_6^{2+}$.

The interaction responsible for hydration is an **ion–dipole interaction**. This interaction is similar to the van der Waals interactions in being electrostatic. In the presence of cations, water molecules will orient themselves so that their negative ends (that is, their oxygen atoms) are directed toward the cations, as shown in Figure 18.17. Ion–dipole interactions are stronger than van der Waals interactions since one of the interacting species—the cation—has at least one full electronic charge. For example, the energy per K–O bond in [K(H$_2$O)$_6$]$^+$ is nearly 70 kJ mol$^{-1}$. On the other hand, ion–dipole interactions are not as strong as ionic bonds because the charge on the H$_2$O dipole is substantially less than a full electronic charge.

The magnitude of the hydration energy of a cation is determined by electrostatic factors. The hydration energy is proportional to the square of the cation charge and inversely proportional to its effective radius:

$$\Delta H_{\text{hyd}} \propto Z^2/r_{\text{eff}} \tag{18.14}$$

The effective radius exceeds the actual ionic radius by a constant value. This value corresponds to the radius of an oxygen atom so that $r_{\text{eff}}$, in essence, is the internuclear distance between the cation and the oxygen end of an attached water molecule. Figure 18.18 shows the correlation between hydration energy and $Z^2/r_{\text{eff}}$ for a number of cations. A linear variation is clearly in evidence.

The hydration of anions generally is of less significance than that of cations. The radii of anions tend to be much larger than those of cations and the hydration energies are consequently smaller. Nonetheless, anions are also hydrated and interact by means of ion–dipole interactions, with the hydrogen atoms of water pointing towards the anion. Hydrogen bonding between water and the anion can also occur.

**b. The Dielectric Constant of Water**  The electrostatic attraction between oppositely charged ions is weaker in solution than it is in vacuum because molecules

**FIGURE 18.18**
Dependence of cation hydration energies on charge and size. The quantity $r_{eff}$ is the sum of the ionic radius of the cation and the atomic radius of oxygen expressed in Å; $Z$ is the ionic charge. Note that $\Delta H_{hyd} < 0$, indicating that hydration is exothermic.

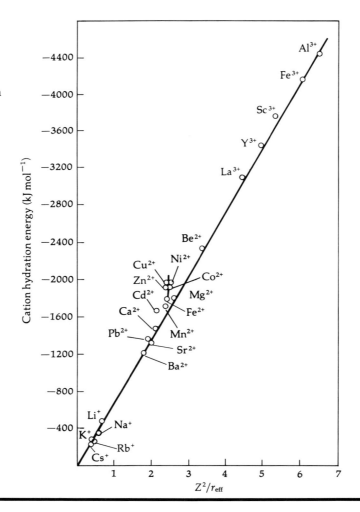

of the solvent fill the space between the ions. Equation 18.6 for the Coulomb energy must be modified as follows for ions in solution:

$$U = -Z_+Z_-/\alpha K d \tag{18.15}$$

where the quantity $K$, the **dielectric constant,** is a measure of the ability of a solvent to screen an electric field. Polar substances tend to have high dielectric constants. They reduce the field between two oppositely charged ions by aligning themselves so that their positive ends point toward the anion and their negative ends point toward the cation (Figure 18.19). In this fashion, the attractive force between the ions is reduced.

The dielectric constant of a vacuum is 1 and that of air is very close to 1. Water has a dielectric constant of approximately 80, one of the highest values of any substance. This high value reflects not only the polarity of the $H_2O$ molecule but also the occurrence of hydrogen bonding. While the dipole moment is a property of a single molecule, the dielectric constant is a property of the liquid as a whole. A hydrogen-bonded liquid has a large number of molecules oriented so that their dipole moments reinforce each other. Such a

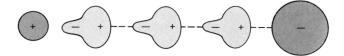

**FIGURE 18.19** Effect of the dielectric constant of water on the electrostatic interaction between oppositely charged ions. Water molecules align themselves between two oppositely charged ions as shown and reduce the attractive interaction between them. Hydrogen bonds (dashed lines) between water molecules enhance the alignment of the dipoles.

liquid can screen an electric field more readily than one consisting of a more randomly oriented collection of polar molecules. For example, water and ethanol have comparable dipole moments, but water, with its stronger hydrogen bonds, has a dielectric constant that is three times larger.

The high dielectric constant of water indicates that the attractive force between two ions is some 80 times weaker when the ions are in aqueous solution than when they are not. This weakening of the attractive force makes it possible for water molecules to squeeze between and separate two oppositely charged ions. It is this separation that makes possible the hydration of the individual ions and thereby makes water such a good solvent for ionic salts.

*c. The Hydrogen Ion in Aqueous Solution* Recall from Section 8.1b that a hydrogen ion—a bare proton—does not exist as such in aqueous solution but rather, in the form of $H_3O^+$, the hydronium ion. The hydronium ion undergoes additional hydration and is believed to be hydrogen bonded to three water molecules to form a structure that can be represented by the formula $H_9O_4^+$ (see Figure 8.2).

The conductivity of a solution of an electrolyte is determined by the **ion mobility**—the speed at which the ions move when an electric field is applied. While solutions of all ionic salts are conducting, the conductivities of acids and bases, that is, solutions containing hydrogen or hydroxide ions, are especially high (see Figure 8.1). However, a species as massive as $H_9O_4^+$ is not expected to have a high enough mobility to account for the unusually high conductivity of acids. This fact suggests that the mechanism by which hydrogen and hydroxide ions migrate through a solution is different than that for other ions, the migration of which involves the actual motion of the hydrated ions.

In addition to the movement of the ions through the solution, the motion of hydrogen and hydroxide ions involves the chain mechanism illustrated in Figure 18.20. The ease with which hydrogen bonds between water molecules break and reform permits the propagation of either positive or negative charge along a chain of water molecules. The mobility of the hydrogen and hydroxide ions is therefore largely determined by the rate at which hydrogen bonds between adjacent water molecules break and reform rather than by the slower rate at which ions migrate.

*d. Hydrates* The ion–dipole or hydrogen bonds between ions and coordinated water molecules often persist when solid salts crystallize out of solution. These solid salts are called **hydrates,** and the attached water is known as **water of hydration** or **water of crystallization.** Some typical examples are copper(II) sulfate

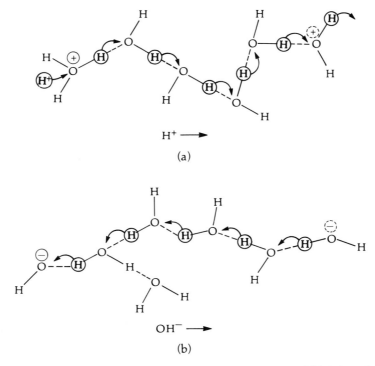

**FIGURE 18.20** Chain mechanism for the migration of (a) H⁺ and (b) OH⁻ ions in aqueous solution. Note that the H⁺ ion entering the chain at one end does not travel along the chain and is *not* the same ion as that leaving the chain at the other end.

---

pentahydrate, CuSO$_4$·5H$_2$O, iron(II) nitrate hexahydrate, Fe(NO$_3$)$_2$·6H$_2$O, and sodium tetraborate decahydrate (borax), Na$_2$B$_4$O$_7$·10H$_2$O.

The structures of hydrates can be fairly complex since the water molecules can be held in various ways in the crystal. The water molecules can be attached to a central cation by coordinate covalent bonds formed by one of the lone pairs on the oxygen atom (**coordination water**), they can be attached to an anion by hydrogen bonds (**anion water**), or they can occupy either random or fixed positions within the crystal structure (**lattice water**). The bonds between the water molecules and the ions are generally not very strong, and the water of hydration can frequently be driven off by gently heating the hydrate.

The structure of CuSO$_4$·5H$_2$O (Figure 18.21) illustrates the presence of both coordination water and anion water. Four of the water molecules are attached to the Cu$^{2+}$ ion by coordinate covalent bonds. The fifth water molecule occupies a bridging position between the [Cu(H$_2$O)$_4$]$^{2+}$ ion and the SO$_4^{2-}$ ion, and is held by hydrogen bonds. The different arrangements of the two types of water molecules are manifested in the thermal behavior of the hydrate. When CuSO$_4$·5H$_2$O is heated, it first loses four molecules of water at 110°C to form the monohydrate, CuSO$_4$·H$_2$O. Further heating to 150°C results in the formation of anhydrous copper sulfate, CuSO$_4$. The process is reversible and can be represented by the equilibrium

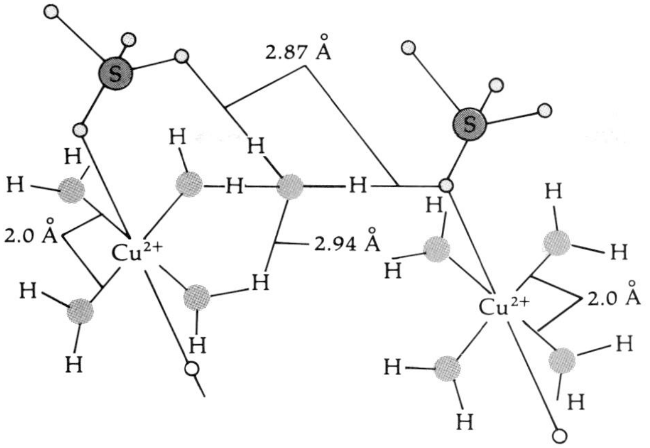

**FIGURE 18.21** The structure of copper(II) sulfate pentahydrate, CuSO$_4$·5H$_2$O. Four of the H$_2$O molecules are attached to each Cu$^{2+}$ ion. The fifth H$_2$O molecule (shown in color) is held by hydrogen bonds between H$_2$O molecules and the SO$_4^{2-}$ ion.

$$\text{CuSO}_4\cdot 5\text{H}_2\text{O(s)} \stackrel{\Delta}{\rightleftharpoons} \text{CuSO}_4\cdot \text{H}_2\text{O(s)} + 4\text{H}_2\text{O(g)} \stackrel{\Delta}{\rightleftharpoons} \text{CuSO}_4\text{(s)} + \text{H}_2\text{O(g)}$$
(blue) (white)

The existence of two or more hydrates at different temperatures is a fairly common occurrence. A solid–solution phase diagram offers a useful summary of the conditions under which different hydrates crystallize, as discussed in Section 12.7.

The presence of lattice water is illustrated by hydrates involving large or highly charged anions, such as FeSiF$_6$·6H$_2$O and Na$_4$XeO$_6$·8H$_2$O. These salts are not stable in anhydrous form, owing to the repulsion between the highly charged anions and because of the disparate sizes of cation and anion. In these instances, the water molecules fill the open spaces in the crystal structure and also reduce the repulsion between the anions.

**e. The Dissolution of Ionic Solids**  A salt will dissolve in water if the free energy of the system decreases upon dissolution,

$$\Delta G = \Delta H - T\,\Delta S < 0$$

The solubility of a salt is determined by both the enthalpy and entropy changes. The sign of the entropy change is generally predictable. If the dissolution is represented by the equation

$$\text{M}^+\text{X}^-(s) + \text{H}_2\text{O} \rightarrow \text{M}^+(aq) + \text{X}^-(aq)$$

the entropy change is determined by the relative magnitudes of the entropies of the hydrated ions and those of the solid salt plus water. Since the hydrated ions represent a more disordered state, the entropy of solution is usually positive. Consequently, if the enthalpy of solution is negative, both entropy and enthalpy factors work in the same direction and dissolution occurs spontaneously. If the enthalpy of solution is positive but relatively small, the increase in entropy may be sufficient to overcome the increase in enthalpy and the free energy change can still be negative, leading to spontaneous dissolution. Since the process is endothermic, the solution will cool as the solute dissolves. Finally, if the enthalpy of solution is positive and large, the increase in entropy will be insufficient to overcome the increase in enthalpy and the salt will not dissolve appreciably.

The dissolution of an ionic compound can be examined by means of a Born–Haber-type cycle in order to see which factors contribute to the enthalpy of solution, as illustrated in Figure 18.22. The enthalpy of solution is given by the sum of three different terms:

$$\Delta H_{solution} = \Delta H_{solute} + \Delta H_{solvent} + \Delta H_{hydration} \tag{18.16}$$

The first term, $\Delta H_{solute}$, represents the effect of the interaction between the ions in the solid. Energy must be supplied to break the ionic bonds in the solid (Section 20.10b), and $\Delta H_{solute}$ is consequently positive. The second term, $\Delta H_{solvent}$, represents the energy required to break apart clusters of water molecules held together by hydrogen bonds and dipole–dipole interactions and is also positive. The third term is the enthalpy of hydration of the ions and, as discussed in Section 18.6a, is negative. The sign of the enthalpy of solution is consequently determined by the magnitude of $\Delta H_{hydration}$ relative to those of the first two terms.

Although the ion–dipole interaction giving rise to hydration is weaker than the ionic bond, hydration energies are generally comparable to or larger than crystal lattice energies because a given ion is attached to several water molecules. We have seen that hydration energies are proportional to the square of the ionic charge and inversely proportional to the ionic radius (Equation 18.14). These same factors, however, also determine the magnitude of the crystal lattice energy, so that large hydration energies are generally accompanied by large crystal lattice energies. The magnitude of the hydration energy consequently is not indicative of the solubility of a salt. For example, highly soluble salts include those with large hydration energies, e.g., $CaI_2$ ($\Delta H_{hydration} = -2180$ kJ mol$^{-1}$), and those with small hydration energies, e.g., KI ($\Delta H_{hydration} = -611$ kJ mol$^{-1}$). Similarly, sparingly soluble salts also include those with large hydration energies, e.g., $CaF_2$ ($\Delta H_{hydration} = -6782$ kJ mol$^{-1}$), and those with small hydration energies, e.g., LiF ($\Delta H_{hydration} = -1004$ kJ mol$^{-1}$). In general, the empirical solubility rules given in Section 9.1 are as adequate a guide to solubility as refined calculations.

*f. The Solubility of Molecular Compounds* The versatility of water as a solvent is manifested in its ability to dissolve a variety of molecular substances includ-

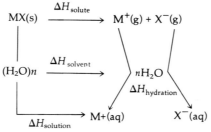

**FIGURE 18.22** Born–Haber-type cycle for the dissolution of an ionic solid. The enthalpies associated with the various steps in the cycle are discussed in the text.

**FIGURE 18.23** The "ammonia fountain", in which a flask filled with ammonia becomes evacuated when the gas dissolves in water drops introduced by the dropper. The resulting decrease in gas pressure permits the higher pressure of the atmosphere to force water into the flask, resulting in fountain action.

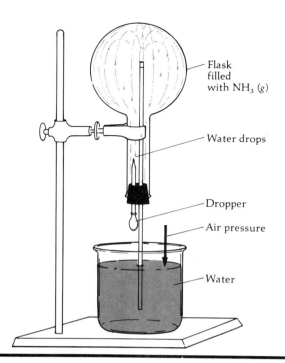

ing various gases and many organic compounds. The solubility of these substances can generally be attributed to their ability to form hydrogen bonds with water. Thus, gases that are highly soluble in water include ammonia and hydrogen chloride, substances that form strong hydrogen bonds with water. The "ammonia fountain," depicted in Figure 18.23, gives a nice visual illustration of the high solubility of ammonia.

The solubility in water of many organic compounds can similarly be attributed to hydrogen bonding, with substances capable of forming hydrogen bonds being fairly soluble and those incapable of forming such bonds being virtually insoluble. For example, dimethyl ether, $(CH_3)_2O$, which forms hydrogen bonds with water (Figure 18.24) is completely miscible, while dimethyl sulfide, $(CH_3)_2S$, which does not form hydrogen bonds, is only slightly soluble in water. Organic substances containing a substantial number of carbon atoms must contain a correspondingly large number of atoms capable of forming hydrogen bonds with water in order to be soluble. When the hydrogen bonds are

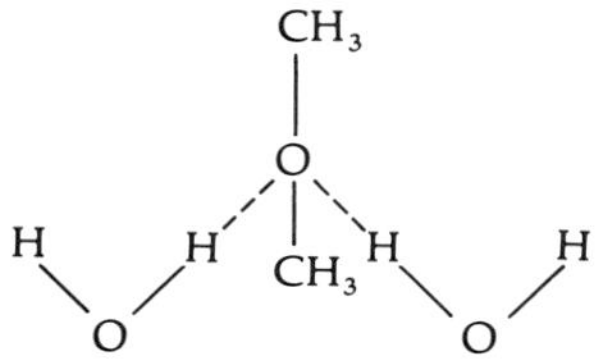

**FIGURE 18.24** Hydrogen bonding between dimethyl ether, $(CH_3)_2O$, and water. Organic compounds which can form hydrogen bonds are far more soluble in water than those which cannot.

formed by oxygen atoms, one oxygen atom is usually required for each three carbon atoms to give a compound that is highly soluble in water. Thus, diethyl ether, $(C_2H_5)_2O$, is only moderately soluble in water while dimethyl ether, $(CH_3)_2O$, is completely miscible, as we have noted.

## 18.7 Natural Waters

*a. The Hydrosphere* The various bodies of water on Earth make up the **hydrosphere**. Water covers over 70% of the surface of the Earth, with all but about 2% of the water being found in the oceans and seas. Water in lakes and streams has an importance that is out of proportion to its relatively small abundance. This fresh water is vital to most living organisms including man. Its use for drinking, irrigation, and many industrial processes is essential, while its many secondary uses, such as for recreation and transportation, contribute immeasurably to the quality of life in those regions that are abundantly endowed with bodies of fresh water.

The composition of natural water is quite variable. At the one extreme, freshly fallen rainwater consists of almost pure $H_2O$ (except where acid rain is a problem); at the other extreme, sea water has a high concentration of solutes that are mostly inorganic ions. Table 18.4 lists the average amounts of the most abundant ions found in sea water and in river water. The ionic content of sea water is approximately 35 parts per thousand (by weight) and is some 300 times larger than the ionic content of river water. The desalination of sea water by one of the techniques mentioned in Section 12.8 has the potential of providing a nearly limitless supply of fresh water. However, the cost of desalination on a large scale is prohibitively expensive and appears to be unfeasible except in special circumstances.

In addition to the naturally occurring impurities in water, most bodies of water contain varying amounts of man-made pollutants. If the concentrations of these substances are sufficiently high, harmful effects to both man and to the indigenous aquatic life forms are sure to follow.

Most pollutants may be grouped into several broad categories. Inorganic pollutants include metal ions such as $Hg^{2+}$, $Pb^{2+}$, and $Cd^{2+}$, nonmetals such as arsenic and selenium, the dilute sulfuric acid that is the chief pollutant in acid rain, and radioactive isotopes. Organic pollutants include pesticides and herbicides, a large variety of industrial compounds, and spilled petroleum.

**TABLE 18.4 Average Abundance of Ions in Natural Waters**

| Ion | Sea water (g ion/kg water) | River water (g ion/kg water) |
|---|---|---|
| $Cl^-$ | 19.4 | 0.008 |
| $Na^+$ | 10.8 | 0.006 |
| $SO_4^{2-}$ | 2.71 | 0.011 |
| $Mg^{2+}$ | 1.29 | 0.004 |
| $Ca^{2+}$ | 0.41 | 0.015 |
| $K^+$ | 0.39 | 0.002 |
| $HCO_3^-$ | 0.14 | 0.059 |
| $Br^-$ | 0.07 | <0.001 |
| $Sr^{2+}$ | 0.008 | <0.001 |

Other substances, although not intrinsically harmful, are problematic because they serve as nutrients for algae, whose rapid growth in fresh waters crowds out other more desirable life forms and fouls the waters, a condition known as **eutriphication.** The substances responsible for this condition are primarily fertilizers, such as nitrates and phosphates, which enter streams and lakes in run-off waters from agricultural lands. Municipal sewage wastes are an additional source of these bionutrients and, in addition, can be a source of harmful microorganisms. While the abatement of the pollution of natural waters is as much a political as a technical problem, scientists and engineers bear a special responsibility because of their technical expertise.

*b. Hard Water* Water that contains appreciable amounts of dipositive metal ions such as $Ca^{2+}$, $Mg^{2+}$, and $Fe^{2+}$ is called **hard water.** These ions give rise to certain practical problems in waters destined for domestic or industrial use since they form a scummy deposit with soap and a scaly precipitate upon boiling.

Hard water can be "softened" by removal of the offending ions. The addition of washing soda ($Na_2CO_3 \cdot 10H_2O$) to hard water leads to the precipitation of the metal carbonate, which can then be filtered off. If the hard water also contains bicarbonate ions, $HCO_3^-$, the metal carbonates can be precipitated by boiling. The reaction that occurs in this process is

$$M^{2+} + 2HCO_3^- \xrightarrow{\Delta} MCO_3(s) + CO_2(g) + H_2O(g)$$

A different method of water softening is **ion exchange,** in which the ions that give rise to hardness are replaced by sodium ions. Ion exchange takes place when hard water is passed through certain naturally occurring minerals called zeolites. **Zeolites** are complex silicates of aluminum and sodium (Figure 18.25a). A typical empirical formula is $NaAlSiO_4$. The characteristic feature of zeolite crystals is that they consist of an open aluminosilicate framework in which the sodium ions are relatively free to move. When hard water flows through a column of zeolite crystals, the dipositive ions in the water displace the $Na^+$ ions from the crystals in a process which can be represented by the equation

$$M^{2+} + 2NaZ(s) \rightarrow 2Na^+ + MZ_2(s)$$

in which Z indicates the anion of zeolite. The water is softened by this process because $Na^+$ ions do not cause the objectionable properties of hard water. When enough water has been passed through a zeolite column to replace most of the $Na^+$ ions with the dipositive cations, the column loses its water softening properties. However, the column can be regenerated by passing a concentrated solution of sodium chloride through it. The $Na^+$ ions now displace the dipositive cations from the column and the latter are washed away with the $Cl^-$ anions (Figure 18.25b). The relatively high concentrations of $Na^+$ ions in water softened by ion exchange are not entirely safe. The presence of significant amounts of sodium in drinking water appears to increase the incidence of heart disease. Consequently, the zeolite process is not recommended for drinking water.

Synthetic ion exchange resins, too, can be used to remove objectionable ions from water. These substances are organic polymers consisting of a hydrocarbon framework with attached ionic groups. Depending on whether these

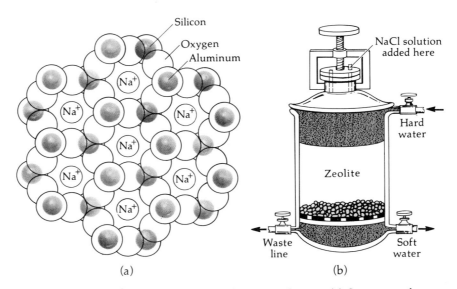

**FIGURE 18.25** The softening of water by ion exchange. (a) Structure of a natural zeolite with the empirical formula NaAlSiO$_4$. (b) Operation of a zeolite ion exchange column. The NaCl solution is added whenever the zeolite column requires regeneration (see text).

groups carry negative or positive charges, the resin can be used for either cation or anion exchange, respectively.

A cation exchange resin contains negatively charged groups such as sulfonate, SO$_3^-$, whose charge is balanced by loosely held cations such as Na$^+$ or H$^+$. When water containing various cations is passed through a cation exchange resin column, cation exchange takes place. Similarly, an anion exchange resin contains positively charged groups such as NR$_3^+$, which are balanced in charge by loosely held anions such as Cl$^-$ or OH$^-$. Objectionable anions can be removed from water by passage through anion exchange resins.

Cation exchange resins containing H$^+$ ions and anion exchange resins containing OH$^-$ ions can be used in tandem to produce **deionized water,** water virtually free of all foreign ions. The cation exchange resin removes cation impurities from water and replaces them with H$^+$ ions. Anions are removed by the anion exchange column and replaced by OH$^-$ ions. Since the concentrations of H$^+$ and OH$^-$ must satisfy the ionization equilibrium of water, neutralization removes most of these ions from the purified water. The cost of deionizing water in this manner is fairly high and this purification process has limited uses.

# Conclusion

The interactions between molecules depend on their polarity. Non-polar molecules interact by means of London dispersion forces. Polar molecules, in addition, can interact by somewhat stronger dipole–dipole interactions. Polar molecules containing hydrogen attached to a very electronegative atom form hydrogen bonds, the strongest of the intermolecular interactions. The energy of the hydrogen bond in water, for example, is about 25 kJ mol$^{-1}$. However, all of these intermolecular interactions are much weaker than the covalent bonds holding atoms in molecules or the ionic bonds between oppositely charged ions.

In spite of their relative weakness, intermolecular forces have readily apparent effects. These forces are responsible for the condensation of gases and their strength in large measure determines the melting and boiling points of molecular substances as well as many of their other properties. The occurrence of hydrogen bonding in water, for example, accounts for water being a liquid rather than a gas at room temperature, a fact of far-reaching practical consequences. Hydrogen bonding in water also accounts for the unusual relative densities of ice and water, and for the high heat capacity and enthalpies of fusion and vaporization of water. The ability of water to interact with other substances by means of hydrogen bonds, dipole–dipole interactions, and ion–dipole interactions accounts for its unique ability to dissolve such varied substances as salts, gases, and certain organic compounds.

# Problems

1. Calculate the repulsion-corrected Coulomb energy of a KF ion pair at the average internuclear distance between the two ions (Table 18.1). For what internuclear distance is this energy equal to the KF bond energy?

2. The bond energy of gaseous KI is 323 kJ mol$^{-1}$, and the bond length is 3.05 Å. Assuming that gaseous KI exists as a K$^+$I$^-$ ion pair, calculate the electron affinity of iodine.

3. Calculate the bond energy in gaseous KCl by means of a Born–Haber-type cycle. Compare with the experimental value in Table 18.1.

4. Arrange the following gas-phase ion pairs in order of increasing energy released when they form from gas-phase ions: Mg$^{2+}$O$^{2-}$, Na$^+$F$^-$, K$^+$Cl$^-$, and Ca$^{2+}$S$^{2-}$.

5. Show that London dispersion forces have much shorter range than Coulomb forces by computing for each interaction the factor by which the attractive potential energy decreases as the distance between the interacting species increases from 1 molecular diameter to 3 molecular diameters.

6. Plot the Lennard–Jones potential energy between two argon atoms in the range of 2 to 20 Å. What is the strength of the interaction between two argon atoms at their most probable internuclear separation in condensed phase?

7. What is the value of the Lennard–Jones potential energy between two molecules separated by a distance $\sigma$? What is the distance between two molecules when the Lennard–Jones potential energy is equal to $\phi$?

8. On the basis of the data in Table 18.2, determine the van der Waals radius of krypton.

9. Use the van der Waals radius of krypton (Problem 8) to determine the fraction of the volume occupied by 1 mole of krypton gas at standard temperature and pressure.

10. Using data tabulated in this chapter, in Chapter 6, and in Chapter 20 (Table 20.4), list the atomic, ionic, and van der Waals radii of fluorine. Which radius has the smallest value and which has the largest? Account for these observations.

11. Use the data in Table 18.2 to calculate the interaction energy between two N$_2$ molecules separated by a distance of (a) 5.00 Å and (b) 1.50 Å. State whether in each case the two molecules attract or repel each other.

12. Which of the following would you expect to show London dispersion forces? dipole–dipole interactions? hydrogen bonding? (a) CH$_4$, (b) CH$_3$Cl, (c) CH$_2$Cl$_2$, (d) CHCl$_3$, and (e) CCl$_4$.

13. Which of the following would you expect to form hydrogen bonds: (a) CH$_3$Br, (b) (CH$_3$)$_2$O, and (c) CH$_3$OH?

14. Which of the following would you expect to show dipole–dipole interactions: (a) H$_2$S, (b) CO$_2$, (c) CF$_4$, and (d) CO?

15. The interhalogen compound ClF has a lower boiling point than $Cl_2$ even though it is polar. Explain.
16. Which of the following substances can form intermolecular hydrogen bonds: (a) $H_2O_2$, (b) $CH_3F$, (c) $H_2SO_4$? Draw hydrogen-bonded structures for each species that can form hydrogen bonds.
17. The H—F hydrogen bond energy is greater than the H—O hydrogen bond energy, yet solid hydrogen fluoride has a much lower melting point than ice. Explain.
18. If the water molecule were linear, hydrogen bonding in water would be less important than it is. Discuss the validity of this statement.
19. Look up the boiling points of the substances in the following series (e.g., in the Handbook of Chemistry and Physics) and explain the trends in terms of the intermolecular interactions involved: (a) C, $N_2$, $O_2$, $F_2$, Ne; (b) He, Ne, Ar, Kr, Xe; and (c) $H_2O$, $H_2S$, $H_2Se$, $H_2Te$.
20. On the basis of their intermolecular interactions, arrange the following substances in order of increasing boiling points: NaF, $Al_2O_3$, Ar, NO, and $CH_4$.
21. Determine which intermolecular forces act in (a) $C_2H_5OH$, (b) $BrF_5$, (c) $(CH_3)_2CO$, and (d) $PH_3$.
22. Which of the following substances are expected to exhibit intermolecular hydrogen bonding: (a) $N_2H_4$, (b) $CH_4$, (c) HI, (d) $H_2Se$, and (e) $BeH_2$?
23. State which intermolecular interactions are important in (a) $SF_6$, (b) $Cl_2CO$, (c) NOCl, (d) $BCl_3$, (e) $AsCl_5$, and (f) $SiH_4$.
24. Arrange the following isomers of pentane, $C_5H_{12}$, in order of increasing boiling point:

    (a) $CH_3CH_2CH_2CH_2CH_3$, n-pentane
    (b) $CH_3CH_2CHCH_3$, isopentane
          |
          $CH_3$

              $CH_3$
              |
    (c) $H_3C—C—CH_3$, neopentane
              |
              $CH_3$

25. Indicate whether positive, negative, or no deviations from Raoult's law might be expected for the following solutions: (a) $C_6H_{12}$ and $C_2H_5OH$, (b) $H_2O$ and $C_7H_{16}$, (c) $(CH_3)_2O$ and HCl, and (d) $C_6H_6$ and $C_6H_5CH_3$.
26. Using the correlation in Figure 18.18, estimate the hydration energies of (a) $Ra^{2+}$, (b) $Tl^+$, and (c) $Tl^{3+}$. (The ionic radii are 1.43 Å, 1.47 Å, and 1.95 Å, respectively.)
27. Silver fluoride is the only silver halide to form a hydrate. Explain.
28. For each of the following pairs of substances, choose the one with the lower value of the property mentioned and justify your choice. (a) Normal boiling point: $CF_4$ or $CCl_4$, (b) melting point: $MgF_2$ or $SiF_4$, and (c) solubility in water: $C_6H_6$ or $C_5H_5N$.
29. The solubility in water of the following isoelectronic salts decreases in the order $KClO_4$, $CaSO_4$, and $ScPO_4$. Explain.
30. Ammonia is very soluble in water but nitrogen trichloride, $NCl_3$, is not. Explain.
31. Methanol, $CH_3OH$, is miscible with water but octanol, $C_8H_{18}OH$, is insoluble in water. Explain.
32. Although oxygen can form much stronger hydrogen bonds with water than chlorine, the chlorides of many metals are much more soluble than their oxides. Explain.
33. Look up the solubilities of the alkali halides in cold water (e.g., in the Handbook of Chemistry and Physics). (a) Convert the data to molalities of the saturated solutions. (b) Write these molal solubilities in a table in which the columns are labelled F, Cl, Br, and I, the rows are labelled Li, Na, K, Rb, and Cs, and the values are written at the intersections of the columns and rows. (c) Comment on any trends that may be seen in this representation of the solubilities.
34. For each of the following pairs of substances, choose the one with the higher value of the property mentioned and justify your choice. (a) Normal boiling point: $CH_3OH$ or $CH_3Cl$, (b) enthalpy of vaporization: $NH_3$ or $PH_3$, and (c) vapor pressure at 0°C: $Br_2$ or $I_2$.
35. Write a chemical equation for the regeneration of a zeolite that has been used to remove calcium ions from water.

# 19

# The Representative Elements

In this chapter we bring to bear on the chemistry of the representative elements many of the principles developed earlier in this book. We shall find that there are many regularities in chemical behavior which can be correlated with the positions of the elements in the periodic table. However, chemical behavior generally does not reflect a single underlying factor, but the combined effect of a number of factors. An apparently simple process, such as the oxidation of a metal in aqueous solution is, as we shall see, affected by the sublimation energy of the metal, its ionization energy, and the hydration energy of the gaseous ion. The influence of so many different factors reduces the regularities that result from the influence of any one of them and results in occasional anomalies and exceptions to otherwise regular trends.

## Acid–Base Chemistry

Acids and bases have been extensively examined in Chapter 8 with an emphasis on their ionization equilibria. The chemical behavior of these substances, as well as that of various related compounds, constitutes a significant component of the chemistry of the representative elements. The substances which exhibit acidic or basic properties include the binary compounds of the representative elements with hydrogen and oxygen as well as the hydroxides and oxoacids. The change of these compounds from basic to acidic occurs primarily from left to right across the periodic table. These trends can, to a significant extent, be correlated with changes in chemical bonding. However, while bonding determines the properties of isolated molecules, additional factors, such as hydration enthalpies, play a substantial role in determining the properties of aqueous solutions of acids and bases.

## 19.1 Binary Compounds with Hydrogen

*a. The Ionic Hydrides* The alkali metals and most of the alkaline earths form binary hydrides by the direct combination of the metals with hydrogen at elevated temperatures, e.g.,

$$2Na(l) + H_2(g) \to 2NaH(s)$$

$$Ca(s) + H_2(g) \to CaH_2(s)$$

The unusual feature of these compounds is that they are ionic. The presence of hydrogen as the negatively charged hydride ion, $H^-$, can readily be demonstrated by electrolysis of the molten hydride (Figure 19.1). Hydrogen gas is liberated at the *anode*, indicating that it is produced in an oxidation half-reaction, i.e.,

$$2H^- \to H_2(g) + 2e^-$$

Recall that the electrolysis of aqueous solutions of electrolytes also liberates hydrogen gas, but at the *cathode*, showing the reduction of $H^+$ to $H_2$ (Section 13.10c).

The formation of ionic hydrides reflects the substantial difference in electronegativity between the highly electropositive alkali or alkaline earth metals and hydrogen. The electronegativities of the representative elements in the higher groups are generally comparable to or larger than the electronegativity of hydrogen. These elements consequently form covalent compounds with hydrogen.

The ionic hydrides are extremely reactive because of the strong tendency of the $H^-$ ion to act as an electron donor, that is, as a Lewis base. The hydrides react vigorously with water, forming hydrogen and the metal hydroxide, for example,

$$KH(s) + H_2O \to K^+ + OH^- + H_2(g)$$

This reaction can be interpreted as a Lewis acid–base reaction in which the $H^-$

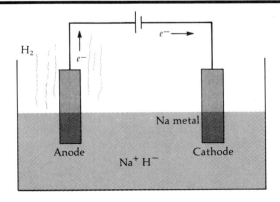

**FIGURE 19.1** The electrolysis of molten sodium hydride, showing the liberation of hydrogen gas at the anode.

ion donates an electron pair to a solvated $H^+$ ion from the ionization of water to form hydrogen:

$$H:^- + H^+ \rightarrow H\text{---}H$$

In their reactions with water, the metal hydrides form strongly basic solutions and at the same time act as powerful reducing agents.

**b. The Basicity of Ammonia** The binary compounds of hydrogen and the elements of Groups IIIA–VIIA are covalently bonded. The compounds with the elements in Groups IIIA and IVA, such as methane, generally have no significant aqueous chemistry. Ammonia is a weak base in aqueous solution and the compounds of Groups VIA and VIIA, such as hydrogen sulfide and the hydrogen halides, form acidic solutions.

Why is it that an aqueous solution of ammonia is basic rather than acidic, that is, why does ammonia react with water to give $NH_4^+$

$$NH_3(g) + H_2O(l) \rightarrow NH_4^+(aq) + OH^-(aq) \tag{19.1}$$

rather than $NH_2^-$?

$$NH_3(g) + H_2O(l) \rightarrow H_3O^+(aq) + NH_2^-(aq) \tag{19.2}$$

This question can be answered by considering the enthalpy changes in reactions 19.1 and 19.2. These reactions are complex processes since they involve the formation of aqueous ions from gaseous ammonia and water. To understand the signs and magnitudes of the reaction enthalpies, the overall reactions can be broken down into a sequence of simpler steps by means of a Born–Haber-type cycle (Section 18.1). It is customary to focus on the enthalpy changes rather than on the free energy changes, which determine the spontaneity of the reactions, since the differences between processes such as reactions 19.1 and 19.2 result primarily from differences in $\Delta H$.

The cycle corresponding to reaction 19.1, shown in Figure 19.2a, may be formulated as follows. First, water is vaporized,

$$H_2O(l) \rightarrow H_2O(g) \qquad \Delta H = \Delta H_{vap}(H_2O) \tag{19.3}$$

**FIGURE 19.2** Born–Haber-type cycles for the reaction of ammonia with water: (a) ammonia acting as a base and (b) ammonia acting as an acid.

$$NH_3(g) + H(g) + OH(g) \xrightarrow{I_H - \varepsilon_{OH}} NH_3(g) + H^+(g) + OH^-(g)$$
$$\uparrow D_{H-O} \qquad \downarrow -P_{NH_3}$$
$$NH_3(g) + H_2O(g) \qquad NH_4^+(g) \qquad + OH^-(g)$$
$$\uparrow \Delta H_{vap} \qquad \downarrow \Delta H_{hyd} \qquad \downarrow \Delta H_{hyd}$$
$$NH_3(g) + H_2O(l) \xrightarrow{\Delta H} NH_4^+(aq) \qquad + OH^-(aq)$$

(a)

$$H(g) + NH_2(g) + H_2O(l) \xrightarrow{I_H - \varepsilon_{NH_2}} H^+(g) + NH_2^-(g) + H_2O(l)$$
$$\uparrow D_{H-N} \qquad \downarrow \Delta H_{hyd} \downarrow \Delta H_{hyd} \downarrow$$
$$NH_3(g) + H_2O(l) \xrightarrow{\Delta H} H_3O^+(aq) + NH_2^-(aq)$$

(b)

The enthalpy change is the known enthalpy of vaporization. Second, water vapor is dissociated into H and OH fragments,

$$H_2O(g) \rightarrow H(g) + OH(g) \qquad \Delta H = D_{H-O} \qquad (19.4)$$

The enthalpy change is given by the H—O bond energy. Third, an electron is transferred from the hydrogen to the hydroxyl radical (i.e., OH),

$$H(g) + OH(g) \rightarrow H^+(g) + OH^-(g) \qquad \Delta H = I_H - \varepsilon_{OH} \qquad (19.5)$$

This process can be pictured as the sum of two separate steps, the ionization of a hydrogen atom, for which $\Delta H$ is given by the ionization energy $I_H$, and the gain of an electron by the hydroxyl radical, for which $\Delta H$ is given by the negative of the electron affinity $-\varepsilon_{OH}$. Fourth, gaseous $NH_3$ and $H^+$ react to form gaseous $NH_4^+$,

$$NH_3(g) + H^+(g) \rightarrow NH_4^+(g), \qquad \Delta H = -p_{NH_3} \qquad (19.6)$$

Since the reverse reaction involves the removal of a proton from a positive ion, the enthalpy change for the reverse reaction is called the proton affinity, $p_{NH_3}$, in analogy to the electron affinity. Consequently, the enthalpy change for the addition of a proton to ammonia is $-p_{NH_3}$. Finally, the gaseous ions are hydrated,

$$NH_4^+(g) + OH^-(g) \rightarrow NH_4^+(aq) + OH^-(aq)$$
$$\Delta H = \Delta H_{hyd}(NH_4^+) + \Delta H_{hyd}(OH^-) \qquad (19.7)$$

and the enthalpy changes are the enthalpies of hydration. The overall reaction, Equation 19.1, is equal to the sum of Equations 19.3–19.7, and the corresponding enthalpy change is equal to the sum of the individual $\Delta H$ values,

$$\Delta H = \Delta H_{vap}(H_2O) + D_{H-O} + I_H - \varepsilon_{OH} - p_{NH_3} + \Delta H_{hyd}(NH_4^+) \qquad (19.8)$$
$$+ \Delta H_{hyd}(OH^-)$$
$$= (44 + 463 + 1312 - 176 - 866 - 377 - 460) \text{ kJ mol}^{-1}$$
$$= -60 \text{ kJ mol}^{-1}$$

The enthalpy change in reaction 19.2 can be evaluated by means of a similar cycle, shown in Figure 19.2b. The sequence of steps involves the dissociation of $NH_3$ to H plus $NH_2$, followed by the transfer of an electron from H to $NH_2$, and concluding with the hydration of the resulting $H^+$ and $NH_2^-$ ions to form the products of reaction 19.2. The result is

$$\Delta H = D_{H-N} + I_H - \varepsilon_{NH_2} + \Delta H_{hyd}(H^+) + \Delta H_{hyd}(NH_2^-) \qquad (19.9)$$
$$= (391 + 1312 - 71 - 1130 - 460) \text{ kJ mol}^{-1}$$
$$= 42 \text{ kJ mol}^{-1}$$

Comparison of Equations 19.8 and 19.9 shows that the ionization of ammonia to form a basic solution is exothermic whereas the ionization of ammonia to form an acidic solution is endothermic. The sign of the free energy change is, of course, determined by a combination of the enthalpy and the entropy terms. Reactions 19.1 and 19.2 are similar in that 1 mole of a gaseous substance reacts to form 2 moles of aqueous ions. Recall from Section 11.5c that a reduction in the number of moles of gaseous substances leads to a decrease in entropy. While this decrease is partially compensated by the increase in the number of ions in solution, the net effect is still a decrease in entropy. Since $\Delta G = \Delta H - T\,\Delta S$, it follows that the combination $\Delta H > 0$ and $\Delta S < 0$, obtained for reaction 19.2, *must* lead to an increase in the free energy. Thus, the formation of an acidic solution from ammonia is not a spontaneous process. On the other hand, the formation of a basic solution *can* occur spontaneously, depending on the relative magnitude of the $\Delta H$ and $T\,\Delta S$ terms. The experimental observation that ammonia is basic indicates that the enthalpy term prevails.

A detailed comparison of Equations 19.8 and 19.9 shows that no one single factor accounts for the basicity of ammonia. We see that the enthalpy change in the base cycle (Equation 19.8) has a large negative contribution from the proton affinity term, $-p_{NH_3} = -866$ kJ mol$^{-1}$, which is not present in the acid cycle. On the other hand, the hydration enthalpy of the cation is much larger in the acid cycle [$\Delta H_{hyd}(H^+) = -1130$ kJ mol$^{-1}$] than in the base cycle [$\Delta H_{hyd}(NH_4^+) = -377$ kJ mol$^{-1}$]—the hydration of a proton is particularly exothermic. This factor, by itself, would make ammonia acidic. However, when all the factors are combined, the net overall enthalpy change favors the base reaction.

---

**Example 19.1**   Construct a Born–Haber-type cycle for the reaction

$$HCl(g) + H_2O(l) \rightarrow H_3O^+(aq) + Cl^-(aq)$$

and show that $\Delta H < 0$. The hydration enthalpy of Cl$^-$ is $-322$ kJ mol$^{-1}$. The additional data may be found elsewhere in this book.

The following cycle, drawn in analogy to Figure 19.2, corresponds to the above reaction:

$$\begin{array}{ccc}
H(g) + Cl(g) + H_2O(l) & \xrightarrow{I_H - \varepsilon_{Cl}} & H^+(g) + Cl^-(g) + H_2O(l) \\
\uparrow D_{HCl} & & \downarrow \Delta H_{hyd}(H^+) \quad \downarrow \Delta H_{hyd}(Cl^-) \\
HCl(g) + H_2O(l) & \xrightarrow{\Delta H} & H_3O^+(aq) + Cl^-(aq)
\end{array}$$

Since the enthalpy change is independent of path, we obtain

$$\Delta H = D_{HCl} + I_H - \varepsilon_{Cl} + \Delta H_{hyd}(H^+) + \Delta H_{hyd}(Cl^-)$$

The bond energy of HCl is given in Table 17.1; the ionization energy and electron affinity are given in Table 6.4; and the hydration energy of H$^+$(g) is given in Equation 19.9. The result is

$$\Delta H = (432 + 1312 - 349 - 1130 - 322) \text{ kJ} = -57 \text{ kJ}$$

The negative value of $\Delta H$ contrasts with the positive $\Delta H$ for the comparable reaction of ammonia (Equation 19.2). Comparison with Equation 19.9 indicates that the difference can be attributed to the much larger electron affinity of Cl than of NH$_2$. ∎

**c. The Hydrohalic Acids**  The acid strength of aqueous solutions of the binary compounds of the nonmetals with hydrogen increases from left to right across a given period. For example, in the third period the acid strength increases in the order H$_3$P < H$_2$S < HCl. The dominant factor responsible for this trend is the increasing electronegativity of the nonmetal atom. The bond between hydrogen and the nonmetal becomes increasingly polar and favors ionization in aqueous solution.

The hydrohalic acids, which are aqueous solutions of the hydrogen halides, are the most widely used of the binary hydrogen acids. The hydrogen halides are colorless gases with sharp odors. They can be made by the direct combination of the elements or, more coveniently, by the action of a nonoxidizing acid on a soluble halide. Sulfuric acid is usually used for the preparation of HF and HCl, as in the reaction

$$2NaCl(aq) + H_2SO_4(aq) \rightarrow 2Na^+ + SO_4^{2-} + 2HCl(g)$$

Phosphoric acid is used for the preparation of HBr and HI since sulfuric acid is too strong an oxidizing agent to permit the formation of these easily oxidized halides. Another preparative method is the hydrolysis of the covalent halides of the nonmetals, for example,

$$PBr_3(l) + 3H_2O \rightarrow H_3PO_3(aq) + 3HBr(g)$$

In addition to the general types of reactions considered in Section 2.5, some of the hydrohalic acids can react in distinctive ways. Concentrated hydrofluoric acid has the unique ability to attack silica and silicates. Glass, which is a mixture of silicates, is etched by this acid:

$$CaSiO_3(s) + 6HF(aq) \rightarrow CaF_2(s) + SiF_4(g) + 3H_2O(l)$$

The escape of gaseous $SiF_4$ from the reaction mixture drives the reaction to completion.

The hydrohalic acids vary in strength in a most unusual way: HF is a moderately weak acid ($K_a = 6.5 \times 10^{-4}$) while HCl, HBr, and HI all are strong acids. The increase in electronegativity of the element bonded to hydrogen, which accounts for the increase in acid strength across a given period, cannot be invoked to explain these observations. Since fluorine is the most electronegative element, HF should on this basis be the strongest rather than the weakest acid in this group.

The ionization of an acid is, in actuality, a rather complicated process. It is helpful to examine the sequence of steps which correspond to ionization by means of a Born–Haber-type cycle. The value of the acid ionization constant is related to the standard free energy change which, in turn, is the resultant of the $\Delta G°$ values associated with the individual steps of the cycle. In comparing the strength of the hydrohalic acids it is appropriate to deal with the free energy changes rather than with the enthalpy changes since the contribution of

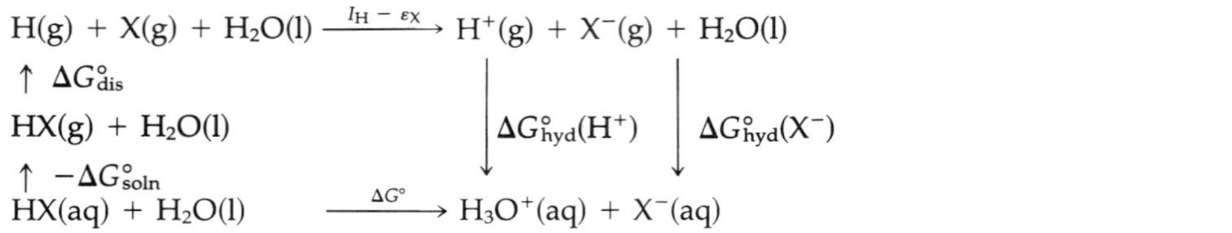

$$H(g) + X(g) + H_2O(l) \xrightarrow{I_H - \varepsilon_X} H^+(g) + X^-(g) + H_2O(l)$$

$\uparrow \Delta G°_{dis}$

$HX(g) + H_2O(l)$ $\qquad\qquad\qquad \Big\downarrow \Delta G°_{hyd}(H^+) \quad \Big\downarrow \Delta G°_{hyd}(X^-)$

$\uparrow -\Delta G°_{soln}$

$HX(aq) + H_2O(l) \xrightarrow{\Delta G°} H_3O^+(aq) + X^-(aq)$

**FIGURE 19.3** Born–Haber-type cycle for the ionization of a hydrohalic acid. The formation of gaseous $H^+$ and $X^-$ from gaseous H and X, respectively, does not change the entropy of the system. Consequently, it is permissible to use $I_H - \varepsilon_X$ for the free energy change in this process.

**TABLE 19.1 Born–Haber Cycle Data for the Ionization of Hydrohalic Acids at 25°C**[a]

| Acid | $-\Delta G°_{soln}$ | $\Delta G°_{dis}$ | $I_H$ | $-\varepsilon_X$ | $\Delta G°_{hyd}(H^+)$ | $\Delta G°_{hyd}(X^-)$ | $\Delta G°_{calc}$ | $K_a$ |
|---|---|---|---|---|---|---|---|---|
| HF  | 25 | 536 | 1312 | −328 | −1100 | −423 | 22  | $1.4 \times 10^{-4}$ |
| HCl | −4 | 402 | 1312 | −349 | −1100 | −305 | −44 | $5.1 \times 10^{7}$ |
| HBr | −4 | 339 | 1312 | −325 | −1100 | −276 | −54 | $2.9 \times 10^{9}$ |
| HI  | −4 | 272 | 1312 | −297 | −1100 | −238 | −55 | $4.3 \times 10^{9}$ |

[a]All data in kJ mol$^{-1}$.

the entropy term is significantly different for HF than for the other acids, as we shall see.

The cycle in Figure 19.3 may be used for the evaluation of $\Delta G°$ for the ionization of the hydrohalic acids. Note that this cycle is essentially the same as that illustrated in Example 19.1, except that we start with the aqueous hydrogen halide instead of with the gas. The formation of HX(g) from HX(aq) is the reverse of the formation of an aqueous solution of the gas, so that the free energy change for this process is the negative of the free energy of solution.

Table 19.1 lists the various contributions to the overall standard free energy changes as well as the calculated values of $K_a$ at 25°C [$\Delta G° = -RT \ln K_a$ (Equation 11.38)]. The calculation is in qualitative agreement with the experimental observations: HF is a moderately weak acid whereas the other acids have such large values of $K_a$ as to be completely ionized. The data show that this difference cannot be attributed to a single factor but to a combination of the following factors. First, the bond energy of HF, and hence $\Delta G°_{dis}$, is especially large. Second, the electron affinity of fluorine is surprisingly small and out of line with the trend for the other halogens. Third, the free energy of hydration of F$^-$, while the most negative of all the halide ions, is actually less negative than expected from the enthalpy of hydration [$\Delta H°_{hyd}(F^-) = -468$ kJ mol$^{-1}$]. The difference reflects the relatively large negative entropy of hydration resulting from hydrogen bonding between the F$^-$ ion and H$_2$O. Hydrogen bonding leads to increased ordering of water molecules about the F$^-$ ions and thereby decreases the entropy.

## 19.2 Binary Oxides

Nearly all the elements form binary compounds with oxygen. Most of these compounds are **oxides,** in which oxygen is assigned an oxidation number of −2. The most active metals—the alkali metals and some of the alkaline earths—also form **peroxides,** such as Na$_2$O$_2$ or BaO$_2$, **superoxides,** such as KO$_2$, and **ozonides,** compounds derived from ozone, such as KO$_3$.

Table 19.2 lists the common oxides of the representative elements, and Table 19.3 summarizes some of the reactions that may be used to make these compounds. Note that the formulas of the Group IA–IIIA oxides change from M$_2$O to MO to M$_2$O$_3$, corresponding to the oxidation numbers of +1, +2, and +3 which characterize the elements in these three groups, respectively (Section 19.4). The elements in Group IVA–VIIA for the most part form two oxides, reflecting the two common oxidation states which characterize these elements.

Recall from Section 16.3 that the type of bond formed between two atoms is correlated with the difference in their electronegativity. Since next to fluorine, oxygen is the most electronegative element, we can expect a gradual change from ionic to covalent bonding for the oxides of elements from left to right across the periodic table. As expected, the oxides of the alkali metals, the

## TABLE 19.2 Common Oxides of the Representative Elements[a]

| IA | IIA | IIIA | IVA | VA | VIA | VIIA | VIIIA |
|---|---|---|---|---|---|---|---|
| $Li_2O$ | BeO | $B_2O_3$ | CO, $CO_2$ | $N_2O$, NO, $NO_2$, $N_2O_4$ | $O_2$, $O_3$ | $OF_2$ | |
| $Na_2O$ | MgO | $Al_2O_3$ | $SiO_2$ | $P_2O_3$, $P_2O_5$ | $SO_2$, $SO_3$ | $Cl_2O$, $ClO_2$, $Cl_2O_7$ | |
| $K_2O$, $Rb_2O$ | CaO, SrO | $Ga_2O_3$, $In_2O_3$ | $GeO_2$, SnO, $SnO_2$ | $As_2O_3$, $Sb_2O_3$, $Sb_2O_5$ | $SeO_2$, $SeO_3$, $TeO_2$, $TeO_3$ | $BrO_2$, $I_2O_5$ | $XeO_3$, $XeO_4$ |
| $Cs_2O$ | BaO | $Tl_2O_3$ | PbO, $PbO_2$ | $Bi_2O_3$ | | | |

[a]Oxides to the left of the black line are ionic. Oxides to the right of the colored line exist as molecules, and oxides to the left and below the colored line exist as polymeric species.

alkaline earths, and some of the Group IIIA metals are ionic. These oxides melt at high temperatures and have a characteristic crystalline structure (Section 20.7).

The elements that lie close to oxygen in the periodic table and have similar electronegativities form discrete, covalently bonded molecular compounds, which are gaseous at ordinary temperatures. Included in this group are the common oxides of carbon, nitrogen, sulfur, and most of the halogens. Since

## TABLE 19.3 Reactions Used for the Preparation of Common Oxides of the Representative Elements

| Group IA | Group IIA |
|---|---|
| $2Li(s) + \frac{1}{2}O_2(g) \rightarrow Li_2O(s)$[a] | $Mg(s) + \frac{1}{2}O_2(g) \xrightarrow{\Delta} MgO(s)$[c] |
| $NaNO_3(s) + 5Na(s) \rightarrow 3Na_2O(s) + \frac{1}{2}N_2(g)$[b] | $MgCO_3(s) \xrightarrow{\Delta} MgO(s) + CO_2(g)$[c] |
| **Group IIIA** | **Group IVA** |
| $2H_3BO_3(s) \xrightarrow{\Delta} B_2O_3(s) + 3H_2O(g)$ | $CaCO_3(s) + 2H_3O^+ \rightarrow Ca^{2+} + CO_2(g) + 3H_2O$ |
| $2Al(s) + \frac{3}{2}O_2(g) \xrightarrow{\Delta} Al_2O_3(s)$ | $SiCl_4(l) + 2H_2O \rightarrow SiO_2(s) + 4HCl(aq)$ |
| **Group VA** | **Group VIA** |
| $NH_4NO_3(s) \xrightarrow{\Delta} N_2O(g) + 2H_2O(g)$ | $SO_3^{2-} + 2H_3O^+ \rightarrow SO_2(g) + 3H_2O$ |
| $2NO_2^- + 2I^- + 4H_3O^+ \rightarrow 2NO(g) + I_2(s) + 6H_2O$ | $Fe_2(SO_4)_3(s) \xrightarrow{\Delta} Fe_2O_3(s) + 3SO_3(g)$ |
| $2NO(g) + O_2(g) \rightarrow 2NO_2(g)$ | |
| $P_4(s) + 5O_2(g) \rightarrow P_4O_{10}(s)$ | |

| Group VIIA |
|---|
| $2F_2(g) + H_2O \xrightarrow{KF} OF_2(g) + 2HF(aq)$[d] |
| $2ClO_3^- + H_2C_2O_4(aq) + 2H_3O^+ \rightarrow 2ClO_2(g) + 2CO_2(g) + 4H_2O$[d] |
| $12HClO_4(aq) + P_4O_{10}(s) \rightarrow 6Cl_2O_7(aq) + 4H_3PO_4(aq)$[d] |
| $Br_2(l) + 4O_3(g) \rightarrow 2BrO_2(l) + 4O_2(g)$[d] |

[a]This reaction forms the oxide for lithium only; the other alkali metals form the peroxide ($Na_2O_2$) or superoxide ($KO_2$).
[b]The analogous reaction may be used to prepare oxides of the heavier alkali metals.
[c]The analogous reaction also occurs for the heavier alkaline earths.
[d]WARNING! Halogen oxides must be handled with care as they tend to explode.

fluorine is more electronegative than oxygen, the electron density in $OF_2$ is concentrated on the fluorine atoms and this compound is more appropriately categorized as a fluoride than as an oxide.

The covalently bonded oxides of the heavier nonmetals as well as those of the semimetals (Table 19.2) do not consist of discrete small molecules but of **polymeric species,** that is, species with a repeating stoichiometric unit. For example, phosphorus pentoxide actually exists as a $P_4O_{10}$ molecule (Figure 19.4a) and silicon dioxide, with the stoichiometric formula $SiO_2$, consists of a very extensive three-dimensional network (Figure 19.4b). The physical properties of molecular and polymeric oxides of similar stoichiometric composition can differ dramatically. For example, carbon dioxide ($CO_2$) is a gas above $-78.5°C$ at atmospheric pressure while silicon dioxide ($SiO_2$) is a solid with a melting point above 1600°C.

The distinction between molecular and polymeric oxides can, at least in part, be attributed to the ease with which the central atom can form pi bonds

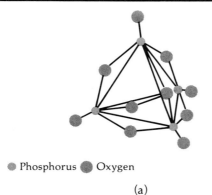

**FIGURE 19.4** The structures of some polymeric oxides: (a) $P_4O_{10}$ and (b) $SiO_2$.

with the attached oxygen atoms. Elements such as carbon, nitrogen, and oxygen readily form pi bonds owing to the presence of half-filled p orbitals. The small size of these atoms permits strong p-orbital overlap to take place and leads to the formation of strong sigma–pi double bonds. By contrast, nonmetallic atoms in the third and higher periods, such as silicon or arsenic, are much larger, and the overlap between their p orbitals and those of oxygen is relatively weak. The formation of double bonds with oxygen is not common. Instead, bonding in these oxides involves a network of single sigma bonds.

The oxides of the elements can be classified by their acid–base properties. Some oxides dissolve in water to form either acidic solutions, e.g.,

$$SO_3(s) + 2H_2O \rightarrow H_3O^+ + HSO_4^-$$

or basic solutions, for example,

$$Na_2O(s) + H_2O \rightarrow 2Na^+ + 2OH^-$$

These oxides are the **anhydrides** of the corresponding acids or bases, meaning that they form the latter when dissolved in water. Other oxides do not dissolve in water but do react with acids or bases. These oxides form weak bases or weak acids, respectively. For example, magnesium oxide is relatively insoluble in water but dissolves readily in acid as a result of the reaction

$$MgO(s) + 2H^+ \rightarrow Mg^{2+} + H_2O$$

Since the acid is neutralized, magnesium oxide is a basic oxide. Similarly, silicon dioxide is insoluble in water but reacts with strong bases:

$$SiO_2(s) + 4OH^- \rightarrow SiO_4^{4-} + 2H_2O$$

Thus, silicon dioxide is an acidic oxide. Finally, some oxides react with both strong acids and strong bases and consequently exhibit both basic and acidic properties. For example, aluminum oxide is insoluble in water but dissolves in either strong acid,

$$Al_2O_3(s) + 6H^+ \rightarrow 2Al^{3+} + 3H_2O$$

or strong base,

$$Al_2O_3(s) + 2OH^- + 3H_2O \rightarrow 2Al(OH)_4^-$$

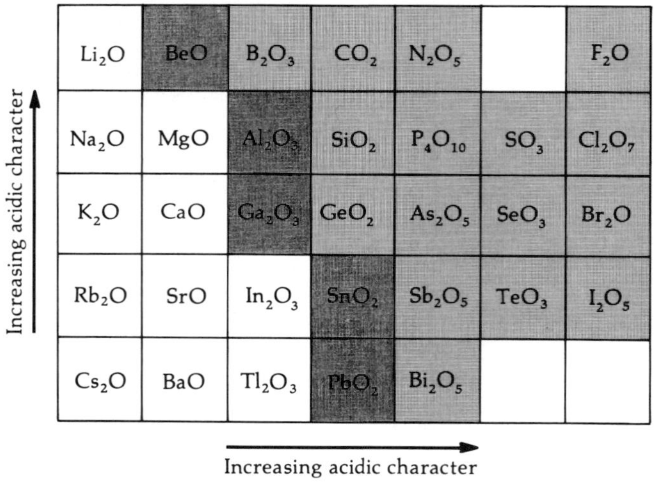

**FIGURE 19.5** Acid–base properties of oxides of the representative elements. The acidic oxides are shown in color and the amphoteric oxides are shaded.

| | | | | | | |
|---|---|---|---|---|---|---|
| Li$_2$O | BeO | B$_2$O$_3$ | CO$_2$ | N$_2$O$_5$ | | F$_2$O |
| Na$_2$O | MgO | Al$_2$O$_3$ | SiO$_2$ | P$_4$O$_{10}$ | SO$_3$ | Cl$_2$O$_7$ |
| K$_2$O | CaO | Ga$_2$O$_3$ | GeO$_2$ | As$_2$O$_5$ | SeO$_3$ | Br$_2$O |
| Rb$_2$O | SrO | In$_2$O$_3$ | SnO$_2$ | Sb$_2$O$_5$ | TeO$_3$ | I$_2$O$_5$ |
| Cs$_2$O | BaO | Tl$_2$O$_3$ | PbO$_2$ | Bi$_2$O$_5$ | | |

Increasing acidic character →

↑ Increasing acidic character

| Oxidation number | +1 | +2 | +3 | +4 | +5 | +6 | +7 |
|---|---|---|---|---|---|---|---|
| Oxide | Na$_2$O | MgO | Al$_2$O$_3$ | SiO$_2$ | P$_4$O$_{10}$ | SO$_3$ | Cl$_2$O$_7$ |
| Acid–base properties | strong base | weak base | amphoteric | weak acid | acid | strong acid | very strong acid |

**FIGURE 19.6** Correlation between acidic properties of oxides and the oxidation number of the element attached to oxygen for the elements of the third period.

Recall that substances exhibiting this behavior are said to be amphoteric (Section 9.10). In all of these cases, basic oxides dissolve or react with the formation of metal cations while acidic oxides form anions.

Figure 19.5 shows the acid–base behavior of various oxides of the representative elements. In analogy to the variation of nonmetallic character (Section 6.9) and to that of electronegativity (Section 6.7), the acidic character of the oxides increases from left to right across the periodic table as well as from bottom to top. These trends lead to a diagonal division, with the basic oxides located in a triangular region on the left side of the table and the acidic oxides located in an inverted triangular region on the right side. The amphoteric oxides lie in a narrow band along the diagonal division, as shown in Figure 19.5. Comparison of this figure with Figure 6.25 indicates that among the representative elements the oxides of the metals generally are basic or amphoteric, those of the semimetals are weakly acidic, and those of the nonmetals are acidic. This correlation is not accidental; it reflects the varying tendency of atoms of these elements to attract electrons.

A different but related way to examine the acidity of oxides is in terms of the oxidation number of the element attached to oxygen. The oxidation numbers of these elements generally increase from left to right across the periodic table. For example, the oxidation number of the elements in the third period increases uniformly from +1 for sodium in Na$_2$O to +7 for chlorine in Cl$_2$O$_7$. The increase in oxidation number is closely correlated with the change from basic to acidic behavior, as shown in Figure 19.6. Moreover, this correlation extends to different oxides of the *same* element. For example, SO$_2$ is less acidic than SO$_3$, As$_2$O$_3$ is less acidic than As$_2$O$_5$, and so on (Figure 19.7).

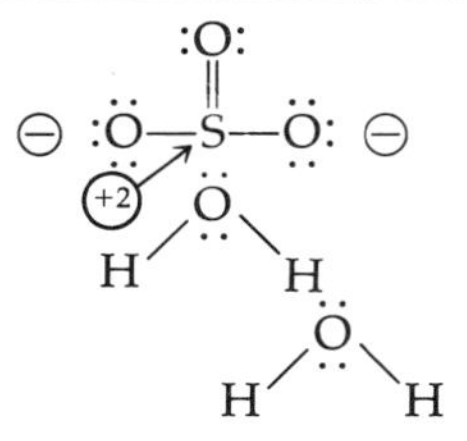

**FIGURE 19.7** Effect of the Lewis structure of SO$_3$ on its acidity in aqueous solution. Because of its high oxidation number (+6) and formal charge, the sulfur atom attracts electrons from an attached H$_2$O molecule. The O—H bonds in this molecule therefore become more polar and one proton (in color) is transferred to a neighboring H$_2$O molecule. SO$_3$ is therefore more acidic than SO$_2$, in which the oxidation number of sulfur is +4.

## 19.3 Hydroxides and Oxoacids

*a. Acidic versus Basic Behavior*  Metal hydroxides and oxoacids can be represented by the general formula $EO_n(OH)_m$, where E is the central atom, $n$ is the number of oxygen atoms without attached hydrogen atoms, and $m$ is the number of hydroxyl groups attached to E. For example, the formulas of aluminum hydroxide and phosphoric acid may be written as $Al(OH)_3$ and $PO(OH)_3$, respectively, reflecting the Lewis structures of the two molecules:

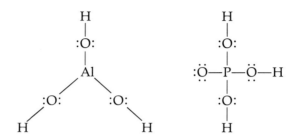

The similarity in structure makes it clear that the behavior of these species as acids or bases depends precisely on which bond is broken in aqueous solution.

Let us examine this question by focusing on the key component of molecules of this type, that is, on the structure

$$E—O—H$$
$$\quad 1 \quad 2$$

If bond 1 is broken in solution, the molecule ionizes with the formation of $E^+$ and $OH^-$ ions and acts as a base. On the other hand, if bond 2 is broken, $H^+$ ions are released, and the molecule acts as an acid. The relative strength of the E—O and O—H bonds depends on the electronegativity of atom E. If the electronegativity difference between E and O is large, that is, for elements of low electronegativity, the E—O bond is ionic. Ionization to $OH^-$ and a cation occurs readily in solution. The alkali and alkaline earth elements consequently form the strongest hydroxides. On the other hand, when the electronegativity of E is relatively high, the electron density is shifted from the O—H to the E—O internuclear region. Ionization then leads to $H^+$ rather than $OH^-$ ions. The strongest oxoacids are consequently formed by the elements in Groups VA–VIIA.

*b. The Common Oxoacids*  The elements in Group VA–VIIA form a number of oxoacids, the more common of which are listed in Table 19.4. The oxoacids of a given element may be categorized by the oxidation state of the nonmetal atom, as indicated. For example, chlorine forms four oxoacids, hypochlorous acid (HClO), chlorous acid ($HClO_2$), chloric acid ($HClO_3$), and perchloric acid ($HClO_4$), in which the oxidation number of chlorine is +1, +3, +5, and +7, respectively.

Oxoacids can be prepared by a variety of reactions. Since many of the nonmetal oxides are anhydrides of oxoacids, the acids may be prepared by dissolving the oxides in water:

**TABLE 19.4 Common Oxoacids of the Elements in Groups VA–VIIA**

| Group VA oxidation state | Acid | Formula |
|---|---|---|
| +1 | Hypophosphorous | $H_3PO_2$ |
| +3 | Nitrous | $HNO_2$ |
|    | Phosphorous | $H_3PO_3$ |
| +5 | Nitric | $HNO_3$ |
|    | Phosphoric | |
|    | (ortho) | $H_3PO_4$ |
|    | (pyro) | $H_4P_2O_7$ |

| Group VIA oxidation state | Acid | Formula |
|---|---|---|
| +4 | Sulfurous | $H_2SO_3$ |
| +6 | Sulfuric | $H_2SO_4$ |

| Group VIIA oxidation state | Acid | Formula |
|---|---|---|
| +1 | Hypohalous[a] | HClO, HBrO, HIO |
| +3 | Halous | $HClO_2$, $HBrO_2$ |
| +5 | Halic | $HClO_3$, $HBrO_3$, $HIO_3$ |
| +7 | Perhalic | $HClO_4$, $HBrO_4$, $H_5IO_6(HIO_4)$[b] |

[a] These are generic names applicable to any member of the group for which a formula is given, e.g., hypochlorous acid, periodic acid, etc.
[b] $HIO_4$ is stable in boiling water.

$$P_4O_{10}(s) + 6H_2O \rightarrow 4H_3PO_4(aq)$$

$$SO_2(g) + H_2O \rightarrow H_2SO_3(aq)$$

$$Cl_2O_7(l) + H_2O \rightarrow 2H^+ + 2ClO_4^-$$

This type of reaction is not completely general since there are instances in which the acid initially formed by dissolution of the oxide is unstable and the overall reaction of the oxide with water is disproportionation. For example, chlorine dioxide, in which chlorine has an oxidation number of +4, reacts with water to form solutions of chlorous and chloric acids, in which the oxidation number of chlorine is +3 and +5, respectively:

$$2ClO_2(g) + H_2O \rightarrow 2H^+ + ClO_2^- + ClO_3^-$$

Some of the oxoacids, notably nitric and sulfuric acids, are among the most widely used chemicals. Nitric acid is generally produced by the oxidation of ammonia (Ostwald process) in the following three-step process:

$$4NH_3(g) + 5O_2(g) \xrightarrow[900°C]{\text{Pt-Rh catalyst}} 4NO(g) + 6H_2O(g) \tag{19.10}$$

$$2NO(g) + O_2(g) \rightarrow 2NO_2(g) \tag{19.11}$$

$$3NO_2(g) + H_2O(l) \rightarrow 2HNO_3(aq) + NO(g) \tag{19.12}$$

The first step, the catalytic oxidation of ammonia, occurs rapidly at high temperature and pressure. Reaction 19.11 is favored at lower temperatures, so the gases from reaction 19.10 are cooled and more air is added to form the reddish-brown nitrogen dioxide. In the final step, nitrogen dioxide is dissolved in water to produce a 70% (by weight) solution of nitric acid in a disproportionation reaction. The nitrogen(II) oxide formed in this step is recycled. Nitric acid is widely used in the manufacture of fertilizers and in the production of organic nitrates which, in turn, are used in plastics, dyes, explosives, and other products.

Sulfuric acid is the most widely used industrial chemical and its production has, in fact, been employed as an index of industrial activity. Among its many uses are the manufacture of fertilizers and other chemicals, petroleum refining, and the production of iron, steel, and other metals. Sulfuric acid is produced by the sequential combustion of sulfur:

$$S(s) + O_2(g) \rightarrow SO_2(g) \tag{19.13}$$

$$SO_2(g) + \tfrac{1}{2}O_2(g) \rightarrow SO_3(g) \tag{19.14}$$

Reaction 19.14 has been discussed in Section 7.6c from the viewpoint of the effect of temperature on the equilibrium point. The reaction is exothermic and the formation of the trioxide is thermodynamically favored at low temperatures. However, the rate of the reaction becomes appreciable only at $\approx 1000°C$, a temperature at which the equilibrium lies far to the left. In practice, the reaction is run at intermediate temperatures and various catalysts are used to increase the reaction rate.

Sulfur trioxide cannot be added directly to water to give sulfuric acid owing to the formation of an acid mist which is not readily absorbed by water. The problem is overcome by first absorbing sulfur trioxide in concentrated sulfuric acid to form pyrosulfuric acid, $H_2S_2O_7$, which is then converted to sulfuric acid of the desired concentration by the addition of water:

$$SO_3(g) + H_2SO_4(l) \rightarrow H_2S_2O_7(l) \tag{19.15}$$

$$H_2S_2O_7(l) + H_2O(l) \rightarrow 2H_2SO_4(aq) \tag{19.16}$$

Concentrated sulfuric acid consists of 98.3% $H_2SO_4$ (by weight) and boils at 338°C. Its dilution is accompanied by the evolution of much heat; the acid must be added to water—very carefully.

### c. Trends in the Strength of Oxoacids

The oxoacids display a remarkably large range of acid ionization constants. Strong acids are completely ionized in aqueous solution and have negative $pK_a$ values. The $pK_a$ values of weak acids can range up to 12 or more for acids such as $HPO_4^{2-}$ or $HS^-$ (Table 8.5).

The values of $pK_a$ can be correlated with the number of nonhydroxylic oxygen atoms in the acid, that is, with the number of oxygen atoms that are not part of an OH group. With the formula of an oxoacid written as $EO_n(OH)_m$, the number of nonhydroxylic oxygen atoms is then equal to $n$. Pauling proposed the following simple correlation between an acid ionization constant and the value of $n$:

$$pK_a = 7 - 5n \tag{19.17}$$

Values calculated from this relation are compared with experimental $pK_a$ values in Table 19.5. It is apparent that the predicted values match the data to within one or two $pK_a$ units. Comparable agreement is obtained for strong and weak acids as well as for monoprotic and polyprotic acids.

## TABLE 19.5 Correlation Between p$K_a$ of Oxoacids and the Value of $n$ in the Formula EO$_n$(OH)$_m$

| Formula | E(OH)$_m$ | | EO(OH)$_m$ | | EO$_2$(OH)$_m$ | | EO$_3$(OH)$_m$ | |
|---|---|---|---|---|---|---|---|---|
| $n$ | 0 | | 1 | | 2 | | 3 | |
| Predicted p$K_a$ (Equation 19.17) | 7 | | 2 | | −3 | | −8 | |
| Acid | | p$K_a$ | Acid | p$K_a$ | Acid[a] | p$K_a$ | Acid[a] | p$K_a$ |
| B(OH)$_3$ (H$_3$BO$_3$) | | 9.2 | NO(OH) (HNO$_2$) | 3.4 | NO$_2$(OH) (HNO$_3$) | −1.4 | ClO$_3$(OH) (HClO$_4$) | −7.3 |
| Si(OH)$_4$ (H$_4$SiO$_4$) | | 9.8 | PO(OH)$_3$ (H$_3$PO$_4$) | 2.1 | SO$_2$(OH)$_2$ (H$_2$SO$_4$) | −3.0 | | |
| Ge(OH)$_4$ (H$_4$GeO$_4$) | | 9.0 | HPO(OH)$_2$ (H$_3$PO$_3$) | 1.8 | ClO$_2$(OH) (HClO$_3$) | −2.7 | | |
| Cl(OH) (HClO) | | 7.5 | H$_2$PO(OH) (H$_3$PO$_2$) | 2.0 | IO$_2$(OH) (HIO$_3$) | 0.8 | | |
| Br(OH) (HBrO) | | 8.7 | SO(OH)$_2$ (H$_2$SO$_3$) | 1.8 | | | | |
| I(OH) (HIO) | | 10.6 | SeO(OH)$_2$ (H$_2$SeO$_3$) | 2.5 | | | | |
| | | | ClO(OH) (HClO$_2$) | 2.0 | | | | |
| PO$_2$(OH)$_2^-$ (H$_2$PO$_4^-$) | | 7.2 | IO(OH) (HIO$_2$) | 1.6 | | | | |
| HPO$_2$(OH)$^-$ (H$_2$PO$_3^-$) | | 6.6 | SO$_3$(OH)$^-$ (HSO$_4^-$) | 1.9 | | | | |
| SO$_2$(OH)$^-$ (HSO$_3^-$) | | 7.2 | SeO$_3$(OH)$^-$ (HSeO$_4^-$) | 1.9 | | | | |

[a] When p$K_a$ < −1, the acid is completely ionized in water. The p$K_a$ values are estimates based on the use of more acidic solvents.

It is clear that such a widely applicable correlation must reflect some general factors about the ability of an oxoacid to donate a proton. Since oxygen is very electronegative, the correlation can be traced to the tendency of the $n$ nonhydroxylic, or terminal, oxygen atoms to attract electrons from the rest of the molecule. A consequence of this attraction is that the O—H bonds are weakened. The tendency of the acid to act as a proton donor therefore increases with the number of electron-withdrawing oxygen atoms, that is, with the value of $n$.

An essentially equivalent interpretation of the Pauling correlation is that the strength of an acid increases with the formal charge of the central atom. Consider, for example, the structure of chloric acid, HClO$_3$:

$$\text{H}-\overset{..}{\underset{..}{\text{O}}}-\overset{+2}{\underset{\underset{\ominus}{\underset{..}{\overset{|}{\text{O}:}}}}{\text{Cl}}}-\overset{..}{\underset{..}{\text{O}}}:\ \ominus$$

Note that the chlorine atom has a formal charge of +2. This unrealistically large positive charge attracts the electron clouds on the oxygen atoms. In particular, the electron density in the O—H bond is reduced as a result of this attraction, making it relatively easy for the molecule to donate the proton. Chloric acid is consequently a strong acid. If the formal charge on Cl were to increase, which happens when an additional oxygen is bonded to chlorine, as in HClO$_4$, the resulting acid loses the proton even more readily. On the other hand, a reduction in the positive formal charge on Cl, which can be achieved by removing an oxygen atom from the molecule, as in HClO$_2$, increases the electron density in the O—H bond and thereby decreases the acid strength. The effects of the number of terminal oxygen atoms and of the positive formal charge on the central atom are equivalent because, so long as only single bonds are used in the Lewis structure of an oxoacid, the value of $n$ is equal to the formal charge of the central atom.

Figure 19.8 shows the equivalence between the number of terminal oxygen atoms and the formal charge on the central atom for the four oxoacids of chlorine and further illustrates the correlation between $n$, or formal charge, and p$K_a$. The oxoacids of phosphorus, whose Lewis structures are depicted in Figure 19.9 provide an additional example. Because some of these acids have one or more P—H bonds, the value of $n$ is equal to 1 and the formal charge on the phosphorus atom is +1 in all cases. Thus, the values of p$K_a$ are in the vicinity of 2, as expected for $n = 1$.

The Pauling correlation expressed in Equation 19.17 has enough general validity that it may be used as a guide to the structure of an oxoacid, as illustrated in the following example.

## Example 19.2

Arsenous acid, H$_3$AsO$_3$, has p$K_a$ = 9.2, in contrast to the corresponding acid of phosphorus, H$_3$PO$_3$, for which p$K_a$ = 1.8. Explain this difference in terms of the structure of H$_3$AsO$_3$.

The small value of p$K_a$ indicates that H$_3$AsO$_3$ is a very weak acid. According to Equation 19.17, we can deduce that $n$ must be zero, for this value yields a p$K_a$ value that is close to the experimental one. The formula should therefore be written as As(OH)$_3$ rather than as HAsO(OH)$_2$, as is the case for phosphorous acid. The Lewis structure of this molecule is therefore

| Acid | p$K_a$ | Single bond Lewis structure | $n$ | Formal charge on Cl |
|---|---|---|---|---|
| Hypochlorous | 7.5 | 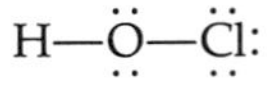 H—Ö—Cl̈: | 0 | 0 |
| Chlorous | 2.0 | 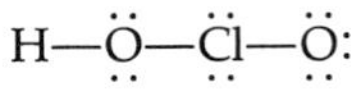 H—Ö—C̈l—Ö: | 1 | +1 |
| Chloric | ~−3 | 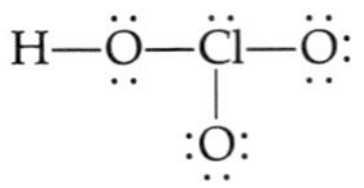 H—Ö—Cl(—Ö:)—Ö: | 2 | +2 |
| Perchloric | ~−8 | 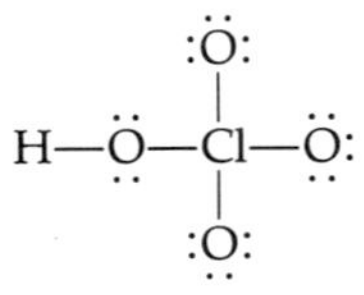 H—Ö—Cl(—Ö:)(—Ö:)—Ö: | 3 | +3 |

**FIGURE 19.8** Correlation between structure and acid strength for the oxoacids of chlorine.

| Acid | $pK_a$ | Single bond Lewis structure | $n$ | Formal charge on P |
|---|---|---|---|---|
| Hypophosphorous ($H_3PO_2$) | 1.9 | 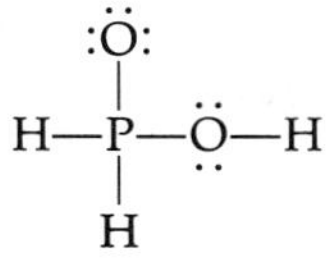 $H-\overset{\overset{\ddot{\text{:\"O:}}}{\vert}}{\underset{\underset{H}{\vert}}{P}}-\ddot{\ddot{O}}-H$ | 1 | +1 |
| Phosphorous ($H_3PO_3$) | 1.5 | 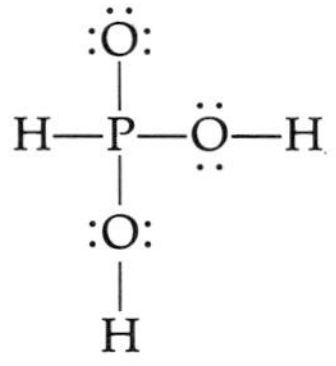 $H-\overset{\overset{\ddot{\text{:\"O:}}}{\vert}}{\underset{\underset{\underset{H}{\vert}}{\underset{\ddot{\text{:\"O:}}}{\vert}}}{P}}-\ddot{\ddot{O}}-H$ | 1 | +1 |
| Phosphoric ($H_3PO_4$) | 2.1 | 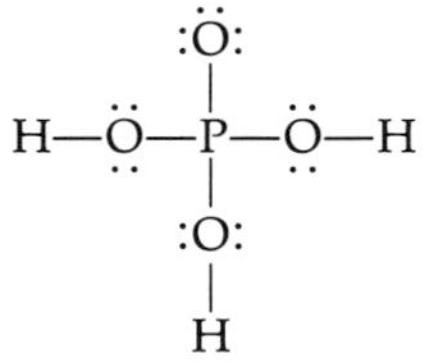 $H-\ddot{\ddot{O}}-\overset{\overset{\ddot{\text{:\"O:}}}{\vert}}{\underset{\underset{\underset{H}{\vert}}{\underset{\ddot{\text{:\"O:}}}{\vert}}}{P}}-\ddot{\ddot{O}}-H$ | 1 | +1 |

**FIGURE 19.9** Correlation between structure and acid strength for the oxoacids of phosphorus.

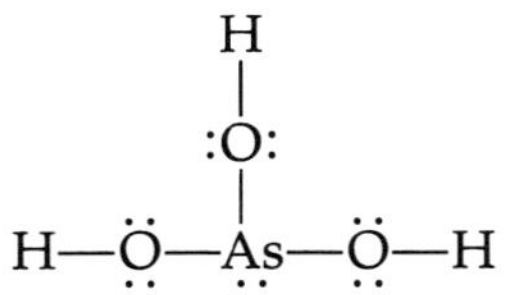

$$H-\ddot{\ddot{O}}-\overset{\overset{\overset{H}{\vert}}{\underset{\ddot{\text{:\"O:}}}{\vert}}}{As}-\ddot{\ddot{O}}-H$$

for which $n = 0$. ∎

# Redox Chemistry

## 19.4 Common Oxidation States of the Representative Elements

The oxidation state of an element in a compound may be determined by means of the rules given in Section 2.7 or by following the procedure discussed in Section 15.5b. The chemical behavior of an element can, to a large measure, be correlated with its oxidation state. Thus, a survey of the important oxidation states of the elements serves a useful role in the examination of similarities and trends in the chemistry of the elements.

The representative elements are arranged by group number in Figure 19.10, and Figure 19.11 shows their common oxidation states. The most striking feature is the periodicity in the oxidation numbers. The maximum oxidation states are generally equal to the group numbers and consequently display the characteristic sawtooth variation between +1 and +7. The metallic elements of Groups IA–IIIA show the group number oxidation states in their simple cations, such as $Na^+$, $Mg^{2+}$, or $Al^{3+}$. Most of the elements in these groups

## 710 The Representative Elements 19.4

| Group | I A | II A | III A | IV A | V A | VI A | VII A | VIII A |
|---|---|---|---|---|---|---|---|---|
| Valence shell configurations | $ns^1$ | $ns^2$ | $ns^2p^1$ | $ns^2p^2$ | $ns^2p^3$ | $ns^2p^4$ | $ns^2p^5$ | $ns^2p^6$ |
| | Li | Be | B | C | N | O | F | Ne |
| | Na | Mg | Al | Si | P | S | Cl | Ar |
| | K | Ca | Ga | Ge | As | Se | Br | Kr |
| | Rb | Sr | In | Sn | Sb | Te | I | Xe |
| | Cs | Ba | Tl | Pb | Bi | Po | At | Rn |

**FIGURE 19.10** The representative elements. Nonmetals are represented by the colored squares, semimetals by the shaded squares, and metals by the open squares.

have only this single oxidation state in their compounds. When the maximum oxidation state of an element is +4 or more, the element is generally covalently bonded in all of its compounds, including those with elements of greatly different electronegativities. Some of the oxoacids examined in Section 19.2 exemplify these high oxidation states, e.g., $HNO_3$ (N, +5), $H_2SO_4$ (S, +6), and $HClO_4$ (Cl, +7).

The minimum oxidation states of the nonmetals in Groups IVA–VIIA are negative and are equal to the group number minus 8. The most electronegative elements display these oxidation states in their common anions, for example, $F^-$, $Cl^-$, and $O^{2-}$. The less electronegative elements show negative oxidation states in their covalent compounds with even less electronegative elements such as hydrogen, as in $CH_4$ (C, −4) and $NH_3$ (N, −3).

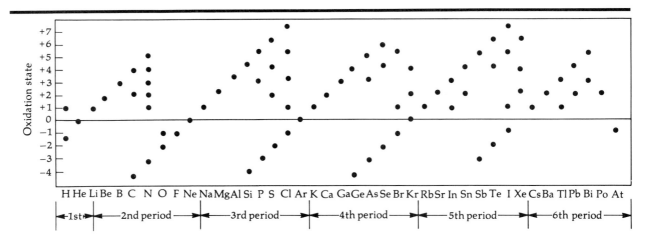

**FIGURE 19.11** Common oxidation states of the representative elements.

Those elements that have a large maximum oxidation state generally also form compounds in which they display one or more oxidation states that are lower than the group number. This tendency is particularly pronounced in the higher periods, where many elements have an oxidation state equal to the group number minus 2. Since the maximum oxidation state corresponds to the withdrawal of *all* the valence electrons (s and p) from the atom towards more electronegative atoms, the state 2 units lower corresponds to the withdrawal of just the p valence electrons. Evidently, the atoms of the heavier elements in a given group are more likely to hold on to their s valence electrons. This happens because the valence shell s orbitals of heavy elements lie at a substantially lower energy than the p orbitals. For example, the most common oxidation state of lead is +2, as in $Pb(NO_3)_2$ or $PbCl_2$, whereas that of carbon, silicon, or germanium is +4.

These generalizations refer to the overall trends, but numerous exceptions occur. For example, fluorine, the most electronegative element, is not assigned any positive oxidation states, oxygen is assigned a positive oxidation state only in its compounds with fluorine, compounds in which bromine has an oxidation state of +7 are very uncommon, and so forth. Nonetheless, the general trends are sufficiently regular to encompass most of the chemistry of the representative elements.

## 19.5 The Metallic Representative Elements

The representative elements include all the nonmetals and semimetals, as well as some of the metals. Figure 19.10 shows the division of the representative elements into these three categories. The metals include all the s-block elements—the alkali and alkaline earth elements—as well as some of the heavier p-block elements.

*a. The s-Block Elements and Their Reactions*  The alkali and alkaline earth elements share several common characteristics: they are excellent conductors of electricity and heat, they are highly reactive, and their compounds are ionic. Beryllium, however, is much less reactive than the others, and its compounds tend to be molecular rather than ionic (Section 19.5b).

Table 19.6 lists several properties of the elements in Groups IA and IIA. The alkali metals have low melting points, and the solids are soft. Both of these properties indicate that the atoms are weakly bonded in the solid state. The densities of the elements are low, reflecting the large atomic radii that characterize the elements at the beginning of each period (Section 6.8b). The melting and boiling points as well as the hardness of the solids decrease in going down the group while the density shows the opposite behavior. The alkaline earth metals are harder, higher melting, and more dense than the alkali metals in the same period.

The alkali and alkaline earth metals are very reactive chemically. They are consequently not found in the elemental state in nature and are difficult to prepare by chemical reduction. Instead, electrolytic reduction of fused salts is commonly used for their preparation (Section 13.10b).

The metals react with nonmetallic elements and with water; these reactions are summarized in Table 19.7. In general, the alkali metals are more reactive than the alkaline earths. Many of the compounds formed in these reactions

## TABLE 19.6 Some Properties of the Metals in Groups IA and IIA

| Group | Element | Atomic number | Atomic radius (Å) | Ionic radius (Å) | Melting point (°C) | Boiling point (°C) | Density (g cm$^{-3}$) |
|---|---|---|---|---|---|---|---|
| IA | Li | 3 | 1.52 | 0.68 | 186 | 1326 | 0.534 |
|    | Na | 11 | 1.86 | 0.97 | 98 | 889 | 0.971 |
|    | K  | 19 | 2.31 | 1.33 | 64 | 774 | 0.862 |
|    | Rb | 37 | 2.44 | 1.47 | 39 | 688 | 1.53 |
|    | Cs | 55 | 2.62 | 1.67 | 28 | 690 | 1.87 |
| IIA | Be | 4  | 1.11 | 0.35 | 1283 | 2970 | 1.85 |
|     | Mg | 12 | 1.60 | 0.66 | 650  | 1120 | 1.74 |
|     | Ca | 20 | 1.97 | 0.99 | 845  | 1420 | 1.55 |
|     | Sr | 38 | 2.15 | 1.12 | 770  | 1380 | 2.60 |
|     | Ba | 56 | 2.17 | 1.34 | 725  | 1640 | 3.51 |

are also very reactive. As already noted in Sections 19.1 and 19.2, the hydrides react with water with the evolution of hydrogen, while the normal oxides are the basic anhydrides of the metal hydroxides. The peroxides, the most common of which are sodium peroxide and barium peroxide, react vigorously with water to form hydrogen peroxide, $H_2O_2$:

$$Na_2O_2(s) + 2H_2O \rightarrow 2Na^+ + 2OH^- + H_2O_2(aq)$$

$$BaO_2(s) + 2H_2O \rightarrow Ba^{2+} + 2OH^- + H_2O_2(aq)$$

Owing to the formation of hydrogen peroxide, the metal peroxides are powerful oxidizing and bleaching agents and are incorporated in commercial bleaching products.

The superoxides, the most common of which is potassium superoxide, disproportionate in water with the liberation of oxygen. This reaction is the ba-

## TABLE 19.7 Common Reactions of the Alkali and Alkaline Earth Metals

| Combining substance | Reaction | Reacting metal |
|---|---|---|
| *Group IA* | | |
| Hydrogen | $2M(s) + H_2(g) \rightarrow 2MH(s)$ | All |
| Nitrogen | $6M(s) + N_2(g) \rightarrow 2M_3N(s)$ | Li |
| Oxygen   | $4M(s) + O_2(g) \rightarrow 2M_2O(s)$ | Li |
|          | $2M(s) + O_2 \rightarrow M_2O_2(s)$ | Na |
|          | $M(s) + O_2(g) \rightarrow MO_2(s)$ | K, Rb, Cs |
| Sulfur   | $2M(s) + S(s) \rightarrow M_2S(s)$ | All |
| Halogens | $2M(s) + X_2 \rightarrow 2MX(s)$ | All |
| Water    | $2M(s) + 2H_2O(l) \rightarrow 2M^+ + 2OH^- + H_2(g)$ | All |
| *Group IIA* | | |
| Hydrogen | $M(s) + H_2(g) \rightarrow MH_2(s)$ | Ca, Sr, Ba |
| Nitrogen | $3M(s) + N_2(g) \rightarrow M_3N_2(s)$ | Mg, Ca, Sr, Ba |
| Oxygen   | $2M(s) + O_2(g) \rightarrow 2MO(s)$ | All |
| Sulfur   | $M(s) + S(s) \rightarrow MS(s)$ | Mg, Ca, Sr, Ba |
| Halogens | $M(s) + X_2 \rightarrow MX_2(s)$ | All |
| Water    | $M(s) + 2H_2O(l) \rightarrow M^{2+} + 2OH^- + H_2(g)$ | Ca, Sr, Ba |
|          | $M(s) + H_2O(g) \rightarrow MO(s) + H_2(g)$ | Mg |

sis of a protective mask used in oxygen-deficient atmospheres, such as mines. Moisture in the exhaled air of the mask wearer continuously releases oxygen by the reaction

$$4KO_2(s) + 2H_2O \rightarrow 4KOH(s) + 3O_2(g)$$

At the same time, the exhaled carbon dioxide is absorbed in the mask by the neutralization reaction

$$2KOH(s) + CO_2(g) \rightarrow K_2CO_3(s) + H_2O$$

One of the most widely used compounds of the alkali metals is sodium carbonate, which forms strongly basic solutions by hydrolysis,

$$CO_3^{2-} + H_2O \rightarrow HCO_3^- + OH^-$$

Consequently, sodium carbonate can be substituted for sodium hydroxide in many applications and is used in the manufacture of glass, soap, paper, and other products. Sodium carbonate has traditionally been made on a large scale by the **Solvay process,** a very efficient process for the conversion of sodium chloride, one of the most widely available substances, to sodium carbonate.

The Solvay process is based on the reaction of carbon dioxide with a concentrated aqueous solution of sodium chloride and ammonia. The reaction involves the following two steps:

$$CO_2(aq) + NH_3(aq) + H_2O \rightarrow NH_4^+ + HCO_3^- \qquad (19.18)$$

$$Na^+ + HCO_3^- \rightarrow NaHCO_3(s) \qquad (19.19)$$

Sodium hydrogen carbonate, the common baking soda, is relatively insoluble in the reaction medium and so precipitates. While it has many uses of its own (Section 8.10), most $NaHCO_3$ is converted to $Na_2CO_3$ by heating:

$$2NaHCO_3(s) \xrightarrow{\Delta} Na_2CO_3(s) + CO_2(g) + H_2O(g) \qquad (19.20)$$

The carbon dioxide used in reaction 19.18 is obtained by heating the common mineral limestone:

$$CaCO_3(s) \xrightarrow{\Delta} CaO(s) + CO_2(g) \qquad (19.21)$$

Carbon dioxide can be used very efficiently because it is also liberated in reaction 19.20 and can be recycled to form additional $NaHCO_3$. The ammonia used in reaction 19.18 is also recovered and recycled. This is done by heating the solution containing ammonium ions with the calcium oxide produced in reaction 19.21:

$$CaO(s) + 2NH_4^+ \rightarrow Ca^{2+} + H_2O + 2NH_3(g) \qquad (19.22)$$

The overall reaction in the Solvay process is

$$2NaCl(s) + CaCO_3(s) \rightarrow Na_2CO_3(s) + CaCl_2(s) \qquad (19.23)$$

showing that calcium chloride is the only byproduct that is not recycled. However, it is not a waste product since it can be used as a drying or a deicing agent.

***b. Diagonal Relationships*** The first elements in Groups IA and IIA more closely resemble in some ways the second elements in the adjacent groups than the second elements in their own groups. This trend gives rise to the diagonal relationship indicated by the direction of the arrows:

Li  Be  B
↘   ↘   ↘
Na  Mg  Al  Si

The diagonal similarities can be ascribed to the small ionic radius and consequent high **charge density** (charge density = ionic charge/ionic radius) of the first element in each group. For example, the charge density of $Be^{2+}$ (5.71 Å$^{-1}$) is closer to that of $Al^{3+}$ (5.88 Å$^{-1}$) than to that of $Mg^{2+}$ (3.03 Å$^{-1}$).

Comparison of lithium with the other alkali metals and with magnesium illustrates the diagonal relationship. Lithium is the only alkali metal to form a normal oxide, $Li_2O$, in its reaction with oxygen as well as the only member of its group to react with nitrogen to form a nitride, $Li_3N$, both reactions undergone by magnesium. Lithium also resembles magnesium in the solubility of some of its salts. For example, the fluorides, carbonates, and phosphates of these two elements are relatively insoluble whereas those of the other alkali metals are quite soluble.

Beryllium resembles aluminum more than the other alkaline earths in the extensive hydrolysis of its salts in solution, in the amphoteric behavior of its oxide, and in the formation of molecular compounds. For example, fused $BeCl_2$ is a fairly poor conductor of electricity, indicating that it does not form ions in the liquid. Boron and silicon are both semimetals and form acidic oxides as well as volatile halides and hydrides.

*c. Reactivity of the Alkali and Alkaline Earth Metals in Solution* The tendency of a species to undergo a redox reaction in aqueous solution at standard conditions is measured by the value of its standard reduction potential. Table 19.8 lists the standard reduction potentials of the alkali and alkaline earth metals. The values of $\epsilon°$ of the alkali metals are very negative, attesting to the strong tendency of the elements to act as reducing agents. Lithium is a stronger reducing agent than sodium, but from sodium down the group,

**TABLE 19.8** Born–Haber Cycle Data for the Reaction $M(s) \rightarrow M^{n+}(aq) + ne^-$ for the Alkali and Alkaline Earth Metals

| Group IA | $\epsilon°$ (V) | $\Delta H°_{sub}$ (kJ mol$^{-1}$) | $I_1$ (kJ mol$^{-1}$) | $\Delta H°_{hyd}(M^+)$ (kJ mol$^{-1}$) | $\Delta H°_{ox}$ (kJ mol$^{-1}$) |
|---|---|---|---|---|---|
| Li | −3.05 | 155 | 520 | −515 | 160 |
| Na | −2.71 | 109 | 496 | −406 | 199 |
| K | −2.92 | 90.0 | 419 | −322 | 187 |
| Rb | −2.93 | 85.5 | 403 | −293 | 195 |
| Cs | −3.08 | 78.8 | 376 | −264 | 190 |

| Group IIA | $\epsilon°$ (V) | $\Delta H°_{sub}$ (kJ mol$^{-1}$) | $I_1$ (kJ mol$^{-1}$) | $I_2$ (kJ mol$^{-1}$) | $\Delta H°_{hyd}(M^{2+})$ (kJ mol$^{-1}$) | $\Delta H°_{ox}$ (kJ mol$^{-1}$) |
|---|---|---|---|---|---|---|
| Be | −1.85 | 319 | 899 | 1757 | −2385 | 590 |
| Mg | −2.37 | 150 | 738 | 1451 | −1940 | 399 |
| Ca | −2.76 | 193 | 590 | 1145 | −1600 | 328 |
| Sr | −2.89 | 164 | 549 | 1064 | −1460 | 317 |
| Ba | −2.90 | 176 | 503 | 965 | −1320 | 324 |

reducing strength increases. The alkaline earths also become increasingly stronger reducing agents with increasing atomic number, and here the first element in the group is not an exception.

An increase in reducing strength in going down a group is expected from the trend in ionization energies. Recall from Section 6.5 that the first ionization energy is a measure of the tendency of a gaseous atom to lose an electron and decreases in going down a group, as shown in Table 19.8. However, electron loss in aqueous solution is a considerably more complicated process than electron loss in the gas phase, and additional factors come into play.

Figure 19.12 shows a Born–Haber-type cycle that may be used to evaluate the enthalpy change for the oxidation of metals in aqueous solution. The enthalpy change in the oxidation half-reaction,

$$M(s) \rightarrow M^+(aq) + e^- \quad \Delta H°_{ox}$$

is equal to the sum of the enthalpy changes in the sublimation of the metal,

$$M(s) \rightarrow M(g) \quad \Delta H°_{sub}$$

the ionization of the gaseous atom,

$$M(g) \rightarrow M^+(g) + e^- \quad I_1$$

and the hydration of the ion,

$$M^+(g) \rightarrow M^+(aq) \quad \Delta H°_{hyd}$$

that is,

$$\Delta H°_{ox} = \Delta H°_{sub} + I_1 + \Delta H°_{hyd} \tag{19.24}$$

Table 19.8 summarizes the various enthalpies involved in Equation 19.24. The quantities that should be compared with the experimental standard reduction potentials actually are the free energies. As is frequently done in comparing group trends, we have used instead the more readily available enthalpies. Since the standard electrode potential refers to the *reduction* of the metal ion while the enthalpy change is that for the *oxidation* of the metal, a large negative $\epsilon°$ value corresponds to a small value of $\Delta H°_{ox}$ (since $\Delta G° = -nF\epsilon°$ and we are approximating $\Delta G°$ by $\Delta H°$).

Comparing the entries in Table 19.8 for $\epsilon°$ and $\Delta H°_{ox}$ we note that they are generally well correlated. In particular, the unusually large negative value of $\epsilon°$ for lithium is mirrored in its small value of $\Delta H°_{ox}$. This small value can directly be attributed to the large negative hydration enthalpy of Li$^+$, for the other components of $\Delta H°_{ox}$, i.e., $\Delta H°_{sub}$ and $I_1$, are larger for Li$^+$ than for the other ions. The large hydration enthalpy of lithium is attributable to its small atomic radius

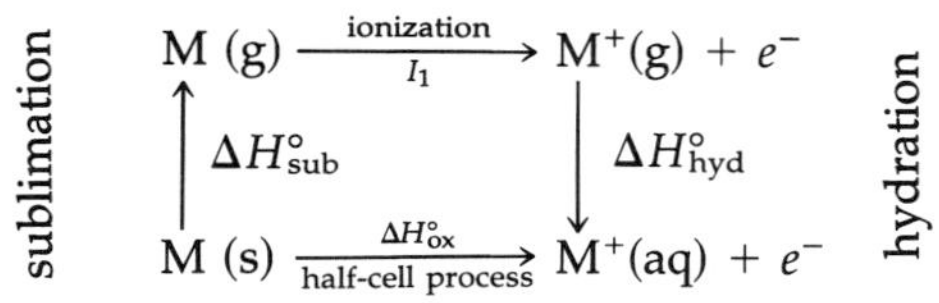

**FIGURE 19.12** Born–Haber-type cycle for the oxidation of the alkali metals to the corresponding ions in aqueous solution.

and consequent high charge density—the same property responsible for the diagonal similarity between lithium and magnesium. The ion–dipole interaction between a cation with high charge density and water is particularly strong and liberates a substantial amount of energy (Section 18.6a).

Table 19.8 also shows that the standard electrode potentials of the alkaline earths correlate well with the calculated oxidation enthalpies. Note that the anomalously strong reducing power of lithium is not duplicated by beryllium, the first element in Group IIA. Although the hydration enthalpy of beryllium is more negative than that of any of the other elements in the group, this effect is counteracted by the large ionization energies (note that we must evaluate $I_1 + I_2$ in the formation of doubly charged ions) and sublimation enthalpy of beryllium. Thus, beryllium is the weakest reducing agent among the alkaline earth elements.

**Example 19.3** Why are the +1 ions of the alkaline earth elements not found in aqueous solution? Answer this question by showing that $\Delta H°$ (and hence $\Delta G°$) for the disproportionation in aqueous solution of any one of the +1 ions, say $Mg^+$, is negative. Approximate the hydration enthalpy of $Mg^+$ by that of $Na^+$.

We wish to show that $\Delta H° < 0$ for the reaction

$$2Mg^+(aq) \rightarrow Mg(s) + Mg^{2+}(aq)$$

The enthalpy change can be evaluated by means of the following cycle:

$$\begin{array}{ccc}
2Mg^+(g) & \xrightarrow{-I_1 + I_2} & Mg(g) + Mg^{2+}(g) \\
\uparrow -2\Delta H°_{hyd} & \downarrow -\Delta H°_{sub} \;\; \downarrow \Delta H°_{hyd} \\
2Mg^+(aq) & \xrightarrow{\Delta H°} & Mg(s) + Mg^{2+}(aq)
\end{array}$$

We see that the desired relation is

$$\Delta H° = -2\Delta H°_{hyd}(Mg^+) - I_1 + I_2 - \Delta H°_{sub} + \Delta H°_{hyd}(Mg^{2+})$$

Using the data in Table 19.8 [including $\Delta H°_{hyd}(Na^+)$ instead of $\Delta H°_{hyd}(Mg^+)$] we obtain

$$\Delta H° = [(-2)(-406) - 738 + 1451 - 150 - 1940] \text{ kJ}$$
$$= -565 \text{ kJ}$$

Since $\Delta H°$ (and hence $\Delta G°$) $< 0$, $Mg^+$ is unstable in aqueous solution with respect to disproportionation. ∎

The strong tendency of the alkali metals to form cations in solution can also be seen when the solvent is ammonia instead of water. All the alkali metals dissolve in liquid ammonia ($T_{bp} = -33°C$) with the formation of characteristic blue solutions which behave as strong electrolytes in their ability to conduct electricity. Dissolution results in the formation of ammoniated cations, e.g., $Na^+(NH_3)_x$, and electrons, $e^-(NH_3)_y$. The surprisingly high conductivity of these solutions results from the ability of the **solvated electrons** to move swiftly through the solution. The formation of these solutions does not permanently alter the metals; they can be recovered by evaporation of the ammonia.

**d. The p-Block Metals**   As noted in Section 6.9, metallic character increases from top to bottom and from right to left in the periodic table. Consistent with this trend, some of the heavier elements in Groups IIIA–VA are metals (Figure 19.10). Table 19.9 lists these metals and summarizes some of their properties.

All the Group IIIA elements except for the first, boron, are metals. Aluminum is a very common and widely used metal. Gallium, indium, and thallium, the heavier members of the group, are less common. Although it has a large standard reduction potential ($\epsilon° = -1.66$ V) and is prepared by electrolytic reduction (Section 13.10b), aluminum is protected against corrosion by a thin and strongly self-adhering layer of aluminum oxide. The relative inertness of aluminum, along with its low density, excellent conductivity, and high tensile strength and malleability, make aluminum useful in the aircraft industry, high-voltage transmission lines, and household utensils.

Aluminum reacts readily with strong acids and bases,

$$2Al(s) + 6H^+ \rightarrow 2Al^{3+} + 3H_2(g)$$

$$2Al(s) + 2OH^- + 6H_2O \rightarrow 2Al(OH)_4^- + 3H_2(g)$$

Like $Be^{2+}$, hydrated $Al^{3+}$ is extensively hydrolyzed,

$$Al^{3+} + H_2O \rightarrow Al(OH)^{2+} + H^+$$

forming an acidic solution.

Aluminum oxide, $Al_2O_3$, is the most stable oxide of the representative elements ($\Delta G_f° = -1582$ kJ mol$^{-1}$). It is formed with the evolution of much heat in the reduction of metal oxides by powdered aluminum. An example of these **aluminothermic reactions** is the **thermite reaction**, used to prepare molten iron for welding,

$$2Al(s) + Fe_2O_3(s) \rightarrow Al_2O_3(s) + 2Fe(l) \qquad \Delta H° = -849 \text{ kJ}$$

The metallic elements of Group IVA are tin (in the allotropic form called white tin) and lead, while bismuth is the only Group VA metal. Tin and lead are soft and malleable, but bismuth is brittle. Tin is a component of various widely used alloys (Section 21.2), and lead is used in large quantitites in the manufacture of storage batteries (Section 13.7b).

As noted in Section 19.4, many of the heavy p-block elements display oxidation states equal to the group number and to the group number minus 2. Tin and lead form compounds in which they display oxidation states of +2 and +4. Many of the tin(IV) compounds, such as the halides, are low-boiling liq-

**TABLE 19.9   Properties of the p-Block Metals**

| Element | Group | Atomic number | Common oxidation states | Melting point (°C) | Boiling point (°C) | Density (g cm$^{-3}$) |
|---|---|---|---|---|---|---|
| Al | IIIA | 13 | +3 | 658 | 1800 | 2.70 |
| Ga | IIIA | 31 | +3 | 29.8 | 1700 | 5.9 |
| In | IIIA | 49 | +3 | 155 | >1450 | 7.30 |
| Tl | IIIA | 81 | +1, +3 | 304 | 1650 | 11.5 |
| Sn | IVA | 50 | +2, +4 | 232 | 2270 | 7.28 |
| Pb | IVA | 82 | +2, +4 | 327 | 1620 | 11.3 |
| Bi | VA | 83 | +3, +5 | 271 | 1560 | 9.80 |

uids, indicating that they consist of molecules. These compounds undergo extensive hydrolysis in water, for example,

$$SnCl_4(l) + 2H_2O \rightarrow SnO_2(s) + 4H^+ + 4Cl^-$$

By contrast, the tin(II) halides are ionic salts. Aqueous solutions are readily oxidized by air and a solution of tin(II) chloride is a moderately strong reducing agent.

## 19.6 The Redox Chemistry of Nitrogen and the Use of Electrode Potential Diagrams

Nitrogen forms compounds in which it displays oxidation numbers ranging from −3 to +5. Table 19.10 lists some compounds of nitrogen with hydrogen or oxygen in which these multiple oxidation states are displayed.

---

**Example 19.4** Draw the Lewis structure of hydroxylamine and show that the oxidation number of nitrogen is −1.

Since the molecule has a hydroxyl group, its structural formula must be

$$\begin{array}{c} H \\ | \\ H-N-O-H \end{array}$$

The total number of valence electrons is 14. As 8 electrons have already been used to form single bonds, the remaining 6 electrons can be disposed as lone pairs about the oxygen and nitrogen, giving each atom an electron octet:

$$\begin{array}{c} H \\ | \\ H-\ddot{N}-\ddot{O}-H \end{array}$$

In determining the oxidation number of nitrogen, we assign the electrons in the H—N bonds to the more electronegative nitrogen and those in the N—O bond to the more electronegative oxygen. Thus, the oxidation number of nitrogen is 5 − 6 = −1. ∎

Owing to the occurrence of so many different oxidation states, the redox chemistry of nitrogen is fairly complicated. It is customary to summarize the

**TABLE 19.10 Oxidation States of Nitrogen in Its Compounds with Hydrogen or Oxygen**

| Nitrogen oxidation number | Compound |
|---|---|
| +5 | HNO$_3$ |
| +4 | NO$_2$, N$_2$O$_4$ |
| +3 | HNO$_2$ |
| +2 | NO |
| +1 | N$_2$O |
| 0 | N$_2$ |
| −1 | NH$_2$OH (hydroxylamine) |
| −2 | N$_2$H$_4$ (hydrazine) |
| −3 | NH$_3$ |

standard electrode potentials connecting the various oxidation states of an element by means of a diagram variously known as an **electrode potential diagram** or a **Latimer diagram**. Figure 19.13 shows such a diagram for the different compounds of nitrogen with oxygen or hydrogen in acidic and in basic solution. Observe that the various species are written from left to right in order of decreasing oxidation number and are connected by lines representing the reduction of the species on the left to the species on the right. Surmounting these lines are the corresponding values of the standard reduction potentials in volts. Remember that these values apply only at standard conditions, with all ionic species present in 1 $M$ concentration and all gases at 1 atm.

The diagram also includes some longer lines connecting non-adjacent species that are commonly formed from each other in redox reactions. The standard reduction potentials listed for these half-reactions can be obtained by appropriate combination of the standard reduction potentials of the intervening adjacent species. Recall from Section 13.5c that in combining the reduction potentials of two half-reactions to obtain the reduction potential of a third half-reaction, we add the corresponding free energies. The reduction potentials are combined according to Equation 13.20,

$$\epsilon_3^\circ = (n_1 \epsilon_1^\circ + n_2 \epsilon_2^\circ)/n_3$$

which expresses the fact that the reduction potential is proportional to the free energy change per transferred electron. For example, the standard reduction potential for the reduction of $NO_3^-$ to $HNO_2$, listed as 0.94 V, is obtained from the values listed for the reduction of $NO_3^-$ to $NO_2$ and that of $NO_2$ to $HNO_2$ as follows:

$$\epsilon^\circ = \frac{[(1)(0.78) + (1)(1.10)]\ V}{2} = 0.94\ V$$

A particularly useful feature of an electrode potential diagram is that we can spot at a glance those species which are thermodynamically unstable with

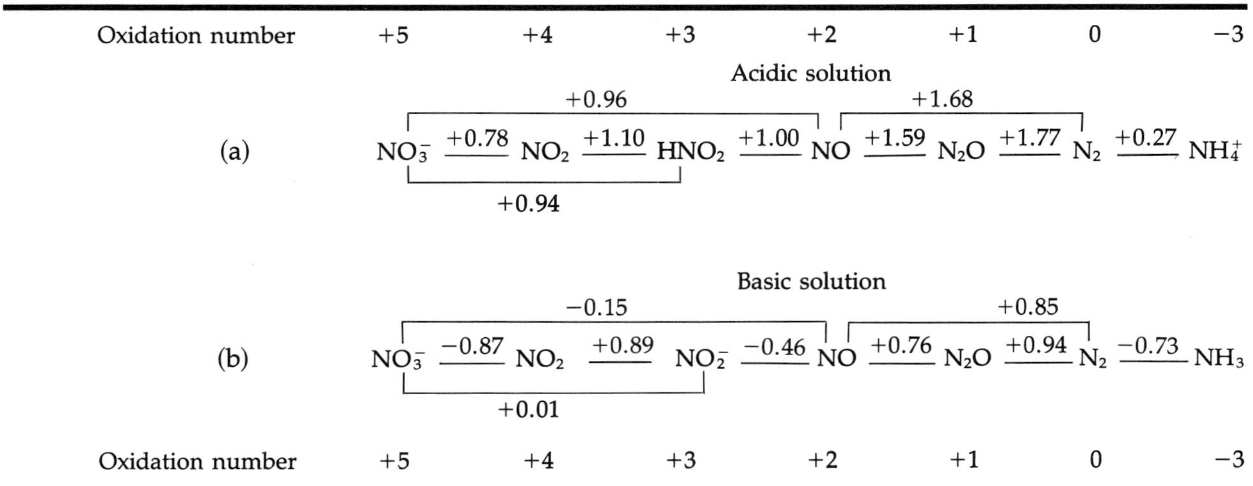

**FIGURE 19.13** Electrode potential diagrams for nitrogen species in (a) acidic solution and (b) basic solution.

respect to disproportionation. This reaction will occur whenever the $\epsilon°$ value immediately to the right of a given species is more positive (or less negative) than that immediately to the left. This criterion indicates, for example, that $NO_2$ should disproportionate in acidic solution. To show that this is indeed so, we write the half-reactions for the disproportionation of this species, taking the appropriate $\epsilon°$ values from the diagram:

$$NO_2 + H^+ + e^- \rightarrow HNO_2 \qquad \epsilon° = 1.10 \text{ V}$$
$$NO_2 + H_2O \rightarrow NO_3^- + 2H^+ + e^- \qquad \epsilon° = -0.78 \text{ V}$$
$$\overline{2NO_2 + H_2O \rightarrow HNO_2 + NO_3^- + H^+ \qquad \epsilon° = 0.32 \text{ V}}$$

Since $\epsilon°$ for the disproportionation is positive, $NO_2$ is thermodynamically unstable in acidic solution. As has been emphasized before, this does not mean that disproportionation will necessarily occur rapidly.

An electrode potential diagram also identifies species that are unstable in aqueous solution because they reduce water or the hydronium ion to hydrogen, or because they oxidize water or the hydroxide ion to oxygen. Figure 19.14a locates on an EMF scale the $\epsilon°$ values for the reduction of the species involved in the redox reactions of water in acidic and basic solutions. This diagram indicates the range of $\epsilon°$ values for which a species is stable in aqueous solution and the range of $\epsilon°$ values for which a species will either oxidize or reduce water. This is also shown in Figure 19.14b, which places these $\epsilon°$ values on the electrode potential diagram of a hypothetical species and indicates the conditions for which the species either is stable or undergoes oxidation or reduction. Applying these criteria to the nitrogen species in Figure 19.13, we find that in acidic solution both NO and $N_2O$ can oxidize $H_2O$ to $O_2$ at standard conditions (the $\epsilon°$ values to the *right* of these species exceed 1.23 V). Similarly, in basic solution NO and $N_2O$ can oxidize water (actually $OH^-$) while $NO_2$ can both oxidize and reduce water (the $\epsilon°$ to the right of $NO_2$ is larger than 0.40 V while that to the left of $NO_2$ is smaller than $-0.83$ V).

Figure 19.13 summarizes a number of additional aspects of the chemistry of nitrogen and Table 19.11 lists some of the more common redox reactions of nitrogen compounds. Elemental nitrogen is unusually stable in aqueous solution; it is not readily oxidized or reduced. The high dissociation energy of the molecule accounts for this stability.

Nitric acid is a strong oxidizing agent, being reduced to $NO_2$ in concentrated solution and to NO in dilute solution, for example, in its reactions with copper:

$$Cu(s) + 4H^+ + 2NO_3^- \rightarrow Cu^{2+} + 2NO_2(g) + 2H_2O$$

and

$$3Cu(s) + 8H^+ + 2NO_3^- \rightarrow 3Cu^{2+} + 2NO(g) + 4H_2O$$

In general, the stronger the reducing agent and the more dilute the acid, the lower will be the oxidation number of nitrogen in the product of the reduction. Thus, zinc reduces dilute nitric acid all the way to ammonium ion:

$$4Zn(s) + 10H^+ + NO_3^- \rightarrow 4Zn^{2+} + NH_4^+ + 3H_2O$$

The oxidizing power of concentrated nitric acid is enhanced by mixing it with three times its volume of concentrated hydrochloric acid, a mixture

19.6 The Representative Elements  721

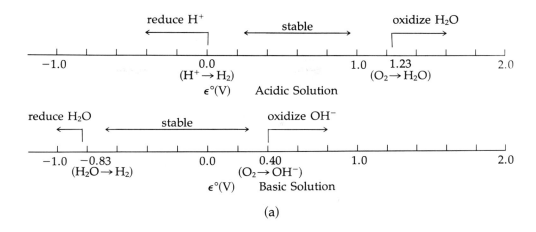

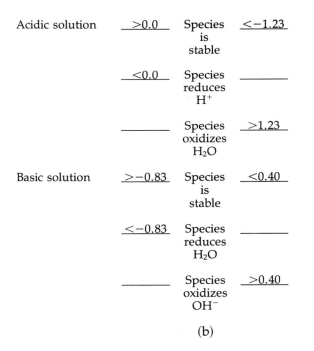

**FIGURE 19.14** Conditions for the stability of a species in aqueous solution. (a) Standard reduction potentials for half-reactions involving water, or its constituent ions, and hydrogen and oxygen. The half-reactions (unbalanced) are written below the corresponding $\epsilon°$ values. The arrows show the range of $\epsilon°$ values for which species in aqueous solution are either stable or can oxidize or reduce water (or one of its ions). The top diagram refers to acidic solution and the bottom diagram to basic solution. (b) Standard electrode potential diagram showing the stability of a species in acidic or basic solution. The $\epsilon°$ values from (a) show the conditions for which the species will either oxidize or reduce water.

TABLE 19.11  Some Redox Reactions of Nitrogen Compounds

| Initial oxidation number | Final oxidation number | Reaction | Comments |
|---|---|---|---|
| +5 | +4 | $Ag(s) + NO_3^- + 2H^+ \rightarrow Ag^+ + NO_2(g) + H_2O$ | Concentrated $HNO_3$ (16 $M$) |
| +5 | +4 | $CuS(s) + 2NO_3^- + 4H^+ \rightarrow Cu^{2+} + S(s) + 2NO_2(g) + H_2O$ | Concentrated $HNO_3$, method of dissolving very insoluble sulfides |
| +5 | +4 | $2Pb(NO_3)_2(s) \xrightarrow{\Delta} 2PbO(s) + 4NO_2(g)$ | For nitrates of less active metals |
| +5 | +3 | $Mg(NO_3)_2(s) \xrightarrow{\Delta} Mg(NO_2)_2(s) + O_2(g)$ | For nitrates of active metals |
| +5, −3 | 0 | $2NH_4NO_3(s) \xrightarrow{>300°C} 2N_2(g) + 4H_2O(g) + O_2(g)$ | DANGER! EXPLOSIVE! |
| +1 | 0 | $2N_2O(g) \xrightarrow{\Delta} 2N_2(g) + O_2(g)$ | DANGER! EXPLOSIVE! |
| −2 | 0 | $N_2H_4(l) + O_2(g) \rightarrow N_2(g) + 2H_2O(g)$ | Rocket fuel reaction, highly exothermic |
| −2, +4 | 0 | $2N_2H_4(l) + N_2O_4(l) \rightarrow 3N_2(g) + 4H_2O(g)$ | Rocket fuel reaction, highly exothermic |

known as **aqua regia** ("royal water"). Noble metals, such as gold or platinum, which are insoluble in either acid alone, will dissolve in aqua regia. The solvent action of aqua regia can be ascribed to the formation of elemental chlorine and nitrosyl chloride (NOCl) in the reaction which occurs when the two acids are mixed:

$$3HCl(aq) + HNO_3(aq) \rightarrow Cl_2(g) + NOCl(g) + 2H_2O$$

The noble metals are attacked by the two gases with the initial formation of chlorides. The metal chlorides are then transformed to stable complex anions by reaction with chloride ion. The overall reaction for gold is

$$Au(s) + 4H^+ + 4Cl^- + NO_3^- \rightarrow AuCl_4^- + NO(g) + 2H_2O$$

In Figure 19.13, contrast the standard reduction potential for the reduction of nitrate ion in acidic solution

$$NO_3^- + 4H^+ + 3e^- \rightarrow NO + 2H_2O \qquad \epsilon° = 0.96 \text{ V}$$

with the corresponding value for the same process in basic solution:

$$NO_3^- + 2H_2O + 3e^- \rightarrow NO + 4OH^- \qquad \epsilon° = -0.15 \text{ V}$$

This difference illustrates the important generalization that oxoanions and oxides are more powerful oxidizing agents in acidic than in basic solution. This rule can be readily understood on the basis of Le Chatelier's principle: since $H^+$ appears as a reactant in the first half-reaction, an increase in its concentration shifts the equilibrium point to the right. Conversely, $OH^-$ appears as a product of the second half-reaction and so an increase in its concentration shifts the equilibrium point to the left.

**Example 19.5** Derive the value of $\epsilon°$ for the reduction of $NO_3^-$ to NO in basic solution from the corresponding value in acidic solution.

The balanced half-reaction in acidic medium is

$$NO_3^- + 4H^+ + 3e^- \rightarrow NO + 2H_2O \qquad \epsilon° = 0.96 \text{ V}$$

The dependence of the reduction potential on pH can be obtained by means of the Nernst equation (Equation 13.16),

$$\epsilon = \epsilon° - \frac{0.0592}{n} \log Q$$

The only species present in variable concentration is $H^+$; all the others can be maintained at 1 $M$ or 1 atm. Thus, the Nernst equation can be written as

$$\epsilon = 0.96 \text{ V} - \frac{0.0592 \text{ V}}{3} \log \frac{1}{[H^+]^4}$$

The value of $\epsilon°$ in basic solution applies when the $OH^-$ concentration is 1 $M$. The $H^+$ concentration at this point must be $10^{-14}$ $M$. Substituting into the Nernst equation, we obtain

$$\epsilon = 0.96 \text{ V} - \frac{0.0592 \text{ V}}{3} \log \frac{1}{(10^{-14})^4} = -0.15 \text{ V}$$

in agreement with the $\epsilon°$ value given in Figure 19.13 for the reduction in basic solution. ∎

---

**Example 19.6** The electrode potential diagram of phosphorus in acidic solution is

$$H_3PO_4 \xrightarrow{-0.28} H_3PO_3 \xrightarrow{-0.50} H_3PO_2 \xrightarrow{-0.51} P_4 \xrightarrow{-0.11} PH_3$$

(a) Which acids of phosphorus, if any, are strong oxidizing agents? (b) Which species undergo disproportionation, reduce $H^+$, or oxidize $H_2O$? Write balanced equations for these reactions and determine the overall $\epsilon°$ values. (c) Determine $\epsilon°$ for the reduction of $H_3PO_4$ to $H_3PO_2$.

(a) The standard reduction potentials of all three oxoacids of phosphorus are negative. None of the acids consequently display a tendency to act as oxidizing agents.

(b) The only species unstable with respect to disproportionation is elemental phosphorus $P_4$. The element undergoes the following reaction:

$$P_4 + 12H^+ + 12e^- \rightarrow 4PH_3 \qquad \epsilon° = -0.11 \text{ V}$$
$$\underline{3 \times (P_4 + 8H_2O \rightarrow 4H_3PO_2 + 4H^+ + 4e^-) \qquad \epsilon° = 0.51 \text{ V}}$$
$$4P_4 + 24H_2O \rightarrow 12H_3PO_2 + 4PH_3$$
$$P_4 + 6H_2O \rightarrow 3H_3PO_2 + PH_3 \qquad \epsilon° = 0.40 \text{ V}$$

Although phosphorus is thermodynamically unstable in aqueous solution, the reaction does not proceed at a significant rate at ordinary temperatures. In fact, white phosphorus is usually stored in water because it ignites spontaneously in air to form $P_4O_{10}$.

Since all the listed $\epsilon°$ values are negative, all the given species excepting $H_3PO_4$ can reduce $H^+$ to $H_2$ but none can oxidize water to oxygen. The reaction of $H_3PO_2$ with $H^+$ illustrates all these reactions:

$$H_3PO_2(aq) + H_2O \rightarrow H_3PO_3(aq) + 2H^+ + 2e^- \quad \epsilon° = 0.50 \text{ V}$$
$$\underline{2H^+ + 2e^- \rightarrow H_2(g) \quad\quad\quad\quad\quad\quad\quad\quad\quad \epsilon° = 0.00 \text{ V}}$$
$$H_3PO_2(aq) + H_2O \rightarrow H_3PO_3(aq) + H_2(g) \quad \epsilon° = 0.50 \text{ V}$$

While the reaction is spontaneous, it actually occurs very slowly.

(c) We apply Equation 13.20,

$$\epsilon_3° = (n_1\epsilon_1° + n_2\epsilon_2°)/n_3$$

to the reduction of $H_3PO_4$, first to $H_3PO_3$ and then to $H_3PO_2$, in order to obtain $\epsilon°$ for the direct reduction to $H_3PO_2$. The oxidation number of phosphorus changes from +5 to +3 to +1 in this process, so that $n_1 = n_2 = 2$ and $n_3 = 4$. We therefore obtain

$$\epsilon° = \frac{(2)(-0.28 \text{ V}) + (2)(-0.50 \text{ V})}{4} = -0.39 \text{ V} \quad \blacksquare$$

## 19.7 The Redox Chemistry of Oxygen and Sulfur

*a. Oxygen* Oxygen is the most abundant element in the earth's crust, where it is present in the form of oxides. The element is a major constituent of the atmosphere (20.9 vol%) and is obtained readily by fractional distillation of liquefied air. Molecular oxygen is a rather strong oxidizing agent but it reacts slowly except at elevated temperatures. Molecular oxygen can be converted to ozone, $O_3$, by the action of ultraviolet radiation or lightning (Section 3.7). Ozone is a powerful and highly reactive oxidizing agent. In aqueous solution it undergoes the half-reaction

$$O_3 + 2H^+ + 2e^- \rightarrow O_2 + H_2O \quad \epsilon° = 2.08 \text{ V}$$

The redox chemistry of oxygen is relatively simple. Oxygen is present in the −2 state in most of its compounds, such as the oxides, the oxoacids, and their salts. In addition, oxygen forms peroxides, in which its oxidation state is −1. Peroxides are characterized by the presence of a single bond between two oxygen atoms. The Lewis structure of hydrogen peroxide, $H_2O_2$, is

$$\begin{array}{c} \ddot{\text{O}}-\text{H} \\ | \\ \text{H}-\ddot{\text{O}}: \end{array}$$

where the HOO bond angles are 93.9°.

Hydrogen peroxide can be prepared in the laboratory by the reaction of barium peroxide with sulfuric acid,

$$BaO_2(s) + 2H^+ + SO_4^{2-} \rightarrow BaSO_4(s) + H_2O_2(aq)$$

Pure hydrogen peroxide is a colorless, syrupy liquid that boils at 151°C and freezes at −0.4°C. While moderately stable in pure form, it readily decomposes in the presence of trace impurities by the reaction

$$2H_2O_2(l) \rightarrow 2H_2O(l) + O_2(g)$$

The reaction is highly exothermic and can occur with explosive violence. Hydrogen peroxide is customarily available in aqueous solution, in which it is more stable. A 3% solution is a common antiseptic.

Hydrogen peroxide is a strong oxidizing agent and a moderately weak reducing agent, as shown by the potential diagram of oxygen in acidic solution:

$$O_2 \xrightarrow{0.68} H_2O_2 \xrightarrow{1.77} H_2O$$

It can oxidize sulfides to sulfates, for example,

$$PbS(s) + 4H_2O_2(aq) \rightarrow PbSO_4(s) + 4H_2O$$

and, in turn, be oxidized by very strong oxidizing agents with the evolution of oxygen:

$$2MnO_4^- + 5H_2O_2(aq) + 6H^+ \rightarrow 2Mn^{2+} + 5O_2(g) + 8H_2O$$

*b. Sulfur* The second elements in Groups VA and VIA, phosphorus and sulfur, resemble each other in their tendency towards self-linkage in the elemental state. We have already noted that white phosphorus, the common form of the element, consists of $P_4$ molecules. Similarly, sulfur forms $S_8$ molecules, which have a characteristic ring structure (Figure 19.15). The two elements are also similar in that they exist in several allotropic forms. White phosphorus can be converted to the polymeric and less reactive red phosphorus by heating to 250°C. It can also be converted to crystalline black phosphorus by heating under extremely high pressure. Similarly, the stable form of sulfur at ordinary temperatures is rhombic sulfur, which can be converted to monoclinic sulfur by slow heating to 96°C (Figure 19.16). In spite of these similarities, sulfur and phosphorus are quite different in their reactivity. Sulfur is less reactive and is found in the elemental state in nature.

Sulfur exists in a number of different oxidation states, as summarized in Table 19.12. Figure 19.17 shows the electrode potential diagrams of sulfur in acidic and basic solution. Note that reduction occurs more readily in acidic solution than in basic solution, where all the standard reduction potentials are negative. Note, too, that some of the intermediate oxidation states, such as sul-

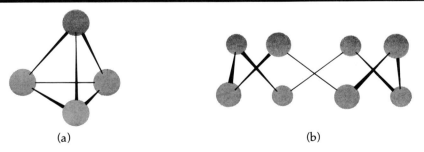

**FIGURE 19.15** The molecular forms of (a) phosphorus, $P_4$, and (b) sulfur, $S_8$.

**FIGURE 19.16** The shape of rhombic and monoclinic sulfur crystals.

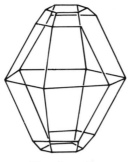
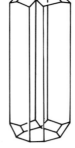

Rhombic sulfur            Monoclinic sulfur

fur dioxide and thiosulfate ion, $S_2O_3^{2-}$, disproportionate in acidic solution. In actuality, $SO_2$ reacts very slowly and its aqueous solution is stable. While sulfur dioxide is nominally the anhydride of sulfurous acid, $H_2SO_3$, it actually appears to be present as hydrated $SO_2$. Stable salts containing both the hydrogen sulfite ion, $HSO_3^-$, and the sulfite ion, $SO_3^{2-}$, are well known. Like sulfur dioxide, these salts act as mild reducing agents.

Figure 19.17 indicates that sulfuric acid is not a particularly strong oxidizing agent. As is true for other oxoacids, the ability to act as an oxidant increases with the hydrogen ion concentration. Concentrated sulfuric acid (18 M) is a moderately strong oxidizing agent capable of oxidizing copper metal:

$$Cu(s) + 4H^+ + SO_4^{2-} \rightarrow Cu^{2+} + SO_2(g) + 2H_2O$$

**Example 19.7** The thiosulfate ion, $S_2O_3^{2-}$, has a structure similar to the sulfate ion, with a sulfur atom replacing one of the oxygen atoms. (a) Draw the Lewis structure

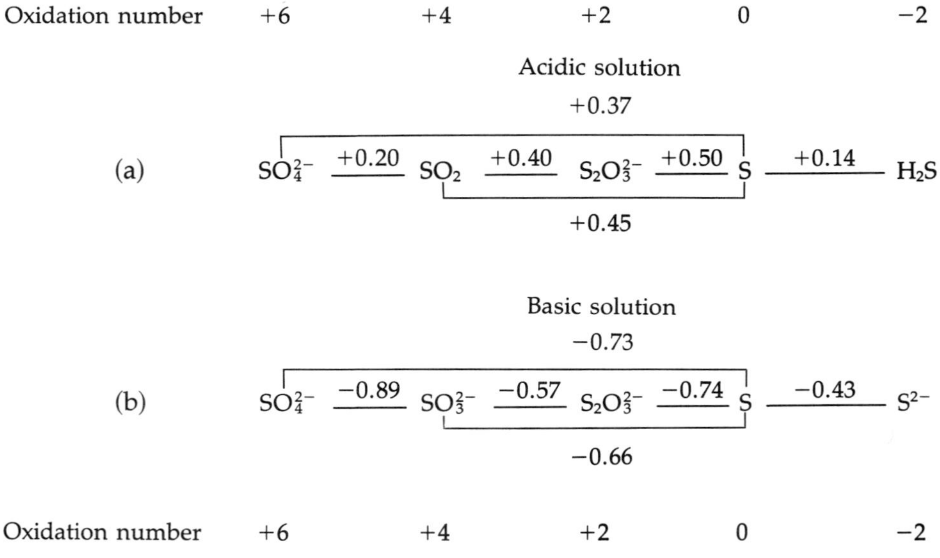

| Oxidation number | +6 | +4 | +2 | 0 | −2 |

Acidic solution
+0.37

(a) $SO_4^{2-}$ —+0.20— $SO_2$ —+0.40— $S_2O_3^{2-}$ —+0.50— $S$ —+0.14— $H_2S$

+0.45

Basic solution
−0.73

(b) $SO_4^{2-}$ —−0.89— $SO_3^{2-}$ —−0.57— $S_2O_3^{2-}$ —−0.74— $S$ —−0.43— $S^{2-}$

−0.66

| Oxidation number | +6 | +4 | +2 | 0 | −2 |

**FIGURE 19.17** Electrode potential diagrams for sulfur species in (a) acidic solution and (b) basic solution.

TABLE 19.12 The Oxidation States of Sulfur

| Oxidation number | Compounds and ions |
|---|---|
| +6 | $SO_3$, $H_2SO_4$, $HSO_4^-$, $SO_4^{2-}$, $SF_6$, $H_2S_2O_7$ (pyrosulfuric acid) $H_2S_2O_8$ (peroxodisulfuric acid) |
| +4 | $SO_2$, $SO_3^{2-}$, $HSO_3^-$ |
| +2 | $S_2O_3^{2-}$ |
| 0 | $S_8$ |
| −2 | $H_2S$, $HS^-$, $S^{2-}$ |

of the ion. (b) Determine the oxidation state of each sulfur atom. (c) Show that the average oxidation state of sulfur is +2.

(a) Recall from Section 15.6 that a large number of resonance structures can be drawn for the sulfate ion. This is also true for the thiosulfate ion. The following structure uses only single bonds and obeys the octet rule:

$$\left[ \begin{array}{c} \ddot{\mathrm{O}}: \\ | \\ :\ddot{\mathrm{O}}-\mathrm{S}-\ddot{\mathrm{O}}: \\ | \\ :\ddot{\mathrm{S}}: \end{array} \right]^{2-}$$

(b) The oxidation number can be determined as in Section 15.5b by assigning the bonded electrons to the more electronegative oxygen atoms. The central sulfur atom is therefore assigned just one electron (from the S—S bond) and its oxidation number is +5. The outer sulfur atom is assigned seven electrons and its oxidation number is −1.

(c) The average of the oxidation numbers of the two sulfur atoms is +2. This value can also be obtained by noting that the oxidation number of each oxygen is −2. Therefore the average oxidation number of sulfur must be +2 if the sum of the oxidation numbers is to equal the ionic charge. ■

Thiosulfate salts can be prepared by boiling aqueous solutions of sulfites with either hydrogen sulfide or sulfur:

$$8Na_2SO_3(aq) + S_8(s) \rightarrow 8Na_2S_2O_3(aq)$$

Thiosulfate ion is a mild reducing agent. It is used in analytical chemistry owing to its ability to reduce iodine quantitatively with the formation of the tetrathionate ion:

$$2S_2O_3^{2-} + I_2 \rightarrow S_4O_6^{2-} + 2I^-$$

The analytical procedure has wide applicability because iodine can, in turn, be generated by the reactions of a variety of oxidizing agents of unknown concentration with excess $I^-$. The resulting $I_2$ is titrated with a thiosulfate solution of known concentration, permitting the determination of the oxidizing agent concentration.

## 19.8 The Halogens

*a. The Elements* The elements of Group VIIA—the halogens—are among the most reactive of the nonmetallic elements. Table 19.13 summarizes some of their properties. At normal temperature and pressure, there is a progression in

**TABLE 19.13  Some Properties of the Halogens**

| Element | Atomic number | Atomic radius (Å) | Melting point (°C) | Boiling point (°C) | Density (g cm$^{-3}$) | Color of gas |
|---|---|---|---|---|---|---|
| Fluorine | 9  | 0.709 | −223   | −187.9 | 1.108[a] | light yellow |
| Chlorine | 17 | 0.994 | −102.4 | −34.0  | 1.57[a]  | greenish yellow |
| Bromine  | 35 | 1.141 | −7.3   | 58.8   | 3.14     | reddish brown |
| Iodine   | 53 | 1.333 | 113.7  | 184.5  | 4.94     | violet |

[a]Density of liquid at its boiling point.

the physical states of the elements from gaseous ($F_2$, $Cl_2$) to liquid ($Br_2$) to solid ($I_2$). This progression is accompanied by a deepening of color from light yellow to deep violet. Bromine and iodine have high vapor pressures and the colored vapors are visibly present in equilibrium with the liquid or solid.

With the exception of iodine, the elements are found in nature in the form of halides. Iodine, which has a greater tendency to form positive oxidation states, is often found in the form of iodates. Since fluorine and chlorine are very strong oxidizing agents, they are difficult to prepare from their corresponding halides by chemical oxidation. They can be made by electrolytic oxidation of either fused salts (KF, NaCl) or concentrated salt solutions (NaCl) (Section 13.10). Bromine can be prepared from bromides by oxidation with chlorine,

$$Cl_2(g) + 2Br^- \rightarrow Br_2(aq) + 2Cl^-$$

and iodine can be made from iodates by sulfite reduction:

$$2IO_3^- + 5SO_3^{2-} + 2H^+ \rightarrow 5SO_4^{2-} + I_2(s) + H_2O$$

Iodine sublimes readily and can be purified by sublimation.

**b. Redox Chemistry**  With the exception of fluorine, the halogens display oxidation states ranging from −1 to +7. Consequently, their redox chemistry is extensive. Figure 19.18 shows the electrode potential diagrams in both acidic and basic solution.

The most general observation based on these diagrams is that the elements as well as the oxoacids or oxoanions are strong oxidizing agents. A more detailed examination reveals several trends within the group. The standard reduction potentials of the elements decrease from fluorine to iodine and the reactivity of the elements parallels the values of $\epsilon°$. Fluorine reacts with practically all the metals and most of the nonmetals to form fluorides while, at the other extreme, iodine reacts to form iodides with many metals but, among the nonmetals, only with hydrogen and phosphorus.

The elements undergo **halogenation** reactions, in which halogen atoms are added to molecular compounds. The reactions with some of the nonmetal oxides are typical, for example,

$$SO_2(g) + X_2(g) \rightarrow SO_2X_2(g) \qquad (X = F, Cl)$$

$$CO(g) + X_2(g) \rightarrow COX_2(g) \qquad (X = Cl, Br)$$

in which, respectively, sulfuryl and carbonyl halides are formed. Both fluorine and chlorine can oxidize water to $O_2$ since their $\epsilon°$ values are larger than 1.23

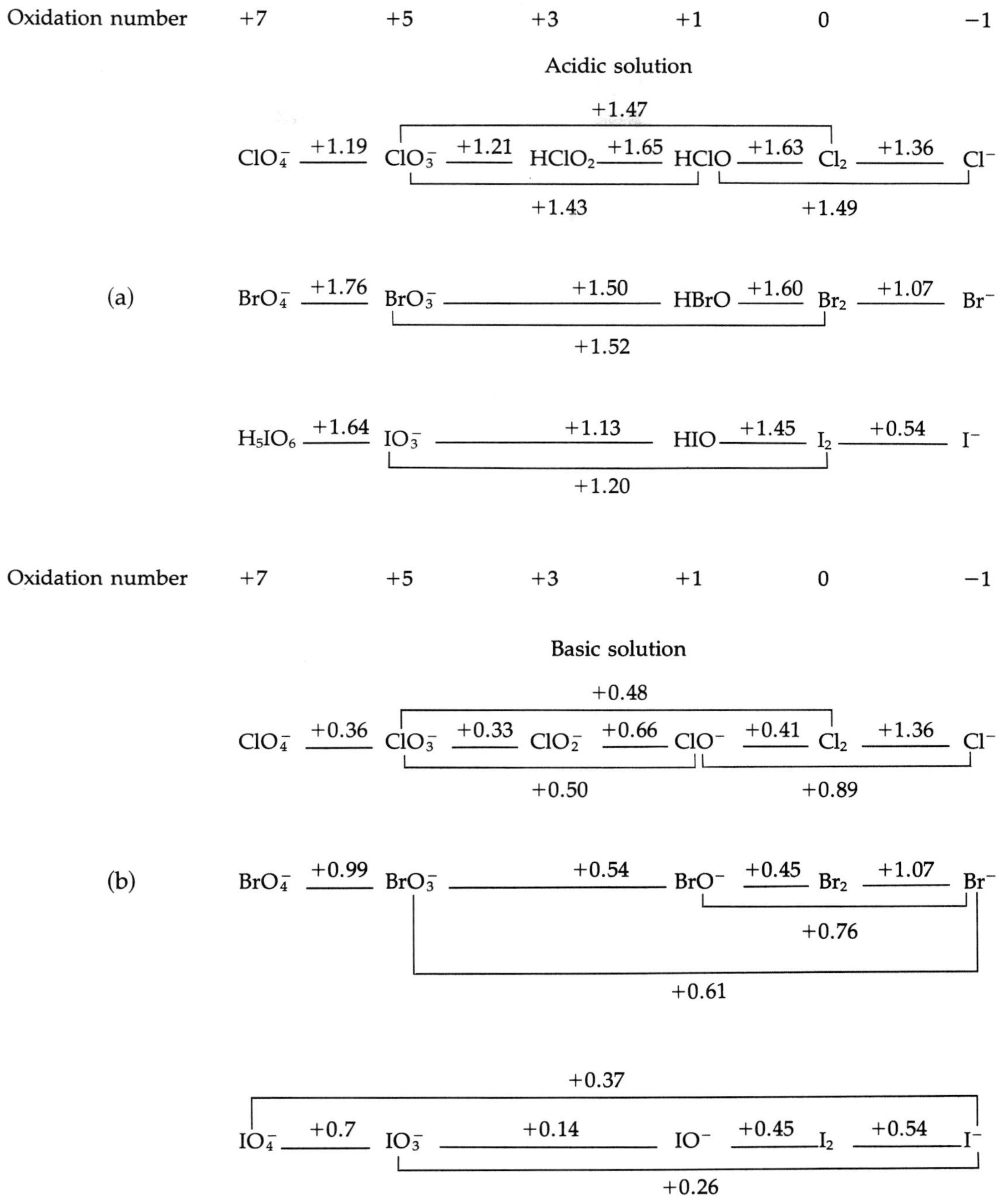

Oxidation number   +7        +5        +3        +1        0        −1

Acidic solution

(a)

$ClO_4^- \xrightarrow{+1.19} ClO_3^- \xrightarrow{+1.21} HClO_2 \xrightarrow{+1.65} HClO \xrightarrow{+1.63} Cl_2 \xrightarrow{+1.36} Cl^-$

ClO₃⁻ → HClO: +1.43
HClO₂ → HClO: (part of +1.47 bracket ClO₃⁻→HClO)
HClO → Cl⁻: +1.49
ClO₃⁻ → HClO (upper): +1.47

$BrO_4^- \xrightarrow{+1.76} BrO_3^- \xrightarrow{+1.50} HBrO \xrightarrow{+1.60} Br_2 \xrightarrow{+1.07} Br^-$

BrO₃⁻ → Br₂: +1.52

$H_5IO_6 \xrightarrow{+1.64} IO_3^- \xrightarrow{+1.13} HIO \xrightarrow{+1.45} I_2 \xrightarrow{+0.54} I^-$

IO₃⁻ → I₂: +1.20

Oxidation number   +7        +5        +3        +1        0        −1

Basic solution

(b)

$ClO_4^- \xrightarrow{+0.36} ClO_3^- \xrightarrow{+0.33} ClO_2^- \xrightarrow{+0.66} ClO^- \xrightarrow{+0.41} Cl_2 \xrightarrow{+1.36} Cl^-$

ClO₃⁻ → ClO⁻: +0.48 (upper)
ClO₃⁻ → ClO⁻: +0.50
ClO⁻ → Cl⁻: +0.89

$BrO_4^- \xrightarrow{+0.99} BrO_3^- \xrightarrow{+0.54} BrO^- \xrightarrow{+0.45} Br_2 \xrightarrow{+1.07} Br^-$

BrO⁻ → Br⁻: +0.76
BrO₃⁻ → Br⁻: +0.61

$IO_4^- \xrightarrow{+0.7} IO_3^- \xrightarrow{+0.14} IO^- \xrightarrow{+0.45} I_2 \xrightarrow{+0.54} I^-$

IO₃⁻ → IO⁻: +0.37 (upper)
IO₃⁻ → I₂: +0.26

**FIGURE 19.18** Electrode potential diagrams for the halogen species in (a) acidic solution and (b) basic solution.

V. Fluorine reacts rapidly with water and chlorine does so more slowly. Both reactions are described by the equation

$$2X_2(g) + 2H_2O \rightarrow 4HX(aq) + O_2(g)$$

The oxoacids and oxoanions generally are strong oxidizing agents in acidic solution but, in accordance with Le Chatelier's principle, tend to lose their oxidizing power in basic solution. A number of the species in intermediate oxidation states are unstable with respect to disproportionation. Figure 19.18 indicates that both $HClO_2$ and $ClO_2^-$ should disproportionate in aqueous solution and, in fact, there is no aqueous chemistry of the +3 state of chlorine.

The halates are fairly strong oxidizing agents. The disproportionation of chlorine in hot base forms $ClO_3^-$ in the reaction

$$3Cl_2(g) + 6OH^- \rightarrow ClO_3^- + 5Cl^- + 3H_2O$$

Bromate and iodate can be made from the corresponding elements by their reaction with chlorine:

$$X_2 + 5Cl_2(g) + 6H_2O \rightarrow 2HXO_3(aq) + 10HCl(aq)$$

The bromate or iodate can be purified by removing the chloride with silver oxide by the reaction

$$2HCl(aq) + Ag_2O(s) \rightarrow 2AgCl(s) + H_2O$$

Perchloric acid is the only oxoacid of chlorine that can be prepared in anhydrous form, for example, by distillation of a solution of a metal perchlorate in sulfuric acid:

$$KClO_4(s) + H_2SO_4(l) \xrightarrow{\Delta} KHSO_4(s) + HClO_4(l)$$

The pure acid as well as the concentrated solution react with explosive violence and are unsafe unless handled properly. In cold, dilute solution, perchloric acid is a very stable and very strong acid.

*c. The Halides and Pseudohalides*  Most elements form at least a few halides. In general, the representative metal halides are ionic salts while the nonmetal halides are covalently bonded. The melting or boiling points of the halides (see Figure 6.26) provide an approximate indication of the type of bond formed. Ionic bonds, being much stronger than the van der Waals interactions between molecules, lead to high melting and boiling points. The representative metal halides have the general formula $MX_n$, where $n$ is usually equal to the group number of the metal, for example, NaCl, $MgCl_2$, and $AlCl_3$. Halide salts are generally very stable.

The most important halides formed by the elements of Group VA are the trihalides and pentahalides of phosphorus, $PX_3$ and $PX_5$, respectively. The trihalides are formed by all the halogens and the pentahalides by all but iodine. Phosphorus pentaiodide presumably cannot be formed owing to **steric hindrance**—the iodine atoms are too large for five of them to squeeze around a single phosphorus atom. The structures of these compounds generally are those predicted by the VSEPR model: the trihalides are trigonal pyramidal and the pentahalides are trigonal bipyramidal in the gas phase (Figure 19.19).

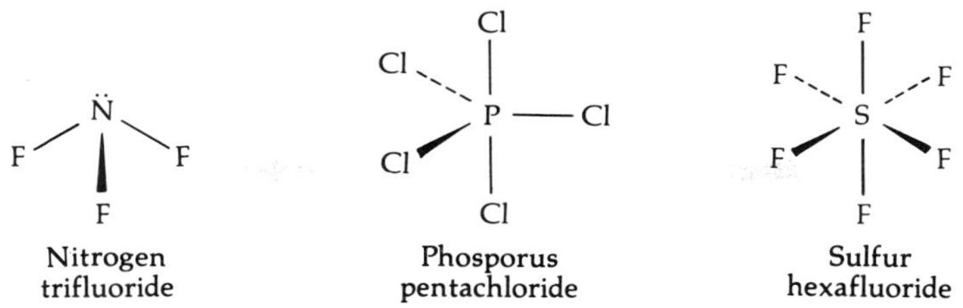

### FIGURE 19.19 The structures of some covalent halides.

Nitrogen trifluoride

Phosporus pentachloride

Sulfur hexafluoride

The halides of phosphorus can be prepared by direct union of the elements. For example, PCl₃ is prepared by passing chlorine over molten phosphorus:

$$P_4(l) + 6Cl_2(g) \xrightarrow{\Delta} 4PCl_3(l)$$

Phosphorus trichloride can be oxidized to the pentachloride with excess chlorine. The trihalides hydrolyze readily, liberating the corresponding hydrogen halide. The reaction of PCl is typical:

$$PCl_3(l) + 3H_2O \rightarrow H_3PO_3(aq) + 3HCl(aq)$$

The pentahalides react with either water or phosphorus pentoxide to form phosphoryl halides:

$$PX_5 + H_2O \rightarrow POX_3 + 2HX$$

$$6PX_5 + P_4O_{10} \rightarrow 10POX_3$$

These compounds hydrolyze with the liberation of the corresponding hydrogen halide:

$$POX_3 + 3H_2O \rightarrow H_3PO_4 + 3HX$$

**Example 19.8** What is the structure of POCl₃?

The structural formula of phosphoryl chloride is

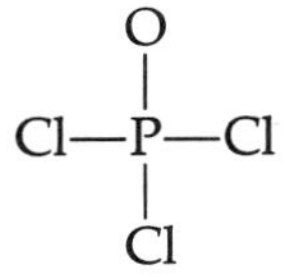

$$\begin{array}{c} O \\ \| \\ Cl-P-Cl \\ | \\ Cl \end{array}$$

The total number of valence electrons is $5 + 6 + 3 \times 7 = 32$ and the Lewis structure can be written as

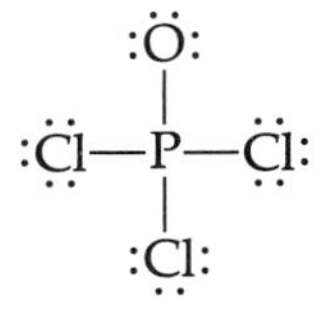

$$\begin{array}{c} :\ddot{O}: \\ \| \\ :\ddot{C}l-P-\ddot{C}l: \\ | \\ :\ddot{C}l: \end{array}$$

According to the VSEPR model, a tetrahedral structure is expected since the phosphorus atom is bonded to four other atoms and has no lone pair electrons. However, since the four atoms bonded to the phosphorus are not identical, some distortion from the tetrahedral shape can be expected. In fact, the Cl—P—Cl bond angle is approximately 103°. ∎

Sulfur is somewhat similar to phosphorus with respect to its halides. The direct reaction between sulfur and fluorine forms primarily the hexafluoride, $SF_6$ (Figure 19.19). This substance is an extremely inert gas that is both thermally stable and an excellent electrical insulator. Therefore, it is used as a gaseous insulator in high-voltage generators.

Sulfur also forms a reactive tetrafluoride $SF_4$, which, in analogy to $PF_3$, is hydrolyzed in water:

$$SF_4(g) + 2H_2O \rightarrow SO_2(g) + 4HF(aq)$$

Nitrogen, too, resembles phosphorus in that it forms trihalides. However, these compounds are generally much less stable than the phosphorus halides. Recall that nitrogen, being in the second period, cannot undergo valence shell expansion. Thus, the nitrogen pentahalides do not exist.

A number of singly charged anions resemble the halides in some of their chemical properties and are consequently known as **pseudohalides**. These species include the azide ion, $N_3^-$, the cyanide ion, $CN^-$, the cyanate ion, $OCN^-$, and the thiocyanate ion, $SCN^-$. Among the similarities to the halides are the insolubility of the salts of silver, lead, and mercury(I), as well as the formation of complex ions, for example,

$$Ag^+ + CN^- \rightarrow AgCN(s)$$

$$AgCN(s) + CN^- \rightarrow Ag(CN)_2^-$$

The corresponding pseudohalogens generally exist as dimeric molecules, for example, cyanogen, $(CN)_2$, oxycyanogen, $(OCN)_2$, and thiocyanogen, $(SCN)_2$.

*d. Interhalogen Compounds* The halogens combine with each other to form

**TABLE 19.14  Interhalogen Compounds and Ions**

| Interhalogen compound | Physical state at room temperature | Anion | Cation |
|---|---|---|---|
| ClF | colorless gas | $ClF_2^-$ | |
| $ClF_3$ | colorless gas | $ClF_4^-$ | $ClF_2^+$ |
| $ClF_5$ | colorless gas | | $ClF_4^+$ |
| BrF | red gas | $BrF_2^-$ | |
| $BrF_3$ | colorless liquid | $BrF_4^-$ | $BrF_2^+$ |
| $BrF_5$ | colorless liquid | $BrF_6^-$ | $BrF_4^+$ |
| BrCl | red gas | $BrCl_2^-$ | |
| $IF_3$ | yellow solid | $IF_4^-$ | $IF_2^+$ |
| $IF_5$ | colorless liquid | $IF_6^-$ | $IF_4^+$ |
| $IF_7$ | colorless gas | $IF_8^-$ | $IF_6^+$ |
| ICl | black solid | $ICl_2^-$ | |
| $ICl_3$ | yellow solid | $ICl_4^-$ | $ICl_2^+$ |
| IBr | brown-black solid | $IBr_2^-$ | |

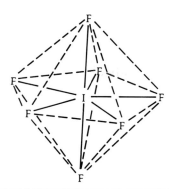

**FIGURE 19.20** The structure of IF$_7$.

molecular compounds known as **interhalogens**. These substances have the general formula XX$'_n$, where $n = 1$, 3, 5, or 7, and the larger halogen atom is surrounded by $n$ atoms of the smaller halogen. Table 19.14, which lists the interhalogens, shows that not all the possible combinations satisfying the general formula are known. For example, iodine is the only halogen atom large enough to accommodate seven surrounding atoms, and then only of fluorine, the smallest of the halogen atoms.

The geometry of the interhalogen molecules is well described by the VSEPR model. The structure of IF$_7$ is of interest since this is one of the relatively few molecules in which the central atom is surrounded by more than six atoms. The Lewis structure indicates that, since iodine has seven valence electrons and is surrounded by seven fluorine atoms, it has no lone pair electrons. The resulting structure is a pentagonal bipyramid (Figure 19.20).

The interhalogens can be regarded as halides of the smaller and more electronegative halogen, which has an oxidation number of $-1$. The other halogen therefore has a positive oxidation number. Consequently, the interhalogens are strong oxidizing agents, and react with most other elements with the formation of mixed halides. The interhalogens hydrolyze in water, for example,

$$BrCl(g) + H_2O \rightarrow HCl(aq) + HBrO(aq)$$

Note that the more electronegative chlorine forms the hydrohalic acid while the less electronegative bromine forms the hypohalous acid, confirming the oxidation number assignments.

Most of the interhalogens react with metal halides to form polyhalide ions, such as ClF$_2^-$, ClF$_4^-$, and BrF$_6^-$. Mixed polyhalide ions such as ICl$_3$F$^-$ and ICIBr$^-$ are also known, as are several polyiodide ions, I$_3^-$, I$_5^-$, and I$_7^-$, and polyhalogen cations.

## 19.9 The Noble Gases

The noble gases were discovered near the end of the nineteenth century. Argon constitutes nearly 1% of the atmosphere and the heavier members of the group are minor constituents of air. Helium is found in varying amounts in natural gas. It has the lowest boiling point of any element (4.2 K) and is widely used as a refrigerant for cryogenic (i.e., low-temperature) studies.

The noble gas atoms have completely filled s and p orbitals in their valence shells. Until 1962, the conventional wisdom held that, owing to the unusually great stability of this electronic configuration, the noble gases could not form stable compounds. There had been some thought that the heavier noble gases might form compounds but various attempts to synthesize xenon fluoride and xenon chloride in the 1930s were unsuccessful, and there was little further interest in the subject. This all changed in 1962 when Neil Bartlett (b. 1932) reported the first preparation of a xenon compound.

It is interesting to examine Bartlett's experiment, not only because its results were in startling contradiction to the accepted dogma, but also because it illustrates how careful consideration of unexpected results can sometimes have consequences of major scientific importance. Bartlett's experiment initially involved an attempt to prepare the then unknown compound $PtF_2$. As one of the steps, fluorine was introduced into a reaction chamber containing $PtF_4$ in a stream of nitrogen gas, which turned out to contain some oxygen impurity. The reaction resulted in the formation of a red solid which was eventually identified as the new compound $O_2^+ PtF_6^-$.

The connection with xenon compounds was made when Bartlett realized that $O_2$ and Xe have virtually identical first ionization energies, 1164 and 1170 kJ $mol^{-1}$, respectively. If $PtF_6$ could remove an electron from oxygen, perhaps it could also remove one from xenon. When xenon was mixed with the deep red $PtF_6$ vapor at room temperature, a yellow solid was obtained immediately. This substance was initially thought to be $XePtF_6$ but is now known to be more complex. The announcement of this result led to renewed interest in noble gas chemistry. Binary compounds of xenon and fluorine were found to be readily preparable and oxocompounds have also been made. The xenon fluorides, $XeF_2$, $XeF_4$, and $XeF_6$ are made sequentially by exposure of xenon to fluorine at elevated temperatures. Xenon tetrafluoride, the best characterized of these compounds, melts at 90°C and is stable at room temperature. It is isoelectronic with $IF_4^-$ and has a similar square-planar structure. The Xe—F bond energy, 126 kJ $mol^{-1}$, is not much smaller than the F—F bond energy.

Hydrolysis of the xenon fluorides results in the formation of oxygenated compounds. For instance, the reaction

$$XeF_6(g) + 3H_2O \rightarrow XeO_3(s) + 6HF(aq)$$

results in the formation of the trioxide. When dry, this solid is violently explosive. In aqueous solution it is relatively stable, but acts as a powerful oxidizing agent:

$$XeO_3(s) + 6H^+ + 6e^- \rightarrow Xe(g) + 3H_2O \qquad \epsilon° = 2.12 \text{ V}$$

While a relatively unstable fluoride of krypton, $KrF_2$, has also been prepared, the lighter noble gases are still thought to be inert, in line with their substantially larger ionization energies.

# Conclusion

In this chapter we have examined some aspects of the acid–base chemistry and redox chemistry of the representative elements. While many regularities may be discerned in such diverse properties as acid strength, oxidation numbers, standard reduction potentials, etc., a significant number of exceptions can also be noted. These exceptions frequently reflect the net effect of a number of factors and can be understood with the help of thermodynamic cycles.

Electrode potential diagrams provide a useful summary of the standard reduction potentials of an element with several oxidation states. These diagrams are of value for an understanding of the redox chemistry of the elements. However, their limitations must be recognized. Since the given electrode potentials apply only at standard conditions, for example to $1 M$ solutions, different $\epsilon$ values, obtainable by means of the Nernst equation, must be used for other conditions. We have seen, for instance, that the oxidizing power of an oxoacid depends strongly on pH. More fundamentally, the standard reduction potentials are thermodynamic quantities and give no indication of the rates of redox reactions. Thus, they tell us what can happen but not necessarily what will happen.

# Problems

1. What mass of sodium chloride will produce enough hydrogen chloride in its reaction with sulfuric acid to make up 300 mL of 40.0% (by weight) hydrochloric acid solution (density = 1.20 g cm$^{-3}$)?

2. An anhydride of an oxoacid is the oxide formed by the removal of water from the acid. Write the formulas of the anhydrides of (a) $HNO_2$, (b) $HNO_3$, and (c) $H_2N_2O_2$.

3. The commercial preparation of chlorine dioxide involves the reaction of sodium chlorate, sulfuric acid, and sulfur dioxide. Sodium hydrogen sulfate is formed as a by-product. Write a chemical equation for this process and indicate the changes in oxidation number which occur.

4. Write the formulas and names of the anhydrides of the following oxoacids: (a) hypobromous acid, (b) chlorous acid, (c) iodic acid, and (d) perchloric acid.

5. Write chemical equations for the formation of each of the following acids from its corresponding anhydride: (a) $H_2SO_4$, (b) $H_3PO_3$, and (c) $H_5IO_6$.

6. Predict the value of $pK_a$ for the ionization of the acid $H_4XeO_6$ in which there are no Xe—H bonds.

7. What mass of nitric acid would be obtained in the Ostwald process from 100 m$^3$ of ammonia measured at 300°C and 3.50 atm if the yield is 91.5% of the maximum obtainable yield?

8. Estimate the $pK_a$ values of the acids (a) $H_5IO_6$ (IO(OH)$_5$) and (b) $H_6TeO_6$ (Te(OH)$_6$).

9. Arrange the following species in order of increasing acid strength: HF, $HClO_4$, $HClO_3$, HCl, $H_2PO_4^-$, and $HPO_4^{2-}$.

10. Two structural isomers with the formula $H_2N_2O_2$ are known. One of these, hyponitrous acid, contains two O—H bonds; the other, nitramide, has no O—H bonds. (a) Draw Lewis structures of these isomers; (b) estimate their standard enthalpies of formation by the bond energy method; (c) estimate $pK_a$ of hyponitrous acid; and (d) determine the oxidation numbers and formal charges of the nitrogen atoms in the two compounds.

11. Potassium superoxide, used as a source of oxygen in breathing masks, has a density of 2.14 g cm$^{-3}$ at 25°C. At what pressure would gaseous oxygen have to be stored at this temperature to have the same density of usable oxygen as potassium superoxide? Assume ideal gas behavior.

12. Listed below are some data relevant to the half-reaction M(s) → M$^{3+}$(aq) + 3$e^-$ for aluminum and gallium. (a) Calculate $\Delta H°$ for this process for each of the elements by means of an appropriate Born–Haber-type cycle. (b) Assuming that the $\Delta G°$ values are comparable to the $\Delta H°$ values, which of the two elements is the stronger reducing agent? Is this trend in accord with that of the $\epsilon°$ values? (c) Identify the factors responsible for the difference.

|  | Aluminum | Gallium |
|---|---|---|
| $I_1$ (kJ mol$^{-1}$) | 578 | 579 |
| $I_2$ (kJ mol$^{-1}$) | 1817 | 1979 |
| $I_3$ (kJ mol$^{-1}$) | 2745 | 2963 |
| $\Delta H°_{sublimation}$ (kJ mol$^{-1}$) | 11,300 | 10,500 |
| $\Delta H°_{hydration}$ (kJ mol$^{-1}$) | −4690 | −4703 |
| $\epsilon°$ (V) (standard reduction potential) | −1.66 | −0.53 |

13. Are Al$^+$ and Al$^{2+}$ stable in aqueous solution? Answer this question by evaluating $\Delta H°$ for the reactions

(a) $3Al^+(aq) \rightarrow 2Al(s) + Al^{3+}(aq)$
(b) $3Al^{2+}(aq) \rightarrow Al(s) + 2Al^{3+}(aq)$

on the assumption that $\Delta G°$ can be approximated by $\Delta H°$. Approximate the enthalpies of hydration of $Al^+$ and $Al^{2+}$ by those of the neighboring Group IA and IIA elements, respectively (Table 19.8).

14. Example 19.3 showed that $Mg^+(aq)$ is unstable with respect to disproportionation into $Mg^{2+}(aq)$ and $Mg(s)$. Is $Mg^+$ stable in the gas phase? Answer this question by computing $\Delta H°$ for
    (a) $2Mg^+(g) \rightarrow Mg(g) + Mg^{2+}(g)$
    (b) $2Mg^+(g) \rightarrow Mg(s) + Mg^{2+}(g)$.
    To what factor can the instability of $Mg^+(aq)$ be attributed?

15. Predict how the following properties of radium, the radioactive alkaline earth element, should compare with those of the other Group IIA elements: first ionization energy, atomic radius, and ionic radius. Note that the lanthanide series falls between barium and radium.

16. The $Be^{2+}$ ion hydrolyzes to $Be(OH)^+$ in aqueous solution. (a) Write a chemical equation for the hydrolysis of $Be(NO_3)_2$. (b) What is the pH of a 0.10 M solution of $Be(NO_3)_2$ given that $Be(OH)^+$ has $pK_b = 5.6$?

17. Using data in Appendix 4 (a) determine $\Delta H°_{298}$ for the decomposition of $NH_4NO_3$. (b) In actuality, the decomposition occurs only at elevated temperatures. Calculate $\Delta H°$ at 300°C.

18. The ONO bond angles are 180° in $NO_2^+$, 134° in $NO_2$, and 115° in $NO_2^-$. Explain this trend in bond angles.

19. Based on the data in Figure 19.13 determine which species containing nitrogen is (a) the strongest oxidizing agent in acidic solution, (b) the strongest reducing agent in acidic solution, (c) the strongest oxidizing agent in basic solution, and (d) the strongest reducing agent in basic solution.

20. Write Lewis structures and determine the oxidation number of nitrogen in each of the following: (a) $NCl_3$, (b) $HN_3$, (c) $N_2O_4$, (d) $N_2O$, (e) $N_2H_4$, and (f) $NOCl$.

21. Use $\epsilon°$ values obtainable from Figure 19.13 for the reduction of $N_2$ to $NH_4^+$ and for the reduction of $N_2$ to $NH_3$ along with the value of $K_w$ to estimate the value of $K_b$ of $NH_3$. Compare with the value given in Table 8.3.

22. Write Lewis structures and determine the oxidation number of phosphorus in each of the following: (a) $PCl_3$, (b) $P_2H_4$, (c) $H_3PO_2$, and (d) $H_4P_2O_7$.

23. Write chemical equations for the following reactions: (a) the hydrolysis of phosphorus trioxide, (b) the hydrolysis of phosphorus pentachloride, and (c) the combustion of white phosphorus in excess oxygen.

24. The electrode potential diagram of phosphorus in basic solution is

$PO_4^{3-} \xrightarrow{-1.12} HPO_3^{2-} \xrightarrow{-1.57}$

$H_2PO_2^- \xrightarrow{-2.05} P_4 \xrightarrow{-0.89} PH_3$

(a) Determine $\epsilon°$ for the reduction of $HPO_3^{2-}$ to $PH_3$. (b) Determine $\epsilon°$ for the reduction of $H_2PO_2^-$ to $PH_3$. (c) Which species, if any, can undergo disproportionation? Write an equation for each disproportionation reaction and determine $\epsilon°$. (d) Which species, if any, react with water or $OH^-$? Write an equation for each reaction and determine $\epsilon°$.

25. Determine the equilibrium constant for the reaction $H_2PO_2^- + OH^- \rightarrow HPO_3^{2-} + H_2$. Use data from problem 24 and Figure 19.14.

26. Find two half-reactions which can be combined to give the overall reaction $2O_3(g) \rightarrow 3O_2(g)$. Use the values of $\epsilon°$ for these half-reactions (found in this chapter or in Appendix 5) to calculate the equilibrium constant for this reaction.

27. A dilute solution of hydrogen peroxide can be prepared by the oxidation of barium metal with oxygen followed by the reaction of the product with dilute sulfuric acid. Write chemical equations for this process.

28. Using data in Appendix 4 calculate the oxygen–oxygen bond energy in (a) $O_2(g)$, (b) $O_3(g)$, (c) $H_2O_2(l)$. Compare with the O—O and O=O bond energy in Table 17.2. Account for the results in terms of the Lewis structures of the three molecules. The enthalpy of vaporization of $H_2O_2(l)$ is 51.47 kJ mol$^{-1}$.

29. Write the Lewis structure of thionyl chloride, $SOCl_2$. Write an equation for the hydrolysis of this compound, showing the formation of two gaseous products.

30. Determine the value of $\epsilon$ for the reduction of $SO_4^{2-}$ to $SO_3^{2-}$ at (a) pH = 0, (b) pH = 7, and (c) pH = 14 (Figure 19.17). Assume that $SO_4^{2-}$ and $SO_3^{2-}$ are present in 1 M concentration.

31. Write chemical equations for the reactions of sulfur with (a) $H_2$, (b) $O_2$, (c) $HNO_3$, and (d) C.

32. Use the data in Figure 19.18 to calculate $\epsilon°$ values for the following half-reactions:
    (a) $BrO_4^- + 7H^+ + 6e^- \rightarrow HBrO + 3H_2O$
    (b) $BrO_4^- + 3H_2O + 6e^- \rightarrow BrO^- + 6OH^-$

33. Write equations for the preparation of (a) $NO_2$ from $HNO_3$, (b) $HClO_4$ from $KClO_4$, and (c) $S_2O_3^{2-}$ from $H_2S$.

34. Write balanced equations for the disproportionation

of (a) $HNO_2$ to $NO_3^-$ and NO (acid), (b) $NO_2^-$ to $NO_3^-$ and NO (base), (c) HClO to $HClO_2$ and $Cl_2$ (acid), and (d) $ClO^-$ to $ClO_2^-$ and $Cl^-$ (base). For each reaction, determine the value of $\epsilon°$. Which of these reactions occur spontaneously at standard conditions?

35. (a) Write a chemical equation for the disproportionation of $ClO_2^-$ to $ClO^-$ and $ClO_3^-$. (b) Calculate the value of $\epsilon$ when the reactant concentration is 0.20 M and the product concentrations are 0.10 M. (c) If the product concentrations are 1 M, what must the concentration of $ClO_2^-$ be for the value of $\epsilon$ to drop to zero?

36. Predict the products of the following reactions and write the appropriate chemical equations: (a) sodium iodide is added to concentrated sulfuric acid; (b) a solution of hypochlorous acid is heated; (c) oxygen difluoride is contacted with powdered aluminum; and (d) sodium sulfite is treated with sodium periodate in dilute sulfuric acid.

37. Nitric oxide (NO) can be made in the reduction of $HNO_2$ by $I^-$ (forming $I_2$). Using data in Figures 19.13 and 19.18 calculate the equilibrium constant for this reaction. Write a balanced chemical equation.

38. The electrode potential diagram in acidic solution of astatine, the radioactive halogen, is

$H_5AtO_6$ —1.6— $AtO_3^-$ —1.5— HAtO —1.0—
$At_2$ —0.3— $At^-$

(a) Which species, if any, are unstable with respect to disproportionation? (b) Which species, if any, decompose water? (c) What is the value of $\epsilon°$ for the reduction of $AtO_3^-$ to $At_2$? (d) What are the values of $\epsilon°$, K, and $\Delta G°$ for the reaction $Br_2(l) + 2At^- \rightarrow 2Br^- + At_2$? (e) Is $H_5AtO_6$ a strong acid? Explain. How does it compare as an oxiding agent with the other perhalic acids?

39. Analyze the energetics of the electrode reaction $\frac{1}{2}X_2(g) + e^- \rightarrow X^-(aq)$ by means of a Born–Haber-type cycle and account for the decrease in the strength of the halogens as oxidizing agents between fluorine and chlorine. Use the data in Tables 19.1 and 17.2.

40. An excess of KI solution is added to 25.0 mL of a solution of $Fe^{3+}$ of unknown concentration, completely reducing it to $Fe^{2+}$. The resulting $I_2$ is titrated with a 0.145 M $S_2O_3^{2-}$ solution and 20.3 mL is required to reach the endpoint. (a) Write chemical equations for this process. (b) Determine the concentration of $Fe^{3+}$ in the original solution.

41. The following halogens and alkali halides are mixed in aqueous solution: $KBr + I_2$, $KI + Cl_2$, $KCl + Br_2$. Use Figure 19.18 to determine $\epsilon°$ values for the halide displacement reactions. Indicate in each case whether a displacement reaction occurs at standard conditions.

42. When iodine is dissolved in NaOH solution at standard conditions it disproportionates to $I^-$ and $IO_3^-$. However, if the solution is acidified, the reaction is reversed. Explain these observations with reference to Figure 19.18.

43. Use the $\epsilon°$ values in Figure 19.18 for the reduction of HClO to $Cl_2$ and for the reduction of $ClO^-$ to $Cl_2$ along with the value of $K_w$ to estimate the value of $pK_a$ of HClO. Compare with the value listed in Table 19.5.

44. The interhalogen compounds react vigorously with water to form oxoacids and hydrogen halides. Write chemical equations for the hydrolysis of (a) ClF, (b) $ClF_5$, (c) $ICl_3$, and (d) $IF_7$.

45. Write chemical equations for the reactions of the following nonmetal fluorides with excess NaOH solution: $PF_5$, $SF_6$, and $IF_7$.

46. Predict the sign (positive, negative, or close to zero) of $\Delta S°$ for each of the following reactions of interhalogen compounds:

(a) $ClF_5(g) \rightarrow ClF_3(g) + F_2(g)$
(b) $3BrF(g) \rightarrow BrF_3(l) + Br_2(g)$
(c) $2ClF(g) \rightarrow Cl_2(g) + F_2(g)$
(d) $IF_5(l) \rightarrow IF_3(s) + F_2(g)$

47. Write the most reasonable Lewis structures you can for (a) SCl and (b) $S_2Cl_2$. On the basis of these structures comment on which molecule is more likely to be stable.

48. Use the data in Figure 19.18 to predict the outcome when the following are mixed: (a) $BrO_3^-$ and $Cl^-$ in 1 M acid and (b) $IO_3^-$ and $Cl_2$ in 1 M base.

49. Draw Lewis structures and predict the geometry of (a) $XeF_4$, (b) $XeO_3$, (c) $XeOF_2$, and (d) $XeOF_4$.

50. (a) Write a chemical equation for the oxidation of bromate ion to perbromate ion by xenon trioxide (which is reduced to the element) in acidic aqueous solution. (b) What is the value of $\epsilon°$ for this reaction? (c) Calculate the equilibrium constant for the reaction.

# 20

# Structure and Bonding in Solids

Most of the substances that we encounter in daily life are solids. These substances display an incredibly large variety of forms. They range from the complex solids that make up the world of plants and animals to the naturally occurring crystalline minerals with their beautiful shapes and colors. Many of the bulk properties of solids reflect the underlying atomic or molecular structure. By investigating the geometric arrangement of atoms in solids and the nature of the forces between them, we can account for the electrical, mechanical, thermal, and optical properties of solids.

## Some Properties of Solids

### 20.1 Crystalline and Amorphous Solids

Many solid substances exist in the form of crystals. A **crystal** is a homogeneous solid in which the atoms, ions, or molecules are arranged in a definite repeating pattern (Figure 20.1). The regularity of crystals on the atomic level gives rise to their distinctive macroscopic shapes, which are clearly in evidence in naturally occurring minerals (Figure 20.2). A crystal is bounded by flat surfaces that intersect at specific angles, the values of which are a characterisic property of a substance. The combined effect of a pleasing color, a brilliance due to the reflection of light by the crystal surfaces, suitable size, and rareness, all enhanced by proper cutting and polishing, makes certain naturally occurring minerals prized as gemstones. For example, the emerald consists of the natural

Structure and Bonding in Solids  20.1  **739**

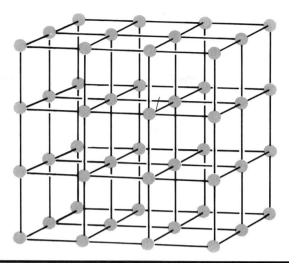

**FIGURE 20.1** Simple cubic crystal. The touching atoms are replaced by small spheres in order to clarify the pattern of this particular arrangement.

mineral beryl, $Be_3Al_2Si_6O_{18}$, in which a small fraction of the aluminum has been replaced by chromium.

Most crystalline solids do not consist of single crystals of the type shown in Figure 20.2 or as exemplified by table salt or sugar. Rather, they are **polycrystalline,** meaning that they are aggregates of many interlocking small crystals, which may be of microscopic size. Solid metals are a typical example.

When a liquid freezes to form a crystalline solid, freezing occurs at a specific temperature. However, many liquids can be supercooled and remain as liquids below their freezing points. Supercooled liquids are thermodynamically unstable with respect to the solid phase (Section 12.3). For many substances, crystallization can be induced by vigorous stirring or by the introduction of a small "seed" crystal of the substance in question. Other liquids, however, do not crystallize under these conditions, particularly if they contain bulky molecules or polymeric species. What happens, instead, is that these species gradually become less mobile and the liquid eventually solidifies as it is cooled. The disordered orientation of the molecules in the liquid persists in the solid. Substances of this type are called **amorphous solids,** and include glass, plastics, and rubber. In contrast to crystals, amorphous solids do not possess

**FIGURE 20.2** Natural quartz mineral.

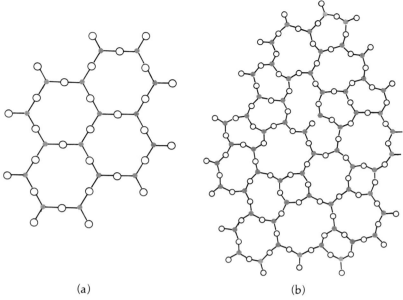

**FIGURE 20.3** The structure of crystalline and amorphous solids. (a) Two-dimensional structure of quartz, a form of crystalline silicon dioxide. (b) Two-dimensional structure of amorphous silicon dioxide, a major constituent of glass. Different glasses are made by fusing SiO$_2$ with CaCO$_3$, Na$_2$CO$_3$, and other compounds.

an ordered internal structure (Figure 20.3) and do not have sharp melting points. When heated, they gradually soften and eventually begin to flow.

In this chapter we will examine the structure and the nature of the interaction in four different types of solid substances: metals, ionic crystals, molecular solids, and covalent solids. Metallic solids consist of an array of cations embedded in a sea of electrons. These electrons are relatively mobile and account for the high electrical conductivity of metals. Ionic solids consist of arrays of oppositely charged ions held together by strong electrostatic forces. Molecular solids consist of discrete molecules held together by the intermolecular forces examined in Chapter 18. Covalent solids, such as diamond, consist of atoms held together by a large network of covalent bonds.

## 20.2 The Heat Capacities of Solids

The atoms of a solid are held rigidly in place and the only motion that they can execute is to vibrate about their equilibrium positions (see Figure 12.1). Since the heat capacity of a body is determined by the change of its internal energy with temperature ($C_v = dE/dT$), the heat capacity of a solid must depend on the variation with temperature of its vibrational energy. At sufficiently high temperatures, the vibrational energy of 1 mole of atoms is $3RT$. Consequently, the heat capacity of an atomic solid is

$$C_v = \frac{dE}{dT} = 3R \approx 25 \text{ J mol}^{-1} \text{ K}^{-1}$$

This relationship, known as the **law of Dulong and Petit,** was found empirically in 1819 and is of historical importance. The predicted value of the molar heat capacity of a metal combined with an experimental measurement of the heat capacity per gram, permits a determination of the atomic weight of the metal (see Problem 20.1), and was one of the early methods used to determine atomic weights. This empirical law can also be applied to many solid compounds: their heat capacity is close to 25 J K$^{-1}$ per mole of atoms.

The law of Dulong and Petit is obeyed at room temperature by all solid elements with atomic weights above about 30. However, solids made up of light atoms, such as boron and carbon, have molar heat capacities that are much smaller than 3R. In fact, measurements performed at low temperatures indicate that the molar heat capacities of *all* solids fall below this limiting value at sufficiently low temperatures, approaching a value of zero at 0 K. A typical curve for the temperature dependence of $C_V$ is shown in Figure 20.4.

The first explanation of the temperature dependence of the heat capacity of solids was given by Einstein in an adaptation of Planck's quantum theory of black-body radiation (Section 5.4a). Following Planck, Einstein assumed that the vibrational energy of an atom was quantized and given by the expression (Equation 5.12)

$$\epsilon = nh\nu$$

where $\nu$ is the vibrational frequency, $h$ is Planck's constant, and $n$ is a positive integer. The total energy of 1 mole of vibrating atoms was obtained by Einstein as

$$E = \frac{3RT(h\nu/kT)}{e^{h\nu/kT} - 1} \tag{20.1}$$

where $k$ is the Boltzmann constant. Note the similarity between this expression and Planck's formula (Equation 5.13) for the intensity of the light emitted by vibrating atoms. The heat capacity is obtained by differentiating Equation 20.1 with respect to $T$:

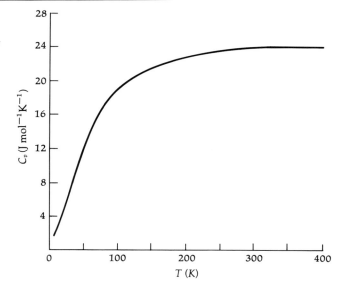

**FIGURE 20.4** Temperature dependence of the molar heat capacity at constant volume, $C_V$, of silver.

$$C_v = \frac{dE}{dT} = 3R\left(\frac{h\nu}{kT}\right)^2 \left\{\frac{\exp(-h\nu/kT)}{[1 - \exp(-h\nu/kT)]^2}\right\} \tag{20.2}$$

The temperature dependence of $C_v$ predicted by this formula can be seen by noting that at low $T$, $\exp(-h\nu/kT)$ drops to zero and so $C_v$ decreases, as shown in Figure 20.4. On the other hand, at sufficiently high $T$, $h\nu/kT \ll 1$ and $\exp(-h\nu/kT)$ can be approximated by $1 - h\nu/kT$ (see Appendix 2, Section A2.4), yielding

$$C_v = 3R\left(\frac{h\nu}{kT}\right)^2 \left(\frac{1 - h\nu/kT}{(h\nu/kT)^2}\right) \approx 3R \tag{20.3}$$

Thus, the theory predicts both the decrease in $C_v$ at low temperatures and the Dulong and Petit limit at high temperatures.

## 20.3 Crystal Systems

*a. Lattices and Unit Cells* The atoms in a crystal are arranged in a regular, repeating pattern. Such a pattern lends itself to a geometric description. Various properties of crystalline substances depend on the geometric arrangement of its atoms. For example, the density of a solid depends on the arrangement of its atoms, as we shall see.

A crystal may be described by specifying all of its points that have identical environments. This collection of points is called a **lattice.** The concept of a lattice is illustrated in Figure 20.5, which shows the crystal structure of sodium chloride. This salt consists of a cubic array of alternating $Na^+$ and $Cl^-$ ions. Various equivalent ways of constructing the lattice corresponding to this structure are possible. The lattice points can be drawn at the centers of the $Na^+$ ions (Figure 20.5a), at the centers of the $Cl^-$ ions (Figure 20.5b), or between the $Na^+$ and $Cl^-$ ions (Figure 20.5c). Comparison of the three sets of lattice points shows that they have the same pattern and form the same lattice, differing only in the exact location of the points.

The periodic repetition of the points in a particular lattice makes it possible to determine the lattice properties by focusing on a small subunit, called the unit cell. The **unit cell** is the smallest part of a lattice that will generate the entire lattice when repeated in three dimensions, which possesses the same symmetry as the lattice. The colored cubes in Figure 20.5 illustrate three ways in which a unit cell of the sodium chloride structure may be drawn. In fact, the three unit cells are identical: they have the same shape and size, and contain the same number of lattice points. The concept of a unit cell is of utmost importance in the study of crystal structure since it reduces the problem of determining the geometry of a large array to that of understanding the properties of a simple figure, containing at most a few points. Since the lattice can be constructed by repeatedly moving the unit cell along the three axes, the unit cell can be regarded as a unique building block exhibiting all the geometrical properties of the entire lattice.

The requirement that the unit cell have the same symmetry as the entire lattice is significant because different types of crystals differ in their symmetry. The significance of this requirement is most easily seen for a two-dimensional lattice, as illustrated in Figure 20.6. The lattice shown in this figure is rectangular—90° angles are formed between the vertical and horizontal axes drawn through the points. The smallest figure that can generate the entire lattice is

Structure and Bonding in Solids 20.3  743

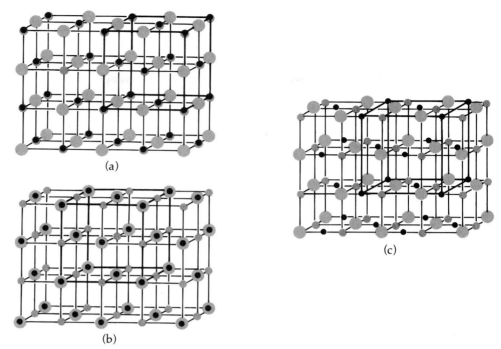

**FIGURE 20.5** The crystal structure of sodium chloride, showing three equivalent lattices. The large spheres represent Cl⁻ ions and the small spheres represent Na⁺ ions. The small colored points define the lattice. (a) The lattice points are placed in the centers of the Na⁺ ions. (b) The lattice points are placed in the centers of the Cl⁻ ions. (c) The lattice points are placed halfway between the Na⁺ and Cl⁻ ions. The colored lines in each structure join the points which define the unit cell of the lattice. For clarity, the ions are drawn as well separated spheres. In actuality, they are in contact.

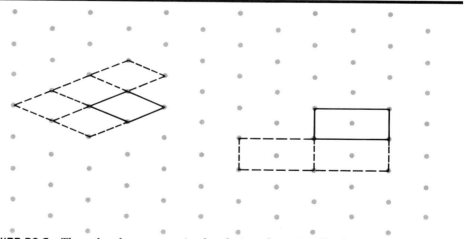

**FIGURE 20.6** The role of symmetry in the choice of a unit cell. The points define a two-dimensional rectangular lattice. The parallelogram drawn on the left is the figure containing the fewest points that can generate the entire lattice when repeated (as shown by the dashed parallelograms). This figure contains no right angles although the occurrence of right angles is one of the important symmetry elements of the lattice. Consequently, the unit cell is the rectangle shown at the right, even though it contains more lattice points than the parallelogram.

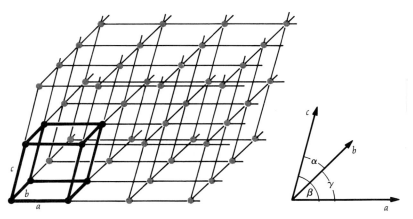

**FIGURE 20.7** The general shape of a unit cell. The diagram on the right identifies the angles between the axes.

the parallelogram shown on the left. However, the sides of this figure do not intersect at right angles and the figure consequently does not have the same symmetry as the lattice. The unit cell is the rectangle shown on the right side of the lattice. Although it contains one more lattice point than the parallelogram, the rectangle is a more satisfactory unit cell because it shows the presence of 90° angles in the lattice.

The unit cell of every known crystal consists of a parallelepiped, a three-dimensional structure whose shape and size are determined by the lengths of its three axes ($a$, $b$, $c$) and the angles between them ($\alpha$, $\beta$, $\gamma$), as shown in Figure 20.7. Seven distinct unit cell shapes, differing in the relative lengths of the three axes or in the angles between them, are sufficient to describe the structures of all known crystals. These figures are known as the seven **crystal systems.** They are shown in Figure 20.8 and their geometric properties are summarized in Table 20.1. The simplest and most symmetric figure is the cube. The

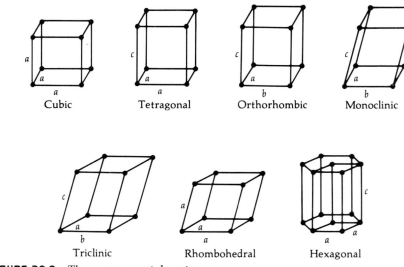

**FIGURE 20.8** The seven crystal systems.

**TABLE 20.1  The Seven Crystal Systems**

| System | Unit cell lengths | Unit cell angles | Examples |
|---|---|---|---|
| Cubic | $a = b = c$ | $\alpha = \beta = \gamma = 90°$ | $CsCl$, $NaCl$, $CaF_2$ |
| Tetragonal | $a = b \neq c$ | $\alpha = \beta = \gamma = 90°$ | $TiO_2$, $Sn$ (white) |
| Orthorhombic | $a \neq b \neq c$ | $\alpha = \beta = \gamma = 90°$ | $HgCl_2$ |
| Monoclinic | $a \neq b \neq c$ | $\alpha = \gamma = 90° \neq \beta$ | $KClO_3$ |
| Triclinic | $a \neq b \neq c$ | $\alpha \neq \beta \neq \gamma \neq 90°$ | $CuSO_4 \cdot 5H_2O$ |
| Rhombohedral | $a = b = c$ | $\alpha = \beta = \gamma \neq 90°$ | $CaCO_3$, $Al_2O_3$ |
| Hexagonal | $a = b \neq c$ | $\alpha = \beta = 90°, \gamma = 120°$ | $ZnS$, $SiO_2$ |

next four systems represent sequential steps in the destruction of the symmetry of the cubic unit cell. First, the sides are made unequal but the angles retain their 90° values (tetragonal and orthorhombic). Next, the angles also become unequal (monoclinic and triclinic). The rhombohedral unit cell can be visualized as a cube whose adjacent sides are tilted with respect to each other. The hexagonal system has the property that when three unit cells are placed side by side, the corners of the cells form regular hexagons.

The chief experimental method for the determination of crystal structure is X-ray diffraction. X rays are a form of electromagnetic radiation and can therefore be diffracted by slits in essentially the same way as visible light (see Figure 5.10). Diffraction occurs when the size of the slit is comparable to the wavelength of the electromagnetic radiation. The wavelength of X rays is of the order of an angstrom, comparable to the internuclear separation of atoms in solids. Consequently, X rays are diffracted by solids. When the atoms in the solid are arranged in an ordered array, i.e., when the solid is crystalline, the diffraction pattern consists of a series of lines from which it is possible to infer the identity of the crystal system and the internuclear distances between neighboring atoms (Figure 20.9).

*b. The Cubic Crystal System*  The unit cells of the three different types of cubic lattices are shown in Figure 20.10. The **simple cubic** (SC) cell is identical to the cell shown in Figure 20.8 and has lattice points at the eight corners of the cube. (Cells whose only lattice points lie at the corners of the unit cell are called **primitive cells**.) The **body-centered cubic** (BCC) cell has an additional lattice point located at the center of the cube. The **face-centered cubic** (FCC) cell has lattice points located at the centers of each of the six faces of the cube in addition to the corner points. Many substances crystallize in these cubic lattices, as we shall see.

A lattice composed of atoms may be characterized by the number of atoms per unit cell, the coordination number, and the distance between nearest neighbors. Table 20.2 lists the values of these properties for the three types of cubic lattices. The number of atoms per unit cell is the ratio of the total number of atoms in the cell to the number of cells among which the atoms are shared. The **coordination number** of a lattice is the number of nearest neighbor atoms surrounding a given atom. In a cubic lattice, the distance between nearest neighbors is expressed in terms of the length of a side of the unit cell, $a$.

The results summarized in Table 20.2 can be deduced readily with the aid of three-dimensional lattice models. To illustrate the procedure, let us derive the results for the BCC lattice. The unit cell is defined by the eight corner

# 746  Structure and Bonding in Solids  20.3

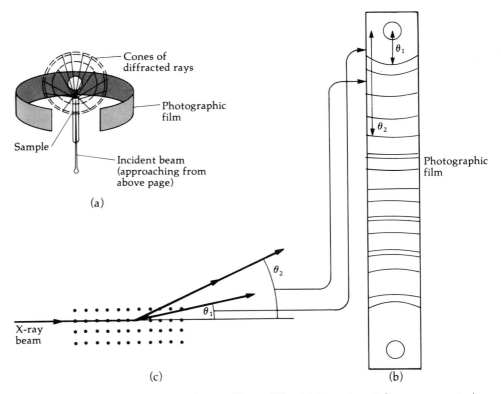

**FIGURE 20.9** X-ray diffraction of crystalline solids. (a) Experimental arrangement. A beam of monochromatic (i.e., single wavelength) X rays strikes a powdered sample of a crystalline solid surrounded by a circular strip of photographic film. The diffracted X rays form a conical pattern whose intersection with the film consists of arcs of circles. The undeflected X-ray beam passes through the hole in the center of the film. (b) X-ray diffraction pattern of NaCl. (c) Relation between the X-ray diffraction pattern and the structure of the crystal. The position of an arc in the spectrum defines the angle by which the X-ray beam is scattered by a plane of atoms. The angle depends on the distance between successive planes of atoms, which can be inferred from the data. The pattern of observed lines permits a determination of the crystal system.

atoms and contains an additional atom in the center (Figure 20.10). As shown in Figure 20.11a, the eight corner atoms are shared among eight unit cells so that there is one corner atom per unit cell. The atom in the center of the unit cell belongs exclusively to that cell. Thus, there are two atoms per unit cell.

The nearest neighbors in a BCC lattice are the atom in the center of the cell and any of the corner atoms. The distance between the nuclei of these

**TABLE 20.2  Properties of Cubic Lattices**

| Lattice | Atoms per unit cell | Coordination number | Distance between nearest neighbors | Examples |
|---|---|---|---|---|
| Simple cubic (SC) | 1 | 6 | $a$[a] | Po |
| Body-centered (BCC) | 2 | 8 | $(\sqrt{3}/2)a$ | Ba, Mo, Fe |
| Face-centered (FCC) | 4 | 12 | $(\sqrt{2}/2)a$ | Ag, Ca, Cu |

[a]The length of a unit cell is equal to $a$.

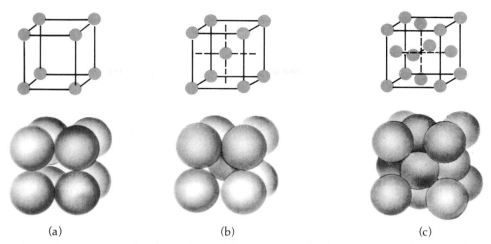

**FIGURE 20.10** Unit cells of the three different types of cubic lattices. (a) Simple cubic, (b) body–centered cubic, and (c) face–centered cubic. The top drawings show the location of the lattice points and the bottom drawings show the locations of identical atoms in crystals of the indicated type.

atoms is half the body diagonal of the cube, $(\sqrt{3}/2)a$ (Figure 20.11b). Since the central atom is surrounded by eight corner atoms, the coordination number is eight. Each corner atom, in turn, is surrounded by the atoms located at the centers of the eight unit cells among which it is shared, showing that each atom in the lattice has a coordination number of eight.

Table 20.2 shows that the number of atoms per unit cell is smallest for an SC lattice and largest for an FCC lattice. It can also be seen that as more atoms are packed into a cubic cell of given size, the number of nearest neighbors (i.e., the coordination number) increases and the distance between them decreases.

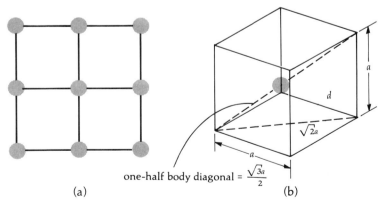

**FIGURE 20.11** Properties of a BCC lattice. (a) Top view of four adjacent simple cubic cells. The atom in the center is shared by the four unit cells in the figure and, in addition, by four other unit cells stacked on top of those in the figure. (b) Distance between nearest neighbors in a BCC lattice.

The length of a side of a unit cell can be obtained from X-ray diffraction measurements. From the value of $a$ and the lattice type, the distance between the nuclei of nearest-neighbor atoms may be obtained. Since these atoms are in contact, the distance between their nuclei is equal to the sum of their radii. The type of radius obtained in this way depends on the interaction between the species present in the crystal. If the crystal consists of nonbonded atoms or molecules, van der Waals radii of the atoms in contact may be obtained. For metallic crystals, the distances correspond to the sum of the atomic radii of the metal atoms. Finally, for ionic crystals, the internuclear distance is equal to the sum of the ionic radii.

**Example 20.1** Xenon crystallizes in a face-centered cubic lattice and its density is 3.52 g cm$^{-3}$. What is the van der Waals radius of xenon?

A face-centered cubic lattice has four atoms per unit cell. Thus, the volume occupied by four atoms is equal to the volume of a unit cell:

$$V = \frac{m}{\rho} = \frac{(131.3 \text{ g mol}^{-1})(4)}{(6.022 \times 10^{23} \text{ mol}^{-1})(3.52 \text{ g cm}^{-3})}$$

$$V = 2.478 \times 10^{-22} \text{ cm}^3$$

Note that division by Avogadro's number converts the mass of a mole to that of a single atom. The volume of a cubic unit cell of length $a$ is equal to $a^3$. The value of $a$ is therefore equal to

$$a = V^{1/3} = (2.478 \times 10^{-22} \text{ cm}^3)^{1/3} = 6.281 \times 10^{-8} \text{ cm}$$

The distance between nearest-neighbor atoms in a face-centered cubic lattice is $(\sqrt{2}/2)a$. Since nearest-neighbor atoms are in contact, this distance is equal to twice the radius. The radius of a xenon atom is therefore equal to

$$r = (\sqrt{2}/4)a = (\sqrt{2}/4)(6.281 \times 10^{-8} \text{ cm})(10^8 \text{ Å cm}^{-1}) = 2.22 \text{ Å}$$

Since xenon atoms interact by the van der Waals interaction, the radius obtained in this manner can be identified with the van der Waals radius. ∎

## Metals

Metals have a number of characteristic physical properties: they are lustrous, malleable, and ductile; they are excellent conductors of heat and electricity, but their electrical conductivity decreases with increasing temperature. These properties can be related to the metal structure and to the interactions between metal atoms.

### 20.4 The Structure of Metals

A majority of metals have a **closest-packed structure:** identical spherical atoms are packed together so that there is a minimum of empty space in the crystal. Closest packing occurs because, as long as the interatomic forces are not strongly directional, this arrangement minimizes the potential energy of the crystal.

A closest-packed arrangement of atoms corresponds to either a hexagonal lattice or to a face-centered cubic lattice. The equivalence of the closest-packed structures and the corresponding crystal system lattices is somewhat difficult to see, and so it is helpful to examine closest packing one layer at a time. Figure 20.12a shows a single layer of closest-packed spherical atoms. Each sphere is surrounded by six nearest-neighbor spheres. The meeting point of any three mutually adjoining spheres defines a small, nearly triangular depression. A

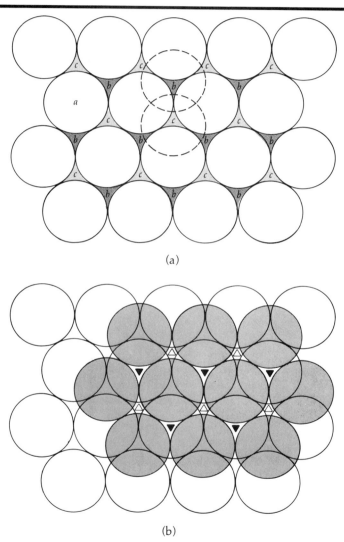

**FIGURE 20.12** The closest packing of spherical atoms. (a) A single layer of a closest-packed structure. Letters a, b, and c define, respectively, the center of a sphere, and the two oppositely oriented types of depressions between spheres. The dashed circles show that adjacent depressions b and c are too close to accommodate a second layer of spheres. (b) A second layer of spheres can be placed in the depressions of type b (marked by the solid triangles) as shown, or in the depressions of type c (marked by the open triangles).

second layer of closest-packed spheres can be added by placing spheres in alternate depressions in the first layer, as shown in Figure 20.12b.

Two choices exist for the placement of spheres in the third layer of a closest-packed structure. First, the spheres can be placed in the depressions which lie above the centers of the spheres in the first layer, i.e., above points $a$. The third layer then lies exactly over the first layer. Succeeding pairs of layers occupy alternate sites giving rise to an ABAB . . . sequence. This arrangement results in the **hexagonal closest-packed** (HCP) structure shown in Figure 20.13, a structure that has the geometrical properties of the hexagonal crystal system (Figure 20.8). The angles between the six atoms surrounding a given atom in layer A are 120°. Each atom in layer A is surrounded by six atoms in the same plane. In addition, three atoms in each of the adjacent layers also are equidistant. Consequently, the coordination number is 12. Some examples of metals with an HCP structure are beryllium, magnesium, and cobalt. Substances with the same crystal structure are said to be **isomorphous.**

The second choice for the placement of the spheres forming the third layer of a closest-packed structure is in the depressions in Figure 20.12a labelled $c$. This arrangement gives rise to an ABCABC . . . sequence of layers, as shown in Figure 20.14a, called **cubic closest packing** (CCP). In order to ascertain that the structure is indeed cubic it must be tilted at an appropriate angle, as shown. The CCP structure is identical to a face-centered cubic lattice (Figure 20.14c). As indicated in Table 20.2, the coordination number is therefore 12. This value is identical to that of an HCP structure, which is to be expected because both configurations involve closest packing. Among the metals crystallizing in the CCP structure are aluminum, calcium, and copper.

The spheres in a closest-packed structure occupy 74% of the available space. This result is readily derived from the parameters for the equivalent face-centered cubic lattice given in Table 20.2. We wish to compare $a^3$, the volume of a unit cell of side $a$, with that of the spheres it contains. Reference to Figure 20.14c shows that the three spheres lying along a face diagonal of a unit cell are tangent to each other and that only half the diameters of the corner spheres lie within the same cell. Therefore, the face diagonal of the unit cell is

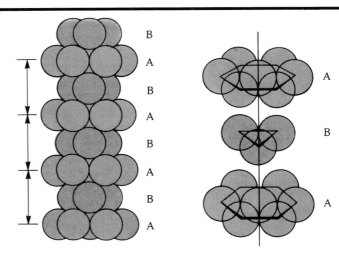

**FIGURE 20.13** Hexagonal closest packing of spheres. The arrows show the height of a unit cell. The letters A and B mark the two different types of layers. An exploded view is shown at the right.

Structure and Bonding in Solids   20.4   **751**

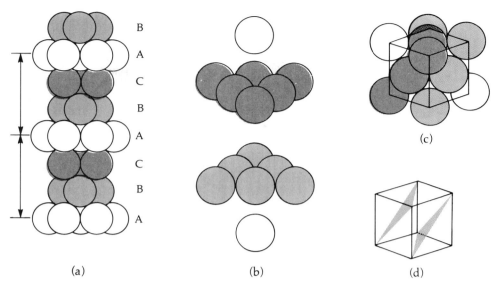

**FIGURE 20.14** Views of cubic closest packing of spheres. (a) Arrangement of layers showing the height of a unit cell (arrows). (b) Exploded view of four layers showing the 14 spheres constituting a face-centered cubic unit cell. (c) Reorientation of (b) showing the face-centered cube. (d) Orientation of the closest-packed layers relative to the axes of a cubic lattice.

equal to four times the radius of the sphere, $r$. The face diagonal is equal to $\sqrt{2}\,a$, and therefore $r$ and $a$ are related to each other by the expression

$$4r = \sqrt{2}\,a$$

Since a unit cell of a face-centered cubic lattice contains four spheres, the fraction of the space occupied by the spheres is given by the ratio

$$\frac{\text{Volume of 4 spheres}}{\text{Volume of cube}} = \frac{4 \times (4/3)\pi r^3}{a^3} = \frac{4 \times (4/3)\pi r^3}{(4r/\sqrt{2})^3} = \frac{16\pi}{3(4/\sqrt{2})^3} = 0.740$$

The same result applies to a hexagonal closest-packed structure.

Approximately two-thirds of all metals crystallize in either the HCP or the CCP structure. Practically all the remaining metals have a body-centered cubic (BCC) structure (Figure 20.10b). Note that since adjacent spheres lying in the same plane do not touch, the BCC arrangement is not closest-packed. Rather, contact between neighboring spheres is achieved along the body diagonal of the cube. The alkali metals are examples of metals with a BCC structure.

**Example 20.2**   What fraction of the available space is occupied by spherical atoms packed as closely as possible in a BCC lattice?

The three spheres lying along a body diagonal of a BCC unit cell are in contact (Figure 20.10b). The length of this diagonal is $4r$, where $r$ is the radius of a sphere, since the center atom contributes a length $2r$ (its diameter) and each end atom contributes a length $r$ (only half the diameter lies in the same unit cell). The length of the body diagonal is $\sqrt{3}\,a$, where $a$ is the side of a unit cell (Figure 20.11b), giving the relation between $r$ and $a$

$$4r = \sqrt{3}\,a$$

As a unit cell of a BCC lattice contains two atoms (Table 20.2), the fraction of the space occupied by the spheres is equal to the ratio

$$\frac{\text{Volume of 2 spheres}}{\text{Volume of cube}} = \frac{2 \times (4/3)\pi r^3}{a^3} = \frac{2 \times (4/3)\pi r^3}{(4r/\sqrt{3})^3}$$

$$= \frac{2 \times (4/3)\pi}{(4/\sqrt{3})^3} = 0.680$$

Thus, while the atoms in a closest-packed structure occupy 74% of the available space, those in a body-centered cubic structure occupy only 68% of the available space. ∎

Many elements can crystallize in more than one structure, a phenomenon known as **polymorphism**. The relative stability of the various possible structures usually depends on temperature and pressure. For example, iron forms a stable BCC structure called $\alpha$-iron at room temperature, a CCP structure called $\gamma$-iron between 906 and 1400°C, and another BCC structure called $\delta$-iron above 1400°C. The change from one structure to another occurs at a definite temperature and involves a phase transition that is similar, in principle, to the more common types of phase transitions examined in Chapter 12.

**Example 20.3** The atomic radius of iron is 1.24 Å. What is the density of (a) $\alpha$-iron and (b) $\gamma$-iron?

(a) The structure of $\alpha$-iron is BCC. Therefore, adjacent atoms are in contact along a body diagonal and the length of the diagonal, $\sqrt{3}\,a$, is equal to 4 atomic radii (see Example 20.2), i.e.,

$$\sqrt{3}\,a = 4r$$

Since there are 2 atoms per unit cell, the density of the iron in a unit cell is

$$\rho = \frac{m}{V} = \frac{m}{a^3} = \frac{m}{(4r/\sqrt{3})^3}$$

where $m$ is the mass of a unit cell, that is, the mass of 2 iron atoms. We obtain

$$\rho = \frac{(2)(55.85 \text{ g mol}^{-1})}{(6.022 \times 10^{23}\text{ mol}^{-1})[(4/\sqrt{3})(1.24\text{ Å})(10^{-8}\text{ cm Å}^{-1})]^3}$$

$$= 7.90 \text{ g cm}^{-3}$$

Since density is an intensive property (Section 1.4c), it is independent of the size of the sample. The result therefore applies to a macroscopic sample.

(b) The structure of $\gamma$-iron is CCP. The density can be obtained by a similar calculation as in (a), the only differences being that the length of a face diagonal, $\sqrt{2}\,a$, is equal to 4 atomic radii and there are 4 atoms per unit cell.

We obtain

$$\rho = \frac{m}{V} = \frac{m}{a^3} = \frac{m}{(4r/\sqrt{2})^3}$$

$$= \frac{(4)(55.85 \text{ g mol}^{-1})}{(6.022 \times 10^{23} \text{ mol}^{-1})[(4/\sqrt{2})(1.24 \text{ Å})(10^{-8} \text{ cm Å}^{-1})]^3}$$

$$= 8.60 \text{ g cm}^{-3}$$

The actual density of iron at room temperature is 7.87 g cm$^{-3}$. This value is in good agreement with the value calculated for $\alpha$-iron. ∎

## 20.5 The Electron-Sea Model of Metallic Bonding

The high melting points of most metals indicate that the interaction holding the atoms of a metal in a crystal—the metallic bond—must be strong. However, the metallic bond differs in important respects from covalent and ionic bonds, the other two strong interactions between atomic species. Recall from Section 6.5a that the metallic elements have low ionization energies and therefore display a tendency to lose electrons. Consequently, the atoms of a metal are unlikely to form electron pair bonds with each other. On the other hand, ionic bonding can be ruled out because there is no driving force for the formation of positive and negative ions from identical atoms.

What, then, is the nature of the metallic bond? A clue is provided by the fact that the number of valence shell electrons of metal atoms is significantly smaller than the number of valence shell orbitals. For example, the alkali elements have only a single valence shell electron but four valence shell orbitals, i.e., one s and three p orbitals. Metal atoms can therefore readily interact with additional electrons. However, this interaction does not take the form of electron sharing or transfer between a single pair of atoms. Rather, the interaction involves a number of atoms simultaneously. A given metal atom is surrounded by eight or twelve nearest neighbors, and it can share its valence electrons with them and, to a lesser extent, with more distant atoms as well. The shared electrons are delocalized over many atoms in a way that is reminiscent of delocalized bonding in conjugated hydrocarbons (Section 16.8b). Figure 20.15 depicts

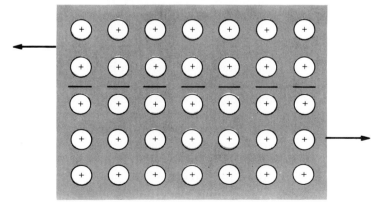

**FIGURE 20.15** Electron-sea model of metals. Positively charged ions occupy fixed lattice sites while the mobile sea of valence electrons is represented by the colored region. The dashed line shows how different planes of atoms can slide by each other, thereby making a metal malleable and ductile.

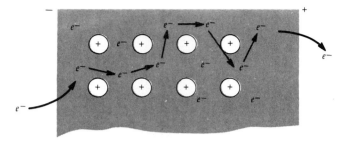

**FIGURE 20.16** Electrical conductivity according to the electron-sea model. As an electron enters one side of the metal, another electron leaves at the other side and moves towards the positive terminal. The vibration of the lattice ions interferes with electron motion and leads to a decrease in conductivity with increasing temperature.

a model of a metallic crystal based on these considerations, the **electron-sea model.** Positively charged ions are fixed in the lattice sites. The valence shell electrons are diffused throughout the entire structure and bind the ions to each other.

The electron-sea model of metals accounts for a number of properties. Normally, the valence electrons are free to move at random throughout the metallic crystal. However, when an electric field is imposed by the application of an electric potential difference, the electrons move in the direction of the positive terminal, and a current flows. Metals are therefore excellent conductors of electricity. Furthermore, the conductivity of metals decreases with increasing temperature. This observation is explained by the interference with the electron motion caused by the vibration of the positive ions in their lattice sites (Figure 20.16). Since the vibrational energy increases with temperature, the conductivity of metals should vary in the opposite way.

The electron-sea model also accounts for two other characteristic properties of metals, their malleability and ductility. In order for metals to be rolled into sheets (malleability) or drawn into wires (ductility) without fracturing, the atoms in one plane of the crystal must be readily displaceable with respect to those in another plane. The sea of electrons makes it possible for planes of positive ions to glide by each other without experiencing the repulsion which, as we shall see, prevents the same motion in ionic crystals.

## 20.6 The Band Theory of Solids

The most successful theory of bonding in metals and various other solids is the band theory, which can be viewed as an extension of molecular orbital theory to a number of atoms large enough to form a macroscopic crystal.

*a. The Molecular Orbital Approach to Metallic Bonding* The presence of delocalized electrons in a molecule has been treated successfully by the molecular orbital approach. This same approach may be extended to the $10^{20}$ or so atoms that make up a macroscopic crystal. However, it should come as no surprise that this enormous jump in the number of atoms in the system leads to some new features.

The changes that occur with the number of atoms can be seen by considering aggregates containing an increasing number of lithium atoms. We focus on the atomic orbitals of these atoms and on the molecular orbitals formed by their overlap (Figure 20.17). We begin by recalling (Section 16.6a) that the formation of $Li_2$ involves the overlap of the 2s atomic orbitals of two Li atoms to form a bonding $\sigma_{2s}$ orbital and an antibonding $\sigma_{2s}^*$ orbital, as shown in Figure 20.17a. These orbitals differ in energy, with the bonding orbital lying at a lower energy. Since each Li atom has one 2s electron, the $\sigma$ orbital is full and the $\sigma^*$ orbital is empty.

The addition of a third Li atom (Figure 20.17b) gives rise to a third molecular orbital. The three orbitals can be arranged in order of increasing energy and change in character from bonding to antibonding. The three electrons occupy the lowest two levels and are delocalized over the entire $Li_3$ molecule. As may be seen in the figure, the energy difference between the levels in $Li_3$ is smaller than that between the levels in $Li_2$. This result can be understood in terms of the particle-in-a-box model (Section 5.9). The valence electrons in $Li_3$ are free to roam over a greater distance than those in the smaller $Li_2$ molecule and the energy levels are therefore spaced more closely.

We are now ready to extrapolate to $Li_N$, a lithium crystal consisting of $N$ atoms ($N \approx 10^{18}$). The crystal has $N$ orbitals which are delocalized over the entire structure. The inverse relation between the energy spacing of the orbitals and their number leads to a distinct difference between the crystal and the molecules, which may be seen in Figure 20.17c: the $N$ orbitals lie so close in en-

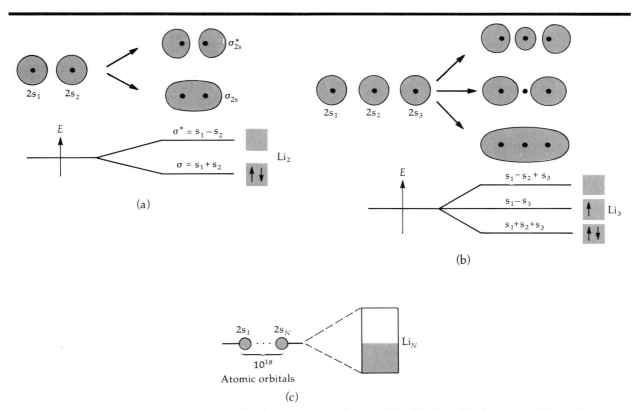

**FIGURE 20.17** The formation of a band of $N$ orbitals in $Li_N$ by the addition of successive Li atoms. (a) $Li_2$, (b) $Li_3$, and (c) $Li_N$ ($N \approx 10^{18}$).

ergy as to form an essentially continuous **energy band.** The orbitals making up this band change in character from strongly bonding to strongly antibonding. As in the case of the well separated molecular orbitals, each orbital can accommodate two spin-paired electrons. Consequently, the $N$ valence electrons of the $N$ Li atoms fill the lower half of the band while the upper half of the band remains empty, as shown.

We have focused so far on the band formed from the overlap of the 2s atomic orbitals. Since these are the valence orbitals of lithium, this band is called the **valence band.** A more complete analysis requires consideration of the bands formed from the overlap of the 1s and the 2p orbitals, as shown in Figure 20.18a. The band structure is determined by the widths of the bands and by their difference in energy. The width of a band depends on the strength of the interaction between the associated atomic orbitals in neighboring atoms. Thus, the band resulting from the overlap of the 1s orbitals is very narrow since these orbitals, being drawn in close to the atomic nuclei, are essentially nonbonding. However, the bands formed by the overlap of the 2s and the 2p orbitals are wide, since these are the strongly interacting valence orbitals. The energy gap between two bands depends on the difference in the energy of the corresponding atomic orbitals. The energy difference between the 2s and 2p orbitals is small. The corresponding energy difference between the resulting bands is also small and, since the bands are broad, they overlap in energy and there is no gap between them. On the other hand, the large difference in the

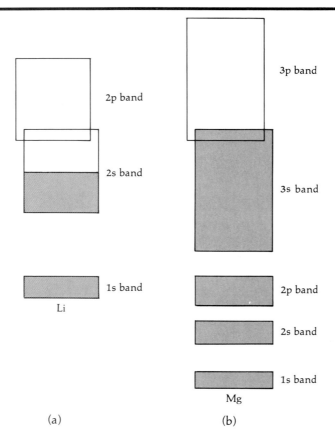

**FIGURE 20.18** Band structure of (a) Li metal and (b) Mg metal.

energy of the 1s and 2s orbitals, coupled with the narrow width of the band formed from the 1s orbitals, leads to a sizeable gap between the lowest two crystal bands.

The crystal band structure of lithium is typical of that of the alkali metals. Figure 20.18b shows the band structure of magnesium. In addition to several narrow bands formed from the filled 1s, 2s, and 2p atomic orbitals, there are two broad, overlapping bands formed from the 3s and 3p atomic orbitals. Since magnesium has two valence electrons, the $N$ molecular orbitals making up the band formed from the 3s atomic orbitals are completely filled. However, the overlap of the bands formed from the 3s and 3p orbitals results in a single partially filled band, as in the case of lithium.

The electrons in the valence band of a metal are free to move through the crystal. However, since there is no preferred direction of motion, there is no flow of current. An electric current in a metal consists of electrons moving in the *same* direction. A current will flow when an electric field is applied provided that the band has vacant orbitals to hold the slightly more energetic moving electrons. As shown in Figure 20.18, the valence band of a metal is only partially filled, so that there are vacant orbitals available. Metals differ in this respect from other solids, as we shall see.

***b. Metals, Insulators, and Semiconductors*** Solids can be divided into three categories on the basis of their ability to conduct electricity. Metals are good conductors, **insulators** are non-conductors, and **semiconductors** have intermediate properties. Figure 20.19 shows an idealized diagram of the band structure of these three types of solids. The key difference between metals and the other two types of solids is that metals have a partially filled valence band while insulators and semiconductors have a completely filled valence band. The next higher energy band of insulators and semiconductors, the **conduction band**, is

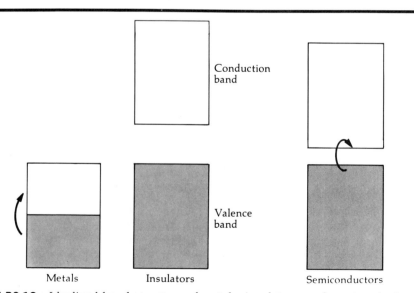

**FIGURE 20.19** Idealized band structure of metals, insulators, and semiconductors. The shaded band is filled with valence electrons and the open band is empty. Metals have a partially filled valence band.

empty and well separated in energy from the valence band.

The energy gap between the valence and conduction bands of an insulator, such as diamond or sulfur, is large, typically well in excess of 5 eV (>500 kJ mol$^{-1}$). At ordinary temperatures, the thermal energy of the valence electrons is much too small to bridge this gap. Consequently, an electric field cannot cause a current to flow, for there are no accessible empty levels.

Semiconductors, such as silicon or germanium, represent an intermediate case. As is true for insulators, semiconductors have a filled valence band and an empty conduction band at low temperatures. However, the gap between these bands is not very large. For example, the gap between the valence and conduction bands in germanium is only 0.60 eV (58 kJ mol$^{-1}$). This gap is sufficiently narrow that, at ordinary temperatures, some electrons have enough thermal energy to jump the gap. The conductivity of these solids is small compared to that of metals but *increases* with temperature as the number of electrons with enough thermal energy to be promoted to the conduction band increases. Recall that this dependence is just the opposite of that displayed by metals.

For each electron that is promoted to the conduction band of a semiconductor, a vacancy, or **hole,** is left in the valence band. The motion of these holes contributes to the conductivity of a semiconductor. As indicated in Figure 20.20, these holes behave like positive charges in the presence of an electric field: electrons hop in and out of holes as they move in one direction while the holes appear to move in the opposite direction. Although the only charged particles in motion are electrons, the electron–hole formalism represents a useful picture of the conduction process in semiconductors.

*c. Doped Semiconductors*  Semiconductors in which the holes are formed by the promotion of electrons of the semiconducting substance are known as **intrinsic** semiconductors. The conductivity of an intrinsic semiconductor depends on its purity. Reproducible results, which are critical to the operation of any device which uses these materials, can be obtained only for ultrapure substances. The presence of random impurities, containing atoms with more or less valence electrons than atoms of the semiconductor, can affect the electrical properties in unpredictable ways. However, if specific impurities are added to

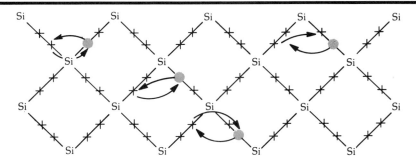

**FIGURE 20.20**  The motion of electrons (+) and holes (○) in a semiconductor such as silicon. Electrons move from filled sites to holes from left to right (black arrows); the holes appear to move from right to left (colored arrows).

semiconductors in controlled amounts, the electrical properties can be tailored to meet specific needs. This powerful technique is called **doping.**

Semiconductors can be doped by the addition of atoms containing either one more or one less valence electron than the host atoms, typically at a level of a few parts per million. This small amount of impurity has a remarkable effect on the conductivity of a semiconductor. For example, the intrinsic conductivity of silicon, the most widely used semiconductor, is some $10^{11}$ times smaller than that of copper at room temperature. The addition of a small amount of appropriate impurity atoms increases its conductivity by many orders of magnitude.

Semiconducting elements such as silicon or germanium crystallize in the diamond structure (see Figure 20.34). Each silicon atom has four valence electrons and forms a covalent bond with each of four neighboring silicon atoms (Figure 20.21a). Suppose we incorporate a small amount of phosphorus or arsenic in a silicon crystal. The atoms of these elements have five valence electrons, one more than silicon. Structural studies indicate that the phosphorus

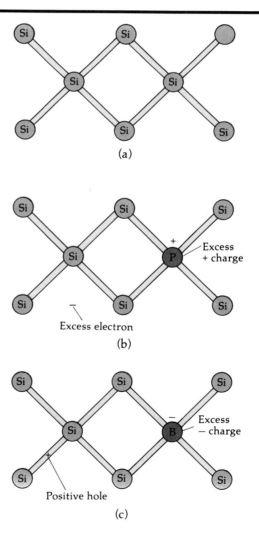

**FIGURE 20.21** The structure of intrinsic and doped semiconductors. (a) Pure silicon, an intrinsic semiconductor. (b) $n$-type silicon containing a P atom as an impurity. The bonded P atom has one extra valence electron which is available for conduction. (c) $p$-type silicon containing a B atom as an impurity. The B atom requires one extra electron to form four bonds. This electron comes from one of the Si atoms, leaving a positive hole in the lattice.

atom replaces a silicon atom in the lattice (Figure 20.21b). The phosphorus atom forms four bonds with the four surrounding silicon atoms. Since there are no additional nearest neighbors, the fifth valence electron of phosphorus cannot be incorporated into a bond and is relatively free to move through the crystal, as shown. Impurity atoms such as phosphorus, which are readily ionized in the host crystal, are called **donors.** The conductivity of a semiconductor doped with a donor impurity is associated with the motion of these donated electrons; the material is said to be an **n-type** semiconductor (n for negative charge).

The incorporation into the silicon lattice of atoms with three valence electrons, such as boron or gallium, is illustrated in Figure 20.21c. In order to form covalent bonds with each of its four nearest-neighbor silicon atoms, a boron atom must attract an electron from a nearby silicon atom. This leaves a positive hole in the lattice. Since the conductivity of the solid is determined by the presence of positive holes, the material is said to be a **p-type** (p for positive charge) semiconductor. Impurities such as boron, which accept electrons from the host material, are called **acceptors.**

The band structure of doped semiconductors is shown in Figure 20.22. In n-type silicon, the extra electrons from the donor impurity occupy energy levels which lie just below the conduction band (Figure 20.22a). At normal temperatures, the electrons in these levels have enough thermal energy to jump into the conduction band and an electric current can flow.

In p-type silicon (Figure 20.22b), the energy levels associated with the acceptor impurity lie just above the valence band. Electrons in the valence band have enough thermal energy to make a transition to the impurity levels. These transitions create holes, whose motion in the presence of an electric field constitutes a current. The development of the **integrated circuit,** a silicon chip con-

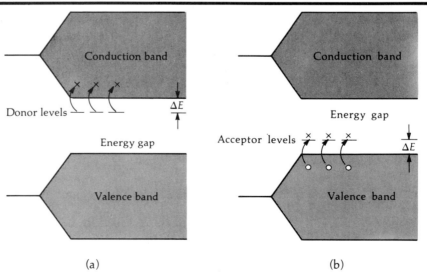

**FIGURE 20.22** The band structure of doped semiconductors. (a) An n-type semiconductor. $\Delta E$ is the very small energy difference between the donor impurity levels and the conduction band. (b) A p-type semiconductor. $\Delta E$ is the very small energy difference between the acceptor impurity levels and the valence band.

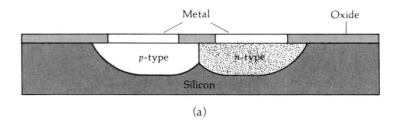

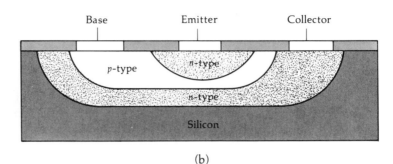

**FIGURE 20.23** Integrated circuit elements. (a) A diode, used to convert an AC voltage to a DC voltage. (b) An *n–p–n* transistor, used to amplify a current.

taining microscopic regions of *n*-type and *p*-type silicon arranged according to specific designs (Figure 20.23), is the basis of the modern electronics and computer industries.

# Ionic Crystals

## 20.7 Some Common Ionic Crystal Structures

Ionic compounds form a number of different crystal structures. The lattices are essentially those examined in Section 20.3 with the added complication that the crystal contains two different types of ions. Some crystals contain an equal number of cations and anions. These are the 1:1 and 2:2 crystals, such as NaCl or ZnS. Other crystals, such as $CaF_2$ or $Na_2O$, have twice as many anions as cations or vice versa.

Figure 20.24 shows some common ionic crystals lattices. The "cesium chloride" structure consists of two interpenetrating simple cubic lattices. The cesium (or chloride) ions occupy the eight corners of the cube and the other ion occupies the center of the cube. Each cation is surrounded by eight anions, and vice versa, so that the coordination number of both ions is 8. The unit cell contains one ion of each type. While Figure 20.24 represents the ions as well separated small spheres, in actuality the ions fill the available space in the manner illustrated in Figure 20.10, with oppositely charged ions generally being in contact.

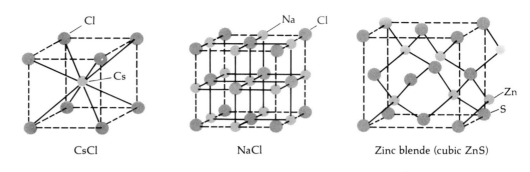

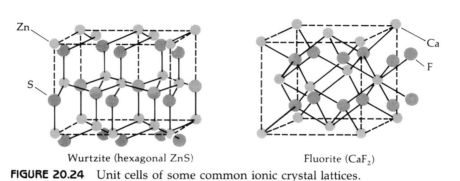

**FIGURE 20.24** Unit cells of some common ionic crystal lattices.

The "sodium chloride" structure consists of two interpenetrating face-centered cubic lattices. Figure 20.24 shows that the ions of one type occupy the corners and the centers of the faces of the cube while the other ions occupy similar sites displaced by half the side of the cube. The coordination number of both ions is six and the unit cell contains four ions of each type.

Zinc sulfide is polymorphic and forms two common types of crystal structures. The "zinc blende" structure has the anions arranged in a face-centered cubic lattice. The location of the cations is slightly more difficult to visualize: they are located near alternate corners of the cube, with an inward displacement in each direction of one-quarter of the side of the cube. Each cation is tetrahedrally coordinated as shown in the figure, and each anion also has a coordination number of 4. A face-centered cubic lattice has four lattice points per unit cell. Thus, the zinc blende structure has four anions per unit cell as well as four cations (the four shown in the figure).

The "wurtzite" structure is the second common form of zinc sulfide. It is similar to the zinc blende structure except that the anions, instead of being in a cubic closest-packed configuration (i.e., the FCC structure), form hexagonal closest-packed planes. The coordination number of each ion is 4.

The "fluorite" structure is a common crystal form for 1:2 salts, such as $CaF_2$, in which there are twice as many anions as cations. The fluorite structure closely resembles the zinc blende structure: the anions are arranged in a face-centered cubic lattice and the cations are displaced inward from each corner of the cube by one-quarter of a side of the cube in each direction. As in the case of zinc blende, the coordination number of each anion is 4. Since there are twice

**TABLE 20.3 Crystal Structures of Ionic Compounds**

| Structure | Examples |
|---|---|
| Cesium chloride | CsCl, CsBr, CsI, NH$_4$Cl, NH$_4$Br, TlCl, TlBr, TlI |
| Sodium chloride | Alkali halides (except the above), Group IIA oxides and sulfides (except Be), AgF, AgCl, AgBr, NH$_4$I, CdO |
| Zinc blende | ZnO, ZnS, CdS, MgS, AlP, GaP, InP, SiC, AgI |
| Wurtzite | BeO, BeS, ZnO, ZnS, CdS, AlN, GaN |
| Fluorite | CaF$_2$, SrF$_2$, BaF$_2$, CdF$_2$, HgF$_2$, PbF$_2$, SrCl$_2$, BaCl$_2$, ZrO$_2$, CeO$_2$, UO$_2$ |
| Antifluorite | Li$_2$O, Na$_2$O, K$_2$O, Li$_2$S, Na$_2$S, K$_2$S |

as many anions as cations, the cation coordination number must be 8. The reversal of the positions and numbers of cations and anions forms the "antifluorite" structure, which is adopted by compounds such as Na$_2$O, and other alkali oxides. Table 20.3 gives some examples of compounds that crystallize in each of these structures.

**Example 20.4** The length of a unit cell of a LiCl crystal is 5.14 Å. Assuming that the Li$^+$ ion is sufficiently small to just fit between touching nearest neighbor Cl$^-$ ions, calculate the ionic radii of Li$^+$ and Cl$^-$.

Table 20.3 indicates that LiCl crystallizes in the sodium chloride structure. The occurrence of anion–anion contact as well as anion–cation contact indicates that the distances between nearest neighbors can be related to the side of the unit cell, $a$, in the manner shown in diagram (a) below (see Table 20.2). We see that the sum of the radii of the adjacent two Cl$^-$ ions is $(\sqrt{2}/2)a$,

$$2r_{Cl^-} = (\sqrt{2}/2)a = (\sqrt{2}/2)(5.14 \text{ Å})$$

$$r_{Cl^-} = 1.82 \text{ Å}$$

Similarly, the sum of the radii of adjacent Li$^+$ and Cl$^-$ ions is $a/2$, giving

$$r_{Li^+} = (\tfrac{1}{2}a - r_{Cl^-}) = \frac{5.14 \text{ Å}}{2} - 1.82 \text{ Å} = 0.75 \text{ Å}$$

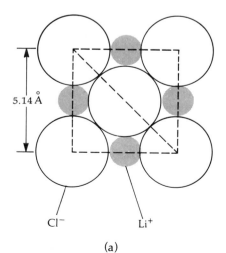

(a)

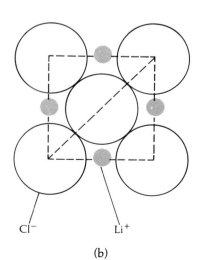

(b)

**TABLE 20.4  Radii of Some Simple Ions**

| Cation | Radius (Å) | Anion | Radius (Å) |
|---|---|---|---|
| Li$^+$ | 0.68 | N$^{3-}$ | 1.71 |
| Be$^{2+}$ | 0.30 | O$^{2-}$ | 1.45 |
| Na$^+$ | 0.98 | F$^-$ | 1.33 |
| Mg$^{2+}$ | 0.65 | S$^{2-}$ | 1.90 |
| Al$^{3+}$ | 0.45 | Cl$^-$ | 1.81 |
| K$^+$ | 1.33 | Se$^{2-}$ | 2.02 |
| Ca$^{2+}$ | 0.94 | Br$^-$ | 1.96 |
| Ga$^{3+}$ | 0.62 | I$^-$ | 2.19 |
| Rb$^+$ | 1.48 | | |
| Sr$^{2+}$ | 1.10 | | |
| In$^{3+}$ | 0.81 | | |
| Cs$^+$ | 1.67 | | |
| Ba$^{2+}$ | 1.29 | | |

The radius of Li$^+$ obtained from an analysis of the crystal structure of a number of lithium salts is only 0.68 Å. This value indicates that Li$^+$ is so small that it is not in contact with adjacent Cl$^-$. Therefore diagram (b) is a more correct representation than (a). ∎

The ionic radii examined in Section 6.8c generally are obtained from X-ray diffraction data in the manner illustrated in Example 20.4. It must be emphasized that this procedure is only approximately valid. The chief problem is that X-ray diffraction yields a value of the internuclear distance between two ions, which must then be apportioned between them to yield the ionic radii. Recall that, owing to the polarizability of ions, ionic bonds are slightly covalent (Section 16.3). This means that two touching ions, instead of having sharply defined surfaces, actually have somewhat overlapping electron clouds. In other words, it is impossible to tell with a high degree of accuracy where the cation ends and the anion begins. The results obtained for many salts have been used to determine average ionic radii; some typical values are listed in Table 20.4.

## 20.8 The Packing of Ions in Crystals

*a. Interstitial Sites* The great stability of ionic crystals is due to the electrostatic Coulomb interaction between the oppositely charged ions. A state of minimum potential energy can be achieved by surrounding each cation with as many anions as possible and vice versa, while maintaining the stoichiometric ratio. This tendency towards closest packing is affected by differences in the ionic radii of anion and cation. Since the radii of anions tend to be larger than those of cations, it is reasonable to let the anions define the lattice and view the cations as occupying the holes in the resulting structure. Different structures can then be regarded as resulting from differences in the relative sizes of the constituent ions.

The structure of ionic crystals can be understood by consideration of the spatial geometry of the holes, or **interstitial sites,** in a closest-packed lattice. There are two types of holes in such a structure, tetrahedral and octahedral. We have encountered the former in our analysis of the closest packing of

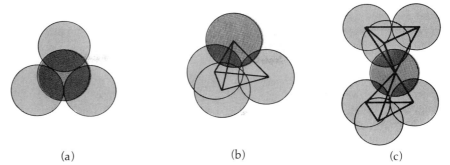

**FIGURE 20.25** Tetrahedral interstitial sites in a closest-packed lattice. (a) Top view. (b) Side view showing the tetrahedral configuration. (c) Two tetrahedrons with a common sphere. Both tetrahedral sites can be associated with this sphere.

spheres. Figure 20.12a showed a single layer of closest-packed spheres with triangle-like depressions between any three mutually touching spheres. A second layer of spheres was built up by placing spheres in alternate depressions in the first layer. Figure 20.25 focuses on three mutually touching spheres defining a depression in which a fourth sphere is placed. The centers of these four identical spheres lie at the corners of a regular tetrahedron. Although the spheres are mutually tangent to each other, a small empty space remains in the middle of the tetrahedron. This space constitutes a tetrahedral site. Two such sites can be associated with each sphere in a closest-packed structure. This is most readily seen by focusing on the single sphere in the second layer (Figure 20.25c).

Compounds crystallizing in the zinc blende, wurtzite, or antifluorite structures, in which at least one of the ions has a coordination number of 4, can be described as closest-packed structures of anions in which the smaller cations occupy tetrahedral sites. Half the tetrahedral sites are filled in compounds with a 1:1 stoichiometry, such as AgI, whereas all the tetrahedral sites are filled in compounds with a 2:1 stoichiometry, such as $Li_2S$.

The octahedral sites in a closest-packed lattice are illustrated in Figure 20.26. Figure 20.26a shows two adjacent layers of spheres arranged so that the

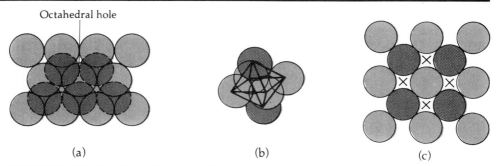

**FIGURE 20.26** Octahedral interstitial sites in a closest–packed lattice. (a) Top view of two layers with an octahedral hole; (b) rotation of the structure showing the octahedral configuration; (c) side view showing the alternation of spheres and octahedral sites.

depression formed by the junction of the three spheres in the first layer is not occupied by any of the spheres in the second layer. The resulting hole is an octahedral site. The hole is equidistantly surrounded by six spheres lying at the corners of an octahedron, as shown in Figure 20.26b. A side view of a face-centered cubic lattice, displayed in Figure 20.26c, shows a regular alternation between spheres and octahedral sites in the lattice, indicating that there is one octahedral site for each sphere.

A comparison of Figures 20.25 and 20.26 shows that an octahedral hole must be larger than a tetrahedral hole, provided, of course, that the spheres constituting the closest-packed structure are of the same size in both cases. The different sizes of the two interstitial sites are closely related to their coordination numbers. The larger the hole, the more spheres can be packed around it. The coordination number of an octahedral site is six, compared to four for a tetrahedral site. Compounds crystallizing in the sodium chloride structure can be described as having the anions in a face-centered cubic lattice in which the cations fill the octahedral holes.

The tetrahedral and octahedral interstitial sites are the only ones present in a closest-packed lattice. However, when the cations and anions are of comparable size, the crystal structure is no longer closest packed. The ions of one type frequently occupy a simple cubic lattice while the ions of the other type occupy cubic interstitial sites in the middle of each cube (Figure 20.27). This structure corresponds to the cesium chloride structure and each ion has a coordination number of 8.

**b. The Radius-Ratio Rule** The crystal structure of an ionic compound depends to a great extent on the relative sizes of the two types of ions present. In general, a cation will seek to maintain contact with as many anions as possible since this arrangement maximizes the electrostatic interaction between oppositely charged ions and leads to a state of minimum potential energy. When a cation is much smaller than an anion, it can fit into a tetrahedral site in a closest–packed anion structure. A somewhat larger cation can also fit into a tetrahedral site but only if the structure expands somewhat, so that the anions are no longer in contact with each other. An even larger cation can simultaneously touch six anions and, since this arrangement has a lower potential energy, the

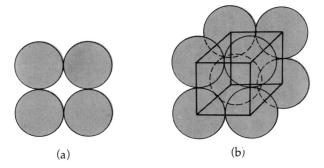

**FIGURE 20.27** Cubic interstitial site. (a) Bottom layer of cube. (b) Two layers define a cube and the interstitial site in its center.

tetrahedral coordination can give way to octahedral coordination. An increase in the size of the cation relative to that of the anion leads to an expansion of the structure until a point is reached when the cation is sufficiently large to fit into a cubic interstitial site, thereby simultaneously contacting eight anions.

The size of a particular type of hole in a closest-packed structure depends on the ratio of the cation and anion radii, $r_+/r_-$. As an example, let us evaluate the range of $r_+/r_-$ values for which an octahedral configuration is expected. The minimum value is that for which a sphere representing the cation can be simultaneously in contact with six closest-packed spheres located at the corners of an octahedron. This value is obtained most readily by reducing the problem to two dimensions. Figure 20.28a shows the equatorial plane of an octahedron. Observe that closest contact is maintained since the cation touches all four anions lying in the plane which, in turn, touch each other. Applying the Pythagorean theorem to the right triangle defined by the centers of the cation and two adjacent anions, we obtain

$$2(r_+ + r_-)^2 = (2r_-)^2$$

which yields

$$r_+/r_- = \sqrt{2} - 1 = 0.414$$

Octahedral coordination gives way to cubic coordination when a cation is large enough to contact eight adjacent anions in a cubic site. The geometry of the cubic arrangement is depicted in Figure 20.28b. The anions lie at the corners of a cube. Since they are in contact, the side of the cube is $2r_-$. The cation lies in the center of the cube and contact with anions occurs along the body diagonal, which is $\sqrt{3}$ times longer than its side. We therefore have

$$2r_- + 2r_+ = \sqrt{3}(2r_-)$$

giving

$$r_+/r_- = \sqrt{3} - 1 = 0.732$$

These results indicate that ionic compounds should display octahedral coordination when $r_+/r_-$ lies between 0.414 and 0.732. Table 20.5 lists similar results for the other structures of interest.

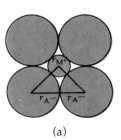 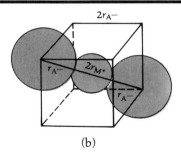

(a)        (b)

**FIGURE 20.28** The radius–ratio rule applied to the stability of the octahedral site structure. (a) Geometry of an octahedral site; (b) geometry of a cubic site.

768   Structure and Bonding in Solids   20.9

TABLE 20.5   Lattice Geometry and the Radius–Ratio Rule

| Coordination number | Geometry | Limiting radius ratio, $r_+/r_-$ | Possible crystal structure |
|---|---|---|---|
| 4 | Tetrahedral | 0.225–0.414 | Zinc blende, wurtzite |
| 6 | Octahedral | 0.414–0.732 | NaCl |
| 8 | Cubic | 0.732–1.0 | CsCl |

While the radius-ratio rule quite often predicts the correct coordination numbers of ions in crystals, it is not a completely reliable guide. Since it assumes that ions can be represented by spheres, it should be used with caution when polarization effects become important. Furthermore, since there is some ambiguity in the values of ionic radii, their ratios cannot be determined very accurately. Thus, the structures of ionic crystals with ionic radius ratios near the limiting values are difficult to predict. Finally, the electrostatic energies of different crystal structures differ by relatively little in certain cases (Section 20.10b) so that small contributions from other sources can tip the balance away from the expected structure.

**Example 20.5**   Use the radius–ratio rule to predict coordination numbers for (a) CsBr and (b) KF. Use the structures in Table 20.3 to determine whether the rule works or not.

(a)   Using the ionic radii in Table 20.4, we obtain

$r_+/r_- = 1.67 \text{ Å}/1.96 \text{ Å} = 0.852$

The $Cs^+$ ions should occupy cubic sites, for which the coordination number is 8. Table 20.3 indicates that CsBr crystallizes in the cesium chloride structure which has a coordination number of 8. In this case, the radius–ratio rule works.

(b)   The ratio of the ionic radii is

$r_+/r_- = 1.33 \text{ Å}/1.33 \text{ Å} = 1.00$

The radius–ratio rule predicts that KF should have the same structure and coordination number as CsBr. However, according to Table 20.3, KF crystallizes in the NaCl structure and so has a coordination number of 6, indicating that the rule fails in this case.   ∎

## 20.9   Defects in Crystals

We have so far emphasized the complete regularity of crystal structures. It would indeed be astonishing if such regularity could be maintained over a volume encompassing the huge number of atoms or ions in a typical crystal. The fact is that real crystals actually contain many defects. These imperfections play an important role in determining the mechanical, optical, and electrical properties of solids. The study of crystal defects is therefore of great importance to such fields as solid state chemistry and physics, and metallurgy.

The most common types of imperfections involve isolated lattice sites and are known as **point defects.** The different kinds of point defects are illustrated in Figure 20.29 as they occur in an ionic crystal. A **Schottky defect** consists of a

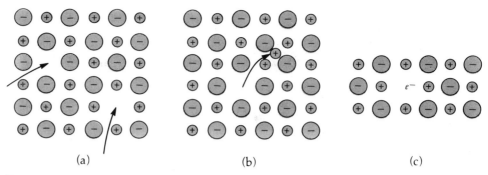

FIGURE 20.29 Point defects in ionic crystals. (a) Schottky defect, (b) Frenkel defect, and (c) F-center.

vacant lattice site. Such vacancies are common in all types of crystals and usually involve the transfer of ions or atoms from the interior to the surface of the crystal. In a **Frenkel defect** an ion is displaced to an interstitial position. The displaced ion is usually a cation, since it tends to be smaller than an anion. Although imperfect crystals have higher internal energy than comparable perfect crystals, their stability relative to the perfectly ordered array is enhanced by the entropy factor. The ions in the imperfect crystal are more randomly arranged and this structure consequently has the greater entropy.

An electron trapped at a vacant site forms an imperfection called an **F-center**. Such imperfections can be formed in a number of ways. Consider, for example, the exposure of an alkali halide crystal to ultraviolet radiation. The exposure causes the loss of free halogen, e.g.,

$$\text{NaCl(s)} \xrightarrow{h\nu} \text{NaCl}_{(1-x)}(s) + xe^- + \tfrac{1}{2}x\ \text{Cl}_2(g)$$

and the electrons are trapped at vacant sites. If the UV radiation is sufficiently intense, the loss of chloride ions can approach the 1% level. The physical appearance of the crystal undergoes some significant changes under these conditions: normally white crystals acquire distinctive colors (the designation F is derived from *Farbe*, German for color). Crystals containing F-centers have a non-stoichiometric composition. If the deviation from perfect stoichiometry is sufficiently large, the electrical and optical properties of the crystal are usually affected.

The sulfides and oxides of the transition metals are examples of non-stoichiometric compounds in which the range of composition is much greater than that displayed by the irradiated alkali halides. Titanium oxide, for instance, can range in composition from $TiO_{0.69}$ to $TiO_{1.32}$. The number of vacancies in the lattice is sufficiently large that the density of this material is substantially lower than expected on the basis of the internuclear spacing in the crystal. The motion of ions from one vacancy to another provides a mechanism by which the conductivity of ionic crystals can be increased.

In addition to point defects, crystals also can have linear, planar, or volume defects. Figure 20.30 shows an example of an edge dislocation, in which an extra plane of atoms is present in part of the crystal. These types of imperfections distort the crystal and thereby reduce its tensile strength. Particularly in the case of metals, such defects can have a severely detrimental effect on the structural properties of the material. The field of metallurgy includes the devel-

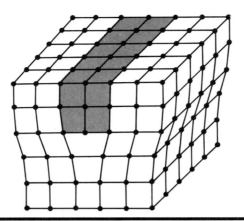

**FIGURE 20.30** Edge dislocation in a crystal. The presence of an extra plane in part of the crystal is shown in color.

opment of methods to strengthen metals and reduce the effects of structural imperfections.

## 20.10 The Stability of Ionic Crystals

*a. Some Properties of Ionic Crystals* Many of the properties of ionic crystals can be related to their crystal structures and to the forces which determine these structures. Ions in crystals generally have high coordination numbers (Section 20.7). The attractive electrostatic interaction between unlike ions is thereby maximized while the repulsive interaction between like ions is minimized. Ionic crystals are consequently very stable and have high melting points.

The melting points of ionic compounds reflect the factors that determine the magnitude of the Coulomb energy, namely, the ionic charges and ionic radii. The effect of ionic charge is seen most readily in a comparison of melting points of isoelectronic compounds. Consider the following pairs of isoelectronic compounds and their melting points:

NaF, $T_{mp}$ = 993°C and MgO, $T_{mp}$ = 2852°C

KF, $T_{mp}$ = 858°C and CaO, $T_{mp}$ = 2614°C

While all four crystals melt at high temperatures, the alkaline earth oxides have much higher melting points because the product of their ionic charges is four times larger than that of the alkali halides.

The dependence of melting point on the sum of the ionic radii is displayed in Figure 20.31 for a number of alkali halides, all of which crystallize in the sodium chloride structure. The inverse relationship expected from the Coulomb energy is clearly in evidence.

Ionic crystals tend to be quite hard and, as in the case of the melting points, the hardness tends to vary with the factors determining the Coulomb energy. In contrast to the metals, ionic crystals are neither malleable nor ductile but, rather, are readily fractured when subjected to mechanical stress. The difference in bonding in metallic and ionic crystals also accounts for the fact that, in contrast to metals, ionic crystals generally are poor conductors. The valence electrons are firmly attached to the anions and are not readily set in motion. The ions themselves are bound tightly in the lattice and, except in certain de-

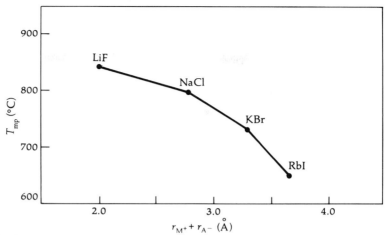

**FIGURE 20.31** Dependence of melting points of alkali halides crystallizing in the sodium chloride structure on the sum of the ionic radii.

fect crystals, are not free to move through the lattice. However, when the ionic lattice is broken down, either by melting or dissolution in water, the resulting melt or solution is a good conductor. The current is carried by the ions as they migrate towards the electrodes, rather than by electrons as in metals.

***b. The Crystal Lattice Energy*** The stability of an ionic crystal can be expressed quantitatively in terms of the **crystal lattice energy** $U_0$, defined as the energy released in the formation of one mole of a crystal from the gaseous ions. The lattice energy of NaCl, for instance, is given by the enthalpy change for the reaction

$$\text{Na}^+(g) + \text{Cl}^-(g) \rightarrow \text{NaCl}(s)$$

Since the energy of an ionic crystal is always lower than that of the separated gas phase ions, $U_0$ is always negative. [In analogy to the dissociation energy of a diatomic molecule (Section 17.1), $U_0$ can also be defined as $\Delta H$ for the reverse reaction—the formation of separate gaseous ions from the crystal. If this convention is used, $U_0$ must be positive.]

Although the lattice energy is difficult to measure directly, it can be calculated readily. The calculation is similar to that described in Section 18.1 for the energy released in the formation of a gaseous ion pair from the separated gaseous ions. Recall that this was just the Coulomb energy between a pair of ions separated by a distance $d_0$, corrected for the repulsion between closely approaching ions (Equation 18.8),

$$U_0 = -\frac{Z_+ Z_-}{\alpha d_0}\left(1 - \frac{1}{n}\right)$$

where $Z_+$ and $Z_-$ are the absolute values of the charges of the cation and anion, respectively, $d_0$ is the equilibrium distance between them, $\alpha$ is a proportionality constant, and $n$ is the exponent in the variation of the repulsive energy with distance.

To apply Equation 18.8 to a pair of ions in a crystal, it must be modified to take into account that each ion interacts with *all* the other ions in the crystal rather than with just a single unlike ion. In order to evaluate this interaction we have to focus on a specific crystal structure. Let us, for example, examine the sodium chloride structure. Figure 20.32 shows the interaction of one particular Na$^+$ ion with some of its neighbors. The ion in question is surrounded by six nearest-neighbor Cl$^-$ ions located at an equilibrium distance $d_0$. The potential energy resulting from this interaction is

$$U_0 = -\frac{6e^2}{\alpha d_0}\left(1 - \frac{1}{n}\right)$$

where $e$ is the electronic charge. The next-nearest neighbors are twelve Na$^+$ ions at a distance $\sqrt{2}\,d_0$ and the potential energy is

$$U_0 = \frac{12e^2}{\alpha\sqrt{2}\,d_0}\left(1 - \frac{1}{n}\right)$$

Since the interaction is between ions of like charge, the force is repulsive and the potential has a positive sign. Figure 20.32 shows that, in order of increasing distance, a given Na$^+$ is next surrounded by eight Cl$^-$ at a distance $\sqrt{3}\,d_0$, 6 Na$^+$ at a distance $2d_0$, 24 Cl$^-$ at $\sqrt{5}\,d_0$, 24 Na$^+$ at $\sqrt{6}\,d_0$, and so on. The potential energy of the Na$^+$ ion resulting from the interaction with all the other ions in the lattice is obtained as the sum of the individual terms:

$$U_0 = -\frac{e^2}{\alpha d_0}\left(1 - \frac{1}{n}\right)\left(6 - \frac{12}{\sqrt{2}} + \frac{8}{\sqrt{3}} - \frac{6}{2} + \frac{24}{\sqrt{5}} - \frac{24}{\sqrt{6}} + \cdots\right) \quad (20.4)$$

The numerical series in Equation 20.4 consists of alternating positive and negative terms and so does not change very drastically as an increasingly larger number of interactions is considered. It is found, in fact, that when a sufficiently large number of terms is included, the sum of the series converges to the value $A = 1.748$, a quantity known as the **Madelung constant** for the NaCl crystal structure. It is customary to report the crystal lattice energy per mole of ion pairs. The result is

$$U_0 = -\frac{N_A A Z_+ Z_-}{\alpha d_0}\left(1 - \frac{1}{n}\right) \quad (20.5)$$

The crystal lattice energy differs from the potential energy of gas-phase ion pairs by the value of the Madelung constant. Thus, the energy holding Na$^+$

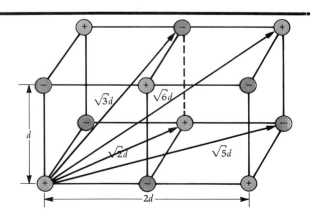

**FIGURE 20.32** Distances between an Na$^+$ ion (at lower left corner) and its neighbors in an NaCl lattice. The figure shows half of a unit cell.

**TABLE 20.6 Parameters of the Crystal Lattice Energy**

| Madelung constant | | Repulsion exponent | |
|---|---|---|---|
| Structure | A | Electronic configuration of ion[a] | n |
| Cesium chloride | 1.763 | [Ne] | 7 |
| Sodium chloride | 1.748 | [Ar], [Ar]3d$^{10}$ | 9 |
| Zinc blende | 1.638 | [Kr], [Kr]4d$^{10}$ | 10 |
| Wurtzite | 1.641 | [Xe], [Xe]4f$^{14}$5d$^{10}$ | 12 |
| Fluorite | 2.519 | | |

[a]If the electronic configuration of the cation and anion differ, an average value of $n$ is used.

and Cl$^-$ ions together in a crystal is 1.75 times greater than that holding the same number of ion pairs together in the gas phase.

The crystal lattice energy depends on the value of the Madelung constant and on that of the repulsion energy exponent $n$. Since the Madelung constant is determined by the number of neighboring ions and the distance between them, it depends on the type of lattice in which a particular salt crystallizes. Table 20.6 lists the values of this constant for a number of the important ionic crystal structures. So long as the stoichiometry of the salts involves an equal number of cations and anions, the Madelung constant is relatively insensitive to the details of the crystal structure. The exponent $n$ ranges between 7 and 12 depending on the electronic configuration of the ions, as indicated in Table 20.6. Since repulsion between ions contributes the term $(1 - 1/n)$ to the crystal lattice energy, the repulsive interaction reduces the magnitude of $U_0$ by approximately 10%.

### Example 20.6

Calculate the crystal lattice energy of sodium chloride.

We use Equation 20.5,

$$U_0 = -\frac{N_A A Z_+ Z_-}{\alpha d_0}\left(1 - \frac{1}{n}\right)$$

The values of the various parameters are

$A = 1.748$ (Table 20.6)

$d_0 = 2.79$ Å (the sum of the ionic radii of Na$^+$ and Cl$^-$, Table 20.4)

$n = 8$ [average of values listed for [Ne] (Na$^+$ configuration) and [Ar] (Cl$^-$ configuration)]

$Z_+ Z_- = e^2$

Inserting these values along with those of the constants $N_A$ and $\alpha$ in Equation 20.5, we obtain

$$U_0 = -\frac{(6.022 \times 10^{23} \text{ mol}^{-1})(1.748)(1.602 \times 10^{-19} \text{ C})^2[1 - (1/8)]}{(1.113 \times 10^{-10} \text{ C}^2 \text{ N}^{-1} \text{ m}^{-2})(1/\text{N m/J})(10^3 \text{ J/kJ})(2.79 \times 10^{-10} \text{ m})}$$

$$= -761 \text{ kJ mol}^{-1} \blacksquare$$

**c. The Born–Haber Cycle** Although the crystal lattice energy is not readily amenable to direct experimental measurement, it can be determined indirectly by examination of an appropriate cyclic process in which the formation of the crystal from the separated gaseous ions is one of the steps. This cycle is similar to the various types of cycles examined in Chapters 18 and 19, but has the added distinction of applying to the first process examined in this manner. It is known by the name of the scientists who performed this analysis as the **Born–Haber cycle.** Figure 20.33 shows the various steps in this cycle for NaCl.

The elements in their standard states are the starting point of the cycle. Two paths are given for the conversion of the elements to crystalline sodium chloride. The first is the direct reaction:

$$Na(s) + \tfrac{1}{2}Cl_2(g) \rightarrow NaCl(s) \qquad \Delta H_f^\circ = -411.2 \text{ kJ mol}^{-1}$$

The energy released in this process can be identified with the standard enthalpy of formation of sodium chloride. The second path is indirect, involving first, the formation of the gaseous atoms, next, that of the gaseous ions, and finally, that of the crystal. Most of these steps have already been considered in Section 18.1. Let us briefly recapitulate what we have learned about the energies involved in these steps.

The energy required to convert a mole of solid sodium to an atomic gas is the standard molar enthalpy of sublimation, $\Delta H_{sub}^\circ$. Similarly, the energy required to convert one-half mole of $Cl_2$ to gaseous atoms is one-half the molar dissociation energy of chlorine, $\tfrac{1}{2}D$. The energies involved in the formation of the gaseous ions from the gaseous atoms are the ionization energy, $I_1$, for the cation, and the negative of the electron affinity, $-\varepsilon$, for the anion. These quantities are tabulated in Table 6.4. The energy released in the final step is the desired crystal lattice energy, $U_0$. Referring to Figure 20.33, we see that, according to Hess's law, these energies obey the relation

$$U_0(NaCl) = \Delta H_f^\circ(NaCl) - \Delta H_{sub}^\circ(Na) - \tfrac{1}{2}D(Cl_2) - I_1(Na) + \varepsilon(Cl) \qquad (20.6)$$

Substituting the appropriate numerical values we obtain

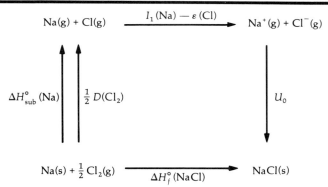

**FIGURE 20.33** The Born–Haber cycle for NaCl. The various quantities used in calculations involving this cycle are $\Delta H_{sub}^\circ(Na)$, enthalpy of sublimation of Na; $D(Cl_2)$, dissociation energy of $Cl_2$; $I_1(Na)$, first ionization energy of sodium; $\varepsilon(Cl)$, electron affinity of chlorine; $U_0$, crystal lattice energy of NaCl; and $\Delta H_f^\circ$, standard enthalpy of formation of NaCl(s).

$$U_0(\text{NaCl}) = (-411.2 - 107.3 - 121.7 - 495.6 + 348.8) \text{ kJ mol}^{-1}$$
$$= -787.0 \text{ kJ mol}^{-1}$$

The value calculated in the preceding section, $-761$ kJ mol$^{-1}$, is in reasonably good agreement with this result. The agreement is not closer because several factors that give rise to small energies have not been included in the calculation of the crystal lattice energy. These include the London dispersion forces between the ions and the zero-point energy of the crystal.

Equation 20.6 may be rearranged in order to permit an examination of the various factors that contribute to the standard enthalpy of formation of a crystalline solid. For a salt of type MX, such as an alkali halide, we obtain

$$\Delta H_f^\circ(\text{MX}) = \Delta H_{\text{sub}}^\circ(\text{M}) + \tfrac{1}{2}D(X_2) + I_1(\text{M}) - \varepsilon(X) + U_0(\text{MX}) \quad (20.7)$$

The sublimation enthalpy of the alkali metal and the dissociation energy of the halogen are always positive, as is the difference between the ionization energy of the alkali metal and the electron affinity of the halogen. Thus, it is only the overwhelmingly strong mutual attraction between the ions and the consequent highly negative crystal lattice energy that makes ionic crystals stable with respect to dissociation into the elements in their standard states.

---

**Example 20.7** The $O^{2-}$ ion is unstable in the gas phase, and the electron affinity of $O^-$ can therefore not be measured directly. Estimate the value of this quantity by application of the Born–Haber cycle to the formation of MgO. Use the data in Tables 6.4, 17.1, 20.4, and Appendix 4. [The sublimation enthalpy of magnesium is equal to $\Delta H_f^\circ$ of Mg(g)].

Let us first sketch the Born–Haber cycle for the formation of MgO from the elements in their standard states:

$$\text{Mg}(g) + \text{O}(g) \xrightarrow{I_1(\text{Mg}) + I_2(\text{Mg}) - \varepsilon(\text{O}) - \varepsilon(\text{O}^-)} \text{Mg}^{2+}(g) + \text{O}^{2-}(g)$$

$$\Delta H_{\text{sub}}^\circ(\text{Mg}) \uparrow \qquad \uparrow \tfrac{1}{2}D(\text{O}_2) \qquad \qquad U_0 \downarrow$$

$$\text{Mg}(s) + \tfrac{1}{2}\text{O}_2(g) \xrightarrow{\Delta H_f^\circ(\text{MgO})} \text{MgO}(s)$$

Note that the only difference between this diagram and the one in Figure 20.33 is the formation of divalent ions and the consequent use of two ionization energies and electron affinities.

All the listed quantities excepting $\varepsilon(\text{O}^-)$ are either tabulated or can be evaluated. We can therefore solve for the electron affinity of $O^-$ by noting that the energy difference between initial and final states is independent of path:

$$\varepsilon(\text{O}^-) = \Delta H_{\text{sub}}^\circ(\text{Mg}) + \tfrac{1}{2}D(\text{O}_2) + I_1(\text{Mg}) + I_2(\text{Mg}) - \varepsilon(\text{O}) + U_0(\text{MgO}) - \Delta H_f^\circ(\text{MgO})$$

The crystal lattice energy of MgO can be evaluated by means of Equation 20.5. We learn from Table 20.3 that MgO crystallizes in the NaCl structure so that the Madelung constant appropriate to this structure must be used. The value of the exponent $n$ is 7 since both $Mg^{2+}$ and $O^{2-}$ have the [Ne] electronic configuration. We obtain

$$U_0 = -\frac{N_A A Z_+ Z_-}{\alpha d_0}\left(1 - \frac{1}{n}\right)$$

$$= -\frac{(6.022 \times 10^{23} \text{ mol}^{-1})(1.748)(2 \times 1.602 \times 10^{-19} \text{ C})^2[1 - (1/7)]}{(1.113 \times 10^{-10} \text{ C}^2 \text{ N}^{-1} \text{ m}^{-2})(1 \text{ N m/J})(10^3 \text{ J kJ}^{-1})(2.10 \times 10^{-10} \text{ m})}$$

$$= -3963 \text{ kJ mol}^{-1}$$

where the value of $d_0$ (2.10 Å) is obtained as the sum of the ionic radii in Table 20.4. All the other quantities needed for the evaluation of $\varepsilon(O^-)$ are tabulated or given and we obtain

$$\varepsilon(O^-) = (147.7 + \tfrac{1}{2}(498.3) + 737.7 + 1451 - 141.1 - 3963 + 601.7) \text{ kJ mol}^{-1}$$

$$= -917 \text{ kJ mol}^{-1}$$

The large negative electron affinity of $O^-$ indicates that the loss of an electron by $O^{2-}$ is highly exothermic. ∎

The crystal lattice energy of MgO, $-3963$ kJ mol$^{-1}$, is much larger in magnitude than that of NaCl, $-761$ kJ mol$^{-1}$. The difference primarily reflects the higher ionic charge of the divalent ions. In ionic solids, the ions have a tendency to acquire the largest possible positive or negative charge because large charges maximize the crystal lattice energy. This tendency is opposed by the increasingly large energy required to form highly charged cations and anions. Thus, the energy required to remove two electrons from a magnesium atom and add them to an oxygen atom, $2.97 \times 10^3$ kJ mol$^{-1}$, is much larger than that required to remove one electron from a sodium atom and add it to a chlorine atom, 147 kJ mol$^{-1}$. In their ionic compounds, the representative elements strike a well-defined optimum balance between these opposing tendencies: they form ions with a noble gas electronic configuration. For example, in their ionic compounds sodium is present as $Na^+$, magnesium as $Mg^{2+}$, and aluminum as $Al^{3+}$. The formation of even more highly charged cations from these elements cannot occur owing to the sharp increase in ionization energies which occurs when a closed shell is broken (Section 6.5b).

# Giant Molecules

## 20.11 Network Covalent Solids

Among the representative elements, atoms of metals are characterized by a smaller number of valence electrons than of valence orbitals as well as by high coordination numbers in the solid. These properties are correlated with the formation of closest-packed structures in which valence electrons are shared among all the atoms in a crystal. The most non-metallic elements are characterized by the opposite properties: their atoms have more valence electrons than valence orbitals, and their coordination numbers in the solid are low. These elements form molecular solids held together by the weak van der Waals forces, and consequently have very low melting and boiling points. Atoms of elements with an equal, or nearly equal, number of valence electrons and valence orbitals often crystallize in structures with intermediate coordination numbers. Each atom has enough valence electrons to form covalent bonds with all of its nearest neighbors. The atoms are therefore linked by a network of electron pair bonds extending throughout the entire crystal in what are called **network cova-**

**lent solids.** In effect, a crystal of such a substance can be regarded as a giant molecule.

Carbon in the form of diamond is a network covalent solid. The structure is the same as the zinc blende structure except that all the sites are occupied by carbon atoms, which are arranged in a face-centered cubic lattice in which alternate tetrahedral holes also are occupied (Figure 20.34a). Each carbon atom is bonded to four others and the bonds form tetrahedral angles. The C—C bonds can therefore be characterized as ordinary sigma bonds resulting from the overlap of hybrid $sp^3$ orbitals. This view is supported by the C—C bond length and bond energy in diamond, which have virtually the same values as they do in ethane and other saturated hydrocarbons.

Since covalent bonds are both strong and directed at specific angles, network covalent solids are hard and incompressible, and have exceedingly high melting points and low vapor pressures. For example, diamond sublimes above 3500°C. As the bonds are localized between pairs of atoms, the valence electrons are not mobile and the solids are poor conductors of electricity. Network covalent solids generally resemble ionic crystals in these properties. However, their electrical conductivity does not increase above the melting point, in contrast to ionic compounds.

Carbon normally exists in the more stable allotropic form of graphite. The structure of graphite consists of two-dimensional networks of covalently bonded atoms (Figure 20.34b). In a given plane, each carbon atom is covalently bonded to three other atoms arranged so as to form the hexagonal array shown in the figure. The bond angles are 120°, indicating that the atoms in a given layer are connected by a framework of sigma bonds formed from the overlap of $sp^2$ hybrid orbitals. Each carbon atom has an additional electron in a p orbital and can therefore form one delocalized pi bond. Figure 20.35 shows the equivalent description of the bonding in terms of resonance structures. The distance between nearest neighbor atoms is 1.415 Å, a distance somewhat shorter than a normal single bond. This shortening is expected given that the bond order of the carbon–carbon bonds is $1\frac{1}{3}$. Since the pi electrons are delocalized

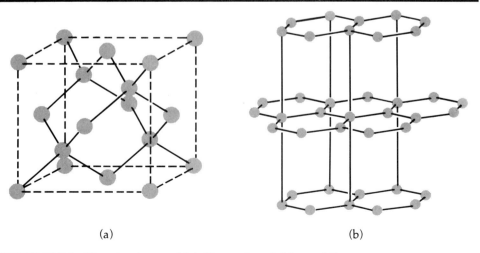

(a)          (b)

**FIGURE 20.34** The structures of (a) diamond and (b) graphite.

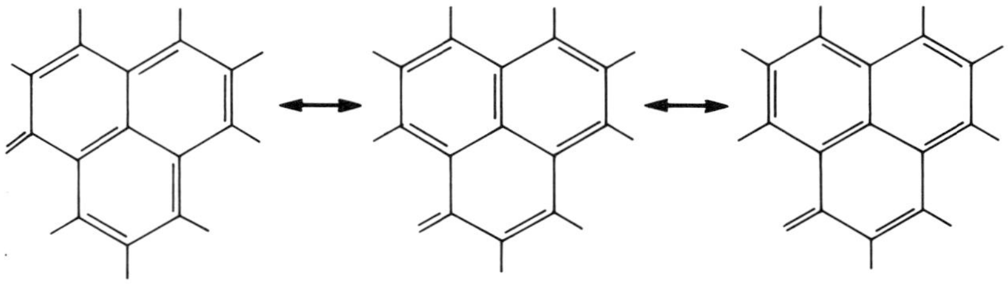

FIGURE 20.35  In-plane bonding in graphite. Three structures contributing to the resonance hybrid are shown.

over an entire plane of atoms, graphite is a moderately good conductor, albeit only in a direction parallel to the hexagonally structured planes.

The distance between adjacent planes of carbon atoms in graphite is 3.41 Å, over twice as large as the in-plane distance between atoms. Adjacent layers are held together by van der Waals forces. Since these forces are weak, the layers can easily slip by each other, contributing to the excellent lubricating properties of graphite.

Diamond is unstable with respect to graphite at ordinary temperatures and pressures, but the rate of transformation is immeasurably slow under these conditions. Figure 20.36 shows the phase diagram of carbon. Diamond, being much denser than graphite, becomes stable at high pressures. Synthetic diamonds can be made by subjecting graphite to high pressures and are widely used in various industrial applications that depend on the hardness and abrasiveness of diamond. However, no one has as yet succeeded in making diamonds of gem quality.

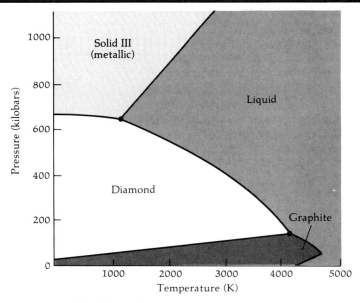

FIGURE 20.36  Phase diagram of carbon.

Silicon and germanium are the two other elements that crystallize in the diamond structure. Some of the allotropic modifications of several of the elements in Groups IIIA, VA, and VIA, such as boron, phosphorus, and sulfur, are solids consisting of atomic chains, rings, layers, and other extended structures.

## 20.12 The Structure of Silicates

Silicates are the most abundant minerals on earth and occur in an uncommonly large variety of forms and combinations with other elements. The basic unit of these substances is the orthosilicate ion, $SiO_4^{4-}$, in which each silicon atom is covalently bonded to four oxygen atoms located at the corners of a tetrahedron. In different types of silicates, from zero to four of the oxygen atoms are shared with adjacent tetrahedra. As we shall see, the number of shared oxygen atoms has a profound effect on the crystal structure and properties of the resulting silicates.

The simplest silicates are the orthosilicates, in which the individual $SiO_4^{4-}$ ions are not connected to each other. Examples of these compounds are zircon, $ZrSiO_4$, willemite, $ZnSiO_4$, and the garnets, $M_3N_2(SiO_4)_3$, where M is a divalent cation such as $Ca^{2+}$, $Fe^{2+}$, or $Mg^{2+}$, and N is a trivalent cation, usually $Al^{3+}$, $Cr^{3+}$, or $Fe^{3+}$. Orthosilicates, particularly $Ca_2SiO_4$, are important constituents of cement.

Orthosilicate ions can join together to form more complex structures. Figure 20.37 shows the structure of two cyclic silicate ions in which either three or six $SiO_4^{4-}$ ions are joined by a common oxygen atom to form the $(Si_3O_9)^{6-}$ and $(Si_6O_{18})^{12-}$ anions, respectively. These ions are found in compounds such as benitoite, $BaTi(Si_3O_9)$, and beryl, $Al_2Be_3(Si_6O_{18})$, an important beryllium mineral. These substances are hard and durable. Many have beautiful colors and are of gem quality. For example, the replacement of $Al^{3+}$ in beryl by a trace of $Cr^{3+}$ or by a trace of $Fe^{3+}$ yields emerald and aquamarine, respectively.

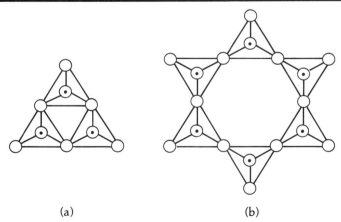

(a)  (b)

**FIGURE 20.37** The structures of cyclic silicate anions in a two-dimensional projection. Three of the oxygen atoms in each $SiO_4$ tetrahedron are shown as open circles (○); the fourth oxygen atom and the silicon atom coincide in this projection and are represented by the dotted circle (⊙): (a) $(Si_3O_9)^{6-}$ with 3 shared oxygen atoms and (b) $(Si_6O_{18})^{12-}$ with 6 shared oxygen atoms.

The silicates considered so far consist of discrete units containing a relatively small number of atoms. Silicate tetrahedra can also form indefinitely large structures extending in one, two, or three dimensions. Figure 20.38 shows the chains formed by single or double strands of tetrahedra. The tetrahedra forming the single strand share two oxygen atoms and the repeating unit therefore is $SiO_3^{2-}$. Minerals of this kind are called pyroxenes; enstatite, $Mg(SiO_3)$, is an example.

The $SiO_4$ tetrahedra forming the double strands alternately, share two or three oxygen atoms to form a characteristic chain of hexagonal rings with $Si_4O_{11}^{6-}$ as the repeating unit. Minerals of this type are known as amphiboles and are closely related to asbestos minerals. Tremolite, $Ca_2(OH)_2Mg_5[Si_4O_{11}]_2$, is a typical example.

Both single- and double-strand chains are held together in the solid by electrostatic interactions with cations located along the chain length. The closeness of the cation packing determines the magnitude of the attraction between adjacent strands. In general, the strength of the covalent bonds forming a particular chain is greater than that of the bonds holding different chains together. As a result, these minerals are fibrous.

The double-stranded structure in Figure 20.38b can be continuously broadened to form a two-dimensional layer such as that displayed in Figure 20.39. Note that three of the four oxygen atoms in each tetrahedron are shared between two tetrahedra. Each layer is tightly held together by covalent bonds while different layers are attached by mutual attraction to cations. Among the various classes of minerals crystallizing in this structure are the micas, the clays, and talc, $Mg_3(OH)_2[Si_4O_{10}]$. Since the bonds holding a given layer together are stronger than those attaching different layers to each other, mica crystals can readily be cleaved into sheets parallel to the silicate layers.

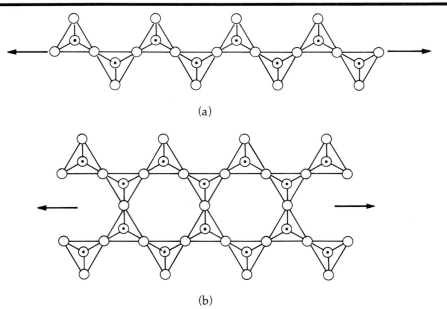

**FIGURE 20.38** Chains of $SiO_4$ tetrahedra: (a) single strand and (b) double strand.

**FIGURE 20.39** Silicate sheet such as those found in mica or clay minerals.

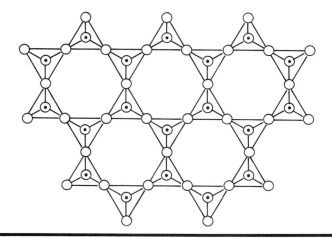

When all four oxygen atoms in a given tetrahedron are shared with neighboring tetrahedra, a three-dimensional network with an $SiO_2$ repeating unit is obtained (see Figure 20.3). Quartz is a typical example of a mineral with this structure. As in the case of other three-dimensional network covalent solids, quartz has a high melting point and is hard and durable. Partial replacement of silicon by aluminum is possible and gives rise to a negatively charged framework in which a variety of cations can be embedded. The feldspars, which have empirical formulas such as $KAlSi_3O_8$, are common minerals with this structure. The zeolites, which also are aluminosilicates, have unusually open structures and absorb water readily. Since the cations needed for charge balance are attached rather loosely, they are readily replaced by cations dissolved in the absorbed water. Because of this property, zeolites are used in water softening (Section 18.7b).

# Conclusion

The bulk properties of solids—electrical conductivity, mechanical strength, melting point—depend on the crystal structure and on the nature of the interaction between the atoms. Metals generally have a closest-packed or nearly closest-packed crystal structure in which the metal cations occupy fixed sites and the valence electrons are free to move throughout the entire crystal. Molecular orbital theory can be used to describe the behavior of these delocalized electrons and the result is the band theory. According to this theory, the electrical properties of metals, insulators, and semiconductors are a consequence of the size of the gap between the valence and conduction bands.

The structure of binary ionic compounds can be described in terms of a closest-packed or nearly closest-packed arrangement of either the cations or the anions, usually the larger anions, with the ions of opposite charge occupying interstitial sites. The specific structure is frequently determined by the relative values of the ionic radii of the cation and anion. The crystal lattice energy is a measure of the stability of ionic crystals. It can be calculated on the basis of the interactions between the ions in a crystal and can also be obtained by means of the Born–Haber cycle.

Network covalent solids contain atoms of elements such as carbon and silicon, which are able to form covalent bonds with all their nearest neighbors. All the atoms in the crystal are thereby connected to form a giant molecule. Covalent bonds are highly directional and, instead of being closest-packed, the atoms have a coordination number determined by the hybridization of their valence orbitals. Since covalent bonds are strong, network covalent solids have very high melting points, comparable to those of the highest melting metals or ionic compounds. They differ in this respect from molecular solids, in which individual molecules are held together by covalent bonds but in which neighboring molecules interact weakly, for example by van der Waals forces. Molecular solids therefore have much lower melting points than network covalent solids, ionic compounds, or metals.

## Problems

1. The heat capacity (per gram) of a certain solid element is 0.237 J g$^{-1}$ K$^{-1}$. The chloride of this element contains 24.74% chlorine by weight. Determine the precise atomic weight of the element.

2. The vibration of an atom in a solid resembles in some respects the vibration of a body attached to a spring (Section 17.3). Using the relationship between the vibrational frequency of such a body and its mass as a guide, explain why the law of Dulong and Petit is not obeyed by light elements.

3. The Einstein expression for the heat capacity of a solid (Equation 20.2) depends on the value of the vibrational frequency $\nu$. The value of $\nu$ for copper is 7.1 × 10$^{12}$ s$^{-1}$. Determine the molar heat capacity of copper at (a) 20 K, (b) 100 K, (c) 300 K, and (d) 800 K. What is the classical value of the heat capacity of copper?

4. Argon crystallizes in a face-centered cubic lattice. The van der Waals radius of argon is 1.90 Å. What is the density of solid argon?

5. What fraction of the available space is occupied by spherical atoms packed as closely as possible in a simple cubic lattice?

6. The atomic radius of copper is 1.28 Å. What is the density of copper metal? Copper crystallizes in the CCP lattice.

7. Aluminum crystallizes in the cubic system with a unit cell volume of 66 Å$^3$. The density of aluminum is 2.7 g cm$^{-3}$. (a) What is the number of atoms per unit cell? (b) In which type of cubic lattice does aluminum crystallize?

8. The density of nickel is 8.902 g cm$^{-3}$, and the metal crystallizes in the CCP lattice. (a) What is the length of an edge of the unit cell? (b) What is the atomic radius of nickel?

9. The length of the edge of a unit cell of a lead crystal is 4.95 Å. What is the size of the largest atom which could fit into an interstitial site in the lead lattice without distortion? Lead crystallizes in the CCP lattice.

10. Germanium containing a small amount of either gallium or arsenic is a doped semiconductor. Which is a $p$-type and which is an $n$-type semiconductor? Which impurity is an electron donor and which is an electron acceptor?

11. The length of a unit cell of a KCl crystal is 6.28 Å. What is the ionic radius of K$^+$ given that the ionic radius of Cl$^-$ is 1.81 Å? Assume that anion-cation contact occurs.

12. Use the radius-ratio rule to predict the coordination numbers of Mg$^{2+}$ and S$^{2-}$ in MgS. Use the structures in Table 20.3 to determine whether the rule works or not.

13. Calculate the distance between adjacent Cs$^+$ and Cl$^-$ ions in cesium chloride given that the density of the solid is 3.97 g cm$^{-3}$. Compare the result with the sum of the ionic radii in Table 20.4.

14. At high pressures, rubidium chloride assumes the CsCl structure. Calculate the ratio of the densities of rubidium chloride at high pressure and at normal pressure. Is the result consistent with Le Chatelier's principle? Explain. Assume that anion-cation contact is maintained in both structures.

15. Consider a planar structure in which three identical

anions in contact form a triangle. A cation sits in the center of the triangle and contacts the three anions. What is the ratio of the radii of cation and anion?

16. Estimate the density of MgO on the basis of data found in this chapter.

17. Crystal structure measurements may be used to evaluate Avogadro's number. The density of $CaF_2$ is 3.180 g cm$^{-3}$, and the length of a side of the unit cell is 5.463Å. Use these data to determine Avogadro's number.

18. Use the radius-ratio rule in conjunction with Tables 20.3 and 20.4 to determine which alkali halides obey this rule and which do not.

19. Lithium hydride crystallizes in the sodium chloride structure, and the length of an edge of the unit cell is 4.09 Å. Evaluate (a) the molar volume and (b) the density of LiH.

20. Show that the minimum value of $r_+/r_-$ for which an ionic compound should have tetrahedral coordination is 0.225. Hint: a tetrahedral site in a closest-packed anion lattice can be generated by placing anions at alternate corners of a cube. The radius of the tetrahedral hole, i.e., $r_+$, is equal to the difference between one-half the body diagonal of the cube and $r_-$.

21. Rhenium oxide crystallizes in a cubic structure. Rhenium ions are located at the corners of the cube and oxide ions at the center of each edge. What is the empirical formula of the oxide?

22. Calcium oxide crystallizes in one of the cubic lattices. The length of an edge of a unit cell is 4.80 Å, and the density is 3.35 g cm$^{-3}$. (a) How many ion pairs are there in a unit cell? (b) Which cubic lattice is formed?

23. The mineral spinel contains 37.9% aluminum, 17.1% magnesium, and 45.0% oxygen. The density is 3.57 g cm$^{-3}$, and the unit cell is a cube of edge 8.09 Å. How many atoms of each element does a unit cell contain?

24. A cubic unit cell contains calcium ions at the corners of the cell, oxide ions in the center of each face, and a titanium ion in the center of the cell. What is the empirical formula of the oxide?

25. For each of the following pairs of compounds, pick the one with the higher melting point: (a) CsF and BaO, (b) LiCl and KCl, (c) MgO and CaS.

26. A certain crystal of iron(II) oxide has the composition $Fe_{0.90}O$ because of incorporation of $Fe^{3+}$ ions in the lattice. What is the ratio of $Fe^{3+}$ to $Fe^{2+}$ in this crystal?

27. The non-stoichiometric compound $TiO_{1.32}$ has 26% of the titanium sites vacant and 2% of the oxygen sites vacant. Calculate the density of this compound relative to that of a TiO oxide in which all the sites are occupied.

28. Calculate the crystal lattice energy of magnesium sulfide on the basis of the interaction between the ions.

29. Determine the first four terms in the crystal lattice energy of cesium chloride (see Equation 20.4).

30. Compare the sum of the terms in the Madelung constant for NaCl (Equation 20.4) with the tabulated value (Table 20.6).

31. Determine the enthalpy of sublimation of potassium by means of a Born–Haber cycle applied to the formation of KCl.

32. Calculate the crystal lattice energy of calcium fluoride on the basis of the interaction between the ions.

33. For the hypothetical compound CaF(s) calculate (a) the crystal lattice energy and (b) the enthalpy of formation. Proceed as follows: first, determine the expected cubic crystal structure on the basis of the radius-ratio rule. Assume that the ionic radius of $Ca^+$ is equal to the average of the radii of $K^+$ and $Ca^{2+}$. Next, calculate the crystal lattice energy. Finally, evaluate the enthalpy of formation by means of a Born–Haber cycle. Obtain the enthalpy of sublimation of calcium and the dissociation energy of fluorine from Appendix 4.

34. Calculate the reaction enthalpy for the disproportionation of CaF(s) to form Ca(s) and $CaF_2$(s). Obtain the needed enthalpies of formation from Problem 20.32 and Problem 20.33 (plus a Born–Haber cycle for $CaF_2$). Comment on the stability of CaF.

35. Calculate the crystal lattice energy of LiF (a) on the basis of the interaction between the ions and (b) by means of a Born–Haber cycle. (Use $n = 7$.)

36. The Born–Haber cycle illustrated in Figure 20.33 is not the only one that can be used to evaluate the crystal lattice energy. Use the following data to construct a Born–Haber cycle and use the cycle to determine the crystal lattice energy of AgCl. For the reaction $Ag(g) + Cl(g) \rightarrow AgCl(g)$, $\Delta H = -301$ kJ. For the vaporization $AgCl(s) \rightarrow AgCl(g)$, $\Delta H = 226$ kJ; $I_1(Ag) = 731.0$ kJ mol$^{-1}$; and $\varepsilon(Cl) = 348.8$ kJ mol$^{-1}$.

37. Some of the alkali and alkaline earth elements form nitrides, e.g., $Li_3N$, in which the nitrogen appears to be present as the nitride ion, $N^{3-}$. Explain the formation of these compounds given that the electron affinity of nitrogen is approximately zero.

38. Calculate the density of diamond given that the C—C bond length is 1.54 Å.

39. Carborundum is a compound of carbon and silicon

which crystallizes in the cubic system. The edge of a unit cell is 4.36 Å in length. The carbon atoms form an FCC structure and the silicon atoms occupy alternate tetrahedral sites within this structure. (a) How many atoms of each kind are there in the unit cell? (b) What is the empirical formula of carborundum? (c) What is the coordination number of each atom? (d) What is the crystal structure of carborundum (see Table 20.3 for structures). (e) Which allotropic form of carbon does carborundum resemble? (f) What is the distance between adjacent carbon and silicon atoms? (g) What is the density of carborundum?

40. The enthalpy of sublimation of graphite is 718 kJ mol$^{-1}$. The energy holding the layers of carbon atoms to each other is 8.2 kJ mol$^{-1}$. (a) From these data, determine the covalent carbon–carbon bond energy in graphite. (b) Use the bond energies in Table 17.2 together with the bond order (Figure 20.35) to calculate the expected carbon–carbon bond energy neglecting resonance. (c) Assuming that the difference between (a) and (b) corresponds to the resonance stabilization energy in graphite, determine its value. How does it compare with the resonance stabilization energy in benzene (Section 17.1)?

41. Determine the repeating unit in a two-dimensional array of silicate ions, such as that found in mica.

# 21

# The Transition Elements and Their Coordination Compounds

The transition elements include most of the industrial metals. Iron has been the most important metal since the early days of civilization. Its present-day use, nearly always in the form of steel, approaches a worldwide yearly total of a billion tons. Many other transition metals are widely used because of their specific properties; the coinage metals—copper, silver, and gold—are familiar to all.

The transition elements have an extensive chemistry. The study of coordination compounds, the complexes they form with diverse ligands, has become a dominant area of inorganic chemistry in the last few decades. In addition to their many practical applications, these compounds have posed many challenges in inorganic synthesis and in the extension of the theories of chemical bonding to large molecules.

## Transition Element Chemistry

### 21.1 General Properties

*a. Electronic Configuration* The transition elements are associated with the filling of the d orbitals. There are three series comprised of ten elements each, corresponding to the filling of the 3d, 4d, and 5d orbitals, and a fourth, incomplete series, consisting of the most unstable artificial elements, in which the 6d orbitals are filling. The two inner transition series—the lanthanides and the actinides—are associated with the filling of the 4f and 5f orbitals, respectively,

# The Transition Elements and Their Coordination Compounds 21.1

| | $(n-1)d^1ns^2$ | $(n-1)d^2ns^2$ | $(n-1)d^3ns^2$ | $(n-1)d^5ns^1$ | $(n-1)d^5ns^2$ | $(n-1)d^6ns^2$ | $(n-1)d^7ns^2$ | $(n-1)d^8ns^2$ | $(n-1)d^{10}ns^1$ | $(n-1)d^{10}ns^2$ |
|---|---|---|---|---|---|---|---|---|---|---|
| $M^{2+a}$ | — | $(n-1)d^2$ | $(n-1)d^3$ | $(n-1)d^4$ | $(n-1)d^5$ | $(n-1)d^6$ | $(n-1)d^7$ | $(n-1)^8$ | $(n-1)d^9$ | $(n-1)d^{10}$ |
| $M^{3+a}$ | [noble gas] | $(n-1)d^1$ | $(n-1)d^2$ | $(n-1)d^3$ | $(n-1)d^4$ | $(n-1)d^5$ | $(n-1)d^6$ | — | $(n-1)d^8$ | — |
| $n=4$ | Sc | Ti | V | Cr | Mn | Fe | Co | Ni | Cu | Zn |
| $n=5$ | Y | Zr | Nb | Mo | Tc | Os | Rh | Pd | Ag | Cd |
| $n=6$ | La | Hf | Ta | W | Re | Ru | Ir | Pt | Au | Hg |
| $n=7$ | Ac | 104 | 105 | 106 | 107 | 108 | 109 | | | |

Lantanides
Actinides

a. Some occasional anomalies are not shown.

**FIGURE 21.1** The transition elements and their electronic configurations. The electronic configurations of the 2+ and 3+ ions also are shown.

and consist of 14 elements each. Figure 21.1 summarizes the electronic configurations of the transition elements and their common ions.

**b. The Elements** All the transition elements are metals. Most of them are hard and have high melting points. The melting points are highest in the middle of each series (Figure 21.2) and tend to be correlated with the number of unpaired d electrons. The lowest melting points are observed at the ends of the series, where the valence d orbitals are completely filled. Thus, mercury is the only metal that is a liquid at room temperature.

In contrast to the representative elements in a given period, the transition elements in each series tend to resemble one another. For example, the atomic radii, although tending to decrease with increasing atomic number, are nearly constant over a large part of any series (Figure 21.3). As noted in Section 6.8, the effective nuclear charge felt by the $ns$ valence electrons increases only slowly across a series owing to the screening by the $(n-1)d$ electrons. This effect also is responsible for the relatively small increase in ionization energies across each series (Section 6.5a). The virtual equality of the atomic radii of corresponding elements in the second and third transition series can be ascribed to the lanthanide contraction (Section 6.8)—the increase in effective nuclear charge due to the filling of the 4f orbitals is sufficiently large to prevent the atomic radii of the elements in the third series from being significantly larger than those of the corresponding elements in the second series.

The reactivity of the transition elements tends to decrease from left to right across a series. The standard reduction potentials for the formation of the elements from their common 2+ or 3+ ions, shown for the elements of the first transition series in Figure 21.4, are indicative of this trend. Scandium, the first element in the series, has the most negative standard reduction potential while copper, near the end of the series, is the only element with a positive $\epsilon°$ value.

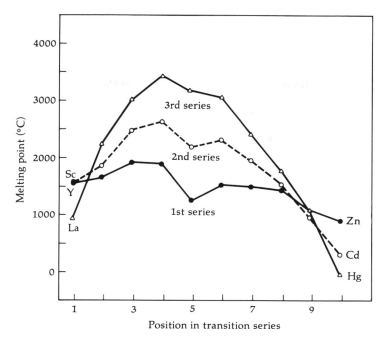

**FIGURE 21.2** Melting points of the transition elements. The lines connect the melting points of the elements in each series.

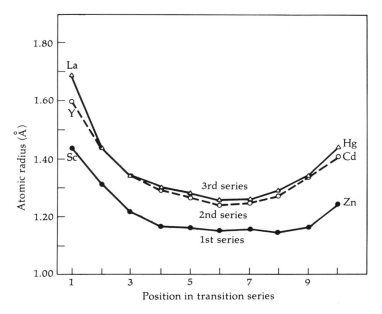

**FIGURE 21.3** Atomic radii of the transition elements.

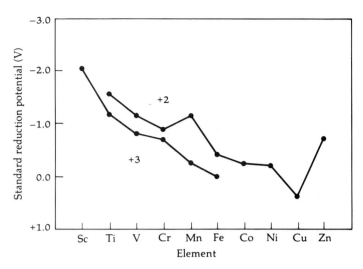

**FIGURE 21.4** Standard reduction potentials of the 2+ and 3+ ions of the elements in the first transition series. The $\epsilon°$ values refer to the reduction to the element.

While the overall trend reflects the increase in ionization energy across the series, the irregularities evident in Figure 21.4 reflect the net effect of *all* the factors which determine the reactivity of the metals in aqueous solution, such as hydration energies and enthalpies of sublimation (Section 19.5c).

Although the $\epsilon°$ values indicate that copper should be the only metal in the first transition series incapable of reacting with a non-oxidizing acid, a number of the metals form a thin protective oxide coating which is resistant to attack by acids. For example, the surface of chromium metal is usually covered by the inert, transparent oxide, $Cr_2O_3$. The presence of this oxide layer makes chromium plating a useful technique for the protection of more active metals against corrosion.

The reactivity of the metals in a particular family generally decreases with increasing atomic weight. Thus, the most inert metals, such as gold and platinum, are found in the third transition series.

*c. Oxidation States* The common oxidation states of the transition elements are shown in Figure 21.5. Although the regularities are less striking than those in the oxidation states of the representative elements (see Figure 19.11), all three series have a number of common features. The first element in each series, scandium, yttrium, or lanthanum, has the single oxidation state of +3. This state is displayed by ions such as $Sc^{3+}$, which have lost all their valence electrons. These elements do not form a +2 oxidation state because once the two $ns$ electrons are lost, the removal of the lone $(n-1)d$ electron requires very little additional energy.

The maximum oxidation state of the next several elements is equal to the group number, indicating the participation in bonding of all the valence shell electrons. These high oxidation states are displayed in oxides or oxoanions, such as $CrO_4^{2-}$ or $MnO_4^-$, in which the metal atom is covalently bonded to oxygen. The lowest oxidation state displayed by these elements is usually +2, corresponding to the loss of the two s electrons in the valence shell, as in $Ti^{2+}$ or

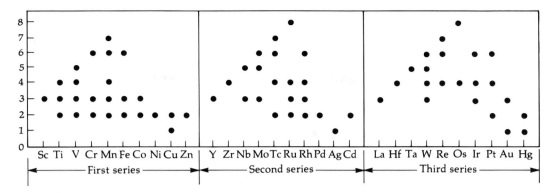

**FIGURE 21.5** Common oxidation states of the transition elements.

$Mn^{2+}$. These elements generally display a number of intermediate oxidation states as well. For the elements in the second and third transition series, the higher oxidation states tend to be favored.

Beyond the middle of the transition series, there is an abrupt decrease in maximum oxidation state and species with low oxidation number are most stable. For example, the common oxidation states of the second five elements in the first transition series are +2 and +3 for iron and cobalt, +2 for nickel and zinc, and +1 and +2 for copper. As may be seen in Figure 6.9, the energy of the $(n-1)d$ orbitals decreases significantly below that of the $ns$ orbitals near the end of the transition series, presumably owing to the gradual increase in effective nuclear charge. Consequently, the $(n-1)d$ electrons are pulled into the inner electron core and are not readily available for bonding.

## 21.2 Redox Chemistry of Transition Elements with Few Oxidation States

The elements near the end of any transition series as well as the first element have only one or two stable oxidation states. Consequently, their redox chemistry is relatively simple. Most of the metals form ions of the type $M^+$, $M^{2+}$, or $M^{3+}$, and their oxides and hydroxides tend to be amphoteric, dissolving in either acid or base. The reactions of zinc hydroxide are typical:

$$Zn(OH)_2(s) + 2H^+ \rightarrow Zn^{2+} + 2H_2O$$

$$Zn(OH)_2(s) + 2OH^- \rightarrow Zn(OH)_4^{2-}$$

*a. The Iron Triad* The three elements in Group VIII of the first transition series, iron, cobalt, and nickel, known collectively as the **iron triad,** have similar properties. The elements are hard metals with relatively high melting points and moderate reactivity, and are often found together in nature. All three metals are **ferromagnetic,** that is, they exhibit magnetism in the absence of an external magnetic field. Iron is of special practical importance because it is by far the most widely used of all metals. This widespread usage reflects its great natural abundance—nearly 5% of the earth's crust, the ease with which it can be produced from its ores, and the wide range of properties that can be obtained by alloying the metal with other substances in various types of steel.

The production of iron in a blast furnace illustrates some of the features of the redox chemistry of this element. The overall reaction,

$$Fe_2O_3(s) + 3CO(g) \rightarrow 2Fe(s) + 3CO_2(g)$$

actually involves a number of steps, as illustrated in Figure 21.6. Iron produced in the blast furnace is known as pig iron. This material is quite impure and has poor structural properties, being hard and very brittle. Its properties can be improved by conversion to steel, a purified alloy of iron, carbon, and other metals, such as chromium, manganese and nickel. (An **alloy** is a metallic substance containing at least two elements. The elements can either form an intermetallic compound, a mixture, or a solid or liquid solution.) The impurities in the pig iron, such as carbon ($\approx 5\%$) are burned away with oxygen at high temperatures and controlled amounts of the alloying materials are added. Worldwide production of steel is close to a billion tons per year, and its properties can be tailored to specific applications.

The elements of the iron triad all form stable 2+ ions that resemble each other in their properties, for example, in the solubility of their salts (Section 9.1). Iron is the only one of these elements to form a stable 3+ ion in aqueous solution. In fact, when solutions containing $Fe^{2+}$ ions are exposed to the atmosphere, oxidation to $Fe^{3+}$ gradually takes place unless the solutions are strongly acidic.

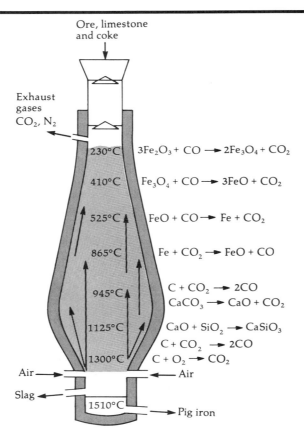

**FIGURE 21.6** The production of iron metal in the blast furnace. The iron ore (mostly impure $Fe_2O_3$), limestone ($CaCO_3$), and coke (carbon) are added from the top and a very strong blast of hot air (or oxygen) is blown in at the bottom. The ore undergoes the indicated reactions as it descends to the hotter regions of the furnace. The reducing agent is carbon monoxide formed in the reaction of carbon with oxygen. The first reduction product, $Fe_3O_4$, is a mixed oxide of Fe(II) and Fe(III). The metallic iron and the slag (largely calcium silicate) accumulate in immiscible layers at the bottom of the furnace.

The elements lying below the iron triad, known as the light (Ru–Rh–Pd) and heavy (Os–Ir–Pt) platinum metals, are rather rare and relatively unreactive. However, they show significant catalytic activity and are used as catalysts. They are often found in the elemental state in nature. Among the interesting compounds formed by these elements are the oxides $RuO_4$ and $OsO_4$ in which the elements display a +8 oxidation state. These compounds are exceptional; oxidation states of +4 or less tend to be more common.

*b. The Coinage Metals*  Copper and the other elements of Group IB (silver and gold), commonly known as the **coinage metals,** are substantially different from the preceding Group VIII elements. Instead of being hard, they are moderately soft as well as ductile and malleable. These properties, as well as their high thermal and electrical conductivity and resistance to chemical attack, have opened new uses for the metals in addition to their traditional uses in coinage and jewelry. Gold, for instance, is currently used as a coating on the surfaces of the more delicate components of space probes and satellites. Due to their relative inertness, the elements of this group are frequently found in the metallic state in nature.

The properties of the coinage metals are readily modified by alloying them with other metals. For example, copper is the chief constituent of brass and bronze, which are harder than pure copper. Table 21.1 gives the composition of a number of common alloys, many of which contain one or more of the coinage metals.

The three elements of the copper family display a +1 oxidation state, but only silver forms $M^+$ ions in aqueous solution. The $Cu^+$ and $Au^+$ ions are unstable in aqueous solution, although they can be stabilized by the formation of insoluble salts or complex ions. These metals are more commonly present in aqueous solution as copper(II) and gold(III).

An important use of silver that depends on its redox chemistry is in photography. Photographic film consists of a suspension of small AgBr crystals in gelatin spread on a cellulose acetate backing. When exposed to light, AgBr is made more sensitive to reduction as a result of changes in its crystal structure. Since more light is focused on the film from the light areas of the object being photographed, this part of the image is more readily reduced on development. The film is developed by exposing it to a mild organic reducing agent such as hydroquinone:

$$AgBr(s) + e^-(\text{reducing agent}) \rightarrow Ag(s) + Br^-$$

**TABLE 21.1  Composition of Some Common Alloys**

| Alloy | Percent composition |
|---|---|
| Brass | 67% Cu, 33% Zn |
| Bronze | 70–95% Cu, 1–25% Zn, 1–18% Sn |
| Sterling silver | 92.5% Ag, 7.5% Cu |
| 18-carat gold | 75% Au, 12.5% Ag, 12.5% Cu |
| 14-carat gold | 58% Au, 4–28% Ag, 14–28% Cu |
| Stainless steel[a] | 82.5% Fe, 16.5% Cr, 0.65% C, 0.35% Mn |
| Type metal | 70% Pb, 18% Sb, 10% Sn, 2% Cu |

[a]Stainless steel does not have a unique composition; elements such as nickel and cobalt are some other typical ingredients.

and the remaining AgBr is then removed by dissolving it in sodium thiosulfate (hypo), thereby forming a negative of the picture:

$$AgBr(s) + 2S_2O_3^{2-} \rightarrow Ag(S_2O_3)_2^{3-} + Br^-$$

A print can be made by illuminating light-sensitive printing paper through the negative.

**Example 21.1** The standard electrode potential diagram of copper is

$$\underset{\underset{0.34}{\underbrace{\qquad\qquad\qquad\qquad}}}{Cu^{2+}\ \underline{\quad 0.16\quad}\ Cu^+\ \underline{\quad 0.52\quad}\ Cu}$$

Calculate the equilibrium constant and $\epsilon°$ for the reaction

$$Cu^{2+} + 2I^- \rightarrow CuI(s) + \tfrac{1}{2}I_2(aq)$$

using data in Appendix 3 (Table A3.3) and Appendix 5.

The reaction can be divided into the following half-reactions:

(1) $Cu^{2+} + I^- + e^- \rightarrow CuI$

(2) $\underline{\qquad\qquad I^- \rightarrow \tfrac{1}{2}I_2 + e^- \qquad\qquad}$

(3) $Cu^{2+} + 2I^- \rightarrow CuI + \tfrac{1}{2}I_2$

Half-reaction (1) can be broken down into the following two steps:

(1a) $Cu^{2+} + e^- \rightarrow Cu^+$

(1b) $\underline{\qquad Cu^+ + I^- \rightarrow CuI(s)\qquad}$

(1) $Cu^{2+} + I^- + e^- \rightarrow CuI(s)$

For (1a), we have $\epsilon° = 0.16$ V. The equilibrium constant of (1b) is the reciprocal of $K_{sp}$ for CuI. From Table A3.3 we obtain

$$K = 1/K_{sp} = (5.1 \times 10^{-12})^{-1} = 2.0 \times 10^{11}$$

Since both $\epsilon°$ and $K$ can be readily converted to $\Delta G°$ values for the respective processes, it is most convenient to first perform this conversion and then add the $\Delta G°$ for (1a), (1b), and (2) to obtain $\Delta G°$ for reaction (3), the desired process. We obtain

(1a) $\Delta G° = -nF\epsilon° = -(1\ \text{mol})(96{,}485\ \text{C mol}^{-1})(0.16\ \text{V})(1\ \text{J/C V})(10^{-3}\ \text{kJ J}^{-1})$
$= -15\ \text{kJ}$

(1b) $\Delta G° = -nRT \ln K = -(1\ \text{mol})(8.314\ \text{J mol}^{-1}\ \text{K}^{-1})(298\ \text{K})(10^{-3}\ \text{kJ J}^{-1}) \ln(2.0 \times 10^{11})$
$= -64\ \text{kJ}$

(2) $\Delta G° = -nF\epsilon° = -(1\ \text{mol})(96{,}485\ \text{C mol}^{-1})(-0.54\ \text{V})(1\ \text{J/C V})(10^{-3}\ \text{kJ J}^{-1})$
$= 52\ \text{kJ}$

where the value of $\epsilon°$ for half-reaction (2) is taken from Appendix 5. The resultant value of the standard free energy change is

(3) $\Delta G° = (-15 - 64 + 52)\ \text{kJ} = -27\ \text{kJ}$

From this we obtain

$$K = \exp\left(\frac{-\Delta G°}{nRT}\right)$$

$$= \exp\left[\frac{-(-27 \text{ kJ})(10^3 \text{ J kJ}^{-1})}{(1 \text{ mol})(8.314 \text{ J mol}^{-1} \text{ K}^{-1})(298 \text{ K})}\right] = 5 \times 10^4$$

and

$$\epsilon° = \frac{-\Delta G°}{nF}$$

$$= \frac{-(-27 \text{ kJ})(10^3 \text{ J kJ}^{-1})}{(1 \text{ mol})(96,485 \text{ C mol}^{-1})(1 \text{ J/C V})} = 0.28 \text{ V}$$

Since $K > 1$, the reaction occurs spontaneously at standard conditions. The addition of a $\approx 1$ M solution of KI to a solution containing copper(II) leads to the formation of a light yellow CuI precipitate while the solution turns violet owing to the release of $I_2$. ∎

*c. The Post-Transition Elements* Zinc, cadmium, and mercury have the electronic configuration $(n - 1)d^{10}ns^2$ and consequently terminate their respective transition series. Since the d orbitals are completely filled, these elements are also known as the **post-transition elements.** All three elements form $M^{2+}$ ions, and mercury also exists in the form of mercury(I). This species is unusual because it is a dimer, $Hg_2^{2+}$, in which the two mercury atoms are held together by a covalent bond.

In contrast to zinc and cadmium, mercury is not readily oxidized, as may be inferred from the standard electrode potential diagram:

$$Hg^{2+} \underline{\quad 0.91 \quad} Hg_2^{2+} \underline{\quad 0.79 \quad} Hg$$
$$\underline{\quad\quad\quad 0.85 \quad\quad\quad}$$

For example, mercury is not attacked by $\approx 1$ M non-oxidizing acids. The standard electrode potential diagram also shows that $Hg_2^{2+}$ is stable against disproportionation. However, disproportionation occurs when $Hg^{2+}$ is removed from solution by precipitation, in accord with Le Chatelier's principle. The reactions of $Hg_2^{2+}$ with strong base and with hydrogen sulfide illustrate this process:

$$Hg_2^{2+} + 2OH^- \rightarrow HgO(s) + Hg(l) + H_2O$$

$$Hg_2^{2+} + H_2S(g) \rightarrow HgS(s) + Hg(l) + 2H^+$$

On the other hand, the halides, nitrate, and sulfate of mercury(I) are stable.

*d. The Scandium Family and the Lanthanides* The elements of the scandium family, scandium, yttrium, and lanthanum, are active metals that readily react with dilute acid to form $M^{3+}$ ions. The fourteen elements following lanthanum comprise the first inner transition series. The properties of these elements are very similar to those of lanthanum and these fifteen elements are usually grouped together as the lanthanides or "rare earths." Since they differ virtually only in the configuration of their $(n - 2)f$ subshell, these elements strongly resemble each other in physical and chemical properties. For example, they all form compounds in which they display the +3 oxidation state. In addition, a few of the elements have stable +2 or +4 states as a result of the extra stability conferred by either the noble gas configuration or by a filled or half-filled f subshell (Section 6.5).

**Example 21.2** What oxidation states other than +3 are likely for cerium and europium?

The electronic configuration of cerium is $[Xe]4f^2 6s^2$. The loss of four electrons leads to the stable [Xe] configuration and cerium(IV) compounds exist.

The electronic configuration of europium is $[Xe]4f^7 6s^2$. The loss of two electrons forms an ion with a half-filled set of f-orbitals, thereby stabilizing the +2 state. ∎

## 21.3 Redox Chemistry of Transition Elements with Multiple Oxidation States

As is evident in Figure 21.5, the elements towards the middle of the transition series exhibit the greatest number of oxidation states, with values typically ranging from +2 to as high as +7. The properties of these elements change in a fairly regular way with increasing oxidation state. In their lower oxidation states, the elements form simple cations such as $V^{2+}$ or $Cr^{3+}$, while in the higher oxidation states they form oxoanions such as $MnO_4^-$, which tend to be strong oxidizing agents in acidic solution. The oxides of these elements change with oxidation state in a similar way as those of the representative elements, the acidity of the oxide increasing with oxidation state. For example, the oxide $VO$ is basic, $VO_2$ is amphoteric, and $V_2O_5$ is acidic.

Chromium and manganese display the most extensive redox chemistry of the transition elements. Figure 21.7 shows their standard electrode potential diagrams in acidic and basic solution.

Chromium forms compounds in which it displays oxidation states ranging from +2 to +6, which change in character from strongly reducing to strongly oxidizing. Chromium(II) compounds, such as $CrCl_2$, can be prepared by reduction of chromium(III) chloride with zinc in the presence of acid. As may be inferred from the standard electrode potential diagram, chromium(II) reduces hydrogen ion and, consequently, is unstable in solution. Chromium(III) is the most stable form of this element. The oxide, $Cr_2O_3$, has a characteristic green color and can be prepared by heating the metal in air. As is true for many other transition metal oxides, $Cr_2O_3$ is amphoteric, dissolving in both acid and base.

Chromium(VI) is present in solution as the deep orange dichromate ion, $Cr_2O_7^{2-}$, or as the yellow chromate ion, $CrO_4^{2-}$. The change of one form to the other proceeds via the hydrogen chromate ion, $HCrO_4^-$, according to the equilibrium

$$Cr_2O_7^{2-} + H_2O \underset{}{\overset{K_1}{\rightleftharpoons}} 2HCrO_4^- \underset{}{\overset{K_2}{\rightleftharpoons}} 2H^+ + 2CrO_4^{2-}$$

where $K_1 = 0.025$ and $K_2 = 1.0 \times 10^{-13}$. The addition of acid shifts the equilibrium point to the left and dichromate is stable in acidic solution while chromate is stable in alkaline medium. Chromium(VI) is a powerful oxidizing agent in acidic solution as witnessed by the standard potential for its reduction to chromium(III):

$$Cr_2O_7^{2-} + 14H^+ + 6e^- \rightarrow 2Cr^{3+} + 7H_2O \qquad \epsilon° = 1.33 \text{ V}$$

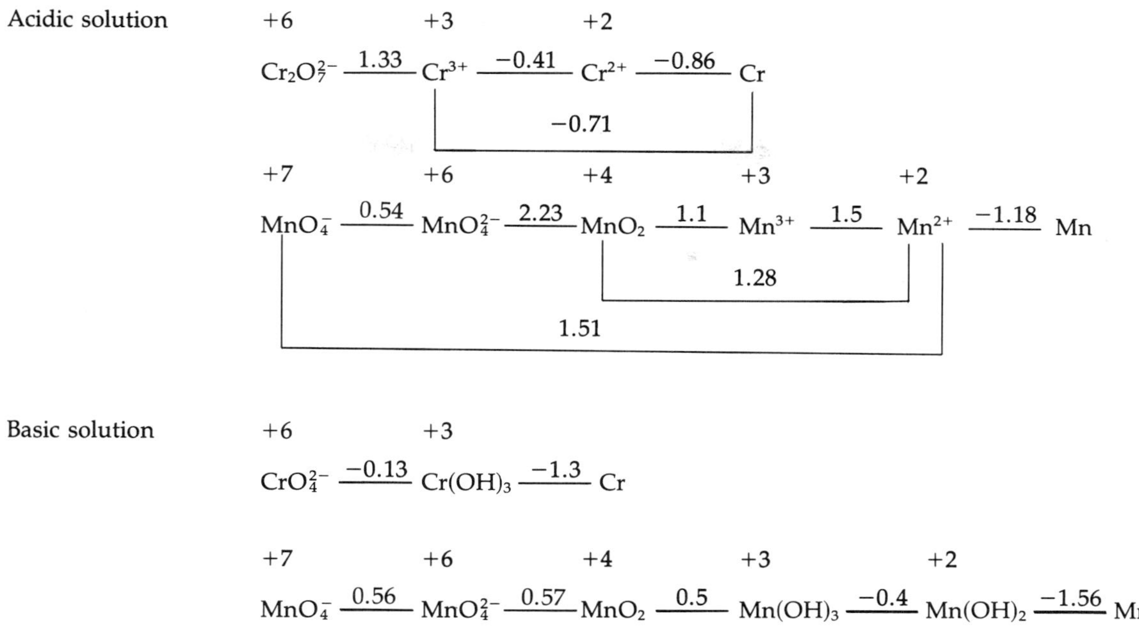

## Figure 21.7

**Acidic solution**

+6      +3      +2

$Cr_2O_7^{2-} \xrightarrow{1.33} Cr^{3+} \xrightarrow{-0.41} Cr^{2+} \xrightarrow{-0.86} Cr$

$\xrightarrow{-0.71}$

+7      +6      +4      +3      +2

$MnO_4^- \xrightarrow{0.54} MnO_4^{2-} \xrightarrow{2.23} MnO_2 \xrightarrow{1.1} Mn^{3+} \xrightarrow{1.5} Mn^{2+} \xrightarrow{-1.18} Mn$

1.28

1.51

**Basic solution**

+6      +3

$CrO_4^{2-} \xrightarrow{-0.13} Cr(OH)_3 \xrightarrow{-1.3} Cr$

+7      +6      +4      +3      +2

$MnO_4^- \xrightarrow{0.56} MnO_4^{2-} \xrightarrow{0.57} MnO_2 \xrightarrow{0.5} Mn(OH)_3 \xrightarrow{-0.4} Mn(OH)_2 \xrightarrow{-1.56} Mn$

**FIGURE 21.7** Standard electrode potential diagrams of chromium and manganese.

---

The large coefficient of the H$^+$ term in this equation indicates that the reduction potential must be strongly dependent on pH, in analogy to some of the oxoanions of the representative elements (Section 19.6). Thus, chromium(VI) does not act as an oxidizing agent in alkaline solution (Figure 21.7). The best way of preparing $Cr_2O_7^{2-}$ from a lower oxidation state of chromium consequently involves oxidation in alkaline solution to form $CrO_4^{2-}$, which can then be converted to $Cr_2O_7^{2-}$ on acidification.

The addition of chromium(VI) salts such as $Na_2CrO_4$ or $K_2Cr_2O_7$ to concentrated sulfuric acid forms a solution of the red oxide $CrO_3$,

$$K_2Cr_2O_7(s) + 3H_2SO_4(conc) \rightarrow 2K^+ + H^+ + 3HSO_4^- + 2CrO_3(aq) + H_2O(l)$$

This solution has strong oxidizing properties and is used to clean laboratory glassware.

Of the elements in the first transition series, manganese exists in the largest number of oxidation states, ranging from manganese(II), as in $MnCl_2$, to manganese(VII), as in $KMnO_4$. In contrast to $Cr^{2+}$, $Mn^{2+}$ is stable in aqueous solution and is not readily oxidized to $Mn^{3+}$. In accord with the trend already considered for other transition elements, MnO is a basic oxide. Trivalent manganese ($Mn^{3+}$) is unstable with respect to disproportionation in acidic solution, and so there is essentially no solution chemistry of this species. However, in basic solution both the hydroxide and the oxide, $Mn_2O_3$, are stable.

Manganese(IV) is commonly found as the dioxide, $MnO_2$, which is a dark brown solid. In acidic medium, this substance is a powerful oxidizing agent, being reduced to $Mn^{2+}$ with $\epsilon° = 1.28$ V. As expected from the pH dependence of the electrode potentials of oxoanions or oxides, the oxidizing power of this substance is much less pronounced in basic solution. Although in line with the increasing acidity of higher-oxidation-state oxides $MnO_2$ is amphoteric, it tends to be relatively inert towards acids and alkalies.

Perhaps the best known compound of manganese is potassium permanganate, KMnO$_4$. This salt has a deep purple color and is a very powerful oxidizing agent in acidic solution:

$$MnO_4^- + 8H^+ + 5e^- \rightarrow Mn^{2+} + 4H_2O \qquad \epsilon° = 1.51 \text{ V}$$

This half-reaction forms the basis of an analytical procedure in which MnO$_4^-$ titrant is used to assay the concentration of a variety of substances present in the reduced state, as already illustrated in Figure 2.9. The procedure works because permanganate is a strong oxidizing agent, because it reacts rapidly and quantitatively to give the single product Mn$^{2+}$, and because it acts as a self-indicator, the endpoint of the titration being marked by the permanent appearance of its purple color. A potentially interfering reaction is the oxidation of water to oxygen, which is thermodynamically allowed. Fortunately, this reaction occurs extremely slowly unless the solution is heated or the acid concentration made very high.

# Coordination Compounds

Coordination chemistry—the study of transition-metal complexes—offers opportunities for the synthesis of novel compounds, for a study of their structures and reactions, and for an application of quantum mechanics to large molecules. Coordination compounds are widely used in various industrial processes, agriculture, and analytical chemistry. Many coordination compounds play significant roles in various biological processes, and their study has spawned the new field of bioinorganic chemistry.

## 21.4 Fundamentals of Coordination Chemistry

*a. Some Basic Definitions* A **coordination compound** consists of a central metal atom or ion attached to a number of groups, or **ligands.** These ligands can be neutral molecules such as NH$_3$ or ions such as CN$^-$, as in [Cu(NH$_3$)$_4$]$^{2+}$ or [Fe(CN)$_6$]$^{3-}$, respectively. These examples illustrate one of the characteristic features of these complexes: the number of ligands (e.g., 6CN$^-$) frequently exceeds the oxidation number of the metal ion (e.g., Fe$^{3+}$). The ligands attached to the central metal atom are species capable of donating electrons and forming coordinate covalent bonds. The specific ligand atom donating the electrons is known as the **donor atom.** Since species such as NH$_3$, CN$^-$, Cl$^-$ and H$_2$O can act as electron donors, it is not surprising that they are frequently encountered ligands. The presence of ligands substantially alters the chemical behavior of the metal ion and also affects its physical properties, most typically by conferring an intense color to the complex in solution. The characteristic equilibria involving complex ions have already been examined (Section 9.6). These equilibria are a source of much information about coordination compounds.

The central atom of a complex can be characterized by its **coordination number,** that is, by the number of attached donor atoms. For example, the coordination number of Pt in [Pt(NH$_3$)$_5$Cl]$^{3+}$ is six while that of Ag in [Ag(NH$_3$)$_2$]$^+$ is two. A complex in which a metal ion is attached to water ligands is a hydrated ion. The coordination number of a metal ion is generally equal to its hydration number (Section 18.6a).

The coordination number of the central metal ion is frequently equal to the number of attached ligands. This is not always the case, however, since there are ligands containing two or more donor atoms. A commonly encountered ligand containing two donor atoms is ethylenediamine,

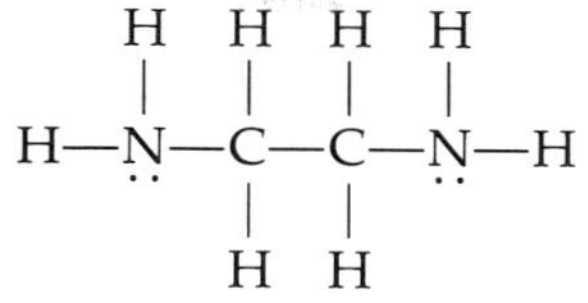

$$\begin{array}{c} \text{H} \quad \text{H} \quad \text{H} \quad \text{H} \\ | \quad | \quad | \quad | \\ \text{H}-\text{N}-\text{C}-\text{C}-\text{N}-\text{H} \\ | \quad | \\ \text{H} \quad \text{H} \end{array}$$

which has two nitrogen atoms that can act as donors. The coordination number of platinum in $[Pt(NH_2CH_2CH_2NH_2)_2Cl_2]^{2+}$ is therefore six even though only four ligands are attached. A ligand attached to a central atom at only one point is said to be **monodentate** while one attached at two points is classified as **bidentate**, and is an example of a **multidentate** ligand. The coordination compounds formed by multidentate ligands are frequently called **chelates** (Greek, *chela* = claw) because of the distinctive manner in which they surround the central metal atom. Figure 21.8 shows the structure of a typical chelate, in which a single ligand, ethylenediaminetetraacetic acid (EDTA), with six donor atoms, is attached to iron. The formation of a ring complex by a chelating ligand is called chelation. Tables 21.2 and 21.3 list a number of common monodentate and multidentate ligands, respectively.

The central metal atom and its surrounding ligands constitute the **coordination sphere** of a complex. These species are usually enclosed in brackets when writing the formula of a complex, e.g., $[Cr(H_2O_5)Cl]^{2+}$. A coordination compound typically consists of a complex ion and enough oppositely charged ions to balance the charge of the complex. These ions are called **counter ions** to distinguish them from similar species that may be part of the complex. For example, two of the $Cl^-$ ions in $[Cr(H_2O)_5Cl]Cl_2$ are counter ions. These ions are not part of the coordination sphere and are not enclosed within brackets in the

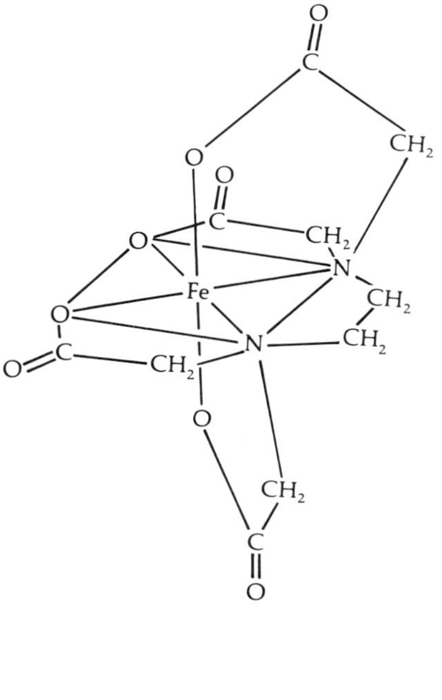

**FIGURE 21.8** The structure of the ethylenediaminetetraacetic acid (EDTA) complex of iron. The oxygen and nitrogen donor atoms occupy the corners of an octahedron at the center of which is the iron atom. This multidentate ligand is called a chelating agent, and the complex is a chelate.

### TABLE 21.2 Common Monodentate Ligands

| Ligand symbol | Donor atom | Ligand name |
|---|---|---|
| F$^-$ | F$^-$ | Fluoro |
| Cl$^-$ | Cl$^-$ | Chloro |
| Br$^-$ | Br$^-$ | Bromo |
| I$^-$ | I$^-$ | Iodo |
| CN$^-$ | C | Cyano |
| OH$^-$ | O | Hydroxo |
| SCN$^-$ | S | Thiocyanato |
| NCS$^-$ | N | Isothiocyanato |
| CH$_3$COO$^-$ | O | Acetato |
| CO$_3^{2-}$ | C | Carbonato |
| SO$_4^{2-}$ | S | Sulfato |
| H$_2$O | O | Aqua |
| NH$_3$ | N | Ammine |
| py (C$_6$H$_5$N) | N | Pyridine |
| CO | C | Carbonyl |
| NO$^+$ | N | Nitrosyl |

formula of the compound. Coordination compounds can also consist of uncharged metal atoms and neutral ligands, for example, [Ni(CO)$_4$].

***b. The Nomenclature of Coordination Compounds*** Although a number of coordination compounds have traditional names, it has become increasingly common to use systematic names for these substances. This nomenclature is sufficiently descriptive to permit the correct formula to be written on the basis

### TABLE 21.3 Common Multidentate Ligands (Chelating Agents)

| Ligand name | Symbol | Formula$^a$ | Bonds |
|---|---|---|---|
| Ethylenediamine | en | N̈H$_2$—CH$_2$—CH$_2$—N̈H$_2$ | 2 |
| Propylenediamine | pn | N̈H$_2$—CH$_2$—CH—N̈H$_2$<br>　　　　　　　　\|<br>　　　　　　　CH$_3$ | 2 |
| Oxalato | ox | $^-$:Ö̤OC—COÖ̤:$^-$ | 2 |
| 8-Hydroxyquinolinato | oxine | (8-hydroxyquinoline structure with N: and :Ö:$^-$) | 2 |
| Diethylenetriamine | dien | N̈H$_2$—(CH$_2$)$_2$—N̈H—(CH$_2$)$_2$—N̈H$_2$ | 3 |
| Triethylenetetraamine | trien | N̈H$_2$—(CH$_2$)$_2$—N̈H—(CH$_2$)$_2$—N̈H—(CH$_2$)$_2$—N̈H$_2$ | 4 |
| Ethylenediaminetetraacetic acid anion | EDTA | $^-$:ÖC(=O)—CH$_2$　　　CH$_2$CÖ:$^-$(=O)<br>　　　　　　\|　　　　　　\|<br>　　　　　　:N—CH$_2$CH$_2$—N:<br>　　　　　　\|　　　　　　\|<br>$^-$:ÖC(=O)—CH$_2$　　　CH$_2$CÖ:$^-$(=O) | 6 |

$^a$The electron pair dots identify the donor atoms.

of the name and vice versa, a procedure that is of value when the formulas are as complicated as those of many coordination compounds. The following rules are used:

1. In naming a coordination compound, the name of the cation precedes that of the anion, in accord with the procedure for naming simple salts.
2. In naming the complex ion proper, the names of the ligands precede that of the central metal atom.
3. The names of a number of ligands are listed in Tables 21.2 and 21.3. Note that negatively charged ligands have the suffix −o added to the stem name, e.g., chloro (Cl$^-$), hydroxo (OH$^-$), oxalato (C$_2$O$_4^{2-}$), and carbonato (CO$_3^{2-}$). Neutral ligands often have names that are the same as those of the non-bonded molecule, e.g., methylamine (CH$_3$NH$_2$) and ethylenediamine (NH$_2$CH$_2$CH$_2$NH$_2$). Some important exceptions to this rule are the names aqua (H$_2$O), ammine (NH$_3$), and carbonyl (CO).
4. When two or more different ligands are attached to the same central atom, their names are given in alphabetical order.
5. If more than one ligand of a given type is present in a single complex, the number is indicated by the Greek prefixes di, tri, tetra, penta, hexa, and so on. For example, the term hexaammine indicates the presence of six NH$_3$ ligands. In those instances where one of these prefixes is already incorporated in the name of the ligand, the prefixes bis- (two) and tris- (three) are used instead. For example, the term tris(ethylenediamine) indicates the presence of three ethylenediamine ligands.
6. The oxidation number of the central metal atom is indicated by a Roman numeral (or by a 0) following the name of the metal.
7. When a complex ion has a negative charge, the suffix -ate is added to the name of the central metal atom.

The following are some examples of the application of these rules:

[Pt(NH$_3$)$_5$Cl]Cl$_3$   Pentaamminechloroplatinum(IV) chloride

[Pt(NH$_3$)$_3$Cl$_3$]$^+$   Triamminetrichloroplatinum(IV) ion

[Co(en)$_3$]$_2$(SO$_4$)$_3$   Tris(ethylenediamine)cobalt(III) sulfate

[Pt(en)$_2$Cl$_2$]Cl$_2$   Dichlorobis(ethylenediamine)platinum(IV) chloride

Note that in all these instances the complex is a cation. In the following examples the complex is an anion:

K$_2$[CuCl$_4$]   Potassium tetrachlorocuprate(II)

[PtCl$_6$]$^{2-}$   Hexachloroplatinate(IV) ion

K$_4$[Fe(CN)$_6$]   Potassium hexacyanoferrate(II) (commonly called potassium ferrocyanide)

Finally, the complex is neutral in the following cases:

[Pt(NH$_3$)$_2$Cl$_4$]   Diamminetetrachloroplatinum(IV)

[Fe(CO)$_5$]   Pentacarbonyliron(0) (commonly called iron carbonyl)

These examples indicate that the oxidation state of the central metal atom in complexes tends to be fairly low, typically +2 or +3 for the elements of the first transition series, and up to +4 for those in subsequent series. When the central metal atom has a high oxidation state, its attraction for the electrons donated by the ligands is so strong as to pull them away completely. In other words, instead of forming a complex, the central atom is reduced. For example, $Fe^{3+}$ oxidizes $I^-$ to $I_2$ instead of forming a complex with this ion. Complexes in which the central metal atom is present in a high oxidation state are formed only with the most electronegative elements, fluorine and oxygen. The atoms of these elements exert a sufficiently strong attraction on their electrons to prevent electron loss. Some of the ions considered in Section 21.3 are examples of complexes in which the metal atom has a high oxidation state, e.g., $TiF_6^{2-}$, $CrO_4^{2-}$, and $MnO_4^-$.

## 21.5 Structure and Isomerism

*a. The Geometry of Metal Complexes*  Recall that the shapes of simple molecules can be related to the coordination numbers of their central atoms (Section 15.9). The same relationship holds true for metal complexes. The most commonly encountered complex structures are shown in Table 21.4. Transition elements in the +1 oxidation state, such as Ag(I), Cu(I), and Au(I), form linear complexes in which their coordination number is two. Fourfold coordination results in either a square-planar or a tetrahedral structure. Square-planar complexes are formed by Ni(II), Cu(II), and Pt(II), among others. Tetrahedral geometry is displayed by some of the complexes of Zn(II), Cd(II), and Hg(II). The most common coordination number of central metal atoms in complexes is six. These complexes are mostly octahedral in shape and are formed by, among other elements, Cr(III), Co(III), Pt(IV), and Mo(V). While other types of coordination, such as fivefold and eightfold, also are known, the number of complexes exhibiting these coordinations is relatively small.

**TABLE 21.4  Structure of Common Coordination Complexes**

| Coordination number | Shape | Examples |
|---|---|---|
| 2 | Linear | $[Ag(NH_3)_2]^+$, $[Cu(CN_2)]^-$ |
| 4 | Tetrahedral | $[Zn(NH_3)_4]^{2+}$, $[MnO_4]^-$, $[Cd(CN)_4]^{2-}$, $[FeCl_4]^-$, $[Ni(CO)_4]$ |
| 4 | Square planar | $[Ni(CN)_4]^{2-}$, $[Pt(NH_3)_4]^{2+}$, $[Cu(NH_3)_4]^{2+}$, $[Au(NH_3)_4]^{3+}$ |
| 6 | Octahedral | $[Co(NH_3)_6]^{3+}$, $[Pt(Cl_6)]^{2-}$, $[CoF_6]^{3-}$, $[MoF_6]^-$ |

**b. Structural Isomerism** Many coordination compounds exist in the form of **isomers**, i.e., substances with the same molecular formula but different arrangements of the atoms in the molecule. Coordination compounds exhibit several types of isomerism. **Structural isomers** have the same molecular formula but differ in the way the atoms are connected. For coordination compounds, differences in structure are often associated with the location of some particular atom or group in either the coordination sphere of the central metal atom or beyond it, as a counter ion.

Hydrated chromium(III) chloride, $CrCl_3 \cdot 6H_2O$, illustrates structural isomerism. Three different compounds with this formula are known. The most apparent differences among them are in their colors, both in the crystalline state and in solution. The structures of these compounds are evident from their structural formulas:

$[Cr(H_2O)_6]Cl_3$          violet

$[Cr(H_2O)_5Cl]Cl_2 \cdot H_2O$      blue-green

$[Cr(H_2O)_4Cl_2]Cl \cdot 2H_2O$      green

The violet isomer is a hexaaqua complex; all three $Cl^-$ ions are counter ions. In the other two isomers one or two $Cl^-$ ions have replaced water molecules as ligands, and the latter are attached to form crystal hydrates. The reactions of the three compounds with excess silver nitrate provide confirmatory evidence for these structures. The amount of silver chloride precipitated for each mole of compound decreases from three moles for the violet isomer, to two moles for the blue-green isomer, to just one mole for the green isomer. Evidently, the $Cl^-$ ions in the coordination sphere of the chromium(III) ion are attached too tightly to be removed by silver ions. However the $Cl^-$ counter ions readily precipitate as silver chloride.

---

**Example 21.3** The compound $Co(NH_3)_5BrSO_4$ forms two structural isomers. Write the formulas and names of the isomers. Suggest a chemical method to distinguish between them.

The given formula and the preceding discussion suggest that the compound in question is a sixfold coordinated complex of Co(III). The incorporation of one bromide and five ammonia ligands in the coordination sphere forms a complex ion with an ionic charge of +2, which therefore requires a sulfate counter ion. This complex is

    $[Co(NH_3)_5Br]SO_4$     Pentaamminebromocobalt(III) sulfate

The second structural isomer involves an exchange of bromide and sulfate and is

    $[Co(NH_3)_5SO_4]Br$     Pentaaminesulfatocobalt(III) bromide

The two isomers can be distinguished by the addition of barium nitrate to solutions of each compound. White barium sulfate precipitates from a solution of the first complex. No precipitate is formed from the second complex since barium bromide is soluble and the $SO_4^{2-}$ ligand is incorporated in the coordination sphere. ∎

*c. Stereoisomerism* Stereoisomers are molecules which differ only in the way their atoms are oriented in space, that is, in their **configuration.** In contrast to structural isomers, stereoisomers do *not* differ in the way different atoms are connected in a molecule. Nonetheless, stereoisomers do differ in chemical and physical properties.

To illustrate stereoisomerism, let us consider the possible structures of diamminedichloroplatinum(II), [Pt(NH$_3$)$_2$Cl$_2$]. Since the Pt atom is bonded to four ligands, it can form either a square-planar or a tetrahedral complex (Figure 21.9). Two stereoisomers are possible for the square-planar complex. The **cis isomer** has two like ligands in close proximity to each other, on the same side of the metal atom. In the **trans isomer**, by contrast, the two like ligands lie on opposite sides of the metal atom. Note that the Pt atom is bonded to two NH$_3$ and two Cl ligands in both cases—there are no structural differences. In contrast to the square-planar structure, isomerism in [Pt(NH$_3$)$_2$Cl$_2$] cannot occur if the structure is tetrahedral. All four ligands are at the corners of a tetrahedron and a simple rotation of the tetrahedron can convert one orientation of the molecule into another.

Two different isomers of [Pt(NH$_3$)$_2$Cl$_2$] are known. It follows that this complex is square-planar. This line of reasoning was first applied to the elucidation of the structure of a complex by Alfred Werner (1866–1919), a pioneer in coordination chemistry. The cis isomer of [Pt(NH$_3$)$_2$Cl$_2$] can be made by treating an aqueous solution of potassium tetrachloroplatinate(II) with ammonia,

$$[\text{PtCl}_4]^{2-} + 2\text{NH}_3 \rightarrow \begin{bmatrix} \text{Cl} & \text{Cl} \\ & \text{Pt} \\ \text{H}_3\text{N} & \text{NH}_3 \end{bmatrix} + 2\text{Cl}^-$$

while the trans isomer is formed in the reaction of tetraammineplatinum(II) chloride solution with HCl:

$$[\text{Pt}(\text{NH}_3)_4]^{2+} + 2\text{HCl} \rightarrow \begin{bmatrix} \text{H}_3\text{N} & \text{Cl} \\ & \text{Pt} \\ \text{Cl} & \text{NH}_3 \end{bmatrix} + 2\text{NH}_4^+$$

Although the two isomeric compounds have a similar creamy white color, they differ in dipole moment, solubility, and chemical properties.

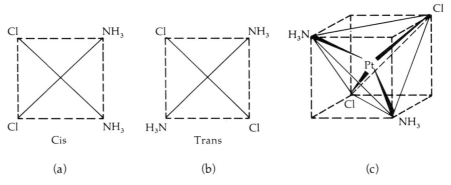

**FIGURE 21.9** The possible structures of [Pt(NH$_3$)$_2$Cl$_2$]. (a) Square-planar structure; cis and trans isomers are possible. (b) Tetrahedral structure; isomerism cannot occur.

### The Transition Elements and Their Coordination Compounds 21.5

The occurrence of stereoisomerism is particularly common for octahedral complexes of Pt(IV), Co(III), and Cr(III). Stereoisomerism is possible when the six ligands consist of at least two different species. Let us consider the general case of an octahedral complex in which a central metal atom M is surrounded by monodenate ligands A and B. The following three different complexes are possible: [MA$_5$B], [MA$_4$B$_2$], and [MA$_3$B$_3$]. (Three additional complexes, in which the number of ligands of type A and B is interchanged, also are possible. However these complexes introduce no additional structural features. For example, [MAB$_5$] has a similar configuration as [MA$_5$B]. ) Figure 21.10 shows the various possible configurations of these complexes. Only a single compound with the formula [MA$_5$B] exists. Two stereoisomers of [MA$_4$B$_2$] are possible. In the cis isomer, the two B ligands occupy adjoining corners of the octahedron while in the trans isomer, they occupy opposite corners. Two isomers of [MA$_3$B$_3$] are possible. The **facial** isomer has all three ligands of a given type in a cis configuration with respect to each other. The **meridional** isomer has one pair of ligands of a given type arranged in the trans configuration. All these types of stereoisomers also are known as **geometrical isomers.**

The complexes of cobalt(III) with ammonia and chlorine are typical examples of this general case. A single complex ion with the formula [Co(NH$_3$)$_5$Cl]$^{2+}$ exists. Its chloride salt has a dark red-violet color. Two complex ions with the formula [Co(NH$_3$)$_4$Cl$_2$]$^+$ exist. The chloride of the cis isomer has a beautiful blue-violet color while the trans isomer is bright green. The neutral complex [Co(NH$_3$)$_3$Cl$_3$] also exists in two forms, corresponding to the facial and meridional isomers.

*d. Optical Stereoisomerism* When two stereoisomers are non-superimposable mirror images of each other, in analogy to the left and right hand (Figure

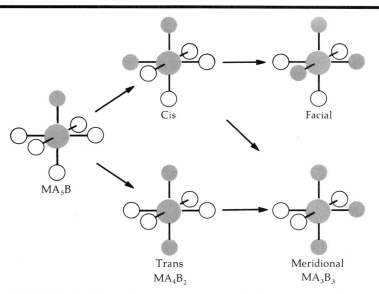

**FIGURE 21.10** Stereoisomerism in octahedral complexes with two different ligands A and B. Starting with [MA$_5$B], the various isomers can be generated by replacing an atom of type A by one of type B, as indicated by the arrows.

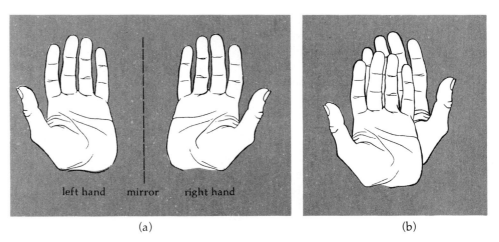

**FIGURE 21.11** An analogy to optical isomerism. (a) The left and right hands are mirror images of each other, i.e., when viewed in a mirror, the left hand looks like a right hand. (b) The two hands are not superimposable, meaning that they are not congruent when laid on top of each other.

21.11), they are said to be **enantiomers.** Isomers of this type respond in different ways to polarized light (Section 22.3). Since this is an optical phenomenon, substances of this type are known as **optical isomers.** In contrast to the geometrical types of stereoisomers considered in Section 21.5c, enantiomers have the same dipole moment, solubility, melting point, and other properties. They differ only in their response to polarized light and in their reactions with other optical isomers.

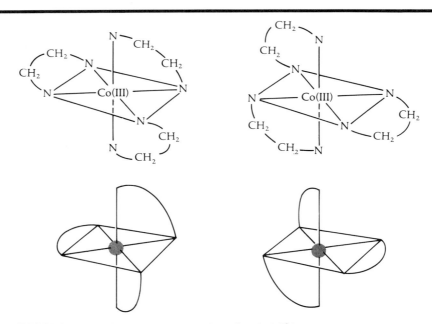

**FIGURE 21.12** The enantiomers of the $[Co(en)_3]^{3+}$ complex ion. A schematic drawing is shown below each structure.

Optical isomerism occurs when a molecule is asymmetric—it lacks a plane or center of symmetry. It is a common occurrence in complexes containing chelate rings. For example, the tris(ethylenediamine)cobalt(III) ion, [Co(en)$_3$]$^{3+}$ forms a pair of enantiomers (Figure 21.12). While the figure shows that the molecule cannot be divided into equal halves, it is somewhat difficult to see on the basis of a drawing that the two structures cannot be superimposed by a suitable rotation. Ths use of three-dimensional models easily shows that this is indeed the case. Optical isomerism is of particular importance in organic chemistry (Chapter 22).

## Example 21.4

Draw the structures of all the isomers of the diamminediaquadichlorocobalt(III) ion, [Co(NH$_3$)$_2$(H$_2$O)$_2$Cl$_2$]$^+$. Identify the enantiomers.

Let us proceed in a systematic manner in order to ensure that all possible structures will be included (and included only once). Since the complex ion has three pairs of identical ligands it is apparent that any two identical ligands must be either cis or trans to each other. Examination of the various combinations of cis and trans pairs provides a scheme that exhausts all the possibilities. The first structure has each ligand trans to one like it:

Note that the same ligands are found on opposite sides of lines drawn through the cobalt atom, indicating that the complex has a center of symmetry at the location of the Co atom. Optical isomerism is consequently not possible for this structure.

Three different stereoisomers can be drawn in which one pair of identical ligands is in the trans configuration while the other pairs are in the cis configuration:

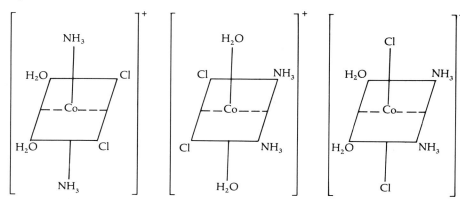

Since each structure is symmetric about a plane drawn through the dashed line and the vertical axis, optical isomerism cannot occur. The final structure is one in which each ligand is cis to one like it. Two enantiomers satisfy this condition:

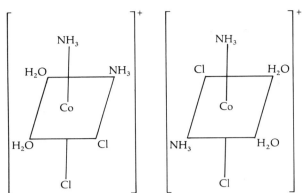

Note that the structures are identical except for the reversal of the Cl and NH₃ ligands. This reversal shows that the two isomers are mirror images and therefore are enantiomers.

The only case that remains to be considered is one in which one ligand is cis to one like it, while the remaining ones are trans to their twin counterparts. An attempt to draw such a structure quickly indicates that this cannot be done. The complex ion in question therefore has six stereoisomers two of which are enantiomers. ■

## 21.6 Color and Magnetism of Complexes

Coordination compounds are frequently colored, and their solutions often have particularly intense and vivid colors. The color is a property of the complex as a whole rather than just of the metal ion. This point is well illustrated by the color of copper(II) in various solvents. Concentrated sulfuric acid is a strong dehydrating agent and copper(II) is present as $Cu^{2+}$ in this medium. The solution has no discernible color. The color of coordinated copper(II) becomes progressively more visible in aqueous solutions, which are light blue due to the presence of $[Cu(H_2O)_4]^{2+}$, and in a concentrated ammonia solution, which has an intense deep blue color associated with the $[Cu(NH_3)_4]^{2+}$ complex.

The occurrence of color is a manifestation of the absorption of electromagnetic radiation in the visible region of the spectrum. Recall from Section 17.4 that the energy of electronic transitions in molecules corresponds to the absorption of radiation in the ultraviolet or visible. Bohr's equation,

$$\Delta E = h\nu = hc/\lambda$$

indicates that the absorption of visible light, which has wavelengths long compared to those of UV radiation, corresponds to relatively low-energy electronic transitions. As discussed in the next section, coordination complexes absorb in the visible because of electronic transitions between close lying d orbitals.

The relationship between the electronic transition energy and the observed color is illustrated schematically in Figure 21.13. When white light is in-

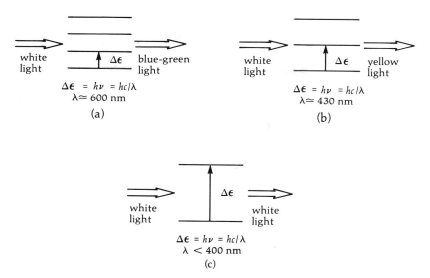

**FIGURE 21.13** The origin of color in coordination complexes and other substances. (a) The spacing of the energy levels corresponds to the absorption of 600 nm radiation (orange light); the transmitted light is blue-green. (b) The spacing of the energy levels corresponds to the absorption of 430 nm radiation (indigo light); the transmitted light is yellow. (c) The spacing of the energy levels corresponds to the absorption of UV radiation; visible light is not absorbed and the complex has no color.

cident upon a substance, the wavelength corresponding to the electronic transition energy is absorbed and the rest of the light is transmitted. The color of a substance is determined by the wavelengths of the light it transmits and is **complementary** to the color of the absorbed light, that is, it is the color that remains in the spectrum after the absorbed wavelengths have been removed. As illustrated in the figure, when a substance absorbs red light, the transmitted radiation has a blue-green color. The observed color gradually shifts towards the yellow as the absorbed wavelengths move towards the violet. If the electronic transition is sufficiently energetic to require the absorption of UV radiation, visible light cannot interact and the substance in question is colorless or white. Table 21.5 summarizes the relationship between the wavelength of the absorbed radiation and the resulting color. It is apparent that the color of a substance provides an approximate indication of the energy of the absorbed radiation and is a guide to its electronic spectroscopy.

A second property of coordination complexes that gives a clue to their electronic structure is their response to a magnetic field. Recall from Section 15.2b that substances with unpaired electrons are attracted by a magnetic field and are said to be paramagnetic. This behavior contrasts with that of diamagnetic substances, which do not possess unpaired electrons and are weakly repelled by a magnetic field. Transition metal complexes display both types of magnetic behavior. For example, $[Co(NO_2)_6]^{3-}$ is diamagnetic while $[CoF_6]^{3-}$ is paramagnetic. The number of unpaired electrons in paramagnetic complexes can be inferred from the strength of the interaction with the magnetic field. The $[CoF_6]^{3-}$ ion, for instance, has four unpaired electrons.

## TABLE 21.5 Relation Between Electronic Transition Energy and Color

| Transition energy (kJ/mol) | Wave number[a] (cm$^{-1}$) | Wavelength (nm) | Spectral color absorbed | Complementary color | Example (Co complex) |
|---|---|---|---|---|---|
| <165 | <13,900 | >720 | Infrared | Colorless | |
| 166 | 13,900 | 720 | Red | Green | trans-[Co(en)$_2$Br$_2$]$^+$ |
| 176 | 14,700 | 680 | Red | Blue-green | trans-[Co(NH$_3$)$_4$Cl$_2$]$^+$ |
| 196 | 16,400 | 610 | Orange | Blue | trans-[Co(en)$_2$Br(NCS)]$^+$ |
| 206 | 17,200 | 580 | Yellow | Indigo | |
| 214 | 17,900 | 560 | Yellow-green | Violet | [Co(EDTA)]$^-$ |
| 226 | 18,900 | 530 | Green | Purple | [Co(NH$_3$)$_5$Cl]$^{2+}$ |
| 239 | 20,000 | 500 | Blue-green | Red | [Co(NH$_3$)$_5$H$_2$O]$^{3+}$ |
| 249 | 20,800 | 480 | Blue | Orange | [Co(NH$_3$)$_5$NCS]$^{2+}$ |
| 279 | 23,300 | 430 | Indigo | Yellow | [Co(NH$_3$)$_6$]$^{3+}$ |
| 292 | 24,400 | 410 | Violet | Lemon yellow | |
| >299 | >25,000 | <400 | Ultraviolet | Colorless | |

[a]Spectroscopic transitions are commonly measured in wave numbers ($\bar{\nu}$), where $\bar{\nu} = \lambda^{-1}$. Bohr's relation in terms of $\bar{\nu}$ is $\Delta E = hc\bar{\nu}$.

## 21.7 Crystal Field Theory

*a. Bonding in Coordination Complexes* A theory of chemical bonding in coordination complexes must be able to account for their electronic structure as revealed, for example, by absorption spectra, for their magnetic behavior, and for other properties as well. The first theoretical approach used to examine bonding in these compounds was the valence bond theory. This theory is an extension of that examined for simple molecules in Chapter 16. According to this theory, the formation of a complex is viewed as a Lewis acid–base reaction in which the ligands act as Lewis bases and form coordinate covalent bonds with the central metal ion, which acts as a Lewis acid. The electron pairs donated by the ligands fill vacant orbitals formed by hybridization of d, s, and p orbitals of the metal atom. The occurrence of octahedral, tetrahedral, and square-planar complexes is readily understandable in terms of the hybridization scheme given in Table 16.1 and applied to complexes in Table 21.6. While these general aspects of valence bond theory provide some useful insights into the nature of bonding in coordination compounds, the theory cannot reliably predict such detailed features as the number of unpaired electrons. Furthermore, the theory cannot account for one of the most striking properties of coordination com-

## TABLE 21.6 Hybridization in Transition Metal Complexes

| Ion | Number of $e^-$ in 3d orbitals | Complex | Number of $e^-$ donated by ligands | Number of $e^-$ in orbitals 3d | 4s | 4p | Hybridization[a] | Geometry |
|---|---|---|---|---|---|---|---|---|
| Cr$^{3+}$ | 3 | [Cr(NH$_3$)$_6$]$^{3+}$ | 12 | 7 | 2 | 6 | d$^2$sp$^3$ | Octahedral |
| Fe$^{2+}$ | 6 | [Fe(CN)$_6$]$^{4-}$ | 12 | 10 | 2 | 6 | d$^2$sp$^3$ | Octahedral |
| Ni$^{2+}$ | 8 | [Ni(CN)$_4$]$^{2-}$ | 8 | 10 | 2 | 4 | dsp$^2$ | Square-planar |
| Zn$^{2+}$ | 10 | [ZnCl$_4$]$^{2-}$ | 8 | 10 | 2 | 6 | sp$^3$ | Tetrahedral |

[a]The hybrid orbitals are formed by the vacant metal orbitals filled by electrons from the ligands. For example, [Ni(CN)$_4$]$^{2-}$ has dsp$^2$ hybridization because the ligands contribute 2$e^-$ to a 3d orbital, 2$e^-$ to a 4s orbital, and 4$e^-$ to two 4p orbitals.

pounds, namely, their vivid colors. Valence bond theory is therefore no longer widely used in coordination chemistry.

The most complete approach to bonding in coordination complexes is molecular orbital theory. As in the case of simple molecules, molecular orbitals are formed by the overlap of atomic orbitals, in this case those of the central atom and the ligand atoms. However, when the molecules become as complex as those of typical coordination compounds, molecular orbital theory becomes difficult to understand in simple terms. Fortunately, there is another approach to bonding in coordination complexes which is conceptually simple and matches the predictions of molecular orbital theory in most respects. This approach is the crystal field theory, so called because it was first applied to ions in crystals. This theory accounts for both the colors and the magnetic properties of coordination compounds, as we shall see.

*b. Octahedral Crystal Field Splitting* **Crystal field theory** focuses on the electrostatic interaction between the central metal ion and the ligands. This interaction involves the attraction between the positively charged ion and either negatively charged ligands or the negative ends of polar ligands such as $NH_3$, and the repulsion between the valence electrons of the metal ion and the ligands. The electrical field resulting from this interaction is called the crystal field. This field affects the energies of the valence orbitals of the central metal atom. Since the orbitals involved in bonding to the ligands are the d orbitals, we must focus on the effect of the crystal field on the energies of the d orbitals. This requires that we review the spatial properties of these orbitals.

Figure 21.14a shows the orientation of the five d orbitals. Note that the orbitals designated $d_{x^2-y^2}$ and $d_{z^2}$ point along the rectilinear axes, indicating that the electron density is highest in that region. The other three orbitals, designated $d_{xy}$, $d_{yz}$, and $d_{xz}$, extend midway between the axes; the electron density along the axes is relatively low.

Recall from Section 6.2 that the energies of the d orbitals of an isolated metal atom or ion are equal; the orbitals are degenerate. Consider what happens in the formation of an octahedral complex. The directions from which the six ligands approach the metal ion may, for convenience, be taken along the axes of the coordinate system, that is, from $x$, $-x$, $y$, $-y$, $z$, and $-z$ (Figure 21.14b). The potential energy of the system increases owing to the repulsion between the negatively charged ligands and the electrons in the d orbitals of the metal ion. However, and this is the key point, the effect is *not* the same on all the d orbitals. The $d_{x^2-y^2}$ and $d_{z^2}$ orbitals point at the approaching ligands while the $d_{xy}$, $d_{yz}$, and $d_{xz}$ orbitals do not. Consequently, the repulsive energy resulting from the interaction of the ligands with the $d_{x^2-y^2}$ and $d_{z^2}$ orbitals is greater than that resulting from the interaction with the other three orbitals, and a splitting of the orbital energies occurs. One group of orbitals, known as the $t_{2g}$ triplet, and consisting of the $d_{xy}$, $d_{yz}$, and $d_{xz}$ orbitals, lies at a lower energy than the other group, consisting of the $d_{x^2-y^2}$ and $d_{z^2}$ orbitals, and known as the $e_g$ doublet.

The effect of the crystal field on the energies of the d orbitals is shown in Figure 21.15. The field increases the average energy of the orbitals and gives rise to the splitting. The overall effect can be depicted as the sum of two separate effects: first, the average energies of the d orbitals increase, as if the crystal field around the metal ion were spherically symmetric, i.e., independent of di-

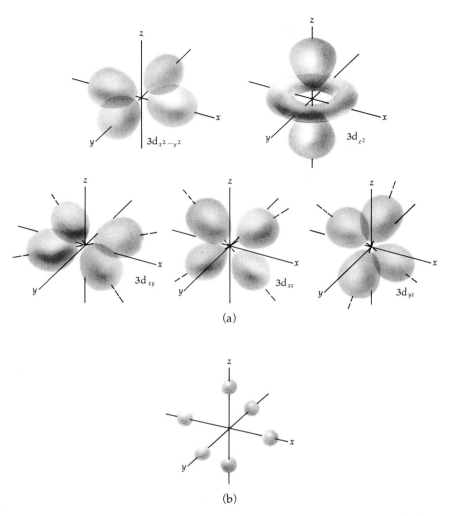

**FIGURE 21.14** Interaction between d orbitals and ligands in an octahedral complex. (a) Shape of the d orbitals relative to a rectilinear coordinate system. (b) Direction of approach of the six ligands.

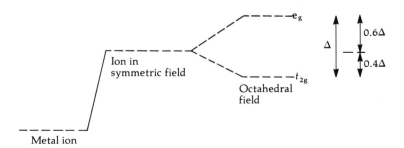

**FIGURE 21.15** Effect of an octahedral crystal field on the d atomic orbitals of the central metal atom. The orbitals lie at a higher average energy and are split. The crystal field stabilization energy, CFSE, is equal to $-0.4\Delta$ when a single electron occupies a $t_{2g}$ orbital.

rection. Second, the two sets of orbitals are split by an energy $\Delta$, called the **crystal field splitting.** The energy of a complex relative to that of the ion in a spherically symmetric field depends on the number of electrons in the $t_{2g}$ and $e_g$ orbitals and on the magnitude of the crystal field splitting. In most cases, the energy of the complex is lower than that of the ion in a spherically symmetric field and the reduction in energy is called the **crystal field stabilization energy,** CFSE. This energy has the value

$$\text{CFSE} = [N(e_g)](0.6\Delta) - [N(t_{2g})](0.4\Delta) \tag{21.1}$$

where $N(e_g)$ and $N(t_{2g})$ are the numbers of electrons in the $e_g$ and $t_{2g}$ orbitals, respectively, $0.6\Delta$ is the upward shift in the energy of the $e_g$ orbitals, and $0.4\Delta$ is the downward shift in the energy of the $t_{2g}$ orbitals (see Figure 21.15). For example, a complex with a single electron in a $t_{2g}$ orbital has CFSE = $-0.4\Delta$. The negative value of CFSE indicates that the complex is more stable than the ion in a spherically symmetric field. When both $e_g$ and $t_{2g}$ orbitals are completely filled, i.e., for ions with a $d^{10}$ configuration,

$$\text{CFSE} = (4)(0.6\Delta) - (6)0.4\Delta = 0$$

and no stabilization occurs. Thus, the CFSE provides a quantitative measure of the stability of a complex.

The energy level diagram in Figure 21.15 can be used to predict the d electron configuration of octahedral complexes. A particular metal ion has a certain number of d electrons (Figure 21.1). These electrons are placed in the $t_{2g}$ and $e_g$ orbitals according to the Pauli exclusion principle and Hund's rule, much in the same way as electrons in atoms fill atomic orbitals (Section 6.3). For example, $Ti^{3+}$ has one d electron, which must occupy one of the degenerate $t_{2g}$ orbitals. Similarly, $Cr^{3+}$ has three d electrons, each of which occupies one of the degenerate $t_{2g}$ orbitals, giving a $t_{2g}^3$ configuration.

The electronic configuration of a complex becomes more difficult to determine once the metal ion has more than three d electrons. To illustrate, consider a complex of $Fe^{3+}$, with five d electrons. As indicated in Figure 21.16, the electronic configuration can be either $t_{2g}^5$, in which the fourth and fifth electron fill two of the $t_{2g}$ orbitals, or $t_{2g}^3 e_g^2$, in which these electrons occupy the higher-energy $e_g$ orbitals. Clearly, it costs energy to occupy an $e_g$ instead of a $t_{2g}$ orbital. However, it also costs energy to add an electron to an orbital which already contains an electron. Since the two electrons are spin paired, this energy is called the **pairing energy** $P$.

The orbital filling scheme is determined by the relative magnitudes of the crystal field splitting and the pairing energy. When $\Delta < P$, the energy required to add an electron to a vacant $e_g$ orbital is lower than that required to form an electron pair in a $t_{2g}$ orbital. This case is called the **weak-field limit** since the crystal field splitting is small. Observe in Figure 21.16a that all the electrons in the weak field configuration have parallel spin so that the complex has high spin. On the other hand, when $\Delta > P$, it is energetically favorable for the $t_{2g}$ orbitals to be filled before any electrons occupy the $e_g$ orbitals (Figure 21.16b). This case is called the **strong-field limit** since the crystal field splitting is large. Note that this configuration gives rise to a low net electron spin, and strong-field complexes have low spin. The magnitude of the spin can be determined experimentally by means of magnetic measurements. Thus, the electronic configuration and the strength of the crystal field can be inferred.

**812** The Transition Elements and Their Coordination Compounds 21.7

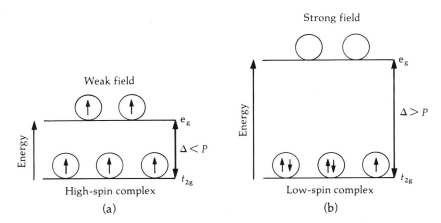

**FIGURE 21.16** Effect of the crystal field splitting and the electron pairing energy on the configuration of an octahedral complex with five d electrons. (a) A weak ligand field ($\Delta < P$) leads to a high-spin complex. (b) A strong ligand field ($\Delta > P$) leads to a low-spin complex.

---

**Example 21.5**  The number of unpaired electrons in the following octahedral complexes is determined by their response to a magnetic field as follows: (a) $[Mn(CN)_6]^{4-}$, 1 unpaired $e^-$; and (b) $[CoF_6]^{3-}$, 4 unpaired $e^-$. For each complex determine whether weak-field or strong-field splitting occurs, determine the d electron configuration, and calculate the magnitude of CFSE in terms of $\Delta$.

(a)  Since the $CN^-$ ligand has an ionic charge of $-1$, manganese is present in $[Mn(CN)_6]^{4-}$ as the dipositive ion. From Figure 21.1, we learn that $Mn^{2+}$ has five d electrons. Since only one of these d electrons is unpaired, all five must occupy the $t_{2g}$ orbitals, four being spin-paired in two orbitals and the fifth occupying the third orbital. The electron configuration is $t_{2g}^5$, corresponding to strong-field splitting. The CFSE can be evaluated by means of Equation 21.1,

$$\text{CFSE} = [N(e_g)](0.6\Delta) - [N(t_{2g})](0.4\Delta)$$
$$= -(5)(0.4\Delta)$$
$$= -2\Delta$$

(We neglect the small reduction in CFSE resulting from the energy required to form two electron pairs.)

(b)  The ionic charge of cobalt in $[CoF_6]^{3-}$ is $+3$ and $Co^{3+}$ has six d electrons, four of which are unpaired. If all the 6 electrons occupied the $t_{2g}$ orbitals, these orbitals would be completely filled and there would be no unpaired electrons. Consequently, the first 3 electrons must occupy the $t_{2g}$ orbitals, the next two electrons must occupy the $e_g$ orbitals, and the sixth electron fills one of the $t_{2g}$ orbitals. The electronic configuration is $t_{2g}^4 e_g^2$ corresponding to weak field splitting. The value of CFSE may be obtained by means of Equation 21.1:

$$\text{CFSE} = [N(e_g)](0.6\Delta) - [N(t_{2g})](0.4\Delta)$$

$$= (2)(0.6\Delta) - (4)(0.4\Delta)$$
$$= -0.4\Delta \quad \blacksquare$$

Figure 21.17, which is a more complete version of Figure 21.16, shows the orbital occupation for weak- and strong-field octahedral complexes as a function of the total number of d electrons. Note that when the ions have less than three or more than seven d electrons, the weak- and strong-field complexes have the same electronic configurations.

***c. Absorption Spectra and the Crystal Field***  The absorption of light by a solution of a coordination compound frequently results in the excitation of electrons from the $t_{2g}$ to the $e_g$ orbitals. This makes it possible to derive values of the crystal field splitting from the results of absorption spectroscopy experiments. For example, solutions of $[Ti(H_2O)_6]^{3+}$ absorb visible light, the strongest absorption occurring at about 20,000 cm$^{-1}$, as shown by the spectrum displayed in Figure 21.18a. This absorption band can be ascribed to the $t_{2g} \rightarrow e_g$ transition shown in Figure 21.18b. The value of $\Delta$ is readily obtained by means

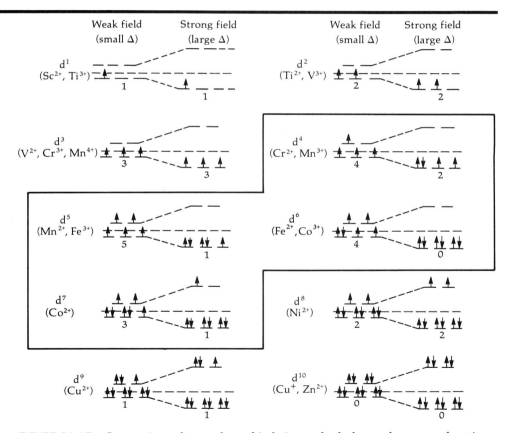

**FIGURE 21.17** Occupation of $t_{2g}$ and $e_g$ orbitals in octahedral complexes as a function of the number of d electrons for the weak-field and strong-field limits. The complexes with 4–7 d electrons (lying within the closed box) have different configurations for the weak and strong field limits.

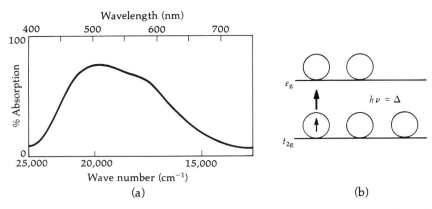

**FIGURE 21.18** Relationship between absorption spectra of solutions of metal complexes and the magnitude of the crystal field splitting. (a) The absorption spectrum of $[Ti(H_2O)_6]^{3+}$ in aqueous solution. (b) Electronic transition giving rise to the observed absorption. Note that $Ti^{3+}$ has one d electron.

of Bohr's formula from the wave number at which the absorption is most intense.

**Example 21.6** What are the values of the crystal field splitting and the CFSE in $[Ti(H_2O)_6]^{3+}$ given that a solution of this complex absorbs at $2.00 \times 10^4$ cm$^{-1}$? What is the color of the solution?

Since the crystal field splitting can be identified with the energy of the absorbed photon, we have

$$\Delta = h\nu = hc\bar{\nu}$$

$$\Delta = (6.626 \times 10^{-34} \text{ J s})(2.998 \times 10^8 \text{ m s}^{-1})(2.00 \times 10^4 \text{ cm}^{-1})$$
$$\times (10^2 \text{ cm m}^{-1})(6.022 \times 10^{23} \text{ mol}^{-1})(10^{-3} \text{ kJ J}^{-1})$$

$$= 239 \text{ kJ mol}^{-1}$$

The value of the CFSE is obtained by means of Equation 21.1,

$$\text{CFSE} = [N(e_g)](0.6\Delta) - [N(t_{2g})](0.4\Delta)$$

Figure 21.18b shows that the complex has a single electron in a $t_{2g}$ orbital. Therefore,

$$\text{CFSE} = -(1)(0.4\Delta) = -(0.4)(239 \text{ kJ mol}^{-1}) = -95.6 \text{ kJ mol}^{-1}$$

The color of the complex is complementary to that of the absorbed radiation. The color of the radiation, as determined by its wavelength,

$$\lambda = 1/\bar{\nu} = (1/2.00 \times 10^4 \text{ cm}^{-1})(10^7 \text{ nm cm}^{-1}) = 500 \text{ nm}$$

is blue-green. According to Table 21.5, the color of the solution must be red. ■

Table 21.7 summarizes the values of $\Delta$ obtained in the manner illustrated in Example 21.6 for a number of octahedral complexes. Typical values range from approximately 80 to 500 kJ mol$^{-1}$. Since the interaction giving rise to crystal field splitting is assumed to be electrostatic, it should be sensitive to the factors determining the magnitude of this interaction. Thus, for a given ion, $\Delta$ should increase with the oxidation state of the metal ion. The more highly charged ion draws the ligands in more closely and thereby increases their interaction with the d orbitals. The results listed for $Fe^{2+}$ and $Fe^{3+}$ confirm the occurrence of this effect.

For a given metal ion, the values of $\Delta$ should, for the same reason, increase with the charge density of the ligands. The results listed for halide ligands are consistent with this expectation, the values of $\Delta$ varying inversely with the size of the halide ion. However, the results for molecular ligands such as $H_2O$ and $NH_3$ yield larger values of $\Delta$ than those for the halide ions even though their dipole moments indicate a small separation of charge. This inconsistency is an indication that the interaction between the metal ion and the ligands cannot be strictly electrostatic and constitutes one of the weaknesses of crystal field theory. Nonetheless, it is possible to arrange the ligands in order of increasing crystal field strength in a way that is essentially independent of the identity of the metal ion. This empirically determined sequence is called the **spectrochemical series.** The following is a partial sequence of ligands:

$$Br^- < SCN^- < Cl^- < F^- < OH^- < H_2O < NH_3 < en < NO_2^- < CN^- < CO$$

weak-field ligands  $\longrightarrow$  strong-field ligands

The magnitude of the crystal field splitting increases by over a factor of two across the series.

Table 21.7 also shows that, for a given ligand, the values of $\Delta$ increase with atomic number within a particular family of elements. For example, the crystal field splitting in $[Ir(NH_3)_6]^{3+}$ is larger than that in $[Rh(NH_3)_6]^{3+}$, which in turn is larger than that in $[Co(NH_3)_6]^{3+}$. This trend is found to be generally valid and its consequence is that complexes of the second and third transition series almost always have low spin (strong field limit). Both low-spin and high-spin complexes are found in the first transition series.

The ability of crystal field theory to correlate the electronic structure, magnetic properties, and colors of transition metal complexes makes it a very useful theory. In addition to relating the color of a complex to the magnitude of the crystal field splitting, the theory also provides an explanation of why certain complexes are either colorless or only weakly colored. The lack of color in complexes can indicate that the $t_{2g} \rightarrow e_g$ transition cannot occur. This is the case

**TABLE 21.7  Crystal Field Splitting $\Delta$ of Octahedral Complexes**[a]

| Metal ion | Ligand | | | | | | |
|---|---|---|---|---|---|---|---|
| | $Br^-$ | $Cl^-$ | $F^-$ | $H_2O$ | $NH_3$ | en | $CN^-$ |
| $Cr^{3+}$ | — | 158 | 182 | 208 | 274 | 262 | 314 |
| $Fe^{2+}$ | — | — | — | 124 | — | — | 393 |
| $Fe^{3+}$ | — | 132 | — | 171 | — | — | 419 |
| $Co^{3+}$ | — | — | — | 218 | 274 | 278 | 401 |
| $Ni^{2+}$ | 84 | 86 | — | 102 | 129 | 138 | — |
| $Rh^{3+}$ | 227 | 243 | — | 323 | 408 | 414 | 544 |
| $Ir^{3+}$ | — | 299 | — | — | 490 | 495 | — |

[a] Values given in kilojoules per mole.

when the metal ion has either an empty or a filled d subshell. Thus, complexes formed by the elements at either end of the transition series—those of the scandium and zinc families—tend to be colorless.

A solution of a complex will be only weakly colored—even when the solution is concentrated—when the probability that the complex will absorb radiation is small. Recall from Section 17.4b that an electronic transition in which an electron undergoes a change of spin is a highly unfavored transition. Consequently, complexes for which a $t_{2g} \rightarrow e_g$ transition involves a spin change absorb radiation only weakly and are therefore only lightly colored. For example, the hexaaqua complexes of manganese(II) and iron(III), as well as other high-spin complexes of these ions are only faintly colored. The reason can be understood with the help of Figure 21.19, which shows the change in electronic configuration in a $t_{2g} \rightarrow e_g$ transition. Since both $Mn^{2+}$ and $Fe^{3+}$ have a $d^5$ configuration, all the $t_{2g}$ and $e_g$ orbitals are half-filled in a high-spin complex. A transition necessarily requires a change of spin. On the other hand, a low-spin complex of these ions (e.g., $Fe(CN)_6^{3-}$) can absorb radiation with no change in electron spin, and solutions of potassium hexacyanoferrate(III) have an intense red color.

The crystal field splitting of the d orbitals of transition elements also affects a number of other properties of these elements. The radii of the metal ions in complexes show such an effect (Figure 21.20). The overall trend across the transition series is one of decreasing radius, which is attributable to the increase in effective nuclear charge. For trivalent ions in low-spin complexes, this trend is rather uniform between $Sc^{3+}$, with no d electrons, and $Co^{3+}$ with 6d electrons and a $t_{2g}^6$ configuration. The next ion in sequence, $Ni^{3+}$, must have a $t_{2g}^6 e_g^1$ configuration. Since the $e_g$ orbital points at the ligands, it repels them. This repulsion leads to an increase in the radius ascribed to the ion.

For high-spin complexes the increase in ionic radius should occur earlier in the series since the first ion with an electron in an $e_g$ orbital has the $t_{2g}^3 e_g^1$ configuration. In the first transition series, this ion is $Mn^{3+}$, and, as shown in Figure 21.20, its radius is larger than that of the $Cr^{3+}$ ion in complexes.

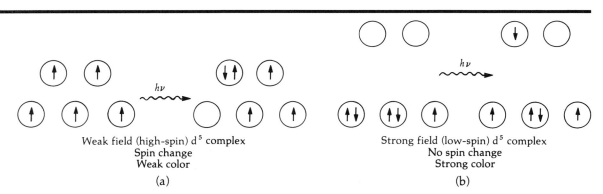

**FIGURE 21.19** Spin change in $t_{2g} \rightarrow e_g$ transitions in $d^5$ complexes, e.g., $Mn^{2+}$ or $Fe^{3+}$. (a) A transition in a high-spin complex is accompanied by a change in electron spin and is unfavored. (b) A transition in a low-spin complex does not require a change in electron spin and is favored. The intensity of the color provides a measure of the transition probability.

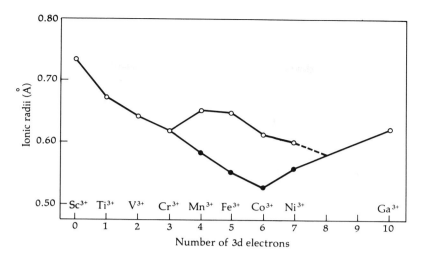

**FIGURE 21.20** Ionic radii of trivalent ions in the first transition series obtained from data for octahedral complexes. For ions with different electronic configurations for low-spin and high-spin complexes, the radii derived from low-spin complexes are shown by the solid circles. Since copper and zinc do not form trivalent ions, no points are shown for these elements.

***

***d. Tetrahedral and Square-Planar Complexes*** Crystal field theory can also be applied to tetrahedral and square-planar complexes. Figure 21.21a shows the structure of a tetrahedral complex relative to the axes of a rectilinear coordinate system. Note that the ligands are placed at alternate corners of a cube whose center defines the location of the central metal ion. A look at Figure 21.14a, which shows the orientation of the d orbitals relative to these axes, indicates that the $d_{xy}$, $d_{xz}$, and $d_{yz}$ orbitals point at the edges of the cube while the $d_{x^2-y^2}$ and $d_{z^2}$ orbitals point toward the centers of its faces. Observe that the distance from a ligand to the center of a face of the cube is greater than that to the center of an edge. Consequently, a tetrahedral crystal field splits the d orbitals so that the $d_{x^2-y^2}$ and $d_{z^2}$ orbitals ($e_g$ orbitals) lie at a lower energy than the other

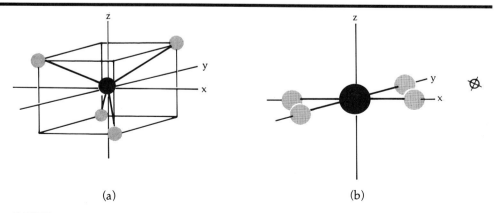

**FIGURE 21.21** Ligand geometries in (a) tetrahedral and (b) square-planar geometries. The metal ion is at the center of the coordinate system.

three d orbitals ($t_{2g}$ orbitals). Figure 21.22 shows a comparison of the effect on the d orbital energies of tetrahedral and octahedral fields. Two features are evident: first, the $t_{2g}$ and $e_g$ orbitals are reversed in energy for the two geometries. Second, the tetrahedral splitting is only about half as large as the octahedral splitting. The reason for this difference is that the number of tetrahedral ligands is smaller (4 versus 6), and these ligands do not approach the metal ion directly along the direction of any of the orbitals.

The square-planar geometry is shown in Figure 21.21b. The crystal field splitting is somewhat more complex than for octahedral and tetrahedral geometries because the d orbitals exhibit four different relationships to the ligands. The $d_{x^2-y^2}$ orbital points directly at the ligands and so this orbital has the highest energy. The $d_{xy}$ orbital lies at the next highest energy because, although it points between the ligands, it is coplanar with them. The $d_{z^2}$ orbital comes next; although its two lobes point out of the plane of the complex, the belt around the center of the orbital, containing about a third of the electron density, lies in the plane. Finally, the $d_{xz}$ and $d_{yz}$ orbitals, which are degenerate, lie at the lowest energy because they point out of the plane of the complex. Figure 21.22c shows the pattern of the orbital energies.

In general, octahedral complexes are more stable than tetrahedral or square-planar complexes because they form a larger number of metal ion–ligand bonds. Nonetheless, the formation of tetrahedral or square-planar complexes is favored in specific cases.

The tetrahedral geometry affords the greatest separation between ligands and consequently minimizes the effects of ligand–ligand repulsion. These effects can be of importance for bulky ligands or for highly charged ligands. Furthermore, owing to the small value of the crystal field splitting, tetrahedral complexes invariably have high spin, that is, all the d orbitals are half-filled be-

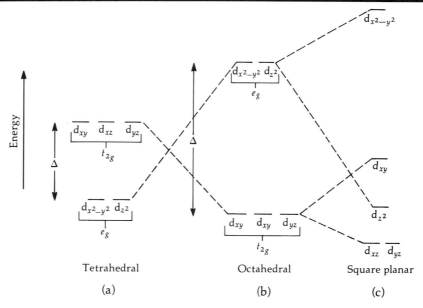

**FIGURE 21.22** The splitting of the d orbitals by ligands in (a) tetrahedral, (b) octahedral, and (c) square-planar geometries.

fore any pairing occurs. Therefore, whenever crystal field splitting stabilizes the low-spin complex, tetrahedral complex formation is energetically unfavored and generally does not occur. This is the case for ions with the $d^3$, $d^6$, and $d^8$ configurations, such as $Cr^{3+}$, $Co^{3+}$, and $Ni^{2+}$, respectively, and these ions generally do not form tetrahedral complexes.

Tetrahedral complexes are most likely when crystal field stabilization cannot occur, either because the d orbitals are completely empty or completely filled. Thus, ions with a $d^0$ or a $d^{10}$ configuration form tetrahedral complexes, some examples being $[CrO_4]^{2-}$, $[MnO_4]^-$, and $[ZnCl_4]^{2-}$. Furthermore, when crystal field stabilization is small, e.g., for ions with a $d^5$ configuration, tetrahedral complexes are formed with ligands that are charged and at least moderately bulky, such as $Cl^-$. Two examples are $[MnCl_4]^{2-}$ and $[FeCl_4]^-$.

The conditions that make square-planar complexes stable can be inferred from Figure 21.22c. As already noted, the square-planar field leads to four low-lying d orbitals and one high-energy orbital. This arrangement suggests that metal ions with a $d^8$ configuration are most likely to form square-planar complexes with strong-field ligands. The eight d electrons can then fill the low-lying d orbitals while the high-energy d orbital remains empty prior to ligand bonding. Most complexes of the $d^8$ ions Pt(II), Pd(II), and Au(III) are square-planar, as are a number of Ni(II) complexes, e.g., $[Ni(CN)_4]^{2-}$. Several $d^9$ ion complexes, notably $[Cu(NH_3)_4]^{2+}$, in which just one electron occupies the high-energy $d_{x^2-y^2}$ orbital, also are square-planar.

## 21.8 Metal Clusters

The complexes considered so far involve a single metal atom coordinated to several ligands. F. Albert Cotton (b. 1930) and others have recently synthesized polynuclear complexes, in which two or more metal atoms are attached to each other, in addition to being coordinated to several ligands. Complexes or clusters with metal–metal bonds possess many interesting structural features.

One of the best studied species is the $Re_2Cl_8^{2-}$ dimer, which can be prepared by the reduction of perrhenate, $ReO_4^-$. Figure 21.23a shows the structure of this dimer. Two unusual features in this structure may be noted. First, the Re—Re bond length is only 2.24 Å, significantly smaller than the 2.75 Å between nearest-neighbor atoms in rhenium metal. This short distance suggests that the bond between the rhenium atoms must be very strong. Second, the chlorine atoms attached to the two rhenium atoms line up with each other when viewed along the Rh—Rh bond axis. Because of this alignment, the two sets of chlorine atoms are said to be in an **eclipsed configuration.** The potential energy of this configuration is higher than that of a **staggered configuration,** in which one set of chlorine atoms is rotated by 45° with respect to the other set (Figure 21.23b). The reason for the difference in potential energy is that, owing to the short Re—Re distance, the distance between eclipsed chlorine atoms is only 3.32 Å, which is less than the sum of the van der Waals radii of two chlorine atoms. The repulsion between the electron clouds of the touching chlorine atoms consequently increases the potential energy of the eclipsed arrangement. Thus, a staggered arrangement, for which the distance between chlorine atoms attached to different rhenium atoms is larger, might have been expected.

The unusual features of the $Re_2Cl_8^{2-}$ structure can be understood by examining the bonds in this species. The electronic configuration of rhenium is $[Xe]4f^{14}5d^56s^2$ and that of rhenium(III), which is the oxidation state of rhenium

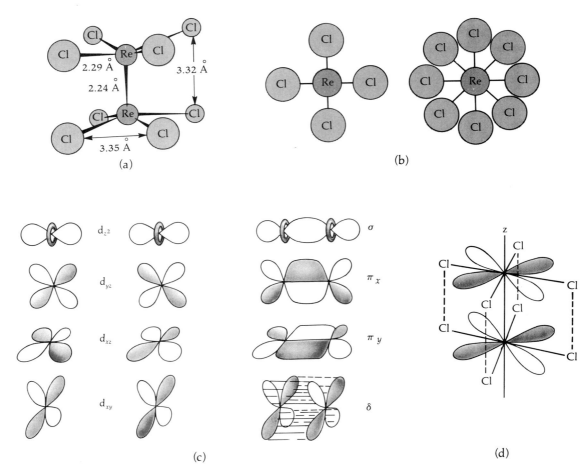

**FIGURE 21.23** The structure of the Re$_2$Cl$_8^{2-}$ dimer. (a) Geometric arrangement and distances. (b) Eclipsed (left) and staggered (right) arrangements of Cl atoms as viewed along the Re–Re axis. (c) Anatomy of the Re–Re quadruple bond. Overlap of the atomic orbitals shown on the left of the diagram forms the molecular orbitals shown on the right. (d) Overlap of the d$_{xy}$ orbitals in the eclipsed configuration.

in Re$_2$Cl$_8^{2-}$, is [Xe]4f$^{14}$5d$^4$. The nature of the Re—Cl bonds can be understood in terms of the description given in Section 21.7. The four Re—Cl bonds formed by each rhenium atom are coordinate covalent bonds in which each chloride ion donates an electron pair to the bond. Since Re$^{3+}$ has one empty 5d orbital as well as empty 6s and 6p orbitals, the eight electrons from the ligands occupy hybridized dsp$^2$ orbitals (Table 21.6). Each ReCl$_4^-$ half of the cluster, therefore, is nearly square-planar. Note that the d orbital involved in the formation of these hybrid orbitals is the d$_{x^2-y^2}$ orbital, which points directly at the ligands (Figure 21.22c).

The electronic configuration of Re$^{3+}$ indicates that each complexed Re$^{3+}$ ion has four d electrons available for bonding. Thus, the two rhenium ions can form a quadruple bond (Figure 21.23c). This bond consists of one sigma bond (from d$_{z^2}$ atomic orbital overlap), two pi bonds (from d$_{xz}$ and d$_{yz}$ overlap), and one delta bond. A delta bond forms from the overlap of the $d_{xy}$ atomic orbitals, which point midway between the chlorine atoms attached to each rhenium atom. Therefore, unless the chlorine atoms are in an eclipsed arrangement, the $d_{xy}$ orbitals cannot overlap, and the delta bond does not form (Figure 21.23d). The energy of the Re—Re quadruple bond is estimated as about 400 kJ mol$^{-1}$. The bond is sufficiently strong that the dirhenium complex is stable in spite of the eclipsed configuration of the chlorine atoms.

Metal–metal bonds are likely when the metal valence shell orbitals can overlap to a substantial extent. Good overlap requires that the metal be in a low oxidation state, for this permits the valence shell orbitals to expand and thereby overlap. Some typical examples of species containing metal–metal bonds are $W_2(C_6H_6)_3$, $Mo_2(O_2CCH_3)_4$, $Fe_2(CO)_9$, and $Re_3Cl_9$. The study of compounds containing metal clusters is an active area of research in inorganic chemistry.

## 21.9 Reactions of Coordination Complexes

As is true for reactions of simple compounds, the reactions of coordination complexes are determined by both thermodynamic and kinetic factors. The stability of complexes towards dissociation is determined by the values of their formation constants (Table 9.3). The values of these constants vary widely. Not surprisingly, some of the factors that determine the magnitude of the crystal field splitting also affect the thermodynamic stability of complexes. For example, the stability of the complexes formed by a particular metal with a given ligand increases with the oxidation state of the metal. Similarly, when the crystal field stabilizes the low-spin complexes of a particular metal, the stability of its complexes varies in a very approximate way with the position of the ligand in the spectrochemical series. For example, formation constants of halide complexes frequently increase from the iodide to the fluoride, and cyanide complexes are particularly stable, as are those of chelating ligands.

In certain cases, complex formation constants are sufficiently large to stabilize certain metals in unusual oxidation states. The complexes of cobalt(III) give perhaps the best illustration of this effect. In aqueous solution, cobalt is usually present as the $Co^{2+}$ ion. The $Co^{3+}$ ion is a powerful oxidizing agent and is unstable in water. The standard reduction potential

$$Co^{3+} + e^- \rightarrow Co^{2+} \qquad \epsilon° = 1.81 \text{ V}$$

indicates that $Co^{3+}$ will oxidize water to oxygen. However, the addition of complexing agents such as ammonia stabilizes cobalt(III) towards reduction:

$$[Co(NH_3)_6]^{3+} + e^- \rightarrow [Co(NH_3)_6]^{2+} \qquad \epsilon° = 0.11 \text{ V}$$

The formation constant of $[Co(NH_3)_6]^{3+}$ is $10^{34}$ while that of $[Co(NH_3)_6]^{2+}$ is only $10^5$, and the cobalt(III) complex forms when the cobalt(II) complex is exposed to oxygen.

One of the most common reactions of coordination complexes is ligand substitution, for example,

$$[Ni(H_2O)_4]^{2+} + 4CN^- \rightarrow [Ni(CN)_4]^{2-} + 4H_2O$$

It is customary to categorize complexes on the basis of the rate at which they undergo such reactions. A complex is said to be **labile** if it undergoes exchange reactions rapidly, in a matter of seconds or less. If the time scale on which exchange takes place is substantially longer, the complex is said to be **inert**. As for other chemical reactions, the rate of these complex exchange reactions depends on the details of the mechanism, the nature of the activated complex, and the magnitude of the activation energy.

To illustrate these points, consider a ligand substitution reaction of an octahedral complex, which can be written in the general form

$$ML_5X + Y \rightarrow ML_5Y + X \tag{21.2}$$

where M is the metal ion, L is a ligand not involved in the reaction, and X and Y are the ligands undergoing exchange. Two of the common mechanisms of this type of reaction involve either the dissociation of the complex,

$$ML_5X \rightarrow ML_5 + X$$
$$ML_5 + Y \rightarrow ML_5Y \tag{21.3}$$

or its association with the entering ligand,

$$ML_5X + Y \rightarrow ML_5XY$$
$$ML_5XY \rightarrow ML_5Y + X \tag{21.4}$$

Observe that the intermediates in reactions (21.3) and (21.4) consist of five-coordinate and seven-coordinate complexes, respectively. To a large extent, the stability of the intermediates determines the magnitude of the activation energy and controls the reaction mechanism. The first step of these reactions frequently consists of a rapid equilibrium and the expected rate law can be derived by means of the steady-state approximation (Section 14.5b).

A second common type of reaction involving coordination complexes is oxidation–reduction. Consider, for example, the reaction

$$[Fe(CN)_6]^{4-} + [Mo(CN)_8]^{3-} \rightarrow [Fe(CN)_6]^{3-} + [Mo(CN)_8]^{4-} \tag{21.5}$$

In this reaction, the oxidation state of iron increases from +2 to +3 while that of molybdenum decreases from +5 to +4. Two different mechanisms are common for reactions of this type. The simpler mechanism, and the one by which reaction 21.5 actually proceeds, is known as the **outer sphere mechanism.** The coordination spheres of the two complexes are not affected in reactions proceeding by this mechanism and the electron transfer occurs between the two stable complexes. This type of a mechanism is generally found when electron transfer occurs at a much faster rate than ligand substitution, for example, when the two complexes are very inert.

The second mechanism of coordination-complex redox reactions is the **inner sphere mechanism**, so called because the ligands are intimately involved in the electron transfer process. The reaction

$$[Co(NH_3)_5Cl]^{2+} + [Cr(H_2O)_6]^{2+} + 5H_2O \rightarrow [Co(H_2O)_6]^{2+} + [Cr(H_2O)_5Cl]^{2+} + 5NH_3 \tag{21.6}$$

in which Co(III) is reduced to Co(II) and Cr(II) is oxidized to Cr(III) illustrates this mechanism. The first step is the formation of a bridged intermediate in which the Cl ligand on the cobalt complex displaces one of the water molecules of the chromium complex:

$$[Co(NH_3)_5Cl]^{2+} + [Cr(H_2O)_6]^{2+} \rightarrow [(NH_3)_5Co-Cl-Cr(H_2O)_5]^{4+} + H_2O \tag{21.7}$$

This reaction can occur because the Co(III) complex is inert while the Cr(II) complex is labile. The second step of the reaction involves electron transfer from Cr(II) to Co(III) and occurs within the intermediate. The resulting Cr(III) forms an inert aquachloro complex and the intermediate breaks up according to the reaction

$$[(NH_3)_5Co-Cl-Cr(H_2O)_5]^{4+} \rightarrow [Cr(H_2O)_5Cl]^{2+} + [Co(NH_3)_5]^{2+} \qquad (21.8)$$

The five-coordinate complex of cobalt quickly picks up a water molecule to form $[Co(NH_3)_5H_2O]^{2+}$ but this complex is so labile that it rapidly undergoes hydrolysis:

$$[Co(NH_3)_5H_2O]^{2+} + 5H_2O \rightarrow [Co(H_2O)_6]^{2+} + 5NH_3 \qquad (21.9)$$

For his studies of the mechanisms of redox reactions of metal complexes, Henry Taube (b. 1915) was awarded the Nobel prize in 1983.

## 21.10 Some Uses of Coordination Compounds

Coordination complexes have a number of applications which depend on their unusual stability. A number of procedures for the separation of specific metal ions by means of complex formation have been developed. Many complexes with organic ligands are both bulky and uncharged. Consequently, they are insoluble in water and can be used for the separation of the ions in question. The bidentate chelating ligand 8-hydroxyquinoline, for example, precipitates a number of metal ions such as zinc, which forms bis(8-hydroxyquinolinato)zinc(II) (Figure 21.24).

EDTA (Table 21.3) is a particularly useful chelating agent because the complex formation constants are very large. Since EDTA is a hexadentate ligand, complex formation occurs in a single step rather than in several steps, as is the case for monodentate ligands (Section 9.6). These features, along with the fast rate of chelation, make titration with EDTA a widely used procedure for the determination of metal ions in solution.

A number of practical applications which depend on the control of metal concentrations achieved by the use of complexing agents can be cited. The electroplating of noble metals such as gold and silver onto baser metals such as copper is a good example. The plating solution usually includes complexing agents such as cyanide, which forms species such as $[Ag(CN)_2]^-$. The formation of this complex drastically reduces the concentration of the free silver ion in solution. The electrodeposition of silver consequently occurs at a slow rate and results in the formation of a smooth coating.

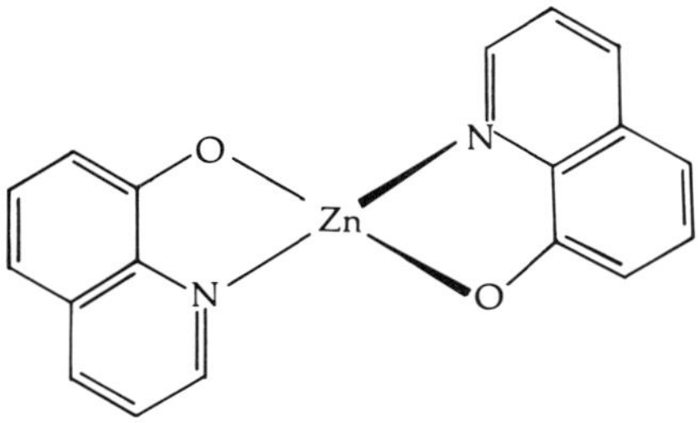

FIGURE 21.24 The structure of bis(8-hydroxyquinolinato)zinc-(II). The formation of this chelate constitutes a method for the removal of $Zn^{2+}$ ions from solution by precipitation.

The control of metal ion concentrations is also important in various biochemical processes. Chelating agents are particularly effective in this respect. Plants, for instance, require trace amounts of iron metal. The addition of iron to plant food in the form of the EDTA complex provides the metal in controlled amounts. Various anticancer drugs that depend on the action of heavy metals such as platinum work in an essentially similar manner. A more mundane practical application involves the addition of suitable complexing agents to laundry detergents. These substances prevent the precipitation of salts by tying up ions such as $Ca^{2+}$, $Mg^{2+}$, and $Fe^{2+}$, which are normally present in hard water (Section 18.7b).

Coordination compounds play a significant role in biological processes. A number of important compounds are based on derivatives of porphin, a planar molecule comprised of four five-membered rings of carbon and nitrogen atoms. The molecule contains four nitrogen atoms which point towards the open center of the molecule and are capable of chelating a metal atom, as shown in Figure 21.25. Naturally occurring complexes having the porphin framework are called porphyrins.

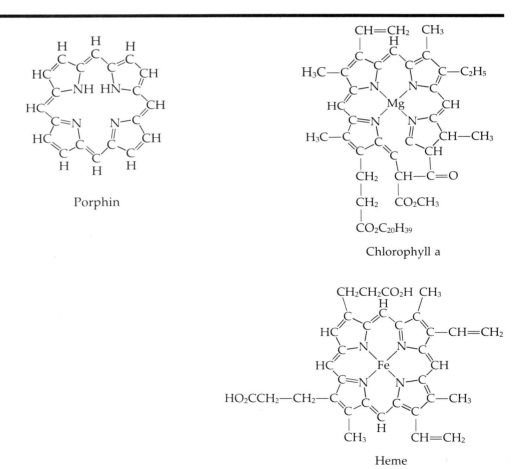

**FIGURE 21.25** Porphin and the naturally occurring porphyrins, chlorophyll *a* and heme.

Chlorophyll is a porphyrin chelate of magnesium. There are two closely related forms of this molecule, labelled a and b, one of which is shown in Figure 21.25. Chlorophyll is the green pigment of plants. The molecule has an absorption minimum in the green region of the spectrum and consequently has this color. The absorption of solar radiation by chlorophyll is the first step in **photosynthesis,** the process by which plants synthesize various essential organic molecules from carbon dioxide and water (Section 22.5c).

Another important porphyrin is heme, a square-planar iron complex whose structure is shown in Figure 21.25. When associated with the protein globin, this complex forms hemoglobin, the substance responsible for the transport of oxygen by the bloodstream. Heme and chlorophyll are just two of the many complexes involved in life processes. It is generally believed that the many trace elements required for the sustenance of life function through coordination chemistry.

# Conclusion

The transition elements display a rich and varied chemistry, much of which can be ascribed to the filling of the d orbitals. Most of the metals, including such important structural metals as iron, are transition elements.

The oxidation states of the transition elements increase both in number and in maximum value in the first half of each series and both of these quantities then decrease rather abruptly near the end of each series. Consequently, the most varied redox chemistry is displayed by the elements near the middle of the series. All the elements generally form either $M^{2+}$ or $M^{3+}$ ions. In their higher oxidation states, the elements in the middle of the series form covalently bonded oxoanions and oxides, many of which are strong oxidizing agents in acidic solution.

One of the characteristic properties of transition elements is their ability to form coordination compounds, in which the central metal ion is surrounded by several ligands, typically four or six, which are attached to it by coordinate covalent bonds. These complexes tend to be colored, are frequently paramagnetic, and in many cases display various types of isomerism. Many of these properties can be explained by crystal field theory, the central feature of which is the splitting of the degenerate d orbitals of the central metal ion by the electric field of the ligands.

# Problems

1. For each of the iron triad elements, state which oxidation state has a $d^7$ configuration. Which of these species form a stable cation in aqueous solution?
2. Write balanced equations for the reactions which occur when each of the following oxides is dissolved in 0.1 $M$ $HClO_4$: (a) $TiO_2$, (b) $V_2O_5$, and (c) MnO.
3. The highest oxidation states of elements with low maximum oxidation states, such as the coinage metals, are found in the fluorides, e.g., $AgF_2$. The highest oxidation states of elements with high maximum oxidation states, such as Mn, are found in the oxides or oxoanions. Explain the difference.
4. The synthesis of compounds in which transition el-

ements are present in high oxidation states is usually carried out in alkaline conditions. Explain.

5. Determine the uncommon oxidation state of the transition element present in each of the following compounds: (a) $SrFeO_4$, (b) $K_2[Ni(CO)_6]$, (c) $Na_4[Ni_2(CN)_6]$, and (d) $K[CrFO_4]$.

6. Write balanced equations for the following reactions: (a) the reaction of $Fe^{2+}$ with excess $CN^-$, (b) the reaction of $Fe_3O_4(s)$ with $CO(g)$ at elevated temperatures, (c) the reaction of $[Au(CN)_2]^-$ with $Zn(s)$, and (d) the reaction of $Fe^{2+}$ with $Cr_2O_7^{2-}$ in acidic solution.

7. What fraction of chromium(VI) is present as $CrO_4^{2-}$ in an aqueous solution of pH = 2.5 if the total concentration of chromium(VI) is 0.0100 M?

8. Using the data in Figure 21.7, determine what reaction occurs when an acidic solution of chromium(III) is treated with excess potassium permanganate. Write a balanced equation for the reaction and determine its equilibrium constant.

9. When carbon dioxide is bubbled into an alkaline solution of manganate, $MnO_4^{2-}$, the color changes from green to purple. Explain this observation with reference to the electrode potential diagram of manganese (Figure 21.7).

10. Manganese(VI) is moderately unstable. (a) Write balanced equations for its disproportionation to manganese(VII) and manganese(IV) in acidic and in basic solution. (b) Determine the equilibrium constants for the two reactions on the basis of the standard electrode potential diagrams (Figure 21.7). (c) Derive the $\epsilon°$ value for the reduction of manganese(VI) to manganese(IV) in base from that in acid.

11. Derive an expression for $\epsilon$ for the half-reaction $MnO_4^{2-} \rightarrow MnO_2(s)$ in terms of $\epsilon°$ for acidic solution as a function of pH.

12. For the half-reaction $MnO_4^- \rightarrow MnO_4^{2-}$, relate the value of $\epsilon°$ in basic solution to the value of $\epsilon°$ in acidic solution.

13. Determine the value of $\epsilon°$ in basic solution for the half-reaction $Mn(OH)_2(s) + 2e^- \rightarrow Mn(s)$ on the basis of $\epsilon°$ for the $Mn^{2+} \rightarrow Mn$ couple and $K_{sp}$ for $Mn(OH)_2$ (Appendix 3).

14. The standard electrode potential for the $[Fe(CN)_6]^{4-} \rightarrow Fe$ couple is $-1.5$ V. Using additional data from Appendix 5, determine the complex formation constant for $[Fe(CN)_6]^{4-}$.

15. Name the following complexes: (a) $[Fe(CN)_6]^{3-}$, (b) $[Fe(H_2O)_6]^{3+}$, (c) $[Co(H_2O)_3F_3]$, and (d) $[Ag(CN)_2]^-$.

16. Write the formulas of the following coordination compounds: (a) hexaaquacobalt(III) chloride, (b) sodium tetrachloronickelate(II), and (c) bromobis(ethylenediamine)cobalt(III) chloride.

17. Write the formulas of the following compounds: (a) triamminetrichloroplatinum(IV) chloride, (b) potassium tricarbonatocobaltate(III), and (c) sodium monochloropentacyanoferrate(III).

18. Name the following coordination compounds: (a) $[Co(NH_3)_4(H_2O)_2]Br_3$, (b) $[Cr(en)_3]Cl_3$, and (c) $[Co(NH_3)_6][Cr(CN)_6]$.

19. What is the coordination number of the transition metal in each of the following complexes: (a) $[Ni(en)_2]^{2+}$, (b) $[Cr(ox)_3]^{3-}$, (c) $[Co(en)Cl_2Br_2]^-$, and (d) triamminetriaquachromium(III) chloride.

20. Draw the structures of all the isomers of the following octahedral complexes and identify those that are optically active:
(a) $[Co(NH_3)_4F_2]^+$
(b) $[Co(NH_3)_2F_4]^-$
(c) $[Co(NH_3)_3Br_2Cl]$

21. Draw the structures of all the isomers of the following complexes and identify those that are optically active: (a) $[Co(en)_2Cl_2]^+$ and (b) $[Co(en)(NH_3)_2Cl_2]^+$.

22. Write the formula and draw the molecular structure of each of the following complexes: (a) *mer*-triamminetrichlorocobalt(III) (*mer* is short for meridional) and (b) *cis*-dichlorotetrafluorochromate(III).

23. Determine the number of isomers of the complex [MABCD], where A, B, C, D are four different ligands, if (a) the complex is square-planar and (b) the complex is tetrahedral. For each isomer, write the structural formula and state whether the isomer is optically active.

24. Chromium(III) forms six possible octahedral isomers with the formula $CrF_3(H_2O)_6$. Draw their structures and write their formulas.

25. Write formulas and names of all the possible structural isomers with the empirical formula $FeFCl·2NH_3·3H_2O$. Write the structural formula of each isomer. Group all the stereoisomers with the same structural formula. Which of these stereoisomers are enantiomers?

26. How many stereoisomers exist for each of the following complexes: (a) $[Pt(NH_3)Cl_3]$ (square-planar), (b) $[Zn(NH_3)_2Cl_2]$ (tetrahedral), (c) $[Pd(H_2O)_2ClF]$ (square-planar), and (d) $[Ru(H_2O)_2(NH_3)_2Cl_2]^+$ (octahedral)?

27. The complex $[Fe(H_2O)_6]^{3+}$ has five unpaired electrons while $[Fe(CN)_6]^{3-}$ has one unpaired electron. Explain these observations by means of crystal field theory.

28. Determine the number of unpaired electrons in the complex $[Cr(NH_3)_6]^{3+}$.

29. The complex [NiCl$_4$]$^{2-}$ has two unpaired electrons while [Ni(CN)$_4$]$^{2-}$ is diamagnetic. Explain these observations by means of crystal field theory.

30. Sketch the occupancy of the d orbitals in octahedral complexes of (a) V(III), (b) Cr(III), (c) Fe(II), and (d) Fe(III), for both the low-field and high-field limits. In those instances for which the occupancy differs in the two limits, identify the high-spin isomer and the low-spin isomer.

31. Sodium hexafluorochromate(III) is paramagnetic. (a) Sketch the occupancy of the $t_{2g}$ and $e_g$ orbitals. (b) Write the d electronic configuration of chromium in the complex. (c) How many unpaired electrons are there?

32. The crystal field splitting of [Mn(H$_2$O)$_6$]$^{3+}$, a high-spin octahedral complex, is 250 kJ mol$^{-1}$, and the crystal field splitting of [Mn(CN)$_6$]$^{3-}$, a low-spin octahedral complex is 460 kJ mol$^{-1}$. (a) Write the d electronic configuration of each complex and give the number of unpaired electrons. (b) Determine the value of CFSE for each complex, neglecting the pairing energy. (c) Would you expect the complex [Mn(OH)$_6$]$^{3-}$ to have high spin or low spin?

33. The diamagnetic complex of cobalt, [Co(NH$_3$)$_6$]$^{3+}$, is orange-yellow while the paramagnetic complex, [CoF$_6$]$_3^-$ is blue. Explain the difference in color.

34. The complex [Cu(NH$_3$)$_4$]$^+$ is colorless while [Cu(NH$_3$)$_4$]$^{2+}$ is intensely blue. Explain this difference in terms of crystal field theory.

35. Which of the following ions might be expected to form colorless aqueous solutions: (a) La$^{3+}$, (b) Ti$^{3+}$, (c) Hf$^{4+}$, (d) Co$^{2+}$, and (e) Ag$^+$? Explain.

36. Using the data in Table 21.7, determine the wavelength absorbed in the $t_{2g} \to e_g$ transition in (a) [Ni(H$_2$O)$_6$]$^{2+}$ and (b) [Ni(NH$_3$)$_6$]$^{2+}$. What would the color of each complex be if no other visible radiation is absorbed?

37. The complex [Co(NH$_3$)$_6$]$^{3+}$ is diamagnetic. (a) State whether this is an example of weak-field or strong-field splitting. (b) Write the electronic configuration. (c) Using data in Table 21.7, calculate the CFSE. (d) What color would the complex have as a result of a $t_{2g} \to e_g$ transition?

38. The effect of octahedral, tetrahedral, and square-planar fields on the d orbitals of a metal atom has been examined in Section 21.7. Apply the same analysis to the p orbitals of a metal atom and determine which of the fields, if any, split the energies of the p orbitals. Draw a diagram such as that in Figure 21.22 for the cases where splitting occurs.

39. For each of the following complexes, sketch the electron occupancy of the d orbitals and write the d electronic configuration of the central metal ion: (a) [Ni(CN)$_4$]$^{2-}$ (square-planar), (b) [NiCl$_4$]$^{2-}$ (tetrahedral), (c) [Zn(NH$_3$)$_4$]$^{2+}$ (tetrahedral), and (d) [Pt(NH$_3$)$_4$]$^{2+}$ (square-planar).

40. The binuclear complex [Mo$_2$Cl$_8$]$^{4-}$ contains a metal–metal quadruple bond. (a) What is the oxidation state of Mo in this complex? (b) Write the corresponding electronic configuration of Mo. (c) Which Mo orbitals are involved in bonding with the ligands and what is their hybridization? (d) Give a molecular orbital description of the Mo—Mo bond.

41. Using data given in Section 21.9, determine the concentration ratio [Co$^{3+}$]/[Co$^{2+}$] in a solution containing 0.5 M NH$_3$, 0.5 M [Co(NH$_3$)$_6$]$^{3+}$, and 0.5 M [Co(NH$_3$)$_6$]$^{2+}$.

42. The mechanism for the ligand substitution reaction in Equation 21.2 is

(1) $\text{ML}_5\text{X} \underset{k_{-1}}{\overset{k_1}{\rightleftharpoons}} \text{ML}_5 + \text{X}$

(2) $\text{ML}_5 + \text{Y} \xrightarrow{k_2} \text{ML}_5\text{Y}$

Use the steady-state approximation to derive the rate law on the assumption that the first step is much faster than the second step.

# 22

# The Chemistry of Carbon

Organic chemistry is the chemistry of carbon compounds. Only a few simple compounds of carbon, such as the oxides, carbonates, and cyanides, are traditionally considered to be inorganic substances. As the names imply, "organic" and "inorganic" chemistry were at one time believed to involve substances obtained exclusively from animate and inanimate sources, respectively. This view, which prevailed at the beginning of the nineteenth century, was first dispelled by Friedrich Woehler (1800–1882), who showed that an inorganic salt, ammonium cyanate, could be converted by heating to urea, known until then only as a constituent of urine and, thus, as "organic":

$$NH_4^+NCO^- \xrightarrow{\Delta} H_2N-\underset{\underset{O}{\|}}{C}-NH_2$$

This experiment showed that "organic" compounds need not come only from living things. To be sure, a lot of very complicated chemistry goes on in living organisms, mostly involving complex carbon compounds. The study of these life processes is the realm of biochemistry.

The bonding properties of carbon—the ability of a carbon atom to form four stable covalent bonds with atoms of carbon and other elements—account for the existence of an enormously large and varied number of carbon compounds. Several million organic compounds are currently known, and the number grows from day to day. Organic compounds are truly ubiquitous. Not only are human beings made up of and nourished by such compounds, but we also use a vast number of industrially produced organic compounds in daily life. We need only think of synthetic fabrics, plastics, wood and paper products, petroleum derivatives, and pharmaceuticals, to name just a few, to appreciate the dominant role played in our society by organic compounds.

# Organic Chemistry

## 22.1 Hydrocarbons

Hydrocarbons are compounds of carbon and hydrogen. Even though they contain only two elements, they can be divided into a number of different categories, as shown in Figure 22.1. **Saturated hydrocarbons** contain only single bonds and may be subdivided into **alkanes,** which have end carbon atoms not bonded to each other, and **cycloalkanes,** which contain a ring of carbon atoms. Hydrocarbons containing one or more C=C double bonds are known as **alkenes,** and those containing C≡C triple bonds are known as **alkynes.** These various types of hydrocarbons belong to the vast class of **aliphatic compounds,** the molecules of which do not contain any bonds of fractional order but only single, double, or triple bonds.

| Type | General Structure | Example |
|---|---|---|
| Alkane | RH ($C_nH_{2n+2}$) | $CH_3CH_3$ (ethane) |
| Cycloalkane | ($C_nH_{2n}$) | 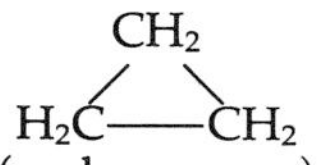 cyclopropane |
| Alkene | 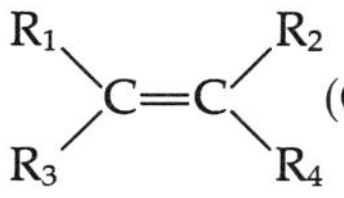 $R_1R_3C=CR_2R_4$  ($C_nH_{2n}$) | $CH_3CH=CH_2$ (propene) |
| Alkyne | $R_1C≡CR_2$ ($C_nH_n$) | $CH_3C≡CH$ (propyne) |
| Aromatic hydrocarbon | 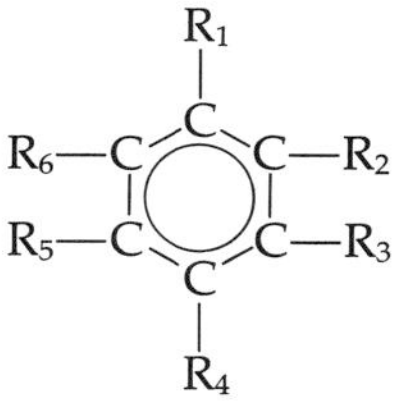 benzene ring with $R_1$–$R_6$ | 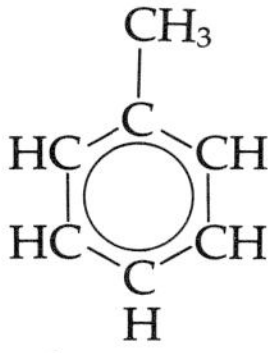 toluene |

**FIGURE 22.1** Types of hydrocarbons. The symbol R stands for an *alkyl group*, derived from an alkane by removal of a hydrogen atom. The simplest alkyl group is the methyl radical ·$CH_3$. The numbered R groups can either be the same or different and can also be hydrogen atoms.

*a. The Alkanes* The structural properties of the simplest alkanes, methane and ethane, were examined in Chapter 15. The carbon atom in methane is located at the center of a tetrahedron with the four hydrogen atoms at the apices. The C—H bonds form tetrahedral angles (109.5°) with respect to each other and can be described as sigma bonds formed by the overlap of hybrid $sp^3$ orbitals of carbon with the hydrogen s orbitals. The same general features are present in ethane. Recall from Section 16.8 that, in addition to forming sigma bonds with the hydrogen atoms, the two carbon atoms form a sigma bond with each other. The geometry of organic molecules is best represented by ball and stick or space-filling models, as shown in Figures 22.2c and 22.2d. Structural formulas (Figure 22.2b) show every covalent bond by a single line, but do not represent the actual molecular geometry. Because of their simplicity, condensed formulas like those in Figure 22.2a are frequently used.

Methane and ethane are the simplest of the alkanes, which have the general formula $C_nH_{2n+2}$, where $n$ is an integer. Each succeeding alkane can be thought of as formed by the replacement of an H atom by a methyl group, —CH$_3$. (A methyl group is just a methyl radical, ·CH$_3$; the line signifies an unpaired electron capable of forming a covalent bond when combined with the unpaired electron on another radical.) For example, propane, $C_3H_8$, the next compound in the sequence, can be thought of as an ethane molecule in which a hydrogen atom has been replaced by a methyl group.

Figure 22.3a shows two representations of propane. It is apparent that the eight hydrogen atoms are not equivalent but can be divided into two groups. The six hydrogen atoms attached to the end carbon atoms are equivalent to each other but not to the two hydrogen atoms bonded to the middle carbon

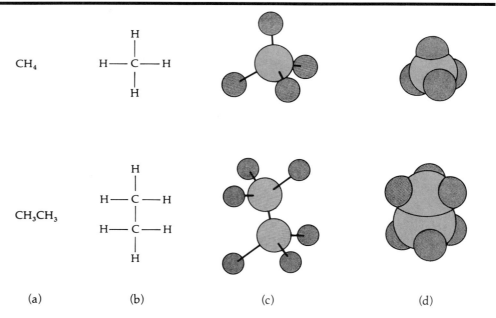

**FIGURE 22.2** Different representations of methane and ethane molecules. (a) Condensed formula, (b) structural formula, (c) ball and stick model, and (d) space-filling model.

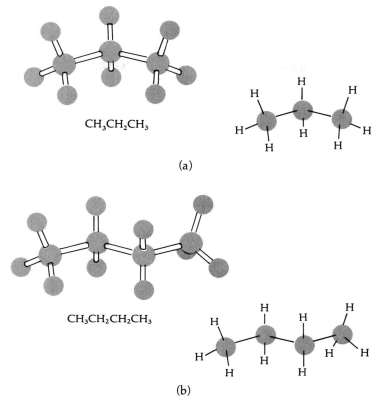

**FIGURE 22.3** The structures of (a) propane and (b) n-butane.

atom. Two structural isomers of butane, the next member of the alkane series, are therefore possible. The first, which has the common name n-butane, is formed by replacement of one of the end hydrogen atoms by a methyl group (Figure 22.3 b):

$$\text{H}-\underset{\underset{\text{H}}{|}}{\overset{\overset{\text{H}}{|}}{\text{C}}}-\underset{\underset{\text{H}}{|}}{\overset{\overset{\text{H}}{|}}{\text{C}}}-\underset{\underset{\text{H}}{|}}{\overset{\overset{\text{H}}{|}}{\text{C}}}-\underset{\underset{\text{H}}{|}}{\overset{\overset{\text{H}}{|}}{\text{C}}}-\text{H} \quad \text{or} \quad \text{CH}_3\text{CH}_2\text{CH}_2\text{CH}_3$$

n-butane

The second isomer, commonly known as isobutane, involves the replacement of one of the middle hydrogen atoms by a methyl group:

$$\text{H}-\underset{\underset{\underset{\underset{\text{H}}{|}}{\overset{\text{H}}{\diagup\diagdown}}}{|}}{\overset{\overset{\text{H}}{|}}{\text{C}}}-\underset{\underset{\text{C}}{|}}{\overset{\overset{\text{H}}{|}}{\text{C}}}-\underset{\underset{\text{H}}{|}}{\overset{\overset{\text{H}}{|}}{\text{C}}}-\text{H} \quad \text{or} \quad \text{CH}_3\underset{\underset{\text{CH}_3}{|}}{\text{CHCH}_3}$$

isobutane

As in the case of previously encountered structural isomers, the two butanes differ in their properties. For example, the boiling point of n-butane is −0.5°C while that of isobutane is −12°C.

The prefix *iso-* is sometimes used to indicate a group branching off the main carbon framework and alkanes such as isobutane are called **branched-chain alkanes**. The prefix *n-*, for normal, indicates that the compound has a **straight chain** of carbon atoms. As may be seen in Figure 22.3, the chain, while unbranched, is actually not straight but, because of the tetrahedral angle between adjacent C—C bonds, zigzags back and forth.

**b. Nomenclature** Alkanes with more than four carbon atoms are named by adding the suffix *-ane* to the Greek root signifying the appropriate number, e.g., pentane and hexane. Table 22.1 summarizes the names and physical properties of some of the straight-chain alkanes. The alkanes have low melting point and boiling points, which increase with the molecular weight, as expected from the increasing van der Waals forces between the molecules.

The number of possible structural isomers of alkanes rapidly increases with the number of carbon atoms. This increase poses a challenge to the task of naming these compounds since it is clear that a few identifying prefixes such as *iso-* will not suffice. The following systematic procedure permits the assignment of a unique name to each alkane and, with some modifications, is used to name a large variety of other types of compounds:

1. Identify the longest continuous chain of carbon atoms and number them sequentially, beginning at the end nearest to the branching. The total number of carbon atoms in the chain determines the base name of the compound. For example, the following molecule is named as a derivative of hexane because the longest chain has six carbon atoms:

$$\overset{1}{C}H_3\overset{2}{C}H-\overset{3}{C}H\overset{4}{C}H_2\overset{5}{C}H_2\overset{6}{C}H_3$$
$$\qquad\quad |\quad\ |$$
$$\qquad\ CH_3\ CH_3$$

2. Designate the location of substituent groups (i.e., groups attached to the longest chain) by the number of the carbon atom to which they are attached. For example, the preceding molecule has methyl groups attached to atoms 2

**TABLE 22.1  Properties of Straight-Chain Alkanes**

| Name | Molecular formula | Molecular weight (g mol$^{-1}$) | Melting point (°C) | Boiling point (°C) | Number of isomers |
|---|---|---|---|---|---|
| Methane | CH$_4$ | 16 | −183 | −162 | 1 |
| Ethane | C$_2$H$_6$ | 30 | −172 | −88 | 1 |
| Propane | C$_3$H$_8$ | 44 | −187 | −42 | 1 |
| Butane | C$_4$H$_{10}$ | 58 | −138 | 0 | 2 |
| Pentane | C$_5$H$_{12}$ | 72 | −130 | 36 | 3 |
| Hexane | C$_6$H$_{14}$ | 86 | −95 | 68 | 5 |
| Heptane | C$_7$H$_{16}$ | 100 | −91 | 98 | 9 |
| Octane | C$_8$H$_{18}$ | 114 | −57 | 126 | 18 |
| Nonane | C$_9$H$_{20}$ | 128 | −54 | 151 | 35 |
| Decane | C$_{10}$H$_{22}$ | 142 | −30 | 174 | 75 |

### TABLE 22.2 The Simplest Alkyl Groups

| CH$_3$—<br>methyl | CH$_3$CH$_2$—<br>ethyl | CH$_3$CH$_2$CH$_2$—<br>n-propyl | CH$_3$$\overset{\mid}{\text{C}}$HCH$_3$<br>isopropyl |
|---|---|---|---|
| CH$_3$CH$_2$CH$_2$CH$_2$—<br>n-butyl | CH$_3$$\overset{\mid}{\text{C}}$HCH$_2$CH$_3$<br>sec-butyl | (CH$_3$)$_2$$\overset{\mid}{\text{C}}$HCH$_2$<br>isobutyl | (CH$_3$)$_3$C—<br>tert-butyl |

and 3. Table 22.2 gives the names of frequently encountered alkyl substituents.

3. Name the compound by designating the location and name of each substituent followed by the base name. As in the case of coordination compounds, different substituents are listed in alphabetical order. The prefixes di, tri, and so on, are used to designate two or more identical substituents. The preceding compound is called 2,3-dimethylhexane.

**Example 22.1** Write condensed formulas for all the isomers of hexane, C$_6$H$_{14}$, and name the compounds.

It is helpful to proceed systematically, starting with the longest possible chain and proceeding to shorter chains, in order to ensure that all isomers are counted. We therefore start with an unbranched six-atom chain,

CH$_3$CH$_2$CH$_2$CH$_2$CH$_2$CH$_3$    hexane (the prefix n- is omitted when the systematic nomenclature is used)

Next, we consider a five-atom chain, which has two possible locations for a methyl substituent:

CH$_3$$\overset{\mid}{\text{C}}$HCH$_2$CH$_2$CH$_3$    2-methylpentane
$\quad\quad$ CH$_3$

CH$_3$CH$_2$$\overset{\mid}{\text{C}}$HCH$_2$CH$_3$    3-methylpentane
$\quad\quad\quad\quad$ CH$_3$

Note that the isomer "4-methylpentane",

CH$_3$CH$_2$CH$_2$$\overset{\mid}{\text{C}}$HCH$_3$
$\quad\quad\quad\quad\quad$ CH$_3$

is identical to 2-methylpentane since it makes no difference at which end of the molecule we start to count carbon atoms. Furthermore, substituents on atoms 1 or 5, the end atoms of the chain, just result in hexane. To find additional isomers we must therefore consider a four-atom chain. The distinguishable possibilities are

$\quad\quad$ CH$_3$
$\quad\quad\mid$
CH$_3$CCH$_2$CH$_3$    2,2-dimethylbutane
$\quad\quad\mid$
$\quad\quad$ CH$_3$

and

$$\underset{\underset{CH_3\ CH_3}{|\quad\;|}}{CH_3CH\ CHCH_3}\quad \text{2,3-dimethylbutane}$$

The preceding guidelines make it clear that any additional isomers with a chain of four atoms must be identical to one of the isomers enumerated already. In particular, the compound with an ethyl group attached to a four-carbon chain, e.g.,

$$\underset{\underset{CH_3}{|}}{\underset{\underset{CH_2}{|}}{CH_3CHCH_2CH_3}}$$

actually is identical to 3-methylpentane.

Let us finally attempt to write a structure based on a three-atom chain, e.g.,

$$\underset{\underset{CH_3}{|}}{\underset{\overset{CH_2CH_3}{|}}{CH_3CCH_3}}$$

We immediately note that in so doing we must necessarily also include a chain longer than three carbon atoms. Thus, this structure is identical with 2,2-dimethylbutane. We therefore conclude that there are five isomers of hexane. ∎

**c. Reactions of Alkanes** Alkanes tend to react very slowly and are relatively inert, which is why they are found naturally as oil and natural gas. For example, alkanes are not attacked by strong acids or bases nor by strong oxidizing agents such as potassium permanganate. Their only important reactions are combustion, dehydrogenation, and halogenation. The combustion of alkanes (Section 10.10) makes these substances exceedingly useful as fuels. A typical reaction is that of butane gas,

$$CH_3CH_2CH_2CH_3(g) + 6\tfrac{1}{2}O_2(g) \rightarrow 4CO_2(g) + 5H_2O(l) \quad \Delta H° = -2878 \text{ kJ}$$

**Dehydrogenation** is the removal of hydrogen. In dehydrogenation reactions, alkanes are converted to alkenes or alkynes. The reactions are endothermic and the position of equilibrium is consequently most favorable at high temperatures. The reactions require the presence of metal oxide catalysts to occur at an appreciable rate. A specific example is the dehydrogenation of ethane,

$$CH_3CH_3(g) \xrightarrow[500°C]{Cr_2O_3} CH_2{=}CH_2(g) + H_2(g)$$

Alkanes undergo **halogenation** when exposed to each of the halogens excepting iodine. In such reactions, successive hydrogen atoms are replaced by halogen atoms. The reaction between methane and chlorine in the presence of light is typical:

$$CH_4(g) + Cl_2(g) \xrightarrow{h\nu} CH_3Cl(g) + HCl(g)$$

As the concentration of chloromethane increases, the following successive reactions can occur:

$$CH_3Cl(g) + Cl_2(g) \xrightarrow{h\nu} CH_2Cl_2(l) + HCl(g)$$

$$CH_2Cl_2(l) + Cl_2(g) \xrightarrow{h\nu} CHCl_3(l) + HCl(g)$$

$$CHCl_3(l) + Cl_2(g) \xrightarrow{h\nu} CCl_4(l) + HCl(g)$$

Halogenation of the heavier alkanes results in a mixture of products since replacement can occur at any carbon atom. Halogenated hydocarbons are useful starting reagents for many reactions.

**Example 22.2**  The reaction of ethane with bromine was allowed to proceed sufficiently far to permit the formation of mono-, di-, and trisubstituted ethanes. Write the condensed formulas and names of all possible products.

There is only one singly substituted ethane:

BrCH$_2$CH$_3$     bromoethane

A second replacement can occur on either the first or second carbon atom, resulting in two structural isomers:

Br$_2$CHCH$_3$     1,1-dibromoethane

BrCH$_2$CH$_2$Br     1,2-dibromoethane

Two distinct tribromoethanes are possible:

Br$_3$CCH$_3$     1,1,1-tribromoethane

Br$_2$CHCH$_2$Br     1,1,2-tribromoethane

Further bromination leads to the formation of tetra-, penta-, and hexabromoethanes. Their condensed formulas can be obtained by exchanging the numbers and positions of the hydrogen and bromine atoms in dibromoethane and bromoethane, respectively. ■

*d. Alkenes and Alkynes*  Alkenes have the general formula $C_nH_{2n}$. The simplest alkene is ethene, H$_2$C=CH$_2$, commonly known as ethylene. Alkenes containing three or more carbon atoms are isomeric with the corresponding cycloalkanes. For example, both propene and cyclopropane (Figure 22.1) have the formula C$_3$H$_6$. Alkynes, which contain triple bonds, have the general formula $C_nH_{2n-2}$. The simplest alkyne is ethyne (acetylene), HC≡CH. As noted in Section 16.8, a C=C double bond consists of a sigma bond and a pi bond while a C≡C triple bond consists of a sigma bond and two pi bonds.

The type of bond formed between two carbon atoms in a molecule determines whether the two parts of the molecule can rotate about this bond. The C—C bond in alkanes is a sigma bond, a bond that is symmetric about the C—C axis (see Figure 16.16). The rotation of one part of the molecule relative to the other does not affect this symmetry, and rotation about a single bond generally is unhindered. This view is supported by the lack of stereoisomerism (Section 21.5c) in disubstituted alkanes such as 1,2-dibromoethane (see Example 22.2). If rotation about the C—C bond were not possible, two different

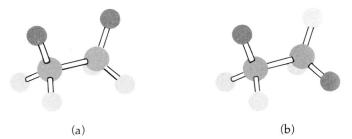

**FIGURE 22.4** The structure of 1,2-dibromoethane (CH$_2$BrCH$_2$Br). The bromine atoms are shown in color. (a) The bromine atoms are on the same side of a plane passing through the two carbon atoms; (b) the bromine atoms are on opposite sides of this plane. Since only one isomer of this compound is known, the two structures must interconvert freely by rotation of the two halves of the molecule about the C—C bond.

stereoisomers of this compound would exist: an isomer in which the Br atoms are on the same side of a plane passing through the two carbon atoms, and an isomer, in which they are on opposite sides of such a plane (Figure 22.4). In actuality, only one stereoisomer of this compound has been isolated.

A C=C double bond, on the other hand, is rigid. The reason is that a pi bond lacks the axial symmetry with respect to the carbon–carbon axis possessed by a sigma bond (see Figure 16.16). Thus, an attempt to rotate one CH$_2$ groups in ethylene relative to the other CH$_2$ group would destroy the $\pi$ orbital as its constituent p orbitals rotate away from each other. The energy required to break a pi bond, $\approx$250 kJ mol$^{-1}$, is so much higher than the thermal energy possessed by molecules at ordinary temperatures that pi bonds are stable and rotation about them does not occur.

The consequence of the rigidity of the double bond in alkenes is the occurrence of cis–trans stereoisomerism. If two identical groups or atoms are on the same side of the double bond between two carbon atoms, the molecule is said to have the cis configuration while, if they are on opposite sides, the molecule has the trans configuration. As in the case of cis–trans isomers of coordination compounds (Section 21.5c), the two isomers differ in physical and

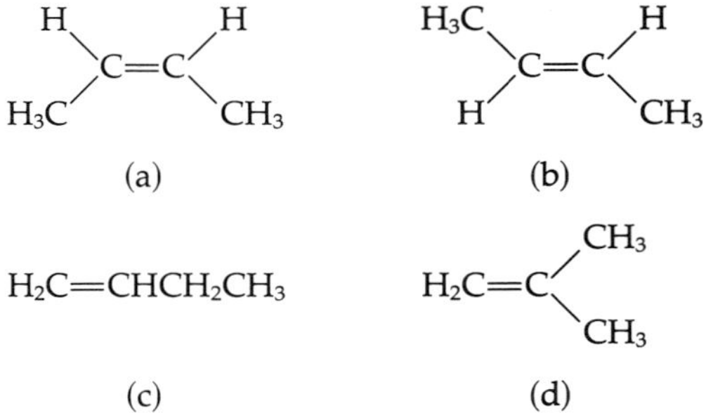

**FIGURE 22.5** Isomerism in alkenes. (a) *cis*-2-butene, (b) *trans*-2-butene, (c) 1-butene, and (d) isobutene (2-methylpropene). Compounds (a) and (b) are cis–trans isomers, a specific type of stereoisomer. Compounds (c) and (d), as well as either (a) or (b), are structural isomers.

chemical properties. Figure 22.5 shows the structures of the cis–trans isomers of 2-butene. Note that the two methyl groups are on the same side of the double bond in the cis isomer whereas they are on opposite sides of the double bond in the trans isomer. Figure 22.5 also shows the structures of 1-butene and isobutene, which, along with the 2-butene isomers, are structural isomers of butene. Since at least one of the double bonded carbon atoms in butene and in isobutene is attached to two identical atoms, cis–trans isomerism cannot occur in these compounds.

**Example 22.3** (a) Write the structures of 1,1-dibromoethylene, cis-1,2-dibromoethylene, and trans-1,2-dibromoethylene. (b) Which of these molecules has a dipole moment? How do you expect the dipole moment of cis-1,2-dibromoethylene to compare with that of cis-1,2-dichloroethylene? (c) Could the compound 1,1-dibromoethylene convey any information about the rigidity of the double bond in the molecule? Explain. (d) Is 2,2-dibromoethylene a different isomer? Explain.

(a) The structures of the dibromoethylenes are

$$\begin{array}{ccc}
\text{Br} \diagdown \quad \diagup \text{H} & \text{Br} \diagdown \quad \diagup \text{Br} & \text{Br} \diagdown \quad \diagup \text{H} \\
\text{C}=\text{C} & \text{C}=\text{C} & \text{C}=\text{C} \\
\text{Br} \diagup \quad \diagdown \text{H} & \text{H} \diagup \quad \diagdown \text{H} & \text{H} \diagup \quad \diagdown \text{Br} \\
\text{1,1-dibromo-} & \text{cis-1,2-dibromo-} & \text{trans-1,2-dibromo-} \\
\text{ethylene} & \text{ethylene} & \text{ethylene}
\end{array}$$

(b) The first two compounds are asymmetric and therefore have dipole moments while the third compound does not. Thus, the C—Br bond moments, which are shown in the structural formulas, reinforce each other in the first two molecules but cancel in the third molecule. Since the C—Cl bond moment is larger than the C—Br bond moment (Table 15.2) cis-1,2-dichloroethylene has a larger dipole moment than the corresponding dibromoethylene.

(c) Since both bromine atoms are attached to the same carbon in 1,1-dibromoethylene, the structure of this molecule would be the same if rotation about the double bond were possible. Thus, this molecule does not convey any information about the rigidity of the double bond.

(d) A molecule is free to assume any orientation in space. Consequently $CBr_2CH_2$ and $CH_2CBr_2$ are identical molecules. ■

The systematic names of the alkenes are based on fairly obvious extensions of the rules given for the naming of alkanes: the base name is that of the longest chain containing the double bond and the positions of substituents are given followed by that of the lowest numbered carbon atom with a double bond. For example, in the compound shown below the double bond is in a six-carbon chain between carbon atoms 2 and 3, and, therefore, the compound is named as a 2-hexene:

$$\underset{\underset{CH_3}{|}}{CH_3C}=CHCH_2\underset{\underset{CH_3}{|}}{CHCH_3} \qquad \text{2,5-dimethyl-2-hexene}$$

The most important reactions of the alkenes involve addition to the double bond with the resulting formation of saturated compounds, which contain only single bonds. The reaction involves the conversion of carbon–carbon pi and sigma bonds into two carbon–carbon sigma bonds. Such a process is usually exothermic since the sigma bond energy is greater than the pi bond energy (see Example 17.2). The addition of hydrogen occurs at ordinary temperatures in the presence of finely divided metal catalysts and is the reverse of alkane dehydrogenation:

$$CH_2=CHCH_3 + H_2 \xrightarrow{Pt} CH_3CH_2CH_3$$
   1-propene                           propane

Other reagents that readily add to the double bond are the halogens, e.g.,

$$CH_2=CHCH_2CH_3 + Br_2 \rightarrow CH_2BrCHBrCH_2CH_3$$
   1-butene                          1,2-dibromobutane

and the hydrogen halides, e.g.,

$$CH_2=CH_2 + HCl \rightarrow CH_3CH_2Cl$$
   ethene                  chloroethane

Alkenes can also undergo **polymerization,** in which very large numbers of small molecules (**monomers**) bond to each other to form polymers. A **polymer** is a very large molecule, a macromolecule, composed of identical repeating units. Some familiar polymers based on alkenes or their derivatives are polyethylene,

$$n\,CH_2=CH_2 \rightarrow -[-CH_2CH_2-]_n-\quad (n \text{ is in the range of } 10^4-10^5)$$

and polytetrafluoroethylene (Teflon),

$$n\,CF_2=CF_2 \rightarrow -[-CF_2CF_2-]_n-$$

The reactions usually require the presence of specific catalysts.

*e. Aromatic Hydrocarbons* **Aromatic compounds** are those based on benzene or on compounds related to benzene. The characteristic feature of these compounds is that their molecules contain delocalized pi electrons (Section 16.8b). The bond order of the carbon–carbon bonds, therefore, is non-integral. Aromatic compounds differ in this respect from aliphatic compounds.

Benzene is usually represented by one of the resonance structures introduced in Section 15.6,

or, perhaps more appropriately, given the complete equivalence of the six C—C bonds, by

In these widely used representations of ring compounds, the carbon atoms comprising the ring and their attached hydrogen atoms are not explicitly

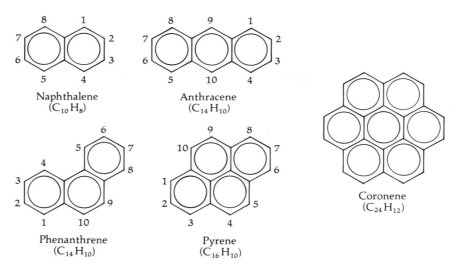

**FIGURE 22.6** Some fused ring aromatic hydrocarbons. The locations of the carbon atoms at which substitution for hydrogen can occur are numbered. Note that the carbon atoms at the points of fusion (unnumbered) are completely bonded to other carbon atoms. The numbering for such fused-ring systems is established by definition.

shown. Benzene, with its one ring, is the simplest of the aromatic hydrocarbons. More complex molecules, in which two or more benzene rings are fused together, are possible. Figure 22.6 shows the structure of some of these fused ring compounds. The ultimate limit to ring fusion, obtained by extending the coronene structure in all directions, is graphite, with its sheets of hexagonal rings (see Figure 20.34).

Recall that benzene is more stable than a $C_6H_6$ molecule containing three single and three double bonds would be (Section 17.1c). The enhanced stability of benzene renders the molecule unusually unreactive when compared to other unsaturated molecules, such as the alkenes. Benzene does not undergo addition reactions such as halogenation, which necessarily would destroy the stabilized multiple bonds. Rather, benzene undergoes a variety of substitution reactions:

## 22.2 The Chemistry of Functional Groups

Most organic compounds contain not only carbon and hydrogen but other elements as well. In many instances, atoms of these elements, present either singly or combined into specific groups, display very similar properties when attached to different carbon skeletons. The presence of an OH group, for example, permits a compound to react with alkali metals with the liberation of hydrogen, independent of the precise identity of the molecule. Groups that confer such a distinctive behavior to a substance are called **functional groups.** A systematic approach to organic chemistry involves a study of the chemistry of functional groups because the chemical behavior of compounds containing such groups is determined largely by the properties of these groups, while the hydrocarbon framework plays a comparatively minor role. Some of the important classes of compounds derived from hydrocarbons by replacement of a hydrogen atom by a functional group are listed in Table 22.3.

*a. Alcohols and Ethers*  The OH group attached to a saturated carbon atom characterizes the **alcohol** family. The simplest and best-known alcohols include the following compounds (common names in parenthesis):

$CH_3OH$   methanol (methyl alcohol, wood alcohol)

$CH_3CH_2OH$   ethanol (ethyl alcohol, grain alcohol)

$CH_3CHCH_3$   isopropanol (isopropyl alcohol, rubbing alcohol)
$\quad\;\,|$
$\quad\,OH$

From a structural viewpoint alcohols can be described as either derivatives of alkanes, in which a hydroxyl group has replaced a hydrogen atom or as derivatives of water, in which the replacement is that of an alkyl group for hydrogen.

Alcohols are classified into three types depending on the number of alkyl groups attached to the carbon atom connected to the hydroxyl radical:

$RCH_2OH$   a primary alcohol

$RCHOH$   a secondary alcohol
$\;\;|$
$\;\;R'$

$\quad\;R'$
$\quad\;|$
$R—COH$   a tertiary alcohol
$\quad\;|$
$\quad\;R''$

where R, R', and R'' may be the same or different. These alcohols undergo different types of reactions, as we shall see.

The systematic names of alcohols are based on those of the alkanes having the same number of carbon atoms in a continuous chain. The alkane name is changed by changing the final *-e* to *-ol*; the chain is numbered so that the hydroxyl substituent has a lower number than any other attached group. Some illustrations of these rules are as follows:

$CH_3CHCH_2CH_3$   2-butanol
$\quad\;|$
$\quad\,OH$

CH₃CHCH₂CHCH₃  4-methyl-2-pentanol
  |       |   (not 2-methyl-4-pentanol)
 CH₃  OH

ClCH₂CHCH₃   3-chloro-2-propanol
    |
   OH

Alcohols have much higher boiling points than alkanes of comparable molecular weight. Methanol, for example, boils at 65°C while ethane, which has nearly the same molecular weight, boils at -89°C. The difference is due to the hydrogen bonding between alcohol molecules. The description of alcohols as derivatives of water emphasizes that hydrogen bonding should occur for alcohols as it does for water (Section 18.3):

$$\underset{H}{\overset{R}{\diagdown}}\ddot{O}:\text{---}H\text{---}\overset{R}{\underset{..}{\ddot{O}}}\diagup \qquad \underset{H}{\overset{H}{\diagdown}}\ddot{O}:\text{---}H\text{---}\overset{H}{\underset{..}{\ddot{O}}}\diagup$$

Pursuing the analogy one step further, alcohols can act as exceedingly weak acids and bases. They accept protons from the very strongest acids to form **ox-**

**TABLE 22.3   Compounds Derived from Alkanes by Functional Group Replacement**

| Compounds | Functional group | General formula | Example |
|---|---|---|---|
| Haloalkanes (alkyl halides) | —X[a] | R—X | CH₂BrCH₂Br<br>1,2–dibromoethane |
| Alcohols | —OH | R—OH | CH₃CHCH₃<br>      \|<br>     OH<br>2–propanol |
| Ethers | —O— | R—O—R′[b] | CH₃OCH₂CH₃<br>methyl ethyl ether |
| Aldehydes | $-\overset{\overset{O}{\|\|}}{C}-H$ | $R-\overset{\overset{O}{\|\|}}{C}-H$ | $CH_3\overset{\overset{O}{\|\|}}{C}H$<br>acetaldehyde |
| Ketones | $-\overset{\overset{O}{\|\|}}{C}-$ | $R-\overset{\overset{O}{\|\|}}{C}-R'^{b}$ | $CH_3\overset{\overset{O}{\|\|}}{C}CH_3$<br>dimethyl ketone<br>(acetone) |
| Carboxylic acids | $-\overset{\overset{O}{\|\|}}{C}-OH$ | $R-\overset{\overset{O}{\|\|}}{C}-OH$ | $CH_3\overset{\overset{O}{\|\|}}{C}OH$<br>acetic acid |
| Esters | $-\overset{\overset{O}{\|\|}}{C}-O-$ | $R-\overset{\overset{O}{\|\|}}{C}-O-R'^{b}$ | $CH_3\overset{\overset{O}{\|\|}}{C}OCH_2CH_3$<br>ethyl acetate |
| Amines | —NH₂ | RNH₂ | CH₃CH₂CH₂NH₂<br>propylamine |

[a] X can be any of the halogens.
[b] R and R′ can be the same or different alkyl groups.

**onium ions,** $ROH_2^+$, which are analogous to the hydronium ion, $H_3O^+$. Alcohols also react with the most reactive metals, e.g., sodium, to produce hydrogen and a salt:

$$ROH(l) + Na(s) \rightarrow RONa(s) + \tfrac{1}{2}H_2(g)$$

The salt RONa contains the alkoxide ion, $RO^-$, which is a stronger base than $OH^-$ and so reacts in aqueous solution to yield the alcohol,

$$RO^- + H_2O \rightarrow ROH + OH^-$$

The similarities between alcohols and water also account for the complete miscibility of the lighter alcohols and water.

Methanol and ethanol are the most widely used alcohols. Methanol, an important starting material for the synthesis of more complex molecules, is currently produced in large quantities by the catalytic hydrogenation of carbon monoxide at high temperature and pressure:

$$CO(g) + 2H_2(g) \xrightarrow[ZnO-Cr_2O_3]{400°C} CH_3OH(l)$$

Methanol is highly toxic; its ingestion in small amounts can cause blindness and even death.

Ethanol has been made since antiquity by the fermentation of naturally occurring sugars and is the alcohol in alcoholic beverages. While ethanol acts as a stimulant in small amounts, the ingestion of sufficiently large amounts depresses the activity of the respiratory center in the brain and can be fatal. Ethanol is an important industrial chemical and is produced by the acid-catalyzed hydration of ethene:

$$CH_2{=}CH_2(g) + H_2O(l) \xrightarrow{acid} CH_3CH_2OH(l)$$

In the presence of sulfuric acid, primary alcohols undergo a condensation reaction in which an ether is formed with the elimination of water, for example,

$$2\,CH_3CH_2OH(l) \xrightarrow{H_2SO_4} CH_3CH_2{-}O{-}CH_2CH_3(l) + H_2O(l)$$
$$\text{diethyl ether}$$

Ethers contain an oxygen atom attached to two organic groups whereas in alcohols one of these groups is replaced by a hydrogen atom. In contrast to the alcohols, ethers are chemically inert. Along with their favorable solvent properties, this feature makes ethers an excellent medium in which to perform organic reactions.

*b. Aldehydes and Ketones* Aldehydes and ketones are compounds containing the **carbonyl group,** $\text{\textgreater}C{=}O$. When this group is attached to two alkyl groups the compound is a **ketone,** and when it is attached to one alkyl group and a hydrogen atom, or to two hydrogen atoms, the compound is an **aldehyde.**

Some examples of aldehydes and ketones are given in Figure 22.7. The lighter aldehydes are generally known by their common names, which are derived from those of the corresponding carboxylic acids (Section 22.2c). The systematic names of aldehydes and ketones are based on those of the alkanes with the same number of carbon atoms, with the ending *-e* replaced by *-al* and *-one,* respectively. The simplest ketone,

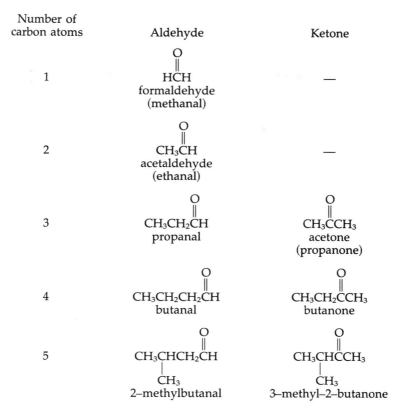

**FIGURE 22.7** Simple aldehydes and ketones. The standard names are given in parentheses for those compounds for which common names are widely used.

$$\underset{\text{CH}_3}{\overset{\overset{\displaystyle O}{\|}}{\text{C}}}\text{CH}_3$$

is universally known by the common name of acetone.

Aldehydes and ketones are polar substances owing to the presence of the polar carbonyl group. Consequently, these compounds have higher boiling points than alkanes of comparable molecular weight, although not as high as the corresponding alcohols which, as already noted, can hydrogen bond. The differences between alkanes, and aldehydes and ketones, are particularly noticeable for the lightest compounds, for which the polar group is a larger part of the molecule.

*c. Carboxylic Acids and Esters* The presence of the carbonyl and hydroxyl groups on the same carbon atom forms the **carboxyl group**. Compounds containing this group are **carboxylic acids**. The simplest carboxylic acids are formic acid and acetic acid, the acidic component of vinegar:

$$\underset{\text{formic acid}}{\overset{\overset{\displaystyle O}{\|}}{\text{HCOH}}} \quad \underset{\text{acetic acid}}{\overset{\overset{\displaystyle O}{\|}}{\text{CH}_3\text{COH}}}$$

Carboxylic acids containing three or more carbon atoms are usually given systematic names in which the *-e* in the alkane corresponding to the longest carbon chain in the acid is replaced by *-oic acid*. The carboxyl carbon is assigned number 1. Some examples of the application of these rules are

$$\underset{\text{propanoic acid}}{CH_3CH_2\overset{\overset{\displaystyle O}{\|}}{C}OH} \qquad \underset{\text{3-methylbutanoic acid}}{CH_3\underset{\underset{\displaystyle CH_3}{|}}{CH}CH_2\overset{\overset{\displaystyle O}{\|}}{C}OH}$$

Carboxylic acids are moderately weak acids. The acid ionization constant for the reaction

$$\underset{\text{carboxylic acid}}{R\overset{\overset{\displaystyle O}{\|}}{C}OH} + H_2O \rightarrow H_3O^+ + \underset{\text{carboxylate ion}}{R\overset{\overset{\displaystyle O}{\|}}{C}O^-}$$

is typically in the range of $10^{-4} - 10^{-5}$. The occurrence of ionization indicates that the acids are polar, extensively hydrogen bonded, and thus have relatively high boiling points. The lowest molecular weight carboxylic acids are completely miscible with water. The ionization of the acids results in the formation of carboxylate ions, and the names of their salts end in *-ate* in the common nomenclature, and in *-oate* in the systematic terminology. Thus $CH_3COO^-Na^+$ is sodium acetate and

$$CH_3\underset{\underset{\displaystyle CH_3}{|}}{CH}CH_2\overset{\overset{\displaystyle O}{\|}}{C}O^-K^+$$

is potassium 3-methylbutanoate. The alkali metal salts of long-chain carboxylic acids are soluble in water and are the major ingredients in soaps.

Although the carboxylic acids are weak acids, they are many orders of magnitude stronger than the alcohols. For example, the acid ionization constant of ethanol is only about $10^{-16}$. Observe that in both types of compounds ionization involves the loss of the proton in the OH group. The difference in acid strengths of alcohols and carboxylic acids can be understood in the same way as the difference in the strengths of various inorganic oxoacids (Section 19.3c). Recall that owing to the large electronegativity of oxygen, nonhydroxylic oxygen atoms attached to an atom reduce the electron density in the O—H bonds of hydroxyl groups attached to the same atom and, thereby, increase the acid strength. Now, the carboxyl group present in carboxylic acids contains one nonhydroxylic oxygen atom. On the other hand, alcohols do not contain any nonhydroxylic oxygen atoms on the C atom attached to the hydroxyl group. Consequently, they are much weaker acids than the carboxylic acids.

The strength of a carboxylic acid can be increased further by the substitution for hydrogen of an electronegative atom or group on the carbon adjacent to the carboxyl group. The presence of an additional electronegative atom further reduces the electron density in the O—H bond and thereby enhances the acid strength. Consider, for example, chloroacetic acid,

$$\overset{\leftrightarrow}{Cl}-CH_2-\overset{\overset{O}{\|}}{C}OH$$

The orientation of the Cl—C dipole indicates that the electron density on the end hydroxyl group must be lower than it is in acetic acid. Consequently, chloroacetic acid ($K_a = 1.4 \times 10^{-3}$) is a stronger acid than acetic acid ($K_a = 1.8 \times 10^{-5}$). The substitution of additional chlorine atoms increases the acid strength further. The effect of a substituent on the electron density distribution in a molecule is called an **inductive effect.** It is one of the ways in which the structure of a molecule affects its reactivity.

**Example 22.4** For each of the following pairs of acids, pick the one that is stronger: (a) FCH$_2$COOH and ClCH$_2$COOH, (b) Cl$_3$CCOOH and Cl$_3$CCH$_2$COOH, (c) CH$_3$COOH and CH$_3$CH$_2$COOH. (Hint: Alkyl groups exert an electron-donating inductive effect.)

(a) Fluorine is more electronegative than chlorine and therefore exerts a stronger electron-withdrawing inductive effect. Consequently, fluoroacetic acid is stronger than chloroacetic acid.

(b) The carbon atom attached to the chlorine atoms in Cl$_3$CCOOH is closer to the hydroxyl group than that in Cl$_3$CCH$_2$COOH. Therefore, Cl$_3$CCOOH is the stronger acid.

(c) Propanoic acid can be described as acetic acid with a methyl group substituted for a hydrogen atom. Since the methyl group increases the electron density in the hydroxyl group, propanoic acid is weaker than acetic acid. ∎

Carboxylic acids react with alcohols to form **esters** plus water. The reaction of acetic acid with ethanol in acid solution, for example, results in the formation of ethyl acetate:

$$CH_3\overset{\overset{O}{\|}}{C}OH(l) + CH_3CH_2OH(l) \underset{}{\overset{H^+}{\rightleftharpoons}} CH_3\overset{\overset{O}{\|}}{C}OCH_2CH_3(l) + H_2O(l)$$
$$\text{ethyl acetate}$$

The removal of water drives the reaction to completion. Since the reaction is that of an acid with a substance containing the hydroxyl group, and results in the formation of water, it is reminiscent of an acid–base reaction. The naming of esters as if they were organic salts of carboxylic acids is consistent with this viewpoint. However, in contrast to the usual salts, esters are not ionic compounds. The reaction of an ester with a base reverses the preceding reaction and can be used for the preparation of carboxylic acids:

$$R\overset{\overset{O}{\|}}{C}OR' + NaOH \overset{H_2O}{\longrightarrow} R\overset{\overset{O}{\|}}{C}ONa + R'OH$$

$$R\overset{\overset{O}{\|}}{C}ONa + H_3O^+ \longrightarrow R\overset{\overset{O}{\|}}{C}OH + Na^+ + H_2O$$

Many esters are volatile liquids having pleasant, fruity odors. The odor of bananas, for example, is due to butyl acetate, while octyl acetate has the odor of oranges. The ester linkage is found in many naturally occurring compounds. Waxes are esters of long-chain carboxylic acids with long-chain alcohols (both chains $C_{16}$ or greater). Fats are esters of long-chain carboxylic acids and glycerol, a propyl alcohol containing three OH groups:

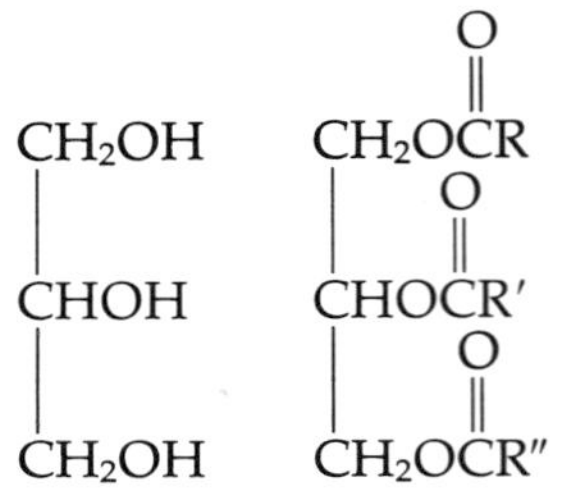

```
                    O
                    ||
CH₂OH      CH₂OCR
 |           |   O
 |           |   ||
CHOH       CHOCR'
 |           |   O
 |           |   ||
CH₂OH      CH₂OCR"
glycerol     a fat
```

***d. Redox Reactions Involving Functional Groups*** While the oxidation state of carbon can be determined by the procedure described in Section 15.5b, it is not customary to use oxidation states in the analysis of organic redox reactions. The functional groups of interest differ in the number of hydrogen and oxygen atoms attached to a given carbon atom. Since the oxidation number of oxygen is −2 and that of hydrogen is +1, the oxidation number of carbon must change with the number of attached oxygen and hydrogen atoms. An organic compound undergoes oxidation when it loses hydrogen or gains oxygen; it undergoes reduction when it gains hydrogen or loses oxygen. Figure 22.8 shows the various functional groups in order of increasing oxidation state of carbon. The correlation with the number of attached hydrogen and oxygen atoms is evi-

| Compound | Formula | Oxidation state of carbon |
|---|---|---|
| Carbon dioxide | Ö=C=Ö | +4 |
| Formic acid | HC—ÖH (with =O above C) | +2 |
| Formaldehyde | HCH (with =O above C) | 0 |
| Methanol | H₃C—ÖH | −2 |
| Methane | CH₄ | −4 |

↑ Oxidation     ↓ Reduction

**FIGURE 22.8** Oxidation and reduction in compounds containing one carbon atom attached to hydrogen or oxygen. Oxidation involves the gain of oxygen or the loss of hydrogen and reduction the opposite. The oxidation states of carbon have been determined by the method of Section 15.5b.

dent: carbon is in its most reduced state (−4) in CH$_4$, a hydrocarbon, and in its most oxidized state (+4) in CO$_2$, an oxide. In order of increasing oxidation state between these limits, we find the alcohols, aldehydes or ketones, and carboxylic acids.

The presence of carbon in different oxidation states suggests that it should be possible to convert compounds with different functional groups into each other by means of redox reactions. As summarized in Figure 22.9, redox reactions are indeed commonly used for the conversion of alcohols to aldehydes, ketones, and carboxylic acids, and vice versa. Primary alcohols are readily oxidized by dichromate in acidic solution to the corresponding aldehydes, for example,

$$3CH_3(CH_2)_2CH_2OH(l) + Cr_2O_7^{2-} + 8H^+ \rightarrow 3CH_3(CH_2)_2\overset{\overset{O}{\|}}{C}H(l) + 2Cr^{3+} + 7H_2O(l)$$
$$\text{1-butanol} \hspace{5cm} \text{butanal}$$

The aldehydes readily undergo further oxidation to carboxylic acids, for example,

$$3CH_3(CH_2)_2\overset{\overset{O}{\|}}{C}H(l) + Cr_2O_7^{2-} + 8H^+ \rightarrow 3CH_3(CH_2)_2\overset{\overset{O}{\|}}{C}OH(l) + 2Cr^{3+} + 4H_2O(l)$$
$$\hspace{7cm} \text{butanoic acid}$$

In order to prevent this further oxidation step, aldehydes must be distilled out of the reaction mixture as they are formed. Since aldehydes form weaker hydrogen bonds than alcohols, their boiling points are lower and such a distillation is practical.

The oxidation of secondary alcohols leads to the formation of ketones under similar conditions as those just described, for example,

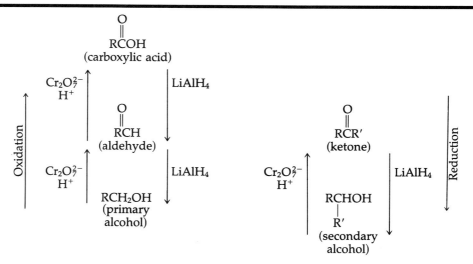

**FIGURE 22.9** Examples of the interconversion of alcohols, aldehydes or ketones, and carboxylic acids by means of redox reactions.

$$3CH_3\underset{\underset{CH_3}{|}}{C}HOH(l) + Cr_2O_7^{2-} + 8H^+ \rightarrow 3CH_3\overset{\overset{O}{\|}}{C}CH_3(l) + 2Cr^{3+} + 7H_2O(l)$$

2-propanol                                           acetone

In contrast to aldehydes, ketones are highly resistant to further oxidation. In a ketone, the carbon atom in the carboxyl group is attached to two other carbon atoms. Consequently, the replacement of a C—H bond by a C—O bond, which occurs in the oxidation of an aldehyde to a carboxylic acid, cannot occur. Oxidation of a ketone requires that the carbon skeleton of the molecule be broken. This happens only under special conditions. For the same reason, tertiary alcohols also are resistant to oxidation.

The oxidation reactions illustrated can be reversed by means of suitable reducing agents, thereby converting compounds with carboxyl and carbonyl groups to alcohols. Carboxylic acids and aldehydes can be reduced to primary alcohols, and ketones can be reduced to secondary alcohols. The mixed metal hydrides, lithium aluminum hydride (LiAlH$_4$) or sodium borohydride (NaBH$_4$), are commonly used as the reducing agent.

## 22.3 Optical Stereoisomerism

Recall from Section 21.5d that two stereoisomers which, in analogy to the left and right hand are nonsuperimposable mirror images of each other (see Figure 21.11), are called **enantiomers** and are examples of **optical isomers**. The term **chirality** is also used to refer to the property of "handedness"; objects that are not superimposable on their mirror images are said to be chiral.

Optical isomerism plays an important role in the chemistry of carbon. Because the four bonds of carbon are directed at tetrahedral angles, a carbon atom bonded to four different groups, called an asymmetric or chiral carbon atom, will form two different enantiomers when incorporated into a molecule.

The alcohols provide a good illustration of stereoisomerism. The simplest chiral alcohol is 2-butanol. The second carbon atom in the chain is attached to four different groups and is thus asymmetric:

$$H_3C - \underset{\underset{OH}{|}}{\overset{\overset{H}{|}}{C}} - CH_2CH_3 \quad \text{or} \quad H_3C - \underset{\underset{H}{|}}{\overset{\overset{OH}{|}}{C}} - CH_2CH_3$$

The two structures are nonsuperimposable mirror images of each other and are a pair of enantiomers. A two-dimensional representation actually does not convincingly show that this is indeed the case. If the molecule were planar, rotation about a C—C single bond would convert the one form into the other. This is not the case, however, for a tetrahedral molecule, as can be seen in the representations in Figure 22.10 or even more clearly by means of actual three-dimensional molecular models. In fact, it was the occurrence of this type of isomerism that led van't Hoff to postulate that the spatial arrangement of the atoms bonded to carbon must be tetrahedral. This brillant insight was confirmed by structural evidence some years later, and van't Hoff was awarded the first Nobel prize for chemistry in 1901.

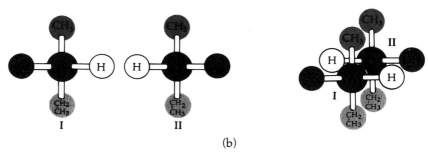

**FIGURE 22.10** Models of the 2-butanol enantiomers. (a) Structural formulas and (b) ball and stick models, showing (right) that the enantiomers are not superimposable.

As noted in Section 21.5d, enantiomers are identical with respect to those properties that have no chiral implications, e.g., boiling point, density, and enthalpy of formation. They also react in an identical way with non-chiral substances. However, enantiomers differ in their interaction with plane-polarized light, as shown in Figure 22.11, and are therefore said to be optically active. The enantiomer which rotates the plane of polarization in the clockwise direction is **dextrorotatory** and is symbolized by a (+)- or a D- in front of the name of the compound. The mirror image enantiomer is **levorotatory** and is symbolized by a (−)- or an L-, e.g., (−)-2-butanol. An equimolar mixture of a pair of enantiomers is called a **racemic mixture.** Since it contains an equal number of dextrorotating and levorotating molecules, such a mixture is optically inactive.

In addition to differing in their optical activity, enantiomers also differ in their reactions with other chiral substances. This difference is of particular importance in biochemical reactions.

**Example 22.5** Which of the unbranched chloropentanes form optical stereoisomers?

The three structural isomers of chloropentane are

$$\underset{\text{1-chloropentane}}{ClCH_2CH_2CH_2CH_2CH_3} \quad \underset{\text{2-chloropentane}}{CH_3\overset{\overset{\displaystyle Cl}{|}}{C}HCH_2CH_2CH_3} \quad \underset{\text{3-chloropentane}}{CH_3CH_2\overset{\overset{\displaystyle Cl}{|}}{C}HCH_2CH_3}$$

Only the second compound has an asymmetric carbon atom (the second carbon atom is attached to CH₃, Cl, H, and CH₂CH₂CH₃ groups) and exhibits optical stereoisomerism. ∎

# 850 The Chemistry of Carbon 22.4

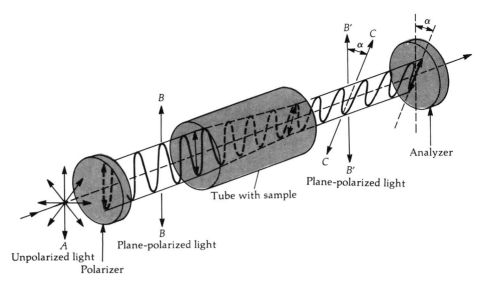

**FIGURE 22.11** The measurement of optical rotation. Ordinary light is unpolarized, i.e., electric and magnetic fields can oscillate in all directions perpendicular to the direction of motion of the wave. A polarizer permits the passage of only a single oscillation component (vertical in the figure). An optically active substance placed in the path of this plane-polarized light rotates the plane of polarization by an angle $\alpha$. If the substance were replaced by an equal amount of its mirror image enantiomer, the plane of polarization would be rotated in the opposite direction, by an angle $-\alpha$. The figure shows a case of clockwise rotation.

## 22.4 Mechanisms of Organic Reactions

*a. Reactive Intermediates* Reaction mechanisms are of special importance in organic chemistry because they help to make sense of the great variety of possible reactions. As in the case of inorganic processes, organic reactions frequently involve the formation of some reactive intermediate. Three general types of intermediates are common; their identity is determined by the manner in which the bond between carbon and some other atom breaks. If each atom retains one of the shared electrons, a **free radical intermediate** is formed:

$$-\overset{|}{\underset{|}{C}}:Z \rightarrow -\overset{|}{\underset{|}{C}}\cdot + \cdot Z$$

If the carbon loses an electron it becomes a positively charged **carbocation,** also known as a **carbonium ion,**

$$-\overset{|}{\underset{|}{C}}:Z \rightarrow -\overset{|}{\underset{|}{C}}^{+} + :Z^{-}$$

Finally, if the carbon retains the second electron, it becomes a negatively charged **carbanion,**

$$-\overset{|}{\underset{|}{C}}:Z \rightarrow -\overset{|}{\underset{|}{C}}:^{-} + Z^{+}$$

Each type of intermediate has specific properties which are of importance in determining the course of the reactions in which they participate. By virtue of their nonbonded electron pair, carbanions are electron-pair donors. Such substances are commonly called **nucleophiles** since they seek some positively charged species to neutralize their charge. Recall from Section 15.3b that electron-pair donors are Lewis bases. As we shall see, the Lewis acid–base definitions turn out to be particularly useful in organic chemistry. The concept of a nucleophile is not confined to an intermediate. Many species with nonbonded electrons pairs, such as :OH$^-$ or :NH$_3$, display the same behavior.

Carbocations display just the opposite behavior. They are electron deficient and are consequently said to be **electrophiles,** species seeking out electrons to complete the octet. Recall that this behavior is characteristic of Lewis acids. Many organic reactions are catalyzed by Lewis acids such as AlCl$_3$ because of the ability of these substances to react with nucleophiles.

Keeping these general concepts in mind, we are now ready to examine some typical organic reaction mechanisms.

***b. Nucleophilic Substitution in Alkyl Halides*** In a nucleophilic substitution reaction, a nucleophile displaces some other substituent in a molecule. Alkyl halides undergo a variety of nucleophilic substitution reactions, examples of which are

$$OH^- + CH_3CH_2Cl \rightarrow CH_3CH_2OH + Cl^-$$

$$CH_3O^- + CH_3Br \rightarrow CH_3OCH_3 + Br^-$$

The first species in each of these equations is a nucleophile or, in acid–base language, a strong Lewis base. The halogen group is, in this context, called a "leaving group." This type of substitution reaction occurs whenever the leaving group is a relatively stable weak Lewis base. The halide ions fulfill this function and these reactions can be used to prepare alcohols and ethers, respectively.

Reactions of this type obey either first-order or second-order kinetics. For example, the reaction

$$CH_3Cl + OH^- \rightarrow CH_3OH + Cl^-$$

obeys the second-order rate law

$$-d[CH_3Cl]/dt = k[CH_3Cl][OH^-]$$

while the reaction

$$(CH_3)_3CCl + OH^- \rightarrow (CH_3)_3COH + Cl^-$$

displays first-order kinetics,

$$-d[(CH_3)_3CCl]/dt = k[(CH_3)_3CCl]$$

As we shall see, the second-order reactions of this type are elementary bimolecular reactions, while the first-order substitutions involve a series of steps, with the rate-determining step being unimolecular. These two types of nucleophilic substitution reactions are commonly designated S$_N$2 and S$_N$1, respectively, where S stands for substitution, N for nucleophilic, and the number for the molecularity of the rate-determining step.

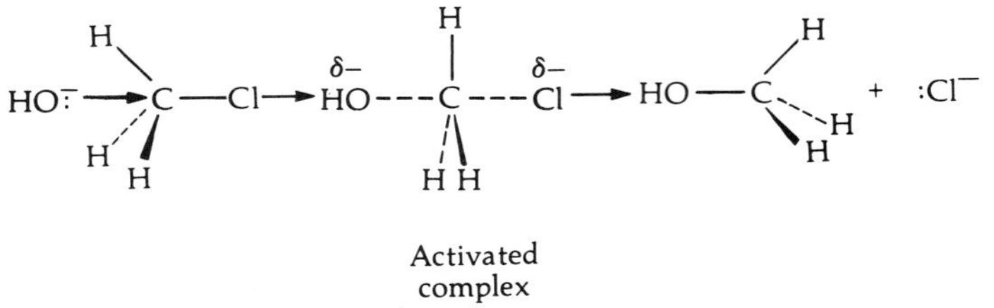

**FIGURE 22.12** One-step displacement mechanism for an S$_N$2 reaction such as CH$_3$Cl + OH$^-$ → CH$_3$OH + Cl$^-$.

We begin with an examination of the S$_N$2 mechanism. The conversion of reactants to products is shown in detail in Figure 22.12. The nucleophilic reactant attacks the alkyl halide molecule from the back side, that is, the side directly opposite the leaving group. A bond between the OH$^-$ ion and the carbon atom begins to form while at the same time the C—Cl bond starts to weaken. In the activated complex, the carbon atom is weakly bonded to both OH and Cl, and the negative charge brought in by the hydroxide ion is equally distributed between the two ends of the complex.

The variation of the potential energy along the reaction coordinate is displayed in Figure 22.13. The curve has a shape similar to those examined in Section 14.9 and can be understood in the same way: the reaction proceeds through an unstable activated complex. The activation energy is 109 kJ mol$^{-1}$, and the overall reaction is exothermic, with an energy release of 75 kJ mol$^{-1}$.

Why does the superficially similar reaction between tert-butyl chloride, (CH$_3$)$_3$CCl, and hydroxide ion follow an S$_N$1 rather than an S$_N$2 mechanism? The main reason for the difference is the occurrence of **steric hindrance,** the name given to the effects of bulky groups in close proximity to each other. We have seen that the formation of the activated complex in S$_N$2 reactions requires that the OH$^-$ approach the chlorine-bearing carbon from the back side. However, the three methyl groups attached to the carbon atom in (CH$_3$)$_3$CCl block the path of the OH$^-$ and effectively prevent the formation of the activated complex. It is apparent that this effect should become increasingly important as the groups attached to the carbon bearing the leaving group increase in size. Figure 22.14 shows some examples of this effect for increasingly bulky sub-

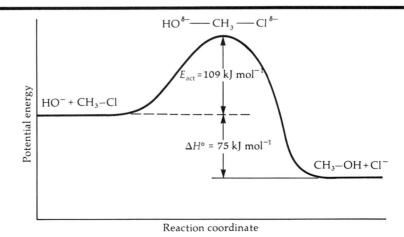

**FIGURE 22.13** Potential energy diagram for the reaction of methyl chloride with hydroxide ion.

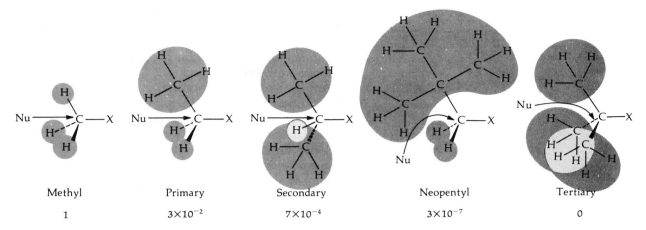

**FIGURE 22.14** Effect of increasing steric hindrance on the rate of nucleophilic $S_N2$ substitution in alkyl halides. Nu is a nucleophilic group such as $OH^-$. The numbers below the figures are the relative rates of reaction.

stituents. The rate of the $S_N2$ reaction is seen to be slowed by many orders of magnitude as a result of steric hindrance.

Although the formation of tert-butyl alcohol from the corresponding chloride cannot involve an $S_N2$ mechanism, the reaction actually occurs readily. As we have already noted, the reaction obeys first-order kinetics and proceeds via an $S_N1$ mechanism. The following steps are involved:

$$(1) \quad CH_3-\underset{\underset{CH_3}{|}}{\overset{\overset{CH_3}{|}}{C}}-Cl \rightarrow CH_3-\underset{\underset{CH_3}{|}}{\overset{\overset{CH_3}{|}}{C^+}} + Cl^- \qquad \text{Slow}$$

$$(2) \quad CH_3-\underset{\underset{CH_3}{|}}{\overset{\overset{CH_3}{|}}{C^+}} + :OH_2 \rightarrow CH_3-\underset{\underset{CH_3}{|}}{\overset{\overset{CH_3}{|}}{C}}-OH_2^+ \qquad \text{Fast}$$

$$(3) \quad CH_3-\underset{\underset{CH_3}{|}}{\overset{\overset{CH_3}{|}}{C}}-OH_2^+ + H_2O \rightarrow CH_3-\underset{\underset{CH_3}{|}}{\overset{\overset{CH_3}{|}}{C}}-OH + H_3O^+ \quad \text{Fast}$$

The first step involves the cleavage of the carbon–chlorine bond with the formation of a carbocation. This species is ordinarily very unstable but when the carbon bearing the charge is attached to three alkyl groups, the charge can be dispersed to some extent over all these groups in the following manner:

$$CH_3-\underset{\underset{CH_3}{|}}{\overset{\overset{CH_3}{|}}{C^+}} \rightarrow CH_3-\underset{\underset{CH_3\,\delta^+}{|}}{\overset{\overset{CH_3\,\delta^+}{|}}{C}}\,\delta^+$$

The resulting decrease in charge density increases the stability of the carbocation with a consequent reduction in the activation energy.

The tert-butyl carbocation formed in the first step is still highly reactive, however, and in step 2 it rapidly combines with one of the solvent water molecules or, occasionally, with a less plentiful hydroxide ion. In the final step, the oxonium ion rapidly transfers a proton to a water molecule with the formation of the alcohol. The first step determines the reaction rate and, since it involves a unimolecular decomposition, the overall reaction obeys first-order kinetics.

*c. Alkene Addition Reactions*  Alkenes readily undergo addition reactions. The pi electrons in a carbon–carbon double bond are localized in a rather exposed position (see Figure 16.26) and the double bond is susceptible to attack by electrophilic reagents. These include positive species such as $H^+$, electron-deficient species such as Lewis acids, and electronegative elements like the halogens. The alkenes undergo addition reactions of the following types with these reagents:

$$\mathrm{C{=}C} + HX \longrightarrow -\underset{H}{\underset{|}{C}}-\underset{X}{\underset{|}{C}}- \quad \text{alkyl halide formation}$$

$$\mathrm{C{=}C} + HOH \xrightarrow{H^+} -\underset{H}{\underset{|}{C}}-\underset{OH}{\underset{|}{C}}- \quad \text{alcohol formation}$$

$$\mathrm{C{=}C} + X_2 \xrightarrow{h\nu} -\underset{X}{\underset{|}{C}}-\underset{X}{\underset{|}{C}}- \quad \text{alkyl dihalide formation}$$

The addition of a hydrogen halide to an asymmetric alkene, that is, one which is not divided into equal halves by the double bond, is of special interest. The addition should result in two isomeric alkyl halides but, suprisingly, only one of them is formed. Consider, for example, the addition of HCl to propene:

$$CH_2{=}CHCH_3 + HCl \begin{cases} \nearrow CH_3CHCH_3 \;\;\; \text{2–chloropropane} \\ \phantom{\nearrow}\;\;\; | \\ \phantom{\nearrow}\;\;\; Cl \\ \searrow\!\!\!\!\!+\!\!+\!\!+\!\!+ ClCH_2CH_2CH_3 \;\;\; \text{1–chloropropane} \end{cases}$$

Only the 2-chloropropane isomer is formed. Many illustrations of this selectivity can be cited. All the examples are consistent with an empirical law, known as **Markovnikov's rule,** which states that in the addition of HX to an alkene, the H atom adds to the carbon with the greater number of hydrogens. Thus, in the addition of HCl to propene, the first carbon atom has two attached hydrogen atoms while the second carbon atom has only one. In agreement with the rule, the hydrogen atom bonds to the first carbon atom.

The reason for Markovnikov's rule becomes apparent when the reaction mechanism is examined. The addition reaction occurs by the following two-step mechanism:

(1) $\displaystyle \mathop{C}\limits^{\diagdown}\!\!=\!\!\mathop{C}\limits^{\diagup} + HX \rightarrow -\underset{+}{C}-\overset{H}{\underset{|}{C}}- + X^-$    Slow

(2) $-\overset{H}{\underset{+}{C}}-\overset{H}{\underset{|}{C}}- + X^- \rightarrow -\overset{H}{\underset{|}{C}}-\overset{H}{\underset{X}{\overset{|}{C}}}-$    Fast

In the rate-determining first step, a proton adds to the double bond with the formation of a carbocation. This highly reactive species combines with a halide ion in the fast second step.

Markovnikov's rule can be understood when we recall from the preceding section that a carbocation becomes increasingly stable as the number of attached alkyl groups increases. These groups are more effective in dispersing the positive charge than attached hydrogen atoms and the carbocation is therefore more stable. The greater stability corresponds to a lower activation energy in the rate-determining step, and the addition reaction therefore occurs in the described manner. For example, in the addition of HCl to propene, the two possible carbocations resulting from proton addition are

(1)  $CH_3CH=CH_2 + H^+ \rightarrow CH_3CH_2\overset{+}{C}H_2$    a primary carbocation

(2)  $CH_3CH=CH_2 + H^+ \rightarrow CH_3\overset{+}{C}HCH_3$    a secondary carbocation

The primary carbocation is less stable than the secondary carbocation, and consequently the latter is formed. Thus, the reaction leads to the eventual formation of 2-chloropropane.

Markovnikov's rule is not restricted to the addition of hydrogen halides but applies to the ionic addition of any polar reagent. The more positive component of the added reagent will attach to a double-bonded carbon so as to form the more stable carbocation. The following example illustrates this rule and shows how the rather general ideas about molecular stability described above make it possible to predict the outcome of a reaction.

---

**Example 22.6**  Show the steps in the acid-catalyzed hydration of 2-methylpropene and indicate which product is formed.

The reaction can be written in the form

$$\underset{\displaystyle CH_3\underset{|}{\overset{|}{C}}=CH_2}{\overset{CH_3}{}} + \overset{\delta+}{H}-\overset{\delta-}{OH} \rightarrow$$

The first step involves the addition of $H^+$ so as to form the more stable carbocation, which in this case has two methyl groups on the charged carbon atom, rather than two hydrogen atoms:

$$\text{(1)} \quad \underset{\underset{CH_3}{|}}{CH_3C}=CH_2 + HOH \xrightarrow{H^+} \underset{\underset{+}{\underset{CH_3}{|}}}{CH_3\overset{CH_3}{\overset{|}{C}}CH_3} + OH^-$$

In the second step the hydroxide ion adds,

$$\text{(2)} \quad \underset{\underset{+}{\underset{CH_3}{|}}}{CH_3\overset{CH_3}{\overset{|}{C}}CH_3} + OH^- \xrightarrow{H^+} \underset{\underset{OH}{|}}{CH_3\overset{CH_3}{\overset{|}{C}}CH_3}$$

forming 2-methyl-2-propanol. (The reaction occurs rapidly only in the presence of an $Hg^{2+}$ catalyst.) ■

---

# Biochemistry

## 22.5 Carbohydrates

With the carbohydrates, which include the various types of sugars, starches, and celluloses, we come to the study of compounds that are of importance in life processes. Carbohydrates are the most abundant constituents of plants, which produce them from inorganic materials by photosynthesis. They are one of the chief ingredients of foodstuffs, and the oxidation of carbohydrates in living organisms serves as an important source of energy. Carbohydrates also are components of such widely used products as paper, wood, and cotton.

The name *carbohydrate* originated in the fact that many of these compounds have the empirical formula $C_n(H_2O)_m$, that is, they appear to be hydrates of carbon. Their actual structures show that they contain both the hydroxyl and carbonyl functional groups and can be regarded as polyhydroxy aldehydes and ketones, or substances that hydrolyze to yield such compounds.

*a. Monosaccharides* The simplest carbohydrates, known as **monosaccharides**, cannot be hydrolyzed to yield smaller carbohydrates, and we begin by examining some of their properties. The most important of these compounds are the sugars containing five or six carbon atoms, known as pentoses and hexoses, respectively. The structure of the hexose D-glucose (D for dextrorotatory), which is both the sugar in blood as well as the repeating unit of starch, illustrates the structure of the monosaccharides. Figure 22.15 shows several representations of this molecule. An equilibrium exists between an open-chain and a cyclic form called α-D-glucose (Figure 22.15a). The former consists of a six-carbon chain containing an aldehyde group at one end and hydroxyl groups on the other carbon atoms. In the conversion to the cyclic form, the carbonyl group of the aldehyde reacts with one of the hydroxyl groups to form a compound called a hemiacetal, which includes a C—O—C linkage in the resulting ring. The cyclic form of the pentoses and hexoses is usually more stable than the open-chain form and the equilibrium strongly favors ring formation.

As may be seen in Figure 22.15a, the open-chain form of a hexose has four asymmetric carbon atoms, each of which is attached to a hydrogen atom

```
         1  CHO                                          1  CHO
      H—C—OH              6 CH₂OH                    HO—C—H                6 CH₂OH
                         5      O                                         5      O
      HO—C—H         H        H                      H—C—OH          H        H    OH
                     4  H                                            4  H
                       OH    1                       HO—H—H              OH    1
      H—C—OH        HO  3   2  OH                                     HO  3   2  H
                       H    OH                       HO—C—H              H    OH
      H—C—OH
                                                        CH₂OH
      6  CH₂OH                                       6

      D—Glucose         α—D—Glucose                 L—Glucose           β—D—Glucose
           (a)                                          (b)                 (c)
```

**FIGURE 22.15** Glucose. (a) The equilibrium between D-glucose and α-D-glucose; (b) the open-chain form of L-glucose; (c) β-D-glucose, which has the opposite configuration from α-D-glucose at C-1.

and an OH group as well as to two unequal portions of the carbon framework. Each of these carbon atoms has two configurations. Consequently, there must be $2^4 = 16$ optical isomers of hexose, which can be grouped into 8 pairs of enantiomers. For example, the enantiomer of D-glucose is L-glucose (L for levorotatory), which has the H and OH groups on each asymmetric carbon atom reversed (Figure 22.15b).

Not all of the optical isomers of the hexoses, or any other compounds that contain more than one asymmetric carbon atom, are enantiomers, or exact mirror images of each other. For example, D-mannose, shown in Figure 22.16, differs from D-glucose in the configuration of one of the four asymmetric carbon atoms. Compounds such as D-glucose and D-mannose, which are optical isomers of one another but are not enantiomers, are called **diastereomers**. In contrast to enantiomers, diastereomers generally differ in their properties.

The cyclic form of glucose (and all other simple sugars) is even more complex than the open-chain form since it contains five asymmetric carbon atoms (labelled 1–5 in Figure 22.15a and c). The additional asymmetric atom is the one which belonged to the aldehyde group in the open-chain form; it is labelled C-1 in the figure. The number of stereoisomers is therefore increased to $2^5 = 32$. In-

**FIGURE 22.16** Example of diastereomers. D-Mannose and D-glucose are optical isomers, but are not enantiomers. L-Glucose (see Figure 22.15), is also a diastereomer but not an enantiomer of D-mannose.

```
         CHO                    CHO
      HO—C—H               H—C—OH
      HO—C—H               HO—C—H
      H—C—OH               H—C—OH
      H—C—OH               H—C—OH
         CH₂OH                  CH₂OH
       D–mannose             D–glucose
```

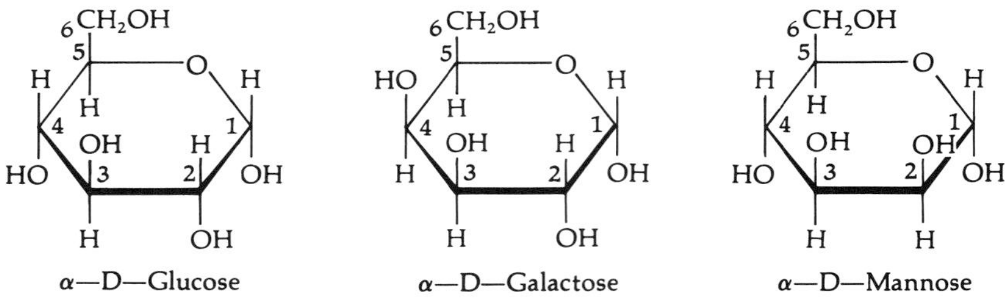

**FIGURE 22.17** The three naturally occurring hexose sugars.

stead of renaming all these compounds, it is customary to use the prefixes $\alpha$ and $\beta$ to designate the configuration of C-1. The structures of $\alpha$-D-glucose and $\beta$-D-glucose are shown in Figures 22.15a and 22.15c and are seen to differ only in the C-1 configuration. The two molecules yield the same open-chain D-glucose when the rings are opened, and can be interconverted by breaking and reforming the ring. There are thus eight types of cyclic hexose sugars, each one having four variants, i.e., $\alpha$-D, $\beta$-D, $\alpha$-L, and $\beta$-L. Only three of these, glucose, galactose, and mannose (Figure 22.17) are commonly found in nature, chiefly in the form of the D-isomers, with glucose having the greatest biochemical importance. It is interesting to note that living organisms generally can only synthesize and use one enantiomeric form of a molecule. Only D-glucose, for instance, is found in human blood. By contrast, glucose synthesized in the laboratory from simpler molecules is a racemic mixture of D- and L-glucose.

**b. Polysaccharides** Polysaccharides consist of two or more monosaccharide rings joined together. The simplest of these compounds are the disaccharides, which consist of two monosaccharide rings. The most common disaccharide is sucrose, ordinary table sugar. As may be seen in Figure 22.18, sucrose is composed of a glucose ring joined by an oxygen bridge (a glycosidic linkage) to fructose, a hexose with a ketone group. Another example of a disaccharide is lactose, the sugar found in milk, which consists of joined glucose and galactose rings. In both instances, the C-1 atom has the $\alpha$ configuration and the bridge is called an $\alpha$ glycosidic linkage.

Some polysaccharides contain hundreds or even thousands of monosaccharide units joined together to form polymers. Three of the most important of these polymers are starch, cellulose, and glycogen. All three are polymers of glucose differing from each other in molecular weight, nature of the glycosidic linkage, and extent of branching of the glucose chains.

A large fraction of the energy required for the sustenance of life is supplied by carbohydrates, chiefly in the form of starch. This substance is a mixture of branched and unbranched polymers of glucose characterized by the $\alpha$ glycosidic linkage shown in Figure 22.19a. Glycogen is a similarly linked, highly branched carbohydrate which serves as an energy reservoir in animals. The first step in the utilization of polysaccharides by animals is the hydrolysis of starch or glycogen to individual glucose molecules. This hydrolysis is catalyzed by enzymes in the digestive tract that have the ability to attack the $\alpha$ glycosidic linkage.

Cellulose, the most abundant component of plants, is a long-chain polymer of glucose in which the glucose rings are joined by a $\beta$ glycosidic linkage

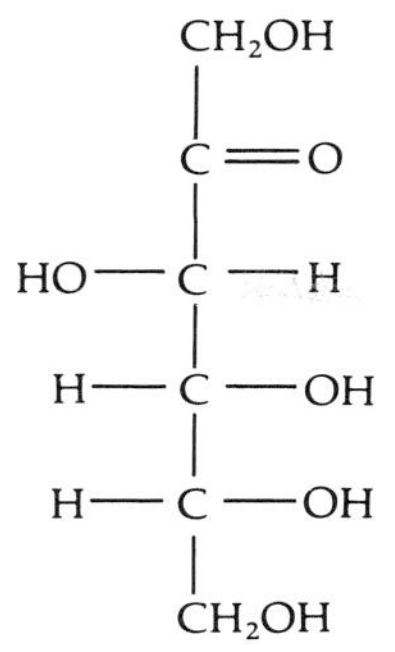

D-Fructose

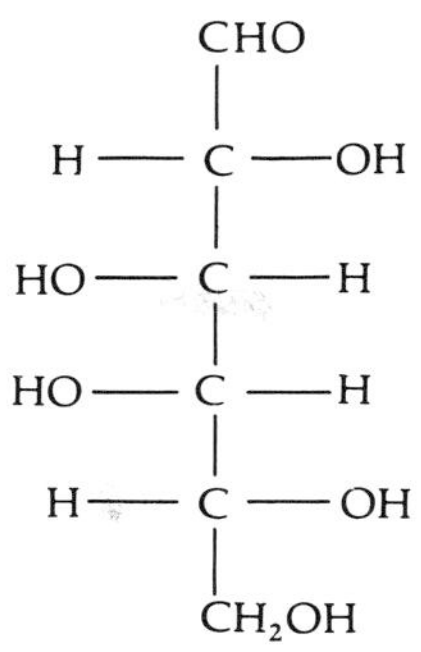

D-Galactose

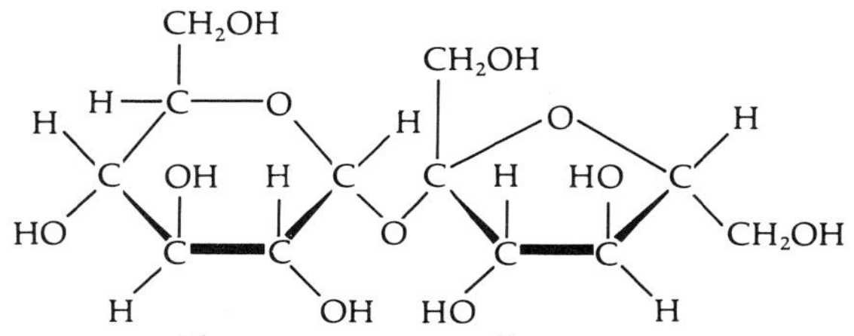

Glucose Fructose

Sucrose

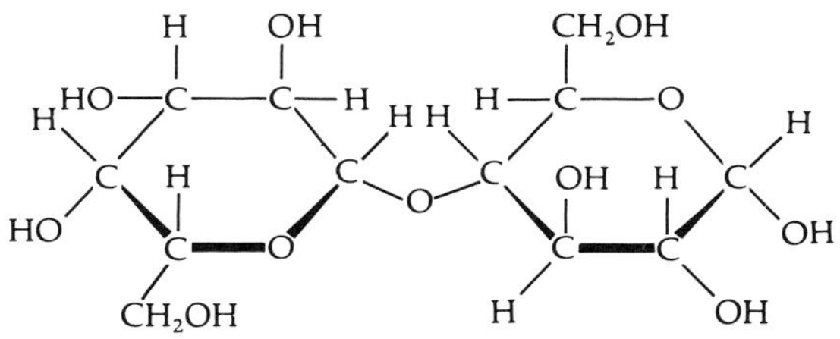

Galactose Glucose

Lactose

**FIGURE 22.18** The structure of sucrose and lactose, two common disaccharides. In these disaccharides, glucose combines with fructose and galactose, respectively.

(Figure 22.19b). This seemingly minor difference from starch has important practical consequences, since many animals, including man, lack the enzymes needed to break down cellulose to glucose. Cellulose can, however, be used indirectly as a foodstuff because grazing animals have bacteria in their intestines that possess the enzymes that can accomplish this task. This difference illustrates the power of enzymes in biological reactions. **Enzymes** are complex naturally occurring molecules, generally proteins (Section 22.6), which function as catalysts in biological reactions. Their chief attributes are high specificity and high catalytic effectiveness. A given enzyme can speed up the rate of a particular reaction by a factor as large as $10^{10}$ without having any discernible effect on other reactions.

*c. **Photosynthesis and Metabolism of Carbohydrates*** Carbohydrates are intimately involved in life processes. Green plants photosynthesize them from car-

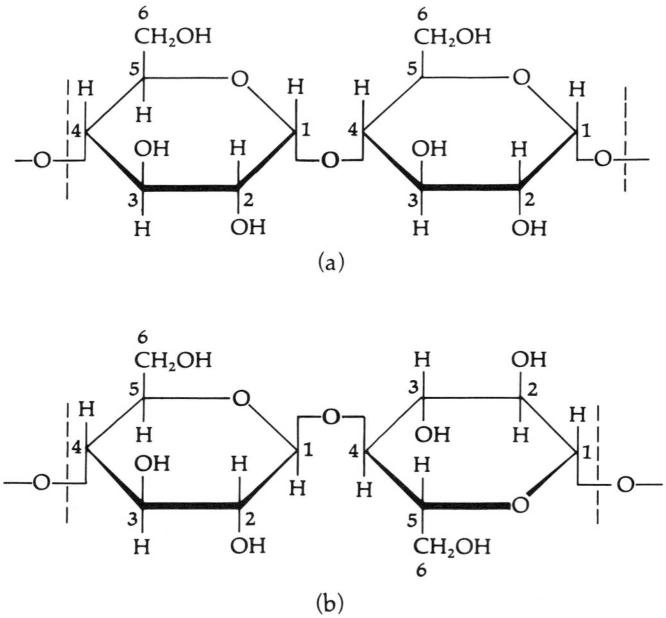

**FIGURE 22.19** Subunits of the chain structure of (a) starch and (b) cellulose, focusing on the important difference between the two: an α-glycosidic linkage in starch and a β glycosidic linkage in cellulose.

bon dioxide and water using solar energy to drive the reaction. In turn, animals use carbohydrates as food in order to obtain the substances and energy needed to sustain life. The complex reactions by which carbohydrates are eventually oxidized back to carbon dioxide and water are an important component of **metabolism,** the general term for the chemical reactions occurring in living organisms. Photosynthesis in plants and metabolism in animals are parts of a vast cyclic process involving the conversion of carbon from inorganic to organic forms and back. Figure 22.20 summarizes some of the essential aspects of this cycle.

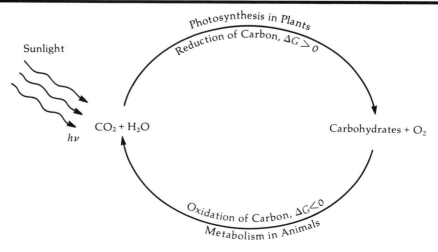

**FIGURE 22.20** Simplified representation of the cyclic interconversion of carbon between inorganic and organic forms by plants and animals.

The overall reaction occurring in photosynthesis is deceptively simple. The formation of glucose, the basic building block of plant carbohydrates, can be represented as

$$6CO_2(g) + 6H_2O(l) \rightarrow C_6H_{12}O_6(s) + 6O_2(g) \qquad \Delta G° = 2870 \text{ kJ}$$

The sign and magnitude of $\Delta G°$ indicate that the overall reaction is very nonspontaneous. It only occurs because of the energy provided by sunlight. Solar radiation is absorbed in plants by molecules of chlorophyll (see Figure 21.26). The ability of chlorophyll to absorb visible radiation is primarily due to its extended system of conjugated double bonds (Section 17.4a). The conversion of solar energy to a usable form and the reduction of carbon dioxide to glucose proceed by a complex series of reactions involving numerous intermediates.

The solar energy stored in carbohydrates is released when these substances undergo metabolic oxidation. The overall reaction is just the reverse of that occurring in photosynthesis. Although the oxygen that animals breathe in is the ultimate oxidant of carbohydrates, the actual process is rather complex and involves several intermediary oxidizing agents and enzymes.

## 22.6 Proteins and Nucleic Acids

Proteins are macromolecules that play a paramount role in practically all biological processes. The proteins known as enzymes catalyze nearly all chemical reactions occurring in living organisms; other proteins are involved in the transport of small molecules, e.g., hemoglobin transports oxygen through the bloodstream; still different proteins are the major components of muscle; others, known as antibodies, help fight foreign organisms. The synthesis of proteins in living organisms is controlled by nucleic acids. Many of the recent advances in biochemistry and molecular biology have been concerned with the structure and function of nucleic acids and proteins.

*a. Amino Acids* The basic units of proteins are amino acids, which have the structural formula

$$\begin{array}{c} \text{H} \\ | \\ \text{R}\!-\!\text{C}\!-\!\text{COOH} \\ | \\ \text{NH}_2 \end{array}$$

The amino ($NH_2$) group and the radical R are attached to the carbon directly bonded to the carboxyl (COOH) group, customarily designated the $\alpha$ carbon. Since this atom is attached to four different groups, the molecule can exist as a pair of enantiomers (except when R = H). As in the case of the sugars, the two isomers are labelled with the prefixes D- and L-. Only the L-isomers are constituents of natural proteins.

The identity of an amino acid is determined by that of the side chain R. Twenty different amino acids are commonly found in proteins. Rather remarkably, the identity of these acids is independent of the protein source, be it bacterium or man. Table 22.4 lists the names, three-letter symbols, and structural formulas of some representative amino acids. The simplest amino acid is glycine, in which the side chain is just a hydrogen atom. Some amino acids contain aliphatic or aromatic side chains, for example valine and tyrosine, re-

TABLE 22.4  Some Representative Amino Acids

| Name | Abbreviation | Structural Formula |
|---|---|---|
| Glycine | Gly | $H-\underset{\underset{NH_2}{\vert}}{\overset{\overset{H}{\vert}}{C}}-C\underset{OH}{\overset{\nearrow O}{}}$ |
| Valine | Val | $\underset{CH_3}{\overset{CH_3}{\diagdown}}CH-\underset{\underset{NH_2}{\vert}}{\overset{\overset{H}{\vert}}{C}}-C\underset{OH}{\overset{\nearrow O}{}}$ |
| Tyrosine | Tyr | $HO-C_6H_4-CH_2-\underset{\underset{NH_2}{\vert}}{\overset{\overset{H}{\vert}}{C}}-C\underset{OH}{\overset{\nearrow O}{}}$ |
| Serine | Ser | $H-\underset{\underset{OH}{\vert}}{\overset{\overset{H}{\vert}}{C}}-\underset{\underset{NH_2}{\vert}}{\overset{\overset{H}{\vert}}{C}}-C\underset{OH}{\overset{\nearrow O}{}}$ |
| Aspartic acid | Asp | $\underset{OH}{\overset{O}{\diagup\diagdown}}C-CH_2-\underset{\underset{NH_2}{\vert}}{\overset{\overset{H}{\vert}}{C}}-C\underset{OH}{\overset{\nearrow O}{}}$ |
| Lysine | Lys | $NH_2-CH_2-CH_2-CH_2-CH_2-\underset{\underset{NH_2}{\vert}}{\overset{\overset{H}{\vert}}{C}}-C\underset{OH}{\overset{\nearrow O}{}}$ |
| Glutamine | Gln | $\underset{NH_2}{\overset{O}{\diagup\diagdown}}C-CH_2-CH_2-\underset{\underset{NH_2}{\vert}}{\overset{\overset{H}{\vert}}{C}}-C\underset{OH}{\overset{\nearrow O}{}}$ |
| Cysteine | Cys | $HS-CH_2-\underset{\underset{NH_2}{\vert}}{\overset{\overset{H}{\vert}}{C}}-C\underset{OH}{\overset{\nearrow O}{}}$ |

spectively. Other amino acids contain specific functional groups capable of undergoing reactions of the type discussed in Section 22.2. For example, serine contains a hydroxyl group, aspartic acid contains an acidic carboxyl side chain, and lysine has a side chain containing a basic amino group. Cysteine is an example of an amino acid containing sulfur. The SH groups in two cysteine molecules can combine to form a disulfide linkage (—S—S—) and thereby link two protein chains together, as we shall see.

Although Table 22.4 shows the amino acids as neutral molecules, they are predominantly present in neutral solution as **zwitterions,** ions containing both a positive and a negative charge. The amino group is protonated while the carboxyl group is ionized, giving rise to the following species:

$$R-\underset{NH_3^+}{\overset{H}{C}}-COO^-$$

The form of an amino acid depends on the pH. In acidic solution the carboxyl group is un-ionized and the molecule has a positive charge. In basic solution the amino group loses the extra proton and the molecule acquires a negative charge. The situation can be represented by the following pH dependent equilibrium:

$$\underset{pH \approx 1}{R-\underset{NH_3^+}{\overset{H}{C}}-COOH} \rightleftarrows \underset{pH \approx 7}{R-\underset{NH_3^+}{\overset{H}{C}}-COO^-} + H^+ \rightleftarrows \underset{pH \approx 11}{R-\underset{NH_2}{\overset{H}{C}}-COO^-} + 2H^+$$

The presence of oppositely charged groups on a single molecule makes zwitterions highly polar. The intermolecular forces in amino acids are consequently strong, and these substances are solid and moderately soluble in water.

Amino acids are linked together to form proteins by means of **peptide bonds.** In forming this bond, the carboxyl group of one amino acid attaches to the amino group of another amino acid with the elimination of $H_2O$:

$$\underset{\text{Amino acid}}{H-N-\underset{H}{\overset{R}{\underset{|}{C}}}-\overset{O}{\overset{\|}{C}}-OH} + \underset{\text{Amino acid}}{H-N-\underset{H}{\overset{R'}{\underset{|}{C}}}-\overset{O}{\overset{\|}{C}}-OH} \longrightarrow$$

$$\underset{\text{Dipeptide}}{H_2NC-\underset{H}{\overset{R}{\underset{|}{C}}}-\underset{H}{\overset{}{\underset{|}{N}}}-\underset{H}{\overset{R'}{\underset{|}{C}}}-\overset{O}{\overset{\|}{C}}-OH} + H_2O$$

Peptide bond

The reaction between two amino acids results in the formation of the simplest condensed structure, a dipeptide. Additional amino acids can link in this fashion to form a **polypeptide,** in which the repeating unit,

$$-\overset{O}{\overset{\|}{C}}-\underset{H}{\overset{}{\underset{|}{N}}}-\underset{H}{\overset{R}{\underset{|}{C}}}-$$

is known as an amino acid residue.

*b. Protein Structure* A **protein** is a naturally occurring polypeptide. Some proteins, such as myoglobin, consist of a single chain while others, such as hemoglobin, consist of several linked chains. The molecular weight of most proteins is between $10^4$ and $10^5$. Since the average molecular weight of an amino acid is about 150, proteins generally contain well over one hundred amino acid residues. The determination of protein structure is consequently a formidable task, but one that must be performed before an understanding of protein function on the molecular level can be attained. The first step is the determination of the number of amino acid residues of each type present and of their precise sequence. This task was accomplished for the first time by Frederick Sanger (b. 1918), who determined the amino acid sequence of insulin, shown in Figure 22.21. Broadly speaking, a protein is first cleaved into smaller chains, and then the amino acid sequence in each chain is found by the use of reagents which degrade the polypeptide by removing one amino acid at a time. Individual amino acids can be identified on the basis of their different properties.

The amino acid sequences of hundreds of proteins are known by now. Each protein has a precisely defined and unique sequence. A change of even a single amino acid in just one protein can have a profound biological effect. For example, the replacement of glutamine by valine at a specific site of one of the polypeptide chains comprising hemoglobin leads to sickle cell anemia, a hereditary disease in which blood circulation is impaired.

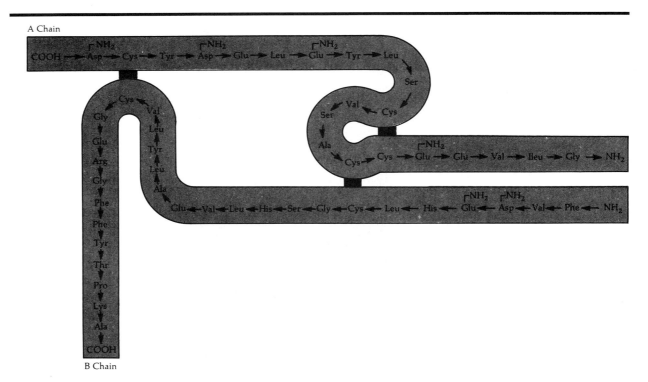

**FIGURE 22.21** Amino acid sequence of insulin. The protein consists of two polypeptide chains joined by disulfide bonds between cysteine molecules (heavy bars).

Proteins have a well-defined three-dimensional structure. The biological activity of proteins depends not only on their composition and sequence but also on their **conformation,** the three-dimensional arrangement of their constituent atoms. The conformation depends, in turn, on the structure of the protein components and on the interactions between different parts of the chain. Figure 22.22 shows the bond lengths and angles in a peptide segment, as determined by X-ray diffraction. The C—C bond length indicates that these are single bonds and rotation about them is possible. On the other hand, the C—N bond length is shorter than the 1.47 Å length of a C—N single bond, indicating that the bond has partial double-bond character. This result is consistent with a resonance description of bonding in what is known as the amide group:

$$\begin{array}{c}:\!\ddot{O}\!: \\ \| \\ C \\ / \quad \backslash \\ \quad \ddot{N}\!\!-\!\! \\ \quad | \\ \quad H\end{array} \quad \longleftrightarrow \quad \begin{array}{c}:\!\ddot{O}\!:^{-} \\ | \\ C \\ / \quad \backslash\!\!\backslash \\ \quad N^{+}\!\!-\!\! \\ \quad | \\ \quad H\end{array}$$

The amide group is planar and relatively rigid, as expected from the partial double bond character of the C—N and C—O bonds implied by the resonance description. The presence of these rigid groups at regular intervals imparts rigidity to proteins.

The three-dimensional shapes of proteins are determined to a large extent by the types of interactions, such as hydrogen bonds and disulfide linkages, that can occur between different parts of a chain or between different chains. The $\alpha$ helix, a rod-like helical structure, is one of the common three-dimensional conformations of proteins. Figure 22.23 shows a model of this structure.

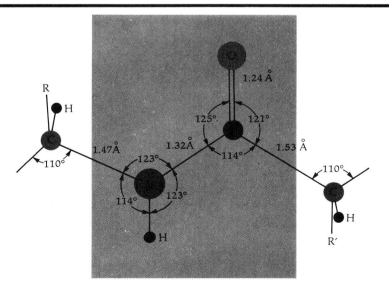

**FIGURE 22.22** Bond lengths and angles in the peptide chain. The amide group (shown in color) is rigid and planar.

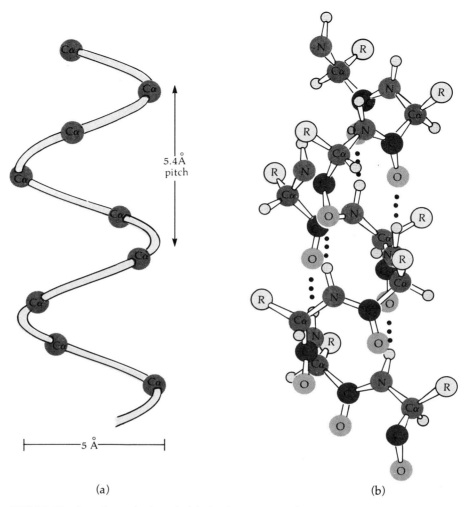

**FIGURE 22.23** The right-handed helical structure of a polypeptide chain. (a) The helix is most clearly seen when only the α carbons are shown. (b) The entire helix, showing hydrogen bonding (dotted lines) between carbonyl and amino groups.

The right-handed helical structure is most clearly seen if only the α carbon atoms are shown. The inclusion of the other groups brings out the important feature that the carbonyl group of each amino acid is hydrogen-bonded to the amino group of the amino acid situated four residues ahead in the sequence. Because of the geometry of the helix, these particular groups end up being in close physical proximity. Hydrogen bonding between these groups stabilizes the structure and, to a large extent, accounts for its occurrence.

If proteins consisted only of α helices, they would all be shaped like long and narrow rods. Most proteins, in fact, have compact globular shapes in which considerable folding and cross linking occur (Figure 22.24). These features are brought about by disulfide linkages between different chains, interchain hydrogen bonding between amide groups, the occurrence of sharp turns in the α helix, and other factors. The determination of all the detailed conformational features of complex proteins is only gradually being achieved.

The Chemistry of Carbon 22.6 **867**

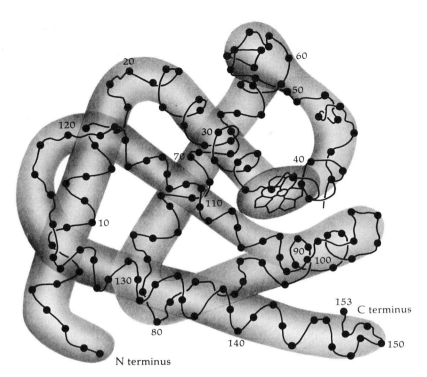

**FIGURE 22.24** Conformation of myoglobin, showing the occurrence of folding. The carbon skeleton is shown by the wiggly line connecting the C atoms.

*c. The Structure of DNA* The information required by an organism to synthesize all its diverse proteins is contained in deoxyribonucleic acid (DNA), the molecule of heredity. One of the great achievements of modern science is the discovery and elucidation of the chemical basis of genetic information and of the mechanisms by which this information is transmitted.

Nucleic acids are polymers of sugar molecules linked by phosphate bridges with one of four different bases attached to each sugar. The sugar in DNA is deoxyribose, while that in ribonucleic acid (RNA) is the closely related ribose. The structures of these sugars as well as the general structure of a nucleic acid molecule are shown in Figure 22.25.

The backbone of DNA consists of an unchanging repetition of the sugar-phosphate combination. The variable part of the molecule is the sequence of bases attached to the sugars. The bases are of types known as purines and pyrimidines. DNA has four different bases: two purines, which are adenine and guanine, and two pyrimidines, which are thymine and cytosine. In RNA thymine is replaced by the closely related uracil. Figure 22.26 shows the structures of these bases. The repeating unit of DNA, called a **nucleotide,** consists of a deoxyribose molecule with its attached base and phosphate group. Figure 22.27 shows four different nucleotides combined to form a short segment of a single strand of DNA.

In 1953, Francis Crick (b. 1916) and James Watson (b. 1928) proposed their famous double-helix model of the three-dimensional structure of DNA. The special significance of this model is that it readily shows how genetic informa-

868  The Chemistry of Carbon  22.6

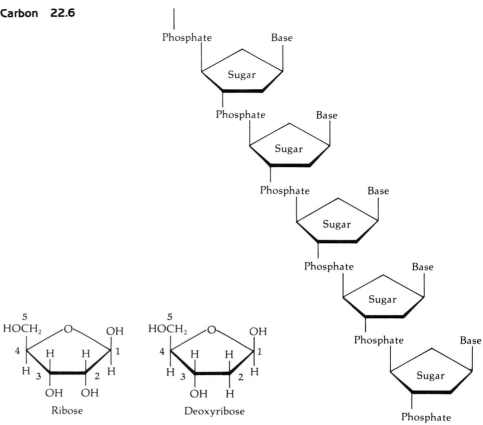

**FIGURE 22.25** Schematic diagram of the nucleic acid structure. The sugar in DNA is deoxyribose while that in RNA is ribose, whose structures are shown at the left.

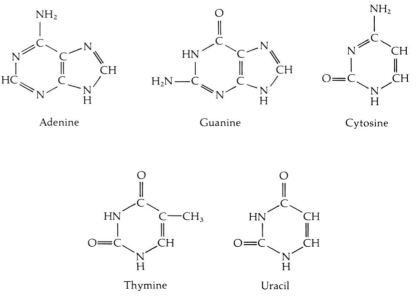

**FIGURE 22.26** The bases in DNA and RNA. The first three bases are common to both DNA and RNA. Thymine is present in DNA and uracil is present in RNA.

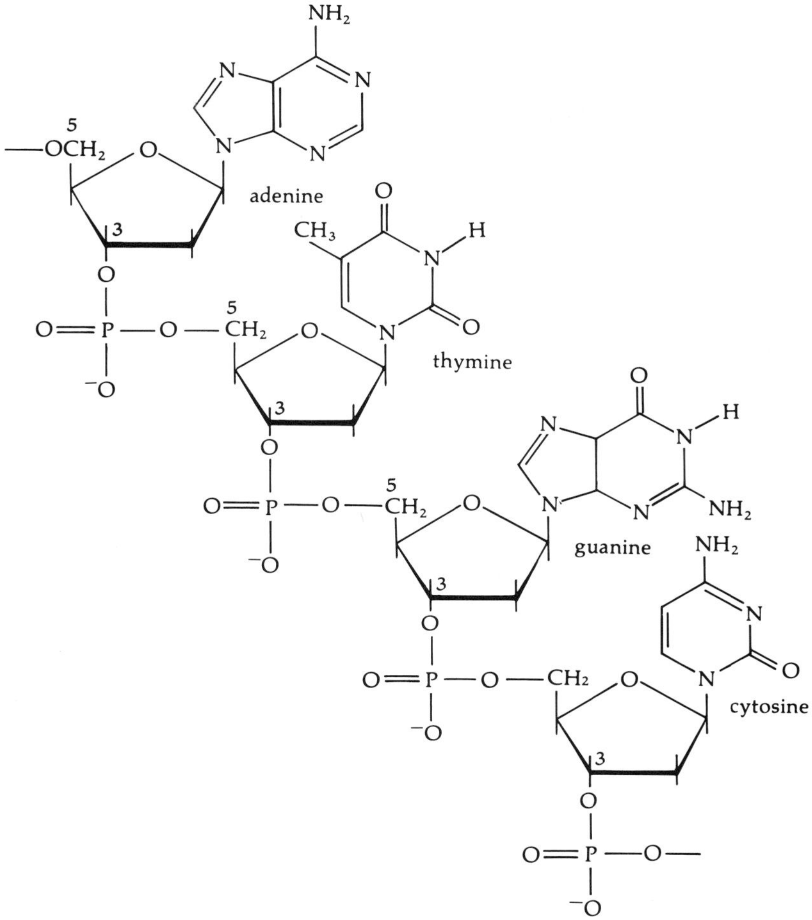

FIGURE 22.27 The four nucleotides combined into a short segment of a single strand of DNA.

tion is preserved, how it can be passed on, and how it can be used to direct the synthesis of specific proteins.

The data on the composition of DNA available prior to 1953 indicated that while the abundance of each of the four bases was variable, the molar ratio of thymine to adenine and that of cytosine to guanine were both equal to unity. These results suggested that thymine and adenine (T–A) as well as cytosine and guanine (C–G) were paired off in some manner in the DNA structure.

The molecular model proposed by Crick and Watson on the basis of X-ray diffraction data on bond lengths and angles revealed the nature of the base pairing. As shown in Figure 22.28, the bases can be arranged side by side in the same plane in such a manner that they are attached and stabilized by hydrogen bonds. At the same time, the atoms connecting the bases to the sugar–phosphate chains lie at opposite ends of this base complex. Remarkably, the horizontal distances between the points of attachment to the chains are nearly identical, about 11 Å, for both T–A and C–G pairs. No other combination of base pairs results in such a unique end-to-end distance while at the same time

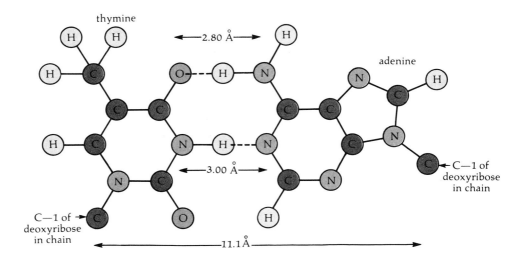

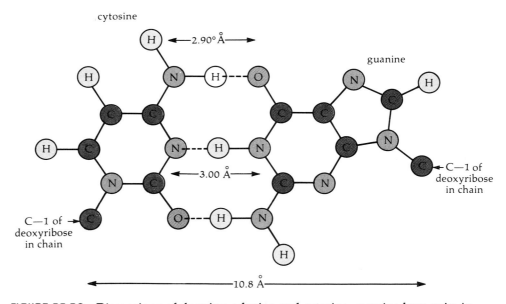

**FIGURE 22.28** Dimensions of thymine–adenine and cytosine–guanine base pairs in DNA. The dashed lines represent hydrogen bonds. The points of attachment to the sugar–phosphate chains are indicated.

permitting effective hydrogen bonding. The specificity of base pairing, cytosine with guanine and thymine with adenine, is the key feature of the DNA structure.

The DNA model resulting form these considerations is shown in Figure 22.29. Two helical polynucleotide strands are coiled about a common axis. The bases are on the inside of the helix and their plane is perpendicular to the helix axis. Successive base pairs are separated by 3.4 Å and a complete turn of the helix occurs every 34 Å. Depending on the organism, a single molecule of DNA can contain between approximately $10^4$ and $10^9$ base pairs, with a corre-

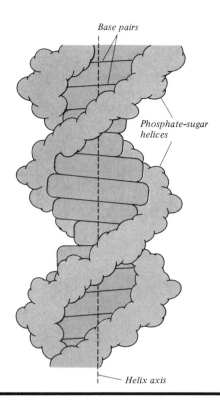

**FIGURE 22.29** The double-helix structure of DNA.

sponding molecular weight ranging from nearly $10^{10}$ to $10^{15}$. The specific way in which the bases pair indicates that every time a particular base, say cytosine, occurs in one strand, the complementary base, guanine, must occur at the same point in the second strand. The two strands are complementary—the base sequence in one strand determines the base sequence in the other strand and vice versa.

The complementarity of the two strands of DNA has profound implications for its biological activity. The genetic information of a cell is carried in the precise sequence of the $10^4$–$10^9$ bases along each strand of a DNA molecule. When a cell divides, this information must be transmitted to the two daughter cells, a process called **replication**. The two strands of a DNA molecule separate and each is incorporated in one of the new cells, where it serves as a template for the assembly of a complementary strand from free nucleotides. In this manner, each cell ends up with a double-stranded DNA molecule, identical in every respect to the DNA molecule in the parent cell.

Living organisms continually synthesize proteins. The **transcription** of DNA is the process by which the sequence of bases in DNA determines the identity of the protein to be synthesized. The precise sequence of amino acids in a given protein is determined by the corresponding sequence of bases in a strand of a particular DNA molecule. For example, the base sequence AAA is a code for the incorporation of the amino acid lysine in a protein. The sequence of bases in DNA is transmitted to the site of protein assembly by various forms of the related molecule RNA. The relationship between the base sequence in DNA and the amino acid sequence in proteins constitutes the **genetic code**. This code can be regarded as a dictionary containing the sequence of bases in

DNA corresponding to each of the twenty amino acids in a protein. Many of the fascinating details of the genetic code and how it functions have been unraveled during the past twenty years, but much remains to be learned.

## Conclusion

The ability of carbon atoms to form four covalent bonds with other carbon atoms and with atoms of other elements in molecules of different sizes and shapes makes organic chemistry an extraordinarily rich subject. Fortunately, certain unifying principles permit the subject to be organized in a manner that emphasizes many of the regularities present. Compounds containing a particular functional group resemble each other in their properties and reactions. Organic compounds can therefore be categorized by functional groups. Certain reactive intermediates, such as carbocations or carbanions, recur in different types of reactions and the mechanisms of these reactions can be understood in terms of the properties of these intermediates.

Organic polymers, such as the carbohydrates, proteins, and nucleic acids, play a vital role in the chemistry of life processes. The gradual understanding of the complex biochemical processes that occur in living organisms surely constitutes one of the great triumphs of modern science. The practical benefits that follow from an understanding of these processes on the molecular level are being felt at an increasing pace in medicine and agriculture. At the same time, biochemical processes reveal a level of chemical sophistication that is unmatched in the laboratory.

## Problems

1. Draw and name the isomers of an alkane with five carbon atoms.
2. Write the structural formulas and give the names of all the isomers with the formula $C_5H_{10}$.
3. Explain why each of the following names is incorrect: (a) methylhexane, (b) 4-methylpentane, (c) 2-ethylpropane, (d) 2,4,8-trimethyloctane.
4. Which of the following compounds exist in the form of cis–trans isomers: (a) 1-butene, (b) 2-pentene, (c) 1,1-dichloropropane, and (d) 1-butyne?
5. Draw and name all the isomers of a three-carbon alkane with two chlorine atoms.
6. How many different isomers can be derived from ethane by replacement of two hydrogen atoms by a fluorine atom and a chlorine atom? Draw their structures.
7. How many different isomers can be derived from ethylene by replacement of two hydrogen atoms by a fluorine atom and a chlorine atom? Draw their structures and give their names. Indicate which are structural isomers and which are stereoisomers.
8. The standard free energies of formation at 25°C of the gaseous isomeric pentanes are: pentane, $\Delta G_f^\circ = -8.37$ kJ mol$^{-1}$; isopentane, $\Delta G_f^\circ = -14.8$ kJ mol$^{-1}$; neopentane, $\Delta G_f^\circ = -15.2$ kJ mol$^{-1}$. Determine the composition of an equilibrium mixture of the three pentanes at 25°C. How does the stability of these alkanes depend on the extent of branching?
9. Name the following compounds:

   (a) BrCH$_2$CHCH=CH$_2$
              |
              Cl

   (b) C(Cl$_3$)CH$_2$CH=CHCH$_3$

   (c) HC≡CBr

          CH(CH$_3$)$_2$
           |
   (d) CH$_3$C=CHBr

10. Write the structures and give the names of all the monochloro substitution products of 2,2,3-trimethylbutane.

11. Write condensed and structural formulas for the following compounds: (a) octane, (b) 3-hexyne, (c) 2-ethyl butanal, (d) 1,2-dibromoethane, (e) 4-ethyl-2-hexanone, (f) methyl pentanoate, and (g) isopropyl t-butyl ether.

12. When a compound C$_3$H$_8$O is exposed to chlorine in the presence of light, three different monochloro derivatives, C$_3$H$_7$OCl, are formed. C$_3$H$_8$O does not react with sodium. Write a structural formula of this compound that is consistent with the experimental observations.

13. Give the names and structural formulas of the products of the following reactions: (a) 3-methyl-2-butanol reacts with formic acid, (b) 2-hexanone is reduced with LiAlH$_4$, (c) 2-ethyl pentanal is oxidized, and (d) 3-ethyl-2-pentanol is treated with potassium dichromate in acidic solution.

14. Starting only with n-propyl alcohol and any needed inorganic reagents, devise a synthesis of n-propyl propanoate.

15. On oxidation, compound A forms compound B. Further oxidation does not occur. On dehydration, A yields isomeric compounds C and D. Addition of H$_2$/Pt to compounds C or D yields butane. Give the names and structures of compounds A–D.

16. Name the following compounds:
    (a) CH$_2$=CCl$_2$
    (b) CH$_3$CH$_2$CH$_2$COOCH$_3$
    (c) CH$_3$CH$_2$CH$_2$OCH$_3$
    (d) CH$_3$COCH(CH$_3$)$_2$

17. Complete and balance the following equations:

(a) CH$_3$CH$_2$CHO $\xrightarrow{Cr_2O_7^{2-}/H^+}$

(b) CH$_3$CH$_2$COCH$_3$ $\xrightarrow[NaOH]{\Delta}$

(c) C$_6$H$_5$—C(=O)—OH + NaOH →

18. What types of compounds are obtained when one or both of the hydrogen atoms in water are replaced by the following groups: (a) one hydrogen atom is replaced by an alkyl group, (b) both hydrogen atoms are replaced by alkyl groups, (c) one hydrogen atom is replaced by the RC=O group.

19. To which class of organic compounds does each of the following compounds belong?

(a) CH$_3$COCH$_2$CH$_3$

(b) C$_6$H$_5$—O—C$_6$H$_5$

(c) CH$_3$CH$_2$CNH$_2$ (with C=O)

(d) C$_6$H$_5$—CH$_2$—CH (with C=O)

(e) CH$_3$CHNH$_2$
    |
    CH$_3$

(f) H$_2$NCHCOH (with C=O)
    |
    CH(CH$_3$)$_2$

(g) HOCH$_2$CHOHCH (with C=O)

(h) HOCH$_2$CCHOHCH$_2$OH (with C=O)

20. For each of the following pairs of acids, pick the one that is stronger: (a) butanoic acid and 2-chlorobutanoic acid, (b) 2-chlorobutanoic acid and 3-chlorobutanoic acid, (c) 4-chlorobutanoic acid and 3-chlorobutanoic acid, and (d) 4-chlorobutanoic acid and butanoic acid.

21. In each of the following pairs, which is the stronger base? Explain.

(a) Cl$^-$ and CH$_3$CO$^-$ (with C=O)

(b) CH$_3$CO$^-$ and CH$_3$CH$_2$CO$^-$ (both with C=O)

(c) FCH$_2$CO$^-$ and F$_2$CHCO$^-$ (both with C=O).

22. Write structural formulas and systematic names of all the substituted pentanes with the formula C$_5$H$_{10}$Cl$_2$. Identify any stereoisomers.

23. Which of the following molecules are chiral: (a) 2-methyl-2-butanol, (b) 2-methylbutane, (c) 2-methylhexane, and (d) 2-methyl-1-butanal?

24. A substance is optically active and the L-isomer rotates the plane of polarization by an angle $\alpha = -14.5°$. What is the percentage of the D-isomer in a mixture of the two enantiomers for which $\alpha = 2.5°$?

25. Which of the following compounds have chiral centers?

(a) CH₃CH₂CHCH₃
         |
         Cl

(b) HOCH₂CH₂CHCH₃
              |
              Br

(c) CH₃COH
      ‖
      O

(d) CH₃CHCH₃
       |
       Cl

26. For all the isomeric alcohols with the formula $C_5H_{11}OH$, (a) write the structural formulas and systematic names, (b) state whether they are primary, secondary, or tertiary alcohols, and (c) identify those that are optically active.

27. Which of the following molecules can form optical isomers: (a) dichlorodifluoromethane, (b) 3-methyloctane, (c) 3-ethyloctane, and (d) trans-2-pentene?

28. The bromination of 3-methylpentane results in a number of monobromo products. (a) Write the structures of all the different monobromo isomers that can be obtained. (b) Which of these isomers are optically active? (c) Which pairs of isomers are enantiomers? (d) Which isomers are diastereomers?

29. Name the compound formed by addition of HBr to (a) propene, (b) 2,5-dimethyl-2-hexene, and (c) 3-methyl-2-pentene.

30. The acid-catalyzed dehydration of acetaldehyde hydrate,

$$CH_3CH(OH)_2 \xrightarrow{HA} CH_3CHO + H_2O$$

involves the following mechanism:

(1) CH₃C(OH)(H)—OH + HA $\underset{k_{-1}}{\overset{k_1}{\rightleftharpoons}}$ CH₃C(OH)(H)—O⁺—H + A⁻   (fast)

(2) CH₃C(OH)(H)—O⁺—H + A⁻ $\xrightarrow{k_2}$ CH₃C(O⁻)(H)—O⁺—H + HA   (slow)

(3) CH₃C(O⁻)(H)—O⁺—H $\xrightarrow{k_3}$ CH₃C=O + H₂O   (fast)

(a) What types of intermediates are formed in steps (1) and (2), respectively? (b) Derive the rate law of this reaction. (c) How does the reaction rate depend on the concentration of the acid HA?

31. Which of the following two reactions is faster? Explain.

(1) $CH_3CH_2CH_2Br \xrightarrow[OH^-]{\Delta} CH_3CH_2CH_2OH + HBr$

(2) $(CH_3)_2CHBr \xrightarrow[OH^-]{\Delta} (CH_3)_2CHOH + HBr$

32. Aldopentoses have the formula

$CH_2OH(CHOH)_3CHO$

(a) Using the notation in Figure 22.15, write the structure of all the isomeric aldopentoses. (b) Number all the isomers and indicate which isomers form pairs of enantiomers.

33. Protonated amino acids are diprotic acids characterized by the ionization equilibria

(1) $H_3\overset{+}{N}CHRCOH \underset{}{\overset{K_1}{\rightleftharpoons}} H_3\overset{+}{N}CHRCO^- + H^+$
              ‖                                    ‖
              O                                    O

(2) $H_3\overset{+}{N}CHRCO^- \underset{}{\overset{K_2}{\rightleftharpoons}} H_2NCHRCO^- + H^+$
              ‖                                    ‖
              O                                    O

The values of $pK_1$ and $pK_2$ of glycine are 2.35 and 9.78, respectively. What is the principal form of glycine present at (a) pH = 2, (b) pH = 4, (c) pH = 8, and (d) pH = 11?

34. Sketch a curve for the titration of glycine hydrochloride with alkali. Let the ordinate be pH and the abscissa the number of equivalents of OH⁻ added. The pK values are given in Problem 33.

35. The pH at which the mean net charge of an amino acid is zero is called its isoelectric point. At this point, the amino acid is predominantly present as the zwitterion. Calculate the isoelectric point of glycine. The pK values are given in Problem 33.

36. Write the structure of the tripeptide, Gly—Val—Ser at neutral pH.

37. The average molecular weight of a pair of nucleotides is approximately 650. What is the length of a DNA molecule with a molecular weight of $3.5 \times 10^9$? Approximately how many base pairs are there in this DNA?

38. A single strand of DNA contains the base sequence T—G—A—A—C—G. What is the corresponding base sequence on the complementary strand?

39. The addition of a particular amino acid in the assembly of a protein is determined by a sequence of 3 bases in DNA. Show that a sequence of 2 bases cannot unambiguously determine the addition of the 20 different amino acids present in protein.

# 23

# Nuclear Chemistry

Nuclear chemistry is the study of the properties and reactions of atomic nuclei and of their chemical applications. The study of the atomic nucleus began at a well defined time: 1896. In that year Henri Becquerel (1852–1908) discovered that uranium compounds emit penetrating rays, capable of darkening photographic plates, at a rate independent of the chemical form of the compounds, and depending only on their uranium content. The particles responsible for this phenomenon were eventually shown to originate in the nucleus. Chemists have been intimately involved in the study of the atomic nucleus from the very beginning. Shortly after Becquerel's discovery, Marie Curie (1867–1935) and Pierre Curie (1859–1906) chemically separated polonium and radium from uranium ores and showed that their radioactivity was far more intense than that of uranium.

Chemists have made major contributions to nuclear science. Their knowledge of the chemistry of the elements has enabled them to study those nuclear phenomena in which the transmutation of elements is of importance. Among the major achievements of nuclear chemists are the discovery of nuclear fission and the synthesis of transuranium elements.

# Nuclear Properties

## 23.1 The Atomic Nucleus

In Sections 1.3 and 1.4 we briefly discussed some of the basic properties of the atomic nucleus such as composition, size, and mass. We begin by recapitulating these features and examining nuclear properties in further detail.

***a. Composition*** Nuclei are composed of neutrons and protons, collectively known as **nucleons**. (Table 1.2 gives the mass and charge of these particles.) The number of protons is equal to the **atomic number** $Z$ and determines the identity of an element. An atom with 20 protons in its nucleus, for example, is a calcium atom, while one with 30 protons is a zinc atom. Since the proton has a positive charge equal in magnitude to the negative charge of the electron, a neutral atom contains equal numbers of protons and electrons.

Atoms of the same element that differ in the number of neutrons, $N$, are called **isotopes**. The **mass number** of an isotope is equal to the sum of the number of protons and neutrons,

$$A = Z + N \tag{23.1}$$

Since isotopes of a given element have the same $Z$ but different $N$, they also differ in mass number $A$.

While chemical behavior is relatively insensitive to neutron number, nuclear properties are as much determined by the number of neutrons as by that of protons. It is customary to use the term "nuclide" instead of "isotope" when dealing with the nuclear properties of different nuclear species. A **nuclide** is a nuclear species characterized by atomic number $Z$, neutron number $N$, and mass number $A$. A nuclide is represented by the chemical symbol corresponding to its atomic number and by its mass number, given as a left superscript, e.g., $^{12}C$, $^{16}O$, $^{40}Ca$. While the neutron number is not explicitly stated, it can be inferred by means of Equation 23.1. Just as nuclides having the same atomic number are called isotopes, those having the same neutron number are called **isotones**, while those having the same mass number are known as **isobars**.

---

**Example 23.1** Of the following pairs of nuclides, which are isotopes, isotones, and isobars? (a) $^{14}N$ and $^{15}N$, (b) $^{13}C$ and $^{14}N$, (c) $^{14}C$ and $^{14}N$, and (d) $^{14}N$ and $^{16}O$.

(a) These are two isotopes of nitrogen.

(b) Since carbon has $Z = 6$, $^{13}C$ has $13-6 = 7$ neutrons. Similarly, $^{14}N$ has $14-7 = 7$ neutrons. Therefore $^{13}C$ and $^{14}N$ are isotones.

(c) These nuclides both have mass number 14 and are therefore isobars.

(d) The neutron numbers of these two nuclides are 7 and 8, respectively. The two nuclides differ in $Z$, $A$, and $N$; they are neither isotopes, isobars, nor isotones. ∎

---

High-energy physicists have obtained considerable evidence that nucleons are complex entities and are made up of other more fundamental particles. However, it is possible to obtain an adequate understanding of nuclear structure without consideration of these subnuclear particles, and we shall not consider them in this chapter.

***b. Nuclear Radii*** While nuclei contain over 99.9% of the atomic mass, they occupy an exceedingly small portion of the volume. This was first demonstrated by Rutherford in his $\alpha$-particle scattering experiment (Section 5.1c). Modern experiments with energetic charged particles have revealed that, to a first approximation, nuclei are spheres of uniform density with a volume proportional to their mass number, $V \propto A$. Since the radius of a sphere is proportional to the cube root of the volume, the nuclear radius must be proportional to the cube root of the mass number. The nuclear radius can be expressed by the relation

$$r = r_0 A^{1/3} \tag{23.2}$$

where $r_0$ is an experimentally determined constant known as the **nuclear radius parameter,** which has a value of approximately 1.4 fm (1 femtometer = 1 fm = $10^{-15}$ m). A nucleus with $A = 64$, for example, is found to have a radius of $5.6 \times 10^{-15}$ m. Comparing this value with the radius of an atom with $A \approx 64$, e.g., copper, which has an atomic radius of $1.16 \times 10^{-10}$ m, we see that nuclei are indeed four to five orders of magnitude smaller in radius than the atoms which contain them.

The proportionality between nuclear volume and mass number implies that a nucleus is composed of nucleons packed closely together to yield a body of constant density, independent of mass number. The density of nuclear matter can be calculated readily. Since $\rho = m/V$ and the mass of an atom in atomic mass units is approximately equal to its mass number (Section 1.4a), we obtain

$$\rho = \frac{A}{V} = \frac{A}{(4/3)\pi(r_0 A^{1/3})^3} = \frac{(1\ \text{u})(1.66 \times 10^{-27}\ \text{kg u}^{-1})(10^3\ \text{g kg}^{-1})}{(4\pi/3)(1.4 \times 10^{-15}\ \text{m})^3 (10^6\ \text{cm}^3\ \text{m}^{-3})}$$

$$= 1.4 \times 10^{14}\ \text{g cm}^{-3}$$

where the denominator in this expression is the volume of a sphere with a radius given by Equation 23.2. The nuclear density is some 13–14 orders of magnitude greater than the density of ordinary matter, so large as to be difficult to imagine. It is believed that neutron stars, the remnants of certain stellar explosions, consist of nuclear matter formed by the collapse of ordinary matter under the action of intense gravitational forces.

**c. Mass and Binding Energy** The masses of individual atoms, known as **nuclidic masses,** are customarily expressed in atomic mass units (Section 1.4a). This mass scale was defined by setting the mass of $^{12}$C equal to exactly 12 u. The masses of nuclides other than $^{12}$C, while close to their respective mass numbers, are generally not equal to them. Since the mass of the neutron or the proton is slightly larger than 1 u (see Table 1.2), the constituent neutrons, protons, and electrons of an atom necessarily have a mass greater than the mass number. It turns out that the mass of an atom is always smaller than the combined mass of its constituents. The missing mass corresponds to the **binding energy** of the nucleus. This energy is given off when the atom is assembled from its constituents. Conversely, this same amount of energy would have to be supplied to the atom in order to break it up into its constituent particles. All that can be said with certainty is that the mass of an atom must be smaller than the mass of its constituents and is generally close to the mass number. However, the actual difference between mass and mass number is variable. This variability is illustrated in Table 23.1, which lists the masses of a number of copper isotopes as well as the sums of the masses of the constituent particles. Note that the integral mass numbers are significantly smaller than the masses of the appropriate numbers of protons, neutrons, and electrons. The nuclidic

TABLE 23.1  Nuclidic Masses and Binding Energies of Some Copper Isotopes

| Nuclide | Mass (u) | Mass of Constituents (u) | $E_b$ (MeV) |
|---|---|---|---|
| $^{58}$Cu | 57.9444 | 58.4782 | 497.2 |
| $^{59}$Cu | 58.9395 | 59.4869 | 509.9 |
| $^{60}$Cu | 59.9374 | 60.4956 | 520.0 |
| $^{61}$Cu | 60.9334 | 61.5042 | 531.7 |
| $^{62}$Cu | 61.9326 | 62.5129 | 540.5 |
| $^{63}$Cu | 62.9296 | 63.5216 | 551.4 |
| $^{64}$Cu | 63.9298 | 64.5302 | 559.3 |
| $^{65}$Cu | 64.9278 | 65.5389 | 569.2 |
| $^{66}$Cu | 65.9289 | 66.5476 | 576.3 |
| $^{67}$Cu | 66.9278 | 67.5562 | 585.4 |

masses are smaller than the masses of their constituents and, for these particular nuclides, are slightly smaller than the mass numbers.

The equivalence of mass and energy follows from Einstein's theory of relativity and is embodied in the famous formula

$$E = mc^2 \tag{23.3}$$

where $c$ is the speed of light. The energy equivalent of one atomic mass unit is

$$E = (1 \text{ u})(1.6606 \times 10^{-24} \text{ g u}^{-1})(10^{-3} \text{ kg g}^{-1})(2.9979 \times 10^8 \text{ m s}^{-1})^2$$
$$\times (1 \text{ J/kg m}^2 \text{ s}^{-2})(1 \text{ eV}/1.6022 \times 10^{-19} \text{ J})$$
$$= 9.315 \times 10^8 \text{ eV} = 931.5 \text{ MeV}$$

where 1 MeV (megaelectron volt) = $10^6$ eV. (Nuclear energies are customarily expressed in eV or one of its multiples.)

The nuclear binding energy $E_b$ can be calculated with the help of this conversion factor as the energy equivalent of the mass difference between its constituents and an atom:

$$E_b = [ZM_H + (A - Z)M_n - M(^AZ)] \text{ u } (931.5 \text{ MeV u}^{-1}) \tag{23.4}$$

where $M(^AZ)$ is the mass of the nuclide in question, $M_H$ is the mass of a hydrogen atom, and $M_n$ is the mass of a neutron. (The shorthand notation $^AZ$, in which the element symbol is omitted and the nuclide represented by its atomic and mass numbers, is used to designate an unspecified nuclide. This notation is also used to designate isotopes of as yet unnamed or undiscovered elements. For example, the unknown isotope of element 114 with mass number 282 is represented by $^{282}114$.) By using the mass of a hydrogen atom rather than that of a proton in Equation 23.4, the mass of the $Z$ electrons that the nuclide $^AZ$ possesses is automatically taken into account. The form of this equation indicates that the binding energy is somewhat analogous to the enthalpy of formation of a compound: both quantities express the difference in energy content of a complex entity and its constituents, an atom in one case and a molecule in the other. However, as noted earlier, the binding energy is *always* positive. Table 23.1 gives the binding energies of the copper isotopes, obtained by taking the differences between the listed masses of the constituents and the nuclidic masses, expressed in MeV.

**Example 23.2** Calculate the binding energy of $^4$He given the following nuclidic masses: $^4$He = 4.002603 u, $^1$H = 1.007825 u, and n = 1.008665 u. What fraction of the mass is converted into energy when $^4$He is assembled from its constituents?

The binding energy is obtained by inserting the given masses into Equation 23.4:

$$E_b = [(2 \times 1.007825) + (2 \times 1.008665) - 4.002603]\ u\ (931.5\ \text{MeV u}^{-1})$$

$$= 28.30\ \text{MeV}$$

The fractional mass loss is

$$\frac{(28.30\ \text{MeV}/931.5\ \text{MeV u}^{-1})}{(2 \times 1.007825\ u + 2 \times 1.008665\ u)} = 0.007533 = 0.7533\% \quad \blacksquare$$

The above example indicates that since nearly 1% of the mass is converted into energy, it is indeed invalid to separate the conservation of mass and energy in nuclear processes. Rather, the conservation of these two important quantities must be replaced by that of mass–energy combined.

Recall from Chapter 2 that in chemical reactions the conservation of mass and energy are expressed as two separate laws. This approximation is valid because the energy change in a chemical reaction is sufficiently small that the effect on the mass is negligible. To see that this is indeed the case, we can compare the binding energy of $^4$He with that of a typical diatomic molecule. For example, the energy binding two chlorine atoms in a $Cl_2$ molecule (i.e., the $Cl_2$ dissociation energy) is 243 kJ mol$^{-1}$. Converting the binding energy of $^4$He to the same units we obtain

$$E_b = (28.30\ \text{MeV})(10^6\ \text{eV MeV}^{-1})(1.602 \times 10^{-19}\ \text{J eV}^{-1})(10^{-3}\ \text{kJ J}^{-1})$$

$$\times\ (6.022 \times 10^{23}\ \text{mol}^{-1})$$

$$= 2.73 \times 10^9\ \text{kJ mol}^{-1}$$

showing that the energy change in a typical nuclear transformation is about 10 million times larger than that in a typical chemical transformation.

## 23.2 Nuclear Stability

Of over 2600 known nuclides, fewer than 300 are stable. The rest are **radioactive**, meaning that they spontaneously decay into other nuclides by the emission of some type of particle. Radioactive decay obeys first-order kinetics (Section 14.3), and radioactive nuclides can consequently be characterized by a half-life—the time it takes for one-half the nuclides present to decay.

***a. Binding Energy per Nucleon*** While the binding energy of a nucleus is a measure of its stability, heavy nuclides have greater binding energies than light ones just because they contain more nucleons. This is also true for molecules, where the total bond energy in, say, $C_5H_{12}$ is greater than it is in $C_2H_6$ because of the larger number of C—H and C—C bonds. The relative stabilities of different nuclides are best compared by examining the binding energy per nucleon, $E_b/A$. Nuclides with large $E_b/A$ values are more stable than nuclides with small $E_b/A$ values. Figure 23.1 is a plot of the mass dependence of this quantity. Over most of the known mass range, $E_b/A$ varies by less than 10%

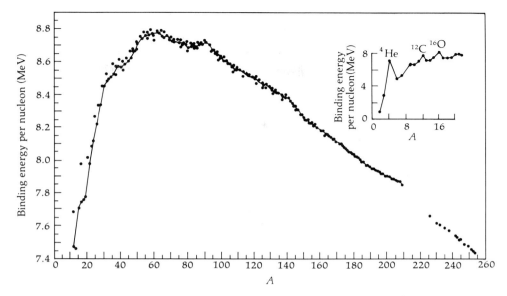

**FIGURE 23.1** Dependence on mass number of the binding energy per nucleon of stable nuclides. The insert shows the values for light nuclides. Note the difference in the ordinate scale of the insert.

from a value of 8 MeV per nucleon. Nonetheless, certain trends are readily apparent. The values of $E_b/A$ increase sharply with A for the lightest nuclides and then increase more slowly toward a maximum in the iron–nickel mass region at $A \approx 60$. Beyond this maximum, the curve decreases rather uniformly with increasing mass number. In the light element region there are a number of irregularities. Subsidiary peaks at $^4$He, $^{12}$C, and $^{16}$O show that these nuclides are particularly stable.

***b. The Liquid Drop Model of the Nucleus*** The experimental values of the binding energy per nucleon and their dependence on mass, displayed in Figure 23.1, can be explained with the help of a simple model of the nucleus, the liquid drop model. As the name implies, this model treats the nucleus as being analogous to a drop of liquid, with the nucleons in the nucleus playing the role of the constituent molecules in a drop of liquid. Just as the molecules in a liquid are held together by van der Waals forces (Section 18.2), the nucleons in a nucleus are bound by the nuclear force. If such a force did not exist, the atomic nucleus would not be stable: the mutual Coulomb repulsion between all the positively charged protons would cause them to fly apart. By pursuing the analogy between a nucleus and a liquid drop it is possible to arrive at the following expression for the nuclear binding energy in terms of nuclear composition as given by A and Z of a nuclide:

$$E_b = 14.10A - 13.00A^{2/3} - 0.6000 \frac{Z^2}{A^{1/3}} - 20.00 \frac{(A-2Z)^2}{A} \pm \frac{130.0}{A} \qquad (23.5)$$

The first term in Equation 23.5 reflects the dominant feature of the mass dependence of the nuclear binding energy displayed in Figure 23.1, that is, the constancy of the binding energy per nucleon. If $E_b/A$ is constant, then $E_b$ must be proportional to $A$, as indicated by the first term in the equation. This proportionality reflects an important property of the force holding the nucleons together in a nucleus. The nuclear force must be able to act only over a very short distance, so that each nucleon is attracted only to its neighboring nucleons in the nucleus (Figure 23.2a). If the nuclear force had sufficiently long range for each nucleon to interact with all the other $(A - 1)$ nucleons present in a nucleus of mass number $A$, $E_b/A$, which is a measure of the nuclear force, would vary as $(A - 1)$. The binding energy itself would then be proportional to $A(A - 1) \approx A^2$ instead of to $A$. Nuclear forces are similar in this respect to chemical forces—a molecule in a liquid or solid is strongly bound to only a small number of neighboring molecules.

Another aspect of the nuclear force implicit in the first term of Equation 23.5 follows from the proportionality between binding energy and mass number, irrespective of how the total number of nucleons is divided between protons and neutrons. This result indicates that the nuclear force is independent of nucleon charge—the nuclear force between two protons is the same as that between two neutrons or as that between a proton and a neutron.

The second term in the liquid drop model formula (Equation 23.5) is proportional to $A^{2/3}$. The negative sign indicates that this term reduces the binding energy from the value given by the first term. Since the nuclear radius is proportional to $A^{1/3}$ (Equation 23.2), the nuclear surface area $(4\pi r^2)$ must be proportional to $A^{2/3}$. Thus, the second term in Equation 23.5 indicates that nucleons occupying the surface region of the nucleus are less tightly bound than those located in the interior of the nucleus. This same effect is observed for molecules located at the surface of a drop and, as shown in Figure 23.2, arises because particles located at the surface are not completely surrounded by other particles. The nuclear surface energy is responsible for the increase in the binding energy per nucleon with mass number which is observed below $A \approx 60$ (Figure 23.1). As shown in Figure 23.2, the nucleons occupying the surface of a nucleus constitute a larger fraction of the total number of nucleons for a light

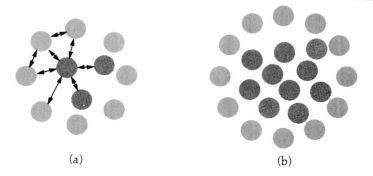

(a)  (b)

**FIGURE 23.2** The liquid drop model of the nucleus. A given nucleon interacts only with its neighboring nucleons, as shown by the arrows in (a). Nucleons occupying the nuclear surface can interact with fewer nucleons than nucleons located in the interior. With increasing mass number, the fraction of nucleons which lie at the surface decreases, as may be seen by comparing the light nucleus in (a) with the heavier nucleus in (b).

nucleus than for a heavy one. Consequently, the binding energy per nucleon increases with mass number.

The third term of Equation 23.5 is proportional to $Z^2/A^{1/3}$ and is negative in sign. This term reflects the repulsive Coulomb force between the positively charged protons. The Coulomb energy between two bodies is proportional to the product of their charges divided by the distance between them. Since each proton repels the other $Z - 1$ protons in the nucleus, the Coulomb energy is proportional to $Z(Z - 1)$, or more simply, to $Z^2$. The average distance between the protons is proportional to the nuclear radius, and the Coulomb energy is therefore also inversely proportional to $A^{1/3}$. Since the Coulomb force is repulsive, the binding energy of the nucleus is lower than it would be in the absence of this effect and the Coulomb term in Equation 23.5 is negative.

Owing to the dependence of the Coulomb energy on $Z^2$, the effect of Coulomb repulsion on the binding energy per nucleon increases in importance with mass number (since $Z \propto A$). This effect works in the opposite direction of the surface energy effect and is responsible for the gradual decrease in the binding energy per nucleon which sets in at $A \approx 60$ (Figure 23.1).

The first two terms of the liquid drop model formula show that the binding energy is independent of the relative number of neutrons and protons in a nucleus and depends only on the total number of nucleons. On the other hand, the third term indicates that the binding energy decreases with increasing nuclear charge. Taken together, the three terms of the formula indicate that the most stable nuclides should have more neutrons than protons. While this is indeed the case for moderately heavy nuclides, it is not so for light nuclides, for which the repulsive Coulomb force is small. The most stable light nuclides contain an equal number of neutrons and protons, for example, $^4$He, $^{12}$C, $^{16}$O, and $^{20}$Ne. Evidently, a factor is at work which increases the binding energy of symmetric nuclei—those with an equal number of neutrons and protons—relative to that of asymmetric nuclei with the same total number of nucleons. The explanation of this factor must be sought in the quantum mechanical description of the nucleus. The result is that the binding energy must be modified by a symmetry energy term. This fourth term in the liquid drop model formula varies as $-(A - 2Z)^2/A$. When $Z = N$, $A - 2Z = 0$ and the binding energy of symmetric nuclides remains unaffected by this term. On the other hand, whenever $Z \neq N$, $(A - 2Z)^2$ is positive, and the binding energy is reduced.

The final term in Equation 23.5 is known as the pairing energy. It reflects the empirical result that, other factors being equal, nuclides are most stable when they contain an even number of protons and an even number of neutrons. Nuclides with an odd number of either neutrons or protons, that is, nuclides with odd mass number, have intermediate stability, and nuclides with an odd number of protons and an odd number of neutrons are least stable. The difference in the stability of these various types of nuclides is manifested in the distribution of the stable isotopes of the elements: there are 163 even–even isotopes (even $Z$, even $N$), 105 odd-$A$ isotopes, and only 4 odd–odd isotopes (odd $Z$, odd $N$).

The pairing energy term is proportional to $\pm A^{-1}$. The positive sign corresponds to even–even nuclides, which have a higher binding energy and the negative sign to odd–odd nuclides, which have a lower binding energy. The pairing energy term is set equal to zero for odd-$A$ nuclides, thereby accounting for their intermediate stability.

While the liquid drop model does an excellent job in predicting the binding energies of most nuclides, it is unable to account for the unusually high stability of nuclides containing certain specific numbers of neutrons or protons. These so-called "magic" numbers occur whenever $N$ or $Z$ is equal to 2, 8, 20, 28, 50, or 82, or when $N$ is equal to 126. This effect is associated with the fact that, in analogy to atomic electrons, nucleons are organized into specific quantum states. Whenever a shell comprised of several closely spaced states is completely filled, an unusually stable configuration is obtained. The nuclear stability of closed-shell nuclides is analogous to the chemical stability of the noble gases.

**Example 23.3** Calculate the nuclidic mass of $^{65}$Cu by means of the liquid drop model. Compare with the experimental mass (Table 23.1).

Equation 23.5 may be used to evaluate the binding energy of $^{65}$Cu:

$$E_b(^{65}\text{Cu}) = 14.10\,A - 13.00\,A^{2/3} - 0.6000\,\frac{Z^2}{A^{1/3}} - 20.00\,\frac{(A-2Z)^2}{A} \pm \frac{130.0}{A}$$

$$= \left[(14.10)(65) - (13.00)(65)^{2/3} - (0.6000)\frac{(29)^2}{65^{1/3}} - (20.00)\frac{(65-58)^2}{65} + 0\right]\text{MeV}$$

$$= (916.5 - 210.2 - 125.5 - 15.1)\text{ MeV}$$

$$= 565.7 \text{ MeV}$$

The mass of $^{65}$Cu can be obtained from the binding energy by means of Equation 23.4:

$$M(^{65}\text{Cu}) = 29 M_H + 36 M_n - E_b(^{65}\text{Cu})$$

Using the masses of the hydrogen atom and the neutron given in Example 23.2 we obtain

$$M(^{65}\text{Cu}) = [(29 \times 1.007825 + (36 \times 1.008665)]\text{u} - (565.7 \text{ MeV}/931.5 \text{ MeV u}^{-1})$$

$$= 64.93 \text{ u}$$

The experimental value given in Table 23.1 is 64.9278 u, which differs from the calculated value by about 0.003%. This result is typical of the accuracy obtainable by means of the liquid drop model. ∎

## 23.3 Beta Decay

Most nuclides are radioactive and decay by the emission of some particle. The most common mode of decay is **beta decay,** in which a nuclide changes in atomic number by one unit without change in mass number. This transformation is accomplished by the emission of a particle such as the electron.

***a. Relative Stabilities of Isobaric Nuclides*** Since beta decay involves isobaric transformations (i.e., constant mass number), we first examine the relative stabilities of isobaric nuclides. Equation 23.5 for the binding energy of a nucleus becomes particularly simple at constant $A$. In this case the only variable is $Z$, and the binding energy is a quadratic in $Z$,

$$E_{b(A\,=\,\text{constant})} = a + bZ + cZ^2 \tag{23.6}$$

where $a$, $b$, and $c$ are $A$-dependent constants. (For even-$A$ nuclides, an additional term corresponding to the pairing energy must be included.) Equation 23.6 is that for a parabola and Figure 23.3 illustrates the parabolic dependence of the binding energy on $Z$ for the nuclides with $A = 131$.

The stability of isobaric nuclides varies in a systematic way with Z. Figure 23.3 shows that $^{131}$Sn, with the smallest binding energy, or equivalently (see Equation 23.4), the largest mass, is the least stable isobar with $A = 131$. As $Z$ increases, the binding energies of the nuclides increase, and their masses decrease. The nuclide $^{131}$Xe lies at the bottom of the parabola and is the only $A = 131$ isobar that is stable. As $Z$ increases beyond 54, the binding energies now decrease, while the nuclidic masses increase, and the nuclear stability decreases.

In beta decay, each nuclide decays to its adjacent isobar of smaller mass (or greater binding energy). The beta decay of the $A = 131$ isobars can thus be represented by the following sequence:

$$^{131}_{50}\text{Sn} \rightarrow {}^{131}_{51}\text{Sb} \rightarrow {}^{131}_{52}\text{Te} \rightarrow {}^{131}_{53}\text{I} \rightarrow {}^{131}_{54}\text{Xe} \leftarrow {}^{131}_{55}\text{Cs} \leftarrow {}^{131}_{56}\text{Ba} \leftarrow {}^{131}_{57}\text{La} \leftarrow {}^{131}_{58}\text{Ce}$$

The nuclides with low $Z$ decay sequentially to $^{131}$Xe by a process in which $Z$ increases by one unit in each decay. On the other hand, the nuclides with high $Z$ decay sequentially to $^{131}$Xe by a process in which $Z$ decreases by one unit in each decay. Both processes are variants of beta decay, as we shall see.

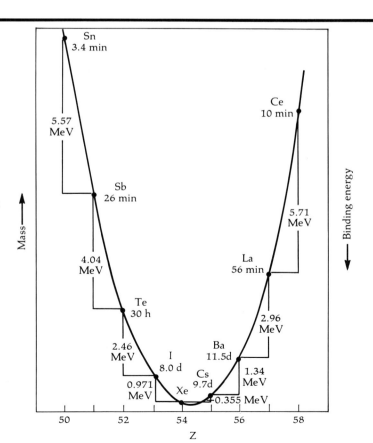

**FIGURE 23.3** Dependence of nuclidic masses (and binding energies) of isobaric nuclides with $A = 131$ on atomic number Z. Each nuclide decays to the adjacent nuclide of lower mass and releases a certain amount of energy when it decays. The decay energies (Section 23.3b) and half-lives of the radioactive nuclides are indicated. The stable nuclide with $A = 131$ is Xe, which lies at the bottom of the mass parabola.

**b. $\beta^-$ Decay** The process by which a nucleus decays isobarically with an increase in atomic number is called $\beta^-$ (**beta minus**) **decay**. This transformation can be represented by the equation

$$^A Z \rightarrow {}^A(Z + 1) + \beta^- + \nu \tag{23.7}$$

For example, the $\beta^-$ decay of $^{131}$Te involves the process

$$^{131}_{52}\text{Te} \rightarrow {}^{131}_{53}\text{I} + \beta^- + \nu$$

(We include the atomic numbers as subscripts for convenience.)

Two particles are emitted in $\beta^-$ decay; they are symbolized as $\beta^-$ and $\nu$ in Equation 23.7. A $\beta^-$ particle is just an electron. However, it is not one of the atomic electrons but, rather, it originates in the nucleus. Since electrons are not found in the nucleus, the electron must be created at the moment the transformation occurs. The emission of the $\beta^-$ particle is accompanied by the conversion of a neutron into a proton with a consequent increase in the atomic number by one unit. The simplest example of $\beta^-$ decay is that of the free neutron. Neutrons are produced in copious numbers in nuclear fission (Section 23.6). Their decay involves the process

$$n \rightarrow p + \beta^- + \nu$$

and occurs with a half-life of 12.8 minutes.

The second particle emitted in $\beta^-$ decay is the neutrino $\nu$. The neutrino is an uncharged particle and is generally believed to have zero mass. Its interaction with matter is so weak that it has a high probability of traversing through the entire Earth without being disturbed. The emission of the neutrino in $\beta^-$ decay was postulated some two decades before this particle was actually observed. The existence of the neutrino was invoked because without it, energy would not have been conserved in beta decay. The conservation of energy has been found to hold true in all other processes, and it was felt preferable to postulate the emission of an unobserved particle rather than to give up this very fundamental law.

To see how neutrino emission is related to energy conservation in beta decay, we must consider the energy released in this process, the **decay energy**. The energy released in $\beta^-$ decay, $Q_{\beta^-}$, is equal to the mass difference between the emitting and product nuclides expressed in energy units,

$$Q_{\beta^-} = (M_{A_Z} - M_{A_{(Z+1)}})(931.5 \text{ MeV u}^{-1}) \tag{23.8}$$

When $\beta^-$ decay occurs, the decay energy is given off in the form of kinetic energy of the decay products. The heavy product nuclide receives practically no energy. Thus, if the electron were the only emitted particle, it would be ejected with a unique energy, $Q_{\beta^-}$. Figure 23.4 shows a typical $\beta^-$ spectrum, that is, a plot of the energy distribution of the electrons emitted when a sample of a particular radioactive nuclide decays. While the maximum kinetic energy of the electrons is indeed equal to $Q_{\beta^-}$, most of the emitted electrons are of lower energy. If the electron were the only emitted particle, energy would not be conserved. On the other hand, if a neutrino also is emitted, it can carry off the missing energy. In any given decay process, the condition

$$Q_{\beta^-} = E_{\beta^-} + E_\nu \tag{23.9}$$

ensures that energy is conserved.

**FIGURE 23.4** Energy spectrum of electrons emitted in the $\beta^-$ decay of $^{32}$P. The maximum electron energy, 1.71 MeV, is equal to the decay energy $Q_{\beta^-}$. The energy of the emitted neutrino is equal to $(Q_{\beta^-}) - (E_{\beta^-})$. For example, if a particular $^{32}$P nuclide decays with the emission of a 1.00-MeV electron, the neutrino energy is 0.71 MeV.

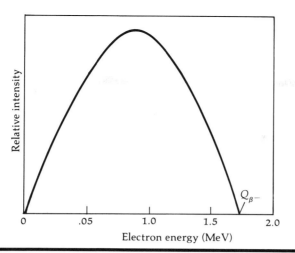

A particular nuclide can undergo $\beta^-$ decay provided it is more massive than its neighboring isobaric nuclide of higher atomic number, that is, so long as Equation 23.8 can be satisfied. As already noted in Figure 23.3, $^{131}$I decays to $^{131}$Xe because its nuclidic mass, 130.90612 u, is greater than that of $^{131}$Xe, 130.90508 u. On the other hand, $^{131}$Xe does not decay to $^{131}$Cs. Rather, $^{131}$Cs decays to $^{131}$Xe because its nuclidic mass (130.90547 u) is greater than that of $^{131}$Xe.

In analogy to atoms and molecules, nuclei have excited states. Frequently, beta decay leads to the formation of nuclei in excited states. These states deexcite to the nuclear ground state by the emission of $\gamma$ rays (Section 5.3), electromagnetic radiation originating in the nucleus. The process is analogous to the emission of radiation in an electronic transition in an atom. On average, $\gamma$ rays have higher energies than other types of photons, such as X rays.

**Example 23.4** The nuclide $^{28}$Al undergoes $\beta^-$ decay. (a) Write an equation for this process. (b) If the masses of $^{28}$Al and $^{28}$Si are 27.981908 u and 27.976927 u, respectively, what is the value of $Q_{\beta^-}$? (c) What is the energy of the neutrino emitted in a decay in which a 0.500-MeV $\beta^-$ is emitted? (d) A fraction of the $\beta^-$ decay of $^{28}$Al populates an excited state of $^{28}$Si at an energy of 1.780 MeV above the ground state. What is the value of $Q_{\beta^-}$ for this transition? What are the energy and wavelength of the $\gamma$ ray emitted when this state deexcites to the $^{28}$Si ground state?

(a) Since Z increases by one unit in $\beta^-$ decay, $^{28}$Al must decay to $^{28}$Si. The equation for this process is

$$^{28}\text{Al} \rightarrow {}^{28}\text{Si} + \beta^- + \nu$$

(b) The value of $Q_{\beta^-}$ may be obtained by means of Equation 23.8:

$$Q_{\beta^-} = [M(^{28}\text{Al}) - M(^{28}\text{Si})](931.5 \text{ MeV u}^{-1})$$

$$= (27.981908 \text{ u} - 27.976927 \text{ u})(931.5 \text{ MeV u}^{-1})$$
$$= 4.640 \text{ MeV}$$

(c) When a particular $^{28}$Al atom decays to $^{28}$Si, the sum of the electron and neutrino energies must be equal to the decay energy (Equation 23.9). Thus,

$$E_\nu = (Q_{\beta^-}) - (E_{\beta^-})$$
$$= (4.640 - 0.500) \text{ MeV} = 4.140 \text{ MeV}$$

(d) The decay of $^{28}$Al to the ground state of $^{28}$Si can proceed via two different paths: the direct path considered above or a path involving the formation of an excited state of $^{28}$Si which then decays to the gound state by $\gamma$-ray emission. The total decay energy must be independent of path. Designating the beta decay energy for the decay to the excited state as $Q^*_{\beta^-}$, we have

$$4.640 \text{ MeV} = Q^*_{\beta^-} + 1.780 \text{ MeV}$$
$$Q^*_{\beta^-} = 2.860 \text{ MeV}$$

Since the energy of the $\gamma$ ray is equal to the energy difference between the initial and final states (Equation 5.18),

$$E_\gamma = 1.780 \text{ MeV}$$

The wavelength of a photon can be obtained from its energy in the usual way:

$$E_\gamma = h\nu = \frac{hc}{\lambda}$$

$$\lambda = \frac{hc}{E_\gamma} = \frac{(6.626 \times 10^{-34} \text{ J s})(2.998 \times 10^8 \text{ m s}^{-1})}{(1.780 \text{ MeV})(10^6 \text{ eV MeV}^{-1})(1.602 \times 10^{-19} \text{ J eV}^{-1})}$$

$$= 6.966 \times 10^{-13} \text{ m}$$

$$= 0.6966 \text{ pm} \quad \blacksquare$$

*c. Positron Decay and Electron Capture Decay* A beta decay process in which the atomic number of the decaying nuclide decreases can occur by one of two different modes. The first mode is analogous to $\beta^-$ decay in that a particle with the mass of an electron is emitted. However, since the atomic number of the nuclide decreases, the ejected particle must be positively charged in order to ensure the conservation of charge. This particle, designated $\beta^+$, is called a **positron**. A positron has the same mass as the electron and a charge of the same magnitude but opposite sign. Modern physics has shown that every known particle must have an antiparticle; the positron is the antiparticle of the electron.

The creation of a positron can be understood as being equivalent to the formation of a "hole" in the electron "sea." As shown in Figure 23.5, the creation of a positron is equivalent to the conversion of an electron of negative energy into one of positive energy. In this process, an amount of energy corresponding to twice the mass of the electron is used up. Since the electron mass expressed in energy units is 0.51 MeV, positron emission requires a minimum energy of 1.02 MeV. This picture of positron creation is somewhat analogous to

**FIGURE 23.5** The creation of a positron. An electron of negative energy is converted to one of positive energy. The positron corresponds to the hole in the electron sea. An energy of 1.02 MeV, equivalent to twice the electron mass, is required for this process.

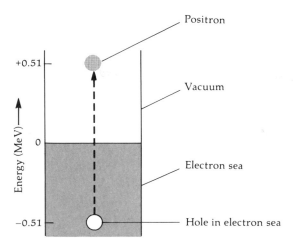

that of the conduction of current in semiconductors by positive "holes" in the electron sea (Section 20.6). In that case, however, the holes corresponded to electron vacancies while, with the much higher energies available in nuclear decay, they correspond to the materialization of a positive antielectron, or positron.

Positron decay can be represented by the equation

$$^AZ \rightarrow {}^A(Z-1) + \beta^+ + \nu \tag{23.10}$$

and the decay energy is given by the relation

$$Q_{\beta^+} = (M_{A_Z} - M_{A(Z-1)})(931.5 \text{ MeV u}^{-1}) \tag{23.11}$$

In analogy to $\beta^-$ decay, $\beta^+$ decay also involves the emission of a neutrino. (The neutrino emitted in $\beta^-$ decay is actually an antineutrino.) The decay energy appears in the form of kinetic energy of the positron and neutrino. In contrast to $\beta^-$ decay, $\beta^+$ decay is a threshold process—it can only occur when the mass difference between the initial and final nuclides, i.e., the decay energy, is larger than 1.02 MeV. For example, $^{131}$Ce, with a decay energy of 5.71 MeV can decay to $^{131}$La by positron emission, as shown in Figure 23.3. On the other hand, $^{131}$Cs, with a decay energy of only 0.36 MeV, cannot decay to $^{131}$Xe by positron emission.

When positron emission is energetically forbidden but when the mass of a nuclide $^AZ$ is still greater than that of nuclide $^A(Z-1)$, as in the case of $^{131}$Cs and $^{131}$Xe, a different decay process, known as electron capture, occurs. **Electron capture** (EC) is a process in which the nucleus captures one of its orbital electrons and thereby loses one unit of positive charge. The captured electron does not retain its identity inside the nucleus. Rather, a proton is converted into a neutron. Electron capture decay can be represented by the equation

$$^AZ + e^- \rightarrow {}^A(Z-1) + \nu \tag{23.12}$$

and the decay energy is given by the relation

$$Q_{EC} = (M_{A_Z} - M_{A(Z-1)})(931.5 \text{ MeV u}^{-1}) \tag{23.13}$$

When the decay energy of a nuclide is large enough for positron decay to be possible, both positron emission and electron capture can occur.

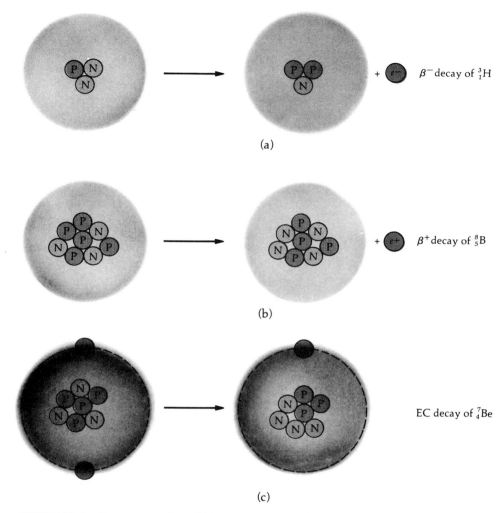

**FIGURE 23.6** Some examples of (a) $\beta^-$ decay, (b) $\beta^+$ decay, and (c) electron capture decay.

When electron capture occurs, the resulting atom is left with an electron vacancy in one of its shells, e.g., the K shell. This vacancy is filled by an electron from a higher shell and is accompanied by X-ray emission (Section 6.5c). Figure 23.6 gives a graphical summary of the three different types of beta decays.

**Example 23.5** The following are the nuclidic masses of the known isobars with $A = 13$:

$^{13}_{5}B$  13.0178 u     $^{13}_{7}N$  13.0057 u

$^{13}_{6}C$  13.0034 u     $^{13}_{8}O$  13.0248 u

Identify the one stable isobar and determine the isobaric decay modes and decay energies of the radioactive isotopes. Write an equation for each decay.

The stable isobar is the nuclide of lowest mass. This nuclide is $^{13}$C. The other nuclides decay into their neighboring isobars of either higher Z or lower Z, depending on which direction leads to a lower mass. The decay of $^{13}$B leads to $^{13}$C formation. Since Z increases, $\beta^-$ emission is involved:

$$^{13}B \rightarrow {}^{13}C + \beta^- + \nu$$

The decay energy is given by Equation 23.8:

$$Q_{\beta^-} = [M(^A Z) - M(^A(Z+1))](931.5 \text{ MeV u}^{-1}) = [M(^{13}B) - M(^{13}C)](931.5 \text{ MeV u}^{-1})$$

$$= (13.0178 \text{ u} - 13.0034 \text{ u})(931.5 \text{ MeV u}^{-1})$$

$$= 13.4 \text{ MeV}$$

The decay of $^{13}$N must form $^{13}$C, which has a lower mass, rather than $^{13}$O, which has a higher mass. The decay involves electron capture, and if the mass difference corresponds to more than 1.02 MeV, positron emission as well. For electron capture we have

$$^{13}N + e^- \rightarrow {}^{13}C + \nu$$

$$Q_{EC} = [M(^A Z) - M(^A(Z-1))](931.5 \text{ MeV u}^{-1}) = [M(^{13}N) - M(^{13}C)](931.5 \text{ MeV u}^{-1})$$

$$= (13.0057 \text{ u} - 13.0034 \text{ u})(931.5 \text{ MeV u}^{-1})$$

$$= 2.1 \text{ MeV}$$

Since the decay energy exceeds 1.02 MeV, positron decay is also possible,

$$^{13}N \rightarrow {}^{13}C + \beta^+ + \nu$$

and

$$Q_{\beta^+} = Q_{EC} = 2.1 \text{ MeV}$$

The decay of $^{13}$O is analogous to that of $^{13}$N:

$$^{13}O + e^- \rightarrow {}^{13}N + \nu$$

$$^{13}O \rightarrow {}^{13}N + \beta^+ + \nu$$

$$Q_{EC} = Q_{\beta^+} = (13.0248 \text{ u} - 13.0057 \text{ u})(931.5 \text{ MeV u}^{-1}) = 17.8 \text{ MeV} \quad \blacksquare$$

**d. The Valley of Nuclear Stability** Isobaric mass parabolas such as that depicted in Figure 23.3 can be drawn for all mass numbers. The locus of the minima of these parabolas gives the location of the stable nuclides in the Z–N (or Z–A) plane. Figure 23.7 shows the distribution in Z and N of the stable isotopes of the elements. By examining the relative numbers of neutrons and protons of these nuclides in the light of the liquid drop model, it is possible to identify some of the factors that contribute to nuclear stability.

In the light-element region the stable isotopes closely follow the $N = Z$ line. This behavior reflects the importance of the symmetry term in the liquid drop formula (Equation 23.5). As already noted, many of the most stable and abundant isotopes of the light elements, such as $^4$He, $^{12}$C, $^{14}$N, $^{16}$O, $^{20}$Ne, $^{28}$Si, $^{32}$S, and $^{40}$Ca lie exactly on this line. This trend does not continue for larger mass numbers. Instead, the band of stable nuclides bends away from the $N = Z$ line in the direction of $N > Z$. For example, the heaviest stable nuclide, $^{209}$Bi, has 83 protons and 126 neutrons. This trend to greater stability for $N > Z$

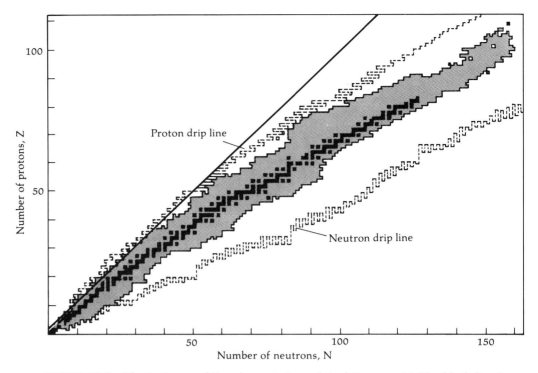

**FIGURE 23.7** The isotopes of the elements in a plot of Z versus N. The black line is the N = Z line. The solid squares represent the stable isotopes. The colored band shows the location of the known radioactive isotopes. The colorless bands show the locations of as yet unknown radioactive isotopes. These bands are terminated by the colored dashed lines, which mark the locations of the proton drip line and the neutron drip line. Nuclides cannot exist outside these lines because they emit protons or neutrons instantaneously.

reflects the increasing importance of Coulomb repulsion with increasing Z. The effect of this repulsion on the nuclear binding energy is minimized by dilution of the protons with an increasing number of neutrons. Eventually, however, Coulomb repulsion becomes so dominant as to prevent the existence of stable nuclides. As we shall see, all nuclides beyond $^{209}$Bi are unstable and undergo radioactive decay.

Surrounding the band of stable nuclides in Figure 23.7 are the radioactive nuclides, most of which undergo one of the three forms of beta decay. The nuclides lying below the stable nuclides are neutron rich and decay by $\beta^-$ emission while the nuclides lying above the stable nuclei are neutron poor and decay by either $\beta^+$ emission or electron capture. The bands of radioactive nuclides do not extend indefinitely away from stability. As a nucleus becomes increasingly neutron rich a point is eventually reached when it can no longer hold an additional neutron. If a neutron were to be added to such a nucleus, it would be ejected immediately. The lower dashed line in Figure 23.7 is the **neutron drip line**. Nuclides lying at this line define the limit of the existence of neutron-rich nuclei. Conversely, as a nucleus becomes increasingly neutron poor, or equivalently, proton rich, the point at which it can no longer hold an additional proton is eventually reached. The upper dashed line in Figure 23.7

shows the location of the **proton drip line,** where spontaneous proton emission marks the limit of the existence of neutron-poor nuclei. As shown in Figure 23.7, only a fraction of the radioactive nuclides lying in the region within the drip lines are presently known.

## 23.4 Alpha Decay

*a. Energetics of Alpha Decay*  An **α particle** is the nucleus of a $^4$He atom and consists of a tightly bound unit of two protons and two neutrons. Many isotopes of heavy elements decay by the emission of α particles. This process occurs in the heavy-element region because Coulomb repulsion between the many protons present in these nuclei lowers their stability. The emission of an α particle reduces the nuclear charge by two units and leads to more stable nuclides.

Alpha decay can be represented by the equation

$$^AZ \rightarrow {}^{(A-4)}(Z-2) + {}^4\text{He} \tag{23.14}$$

As is true for any decay process, the α-decay energy is equal to the mass difference between the initial and final nuclides expressed in energy units:

$$Q_\alpha = [M(^AZ) - M(^{(A-4)}(Z-2)) - M(^4\text{He})](931.5 \text{ MeV u}^{-1}) \tag{23.15}$$

When $Q_\alpha > 0$, the mass of the decaying nuclide is greater than that of the products, and mass can be converted to energy. Consequently, the process is allowed energetically. When $Q_\alpha < 0$, the opposite condition holds true, and the process is energetically forbidden.

---

**Example 23.6**  The α decay of $^{210}$Po results in the formation of $^{206}$Pb. Calculate the α-decay energy. Can $^{210}$Po decay by proton emission? The relevant nuclidic masses are

$^{210}_{84}$Po  209.98287 u     $^{209}_{83}$Bi  208.98042 u

$^{206}_{82}$Pb  205.97446 u     $^{1}_{1}$H   1.0078252 u

$^{4}_{2}$He  4.0026036 u

The α-decay energy is given by Equation 23.15:

$$Q_\alpha = [M(^{210}\text{Po}) - M(^{206}\text{Pb}) - M(^4\text{He})](931.5 \text{ MeV u}^{-1})$$

$$= (209.98287 \text{ u} - 205.97446 \text{ u} - 4.0026036 \text{ u})(931.5 \text{ MeV u}^{-1})$$

$$= 5.41 \text{ MeV}$$

The proton decay energy is given by an analogous equation:

$$Q_p = [M(^AZ) - M(^{(A-1)}(Z-1)) - M(^1\text{H})](931.5 \text{ MeV u}^{-1})$$

$$= [M(^{210}\text{Po}) - M(^{209}\text{Bi}) - M(^1\text{H})](931.5 \text{ MeV u}^{-1})$$

$$= (209.98287 \text{ u} - 208.98042 \text{ u} - 1.0078252 \text{ u})(931.5 \text{ MeV u}^{-1})$$

$$= -5.01 \text{ MeV}$$

The negative value of the proton decay energy means that $^{210}$Po is lighter than its decay products. Thus, there is no extra mass that can be converted to energy, and proton decay cannot occur. ∎

The results of this example are typical for radioactive isotopes of heavy elements: $\alpha$ decay is energetically allowed and is a common decay mode; proton decay is energetically forbidden and does not occur.

**b. The Effect of the Coulomb Barrier** Recall that in Rutherford's $\alpha$-particle scattering experiment (Section 5.1c), the $\alpha$ particles were deflected by the nucleus owing to the Coulomb repulsion between these two positively charged bodies. We say that a Coulomb barrier prevents the $\alpha$ particle from penetrating the nucleus unless it has enough energy to surmount the barrier. This same barrier also faces an $\alpha$ particle attempting to leave the nucleus and has a large effect on the rate at which $\alpha$ decay occurs.

The height of the Coulomb barrier is equal to the product of the charges of the nucleus and the $\alpha$ particle divided by the sum of their radii. It turns out, fortuitously, that the various constants and conversion factors cancel out and the barrier, expressed in MeV, is given by the simple relation

$$B = \frac{Z_1 Z_2}{A_1^{1/3} + A_2^{1/3}} \text{ MeV} \tag{23.16}$$

where $Z_1$ and $A_1$ are the atomic and mass numbers of a particle attempting to penetrate or escape from a nucleus. The quantities $Z_2$ and $A_2$ are the atomic and mass numbers of the nucleus either before the capture of the particle or after the escape of the particle, depending on whether the process involves the capture or emission of a particle, respectively. For example, the same Coulomb barrier exists between an $\alpha$ particle incident on $^{206}$Pb as between an $\alpha$ particle and a $^{206}$Pb nucleus formed in the $\alpha$ decay of $^{210}$Po. Its magnitude is

$$B = \frac{(2)(82)}{(4)^{1/3} + (206)^{1/3}} = 21.9 \text{ MeV}$$

Figure 23.8 shows the potential energy between an $\alpha$ particle and $^{206}$Pb. The potential energy inside the nucleus is negative as a consequence of the presence of the attractive nuclear force. At the nuclear radius the potential changes sign as the nuclear force gives way to the repulsive Coulomb force. The value of the potential energy at this point is equal to the Coulomb barrier. With increasing distance from the nucleus, the potential energy decreases according to the well-known $1/r$ dependence.

Consider the emission of an $\alpha$ particle from $^{210}$Po. Since the $\alpha$-decay energy is 5.4 MeV (Example 23.6) the $\alpha$ particle is shown in Figure 23.8 at a positive energy of 5.4 MeV. In the absence of the Coulomb barrier, $^{210}$Po would decay essentially instantaneously because the $\alpha$ particle has enough energy to escape from the nucleus. However, the Coulomb barrier is 21.9 MeV high and the 5.4-MeV $\alpha$ particle has much too low an energy to surmount the barrier. What happens, instead, is that the $\alpha$ particle "tunnels" through the barrier. (Tunneling is not allowed by classical physics. It is a quantum mechanical process that occurs because the wave function representing the $\alpha$ particle does not go abruptly to zero at the barrier, but has a small value outside the nucleus.) While the tunneling probability is very low, the $\alpha$ particle makes so many collisions with the barrier that it eventually tunnels through. The tunneling probability increases as the barrier becomes narrower. As can be seen in Figure 23.8, the width of the barrier decreases with increasing $\alpha$-decay energy. Consequently, $\alpha$ decay is most probable for nuclides having a large $\alpha$-decay energy

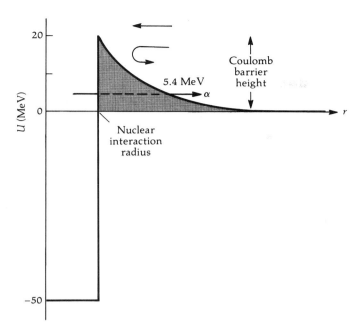

**FIGURE 23.8** Potential energy of a positively charged particle approaching or leaving a nucleus. The height of the Coulomb barrier shown corresponds to the case of an $\alpha$ particle incident on $^{206}$Pb or leaving $^{210}$Po (Equation 23.16). The width of the barrier is shown by the colored region. An $\alpha$ particle with a positive energy of 5.4 MeV (corresponding to $Q_\alpha$ for $^{210}$Po) cannot surmount the barrier but has a small probability of tunneling through it and thereby be emitted (arrow pointing right). The straight arrow pointing left represents an $\alpha$ particle with kinetic energy greater than the barrier approaching the nucleus. Such a particle can penetrate the nucleus. The curved arrow represents an incident $\alpha$ particle with an energy less than the barrier. Since the tunneling probability is low, such a particle cannot penetrate the nucleus with significant probability.

and the half-lives of $\alpha$ emitters are a strong inverse function of their decay energies. For example, $^{210}$Po, with $Q_\alpha = 5.4$ MeV, has a half-life of 138.4 d while $^{214}$Po, with $Q_\alpha = 7.71$ MeV has a half-life of only 164 $\mu$s.

*c. The Decay Series of the Naturally Occurring Heavy-Element Isotopes* All isotopes of the elements beyond bismuth are unstable and many of them undergo $\alpha$ decay. To be sure, some heavy-element isotopes are sufficiently long lived to be found on Earth. These include $^{232}$Th ($t_{1/2} = 1.4 \times 10^{10}$ y), $^{238}$U ($t_{1/2} = 4.5 \times 10^9$ y) and $^{235}$U ($t_{1/2} = 7.0 \times 10^8$ y). Since the Earth is believed to have been formed nearly five billion years ago, nuclides with significantly shorter half-lives than that of $^{235}$U which might have been present at that time would have undergone virtually complete decay in the intervening period.

The decay of the naturally occurring heavy-element isotopes occurs by the sequential emission of $\alpha$ and $\beta^-$ particles, ultimately leading to the formation of stable isotopes of lead. Each of the isotopes $^{232}$Th, $^{235}$U, and $^{238}$U, is the starting member of a series of radioactive isotopes formed at different stages of the decay chain. Figure 23.9 shows the uranium series, in which $^{238}$U ultimately decays to $^{206}$Pb. The series encompasses a number of different radioactive ele-

**896** Nuclear Chemistry 23.5

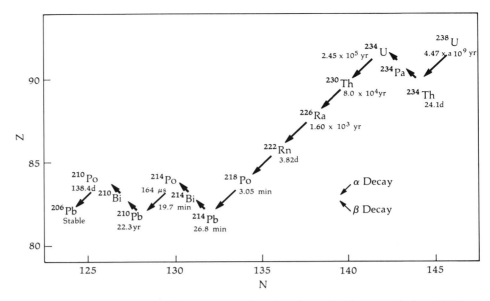

**FIGURE 23.9** The uranium decay series, showing the main decay path from $^{238}$U to $^{206}$Pb. The half-lives of the various nuclides in the series are given.

ments including radium, the separation of which from naturally occurring ores by the Curies at the beginning of the twentieth century may be said to mark the beginning of nuclear chemistry.

**Example 23.7** Uranium-235 decays by $\alpha$ and $\beta^-$ decay to $^{207}$Pb. How many $\alpha$ and $\beta^-$ particles are emitted in the conversion of one $^{235}$U atom to $^{207}$Pb?

The emission of an $\alpha$ particle reduces the nuclear charge by two units and the mass number by four units. On the other hand, the emission of a $\beta^-$ particle increases the nuclear charge by one.

The overall change in mass number is

$$A = 235 - 207 = 28$$

Dividing by four, we see that seven $\alpha$ particles must be emitted. The nuclear charge is therefore reduced by fourteen units. However, the difference in nuclear charge of uranium ($Z = 92$) and lead ($Z = 82$) is only 10. Therefore, 4 $\beta^-$ particles must be emitted to counteract the loss of the extra nuclear charge caused by $\alpha$-particle emission. ∎

# Nuclear Reactions

## 23.5 Fundamentals of Nuclear Reactions

A **nuclear reaction** is a process in which a nucleus reacts with a particle, a photon, or another nucleus to produce one or more other nuclei and, usually, some light particles as well. Typically, a nuclear reaction occurs between an energetic reaction partner, called the projectile or bombarding particle, and one that is at rest, called the target. The projectile strikes the target, some particles are emitted, and one or more nuclei are formed. Commonly used projectiles include protons, neutrons, $\alpha$ particles, and heavy ions ranging all the way up to $^{238}$U$^{n+}$ ($n$ depends on the projectile energy). Any stable isotope can serve as the target, and a number of radioisotopes have also been used in this manner. Nuclear reactions occur on an exceedingly fast time scale and are over in $10^{-12}$ s or less.

The first study of a nuclear reaction was performed by Rutherford in 1919. He used $\alpha$ particles emitted in the decay of $^{214}$Po to transform $^{14}$N to $^{17}$O by the reaction

$$^{14}_{7}\text{N} + ^{4}_{2}\text{He} \rightarrow ^{17}_{8}\text{O} + ^{1}_{1}\text{H} \tag{23.17}$$

The observation of energetic protons signalled the occurrence of the reaction. The development of particle accelerators has made it possible to study a large variety of nuclear reactions since the days of Rutherford's pioneering work.

It is customary to represent a nuclear reaction by means of a shorthand notation. The target nuclide is indicated first, and then, within parentheses, the projectile, followed after a comma by the light emitted particles. The identity of the massive residual nucleus follows outside the parentheses. The abbreviations used for $^{1}$H, $^{2}$H, and $^{4}$He are p, d, and $\alpha$, respectively. Reaction 23.17, for example, is written as

$^{14}$N($\alpha$, p)$^{17}$O

in this notation.

Nuclear reactions obey the same conservation laws as chemical reactions. The most important quantities conserved are nuclear charge, mass number, and mass–energy. Let us apply these conservation laws to reaction 23.17. The total nuclear charge of the nitrogen target and the $^{4}$He projectile is $7 + 2 = 9$. The nuclear charge of the reaction products is $8 + 1 = 9$, indicating that nuclear charge is conserved. Similarly, the total mass number of both reactants and products is 18. Viewed in these terms, a nuclear reaction involves a rearrangement in the number of neutrons and protons present in the nucleus of each species participating in the reaction.

The conservation of mass–energy can be understood in the same way as in the case of $\alpha$ decay. The energy released or absorbed in a nuclear reaction, called its **Q value,** is equal to the mass difference between reactants and products expressed in energy units:

$$Q = (\Sigma\, M_{\text{reactants}} - \Sigma\, M_{\text{products}})(931.5 \text{ MeV u}^{-1}) \tag{23.18}$$

There is one important difference between a reaction $Q$ value and a decay $Q$ value, such as $Q_\alpha$. Radioactive decay necessarily is exothermic, i.e., $Q_\alpha > 0$. On the other hand, a nuclear reaction can be either exothermic ($Q > 0$) or endothermic ($Q < 0$). In an endothermic reaction, the reactants are less massive than the products. The mass deficit is made up by the kinetic energy of the projectile. In addition to having enough kinetic energy to make up for a possible mass deficit, a charged projectile must be sufficiently energetic to surmount the Coulomb barrier between it and a target nucleus (see Equation 23.16 and Figure 23.8). The probability of a nuclear reaction induced by a projectile with an energy lower than that of the Coulomb barrier is low.

**Example 23.8** (a) Calculate the $Q$ value of the $^{65}$Cu(p, p2n)$^{63}$Cu reaction. Use the data in Table 23.1 and the neutron mass, $M_n = 1.0086654$ u. (b) What is the Coulomb barrier for protons incident on $^{65}$Cu? Does the barrier inhibit the occurrence of the (p, p2n) reaction?

(a) The $Q$ value is obtained by means of Equation 23.18:

$$Q = [M(^{65}\text{Cu}) + M(^1\text{H}) - M(^1\text{H}) - 2M(\text{n}) - M(^{63}\text{Cu})](931.5 \text{ MeV u}^{-1})$$

$$= [(64.9278 \text{ u}) - (2)(1.0086654 \text{ u}) - (62.9296 \text{ u})](931.5 \text{ MeV u}^{-1})$$

$$= -17.8 \text{ MeV}$$

(b) The Coulomb barrier is given by Equation 23.16:

$$B = \frac{Z_1 Z_2}{A_1^{1/3} + A_2^{1/3}}$$

$$= \frac{(1)(29)}{(1)^{1/3} + (65)^{1/3}} \text{ MeV}$$

$$= 5.8 \text{ MeV}$$

Since $Q = -17.8$ MeV, the reaction is endothermic and can occur only when the energy of the incident protons is higher than 17.8 MeV. As this value is much larger than that of the Coulomb barrier, the barrier does not affect the reaction. ■

## 23.6 Types of Nuclear Reactions

The probability of a nuclear reaction depends on a variety of factors, including the identity and energy of the projectile and the nuclear composition of the target.

*a. The (n, γ) Reaction* Neutrons are uncharged and consequently do not feel a Coulomb barrier. This makes possible the capture by a nucleus of very low-energy neutrons. In particular, neutrons can be slowed down until they are in thermal equilibrium with their surroundings. The velocity of such thermal neutrons obeys the Maxwell–Boltzmann distribution (Section 4.6), just as is true for the molecules of any other gas. Thermal neutrons have an unusually large probability of being captured by nuclei. The de Broglie wavelength of a thermal neutron, $\lambda = h/mv$, is much larger than that of an energetic projectile. Because of its wave properties, a thermal neutron can be captured even when its trajectory does not bring it in contact with the nucleus, as shown in Figure 23.10.

**Example 23.9** Calculate the average kinetic energy in electron volts and the de Broglie wavelength of a thermal neutron at 25°C.

We know from Section 4.2 that the average kinetic energy per mole is $\frac{3}{2}RT$. To obtain the average energy of a single neutron we must divide by Avogadro's number and convert from joules to eV:

$$\varepsilon_n = \frac{(3/2)RT}{N_A}$$

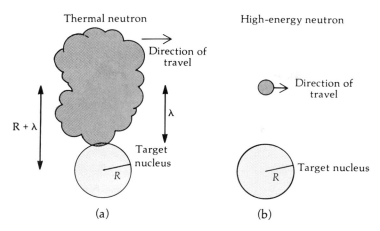

**FIGURE 23.10** Schematic illustration of the large capture probability of a thermal neutron. (a) A thermal neutron has a large de Broglie wavelength λ and can be captured by a nucleus so long as the distance of closest approach is less than $R + \lambda$, where $R$ is the radius of the target nucleus. (b) A high-energy neutron has a negligibly small de Broglie wavelength and must collide with the nucleus in order to react.

$$= \left(\frac{(1.5)(8.314 \text{ J mol}^{-1} \text{ K}^{-1})(298 \text{ K})}{6.022 \times 10^{23} \text{ mol}^{-1}}\right)\left(\frac{1 \text{ eV}}{1.602 \times 10^{-19} \text{ J}}\right)$$

$$= 0.0385 \text{ eV}$$

The de Broglie wavelength is $\lambda = h/mv$. The neutron speed can be obtained from the calculated average neutron energy (expressed in joules) and the neutron mass:

$$v = \left(\frac{2\varepsilon_n}{m}\right)^{1/2}$$

$$= \frac{(2)(6.17 \times 10^{-21} \text{ J})(1 \text{ kg m}^2 \text{ s}^{-2}/\text{J})^{1/2}}{(1.0087 \text{ u})(1.661 \times 10^{-27} \text{ kg u}^{-1})}$$

$$= 2714 \text{ m s}^{-1}$$

$$\lambda = \frac{h}{mv}$$

$$= \frac{(6.626 \times 10^{-34} \text{ J s})(1 \text{ kg m}^2 \text{ s}^{-2}/\text{J})}{(1.0087 \text{ u})(1.661 \times 10^{-27} \text{ kg u}^{-1})(2714 \text{ m s}^{-1})} = 1.46 \times 10^{-10} \text{ m}$$

$$= 1.46 \text{ Å} \quad \blacksquare$$

This example shows that the wavelength of a thermal neutron is of atomic rather than of nuclear dimensions, and its interaction probability is correspondingly large. Furthermore, because of their long wavelength, thermal neutrons can be used to study the structure of solids. This technique, **neutron diffraction,** is analogous to X-ray diffraction (Section 20.3).

The capture of a thermal neutron by a nucleus generally does not transfer enough energy to permit the emission of a particle from the nucleus (other than the neutron). The neutron is often incorporated into the nucleus, and the excess energy is dissipated by the emission of $\gamma$ rays. The result is the (n, $\gamma$) reaction, resulting in the formation of an isotope of the target element containing one additional neutron. Some examples of this reaction are $^{59}$Co(n, $\gamma$)$^{60}$Co and $^{75}$As(n, $\gamma$)$^{76}$As. Many of the resulting isotopes are radioactive and the (n, $\gamma$) reaction can be used for their preparation.

The (n, $\gamma$) reaction also is widely used in **activation analysis,** a sensitive method for the analysis of the elemental composition of materials. A sample is irradiated with neutrons from a nuclear reactor, and the radioisotopes produced in the (n, $\gamma$) reaction are detected by assay of their characteristic radiations. The technique is very sensitive because the reaction probability is high, the neutron source is intense, and a small number of radioactive atoms can be detected. The detection of amounts as small as $10^{-12}$ g make the technique particularly useful for the determination of trace amounts of elements.

*b. Nuclear Fission*  When certain isotopes of heavy elements capture thermal neutrons, they undergo **fission.** In this process the nucleus splits into two roughly equal parts with the release of a large amount of energy, approximately 200 MeV. Several neutrons are emitted, each of which is capable of inducing another nucleus to fission, thereby making a self-sustaining chain reaction (Section 14.6) possible. The combination of the large amount of energy released and the possibility of a chain reaction are the two factors that endow fission with an awesome power. The slow controlled release of fission energy in nuclear reactors is a valuable energy source for the generation of electrical power; the rapid uncontrolled release of fission energy forms the basis of nuclear weapons capable of destroying civilization.

Figure 23.11 shows the sequence of events occurring in nuclear fission. The capture of a neutron gives the nucleus some excess energy. As a result, the nucleus can distort from the spherical shape. This distortion is favored by the resulting increase in the distance between the protons and the consequent decrease in Coulomb repulsion energy. However, the distortion of the nucleus also results in an increase in its surface area and, consequently, in its surface energy. Since an increase in surface energy reduces the nuclear binding energy (Equation 23.5), the distortion is disfavored. The opposing effects of a reduction in Coulomb energy and an increase in surface energy are nearly in balance for heavy elements. It turns out that for two isotopes, $^{235}$U and $^{239}$Pu, the energy available from neutron capture is sufficiently large to permit the distortion to continue to the point of no return (labelled (4) in the figure) and fission rapidly follows. The change in potential energy accompanying this distortion is similar to that occurring in the passage of chemical reactants through the activated complex (Section 14.9).

After the fission fragments are born (step 5), their mutual Coulomb repulsion rapidly drives them apart (step 6). Most of the energy released in fission is in the form of the kinetic energy of the fission fragments. These fragments collide with the atoms of the surrounding environment and dissipate their energy in the form of heat, eventually slowing down and coming to rest.

The moving fragments emit, on the average, 2–3 neutrons per fission (step 7). Neutron emission occurs because the fragments have high internal energy and more than the optimum number of neutrons. As shown in Figure 23.7, the stable isotopes in the mass region where the fission products lie, $A \approx 85$–150, have a lower neutron-to-proton ratio than $^{235}$U. Neutron emission partially corrects this imbalance. However, the resulting products are still too

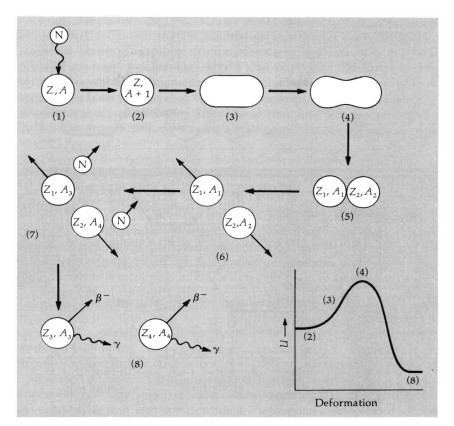

**FIGURE 23.11** Steps in thermal neutron fission. A $^{235}$U nucleus captures a thermal neutron (1). The excess energy is shared among all the nucleons (2) and the nucleus distorts (3). If the distortion reaches the configuration labelled (4), then the nucleus rapidly fissions (5). The fission fragments repel each other (6) and emit neutrons (7). The resulting fission products are radioactive and decay towards stability by $\beta^-$ decay (8). The potential energy change is plotted as a function of deformation in the lower right diagram. The numbers identify the configurations corresponding to specific regions of the curve.

---

neutron rich. Consequently, they are radioactive and decay to stability by a sequence of $\beta^-$ decays (step 8).

*c. Reactions Induced by Charged Particles* By virtue of their charge, charged particles can be accelerated to well-defined energies. This makes it possible to study the yields of specific reactions as a function of projectile energy. As long as the projectile energy exceeds the Coulomb barrier, all energetically possible reactions will in general occur. For example, the interaction of $^{64}$Zn with 30 MeV $\alpha$ particles can lead to the following reactions:

$$^{64}\text{Zn} + {}^{4}\text{He} \rightarrow [{}^{68}\text{Ge}]^* \begin{cases} \rightarrow {}^{67}\text{Ge} + \text{n} & (\alpha, \text{n}) \\ \rightarrow {}^{67}\text{Ga} + \text{p} & (\alpha, \text{p}) \\ \rightarrow {}^{66}\text{Ge} + 2\text{n} & (\alpha, 2\text{n}) \\ \rightarrow {}^{66}\text{Ga} + \text{p} + \text{n} & (\alpha, \text{pn}) \\ \rightarrow {}^{63}\text{Zn} + \alpha + \text{n} & (\alpha, \alpha\text{n}) \end{cases}$$

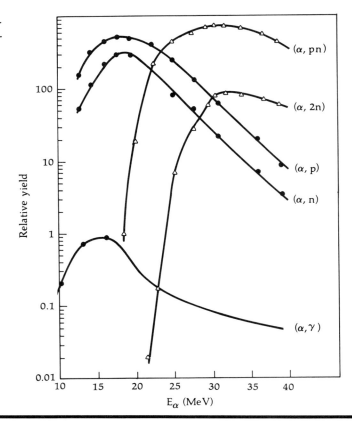

**FIGURE 23.12** Dependence on the bombarding energy of the yields of various reactions of $^{64}$Zn with α-particles. Such curves are called *excitation functions*.

The reaction involves, as a first step, the amalgamation of projectile and target to form an excited compound nucleus, which is $^{68}$Ge. This nucleus has a very large amount of excess energy and rapidly decays by the emission of one or more neutrons and protons, or even by the emission of more complex particles. The relative yields of the different reactions vary widely, as shown in Figure 23.12, where the variation of yield with bombarding energy is plotted. It is apparent that each reaction occurs most readily at some optimum energy. For example, the maximum yield of the radioisotope $^{67}$Ge produced by the $^{64}$Zn(α, n) reaction is obtained at an energy of ≈20 MeV.

With increasing projectile energy, it is possible to make products whose atomic and mass numbers are increasingly smaller than those of the target nuclide. One useful way to display the data is by means of a plot of product yield versus mass number. Mass yield curves for the interaction of protons of several energies with lead are shown in Figure 23.13. At a relatively low energy, 40 MeV, the entire yield is concentrated in just a few mass numbers close to that of the target. With increasing energy the curve broadens, and in the GeV energy range (1 GeV = $10^9$ eV) all isotopes in the periodic table lighter than the target nuclide are produced. In particular, the conversion of lead to gold (A = 197) occurs readily and is optimal at an energy of several hundred MeV. This ancient dream of the alchemists is just one of the many kinds of nuclear transformations made possible by the advent of powerful particle accelerators. Unfortunately, the number of particles that can be accelerated to high energies

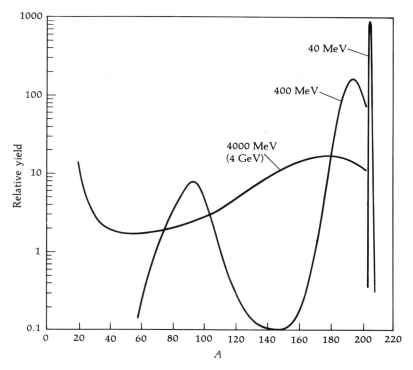

**FIGURE 23.13** Mass yield curves for the interaction of lead with 40-MeV, 400-MeV, and 4-GeV (4000-MeV) protons.

by any accelerator built to date add up to less than a mole per year. Thus, this method cannot yield practical amounts of gold.

## 23.7 Artificially Produced Elements

The heaviest naturally occurring element found on Earth is uranium ($Z = 92$). One of the great triumphs of modern nuclear chemistry and physics has been the production and characterization of elements beyond uranium. Much of this work has been done by Glenn Seaborg (b. 1912) and his collaborators at Berkeley. Isotopes of 17 elements with $Z = 93$–109 have been made to date. Their half-lives generally decrease with increasing atomic number to the point that the heaviest elements decay in a fraction of a second. Owing to the large magnitude of the Coulomb repulsion energy at these high Z values, a decay mode not seen for elements of lower atomic numbers becomes important: **spontaneous fission.** As in induced fission (Section 23.6b) the nucleus splits into two fragments of comparable mass and charge. Spontaneous fission becomes increasingly important with increasing atomic number and is the process which limits the production of heavier elements.

Figure 23.14 shows the location of the heaviest elements in the periodic table and Table 23.2 summarizes some of their properties. The elements through lawrencium ($Z = 103$) complete the actinide series, in which the 5f subshell is filled. The subsequent elements belong to the fourth transition series, in which electrons are added to the 6d subshell.

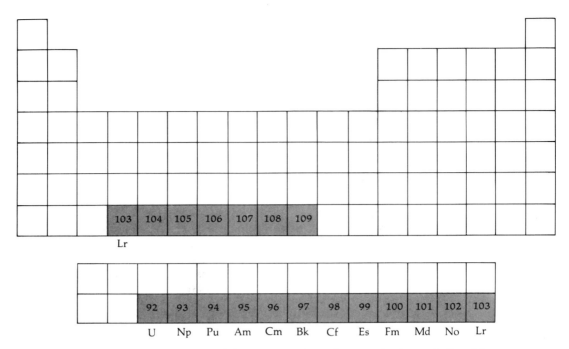

**FIGURE 23.14** Location of the transuranium elements in the periodic table.

The transuranium elements can be produced by a variety of nuclear reactions. Neptunium and plutonium can be made by thermal neutron capture in $^{238}$U followed by $\beta^-$ decay of the resulting products:

$$^{238}_{92}\text{U}(n, \gamma)\ ^{239}\text{U} \xrightarrow[t_{1/2}\ =\ 23.5\text{ m}]{\beta^-}\ ^{239}_{93}\text{Np} \xrightarrow[t_{1/2}\ =\ 2.35\text{ d}]{\beta^-}\ ^{239}_{94}\text{Pu} \rightarrow$$

or by charged particle bombardment of uranium, e.g.,

$$^{238}_{92}\text{U}(d, 2n)\ ^{238}_{93}\text{Np} \xrightarrow[t_{1/2}\ =\ 2.1\text{ d}]{\beta^-}\ ^{238}_{94}\text{Pu} \rightarrow$$

Special attention has been focused on $^{239}$Pu because, along with $^{235}$U, it undergoes thermal neutron-induced fission and, with a half-life of $2.4 \times 10^4$ y, is sufficiently long-lived to be useful.

A variety of reactions have been used to prepare the heavier transuranium elements. For example, mendelevium (Md) was first made in 1955 by the reaction

$$^{253}_{99}\text{Es }(\alpha, n)^{256}_{101}\text{Md}$$

The target consisted of approximately $10^9$ atoms of $^{253}$Es, an isotope whose half-life is only 20 days. Less than 100 atoms of Md were produced. In spite of this small number, a chemical separation from the other transuranium elements proved possible. The development of chemical methods to deal with such trace amounts of materials has been of great value in elucidating the chemistry of the transuranium elements.

In recent years, macroscopic amounts, i.e., micrograms or greater, of the elements through einsteinium have become available. The chemistry of these elements is actively being studied. For example, the principal oxidation states displayed by these elements in aqueous solution are known (Table 23.2). While

### TABLE 23.2 The Actinide and Transactinide Elements

| Z | Element | Longest-Lived Isotope | Half-Life | Principal Oxidation States in Solution[a] |
|---|---|---|---|---|
| 90 | Thorium | $^{232}$Th | $1.4 \times 10^{10}$ y | +4 |
| 91 | Protoactinium | $^{231}$Pa | $3.3 \times 10^{4}$ y | +5,+4 |
| 92 | Uranium | $^{238}$U | $4.5 \times 10^{9}$ y | +6,+3,+4,+5 |
| 93 | Neptunium | $^{237}$Np | $2.1 \times 10^{6}$ y | +5,+3,+4,+6,+7 |
| 94 | Plutonium | $^{244}$Pu | $8.1 \times 10^{7}$ y | +4,+3,+5,+6,+7 |
| 95 | Americium | $^{243}$Am | $7.4 \times 10^{3}$ y | +3,+4,+5,+6 |
| 96 | Curium | $^{247}$Cm | $1.6 \times 10^{7}$ y | +3,+4 |
| 97 | Berkelium | $^{247}$Bk | $1.4 \times 10^{3}$ y | +3,+4 |
| 98 | Californium | $^{251}$Cf | $9.0 \times 10^{2}$ y | +3 |
| 99 | Einsteinium | $^{252}$Es | 1.3 y | +3,+2 |
| 100 | Fermium | $^{257}$Fm | 100 d | +3,+2 |
| 101 | Mendelvium | $^{258}$Md | 56 d | +3,+2 |
| 102 | Nobelium | $^{259}$No | 58 m | +2,+3 |
| 103 | Lawrencium | $^{260}$Lr | 3.0 m | +3 |
| 104 | (Unnilquadium)[b] | $^{261}$104 (Unq)[b] | 1.1 m | |
| 105 | (Unnilpentium) | $^{262}$105 (Unp) | 0.7 m | |
| 106 | (Unnilhexium) | $^{263}$106 (Unh) | 0.9 s | |
| 107 | (Unnilseptium) | $^{261}$107 (Uns) | 2 ms | |
| 108 | (Unniloctium) | $^{265}$108 (Uno) | short | |
| 109 | (Unnilenium) | $^{266}$109 (Une) | short | |

[a]The most stable oxidation state is listed first.
[b]IUPAC proposed names and symbols for elements with Z ≥ 104. These names are not gaining wide acceptance among nuclear chemists, who prefer to refer to unnamed elements by their atomic numbers.

in analogy to the lanthanides the +3 state is important, the lighter elements are more commonly found in higher oxidation states. The chemical behavior of the transeinsteinium (Z > 99) elements is being studied using the available trace amounts of rapidly decaying isotopes. Although this work poses immense experimental challenges it is of importance in view of the transition between the actinide series and the fourth transition series that occurs in this region.

---

**Example 23.10** Element 109 was discovered at the GSI laboratory in the Federal Republic of Germany in 1982 in heavy ion bombardment of $^{209}$Bi. The $^{267}$109 compound nucleus was formed and emission of one neutron resulted in the formation of $^{266}$109. While only a single atom of this isotope was formed, its identity could be established by the observation of its sequential decay:

$$^{266}109 \xrightarrow[5 \text{ ms}]{\alpha} {}^{262}107 \xrightarrow[22 \text{ ms}]{\alpha} {}^{258}105 \xrightarrow{EC} {}^{258}104 \rightarrow$$

(a) Determine the identity of the projectile and write the reaction leading to the formation of $^{266}$109 in shorthand notation. (b) Estimate the $Q$ value of the reaction and the Coulomb barrier for the interaction of the projectile and target.

(a) Since the compound nucleus involves amalgamation of target and projectile, the identity of the latter can be determined by subtracting the atomic and mass numbers of the target from the corresponding values for the compound nucleus:

$$Z = 109 - 83 = 26$$
$$A = 267 - 209 = 58$$

The projectile is $^{58}$Fe and the reaction is $^{209}$Bi($^{58}$Fe, n)$^{266}$109.

(b) The $Q$ value of the reaction can be estimated from masses obtained by means of the liquid drop model, Equation 23.5. Using Equation 23.18 for the $Q$ value we have

$$Q = [M(^{209}\text{Bi}) + M(^{58}\text{Fe}) - M(^{266}109) - M(n)] (931.5) \text{ MeV}$$

Since the nuclidic mass is related to the nuclear binding energy by the relation (Equation 23.4)

$$M(Z, A) = ZM_H + (A-Z)M_n - E_b(Z, A)$$

and

$$E_b = 14.1A - 13.0A^{2/3} - 0.600\frac{Z^2}{A^{1/3}} - 20.0\frac{(A-2Z)^2}{A} \pm \frac{130}{A}$$

we obtain

$$Q = [83M_H + 126M_n - E_b(^{209}\text{Bi})/931.5 + 26M_H + 32M_n - E_b(^{58}\text{Fe})/931.5$$
$$- 109M_H - 157M_n + E_b(^{266}109)/931.5 - M_n]931.5$$
$$= E_b(^{266}109) - E_b(^{209}\text{Bi}) - E_b(^{58}\text{Fe})$$
$$= [14.1(266 - 209 - 58) - 13.0[(266)^{2/3} - (209)^{2/3} - (58)^{2/3}]$$
$$- 0.600\left[\frac{(109)^2}{(266)^{1/3}} - \frac{(83)^2}{(209)^{1/3}} - \frac{(26)^2}{(58)^{1/3}}\right]$$
$$- 20.0\left[\frac{(266-218)^2}{266} - \frac{(209-166)^2}{209} - \frac{(58-52)^2}{26}\right] - \frac{130}{266} + \frac{130}{58} \text{ (MeV)}$$

$$Q = -173 \text{ MeV}$$

The Coulomb barrier is given by Equation 23.16,

$$B = \frac{Z_1 Z_2}{A_1^{1/3} + A_2^{1/3}}$$
$$= \frac{26 \times 83}{(58)^{1/3} + (209)^{1/3}}$$
$$= 220 \text{ MeV}$$

The $^{58}$Fe projectile should have a kinetic energy above 220 MeV in order to permit the product to be formed. In practice, an energy of 299 MeV was used. ∎

Nuclear theory has indicated that a closed proton shell should occur at $Z = 114$. Shell closure stabilizes a nucleus, and it has been estimated that the half-lives of nuclides in the vicinity of $Z = 114$ could be as long as days, years, or even longer. Since, as noted in Table 23.2, the half-lives of the longest-lived isotopes of the heaviest known elements are only in the range of milliseconds, the observation of such an abrupt reversal in the trend of decreasing half-lives with increasing atomic number would be dramatic. We appear to be nearing the end of a period of intense worldwide experimental attempts to produce these so-called superheavy elements. Because of the instability of the intervening elements with $Z \geq 102$, it is apparent that superheavy elements can be made only in reactions induced by heavy ions on target elements with $Z \leq 101$, e.g., $^{244}\text{Pu}(^{48}\text{Ca},n)^{291}114$. While a large variety of approaches have been tried, none have been successful. Nuclear scientists are coming to the conclusion that superheavy elements are either not sufficiently stable to be observed or cannot be produced in measurable yields by nuclear reactions.

# Applications of Nuclear Chemistry

## 23.8 Radioactive Decay

Radioactive decay obeys first-order kinetics because the decay of a particular radioactive atom is independent of the presence of other atoms. Thus, in an ensemble of radioactive atoms of a given isotope, the **disintegration rate** $D$—the number of radioactive atoms decaying per unit time—is proportional to the number of atoms $N$ present,

$$D = -dN/dt = \lambda N \tag{23.19}$$

The disintegration rate is commonly expressed in disintegrations per minute (dpm) or some other convenient time unit. The decay constant $\lambda$ is analogous to the first-order rate constant $k$ in a chemical reaction. The half-life of a radioisotope is related to the decay constant by

$$t_{1/2} = \ln 2 / \lambda \tag{23.20}$$

in analogy to the similar relation for first-order kinetics (Equation 14.20). The differential rate law can be integrated following the procedure described in Section 14.3a to yield the variation with time of either the number of atoms $N$ or the disintegration rate $D$,

$$N = N° \, e^{-\lambda t}$$
$$D = D° \, e^{-\lambda t} \tag{23.21}$$

where the quantities with the superscript ° denote the values at $t = 0$.

The half-life of a radioisotope is usually determined by observation of the exponential decrease in its disintegration rate. This technique becomes impractical for long-lived radioisotopes, e.g., $^{41}\text{Ca}$, $t_{1/2} = 1.2 \times 10^5$ y, since their disintegration rates do not perceptibly decrease during a reasonable observation time. In these instances the half-life can be determined by means of the differential rate law (Equation 23.19). The disintegration rate is measured, and the number of atoms present is determined by mass spectrometry, thereby permitting Equation 23.19 to be solved for $\lambda$.

**Example 23.11** The radioisotope $^{68}\text{Ga}(t_{1/2} = 68.3 \text{ min})$ is produced in the bombardment of $^{65}\text{Cu}$ by $\alpha$ particles. The disintegration rate of a $^{68}\text{Ga}$ sample 30.0 min after the end of bombardment was $6.05 \times 10^6$ dpm. Determine the disintegration rate and the number of atoms of $^{68}\text{Ga}$ present at the end of bombardment.

## 908 Nuclear Chemistry 23.8

Let $t = 0$ correspond to the time at the end of bombardment. The disintegration rate at $t = 0$ is $D°$. Solving Equation 23.21 for $D°$, we obtain

$$D° = D\, e^{\lambda t}$$

$$= (6.05 \times 10^6 \text{ dpm}) \exp\left[\frac{(\ln 2)(30.0 \text{ min})}{68.3 \text{ min}}\right]$$

$$= 8.20 \times 10^6 \text{ dpm}$$

where the decay constant $\lambda$ has been expressed in terms of the half-life by means of Equation 23.20.

The number of atoms of $^{68}$Ga is obtained from the disintegration rate by means of Equation 23.19:

$$N = \frac{D}{\lambda} = \frac{D\, t_{1/2}}{\ln 2}$$

$$= \frac{(8.20 \times 10^6 \text{ dpm})(68.3 \text{ m})}{\ln 2} = 8.08 \times 10^8 \quad \blacksquare$$

Since radioactive decay provides a measure of the number of atoms present in a sample, it can be used to determine Avogadro's number, as shown in the following example.

---

**Example 23.12** The isotope $^{250}$Cf, of half-life 13.1 y, decays by $\alpha$ decay. A 1.00 mg sample of $^{250}$Cf undergoes $2.424 \times 10^{11}$ disintegrations per minute. The emitted $\alpha$ particles are collected in an evacuated 1.00 cm³ container for 36.5 days. The pressure of the resulting helium gas at 25°C is 0.3933 torr. Determine Avogadro's number from these data.

The data on the accumulated helium gas can be used in conjunction with the ideal gas law to determine the number of moles of helium produced in the decay of $^{250}$Cf:

$$PV = nRT$$

$$n = \frac{(0.3933 \text{ torr})(1 \text{ atm}/760 \text{ torr})(1.00 \text{ cm}^3)(10^{-3} \text{ L cm}^{-3})}{(0.08206 \text{ L atm mol}^{-1} \text{ K}^{-1})(298 \text{ K})}$$

$$= 2.116 \times 10^{-8} \text{ mol}$$

Since $^{250}$Cf decays by $\alpha$ emission, one $^4$He atom is produced in each decay. The number of $^4$He atoms produced per minute is therefore equal to the number of disintegrations per minute of $^{250}$Cf, $2.424 \times 10^{11}$. The number of $^4$He atoms produced in 36.5 d is

$$N_{^4\text{He}} = (2.424 \times 10^{11} \text{ m}^{-1})(60 \text{ m h}^{-1})(24 \text{ h d}^{-1})(36.5 \text{ d})$$

$$= 1.274 \times 10^{16}$$

Avogadro's number is just the ratio of the number of $^4$He atoms to the corresponding number of moles:

$$N_A = \frac{1.274 \times 10^{16}}{2.116 \times 10^{-8} \text{ mol}} = 6.02 \times 10^{23} \text{ mol}^{-1} \quad \blacksquare$$

## 23.9 Age Determinations by Radioactivity Measurements

The presence on Earth of naturally occurring long-lived radioisotopes has led to their use as "clocks" to determine the age of geological materials. The age determination is based on the measurement of two related quantities: the disintegration rate of some long-lived radioisotope embedded in a geological sample and the amount of decay product that has accumulated in the sample. The combination of these two quantities yields the time elapsed since the sample was formed, i.e., its age. This technique convincingly showed for the first time that terrestrial rocks underwent solidification several billion years ago.

We shall illustrate the method by consideration of the decay of $^{238}U$ ($t_{1/2} = 4.47 \times 10^9$ y) to $^{206}Pb$ in a sample of pitchblende, a uranium ore. The quantities measured are the number of atoms of $^{238}U$ and $^{206}Pb$ present in the rock, designated $N_{238}$ and $N_{206}$, respectively. Assuming that the $^{206}Pb$ present was exclusively formed in the decay of $^{238}U$, we start with the relation

$$N_{206} = N°_{238} - N_{238} \tag{23.22}$$

which states that the number of $^{206}Pb$ atoms is equal to the difference between the number of $^{238}U$ atoms present initially ($N°_{238}$) and the number present now, that is, to the number of $^{238}U$ atoms that have decayed. Relating $N°_{238}$ to $N_{238}$ by means of Equation 23.21 we obtain

$$N_{206} = N_{238}\, e^{\lambda t} - N_{238} \tag{23.23}$$

and solving for $t$ we get

$$t = \frac{1}{\lambda} \ln\left(1 + \frac{N_{206}}{N_{238}}\right) \tag{23.24}$$

Let us evaluate this expression for some realistic values of $N_{206}$ and $N_{238}$. The $^{238}U$ content of pitchblende is typically $2.5 \times 10^{20}$ atoms per gram. If the value of $N_{206}$ is $1.0 \times 10^{20}$ atoms per gram of pitchblende, the value of $t$ is $2.2 \times 10^9$ y. This is the time elapsed since the particular pitchblende sample solidified, thereby preventing any subsequent separation between uranium and its decay products.

Such an analysis assumes that the particular sample has been a closed system since its formation, i.e., that there has been no loss or gain of either $^{238}U$ or $^{206}Pb$ since solidification, all the $^{206}Pb$ product being due to uranium decay. It is this last point that is most subject to error since lead is generally more common than uranium. Fortunately, lead consists of several stable isotopes, one of which, $^{204}Pb$, is not formed in the radioactive decay of heavy elements and is primordial. If the pitchblende sample is found to be free of $^{204}Pb$ it is safe to assume that all the $^{206}Pb$ resulted from the decay of $^{238}U$; if not, a correction can be applied on the basis of the natural isotopic composition of lead.

The $^{238}U$–$^{206}Pb$ method is not the only one that can be used to determine the age of geological samples. Table 23.3 lists some other pairs of nuclides that have been used for this purpose. As in the preceding case, one nuclide is a long-lived radioisotope and the other is its stable decay product. Since experimental conditions in geological samples obviously cannot be controlled in the same way that they can in the laboratory, any given measurement can yield erroneous results. It is the general concordance between different measurements performed on various samples using different pairs of isotopes that indicates that the method is valid.

TABLE 23.3 Long-Lived Radioisotopes Used for Geological Age Determinations

| Radioisotope | Half-Life (years) | Stable product |
|---|---|---|
| $^{87}$Rb | $4.7 \times 10^{10}$ | $^{87}$Sr |
| $^{238}$U | $4.5 \times 10^{9}$ | $^{206}$Pb |
| $^{40}$K | $1.3 \times 10^{9}$ | $^{40}$Ar |
| $^{235}$U | $7.1 \times 10^{8}$ | $^{207}$Pb |

## 23.10 Radiocarbon Dating

Radiocarbon dating is commonly used to determine the age of materials of biological origin. This technique, which has proved to be extremely valuable in archeology, is based on a rather different principle than the method discussed in Section 23.9.

As a result of the action of cosmic rays, which are energetic particles from outer space continually bombarding the Earth, the most abundant isotope of nitrogen is converted to radioactive $^{14}$C in the upper atmosphere by the reaction $^{14}$N(n, p)$^{14}$C. Carbon-14, commonly called radiocarbon, is incorporated in atmospheric carbon dioxide, and in this form eventually makes its way down to the surface of the Earth. Owing to the cyclical conversion between inorganic and organic carbon (Section 22.5c), some of the $^{14}$C ultimately becomes part of all living organisms. Since the half-life of $^{14}$C is rather long, $5.7 \times 10^3$ y, there is ample time for equilibrium to be established. As a result, all living matter has approximately 15 dpm $^{14}$C per gram of carbon. Once an organism dies, it ceases to incorporate $^{14}$C. Consequently, the disintegration rate of $^{14}$C decreases exponentially at a rate determined by its half-life. If some artifact containing organic material is found, for example in an archeological excavation, the disintegration rate of $^{14}$C per gram of carbon permits a determination of the time elapsed since the organism from which the material was obtained was alive. For instance, if a recently excavated piece of cloth has 7.5 dpm $^{14}$C per gram of carbon, some $5.7 \times 10^3$ years must have elapsed since the cotton that went into the cloth was grown.

The validity of radiocarbon dating is based on the assumption that the amount of $^{14}$C present in a living organism has yielded 15 dpm per gram of carbon over the entire time range over which the method can be used. This assumes, in turn, that the cosmic ray intensity has been constant in time. Independent evidence based on the known ages of various objects permits a check of the method and indicates that the cosmic ray intensity has indeed been nearly constant.

Figure 23.15 shows the ages of various objects that have been determined in this fashion superimposed on the exponential decay curve of $^{14}$C. It can be seen that the disintegration rates of objects much older than some 25,000 years are very low and this fact establishes the practical limits of the technique. Fortunately, the period of the past 10,000 years, for which the technique yields the most accurate results, is just the time that is of greatest interest from the historical viewpoint.

## 23.11 Isotopic Tracers in Chemistry

Isotopes are useful in studying the mechanisms of chemical reactions because their presence in the natural mixture of isotopes of a particular element can be

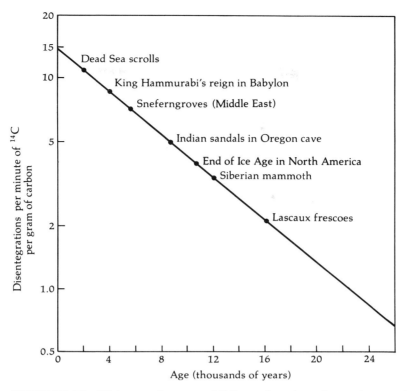

**FIGURE 23.15** Disintegration rate of $^{14}$C as a function of age of some objects assayed by radiocarbon dating.

identified readily. The behavior of an element in a chemical reaction is in most instances essentially independent of its isotopic composition. Consequently, isotopes can be used as tracers or indicators, in the sense that their behavior will be the same as that of the other atoms of the same element originally present in a particular chemical species. Because of the ease and sensitivity with which they can be detected, radioisotopes make ideal tracers. However, for elements such as oxygen, all of whose radioisotopes are short-lived, stable isotopes can sometimes be used. For example, the rare but stable $^{18}$O isotope is used as an isotopic tracer for oxygen.

As an illustration of the usefulness of isotopic tracers consider the formation of an ester in the reaction between a carboxylic acid and an alcohol, e.g.,

$$C_2H_5COOH + CH_3OH \rightarrow C_2H_5CO\overset{*}{O}CH_3 + H_2O$$

We are interested here in the source of the oxygen atom labelled with the star. Does this atom originate in the acid or the alcohol? Isotopic labelling provides a ready answer to this question. The alcohol is labelled with $^{18}$O, CH$_3^{18}$OH, and the reaction is carried out. It is found that the isotope appears in the ester, $C_2H_5CO^{18}OCH_3$, unambiguously indicating that the alcohol is the source of the oxygen atom in question.

An example drawn from inorganic chemistry concerns the source of oxygen in the reaction of aqueous hydrogen peroxide with strong oxidizing agents such as $PbO_2$ or $MnO_4^-$. Consider the reaction

$$PbO_2(s) + H_2O_2(aq) \rightarrow PbO(s) + H_2O(l) + O_2(g)$$

The most plausible hypothesis for the formation of $O_2$ is that each of the two reacting species provides one oxygen atom, as suggested by the formation of the following possible intermediate:

$$PbO-\boxed{O-O}-OH_2$$

However, experiments in which $PbO_2$ and $H_2O_2$ were separately labelled with two $^{18}O$ atoms apiece indicated that both oxygen atoms originated in $H_2O_2$. A different reaction mechanism was therefore indicated.

Isotopic tracers have been of great value in the study of the kinetics of electron transfer reactions. When the transfer process involves two different elements, e.g.,

$$Fe^{2+} + Ce^{4+} \rightarrow Fe^{3+} + Ce^{3+}$$

the kinetics can be studied by conventional means. However, when only a single element is involved, as in

$$Fe^{2+} + Fe^{3+} \rightarrow Fe^{3+} + Fe^{2+}$$

it is impossible to determine the rate of electron transfer without isotopic labelling. The use of a radioactive tracer such as $^{59}Fe$ incorporated in, say, $Fe^{2+}$, and a measurement of the rate at which the tracer appears in the $Fe^{3+}$, permits a determination of the kinetics.

## 23.12 Nuclear Magnetic Resonance Spectroscopy

Like atomic electrons, nucleons possess angular momentum as a result of their intrinsic spin and orbital motion. When several nucleons are present in a single nucleus, their angular momenta combine to yield a net angular momentum characteristic of the nucleus as a whole. This angular momentum is known as the **nuclear spin.**

Nuclear spin is the basis of **nuclear magnetic resonance spectroscopy,** commonly known as **NMR.** This technique is a powerful tool for structural studies in chemistry, particularly in organic chemistry. It has been employed largely to elucidate the nature of the chemical environment of hydrogen atoms in organic compounds, and it is on this application that we shall focus.

To understand the phenomenon that makes NMR possible, recall from Section 5.11 that a particle possessing angular momentum is affected by an external magnetic field. Since the nucleus of a hydrogen atom is just a proton, which has spin $\frac{1}{2}$, the application of a magnetic field causes a hydrogen atom to assume one of two different orientations with respect to the field. As shown in Figure 23.16, the spin axis may be aligned parallel to the magnetic field direction or antiparallel to it. The atom has a different energy corresponding to each of these alignments, and the application of a magnetic field gives rise to two different quantum states, as shown. If electromagnetic radiation with a frequency corresponding to the energy difference between these states is incident on a sample placed in a magnetic field, the radiation can be absorbed. The absorption associated with the transition between the two states constitutes the nuclear magnetic resonance. Figure 23.17 shows a schematic diagram of an NMR spectrometer.

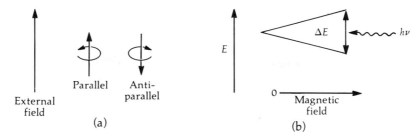

**FIGURE 23.16** Effect of external magnetic field on a proton. (a) Parallel and antiparallel alignments of the proton spin axis relative to the direction of the field. (b) The proton can occupy one of two quantum states depending on the direction of its alignment. The energy difference $\Delta E$ between the states increases with the magnitude of the applied field. In the presence of a magnetic field, electromagnetic radiation with a frequency satisfying the relation $\Delta E = h\nu$ is absorbed and induces a transition between the states.

If the magnetic field in the vicinity of each proton were just equal to the external magnetic field, all hydrogen atoms in a sample would absorb at exactly the same value of the field. Under these circumstances NMR would not provide any useful chemical information. The technique is valuable from a chemist's perspective because the external magnetic field is modified by the presence of electrons in the vicinity of the absorbing hydrogen atom. The magnitude of the field required to induce absorption consequently depends on the specific chemical environment of each hydrogen atom present in a molecule. Figure 23.18 shows the NMR spectrum of tert-butyl acetate, a compound that has a simple spectrum. The two distinct peaks can be associated with the nine equivalent tert-butyl hydrogen atoms and the three equivalent methyl hydrogen atoms, respectively. The peak intensities are in the ratio of 3 : 1 as expected from the relative numbers of these two types of hydrogen atoms. The peaks appear at different locations because the electron density differs at the two

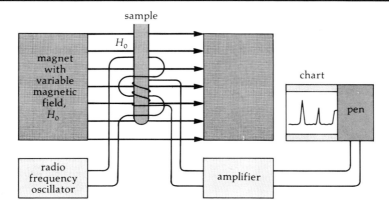

**FIGURE 23.17** Schematic diagram of an NMR spectrometer. The radiation required to induce a transition lies in the radio band and is produced by a radio frequency oscillator. The magnitude of the magnetic field is varied until the energy of the transition matches that of the radiowave and resonant absorption occurs.

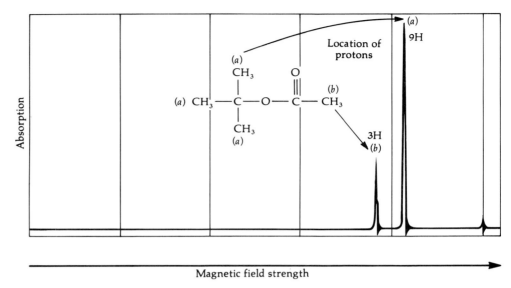

**FIGURE 23.18** The NMR spectrum of tert-butyl acetate.

sites. The methyl hydrogens (b in Figure 23.18) are in a region of lower electron density because the $\alpha$ carbon is attached to two electronegative oxygen atoms. For a more complicated molecule, the number, location, and relative intensity of the peaks in the NMR spectrum provide an indication of the types and relative numbers of groups present, and thus provide a detailed picture of molecular structure. Nuclear magnetic resonance spectroscopy is undoubtedly the outstanding application to chemistry of a nuclear pheonomenon.

# Conclusion

Nuclear processes obey the same general laws as ordinary chemical processes, for example, conservation of charge and mass–energy. However, the scale is different—nuclei are much smaller than molecules and the energies involved in nuclear processes are much larger than those involved in chemical processes. Owing to their much smaller size, nuclei collide and react much less frequently than molecules do. Consequently, the energy changes occurring in nuclear reactions are expressed on a per nucleus rather than on a per mole basis.

The relative stability of nuclei depends on a number of factors, such as the surface energy, Coulomb energy, symmetry energy, and pairing energy. The interplay between these factors determines the dependence of the stability line on the number of protons and neutrons. Nuclides lying off the stability line undergo radioactive decay. Most radionuclides undergo isobaric decay, either $\beta^-$, $\beta^+$, or electron capture decay. Radioisotopes of heavy elements can undergo $\alpha$ decay, while those of the heaviest known elements also can decay by spontaneous fission.

The advent of accelerators and nuclear reactors has made possible the study of a large variety of nuclear reactions. Different reactions occur depending on the identity of target and projectile and the bombarding energy. In general, any reaction that is energetically allowed will occur with some probability. However, reactions induced by charged particles with less energy than the Coulomb barrier are severely hindered. Many nuclear reactions result in the formation of radioisotopes. These isotopes are useful in a number of applications, such as age determinations and tracer studies.

## Problems

1. Which of the following nuclides are (a) isotopic, (b) isotonic, and (c) isobaric with $^{95}$Mo, $^{95}$Ru, $^{96}$Tc, $^{97}$Rh, $^{96}$Mo, $^{94}$Nb, $^{95}$Nb?

2. The nuclidic mass of $^{232}$Th is 232.03821 u. (a) What are the binding energy and the binding energy per nucleon of $^{232}$Th? (b) What fraction of the mass is converted into energy when $^{232}$Th is assembled from its constituents? Use the masses of the hydrogen atom and the neutron given in this chapter.

3. Calculate the nuclidic mass of $^{232}$Th by means of the liquid drop model. Compare with the experimental mass given in Problem 2.

4. The nuclide $^{115}$Sb decays by both positron emission and electron capture. (a) Write equations for these two processes. (b) The nuclidic masses of $^{115}$Sb and $^{115}$Sn are 114.90680 u and 114.90335 u, respectively. What is the value of $Q_{EC}$? (c) What is the energy of the neutrino emitted when $^{115}$Sb decays by electron capture? (d) What is the energy of the neutrino emitted when $^{115}$Sb decays by the emission of a 0.30 MeV positron? (e) Some of the decays of $^{115}$Sb populate an excited state of $^{115}$Sn at an energy of 0.50 MeV above the ground state. What is the value of $Q_{EC}$ for this process? What are the energy and wavelength of the $\gamma$ ray emitted when this state de-excites to the $^{115}$Sn ground state?

5. The nuclidic masses of some isobaric nuclides with $A = 121$ are as follows: $^{121}$In, 120.90785 u; $^{121}$Sn, 120.90424 u; $^{121}$Sb, 120.90381 u; $^{121}$Te, 120.90499 u; and $^{121}$I, 120.90753 u. Identify the stable nuclide and determine the isobaric decay modes and decay energies of the radioactive nuclides. Write an equation for each decay.

6. The nuclidic masses of the $A = 96$ isobars are as follows:

| Nuclide | Mass |
|---|---|
| $^{96}$Sr | 95.92031 u |
| $^{96}$Y | 95.91564 u |
| $^{96}$Zr | 95.90810 u |
| $^{96}$Nb | 95.90823 u |
| $^{96}$Mo | 95.90474 u |
| $^{96}$Tc | 95.90783 u |
| $^{96}$Ru | 95.90760 u |
| $^{96}$Rh | 95.91451 u |

(a) Plot the masses of these nuclides versus atomic number. Discuss the shape of this plot in terms of the contribution of the pairing energy to the liquid drop formula (Equation 23.5). Compare with a similar plot for odd-$A$ isobaric nuclides (e.g., Figure 23.3). (b) Identify all the stable nuclides with $A = 96$ on the assumption that a nuclide can decay only to an *adjacent* lighter nuclide. (c) For each radioactive nuclide, list all the possible isobaric decay modes and decay energies. Write an equation for each decay process.

7. Using the following mass table, determine which of the following nuclides with $A = 138$ are stable and which undergo $\beta^-$, $\beta^+$, or EC decay. For each nuclide undergoing decay, determine the value of $Q_{\beta^-}$, $Q_{\beta^+}$, or $Q_{EC}$, and write an equation for the decay process.

| Nuclide | Mass |
|---|---|
| $^{138}$Xe | 137.91404 u |
| $^{138}$Cs | 137.91104 u |
| $^{138}$Ba | 137.90524 u |
| $^{138}$La | 137.90716 u |
| $^{138}$Ce | 137.90602 u |
| $^{138}$Pr | 137.91079 u |
| $^{138}$Nd | 137.91111 u |

*8. Use the liquid drop model expression for the binding energy (Equation 23.5) to derive the relation

$$Z_A = \frac{A}{2.00 + 0.0150\, A^{2/3}}$$

for the most stable nuclear charge at mass number $A$.

9. The most stable nuclear charge at mass number $A$, $Z_A$, is given by the expression (see Problem 8)

$$Z_A = \frac{A}{2.00 + 0.0150 A^{2/3}}$$

(a) Show that the most stable light nuclides have equal numbers of neutrons and protons. To what factor can this effect be ascribed? (b) Show that the most stable heavy nuclides have more neutrons than protons. To what factor can this effect be ascribed?

10. The masses of the known technetium isotopes and those of their neighboring ruthenium and molybdenum isobars are given below. Use these data to predict the isobaric decay mode and decay energy of each Tc isotope. Write an equation for each decay. Are there any stable isotopes of technetium?

| A | Mass (Mo) | Mass (Tc) | Mass (Ru) |
|---|---|---|---|
| 95 | 94.90584 u | 94.90767 u | 94.91041 u |
| 96 | 95.90474 u | 95.90783 u | 95.90760 u |
| 97 | 96.90602 u | 96.90640 u | 96.90763 u |
| 98 | 97.90541 u | 97.90712 u | 97.90529 u |
| 99 | 98.90772 u | 98.90626 u | 98.90594 u |
| 100 | 99.90748 u | 99.90784 u | 99.90422 u |

11. Which of the following decay processes are energetically possible for $^{220}$Rn: (a) $\alpha$ decay, (b) proton decay, (c) $^{14}$C decay, (d) $\beta^-$ decay. Calculate the decay energy for each allowed process and identify the decay products. Use the following nuclidic masses: $^{220}$Rn, 220.01140 u; $^{216}$Po, 216.00192 u; $^{219}$At, 219.0114 u; $^{206}$Hg, 205.97747 u; $^{220}$Fr, 220.01233 u; $^{4}$He, 4.0026036 u; $^{1}$H, 1.0078252 u; and $^{14}$C, 14.0032419 u.

12. Calculate the Coulomb barrier against the emission from $^{220}$Rn of (a) an $\alpha$ particle, (b) a proton, and (c) a $^{14}$C fragment.

13. Thorium-232 decays to an isotope of lead by a sequence of $\alpha$ and $\beta^-$ decays. (a) Which isotope of lead is formed? (b) How many $\alpha$ particles and $\beta^-$ particles are emitted in this process?

14. Calculate the amount of energy required to dissociate one atom of $^{12}$C into 3 atoms of $^{4}$He.

15. The nuclide $^{256}$Es decays to $^{252}$Bk by $\alpha$ decay and to $^{256}$Fm by $\beta^-$ decay. The values $Q_\alpha$ and $Q_{\beta^-}$ are 6.300 MeV and 1.770 MeV, respectively. In turn, $^{256}$Fm decays by $\alpha$ decay to $^{252}$Cf and $Q_\alpha = 7.034$ MeV. Which is heavier, $^{252}$Cf or $^{252}$Bk? Which of these two nuclides can decay into the other and by what mode? What is the decay energy of this process? Answer this question by setting up a Born–Haber-type cycle involving these four nuclides.

16. There is no stable nuclide with $A = 8$, although $^{8}$Be has the largest binding energy and thus comes closest to being stable. Make a list of all the possible decay modes of $^{8}$Be and use the following table of nuclidic masses to determine which decay mode releases the most energy. Restrict your list to decays involving the formation of two fragments. From the largest energy value, obtain a lower limit for the mass of $^{8}$Be.

| Nuclide | Mass |
|---|---|
| $^{7}$Be | 7.016931 u |
| $^{7}$Li | 7.016005 u |
| $^{6}$Li | 6.015126 u |
| $^{4}$He | 4.0026036 u |
| $^{2}$H | 2.0141022 u |
| $^{1}$H | 1.0078252 u |
| n | 1.0086654 u |

17. Complete each of the following nuclear reactions and write the reaction in shorthand notation.
(a) $^{63}$Cu + p → $^{60}$Ni + ?
(b) $^{24}$Mg + $\alpha$ → ? + $\alpha$ + n
(c) $^{36}$S + d → $^{37}$S + ?
(d) ? + n → $^{66}$Cu + $\gamma$

18. (a) Calculate the Q value of the $^{56}$Fe ($\alpha$, 2n) reaction. (b) What is the Coulomb barrier for $\alpha$ particles incident on $^{56}$Fe? Does the barrier inhibit the occurrence of this reaction? Use masses given in this chapter plus $M(^{56}$Fe$) = 55.93493$ u and $M(^{58}$Ni$) = 57.93534$ u.

19. A beam of 10.0-MeV protons is used to bombard a nickel foil. Which of the following reactions are energetically possible? (a) $^{58}$Ni(p, 2n), (b) $^{58}$Ni(p, n), (c) $^{60}$Ni(p, $\alpha$), (d) $^{64}$Ni(p, 2n). Use the following nuclidic masses: $^{58}$Ni, 57.93534 u; $^{60}$Ni, 59.93078 u; $^{64}$Ni, 63.92796 u; $^{57}$Cu, 56.9489 u; $^{58}$Cu, 57.9447 u; $^{57}$Co, 56.93629 u; and $^{63}$Cu, 62.92959 u.

20. The compound nucleus [$^{64}$Zn]* is formed in a level lying 20.0 MeV above the $^{64}$Zn ground state when $^{60}$Ni is bombarded by $\alpha$ particles. What energy $\alpha$ particles must be used? Assume that all the $\alpha$-particle energy can be converted to internal energy of $^{64}$Zn. The nuclidic masses of $^{60}$Ni and $^{64}$Zn are 59.93078 u and 63.929145 u, respectively.

21. The [$^{64}$Zn]* compound nucleus (see Problem 20) can also be formed when $^{63}$Cu is bombarded by particles of a certain type. (a) What is this particle? (b) What must its energy be if the compound nucleus is to be formed in the same energy level as when $^{60}$Ni is bombarded by 25.0-MeV $\alpha$ particles? The nuclidic mass of $^{63}$Cu is 62.92959 u.

22. The Q value of the $^{33}$S(n, p)$^{33}$P reaction is 0.533 MeV. The mass of $^{33}$S is 32.971458 u. What is the mass of $^{33}$P?

23. The nuclide $^{33}$Cl decays by $\beta^+$ emission and the maximum kinetic energy of the emitted positrons is 4.5 MeV. What is the Q value of the $^{33}$S(p, n)$^{33}$Cl reaction?

24. Uranium-235 undergoes thermal neutron fission. In one particular fission event, 3 neutrons are emitted and $^{91}$Rb and $^{142}$Cs fission products are formed. Calculate the energy released in this process. The nuclidic mass of $^{235}$U is 235.04393 u. The nuclidic masses of the fission products can be obtained by means of the liquid drop formula.

25. Californium-252 ($t_{1/2} = 2.55$ y) decays by $\alpha$ decay in 96.9% of the cases and by spontaneous fission in the remaining 3.1%. (a) How many fission fragments are emitted in 1 second by a 1.00-$\mu$g source of $^{252}$Cf? (b) If an average of 3.8 neutrons are emitted in each spontaneous fission decay, what is the rate of neutron emission?

26. The Coulomb repulsion between two newly formed fission fragments causes them to fly apart with a certain kinetic energy. The total kinetic energy of $^{138}$I and $^{98}$Y fragments formed in fission is 167 MeV. (a) What is the distance between the centers of the two nuclides at the time of fragment formation? (b) How does this distance compare with the sum of the nuclear radii of spherical fragments? (c) What can you conclude about the shapes of newly formed fission fragments?

27. Use the liquid drop formula (Equation 23.5) to derive the expression

$$Q_{SF} = 0.222(Z^2/A^{1/3}) - 3.38A^{2/3}$$

for the energy released in the spontaneous fission of nuclide $(Z, A)$ to form two identical fission fragments, $(Z/2, A/2)$. Neglect the pairing energy term in Equation 23.5.

28. Use the expression

$$Q_{SF} = 0.222(Z^2/A^{1/3}) - 3.38A^{2/3}$$

for the energy released in spontaneous fission (see Problem 27) to derive an expression for the minimum value of $Z^2/A$ for which spontaneous fission is energetically possible. Assuming that in this mass range $Z = 0.43 A$, what is the identity of the nuclide which meets this condition?

29. Spontaneous fission is expected to occur instantaneously when the energy released in fission becomes as large as the Coulomb barrier between the two fission fragments. For spontaneous fission of nuclide $(Z, A)$ into two identical fragments $(Z/2, A/2)$, determine the value of $Z^2/A$ at which this condition is met. Use the expression (see Problem 27)

$$Q_{SF} = 0.222(Z^2/A^{1/3}) - 3.38A^{2/3}$$

Assuming that $Z = 0.36A$ in the mass range of interest, what is the atomic number of the heaviest element that does not decay instantaneously by spontaneous fission?

30. The disintegration rate of $^{48}$V ($t_{1/2} = 16.1$ d) measured 10.5 d after its production in the $^{45}$Sc($\alpha$, n)$^{48}$V reaction is $1.4 \times 10^3$ dpm. (a) What was the disintegration rate of $^{48}$V at the end of the irradiation? (b) How many atoms of $^{48}$V were present at the end of the irradiation?

31. In an experiment designed to produce a new nuclide, 10 dpm of a new $\beta^-$ emitter were obtained. To how many grams of isotope does this correspond if the half-life is 1.1 h and the mass number is 120?

32. The half-life of a long-lived nuclide is measured by assay of its radioactivity and mass spectrometric determination of the number of atoms present. It is found that the sample contains $1.50 \times 10^{18}$ atoms of the nuclide and decays at a rate of 1500 disintegrations per hour. What is the half-life in years?

33. A certain radioactive isotope has a half-life of 4.0 days. What fraction of the initial radioactivity is present after (a) 2.0 days, (b) 4.0 days, (c) 8.0 days, and (d) 40 days?

34. What volume of helium at STP is collected when 1.00 g of pure radium ($^{226}$Ra) is stored in a container for 10.0 y. (See Figure 23.9 for data on the decay of $^{226}$Ra. Neglect the decay of $^{210}$Pb.)

35. Natural potassium contains 1.18% of the radioactive isotope $^{40}$K ($t_{1/2} = 1.27 \times 10^9$ y). What is the natural radioactivity in disintegrations per minute of one gram KCl?

36. The age of a certain rock is to be determined by the $^{238}$U–$^{206}$Pb method. The disintegration rate of $^{238}$U per gram of rock is $7.095 \times 10^4$ dpm. Mass spectrometric assay indicates that one gram of rock contains $2.04 \times 10^{20}$ atoms of $^{206}$Pb and that the ratio of $^{206}$Pb to $^{204}$Pb atoms is 40.0, compared to a value of 17.2 in primordial lead. The half-life of $^{238}$U is $4.47 \times 10^9$ y. What is the age of the rock?

37. The radioactivity of equal amounts of $CO_2$ from different samples of carbon is measured with the same counter. Sample A, which consists of $CO_2$ from ancient coal, gives 12.3 dpm (this is the natural background of the counter). Sample B, which consists of $CO_2$ from the burning of freshly grown wood, gives 121 dpm. Sample C, which consists of $CO_2$ from an unknown source of carbon, gives 42.9 dpm. What is the age of sample C? (The half-life of $^{14}$C is 5720 y.)

# Appendix 1: Units and Conversion Factors

The International System of Units (SI units) was introduced in 1960 to provide a common, internationally accepted set of scientific units. The SI system is based on the metric system and consists of the seven fundamental units listed in Table A1.1. In this book we use all of these units except for the candela, the unit of luminous intensity.

The magnitudes of the fundamental units often are either much larger or much smaller than the values of the quantities they measure. For example, the meter is much larger than the size of an atom but much smaller than the distances between the planets. To deal with such scale factors, the prefixes listed in Table A1.2 are used to denote fractions or multiples of the SI units differing from each other by powers of 10. A micrometer, $\mu$m, for example, is $10^{-6}$ meters, and a kiloampere, kA, is 1000 amperes. The kilogram is anomalous since the prefix *kilo* is incorporated in the fundamental unit of mass. The SI system emphasizes the use of prefixes that differ by consecutive factors of $10^3$ from the fundamental units. Only a few prefixes that differ from the fundamental units by factors of 10 or 100, such as *deca* or *centi*, are used.

We continue to use several traditional units that are decimal fractions of SI units, but do not correspond to any of the prefixes in Table A1.2. Since the sizes of atoms and molecules are in the range of $10^{-10}$ m, we use the traditional angstrom (Å)

$$1 \text{ Å} = 10^{-10} \text{ m}$$

to express chemical bond lengths and other quantities related to the sizes of atoms and molecules. A strict SI system expresses these lengths in either nanometers ($10^{-9}$ m) or picometers ($10^{-12}$ m).

The SI unit of volume is the cubic meter, m³. However, it is customary to express the volumes of laboratory amounts of liquids or gases in liters (L), where

$$1 \text{ L} = 10^{-3} \text{ m}^3 = 1 \text{ dm}^3$$

The milliliter (mL),

$$1 \text{ mL} = 10^{-3} \text{ L} = 1 \text{ cm}^3$$

is also commonly used.

Density, the mass per unit volume, can be expressed in SI units as

$$\rho = m/V$$
(kg m$^{-3}$)

It is customary, instead, to express density in units of g cm$^{-3}$. The numerical values of densities expressed in these units are $10^3$ times smaller than in units of kg m$^{-3}$. For example, the density of water is 1 g cm$^{-3}$, a more convenient value

**TABLE A1.1  The Seven Fundamental SI Units**

| Physical quantity | Unit | Abbreviation |
|---|---|---|
| Length | meter | m |
| Mass | kilogram | kg |
| Time | second | s |
| Electric current | ampere | A |
| Temperature | kelvin | K |
| Amount of substance | mole | mol |
| Luminous intensity | candela | cd |

**TABLE A1.2  Prefixes Denoting Power of 10 Fractions or Multiples of SI Units**

| Fraction | Prefix | Symbol | Multiple | Prefix | Symbol |
|---|---|---|---|---|---|
| $10^{-1}$ | deci | d | 10 | deca | da |
| $10^{-2}$ | centi | c | $10^2$ | hecto | h |
| $10^{-3}$ | milli | m | $10^3$ | kilo | k |
| $10^{-6}$ | micro | $\mu$ | $10^6$ | mega | M |
| $10^{-9}$ | nano | n | $10^9$ | giga | G |
| $10^{-12}$ | pico | p | $10^{12}$ | tera | T |
| $10^{-15}$ | femto | f | | | |
| $10^{-18}$ | atto | a | | | |

than the equivalent $10^3$ kg m$^{-3}$. The densities of gases are much smaller than those of liquids or solids and are customarily expressed in g L$^{-1}$.

The fundamental SI units are combined in appropriate ways to obtain the units of derived quantities, such as energy, pressure, or force. For example, force is defined by means of Newton's second law, $F = ma$, where the acceleration $a$ is expressed in units of m s$^{-2}$. The SI units of force are obtained by combining the SI units of mass and acceleration according to this law; the unit is kg m s$^{-2}$. Since this is an unwieldy unit, it is given a special name, a *newton*, and a special symbol, N, so that

$$1 \text{ N} = 1 \text{ kg m s}^{-2}$$

Table A1.3 summarizes the SI units, special names, and symbols of a number of widely used physical quantities.

In addition to being expressed in SI units, a number of physical quantities are commonly given in traditional units. For example, pressure is frequently expressed in terms of the pressure exerted by the atmosphere, the units of which are atmospheres, torr, or millimeters of mercury. At sea level at 0°C the average height of a column of mercury supported by the atmosphere is 760 mm Hg. The torr is defined as equal to 1 mm Hg, and

$$1 \text{ atm} = 760 \text{ torr} = 760 \text{ mm Hg}$$

Energy is customarily expressed in a number of different units. The SI unit of energy is the joule (1J = 1 kg m$^2$ s$^{-2}$). The energy of chemical reactions and other processes is usually given in kilojoules per mole of substance, kJ mol$^{-1}$. In the older scientific literature, energy instead is expressed in kilocalories per mole, kcal mol$^{-1}$, the calorie being a traditional unit of heat. The energies of single particles, such as the electron, are exceedingly small when expressed in joules. It is customary to give the energies of single particles in electron volts, eV, where

**TABLE A1.3  Some Derived SI Units**

| Physical quantity | Unit | Symbol | Fundamental Units |
|---|---|---|---|
| Force | newton | N | N = kg m s$^{-2}$ |
| Energy | joule | J | J = Nm = kg m$^2$ s$^{-2}$ |
| Pressure | pascal | Pa | Pa = Nm$^{-2}$ = kg m$^{-1}$ s$^{-2}$ |
| Frequency | hertz | Hz | Hz = s$^{-1}$ |
| Power | watt | W | W = Js$^{-1}$ = kg m$^2$ s$^{-3}$ |
| Electric charge | coulomb | C | C = A s |
| Electric potential | volt | V | V = WA$^{-1}$ = kg m$^2$ A$^{-1}$ s$^{-3}$ |
| Electric resistance | ohm | $\Omega$ | $\Omega$ = VA$^{-1}$ = kg m$^2$ A$^{-2}$ s$^{-3}$ |

one electron volt is the energy acquired by an electron accelerated through a potential difference of 1 volt. Table A1.4 summarizes the traditional units used in this book and gives their equivalent values in SI units.

**TABLE A1.4  Traditional Units and Their SI Equivalents**

| Physical quantity | Unit | Symbol | SI equivalent |
|---|---|---|---|
| Length | angstrom | Å | $1 \text{ Å} = 100 \text{ pm}$ |
| Mass of atoms or molecules | atomic mass unit | u | $1 \text{ u} = 1.66053 \times 10^{-27} \text{ kg}$ |
| Temperature | degree Celsius | °C | $T(°C) = T(K) - 273.15 \text{ K}$ |
| Pressure | atmosphere | atm | $1 \text{ atm} = 1.01325 \times 10^5 \text{ Pa}$ |
|  | torr, millimeter of mercury | torr, mm Hg | $1 \text{ torr} = 1 \text{ mm Hg} = \frac{1}{760} \text{ atm}$ $= 1.3332 \times 10^2 \text{ Pa}$ |
| Volume | liter | L | $1 \text{ L} = 10^{-3} \text{ m}^3$ |
| Energy | calorie | cal | $1 \text{ cal} = 4.1840 \text{ J}$ $1 \text{ kcal mol}^{-1} = 4.1840 \text{ kJ mol}^{-1}$ |
|  | electron volt | eV | $1 \text{ eV} = 1.6022 \times 10^{-19} \text{ J}$ $= 96.48 \text{ kJ mol}^{-1}$ |
|  | liter atmosphere | L atm | $1 \text{ L atm} = 101.325 \text{ J}$ |
| Dipole moment | debye | D | $1 \text{ D} = 3.338 \times 10^{-30} \text{ C m}$ |

# Appendix 2: Mathematical Operations

## A2.1 Exponentials

We generally express numbers in scientific (exponential) notation as $N \times 10^n$, where $N$, the *preexponential*, is a decimal number between 1 and 10 and the exponent $n$ is a positive or negative integer. For example,

$$6{,}543 = 6.543 \times 10^3$$
$$0.00018 = 1.8 \times 10^{-4}$$

Observe that in making these conversions, the decimal point must be moved $n$ places to the left (for $n > 0$) or $n$ places to the right (for $n < 0$) in order to obtain the number $N$.

In order to add or subtract exponential numbers, it is first necessary to express them in terms of the same power of 10. (Calculators do this automatically.) The values of $N$ are then added or subtracted and the result expressed in scientific notation. For example,

$$3.46 \times 10^3 - 2.14 \times 10^2 = (3.46 - 0.214) \times 10^3 = 3.25 \times 10^3$$

The product of two exponential numbers is obtained by adding the values of the exponents and forming the product of the preexponential terms. Thus, the product of $(A \times 10^a)$ and $(B \times 10^b)$ is $AB \times 10^{a+b}$. For example,

$$(5.78 \times 10^4)(4.83 \times 10^{-2}) = 27.92 \times 10^{(4-2)} = 2.79 \times 10^3$$

Similarly, the quotient of two exponentials is formed by subtracting the exponent of the denominator from that of the numerator and dividing the preexponential term of the numerator by that of the denominator, i.e.,

$$\frac{A \times 10^a}{B \times 10^b} = \left(\frac{A}{B}\right) 10^{(a-b)}$$

For example,

$$\frac{2.38 \times 10^2}{6.022 \times 10^{23}} = 0.3952 \times 10^{(2-23)} = 3.95 \times 10^{-22}$$

When an exponential number is raised to a power, the preexponential is raised to that power while the exponent is multiplied by the power in question:

$$(N \times 10^n)^b = N^b \times 10^{bn}$$

The power $b$ can be either larger or smaller than unity, as well as positive or negative. When $b$ is negative, the procedure is equivalent to raising the reciprocal of the exponential number to the power $+b$. Some examples of these operations are

$$(6.626 \times 10^{-34})^2 = (6.626)^2 \times 10^{-34 \times 2} = 43.904 \times 10^{-68} = 4.390 \times 10^{-67}$$
$$(6.022 \times 10^{23})^{1/2} = (60.22 \times 10^{22})^{1/2} = (60.22)^{1/2} \times 10^{22/2} = 7.760 \times 10^{11}$$
$$(2.998 \times 10^8)^{-1/3} = (0.2998 \times 10^9)^{-1/3} = (0.2998)^{-1/3} \times 10^{-9/3} = 1.494 \times 10^{-3}$$

Note that in the last two cases the number must first be changed so that the exponent gives an integral power of 10 when multiplied by the fractional power. Calculators do this automatically.

## A2.2 Logarithms

The common logarithm (base 10) of a number is the power to which 10 must be raised to give that number. Thus, if $A = 10^a$, $\log A = a$. A logarithm consists of two parts: an integer called the *characteristic*, followed by a decimal fraction called the *mantissa*.

Suppose we wish to find the logarithm of $4.760 \times 10^5$. The log of $10^5$ is 5, and this number is the characteristic. The log of 4.760 is 0.6776, and this number is the mantissa. The log of $4.760 \times 10^5$ is 5.6776. When a number changes by powers of 10 its logarithm changes in the characteristic, but the mantissa remains unchanged. For example, the log of $4.760 \times 10^{10}$ is 10.6776 and the log of $4.760 \times 10^{-3}$ is $(-3 + 0.6776) = -2.3224$. Mantissa values are tabulated in log tables but scientific calculators give the logs of numbers in a single step.

The procedure for obtaining the logarithm of a number can be reversed to obtain its antilogarithm (antilog). For example, in an equation of the type

$$\log K = -8.7152$$

the antilog of $\log K$ is $K$. The value of $K$ is

$$K = 10^{-8.7152} = 1.927 \times 10^{-9}$$

In addition to using common logarithms, we use natural logarithms (represented as ln), that is, logarithms that are exponents of the base $e$, where

$$e = 2.718281828\ldots$$

For example, $e^2 = 7.3891$; therefore, $\ln 7.3891 = 2$. Natural logarithms are important in chemistry because of the widespread use of the exponential function $e^x$ in various areas of chemistry. Furthermore, various mathematical procedures (e.g., integration of $x^{-1}\, dx$) result in the occurrence of natural logarithms.

The relation between the natural and common logarithms of a number is obtained readily. Take the log of a number $x$ with respect to each base:

$$\log x = a \qquad x = 10^a$$
$$\ln x = b \qquad x = e^b$$

Therefore, $10^a = e^b$ and $10 = e^{b/a}$. Taking the natural log of both sides, we get

$$\ln 10 = b/a$$
$$b = a \ln 10 = 2.3026a$$

Substituting for $a$ and $b$, we obtain

$$\ln x = 2.3026 \log x$$
$$\log x = 0.43429 \ln x$$

Since the logarithm of a number is the exponent to which the base must be raised to yield that number, mathematical operations involving logarithms obey the same rules as those involving exponents. Thus, the log of a product is the sum of the logs of the factors,

$$\log(ab) = \log a + \log b$$

The log of a quotient is the difference between the log of the numerator and that of the denominator,

$$\log(a/b) = \log a - \log b$$

A special case of this rule is that $\log(1/b) = -\log b$, since $\log 1 = 0$. The logarithm of the $n$th power of a number is $n$ times the log of the number,

$$\log(a^n) = n \log a$$

As an illustration of some of these rules, we compute the logarithms of the following numbers, given only that $\log 2 = 0.3010$:

$$\log 4 = \log(2)^2 = 2 \log 2 = 0.6020$$

$$\log 8 = \log(2)^3 = 3 \log 2 = 0.9030$$

$$\log 0.5 = \log \tfrac{1}{2} = -\log 2 = -0.3010$$

$$\log 20 = \log(10 \times 2) = \log 10 + \log 2 = 1.3010$$

$$\log 0.2 = \log \tfrac{2}{10} = \log 2 - \log 10 = -0.6990$$

## A2.3 The Quadratic Formula

The quadratic formula gives the solutions to a quadratic equation,

$$ax^2 + bx + c = 0$$

where $x$ is a variable and $a$, $b$, and $c$ are numerical coefficients. The general solutions to this equation are

$$x = \frac{-b \pm (b^2 - 4ac)^{1/2}}{2a}$$

While the equation has two solutions, only one of them may be physically meaningful. For example, if $x$ is the hydrogen ion concentration of a solution of a weak acid, $x$ cannot be negative nor can it be larger than the total acid concentration.

Not every quadratic equation has a real solution. In order for an equation to be solvable, the condition $b^2 - 4ac \geq 0$ must be satisfied. In the special case when $b^2 - 4ac = 0$, the equation has only one solution, $x = -b/2a$.

## A2.4 Mathematical Approximations

The description of physical phenomena sometimes leads to complicated formulas (e.g., Equation 20.2 for the heat capacities of solids). Under certain conditions formulas can be simplified, thereby making the nature of the relationship between two physical quantities more readily apparent.

A useful procedure for the simplification of complicated functions is the use of infinite series expansions, in which some function of a variable $x$ is replaced by an infinite series of powers of $x$. The exponential function $e^x$ can be expanded in the infinite series

$$e^x = 1 + x + \frac{x^2}{2!} + \frac{x^3}{3!} + \cdots$$

where 2! and 3! are factorials [$N! = N(N-1)(N-2) \cdots (1)$]. The expansion becomes particularly useful when $x \ll 1$, for then only the first power of $x$ contributes significantly. Under these conditions the series reduces to

$$e^x = 1 + x \qquad -1 \ll x \ll 1$$

For example, when $x = 0.100$, $1 + x = 1.100$ while $e^x = 1.105$, a difference of less than 1%.

Some other useful expansions are

$$\ln(1 + x) = x - \tfrac{1}{2}x^2 + \tfrac{1}{3}x^3 - \tfrac{1}{4}x^4 + \cdots \qquad (-1 < x < 1)$$

$$\frac{1}{1 \pm x} = 1 \mp x + x^2 \mp x^3 + x^4 \mp x^5 + \cdots \qquad (x^2 < 1)$$

[the minus signs in the series apply to the expansion of $(1 + x)^{-1}$ and the plus signs to the expansion of $(1 - x)^{-1}$], and

$$(1 + x)^{1/2} = 1 + \left(\frac{1}{2}\right)x - \left(\frac{1}{(2)(4)}\right)x^2 + \left(\frac{(1)(3)}{(2)(4)(6)}\right)x^3 - \left(\frac{(1)(3)(5)}{(2)(4)(6)(8)}\right)x^4 + \cdots \quad (x^2 < 1)$$

Table A2.1 summarizes the approximations that result when only the first power of $x$ is kept and indicates the range of $x$ values for which the error in the approximation is less than 1%.

## A2.5 Differentiation

Many types of experimental observations can be summarized in the form of laws that state the dependence of one observable on another. For example, at constant temperature, the pressure of a gas can be expressed as a function of its volume, $P = f(V)$. When written in this form, the volume is the *independent variable*, and the pressure is the *dependent variable*, since its value depends on that of the volume. It is customary to express such a relation in the form $y = f(x)$.

We are frequently interested in the magnitude of the change in $y$, $\Delta y$, produced by a given change in $x$, $\Delta x$. In order to determine the relationship between $\Delta y$ and $\Delta x$, the form of the function $f(x)$ must be specified. Suppose that $y$ is a linear function of $x$,

$$y = a + bx \tag{A2.1}$$

where $a$ and $b$ are constants. If $x$ changes by an amount $\Delta x$, the relation between $y$ and $x$ becomes

$$y + \Delta y = a + b(x + \Delta x) \tag{A2.2}$$

**TABLE A2.1  Some Useful Mathematical Approximations**

| Function | Approximation | Limits of validity[a] |
|---|---|---|
| $e^x$ | $1 + x$ | $-0.13 \leq x \leq 0.14$ |
| $\ln(1 + x)$ | $x$ | $-0.02 \leq x \leq 0.02$ |
| $(1 \pm x)^{-1}$ | $1 \mp x$ | $-0.10 \leq x \leq 0.10$ |
| $(1 + x)^{1/2}$ | $1 + x/2$ | $-0.24 \leq x \leq 0.32$ |
| $(1 + x)^2$ | $1 + 2x$ | $-0.09 \leq x \leq 0.11$ |

[a] For this range of $x$ values, the answers obtained from the approximation differ from the true answer by less than 1%.

Subtracting Equation A2.1 from A2.2, we get

$$\Delta y = b \, \Delta x$$

and the change in $y$ per unit change in $x$ is

$$\Delta y / \Delta x = b$$

The constant $b$ is the slope of the straight line represented by Equation A2.1.

A straight line has a constant slope and so the value of $\Delta y / \Delta x$ is independent of the width of the interval $\Delta x$ (Figure A2.1a). This is not true for a more complex function, such as $y = ax^2$. The plot of this function is a parabola (Figure A2.1b). It is evident that the value of $\Delta y / \Delta x$ increases with $x$ and consequently depends on the value of $\Delta x$. In order to obtain an unambiguous value for the change in $y$ with the change in $x$, the width of the interval $\Delta x$ must be reduced until it is infinitesimally small. Such an infinitesimally wide interval is called a differential and is symbolized by $dx$. The rate of change of $y$ with respect to $x$, $dy/dx$, is the derivative of $y$ with respect to $x$.

As may be seen in Figure A2.1b, the derivative can be interpreted geometrically as the slope of a line which is tangent to the curve representing the

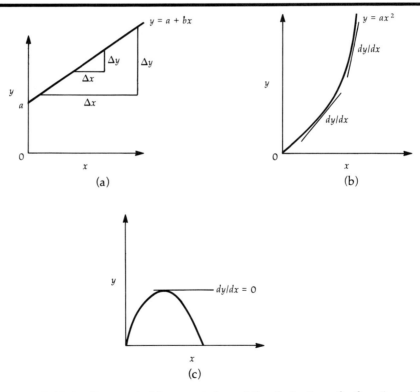

**FIGURE A2.1** Geometrical interpretation of the derivative of a function. (a) The derivative of a linear function $y = a + bx$ is a constant equal to the slope $b$. (b) The derivative of a more complicated function, such as the parabola $y = ax^2$, is given by the slope of a tangent to the curve. Note that the slope is not constant but varies with $x$. (c) When a curve has a maximum or a minimum, the tangent to the curve at this point is horizontal and the derivative is zero.

function. This interpretation permits us to infer that when a function goes through a maximum or a minimum, its derivative must be zero (Figure A2.1c). This conclusion follows since at a maximum or a minimum the tangent is horizontal and its slope is zero.

Table A2.2 summarizes the derivatives of some widely used functions. The results can be generalized to more complicated functions. If $x$ in the tabulated functions is replaced by some function of $x$, $u(x)$, then $x$ is replaced by $u$ in the expression for $dy/dx$, and the resulting expression is multiplied by the derivative $du/dx$. For example, if

$$y = e^u$$

then

$$dy/dx = e^u(du/dx)$$

If we let

$$u = ax^2$$

then

$$y = e^{ax^2}$$

and, using the derivative of $ax^n$ from Table A2.2, we obtain

$$dy/dx = e^{ax^2} d(ax^2)/dx$$
$$= 2axe^{ax^2}$$

Complicated functions of $x$ can frequently be expressed as a sum, a product, or a ratio of separate functions of $x$. If $u$ and $v$ are functions of $x$,

$$\frac{d(u+y)}{dx} = \frac{du}{dx} + \frac{dv}{dx} \tag{A2.3}$$

$$\frac{d(uv)}{dx} = u\frac{dv}{dx} + v\frac{du}{dx} \tag{A2.4}$$

$$\frac{d}{dx}\left(\frac{u}{v}\right) = \left(\frac{1}{v}\right)\left(\frac{du}{dx}\right) - \left(\frac{u}{v^2}\right)\left(\frac{dv}{dx}\right) \tag{A2.5}$$

To illustrate some of these relationships, we derive the expression for the most probable speed of the molecules in a gas (Equation 4.27),

**TABLE A2.2 Derivatives of Some Important Functions**

| Function[a] | Derivative, $dy/dx$ |
|---|---|
| $y = a$ | 0 |
| $y = ax^n$ | $nax^{n-1}$ |
| $y = e^{ax}$ | $ae^{ax}$ |
| $y = \ln x$ | $1/x$ |
| $y = \sin ax$ | $a \cos ax$ |
| $y = \cos ax$ | $-a \sin ax$ |

[a] In these functions $a$ and $n$ represent constant numbers.

$$v_{mp} = (2kT/m)^{1/2}$$

from the Maxwell–Boltzmann distribution of molecular speeds (Equation 4.26),

$$N(v)/N_{tot} = 4\pi(m/2\pi kT)^{3/2} v^2 \exp(-mv^2/2kT)$$

A function goes through a maximum when it attains its most probable value. Thus, we must evaluate the derivative of the function with respect to the speed,

$$\frac{dN(v)/N_{tot}}{dv}$$

and set the result equal to zero. The value of $v$ which obeys the resulting equation is the most probable speed $v_{mp}$.

The Maxwell–Boltzmann distribution has two different factors that depend on $v$. Consequently its derivative can be obtained by means of Equation A2.4. The separate factors are

$$4\pi(m/2\pi kT)^{3/2} v^2 \quad \text{and} \quad \exp(-mv^2/2kT)$$

The derivative of the function is

$$\frac{dN(v)/N_{tot}}{dv} = 4\pi\left(\frac{m}{2\pi kT}\right)^{3/2} v^2 \frac{d[\exp(-mv^2/2kT)]}{dv} + \exp\left(-\frac{mv^2}{2kT}\right)\left(\frac{d[4\pi(m/2\pi kT)^{3/2}v^2]}{dv}\right)$$

$$= 4\pi\left(\frac{m}{2\pi kT}\right)^{3/2} v^2 \left[\exp\left(-\frac{mv^2}{2kT}\right)\right]\left(\frac{-2mv}{2kT}\right) + \exp\left(\frac{-mv^2}{2kT}\right) 4\pi\left(\frac{m}{2\pi kT}\right)^{3/2} 2v$$

where the derivatives have been evaluated by means of the formulas in Table A2.2. Setting this result equal to zero, cancelling common factors, and specifying that $v$ now has its most probable value, $v_{mp}$, we obtain

$$(v_{mp})^2(-2mv_{mp}/2kT) + 2v_{mp} = 0$$

Solving for $v_{mp}$, we get

$$(v_{mp})^2 = 2kT/m$$

$$v_{mp} = (2kT/m)^{1/2}$$

## A2.6 Integration

Integration is a procedure for obtaining the area under a curve represented by the function $y = f(x)$. As illustrated in Figure A2.2, a curve can be approximated by a series of horizontal segments of width $\Delta x$. The area under each segment is a rectangle of height $y$ and width $\Delta x$,

$$\Delta A = y \, \Delta x = f(x) \, \Delta x \tag{A2.6}$$

The area under the curve can be approximated by the sum of the areas of the rectangles,

$$A \simeq \sum_{i=1}^{n} \Delta A_i = \sum_{i=1}^{n} y_i \, \Delta x \tag{A2.7}$$

It is apparent that the approximation given by Equation A2.7 becomes increasingly good as the width of the interval $\Delta x$ decreases. In the limit that $\Delta x$ approaches zero, it can be replaced by its differential $dx$, and Equation A2.6 can be written in the form

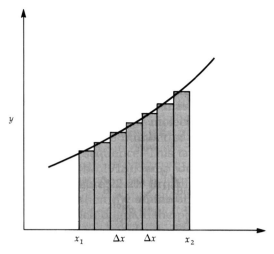

**FIGURE A2.2** Geometrical interpretation of the integral of a function. The function is approximated by a series of horizontal segments of width $\Delta x$. The integral of the function between $x_1$ and $x_2$ is equal to the sum of the areas of the rectangles in the limit as $\Delta x$ approaches zero.

$$\lim_{\Delta x = 0} \Delta A/\Delta x = dA/dx = f(x)$$

$$dA = f(x)\, dx$$

The differential element of area, $dA$, is the area of a rectangle of height $f(x)$ and infinitesimal width $dx$. The total area under the curve is given by the integral of all the infinitesimal areas

$$A = \int dA = \int_{x_1}^{x_2} f(x)\, dx \qquad (A2.8)$$

Thus, integration of the function $f(x)$ between the limits $x = x_1$ and $x = x_2$ can be understood as a summation of the areas of the rectangles under the curve in the limit in which the width of each rectangle approaches zero.

In order to evaluate the integral of a function $f(x)$, we note from Equation A2.8 that

$$A = \int dA$$

In other words, since $dA$ is the differential of $A$ and its integral is equal to $A$, integration converts the differential of a function into the function. Consequently, integration corresponds to the reverse of differentiation. To integrate a function $f(x)$, we must therefore represent it as the derivative with respect to $x$ of some other function $F(x)$. That is, if

$$f(x) = (d/dx)\, F(x)$$

then

$$A = \int f(x)\, dx = \int \frac{dF(x)\, dx}{dx} = \int dF(x) = F(x)$$

Let us pick the simple function $f(x) = x$ to illustrate this procedure. We know from Section A2.5 that $x$ is the derivative of the function $x^2/2$, that is,

$$\frac{d}{dx}\left(\frac{x^2}{2}\right) = \frac{2x}{2} = x$$

Therefore,

$$\int x \, dx = \frac{x^2}{2}$$

This answer, while essentially correct, is actually not the most general answer. The desired answer is

$$\int x \, dx = \frac{x^2}{2} + c$$

where $c$ is a constant called an integration constant. To see that this is true, let us differentiate this result with respect to $x$. We obtain

$$\frac{d(x^2/2 + c)}{dx} = \frac{d(x^2/2)}{dx} + \frac{dc}{dx} = x + 0 = x$$

where, according to Table A.2.2, the derivative of a constant is zero. Thus, the derivative of $(x^2/2 + c)$ with respect to $x$ is the same as that of $x^2/2$. The integral of $x \, dx$ is therefore $(x^2/2 + c)$, a result that is more general than $x^2/2$. More generally, for a function $f(x)$,

$$A(x) = \int f(x) \, dx = F(x) + c \tag{A2.9}$$

Whenever a constant such as $c$ appears in the integration of some function, the integral is called an *indefinite* integral. Note that in contrast to Equation A2.8, an indefinite integral is not integrated between specific limits $x_1$ and $x_2$. It is customary to tabulate values of indefinite integrals since they are more general than those of *definite* integrals, that is, those for which the integration is limited to a specific interval, say, between $x_1$ and $x_2$.

Table A2.3 lists the values of some common indefinite integrals. The area between specific limits under the curve given by some function, that is, the value of the definite integral, can be obtained readily from the indefinite integral. Let $A(x_2)$ be the area under the curve up to $x = x_2$ and $A(x_1)$ be the area under the curve up to $x = x_1$, where $x_2 > x_1$. The area under the curve between $x_1$ and $x_2$ can be obtained from Equation A2.9 as

$$A(x_2) - A(x_1) = A = F(x_2) + c - (F(x_1) + c) = F(x_2) - F(x_1)$$

In other words, this area is just the difference between the integral evaluated at the upper limit $x_2$ and the lower limit $x_1$ and can be represented as

$$A = \int_{x_1}^{x_2} f(x) \, dx = F(x) \Big|_{x_1}^{x_2} = F(x_2) - F(x_1) \tag{A2.10}$$

where $F(x)\big|_{x_1}^{x_2}$ specifies the functional form of the integral and the integration limits, and $F(x_2) - \overline{F(x_1)}$ gives the result of the integration between these limits.

**TABLE A2.3  Some Important Indefinite Integrals**[a]

$$\int a \, dx = ax + c$$

$$\int af(x) = a \int f(x)$$

$$\int x^n \, dx = \frac{x^{n+1}}{n+1} + c \text{ (except when } n = -1\text{)}$$

$$\int \frac{dx}{x} = \ln x + c$$

$$\int e^{ax} \, dx = \frac{e^{ax}}{a} + c$$

$$\int \sin x \, dx = -\cos x + c$$

$$\int \cos x \, dx = \sin x + c$$

$$\int (u + v) \, dx = \int u \, dx + \int v \, dx$$

[a] $a$, $c$, and $n$ are constants, and $f$, $u$, and $v$ are functions of $x$.

For example, the integral of $x \, dx$ between $x = 1$ and $x = 2$ is

$$\int_{x=1}^{x=2} x \, dx = \left. \frac{x^2}{2} \right|_1^2 = \frac{(2)^2}{2} - \frac{1^2}{2} = 2 - \frac{1}{2} = 1.5$$

# Appendix 3: Equilibrium Constants

## TABLE A3.1  Ionization Constants of Acids at 25°C[a]

| Name | Formula | $K_a$ | Name | Formula | $K_a$ |
|---|---|---|---|---|---|
| **Inorganic and organic acids** | | | | | |
| *Arsenic* | | | *Oxygen* | | |
| Arsenic | $H_3AsO_4$ | $6.5 \times 10^{-3}$ | Hydrogen peroxide | $H_2O_2$ | $2.2 \times 10^{-12}$ |
| | $H_2AsO_4^-$ | $1.1 \times 10^{-7}$ | *Phosphorus* | | |
| | $HAsO_4^{2-}$ | $3 \times 10^{-12}$ | Phosphoric | $H_3PO_4$ | $7.5 \times 10^{-3}$ |
| *Boron* | | | | $H_2PO_4^-$ | $6.6 \times 10^{-8}$ |
| Boric | $H_3BO_3$ | $6.0 \times 10^{-10}$ | | $HPO_4^{2-}$ | $1 \times 10^{-12}$ |
| *Bromine* | | | Phosphorous | $H_3PO_3$ | $3 \times 10^{-2}$ |
| Hydrobromic | HBr | large | | $H_2PO_3^-$ | $1.6 \times 10^{-7}$ |
| Hypobromous | HBrO | $2.2 \times 10^{-9}$ | Hypophosphorous | $H_3PO_2$ | $1.23 \times 10^{-2}$ |
| *Carbon* | | | Pyrophosphoric | $H_4P_2O_7$ | $1.2 \times 10^{-1}$ |
| Acetic | $CH_3COOH$ | $1.76 \times 10^{-5}$ | | $H_3P_2O_7^-$ | $7.9 \times 10^{-2}$ |
| Benzoic | $C_6H_5COOH$ | $6.6 \times 10^{-5}$ | | $H_2P_2O_7^{2-}$ | $2.0 \times 10^{-7}$ |
| Carbonic | $H_2CO_3[CO_2(aq)]$ | $4.5 \times 10^{-7}$ | | $HP_2O_7^{3-}$ | $4.8 \times 10^{-10}$ |
| | $HCO_3^-$ | $4.8 \times 10^{-11}$ | *Silicon* | | |
| Chloroacetic | $CH_2ClCOOH$ | $1.40 \times 10^{-3}$ | Metasilicic | $H_2SiO_3$ | $3.2 \times 10^{-10}$ |
| Cyanic | HOCN | $3.3 \times 10^{-4}$ | | $HSiO_3^-$ | $1.5 \times 10^{-12}$ |
| Dichloroacetic | $CHCl_2COOH$ | $3.32 \times 10^{-2}$ | *Sulfur* | | |
| Formic | HCOOH | $1.772 \times 10^{-4}$ | Hydrosulfuric | $H_2S$ | $1.0 \times 10^{-7}$ |
| Hydrocyanic | HCN | $6.2 \times 10^{-10}$ | | $HS^-$ | $3 \times 10^{-13}$ |
| Oxalic | $H_2C_2O_4$ | $5.60 \times 10^{-2}$ | Sulfuric | $H_2SO_4$ | large |
| | $HC_2O_4^-$ | $6.2 \times 10^{-5}$ | | $HSO_4^-$ | $1.0 \times 10^{-2}$ |
| Propionic | $CH_3CH_2COOH$ | $1.34 \times 10^{-5}$ | Sulfurous | $H_2SO_3$ | $1.43 \times 10^{-2}$ |
| Thiocyanic | HNCS | large | | $HSO_3^-$ | $5.0 \times 10^{-8}$ |
| Trichloroacetic | $CCl_3COOH$ | $2 \times 10^{-1}$ | Thiosulfuric | $H_2S_2O_3$ | $2.0 \times 10^{-2}$ |
| *Chlorine* | | | | $HS_2O_3^-$ | $3.2 \times 10^{-3}$ |
| Hydrochloric | HCl | large | **Amphoteric hydroxides** | | |
| Perchloric | $HClO_4$ | large | Aluminum hydroxide | $Al(OH)_3$ | $4 \times 10^{-13}$ |
| Chloric | $HClO_3$ | large | Antimony(III) hydroxide | $SbO(OH)$ | $1 \times 10^{-11}$ |
| Chlorous | $HClO_2$ | $1.1 \times 10^{-2}$ | Chromium(III) hydroxide | $Cr(OH)_3$ | $9 \times 10^{-17}$ |
| Hypochlorous | HClO | $2.90 \times 10^{-8}$ | Copper(II) hydroxide | $Cu(OH)_2$ | $1 \times 10^{-19}$ |
| *Chromium* | | | | $HCuO_2^-$ | $7.0 \times 10^{-14}$ |
| Chromic | $H_2CrO_4$ | $1.8 \times 10^{-1}$ | Lead(II) hydroxide | $Pb(OH)_2$ | $4.6 \times 10^{-16}$ |
| | $HCrO_4^-$ | $3.2 \times 10^{-7}$ | Tin(IV) hydroxide | $Sn(OH)_4$ | $\times 10^{-32}$ |
| *Fluorine* | | | Tin(II) hydroxide | $Sn(OH)_2$ | $3.8 \times 10^{-15}$ |
| Hydrofluoric | HF | $6.5 \times 10^{-4}$ | Zinc hydroxide | $Zn(OH)_2$ | $1.0 \times 10^{-29}$ |
| *Iodine* | | | **Metal cations** | | |
| Hydroiodic | HI | large | Aluminum ion | $Al^{3+}$ | $1.4 \times 10^{-5}$ |
| Periodic | $HIO_4$ | $5.6 \times 10^{-9}$ | Ammonium ion | $NH_4^+$ | $6.3 \times 10^{-10}$ |
| Iodic | $HIO_3$ | $1.6 \times 10^{-1}$ | Bismuth(III) ion | $Bi^{3+}$ | $1 \times 10^{-2}$ |
| Hypoiodous | HIO | $2.3 \times 10^{-11}$ | Chromium(III) ion | $Cr^{3+}$ | $1 \times 10^{-4}$ |
| *Manganese* | | | Copper(II) ion | $Cu^{2+}$ | $1 \times 10^{-8}$ |
| Permanganic | $HMnO_4$ | large | Iron(III) ion | $Fe^{3+}$ | $4.0 \times 10^{-3}$ |
| *Nitrogen* | | | Iron(II) ion | $Fe^{2+}$ | $1.2 \times 10^{-6}$ |
| Nitric | $HNO_3$ | large | Magnesium ion | $Mg^{2+}$ | $2 \times 10^{-12}$ |
| Nitrous | $HNO_2$ | $7.2 \times 10^{-4}$ | Mercury(II) ion | $Hg^{2+}$ | $2 \times 10^{-3}$ |
| | | | Zinc ion | $Zn^{2+}$ | $2.5 \times 10^{-10}$ |

[a] Adapted from Bailar, Jr., J. C., Moeller, Th., Kleinberg, J., Guss, C. O., Castellion, M. E., and Metz, C. (1984). "Chemistry: Second Edition." Academic Press, Orlando, Florida.

## TABLE A3.2  Ionization Constants of Bases at 25°C[a]

| Name | Formula | $K_b$ | Name | Formula | $K_b$ |
|---|---|---|---|---|---|
| **Inorganic and organic bases** | | | Cyanide ion | $CN^-$ | $1.6 \times 10^{-5}$ |
| Ammonia | $NH_3$ | $1.6 \times 10^{-5}$ | Fluoride ion | $F^-$ | $1.5 \times 10^{-11}$ |
| Aniline | $C_6H_5NH_2$ | $4.2 \times 10^{-10}$ | Formate ion | $HCOO^-$ | $5.643 \times 10^{-11}$ |
| Diethylamine | $(C_2H_5)_2NH$ | $9.5 \times 10^{-4}$ | Nitrite ion | $NO_2^-$ | $1.4 \times 10^{-11}$ |
| Dimethylamine | $(CH_3)_2NH$ | $5.9 \times 10^{-4}$ | Oxalate ion | $C_2O_4^{2-}$ | $1.6 \times 10^{-10}$ |
| Ethylamine | $C_2H_5NH_2$ | $4.7 \times 10^{-4}$ | | $HC_2O_4^-$ | $1.79 \times 10^{-13}$ |
| Methylamine | $CH_3NH_2$ | $3.9 \times 10^{-4}$ | Phosphate ion | $PO_4^{3-}$ | $1 \times 10^{-2}$ |
| Triethylamine | $(C_2H_5)_3N$ | $5.2 \times 10^{-4}$ | | $HPO_4^{2-}$ | $1.5 \times 10^{-7}$ |
| Trimethylamine | $(CH_3)_3N$ | $6.3 \times 10^{-5}$ | | $H_2PO_4^-$ | $1.3 \times 10^{-12}$ |
| | | | Phosphite ion | $HPO_3^{2-}$ | $6.3 \times 10^{-8}$ |
| **Anions** | | | | $H_2PO_3^-$ | $3 \times 10^{-13}$ |
| Acetate ion | $CH_3COO^-$ | $5.68 \times 10^{-10}$ | Metasilicate ion | $SiO_3^{2-}$ | $6.7 \times 10^{-3}$ |
| Arsenate ion | $AsO_4^{3-}$ | $3.3 \times 10^{-3}$ | | $HSiO_3^-$ | $3.1 \times 10^{-5}$ |
| | $HAsO_4^{2-}$ | $9.1 \times 10^{-8}$ | Sulfate ion | $SO_4^{2-}$ | $1.0 \times 10^{-12}$ |
| | $H_2AsO_4^-$ | $1.5 \times 10^{-12}$ | Sulfite ion | $SO_3^{2-}$ | $2.0 \times 10^{-7}$ |
| Borate ion | $H_2BO_3^-$ | $1.6 \times 10^{-5}$ | | $HSO_3^-$ | $6.99 \times 10^{-13}$ |
| | $B_4O_7^{2-}$ | $\times 10^{-3}$ | Sulfide ion | $S^{2-}$ | $3 \times 10^{-2}$ |
| Carbonate ion | $CO_3^{2-}$ | $2.1 \times 10^{-4}$ | | $HS^-$ | $1.0 \times 10^{-7}$ |
| | $HCO_3^-$ | $2.2 \times 10^{-8}$ | Thiocyanate ion | $NCS^-$ | $1.4 \times 10^{-11}$ |
| Chromate ion | $CrO_4^{2-}$ | $3.1 \times 10^{-8}$ | Thiosulfate ion | $S_2O_3^{2-}$ | $3.1 \times 10^{-12}$ |

[a] Adapted from Bailar, Jr., J. C., Moeller, Th., Kleinberg, J., Guss, C. O., Castellion, M. E., and Metz, C. (1984). "Chemistry: Second Edition." Academic Press, Orlando, Florida.

## TABLE A3.3  Solubility Products of Sparingly Soluble Salts and Bases at 25°C[a]

| Salt | $K_{sp}$ | Salt | $K_{sp}$ | Salt | $K_{sp}$ |
|---|---|---|---|---|---|
| *Acetates* | | $FeCO_3$ | $2.1 \times 10^{-11}$ | *Ferrocyanides* | |
| $Ag(CH_3COO)$ | $2.3 \times 10^{-3}$ | $ZnCO_3$ | $1.7 \times 10^{-11}$ | $KFe[Fe(CN)_6]$ | $3 \times 10^{-41}$ |
| $Hg_2(CH_3COO)_2$ | $4 \times 10^{-10}$ | $Ag_2CO_3$ | $8.1 \times 10^{-12}$ | $Ag_4[Fe(CN)_6]$ | $2 \times 10^{-41}$ |
| *Arsenates* | | $CdCO_3$ | $1.0 \times 10^{-12}$ | $K_2Zn_3[Fe(CN)_6]_2$ | $1 \times 10^{-95}$ |
| $Ag_3AsO_4$ | $1 \times 10^{-20}$ | $PbCO_3$ | $7.4 \times 10^{-14}$ | *Fluorides* | |
| *Bromides* | | *Chlorides* | | $BaF_2$ | $1.0 \times 10^{-6}$ |
| $PbBr_2$ | $6.2 \times 10^{-6}$ | $PbCl_2$ | $1.6 \times 10^{-5}$ | $MgF_2$ | $6.8 \times 10^{-9}$ |
| $CuBr$ | $5.2 \times 10^{-9}$ | $CuCl$ | $2 \times 10^{-7}$ | $SrF_2$ | $7.9 \times 10^{-10}$ |
| $AgBr$ | $5.0 \times 10^{-13}$ | $AgCl$ | $1.8 \times 10^{-10}$ | $CaF_2$ | $3.9 \times 10^{-11}$ |
| $Hg_2Br_2$ | $1.3 \times 10^{-22}$ | $Hg_2Cl_2$ | $1.3 \times 10^{-18}$ | $ThF_4$ | $4 \times 10^{-28}$ |
| *Carbonates* | | *Chromates* | | *Hydroxides* | |
| $MgCO_3$ | $1 \times 10^{-5}$ | $CaCrO_4$ | $7.1 \times 10^{-4}$ | $Ba(OH)_2$ | $1.3 \times 10^{-2}$ |
| $NiCO_3$ | $1.3 \times 10^{-7}$ | $SrCrO_4$ | $3.6 \times 10^{-5}$ | $Sr(OH)_2$ | $6.4 \times 10^{-3}$ |
| $CaCO_3$ | $4.8 \times 10^{-9}$ | $Hg_2CrO_4$ | $2.0 \times 10^{-9}$ | $Ca(OH)_2$ | $7.9 \times 10^{-6}$ |
| $BaCO_3$ | $8.1 \times 10^{-9}$ | $BaCrO_4$ | $2.0 \times 10^{-10}$ | $Ag_2O$ | $2 \times 10^{-8}$ |
| $SrCO_3$ | $9.4 \times 10^{-10}$ | $Ag_2CrO_4$ | $2.5 \times 10^{-12}$ | $Mg(OH)_2$ | $7.1 \times 10^{-12}$ |
| $MnCO_3$ | $8.8 \times 10^{-11}$ | $PbCrO_4$ | $1.8 \times 10^{-14}$ | $BiO(OH)$ | $1 \times 10^{-12}$ |
| $CuCO_3$ | $2.3 \times 10^{-10}$ | *Cyanides* | | $Be(OH)_2$ | $4 \times 10^{-13}$ |
| $CoCO_3$ | $1.0 \times 10^{-10}$ | $AgCN$ | $1.2 \times 10^{-16}$ | $Mn(OH)_2$ | $2 \times 10^{-13}$ |

## TABLE A3.3 Solubility Products of Sparingly Soluble Salts and Bases at 25°C[a]

| Salt | $K_{sp}$ | Salt | $K_{sp}$ | Salt | $K_{sp}$ |
|---|---|---|---|---|---|
| $Cd(OH)_2$ | $8.1 \times 10^{-15}$ | Oxalates | | $Sr_3(PO_4)_2$ | $4 \times 10^{-28}$ |
| $Pb(OH)_2$ | $1.2 \times 10^{-15}$ | $MgC_2O_4$ | $8 \times 10^{-5}$ | $Mg_3(PO_4)_2$ | $1 \times 10^{-32}$ |
| $Fe(OH)_2$ | $8 \times 10^{-16}$ | $CoC_2O_4$ | $4 \times 10^{-6}$ | $Pb_3(PO_4)_2$ | $7.9 \times 10^{-43}$ |
| $Ni(OH)_2$ | $3 \times 10^{-16}$ | $FeC_2O_4$ | $2 \times 10^{-7}$ | Sulfates | |
| $Co(OH)_2$ | $2 \times 10^{-16}$ | $NiC_2O_4$ | $1 \times 10^{-7}$ | $CaSO_4$ | $2.5 \times 10^{-5}$ |
| $Zn(OH)_2$ | $1.2 \times 10^{-17}$ | $SrC_2O_4$ | $5 \times 10^{-8}$ | $Ag_2SO_4$ | $1.5 \times 10^{-5}$ |
| $SbO(OH)$ | $1 \times 10^{-17}$ | $CuC_2O_4$ | $3 \times 10^{-8}$ | $Hg_2SO_4$ | $6.8 \times 10^{-7}$ |
| $Cu(OH)_2$ | $1.3 \times 10^{-20}$ | $BaC_2O_4$ | $2 \times 10^{-8}$ | $SrSO_4$ | $2.8 \times 10^{-7}$ |
| $Hg(OH)_2$ | $4 \times 10^{-26}$ | $CdC_2O_4$ | $2 \times 10^{-8}$ | $PbSO_4$ | $2.2 \times 10^{-8}$ |
| $Sn(OH)_2$ | $6 \times 10^{-27}$ | $ZnC_2O_4$ | $2 \times 10^{-9}$ | $BaSO_4$ | $1.1 \times 10^{-10}$ |
| $Cr(OH)_3$ | $6 \times 10^{-31}$ | $CaC_2O_4$ | $1 \times 10^{-9}$ | | |
| $Al(OH)_3$ | $1.9 \times 10^{-33}$ | $Ag_2C_2O_4$ | $3.5 \times 10^{-11}$ | Sulfides | |
| $Fe(OH)_3$ | $6.3 \times 10^{-38}$ | $PbC_2O_4$ | $8 \times 10^{-12}$ | $MnS$ | $2.3 \times 10^{-13}$ |
| $Sn(OH)_4$ | $10^{-57}$ | $Hg_2C_2O_4$ | $2 \times 10^{-13}$ | $FeS$ | $4.2 \times 10^{-17}$ |
| Iodates | | $MnC_2O_4$ | $1 \times 10^{-15}$ | $NiS$ | $3 \times 10^{-19}$ |
| $AgIO_3$ | $3.0 \times 10^{-8}$ | $La_2(C_2O_4)_3$ | $2 \times 10^{-28}$ | $CoS$ | $4 \times 10^{-21}$ |
| Iodides | | | | $ZnS$ | $1.1 \times 10^{-21}$ |
| $PbI_2$ | $7.1 \times 10^{-9}$ | Phosphates | | $SnS$ | $3 \times 10^{-27}$ |
| $CuI$ | $5.1 \times 10^{-12}$ | $Li_3PO_4$ | $3 \times 10^{-13}$ | $CdS$ | $2 \times 10^{-28}$ |
| $AgI$ | $8.3 \times 10^{-17}$ | $Mg(NH_4)PO_4$ | $3 \times 10^{-13}$ | $PbS$ | $1 \times 10^{-28}$ |
| $HgI_2$ | $3 \times 10^{-26}$ | $Ag_3PO_4$ | $1.4 \times 10^{-16}$ | $CuS$ | $6 \times 10^{-36}$ |
| $Hg_2I_2$ | $4.5 \times 10^{-29}$ | $AlPO_4$ | $5.8 \times 10^{-19}$ | $Cu_2S$ | $3 \times 10^{-48}$ |
| Nitrates | | $Mn_3(PO_4)_2$ | $1 \times 10^{-22}$ | $Ag_2S$ | $7.1 \times 10^{-50}$ |
| $BiO(NO_3)$ | $2.8 \times 10^{-3}$ | $Ba_3(PO_4)_2$ | $3 \times 10^{-23}$ | $HgS$ | $4 \times 10^{-53}$ |
| Nitrites | | $BiPO_4$ | $1.3 \times 10^{-23}$ | $Fe_2S_3$ | $1 \times 10^{-88}$ |
| $Ag(NO_2)$ | $6.0 \times 10^{-4}$ | $Ca_3(PO_4)_2$ | $10^{-26}$ | $Bi_2S_3$ | $1 \times 10^{-96}$ |

[a] Adapted from Bailar, Jr., J. C., Moeller, Th., Kleinberg, J., Guss, C. O., Castellion, M. E., and Metz, C. (1984). "Chemistry: Second Edition." Academic Press, Orlando, Florida.

## TABLE A3.4 Dissociation Constants of Complexes at 25°C.[a]

| Complex | $K_d$ | Complex | $K_d$ | Complex | $K_d$ |
|---|---|---|---|---|---|
| Aluminum | | Cobalt | | $[Au(CN)_2]^-$ | $5 \times 10^{-39}$ |
| $[AlF_6]^{3-}$ | $3 \times 10^{-20}$ | $[Co(NH_3)_6]^{2+}$ | $9 \times 10^{-6}$ | Iron | |
| Calcium | | $[Co(C_2O_4)_3]^{4-}$ | $2.2 \times 10^{-7}$ | $[Fe(C_2O_4)_3]^{4-}$ | $6 \times 10^{-6}$ |
| $[Ca(P_2O_7)]^{2-}$ | $1 \times 10^{-5}$ | $[Co(en)_3]^{2+}$ | $1.52 \times 10^{-14}$ | $[Fe(SCN)_3]$ | $5 \times 10^{-7}$ |
| $[Ca(nta)_2]^{4-}$ | $2.44 \times 10^{-12}$ | $[Co(en)_3]^{3+}$ | $2.04 \times 10^{-49}$ | $[Fe(C_2O_4)_2]^{2-}$ | $2 \times 10^{-8}$ |
| Cadmium | | Copper | | $[Fe(C_2O_4)_3]^{3-}$ | $3 \times 10^{-21}$ |
| $[CdCl_4]^{2-}$ | $9.3 \times 10^{-3}$ | $[Cu(SCN)_2]$ | $1.8 \times 10^{-4}$ | $[Fe(CN)_6]^{4-}$ | $1.3 \times 10^{-37}$ |
| $[Cd(SCN)_4]^{2-}$ | $1 \times 10^{-3}$ | $[CuCl_2]^-$ | $1.15 \times 10^{-5}$ | $[Fe(CN)_6]^{3-}$ | $1.3 \times 10^{-44}$ |
| $[CdBr_4]^{2-}$ | $2 \times 10^{-4}$ | $[Cu(P_2O_7)]^{2-}$ | $2.0 \times 10^{-7}$ | Lead | |
| $[Cd(NH_3)_6]^{2+}$ | $1 \times 10^{-5}$ | $[Cu(C_2O_4)_2]^{2-}$ | $6 \times 10^{-11}$ | $[Pb(SCN)_2]$ | $3 \times 10^{-3}$ |
| $[CdI_4]^{2-}$ | $8 \times 10^{-7}$ | $[Cu(NH_3)_4]^{2+}$ | $9.1 \times 10^{-14}$ | Magnesium | |
| $[Cd(CH_3NH_2)_4]^{2+}$ | $2.82 \times 10^{-7}$ | $[Cu(gly)_2]$ | $5.6 \times 10^{-16}$ | $[Mg(P_2O_7)]^{2-}$ | $2 \times 10^{-6}$ |
| $[Cd(NH_3)_4]^{2+}$ | $1 \times 10^{-7}$ | $[Cu(OH)_4]^{2-}$ | $7.6 \times 10^{-17}$ | $[Mg(nta)_2]^{4-}$ | $6.3 \times 10^{-11}$ |
| $[Cd(en)_4]^{2+}$ | $2.60 \times 10^{-11}$ | $[Cu(eta)]^{2-}$ | $1.38 \times 10^{-19}$ | Mercury | |
| $[Cd(CN)_4]^{2-}$ | $8.3 \times 10^{-18}$ | Gold | | $[HgCl_4]^{2-}$ | $2 \times 10^{-16}$ |

## TABLE A3.4 Dissociation Constants of Complexes at 25°C.[a]

| Complex | $K_d$ | Complex | $K_d$ | Complex | $K_d$ |
|---|---|---|---|---|---|
| $[Hg(SCN)_4]^{2-}$ | $2.0 \times 10^{-22}$ | $[Ag(en)]^+$ | $1 \times 10^{-5}$ | Zinc | |
| Nickel | | $[AgCl_2]^-$ | $9 \times 10^{-6}$ | $[Zn(NH_3)_4]^{2+}$ | $3.4 \times 10^{-10}$ |
| $[Ni(NH_3)_6]^{2+}$ | $1 \times 10^{-9}$ | $[AgCl_4]^{3-}$ | $5 \times 10^{-6}$ | $[Zn(gly)_2]$ | $1.1 \times 10^{-10}$ |
| Palladium | | $[AgBr_2]^-$ | $7.8 \times 10^{-8}$ | $[Zn(edta)]^{2-}$ | $2.63 \times 10^{-17}$ |
| $[PdBr_4]^{2-}$ | $8.0 \times 10^{-14}$ | $[Ag(SCN)_4]^{3-}$ | $1.0 \times 10^{-8}$ | $[Zn(CN_4)]^{2-}$ | $2.4 \times 10^{-20}$ |
| $[PdCl_4]^{2-}$ | $6 \times 10^{-14}$ | $[Ag(CN)_2]^-$ | $2.1 \times 10^{-10}$ | $[Zn(OH)_4]^{2-}$ | $5 \times 10^{-21}$ |
| Silver | | $[Au(CN)_2]^-$ | $1 \times 10^{-22}$ | | |
| $[Ag(OH)_3]^{2-}$ | $1.7 \times 10^{-5}$ | | | | |

[a] Adapted from Bailar, Jr., J. C., Moeller, Th., Kleinberg, J., Guss, C. O., Castellion, M. E., and Metz, C. (1984). "Chemistry: Second Edition." Academic Press, Orlando, Florida. The complex formation constants discussed in Chapter 9 are the reciprocals of the dissociation constants. The following abbreviations are used to represent ligands: (en) is the ethylenediamine molecule, $H_2NCH_2CH_2NH_2$; (nta) is the nitrilotriacetate ion, $N(CH_2COO)_2^{3-}$; (gly) is the glycine ion, $H_2NCH_2COO^-$; (edta) is the ethylenediaminetetraacetate ion, $(OOCCH_2)_2NCH_2CH_2N(CH_2COO)_2^{4-}$.

# Appendix 4: Thermodynamic Data Tables[a]

| Substance | $\Delta H_f^\circ$ (kJ mol$^{-1}$) | $\Delta G_f^\circ$ (kJ mol$^{-1}$) | $S^\circ$ (J mol$^{-1}$ K$^{-1}$) | $C_p^\circ$ (J mol$^{-1}$ K$^{-1}$) |
|---|---|---|---|---|
| **Aluminum** | | | | |
| Al(s) | 0.0 | 0.0 | 28.33 | 24.35 |
| Al(g) | 326.4 | 285.7 | 164.54 | 21.38 |
| Al$^{3+}$(g) | 5483.17 | | | |
| Al$^{3+}$(aq) | −531 | −485 | −321.7 | |
| Al$_2$O$_3$($\alpha$-solid) | −1675.7 | −1582.3 | 50.92 | 79.04 |
| AlF$_3$(s) | −1504.1 | −1425.0 | 66.44 | 75.10 |
| AlCl$_3$(s) | −704.2 | −628.8 | 110.67 | 91.84 |
| Al$_2$Cl$_6$(g) | −1290.8 | −1220.4 | 490 | |
| AlBr$_3$(s) | −527.2 | | | 101.7 |
| Al$_2$(SO$_4$)$_3$(s) | −3440.84 | −3099.94 | 239.3 | 259.41 |
| AlPO$_4$(s) | −1733.8 | −1617.9 | 90.79 | 93.18 |
| **Antimony** | | | | |
| Sb(solid III) | 0.0 | 0.0 | 45.69 | 25.23 |
| Sb(g) | 262.3 | 222.1 | 180.27 | 20.79 |
| Sb$_4$O$_6$(cubic) | −1440.6 | −1268.2 | 220.9 | |
| Sb$_4$O$_6$(orthorhombic) | −1417.1 | −1253.0 | 246.0 | 202.76 |
| Sb$_2$O$_5$(s) | −971.9 | −829.2 | 125.1 | |
| SbCl$_5$(l) | −440.2 | −350.1 | 301 | |
| SbCl$_3$(s) | −382.17 | −323.67 | 184.1 | 107.9 |
| SbOCl(s) | −374.0 | | | |
| **Argon** | | | | |
| Ar(g) | 0.0 | 0.0 | 154.843 | 20.786 |
| **Arsenic** | | | | |
| As($\alpha$-solid) | 0.0 | 0.0 | 35.1 | 24.64 |
| As(g) | 302.5 | 261.0 | 174.21 | 20.786 |
| As$_4$(g) | 143.9 | 92.4 | 314 | |
| H$_3$As(g) | 66.44 | 68.93 | 222.78 | 38.07 |
| As$_2$O$_5$(s) | −924.87 | −782.3 | 105.4 | 116.52 |
| As$_4$O$_6$(monoclinic) | −1309.6 | −1153.93 | 234 | |
| H$_3$AsO$_3$(aq) | −742.2 | −639.80 | 195 | |
| H$_3$AsO$_4$(aq) | −904.6 | | | |
| AsCl$_3$(l) | −305.0 | −259.4 | 216.3 | |
| **Barium** | | | | |
| Ba(s) | 0.0 | 0.0 | 62.8 | 28.07 |
| Ba(g) | 180 | 146 | 170.243 | 20.786 |
| Ba$^{2+}$(g) | 1660.38 | | | |
| Ba$^{2+}$(aq) | −537.64 | −560.77 | 9.6 | |
| BaO(s) | −553.5 | −525.1 | 70.42 | 47.78 |
| Ba(OH)$_2$(s) | −944.7 | | | |
| BaF$_2$(s) | −1207.1 | −1156.8 | 96.36 | 71.21 |
| BaCl$_2$(s) | −858.6 | −810.4 | 123.68 | 75.14 |
| BaCl$_2\cdot$2H$_2$O(s) | −1160.6 | −1055.63 | 166.9 | |
| BaCO$_3$(s) | −1216.3 | −1137.6 | 112.1 | 85.35 |
| Ba(NO$_3$)$_2$(s) | −992.07 | −796.59 | 213.8 | 151.38 |
| BaSO$_4$(s) | −1473.2 | −1362.2 | 132.2 | 101.75 |
| BaCrO$_4$(s) | −1446.0 | −1345.22 | 158.6 | |
| **Beryllium** | | | | |
| Be(s) | 0.0 | 0.0 | 9.50 | 16.44 |
| Be(g) | 324.3 | 286.6 | 136.269 | 20.786 |

| Substance | $\Delta H_f^\circ$ (kJ mol$^{-1}$) | $\Delta G_f^\circ$ (kJ mol$^{-1}$) | $S^\circ$ (J mol$^{-1}$ K$^{-1}$) | $C_p^\circ$ (J mol$^{-1}$ K$^{-1}$) |
|---|---|---|---|---|
| BeO(s) | −609.6 | −580.3 | 14.14 | 25.52 |
| BeF$_2$(s) | −1026.8 | −979.4 | 53.35 | 51.84 |
| BeCl$_2$(s) | −490.4 | −445.6 | 82.68 | 64.85 |
| BeSO$_4$(s) | −1205.20 | −1093.80 | 77.91 | 85.69 |
| **Bismuth** | | | | |
| Bi(s) | 0.0 | 0.0 | 56.74 | 25.52 |
| Bi(g) | 207.1 | 168.2 | 187.005 | 20.786 |
| Bi$_2$O$_3$(s) | −573.88 | −493.7 | 151.5 | 113.51 |
| BiCl$_3$(s) | −379.1 | −315.0 | 177.0 | 105 |
| BiOCl(s) | −366.9 | −322.1 | 120.5 | |
| Bi$_2$S$_3$(s) | −143.1 | −140.6 | 200.4 | 122.2 |
| **Boron** | | | | |
| B(s) | 0.0 | 0.0 | 5.86 | 11.09 |
| B(g) | 562.7 | 518.8 | 153.45 | 20.799 |
| B$_2$H$_6$(g) | 35.6 | 86.7 | 232.11 | 56.90 |
| B$_2$O$_3$(s) | −1272.77 | −1193.65 | 53.97 | 62.93 |
| H$_3$BO$_3$(s) | −1094.33 | −968.92 | 88.83 | 81.38 |
| [B(OH)$_4$]$^-$(aq) | −1344.03 | −1153.17 | 102.5 | |
| BN(s) | −254.4 | −228.4 | 14.81 | 19.71 |
| BF$_3$(g) | −1137.00 | −1120.33 | 254.12 | 50.46 |
| BCl$_3$(l) | −427.2 | −387.4 | 206.3 | 106.7 |
| NaBH$_4$(s) | −183.34 | −119.54 | 104.68 | 86.6 |
| **Bromine** | | | | |
| Br$_2$(l) | 0.0 | 0.0 | 152.231 | 75.689 |
| Br(g) | 111.884 | 82.396 | 175.022 | 20.786 |
| Br$_2$(g) | 30.907 | 3.110 | 245.463 | 36.02 |
| Br$^-$(g) | −219.07 | | | |
| Br$^-$(aq) | −121.55 | −103.96 | 82.4 | −141.8 |
| HBr(g) | −36.40 | −53.45 | 198.695 | 29.142 |
| BrO$^-$(aq) | −94.1 | −33.4 | 42 | |
| BrO$_3^-$(aq) | −67.07 | 18.60 | 161.71 | |
| BrF(g) | −93.85 | −109.18 | 228.97 | 32.97 |
| BrCl(g) | 14.64 | −0.98 | 240.10 | 34.98 |
| **Cadmium** | | | | |
| Cd($\gamma$-solid) | 0.0 | 0.0 | 51.76 | 25.98 |
| Cd(g) | 112.01 | 77.41 | 167.746 | 20.786 |
| Cd$^{2+}$(g) | 2623.54 | | | |
| Cd$^{2+}$(aq) | −75.90 | −77.612 | −73.2 | |
| CdO(s) | −258.2 | −228.4 | 54.8 | 43.43 |
| CdCl$_2$(s) | −391.50 | −343.93 | 115.27 | 74.68 |
| CdS(s) | −161.9 | −156.5 | 64.9 | |
| **Calcium** | | | | |
| Ca(s) | 0.0 | 0.0 | 41.42 | 25.31 |
| Ca(g) | 178.2 | 144.3 | 154.884 | 20.786 |
| Ca$^{2+}$(g) | 1925.90 | | | |
| Ca$^{2+}$(aq) | −542.83 | −553.58 | −53.1 | |
| CaO(s) | −635.09 | −604.03 | 39.75 | 42.80 |
| Ca(OH)$_2$(s) | −986.09 | −896.49 | 83.39 | 87.49 |
| CaC$_2$(s) | −59.8 | −64.9 | 69.96 | 62.72 |
| CaCO$_3$(calcite) | −1206.87 | −1128.76 | 92.9 | 81.88 |
| CaCO$_3$(aragonite) | −1207.04 | −1127.72 | 88.7 | 81.25 |
| CaSO$_4$(s) | −1434.11 | −1321.79 | 106.7 | 99.66 |

| Substance | $\Delta H_f^\circ$ (kJ mol$^{-1}$) | $\Delta G_f^\circ$ (kJ mol$^{-1}$) | $S^\circ$ (J mol$^{-1}$ K$^{-1}$) | $C_p^\circ$ (J mol$^{-1}$ K$^{-1}$) |
|---|---|---|---|---|
| CaSO$_4 \cdot$2H$_2$O(s) | −2022.63 | −1797.28 | 194.1 | 186.02 |
| CaF$_2$(s) | −1219.6 | −1167.3 | 68.87 | 67.03 |
| CaCl$_2$(s) | −795.8 | −748.1 | 104.6 | 72.59 |
| CaCl$_2 \cdot$6H$_2$O(s) | −2607.9 | | | |
| CaBr$_2$(s) | −682.8 | −663.6 | 130 | |
| Ca(NO$_3$)$_2$(s) | −938.39 | −743.07 | 193.3 | 149.37 |
| Ca$_3$(PO$_4$)$_2$(s) | −4120.8 | −3884.7 | 236.0 | 227.82 |
| Carbon | | | | |
| C(graphite) | 0.0 | 0.0 | 5.740 | 8.527 |
| C(diamond) | 1.895 | 2.900 | 2.377 | 6.113 |
| C(g) | 716.682 | 671.257 | 158.096 | 20.838 |
| CH$_4$(g) | −74.81 | −50.72 | 186.264 | 35.309 |
| C$_2$H$_2$(g) | 226.73 | 209.20 | 200.94 | 43.93 |
| C$_2$H$_4$(g) | 52.26 | 68.15 | 219.56 | 43.56 |
| C$_2$H$_6$(g) | −84.68 | −32.82 | 229.60 | 52.63 |
| CO(g) | −110.525 | −137.168 | 197.674 | 29.142 |
| CO$_2$(g) | −393.509 | −394.359 | 213.74 | 37.11 |
| HCN(g) | 135.1 | 124.7 | 201.78 | 35.86 |
| NH$_4$CN(s) | 0.42 | | | |
| CS$_2$(l) | 89.70 | 65.27 | 151.34 | 75.7 |
| CF$_4$(g) | −925 | −879 | 261.61 | 61.09 |
| CCl$_4$(l) | −135.44 | −65.21 | 216.40 | 131.75 |
| CCl$_4$(g) | −102.9 | −60.59 | 309.85 | 83.30 |
| CH$_3$Cl(g) | −80.83 | −57.37 | 234.58 | 40.75 |
| CH$_2$Cl$_2$(l) | −121.46 | −67.26 | 177.8 | 100.0 |
| CHCl$_3$(l) | −134.47 | −73.66 | 201.7 | 113.8 |
| CHCl$_3$(g) | −103.14 | −70.34 | 295.71 | 65.69 |
| HCOOH(l) | −424.72 | −361.35 | 128.95 | 99.04 |
| CH$_3$COOH(l) | −484.5 | −389.9 | 159.8 | 124.3 |
| CH$_3$COOH(aq) | −485.76 | −396.46 | 178.7 | |
| CH$_3$OH(l) | −238.66 | −166.27 | 126.8 | 81.6 |
| CH$_3$CH$_2$OH(l) | −277.69 | −174.78 | 160.7 | 111.46 |
| cis-CHClCHCl(l) | −27.6 | 22.05 | 198.41 | 113 |
| trans-CHClCHCl(l) | −23.14 | 27.28 | 195.85 | 113 |
| CH$_3$NH$_2$(l) | −47.3 | 35.6 | 150.21 | |
| HCHO(g) | −108.57 | −102.53 | 218.77 | 35.40 |
| CH$_3$CHO(g) | −166.19 | −128.86 | 250.3 | 57.3 |
| C$_6$H$_6$(l) | 49.028 | 124.50 | 172.8 | |
| C$_6$H$_6$(g) | 82.927 | 129.66 | 269.2 | |
| C$_3$H$_6$(g) (cyclopropane) | 54 | 100 | | |
| C$_5$H$_{10}$(g) (cyclopentane) | −77.2 | 38.6 | 293.9 | |
| Chlorine | | | | |
| Cl$_2$(g) | 0.0 | 0.0 | 223.066 | 33.907 |
| Cl(g) | 121.679 | 105.680 | 165.198 | 21.840 |
| Cl$^-$(g) | −233.13 | | | |
| Cl$^-$(aq) | −167.59 | −131.228 | 56.5 | −136.4 |
| HCl(g) | −92.307 | −95.299 | 186.908 | 29.12 |
| ClO$^-$(aq) | −107.1 | −36.8 | 42 | |
| ClO$_4^-$(aq) | −129.33 | −8.52 | 182.0 | |
| ClO$_2$(g) | 102.5 | 120.5 | 256.84 | 41.97 |

Appendix 937

| Substance | $\Delta H_f^\circ$ (kJ mol$^{-1}$) | $\Delta G_f^\circ$ (kJ mol$^{-1}$) | $S^\circ$ (J mol$^{-1}$ K$^{-1}$) | $C_p^\circ$ (J mol$^{-1}$ K$^{-1}$) |
|---|---|---|---|---|
| ClF(g) | −54.48 | −55.94 | 217.89 | 32.05 |
| ClF$_3$(g) | −163.2 | −123.0 | 281.61 | 63.85 |
| **Chromium** | | | | |
| Cr(s) | 0.0 | 0.0 | 23.77 | 23.35 |
| Cr(g) | 396.6 | 351.8 | 174.50 | 20.79 |
| Cr$_2$O$_3$(s) | −1139.7 | −1058.1 | 81.2 | 118.74 |
| CrO$_3$(s) | −589.5 | | | |
| Cr(OH)$_3$(s) | −1064.0 | | | |
| [Cr(H$_2$O)$_6$]$^{3+}$ | −1999.1 | | | |
| HCrO$_4^-$(aq) | −878.2 | −764.7 | 184.1 | |
| CrO$_4^{2-}$(aq) | −881.15 | −727.75 | 50.21 | |
| Cr$_2$O$_7^{2-}$(aq) | −1490.3 | −1301.1 | 261.9 | |
| CrCl$_3$(s) | −556.5 | −486.1 | 123.0 | 91.80 |
| **Cobalt** | | | | |
| Co(hexagonal) | 0.0 | 0.0 | 30.04 | 24.81 |
| Co(g) | 424.7 | 380.3 | 179.515 | 23.020 |
| Co$^{2+}$(g) | 2844.20 | | | |
| Co$^{2+}$(aq) | −58.2 | −54.4 | −113 | |
| CoO(s) | −237.94 | −214.20 | 52.97 | 55.23 |
| Co$_3$O$_4$(s) | −891 | −774 | 102.5 | 123.4 |
| CoCl$_2$(s) | −312.5 | −269.8 | 109.16 | 78.49 |
| CoSO$_4$(s) | −888.3 | −782.3 | 118.0 | |
| [Co(NH$_3$)$_6$]$^{3+}$(aq) | −584.9 | −157.0 | 146 | |
| **Copper** | | | | |
| Cu(s) | 0.0 | 0.0 | 33.150 | 24.435 |
| Cu(g) | 338.32 | 298.58 | 166.38 | 20.786 |
| Cu$^{2+}$(g) | 1089.986 | | | |
| Cu$^{2+}$(aq) | 64.77 | 65.49 | −99.6 | |
| CuO(s) | −157.3 | −129.7 | 42.63 | 42.30 |
| Cu$_2$O(s) | −168.6 | −146.0 | 93.14 | 63.64 |
| CuS(s) | −53.1 | −53.6 | 66.5 | 47.82 |
| Cu$_2$S(s) | −79.5 | −86.2 | 120.9 | 76.32 |
| CuF$_2$(s) | −542.7 | | | |
| CuCl$_2$(s) | −220.1 | −175.7 | 108.07 | 71.88 |
| CuCl(s) | −137.2 | −119.86 | 86.2 | 48.5 |
| CuBr(s) | −104.6 | −100.8 | 96.11 | 54.73 |
| CuI(s) | −67.8 | −69.5 | 96.7 | 54.06 |
| CuSO$_4$(s) | −771.36 | −661.8 | 109 | 100.0 |
| CuSO$_4$·5H$_2$O(s) | −2279.65 | −1879.745 | 300.4 | 280 |
| Cu(NO$_3$)$_2$(s) | −302.9 | | | |
| [Cu(NH$_3$)$_4$]$^{2+}$(aq) | −348.5 | −111.07 | 273.6 | |
| **Fluorine** | | | | |
| F$_2$(g) | 0.0 | 0.0 | 202.78 | 31.30 |
| F(g) | 78.99 | 61.91 | 158.754 | 22.744 |
| F$^-$(g) | −255.39 | | | |
| F$^-$(aq) | −332.63 | −278.79 | −13.8 | −106.7 |
| HF(g) | −271.1 | −273.2 | 173.669 | 29.133 |
| HF(aq) | −320.08 | −296.82 | 88.7 | |
| **Helium** | | | | |
| He(g) | 0.0 | 0.0 | 126.150 | 20.786 |
| **Hydrogen** | | | | |
| H$_2$(g) | 0.0 | 0.0 | 130.684 | 28.824 |

| Substance | $\Delta H_f^\circ$ (kJ mol$^{-1}$) | $\Delta G_f^\circ$ (kJ mol$^{-1}$) | $S^\circ$ (J mol$^{-1}$ K$^{-1}$) | $C_p^\circ$ (J mol$^{-1}$ K$^{-1}$) |
|---|---|---|---|---|
| H(g) | 217.965 | 203.247 | 114.713 | 20.784 |
| H$^+$(g) | 1536.20 | | | |
| H$^+$(aq) | 0.0 | 0.0 | 0.0 | 0.0 |
| Iodine | | | | |
| I$_2$(s) | 0.0 | 0.0 | 116.135 | 54.438 |
| I(g) | 106.838 | 70.250 | 180.791 | 20.786 |
| I$_2$(g) | 62.438 | 19.327 | 260.69 | 36.90 |
| I$^-$(g) | −197 | | | |
| I$^-$(aq) | −55.19 | −51.57 | 111.3 | −142.3 |
| [I$_3$]$^-$(aq) | −51.5 | −51.4 | 239.3 | |
| HI(g) | 26.48 | 1.70 | 206.594 | 29.158 |
| HIO(aq) | −138.1 | −99.1 | 95.4 | |
| IO$^-$(aq) | −107.5 | −38.5 | −5.4 | |
| IO$_3^-$(aq) | −221.3 | −128.0 | 118.4 | |
| IF$_7$(g) | −943.9 | −818.3 | 346.5 | 136.4 |
| IF$_5$(l) | −864.8 | | | |
| ICl$_3$(s) | −89.5 | −22.29 | 167.4 | |
| IBr(s) | −10.5 | | | |
| Iron | | | | |
| Fe($\alpha$-solid) | 0.0 | 0.0 | 27.28 | 25.10 |
| Fe(g) | 416.3 | 370.7 | 180.490 | 25.677 |
| Fe$^{2+}$(g) | 2749.93 | | | |
| Fe$^{3+}$(g) | 5712.8 | | | |
| Fe$^{2+}$(aq) | −89.1 | −78.90 | −137.7 | |
| Fe$^{3+}$(aq) | −48.5 | −4.7 | −315.9 | |
| FeO(s) | −272.0 | | | |
| Fe$_2$O$_3$(s) | −824.2 | −742.2 | 87.40 | 103.85 |
| Fe$_3$O$_4$(s) | −1118.4 | −1015.4 | 146.4 | 143.43 |
| Fe(OH)$_2$(s) | −569.0 | −486.5 | 88 | |
| Fe(OH)$_3$(s) | −823.0 | −696.5 | 106.7 | |
| FeCl$_2$(s) | −341.79 | −302.30 | 117.95 | 76.65 |
| FeCl$_3$(s) | −399.49 | −334.00 | 142.3 | 96.65 |
| FeSO$_4$(s) | −928.4 | −820.8 | 107.5 | 100.58 |
| FeSO$_4\cdot$7H$_2$O(s) | −3014.57 | −2509.87 | 409.2 | 394.47 |
| Fe$_2$(SO$_4$)$_3$(s) | −2581.5 | | | |
| K$_3$[Fe(CN)$_6$](s) | −173.2 | | | |
| K$_4$[Fe(CN)$_6$](s) | −523.4 | | | |
| Lead | | | | |
| Pb(s) | 0.0 | 0.0 | 64.81 | 26.44 |
| Pb(g) | 195.0 | 161.9 | 175.373 | 20.786 |
| Pb$^{2+}$(g) | 2373.33 | | | |
| Pb$^{2+}$(aq) | −1.7 | −24.43 | 10.5 | |
| PbO(yellow) | −217.32 | −187.89 | 68.70 | 45.77 |
| PbO(red) | −218.99 | −188.93 | 66.5 | 45.81 |
| PbO$_2$(s) | −277.4 | −217.33 | 68.6 | 64.64 |
| Pb$_3$O$_4$(s) | −718.4 | −601.2 | 211.3 | 146.9 |
| PbS(s) | −100.4 | −98.7 | 91.2 | 49.50 |
| PbCl$_2$(s) | −359.41 | −314.10 | 136.0 | |
| Pb(NO$_3$)$_2$(s) | −451.9 | | | |
| PbSO$_4$(s) | −919.94 | −813.14 | 148.57 | 103.207 |
| Lithium | | | | |
| Li(s) | 0.0 | 0.0 | 29.12 | 24.77 |

| Substance | $\Delta H_f^\circ$ (kJ mol$^{-1}$) | $\Delta G_f^\circ$ (kJ mol$^{-1}$) | $S^\circ$ (J mol$^{-1}$ K$^{-1}$) | $C_p^\circ$ (J mol$^{-1}$ K$^{-1}$) |
|---|---|---|---|---|
| Li(g) | 159.37 | 126.66 | 138.77 | 20.786 |
| Li$^+$(g) | 685.783 | | | |
| Li$^+$(aq) | −278.49 | −293.31 | 13.4 | 68.6 |
| Li$_2$O(s) | −597.94 | −561.18 | 37.57 | 54.10 |
| LiH(s) | −90.54 | −68.35 | 20.008 | 27.87 |
| LiOH(s) | −484.93 | −438.95 | 42.80 | 49.66 |
| LiF(s) | −615.97 | −587.71 | 35.65 | 41.59 |
| LiCl(s) | −408.61 | −384.37 | 59.33 | 47.99 |
| LiNO$_3$(s) | −483.13 | −381.1 | 90.0 | |
| LiAlH$_4$(s) | −116.3 | −44.7 | 78.74 | 83.18 |
| **Magnesium** | | | | |
| Mg(s) | 0.0 | 0.0 | 32.68 | 24.89 |
| Mg(g) | 147.70 | 113.0 | 148.650 | 20.786 |
| Mg$^{2+}$(g) | 2348.504 | | | |
| Mg$^{2+}$(aq) | −466.85 | −454.8 | −138.1 | |
| MgO(s) | −601.70 | −569.43 | 26.94 | 37.15 |
| Mg(OH)$_2$(s) | −924.54 | −833.51 | 63.18 | 77.03 |
| Mg$_3$N$_2$(s) | −461.24 | | | 104.56 |
| MgF$_2$(s) | −1123.4 | −1070.2 | 57.24 | 61.59 |
| MgCl$_2$(s) | −641.32 | −591.79 | 89.62 | 71.38 |
| MgBr$_2$(s) | −524.3 | −503.8 | 117.2 | |
| MgI$_2$(s) | −364.0 | −358.2 | 129.7 | |
| MgCO$_3$(s) | −1095.8 | −1012.1 | 65.7 | 75.52 |
| Mg(NO$_3$)$_2$(s) | −790.65 | −589.4 | 164.0 | 141.92 |
| Mg$_3$(PO$_4$)$_2$(s) | −3780.7 | −3538.7 | 189.20 | 213.47 |
| MgSO$_4$(s) | −1284.9 | −1170.6 | 91.6 | 96.48 |
| **Manganese** | | | | |
| Mn($\alpha$-solid) | 0.0 | 0.0 | 32.01 | 26.32 |
| Mn(g) | 280.7 | 238.5 | 173.70 | 20.79 |
| Mn$^{2+}$(g) | 2519.69 | | | |
| Mn$^{2+}$(aq) | −220.75 | −228.1 | −73.6 | 50 |
| MnO(s) | −385.22 | −362.90 | 59.71 | 45.44 |
| MnO$_2$(s) | −520.03 | −465.14 | 53.05 | 54.14 |
| Mn$_2$O$_3$(s) | −959.0 | −881.1 | 110.5 | 107.65 |
| Mn$_3$O$_4$(s) | −1387.8 | −1283.2 | 155.6 | 139.66 |
| MnCl$_2$(s) | −481.29 | −440.50 | 118.24 | 72.93 |
| MnSO$_4$(s) | −1065.25 | −957.36 | 112.1 | 100.50 |
| MnO$_4^-$(aq) | −541.4 | −447.2 | 191.2 | −82.0 |
| MnO$_4^{2-}$(aq) | −653 | −500.7 | 59 | |
| **Mercury** | | | | |
| Hg(l) | 0.0 | 0.0 | 76.02 | 27.983 |
| Hg(g) | 61.317 | 31.820 | 174.96 | 20.786 |
| Hg$^+$(g) | 1074.58 | | | |
| Hg$^{2+}$(g) | 2890.47 | | | |
| Hg$^{2+}$(aq) | 171.1 | 164.40 | −32.2 | |
| Hg$_2^{2+}$(aq) | 172.4 | 153.52 | 84.5 | |
| HgO(red) | −90.83 | −58.539 | 70.29 | 44.06 |
| HgO(yellow) | −90.46 | −58.409 | 71.1 | |
| HgCl$_2$(s) | −224.3 | −178.6 | 146.0 | |
| Hg$_2$Cl$_2$(s) | −265.22 | −210.745 | 192.5 | |
| HgS(red) | −58.2 | −50.6 | 82.4 | 48.41 |
| HgS(black) | −53.6 | −47.7 | 88.3 | |

| Substance | $\Delta H_f^\circ$ (kJ mol$^{-1}$) | $\Delta G_f^\circ$ (kJ mol$^{-1}$) | $S^\circ$ (J mol$^{-1}$ K$^{-1}$) | $C_p^\circ$ (J mol$^{-1}$ K$^{-1}$) |
|---|---|---|---|---|
| [Hg(NH$_3$)$_4$]$^{2+}$ | −282.8 | −51.7 | 335 | |
| **Neon** | | | | |
| Ne(g) | 0.0 | 0.0 | 146.328 | 20.786 |
| **Nickel** | | | | |
| Ni(s) | 0.0 | 0.0 | 29.87 | 26.07 |
| Ni(g) | 429.7 | 384.5 | 182.193 | 23.359 |
| Ni$^{2+}$(g) | 2931.390 | | | |
| Ni$^{2+}$(aq) | −54.0 | −45.6 | −128.9 | |
| NiO(s) | −239.7 | −211.7 | 37.99 | 44.31 |
| NiS(s) | −82.0 | −79.5 | 52.97 | 47.11 |
| NiCl$_2$(s) | −305.332 | −259.032 | 97.65 | 71.67 |
| NiSO$_4$(s) | −872.91 | −759.7 | 92 | 138 |
| [Ni(NH$_3$)$_6$]$^{2+}$(aq) | −630.1 | −255.7 | 394.6 | |
| [Ni(CN)$_4$]$^{2-}$(aq) | 367.8 | 472.1 | 218 | |
| **Nitrogen** | | | | |
| N$_2$(g) | 0.0 | 0.0 | 191.61 | 29.125 |
| N(g) | 472.704 | 455.563 | 153.298 | 20.786 |
| NH$_3$(g) | −46.11 | −16.45 | 192.45 | 35.06 |
| NH$_3$(aq) | −80.29 | −26.50 | 111.3 | |
| NH$_4^+$(aq) | −132.51 | −79.31 | 113.4 | 79.9 |
| N$_2$H$_4$(l) | 50.63 | 149.34 | 121.21 | 98.87 |
| NO(g) | 90.25 | 86.55 | 210.761 | 29.844 |
| NO$_2$(g) | 33.18 | 51.31 | 240.06 | 37.20 |
| N$_2$O(g) | 82.05 | 104.20 | 219.85 | 38.45 |
| N$_2$O$_3$(g) | 83.72 | 139.46 | 312.28 | 65.61 |
| N$_2$O$_4$(g) | 9.16 | 97.89 | 304.29 | 77.28 |
| N$_2$O$_5$(s) | −43.1 | 113.9 | 178.2 | 143.1 |
| N$_2$O$_5$(g) | 11.3 | 115.1 | 355.7 | 84.5 |
| HNO$_3$(l) | −174.10 | −80.71 | 155.60 | 109.87 |
| HNO$_3$(g) | −135.06 | −74.72 | 266.38 | 53.35 |
| NO$_3^-$(aq) | −205.0 | −108.74 | 146.4 | −86.6 |
| NH$_4$NO$_3$(s) | −365.56 | −183.87 | 151.08 | 139.3 |
| NH$_4$NO$_2$(s) | −256.5 | | | |
| NH$_4$F(s) | −463.96 | −348.68 | 71.96 | 65.27 |
| NH$_4$Cl(s) | −314.43 | −202.87 | 94.6 | 84.1 |
| NH$_4$Br(s) | −270.83 | −175.2 | 113 | 96 |
| NH$_4$I(s) | −201.42 | −112.5 | 117 | |
| (NH$_4$)$_2$SO$_4$(s) | −1180.85 | −901.67 | 220.1 | 187.49 |
| NOCl(g) | 51.71 | 66.08 | 261.69 | 44.69 |
| NOBr(g) | 82.17 | 82.42 | 273.66 | 45.48 |
| **Oxygen** | | | | |
| O$_2$(g) | 0.0 | 0.0 | 205.138 | 29.355 |
| O(g) | 249.170 | 231.731 | 161.055 | 21.912 |
| O$_3$(g) | 142.7 | 163.2 | 238.93 | 39.20 |
| OH$^-$(g) | −143.5 | | | |
| OH$^-$(aq) | −229.994 | −157.244 | −10.75 | −148.5 |
| H$_2$O(l) | −285.830 | −237.129 | 69.91 | 75.291 |
| H$_2$O(g) | −241.818 | −228.72 | 188.825 | 33.577 |
| H$_2$O$_2$(l) | −187.78 | −120.35 | 109.6 | 89.1 |
| **Phosphorus** | | | | |
| P(white) | 0.0 | 0.0 | 41.09 | 23.840 |
| P(g) | 314.64 | 278.25 | 163.193 | 20.786 |

| Substance | $\Delta H_f^\circ$ (kJ mol$^{-1}$) | $\Delta G_f^\circ$ (kJ mol$^{-1}$) | $S^\circ$ (J mol$^{-1}$ K$^{-1}$) | $C_p^\circ$ (J mol$^{-1}$ K$^{-1}$) |
|---|---|---|---|---|
| P(red) | −17.6 | −12.1 | 22.80 | 21.21 |
| P(black) | −39.3 | | | |
| P$_4$(g) | 58.91 | 24.44 | 279.98 | 67.15 |
| PH$_3$(g) | 5.4 | 13.4 | 210.23 | 37.11 |
| P$_4$O$_6$(s) | −1640.1 | | | |
| P$_4$O$_{10}$(s) | −2984.0 | −2697.7 | 228.86 | 211.71 |
| HPO$_3$(s) | −948.5 | | | |
| H$_3$PO$_2$(s) | −604.6 | | | |
| H$_3$PO$_3$(s) | −964.4 | | | |
| H$_3$PO$_4$(s) | −1279.0 | −1119.1 | 110.50 | 106.06 |
| H$_3$PO$_4$(aq) | −1288.34 | −1142.54 | 158.2 | |
| H$_2$PO$_4^-$(aq) | −1296.29 | −1130.28 | 90.4 | |
| HPO$_4^{2-}$(aq) | −1292.14 | −1089.15 | −33.5 | |
| PO$_4^{3-}$(aq) | −1277.4 | −1018.7 | −222 | |
| H$_4$P$_2$O$_7$(s) | −2241.0 | | | |
| PF$_5$(g) | −1595.8 | | | |
| PF$_3$(g) | −918.8 | −897.5 | 273.24 | 58.70 |
| PCl$_5$(g) | −374.9 | −305.0 | 364.58 | 112.80 |
| PCl$_5$(s) | −443.5 | | | |
| PCl$_3$(l) | −319.7 | −272.3 | 217.1 | |
| PCl$_3$(g) | −287.0 | −267.8 | 311.78 | 71.84 |
| **Potassium** | | | | |
| K(s) | 0.0 | 0.0 | 64.18 | 29.58 |
| K(g) | 89.24 | 60.59 | 160.336 | 20.786 |
| K$^+$(g) | 514.26 | | | |
| K$^+$(aq) | −252.38 | −283.27 | 102.5 | 21.8 |
| KO$_2$(s) | −284.93 | −239.4 | 116.7 | 77.53 |
| K$_2$O$_2$(s) | −494.1 | −425.1 | 102.1 | |
| KOH(s) | −424.764 | −379.08 | 78.9 | 64.9 |
| KF(s) | −567.27 | −537.75 | 66.57 | 49.04 |
| KCl(s) | −436.747 | −409.14 | 82.59 | 51.30 |
| KBr(s) | −393.798 | −380.66 | 95.90 | 52.30 |
| KI(s) | −327.900 | −324.892 | 106.32 | 52.93 |
| KClO$_4$(s) | −432.75 | −303.09 | 151.0 | 112.38 |
| K$_2$CO$_3$(s) | −1151.02 | −1063.5 | 155.52 | 114.43 |
| KNO$_3$(s) | −494.63 | −394.86 | 133.05 | 96.40 |
| K$_2$SO$_4$(s) | −1437.79 | −1321.37 | 175.56 | 131.46 |
| KMnO$_4$(s) | −837.2 | −737.6 | 171.71 | 117.57 |
| K$_2$CrO$_4$(s) | −1403.7 | −1295.7 | 200.12 | 145.98 |
| K$_2$Cr$_2$O$_7$(s) | −2061.5 | −1881.8 | 291.2 | 219.24 |
| K$_4$Fe(CN)$_6$(s) | −594.1 | −453.0 | 418.8 | 332.21 |
| K$_3$Fe(CN)$_6$(s) | −249.8 | −129.6 | 426.06 | |
| **Selenium** | | | | |
| Se(black) | 0.0 | 0.0 | 42.442 | 25.363 |
| Se(red) | 6.7 | | | |
| Se(g) | 227.07 | 187.03 | 176.72 | 20.820 |
| H$_2$Se(g) | 29.7 | 15.9 | 219.02 | 34.73 |
| SeO$_2$(s) | −225.35 | | | |
| SeO$_3$(s) | −166.9 | | | |
| H$_2$SeO$_4$(s) | −530.1 | | | |
| SeF$_6$(g) | −1117 | −1017 | 313.87 | 110.5 |
| SeCl$_4$(s) | −183.3 | | | |

| Substance | $\Delta H_f^\circ$ (kJ mol$^{-1}$) | $\Delta G_f^\circ$ (kJ mol$^{-1}$) | $S^\circ$ (J mol$^{-1}$ K$^{-1}$) | $C_p^\circ$ (J mol$^{-1}$ K$^{-1}$) |
|---|---|---|---|---|
| **Silicon** | | | | |
| Si(s) | 0.0 | 0.0 | 18.83 | 20.00 |
| Si(g) | 455.6 | 411.3 | 167.97 | 22.251 |
| SiH$_4$(g) | 34.3 | 56.9 | 204.62 | 42.84 |
| SiO$_2$($\alpha$-quartz) | −910.94 | −856.64 | 41.84 | 44.43 |
| H$_4$SiO$_4$(s) | −1481.1 | −1332.9 | 192 | |
| H$_2$SiO$_3$(s) | −1188.7 | −1092.4 | 134 | |
| SiF$_4$(g) | −1614.94 | −1572.65 | 282.49 | 73.64 |
| SiCl$_4$(l) | −687.0 | −619.84 | 239.7 | 145.31 |
| SiBr$_4$(l) | −457.3 | −443.9 | 277.8 | |
| SiC(cubic) | −65.3 | −62.8 | 16.61 | 26.86 |
| **Silver** | | | | |
| Ag(s) | 0.0 | 0.0 | 42.55 | 25.351 |
| Ag(g) | 284.55 | 245.65 | 172.997 | 20.786 |
| Ag$^+$(g) | 1021.73 | | | |
| Ag$^+$(aq) | 105.579 | 77.107 | 72.68 | 21.8 |
| Ag$_2$O(s) | −31.05 | −11.20 | 121.3 | 65.86 |
| Ag$_2$S(orthorhombic) | −32.59 | −40.67 | 144.01 | 76.53 |
| AgF(s) | −204.6 | | | |
| AgCl(s) | −127.068 | −109.789 | 96.2 | 50.79 |
| [AgCl$_2$]$^-$(aq) | −245.2 | −215.4 | 231.4 | |
| AgBr(s) | −100.37 | −96.90 | 107.1 | 52.38 |
| AgI(s) | −61.84 | −66.19 | 115.5 | 56.82 |
| AgNO$_3$(s) | −124.39 | −33.41 | 140.92 | 93.05 |
| Ag$_2$SO$_4$(s) | −715.88 | −618.41 | 200.4 | 131.38 |
| AgSCN(s) | 87.9 | 101.39 | 131.0 | 63 |
| [Ag(NH$_3$)$_2$]$^+$(aq) | −111.29 | −17.12 | 245.2 | |
| [Ag(CN)$_2$]$^-$(aq) | 270.3 | 305.5 | 192 | |
| **Sodium** | | | | |
| Na(s) | 0.0 | 0.0 | 51.21 | 28.24 |
| Na(g) | 107.32 | 76.761 | 153.712 | 20.786 |
| Na$^+$(g) | 609.358 | | | |
| Na$^+$(aq) | −240.12 | −261.905 | 59.0 | 46.4 |
| Na$_2$O(s) | −414.22 | −375.46 | 75.06 | 69.12 |
| Na$_2$O$_2$(s) | −510.87 | −447.7 | 95.0 | 89.24 |
| NaH(s) | −56.275 | −33.46 | 40.016 | 36.401 |
| NaOH(s) | −425.609 | −379.494 | 64.455 | 59.54 |
| Na$_2$S(s) | −364.8 | −349.8 | 83.7 | |
| NaF(s) | −573.647 | −543.494 | 51.46 | 46.86 |
| NaCl(s) | −411.153 | −384.138 | 72.13 | 50.50 |
| NaBr(s) | −361.062 | −348.983 | 86.82 | 51.38 |
| NaI(s) | −287.78 | −286.06 | 98.53 | 52.09 |
| NaClO$_4$(s) | −383.30 | −254.85 | 142.3 | |
| Na$_2$CO$_3$(s) | −1130.68 | −1044.44 | 134.98 | 112.30 |
| NaHCO$_3$(s) | −950.81 | −851.0 | 101.7 | 87.61 |
| Na$_2$SO$_4$(s) | −1387.08 | −1270.16 | 149.58 | 128.20 |
| NaHSO$_4$(s) | −1125.5 | −992.8 | 113.0 | |
| Na$_2$S$_2$O$_3$(s) | −1123.0 | −1028.0 | 155 | 146.0 |
| NaNO$_3$(s) | −467.85 | −367.00 | 116.52 | 92.88 |
| Na$_3$PO$_4$(s) | −1917.40 | −1788.80 | 173.80 | 153.47 |
| NaCH$_3$COO(s) | −708.81 | −607.18 | 123.0 | 79.9 |
| NaCN(s) | −87.49 | −76.43 | 115.60 | 70.37 |

| Substance | $\Delta H_f^\circ$ (kJ mol$^{-1}$) | $\Delta G_f^\circ$ (kJ mol$^{-1}$) | $S^\circ$ (J mol$^{-1}$ K$^{-1}$) | $C_p^\circ$ (J mol$^{-1}$ K$^{-1}$) |
|---|---|---|---|---|
| **Strontium** | | | | |
| Sr(s) | 0.0 | 0.0 | 52.3 | 26.4 |
| Sr(g) | 164.4 | 130.9 | 164.62 | 20.786 |
| Sr$^{2+}$(g) | 1790.54 | | | |
| Sr$^{2+}$(aq) | −545.80 | −559.48 | −32.6 | |
| SrO(s) | −592.0 | −561.9 | 54.4 | 45.02 |
| SrCO$_3$(s) | −1220.1 | −1140.1 | 97.1 | 81.42 |
| SrCl$_2$(s) | −828.9 | −781.1 | 114.85 | 75.60 |
| **Sulfur** | | | | |
| S(rhombic) | 0.0 | 0.0 | 31.80 | 22.64 |
| S(monoclinic) | 0.33 | | | |
| S(g) | 278.805 | 238.250 | 167.821 | 23.673 |
| S$_8$(g) | 102.30 | 49.63 | 430.98 | 156.44 |
| H$_2$S(g) | −20.63 | −33.56 | 205.79 | 34.23 |
| SO$_2$(g) | −296.830 | −300.194 | 248.22 | 39.87 |
| SO$_3$(s) | −454.51 | −374.21 | 70.7 | |
| SO$_3$(g) | −395.72 | −371.06 | 256.76 | 50.67 |
| H$_2$SO$_4$(l) | −813.989 | −690.003 | 156.904 | 138.91 |
| H$_2$SO$_4$(aq) | −909.27 | −744.53 | 20.1 | −293 |
| HSO$_4^-$(aq) | −887.34 | −755.91 | 131.8 | −84 |
| SO$_4^{2-}$(aq) | −909.27 | −774.53 | 20.1 | −293 |
| SF$_6$(g) | −1209 | −1105.3 | 291.82 | 97.28 |
| SCl$_2$(l) | −50 | | | |
| S$_2$Cl$_2$(l) | −59.4 | | | |
| **Tellurium** | | | | |
| Te(s) | 0.0 | 0.0 | 49.71 | 25.73 |
| Te(g) | 196.73 | 157.08 | 182.74 | 20.786 |
| H$_2$Te(g) | 99.6 | | | |
| TeO$_2$(s) | −322.6 | −270.3 | 79.5 | |
| **Tin** | | | | |
| Sn(white) | 0.0 | 0.0 | 51.55 | 26.99 |
| Sn(gray) | −2.09 | 0.13 | 44.14 | 25.77 |
| Sn(g) | 302.1 | 267.3 | 168.486 | 21.259 |
| SnO(s) | −285.8 | −256.9 | 56.5 | 44.31 |
| SnO$_2$(s) | −580.7 | −519.6 | 52.3 | 52.59 |
| SnCl$_4$(l) | −511.3 | −440.1 | 258.6 | 165.3 |
| SnCl$_2$(s) | −325.1 | | | |
| SnCl$_2 \cdot$2H$_2$O(s) | −921.3 | | | |
| **Uranium** | | | | |
| U(s) | 0.0 | 0.0 | 50.21 | 27.665 |
| U(g) | 535.6 | 491.9 | 199.77 | 23.694 |
| UO$_2$(s) | −1084.9 | −1031.7 | 77.03 | 63.60 |
| UO$_3$(s) | −1223.8 | −1145.9 | 96.11 | 81.67 |
| UF$_6$(g) | −2147.4 | −2063.7 | 377.9 | 129.62 |
| **Xenon** | | | | |
| Xe(g) | 0.0 | 0.0 | 169.683 | 20.786 |
| XeF$_4$(s) | −261.5 | | | |
| **Zinc** | | | | |
| Zn(s) | 0.0 | 0.0 | 41.63 | 25.40 |
| Zn(g) | 130.79 | 95.145 | 160.984 | 20.786 |
| Zn$^{2+}$(g) | 2782.78 | | | |
| Zn$^{2+}$(aq) | −153.89 | −147.06 | −112.1 | 46 |

| Substance | $\Delta H_f^\circ$ (kJ mol$^{-1}$) | $\Delta G_f^\circ$ (kJ mol$^{-1}$) | $S^\circ$ (J mol$^{-1}$ K$^{-1}$) | $C_p^\circ$ (J mol$^{-1}$ K$^{-1}$) |
|---|---|---|---|---|
| ZnO(s) | −348.28 | −318.30 | 43.64 | 40.25 |
| Zn(OH)$_2$(s) | −643.25 | −555.07 | 81.6 | 72.4 |
| ZnS(wurtzite) | −192.63 | | | |
| ZnS(sphalerite) | −205.98 | −201.29 | 57.7 | 46.0 |
| ZnCl$_2$(s) | −415.05 | −369.398 | 111.46 | 71.34 |
| ZnSO$_4$(s) | −982.8 | −871.5 | 110.5 | 99.2 |
| Zn(NO$_3$)$_2$(s) | −483.7 | | | |
| Zn(NO$_3$)$_2$·6H$_2$O(s) | −2306.64 | −1772.71 | 456.9 | 323.0 |
| ZnCO$_3$(s) | −812.78 | −731.52 | 82.4 | 79.71 |

[a] Adapted from the National Bureau of Standards tables of chemical thermodynamic properties, *J Phys. Chem. Ref. Data* **11**, Suppl. 2 (1982).

# Appendix 5: Standard Reduction Potentials at 25°C

**TABLE A5.1  Acidic Media**

| Reduction half-reaction | $\epsilon°(V)$ |
|---|---|
| $Li^+ + e^- \rightarrow Li(s)$ | −3.045 |
| $Rb^+ + e^- \rightarrow Rb(s)$ | −2.925 |
| $K^+ + e^- \rightarrow K(s)$ | −2.925 |
| $Cs^+ + e^- \rightarrow Cs(s)$ | −2.923 |
| $Ra^{2+} + 2e^- \rightarrow Ra(s)$ | −2.916 |
| $Ba^{2+} + 2e^- \rightarrow Ba(s)$ | −2.906 |
| $Sr^{2+} + 2e^- \rightarrow Sr(s)$ | −2.888 |
| $Ca^{2+} + 2e^- \rightarrow Ca(s)$ | −2.866 |
| $Na^+ + e^- \rightarrow Na(s)$ | −2.714 |
| $La^{3+} + 3e^- \rightarrow La(s)$ | −2.522 |
| $Ce^{3+} + 3e^- \rightarrow Ce(s)$ | −2.483 |
| $Mg^{2+} + 2e^- \rightarrow Mg(s)$ | −2.363 |
| $H_2(g) + 2e^- \rightarrow 2H^-$ | −2.25 |
| $Sc^{3+} + 3e^- \rightarrow Sc(s)$ | −2.077 |
| $[AlF_6]^{3-} + 3e^- \rightarrow Al(s) + 6F^-$ | −2.069 |
| $Be^{2+} + 2e^- \rightarrow Be(s)$ | −1.847 |
| $V^{3+} + 3e^- \rightarrow V(s)$ | −1.798 |
| $Hf^{4+} + 4e^- \rightarrow Hf(s)$ | −1.700 |
| $Al^{3+} + 3e^- \rightarrow Al(s)$ | −1.662 |
| $Ti^{2+} + 2e^- \rightarrow Ti(s)$ | −1.628 |
| $Zr^{4+} + 4e^- \rightarrow Zr(s)$ | −1.529 |
| $V^{4+} + 4e^- \rightarrow V(s)$ | −1.50 |
| $[SiF_6]^{2-} + 4e^- \rightarrow Si(s) + 6F^-$ | −1.24 |
| $[TiF_6]^{2-} + 4e^- \rightarrow Ti(s) + 6F^-$ | −1.191 |
| $Mn^{2+} + 2e^- \rightarrow Mn(s)$ | −1.185 |
| $V^{2+} + 2e^- \rightarrow V(s)$ | −1.175 |
| $Cr^{2+} + 2e^- \rightarrow Cr(s)$ | −0.913 |
| $H_3BO_3(s) + 3H^+ + 3e^- \rightarrow B(s) + 3H_2O(l)$ | −0.869 |
| $SiO_2(s) + 4H^+ + 4e^- \rightarrow Si(s) + 2H_2O(l)$ | −0.857 |
| $Zn^{2+} + 2e^- \rightarrow Zn(s)$ | −0.7628 |
| $Cr^{3+} + 3e^- \rightarrow Cr(s)$ | −0.744 |
| $Te(s) + 2H^+ + 2e^- \rightarrow H_2Te(aq)$ | −0.739 |
| $U^{4+} + e^- \rightarrow U^{3+}$ | −0.607 |
| $As(s) + 3H^+ + 3e^- \rightarrow AsH_3(g)$ | −0.607 |
| $Ga^{3+} + 3e^- \rightarrow Ga(s)$ | −0.529 |
| $Fe^{2+} + 2e^- \rightarrow Fe(s)$ | −0.4402 |
| $Cr^{3+} + e^- \rightarrow Cr^{2+}$ | −0.408 |
| $Cd^{2+} + 2e^- \rightarrow Cd(s)$ | −0.4029 |
| $Se(s) + 2H^+ + 2e^- \rightarrow H_2Se(aq)$ | −0.399 |
| $Ti^{3+} + e^- \rightarrow Ti^{2+}$ | −0.369 |
| $PbI_2(s) + 2e^- \rightarrow Pb(s) + 2I^-$ | −0.365 |
| $PbSO_4(s) + 2e^- \rightarrow Pb(s) + SO_4^{2-}$ | −0.3588 |
| $In^{3+} + 3e^- \rightarrow In(s)$ | −0.343 |
| $Tl^+ + e^- \rightarrow Tl(s)$ | −0.3363 |
| $PbBr_2(s) + 2e^- \rightarrow Pb(s) + 2Br^-$ | −0.284 |
| $Co^{2+} + 2e^- \rightarrow Co(s)$ | −0.277 |
| $PbCl_2(s) + 2e^- \rightarrow Pb(s) + 2Cl^-$ | −0.268 |
| $V^{3+} + e^- \rightarrow V^{2+}$ | −0.256 |
| $Ni^{2+} + 2e^- \rightarrow Ni(s)$ | −0.250 |
| $AgI(s) + e^- \rightarrow Ag(s) + I^-$ | −0.1518 |
| $Sn^{2+} + 2e^- \rightarrow Sn(s)$ | −0.136 |

| Reduction half-reaction | $\epsilon°(V)$ |
|---|---|
| $Pb^{2+} + 2e^- \rightarrow Pb(s)$ | −0.126 |
| $P(s) + 3H^+ + 3e^- \rightarrow PH_3(g)$ | −0.063 |
| $Fe^{3+} + 3e^- \rightarrow Fe(s)$ | −0.036 |
| $2H^+ + 2e^- \rightarrow H_2(g)$ | 0.000 |
| $AgBr(s) + e^- \rightarrow Ag(s) + Br^-$ | +0.0713 |
| $Si(s) + 4H^+ + 4e^- \rightarrow SiH_4(g)$ | +0.102 |
| $Hg_2Br_2(s) + 2e^- \rightarrow 2Hg(l) + 2Br^-$ | +0.1397 |
| $S(s) + 2H^+ + 2e^- \rightarrow H_2S(aq)$ | +0.142 |
| $Sn^{4+} + 2e^- \rightarrow Sn^{2+}$ | +0.15 |
| $Sb_2O_3(s) + 6H^+ + 6e^- \rightarrow 2Sb(s) + 3H_2O(l)$ | +0.152 |
| $Cu^{2+} + e^- \rightarrow Cu^+$ | +0.153 |
| $SO_4^{2-} + 4H^+ + 2e^- \rightarrow H_2SO_3(aq) + H_2O(l)$ | +0.172 |
| $AgCl(s) + e^- \rightarrow Ag(s) + Cl^-$ | +0.2222 |
| $[Hg_2Br_4]^{2-} + 2e^- \rightarrow Hg(l) + 4Br^-$ | +0.223 |
| $Hg_2Cl_2(s) + 2e^- \rightarrow 2Hg(l) + 2Cl^-$ | +0.2676 |
| $Cu^{2+} + 2e^- \rightarrow Cu(s)$ | +0.337 |
| $SO_4^{2-} + 8H^+ + 6e^- \rightarrow S(s) + 4H_2O(l)$ | +0.3572 |
| $VO^{2+} + 2H^+ + e^- \rightarrow V^{3+} + H_2O(l)$ | +0.359 |
| $[Fe(CN)_6]^{3-} + e^- \rightarrow [Fe(CN)_6]^{4-}$ | +0.36 |
| $H_2SO_3(aq) + 4H^+ + 4e^- \rightarrow S(s) + 3H_2O(l)$ | +0.450 |
| $Cu^+ + e^- \rightarrow Cu(s)$ | +0.521 |
| $I_2(s) + 2e^- \rightarrow 2I^-$ | +0.5355 |
| $MnO_4^- + e^- \rightarrow MnO_4^{2-}$ | +0.564 |
| $Hg_2SO_4(s) + 2e^- \rightarrow 2Hg(l) + SO_4^{2-}$ | +0.6151 |
| $Cu^{2+} + Br^- + e^- \rightarrow CuBr(s)$ | +0.640 |
| $Po^{2+} + 2e^- \rightarrow Po(s)$ | +0.65 |
| $[PtCl_6]^{2-} + 2e^- \rightarrow [PtCl_4]^{2-} + 2Cl^-$ | +0.68 |
| $O_2(g) + 2H^+ + 2e^- \rightarrow H_2O_2(aq)$ | +0.6824 |
| $[PtCl_4]^{2-} + 2e^- \rightarrow Pt(s) + 4Cl^-$ | +0.73 |
| $Fe^{3+} + e^- \rightarrow Fe^{2+}$ | +0.771 |
| $Hg_2^{2+} + 2e^- \rightarrow 2Hg(l)$ | +0.788 |
| $Ag^+ + e^- \rightarrow Ag(s)$ | +0.7991 |
| $Rh^{3+} + 3e^- \rightarrow Rh(s)$ | +0.80 |
| $2NO_3^- + 4H^+ + 2e^- \rightarrow N_2O_4(g) + 2H_2O(l)$ | +0.803 |
| $Cu^{2+} + I^- + e^- \rightarrow CuI(s)$ | +0.86 |
| $2Hg^{2+} + 2e^- \rightarrow Hg_2^{2+}$ | +0.920 |
| $NO_3^- + 3H^+ + 2e^- \rightarrow HNO_2(aq) + H_2O(l)$ | +0.94 |
| $NO_3^- + 4H^+ + 3e^- \rightarrow NO(g) + 2H_2O(l)$ | +0.96 |
| $Pd^{2+} + 2e^- \rightarrow Pd(s)$ | +0.987 |
| $[AuCl_4]^- + 3e^- \rightarrow Au(s) + 4Cl^-$ | +1.00 |
| $Br_2(l) + 2e^- \rightarrow 2Br^-$ | +1.0652 |
| $Br_2(aq) + 2e^- \rightarrow 2Br^-$ | +1.087 |
| $SeO_4^{2-} + 4H^+ + 2e^- \rightarrow H_2SeO_3(aq) + H_2O(l)$ | +1.15 |
| $ClO_4^- + 2H^+ + 2e^- \rightarrow ClO_3^- + H_2O(l)$ | +1.19 |
| $2IO_3^- + 12H^+ + 10e^- \rightarrow I_2(s) + 6H_2O(l)$ | +1.195 |
| $Pt^{2+} + 2e^- \rightarrow Pt(s)$ | ~1.2 |
| $ClO_3^- + 3H^+ + 2e^- \rightarrow HClO_2(aq) + H_2O(l)$ | +1.21 |
| $O_2(g) + 4H^+ + 4e^- \rightarrow 2H_2O(l)$ | +1.229 |
| $MnO_2(s) + 4H^+ + 2e^- \rightarrow Mn^{2+} + 2H_2O(l)$ | +1.23 |
| $2HNO_2(aq) + 4H^+ + 4e^- \rightarrow N_2O(g) + 3H_2O(l)$ | +1.29 |
| $Cr_2O_7^{2-} + 14H^+ + 6e^- \rightarrow 2Cr^{3+} + 7H_2O(l)$ | +1.33 |
| $Cl_2(g) + 2e^- \rightarrow 2Cl^-$ | +1.3595 |

| Reduction half-reaction | $\epsilon°$ (V) |
|---|---|
| $PbO_2(s) + 4H^+ + 2e^- \to Pb^{2+} + 4H_2O(l)$ | +1.455 |
| $Au^{3+} + 3e^- \to Au(s)$ | +1.498 |
| $MnO_4^- + 8H^+ + 5e^- \to Mn^{2+} + 4H_2O(l)$ | +1.51 |
| $2BrO_3^- + 12H^+ + 10e^- \to Br_2(l) + 6H_2O(l)$ | +1.52 |
| $Ce^{4+} + e^- \to Ce^{3+}$ | +1.61 |
| $2HClO(aq) + 2H^+ + 2e^- \to Cl_2(g) + 2H_2O(l)$ | +1.63 |
| $HClO_2(aq) + 2H^+ + 2e^- \to HClO(aq) + H_2O(l)$ | +1.645 |
| $Au^+ + e^- \to Au(s)$ | +1.691 |
| $H_2O_2(aq) + 2H^+ + 2e^- \to 2H_2O(l)$ | +1.776 |
| $Co^{3+} + e^- \to Co^{2+}$ | +1.808 |
| $Ag^{2+} + e^- \to Ag^+$ | +1.980 |
| $S_2O_8^{2-} + 2e^- \to 2SO_4^{2-}$ | +2.01 |
| $O_3(g) + 2H^+ + 2e^- \to O_2(g) + H_2O(l)$ | +2.07 |
| $F_2(g) + 2e^- \to 2F^-$ | +2.87 |
| $F_2(g) + 2H^+ + 2e^- \to 2HF(aq)$ | +3.06 |

### TABLE A5.2  Alkaline Media

| Reduction half-reaction | $\epsilon°$ (V) |
|---|---|
| $Ca(OH)_2(s) + 2e^- \to Ca(s) + 2OH^-$ | −3.02 |
| $Sr(OH)_2(s) + 2e^- \to Sr(s) + 2OH^-$ | −2.88 |
| $Ce(OH)_2(s) + 3e^- \to Ce(s) + 3OH^-$ | −2.87 |
| $Mg(OH)_2(s) + 2e^- \to Mg(s) + 2OH^-$ | −2.690 |
| $BeO(s) + H_2O(l) + 2e^- \to Be(s) + 2OH^-$ | −2.613 |
| $Al(OH)_3(s) + 3e^- \to Al(s) + 3OH^-$ | −2.30 |
| $U(OH)_4(s) + e^- \to U(OH)_3(s) + OH^-$ | −2.20 |
| $U(OH)_3(s) + 3e^- \to U(s) + 3OH^-$ | −2.17 |
| $H_2PO_2^- + e^- \to P(s) + 2OH^-$ | −2.05 |
| $SiO_3^{2-} + 3H_2O(l) + 4e^- \to Si(s) + 6OH^-$ | −1.697 |
| $Mn(OH)_2(s) + 2e^- \to Mn(s) + 2OH^-$ | −1.55 |
| $Zn(OH)_4^{2-} + 2e^- \to Zn + 4OH^-$ | −1.36 |
| $Cr(OH)_3(s) + 3e^- \to Cr(s) + 3OH^-$ | −1.34 |
| $Zn(OH)_2(s) + 2e^- \to Zn(s) + 2OH^-$ | −1.245 |
| $Te(s) + 2e^- \to Te^{2-}$ | −1.143 |
| $PO_4^{3-} + 2H_2O(l) + 2e^- \to HPO_3^{2-} + 3OH^-$ | −1.12 |
| $WO_4^{2-} + 4H_2O(l) + 6e^- \to W(s) + 8OH^-$ | −1.05 |
| $MoO_4^{2-} + 4H_2O(l) + 6e^- \to Mo(s) + 8OH^-$ | −1.05 |
| $In(OH)_3(s) + 3e^- \to In(s) + 3OH^-$ | −1.00 |
| $PbS(s) + 2e^- \to Pb(s) + S^{2-}$ | −0.93 |
| $SO_4^{2-} + H_2O(l) + 2e^- \to SO_3^{2-} + 2OH^-$ | −0.93 |
| $Se(s) + 2e^- \to Se^{2-}$ | −0.92 |
| $P(s) + 3H_2O(l) + 3e^- \to PH_3(g) + 3OH^-$ | −0.89 |
| $Fe(OH)_2(s) + 2e^- \to Fe(s) + 2OH^-$ | −0.877 |
| $2H_2O(l) + 2e^- \to H_2(g) + 2OH^-$ | −0.8281 |
| $Cd(OH)_2(s) + 2e^- \to Cd(s) + 2OH^-$ | −0.809 |
| $Co(OH)_2(s) + 2e^- \to Co(s) + 2OH^-$ | −0.73 |
| $Ni(OH)_2(s) + 2e^- \to Ni(s) + 2OH^-$ | −0.72 |
| $SbO_2^- + 2H_2O(l) + 3e^- \to Sb(s) + 4OH^-$ | −0.66 |
| $PbO(s) + H_2O(l) + 2e^- \to Pb(s) + 2OH^-$ | −0.580 |
| $TeO_3^{2-} + 3H_2O(l) + 4e^- \to Te(s) + 6OH^-$ | −0.57 |
| $Fe(OH)_3(s) + e^- \to Fe(OH)_2(s) + OH^-$ | −0.56 |

| Reduction half-reaction | $\epsilon°(V)$ |
|---|---|
| $S(s) + 2e^- \rightarrow S^{2-}$ | $-0.447$ |
| $Cu_2O(s) + H_2O(l) + 2e^- \rightarrow 2Cu(s) + 2OH^-$ | $-0.358$ |
| $TlOH(s) + e^- \rightarrow Tl(s) + OH^-$ | $-0.343$ |
| $CrO_4^{2-} + 4H_2O(l) + 3e^- \rightarrow Cr(OH)_3(s) + 5OH^-$ | $-0.13$ |
| $2Cu(OH)_2(s) + 2e^- \rightarrow Cu_2O(s) + H_2O(l) + 2OH^-$ | $-0.080$ |
| $Tl(OH)_3(s) + 2e^- \rightarrow TlOH(s) + 2OH^-$ | $-0.05$ |
| $MnO_2(s) + 2H_2O(l) + 2e^- \rightarrow Mn(OH)_2(s) + 2OH^-$ | $-0.05$ |
| $NO_3^- + H_2O(l) + 2e^- \rightarrow NO_2^- + 2OH^-$ | $+0.01$ |
| $SeO_4^{2-} + H_2O(l) + 2e^- \rightarrow SeO_3^{2-} + 2OH^-$ | $+0.05$ |
| $HgO(s) + H_2O(l) + 2e^- \rightarrow Hg(l) + 2OH^-$ | $+0.098$ |
| $PbO_2(s) + H_2O(l) + 2e^- \rightarrow PbO(s) + 2OH^-$ | $+0.247$ |
| $IO_3^- + 3H_2O(l) + 6e^- \rightarrow I^- + 6OH^-$ | $+0.26$ |
| $ClO_3^- + H_2O(l) + 2e^- \rightarrow ClO_2^- + 2OH^-$ | $+0.33$ |
| $Ag_2O(s) + H_2O(l) + 2e^- \rightarrow 2Ag(s) + 2OH^-$ | $+0.345$ |
| $ClO_4^- + H_2O(l) + 2e^- \rightarrow ClO_3^- + 2OH^-$ | $+0.36$ |
| $O_2(g) + 2H_2O(l) + 4e^- \rightarrow 4OH^-$ | $+0.401$ |
| $IO^- + H_2O(l) + 2e^- \rightarrow I^- + 2OH^-$ | $+0.485$ |
| $NiO_2(s) + 2H_2O(l) + 2e^- \rightarrow Ni(OH)_2(s) + 2OH^-$ | $+0.490$ |
| $MnO_4^- + 2H_2O(l) + 3e^- \rightarrow MnO_2(s) + 4OH^-$ | $+0.588$ |
| $BrO_3^- + 3H_2O(l) + 6e^- \rightarrow Br^- + 6OH^-$ | $+0.61$ |
| $BrO^- + H_2O(l) + 2e^- \rightarrow Br^- + 2OH^-$ | $+0.761$ |
| $ClO^- + H_2O(l) + 2e^- \rightarrow Cl^- + 2OH^-$ | $+0.89$ |
| $O_3(g) + H_2O(l) + 2e^- \rightarrow O_2(g) + 2OH^-$ | $+1.24$ |

# Answers to Selected Problems

**Chapter 1:** **2.** 0.1588 u; 0.8289%. **3.** 14.01 u. **5.** $^{35}$Cl, 75.9%; $^{37}$Cl, 24.1%. **10.** 1.3 ppt; CH$_3$OH. **12.** $2.56 \times 10^{-9}$ N. **15.** (a) 2.68; (b) 9.81; (c) $5.06 \times 10^{-3}$; (d) 0.114. **16.** (a) 70.0 g; (b) 77.0 g; (c) 3.59 g; (d) 0.621 g. **18.** $6.024 \times 10^{23}$. **20.** (a) 307 cm$^3$; (b) 13.5 g; (c) 55.7 g; (d) 52.9 g. **22.** 7. **23.** 9.16 g. **24.** $1.4 \times 10^2$ g/mol. **26.** CH$_2$O; C$_6$H$_{12}$O$_6$; 180.157 g/mol. **28.** C$_2$H$_5$Cl. **29.** N; NO$_2$Cl; 81.456 g/mol. **31.** 51.996 g/mol; Cr. **33.** KNO$_3$, 76.9%; (NH$_4$)$_3$PO$_4$, 23.1%. **34.** NaNO$_3$, 43.6%; Na$_2$SO$_4$, 56.4%. **35.** AgNO$_3$, 23.6%; Ag$_2$SO$_4$, 31.0%; AgClO$_4$, 45.4%

**Chapter 2:** **1.** 131 g. **3.** 10.6 g. **4.** 0.103 g. **6.** 2.53 g. **8.** 5.74 g. **9.** 1.97 g; 4.25 g NaCl. **10.** 33.1 g ZnI$_2$, 18.2 g Zn. **11.** 60.7 g. **12.** 9 g. **14.** 59%. **16.** 3.8 g. **18.** 13.7 g Ag, 11.3 g Ni. **20.** 7.922 g; 41.78%. **21.** 0.898 M. **22.** 0.319 g. **23.** 0.90 mL. **24.** 6.147 L. **26.** 0.992 g; [Na$^+$] = 0.140 M, [NO$_3^-$] = 0.0567 M, [SO$_4^{2-}$] = 0.042 M. **28.** 0.390 M. **29.** 5.19 mL. **30.** 41.4 mL. **32.** [Na$^+$] = 0.141 M, [NO$_3^-$] = 0.0820 M, [OH$^-$] = 0.0590 M. **33.** 0.506 M; 0.215 g. **35.** (b) 14ClO$_3^-$ + 3As$_2$S$_3$ + 18H$_2$O $\rightarrow$ 14Cl$^-$ + 6H$_2$AsO$_4^-$ + 9HSO$_4^-$ + 15H$^+$; (h) 3CuS + 8NO$_3^-$ + 11H$^+$ $\rightarrow$ 8NO + 3Cu$^{2+}$ + 3HSO$_4^-$ + 4H$_2$O. **36.** (d) Cr(OH)$_3^-$ + 2HO$_2^-$ $\rightarrow$ CrO$_4^{2-}$ + 2H$_2$O + OH$^-$; (h) 2MnO$_4^-$ + 3IO$_3^-$ + H$_2$O $\rightarrow$ 3IO$_4^-$ + 2MnO$_2$ + 2OH$^-$.

**Chapter 3:** **1.** 0.35. **2.** 0.75. **4.** 185 L. **6.** (a) CH$_2$; (b) C$_2$H$_4$, 28.054 g/mol. **8.** 628 K. **9.** 145 atm. **10.** $1.43 \times 10^{-3}$ mol. **13.** 5.83 g. **14.** 1.7 m. **15.** $2.34 \times 10^3$ g. **16.** $6.3 \times 10^{-3}$ mol/L. **18.** (a) 2.40 atm; (b) 3.8 atm, 1.0 atm. **20.** 46.0 g/mol, NO$_2$. **22.** C$_6$H$_4$Cl$_2$. **23.** 0.76. **24.** $P_{NO_2}$ = 1.02 atm, $P_{N_2O_4}$ = 0.18 atm. **26.** (a) $7.86 \times 10^{-6}$ atm; (b) $7.86 \times 10^{-6}$. **27.** 611 torr. **29.** $X_{O_2}$ = 0.207, $X_{CO_2}$ = 0.793; P = 3.77 atm. **30.** 0.0197 mol. **31.** 0.058 atm. **33.** (a) $1.8 \times 10^{-5}$; (b) 0.62

**Chapter 4:** **2.** (a) 1.86 kJ; (b) 0.78 kJ. **3.** 515 m/s; 3.72 kJ/mol. **4.** CO. **6.** (a) $5 \times 10^{21}$/s; (b) 2 min. **7.** 4.00 g/mol. **8.** (a) $Z_1 = 7.7 \times 10^9$/s, $\Lambda = 6.6 \times 10^{-8}$ m; (b) $Z_1 = 1.0 \times 10^5$/s, $\Lambda = 5.0 \times 10^{-3}$ m. **11.** $1.73 \times 10^{-9}$. **12.** $2.0 \times 10^{-4}$. **15.** $(kT/m)^{1/2}$. **18.** (a) 25.6 L; (b) 22.5 L. **19.** 26.3%. **20.** 0.0459 L. **21.** (a) 611 atm; (b) 427 atm. **24.** 0.375.

**Chapter 5:** **2.** (a) $5.0 \times 10^{-8}$ s; (b) 0.053 W/m$^2$; (c) $1.5 \times 10^{16}$. **4.** $6.0 \times 10^{-19}$ J **5.** $\nu = 1.30 \times 10^{15}$/s, $\lambda = 231$ nm. **6.** 88. **8.** $6.06 \times 10^3$ K. **9.** 17.0 keV. **10.** 3329, 3.865, 1. **11.** $\lambda_{3 \to 1} = 25.64$ nm, $\lambda_{2 \to 1} = 30.39$ nm, $\lambda_{3 \to 2} = 164.1$ nm. **12.** 54.5 eV, 123 eV, 218 eV. **14.** 3; n = 3, 4, 5. **16.** 2.1050 eV; 2.1029 eV; $\Delta E$ = 0.0021 eV. **19.** (a) $3.9 \times 10^{-11}$ m; (b) $9.1 \times 10^{-13}$ m. **20.** $\lambda = 3.26 \times 10^{-11}$ m. **21.** $2.4 \times 10^{-5}$ eV. **24.** (a) $1.385 \times 10^3$ m/s; (b) $5.3 \times 10^{-26}$ kg m/s; (c) 0.57%. **26.** 0.66 eV; 52 keV. **28.** (a) $4.2 \times 10^9$; (b) $4.8 \times 10^{-10}$. **29.** 2. **30.** 0.609, 0.196, 0.333. **34.** (a) 2a$_0$; (b) 6a$_0$. **38.** (a) 18; (b) 50.

**Chapter 6:** **4.** 0, 2, 0, 1, 2, 2, 5, 0, 0, 0. **7.** Mg$^{2+}$, Ca$^{2+}$, Mn$^{2+}$, Fe$^{2+}$, Ni$^{2+}$, Eu$^{2+}$. **17.** $1 \times 10^3$ kJ/mol. **18.** (a) 2.260; (b) 2.682. **19.** (a) 56.37 pm; (b) 49.76 pm. **21.** 87.528 pm. **22.** $4.2 \times 10^5$ kJ/mol. **23.** 35.3, 36.3. **24.** 0.17434 nm. **25.** Gd. **26.** 981.3 nm, 362.5 nm, 596.9 nm. **28.** 146.8 kJ/mol.

**Chapter 7:** **2.** 0.135. **3.** K = $K_1^{-1/2}K_2^2K_3^{3/2}$. **4.** (a) $1.1 \times 10^6$; (b) $9.5 \times 10^{-4}$. **7.** $K_p = 1.43 \times 10^{-2}$, K = $5.81 \times 10^{-4}$. **8.** 2.5 atm. **9.** $8.95 \times 10^{-3}$ atm. **10.** (a) 4.07, (b) 0.588. **12.** 1.03 atm. **14.** 0.56 atm. **15.** $X_{CO_2} = X_{H_2} = 0.400$, $X_{CO} = X_{H_2O} = 0.100$. **16.** $P_{N_2O_4} = 0.11$ atm, $P_{NO_2} = 0.72$ atm. **17.** 0.88 atm. **18.** (a) $3.33 \times 10^{-16}$; (b) $X_{SO_2} = 0.665$, $X_{O_2} = 0.332$, $X_{SO_3} = 1.97 \times 10^{-3}$, P = 7.91 atm. **20.** (a) 3.9 g/L; (b) 77 g/mol. **21.** (a) 0.0058; (b) $4.2 \times 10^{-4}$, (c) 2.8 atm. **22.** 0.24 atm. **24.** (a) 1.35 atm, (b) 0.13 atm, (c) 2.57 atm, (d) 0.904. **25.** $P_{PCl_5} = 0.79$ atm, $P_{PCl_3} = 1.20$ atm, $P_{Cl_2} = 0.00979$ atm, f = 0.00979. **32.** (a) $n_{NH_3} = 0.2$ mol, $n_{N_2} = 0.9$ mol, $n_{H_2} = 2.7$ mol; (b) $n_{NH_3} = 0.2$ mol, $n_{N_2} = 0.8$ mol, $n_{H_2} = 2.5$ mol. **33.** $2 \times 10^1$ atm.

**Chapter 8:** **4.** 2.456, 12.176, 11.431. **5.** 1.49. **6.** 148 mL. **7.** (a) 7.435; (b) pD < 7.435; (c) pD + pOD = 14.870. **10.** $[H_3O^+] = [OAc^-] = 6.54 \times 10^{-4}$ M, $[HOAc] = 0.0243$ M, $[OH^-] = 1.53 \times 10^{-11}$ M; 3.184; 2.62%. **12.** $2.64 \times 10^{-5}$ M. **14.** 34 g. **15.** (a) $4.6 \times 10^{-4}$; (b) 3.0%. **16.** 3.59. **17.** 2.77. **18.** (a) 0.036; (b) 55%. **20.** 2% low. **22.** 6.791. **24.** $[H_3O^+] = 3.0 \times 10^{-3}$ M, $[CH_3CH_2COO^-] = 4.5 \times 10^{-4}$ M, $[C_6H_5COO^-] = 2.6 \times 10^{-3}$ M, $[CH_3CH_2COOH] = 0.100$ M, $[C_6H_5COOH] = 0.117$ M, $[OH^-] = 3.3 \times 10^{-12}$ M. **27.** $2.3 \times 10^{-4}$ M. **28.** $9 \times 10^{-11}$. **30.** (a) 4.879; (b) 4.903; (c) 0.22 g. **31.** (a) no change; (b) 0.03; (c) −0.05. **32.** 79.0 mL; 8.781. **34.** 2.18-4.74. **35.** $4 \times 10^{-5}$. **37.** $1.3 \times 10^{-5}$. **39.** $[CO_2] = 9.23 \times 10^{-7}$ M, $[CO_3^{2-}] = 0.133$ M, $[HCO_3^-] = 0.0334$ M, $[OH^-] = 8.4 \times 10^{-4}$ M, $[H^+] = 1.2 \times 10^{-11}$ M, $[Na^+] = 0.333$ M, $[Cl^-] = 0.0333$ M. **40.** 9.7 g. **42.** 0.1319 M. **43.** (a) 9.23; (b) 6.27; (c) 4.754. **46.** $[H_3O^+] = 1.0 \times 10^{-5}$ M, $[C_2O_4^{2-}] = 0.129$ M, $[HC_2O_4^-] = 0.021$ M, $[H_2C_2O_4] = 3.7 \times 10^{-6}$ M, $[OH^-] = 1.0 \times 10^{-9}$ M. **47.** 0.77 L, 0.23 L. **49.** (a) 6.98, $5.9 \times 10^{-3}$; (b) 9.21, 0.50.

**Chapter 9:** **2.** $1.7 \times 10^{-4}$. **3.** (a) $5.3 \times 10^{-4}$ M; (b) $1.9 \times 10^{-6}$ M. **6.** $[Ag^+] = 3 \times 10^{-9}$ M, $[Cl^-] = 0.07$ M, $[Ba^{2+}] = 0.060$ M, $[NO_3^-] = 0.050$ M. **7.** $4 \times 10^{-4}$. **8.** $[Mg^{2+}] = 0.030$ M, $[PO_4^{3-}] = 2 \times 10^{-14}$ M. **9.** (a) $8.8 \times 10^{-10}$ M; (b) $4.7 \times 10^{-6}$ M. **11.** $3.3 \times 10^{-9}$ M. **12.** 0.075 M. **16.** (a) $2.0 \times 10^{-14}$ M; (b) 0.010 M. **18.** $1.4 \times 10^{-3}$ M. **20.** 0.12 M. **21.** (a) 3.3; (b) 8.5. **23.** (a) −6.8 to 0.3; (b) 0.5 to 1.5; (c) no. **24.** $2 \times 10^{-15}$ M; solubility increases by $10^3$. **25.** $5 \times 10^{-5}$ M. **26.** (a) $2.2 \times 10^{-10}$ M; (b) $6.3 \times 10^{-17}$ M. **27.** 4.7 mL. **29.** 7.18, 14.00.

**Chapter 10:** **2.** (a) $-1.63 \times 10^3$ J; (b) $-8.1 \times 10^2$ J. **4.** (a) $-5.6 \times 10^2$ J; (b) $-2.3 \times 10^3$ J. **6.** 36°C. **7.** 25.02°C. **8.** $w = -2.74 \times 10^3$ J, $q = 2.74 \times 10^3$ J, $\Delta E = 0$, $\Delta H = 0$. **9.** (a) 58.4 L; (b) 285 K; (c) $1.76 \times 10^3$ J, work on gas; (d) $-2.64 \times 10^3$ J; (e) −4400 J; (f) 20.75 J/mol K. **10.** q = 374 J, w = 0, $\Delta E$ = 374 J, $\Delta H$ = 623 J, $\Delta P$ = 0.11 atm. **12.** (a) $q = 2.5 \times 10^2$ J, w = 0; (b) $q = 4.2 \times 10^2$ J, $w = -1.7 \times 10^2$ J. **13.** $-9.5 \times 10^2$ J. **16.** 40°C. **17.** $5 \times 10^1$ g. **18.** $q = \Delta H = 61.8$ kJ; $\Delta E$ = 58.1 kJ, w = −3.7 kJ. **19.** −277.5 kJ/mol. **20.** −1192.49 kJ. **22.** −79.14 kJ/mol. **24.** (a) 55 kJ/mol, 232 kJ/mol; (b) −641 kJ. **25.** 5.4 g $H_2$, 4.6 g $CH_4$. **26.** −1266.0 kJ. **28.** (a) −14 kJ; (b) 28°C **29.** −33 kJ; 8.4 m. **31.** (a) 131.29 kJ; (b) 132.88 kJ; (c) −284.41 kJ. **32.** −52.98 kJ; −47.99 kJ. **34.** (a) 498.34 kJ; (b) 506.66 kJ.

**Chapter 11:** **1.** $\Delta S_{sys} = -2.72$ J/K, $\Delta S_{surr} = 6.76$ J/K, $\Delta S_{total} = 4.04$ J/K. **2.** (a) 0.50; (b) 0.22; (c) $2 \times 10^{-23}$. **4.** (a) $\Delta S_{sys} = -11.66$ J/K, $\Delta S_{surr} = 11.97$ J/K, $\Delta S_{tot} = 0.31$ J/K; (b) $\Delta S_{sys} = -11.66$ J/K, $\Delta S_{surr} = 11.7$ J/K, $\Delta S_{tot} = 0.0$ J/K; (c) $\Delta S_{sys} = -11.66$ J/K, $\Delta S_{surr} = 11.66$ J/K, $\Delta S_{tot} = 0$. **5.** 180.6 J/K. **6.** $\Delta S_{sys} = -22.0$ J/K, $\Delta S_{surr} = 22.0$ J/K, $\Delta S_{tot} = 0$; yes. **7.** $\Delta S_{sys} = -20.6$ J/K, $\Delta S_{surr} = 21.4$ J/K, $\Delta S_{tot} = 0.8$ J/K; no. **8.** 10.3 J/K. **15.** 87.11 J/mol K, 0. **17.** 11.5 kJ. **18.** $\Delta E$ = 529 J, $\Delta H$ = 731 J, $\Delta S$ = 1.69 J/K, $\Delta G = -3.93 \times 10^4$ J, q = 731 J, w = −202 J. **19.** (a) 40°C; (b) 0; (c) 0.24 J/K; (d) 0; (e) 0.24 J/K; (f) 73.0 J. **22.** $\Delta G° = 37.2$ J/K, $K = 3 \times 10^{-7}$. **24.** $\Delta G° = 74.4$ kJ, $K = 9 \times 10^{-14}$, P = 0.002 atm. **25.** $1.77 \times 10^{-10}$. **26.** $6.94 \times 10^{-4}$. **30.** 40 kJ. **31.** (a) $1.3 \times 10^5$, 38; (b) $9.1 \times 10^4$, $1.3 \times 10^{-11}$. **32.** 209.8 K; $3.6 \times 10^{14}$. **34.** −14.3 kJ, 17. **38.** $4.2 \times 10^{-6}$; $9.3 \times 10^{-6}$.

**Chapter 12:** **1.** 113.5 J/mol K, 0. **4.** (a) 0.26 atm; (b) 0.70 atm. **5.** 71°C. **6.** 1.6 kg. **8.** 58.1 torr. **9.** 35°C. **10.** 24.35 kJ/mol, −33.4°C. **12.** (a) 5.16 kJ/mol, 0.12 atm; (b) 79°C. **14.** 432 torr, 502 torr. **15.** 31.9 torr. **16.** 196 torr. **20.** M = 1.498 mol/L, X = 0.02857, m = 1.633 mol/kg. **21.** $X_{C_6H_5Cl} = 0.724$, $X_{C_6H_5Cl}^{vap} = 0.833$. **22.** $X_{tot} = 0.920$, $X_{tot}^{vap} = 0.968$. **24.** (a) $X_A = 0.667$; (b) $X_A = 0.800$. **25.** X = 0.0030, m = 0.166 mol/kg. **27.** $m_{O_2} = 2.7 \times 10^{-4}$ mol/kg, $m_{N_2} = 5.1 \times 10^{-4}$ mol/kg. **33.** (a) 0.735 torr; (b) 0.856 torr. **34.** $4.3 \times 10^2$ g/mol. **36.** $2.0 \times 10^3$ g/mol. **37.** 27 torr; 0.37 m. **38.** −0.136°C. **40.** (a) 81 g/mol; (b) 81.1°C. **41.** $3.69 \times 10^4$ g/mol; $2.95 \times 10^{-6}$. **42.** $4.2 \times 10^2$ torr. **44.** 156 g/mol, 1.95 K kg/mol.

**Chapter 13:** **4.** (b) 1.21 V. **7.** $2.0 \times 10^3$ atm. **8.** $\epsilon° = 0.462$ V, $\epsilon = 0.41$ V, $K = 4 \times 10^{15}$. **10.** 0.09 M. **12.** (b) 0.36 V; (c) $10^{16}$. **15.** (a) $\epsilon° = 0.70$ V, $\epsilon° = -0.40$ V; (b) $A + B^{2+} \to A^{2+} + B$, $\epsilon° = 1.10$ V. **17.** $\epsilon° = 1.10$ V, no. **20.** $1.78 \times 10^{-10}$. **21.** $10^{20}$. **22.** (a) 0.33 V; (b) $-0.21$ V. **23.** $9 \times 10^{-12}$. **24.** 0.854 V. **25.** $-0.34$. **26.** $-0.756$ V. **28.** (b) 2.433 V; (c) $\Delta G° = -7.042 \times 10^5$ J, $K = 3 \times 10^{123}$. **30.** 2.8, $Cl^-$, $-131.17$ kJ/mol; $Mg^{2+}$, $-456.0$ kJ/mol, $Al^{3+}$, $-481.1$ kJ/mol. **33.** 22.8 h. **34.** 2.74 g, 1.74 g; 17.0 min. **35.** (a) 0.0384 L; (b) 0.735 A. **36.** (a) 0.0715 F; (b) 19.2 min; (c) $[Ca^{2+}] = 0.150$ M, $[I^-] = 0.211$ M, $[OH^-] = 0.0894$ M.

**Chapter 14:** **2.** $d[HNO_2]/dt = -4.5 \times 10^{-4}$ mol/L s. **5.** $3.4 \times 10^6$ mol/L s. **7.** (a) NO order = 2, $Cl_2$ order = 1, overall order = 3; (b) 10.0 $L^2/mol^2$ s; (c) $8.00 \times 10^{-7}$ mol. **8.** $l = 1$, $m = 1.5$, $n = -1$. **9.** 8.65. **11.** 7.0 h. **12.** 0.160 mol/L. **13.** 58 min, 0.012/min. **14.** 12 h. **16.** $2^{nd}$ order, $1.00 \times 10^{-4}$/torr s, 3 h. **18.** 12 min. **20.** $R = k[N_2O_3]^2$, $k = 4.87 \times 10^{-4}$ L/mol s. **21.** (a) second order, $4.50 \times 10^{-3}$ L/mol s; (b) 0.3 y. **22.** (a) 0.10 mol/L; (b) 0.13 mol/L; (c) 0.00 mol/L. **23.** 38.0 min. **24.** (a) 0.98 L/mol s; (b) 0.011 mol/L. **27.** $R = k[H_2][NO]^2$, $k = 0.012$ $L^2/mol^2$ s. **36.** $1.17 \times 10^3$ L/mol s. **37.** $7.1 \times 10^9$ L/mol s. **38.** 53 kJ/mol. **39.** $1.8 \times 10^5$ J/mol, $6.2 \times 10^{10}$ L/mol s. **40.** (a) $1.9 \times 10^{-3}$; (b) $2.2 \times 10^{-2}$. **41.** 1.4 y. **42.** 2.62. **44.** (a) 388 K; (b) 319 K. **46.** $1.39 \times 10^{11}$ L/mol s. **48.** (a) $2.4 \times 10^4$; (b) $3.1 \times 10^2$.

**Chapter 15:** **9.** (a) acceptor; $FC_P = -1$, $FC_N = 1$; (b) donor; $FC_P = 1$, $FC_B = -1$. **11.** $ON_{Cl} = +3$, $ON_O = -2$; $FC_{Cl} = +1$, $FC_O = -1$. **24.** 0.94 D. **26.** 0, 2.72 D; m, 1.57 D; P, O.

**Chapter 16:** **5.** 64.4%. **10.** (a) 2.2%; (b) 22%. **12.** (a) 22%; (b) 1.5 D; (c) 115°. **16.** (a) 0.78; (b) 0.17; (c) 2.3.

**Chapter 17:** **1.** $-243$ kJ/mol. **2.** (a) 9 kJ; (b) $-146$ kJ; (c) $-34$ kJ; (d) $-56$ kJ. **4.** (a) 92 kJ; (b) $-103$ kJ; (c) 492 kJ; (d) $-104$ kJ. **6.** $-278$ kJ; $-282.984$ kJ. **7.** 192 kJ/mol. **8.** (a) $-810$ kJ; (b) $-1251$ kJ; (c) $-657$ kJ; (d) $-243$ kJ; $H_2$. **9.** 170 kJ. **11.** (a) $8.3 \times 10^{-25}$ g; (b) $1.11 \times 10^{-24}$ g; (c) $1.58 \times 10^{-24}$ g; (d) $1.61 \times 10^{-24}$ g. **12.** 1.604 Å. **14.** 10.5 $cm^{-1}$. **15.** $3.04 \times 10^{-3}$ m, $1.52 \times 10^{-3}$ m, $1.01 \times 10^{-3}$ m. **16.** (a) $2.18320 \times 10^{-45}$ kg $m^2$. (b) 2.66974 Å. **19.** 408.7 N/m, 0.1643 eV. **20.** $5.652 \times 10^{13}$/s. **21.** (a) 432.0 kJ/mol; (b) 552 N/m; (c) 439.6 kJ/mol. **22.** (a) 557 $cm^{-1}$; (b) 3.24 kJ/mol; (c) 239 kJ/mol. **24.** 1903 N/m. **25.** 4.27 Å. **26.** 328 nm. **27.** 4; $C_8H_{10}$.

**Chapter 18:** **1.** 2.51 Å. **2.** 332 kJ/mol. **3.** 398 kJ/mol. **8.** 1.80 Å. **9.** $6.5 \times 10^{-4}$. **11.** (a) $-0.0045$ eV, attract; (b) 1.66 keV, repel. **26.** (a) $-1100$ kJ/mol; (b) $-250$ kJ/mol; (c) $-3600$ kJ/mol.

**Chapter 19:** **1.** 231 g. **6.** $-3$. **7.** 429 kg. **8.** 2, 7. **11.** 736 atm. **13.** (a) $-22,700$ kJ; (b) $-11,800$ kJ. **16.** (b) 4.7. **17.** (a) $-118.08$ kJ/mol; (b) $-125.9$ kJ/mol. **21.** $10^{-5}$. **24.** (a) $-1.31$ V; (b) $-1.18$ V; $P_4$, 1.16 V; $H_2PO_2^-$ $\epsilon° = 0.39$ V; (d) $HPO_3^{2-}$, $\epsilon° = 0.29$ V; $H_2PO_2^-$, $\epsilon° = 0.74$ V; $P_4$, $\epsilon° = 1.22$ V; $PH_3$, $\epsilon° = 0.06$ V. **25.** $10^{25}$. **26.** $10^{57}$. **30.** (a) $-0.06$ V; (b) $-0.48$ V; (c) $-0.89$ V. **32.** (a) 1.59 V; (b) 0.69 V. **37.** $10^{15}$. **38.** (d) $\epsilon° = 0.8$ V, $K = 10^{27}$. **40.** (b) 0.118 M. **41.** (a) $-0.53$ V, 0.82 V, $-0.29$ V. **43.** 8. **50.** (b) 0.36 V; (c) $10^{36}$.

**Chapter 20:** **1.** 107.8 g/mol. **3.** (a) $2.86 \times 10^{-4}$ J/mol K; (b) 10.3 J/mol K; (c) 22.4 J/mol K; (d) 24.6 J/mol K; 24.942 J/mol K. **4.** 1.71 g/$cm^3$. **5.** 0.524. **6.** 8.90 g/$cm^3$. **8.** (a) $3.525 \times 10^{-8}$ cm; (b) 1.246 Å. **9.** 0.725 Å. **11.** 1.33 Å. **13.** 3.58 Å. **15.** 0.155. **19.** (a) 10.3 $cm^3$/mol; (b) 0.772 g/$cm^3$. **23.** $Al_{16}Mg_8O_{32}$. **24.** $CaTiO_3$. **26.** 0.29. **27.** 0.80. **28.** $-3.12 \times 10^3$ kJ/mol. **31.** 59 kJ/mol. **32.** $-2.70 \times 10^3$ kJ/mol. **33.** (a) $-867$ kJ/mol; (b) $-348$ kJ/mol. **35.** (a) $-1.04 \times 10^3$ kJ/mol; (b) $-1046.6$ kJ. **36.** $-909$ kJ. **38.** 3.55 g/$cm^3$. **40.** (a) 237 kJ/mol; (b) 437 kJ/mol; (c) 200 kJ/mol.

**Chapter 21:** **7.** $2 \times 10^{-6}$ M. **8.** $10 Cr^{3+} + 6 MnO_4^- + 11 H_2O \to 5 Cr_2O_7^{2-} + 6 Mn^{2+} + 22 H^+$; $10^{91}$. **10.** (b) $1 \times 10^{57}$, 2; (c) 0.57 V. **13.** $-1.56$ V. **14.** $10^{36}$. **19.** (a) 4; (b) 6; (c) 6; (d) 6. **26.** (a) 1; (b) 1; (c) 2; (d) 6. **28.** 3. **32.** (b) $-150$ kJ/mol. **36.** (a) $1.17 \times 10^3$ nm, colorless; (b) 927 nm, colorless. **41.** $2 \times 10^{-29}$.

**Chapter 22:** **8.** 3 mol % p, 43 mol % iso-p, 54 mol % neo-p. **23.** (a) no; (b) no; (c) no; (d) yes. **24.** 59% (a) yes; (b) yes; (c) no; (d) no. **30.** (b) $R = Kk_2 [CH_3CH(OH)_2][HA]$. **33.** (a) $H_3NCHRCOOH^+$; (b) and (c) $H_3NCHRCOO$; (d) $H_2NCHRCOO^-$. **35.** 6.07 **37.** 1.8 nm, $5.4 \times 10^6$ base pairs.

**Chapter 23:** **2.** (a) 7.614 MeV/nucleon; (b) 0.8107%. **3.** 232.048 u. **4.** (b) 3.2 MeV; (c) 3.2 MeV; (d) 1.9 MeV; (e) 2.7 MeV; 0.50 MeV; 2.5 pm. **11.** (a) 6.41 MeV, yes; (b) −7.3 MeV, no; (c) 28.59 MeV, yes; (d) −0.87 MeV, no. **12.** (a) 22.14 MeV; (b) 12.10 MeV; (c) 57.72 MeV. **14.** 7.276 MeV. **16.** 8.0052072 u. **18.** (a) −14.10 MeV; (b) 9.61 MeV, no. **20.** 16.1 MeV. **21.** (a) proton; (b) 21.3 MeV. **22.** 32.971726 u. **23.** −6.3 MeV. **24.** 146.2 MeV. **25.** (a) $1.3 \times 10^6$; (b) $2.4 \times 10^6$/s. **26.** (a) $1.78 \times 10^{-14}$ m; (b) $1.37 \times 10^{-14}$ m. **28.** $^{82}$Br. **30.** (a) $2.2 \times 10^3$ dpm; (b) $7.4 \times 10^7$. **31.** $1.9 \times 10^{-19}$ g. **32.** $7.91 \times 10^{10}$ y. **34.** 1.71 mL. **35.** $9.90 \times 10^4$ dpm. **36.** $2.53 \times 10^9$ y. **37.** $1.05 \times 10^4$ y.

# Index

Absolute temperature, 73
Absolute zero, 72
Absorption spectrum, 635
Acceleration of gravity, 10, 68
Acceptor impurities, 760
Accuracy vs. precision, 27
Acetylene, bonding in, 563, 620
Acid ionization constants, 255–257
  table, 255, 931
Acid-base conjugate pairs, 253
Acid-base indicators, 278–279
Acid-base titrations, 276–285
Acid rain, 87, 259
Acid strength, 255
Acidic and basic oxides, 699–703
Acidic and basic solvents, 256–257
Acids,
  Arrhenius model of, 41, 250–252
  Brønsted-Lowry model of, 252–254
  leveling effect of water on strength of, 257
  Lewis model of, 254, 262
  polyprotic, 286–295
  polyprotic, ionization constants, table, 287
  strength of, 255
  strong distinguished from weak, 256
  stronger than hydronium ion, 256
  weak, 41–42
Actinide and transactinide elements, table, 905
Actinides, 191
Activated complex, 536–537
Activation analysis, 900
Activation-controlled reactions, 542
Activation energy, 530, 532–533
  table, 533
Adabatic process, 335
Addition compounds (adducts), 560
Air, composition of, 82
  composition of, table, 82
Air pollution, 86–88
Alcohols, 840–842
  hydrogen bonding in, 841
  nomenclature of, 840
Aldehydes, 842–843
Aliphatic compounds, 829
Alkali metal hydrides, 694–695
Alkali metals, 192, 215
  chemistry of, 711–716
Alkaline dry cell, 481–482
Alkaline earth hydrides, 694–695
Alkaline earth metals, 192, 215
  chemistry of, 711–716

Alkanes, 364, 829, 830–832
  branched chain isomers of, 832
  dehydrogenation of, 834
  halogenation of, 834–835
  nomenclature of, 832–834
  reactions of, 834–835
  straight chain isomers of, 832
Alkenes, 829, 835–837
  polymerization of, 838
  reactions of, 838
Alkoxide ion, 842
Alkyl groups, table, 833
Alkynes, 829, 835–837
Allotrope, 85
Alloys, 441, 790, 791
  table, 791
Alpha decay, 893
Alpha helix, in proteins, 865–866
Alpha particles, 127, 893
  scattering by metal foils, 127–129
Aluminothermic reactions, 717
Aluminum, 215, 717
  anodized, 488
  electrolytic production of, 492
Amalgam, 482
Amino acids, 861–863
  table, 862
Ammonia,
  as base, 695–697
  basicity constant of, 260–261
  bond angles of, 580
  complexes with metal ions, 308–311
  dipole moment of, 575
  production of, 245
  reaction with boron trifluoride, 560
Ammonia fountain, 687
Amorphous solids, 739–740
Amphiprotic species, 253
Amphoteric oxides and hydroxides, 327–328
Amplitude of wave, 130
Angstrom unit, 4
Angular function, 162, 169–171
Angular momentum, quantized, 163, 638
Angular nodes, 171
Anhydrides, 702
Anion exchange column, 690
Anion water, in hydrates, 684
Anions, 16, 458
Anode, 457
Antibonding molecular orbitals, 601–603
Antifluorite crystal structure, 763
Antiparticles, 888

Approximations in equilibrium calculations, 230–236
  method of successive, 234
Approximations, mathematical, 923–924
Aqua regia, 722
Aromatic compounds, 838–839
Arrhenius equation, 530–533
  parameters of, table, 542
Arrhenius model of acids and bases, 41, 250–252
Arrhenius, Svante, 250
Atmosphere, 68–69, 82–85
Atom, 3
  ground state of, 143
  excited states of, 143
  structure of, 4–6
Atomic constituents, mass and charge of, table, 6
Atomic emission spectra, 141–148
Atomic mass unit, 7
Atomic number, 6, 877
Atomic orbitals, 165–171
Atomic radii, 209
  table, 210
Atomic radius, 4
Atomic spectrum, of hydrogen, 143–148
Atomic volume per mole, 208–209
Atomic weight, 8–9
Atoms, multielectron, energy levels of, 179–182
Atomization, enthalpy of, 630
Aufbau of periodic table, 184–190
Average translational kinetic energy per molecule, 99
Avogadro, Amedeo, 15, 66
Avogadro's law, 66–67, 77
Avogadro's number, 20–21
Azeotrope, 438–439
Azimuthal quantum number, 163

Baking soda, 293–294
Balance, 9
Balancing chemical equations, 35, 45–46, 55–59
Balmer series, 144
Band spectra, 636
Band theory of solids, 754–757
Bar, unit of pressure, 70
Barometer, 68–70
Barometric formula, 83–84
Barrier,
  Coulomb, 894–895
  potential energy, 536
Bartlett, Neil, 734

# Index

Base ionization constants, 259–261
  table, 261, 932
Bases,
  Arrhenius model of, 41, 250–252
  Brønsted-Lowry model of, 252–254
  leveling effect of water on strength of, 257
  Lewis model of, 254, 562
  weak, 41–42
Batteries, 480–485
Becquerel, Henri, 876
Bends, from deep-sea diving, 438
Bent molecules, 571
Benzene,
  bonding in, 568–569, 620–623
  molecular orbitals of, 621–622
  resonance energy of, 633–634
  resonance in, 568–569
Beta decay, 884
Beta minus decay, 886–888
Beta minus particle, 886
Bimolecular reactions, 518
Binding energy, of nucleus, 878–879
Binding energy per nucleon, 880–881
Black-body radiation, 135–137
Body-centered cubic (bcc) unit cell, 745–747
Bohr frequency condition, 143
Bohr, Niels, 143
Bohr radius, 165
Bohr theory of atomic hydrogen, 145–147
Boiling point,
  effect of dipole-dipole forces on, 669–670
  effect of hydrogen bonding on, 673–674
  effect of London dispersion forces on, 669–670
Boiling point elevation, 444–448
  for electrolytic solutions, 448
Boiling point elevation constant, 447
  table, 447
Boltzmann, Ludwig, 92, 380
Boltzmann's constant, 99
Bomb calorimeter, 351
Bond angle, 571
Bond energies, 553, 612, 628–634
  and bond order, 563
  and reaction enthalpy, 632–633
  average, 630–631
  average, table, 631
  of diatomic molecules, 628–629
  of polyatomic molecules, 630–632
  table, 629, 662
Bond force constant, 642–645
  table, 645

Bond lengths, 553, 612, 640–641
  and bond order, 563
  average, table, 640
  table, 629, 662
Bond moment, 575
  table, 576
  vector addition of, 577–578
Bond order, 563, 568, 603, 612
  and bond lengths and energies, 563
Bonding molecular orbitals, 601–603
Boranes, 623–625
Born, Max, 661
Born-Haber cycle, 661, 774–776
  data, for hydrohalic acids, table, 699
  data, for metals, table, 714
Boron, 215
Boron hydrides, bonding in, 623–625
Boundary conditions of Schrodinger equation, 153
Boyle, Robert, 70–71
Boyle's law, 70–72
  derivation from kinetic theory, 94–98
Brackett series, 144
Brønsted-Lowry model of acids and bases, 252–254
Brownian motion, 451
Buffers, pH range of, 274–276
  figure, 275
Buffer solutions, 269–276
Build-up (Aufbau) of periodic table, 184–190
Buret, 44

Calorie, 342–343
Calorimeter, 350–351, 353
Cannizzaro, Stanislao, 15
Carbanion, 850
Carbocation (carbonium ion), 850
Carbohydrates, 856
Carbon-carbon bond lengths, 563
Carbon dioxide,
  aqueous solution of, 287–290
  dipole moment of, 576
  phase diagram of, 424
Carbon, phase diagram of, 778
Carbonyl group, 842
Carboxyl group, 843
Carboxylate ion, 844
Carboxylic acids, 843–845
Catalysis, 543–546
  heterogeneous, 544–546
  homogeneous, 544–545
Catalyst, 54, 244
Cathode, 457
Cathode rays, 124–125
Cathodic protection, 487–488

Cation exchange column, 690
Cation hydrolysis, 325–326
Cations, 16, 458
Cell voltage (potential), 460
Cellulose, 858–859
Celsius temperature scale, 73
Central atom, and Lewis structures, 556
Cesium chloride crystal structure, 761
Chain reaction, 85, 526–528
  explosive, 527–528
Chalcogens, 192
Charge and mass of atomic constituents, table, 6
Charge balance equation, 268, 292, 324
Charge conservation, in balanced equations, 45–46
Charge density, of ions, 325–326, 714
Charge-to-mass ratio of electron, 124–125
Charles' (Gay Lussac's) law, 72–73
Chelate, 797
Chelating agents, table, 798
Chemical bond, 11, 553
Chemical bonding and melting points, 217–218
Chemical equations, balancing of, 35, 55–59
Chemical equilibrium, 222–224
  approximations in calculations of, 230–236
  factors affecting position of, 240–245
  for real gases, 245–246
  involving acids and bases, 261–295
  involving amphiprotic species, 291–295
Chemical properties, 14
Chemical reactions,
  classification of, 48–51
  conservation of charge in, 45–46
  conservation of energy in, 47
  conservation of mass in, 33–35
  rate of, 47, 502
Chemiluminescence, 653
Chemisorption, 545
Chirality, 848
Chlorine,
  chemistry of, 727–733
  production of by electrolysis, 492
Chlorofluorocarbons, 85
Chlorophyll, 825
Chromium, chemistry of, 794–795
Cis-trans isomerism, in alkenes, 836–837
Cis-trans isomers, 802
Clausius, Rudolf, 92, 379

Clausius-Clapeyron equation, 420
Closed electron shell, 184
Closed-shell nuclides, 884
Closest-packed structures, 748–752
Coal, 364–365
Coherent light, 653
Coinage metals, 791–793
Colligative properties, 444–450
Collision frequency, 103, 528–529
Collision rate constant, 529–530
Colloid, 451
Color, and electronic transition energy, table, 808
Color of transition metal complexes, 806–808
Combination reaction, 48
Combined gas law, 73–74
Combining power, 216
Combustion, 48
  enthalpy of, 364–366
Common ion effect, 304–305
Complex formation constants, table, 311
Complex ions, 307
  equilibria of, 307–312
Composition of air, table, 82
Compound nucleus, 902
Compound, 13
Compressibility factor, 110
Compton effect, 140–141
Concentration units for solutions, 42–43, 426
Concentrations cells, 468–470
Condensation, 113–115
Condensed formula, 830
Conduction band, 757–758
Conformation, of proteins, 865
Conjugate acid-base pair, 253–254
Conjugated molecules, 620
Conservation of charge in chemical reactions, 45–46
Conservation of energy, 47
Conservation of mass-energy, 880
Conservation of mass in chemical reactions, 33–35
Constant volume process, 350–352
Constructive interference, 133
  of atomic wavefunctions, 593–594, 600
Conversion factors and units, 918–920
Cooling curve, 442
Coordinate covalent (or dative) bonds, 560–562, 796
Coordination complexes,
  reactions of, 821–823
  structure of, table, 800
  valence bond theory of, 808–809
Coordination compounds, rules for naming, 798–800
Coordination number, 796

  of lattice, 745
Coordination sphere, 797
Coordination water, in hydrates, 684
Copper sulfate pentahydrate, structure of, 684–685
Corresponding states, law of, 115–117
Corrosion, 485–488
Cosmic rays, 910
Cotton, F. A., 820
Coulomb, unit of charge, 6
Coulomb barrier, 894–895
Coulomb, Charles, 552
Coulomb force, 17, 101, 660
Coulomb's law, 17
Counter ions, in coordination compounds, 797
Covalent bond, 11, 553
Covalent radii, 640
Crewe, Albert, 4
Crick, Francis, 867
Critical point, 115
Critical point parameters, table, 115
Crystal defects, 768–770
Crystal field splitting,
  of octahedral complexes, 809–813
  of square planar complexes, 818
  of tetrahedral complexes, 817–818
  table, 815
Crystal field stabilization energy, 811
Crystal field theory, 809
Crystal lattice energy, 771–773
  table of parameters, 773
Crystal structure,
  of cesium chloride, 761
  of ionic compounds, table, 763
  of sodium chloride, 762
Crystal systems, 744–745
  table, 745
Crystalline solids, 738–739
Cubic closest packing, 750–751
Cubic crystal system, 745–748
Cubic lattices, table of properties, 746
Cubic sites, in ionic crystals, 766
Cubic unit cell, 745–748
Curie, Marie, 876
Curie, Pierre, 876
Cyclic process, 347
Cycloalkanes, 829

Dalton, John, 3, 14
Dalton's law of partial pressures, 77–82, 427–429
Daniell cell, 457–458
Davisson, Clinton J., 149
De Broglie, Louis, 148
De Broglie wavelength, 148–150

Debye (D), unit of dipole moment, 574
Debye, Peter, 574
Decay constant, 907
Decay energy, 886
Decomposition reaction, 48
Defects in crystals, 768–770
Definite proportions, law of, 14
Deionized water, 690
Degeneracy, 165
Degree of dissociation, 237
Degree of ionization, of weak acid or base, 261–262
Degrees of freedom of a system, 425
Dehydrogenation, of alkanes, 834
Delocalized electrons, 568
Delocalized orbitals, 620–623
Delta bond, 820–821
Democritus, 3
Dempster, Arthur, 16
Density, 10
  nuclear, 878
  of elements, 10
  of elements, table, 10
Derivatives, table, 926
Derived SI units, table, 919
Destructive interference, 133
  of atomic wave functions, 593–594, 600
Deuterium, 6
Dextrorotatory enantiomer, 849
Diagonal relationships, in periodic table, 713–714
Dialysis, 451
Diamagnetism, 558
Diamond,
  bonding in, 777
  structure of, 777
Diastereomers, 857
Diatomic molecules (homonuclear), electronic configuration of, table, 610
Diborane, 623–625
Dielectric constant, 682
Differentiation, 924–927
Diffraction, 133
Diffusion, 102
Diffusion-controlled reactions, 541
Dipole-dipole interactions, 665–666
  effect on boiling point, 669–670
Dipole-induced dipole interaction, 666
Dipole moments, 573–578
  induced, 664–665
  of diatomic molecules, 573–575
  of polyatomic molecules, 575–578
  table, 575
  units of, 574
Disintegration rate, 907
Displacement reaction, 49

Disproportionation, 59
Dissociation and formation constants, of complexes, 308
Dissociation constants of complexes, table, 933–934
Dissociation energy, of diatomic molecule, 628
Dissociation equilibria, 237–240
Distillation, 433
 fractional, 433–434
DNA (deoxyribornucleic acid), 867–872
 replication of, 871
 transcription of, 871
Dobereiner, J., 176
Donor atoms, in coordination compounds, 796
Donor impurities, 760
Doping, of semiconductors, 759
Double bond, 562–563
Double displacement reaction, 50
Downs cell, 491–492
Drip lines, neutron and proton, 892–893
Dry cell, 481
 alkaline, 481–482
Dry ice, 416
Ductility, of metals, 754
Dulong and Petit, law of, 344, 740–741

Eclipsed versus staggered configuration of atoms, 820
EDTA titrations, 823
Effective nuclear charge, 181–182, 194–195
Effusion of gases, 101–103
Einstein, Albert, 138, 344, 741, 879
Elastic collisions, 94
Electric current, 460
Electrical potential difference, 460
Electrical work, 460
Electrochemical cell, 489
Electrode, 457
 calomel, 485
 gas, 461–462
 glass, 486
 inert, 461
 ion-selective, 485
 metal/insoluble metal salt, 462–463
 reference, 470
Electrode potential diagram, 719–721
 of chromium, 795
 of copper, 792
 of halogens, 729
 of manganese, 795
 of mercury, 793
 of nitrogen, 719

 of phosphorus, 723
 of sulfur, 726
Electrolysis, 488–495
 Faraday's laws of, 490–491
 of aqueous solutions, 493–494
 of water, 490
 table of products, 494
Electrolytes, 42
 distinction between strong and weak, 250–251
Electrolytic cell, 489
Electrolytic refining, 494–495
Electromagnetic radiation, 129–133
Electromagnetic spectrum, 133–135
 optical region of, 134
 table, 135
Electromotive force (EMF), 460–461
Electron, 5–6, 124–126
 charge/mass ratio, 124–125
 delocalized, 568
 mass, 126
 solvated, 716
Electron affinity, 204–206
 table, 193
Electron capture decay, 889
Electron deficient molecules, 557–558
Electron diffraction, 149
Electron exchange, in valence bond theory of $H_2$, 590–591
Electron octet, 554
Electron pairing, 182
Electron-sea model, of metals, 753–754
Electron shells, 184
 closed, 184
 inner, 184
Electron subshells, 184
Electron volt (eV), 139–140
Electron-pair acceptor, 254, 266
Electron-pair donor, 254
Electronegativity, 40, 206–208, 213, 553, 597
 and bond type, 218, 597–598
 table, 597
Electronic charge, 125–126
Electronic configuration,
 of elements, 184–190
 of elements, table, 187
 of homonuclear diatomic molecules, 608–612
 of homonuclear diatomic molecules, table, 610
Electronic transition energy, and color, table, 808
Electrophile, 851
Electroplating, 494
Element, 2
Elementary reaction, 500, 517, 519
Elemental composition of Earth, table, 3

Elements, electronic configuration of, table, 187
EMF (electromotive force, cell potential), 460–461
EMF series, 472–475
Emission spectrum, 635
Empedocles, 2
Empirical formula, 24
Enantiomers, 804, 848
 dextrorotatory, 849
 levorotatory, 849
Encounter (between molecules), 540
Endothermic reaction, 47, 359
Endpoint, 60, 227–228
Energy band, 756
Energy levels,
 of hydrogen atom, 146
 of molecules, 636
 of multielectron atoms, 179–182
 population inversion of, 654
Energy, types of, 332
Enthalpy, 352
 of atomization, 630
 of combustion, table, 365
 of formation, standard molar, table, 362
 of fusion, 359, 415–416
 of phase transitions, table, 360, 414, 415
 of vaporization, 359, 415–416
Entropy, 374
 molecular interpretation of, 379–381
 of fusion, 415–416
 of vaporization, 415–416
 of vaporization and fusion, table, 415
 standard, table, 388
Entropy change,
 in chemical reaction, 389–390
 in gas expansion, 374–379
 in phase transition, 382–385
 with change in temperature, 382
Enzymes, 543, 859
Equation of state, 75
Equatorial and axial sites, in trigonal bipyramidal geometry, 572–573
 placement of lone pair electrons, 580–582
Equilibrium constant, 224–230
 and rate constants, 519
 and standard cell potential, 465
 for gas-phase reactions, 226–227
 for heterogeneous reactions, 225
 Kp, table, 230
 magnitude of, 229–230
 temperature dependence of, 400–403
 units of, 227
Equivalence point, 60, 276

ESCA, 204
Esters, 845
Ethanol, 365, 842
Ethers, 842
Eutectic point, 441
Eutriphication, 689
Even-even nuclides, 883
Excitation functions, 902
Excited states of atom, 143
Excluded volume of gas molecules, 117–118
Exclusion principle, 182–183, 608
Exothermic reaction, 47, 359
Expanded valence shells, 558–560
Exponentials, 921
Extensive properties, 10

F-center, 769
Face-centered cubic (fcc) unit cell, 745–747
Facial isomers, 803
Faraday, Michael, 490
Faraday, unit of charge, 460
Faraday's laws of electrolysis, 490–491
Ferromagnetic metals, 789
First ionization energy, 192–197
First law of thermodynamics, 344–350
First-order kinetics, 508–511
Fission, nuclear, 900–901
Fluorescence, 650
Fluorescent dyes, 650
Fluorite crystal structure, 762
Force constant, 642
Formal charge, 564–566
  and strength of oxoacids, 707–709
Formation and dissociation constants, of complexes, 308
Formation constants, of complexes, table, 311
Formula weight, 17
Fractional distillation, 433–434
Free electron model, 649–650
Free energy (Gibbs), 390–392
Free energy change,
  and cell potential, 464–465
  as criterion of spontaneity, 390–391
  in chemical reactions, 392–394
  in phase changes, 391–392
  makeup of, 392–393
  variation with temperature, 403–405
Free radical, 86
Free-radical chain reaction, 526–527
Free radical intermediate, 850
Freezing point depression, 444–448
Freezing point depression constant, 447

Freon, 85
Frenkel defect, 769
Frequency factor, 530
Frequency of wave, 131
Fuel cells, 480, 484–485
Functional groups, 840
  table of, 841
  redox reactions of, 846–848
Fundamental SI units, table, 918

Galvani, Luigi, 456
Galvanic (voltaic) cells, 456–463
  line notation for, 458
Gamma-rays, 134
Gas constant, 74–75
Gases, molar volume at STP, table, 66
Gay-Lussac, Joseph, 66
Genetic code, 871–872
Geometrical isomers, 802–803
Germer, L. H., 149
Gibbs, J. Willard, 425
Glass, 740
Glucose, 856
Graham, Thomas, 102
Graham's law of effusion, 102
Graphite,
  bonding in, 777–778
  structure of, 777–778
Gravity, acceleration of, 10
Greenhouse effect, 87–88
Ground state of atom, 143

Haber, Fritz, 661
Half-life, 509
  of first-order reaction, 509
  of radioactive nuclei, 511
  of radioisotopes, 907
  of second-order reaction, 511–512
Half-reactions, 55–58
Halides, 730–731
Hall-process electrolysis cell, 492
Halogenation, 728
  of alkanes, 834–835
Halogens, 192, 216
  chemistry of, 727–733
  table of properties, 728
Hard water, 688–689
Harmonic oscillator potential, 642–643
Heat, 332–335
Heat capacity, 343–344
  at constant pressure, 353
  at constant volume, 352
  molar, 344
  molar, table, 354
  of solids, 740–742
  relationship between $C_p$ and $C_v$, 356–357

Heavy elements, naturally occurring isotopes of, 895–896
Heisenberg uncertainty principle, 150–152
Henderson-Hasselbalch equation, 273–274
Henry's law, 435–438
Henry's law constant, table, 437
Hertz, Heinrich, 134
Hertz (Hz), unit of frequency, 131
Hess's law, 360–361
Heterogeneous equilibria, 225, 239–240
Heteronuclear diatomic molecules, 553
  molecular orbital configuration of, 613–617
Hexagonal close packing, 750
Hexose, 856
High-spin vs low-spin octahedral complexes, 811–812
Homonuclear diatomic molecules, 553
  electronic configuration of, 608–612
Hooke's law, 642
Hund's rule, 182–183, 189, 196, 608
Hybrid atomic orbitals, 592–596
  table, 595
Hydrated ions, 40–41, 252, 680–681
Hydrates, 683–685
  phase diagram of, 441–442
Hydration energy, 680–681
Hydration number, 680
Hydride ion, 215, 694
Hydrides, 694–695
  of alkali metals, 694–695
  of alkaline earths, 694–695
Hydrocarbons, 364, 829
  delocalized bonding in, 620–623
  localized bonding in, 617–620
  saturated, 829–835
  unsaturated, 835
Hydrochloric acid, 698
Hydrofluoric acid, 698
Hydrogen, 215
  atom orbitals, table, 164
  atomic spectrum of, 143–148
  atomic spectrum of, table of series, 144
  Bohr theory of atom, 145–147
  isotopes of, 6
  laboratory preparation of, 50
  molecular orbital description of, 603
  position in periodic table, 215
  properties of, 215
  valence bond description of, 589–592
Hydrogen and oxygen, reaction between, mechanism of, 527

# Index

Hydrogen bond, 670–673
Hydrogen bonding,
  and deviations from ideal solution behavior, 435
  and dielectric constant, 682–683
  and ion mobility, 683
  and solubility, 686–688
  and the structure of ice, 677–678
  and Trouton's rule, 674
  effects on nonmetal hydrides, 673–676
  in alcohols, 841
  in DNA, 869–870
  in proteins, 866
  in water, 678–679
Hydrogen bridge bonds, in boranes, 623–625
Hydrogen electrode, 471
Hydrogen halides, 215, 698
Hydrogen-molecule ion, 599–603
Hydrogen peroxide, 724–725
Hydrogenic wave functions, table, 166
Hydrohalic acids, 215, 697–699
  Born-Haber cycle data for, table, 699
Hydrolysis, 265–266
Hydronium ion, 252
Hydrosphere, 688
Hydroxides, 704

Ice, structure of, 677–678
Ideal gas, 74
Ideal gas law, 74–77
  combined, 73–74
  deviations from, 110–113
Ideal solution, 426–429
Impurities in semiconductors, donor and acceptor, 760
Indefinite integrals, table, 930
Indicators,
  acid-base, 278–279
  of pH, figure, 278
Induced dipole moment, 664–665
Inductive effect, and strength of carboxylic acids, 845
Inert complexes, 822
Inexact differentials, 348
Infrared,
  far, 135
  near, 135
Infrared absorption bands, table, 646
Infrared (IR) region, 134–135
Infrared (IR) spectroscopy, 644, 645–647
Infrared spectrophotometer, 645
Inner electron shells, 184
Inner transition series, 191
Inorganic nomenclature, table, 18, 19

Inorganic qualitative analysis, 314
  of insoluble metal chlorides, 314–316
  of insoluble metal sulfides, 320–322
Insoluble metal chlorides, flow diagram for separation, 316
Insulators, 757
Integrated circuit, 760–761
Integration, 927–930
Intensity of wave, 131
Intensive properties, 10
Interhalogen compounds, 732–733
  table, 732
Intermediate species in mechanisms, 500, 518
Intermolecular forces, 111–113, 664–670
  compared with interatomic bonds, 673
Intermolecular potential energy function, 113, 666–668
Internal energy, 47, 345
Internal pressure of a gas, 117–118
Interstitial sites, in ionic crystals, 764–766
Ion-dipole interaction, 681
Ion exchange, 689–690
Ion mobility, 683
Ion-pairs, 658–664
  Lewis structure of, 560
Ion product of water, 257
Ionic bonds, 17, 553, 658–664
Ionic character of covalent bonds, 596–599
  and dipole moments, 596–597
  and electronegativity, 597–598
Ionic charge density, 714
Ionic compounds,
  crystal structures of, table, 763
  nomenclature, 17–20
Ionic crystalline solids,
  interstitial sites in, 764–766
  properties of, 770–771
Ionic radii, 209–213, 764
  table, 764
Ionization, 16
Ionization constants,
  of acids, 255–257
  of acids, table, 931
  of bases, 259–261
  of bases, table, 932
Ionization energy, 192–204
  first, 192–197
  higher than first, 197–199
  of hydrogen atom, 147
  table, 193
Ions, 16–17
  charge density of, 325–326, 714
  hydrated, 40–41, 252, 680–681
  mobility of, 683

  names and formulas, table, 18, 19
Iron,
  galvanized, 488
  polymorphism in, 752
  production of metal, 790
Iron triad, elements of, 789–791
Irreversible process, 339
Isobaric nuclides, relative stability of, 884–885
Isobaric process, 352–353
Isobars, 877
Isoelectronic species, 198–199, 564
Isolated system, 332
Isolation method (in kinetics), 514–517
Isomerism, 801–806
  cis-trans, 836–837
Isomers,
  facial, 803
  geometrical, 802–803
  meridional, 803
  of alkanes, 832
  optical, 803–805, 848
  structural, 553–554
Isomorphous substances, 750
Isoteniscope, 417–418
Isotherm, 73
Isothermal process involving ideal gas, 357
Isotones, 877
Isotope separation, 102
Isotopes, 6, 877
  of heavy elements, naturally occurring, 895–896
Isotopic anomalies, 9
Isotopic tracers, 910–912

Joule, James, 353

Kelvin, Lord (William Thomson), 72–73
Kelvin temperature scale, relative to Celsius, 73
Ketones, 842–843
Kinetic energy, 94–95
Kinetic theory of gases, 92–110
  derivation of Boyle's law, 94–98
  postulates, 92–94
Kinetics, 499

Labile complexes, 821–822
Lanthanide contraction, 212–213
Lanthanides, 190, 191, 216, 793–794
Laser, 653–655
Latimer diagram, see electrode potential diagram
Lattice, crystal, 742–744
Lattice water, in hydrates, 684

# 960 Index

Lattices, cubic, table of properties, 746
Lavoisier, Antoine, 33–34, 249, 342
Law of corresponding states, 115–117
Law of definite proportions, 14
Law of multiple proportions, 14
LCAO-MO method, 599–604
Lead storage battery, 482–483
Le Chatelier, Henri, 241
Le Chatelier's principle, 241
 and buffers, 270
 and cell potential, 468, 479
 and common ion effect, 304
 and complex ion formation, 309, 313
 and concentration changes, 241–243
 and degree of ionization, 262
 and effect of pressure on melting point, 423
 and ionization of polyprotic acids, 289
 and pH dependence of oxidizing strength of oxoacids, 722
 and pressure changes, 243–244
 and repression of self-ionization, 266
 and solubility, 323
 and temperature changes, 244–245
 and temperature dependence of solubility, 299
Leclanché cell, 480–481
Lennard-Jones parameters, table, 668
Lennard-Jones potential, 666–668
Leveling effect of water, 257
 on strong acids and bases, 257
Levorotatory enantiomer, 849
Lewis acid and Lewis base, 254, 562
Lewis formula, 13
Lewis, G. N., 254, 562
Lewis structures, 554–560
Lewis theory of acids and bases, 254
Ligand, 307, 796
 monodentate, bidentate, and polydentate, 797–798
 monodentate, table, 798
 multidentate, table, 798
Ligand substitution reactions, 821–822
Light,
 coherent, 563
 plane-polarized, 849–850
 speed of, 131
Limiting reagent, 39
Linear combinations of atomic orbitals, as molecular orbitals, 599–600
Linear molecules, 571
Lipscomb, William N., 625
Liquid drop model, of nucleus, 881–884
Liquid-vapor equilibrium, 423
Liquids, vapor pressure of, table, 418
Lithium aluminum hydride, as reducing agent, 848
Localized bonding in hydrocarbons, 617–620
Logarithms, 922–923
London (dispersion) forces, 665
 effect on boiling point, 669–670
London, F., 665
Lone pair electrons, 554
 effect on molecular geometry, 579–584
Low-spin vs high-spin octahedral complexes, 811–812
LP gas, 364
Lyman series, 144

Madelung constant, 772–773
Magic numbers, in nuclei, 884
Magnetic properties of transition metal complexes, 807–808
Magnetic quantum number, 163
Maiman, Theodore, 655
Malleability, of metals, 754
Manganese,
 chemistry of, 795–796
 redox chemistry of, 795–796
Manometer, 71
Markovnikov's rule, 854–855
Mass, 9
 conservation of, in balanced equations, 35
 conservation of, in chemical reactions, 33–35
Mass and charge of atomic constituents, table, 6
Mass balance equation, 263, 268, 292, 324
Mass-energy,
 conservation of, 880
 equivalence of, 879
Mass number, 6, 877
Mass spectrometry, 15
Mass yield curve, 902
Mathematical approximations, 923–924
 table, 924
Maxwell, James, C., 92, 130
Maxwell-Boltzmann distribution, 106–109
Mean free path, 104

Mean square speed, 97
Mechanical equivalent of heat, 342–343
Mechanical work, 335–336
Mechanisms, of organic reactions, 850–856
Melting points, and chemical bonding, 217–218
Mendeleev, Dmitri, 177–179
Mercury cell, 482
Meridional isomers, 803
Metabolism, 860–861
Metal chlorides insoluble, flow diagram for separation, 316
Metal clusters, 820–821
Metal complexes, structure of, 800
Metal hydroxides, solubility of, 326–328
Metal-metal bonds, 820–821
Metal salts, solubility of, table, 299
Metal sulfides,
 highly insoluble, flow diagram for separation, 321
 moderately insoluble, flow diagram for separation, 322
Metals, 213–216, 748–757
 band theory of, 754–757
 bonding of, 753–757
 Born-Haber cycle data for, table, 714
 ductility of, 754
 electron sea model of, 753–754
 in groups IA and IIA, table of properties, 712
 location in periodic table, 213–214
 malleability of, 754
 p-block, table of properties, 717
 reactions of, table, 712
 structure of, 748–753
 work function of, 138
Methanol, 365, 842
Meyer, Lothar, 177
Microwave spectroscopy, 638
Millikan, Robert, 125
Mixture, 13
Mobility, of ions, 683
Molality, 426
Molar conductivity of electrolytic solutions, 250
Molar heat capacities, table, 354
Molar mass, 21
Molar volume of gas at STP, 67
 table, 66
Molarity, 42–43
Mole, 20–25
Mole fraction, 78
Molecular beams, 539
Molecular energy levels, 635–637
 electronic, 635
 rotational, 635

# Index

vibrational, 635
Molecular formula, 11, 24
Molecular orbital theory, 599–617
Molecular spectroscopy, 634–637
Molecular speeds, 99–101
Molecular weight, 15
Molecularity (of a reaction), 518
Molecule, 11
Moment of inertia, 637
Momentum, 94
Monodentate ligands, table, 798
Monosaccharides, 856–858
Moseley, Henry, 199
Most probable speed of gas molecules, 107
Mulliken, R. S., 206
Multidentate ligands, table, 798
Multielectron atoms, energy levels of, 179–182
Multiple bonds, 562–564
Multiple proportions, law of, 14

n-Type semiconductors, 760
Natural gas, 364–365
Naturally occurring isotopes of heavy elements, 895–896
Nearest neighbors, number of, in cubic lattices, 746–747
Negative deviations from Raoult's law, 434
Nernst equation, 467–468, 478–479
Net ionic equations, 46
Network covalent solids, 776–779
Neutralization, 51, 276–277
Neutrino, 886
Neutron, 5
Neutron diffraction, 149, 899
Neutron drip line, 892
Neutrons, thermal, 898
Newlands, John, 176
Newton, Isaac, 123
Newton's second law, 9–10, 68
Nickel-cadmium battery, 483
Nitrate ion, resonance structures of, 567–568
Nitric acid, 705–706, 720
  as oxidizing agent, 720
  Ostwald process for synthesis of, 705–706
Nitrogen, and its compounds, 718–724
  Lewis structure of, 563
  molecular orbitals of, 611
  oxidation states of, 718
  redox reactions, table, 722
Noble gases, 185, 191, 192, 733
Node, 159, 171, 600
Nomenclature,
  inorganic, 18–20

of alcohols, 840
of alkanes, 832–833
of coordination compounds, 798–800
Nonbonding atomic orbitals, 604
Nonideal solutions, 434–435
Nonmetals, 213–216
  location in periodic table, 213–214
Nonstoichiometric compounds, 769
Normal boiling point, 72, 419
Normal freezing point, 72
Normalization constant of wave function, 157
Nuclear binding energy, 878–879
Nuclear density, 878
Nuclear fission, 900–901
Nuclear force, 881–882
Nuclear magnetic resonance spectroscopy (NMR), 912–914
Nuclear radii, 877–878
Nuclear radius parameter, 878
Nuclear reactions, 896–898
Nuclear spin, 912
Nucleic acid, 867
Nucleon, 877
Nucleophile, 851
Nucleophilic substitution reactions ($S_N1$, $S_N2$), 851–854
Nucleotide, 867
Nucleus, 4, 129
  liquid drop model of, 881–884
Nuclides, 877
  radioactive, 880
  stable, 880
  with closed shells, 884
Nuclidic mass, 878

Octahedral complexes,
  crystal field splitting in, table, 815
  crystal field theory of, 809–813
  high spin vs low spin, 811–812
Octahedral molecules, 573
Octahedral sites, in ionic crystals, 765–766
Octahedral vs tetrahedral vs square-planar complexes, 818–820
Octet rule, 555
  exceptions to, 557–560
Odd-odd nuclides, 883
Oil drop experiment, 125–126
Oklo mine, 9
Optical isomers, 803–805, 848
Optical region of electromagnetic spectrum, 134
Orbitals, 162–163
  delocalized, 620–623

hybrid, table, 595
penetration and screening of, 181, 195–196
Organic chemistry, 828
Organic reactions,
  mechanisms of, 850–856
  reactive intermediates in, 850–851
  redox, 846–847
Orientation effects (in reaction rates), 533–534
Osmosis and osmotic pressure, 448–450
Ostwald process for making nitric acid, 705–706
Overpotential, 494
Oxidation, 52, 54
Oxidation-reduction equations, methods of balancing, 55–59
Oxidation state (oxidation number) 53–54, 566–567
  and formal charge, 564–567
  as periodic property, 216–217, 710
  of representative elements, 709–711
  of transition elements, 788–789
Oxides,
  acidic and basic, table, 702
  acidic vs basic, 702–703
  amphoteric, 702–703
  ionic, 699–700
  molecular, 700–701
  of representative elements, 699–703
  of representative elements, table, 700
  polymeric, 701–702
  preparation of, table, 700
Oxidizing agent, 52, 472
Oxoacids, 704–709
  preparation of, 705–706
  table, 705
  trends in strength of, 706–709
Oxoanions, 18
Oxonium ion, 841–842
Oxygen,
  allotropes of, 85
  chemistry of, 724–725
  molecular orbital theory of, 611–612
  paramagnetism of, 564, 611–612
  phosphorescence of, 652–653
Oxygen and hydrogen, reaction between, mechanism of, 527
Ozone, 85, 724
Ozonides, 699

p-Type semiconductor, 760
P-V isotherm, 73

## 962 Index

Pairing energy,
  of electrons, 811
  of nuclei, 883
Paramagnetism, 558
  of diatomic oxygen, 564
  of transition metal complexes, 807
Partial pressure, 77
  Dalton's law of, 77–82
Particle in a box, Schrodinger equation, 154–161
Pascal (Pa), 68
Paschen series, 144
Path function, 345
Pauli exclusion principle, 182–183, 608
  and bonding in $H_2$, 592
Pauli, Wolfgang, 182
Pauling, Linus, 592, 706
Penetration and screening of orbitals, 181, 195–196
Peptide bond, 863
Perchloric acid, 730
Periodic properties, 192
  atomic volume per mole, 208–209
  atomic radii, 209
  ionic radii, 210–211
  electron affinity, 204–206
  electronegativity, 206–208
  ionization energy, 192–197
  oxidation state, 710–711, 788–789
Periodic table, 175–179, 190–192
  aufbau of, 184–190
  diagonal relationships in, 713–714
  divided into blocks, 191
  divided into metals, nonmetals, and semimetals, 213–214
  figure, 191
  groups of, 191–192
  history of, 175–179
  periods of, 190
  short form, 214
Peroxides, 699, 712
Perpetual motion machines,
  of the first kind, 348
  of the second kind, 379
Petroleum, 364–365
Pfund series, 144
pH, electrochemical determination of, 485–486
pH indicators, 278–279
  figure of, 278
pH meter, 486
pH of weak acids and bases, 261–265
pH range of buffers, 274–276
pH scale, 257–259
Phase diagram,
  liquid-vapor, 429–431
  of carbon, 778
  of carbon dioxide, 424
  of hydrates, 441–442
  of water, 422–424
  one-component, 422–425
  solid-liquid, 439–443
  two components, liquid-vapor, 426–439
Phase of wave, 133
Phase rule, 425, 432–433
Phase transitions, 410–416
  and enthalpy and entropy changes, 413–416
Phosphorescence, 651
Phosphorus, structure of, 725
Photochemical reaction, 85
Photochemical smog, 86–87
Photochemistry, 650
Photoelectric effect, 137–140
Photoelectron spectroscopy, 203–204
Photon, 138
Photosynthesis, 825, 859–861
Physical properties, 13
Pi ($\pi$) bond, 608, 619–620
Pi ($\pi$) molecular orbitals, 607–608
Pi to pi star transition ($\pi - \pi^*$), 647
$pK_a$ values, 256
Planck, Max, 136, 741
Planck's constant, 136
Plane-polarized light, 849–850
pOH scale, 258
Point defects, in crystals, 768–769
Polar covalent bonds, 40, 553
Polarizability, 598–599, 665
Polarization, 458
Polycrystalline solids, 739
Polyethylene, 838
Polymer, 838
Polymerization, of alkenes, 838
Polymorphism, 752
Polynuclear complexes, 820–821
Polypeptide, 863
Polyprotic acids, 286–295
  ionization constants, table, 287
Polysaccharides, 858–859
Population inversion, of energy levels, 654
Porphyrins, 824–825
Positive deviations from Raoult's law, 435
Positron, 888
Positron decay, 889
Post-transition elements, 793
Potassium permanganate, 60, 796
Potential energy barrier, 536
Potential energy contour plot, 537–538
Potential energy, intermolecular, 112–113
Power, 131

Precision vs accuracy, 27
Prefixes of SI units, table, 919
Pressure, 68–70
  of a gas, internal, 117–118
  units of, table, 70
Primary cells, 480–482
Primitive cell, 745
Principal quantum number, 163
Proteins, 864–866
  conformation of, 865
Proton, 5
Proton affinity, 696
Proton drip line, 893
Pseudo-first-order reactions, 514–515
Pseudohalides, 732

Q value, 897
Quadratic formula, 923
Quantized energy, 142
Quantum mechanics, 123, 152
Quantum numbers,
  electronic, 146, 163–165
  of H orbitals, table, 164
  rotational, 638
  vibrational, 643
Quartz, 781

Racemic mixture, 849
Radial function, 162
Radial nodes, 168
Radial probability function, 165–168
Radiative decay, 650
Radical, 830
Radii,
  ionic, table, 764
  nuclear, 877–878
  of atoms, table, 210
  of isoelectronic ions, 211
Radio waves, 135
Radioactive clocks, 909
Radioactive decay, 907
Radioactive isotopes, 6, 223, 880
Radioactivity, 127
Radiocarbon dating, 910
Radioisotopes used for age determinations, table, 910
Radium, 896
Radius-ratio rule, 766–768
  table of values, 768
Random walk, 98
Raoult's law, 427–429, 435–436
Rare earths, see lanthanides
Rate constant, 503
  and equilibrium constant, 519
  table, 530
Rate-determining step, 520–521
Rate laws, 500, 503–505
  initial rate method for the deter-

mination of, 505–508
  integrated, 508–517
  table, 517
Rate of a chemical reaction, 47, 502
Reaction coordinate, 535
Reaction enthalpy, 358–359
  temperature dependence of, 366–367
Reaction mechanism, 500, 517, 519–525
Reaction order, 503
Reaction quotient, 230–231
  equilibrium constant, relative to, 230
  solubility product, compared to, 303
Reactive intermediates, in organic reactions, 850–851
Real gases, chemical equilibrium of, 245–246
Rechargeable battery, 482
Redox couple, 461
Redox equations, rules for balancing, 55–59
Redox reactions, 52
  inner sphere mechanism in, 822–823
  involving functional groups, 846–848
  of coordination complexes, 822–823
  of nitrogen compounds, table, 722
  organic, 846–847
  outer sphere mechanism in, 822
Redox titration, 60
Reduced mass, 637
Reduced variables, 115
Reducing agent, 52, 472
Reduction, 52, 54
Reduction potentials, standard, table, 473
Renewable fuels, 365
Replication, of DNA, 871
Representative element oxides, table, 700
Representative elements, 191–192
  oxidation numbers of, 709–711
Resolution, 16
Resonance, 567–571
  in benzene, 568–569
  in nitrate ion, 567–568
Resonance energy, 633–634
Resonance hybrid, 568
Reverse osmosis, 450
Reversible process, 339, 372–373
Reversible reaction, 223
RNA, 867
Roentgen, Wilhelm, 199
Root-mean-square speed, 100–101

Rotational energy, 358
Rotational quantum number, 638
Rotational spectra, of diatomic molecules, 637–640
Rounding off numbers, 29
Rutherford, Ernest, 127, 897
Rydberg constant, 144

Salt bridge, 457–458
Salts, 42
  dissolution in water of, 685–686
  of weak acid or weak base, 265–266
Sanger, Frederick, 864
Saturated solutions, 300–301
Scandium family, elements of, 793–794
Schottky defect, 769
Schrödinger, Erwin, 153
Schrödinger wave equation, 153–154
  boundary conditions, 153
  normalization constant of, 157
  of hydrogen atom, 161–163
  of particle in a box, 154–161
Seaborg, Glenn, 903
Second-order kinetics, 511–514
Self-ionization of water, 257
Semiconductors, 757
  doped, 758–761
  intrinsic, 757–758
  n-type and p-type, 760
Semimetals, 213–214
  location in periodic table, 213–214
Semipermeable membrane, 448
SI prefixes, table, 919
SI units,
  derived, table, 919
  fundamental, table, 918
Sigma bond, 608, 618–620
Sigma molecular orbitals, 600–606
Significant figures, 28–30
Silicates, structure of, 779–781
Silver, use in photography, 791–792
Silverplating, 494, 823
Simple cubic (sc) cell, 745–747
Simple harmonic motion, 642
Simultaneous equilibria, 312–328
Singlet state, 652
Six-twelve potential, 667
Soap, 844
Sodium borohydride, as reducing agent, 848
Sodium carbonate, 713
Sodium chloride,
  crystal lattice energy of, 771–773
  crystal structure of, 762
  electrolysis of aqueous, 493–494
  electrolysis of molten, 491

Sodium, electrolytic production of, 491–492
Sodium hydrogen carbonate, 293–294, 713
Sodium hydroxide, electrolytic production of, 494
Sodium-sulfur battery, 484
Solids,
  amorphous, 739–740
  band theory of, 754–757
  categorized as metals, insulators, and semiconductors, 757
  crystalline, 738–739
  heat capacity of, 740–742
  polycrystalline, 739
Solubility,
  and complex formation, 313–319
  effect of hydrolysis on, 322–326
  effect of pH on, 319–320
  effect of strong electrolytes on, 318–319
  of electrolytes in water, 298–301
  of metal hydroxides, 326–328
  of metal salts, table, 299
Solubility product, 300–304
  and standard cell potential, 478–480
  table, 302, 932–933
Solute, 40
Solutions, 40, 426
  concentration units, 42–43, 426
  ideal, 426–429
  neutral acidic, and basic, 257
  non-ideal, 434–439
  saturated, 300–301
  supersaturated, 303
Solvated electrons, 716
Solvation, 680
Solvay process, 713
Solvent, 40
  acidic and basic, 256–257
Solvent cage, 540
Spectator ions, 46
Spectrochemical series, 815
Spectroscopy,
  atomic, 141–148
  infrared, 644, 645–647
  microwave, 638
  molecular, 634–637
  nuclear magnetic resonance, 912–914
  photoelectron, 203–204
  ultraviolet, 647
  visible, 647
Spectrum,
  absorption, 635
  emission, 635
  of hydrogen atom, 143–148
  of hydrogen atom, table, 144
Speed of light, 131

Spherical polar coordinates, 162
Spin quantum number, 163
Spontaneous emission, 653
Spontaneous fission, 903
Spontaneous processes, 370–374
Square-planar complexes, 818
Staggered vs eclipsed configuration of atoms, 820
Standard cell potential (EMF), 460
 and equilibrium constant, 477–480
 and solubility products, 478–480
Standard electrode potentials, 470–472
 of elements with three or more oxidation states, 475–477
Standard enthalpies of formation, 362–364
Standard entropies, 385, 386–389
 table, 388
Standard free energy change,
 and standard cell potential, 465
 and the equilibrium constant, 394–400
Standard free energy of formation, 393–394
Standard reduction potentials, 472–473
 table, 473, 946–949
Standard state, 362
Standard temperature and pressure (STP), 66–67
Standing waves, 154
Starch, 858
State functions, 345–350
Statistical mechanics, 358
Steady-state approximation, 521–525
Stereoisomers, 802–803
Steric factor, 534
Steric hindrance, 730, 852
Stimulated emission, 654
Stoichiometric calculations, 35–40
Stoichiometry, 36
Storage (secondary) cells, 480, 482–484
Straight-chain alkanes, table, 832
Strong acids, 256
Strong ligand field limit, 811
Structural formula, 13, 556, 830
Structural isomerism, 553–554, 801
 in alkenes, 837
Sublimation, 416
Subshells of atomic orbitals, 184
Successive approximations, method of, 234
Sucrose, 858
Sulfate ion, structure of, 569–571
Sulfur,
 chemistry of, 725–727
 monoclinic and rhombic, 726

Sulfuric acid, 706
Supercooling, 442
Superheavy elements, 906–907
Superoxides, 699, 712–713
Supersaturated and unsaturated solutions, 303
Surroundings (in thermodynamics), 332
Symmetric nuclei, 883
Symmetry energy, of nuclei, 883
System, 331–335
 and surroundings, 331–335
 closed, 332
 degrees of freedom of, 425
 isolated, 332
 open, 332
 state of, 332
Systematic errors, 26

Taube, Henry, 823
Teflon, 838
Temperature, kinetic theory of, 98–99
 absolute or Kelvin, 73
 Celsius or centigrade, 73
Termolecular reactions, 518
Tetrahedral bond angle, 572
Tetrahedral complexes, 817–818
Tetrahedral molecules, 571–572
Tetrahedral sites, in ionic crystals, 765
Thermal analysis, 442–443
Thermal neutrons, 898
Thermite process, 717
Thermochemical equation, 358–359
Thermochemistry, 47, 358–367
Thermodynamic data, table, 935–945
Thermodynamic properties of substances, 394
Thermodynamics,
 first law of, 344–350
 second law of, 374–379
 third law of, 386
Thiosulfate ion, 727
Third-law entropies, 385–386
Third law of thermodynamics, 386
Thomson, Joseph J., 124, 127
Three-center bonds in boron hydrides, 623–625
Time-of-flight measurement, 109
Titration curve,
 diprotic acid vs strong base, 294–295
 strong acid vs strong base, 280–281
 strong base vs strong acid, 281
 weak acid vs strong base, 281–284
Titration,

 acid-base, 276–285
 graphical representation of concentration changes, 285–286
 redox, 60
 with EDTA, 823
Torr, 69
Torricelli, Evangelista, 68
Townes, Charles, 655
Traditional units, table, 920
Transcription, of DNA, 871
Transition elements, 191–192, 216, 785–789
 electronic configurations of, 785–786
 oxidation states of, 788–789
 properties of, 786–788
Transition metal complexes,
 color of, 806–808, 815–816
 hybridization in, table, 808
 magnetic properties of, 807–808
Transition series, 189
Transition state, 536–537
Translational kinetic energy, 99, 357
Transuranium elements, 903–905
Trigonal bipyramidal molecules, 571
 equatorial and axial sites in, 572–573
Trigonal planar molecules, 571
Trigonal pyramidal molecules, 571
Triple bond, 563
Triple point, 424
Triplet state, 652
Tritium, 6
Trouton's rule, 416
Tunneling, 894
Tyndall effect, 451

Ultraviolet radiation, 85
Ultraviolet (UV) spectroscopy, 647
Uncertainty principle, 150–152
Unimolecular reactions, 518
Unit cell, 742–744
 body centered cubic, 745–747
 face centered cubic, 745–747
 of the seven crystal systems, 744–745
 primitive, 745
 simple cubic, 745–747
Units,
 and conversion factors, 918–920
 conversion of, 22–23
 SI, table, 918–919
 traditional, table, 920

Valence band, 756
Valence bond theory, 589
 applied to $H_2$, 589–592
 of coordination complexes, 808–809

# Index

Valence electrons, 11
Valence shell, 184
Van der Waals constants, 118
   and Lennard-Jones parameters, 668
   table, 118
Van der Waals equation, 117–121
Van der Waals forces, 118, 664–670
   and boiling points, 670
Van der Waals, Johannes, 117
Van der Waals radii, 668, 748
Van't Hoff equation, 401
Van't Hoff factor, 448
Van't Hoff, Jacobus, 401, 841
Vapor pressure, 416–422
   measurement of, 418
   of liquids, table, 418
   of solution of a non-volatile solute, 444
   of water, 80
   of water, table, 80
   temperature dependence of, 419–423
Vaporization, enthalpy of, 415–416
Vector, resolution into components, 577
Velocity, distinction from speed, 94
Vibrational energy, 358
Vibrational frequency, 642
Vibrational quantum number, 643
Vibrational spectra, of diatomic molecules, 641–647
Visible region of electromagnetic spectrum, 134–135
Visible spectroscopy, 647

Volt, 460
Volta, Alessandro, 456
VSEPR model, 578–584

Water,
   abundance of ions in, table, 688
   as solvent, 685–688
   critical point of, 423
   deionized, 690
   density of, 678
   dielectric constant of, 681–683
   electrolysis of, 490
   hard, 689–690
   ion product of, 257–258
   leveling effect of, 257
   natural, 688–689
   of crystallization, 683
   phase diagram of, 422–424
   self-ionization of, 254, 266–269
   solubility of gases in, 437–438
   supercooled, 423
   triple point of, 424
   vapor pressure of, table, 80
Watson, James, 867
Wave,
   frequency of, 131
   intensity of, 131
   phase of, 133
Wavelength, 130
   de Broglie, 148–150
Wave number, 131
Weak acids and bases,
   degree of ionization of, 261–262
   effect of ionization of water on
      pH of, 267–269
   pH of, 261–265
Weak ligand field limit, 811
Weight, 9
Werner, Alfred, 802
Woehler, Friederich, 828
Work, 332–335
   electrical, 460
   mechanical, 335–336
   of expansion, 336–337
   of expansion against variable external pressure, 337–339
   reversible, ideal gas expansion, 339–342
   sign convention for, 336–337
Work function of metals, 138
Wurtzite crystal structure, 762

Xenon compounds, 724
X-ray diffraction, 134, 745
X-ray emission, 199–203
X-rays, 134, 199

Yield,
   maximum, 39
   percentage, 39

Zeolite, 689
Zero point energy, 157, 643
Zinc blende crystal structure, 762
Zwitterions, 863

# Physical Constants

| | |
|---|---|
| Avogadro's number | $N_A = 6.022045 \times 10^{23}$ mol$^{-1}$ |
| Bohr radius | $a_o = 0.5291771$ Å $= 5.291771 \times 10^{-11}$ m |
| Boltzmann constant | $k = 1.38066 \times 10^{-23}$ J K$^{-1}$ |
| Coulomb's law constant | $\alpha = 1.1126501 \times 10^{-10}$ C$^2$ N$^{-1}$ m$^{-2}$ |
| Electron charge | $e = 1.602189 \times 10^{-19}$ C |
| Electron mass | $m_e = 9.10953 \times 10^{-31}$ kg |
| Faraday's constant | $F = 96{,}485$ C mol$^{-1}$ |
| Gas constant | $R = 8.3144$ J mol$^{-1}$ K$^{-1}$ |
| | $= 0.082057$ L atm mol$^{-1}$ K$^{-1}$ |
| Gravitational acceleration | $g = 9.80665$ m s$^{-2}$ |
| Ideal gas molar volume at STP | $V_{id} = 22.4138$ L mol$^{-1}$ |
| Neutron mass | $m_n = 1.674954 \times 10^{-27}$ kg |
| Planck's constant | $h = 6.62618 \times 10^{-34}$ J s |
| Proton mass | $m_p = 1.672648 \times 10^{-27}$ kg |
| Rydberg constant | $\mathcal{R} = 1.09737318 \times 10^7$ m$^{-1}$ |
| Speed of light (in vacuum) | $c = 2.9979246 \times 10^8$ m s$^{-1}$ |

# Conversion Factors

1 electron volt (eV) $= 1.602189 \times 10^{-19}$ J
1 calorie $= 4.184$ J
1 u $= 1.660566 \times 10^{-27}$ kg $= 931.502$ MeV
1 L atm $= 101.325$ J